Flora of North America

Contributors to Volume 5

Mihai Costea
Garrett E. Crow
Craig C. Freeman
Ronald L. Hartman
Harold R. Hinds†
Walter C. Holmes
Claude Lefèbvre

Nancy R. Morin
John K. Morton
Sergei L. Mosyakin
Mark A. Nienaber
John G. Packer
James S. Pringle
Richard K. Rabeler
James L. Reveal

Alan R. Smith
Ann Swanson
François J. Tardif
John W. Thieret
Frederick H. Utech
Xavier Vekemans
Warren L. Wagner

Editors for Volume 5

Craig C. Freeman,
Taxon Editor for Polygonaceae
Kanchi Gandhi,
Nomenclatural Editor

Ronald L. Hartman,
Co-Taxon Editor for
Caryophyllaceae
Robert W. Kiger,
Lead and Bibliographic Editor
Nancy R. Morin,
Taxon Editor for Plumbaginaceae

Richard K. Rabeler,
Co-Taxon Editor for
Caryophyllaceae
John L. Strother,
Reviewing Editor

James L. Zarucchi,
Editorial Director

Volume Composition

Pat Harris,
Compositor
and Editorial Assistant

Claire A. Hemingway,
Compositor
and Technical Editor

Kristin Pierce,
Compositor
and Herbarium Assistant

Kay Yatskievych,
Production Coordinator
and Compositor

Silene petersonii

Flora of North America

North of Mexico

Edited by FLORA OF NORTH AMERICA EDITORIAL COMMITTEE

VOLUME 5

Magnoliophyta: Caryophyllidae, part 2

CARYOPHYLLALES, part 2 (Pink order)

POLYGONALES (Buckwheat order)

PLUMBAGINALES (Leadwort order)

NEW YORK OXFORD · OXFORD UNIVERSITY PRESS · 2005

Oxford University Press, Inc., publishes works that further
Oxford University's objective of excellence
in research, scholarship, and education.

Oxford New York
Auckland Cape Town Dar es Salaam Hong Kong Karachi
Kuala Lumpur Madrid Melbourne Mexico City Nairobi
New Delhi Shanghai Taipei Toronto

With offices in
Argentina Austria Brazil Chile Czech Republic France Greece
Guatemala Hungary Italy Japan Poland Portugal Singapore
South Korea Switzerland Thailand Turkey Ukraine Vietnam

Published by Oxford University Press, Inc.
198 Madison Avenue, New York, New York 10016
http://www.oup.com

Oxford is a registered trademark of Oxford University Press

Library of Congress Cataloging-in-Publication Data
(Revised for volume 5)
Flora of North America north of Mexico
edited by Flora of North America Editorial Committee.
Includes bibliographical references and indexes.
Contents: v. 1. Introduction—v. 2. Pteridophytes and gymnosperms—
v. 3. Magnoliophyta: Magnoliidae and Hamamelidae—
v. 22. Magnoliophyta: Alismatidae, Arecidae, Commelinidae (in part), and Zingiberidae—
v. 26. Magnoliophyta: Liliidae: Liliales and Orchidales—
v. 23. Magnoliophyta: Commelinidae (in part): Cyperaceae—
v. 25. Magnoliophyta: Commelinidae (in part): Poaceae, part 2—
v. 4. Magnoliophyta: Caryophyllidae (in part): part 1—
v. 5. Magnoliophyta: Caryophyllidae (in part): part 2

ISBN-13: 978-0-19-522211-1 (v. 5)
ISBN 0-19-522211-3 (v. 5)
1. Botany—North America.
2. Botany—United States.
3. Botany—Canada.
I. Flora of North America Editorial Committee.
QK110.F55 2002 581.97 92-30459

Printing number: 9 8 7 6 5 4 3 2 1
Printed in the United States of America
on acid-free paper

Contents

*For their support of the preparation of this volume,
we gratefully acknowledge and thank:*

Chanticleer Foundation
National Science Foundation
The David and Lucile Packard Foundation
National Fish and Wildlife Foundation
The William and Flora Hewlett Foundation
The Andrew W. Mellon Foundation
The Fairweather Foundation
ChevronTexaco
ESRI

FOUNDING MEMBER INSTITUTIONS

Flora of North America Association

Arnold Arboretum
Jamaica Plain, Massachusetts

Biosystematics Research Institute
Ottawa, Ontario

Canadian Museum of Nature
Ottawa, Ontario

Carnegie Museum of Natural
 History
Pittsburgh, Pennsylvania

Field Museum of Natural History
Chicago, Illinois

Fish and Wildlife Service
United States Department of the
 Interior
Washington, D.C.

Harvard University Herbaria
Cambridge, Massachusetts

Hunt Institute for Botanical
 Documentation
Carnegie Mellon University
Pittsburgh, Pennsylvania

Jacksonville State University
Jacksonville, Alabama

Jardin Botanique de Montréal
Montréal, Québec

Kansas State University
Manhattan, Kansas

Missouri Botanical Garden
St. Louis, Missouri

New Mexico State University
Las Cruces, New Mexico

New York State Museum
Albany, New York

Northern Kentucky University
Highland Heights, Kentucky

The New York Botanical Garden
Bronx, New York

The University of British Columbia
Vancouver, British Columbia

The University of Texas
Austin, Texas

Université de Montréal
Montréal, Québec

University of Alaska
Fairbanks, Alaska

University of Alberta
Edmonton, Alberta

University of California
Berkeley, California

University of California
Davis, California

University of Idaho
Moscow, Idaho

University of Illinois
Urbana-Champaign, Illinois

University of Iowa
Iowa City, Iowa

The University of Kansas
Lawrence, Kansas

University of Michigan
Ann Arbor, Michigan

University of Oklahoma
Norman, Oklahoma

University of Ottawa
Ottawa, Ontario

University of Southwestern
 Louisiana
Lafayette, Louisiana

University of Western Ontario
London, Ontario

University of Wyoming
Laramie, Wyoming

Utah State University
Logan, Utah

Project Staff — past and present
involved with the preparation of this volume

Barbara Alongi, *Illustrator*
Michael Blomberg, *Scanning Specialist*
Trisha K. Consiglio, *GIS Analyst*
Bee F. Gunn, *Illustrator*
Pat Harris, *Editorial Assistant and Compositor*
Claire A. Hemingway, *Technical Editor and Compositor*
Fred Keusenkothen, *Scanning Supervisor*
John Myers, *Illustrator and Illustration Compositor*
Kristin Pierce, *Editorial Assistant and Compositor*
Elizabeth A. Polen, *Technical Editor*
Mary Ann Schmidt, *Technical Editor*
Hong Song, *Programmer*
Yevonn Wilson-Ramsey, *Art Director and Illustrator*
George Yatskievych, *Technical Adviser*
Kay Yatskievych, *Production Coordinator and Compositor*

Contributors

Mihai Costea
University of Guelph
Guelph, Ontario

Garrett E. Crow
University of New Hampshire
Durham, New Hampshire

Craig C. Freeman
The University of Kansas
Lawrence, Kansas

Ronald L. Hartman
University of Wyoming
Laramie, Wyoming

Harold R. Hinds†
University of New Brunswick
Fredericton, New Brunswick

Walter C. Holmes
Baylor University
Waco, Texas

Claude Lefèbvre
Université Libre de Bruxelles
Brussels, Belgium

Nancy R. Morin
The Arboretum at Flagstaff
Flagstaff, Arizona

John K. Morton
University of Waterloo
Waterloo, Ontario

Sergei L. Mosyakin
N. G. Kholodny Institute of
  Botany
Kiev, Ukraine

Mark A. Nienaber
Northern Kentucky University
Highland Heights, Kentucky

John G. Packer
University of Alberta
Edmonton, Alberta

James S. Pringle
Royal Botanical Gardens
Hamilton, Ontario

Richard K. Rabeler
University of Michigan
Ann Arbor, Michigan

James L. Reveal
University of Maryland
College Park, Maryland

Alan R. Smith
University of California
Berkeley, California

Ann Swanson
Northern Kentucky University
Highland Heights, Kentucky

François J. Tardif
University of Guelph
Guelph, Ontario

John W. Thieret
Northern Kentucky University
Highland Heights, Kentucky

Frederick H. Utech
Hunt Institute for Botanical
  Documentation
Pittsburgh, Pennsylvania

Xavier Vekemans
Université Libre de Bruxelles
Brussels, Belgium

Warren L. Wagner
Smithsonian Institution
Washington, D.C.

Taxonomic Reviewers

Michael F. Baad
California State University
Sacramento, California

John R. Edmondson
National Museum Liverpool
Liverpool, England

Barbara J. Ertter
University of California
Berkeley, California

James L. Luteyn
The New York Botanical Garden
Bronx, New York

Kimberlie A. McCue
Missouri Botanical Garden
St. Louis, Missouri

Rob Smissen
Landcare Research
Lincoln, New Zealand

Dieter H. Wilken
Santa Barbara Botanic Garden
Santa Barbara, California

Peter F. Zika
Seattle, Washington

Regional Reviewers

ALASKA

Alan Batten
University of Alaska Museum
Fairbanks, Alaska

Robert Lipkin
Alaska Natural Heritage Program
University of Alaska
Anchorage, Alaska

Mary Stensvold
Forest Service, Alaska Region
United States Department of
Agriculture
Sitka, Alaska

PACIFIC NORTHWEST

Geraldine Allen
University of Victoria
Victoria, British Columbia

Edward R. Alverson
The Nature Conservancy
Eugene, Oregon

Adolf Ceska
British Columbia Conservation
Data Centre
Victoria, British Columbia

Kenton L. Chambers
Oregon State University
Corvallis, Oregon

George W. Douglas
British Columbia Conservation
Data Centre
Victoria, British Columbia

Richard R. Old
XID Services, Inc.
Pullman, Washington

Jim Pojar
British Columbia Forest Service
Smithers, British Columbia

Scott Sundberg
Oregon State University
Corvallis, Oregon

Peter F. Zika
Seattle, Washington

SOUTHWESTERN UNITED STATES

Roxanne L. Bittman
California Department of Fish and
Game
Sacramento, California

Steve Boyd
Rancho Santa Ana Botanic
Garden
Claremont, California

Barbara J. Ertter
University of California
Berkeley, California

H. David Hammond
Northern Arizona University
Flagstaff, Arizona

G. Frederic Hrusa
Courtland, California

Charles T. Mason Jr.
University of Arizona
Tucson, Arizona

James D. Morefield
Nevada Natural Heritage Program
Carson City, Nevada

Donald J. Pinkava
Arizona State University
Tempe, Arizona

WESTERN CANADA

Bruce Bennett
NatureServe Yukon-V5N
Whitehorse, Yukon

William T. Cody
Agriculture and Agri-food Canada
Crop Protection Program
Central Experimental Farm
Ottawa, Ontario

Bruce A. Ford
University of Manitoba
Winnepeg, Manitoba

Vernon L. Harms
University of Saskatchewan
Saskatoon, Saskatchewan

ROCKY MOUNTAINS

Bonnie Heidel
Wyoming Natural Diversity
  Database
University of Wyoming

Tim Hogan
University of Colorado Museum
Boulder, Colorado

B. E. Nelson
Rocky Mountain Herbarium
Department of Botany
University of Wyoming

J. Stephen Shelly
U.S.D.A. Forest Service, Northern
  Region
Missoula, Montana

NORTH-CENTRAL UNITED STATES

William T. Barker
North Dakota State University
Fargo, North Dakota

Anita F. Cholewa
University of Minnesota
St. Paul, Minnesota

Theodore S. Cochrane
University of Wisconsin
Madison, Wisconsin

Patricia Folley
Noble, Oklahoma

Neil A. Harriman
University of Wisconsin
Oshkosh, Wisconsin

Bruce W. Hoagland
University of Oklahoma
Norman, Oklahoma

Robert B. Kaul
University of Nebraska
Lincoln, Nebraska

Gary E. Larson
South Dakota State University
Brookings, South Dakota

Deborah Q. Lewis
Iowa State University
Ames, Iowa

Ronald L. McGregor
University of Kansas
Lawrence, Kansas

Lawrence R. Stritch
U.S.D.A. Forest Service
Shepherdstown, West Virginia

Connie Taylor
Southeastern Oklahoma State
  University
Durant, Oklahoma

George Yatskievych
Missouri Botanical Garden
St. Louis, Missouri

SOUTH-CENTRAL UNITED STATES

David E. Lemke
Southwest Texas State University
San Marcos, Texas

Jackie M. Poole
Texas Parks and Wildlife
  Department
Austin, Texas

Robert C. Sivinski
New Mexico Energy, Mineral, and
  Natural Resources Department
Santa Fe, New Mexico

EASTERN CANADA

Jacques Cayouette
Agriculture and Agri-Food Canada
Ottawa, Ontario

William J. Crins
Ontario Ministry of Natural
  Resources
Peterborough, Ontario

Stuart G. Hay
Herbier Marie-Victorin
Université de Montréal
Montréal, Québec

John K. Morton
University of Waterloo
Waterloo, Ontario

Marian Munro
Nova Scotia Museum of Natural
  History
Halifax, Nova Scotia

Michael J. Oldham
Natural Heritage Information
  Centre
Peterborough, Ontario

Claude Roy
Université Laval
Québec, Québec

GREENLAND

Bent Fredskild
University of Copenhagen
Copenhagen, Denmark

Geoffrey Halliday
University of Lancaster
Lancaster, England

NORTHEASTERN UNITED STATES

Ray Angelo
New England Botanical Club
Cambridge, Massachusetts

Steven E. Clemants
Brooklyn Botanic Garden
Brooklyn, New York

Tom S. Cooperrider
Kent State University
Kent, Ohio

Arthur Haines
Bowdoin, Maine

Les Mehrhoff
University of Connecticut
Storrs, Connecticut

Edward G. Voss
University of Michigan
Ann Arbor, Michigan

SOUTHEASTERN UNITED STATES

Raymond Cranfill
University of California
Berkeley, California

John B. Nelson
University of South Carolina
Columbia, South Carolina

Bruce A. Sorrie
Longleaf Ecological
Whispering Pines, North Carolina

R. Dale Thomas
University of Louisiana at Monroe
Monroe, Louisiana

Lowell E. Urbatsch
Louisiana State University
Baton Rouge, Louisiana

Alan S. Weakley
Association for Biodiversity
Information/The Nature
Conservancy
Durham, North Carolina

Thomas F. Wieboldt
Virginia Polytechnic Institute and
State University
Blacksburg, Virginia

B. Eugene Wofford
University of Tennessee
Knoxville, Tennessee

FLORIDA

Loran C. Anderson
Florida State University
Tallahassee, Florida

Edwin L. Bridges
Botanical and Ecological
Consultant
Bremerton, Washington

Bruce F. Hansen
University of South Florida
Tampa, Florida

Walter S. Judd
University of Florida
Gainesville, Florida

Richard P. Wunderlin
University of South Florida
Tampa, Florida

Preface

During 2002 and 2003 the Flora of North America Association [FNAA] underwent a reorganization in which the former Editorial Committee has been succeeded by a Board of Directors. An Executive Committee and an Editorial Management Committee have been created to focus on the project's policymaking and production efforts, respectively. For the sake of continuity of citation, authorship of *Flora* volumes is still to be cited as "Flora of North America Editorial Committee, eds."

Since the publication of volume 4, the eighth volume in the *Flora* series in 2003, Wayne J. Elisens (Taxon Editor) and Jackie M. Poole (Regional Coordinator) have become members of the FNAA Board of Directors.

Editorial processing for this volume of the *Flora* was undertaken at the Hunt Institute for Botanical Documentation at Carnegie Mellon University in Pittsburgh and the Missouri Botanical Garden in St. Louis. The maps were prepared at the St. Louis center based on taxon distribution statements found in the treatments along with additional data for the indicators showing occurrence in Alaska, Greenland, and the larger Canadian provinces and territories. Pre-press production, including typesetting, layout, and scanning and labeling of illustrations, was done at the Missouri Botanical Garden. The St. Louis center also provided coordination for all aspects of planning and executing the illustrations included in this volume.

In addition to her duties as the project's Art Director, Yevonn Wilson-Ramsey prepared the illustrations for numerous taxa in genera of Caryophyllaceae (*Achyronychia, Agrostemma, Cardionema, Cerastium, Corrigiola, Lepyrodiclis, Scopulophila,* and *Silene*), including the color frontispiece depicting *Silene petersonii,* and Polygonaceae (*Aconogonon, Antigonon, Bistorta, Brunnichia, Fagopyrum, Fallopia, Koenigia, Persicaria, Polygonum, Rheum,* and *Rumex*). Other illustrations were prepared by John Myers, for Plumbaginaceae and eriogonoid genera of Polygonaceae (except *Eriogonum* subgenera *Eucycla* and *Micrantha*); by Barbara Alongi, for various taxa of Caryophyllaceae (*Dianthus, Drymaria, Eremogone, Geocarpon, Gypsophila, Herniaria, Holosteum, Loeflingia, Moehringia, Moenchia, Myosoton, Paronychia, Petrorhagia, Polycarpaea, Polycarpon, Saponaria, Scleranthus, Spergula, Spergularia, Stipulicida, Vaccaria,* and *Velezia*); and by Bee F. Gunn, for Caryophyllaceae (*Honckenya, Minuartia, Pseudostellaria, Sagina, Stellaria,* and *Wilhelmsia*) and Polygonaceae (*Coccoloba, Emex,* part of *Eriogonum* [subgenera *Eucycla* and *Micrantha*], *Muehlenbeckia, Oxyria,* and *Polygonella*). Illustrations were scanned at the Missouri Botanical Garden by Michael Blomberg and Fred Keusenkothen. Composition and labeling of all artwork was completed by John Myers.

The Flora of North America Association remains deeply grateful to the many people who continue to help create and sustain the *Flora*.

Introduction

Scope of the Work

Flora of North America North of Mexico is a synoptic account of the plants of North America north of Mexico: the continental United States of America (including the Florida Keys and Aleutian Islands), Canada, Greenland (Kalâtdlit-Nunât), and St. Pierre and Miquelon. The *Flora* is intended to serve both as a means of identifying plants within the region and as a systematic conspectus of the North American flora.

The *Flora* will be published in 30 volumes. Volume 1 contains background information that is useful for understanding patterns in the flora. Volume 2 contains treatments of ferns and gymnosperms. Families in volumes 3–26, the angiosperms, are arranged according to the classification system of A. Cronquist (1981). Bryophytes will be covered in volumes 27–29. Volume 30 will contain the cumulative bibliography and index.

The first two volumes were published in 1993, Volume 3 in 1997, and Volumes 22, 23, and 26, the first three of five covering the monocotyledons, appeared in 2000, 2002, and 2002, respectively. Volume 4, the first part of the Caryophyllales, was published in late 2003. Volume 25, the second part of the Poaceae, was published in mid-2003. The correct bibliographic citation for the *Flora* is: Flora of North America Editorial Committee, eds. 1993+. *Flora of North America North of Mexico*. 9+ vols. New York and Oxford.

Volume 5 treats 739 species in 75 genera contained in three families. For additional statistics, please refer to Table 1.

Contents · General

The *Flora* includes accepted names, selected synonyms, literature citations, identification keys, descriptions, chromosome numbers, phenological information, summaries of habitats and geographic ranges, and other biological observations. Economic uses, weed status, and conservation status are provided from specified sources. Each volume contains a bibliography and an index to the taxa included in that volume. The treatments, written and reviewed by experts from throughout the systematic botanical community, are based on original observations of herbarium specimens and, whenever possible, on living plants. These observations are supplemented by critical reviews of the literature.

Table 1. *Statistics for Volume 5 of Flora of North America.*

Family	Total Genera	Total Species	Endemic Genera	Endemic Species	Introduced Genera	Introduced Species	Conservation Taxa
Caryophyllaceae	37	286	1	120	15	82	34
Polygonaceae	35	442	12	310	5	54	111
Plumbaginaceae	3	11	0	1	0	6	0
Totals	75	739	13	431	20	142	145

Basic Concepts

Our goal is to make the *Flora* as clear, concise, and informative as practicable so that it can be an important resource for both botanists and nonbotanists. To this end, we are attempting to be consistent in style and content from the first volume to the last. Readers may assume that a term has the same meaning each time it appears and that, within groups, descriptions may be compared directly with one another. Any departures from consistent usage will be explicitly noted in the treatments (see also References).

Treatments are intended to reflect current knowledge of taxa throughout their ranges worldwide, and classifications are therefore based on all available evidence. Where notable differences of opinion about the classification of a group occur, appropriate references are mentioned in the discussion of the group.

Documentation and arguments supporting significantly revised classifications are published separately in botanical journals before publication of the pertinent volume of the *Flora*. Similarly, all new names and new combinations are published elsewhere prior to their use in the *Flora*. No nomenclatural innovations will be published intentionally in the *Flora*.

Taxa treated in full include extant and recently extinct native species, hybrids that are well established (or frequent), and waifs or cultivated plants that are found frequently outside cultivation and give the appearance of being naturalized. Taxa mentioned only in discussions include waifs or naturalized plants now known only from isolated old records and some nonnative, economically important or extensively cultivated plants, particularly when they are relatives of native species. Excluded names and taxa are listed at the ends of appropriate sections, e.g., species at the end of genus, genera at the end of family.

Treatments are intended to be succinct and diagnostic but adequately descriptive. Characters and character states used in the keys are repeated in the descriptions. Descriptions of related taxa at the same rank are directly comparable.

With few exceptions, taxa are presented in taxonomic sequence. If an author is unable to produce a classification, the taxa are arranged alphabetically, and the reasons are given in the discussion.

Treatments of hybrids follow that of one of the putative parents. Hybrid complexes are treated at the ends of their genera, after the descriptions of species.

We have attempted to keep terminology as simple as accuracy permits. Common English equivalents usually have been used in place of Latin or Latinized terms or other specialized terminology, whenever the correct meaning could be conveyed in approximately the same space, e.g., "pitted" rather than "foveolate," but "striate" rather than "with fine longitudinal lines." See *Categorical Glossary for the Flora of North America Project* (R. W. Kiger and D. M. Porter 2001; also available online at http://huntbot.andrew.cmu.edu) for standard definitions of generally used terms. Very specialized terms are defined, and sometimes illustrated, in the relevant family or generic treatments.

References

Authoritative general reference works used for style are *The Chicago Manual of Style*, ed. 14 (University of Chicago Press 1993); *Webster's New Geographical Dictionary* (Merriam-Webster 1988); and *The Random House Dictionary of the English Language*, ed. 2, unabridged (S. B. Flexner and L. C. Hauck 1987). *B-P-H/S. Botanico-Periodicum-Huntianum/Supplementum* (G. D. R. Bridson and E. R. Smith 1991) has been used for abbreviations of serial titles, and *Taxonomic Literature*, ed. 2 (F. A. Stafleu and R. S. Cowan 1976–1988) and its supplements by F. A. Stafleu and E. A. Mennega (1992+) have been used for abbreviations of book titles.

Graphic Elements

All genera and approximately 30 percent of the species in this volume are illustrated. Illustration panels have been enlarged for this and subsequent volumes in the series. The illustrations may show diagnostic traits or complex structures. Most illustrations have been drawn from herbarium specimens selected by the authors. In some cases living material or photographs have been used. Data on specimens that were used and parts that were illustrated have been recorded. This information, together with the archivally preserved original drawings, is deposited in the Missouri Botanical Garden Library and is available for scholarly study.

Specific Information in Treatments

Keys

Dichotomous keys are included for all ranks below family if two or more taxa are treated. For dioecious species, keys are designed for use with either staminate or pistillate plants. Keys are designed also to facilitate identification of taxa that flower before leaves appear. More than one key may be given, and for some groups tabular comparisons may be presented in addition to keys.

Nomenclatural Information

Basionyms of accepted names, with author and bibliographic citations, are listed first in synonymy, followed by any other synonyms in common recent use, listed in alphabetical order, without bibliographic citations.

Vernacular names in common use are given in the appropriate language. In general, such names have not been coined for use in the *Flora*. Those preferred by governmental or conservation agencies are listed if known.

The last names of authors of taxonomic names have been spelled out. The conventions of *Authors of Plant Names* (R. K. Brummitt and C. E. Powell 1992) have been used as a guide for including first initials to discriminate individuals who share surnames. An exception is "Alph. Wood" instead of "A. W. Wood" (Brummitt, pers. comm.).

If only one infraspecific taxon within a species occurs in the flora area, nomenclatural information (literature citation, basionym with literature citation, relevant other synonyms) is given for the species, as is information on the number of infraspecific taxa in the species and their distribution worldwide, if known. A description and detailed distributional information are given only for the infraspecific taxon.

Descriptions

Character states common to all taxa are noted in the description of the taxon at the next higher rank. For example, if flowers are unisexual for all species treated within a genus, that character state is given in the generic description. Characters used in keys are repeated in the descriptions. Characteristics are given as they occur in plants from the flora area. Characteristics that occur only in plants from outside the flora area may be given within square brackets, or instead may be noted in the discussion following the description. In families with one genus and one or more species, the family description is given as usual, the genus description is condensed, and the species are described as usual. Any special terms that may be used when describing members of a genus are presented and explained in the genus description.

In reading descriptions, the reader may assume, unless otherwise noted, that: the plants are green, photosynthetic, and reproductively mature; woody plants are perennial; stems are erect; roots are fibrous; leaves are simple and petiolate; flowers are bisexual, radially symmetric, and pediceled; perianth parts are hypogynous, distinct, and free; and ovaries are superior. Because measurements and elevations are almost always approximate, modifiers such as "about," "circa," or "±" are usually omitted.

Unless otherwise noted, dimensions are length × width. If only one dimension is given, it is length or height. All measurements are given in metric units. Measurements usually are based on dried specimens but these should not differ significantly from the measurements actually found in fresh or living material.

Chromosome numbers generally are given only if published, documented counts are available from North American material or from an adjacent region. No new counts are published intentionally in the *Flora*. Chromosome counts from nonsporophyte tissue have been converted to the **2n** form. The base number (**x** =) is given for each genus. This represents the lowest known haploid count for the genus unless evidence is available that the base number differs.

Flowering time and often fruiting time are given by season, sometimes qualified by early, mid, or late, or by months. Elevations over 50 m generally are rounded to the nearest 100 m; those 50 m and under are rounded to the nearest 10 m. Mean sea level is shown as 0 m, with the understanding that this is approximate. Elevation often is omitted from herbarium specimen labels, particularly for collections made where the topography is not remarkable, and therefore precise elevation is sometimes not known for a given taxon.

The term "introduced" is defined broadly to refer to plants that were released deliberately or accidentally into the flora and that now exist as wild plants in areas in which they were not recorded as native in the past. The distribution of non-native plants is often poorly documented and may be ephemeral.

If a taxon is globally rare or if its continued existence is threatened in some way, the words "of conservation concern" appear before the statements of elevation and geographic range.

Criteria for taxa of conservation concern are based on NatureServe's (formerly The Nature Conservancy)—see http://www.natureserve.org—designations of global rank (G-rank) G1 and G2:

G1 Critically imperiled globally because of extreme rarity (5 or fewer occurrences or fewer than 1000 individuals or acres) or because of some factor(s) making it especially vulnerable to extinction.

G2 Imperiled globally because of rarity (5–20 occurrences or fewer than 3000 individuals or acres) or because of some factor(s) making it very vulnerable to extinction throughout its range.

The occurrence of species and infraspecific taxa within political subunits of the *Flora* area is depicted by dots placed on the outline map to indicate occurrence in a state or province. For the 48 contiguous states of the United States and the smaller Canadian provinces, a single dot is used in those units where a taxon is known to occur. In the case of Greenland, the larger Canadian provinces and territories, and the main area of Alaska, a dot's position can vary to indicate more northern, southern, or central/scattered distributions (also western or eastern only for Alaska). In the case of Alaska, the occurrence of a taxon in the Aleutian Islands and/or the panhandle area adjacent to British Columbia may also be indicated. The Nunavut boundary on the maps has been provided by the GeoAccess Division, Canada Centre for Remote Sensing, Earth Science. Authors are expected to have seen at least one specimen documenting each geographic unit record and have been urged to examine as many specimens as possible from throughout the range of each taxon. Additional information about taxon distribution may be presented in the discussion.

Distributions are stated in the following order: Greenland; St. Pierre and Miquelon; Canada (provinces and territories in alphabetic order); United States (states in alphabetic order); Mexico (11 northern states may be listed specifically, in alphabetic order); West Indies; Bermuda; Central América (Belize, Costa Rica, El Salvador, Guatemala, Honduras, Nicaragua, Panama); South America; Europe, or Eurasia; Asia (including Indonesia); Africa; Pacific Islands; Australia; Antarctica.

Discussion

The discussion section may include information on taxonomic problems, distributional and ecological details, interesting biological phenomena, economic uses, and toxicity (see "Caution," below).

Selected References

Major references used in preparation of a treatment or containing critical information about a taxon are cited following the discussion. These, and other works that are referred to in discussion or elsewhere, are included in Literature Cited at the end of the volume.

CAUTION

The Flora of North America Editorial Committee **does not encourage, recommend, promote, or endorse** any of the folk remedies, culinary practices, or various utilizations of any plant described within these volumes. Information about medicinal practices and/or ingestion of plants, or of any part or preparation thereof, has been included only for historical background and as a matter of interest. Under no circumstances should the information contained in these volumes be used in connection with medical treatment. Readers are strongly cautioned to remember that many plants in the flora are toxic or can cause unpleasant or adverse reactions if used or encountered carelessly.

Key to boxed codes following accepted names:

- C of conservation concern
- E endemic to the flora area
- F illustrated
- I introduced to the flora area
- W weedy, based mostly on R. H. Callihan et al. (1995) and/or D. T. Patterson et al. (1989)

Flora of North America

43. CARYOPHYLLACEAE Jussieu

- Pink Family

Richard K. Rabeler

Ronald L. Hartman

Herbs [small trees, shrubs, or vines], winter annual, annual, biennial, or perennial, glabrous or pubescence of simple hairs or stalked glands; taprooted and/or rhizomatous with fibrous roots, sometimes from woody caudex, rhizomes rarely with tuberous thickenings. **Stems** erect to prostrate, often with swollen nodes, herbaceous. **Leaves** opposite, pseudoverticillate, whorled, or rarely alternate, distinct or connate proximally, simple; petiole often present; stipules present or absent; blade subulate to linear, spatulate to broadly ovate or suborbiculate, succulent or not, margins entire. **Inflorescences** terminal or axillary cymes, thyrses, or capitula, or flowers solitary; bracts usually paired, foliaceous or reduced, herbaceous to scarious, or absent; involucel bracteoles (epicalyces) immediately subtending calyx occasionally present. **Pedicels** present, or flowers sessile. **Flowers** bisexual or occasionally unisexual, radially symmetric; perianth and androecium hypogynous or perigynous; hypanthium, when present, urceolate, cup-, disc-, or dish-shaped, sometimes abruptly expanded distally; sepals persistent in fruit, (3–)4–5, distinct or connate proximally into cup or tube, herbaceous or scarious, apex sometimes hooded or with apical or subapical spine; petals absent or (1–)4–5, often fugacious in *Polycarpon*, distinct, often clawed, auricles present or absent, coronal appendages present or absent, blade apex entire, notched, or 2(–4)-fid, sometimes dentate or laciniate; stamens 1–10, in 1 or 2 whorls, arising from base of ovary, nectariferous disc, or hypanthium rim, absent in pistillate flowers; staminodes usually absent, or 1–10 or 16–19; ovary 1, superior, 1-locular, rarely 2-locular proximally, or 3–5 locular, placentation free-central, basal, or axile in proximal half; ovules mostly campylotropous, bitegmic, crassinucellate; styles 1–5(–6), distinct or connate proximally, absent in staminate flowers; stigmas 2–5(–6), linear along adaxial surface of styles (or style branches), subcapitate, or terminal, papillate or obscurely so, absent in staminate flowers. **Fruits** capsules, carpels opening into entire valves or valves split axially into teeth to divided to base, or a usually indehiscent utricle; carpophore sometimes present. **Seeds** 1–150(–500+), often brown or black, sometimes white or yellowish to tan, reniform or triangular to globose and often laterally compressed, sometimes shield-shaped or oblong and dorsiventrally compressed, horizontal wing sometimes present, spongy appendage (strophiole) rarely present (*Moehringia*);

embryo often peripheral, curved, surrounding the perisperm, rarely annular or central and straight; endosperm absent.

Genera 83 or 89, species ca. 3000 (37 genera, 286 species in the flora): worldwide, especially north-temperate, montane and alpine, and Mediterranean areas.

In addition to the species treated here, *Illecebrum verticillatum* Linnaeus was collected in 1912 as a weed in a plant nursery in Reading, Massachusetts. This species of Europe and the Mediterranean region resembles species of *Herniaria* or *Paronychia*; it is distinguished by minute stipules and white, spongy, awned sepals.

Caryophyllaceae includes 54 locally endemic genera (many of them in the eastern Mediterranean region of Europe, Asia, and Africa), cultivated taxa (especially *Dianthus*, *Gypsophila*, and *Silene*), and weedy taxa (mostly from Eurasia). Of the 37 genera in the flora area, 15 are entirely non-native: *Agrostemma*, *Corrigiola*, *Gypsophila*, *Holosteum*, *Lepyrodiclis*, *Moenchia*, *Myosoton*, *Petrorhagia*, *Polycarpaea*, *Polycarpon*, *Saponaria*, *Scleranthus*, *Spergula*, *Vaccaria*, and *Velezia*.

Caryophyllaceae and Molluginaceae are exceptional in Caryophyllales in possessing anthocyanin pigments rather than betalains. Evidence from morphology (e.g., pollen type, sieve-tube plastids) and molecular studies confirms this alignment. S. R. Downie et al. (1997) studied the ORF2280 plastid gene and noted that Caryophyllaceae is monophyletic and a sister group to a clade containing Chenopodiaceae + Amaranthaceae.

The number of genera recognized within Caryophyllaceae depends on adoption of a broad (as here) or narrow (see B. Oxelman et al. 2000) circumscription of *Silene*, the largest genus in the family. The number of genera increases to over 120 if, as done by some Old World authors, segregates of *Silene* and other large genera (*Arenaria*, *Gypsophila*, *Minuartia*, *Saponaria*, and *Stellaria*) are recognized. Molecular evidence reinforces recognition of six monophyletic segregates from *Silene* in the broad sense and one from *Arenaria* (*Eremogone*); further studies may provide evidence that additional groups within the larger genera are monophyletic and deserve separate generic recognition.

Interpretation of appendages, sometimes petal-like, in some members of the Paronychioideae has been debated. Some workers (e.g., V. Bittrich 1993; A. Cronquist 1981) have considered them petals; we follow L. Petrusson and M. Thulin (1996) in considering these structures to be staminodes that replace the inner whorl of stamens. When present, these structures alternate with the fertile stamens and arise together from the same hypanthial rim.

Depending on the characters emphasized, the Caryophyllaceae can be divided into three, four, or five subfamilies; we accept four in this treatment. The "traditional" division recognizes three subfamilies: Paronychioideae (stipules present, sepals distinct; genera 1–13), Alsinoideae (stipules absent, sepals mostly distinct; genera 14–29), and Caryophylloideae (stipules absent, sepals connate proximally into a cup or tube; genera 30–37). We recognize also Polycarpoideae, moving to it the paronychioid species having capsules (genera 1–7), and retaining in Paronychioideae those species with utricles (genera 8–13). This alignment, suggested by M. G. Gilbert (1987), is also consistent with the concept that some authors (e.g., J. Hutchinson 1973) have followed in segregating Paronychioideae in the narrow sense as the Illecebraceae. A. Takhtajan (1997) segregated the Scleranthoideae (*Pentastemonodiscus* and *Scleranthus*) from the Alsinoideae on the basis of their having connate sepals and a utricle but lacking stipules. While not addressing Takhtajan's concept directly, R. D. Smissen et al. (2003) showed that molecular data favor continued placement of *Scleranthus* within the Alsinoideae. Results from molecular studies (M. Nepokroeff et al. 2002; Smissen et al. 2002) have suggested that the traditional morphological characters may not reflect the phylogeny within the family.

SELECTED REFERENCES   Behnke, H.-D. and T. J. Mabry, eds. 1994. Caryophyllales: Evolution and Systematics. Berlin. Bittrich, V. 1993. Caryophyllaceae. In: K. Kubitzki et al., eds. 1990+. The Families and Genera of Vascular Plants. 4+ vols. Berlin, etc. Vol. 2, pp. 206–236. Hartman, R. L. 1971. The Family Caryophyllaceae in Wyoming. M.S. thesis. University of Wyoming. Hartman, R. L. 1972. [Flora of Wyoming] Caryophyllaceae. Res. J. Wyoming Agric. Exp. Sta. 64: 14–45. Rabeler, R. K. 2004. Caryophyllaceae (pink family). In: N. P. Smith et al., eds. 2004. Flowering Plants of the Neotropics. Princeton. Pp. 88–90.   Rabeler, R. K. and J. W. Thieret. 1988. Comments on the Caryophyllaceae of the southeastern United States. Sida 13: 149–156.

1. Leaves with stipules present, stipules ovate to spatulate, triangular, or bristlelike, mostly scarious.
    2. Fruits utricles (± modified in *Achyronychia*, *Scopulophila*); seed 1; petals absent; staminodes petaloid or scalelike; flowers perigynous; stamens arising from hypanthium rim [43b. subfam. Paronychioideae].
        3. Leaves alternate or rarely subopposite; staminodes petaloid, ovate to oblong . . . . 13. *Corrigiola*, p. 48
        3. Leaves opposite (or occasionally distalmost alternate in *Herniaria*); staminodes not petaloid (narrowly lanceolate in *Scopulophila*), subulate to narrowly oblong.
            4. Sepals broadly ovate to reniform, margins white, scarious, awns absent; staminodes 14–19 and filiform or 5 and petaloid, lanceolate.
                5. Plants annual, base ± glabrous; stems prostrate to ascending; staminodes 14–19, filiform; styles 2-branched . . . . 11. *Achyronychia*, p. 46
                5. Plants perennial, base densely woolly; stems ascending to erect; staminodes 5, oblong, petaloid; styles 3-branched . . . . 12. *Scopulophila*, p. 47
            4. Sepals lanceolate to oblong (if ovate or obovate, then margins scarious and awns present or margins herbaceous and awns absent), margins herbaceous or scarious; staminodes 4–5, filiform or subulate, or absent.
                6. Stipules inconspicuous; sepals obtuse, awnless, margins green, herbaceous . . . . 9. *Herniaria*, p. 43
                6. Stipules often conspicuous; sepals hooded or awned, margins white, scarious.
                    7. Sepal awns very stout, spinose, 1.5–4 mm; flowers densely woolly . . . . 10. *Cardionema*, p. 45
                    7. Sepals awnless or awns threadlike to somewhat stout, not especially spiny, 0.3–1.5(–2.1) mm; flowers not woolly, hairs ± straight or tips coiled . . . . 8. *Paronychia*, p. 30
    2. Fruits capsules; seeds 3–150+; petals rarely absent or replaced by staminodes; flowers hypogynous (stamens arising from ovary base) or sometimes perigynous (stamens arising from rim of hypanthium or disc surrounding ovary base) [43 a. subfam. Polycarpoideae, in part].
        8. Basal leaf blades spatulate to obovate, stipules in tuft of 2–14+, filiform, scarious; cauline leaf blades subulate to triangular, scalelike, stipules forming incised or notched nodal fringe . . . . 7. *Stipulicida*, p. 27
        8. Basal leaves present or absent, if blade spatulate to ovate, then stipules 2 or 4 per node, not as a tuft; cauline leaf blades subulate to linear, lanceolate, ovate, or orbiculate, not scalelike, stipules bristlelike to triangular, not forming nodal fringe.
            9. Styles 1, or (2–)3, connate proximally for ½ or more of length or rarely distinct.
                10. Petal blades divided into 2–4 lobes; styles 2(–3), connate proximally for ½ or more of length or rarely distinct . . . . 1. *Drymaria* (in part), p. 9
                10. Petal blades entire, emarginate, or irregularly toothed; style 1.
                    11. Sepals herbaceous, margins white, apex acuminate to aristate, obscurely to distinctly keeled . . . . 5. *Polycarpon*, p. 25
                    11. Sepals scarious, central portion silvery with brown midrib, margins silvery, apex acute, not keeled . . . . 4. *Polycarpaea*, p. 23
            9. Styles 3 or 5, occasionally 2, distinct or nearly so.
                12. Leaf blades broadly ovate to orbiculate; petal blades with apex divided into 2 linear lobes . . . . 1. *Drymaria* (in part), p. 9

12. Leaf blades linear or subulate to oblong; petal blades with apex entire or petals rudimentary or absent.

    13. Leaves opposite but appearing whorled, as 8–15 per axillary cluster, 2 clusters per node; styles and capsule valves 5 . . . . 2. *Spergula*, p. 14

    13. Leaves opposite, axillary leaf clusters often present; styles and capsule valves 3.

        14. Flowers axillary, 1(–2); sepals acute to spinose, usually with 2 filamentous to stiff lateral spurs; petals absent or rudimentary; stipules bristlelike . . . . 6. *Loeflingia*, p. 26

        14. Flowers in terminal cymes, or sometimes solitary, axillary; sepals acute to obtuse, lateral spurs absent; petals present; stipules lanceolate to widely triangular . . . . 3. *Spergularia*, p. 16

1. Leaves with stipules absent.

  15. Sepals connate ($^1/_4$–)$^1/_2$+ of length, forming cup or usually tube; petals white to pink, red, or purple (rarely absent); perianth hypogynous [43d. subfam. Caryophylloideae].

    16. Flowers immediately subtended, or sometimes partially concealed, by 2–6 appressed involucel bracteoles.

      17. Calyx veins 15–40, commissure between sepals absent. . . . . 33. *Dianthus*, p. 159

      17. Calyx veins 15, commissure between sepals scarious . . . . 34. *Petrorhagia*, p. 162

    16. Flowers not subtended by appressed involucel bracteoles.

      18. Sepals 25–62 mm, lobes longer than calyx tube, linear-lanceolate, often equaling or longer than petals . . . . 37. *Agrostemma*, p. 214

      18. Sepals (1–)10–28(–40) mm, lobes usually shorter than calyx tube or equaling or exceeding 1–5 mm cup (*Gypsophila*), subulate to triangular, lanceolate-acuminate, ovate, or elliptic, shorter (very rarely longer) than petals.

        19. Styles 3 or 5, occasionally 4 (absent in staminate flowers); fruit valves 3–5 or splitting into 6–10 teeth . . . . 36. *Silene*, p. 166

        19. Styles 2(–3); fruit valves usually 4; plants not dioecious.

          20. Stamens 5; sepals 11–15 mm; calyx tubes narrowly cylindric; flowers usually axillary, solitary. . . . . 35. *Velezia*, p. 166

          20. Stamens 10; sepals 1–25 mm, calyx tubes campanulate to cylindric; flowers in open to subcapitate, terminal cymes.

            21. Sepals 1–5 mm, forming obconic to campanulate cup, commissures between sepals scarious. . . . . 30. *Gypsophila*, p. 153

            21. Sepals 7–25 mm, forming cylindric to ovoid tube, commissures between sepals absent.

              22. Plants perennial; inflorescences congested or open; pedicels 1–6 mm; calyx tubes terete; coronal appendages 2 . . . . . . 32. *Saponaria*, p. 157

              22. Plants annual; inflorescences open; pedicels (5–)10–30(–55) mm; calyx tubes 5-angled or keeled; coronal appendages absent . . . . 31. *Vaccaria*, p. 156

  15. Sepals distinct or seldom basally connate; petals usually white, sometimes yellowish, pink, or brown (sometimes absent); perianth hypogynous or sometimes perigynous [43c. subfam. Alsinoideae].

    23. Fruits utricles enclosed by persistent hypanthium; flowers inconspicuous, sessile or subsessile; petals absent . . . . 29. *Scleranthus*, p. 149

    23. Fruits capsules; hypanthium if present, not enclosing fruit; flowers mostly conspicuous, mostly pedicellate; petals present or absent.

      24. Petals absent or rudimentary.

        25. Capsules cylindric, opening by 8 or 10 teeth.

          26. Capsules often curved, usually opening by 10 teeth, longer than sepals; styles usually 5 . . . . 18. *Cerastium* (in part), p. 74

          26. Capsules straight, usually opening by 8 teeth, equaling or shorter than sepals; styles usually 4 . . . . 19. *Moenchia* (in part), p. 93

25. Capsules ovoid to globose, opening by valves or teeth.
    27. Capsule valves or teeth 2 times number of styles.
        28. Styles 3(–5); capsule valves (3, 4) 6 (8 or 10); seeds yellowish to
            brown, rarely black, if shiny, then surface tuberculate ... 22. *Stellaria* (in part), p. 96
        28. Styles 3; capsule valves or teeth 6; seeds dark brown or black,
            shiny, smooth or at most obscurely tuberculate .... 14. *Arenaria* (in part), p. 51
    27. Capsule valves equal in number to styles.
        29. Sepals 4 or 5; styles 4 or 5; capsule valves 4 or 5 ..... 27. *Sagina* (in part), p. 140
        29. Sepals 5; styles 3(–4); capsule valves 3.
            30. Flowers pedicellate; stamens 10; staminodes absent ......
                .... 24. *Minuartia* (in part), p. 116
            30. Flowers sessile; stamens 5, alternating with sepals; staminodes
                5, scalelike .... 28. *Geocarpon*, p. 148
    [24. Shifted to left margin.—Ed.]
24. Petals present.
  31. Petal apices 2-lobed or 2-fid, often divided nearly to base, or 4-fid.
    32. Plants prostrate or sprawling, glabrous, glaucous, and succulent; leaves appearing
      whorled; petals 4-fid; inflorescences umbelliform [43a. subfam. Polycarpoideae, in
      part] .... 1. *Drymaria* (in part), p. 9
    32. Plants, if prostrate, not glabrous, glaucous, succulent; leaves opposite; petals 2-
      lobed or 2-fid; inflorescences rarely umbelliform.
      33. Petals 2-lobed for $^1/_{10}$–$^1/_5$ their length; fruits ovoid, opening by 6, ca. 2–3 times
        recoiled valves; seeds 1–3; rhizomes usually with tuberous thickenings or verti-
        cal fleshy root.................................... 23. *Pseudostellaria*, p. 114
      33. Petals 2-lobed or 2-fid nearly to base; fruits ovoid or globose to cylindric, open-
        ing by teeth or valves, valves not recoiled; seeds 1–100; rhizomes, when present,
        without tuberous thickenings or vertical fleshy roots.
        34. Capsules cylindric, often curved, opening by 6, 8, or 10 teeth .... 18. *Cerastium* (in part), p. 74
        34. Capsules ovoid to globose, opening by valves.
            35. Leaf blades filiform to subulate, congested at or near base of flowering
              stems, apex apiculate or sharp-spinose .... 15. *Eremogone* (in part), p. 56
            35. Leaf blades linear or lanceolate to ovate or deltate; apex acute to ob-
              tuse.
                36. Pedicels minutely glandular-pilose; styles 5(–6); capsules opening
                  by 5(–6) shallowly 2-fid valves .... 21. *Myosoton*, p. 95
                36. Pedicels glabrous or pubescent, not glandular; styles 3(–5); cap-
                  sules opening by (3, 4) 6 (8 or 10) valves .... 22. *Stellaria* (in part), p. 96
  31. Petal apices entire, emarginate, jagged, or notched.
    37. Leaf blades conspicuously fleshy; plants maritime, found on coastal beaches
      .... 26. *Honckenya*, p. 137
    37. Leaf blades herbaceous to slightly succulent; plants of various nonmaritime habi-
      tats.
      38. Styles and capsule valves equal in number.
        39. Styles 2; capsule valves 2; stems ascending to sprawling, 40–100 cm; dis-
          turbed areas, agricultural fields .... 17. *Lepyrodiclis*, p. 72
        39. Styles 3–5; capsule valves 3–5; stems erect to ascending, 3–30(–55) cm;
          rarely in disturbed areas.
            40. Sepals 4 or 5; styles 4 or 5; capsule valves 4 or 5 .... 27. *Sagina* (in part), p. 140
            40. Sepals 5; styles 3, occasionally 4; capsule valves or teeth 3.

41. Leaf blades filiform-linear to subulate, lanceolate, or oblanceolate, rarely to ovate; capsules not inflated, 1-locular, opening by 3 teeth or valves . . . . 24. *Minuartia* (in part), p. 116

41. Leaf blades ovate to elliptic; capsules inflated, 3-locular below, opening along false septa, sometimes irregularly, into 3, 2-toothed parts . . . . 25. *Wilhelmsia*, p. 136

[38. Shifted to left margin.—Ed.]

38. Capsule valves or teeth 2 times number of styles.

42. Capsules cylindric, opening by 6, 8, or 10 teeth.

43. Cymes umbellate (sometimes not yet developed in young individuals, then with 2–4 bracts and often buds at summit of peduncle); petal blades with apex jagged; seeds shield-shaped . . . . 20. *Holosteum*, p. 94

43. Cymes repeatedly branched; petal blades with apex entire, emarginate, or notched; seeds angular-obovate or reniform.

44. Petals (4–)5, blade apex shallowly emarginate or notched; capsules longer than sepals . . . . 18. *Cerastium* (in part), p. 74

44. Petals 4, blade apex entire; capsules shorter than or equaling sepals . . . . 19. *Moenchia* (in part), p. 93

42. Capsules ovoid to urceolate or globose, opening by 6 valves or teeth.

45. Leaf blades needlelike or filiform to subulate or narrowly linear, usually congested at or near base of flowering stems, apex blunt to usually apiculate or spinose . . . . 15. *Eremogone* (in part), p. 56

45. Leaf blades ovate to lanceolate (sometimes narrowly so), not congested at or near base of flowering stems, apex blunt to acute or apiculate (rarely spinose).

46. Seeds with appendage present, elliptic, white, spongy . . . . 16. *Moehringia*, p. 70

46. Seeds with appendage absent . . . . 14. *Arenaria* (in part), p. 51

43a. Caryophyllaceae Jussieu subfam. Polycarpoideae Tanfani in F. Parlatore, Fl. Ital. 9: 623. 1892 (as Polycarpinee)

Richard K. Rabeler

Ronald L. Hartman

Herbs [**small shrubs**], winter annual, annual, or perennial; taprooted, not rhizomatous. **Stems** prostrate to erect, simple or branched. **Leaves** opposite or sometimes appearing whorled, bases connate or not, sometimes petiolate or often sessile, stipulate; stipules ovate or deltate to lanceolate or bristlelike, scarious; blade subulate or subtriangular to linear and threadlike or spatulate to ovate or orbiculate, seldom succulent. **Inflorescences** terminal or axillary cymes, or flowers solitary; bracts scarious or absent; involucel bracteoles absent. **Pedicels** present or flowers sessile. **Flowers** bisexual or rarely unisexual; perianth and androecium hypogynous or perigynous; hypanthium absent or dish- or cup-shaped; sepals 5, distinct or sometimes connate proximally, hooded (*Drymaria, Polycarpon*) or not, awned (*Polycarpon*) or not; petals absent or (3–)5, blade clawed (*Drymaria*) or not, auricles absent, coronal appendages absent, blade apex entire, erose, or 2(–4)-fid, sometimes emarginate; stamens (1–)3–5(–10), in 1 or 2 whorls, usually arising from base of ovary or from rim of hypanthium (*Spergula, Spergularia*); staminodes absent; ovary 1-locular; styles 1 or 3, occasionally 2 or 5, distinct or sometimes connate proximally; stigmas 3, occasionally 2 or 5. **Fruits** capsules, opening by 3 or 5, occasionally 4 valves; carpophore sometimes present. **Seeds** 3–150+, whitish or tan to often brown or black, ± triangular, pyriform, or reniform to circular, subglobose or laterally compressed to angular (*Polycarpon*); embryo peripheral, curved or rarely annular to spirally curved (*Spergula*). x = [7], 8, 9, (11), 12.

Genera 16, species ca. 210 (7 genera, 28 species in the flora): w North America (including Mexico), w South America, Europe (Mediterranean region), Asia (Mediterranean region), Africa (Mediterranean and tropical regions), less diverse in temperate areas.

More commonly recognized as a tribe, Polycarpoideae is characterized by the presence of stipules, well-developed petals, and capsules. It is a relatively small group, with about four-fifths of the species in *Drymaria*, *Polycarpaea*, and *Spergularia*. Although clearly stipulate, *Spergularia* clusters with members of Alsinoideae in recent molecular studies (M. Nepokroeff et al. 2002; R. D. Smissen et al. 2002).

1. DRYMARIA Willdenow ex Schultes in J. J. Roemer et al., Syst. Veg. 5: xxxi, 406. 1819 • Drymary [Greek *drymos*, forest, alluding to habitat of at least one species]

Ronald L. Hartman

Herbs, annual or perennial, caudices often branched. **Taproots** slender, elongate. **Stems** sprawling to erect, simple or branching proximally or throughout, terete. **Leaves** opposite or appearing whorled, connate by membranous to thickened line, petiolate or sessile, stipulate (*D. pachyphylla* not stipulate); stipules 2 per node, white to tan, simple or divided into segments, subulate to filiform, often minute, margins entire, apex acute to acuminate; blade 1–5-veined, linear to lanceolate, spatulate, ovate, reniform, or orbiculate, not succulent, apex rounded to acuminate. **Inflorescences** terminal or axillary, open to congested, bracteate cymes or umbelliform clusters or flowers solitary, axillary; bracts paired, scarious or central portion herbaceous. **Pedicels** erect to spreading or reflexed. **Flowers:** perianth and androecium hypogynous; sepals 5, distinct, white, lanceolate to oblong, ovate, or orbiculate, 1.5–4.8(–5) mm, herbaceous, margins white to purple, scarious, apex acuminate to rounded, hooded or not; petals (3–)5, sometimes absent, white, claw narrow, tapering distally or with oblong or expanded, sessile or short-clawed trunk, auricles absent, blade apex divided into 2 or 4 lobes; nectaries at base of filaments opposite sepals; stamens 5; filaments distinct or briefly connate proximally; styles 3, occasionally 2, connate proximally for 1/2 of length, rarely to nearly distinct (*D. cordata*), filiform, 0.1–0.3 mm, glabrous proximally; stigmas 3, occasionally 2, linear along adaxial surfaces of styles (or branches), obscurely papillate (30×). **Capsules** ellipsoid to globose, opening by (2–)3 spreading to recurved valves; carpophore absent. **Seeds** 3–25, tan, reddish brown, dark brown, black, or transparent (white embryo visible), horseshoe-, snail-shell- or teardrop-shaped, compressed laterally, at least somewhat, tuberculate, marginal wing absent, appendage absent. x = (11), 12.

Species 48 (9 in the flora): sw United States, Mexico, West Indies, Central America, South America; introduced in Asia (Indonesia), e, s Africa, Australia, Pacific Islands.

Drymaria arenarioides Willdenow ex Schultes, alfombrilla, is on the Federal Noxious Weed List. It is highly toxic to livestock and is native in northwestern Mexico, where it has been reported within a few miles of the United States border.

J. A. Duke (1961) proposed an infraspecific classification for *Drymaria* consisting of 17 "informal" series that he did not validly publish. Some of Duke's series were cited by M. Escamilla and V. Sosa (2000), apparently assuming that they were "real" series.

SELECTED REFERENCE Duke, J. A. 1961. Preliminary revision of the genus *Drymaria*. Ann. Missouri Bot. Gard. 48: 173–268.

1. Blades of cauline leaves ovate, deltate, cordate, reniform, orbiculate, or suborbiculate, 4–25 mm wide, base obtuse or cordate to rounded.
    2. Stems erect or ascending; petals 2-fid, lobe apex not notched; seeds 0.5–0.7 mm .... 4. *Drymaria glandulosa*
    2. Stems prostrate or sprawling; petals 2- or 4-fid, if 2-fid, lobe apex distinctly notched or seeds 1–1.5 mm.
        3. Plants glabrous, glaucous, succulent; petals 4-fid; leaves appearing whorled, not stipulate; inflorescences umbelliform clusters .... 8. *Drymaria pachyphylla*
        3. Plants often glandular in inflorescence, not glaucous, herbaceous; petals 2-fid; leaves opposite, stipules present or deciduous in part; inflorescences open cymes.
            4. Petals with lobe apex rounded to acute, 1-veined, vein not branched; seeds 1–1.5 mm, tubercles rounded .... 1. *Drymaria cordata*
            4. Petals with lobe apex distinctly notched, 1-veined, vein dichotomously branched; seeds 0.5–0.7 mm, tubercles conic .... 5. *Drymaria laxiflora*
1. Blades of cauline leaves linear to oblong or lanceolate, 0.2–3 mm wide, base briefly obtuse to attenuate.
    5. Cauline leaves appearing whorled, at least in part.
        6. Plants viscid; stems prostrate; sepals stipitate-glandular; inflorescences 4–7-flowered, axillary and terminal cymes .... 9. *Drymaria viscosa*
        6. Plants glabrous or sparsely glandular; stems erect; sepals glabrous; inflorescences 3–30+-flowered, terminal cymes initially becoming racemose .... 7. *Drymaria molluginea*
    5. Cauline leaves opposite.
        7. Petals 1 1/4–1 1/2 times as long as sepals .... 3. *Drymaria effusa*
        7. Petals 1/2–1 times as long as sepals.
            8. Herbaceous portion of sepals ± oblong, apex blunt or rounded, veins ± parallel, apically confluent .... 2. *Drymaria depressa*
            8. Herbaceous portion of sepals lanceolate, apex acute to acuminate, veins with lateral pair distinctly arcing outward .... 6. *Drymaria leptophylla*

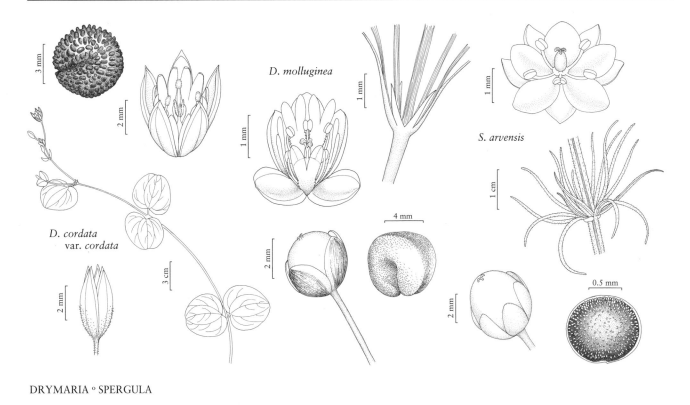

D. molluginea

S. arvensis

D. cordata
var. *cordata*

DRYMARIA ° SPERGULA

1. Drymaria cordata (Linnaeus) Willdenow ex Schultes in J. J. Roemer et al., Syst. Veg. 5: 406. 1819 F I W

Holosteum cordatum Linnaeus, Sp. Pl. 1: 88. 1753

Varieties 2 (1 in the flora): introduced; Mexico, West Indies, Central America, South America (to Argentina); introduced in Africa, Asia, Pacific Islands.

Contrary to J. A. Duke (1961), it appears best to consider *Drymaria cordata* as introduced in North America. The earliest collections were made in Florida in the early 1900s.

1a. Drymaria cordata (Linnaeus) Willdenow ex Schultes var. **cordata** F I W

Plants annual, herbaceous, glabrate to glandular-puberulent, often densely so in inflorescences, not glaucous. Stems prostrate and spreading, branched, rooting at some nodes, 20–60+ cm. Leaves opposite; stipules persistent, divided into 3–5 subulate to fili-form segments, 1–2 mm; petiole 2–15 mm; blade orbiculate to reniform, 0.5–2.5 cm × 5–25 mm, base cordate to rounded, apex rounded to acute. Inflorescences axillary and terminal, mostly open, 5–15+-flowered cymes. Pedicels shorter to longer than subtending bracts at maturity. Flowers: sepals with weak veins arcing outward at midsection and ± confluent

apically, lanceolate to ovate (herbaceous portion similar), 2–4(–5) mm, ± equal, apex acute to acuminate (herbaceous portion similar), not hooded, glabrous or with ± sessile glands; petals 2-fid for $^3/_4$+ their total length, 2–3 mm, $^1/_2$–$^2/_3$ times as long as sepals, lobes 1-veined, vein unbranched, linear, trunk absent, base abruptly tapered, apex rounded to acute. Seeds dark reddish brown, snail-shell-shaped, 1–1.5 mm; tubercles prominent, rounded. $2n$ = 24 (Africa, India), 36 (South America, Taiwan).

Flowering year-round. Lawns, gardens, disturbed areas; 10–30 m; introduced; Fla., Ga., La., Miss.; Mexico; West Indies; Central America; South America; introduced in tropical areas of the Old World.

2. Drymaria depressa Greene, Leafl. Bot. Observ. Crit. 1: 153. 1905

Drymaria effusa A. Gray var. *depressa* (Greene) J. A. Duke

Plants annual, herbaceous, gla-brous or minutely puberulent, not glaucous. Stems ascending to erect, generally branching at base, 0.5–5 cm. Leaves opposite; stipules ± deciduous, simple, subulate, 0.5–1.2 mm; petiole absent or nearly so; blade orbiculate to spatulate (basal leaves) or oblong (cauline leaves), 0.3–1 cm × 0.2–3 mm, base attenuate, apex rounded to acute. **Inflorescences** terminal, congested to open, 3–25+-flowered cymes. **Pedicels** mostly shorter than subtending bracts at maturity. **Flowers:** sepals with 3 prominent, ± parallel, apically confluent veins, lanceolate, oblong, or ovate (herbaceous portion ± oblong), 1.8–2.3 mm, ± equal, apex blunt or rounded, hood at ± right angle to apex, formed in part by scarious margins, glabrous; petals 2-fid for ca. ¹/₂ their length, 1.5–2.8 mm, ³/₄–1 times as long as sepals, lobes 1-veined, vein unbranched, linear, trunk absent, base gradually tapered, apex rounded. **Seeds** light reddish brown to tan, snail-shell- to teardrop-shaped, 0.5–0.6 mm; tubercles minute, rounded.

Flowering late summer–early fall. Rocky or gravelly slopes, disturbed areas in pine or aspen woodland; 2200–3100 m; Ariz., Colo., N.Mex.; Mexico; Central America.

Drymaria depressa has been treated as a variety of *D. effusa*, from which it differs in a number of respects, as indicated in the key.

3. Drymaria effusa   A. Gray, Smithsonian Contr. Knowl. 5(6): 19. 1853

Varieties 2 (1 in the flora): Arizona, Mexico.

3a. Drymaria effusa   A. Gray var. **effusa**

Plants annual, herbaceous, minutely glandular-puberulent or glabrate, not glaucous. **Stems** mostly erect, usually unbranched at base, 0.8–25 cm. **Leaves** sometimes appearing whorled proximally, opposite distally; stipules ± deciduous, simple, subulate, 0.5–1.2 mm; petiole absent or nearly so; blade orbiculate or spatulate (basal leaves) or linear to oblong (cauline leaves), 0.5–2.5 cm × 0.2–1.2 mm, base briefly obtuse or attenuate, apex rounded to apiculate. **Inflorescences** terminal, congested or open, 7–50+-flowered cymes. **Pedicels** shorter to longer than subtending bracts at maturity. **Flowers:** sepals with 3 prominent veins arcing outward at midsection and confluent apically, broadly ovate to elliptic (herbaceous portion narrowly elliptic to ovate), 1.5–2.3 mm, unequal, outer 2 shorter (± 0.5 mm) than inner, apex acute or obtuse (herbaceous portion acute to acuminate), hood oblique to apex, on outer 2 or all 5 sepals formed in part by scarious margins, glabrous or glandular; petals 2-fid for ca. ¹/₂+ their length, 1.5–4 mm, 1¹/₄–1¹/₂ times as long as sepals, lobes 1-veined, vein unbranched, linear, trunk absent, base gradually tapered, apex rounded. **Seeds** brown, snail-shell- to teardrop-shaped, 0.6–0.7 mm; tubercles minute, rounded.

Flowering late summer–early fall. Rocky or gravelly flats or slopes, disturbed areas in pine or oak woodland; 1400–1900 m; Ariz.; n Mexico (Sinaloa, Sonora).

4. Drymaria glandulosa   Bartling in C. B. Presl, Reliq. Haenk. 2: 9. 1831

Varieties 2 (1 in the flora): sw United States, Mexico.

4a. Drymaria glandulosa   Bartling var. **glandulosa**

Drymaria fendleri S. Watson

Plants annual or perennial, herbaceous, pubescent to stipitate-glandular (especially in inflorescences), not glaucous. **Stems** erect or ascending, simple or sparingly branched throughout, 5–35 cm. **Leaves** opposite; stipules persistent, divided into 2 filiform segments, 0.5–1 mm; petiole 1–8 mm; blade ovate to reniform, 0.5–1.5(–2) cm × 4–17(–25) mm, base truncate to cordate, apex acute to cuspidate. **Inflorescences** terminal, usually congested, 3–15-flowered clusters terminating branches of cymes. **Pedicels** shorter to longer than subtending bracts at maturity. **Flowers:** sepals with 3 distinct, usually prominent veins arcing outward at midsection and ± confluent apically, lanceolate (herbaceous portion similar), 3–4.8 mm, subequal or outer sepals shorter than inner, apex acute to setaceous-acuminate (herbaceous portion similar), not hooded, glabrous or stipitate-glandular; petals 2-fid for ¹/₂+ their length, 1.2–3.2 mm, equaling or shorter than sepals, lobes 1-veined, vein unbranched, lanceolate, trunk absent, base gradually tapered, apex ± rounded. **Seeds** tan to reddish brown, snail-shell-shaped, 0.5–0.7 mm; tubercles prominent, rounded.

Flowering late summer–early fall. Woodlands of mountainous areas; 1500–2300 m; Ariz., N.Mex.; Mexico (Chihuahua, Coahuila, Nuevo León, San Luis Potosí, Sonora).

5. Drymaria laxiflora   Bentham, Pl. Hartw., 73. 1839

Drymaria chihuahuensis Briquet

Plants annual or perennial, herbaceous, glabrous or densely stipitate-glandular, not glaucous. **Stems** sprawling, branching proximally, 10–30+ cm. **Leaves** opposite; stipules ± persistent, divided into 2 or 3 filiform segments, 0.5–3.5 mm; petiole (1–)3–6 mm; blade broadly ovate, deltate, cordate, or reniform, (0.2–)0.5–1.6 cm × 4–12 mm, base truncate to rounded, apex acute to cuspidate. **Inflorescences** terminal, open, 3–15-flowered cymes. **Pedicels** shorter to longer than subtending bracts at maturity. **Flowers:** sepals with 3 distinct veins arcing outward at midsection and ± confluent apically, lance-ovate to elliptic (herbaceous portion linear-lanceolate to narrowly ovate), 3–4(–5) mm, subequal, apex acute to acuminate (herbaceous portion often obtuse), not hooded, glabrous; petals 2-fid for $1/2$–$3/4$ their length, 2.7–6 mm, subequal to sepals, lobes 1-veined, vein dichotomously branched distally, oblong, trunk entire, base abruptly tapered, apex deeply notched. **Seeds** dark brown to blackish, snail-shell-shaped, 0.5–0.7 mm; tubercles prominent, conic.

Flowering spring–late fall. Montane chaparral and woodlands, igneous-rock mountains; 1400–2500 m; Tex.; Mexico (Chihuahua, Coahuila, San Luis Potosí, Zacatecas).

6. **Drymaria leptophylla** (Chamisso & Schlechtendal) Fenzl ex Rohrbach, Linnaea 37: 195. 1871

Arenaria leptophylla Chamisso & Schlechtendal, Linnaea 5: 233. 1830

Varieties 3 (1 in the flora): sw United States, Mexico.

6a. **Drymaria leptophylla** (Chamisso & Schlechtendal) Fenzl ex Rohrbach var. **leptophylla**

Drymaria tenella A. Gray

Plants annual, herbaceous, glabrous, not glaucous. **Stems** erect, sparingly branched, 8–25 cm. **Leaves** mostly opposite; stipules ± deciduous, divided into 2 filiform segments, 0.3–1 mm; petiole absent or nearly so; blade linear to narrowly oblong, 0.5–2.5 cm × 0.2–1.2 mm, base attenuate, apex rounded to apiculate. **Inflorescences** terminal, open, (5–)15–75-flowered cymes. **Pedicels** shorter to longer than subtending bracts at maturity. **Flowers:** sepals with 3 prominent veins arcing outward at midsection and confluent apically, lanceolate to ovate (herbaceous portion lanceolate), 1.8–3.5 mm, unequal, with outer 2 often shorter (± 0.5 mm) than inner, apex acute to acuminate (herbaceous portion similar), hood oblique to apex, at least weakly developed on outer 2 sepals, glabrous or sessile-glandular; petals 2-fid for $1/2$ or less their length, 1.3–2.6 mm, $1/2$–1 times as long as sepals, lobes 1-veined, vein unbranched, linear, trunk absent, base gradually tapered, apex rounded. **Seeds** tan, snail-shell-shaped, lateral thickenings absent, 0.6–0.7 mm; tubercles minute, rounded. *2n* = 36.

Flowering late summer–early fall. Rocky or gravelly flats or slopes, disturbed areas in pine or oak woodland; 1700–2400 m; Ariz., Colo., N.Mex., Tex.; Mexico (Baja California, Chihuahua, San Luis Potosí, Sonora).

7. **Drymaria mulluginea** (Seringe) Didrichsen, Linnaea 29: 738. 1858 F

Arenaria mulluginea Seringe in A. P. de Candolle and A. L. P. P. de Candolle, Prodr. 1: 400. 1824; *Alsine mulluginea* Lagasca, Gen. Sp. Pl., 13. 1816, not Hornemann 1813; *Drymaria sperguloides* A. Gray

Plants annual, herbaceous, glabrous or sparsely glandular, not glaucous. **Stems** erect, simple or dichotomously branched, 3–25(–30) cm. **Leaves** mostly appearing whorled; stipules ± persistent, simple, filiform to subulate, 0.5–2 mm; petiole absent or nearly so; blade linear, 1–3(–3.5) cm × 0.5–1.5(–2) mm, base attenuate, apex rounded, sometimes apiculate. **Inflorescences** terminal or racemose, open, 3–30+-flowered cymes. **Pedicels** shorter to longer than subtending bracts at maturity. **Flowers:** sepals with midvein prominent, lateral pair often evident, then arcing outward at midsection and confluent apically, broadly ovate to ± orbiculate (herbaceous portion elliptic to oblong or lanceolate), 2–3.5 mm, subequal, apex obtuse (herbaceous portion similar), hood oblique or at right angles to apex, formed in part by scarious margins, glabrous; petals 4-fid for $1/2$ or less their length, 1.7–2.5 mm, subequal to sepals, lobes 1-veined, vein unbranched, linear, outer pair $1/3$–$1/2$ as long as petal trunk, inner pair slightly shorter or, sometimes, greatly reduced or absent, trunk laterally denticulate, base abruptly tapered, apex rounded. **Seeds** dark brown to purplish, horseshoe-shaped, 0.8–0.9 mm; tubercles minute, rounded.

Flowering late summer–early fall. Rocky to sandy soil; 2100–2400 m; Ariz., N.Mex., Tex.; Mexico.

8. **Drymaria pachyphylla** Wooton & Standley, Contr. U.S. Natl. Herb. 16: 121. 1913 • Thickleaf drymary, inkweed

Plants annual, succulent, glabrous, glaucous. **Stems** nearly prostrate, radiating pseudo-verticillately from base, 10–20 cm. **Leaves** appearing whorled; not stipulate; petiole 2–8 mm; blade ovate to suborbiculate, (0.2–)0.5–1.3 cm × 4–10 mm, base obtuse to rounded, apex ± obtuse. **Inflor-**es axillary, congested, 3–12-flowered umbelliform clusters. **Pedicels** shorter to longer than subtending bracts at maturity. **Flowers:** sepals with 3 or 5 obscure veins usually not confluent apically, oblong to broadly elliptic (herbaceous portion similar), 2–3.5 mm, subequal, apex obtuse (herbaceous portion generally acute), not hooded, glabrous; petals 4-fid for $1/2$ or less their length, 2.5–3 mm, $^2/_3$–1 times as long as sepals, lobes 1-veined, vein unbranched, linear, outer pair $^1/_2$ length of petal, each with narrower, slightly shorter lobe on inner flank, trunk laterally denticulate, base abruptly tapered, apex ± rounded. **Seeds** olive green to black, teardrop-shaped (with elongate or crescent-shaped lateral thickening), 1.1–1.3 mm; tubercles marginal, minute, elongate.

Flowering spring–late summer. Heavy, saline soils, desert flats, river bottoms, playa margins; 1200–1500 m; Ariz., N.Mex., Tex.; Mexico (Chihuahua, Coahuila, Durango, Nuevo León).

Drymaria pachyphylla is highly toxic to livestock.

9. Drymaria viscosa S. Watson, Proc. Amer. Acad. Arts 22: 469. 1887

Plants annual, herbaceous, viscid, not glaucous. **Stems** prostrate, diffusely branched proximally, 5–20 cm. **Leaves** appearing whorled or opposite; stipules deciduous, simple, filiform, 1–1.5 mm; petiole absent or nearly so; blade linear to narrowly lanceolate, 0.3–1.5 cm × 0.5–1.5 mm, base attenuate, apex rounded to acute. **Inflorescences** axillary and terminal, somewhat congested, 4–7-flowered cymes.

Pedicels shorter to longer than subtending bracts at maturity. **Flowers:** sepals with 3 obscure veins arcing outward at midsection and ± confluent apically, lanceolate to narrowly ovate (herbaceous portion similar), 2.3–3 mm, subequal, apex obtuse (herbaceous portion acute to acuminate), not hooded, stipitate-glandular; petals 2-fid for $^1/_2$+ their length, 1.8–2.2 mm, $^2/_3$–1 times sepals, lobes 1-veined, vein unbranched, spatulate, trunk absent, base gradually tapered, apex rounded. **Seeds** brown abaxially, transparent (or white embryo visible) adaxially, snail-shell- or teardrop-shaped, 0.6–0.7 mm; tubercles minute, rounded.

Flowering spring. Stabilized sand dunes; 200–300 m; Ariz.; Mexico (Baja California, Sonora).

Drymaria viscosa is often attributed to S. Watson ex Orcutt; however, as B. D. Parfitt and W. C. Hodgson (1985) correctly stated, C. R. Orcutt (1886) merely mentioned the name and did not publish it validly.

2. SPERGULA Linnaeus, Sp. Pl. 1: 440. 1753; Gen. Pl. ed. 5, 199. 1754 • Spurrey, spargoute [Latin *spargo*, scatter or sow, alluding to discharge of seeds] [I]

Ronald L. Hartman

Richard K. Rabeler

Herbs, annual or winter annual. **Taproots** slender to ± stout, especially proximally. **Stems** spreading or ascending to erect, simple or branched, terete to somewhat angular. **Leaves** opposite but appearing whorled, as 8–15 per axillary cluster, 2 clusters per node, connate proximally by often-prominent ridge from which stipules arise, sessile; stipules 4 per node, white, ovate to triangular, margins entire but splitting variously with age, apex obtuse to acuminate; blade 1-veined, linear or filiform, sometimes succulent, apex blunt to apiculate. **Inflorescences** terminal, open to diffuse cymes; bracts paired, minute. **Pedicels** erect to ascending, spreading or usually reflexed and sometimes secund in fruit. **Flowers** usually bisexual, sometimes pistillate by stamen abortion; perianth and androecium briefly perigynous; hypanthium dish- or cup-shaped, not abruptly expanded distally; sepals distinct, silvery, elliptic to nearly ovate, 2.5–5 mm, herbaceous, margins scarious, apex acute to obtuse; petals 5, white, blade apex entire; nectaries at adaxial base of broader filaments opposite sepals; stamens 5 and opposite sepals, or 10 and arising from distally tapered rim of hypanthium; filaments distinct; styles 5, distinct, filiform, 0.4–0.6 mm, glabrous proximally; stigmas 5, linear along adaxial surface of styles, obscurely papillate (30×). **Capsules** ovoid, opening by 5 spreading to somewhat recurved valves;

carpophore absent. **Seeds** 5–25, blackish, circular, subglobose or lenticular and laterally compressed, nearly smooth or finely papillate, membranous, entire marginal wing often present, appendage absent; embryo peripheral, annular to spirally curved. $x = 9$.

Species 5 (3 in the flora): introduced; Eurasia (esp. Mediterranean region, Europe).

Spergula arvensis is the only species of the genus that has been introduced extensively outside of Eurasia.

1. Leaf blades usually appearing terete, 1.5–3(–5) cm, margins often revolute, forming abaxial channel; seeds subglobose, sometimes with keel or wing ± 0.1 mm wide . . . . 1. *Spergula arvensis*
1. Leaf blades usually flat, 0.3–1.5(–2) cm, usually not channeled abaxially; seeds lenticular, laterally compressed, wings 0.2–0.6 mm wide.
  2. Seed wings light brown to brownish black, 0.2–0.3 mm wide; stamens usually 10 . . . 2. *Spergula morisonii*
  2. Seed wings white or sometimes slightly tannish, 0.4–0.6 mm wide; stamens usually 5
  . . . . 3. *Spergula pentandra*

1. **Spergula arvensis** Linnaeus, Sp. Pl. 1: 440. 1753 • Corn spurrey, stickwort, starwort, spargoute des champs F I W

Spergula arvensis var. *sativa* (Boenninghausen) Mertens & W. D. J. Koch

Plants glabrous or, often, glandular. **Stems** usually branched proximally, 10–50+ cm. **Leaf blades** usually appearing terete, 1.5–3(–5) cm, margins often revolute, forming abaxial channel. **Pedicels** erect to ascending, reflexed, secund in fruit. **Flowers:** sepals 3.5–5 mm; petals ovate, ³/₄–1 times as long as sepals in flower, apex obtuse; stamens usually 10. **Capsule valves** 3.5–5 mm. **Seeds** sometimes keeled or winged, subglobose, 1–1.1 mm wide, surface minutely roughened or obscurely low-tuberculate (50×), covered with white, club-shaped papillae in part or throughout (packing of seeds in capsule may prevent papillae development in spots), wings white, ± 0.1 mm wide. $2n = 18, 36$ (both Europe).

Flowering spring–early summer. Sandy roadsides, cultivated fields, other disturbed areas; 10–2000 m; introduced; Greenland; St. Pierre and Miquelon; Alta., B.C., N.B., Nfld. and Labr. (Nfld.), N.W.T., N.S., Ont., P.E.I., Que., Sask., Yukon; Ala., Alaska, Ark., Calif., Colo., Conn., Del., D.C., Fla., Ga., Idaho, Ill., Ind., Ky., La., Maine, Md., Mass., Mich., Minn., Miss., Mo., Mont., N.H., N.J., N.Y., N.C., Ohio, Oreg., Pa., R.I., S.C., Tex., Vt., Va., Wash., W.Va., Wis., Wyo.; Eurasia; introduced in Central America, South America, Asia (Korea), Africa, Australia.

Spergula arvensis is often a significant weed in sandy crop lands, but it is sometimes used as a forage crop in areas with poor, sandy soils; it was intentionally introduced to Crawford County, Michigan, in 1888 (O. Clute and O. Palmer 1893). Historical collections are known also from Kansas, Nebraska, North Dakota, and South Dakota, where *Spergula arvensis* may have been introduced but apparently did not persist.

SELECTED REFERENCE New, J. K. 1961. Biological flora of the British Isles. *Spergula arvensis* L. (*S. sativa* Boenn., *S. vulgaris* Boenn.). J. Ecol. 49: 205–215.

2. **Spergula morisonii** Boreau, Rev. Bot. Recueil Mens. 2: 424. 1847 • Morison's spurrey I

Plants glabrous or densely pubescent or glandular. **Stems** often branched proximally, 5–35 cm. **Leaf blades** usually flat, 0.3–1.5 (–2) cm, usually not channeled abaxially. **Pedicels** erect to ascending, spreading or sometimes reflexed in fruit, sometimes secund. **Flowers:** sepals 3–4 mm; petals ovate, ²/₃–⁷/₈ times as long as sepals in flower, apex obtuse; stamens usually 10. **Capsule valves** 3.5–6 mm. **Seeds** winged, lenticular, 0.9–1 mm wide, surface minutely roughened or low-tuberculate (50×), with marginal ring of tan, club-shaped papillae; wings light brown to brownish black, 0.2–0.3 mm wide. $2n = 18$ (Europe).

Flowering spring–early summer. Sandy roadsides, disturbed areas; 10–100 m; introduced; Md., Mass., N.J.; Europe.

Spergula morisonii was first reported for North America from New Jersey in 1966; the earliest collections date from 1917 (D. B. Snyder 1987). It should be expected elsewhere in the flora area; the collections from Maryland and Massachusetts date from 2002 and 2000 respectively, with the Maryland population described as including "thousands of plants" (B. W. Steury 2004).

3. Spergula pentandra Linnaeus, Sp. Pl. 1: 440.
1753 • Wingstem spurrey [I]

Plants glabrous or moderately pubescent. **Stems** often branched proximally, 5–30 cm. **Leaf blades** usually flat, 0.5–1.5 cm, usually not channeled abaxially. **Pedicels** erect to ascending, spreading or, sometimes, reflexed in fruit, not secund. **Flowers:** sepals 2.5–4 mm; petals ± lanceolate, ³/₄–⁷/₈ times as long as sepals in flower, apex acute to acuminate; stamens usually 5. **Capsule** valves 4–5 mm. **Seeds** winged, lenticular, 0.6–0.9 mm wide, surface minutely roughened or obscurely tuberculate (50×), papillae absent or relatively few in marginal ring; wings usually white, sometimes slightly tannish, 0.4–0.6 mm wide. $2n$ = 18 (Europe).

Flowering spring–early summer. Sandy fields, other disturbed areas; 0–100 m; introduced; Conn., N.J., N.C., Va.; Europe; sw Asia; nw Africa; introduced in Australia.

Spergula pentandra was first collected in North America in 1956; see D. B. Snyder (1987) for an account of earlier confusion of this taxon with *S. morisonii* in New Jersey.

3. SPERGULARIA (Persoon) J. Presl & C. Presl, Fl. Čech., 94. 1819, name conserved

• Sand-spurrey, spergulaire [genus name *Spergula*, and Latin -*aria*, pertaining to]

Ronald L. Hartman

Richard K. Rabeler

Arenaria Linnaeus subg. *Spergularia* Persoon, Syn. Pl. 1: 504. 1805; *Delia* Dumortier; *Lepigonum* Wahlenberg

Herbs, annual or strongly perennial with branched, woody caudex. **Taproots** filiform to stout. **Stems** erect to sprawling, simple to freely branching distally or throughout, terete, sometimes woody. **Leaves** opposite, axillary clusters of leaves often present, distinct, sessile; stipules 2 per node, white to tan, lanceolate and acuminate to widely triangular, margins entire, apex entire to variously split; blade 1-veined, threadlike to linear, mostly succulent, apex acute to acuminate. **Inflorescences** terminal cymes, branching symmetrically or to 1 side (monochasium), simple to 8+-compound or sometimes flowers solitary and axillary; bracts usually paired or sometimes single, smaller, foliaceous, distalmost sometimes with scarious margins. **Pedicels** ascending to erect, divergently spreading, reflexed, or arching downward in fruit. **Flowers:** perianth and androecium hypogynous, briefly perigynous; hypanthium dish- or cup-shaped, not abruptly expanded distally; sepals connate in proximal ¹/₅, green, lanceolate to ovate, 0.9–8 mm, herbaceous, margins scarious, apex acute to obtuse; petals 5, white to pink, blade apex entire; nectaries as lateral expansion of bases of filaments opposite sepals; stamens 1–10, arising from rim of hypanthium; filaments distinct; styles 3, distinct or nearly so, filiform, 0.2–3 mm, glabrous proximally; stigmas 3, linear along adaxial surface of styles, obscurely papillate (30×). **Capsules** ovoid, opening by 3 spreading valves with recurved tips; carpophore absent. **Seeds** 30–150+, light to dark brown, reddish brown, or black, circular to angular, plump or laterally compressed, smooth to variously sculptured to papillate, complete or partial, membranous, laciniate, marginal wing often present, appendage absent. x = 9.

Species ca. 60 (11 in the flora): coastal and saline areas, w North America (including Mexico), Central America, w South America, Europe (Mediterranean region), Africa (Mediterranean region).

We follow the species order of R. P. Rossbach (1940). While there have been attempts to define species groups, and sections have been described, all such efforts focus only on European-Mediterranean taxa, ignoring the many taxa endemic to South America. Likewise, the major breaks in the key to species follow Rossbach's revision. Attempts to construct a key independent of habit were not successful; *Spergularia* is a difficult genus in which emphasis on seed characters appears necessary.

Spergularia is included in the first list of *nomina genericorum conservanda* published in 1905, conserved against *Buda* and *Tissa*. This action most likely occurred as a result of bickering between J. Britten and N. L. Britton in 1888–1890 about usage of the genera that Adanson had erected for *Spergularia* species with ten and five stamens respectively; see J. Britten (1890) for a summary of the arguments.

SELECTED REFERENCE Rossbach, R. P. 1940. *Spergularia* in North and South America. Rhodora 42: 57–83, 105–143, 158–193, 203–213.

1. Plants strongly perennial, base woody; stamens 7–10.
  2. Seeds 0.4–0.5 mm; sepal lobes 2.5–4 mm, to 5 mm in fruit; styles 0.4–0.6 mm . . . . 10. *Spergularia villosa*
  2. Seeds (0.6–)0.7–1.2 mm; sepal lobes 4.5–7 mm, to 8 mm in fruit; styles 0.6–3 mm
  . . . . 1. *Spergularia macrotheca*
1. Plants annual or short-lived perennial, base not woody; stamens 1–10.
  3. Seeds black, not papillate.
    4. Seeds 0.6–0.8 mm, broadly ovate, often with iridescent tinge, sculpturing of parallel, wavy lines; stipules 1–2.5 mm. . . . . 2. *Spergularia atrosperma*
    4. Seeds 0.4–0.6 mm, pyriform, often with silvery, not iridescent, tinge, sculpturing of low, elongate tubercles; stipules 1.5–2 mm . . . . 3. *Spergularia diandra*
  3. Seeds light to dark brown or reddish brown, if nearly black, seeds covered with gland-tipped papillae.
    5. Stamens 6–10.
      6. Seeds 0.8–1.1 mm, ± smooth, not papillate, usually winged; capsules usually (4.5–)5.5–8 mm; plants stout; main stems usually 1–4 mm diam. proximally; leaf blades markedly fleshy. . . . . 7. *Spergularia media*
      6. Seeds 0.4–0.6 mm, roughened or variously sculptured, papillate, not winged; capsules 2.8–5.4 mm; plants ± delicate; main stems usually 0.3–1 mm diam. proximally; leaf blades scarcely to somewhat fleshy.
        7. Axillary leaves 2–4+ per cluster, blade scarcely fleshy; seeds reddish brown to dark brown; stipules lanceolate, apex ± long-acuminate, shiny white, conspicuous . . . . 4. *Spergularia rubra*
        7. Axillary leaves absent or 1–2 per cluster, blade at least moderately fleshy; seeds light brown; stipules mostly deltate, apex acute to short-acuminate, dull white to tan, usually inconspicuous . . . . 5. *Spergularia bocconi*
    5. Stamens 1–5.
      8. Seeds 0.9–1.4 mm; capsules (1.2–)1.5–2 times as long as sepals . . . . . . 6. *Spergularia canadensis*
      8. Seeds 0.3–0.7(–0.8) mm; capsules 0.9–1.5 times as long as sepals.
        9. Plants glabrous; capsules 1.4–2.6 mm; sepal lobes 0.9–1.6 mm, to 2 mm in fruit; seeds 0.3–0.4 mm . . . . 11. *Spergularia platensis*
        9. Plants with stalked glands in inflorescence or throughout; capsules 2.8–6.4 mm; sepals lobes 1.8–4.5 mm, to 4.8 mm in fruit; seeds 0.5–0.7(–0.8) mm.
          10. Seeds dull, not silver tinged, ± smooth; stipules longer than wide; styles 0.4–0.7 mm . . . . 8. *Spergularia salina*
          10. Seeds shiny, silver tinged, slightly roughened; stipules shorter than wide; styles 0.3–0.4 mm . . . . 9. *Spergularia echinosperma*

1. **Spergularia macrotheca** (Hornemann ex Chamisso & Schlechtendal) Heynhold, Alph. Aufz. Gew., 689. 1846 • Sticky sand-spurrey

Arenaria macrotheca Hornemann ex Chamisso & Schlechtendal, Linnaea 1: 53. 1826

Plants strongly perennial with branched, woody base, stout, 5–40 cm, densely stipitate-glandular in inflorescence or throughout. **Taproots** becoming stout, woody. **Stems** erect or ascending to prostrate, usually branched proximally; main stem 0.8–3 mm diam. proximally. **Leaves:** stipules conspicuous, dull white to tan, narrowly triangular, 4.5–11 mm, apex long-acuminate; blade linear, (0.6–)1–5.5 cm, fleshy, apex apiculate to spine-tipped; axillary leaves 1–2+ per cluster. **Cymes** simple to 3-compound or flowers solitary and axillary. **Pedicels** erect, divergent, or reflexed in fruit. **Flowers:** sepals connate 0.5–1.8 mm proximally, lobes often 3-veined, ovate to lanceolate, 4.5–7 mm, to 8 mm in fruit, margins 0.3–0.7 mm wide, apex blunt to rounded; petals white or pink to rosy, elliptic to obovate, 0.9–1.3 times as long as sepals in flower; stamens 9–10; styles 0.6–3 mm. **Capsules** tan, 4.6–10 mm, 0.8–1.4 times as long as sepals. **Seeds** ± red-brown, often with submarginal groove or depression, suborbiculate to pyriform, compressed, (0.6–)0.7–1.2 mm, smooth to faintly or prominently tuberculate or sculpturing of parallel, wavy lines or of low, rounded mounds, not papillate; wing absent or white to reddish brown proximally, 0.1–0.3 mm wide, margins irregular.

Varieties 3 (3 in the flora): w North America, including nw Mexico.

1. Petals pink to rosy; styles 0.6–1.2 mm . . . . . .
. . . . 1a. *Spergularia macrotheca* var. *macrotheca*
1. Petals white; styles 1.2–3 mm.
   2. Capsules 1.2–1.4 times as long as sepals; styles 1.2–1.9 mm . . . . 1b. *Spergularia macrotheca* var. *leucantha*
   2. Capsules 0.8–1 times as long as sepals; styles 2–3 mm . . . . 1c. *Spergularia macrotheca* var. *longistyla*

1a. **Spergularia macrotheca** (Hornemann ex Chamisso & Schlechtendal) Heynhold var. **macrotheca**

Plants 5–35 cm. **Stems** erect to prostrate. **Leaves:** stipules 5–11 mm; blade (0.6–)1–4 cm. **Flowers:** sepal lobes ovate to lanceolate, (4.5–)5–7 mm, to 8 mm in fruit; petals pink to rosy, 0.9–1.1 times as long as sepals; styles 0.6–1.2 mm. **Capsules** 0.8–1.2 times as long as calyx. **Seeds** red-brown, often with submarginal groove or depression, suborbiculate to ovate, (0.6–)0.7–0.8 mm, smooth (40×); wing absent or 0.1–0.2 mm wide. **2*n*** = 36, 72.

Flowering spring–fall. Salt flats and marshes, dunes, rocky outcrops, sandy or rocky coastal bluffs, gravelly ridges, alkaline fields; less than 300 m; B.C.; Calif., Oreg., Wash.; Mexico (Baja California).

1b. **Spergularia macrotheca** (Hornemann ex Chamisso & Schlechtendal) Heynhold var. **leucantha** (Greene) B. L. Robinson, Proc. Amer. Acad. Arts 29: 313. 1894 E

Tissa leucantha Greene, Pittonia 1: 301. 1889

Plants 10–40 cm. **Stems** usually ascending to erect, sometimes prostrate. **Leaves:** stipules 6–9 mm; blade 1.5–5.5 cm. **Flowers:** sepal lobes ± lanceolate, 4.5–5.5 mm, to 6.5 mm in fruit; petals white, 1–1.1 times as long as sepals; styles 1.2–1.9 mm. **Capsules** 1.2–1.4 times as long as sepals. **Seeds** dark red-brown, not grooved, suborbiculate, 0.7–0.8 mm, smooth to faintly tuberculate (40×); wing absent or 0.1–0.3 mm wide.

Flowering spring. Alkaline soils, floodplains, vernal pools, meadows, marshy ground; 0–800 m; Calif.

1c. **Spergularia macrotheca** (Hornemann ex Chamisso & Schlechtendal) Heynhold var. **longistyla** R. Rossbach, Rhodora 42: 78, plate 589, fig. 1e,f. 1940 E

Plants 10–30 cm. **Stems** ascending or erect, sometimes prostrate. **Leaves:** stipules 4.5–8 mm; blade 1.5–4 cm. **Flowers:** sepal lobes ovate to lanceolate, (4.5–)5–5.5 mm, to 7 mm in fruit; petals white, 1–1.3 times as long as sepals; styles 2–3 mm. **Capsules** 0.8–1 times as long as sepals. **Seeds** dark red-brown, often with submarginal groove, suborbiculate to pyriform, 0.8–1.2 mm, covered with rounded, elongate tubercles; wing sometimes absent, 0.2–0.3 mm wide.

Flowering spring. Alkaline marshes, mud flats, meadows, hot springs; 20–200 m; Calif.

2. Spergularia atrosperma R. Rossbach, Rhodora 42: 80, plate 589, fig. 3a–c. 1940 • Black-seed sand-spurrey E

Plants annual, delicate, 6–15 cm, glabrous or stipitate-glandular distally. **Taproots** filiform. **Stems** erect to ascending or arcuate-spreading, simple or branched proximally; main stem 0.3–1 mm diam. proximally. **Leaves:** stipules inconspicuous, dull white to tan, broadly triangular, 1–2.5 mm, apex acute to short-acuminate; blade filiform to linear, 0.5–3.5 cm, fleshy, apex blunt to apiculate; axillary leaf clusters absent. **Cymes** simple to 2-compound. **Pedicels** divergently spreading to reflexed in fruit. **Flowers:** sepals connate 0.2–0.5 mm proximally, lobes ± weakly 1-veined, elliptic to ovate, 1.8–2.7 mm, to 3.5 mm in fruit, margins 0.1–0.5 mm wide, apex broadly acute to rounded; petals white to rosy, ± ovate, 0.7–0.8 times as long as sepals; stamens 4–8; styles 0.4–0.8 mm. **Capsules** greenish, 3.3–5 mm, 1–1.3 times as long as sepals. **Seeds** black, often iridescent, submarginal groove usually absent, broadly ovate, compressed, 0.6–0.8 mm, shiny, sculpturing of parallel, wavy lines, not papillate (40×); wing absent or, rarely, incomplete, brownish, 0.1–0.2 mm, margins irregular.

Flowering summer. Alkaline places, mud flats, stream beds, sandy places; 30–1500 m; Calif., Nev.

3. Spergularia diandra (Gussone) Heldreich, Pl. Atticae, unnumbered. 1851 • Alkali sand-spurrey I

Arenaria diandra Gussone, Fl. Sicul. Prodr. 1: 515. 1827; *Spergularia salsuginea* Fenzl

Plants annual, delicate, 5–15 cm, stipitate-glandular throughout or nearly so. **Taproots** slender. **Stems** erect to diffusely spreading, much-branched proximally and distally; main stem 0.3–0.7 mm diam. proximally. **Leaves:** stipules inconspicuous, silvery to dull tan, broadly triangular, 1.5–2 mm, apex acuminate; blade linear, 0.6–2.3 cm, somewhat fleshy, apex blunt to apiculate; axillary leaf cluster usually absent. **Cymes** simple but commonly 4–8+-compound. **Pedicels** erect to reflexed in fruit. **Flowers:** sepals connate 0.2–0.5 mm proximally, lobes 1-veined or not, lanceolate to ovate, 2.3–3.1 mm, enlarging little in fruit, margins 0.1–0.3 mm wide, apex obtuse to rounded; petals white, elliptic to ovate, 0.7–0.8 times sepals; stamens 4–7; styles 0.4–0.6 mm. **Capsules** greenish tan, 2.5–2.8 mm, 0.9–1.2 times sepals. **Seeds** black, often with silvery, not iridescent tinge, with submarginal groove, pyriform, somewhat compressed, 0.4–0.6 mm, shiny, sculpturing

of low, elongate tubercles, not papillate (40×); wing absent. $2n = 18$ (Europe).

Flowering spring–fall. Sandy beaches, river shores; 600–700 m; introduced; Alta., B.C., Sask.; Idaho, Mass., Oreg., Wash.; Europe (Mediterranean region); sw, c Asia; n Africa; introduced in Australia.

R. P. Rossbach's (1940) report of *Spergularia diandra* from Georgia is referred to 9. *S. echinosperma*.

The name *Spergularia diandra* was effectively and validly published via an autographic label distributed in 1851, predating the other commonly seen attributions of this combination.

4. Spergularia rubra (Linnaeus) J. Presl & C. Presl, Fl. Čech., 94. 1819 • Red sand-spurrey, spergulaire rouge I W

Arenaria rubra Linnaeus, Sp. Pl. 1: 423. 1753; *Tissa rubra* (Linnaeus) Britton

Plants annual or short-lived perennial, delicate, 4–25 cm, stipitate-glandular in inflorescence. **Taproots** slender to ± stout. **Stems** erect to ascending or prostrate, usually much-branched proximally; main stem 0.3–0.5 mm diam. proximally. **Leaves:** stipules conspicuous, shiny white, lanceolate, 3.5–5 mm, apex long-acuminate; blade filiform to linear, 0.4–1.5 cm, scarcely fleshy, apex apiculate to spine-tipped; axillary leaves 2–4+ per cluster. **Cymes** simple to 3+-compound or flowers solitary and axillary. **Pedicels** ascending to reflexed. **Flowers:** sepals connate 0.5–0.7 mm proximally, lobes often 3-veined, lanceolate, (2–)2.5–3.2 mm, to 4 mm in fruit, margins 0.1–0.3 mm wide, apex obtuse to acute; petals pink, obovate to ovate, 0.9–1 times as long as sepals; stamens 6–10; styles 0.6–0.8 mm. **Capsules** greenish to tan, 3.5–5 mm, 1–1.2 times as long as sepals. **Seeds** red-brown to dark brown, with submarginal groove, broadly ovate or ± truncate, angular at broad end, plump, 0.4–0.6 mm, sculpturing of parallel, wavy lines, margins with peglike papillae (30×); wing absent. $2n = 18, 27, 36, 54$ (all Europe).

Flowering spring–fall. Open forests, gravelly glades, meadows, mud flats, roadsides, disturbed places; 0–2400 m; introduced; St. Pierre and Miquelon; B.C., N.B., Nfld. and Labr. (Nfld.), N.S., Ont., P.E.I., Que., Yukon; Alaska, Calif., Colo., Conn., Del., D.C., Idaho, Ill., Ind., Ky., Maine, Md., Mass., Mich., Minn., Mont., Nev., N.H., N.J., N.Mex., N.Y., Ohio, Oreg., Pa., R.I., Utah, Vt., Va., Wash., Wis., Wyo.; Europe; Asia; introduced in South America, Australia.

Spergularia rubra was collected in 1901 on ballast in Alabama (*Mohr*, DS), the only record in the southeastern United States. It is the most widely distributed *Spergularia* species found outside of saline areas in the flora and has been in North America since at least the 1860s.

5. Spergularia bocconi (Scheele) Graebner in P. F. A. Ascherson et al., Syn. Mitteleur. Fl. 5(1): 849. 1919 (as bocconei) • Boccone's sand-spurrey [I]

Alsine bocconi Scheele, Flora 26: 431. 1843

Plants annual, ± delicate, 6–25+ cm, often densely stipitate-glandular, at least in inflorescence. **Taproots** ± filiform. **Stems** erect to spreading or sprawling, usually much-branched proximally; main stem 0.5–1 mm diam. proximally. **Leaves:** stipules usually inconspicuous, dull white to tan, broadly triangular, 1.5–4.5 mm, apex acute to short-acuminate; blade ± linear, 1–4.2 cm, at least moderately fleshy, apex apiculate to spine-tipped; axillary leaves absent or 1–2 per cluster. **Cymes** simple to 6+-compound. **Pedicels** often oriented to 1 side in fruit. **Flowers:** sepals connate 0.3–0.6 mm proximally, lobes often 3-veined, ovate to elliptic-oblong, 2.2–3.5 mm, to 4.5 mm in fruit, margins 0.2–0.5 mm wide, apex acute to rounded; petals white or pink to rosy, ovate to obovate, 0.8–1 times as long as sepals; stamens 8–10; styles 0.4–0.6 mm. **Capsules** greenish, 3(–4) mm, 1–1.2 times as long as sepals. **Seeds** light brown, with submarginal groove, broadly ovate, plump, 0.4–0.6 mm, somewhat shiny, smooth to minutely roughened, margins with peglike papillae (40×); wing absent. $2n$ = 18? (Africa), 36 (Europe).

Flowering spring. Salt marshes, alkaline places, sandy soils; 0–400 m; introduced; Calif., Oreg.; sw Europe (Mediterranean region).

The spelling of the epithet *bocconi*, often "corrected" to *bocconii*, is debatable. It commemorates Paolo Boccone, suggesting a correction to *bocconei*, but he also used the Latinized form Bocconus, allowing *bocconi*. We have used *bocconi*, following the first usage by Scheele.

6. Spergularia canadensis (Persoon) G. Don, Gen. Hist. 1: 426. 1831 [E]

Arenaria canadensis Persoon, Syn. Pl. 1: 504. 1805

Plants annual, delicate to stout, 3–25 cm, glabrous or stipitate-glandular throughout or only in inflorescence. **Taproots** slender. **Stems** prostrate or decumbent to erect, usually branched proximally; main stem usually 0.4–1.8 mm diam. proximally. **Leaves:** stipules inconspicuous, dull white, broadly triangular, 1–2.8 mm, apex obtuse to acute; blade linear, 1.5–4.5 cm, fleshy, apex ± blunt; axillary leaf clusters mostly absent. **Cymes** simple to 2+-compound or flowers solitary and axillary. **Pedicels**

reflexed in fruit. **Flowers:** sepals connate 0.5–0.6 mm proximally, lobes weakly 1-veined, ovate to elliptic-oblong, 2.2–3.5 mm, to 4.3–4.5 mm in fruit, margins 0.2–0.5 mm wide, apex ± acute to rounded; petals white or pink, narrowly ovate, 0.9–1 times as long as sepals; stamens 2–4; styles 0.3–0.7 mm. **Capsules** greenish, 3.5–5.3 mm, 1.2–2 times as long as sepals. **Seeds** reddish brown, submarginal groove absent, broadly ovate, compressed, 0.9–1.4 mm, shiny, ± smooth, papillate; wing absent or often present, whitish, 0.2–0.3 mm wide, margins irregular.

Varieties 2 (2 in the flora): coastal North America.

1. Capsules 1.5–2 times as long as sepals; sepals 2.2–3.2 mm, to 4.5 mm in fruit, apex broadly rounded; stems prostrate or decumbent; plants glabrous except pedicels rarely stipitate-glandular . . . . 6a. *Spergularia canadensis* var. *canadensis*
1. Capsules 1.2–1.3 times as long as sepals; sepals 2.5–3.5 mm, to 4.3 mm in fruit, apex ± acute to narrowly obtuse; stems widely spreading to erect; plants glabrous or stipitate-glandular . . . . 6b. *Spergularia canadensis* var. *occidentalis*

6a. Spergularia canadensis (Persoon) G. Don var. **canadensis** • Canadian sand-spurrey, spergulaire du Canada [E]

Plants 3–25 cm, glabrous except pedicels rarely stipitate-glandular. **Stems** prostrate or decumbent, widely branched. **Stipules** 1–2.8 mm. **Sepals** 2.2–3.2 mm, to 4.5 mm in fruit, apex broadly rounded or blunt. **Capsules** 3.6–5.2 mm, 1.5–2 times as long as sepals. **Seeds** 0.9–1.4 mm. $2n$ = 36.

Flowering summer. Coastal salt marshes, brackish river margins, moist beach sand; 0–50 m; St. Pierre and Miquelon; N.B., Nfld. and Labr. (Nfld.), N.S., Nunavut, Ont., P.E.I., Que.; Conn., Maine, Mass., N.H., N.Y., R.I.

6b. Spergularia canadensis (Persoon) G. Don var. **occidentalis** R. Rossbach, Rhodora 42: 116. 1940 [E]

Plants 7–25 cm, glabrous or stipitate-glandular. **Stems** widely spread-ing to erect, branching proximally. **Stipules** 2–2.7 mm. **Sepals** 2.5–3.5 mm, to 4.3 mm in fruit, apex ± acute to narrowly obtuse. **Capsules** 3.5–5.3 mm, 1.2–1.3 as long as times sepals. **Seeds** 0.9–1.1 mm. $2n$ = 36.

Flowering summer. Disturbed marshy areas; 0–50 m; B.C.; Alaska, Calif., Oreg., Wash.

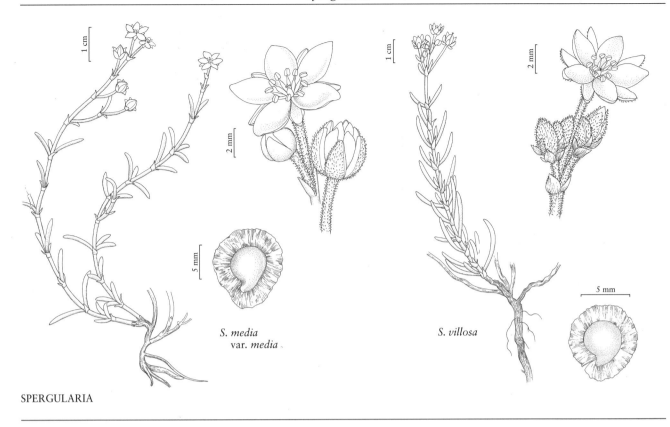

S. media
var. *media*

S. villosa

SPERGULARIA

7. Spergularia media (Linnaeus) C. Presl, Fl. Sicul., 161. 1826 • Greater sea-spurrey, spergulaire moyenne, spergulaire marginée F I

Arenaria media Linnaeus, Sp. Pl. ed. 2, 1: 606. 1762

Varieties 5 (1 in the flora): introduced; coastal Europe, Asia, Africa (Mediterranean region); introduced widely.

Although a proposal to reject *Spergularia media* as a confused name (J. Lambinon 1981) was rejected, some authors still favor that argument, preferring to use the name *S. maritima* for this species.

Spergularia media is one of the "highway halophytes" (A. A. Reznicek 1980) that have spread along highways that are heavily salted during the winter, where saline areas have been created. *Spergularia* distribution has been investigated in Ohio (A. W. Cusick 1983), where such records were first noted in the early 1970s.

7a. Spergularia media (Linnaeus) C. Presl var. media F I

Spergularia marginata Kittel; *S. maritima* (Linnaeus) Chiovenda

Plants annual or short-lived perennial with branching caudex, stout, 7–30 cm, glabrous or sparsely stipitate-glandular distally. **Taproots** slender to moderately stout. **Stems** erect to ascending or prostrate, much-branched proxi-

mally; main stem 1–4 mm diam. proximally. **Leaves:** stipules inconspicuous, dull white, broadly triangular, 2.6–6 mm, apex acuminate to spine-tipped; blade linear to broadly so, 0.5–3.5 cm, distinctly fleshy, apex apiculate to spine-tipped; axillary leaf clusters mostly absent, sometimes of 2–4 leaves. **Cymes** simple or 3+-compound. **Pedicels** reflexed and oriented to 1 side in fruit. **Flowers:** sepals connate 0.3–1 mm proximally, lobes often 3-veined, lanceolate to ovate, 2.5–5 mm, to 7 mm in fruit, margins 0.1–0.5 mm wide, apex obtuse to rounded; petals white to apically pink, elliptic to obovate, 0.8–1 times as long as sepals; stamens 9–10; styles 0.5–1 mm. **Capsules** tan, (4.5–)5.5–7(–8) mm, 1.2–1.4 times as long as sepals. **Seeds** dark reddish brown, submarginal groove absent, broadly ovate, compressed, 0.8–1.1 mm, ± smooth, not papillate; wings usually present, white to reddish brown, 0.3–0.5 mm wide, margins irregular. $2n$ = 18, 36? (both Europe).

Flowering summer–fall. Salt flats, salt marshes, sandy beaches, saline highway edges (Great Lakes region); 0–1700 m; introduced; Ont., Que.; Calif., Colo., Ill., Ind., Mass., Mich., Mont., N.Mex., N.Y., N.Dak., Ohio, Oreg., Pa., Utah, Wyo.; coastal Europe except ne; Asia; Africa (Mediterranean region); introduced widely elsewhere including South America, e Asia, Australia.

8. **Spergularia salina** J. Presl & C. Presl, Fl. Čech., 95. 1819 • Salt-marsh sand-spurrey, lesser sea-spurrey, spergulaire des arais salés [I]

Spergularia marina (Linnaeus) Grisebach; *S. marina* var. *tenuis* (Greene) R. Rossbach; *S. salina* var. *tenuis* (Greene) Jepson; *S. tenuis* Greene; *Tissa marina* (Linnaeus) Britton

Plants annual, delicate, 8–25(–30) cm, stipitate-glandular, at least in inflorescence. **Taproots** ± slender. **Stems** erect to ascending or prostrate, usually much-branched proximally; main stem 0.6–2(–3) mm diam. proximally. **Leaves:** stipules inconspicuous, dull white, broadly triangular, 1.2–3.5 mm, longer than wide, apex acute to short-acuminate; blade linear, (0.8–)1.5–4 cm, fleshy, apex blunt to apiculate; axillary leaf clusters usually absent. **Cymes** simple to 3+-compound or flowers solitary and axillary. **Pedicels** reflexed and oriented to 1 side in fruit. **Flowers:** sepals connate 0.5–1 mm proximally, lobes often 3-veined, ovate to elliptic, 2.5–4.5 mm, to 4.8 mm in fruit, margins 0.1–0.5 mm wide, apex acute to rounded; petals white or pink to rosy, ovate to elliptic-oblong, 0.8–1 times as long as sepals; stamens (1–)2–3(–5); styles 0.4–0.7 mm. **Capsules** greenish to tan, 2.8–6.4 mm, 1–1.5 times as long as sepals. **Seeds** light brown to reddish brown, with submarginal groove, broadly ovate, ± plump, 0.5–0.7 (–0.8) mm, dull, ± smooth, often with gland-tipped papillae (30×); wing usually absent or incomplete. $2n$ = 18? (Asia), 36 (Europe).

Flowering summer–early fall. Mud flats, alkaline fields, sandy river bottoms, sandy coasts, salt marshes, saline highway edges (Great Lakes region); 0–1400 m; introduced and native; St. Pierre and Miquelon; Alta., B.C., Man., N.B., N.W.T., N.S., Ont., P.E.I., Que., Sask.; Ala., Ariz., Calif., Colo., Conn., Del., D.C., Fla., Ga., Idaho, Ill., Iowa, Ky., La., Maine, Md., Mass., Mich., Miss., Mo., Mont., Nebr., Nev., N.H., N.J., N.Mex., N.Y., N.C., N.Dak., Ohio, Okla., Oreg., Pa., R.I., S.C., Tex., Utah, Va., Wash., Wis., Wyo.; Eurasia; almost cosmopolitan via introduction.

While *Spergularia salina* may be native in coastal areas and some inland saline sites in much of the cited range, populations in the Great Lakes region are introduced where, as in *S. media*, highway and sidewalk salt runoff has created favorable habitats.

Variety *tenuis* has been distinguished from var. *salina* by some authors as follows: cyme crowded versus lax, sepals 1.6–3.8 mm versus 2.4–5 mm, mature capsules 3–4.4 mm versus 3.6–6.4 mm, respectively. Due to the extreme overlap in morphologic features as well as geographic ranges, var. *tenuis* is not recognized here.

The name *Spergularia marina* var. *leiosperma* (Kindberg) Gurke has been applied to plants with smooth seeds but, as pointed out by R. P. Rossbach (1940), separation of plants with smooth versus papillose seeds is not practical.

Some authors believe that the correct name for this species is *Spergularia marina*.

9. **Spergularia echinosperma** Čelakovský, Arch. Naturwiss. Landesdurchf. Böhmen 4: 867. 1881 • Bristle-seed sand-spurrey [I]

Plants annual, delicate, 5–15 cm, stipitate-glandular throughout. **Taproots** slender. **Stems** erect to ascending, simple to diffusely branched proximally and distally; main stem occasionally prostrate, 0.2–1.5 mm diam. proximally. **Leaves:** stipules inconspicuous, silvery to dull tan, broadly triangular, 1.4–2.4 mm, shorter than wide, apex acuminate; blade linear, 0.5–3.5 cm, somewhat fleshy, apex blunt to apiculate; axillary leaf cluster usually absent. **Cymes** commonly 4–8+-compound. **Pedicels** reflexed and oriented to 1 side in fruit. **Flowers:** sepals connate 0.2–0.3 mm proximally, lobes 1–3-veined, lanceolate to ovate, 2.5–3.6 mm, to 4 mm in fruit, margins 0.1–0.5 mm wide, apex rounded; petals white to pink or rosy, lanceolate, 0.4–0.6 times as long as sepals; stamens 1–4(–5); styles 0.3–0.4 mm. **Capsules** greenish to tan, 2.8–4 mm, 0.9–1.4 times as long as sepals. **Seeds** reddish brown to blackish, silver tinged, with submarginal groove, pyriform, ± compressed, 0.5–0.7(–0.8) mm, shiny, slightly roughened, with dense, gland-tipped papillae and appearing echinate (30×); wing usually absent, whitish to reddish brown, 0.1–0.2 mm wide, margins irregular and not papillate. $2n$ = 18, 36 (both Europe).

Flowering spring. Dunes, clay flats, sandy river banks; 0–20+ m; introduced; Ala., Ga., La., Tex.; Europe.

10. **Spergularia villosa** (Persoon) Cambessèdes in A. St.-Hilaire et al., Fl. Bras. Merid. 2: 129. 1830 • Hairy sand-spurrey [F] [I]

Spergula villosa Persoon, Syn. Pl. 1: 522. 1805

Plants strongly perennial with branched, woody base, stout, 11–30 cm, stipitate-glandular in inflorescence or throughout. **Taproots** becoming stout, woody. **Stems** erect to ascending, often arcuately so, much-branched proximally; main stem 0.4–1.3 mm diam. proximally. **Leaves:** stipules ± conspicuous, dull white, broadly lanceacuminate, 3–8 mm, apex mucronate; blade filiform to linear, 1–4.2 cm, somewhat fleshy, apex apiculate or spine-tipped; axillary leaves 2–4 per cluster. **Cymes** simple to 3-compound. **Pedicels** spreading to reflexed in fruit. **Flowers:** sepals connate 0.5–0.7 mm proximally, lobes 1- or 3-veined, lance-ovate to lanceolate, 2.5–4 mm, to 5 mm in fruit, margins 0.1–0.6 mm wide, apex acute to acuminate but often briefly rounded at tip; petals white, ± elliptic, 0.7–0.8 times as long as sepal; stamens 7–10; styles 0.4–0.6 mm. **Capsules** greenish to tan, (4–)5–6.5 mm, 1.1–1.3 times as long as sepals. **Seeds** reddish brown to dark brown, often with submarginal groove, broadly ovate, plump, 0.4–0.5 mm, smooth, often sculptured with parallel, wavy lines, papillae often present; wing often present, white, 0.1–0.2 mm wide, margins irregular.

Flowering spring. Sandy slopes and bluffs, clay ridges and plains, disturbed areas; 0–500 m; introduced; Calif., Oreg.; Mexico (Baja California); South America (Chile?).

11. Spergularia platensis (Cambessèdes) Fenzl, Ann. Wiener Mus. Naturgesch. 2: 272. 1839 • La Plata sand-spurrey [I]

Balardia platensis Cambessèdes in A. St.-Hilaire et al., Fl. Bras. Merid. 2: 130, plate 111. 1830

Varieties 2 (1 in the flora): introduced; South America (Argentina); introduced in South America (Chile).

11a. Spergularia platensis (Cambessèdes) Fenzl var. **platensis** [I]

Plants annual, delicate, 6–18 cm, glabrous. **Taproots** filiform. **Stems** erect to diffusely spreading, much-branched proximally and distally; main stem 0.3–1 mm diam. proximally. **Leaves:** stipules inconspicuous, dull white, broadly triangular, 1.5–3.5 mm, apex acute to acuminate; blade filiform to linear, 0.5–3 cm, somewhat fleshy, apex acute to spine-tipped; axillary leaf clusters absent. **Cymes** branching symmetrically or mostly to 1 side, usually 4–7+-compound. **Pedicels** erect to spreading in fruit. **Flowers:** sepals connate 0.1–0.3 mm proximally, lobes weakly 1-veined, ovate to broadly elliptic, 0.9–1.6 mm, to 2 mm in fruit, margins 0.1–0.3 mm wide, apex obtuse to broadly rounded; petals white, elliptic, 0.6–0.8 times as long as sepals; stamens 3–5; styles 0.3–0.4 mm. **Capsules** greenish to tan, 1.4–2.6 mm, 1.3–1.5 times as long as sepals. **Seeds** dark brown or reddish brown to black, submarginal groove usually absent, pyriform, plump, 0.3–0.4 mm, shiny, with low, elongate tubercles, gland-tipped papillae often present; wing absent.

Flowering early spring. Dried or brackish mud flats with adobe mesas; 0–400 m; introduced; Calif., Tex.; Mexico (Baja California); probably South America (Argentina).

Although var. *platensis* has been known from North America since at least the 1880s, R. P. Rossbach (1940) thought it likely that the taxon was introduced here, and that the only native occurrences are in Argentina.

4. POLYCARPAEA Lamarck, J. Hist. Nat. 2: 3, 5, plate 25. 1792, name conserved (as Polycarpea) • [Greek *poly*-, numerous, and *karpos*, fruit, alluding to the numerous capsules] [I]

John W. Thieret

Richard K. Rabeler

Herbs [small shrubs], annual [perennial]. **Taproots** slender to stout. **Stems** erect, branched, terete. **Leaves** opposite, sometimes appearing whorled, not connate, short-petiolate (basal leaves) or sessile (cauline leaves); stipules 2 per node, white, lanceolate, margins entire, apex hairlike; blade 1-veined, oblong-ovate to suborbiculate (basal leaves), subulate to linear (cauline leaves), not succulent, apex acute or hairlike. **Inflorescences** terminal, compact to loose cymes; bracts paired, scarious. **Pedicels** erect or spreading. **Flowers:** perianth and androecium perigynous; hypanthium cup-shaped, not abruptly expanded distally; sepals distinct, silvery with brown midrib, ovate to lanceolate, not keeled, 2.2–3.1 mm, scarious, margins scarious, apex acute;

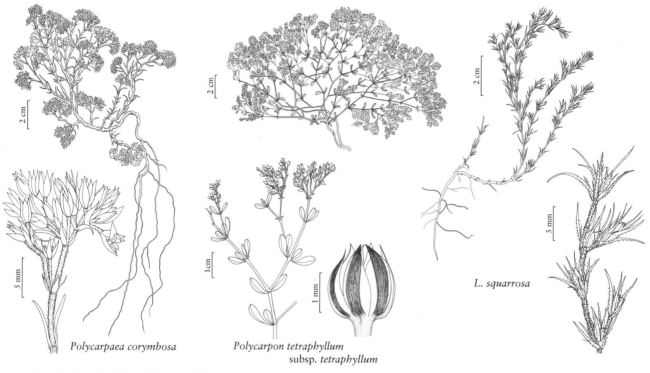

Polycarpaea corymbosa   *Polycarpon tetraphyllum*
                              subsp. *tetraphyllum*

L. squarrosa

POLYCARPAEA ∘ POLYCARPON ∘ LOEFLINGIA

petals 5, pink, blade apex irregularly toothed or entire; nectaries between filament bases; stamens 5; filaments distinct; style 1, filiform, minute, less than 0.1 mm, glabrous proximally; stigmas 3, subcapitate, smooth to obscurely papillate (50×). **Capsules** ellipsoid, opening by 3 recurved valves; carpophore present. **Seeds** 5–8, tan, translucent, pyriform, not compressed, rugulose, marginal wing absent, appendage absent. *x* = 9.

Species ca. 50 (1 in the flora): introduced; Florida, tropics and subtropics, chiefly Old World.

SELECTED REFERENCES   Bakker, K. 1957. Revision of the genus *Polycarpaea* (Caryoph.) in Malaysia. Acta Bot. Neerl. 6: 48–53. Gagnepain, F. 1908. Contribution à la connaissance du genre *Polycarpaea* Lamk. J. Bot. (Morot), sér. 2, 1: 275–280.   Lakela, O. 1962. Occurrence of species of *Polycarpaea* Lam. (Caryophyllaceae) in North America. Rhodora 64: 179–182.   Lakela, O. 1963. Annotation of North American *Polycarpaea*. Rhodora 65: 35–44.   Sundaramoorthy, S. and D. N. Sen. 1988. Ecology of Indian arid zone weeds. XI. *Polycarpaea corymbosa* (Linn.) Lamk. Geobios (Jodhpur) 15: 235–237. .

1. **Polycarpaea corymbosa**   (Linnaeus) Lamarck in J. Lamarck and J. Poiret, Tabl. Encycl. 2: 129. 1797 (as Policarpaea)   F I

Achyranthes corymbosa Linnaeus, Sp. Pl. 1: 205. 1753; *Polycarpaea nebulosa* Lakela

Stems 6–18 cm, ± pilose. **Leaves** dimorphic; basal (sometimes absent in older specimens) with petiole 2.5–4 mm, blade oblong-ovate to suborbiculate, 8–12 mm, margins flat, glabrous; cauline subtending fascicles of usually smaller leaves, blade linear, 1–2.5 cm, margins revolute, glabrous or ciliate.

Cymes dense. **Pedicels** 0.8–3 mm. **Flowers:** sepals 2.2–3.1 mm; petals marcescent, elliptic to ovate, 0.7–1 mm; stigmas nearly sessile. **Capsules** 1.5 mm. **Seeds** 0.4–0.5 mm. **2n** = 36.

Flowering summer–winter. Sandhills, grassy areas, disturbed sites; 10 m; introduced; Fla.; South America; Asia; Africa; Australia.

Polycarpaea corymbosa was first found in the sandhills of Tampa, Florida, in 1960 by O. Lakela (1962, 1963). Although she eventually described her plants as a new species, subsequent authors have concluded that the features supposedly distinguishing it lie within the range of *P. corymbosa*, a widely distributed paleotropical taxon. It is now known from five counties in central Florida.

Polycarpaea corymbosa is morphologically diverse, especially in Africa. While at least five varieties have been described (four since 1975) for distinctive populations in Asia and South America, a thorough study across the range of the species is yet to be undertaken.

K. R. Beena et al. (2000) demonstrated the presence of endophytic fungi in *Polycarpaea corymbosa*.

5. POLYCARPON Loefling ex Linnaeus, Syst. Nat. ed. 10, 2: 859 (as Polycarpa), 881, 1360. 1759 • Manyseed [Greek *polys*, many, and *karpos*, fruit, alluding to numerous capsules]

John W. Thieret

Richard K. Rabeler

Herbs, annual. **Taproots** slender. **Stems** prostrate to erect, branched, terete to finely ridged. **Leaves** opposite or in whorls of 4, not connate, petiolate; stipules 2 per node, silvery, lanceolate to triangular-ovate, margins entire or irregularly cut, apex acuminate to aristate; blade 1-veined, spatulate or oblanceolate to ovate or elliptic, not succulent, apex obtuse, sometimes mucronate. **Inflorescences** terminal or axillary, dense or lax cymes; bracts paired or absent. **Pedicels** erect. **Flowers:** perianth and androecium perigynous; hypanthium minute, cup-shaped, not abruptly expanded distally; sepals distinct, green, lanceolate to elliptic or ovate, often keeled, 1–2.5 mm, herbaceous, margins white, scarious, apex acute, ± hooded, ± awned; petals often fugacious, 5, white, blade apex emarginate; nectaries between filament bases; stamens 3–5; filaments shortly connate distally around ovary; style 1, obscurely 3-branched, filiform, 0.1–0.3 mm, glabrous proximally; stigmas 3, linear along adaxial surface of style branches, papillate (30×). **Capsules** ovoid to spherical, opening by 3 incurved or twisting valves; carpophore present. **Seeds** ca. 8–15, whitish, ovoid to lenticular or triangular, laterally compressed to angular, papillate or granular, marginal wing absent, appendage absent. $x = [7], 8, 9$.

Species ca. 9 or 15 (2 in the flora): w North America, South America, Europe (including Mediterranean region), Asia, Africa; *Polycarpon tetraphyllum* introduced widely, including e North America, Australia.

C. Mohr's (1901) report of *Polycarpon alsinifolium* (Bivona-Bernardi) de Candolle from Alabama was based on misidentification of *P. tetraphyllum* subsp. *tetraphyllum*.

The shape of the sepal apex in *Polycarpon* is varied, with the herbaceous central portion sometimes prolonged beyond the scarious margins into a brief, narrow hood or awn, suggesting an aristate apex.

SELECTED REFERENCE Howell, J. T. 1941. Notes on *Polycarpon*. Leafl. W. Bot. 3: 80.

1. Sepals 1.5–2.5 mm, keeled; stipules 1.8–2.8 mm; leaves in whorls of 4 or opposite . . . . 1. *Polycarpon tetraphyllum*
1. Sepals 1–1.5 mm, flat or sometimes barely keeled; stipules 0.4–1.2 mm; leaves opposite, not whorled . . . . 2. *Polycarpon depressum*

1. Polycarpon tetraphyllum (Linnaeus) Linnaeus, Syst. Nat. ed. 10, 2: 881. 1759 • Fourleaf manyseed ⬚F⬚ ⬚I⬚

Mollugo tetraphylla Linnaeus, Sp. Pl. 1: 89. 1753

Subspecies 3 (1 in the flora): introduced; s Europe (Mediterranean region); considered cosmopolitan via introduction.

The flowers of *Polycarpon tetraphyllum* are said to be homogamous, self-pollinated, and often cleistogamous.

1a. Polycarpon tetraphyllum (Linnaeus) Linnaeus subsp. **tetraphyllum** ⬚F⬚ ⬚I⬚

Stems prostrate to erect, 3–15 cm. **Leaves** in whorls of 4 or opposite; stipules 1.8–2.8 mm; blade elliptic, ovate, or obovate, 2–12 mm. **Inflorescence bracts** absent. **Flowers:** sepals with keel and apex sometimes purplish, keeled, 1.5–2.5 mm; petals linear to oblong, 0.5–0.8 mm, ca. 1/2 times as long as sepals; stamens 3(–5). **Capsules** clasped by calyx, broadly ovoid, valve margins rolled inward after dehiscence. **Seeds** ovoid to lenticular, 0.4–0.8 mm, finely papillate. $2n$ = 32, 48, 64 (all Europe).

Flowering spring–fall. Waste places, cultivated land; 0–1400 m; introduced; B.C.; Ala., Calif., Fla., Ga., S.C., Tex.; s Europe (Mediterranean region); considered cosmopolitan via introduction, at least in part from ships' ballast.

Historic records are known from Massachusetts [wool waste, North Worcester (*Woodward* in 1917, GH)] and Pennsylvania [ballast at Philadelphia (*Burk* in 1880, PH), South Bethlehem, 1890s (T. C. Porter 1892)].

2. Polycarpon depressum Nuttall in J. Torrey and A. Gray, Fl. N. Amer. 1: 174. 1838 • California manyseed

Stems prostrate, 1–6 cm. **Leaves** opposite, not whorled; stipules 0.4–1.2 mm; blade spatulate to obovate or oblanceolate, 3–9 mm. **Inflorescence bracts** mostly present. **Flowers:** sepals flat or sometimes barely keeled, 1–1.5 mm; petals linear to elliptic, 0.8–1 mm, 1/2 times as long as sepals; stamens 3–5. **Capsules** not clasped by calyx, spherical, valve margins straight or only slightly rolled inward after dehiscence. **Seeds** obliquely triangular, 0.4–0.5 mm, finely granular.

Flowering spring–summer. Sandy places, fields, weedy areas; 0–500 m; Calif.; Mexico (Baja California).

6. LOEFLINGIA Linnaeus, Sp. Pl. 1: 35. 1753; Gen. Pl. ed. 5, 22. 1754 • [For P. Loefling, 1729–1756, Swedish botanist and explorer]

Ronald L. Hartman

Richard K. Rabeler

Herbs, annual. **Taproots** slender. **Stems** prostrate to erect, usually dichotomously branched at or near base, terete. **Leaves** opposite, basal leaves absent, axillary leaf clusters often present, weakly connate, sessile; stipules 2 per node, ± white to silvery, bristlelike, margins entire, apex acuminate; blade 1-veined or obscurely so, subulate to oblong, not succulent, apex blunt to spine-tipped. **Inflorescences** axillary, 1(–2)-flowered; bracts absent. **Pedicels:** flowers sessile. **Flowers:** perianth and androecium hypogynous; sepals ± distinct, green, linear to lanceolate, 1.8–6.5 mm, herbaceous, margins silvery, often scarious, apex acute to spinose; petals absent or rudimentary; nectaries not apparent; stamens 3–5, arising from base of ovary; filaments distinct; styles 3, distinct, filiform, shorter than 0.1 mm, glabrous proximally; stigmas 3, linear along adaxial surface of styles, papillate (30×). **Capsules** lanceoloid to ovoid, opening by 3 slightly recurved valves; carpophore absent. **Seeds** 25–35, tan with reddish brown band on curved edge, tear-shaped, plump, minutely papillate on broadly grooved edge, marginal wing absent, appendage absent.

Species 7 (1 in the flora): North America, Europe (w Mediterranean region), sw Asia, Africa (w Mediterranean region, Sahara).

SELECTED REFERENCE Barneby, R. C. and E. C. Twisselmann. 1970. Notes on *Loeflingia* (Caryophyllaceae). Madroño 20: 398–408.

1. Loeflingia squarrosa Nuttall in J. Torrey and
A. Gray, Fl. N. Amer. 1: 174. 1838   F

Loeflingia pusilla Curran;
L. squarrosa subsp. *artemisiarum*
Barneby & Twisselmann;
L. squarrosa var. *artemisiarum*
(Barneby & Twisselmann) Dorn;
L. squarrosa subsp. *cactorum*
Barneby & Twisselmann;
L. squarrosa subsp. *texana*
(Hooker) Barneby & Twisselmann;
L. texana Hooker

Plants 1–12 cm, covered with stalked glands, somewhat fleshy. **Stems** stiff, usually dichotomously branched at or near base, variously branched or not distally. **Leaves** usually connate proximally into short, scarious sheath; stipules filamentous to spinose, 0.4–1.5 mm; blade erect to ± recurved, 0.4–5.5 mm, apex blunt to spine-tipped. **Inflorescences** often secund. **Flowers** cleistogamous; sepals erect to squarrose, resembling leaves (especially outer pair), usually with 2 filamentous to stiff lateral spurs, 1.8–6.5 mm, becoming hardened, margins often scarious. **Capsules** 3-angled, 1.5–3.7 mm, $^{1}/_{2}$–$^{4}/_{5}$ times as long as sepals. **Seeds** 0.4–0.7 mm.

Flowering spring–summer. Sandy, gravelly areas; 0–2100 m; Ariz., Ark., Calif., Kans., Nebr., Nev., Okla., Oreg., Tex., Utah, Wash., Wyo.; Mexico (Sonora).

R. C. Barneby and E. C. Twisselmann (1970) recognized four subspecies of *Loeflingia squarrosa*, for the most part allopatric. After a reevaluation of the characters used in their key, we feel that those entities are best regarded as geographical races of the species. This is justified largely by both the overlap in expressions of and the lack of correlation of the characters. Barneby and Twisselmann placed major emphasis on habit and stature, including the position of the primary dichotomy of the plant and the location of intermediate monochasial branching subsequent to last branching, if present. In their words, "We have developed an objective formula for describing the permutations of branching, but believe that the intangible quality of habit permits intuitive sorting of material into categories that coincide with comprehensive dispersal patterns." They used, in addition to habit, the size and stiffness of stipules and sepal spurs, size of sepals and their orientation, size and shape of capsules, and size of seeds to discriminate between subspecies.

7. STIPULICIDA Michaux, Fl. Bor.-Amer. 1: 26, plate 6. 1803 • [Latin *stipula*, diminutive of *stipes*, stalk, and *-cida*, cut, alluding to the incised stipules)

Ann Swanson

Richard K. Rabeler

Herbs, annual (or short-lived perennial?). **Taproots** filiform to stout. **Stems** diffuse to erect, repeatedly dichotomous, terete. **Leaves** opposite (cauline) or rosulate (basal), connate (distally) or not (proximally), petiolate (basal) or sessile (cauline); stipules 2 per node (cauline leaves) or forming tuft of to 14+ per node (basal leaves), white to tan, filiform, forming incised or notched nodal fringe; blade 1-veined, spatulate to suborbiculate (basal) or scalelike, subulate to triangular (cauline), not succulent, apex obtuse. **Inflorescences** terminal, compact, few-flowered cymes; bracts paired, scalelike. **Pedicels** erect. **Flowers:** perianth and androecium hypogynous; sepals distinct, reddish brown, elliptic to obovate, 0.8–2 mm, scarious, margins scarious, apex acute to obtuse or mucronate; petals 5, white, blade apex entire to erose; nectaries as minute, rounded lobes flanking filament bases; stamens 3–5; filaments distinct; styles 3, distinct or nearly so, capitate, ca. 0.2 mm, glabrous proximally; stigmas 3, terminal, obscurely papillate (30×). **Capsules** ellipsoid to globose, opening by 3 recurved valves; carpophore absent. **Seeds** ca. 20, golden chestnut to reddish brown, ± triangular, laterally compressed, lustrous, reticulate, marginal wing absent, appendage absent.

Species 1: se United States, West Indies (Cuba).

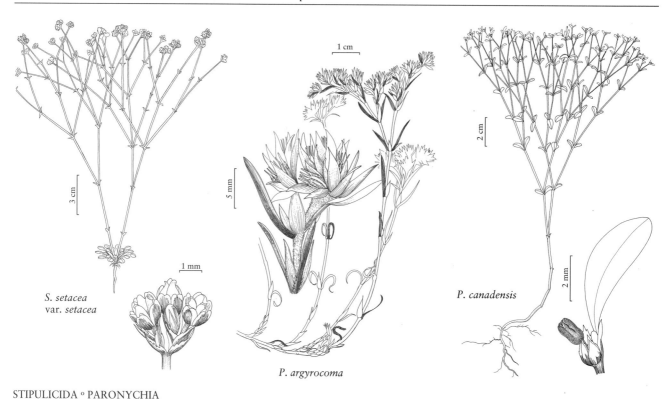

S. setacea
var. setacea

P. argyrocoma

P. canadensis

STIPULICIDA ○ PARONYCHIA

1. Stipulicida setacea Michaux, Fl. Bor.-Amer. 1: 26, plate 6. 1803 • Pineland scaly-pink F

Plants 5–25 cm. **Leaf blades:** basal 5–20 mm, cauline 0.5–2.5 mm, margins shallowly to deeply incised, especially toward base. **Flowers:** petals 1–2.5 mm, equaling or slightly longer than sepals, margins minutely 2–5-toothed near base. **Seeds** 0.5 mm.

Varieties 2 (2 in the flora): se United States, West Indies (Cuba).

1. Sepal margins ± entire; outer sepal apices acute to obtuse . . . . 1a. *Stipulicida setacea* var. *setacea*
1. Sepal margins lacerate; outer sepal apices mucronate . . . . 1b. *Stipulicida setacea* var. *lacerata*

1a. Stipulicida setacea Michaux var. **setacea** E F

Stipulicida filiformis Nash; *S. setacea* var. *filiformis* (Nash) D. B. Ward

Sepals with margins ± entire, outer sepals with apex acute to obtuse.

Flowering spring–fall. Dry, sandy soil of pinelands, scrub, fields; 0–200 m; Ala., Fla., Ga., La., Miss., N.C., S.C., Va.

We follow W. S. Judd (1983) in considering *Stipulicida filiformis*, described to account for plants with fewer flowers and thinner stems, as a morphological/ecological extreme of *S. setacea*; for an alternative view, see D. B. Ward (2001).

1b. Stipulicida setacea Michaux var. **lacerata** C. W. James, Rhodora 59: 98. 1957

Sepals with margins lacerate, outer sepals with apex mucronate.

Flowering spring-summer. Dry, sandy soil of pinelands, scrub, fields; 0–100 m; Fla.; West Indies (Cuba).

43b. CARYOPHYLLACEAE Jussieu subfam. PARONYCHIOIDEAE Meisner, Pl. Vasc. Gen. 1: 132; 2: 96. 1838 (as Paronychieae)

Richard K. Rabeler

Ronald L. Hartman

Herbs [small shrubs], annual, biennial, or perennial; taprooted, not rhizomatous. **Stems** prostrate to ascending or erect, simple or branched. **Leaves** opposite, distalmost or all sometimes alternate, bases connate or not, sometimes petiolate, stipulate; stipules ovate or deltate to lanceolate or spatulate, scarious; blade needlelike or often spatulate to elliptic or suborbiculate, seldom succulent. **Inflorescences** terminal or axillary cymes or flowers solitary; bracts foliaceous or usually scarious; involucel bracteoles absent. **Pedicels** present or flowers sessile. **Flowers** bisexual or sometimes unisexual (the plant then dioecious or polygamodioecious); perianth and androecium perigynous; hypanthium usually cup-shaped or cylindric or conic to urceolate; sepals (3–)5, distinct or rarely connate proximally, apex often hooded or awned (awn often subapical); petals absent; stamens absent or 1–5, in 1 whorl arising from hypanthium rim; staminodes absent or 5 (16–19 in *Achyronychia*); ovary 1-locular; styles 1–3, distinct or sometimes connate proximally; stigmas 2 or 3. **Fruits** utricles, indehiscent or sometimes opening by 3 or 8–10 valves; carpophore absent. **Seeds** 1, white to tan or brown to black, ovoid to reniform, not or slightly laterally compressed; embryo peripheral or central, curved or straight. $x = 7, 8, 9$.

Genera 17, species ca. 200 (6 genera, 32 species in the flora): s North America, South America (Andean region), Europe (Mediterranean region), Asia (Mediterranean region, e to India), Africa (Mediterranean region).

Paronychioideae is characterized by the presence of stipules, petaloid staminodes, and usually indehiscent utricles. It is of similar size to Polycarpoideae; about two-thirds of the species are found in *Paronychia* and *Herniaria*. Paronychioideae is sometimes segregated from Caryophyllaceae as Illecebraceae, due to emphasis on the utricle; molecular data does not support recognition of Illecebraceae (M. Nepokroeff et al. 2002; R. D. Smissen et al. 2002). While there are some features shared with Polycarpoideae (stipules, solanad type of embryogeny), floral reduction is more pronounced in this group.

Tentatively, Corrigioleae (*Telephium* and *Corrigiola*) is included here. M. G. Gilbert (1987) proposed transferring this tribe to Molluginaceae, noting that the morphological anomalies within Caryophyllaceae, including alternate leaves, exhibited in these plants were reduced under such an alignment. M. Nepokroeff et al. (2002) retained the tribe within Caryophyllaceae, placed as a sister group to the rest of the family.

SELECTED REFERENCE Chaudhri, M. N. 1968. A revision of the Paronychiinae. Meded. Bot. Mus. Utrecht 285: 1–440.

8. PARONYCHIA Miller, Gard. Dict. Abr. ed. 4, vol. 3. 1754 • Nailwort, whitlow-wort

[Greek *para*-, beside, and *onyx* or *onychos*, fingernail, alluding to use for treating whitlow or felon, a disease of the fingernails]

Ronald L. Hartman

John W. Thieret

Richard K. Rabeler

Anychia Michaux; *Anychiastrum* Small; *Gastronychia* Small; *Gibbesia* Small; *Nyachia* Small; *Odontonychia* Small; *Siphonychia* Torrey & A. Gray

Herbs, annual, biennial, or perennial, sometimes with woody base. **Taproots** filiform to stout. **Stems** prostrate, ascending, or erect, simple or branched, terete to angular. **Leaves** opposite, connate by stipules from adjacent leaves, petiolate (basal) or sessile (cauline); stipules 2 per node, often conspicuous, white or silvery, subulate to lanceolate or ovate, margins entire or fimbriate, apex subobtuse or acute to acuminate, unlobed or sometimes deeply 2-fid; blade 1-veined, linear to elliptic, oblanceolate, or spatulate, sometimes thickened and succulent, apex obtuse or acute to acuminate or spinose. **Inflorescences** terminal or sometimes axillary, frequently much-branched or congested cymes, or flowers solitary; bracts paired, dimorphic (resembling leaf blades and stipules), often concealing flowers. **Pedicels** erect in fruit. **Flowers** bisexual or rarely unisexual, some plants also having staminate unisexual flowers, others also having pistillate unisexual flowers, not woolly, with hairs ± straight or tips coiled, 0.1–0.3 mm; hypanthium cup-shaped, tapering or expanded distally; sepals (3–)5, connate proximally, white or yellowish to green or reddish or purplish brown, subulate to linear-oblong, lanceolate, spatulate, or ovate, 0.4–4.5 mm, margins translucent to white, scarious or papery, apex defined by a usually prominent adaxial hood, ascending to slightly descending, rounded to triangular, sometimes absent (*P. americana*, *P. erecta*), apex obtuse or rounded, usually with terminal or subterminal cusp, crest, mucro, or prominent awn (often thickened-conic proximally, spinose distally); nectar secreted from within hypanthium; stamens usually 5; filaments distinct or connate proximally with alternating staminodes; staminodes absent or 5, arising from hypanthium rim, subulate to narrowly triangular, filiform, or oblong; styles 1–2(–3), distinct or often connate proximally $^{1}/_{10}$–$^{9}/_{10}$ of length, subcapitate to filiform, 0.07–3.2 mm, glabrous proximally; stigmas 2(–3), subterminal or linear along adaxial surface of style branches, obscurely papillate (50×). **Utricles** ovoid to globose or rarely 4-angular, membranous, indehiscent. **Seeds** brown, subglobose to ellipsoid, laterally compressed, smooth, marginal wing absent, appendage absent; embryo peripheral, curved. $x = 7, 8, 9$.

Species ca. 110 (26 in the flora): warm-temperate North America, South America, Eurasia, Africa.

Much of this treatment follows M. N. Chaudhri (1968), the only recent monograph of the genus; we agree with B. L. Turner (1983b) in not recognizing the infraspecific taxa that Chaudhri proposed for North American taxa. We follow L. H. Shinners (1962c) and Chaudhri in including the genus *Siphonychia* within *Paronychia*. This disposition is supported by results of the molecular study by B. Oxleman et al. (2002), which indicated that subg. *Paronychia* and subg. *Siphonychia* (Torrey & A. Gray) Chaudhri are sister taxa within a strongly supported clade.

We have chosen not to utilize an infrageneric classification of *Paronychia*. M. N. Chaudhri (1968) recognized three subgenera within *Paronychia*. Most of our species are included in one of the two sections of subg. *Paronychia*; within sect. *Paronychia*, North American taxa fall into three subsections (including five different series within subsect. *Paronychia*). B. Oxelman et al. (2002) found that *Paronychia* as presently accepted is polyphyletic, with subg. *Anoplonychia* (Fenzl) Chaudhri (48 Old World species) clustering with *Herniaria*. Since their investigation included only eight species of *Paronychia*, additional study is warranted before splitting *Paronychia* and, as a logical consequence, revising the infrageneric classification.

Flowers are described here in the closed state; the vast majority seen on *Paronychia* specimens examined are closed. The flower lengths given include awns (cusps, mucros, crests); sepal measurements do not include the awns; awns are interpreted as originating at the point of the departure of the usually ± horizontal, adaxial hood.

SELECTED REFERENCES Core, E. L. 1939. A taxonomic revision of the genus *Siphonychia*. J. Elisha Mitchell Sci. Soc. 55: 339–345. Core, E. L. 1940. Notes on the mid-Appalachian species of *Paronychia*. Virginia J. Sci. 1: 110–116. Core, E. L. 1941. The North American species of *Paronychia*. Amer. Midl. Naturalist 26: 369–397. Fernald, M. L. 1936c. Notes on *Paronychia*, *Anychia*. Rhodora 38: 416–421. Hartman, R. L. 1974. Rocky Mountain species of *Paronychia* (Caryophyllaceae): A morphological, cytological, and chemical study. Brittonia 26: 256–263. Shinners, L. H. 1962c. *Siphonychia* transferred to *Paronychia* (Caryophyllaceae). Sida 1: 101–103. Turner, B. L. 1983b. The Texas species of *Paronychia* (Caryophyllaceae). Phytologia 54: 9–23. Ward, D. B. 1977c. Keys to the flora of Florida. 2. *Paronychia* (Caryophyllaceae). Phytologia 35: 414–418.

1. Sepals rounded or with conic to triangular cusp, crest, or mucro, 0.1–0.3 mm.
   - 2. Sepals papery, typically red-brown proximally, white or whitish in distal $^1/_3$–$^1/_4$, mucronate or not.
     - 3. Sepals not mucronate; plant cespitose perennial (occasionally biennial?); stems typically dimorphic, sterile ones with leaves more closely imbricate than fertile . . . . 11. *Paronychia erecta*
     - 3. Sepals mucronate; plants annual or rarely short-lived, noncespitose perennial; stems only fertile.
       - 4. Flowers 1.2–2.2 mm, in clusters 2–5 mm wide; sepals 0.6–1.2 mm; styles 1.3 mm or less . . . . 20. *Paronychia patula*
       - 4. Flowers 2.3–3.5 mm, in clusters often 5–30 mm wide; sepals 1.5–2 mm; styles 1.5 mm or longer . . . . 22. *Paronychia rugelii*
   - 2. Sepals herbaceous to leathery or rigid, mucronate.
     - 5. Leaf blade margins revolute . . . . 6. *Paronychia chartacea*
     - 5. Leaf blade margins flat.
       - 6. Plants glabrous. . . . . 5. *Paronychia canadensis*
       - 6. Plants, at least stems and often hypanthia, pubescent, sometimes sparsely so.
         - 7. Flowers briefly constricted above hypanthium; sepal apices broadly rounded, with cusps absent; sepals obovate, 0.4–0.8 mm. . . . . 2. *Paronychia americana*
         - 7. Flowers not constricted between hypanthium and calyx; sepal apices with cusps 0.1–0.3 mm; sepals linear-oblong to ovate, 0.8–1.3 mm.
           - 8. Leaf margins ciliate, at least proximally; staminodes subulate, 0.2–0.3 mm . . . . 4. *Paronychia baldwinii*
           - 8. Leaf margins glabrous; staminodes absent . . . . 12. *Paronychia fastigiata*
1. Sepals with distinct apical or subapical awn or spine, 0.5–2.1 mm, proximal $^1/_3$–$^2/_3$ (measured from point of departure of hood) often thickened and ± conic or awn ± lanceoloid and 0.4–2.1+ mm.
   - 9. Awns and often margins of calyx lobes white, often prominent.
     - 10. Awns of calyx lobes ± lanceoloid, often flattened, distinct spine absent; s, w Texas.
       - 11. Leaves glabrous; hypanthia with numerous hooked hairs. . . . . 18. *Paronychia maccartii*
       - 11. Leaves hirtulose; hypanthia with straight hairs . . . . 26. *Paronychia wilkinsonii*
     - 10. Awns of calyx lobes spinose or ± conic with spinose tip.
       - 12. Awns of calyx lobes spinose, ± erect; plants perennial; Kentucky, Georgia, Tennessee, and east . . . . 3. *Paronychia argyrocoma*
       - 12. Awns of calyx lobes ± broadly conic, divergent; plants annual or biennial.
         - 13. Stems erect; calyx rather uniformly pubescent with hooked hairs; Louisiana, Oklahoma, Texas. . . . . 9. *Paronychia drummondii*

13. Stems prostrate; calyx with at least a few decidedly enlarged, hooked hairs; s Texas . . . . 16. *Paronychia jonesii*

[9. Shifted to left margin.—Ed.]

9. Awns and margins green to brown or yellowish (awns white but inconspicuous and threadlike in *P. ahartii*).

14. Plants diminutive, annual; stems erect, 0.5–1.2 cm; calyx lobes with linear, herbaceous midrib terminating in white, wavy, threadlike, spreading awn 1.5–2 mm, margins scarious, resembling stipule with erect, split apex apparently representing hood . . . . . . 1. *Paronychia ahartii*

14. Plants if annual, 3–38 cm; calyx lobes lanceolate to ovate with relatively stout awn or spine, margins if scarious, not stipulelike, hoods ± horizontal, adaxial.

15. Leaves relatively broad (ovate to elliptic or oblong).

16. Plants erect to prostrate, annual.

17. Flowers 3–12 per node along stems and branches, moderately pubescent with hooked hairs proximally; awns of calyx lobes widely divergent, 0.6–0.9 mm; introduced; California . . . . 10. *Paronychia echinulata*

17. Flowers 1 or 2 per node in diffuse cymes, glabrous to strigose; awns of calyx lobes straight to slightly spreading, 0.3–0.4 mm; native; Florida, Georgia, South Carolina . . . . 14. *Paronychia herniarioides*

16. Plants cespitose or prostrate, strongly perennial.

18. Plants mat-forming; stems mostly 20–50 cm; leaves ± moderately antrorsely appressed-pubescent; flowers 3–6 per leaf axil along stem, 1.9–2.4 mm; coastal hills; California . . . . 13. *Paronychia franciscana*

18. Plants densely cespitose, forming cushions; stems 5–10 cm; leaves glabrous (or minutely ciliolate); flowers solitary, terminal, 2.5–2.8 mm; alpine; Colorado, New Mexico, Utah, Wyoming . . . . 21. *Paronychia pulvinata*

15. Leaves needlelike (linear to subulate or filiform).

19. Plants annual, biennial, or weak perennial, without vegetative shoots or mats or numerous old stems.

20. Calyces of distalmost flowers noticeably exceeded by subtending bracts . . . . 24. *Paronychia setacea*

20. Calyces of distalmost flowers longer than subtending bracts.

21. Flowers mostly solitary and separate; flowers 1.6–2 mm; c Texas . . . . 17. *Paronychia lindheimeri*

21. Flowers mostly clustered; flowers 2.3–3.2 mm; Trans-Pecos Texas . . . . 19. *Paronychia monticola* (in part)

19. Plants strongly perennial with vegetative shoots and remnants of old flowering stems, caudex woody, branched.

22. Flowers 1–6, sessile, terminating stems; leaves of sterile and flowering stems appressed, tightly overlapping ("juniperlike"), blades mostly with rounded apex; awns of sepals erect to somewhat spreadin . . . . 23. *Paronychia sessiliflora*

22. Flowers 7–80+, pedicellate, in diffuse to compact cymes; leaves of stems spreading, at least moderately spaced, blades with rounded to mucronate or cuspidate apex.

23. Sepals prominently 3-veined throughout . . . . 25. *Paronychia virginica*

23. Sepals not veined or obscurely so.

24. Stems and leaf blades glabrous; awns 1–1.5(–2) mm . . . . 19. *Paronychia monticola* (in part)

24. Stems and leaf blades puberulent to pubescent; awns 0.4–0.9 mm.

25. Calyces gradually narrowing distally; leaf blades 4–7 mm . . . . 7. *Paronychia congesta*

25. Calyces ± cylindric to expanding distally; leaf blades 7–25 (–34) mm.

26. Plants mat-forming with prostrate stems seldom more than 8 cm above ground surface; flowers 2.3–3.5 mm . . . 8. *Paronychia depressa*

26. Plants with erect to ascending stems, 10–30 cm; flowers 1.8–2.8 mm . . . . 15. *Paronychia jamesii*

1. Paronychia ahartii Ertter, Madroño 32: 87, fig. 1. 1985 • Ahart's nailwort C E

Plants annual; taproot filiform to stout. **Stems** erect, tightly branched, 0.5–1.2 cm, glabrous. **Leaves:** stipules broadly ovate, 3–6 mm, apex acute to acuminate, entire; blade linear to oblanceolate, 2.5–7.5 × 0.5–1.2 mm, leathery, apex spinulose, glabrous. **Inflorescences:** flowers axillary, solitary. **Flowers** 5-merous, cylindric, with enlarged hypanthium and calyx cylindric to somewhat tapering distally, 4.2–5 mm, moderately hairy in proximal $1/2$ with hooked to coiled hairs; sepals green to tan, veins absent, lanceolate to elliptic, 3.5–4.5 mm, margins translucent, 0.7–1+ mm wide, scarious (resembling stipules), apex (of herbaceous midrib) terminated by awn, hood apparently consisting of prominent, erect, scarious extension of margins split at apex, awn ± spreading, 1.5–2 mm, oblong extension of midrib in proximal $1/4$, with white, wavy, threadlike spine; staminodes filiform, ca. 1 mm; style 1, cleft in distal $3/4$, ca. 0.5 mm. **Utricles** ovoid, ca. 1.3 mm, papillose distally.

Flowering spring. Well-drained rocky outcrops, vernal pool edges, volcanic uplands; of conservation concern; 0–500 m; Calif.

Paronychia ahartii, first collected in 1938, is known from three counties in north-central California. It most closely resembles *P. arabica* (Linnaeus) de Candolle, a species of northern African and Arabian deserts.

2. Paronychia americana (Nuttall) Fenzl ex Walpers, Repert. Bot. Syst. 1: 262. 1842 • American nailwort E

Herniaria americana Nuttall, Amer. J. Sci. Arts 5: 291. 1822; *Paronychia americana* subsp. *pauciflora* (Small) Chaudhri; *Siphonychia americana* (Nuttall) Torrey & A. Gray; *S. pauciflora* Small

Plants annual or biennial; taproot slender. **Stems** erect or ascending to prostrate, branched, 5–60 cm, often retrorsely pubescent on 1 side. **Leaves:** stipules ovate-lanceolate, 1.5–7 mm, apex acuminate, fimbriate; blade spatulate to oblanceolate or linear-oblanceolate, 3–20 × 1–4 mm, herbaceous, apex acute to obtuse or rounded, glabrous to scaberulous. **Cymes** terminal, 9–25-flowered, ± compact, forming spheroid glomerules 2–6 mm wide. **Flowers** 5-merous, ± short-cylindric, with slightly enlarged hypanthium and calyx widening somewhat distally, 1–1.8 mm, sparsely to moderately pubescent proximally with hooked to straight hairs; sepals red-

brown (sometimes finely striped or mottled), white distally, midrib obscure, obovate, 0.4–0.8 mm, leathery to rigid, margins white, 0.03–0.1 mm wide, thinly herbaceous, apex broadly rounded, hood broadly rounded, awn or mucro usually absent or minute; staminodes narrowly triangular, 0.3–0.4 mm; style 1, cleft in distal $1/6$, 0.6–0.8 mm. **Utricles** ovoid to ellipsoid, 0.6–0.8 mm, smooth, glabrous.

Flowering year round. Dunes, pine/oak woodland, fields, clearings, roadsides; 0–300 m; Ala., Fla., Ga., S.C.

Plants of *Paronychia americana* with fewer flowers in lax cymes from Florida and Georgia were named subsp. *pauciflora*, a distinction that we do not feel is worth recognition.

3. Paronychia argyrocoma (Michaux) Nuttall, Gen. N. Amer. Pl. 1: 160. 1818 • Silvery nailwort, silverling E F

Anychia argyrocoma Michaux, Fl. Bor.-Amer. 1: 113. 1803; *Paronychia argyrocoma* subsp. *albimontana* (Fernald) Maguire; *P. argyrocoma* var. *albimontana* Fernald

Plants perennial, matted; caudex woody. **Stems** prostrate to ascending, much-branched, 5–60 cm, often retrorsely pubescent on 1 side. **Leaves:** stipules lanceolate, 2.5–8 mm, apex acute, entire; blade linear to linear-lanceolate or -oblanceolate, 5–30 × 0.5–2 mm, leathery, apex acute, often mucronate, sparsely appressed-pubescent. **Cymes** terminal, 15–25+-flowered, very compact, forming conspicuous glomerules 10–20 mm wide. **Flowers** 5-merous, short-cylindric to ovoid, with enlarged hypanthium and calyx cylindric to tapering distally, 3.5–6.5 mm, pubescent with antrorse, slightly spreading, silky hairs; sepals greenish to brownish, veins 3, obscure, ribs absent, narrowly lanceolate, 2–3.2 mm, leathery to rigid, margins translucent, ca. 0.1 mm wide, scarious, apex terminated by awn, hood narrowly triangular, awn straight to slightly divergent, white, 0.9–2 mm, scabrous, spinose; staminodes narrowly triangular, 0.4–0.6 mm; style 1, cleft in distal $1/6$, 1.4–2 mm. **Utricles** oblong, 1.5–1.8 mm, smooth, pubescent distally.

Flowering spring–early fall. On or among rocks; 200–1800 m; Ga., Ky., Maine, Md., Mass., N.H., N.C., Tenn., Vt., Va., W.Va.

Plants of *Paronychia argyrocoma* with glabrous or barely scabrous leaves and glabrous sepal awns have sometimes been recognized as var. *albimontana*. They are found in both the southern and northern areas of the species range, but, curiously, not in the central portion (M. N. Chaudhri 1968).

4. Paronychia baldwinii (Torrey & A. Gray) Fenzl ex Walpers, Repert. Bot. Syst. 1: 262. 1842 • Baldwin's nailwort E

Anychia baldwinii Torrey & A. Gray, Fl. N. Amer. 1: 172. 1838; *Anychiasatrum baldwinii* (Torrey & A. Gray) Small; *A. riparium* (Chapman) Small; *Paronychia baldwinii* var. *ciliata* Chaudhri; *P. baldwinii* subsp. *riparia* (Chapman) Chaudhri; *P. riparia* Chapman

Plants annual, biennial, or perennial, often matted; taproot slender. **Stems** prostrate to erect, branched, 5–70 cm, mostly retrorsely to spreading-pubescent on 1 side or throughout. **Leaves:** stipules lanceolate, 2–6 mm, apex acuminate, entire; blade oblong to elliptic or oblanceolate, 3–25 × 1–6 mm, herbaceous, apex acute and briefly cuspidate, glabrous. **Cymes** terminal, 20–40+-flowered, diffuse, lax, repeatedly forked or dichotomous. **Flowers** 5-merous, ± short-cylindric, with enlarged hypanthium and calyx cylindric, 1–1.7 mm, glabrous to pubescent with short hairs, often minutely ciliate, sometimes glaucous; sepals greenish or greenish white to brownish, veins absent, ovate to oblong, 0.8–1.3 mm, herbaceous, margins white, 0.05–0.1 mm wide, scarious to papery, apex terminated by minute cusp, hood narrowly rounded, cusp light green to whitish, straight, short-conic, 0.1–0.15 mm, minutely scabrous; staminodes subulate, 0.2–0.3 mm; style 1, cleft in distal ⁴/₅+, 0.2–0.4 mm. **Utricles** ellipsoid, 1–1.3 mm, papillate distally.

Flowering summer–fall. Dunes, woodlands, fields, clearings, roadsides, riverbanks, hummocks, waste places; 0–200 m; Ala., Fla., Ga., N.C., S.C., Va.

Chaudhri used duration and pubescence to recognize two subspecies of *Paronychia baldwinii*, characters that L. H. Shinners (1962c) found to vary independently in this species.

5. Paronychia canadensis (Linnaeus) Alph. Wood, Class-Book Bot. ed. s.n.(b), 262. 1861 • Smooth-forked nailwort, forked-chickweed E F

Queria canadensis Linnaeus, Sp. Pl. 1: 90. 1753; *Anychia canadensis* (Linnaeus) Elliott

Plants annual; taproot filiform to slender. **Stems** erect, dichotomously branched, 3–40 cm, ultimate branches filiform, glabrous. **Leaves:** stipules subulate to lanceolate, 0.5–4 mm, apex acuminate, entire; blade usually dotted or blotched, elliptic to obovate, 3–30 × 1–11 mm, herbaceous, apex acute to obtuse or cuspidate, glabrous. **Cymes** diffuse,

flowers solitary (paired) in axils of leafy bracts. **Flowers** 5-merous, ± short-cylindric, with enlarged hypanthium and calyx cylindric, 0.9–1.3 mm, glabrous; sepals greenish to brownish, midrib and lateral pair of veins prominent, ovate to oblong, 0.5–1 mm, herbaceous, margins white to translucent, 0.03–0.1 mm wide, scarious, apex terminated by mucro, hood rounded-triangular, mucro triangular, 0.07–0.15 mm, glabrous; staminodes absent or indistinct; styles 2, 0.2–0.3 mm. **Utricles** ovoid to globose, ca. 0.5 mm, granular in distal ¹/₂, glabrous.

Flowering spring–fall. Woodlands, thickets, fields, clearings, roadsides, waste places; 0–1300 m; Ont.; Ala., Ark., Conn., Del., D.C., Ga., Ill., Ind., Iowa, Kans., Ky., La., Md., Mass., Mich., Minn., Mo., Nebr., N.H., N.J., N.Y., N.C., Ohio, Okla., Pa., R.I., S.C., Tenn., Vt., Va., W.Va., Wis.

6. Paronychia chartacea Fernald, Rhodora 38: 418. 1936 • Paper nailwort, papery whitlow-wort C E F

Nyachia pulvinata Small, Torreya 25: 12. 1925, not *Paronychia pulvinata* A. Gray 1864

Plants annual or short-lived perennial, often densely matted; taproot filiform to slender. **Stems** prostrate to ascending, dichotomously branched, 5–20 cm, often retrorsely pubescent on 1 side. **Leaves:** stipules ovate-lanceolate, 1–1.5 mm, apex obtuse to acuminate, fimbriate; blade oblong to triangular-ovate, 1.5–6 × 0.5–1.5 mm, leathery, apex acute to obtuse, glabrous. **Cymes** terminal, 7–25+-flowered, much-branched and very profusely developed along all branches, 1.5–20 mm wide. **Flowers** 5-merous, occasionally 3- or 4-merous, obconic to short-campanulate, with gradually expanded hypanthium and calyx widening distally, 0.5–0.8 mm, pubescent proximally with straight to hooked hairs; sepals brown-purple, white to greenish at apex, sometimes proximally, veins absent, ± oblong, 0.5–0.7 mm, leathery to rigid, margins whitish to translucent, 0.1–0.2 mm wide, scarious, apex terminated by crest or mucro, hood broadly rounded, mucro rounded-triangular, ca. 0.1 mm, glabrous; staminodes absent or rudimentary; styles 2 (–3), ca. 0.3 mm. **Utricles** ovoid or ellipsoid to subglobose, 0.4–0.5 mm, smooth, glabrous.

Varieties 2 (2 in the flora): Florida.

The occasional occurrence of 3- or 4-merous flowers is unique to *Paronychia chartacea* among North American species of the genus.

Paronychia chartacea is in the Center for Plant Conservation's National Collection of Endangered Plants.

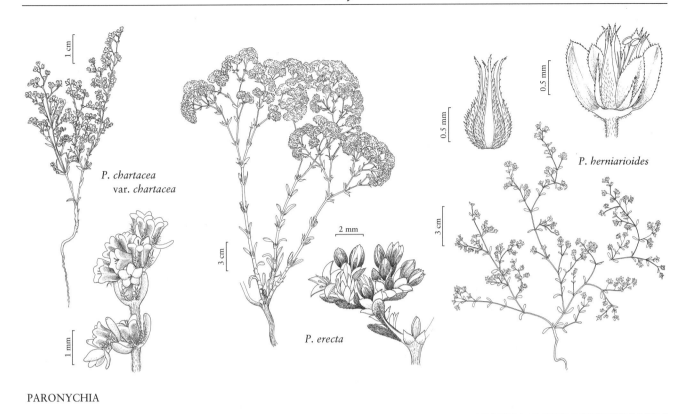

P. chartacea
var. *chartacea*

P. erecta

P. herniarioides

PARONYCHIA

The following key is adapted from L. C. Anderson (1991).

1. Caudex (stem base) (1–)1.5–3.5(–4.2) mm; leaf blades 0.5–1.5 mm wide; flower clusters 3–20 mm wide; c peninsular Florida . . . .
. . . . 6a. *Paronychia chartacea* var. *chartacea*
1. Caudex (stem base) (0.4–)0.7–1(–1.5) mm; leaf blades 1.2–3 mm wide; flower clusters 1.5–4 mm wide; Florida panhandle . . . .
. . . . 6b. *Paronychia chartacea* var. *minima*

6a. Paronychia chartacea Fernald var. **chartacea**
C E F

Plants annual or short-lived perennial; caudex (stem base) (1–)1.5–3.5(–4.2) mm. **Stems** without purple epidermal inclusions. **Leaf blades** 0.5–1.5 mm wide. **Flower clusters** 3–20 mm wide.

Flowering year-round (mainly Aug–Mar). Sandhills, pine/oak woodland, open scrub; of conservation concern; 30–50 m; Fla.

The diminutive var. *chartacea* is threatened by habitat destruction via both housing developments and citrus grove expansion in central Florida. R. Kral (1983) suggested that it is an early successional species in sand scrub areas since it thrives on bare sands but disappears as taller herbs become established.

6b. Paronychia chartacea Fernald var. **minima** (L. C. Anderson) R. L. Hartman, Sida 21: 754. 2004   C E

Paronychia chartacea subsp. *minima* L. C. Anderson, Sida 14: 436, fig. 1. 1991

Plants annual; caudex (stem base) (0.4–)0.7–1(–1.5) mm. **Stems** purple-spotted with vertically elongate epidermal inclusions. **Leaf blades** 1.2–3 mm wide. **Flower clusters** 1.5–4 mm wide.

Flowering summer–fall (mainly Jul–Oct). Coarse, white sand; of conservation concern; 100 m; Fla.

Variety *minima* is known from sandy lake margins on the panhandle of Florida in two counties. It exhibits pronounced sexual dimorphism. "Predominantly male plants are more openly branched, usually with two successive dichotomies that result in a spindly, cruciform prostrate plant, whereas plants with predominantly bisexual or rarely pistillate flowers are more densely matted and have more numerous, shorter branches...Sexual dimorphism is present, but much less pronounced, in ssp. *chartacea*" (L. C. Anderson 1991).

7. Paronychia congesta Correll, Brittonia 18: 307. 1967 • Rio Grande nailwort C E

Plants perennial; caudex woody, branched. **Stems** erect, 6–10 cm, evenly puberulent with spreading hairs. **Leaves:** stipules lanceolate, 4–7 mm, apex acuminate, entire; blade linear or needlelike, 4–7 × ca. 0.5 mm, leathery, apex spinulose, puberulous. **Cymes** terminal, 7–28+-flowered, branched, mostly congested in clusters 5–15 mm wide. **Flowers** 5-merous, extended-urecolate, with prominent hypanthial bulge and calyx tapering somewhat distally, 2.8–3.1 mm, moderately pubescent with straight, spreading to appressed hairs generally throughout; sepals green to red-brown, veins absent or midrib and lateral pair of veins evident in fruit, lanceolate to oblong, 1.3–1.5 mm, leathery to rigid, margins translucent, less than 0.05 mm wide, scarious, apex terminated by awn, hood prominent, narrowly rounded, awn widely divergent, 0.5–0.7 mm, broadly conic in proximal $^1/_2$–$^2/_3$ with yellowish, ± glabrous spine; staminodes subulate, ca. 0.5 mm; style 1, 2-lobed in distal $^1/_3$–$^3/_5$, 0.5–0.7 mm. **Utricles** ± ovoid, 0.8–0.9 mm, ± smooth, glabrous.

Flowering spring-summer. Thorn scrubland openings; of conservation concern; 200-300 m; Tex.

Paronychia congesta is known only from two collections in Jim Hogg County.

8. Paronychia depressa (Torrey & A. Gray) Nuttall ex A. Nelson, Bull. Torrey Bot. Club 26: 236. 1899 • Spreading nailwort E

Paronychia jamesii Torrey & A. Gray var. *depressa* Torrey & A. Gray, Fl. N. Amer. 1: 171. 1838; *P. depressa* var. *brevicuspis* (A. Nelson) Chaudhri; *P. depressa* var. *diffusa* (A. Nelson) Chaudhri

Plants perennial, often matted; caudex branched, woody. **Stems** prostrate to sprawling, much-branched, 8–15 cm, scabrous-puberulent to puberulent throughout. **Leaves:** stipules lanceolate, 2–8 mm, apex acuminate, entire; blade linear, 8–15(–23) × 0.5–1 mm, leathery, apex shortly cuspidate, puberulent. **Cymes** terminal, 3–7-flowered, branched, congested, in clusters 7–25 mm wide. **Flowers** 5-merous, ± ovate, with prominently enlarged hypanthium and calyx tapering only slightly distally, 2.3–3.5 mm, puberulent; sepals green to purple-brown, midrib and lateral pair of veins ± obscure, oblong to lanceolate-oblong, 1.7–2 mm, leathery to rigid, margins whitish to translucent, 0.05–0.1 mm wide, scarious, apex terminated by awn, hood prominent, rounded-triangular, awn divergent, conic in

proximal $^1/_2$–$^2/_3$ with yellowish, scabrous spine 0.7–0.9 mm; staminodes filiform, 0.5–0.8 mm; style 1, cleft in distal $^1/_3$–$^2/_3$, 0.8–1.4 mm. **Utricles** ± ovoid, 0.8–0.9 mm, smooth, glabrous. **2n** = 32.

Flowering summer. Dry plains, rocky ridges, hillsides; 800–3000 m; Colo., S.Dak., Wyo.

Paronychia depressa is considered to be closely related to the nearly allopatric *P. jamesii*.

9. Paronychia drummondii Torrey & A. Gray, Fl. N. Amer. 1: 170. 1838 • Drummond's nailwort E

Paronychia drummondii subsp. *parviflora* Chaudhri

Plants annual or biennial; taproot filiform to slender. **Stems** sprawling t2o erect, nearly simple to much-branched especially distally, 7–35 cm, retrorsely pubescent on 1 side or throughout. **Leaves:** stipules lanceolate to ovate, 5–10 mm, apex acuminate, entire; blade linear-oblong to oblanceolate, 5–30 × 1–7 mm, leathery, apex acute to cuspidate, moderately antrorsely pubescent. **Cymes** terminal, 25+-flowered, much-branched but congested, clusters 5–20 mm wide. **Flowers** 5-merous, ± short-campanulate, with prominently enlarged hypanthium and calyx flaring distally, (1.5–)2–2.3 mm, pubescent proximally with short, hooked hairs; sepals greenish to brownish or red-brown, white distally, veins absent, oblong to obovate, 1–1.5 mm, leathery to rigid, margins white, 0.2–0.3 mm wide, papery, apex terminating in divergent awn, hood broadly rounded, awn white, stout-conic, 0.5–0.6 mm, glabrous; staminodes filiform, ± 0.3 mm; style 1, cleft in distal $^1/_5$, 0.3–0.6 mm. **Utricles** ellipsoid to subglobose, 0.5–0.8 mm, minutely papillate distally.

Flowering spring–fall. Sandy woodlands, clearings, roadsides; 0–100 m; La., Okla., Tex.

Chaudhri described two subspecies of *Paronychia drummondii* based on differences in leaf pubescence, flower size, and style length. We follow B. L. Turner (1983b), who noted that many specimens demonstrate intermediate characteristics, suggesting that taxonomic recognition not be given to the extremes.

10. Paronychia echinulata Chater, Feddes Repert. Spec. Nov. Regni Veg. 69: 52. 1964 • Eurasian nailwort I

Varieties 2 (1 in the flora): introduced; Europe (Mediterranean region), sw Asia, Africa (nw Mediterranean region).

Paronychia echinulata was for many years known as *P. echinata* Lamarck. The latter cannot be used since, as Chater noted, it is a superfluous name for *Illecebrum cymosa* Linnaeus [= *P. cymosa* (Linnaeus) de Candolle].

10a. Paronychia echinulata Chater var. echinulata

Plants annual; taproot filiform to slender. **Stems** erect or ascending, often basally branched, 3–20(–38) cm, retrorsely pubescent throughout. **Leaves:** stipules lanceolate, 1–3 mm, apex acuminate, entire; blade elliptic to oblong, 3–7 × 1.5–3.5 mm, granulose, leathery, apex shortly mucronate, glabrous. **Cymes** axillary, 3–12-flowered, subsessile, congested, in clusters 3–5 mm wide. **Flowers** 5-merous, short-campanulate, with prominent hypanthial bulge and calyx somewhat flaring distally, 2–2.5 mm, moderately pubescent with hooked hairs proximally; sepals red-brown, often finely striped or mottled, white distally, veins absent, obovate to oblong, 0.7–1 mm, leathery to rigid, margins white, 0.1–0.2 mm wide, papery to scarious, apex terminated by awn, hood prominent, broadly rounded, awn widely divergent, 0.6–0.9 mm, conic in proximal 1/3 with yellow, ± scabrous spine; staminodes filiform, 0.5–0.6 mm; style 1, split in distal 1/5, 0.4–0.5 mm. **Utricles** ± globose, 0.8–0.9 mm, minutely papillose distally. *2n* = 10, 14, 24, 28 (all Europe).

Flowering spring. Disturbed clay soil; 20–30 m; introduced; Calif.; s Europe; w Asia; nw Africa.

Variety *echinulata* is the only *Paronychia* introduced to North America from the Old World and is known from a single collection in Stanislaus County (*P. Allen*, in 1968, UC).

11. Paronychia erecta (Chapman) Shinners, Sida 1: 102. 1962 • Squareflower E F

Siphonychia erecta Chapman, Fl. South. U.S., 47. 1860; *Odontonychia corymbosa* (Small) Small; *O. erecta* (Chapman) Small; *Paronychia erecta* var. *corymbosa* (Small) Chaudhri

Plants perennial (occasionally biennial), often matted; taproot stout. **Stems** prostrate to ascending, branched especially distally, retrorsely to spreading-pubescent throughout (when pubescent); flowering stems 8–48 cm; sterile stems 2–10 cm. **Leaves:** stipules ovate-lanceolate, 2–12 mm, apex acuminate, entire; blade linear to spatulate-oblanceolate, 4–40 × 1–4 mm, leathery, apex obtuse to acute, moderately antrorsely pubescent. **Cymes** terminal, 15–200+-flowered, branched, densely to loosely grouped to form subcorymbose clusters 5–50 mm wide. **Flowers** 5-merous, narrowly ellipsoid, with slightly enlarged hypanthium and calyx tapering distally, 2.3–3.5 mm,

glabrous to slightly puberulent proximally with straight to hooked hairs; sepals red-brown, white or whitish distally, veins absent, narrowly oblong to lanceolate-oblong, 1.4–2 mm, papery, margins white, ca. 0.1 mm wide, papery, apex rounded, hood formed from slight incurving, awn or mucro absent; staminodes narrowly oblong, 0.4 mm; style 1, cleft in distal 1/10, 1.3–1.7 mm. **Utricles** ovoid to ellipsoid, 1–1.2 mm, rugulose, glabrous.

Flowering spring–fall. Coastal dunes, sandflats, pine/oak woodlands; 0–100 m; Ala., Fla., La., Miss.

Plants with pubescent stems, strigose leaves, and pubescent receptacles have been recognized as var. *corymbosa*.

12. Paronychia fastigiata (Rafinesque) Fernald, Rhodora 38: 421. 1936 • Hairy forked nailwort, forked-chickweed E

Anychia fastigiata Rafinesque, Atlantic J. 1: 16. 1832

Plants annual; taproot filiform to slender. **Stems** erect, usually much-branched, 4–30 cm, retrorsely to spreading-pubescent mostly on 1 side. **Leaves:** stipules subulate to lanceolate, 0.5–4.5 mm, apex acuminate, sometimes fimbriate; blade usually dotted or blotched, oblanceolate to elliptic or obovate, 2–25 × 0.5–7(–10) mm, herbaceous to leathery, apex subobtuse to acute or cuspidate, glabrous or occasionally with very few scattered trichomes along midrib. **Cymes** terminal, 25–70+-flowered, branched, loose to compact, forming clusters 3–10 mm wide. **Flowers** 5-merous, short-cylindric, with enlarged hypanthium and calyx ± cylindric, 1.1–1.6 mm, glabrous or with few scattered hairs; sepals greenish to brownish, veins absent, linear-oblong, 1–1.2 mm, leathery to rigid, margins white to translucent, 0.05–0.1 mm wide, scarious, apex terminated by mucro, hood narrowly rounded, mucro short-conic, 0.05–0.3 mm, ± minutely scabrous; staminodes absent; styles 2, 0.07–0.6(–0.7) mm. **Utricles** obovoid to obconic, 0.7–1 mm, minutely papillate. *2n* = 32, 36.

Varieties 3 (3 in the flora): e North America.

The varieties here recognized are sympatric over at least portions of their ranges. A detailed study on infraspecific variation within *Paronychia fastigiata* is warranted.

1. Styles 0.3–0.6(–0.7) mm, often bent or contorted
. . . . 12c. *Paronychia fastigiata* var. *pumila*
1. Styles 0.07–0.25(–0.3) mm, straight.
  2. Sepal cusps 0.2–0.3 mm . . . .
. . . . 12b. *Paronychia fastigiata* var. *nuttallii*
  2. Sepal cusps 0.05–0.15 mm . . . .
. . . . 12a. *Paronychia fastigiata* var. *fastigiata*

12a. Paronychia fastigiata (Rafinesque) Fernald var. **fastigiata** E

Anychia polygonoides Rafinesque; *Paronychia fastigiata* var. *paleacea* Fernald

Flowers: sepal cusps 0.05–0.15 mm; styles straight, 0.1–0.25 (–0.3) mm.

Flowering spring–fall. Woodlands, fields, clearings, rocky areas, roadsides, waste places; 0–1200 m; Ont.,Que.; Ala., Ark., Conn., Del., D.C., Ga., Ill., Ind., Iowa, Kans., Ky., La., Md., Mass., Mich., Minn., Miss., Mo., N.J., N.Y., N.C., Ohio, Okla., Pa., R.I., S.C., Tenn., Tex., Va., W.Va., Wis.

12b. Paronychia fastigiata (Rafinesque) Fernald var. **nuttallii** (Small) Fernald, Rhodora 38: 421. 1936 E

Anychia nuttallii Small, Torreya 25: 60. 1925

Flowers: sepal cusps 0.2–0.3 mm; styles straight, 0.07–0.25(–0.3) mm.

Flowering summer–fall. Woodlands, thickets, rocky areas, waste places; 0–2000 m; Maine, Md., N.C., Pa., Tenn., Va., W.Va.

Variety *nuttallii* is perhaps better considered as a form of the species.

12c. Paronychia fastigiata (Rafinesque) Fernald var. **pumila** (Alph. Wood) Fernald, Rhodora 38: 421. 1936 E

Paronychia canadensis (Linnaeus) Alph. Wood var. *pumila* Alph. Wood, Class-Book Bot. ed. s.n.(b), 263. 1861; *Anychiastrum montanum* Small; *Paronychia montana* (Small) Pax & Hoffmann; *P. pumila* (Alph. Wood) Core

Flowers: sepal cusps 0.1–0.3 mm; styles often bent or contorted, 0.3–0.6(–0.7) mm.

Flowering late spring–fall. Rocky places and woodlands, especially on shales; 0–2000 m; Ala., Ga., Md., N.J., N.C., Pa., S.C., Tenn., Va., W.Va.

13. Paronychia franciscana Eastwood, Bull. Torrey Bot. Club 28: 288. 1901 • San Francisco nailwort I

Plants perennial, mat-forming; caudex branched, woody. **Stems** prostrate, usually much-branched throughout, 5–50 cm, pubescent. **Leaves:** stipules ovate-lanceolate, 3–6 mm, apex narrowly acute to long-acuminate, entire; blade elliptic to oblanceolate, 5–10 × 1.5–2.5 mm, ± fleshy, apex spinulous, entire, ± moderately antrorsely appressed-pubescent. **Cymes** axillary, inconspicuous, 2–6-flowered, tightly congested. **Flowers** 5-merous, short-cylindric, with enlarged hypanthium and calyx cylindric to slightly tapering distally, 1.9–2.4 mm, glabrous, sepals puberulent distally; sepals greenish, becoming reddish brown, midrib and lateral pair of veins often apparent, oblong to ovate, 1.2–1.3 mm, herbaceous, margins translucent, ca. 0.1 mm wide, scarious, apex terminated by awn, hood broadly rounded, awn erect, 0.5–0.7 mm, conic in proximal $^1/_6$ with whitish, smooth spine; staminodes absent; styles 2, 0.2–0.3 mm. **Utricles** ± globose to 4-angled, 1.2–1.3 mm, papillate distally.

Flowering spring. Grassy hills; 20–300 m; introduced; Calif.; South America (Chile).

Although *Paronychia franciscana* was described from California, where it has been known from the San Francisco area since 1887, the species is native in Chile.

14. Paronychia herniarioides (Michaux) Nuttall, Gen. N. Amer. Pl. 1: 159. 1818 (as herniaroides) • Coastal-plain nailwort E F

Anychia herniarioides Michaux, Fl. Bor.-Amer. 1: 113. 1803; *Anychiastrum herniarioides* (Michaux) Small; *Gastronychia herniarioides* (Michaux) Small

Plants annual, matted; taproot filiform. **Stems** prostrate to erect, dichotomously branched, 3–20 cm, retrorsely to spreading-pubescent throughout. **Leaves:** stipules ovate-lanceolate, 0.5–7 mm, apex acuminate, entire; blade elliptic to oblong or ovate, or basal ones spatulate, 2–18 × 1–6 mm, herbaceous to leathery, apex mucronate to short-spine-tipped, moderately antrorsely pubescent. **Inflorescences** diffuse, flowers solitary (paired) in axils of leafy bracts. **Flowers** 5-merous, broadly ovoid, with enlarged hypanthium and calyx tapering distally, 2–2.3 mm, glabrous to strigose; sepals brownish to yellowish, veins absent, subulate to lanceolate, 1–1.7 mm, leathery to rigid, margins whitish, 0.03–0.1 mm wide, papery, apex terminated by awn, hood obscure, small flange, awn straight to slightly divergent, yellow, 0.3–0.4 mm, scabrous to antrorsely strigose, spinose; staminodes narrowly triangular, 0.2–0.3 mm; style 1, cleft in distal ¹/₈, 0.7–0.9 mm. **Utricles** ellipsoid to subglobose, 0.5–0.8 mm, essentially smooth, glabrous.

Flowering spring–fall. Sandy ridges, dunes, pine/oak woodland; 0–300 m; Fla., Ga., N.C., S.C.

15. **Paronychia jamesii** Torrey & A. Gray, Fl. N. Amer. 1: 170. 1838 • James' nailwort

Paronychia jamesii var. *hirsuta* Chaudhri, *P. jamesii* var. *parviflora* Chaudhri; *P. jamesii* var. *praelongifolia* Correll

Plants perennial; caudex branched, woody. **Stems** erect to ascending, much-branched, 10–35 cm, scabrous, puberulent to pubescent, sometimes glabrous with age. **Leaves:** stipules lanceolate, 5–15 mm, apex acuminate, entire; blade linear, 7–25(–34) × 0.5–1 mm, leathery, apex obtuse to subacute or submucronate, pubescent to puberulent. **Cymes** terminal, 20–70-flowered, open, clusters 1–2 cm wide. **Flowers** 5-merous, short-campanulate, with enlarged hypanthium and calyx widening somewhat distally, 1.8–2.8 mm, puberulent, glabrous to hirtellous distally; sepals green to red-brown, veins absent, oblong, 1.3–1.8 mm, leathery to rigid, margins whitish to translucent, 0.05–0.1 mm wide, scarious, apex terminated by awn, broadly rounded, awn widely divergent, 0.4–0.8 mm, conic in proximal ¹/₂–²/₃ with yellowish, scabrous spine; staminodes filiform, 0.6–1 mm; style 1, cleft in distal ¹/₃–¹/₆, 0.8–1.2 mm. **Utricles** ellipsoid-ovoid, 0.8–1 mm, smooth, glabrous.

Flowering summer–fall. Limestone rocky ledges, slopes, hilltops, grasslands; 500–2500 m; Colo., Kans., Nebr., N.Mex., Okla., Tex., Wyo.; Mexico (Chihuahua, Coahuila).

We agree with B. L. Turner (1983b) in not adopting the four varieties of *Paronychia jamesii* that Chaudhri recognized.

16. **Paronychia jonesii** M. C. Johnston, Wrightia 2: 250. 1963 • Jones' nailwort E

Plants annual; taproot slender. **Stems** prostrate, sprawling-spreading, much-branched, 10–35 cm, short-pubescent. **Leaves:** stipules ovate, 3–7.5 mm, apex acuminate, entire; blade oblanceolate to spatulate, 5–18 × 1.5–3.5 mm, leathery, apex obtuse to very short-cuspidate, densely appressed-pubescent. **Cymes** terminal and subterminal, occasionally axillary, 3–7-flowered, congested, clusters 3–8

mm wide. **Flowers** 5-merous, ± extended-urceolate, with enlarged hypanthium and calyx constricted proximally, flaring distally, 1.8–2 mm, pubescent with short, hooked hairs proximally; sepals red-brown, often finely striped or mottled, veins absent, spatulate-obovate to suboblong, 0.8–1.3 mm, leathery to rigid, margins white, 0.1–0.2 mm wide, papery, apex terminated by awn, broadly rounded, awn divergent, 0.3–0.4 mm, broadly conic in proximal ¹/₃–¹/₂ with white, scabrous spine; staminodes filiform, 0.5–0.6 mm; style 1, 0.7–0.8 mm, cleft in distal ¹/₅. **Utricles** globose, 0.8–1 mm, smooth, glabrous.

Flowering spring–summer. Sandy, open grounds; 10–40 m; Tex.

Paronychia jonesii most closely resembles *P. drummondii* and is known only from southeastern coastal Texas.

17. **Paronychia lindheimeri** Engelmann ex A. Gray, Boston J. Nat. Hist. 6: 152. 1850 • Forked nailwort E

Paronychia chorizanthoides Small; *P. lindheimeri* var. *longibracteata* Chaudhri

Plants annual; taproot filiform to slender. **Stems** erect to ascending, much-branched, 15–33 cm, glabrous to minutely puberulent. **Leaves:** stipules lanceolate to ovate, 2.5–5 mm, apex acuminate to long-acuminate, entire; blade linear to filiform, 8–15 × 0.4–0.8 mm, leathery, apex acute to submucronate, glabrous. **Cymes** terminal and axillary, 5–20+-flowered, dichotomous, repeatedly branching and diffusely spreading, open, clusters 0.5–1.5 mm wide. **Flowers** 5-merous, short-campanulate, with enlarged hypanthium and calyx slightly constricted proximally, 1.6–2 mm, appressed-puberulent proximally; sepals red-brown, midrib and lateral pair of veins absent to evident, oblong, 0.8–1.1 mm, leathery to rigid, margins whitish to translucent, 0.05–0.1 mm wide, scarious, apex terminated by awn, hood present, broadly rounded, awn divergently spreading, 0.4–0.6 mm, ± broadly conic in proximal ¹/₂ with pale-yellow, scabrous spine; staminodes filiform, 0.5–0.7 mm; style 1, cleft in distal ¹/₆, 0.6–0.7 mm. **Utricles** ovoid-globose, 0.8–0.9 mm, smooth, glabrous.

Flowering summer–fall. Rocky places on limestone hills and high prairies; 100–600 m; Tex.

18. Paronychia maccartii Correll, Brittonia 18: 307. 1967 • McCart's nailwort C E

Plants perennial; taproot ± stout. **Stems** sprawling, branched from base, 3–10 cm, puberulent. **Leaves:** stipules lanceolate, 2.5–5 mm, apex narrowly acuminate, entire; blade linear-oblong to linear-lanceolate, 2–10 × 0.7–1.5 mm, leathery, apex spinose, minutely puberulent. **Cymes** terminal, 10–20-flowered, congested, in clusters 0.5–1.5 mm wide. **Flowers** 5-merous, extended-urceolate, with enlarged hypanthial bulge and calyx cylindric to slightly tapering distally, 3.5-4 mm, moderately pubescent proximally, with minute, hooked hairs on hypanthial bulge, sparsely pubescent on calyx, densely so with hooked to coiled hairs on hood; sepals green to red-purple, midrib apparent, oblong, 1.9–2.1 mm, herbaceous to leathery, margins white, 0.2–0.3 mm wide, papery, apex terminated by awn, hood triangular-rounded, awn erect, becoming widely divergent, white, ± lanceoloid from proximal constriction, 0.8–1.3 mm, slightly scarious, distinct spine absent; staminodes narrowly subulate, 0.6–0.8 mm; style 1, cleft in distal $^7/_8$, 0.7–0.8 mm. **Utricles** unknown.

Flowering spring. Dense red sands; of conservation concern; 200-300 m; Tex.

Paronychia maccartii is still known only from the type collection in Webb County.

19. Paronychia monticola Cory, Rhodora 46: 279. 1944 • Livermore nailwort

Paronychia nudata Correll

Plants biennial but usually perennial; taproot stout. **Stems** erect to ascending, much-branched throughout, 10–30 cm, glabrous. **Leaves:** stipules lanceolate, 2–5 mm, apex acuminate to long-acuminate, entire; blade linear, 7–15(–19) × 0.4–0.8(–1) mm, leathery, apex subobtuse or minutely cuspidate, glabrous. **Cymes** terminating branchlets, 10–40+-flowered, repeatedly forked and diffuse, clusters 1–3 cm wide. **Flowers** 5-merous, ± short-campanulate, with enlarged hypanthium and calyx constricted proximally, 2.3–3.2 mm, hirtellous, glabrous distally; sepals red-brown, veins absent to obscure, narrowly oblong, 1.8–2.3 mm, leathery to rigid, margins translucent, 0.05–0.1 mm wide, scarious, apex terminated by awn, hood conspicuously rounded, awn divergently spreading, 1–1.5(–2) mm, ± broadly conic in proximal $^1/_3$ with yellowish, glabrous spine; staminodes filiform, 0.7–0.9 mm; style 1, cleft in

distal $^1/_5$, 0.6–1.3 mm. **Utricles** ovoid, 1.2–1.5 mm, minutely papillose distally.

Flowering summer–fall. Mountain tops, rocky slopes and ledges, gravel beds of mountain streams; ca 1200 m; Tex.; Mexico (Coahuila).

20. Paronychia patula Shinners, Sida 1: 102. 1962 • Pineland nailwort E

Siphonychia diffusa Chapman, Fl. South. U.S., 47. 1860, not *Paronychia diffusa* A. Nelson 1899

Plants annual or rarely short-lived perennial; taproot filiform to slender. **Stems** prostrate to erect, much-branched, 5–50 cm, retrorsely pubescent throughout. **Leaves:** stipules lanceolate, 1.5–7 mm, apex acuminate, entire; blade linear-oblong to oblanceolate, 3–20 × 0.5–3 mm, herbaceous, apex obtuse to subacute, sparsely antrorsely pubescent. **Cymes** terminal and lateral, 15–50+-flowered, densely congested, ± spheroid clusters 3–5 mm wide. **Flowers** 5-merous, cylindric, with enlarged hypanthium and calyx constricted then widening distally, 1.2–2.2 mm, pubescent, sometimes only slightly so, mostly proximally with straight or usually hooked hairs; sepals red-brown, white distally, midrib and lateral pair of veins becoming prominent, linear to oblong, 0.6–1.2 mm, papery, margins white, 0.03–0.1 mm wide, papery, apex terminated by mucro, hood obscure, narrowly rounded, mucro short-conic, 0.05–0.1 mm, glabrous; staminodes triangular, 0.2–0.4 mm; style 1, cleft (2- or 3-lobed) in distal $^1/_5$, 0.5–1.3 mm. **Utricles** ovoid to subglobose, 0.4–0.6 mm, ± smooth, glabrous.

Flowering spring–fall. Sandhills, woodlands, fields, clearings, roadsides; 0–100 m; Ala., Fla., Ga., La.

The Louisiana record of *Paronychia patula*, based on a specimen at NY, is questionable.

21. Paronychia pulvinata A. Gray, Proc. Acad. Nat. Sci. Philadelphia 15: 58. 1864 • Rocky Mountain nailwort E F

Paronychia pulvinata A. Gray var. *longiaristata* Chaudhri

Plants perennial, densely cespitose, cushion-forming; caudex much-branched, woody. **Stems** prostrate, much-branched, 5–10 cm, puberulent. **Leaves:** stipules ovate, 3–6 mm, apex subobtuse, entire; blade narrowly elliptic-oblong or oblong to narrowly elliptic-oblanceolate, 2–5 × 0.2–1.8(–2) mm, fleshy, apex obtuse to subacute,

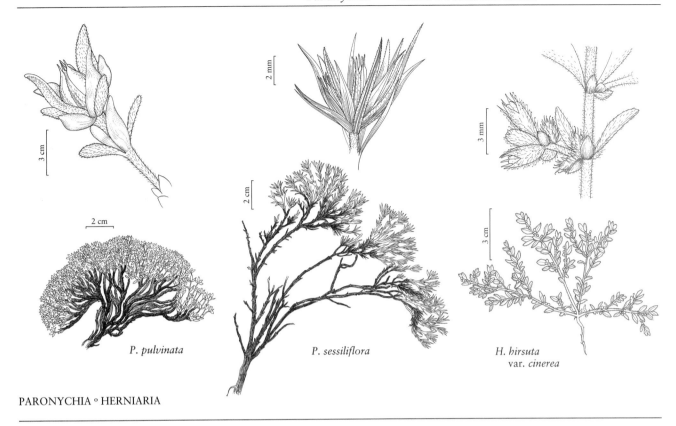

P. pulvinata

P. sessiliflora

H. hirsuta
var. *cinerea*

PARONYCHIA ∘ HERNIARIA

glabrous. **Flowers** mostly solitary at end of shoots, almost concealed by leaves. **Flowers** 5-merous, elliptic-oblong, with enlarged hypanthium and calyx straight to tapering distally, 2.5–2.8 mm, densely appressed-puberulent, sparsely so distally; sepals whitish to green, veins absent, narrowly oblong to ovate-oblong, 1.5–1.7 mm, papery to herbaceous, margins white, 0.2–0.3 mm wide, papery, apex with subterminal awn, hood ascending as continuation of sepal, broadly rounded to notched, awn erect, white, 0.3–0.6(–1) mm, ± glabrous spine; staminodes filiform, 0.7–0.9 mm; style 1, cleft in distal ⅓, 0.8–1 mm. **Utricles** ovoid, 1.3–1.5 mm, smooth, glabrous. $2n = 32$.

Flowering summer. Rocky slopes and summit screes in alpine regions; 3300–4200 m; Colo., Utah, Wyo.

22. **Paronychia rugelii** (Chapman) Shuttleworth ex Chapman, Fl. South. U.S. ed. 3, 397. 1897
- Rugel's nailwort C E

Siphonychia rugelii Chapman, Fl. South. U.S., 47. 1860; *Gibbesia rugelii* (Chapman) Small; *Odontonychia interior* Small; *Paronychia rugelii* var. *interior* (Small) Chaudhri; *Siphonychia interior* (Small) Core

Plants annual; taproot filiform to slender. **Stems** erect, dichotomously branched, 5–50 cm, retrorsely to spreading-

pubescent throughout. **Leaves:** stipules ovate-lanceolate, 2–6 mm, apex acuminate, entire; blade linear to linear-oblanceolate or linear-oblong, 3–25 × 1–2.3 mm, herbaceous, apex ± acute, moderately antrorsely puberulent. **Cymes** terminal and lateral, 15–50+-flowered, often forming large, congested clusters 5–30 mm wide. **Flowers** 5-merous, cylindric to narrowly lanceolate, with enlarged hypanthium and calyx cylindric to tapering distally, 2.3–3.5 mm, pubescent mostly proximal with hooked or usually straight hairs, sometimes with scattered straight hairs distally; sepals flecked or solid red-brown, white distally, midrib obscure, linear-lanceolate, 1.5–2 mm, papery, margins white, 0.1–0.2 mm wide, papery, apex terminated by mucro, hood obscure, narrowly rounded, mucro short-conic, ca. 0.1 mm, glabrous; staminodes filiform, widening proximally, 0.5–0.6 mm; style 1, cleft in distal ¹/₁₀. **Utricles** broadly ellipsoid to ± globose, 0.5–0.7 mm, smooth, glabrous.

Flowering late spring–fall. Woodlands, fields, clearings, roadsides, waste places; of conservation concern; 0–200 m; Fla., Ga.

23. Paronychia sessiliflora Nuttall, Gen. N. Amer. Pl. 1: 160. 1818 • Creeping nailwort E F

Plants perennial, densely cespitose and mat-forming; caudex branched, woody. **Stems** erect to ascending, branched proximally, 5–25 cm, hirtellous. **Leaves:** stipules lanceolate to subulate, 2–3 mm, apex long-acuminate, often deeply cleft; blade linear-subulate, 4–7.5 × 0.5–0.8 mm, leathery, apex acute or shortly cuspidate-mucronate, very finely puberulent to glabrous. **Cymes** terminal, 3–6-flowered, congested, or flowers solitary. **Flowers** 5-merous, ± ovoid, with enlarged hypanthium and calyx narrowing distally, 3.6–5 mm, moderately pubescent with silky to stiff, antrorse to somewhat spreading hairs; sepals green to red-brown, midrib and lateral pair of veins prominent, lanceolate-oblong, 1.5–2 mm, leathery, margins whitish to translucent, 0.1–0.2 mm wide, scarious, apex terminated by awn, hood ± obscure, narrowly rounded, awn erect to somewhat spreading, 1–1.5(–2) mm, narrowly conic in proximal 1/2 with white, scabrous spine; staminodes filiform, 0.7–1 mm; style 1, cleft in distal 1/5, 1.4–1.5 mm. **Utricles** ovoid-oblong, 1.3–1.4 mm, densely pubescent in distal 1/2.

Flowering summer. Dry, stony hillsides, summits, and sandstone mesas; 700–3100 m; Alta., Sask.; Colo., Mont., Nebr., Nev., N.Mex., N.Dak., Okla., Tex., Utah, Wyo.

24. Paronychia setacea Torrey & A. Gray, Fl. N. Amer. 1: 170. 1838 • Bristle nailwort E

Paronychia lundellorum B. L. Turner; *P. setacea* var. *longibracteata* Chaudhri

Plants annual, biennial, or perennial; taproot filiform to slender. **Stems** erect to ascending, much-branched distally, 5–20 cm, finely pubescent to puberulent. **Leaves:** stipules lanceolate, 3–8 mm, apex acuminate, entire; blade linear, 5–20 × 0.3–1 mm, leathery, apex minutely cuspidate, glabrous. **Cymes** terminal and lateral, 5–20+-flowered, dichasial and rather diffuse, forming clusters 3–15 mm wide. **Flowers** 5-merous, ± short-campanulate, with enlarged hypanthium and calyx widening distally, 2.4–3.2 mm, moderately pubescent proximately with antrorsely appressed to spreading hairs; sepals green to tan, midrib and lateral pair of veins absent to evident, oblong, 1.2–1.3 mm, leathery to rigid, margins whitish to translucent, 0.05–0.1 mm wide, scarious, apex terminated by awn, hood ± obscure, narrowly rounded, awn widely divergent, 0.6–1.3 mm, broadly conic in proximal 1/3

with white to yellowish, scabrous spine; staminodes subulate, 0.6–0.7 mm; style 1, cleft in distal 1/2, 0.4–0.6 mm. **Utricles** ovoid-oblong, 1 mm, smooth, glabrous. $2n = 64$.

Flowering spring–fall. Limestone barrens, gravelly or sandy slopes and grasslands; 30–200 m; Tex.

We could not distinguish *Paronychia lundellorum* from *P. setacea* based on features that Turner used for differentiation, nor did we discover any other discriminating characters. The former was described as perennial (versus annual for the latter) and as having "decided pedicellate" (versus sessile) and longer flowers. Recent data (W. Carr, pers. comm.) indicate that it does bloom in the first year. Furthermore, we have observed that the pedicel character is inconsistent and the flower lengths overlap extensively. Further study of the allopatric populations named *P. lundellorum* is warranted.

25. Paronychia virginica Sprengel, Syst. Veg. 1: 822. 1824 • Yellow nailwort, Appalachian nailwort E

Paronychia parksii Cory; *P. scoparia* Small; *P. virginica* var. *parksii* (Cory) Chaudhri; *P. virginica* var. *scoparia* (Small) Cory

Plants perennial; caudex branched, woody. **Stems** procumbent, branched from base, hirtellose; flowering stems 7–45 cm; sterile stems 3–10 cm. **Leaves:** stipules narrowly lanceolate, 6–13 mm, apex acuminate, often deeply cleft; blade linear, 10–30 × 0.4–1 mm, leathery, apex short-spinose, minutely hirtellous to puberulent. **Cymes** terminal, 3–10+-flowered, somewhat open to compact, often forming clusters 6–30 mm wide. **Flowers** 5-merous, narrowly ovoid, with enlarged hypanthium and calyx tapering gradually distally, 2.8–5.1 mm, glabrous to puberulent, especially proximally; sepals brown to yellowish, midrib and lateral pair of veins prominent, lanceolate to oblong-lanceolate, 2–2.9 mm, leathery to rigid, margins whitish, 0.1–0.2 mm wide, papery, apex terminated by awn, hood rounded-triangular, awn curved outward, green to red-brown, ± conic, 0.4–1.1 mm, scabrous, distinct spine absent; staminodes filiform, 0.8–1 mm; style 1, cleft in distal 1/5, 1.2–2 mm. **Utricles** ovoid to obovoid, 1.8–2 mm, smooth, glabrous.

Flowering summer–fall. On or among rocks; 700–1300 m; Ala., Ark., D.C., Md., Mo., N.C., Okla., Tenn., Tex., Va., W.Va.

Cory established var. *scoparia* for the western populations to reflect the disjunct distribution. *Paronychia dichotoma* (Linnaeus) Nuttall (1818), sometimes applied to this species, is a later homonym of *P. dichotoma* de Candolle (1805); see E. L. Core (1940).

26. Paronychia wilkinsonii S. Watson, Proc. Amer. Acad. Arts 21: 454. 1886 (as wilkinsoni)

• Wilkinson's nailwort [C]

Plants perennial, cushion-forming, hirtellous; caudex much-branched, woody. **Stems** erect to ascending, much-branched, 4–10 cm, hirtellous. **Leaves:** stipules lanceolate, 5–8 mm, apex acuminate, often deeply cleft; blade lanceolate, 5–9 × 0.75–1.1 mm, leathery, apex pungently mucronate, minutely hirtellous to puberulent. **Cymes** terminal and subterminal, 3–7(–10)-flowered, densely congested, forming large and conspicuous glomerules 5–15 mm wide. **Flowers** 5-merous, ± short-campanulate, with enlarged hypanthium and calyx slightly constricted then widening distally, 3.3–4.5 mm, puberulent with mostly ascending hairs; sepals red-brown, midrib and lateral pair of veins absent to evident, oblong, 1.7–2 mm, leathery to rigid, margins white, 0.2–0.3 mm wide, papery to scarious, apex terminated by awn, hood ± obscure, broadly rounded, awn moderately divergent, white, ± lanceoloid, 1.4–2.1 mm, scabrous, distinct spine absent; staminodes ± triangular, 0.2–0.4 mm; styles 2, 0.5–0.7 mm. **Utricles** subglobose to ± 4-angular, 1.2–1.3 mm, papillate distally.

Flowering spring–fall. Dry, rocky hills, gravelly slopes and summits; of conservation concern; 1200–1600 m.; Tex.; Mexico (Chihuahua, Coahuila).

9. HERNIARIA Linnaeus, Sp. Pl. 1: 218. 1753; Gen. Pl. ed. 5, 103. 1754

• Rupturewort, herniary, herniaire [Latin *hernia*, rupture, and *-aria*, pertaining to, alluding to use in treatment of hernias] [T]

John W. Thieret

Ronald L. Hartman

Richard K. Rabeler

Herbs, annual, biennial, or perennial. **Taproots** slender. **Stems** ascending or spreading to often prostrate, much-branched from base, mat-forming, terete. **Leaves** opposite, or distalmost alternate (from reduction of 1 member of a pair), connate by a line of tissue between adjacent stipules, sessile or virtually so; stipules 2 per node, inconspicuous, white, ovate to deltate, margins ciliate, apex acute; blade 1-veined, oblanceolate to elliptic or suborbiculate, not succulent, apex acute to rounded. **Inflorescences** densely clustered cymes usually on short lateral branches opposite a leaf; bracts paired, resembling stipules, smaller. **Pedicels:** flowers sessile. **Flowers:** hypanthium cup-shaped, not abruptly expanded distally; sepals 5, distinct, greenish to whitish green, lanceolate to oblong, 0.5–1.5 mm, herbaceous, margins green, herbaceous, apex acute to subobtuse, not hooded, not awned; nectaries near inner surface of filament bases; stamens (2–)4–5; filaments distinct; staminodes 5, arising from hypanthium rim, subulate-filiform, inconspicuous; styles 2, connate in proximal 1/3, filiform, 0.1–0.4 mm, glabrous proximally; stigmas 2, linear along adaxial surfaces of style branches, papillate (100×). **Utricles** at least partly enclosed by hypanthium, opening irregularly. **Seeds** dark brown or black, ovoid to lenticular, slightly laterally compressed, shiny, smooth, marginal wing absent, appendage absent; embryo peripheral, curved. *x* = 9.

Species ca. 45 (2 in the flora): introduced; South America (Andes), Europe, c, w Asia, Africa; introduced elsewhere.

SELECTED REFERENCES Akeroyd, J. R. 1993. *Herniaria*. In: T. G. Tutin et al., eds. 1993+. Flora Europaea, ed. 2. 1+ vol. Cambridge and New York. Vol. 1, pp. 182–184. Hermann, F. 1937. Übersicht über die *Herniaria*-Arten des Berliner Herbars. Repert. Spec. Nov. Regni Veg. 42: 203–224. Williams, F. N. 1896. A revision of the genus *Herniaria*. Bull. Herb. Boissier 4: 556–570.

1. Flowers glabrous or minutely ciliate; leaf blades glabrous or sometimes minutely ciliate; plants green, glabrous or pubescent . . . . 1. *Herniaria glabra*
1. Flowers densely pubescent; leaf blades hirsute or ciliate; plants gray-green, densely pubescent . . . . 2. *Herniaria hirsuta*

1. Herniaria glabra Linnaeus, Sp. Pl. 1: 218. 1753

• Smooth rupturewort, green-carpet, herniaire glabre

Plants annual, biennial, or perennial, light to yellowish green, glabrous or puberulent, sometimes with woody caudex. **Stems** spreading to prostrate, 5–35 cm. **Leaves** opposite, or distalmost alternate; stipules 0.5–1.5 mm; blade obovate-elliptic to suborbiculate, 3–7(–10) mm, glabrous or sometimes short-ciliate. **Inflorescences** mostly leaf-opposed, 6–10-flowered. **Flowers** 1–1.5 (–1.8) mm, usually glabrous or sometimes short-ciliate; calyx not burlike; sepals equal or sometimes unequal, 0.5–0.6 mm, glabrous; stamens 5; staminodes petaloid, 0.5 mm; styles connate in proximal ⅓. **Utricles** 1–1.3 mm, usually longer than sepals. *2n* = 18, 36, 72 (Europe), 54 (Africa).

Flowering spring–summer. Roadsides, dry or rocky, sandy places; 0–1200 m; introduced; Ont., Que.; Md., Mich., N.J., Pa., Utah; Europe; Asia (Turkey); introduced elsewhere.

Historical collections are known also from Maine (1903) and New York (1943).

Herniaria glabra, variable in habit, vesture, flower size, and fruit length (H. W. Pugsley 1930), has been reported to hybridize naturally with *H. hirsuta* (M. N. Chaudhri 1968). It makes a dense mat of foliage, being occasionally planted as a ground or grave cover.

2. Herniaria hirsuta Linnaeus, Sp. Pl. 1: 218. 1753

• Hairy rupturewort [F] [I]

Plants annual, gray-green, densely pubescent. **Stems** prostrate to ascending, 4–20 cm. **Leaves** opposite proximally, often alternate distally; stipules 0.5–1.3 mm; blade elliptic to oblanceolate, 3–12 mm, hirsute or ciliate, adaxial surface sometimes glabrescent. **Inflorescences** axillary, leaf-opposed or on short branches, mostly 3–8-flowered. **Flowers** 0.9–1.8 mm, densely pubescent; calyx burlike; sepals equal or somewhat unequal, 0.8–1.5 mm, hirsute, hairs of perigynous zone hooked or tightly coiled, each sepal with 1–2 spinelike hairs at apex; stamens 2–3 or 5; staminodes petaloid, 0.4–0.6 mm; styles distinct or connate in proximal ⅓. **Utricles** 0.7–0.9 mm, ca. equaling sepals. *2n* = 18, 36 (Europe).

Varieties 4+ (2 in the flora): introduced; Eurasia; Africa.

We are following J. R. Akeroyd (1993) rather than M. N. Chaudhri (1968) in treating *Herniaria cinerea* as an infraspecific taxon of *H. hirsuta*. We believe it more appropriate to recognize the differences at varietal level; intermediate conditions found in both European and North American populations weaken the distinctions.

1. Sepals in fruit of ± equal lengths; hairs on flowers of 1 size, ⅕–⅓ times sepals, tips of hairs ± straight, reduced or absent on hypanthium area; stamens usually 5 . . . . 2a. *Herniaria hirsuta* var. *hirsuta*
1. Sepals in fruit of 2 ± unequal lengths; hairs on flowers of 2 sizes, long one ½–⅔ times sepals, short ones ¼–⅓ times sepals, tips of some or all hairs hooked or tightly coiled; hypanthium area pubescent, hairs with hooked or tightly coiled tips; stamens 2–3 . . . . 2b. *Herniaria hirsuta* var. *cinerea*

2a. Herniaria hirsuta Linnaeus var. **hirsuta** [I]

Stems 4–15 cm. **Inflorescences** 3–6-flowered. **Flowers** 0.9–1 mm, hairs mostly of 1 size, ⅕–⅓ as long as sepals, tips of hairs ± straight; hypanthium glabrous or pubescent, tips of hairs ± straight; sepals in fruit ± equal; stamens usually 5; styles connate in proximal ⅓, 2-branched, shorter than 0.1 mm. **Seeds** 0.6–0.7 mm. *2n* = 18, 36 (Europe).

Flowering spring–fall. Sandy flats, roadsides, woodlands; 200–1800 m; introduced; Calif., Md., Mass.; s Europe; sw Asia; n Africa.

Variety *hirsuta* is apparently much less common than var. *cinerea* in North America. A historical record exists from Pennsylvania (1915).

2b. Herniaria hirsuta Linnaeus var. **cinerea** (de Candolle) Loret & Barrandon, Fl. Montpellier, 243. 1876 [F] [I]

Herniaria cinerea de Candolle in J. Lamarck and A. P. de Candolle, Fl. Franç. ed. 3, 5: 375. 1815; *H. hirsuta* subsp. *cinerea* (de Candolle) Coutinho

Stems 5–20 cm. **Inflorescences** 3–8-flowered. **Flowers** 1.2–1.8 mm, hairs of 2 sizes, long hairs ½–⅔ as long as sepals, short hairs

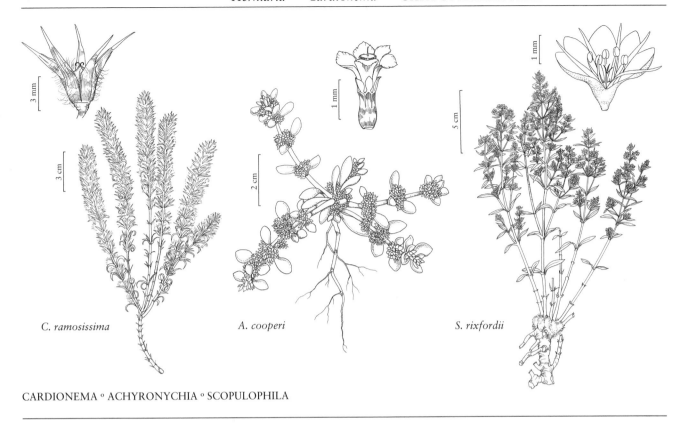

C. ramosissima   A. cooperi   S. rixfordii

CARDIONEMA ° ACHYRONYCHIA ° SCOPULOPHILA

¹/₄–¹/₃ as long as sepals, tips of some or all hooked or tightly coiled; hypanthium area pubescent, tips of hairs hooked or tightly coiled; sepals in fruit ± unequal, of 2 lengths; stamens 2–3; styles distinct, 0.2–0.4 mm. **Seeds** 0.5–0.6 mm. $2n$ = 36 (Europe).

Flowering spring–fall. Disturbed areas, alkaline hills, clay flats; 40–800 m; introduced; Ariz., Calif., Md., Oreg.; s Europe; sw Asia; n Africa.

Historical collections of var. *cinerea* are known from New York (1890s) and Wisconsin (1870). It was also collected once in Washington in 1979 (C. T. Roché 1991) but did not persist.

10. CARDIONEMA de Candolle in A. P. de Candolle and A. L. P. P. de Candolle, Prodr. 3: 372. 1828 • Sandcarpet [Greek *kardio*, heart, and *nema*, thread, alluding to the obcordate anthers and slender filaments]

Ronald L. Hartman

Herbs, perennial. **Taproots** often stout. **Stems** prostrate, much-branched proximally, terete to angular. **Leaves** opposite, not connate, connected by thickened margins from which stipules arise, sessile; stipules 2 per node, silvery, lanceolate to ovate, margins entire to irregularly fimbriate, apex 2-lobed; blade prominently 1-veined, needlelike, not succulent, apex spinose. **Inflorescences** axillary, 1–3-flowered clusters; bracts paired, resembling stipules, smaller. **Flowers** sessile; hypanthium cup-shaped, not abruptly expanded distally; sepals 5, distinct, olive-green, oblong to obovate, 1.2–2.8 mm (excluding awns), indurate, herbaceous, margins whitish, scarious, apex spinose, prominently hooded, awned, hoods projecting inward and enclosing developing fruit, awns arising from near apex, widely divergent, very stout, spinose, 3 awns and associated sepals well developed, alternating with 2 reduced sepals; nectaries not apparent; stamens 3–5; filaments distinct to base; staminodes 5, adnate basally with alternating filaments, ovate-triangular, apex acuminate; styles 2, distinct, subcapitate, 0.2 mm, glabrous proximally; stigmas 2, subterminal, obscurely papillate (50×). **Utricles** ellipsoid to cylindric, teeth absent, indehiscent. **Seeds** tan with red spot at hilum, narrowly ovate, not compressed, smooth, marginal wing absent, appendage absent; embryo peripheral, straight to slightly curved.

Species 6 (1 in the flora): coastal w North America, South America (Chile, Ecuador, Peru).

1. Cardionema ramosissima (Weinmann) A. Nelson & J. F. Macbride, Bot. Gaz. 56: 473. 1913 F

Loeflingia ramosissima Weinmann, Flora 3: 608. 1820 (as Loefflingia).

Stems often forming dense mats, 5–30+ cm, often obscured by stipules and dense pubescence. **Leaves:** stipules 1–8 mm, often nearly as long as leaves, lobes acuminate; blade 2-grooved, 5–13 mm, apex finely spinose, glabrous. **Flowers** densely woolly, hairs 0.5–1.5 mm; sepals 5, margins fimbriate, awn 1.5–4 mm; staminodes white, 0.3–0.5 mm. **Utricles:** apex apiculate. **Seeds** 1.4–1.6 mm.

Flowering spring–early summer. Sandy beaches, grassy bluffs, sand dunes; 0–200 m; Calif., Oreg., Wash.; Mexico; South America (Chile, Ecuador, Peru).

The flowers are burlike in fruit with the utricle enclosed in the rigid, persistent calyx, the presumed unit of seed dispersal.

11. ACHYRONYCHIA Torrey & A. Gray, Proc. Amer. Acad. Arts 7: 330. 1868 • Onyx flower, frost-mat [Greek *achuron*, chaff, and *onyx*, *onychos*, nail or fingernail, alluding to the chaffy sepals]

Ronald L. Hartman

Herbs, annual. **Taproots** slender. **Stems** prostrate to ascending, branched, terete to angular, base glabrous. **Leaves** opposite, connate or connected by thickened ridge or transverse wing from which stipules arise, sessile; stipules 2 per node, white, ovate to spatulate, margins fringed to ciliate, apex ± entire; blade obscurely 1-veined, spatulate, somewhat succulent, apex rounded. **Inflorescences** axillary cymes proliferating throughout length of stem; bracts paired, resembling stipules, smaller. **Flowers** sessile; hypanthium cylindric to urceolate, abruptly expanded proximally and distally in fruit; sepals 5, distinct, green, broadly ovate to orbiculate or reniform, 1.2–1.5 mm, herbaceous, margins white, scarious, apex broadly rounded, not hooded, not awned; nectaries at filament bases subtended adaxially by flaps of tissue; stamens 1–4; filaments distinct to base; staminodes 14–19, arising from hypanthium rim, filiform; styles 2, connate in proximal 1/2, filiform, 0.2–0.4 mm, glabrous proximally; stigmas 2, linear along adaxial surface of styles, obscurely papillate (50×). **Modified utricles** obconic, opening by 8 or 10, not spreading, toothlike valves. **Seeds** tan with red spot near one end, ovoid, slightly laterally compressed, smooth, marginal wing absent, appendage absent; embryo peripheral, straight to slightly curved.

Species 1: sw United States, Mexico.

Based on characters of the seeds and flowers, *Achyronychia* is closely related to and possibly congeneric with *Scopulophila*.

1. Achyronychia cooperi Torrey & A. Gray, Proc. Amer. Acad. Arts 7: 331. 1868 F

Stems 3–17 cm, mostly glabrous. **Leaves** pairs unequal; stipules 0.1–0.4 mm; blade 3–20 mm. **Inflorescences** 20–60+-flowered. **Flowers:** perianth 2.5–3 mm; hypanthium green, becoming brown and hard, 10-ribbed distally; sepals erect to spreading, concave, central portion green, fleshy, white, scarious portion forming an ovate to orbiculate apex, 1–1.3+ mm wide, becoming deciduous with age; staminodes ca. 0.5 mm. **Seeds** 0.9–1.1 mm.

Flowering spring. Dry, sandy areas, sand dunes, desert washes; 50–700 m; Ariz., Calif.; Mexico.

12. SCOPULOPHILA M. E. Jones, Contr. W. Bot. 12: 5. 1908 • [Greek *scopulus*, rock or crag, and *phil*, fond of, alluding to habitat]

Ronald L. Hartman

Herbs, perennial, dioecious, with thick, woody crown. **Taproots** stout. **Stems** ascending to erect, branched distally, terete to angular, base densely woolly. **Leaves** opposite, connected by thickened ridge or transverse wing from which stipules arise, sessile; stipules 2 per node, white, triangular, margins jagged to ciliate, apex entire to 2-lobed; blade obscurely 1-veined, linear to lanceolate, somewhat succulent, apex acute to apiculate. **Inflorescences** reduced axillary cymes, proliferating with age; bracts paired, resembling stipules, smaller, scarious. **Flowers** functionally unisexual, sessile; hypanthium conic to urceolate, abruptly expanded distally; sepals 5, distinct, green, linear-elliptic to orbiculate, 1.1–2.1 mm, herbaceous, margins white, scarious, apex broadly rounded, not hooded, not awned; nectaries minute, reddish thickenings adaxially subtending filament bases; stamens 5, abortive or sterile in pistillate flowers; filaments distinct to base; staminodes 5, arising from hypanthium rim, alternating with stamens, petaloid, oblong; ovary sterile in staminate flowers; styles 3, connate in proximal $1/3$–$2/3$, filiform, ca. 1.5 mm, glabrous proximally; stigmas 3, linear along adaxial surface of styles, obscurely papillate (50×). **Modified utricles** ellipsoid, opening by 3 spreading, minute, toothlike valves. **Seeds** tan with red spot near one end, ovoid, slightly laterally compressed, minutely roughened, marginal wing absent, appendage absent; embryo central, straight to slightly curved.

Species 2 (1 in the flora): sw United States, Mexico.

The second species, *Scopulophila parryi* (Hemsley) I. M. Johnston, is restricted to central Mexico north to the southern margin of the Chihuahuan Desert.

Based on characters of the seeds and flowers, *Scopulophila* is closely related to and possibly congeneric with *Achyronychia*.

1. Scopulophila rixfordii (Brandegee) Munz & I. M. Johnston, Bull. Torrey Bot. Club 49: 351. 1923

 • Rixford's rockwort　E　F

Achyronychia rixfordii Brandegee, Zoë 1: 230. 1890

Stems 10–30 cm, glabrous except densely woolly at base. **Leaves:** stipules 0.8–3.5 mm; leaf pairs equal; blade 8–25 mm, fleshy. **Inflorescences** 2–4-flowered. **Flowers:** perianth 2.2–4.2 mm; hypanthium green, becoming brown and hard, terete to angular; sepals erect to spreading, central portion green, linear to oblong, concave, often unequal, fleshy, scarious portion becoming deciduous with age; staminodes slightly longer than sepals. **Utricles** ovoid. **Seeds** 0.9–1.1 mm.

Flowering spring–early summer. Limestone and quartzite outcrops; 1200–1600 m; Ariz., Calif., Nev.

13. CORRIGIOLA　Linnaeus, Sp. Pl. 1: 271. 1753; Gen. Pl. ed. 5, 132. 1754

 • Strapwort [Latin *corrigia*, shoelace, perhaps alluding to the slender stems]　I

John W. Thieret

Richard K. Rabeler

Herbs, annual or biennial [perennial]. **Taproots** slender. **Stems** decumbent to ascending, branched, terete. **Leaves** alternate or rarely subopposite, sometimes forming basal rosette, not connate, petiolate or sessile; stipules 2 per node, white, ovate, margins mostly entire, apex acuminate; petiole minute (subopposite leaves); blade 1-veined, narrowly oblanceolate to spatulate, not succulent, apex obtuse. **Inflorescences** terminal or axillary, dense cymes; bracts paired, smaller, scarious. **Pedicels** erect. **Flowers:** perianth and androecium slightly perigynous; hypanthium cup-shaped, not abruptly expanded distally; sepals 5, distinct, green or reddish brown, 1–1.5 mm, herbaceous, margins white-scarious, apex blunt, not hooded, not awned; nectaries not apparent; stamens 5; filaments distinct; staminodes 5, arising from hypanthium rim, ovate to oblong; styles 3, distinct, capitate, 0.1–0.2 mm, glabrous proximally; stigmas 3, capitate, obscurely papillate (at 50×). **Utricles** at least partly enclosed by persistent calyx/hypanthium, obscurely 3-gonous, indehiscent. **Seeds** white with tan band, reniform or subglobose, not compressed, finely papillose, marginal wing absent, appendage absent; embryo peripheral, curved. $x = 8, 9$.

Species ca. 10 (1 in the flora): introduced; South America, Eurasia, Africa; introduced elsewhere.

SELECTED REFERENCES　Coker, P. D. 1962. *Corrigiola litoralis* L. J. Ecol. 50: 833–840.　Gilbert, M. G. 1987. The taxonomic position of the genera *Telephium* and *Corrigiola*. Taxon 36: 47–49.　Vogel, A. 1997. Die Verbreitung, Vergesellschaftung und Populationsökologie von *Corrigiola litoralis*, *Illecebrum verticillatum* und *Herniaria glabra* (Illecebraceae). Berlin.

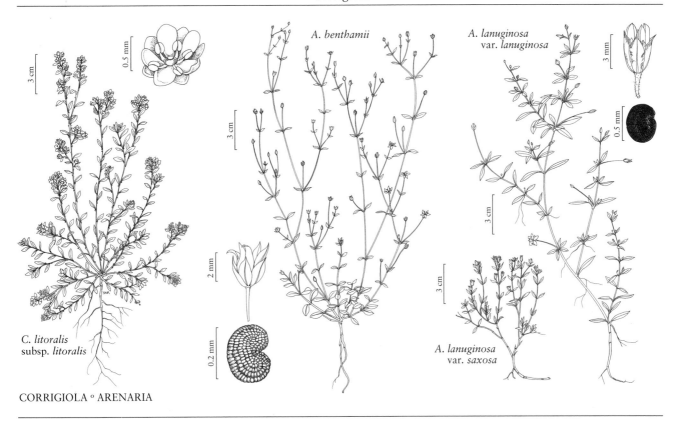

A. benthamii

A. lanuginosa
var. *lanuginosa*

C. *litoralis*
subsp. *litoralis*

A. *lanuginosa*
var. *saxosa*

CORRIGIOLA ° ARENARIA

1. Corrigiola litoralis Linnaeus, Sp. Pl. 1: 271. 1753

F I

Subspecies 2 (1 in the flora): introduced; South America, Eurasia, Africa.

1a. Corrigiola litoralis Linnaeus subsp. **litoralis**

F I

Plants annual or biennial, gla-brous, glaucous. **Stems** 5–50 cm. **Leaves** mostly alternate; blade lance-oblong to linear-oblance-olate, 5–30 mm, base narrowed to obscure petiole. **Pedicels** articulated, to 1 mm. **Flowers:** sepals 1–1.5 mm; staminodes white to pink, sometimes red-tipped. **Utricles** 1–2 mm. $2n$ = 16, 18, 32, 54 (Europe).

Flowering summer–fall. Waste places, ballasts, grassy areas; 0–200 m; introduced; B.C.; Md., Oreg.; Eurasia.

Historical records from ballast grounds at Camden, New Jersey (1870s), and Philadelphia, Pennsylvania, exist, as well as a single report as a garden weed from Massachusetts (1900?; *Manning s.n.*, GH).

43c. CARYOPHYLLACEAE Jussieu subfam. ALSINOIDEAE Fenzl in S. L. Endlicher, Gen. Pl. 13: 963. 1840 (as Alsineae)

Richard K. Rabeler

Ronald L. Hartman

Herbs, winter annual, annual, biennial, or perennial; taprooted and/or rhizomatous, rarely with tuberous thickenings (*Pseudostellaria*). **Stems** prostrate to ascending or erect, simple or branched. **Leaves** opposite, connate proximally or not, often petiolate (basal leaves), not stipulate; blade subulate or linear to spatulate, lanceolate, or broadly ovate, seldom succulent. **Inflorescences** terminal or axillary cymes, or flowers solitary; bracts foliaceous or reduced, herbaceous to scarious (or rarely absent); involucel bracteoles absent. **Pedicels** present or rarely flowers sessile. **Flowers** bisexual or seldom unisexual, sometimes inconspicuous; perianth and androecium hypogynous or perigynous, often slightly; hypanthium cup-, dish-, or disc-shaped; sepals (4–)5, distinct or seldom connate basally, sometimes hooded, not awned; petals absent or (1–)4–5, usually white, sometimes translucent, yellowish white, pink, or brownish, seldom clawed, auricles absent, coronal appendages absent, blade apex entire or 2-fid, sometimes jagged or emarginate, rarely laciniate; stamens absent or (1–)5(–10), in 1 or 2 whorls, arising from base of ovary, a nectariferous disc, or sometimes the hypanthium or hypanthium rim; staminodes absent or 1–5(–8); ovary 1- or rarely 3-locular (*Wilhelmsia*); styles (2–)3–5(–6), distinct; stigmas (2–)3–5(–6). **Fruits** capsules, or rarely utricles (*Scleranthus*), opening by (2–)3–6, occasionally 8 or 10 valves or (3 or) 6–10 teeth; carpophore present or often absent. **Seeds** 1–60+, yellowish or tan to dark red or often brown or black, usually reniform or triangular to circular and laterally compressed or ovoid to globose, rarely oblong and dorsiventrally compressed (*Holosteum*); embryo usually peripheral and curved, rarely central and straight (*Holosteum*). x = 6–15, 17–19, 23.

Genera 30, species ca. 1040 (16 genera, 137 species in the flora): north-temperate regions, South America (Andean region), Europe (Mediterranean region), w, c Asia (Himalayas, Mediterranean region), Africa (Mediterranean region).

Alsinoideae, often considered basal in the family and the least specialized, is in some ways the most heterogeneous of the subfamilies. Members of its largest tribe (Alsineae) share the following characteristics: stipules absent, sepals free or at most basally connate, and capsular fruits. Indehiscent fruits, relatively short hypanthia, and other floral reductions occur in varying combinations in the approximately 30 species placed in four other tribes. A broad molecular survey of Alsinoideae has revealed two major lineages and lack of support for the existing tribal circumscriptions (M. Nepokroeff et al. 2002). About three-fourths of the species are members of *Arenaria*, *Cerastium*, *Minuartia*, and *Stellaria*.

Attempts have been made to move *Scleranthus* (fruit a utricle surrounded by an enlarged hypanthium) from Alsinoideae to either Paronychioideae (J. Hutchinson 1973, as Illecebraceae) or Scleranthaceae (A. Takhtajan 1997). Recent molecular and morphological studies by R. D. Smissen et. al. (2002, 2003) supported its retention in the Alsinoideae.

SELECTED REFERENCES Fernald, M. L. 1919. The unity of the genus *Arenaria*. Rhodora 21: 1–7. Maguire, B. 1951. Studies in the Caryophyllaceae. V. *Arenaria* in America north of Mexico. Amer. Midl. Naturalist 46: 493–511. McNeill, J. 1962. Taxonomic studies in the Alsinoideae: I. Generic and infra-generic groups. Notes Roy. Bot. Gard. Edinburgh 24: 79–155. McNeill, J. 1980b. The delimitation of *Arenaria* (Caryophyllaceae) and related genera in North America, with 11 new combinations in *Minuartia*. Rhodora 82: 495–502. Wofford, B. E. 1981. External seed morphology of *Arenaria* (Caryophyllaceae) of the southeastern United States. Syst. Bot. 6: 126–135.

14. ARENARIA Linnaeus, Sp. Pl. 1: 423. 1753; Gen. Pl. ed. 5, 193. 1754 • Sandwort, sabline [Latin *arena*, sand, a common habitat]

Ronald L. Hartman

Richard K. Rabeler

Frederick H. Utech

Spergulastrum Michaux

Plants annual or perennial, sometimes densely matted. **Taproots** filiform to moderately thickened; rhizomes slender. **Stems** prostrate to ascending or erect, simple or branched, terete to ellipsoid, angular or grooved. **Leaves** mostly connate, mostly sessile, not congested at or near base of flowering stem; blade 1- or 3–5-veined, rarely linear or linear-lanceolate to usually elliptic to ovate or rarely orbiculate, sometimes subsucculent or succulent, apex blunt or obtuse to acute, acuminate, or apiculate. **Inflorescences** terminal or axillary open cymes, or flowers solitary; bracts paired, smaller, foliaceous. **Pedicels** erect or ascending to reflexed in fruit. **Flowers:** perianth and androecium hypogynous; sepals 5, distinct or barely connate proximally, green or rarely distally purple, lanceolate to broadly ovate, 2–5 mm, margins foliaceous or white and scarious, apex obtuse to rounded or acute to acuminate, not hooded; petals 5 or absent, white, not clawed, blade apex entire; nectaries at base of filaments opposite sepals; stamens 10 (ca. 8 in *A. livermorensis*), arising from base of ovary; filaments distinct; staminodes absent; styles 3, filiform, 0.5–2 mm, glabrous proximally; stigmas 3, linear along adaxial surfaces of styles, papillate (30×). **Capsules** broadly ellipsoid or ovoid to cylindric, opening by 6 ascending to recurved teeth; carpophore absent. **Seeds** [1–]5–35, brown to dark brown or black, reniform or suborbicular, laterally compressed or not, shiny or dull, smooth, rugulose, or tuberculate, marginal wing absent, appendage absent. x = [7, 8?] 10, 11.

Species 210 (9 in the flora): north-temperate regions, w North America (including Mexico), Central America, South America (Andes), Eurasian mountains, Mediterranean region, Asia Minor.

Our treatment of *Arenaria* follows J. McNeill (1962, 1980b) in segregating *Minuartia* from *Arenaria*, while we take an additional step in recognizing *Eremogone*. While a broad concept of *Arenaria* is often used in North American works (e.g., B. E. Wofford 1981), splitting *Arenaria* is supported by morphological and molecular evidence (M. Nepokroeff et al. 2001).

J. McNeill (1962) outlined an infrageneric classification of *Arenaria* consisting of ten subgenera. The nine species of *Arenaria* occurring in North America belong to two subgenera (*Leiosperma* McNeill and *Arenaria*) and two sections within subg. *Arenaria* (*Rariflorae* F. Williams and *Arenaria*). Based in part on recent molecular evidence (M. Nepokroeff et al. 2001), we are here treating two of McNeill's subgenera as belonging to the genus *Eremogone*.

1. Plants annual, not mat-forming; stems erect to ascending.
    2. Stems uniformly puberulent; leaf blades 3–5-veined; sepals 3-veined, stipitate-glandular . . . . 9. *Arenaria serpyllifolia*
    2. Stems puberulent in 2 lines; leaf blades 1-veined; sepals 1-veined, glabrous.
        3. Sepals strongly keeled proximally; seeds dark brown to black, shiny, suborbicular, obscurely tuberculate . . . . 6. *Arenaria ludens*
        3. Sepals not keeled; seeds ashy brown, dull, broadly reniform, minutely/prominently rounded-tuberculate . . . . 1. *Arenaria benthamii*

1. Plants perennial, often mat-forming, usually rhizomatous; stems prostrate, seldom erect or
   ascending.
   4. Plants cespitose or pulvinate; stems usually shorter than 10 cm; seeds brown to black,
      rugulose or tuberculate; boreal.
      5. Rhizomes absent; sepals narrowly lanceolate; petals obovate, $1^2/_3$–2 times as long
         as sepals; seeds black to reddish brown, low-tuberculate . . . . 8. *Arenaria pseudofrigida*
      5. Rhizomes present, sepals ovate; petals broadly elliptic or oblanceolate, ± 1–$1^1/_2$
         times as long as sepals; seeds brown, rugulose or faintly so.
         6. Leaf blade margins ciliate proximally; pedicels 10–20 mm; capsules ellipsoid
            . . . . 5. *Arenaria longipedunculata*
         6. Leaf blade margins smooth; pedicels 1–10 mm; capsules broadly ellipsoid
            . . . . 2. *Arenaria humifusa*
   4. Plants with stems mostly longer than 10 cm (if matted, then to 4 cm); seeds dark
      brown or black, shiny, smooth; s, sw United States.
      7. Plants matted, mosslike; leaf blade margins with peglike cilia . . . . 4. *Arenaria livermorensis*
      7. Plants not matted; leaf blade margins glabrous or with hairlike cilia.
         8. Stem internodes and pedicels glabrous; sepals obscurely 1-veined, apices ob-
            tuse to rounded; petals $1^1/_2$–2 times as long as sepals . . . . 7. *Arenaria paludicola*
         8. Stem internodes and pedicels retrorsely-pubescent; sepals 1–3-veined, apices
            acute to acuminate; petals $^1/_2$–$1^2/_5$ times as long as sepals . . . . 3. *Arenaria lanuginosa*

1. **Arenaria benthamii** Fenzl ex Torrey & A. Gray, Fl.
N. Amer. 1: 675. 1840 • Hilly sandwort [F]

Plants annual. **Taproots** filiform.
Stems 1–40+, erect to ascending,
branching, green, 10–30(–50) cm;
internodes ± terete, mostly 1–10
times as long as leaves, dull,
retrosely puberulent in 2 lines.
Leaves usually connate proximally, with scarious sheath 0.2–
0.4 mm, petiolate (proximal
leaves) or sessile; petiole 2–4 mm; blade 1-veined, vein ±
weak abaxially, ovate (petiolate blades) or elliptic-
lanceolate to oblanceolate, 5–15(–20) × 2–4 mm,
herbaceous, margins undulate, scarious, shiny, sparsely
ciliate especially proximally, apex acute to apiculate or
cuspidate, often pustulate, glabrous; axillary leaf clusters
absent. **Inflorescences** terminal, open, leafy, 3–50+-
flowered cymes. **Pedicels** erect to reflexed in fruit, 10–
40 mm, glabrous. **Flowers:** sepals green, 1-veined, vein
obscure or prominent proximally, not keeled, elliptic to
broadly ovate (herbaceous portion light green, ovate),
2–3.5 mm, to 4.5 mm in fruit, apex acute to acuminate,
often pustulate, glabrous; petals elliptic, 1.5–3 mm, $^1/_2$–
$^3/_4$ times as long as sepals, apex obtuse. **Capsules** loosely
to tightly enclosed by calyx, ovoid, 3–4 mm, $1^1/_5$–2 times
as long as sepals. **Seeds** 20–30, ashy brown, broadly
reniform, compressed, 0.5–0.6 mm, dull, minutely or
prominently rounded-tuberculate.
  Flowering late spring. Open woodlands, limestone
slopes and outcrops; 0–400 m; La., N.Mex., Okla., Tex.;
Mexico (Coahuila, Nuevo León).

2. **Arenaria humifusa** Wahlenberg, Fl. Lapp., 129.
1812 • Creeping sandwort, sabline rampante

Plants perennial, matted. **Taproots** filiform; rhizomes weak,
slender, 0.5–15+ cm. **Stems** 5–
40+, flowering stems erect to
ascending, green, 2–8(–10) cm;
internodes terete to angular, $^1/_2$–4
times as long as leaves, shiny,
glabrous or occasionally sparsely
stipitate-glandular at distalmost
internodes. **Leaves** connate basally, with scarious sheath
0.2–0.3 mm, sessile; blade weakly to prominently 1-
veined, narrowly oblong or elliptic to ovate-lanceolate,
2–12 × 1–3 mm, subsucculent, margins thickened,
herbaceous, shiny, smooth, apex blunt to acute,
prominently pustulate or not, glabrous; axillary leaf
clusters absent. **Inflorescences** terminal, open, leafy, 1–
3-flowered cymes or often flowers solitary. **Pedicels** erect
in fruit, 1–10 mm, minutely retrorsely pubescent to
stipitate-glandular. **Flowers:** sepals green, 3-veined, 2
lateral veins $^1/_2$–$^3/_4$ times as long as midvein, more
prominent in fruit, not keeled, ovate (herbaceous portion
ovate), 2.5–4 mm, not enlarging in fruit, apex obtuse or
somewhat acute, not pustulate, glabrous; petals
oblanceolate, 3–4 mm, ± $1^1/_2$ times as long as sepals,
apex acute. **Capsules** tightly enclosed by calyx, broadly
ellipsoid, 4–4.5 mm, ca. $1^1/_4$ times as long as sepals. **Seeds**
12–18, brown, reniform, compressed, 0.5–0.7 mm, shiny,
faintly rugulose. $2n$ = 40, 44 (both Europe).

Flowering spring–summer. Moist, calcareous gravels and rock crevices; 0–100 m; Greenland; St. Pierre and Miquelon; Man., Nfld. and Labr., N.W.T., N.S., Nunavut, Ont., P.E.I., Que.; Europe.

Arenaria humifusa is similar to *A. longipedunculata*, but is smaller in some dimensions than that species.

3. **Arenaria lanuginosa** (Michaux) Rohrbach in C. F. P. von Martius et al., Fl. Bras. 14(2): 274. 1872

• Spreading sandwort F

Spergulastrum lanuginosum Michaux, Fl. Bor.-Amer. 1: 275. 1803

Plants ± strongly perennial, possibly blooming first year, not matted. **Taproots** filiform to moderately thickened; rhizomes often present, slender, 2–15+ cm. **Stems** 1–80+, erect or ascending to procumbent or prostrate to trailing, green, 5–60 cm; internodes terete to angular, $^{1}/_{3}$–8+ times as long as leaves, dull, retrorsely pubescent throughout or in lines, hairs minute. **Leaves** usually connate basally, with scarious sheath 0.1–0.5 mm, occasionally petiolate (proximal leaves) or sessile; petiole 2–5 mm; blade 1-veined, vein prominent abaxially, linear-lanceolate to narrowly elliptic or oblanceolate, 3–35 × 2–14 mm, herbaceous, margins thickened, scarious, shiny, ciliate proximally or throughout, apex obtuse or acute to apiculate, often minutely pustulate, ciliate on margins and adaxial midrib; axillary leaf clusters absent. **Inflorescences** axillary, solitary flowers or in proliferating, mostly terminal, leafy, 1–80+-flowered cymes. **Pedicels** erect to ascending (often arcuately so), or straight to widely divergent, often hooked distally in fruit, 2–40 mm, retrorsely pubescent. **Flowers:** sepals green, 1–3-veined, 2 lateral veins $^{1}/_{4}$–$^{3}/_{4}$ times as long as midvein, often appearing prominently keeled proximally, lanceolate to ovate (herbaceous portion oblong or lanceolate to ovate), 2–5 mm, to 5.5 mm in fruit, apex acute to acuminate, not pustulate, glabrous; petals narrowly spatulate to obovate, 1.5–6 mm, $^{1}/_{2}$–1$^{2}/_{5}$ times as long as sepals or absent, apex obtuse to rounded, petals sometimes absent. **Capsules** ± loosely to tightly enclosed by calyx, ovoid, 3–6 mm, $^{4}/_{5}$–1$^{1}/_{2}$ times as long as sepals. **Seeds** 8–35, black, suborbicular, slightly compressed, 0.7–0.8 mm, shiny, smooth. $2n$ = 40, 44.

Varieties 4+ (2 in the flora): North America (including Mexico), Central America, South America.

Arenaria lanuginosa is morphologically diverse, both in our area and southward into northern South America, and is in serious need of comprehensive study. Other species in subg. *Leiosperma* (e.g., *A. gypsostrata* B. L. Turner) that occur in Mexico resemble *A. lanuginosa*; the nature of those relationships also requires study. We

have taken the "conservative approach" of treating the two taxa that occur in the flora area as varieties.

1. Stems often 1–10, prostrate to trailing; inflorescences of solitary, axillary flowers; petals absent or $^{1}/_{2}$–$^{3}/_{4}$ times as long as sepals . . . . 3a. *Arenaria lanuginosa* var. *lanuginosa*
1. Stems 1–80+, erect or ascending to procumbent; inflorescences of proliferating, leafy, 1–80+-flowered cymes; petals $^{3}/_{4}$–1$^{2}/_{5}$ times as long as sepals . . . . 3b. *Arenaria lanuginosa* var. *saxosa*

3a. **Arenaria lanuginosa** (Michaux) Rohrbach var. **lanuginosa** F

Arenaria lanuginosa var. *longipedunculata* W. H. Duncan

Stems often 1–10, prostrate to trailing. **Inflorescences** axillary, solitary flowers. **Pedicels** straight to widely divergent, often hooked distally in fruit, 25–40 mm. **Petals** $^{1}/_{2}$–$^{3}/_{4}$ times as long as sepals or absent. $2n$ = 44 (South America).

Flowering spring–summer. Forests, limestone outcrops, moist hummocks, dunes; 0–200 m; Ala., Fla., Ga., La., Miss., N.C., S.C., Tenn., Tex., Va.; Mexico; Central America; South America.

Variety *lanuginosa* occurs through much of Mexico, mostly in the mountainous areas. We find material from the southeastern United States to be indistinguishable from many Mexican specimens.

Variety *longipedunculata* was described from two populations in Georgia and represents plants with particularly long pedicels, 40–50 mm, but this range overlaps that found otherwise in var. *lanuginosa*.

3b. **Arenaria lanuginosa** (Michaux) Rohrbach var. **saxosa** (A. Gray) Zarucchi, R. L. Hartman & Rabeler, Sida 21: 753. 2004 F

Arenaria saxosa A. Gray, Smithsonian Contr. Knowl. 5(6): 18. 1853; *A. confusa* Rydberg; *A. lanuginosa* subsp. *saxosa* (A. Gray) Maguire; *A. mearnsii* Wooton & Standley; *A. saxosa* A. Gray var. *cinerascens* B. L. Robinson; *A. saxosa* A. Gray var. *mearnsii* (Wooton & Standley) Kearney & Peebles; *Sperulastrum lanuginosum* Michaux subsp. *saxosum* (A. Gray) W. A. Weber

Stems 1–80+, erect or ascending to procumbent. **Inflorescences** proliferating, mostly terminal, leafy, 1–80+-flowered cymes. **Pedicels** erect to ascending (often arcuately so) in fruit, 2–25 mm. **Petals** $^{3}/_{4}$–1$^{2}/_{5}$ times as long as sepals. $2n$ = 40, 44.

Flowering spring–summer. Forests, stream banks; 1000–3100 m; Ariz., Calif., Colo., N.Mex., Tex., Utah; Mexico (Baja California, Chihuahua, Coahuila).

The varietal combination *Arenaria lanuginosa* var. *cinerascens* (B. L. Robinson) Shinners, often applied to this taxon, is not correct because the epithet of the earlier autonym *A. saxosa* var. *saxosa* has priority.

4. **Arenaria livermorensis** Correll, Brittonia 18: 308. 1967 • Livermore sandwort C E

Plants perennial, matted, moss-like. **Taproots** filiform. **Stems** 15–25, decumbent, trailing, green, 1–4 cm; internodes angular to grooved, 1–1 1/2 times as long as leaves, minute, shiny, pubescence of widely spreading to slightly retrorse hairs in 2 lines. **Leaves** connate basally, with herbaceous sheath 0.1–0.2 mm, sessile; blade 1-veined, vein prominent abaxially, linear, 4–6 × 1 mm, herbaceous, rigid, margins thickened, herbaceous, shiny, lined with peglike cilia, apex acute, pungent-tipped, not pustulate, glabrous; axillary leaf clusters present. **Inflorescences** solitary flowers in distal leaf axils. **Pedicels** erect or ascending in fruit, 6–10 mm, minutely pubescent. **Flowers:** sepals green, 1-veined, slightly keeled proximally, broadly lanceolate (herbaceous portion narrowly lanceolate), 3–4 mm, not enlarging in fruit, apex acute to acuminate, not pustulate, glabrous; petals absent; stamens ca. 8. **Capsules** very loosely enclosed by calyx, ovoid, 2–3 mm, 2/3–4/5 times as long as sepals. **Seeds** 5–7, dark brown to black, subglobose, slightly compressed, ca. 1 mm, shiny, smooth.

Flowering spring–summer. Crevices of cliffs and bare walls; of conservation concern; 2100–2500 m; Tex.

Arenaria livermorensis is known only from the Davis Mountains in the Trans-Pecos region of western Texas. It may be related to *A. lycopodioides* Willdenow ex Schlechtendal, a species found in the mountains of central Mexico.

5. **Arenaria longipedunculata** Hultén, Bot. Not. 119: 313, fig. 1. 1966 • Longstem sandwort

Plants perennial, tufted to mat-forming. **Taproots** filiform; rhizomes slender, 0.5–3 cm. **Stems** 25–60+, erect to ascending, green, 2–4 cm; internodes terete, 1–4 times as long as leaves, shiny, minutely glandular-villous. **Leaves** briefly connate basally, with herbaceous sheath 0.1–0.3 mm, sessile; blade obscurely 1-veined, narrowly elliptic to lanceolate or ovate, 2–6 × 1–1.5 mm, subsucculent, margins thickened, herbaceous, shiny, ciliate proximally, apex acute to apiculate, pustulate, glabrous; axillary leaf clusters absent. **Inflorescences** terminal, solitary flowers. **Pedicels** erect in fruit, 10–20 mm, densely glandular-villous. **Flowers:** sepals green, obscurely veined, not keeled, ovate (herbaceous portion ovate), 3.5–4 mm, not enlarging in fruit, apex acute to acuminate, not pustulate, minutely stipitate-glandular proximally; petals broadly elliptic, 2.5–3.5 mm, 1 1/4–1 1/2 times as long as sepals, apex rounded. **Capsules** tightly enclosed by calyx, ellipsoid, 4–5 mm, 1–1 1/5 times as long as sepals. **Seeds** 13–20, brown, reniform, compressed, 0.7–0.9 mm, shiny, rugulose (30×). $2n = 40, 80$.

Flowering spring–summer. Gravelly, moist, montane areas, open alpine woods; 50–1000 m; Alta., B.C., N.W.T., Yukon; Alaska; Asia.

Arenaria longipedunculata was for many years included in *A. humifusa*.

6. **Arenaria ludens** Shinners, Sida 1: 51. 1962 • Trans-Pecos sandwort

Plants annual. **Taproots** filiform. **Stems** 1–18+, erect to ascending, green or reddish, 15–30(–45) cm; internodes ± terete, 1–10 times as long as leaves, dull, retrorsely pubescent in 2 lines. **Leaves** usually connate basally, with narrow, scarious sheath 0.2–0.5 mm, petiolate (proximal leaves) or sessile; petiole 2–4 mm; blade 1-veined, vein prominent abaxially, usually narrowly lanceolate, elliptic, or oblanceolate, 10–17 × 2–4 mm, herbaceous, margins ± flat, herbaceous, dull, ciliate in proximal 1/2, apex acute to acuminate, not pustulate, glabrous; axillary leaf clusters absent. **Inflorescences** terminal, open, minutely bracteate, 3–45+-flowered cymes. **Pedicels** erect or ascending in fruit, 3–20 mm, retrorsely pubescent in 2 lines. **Flowers:** sepals green or often purple, 1-veined, strongly keeled proximally, ovate-lanceolate to narrowly lanceolate (herbaceous portion pale, narrowly lanceolate to linear), 3–4 mm, not enlarging in fruit, apex acuminate, not pustulate, glabrous; petals oblong to broadly elliptic, 2.8–4 mm, 1–1 1/3 times as long as sepals, apex rounded. **Capsules** loosely enclosed by calyx, ovoid, 3–3.5 mm, 4/5–1 times as long as sepals. **Seeds** 7–15, dark brown to black, suborbicular, slightly compressed, 0.6–0.7 mm, shiny, obscurely tuberculate (20×).

Flowering late summer–early autumn. Igneous soil on cliffs and ledges; 1000–2000 m; Tex.; n Mexico (Coahuila).

Arenaria ludens may be more closely related to *A. lanuginosa* than to *A. benthamii*, the taxon with which it is often confused, if seed morphology is any indication.

7. Arenaria paludicola B. L. Robinson, Proc. Amer. Acad. Arts 29: 298. 1894 • Marsh sandwort C

Alsine palustris Kellogg, Proc. Calif. Acad. Sci. 3: 61. 1863 (as palustre), not *Arenaria palustre* Gay 1845; *Minuartia paludicola* (B. L. Robinson) House

Plants perennial, not matted. **Taproots** filiform; rhizomes slender, 15+ cm. **Stems** 5–15, weakly erect, green, 25–90 cm; internodes angular to grooved, 2/3–3 times as long as leaves, shiny, glabrous except for fine hairs at nodes. **Leaves** connate basally, with scarious sheath 0.2–0.5 mm, sessile; blade 1-veined (vein prominent abaxially), linear to linear-lanceolate, 20–55 × 2–7 mm, herbaceous to subsucculent, margins ± flat, scarious, shiny, ciliate, apex acute or acuminate to cuspidate, sometimes pustulate, sparsely pubescent on adaxial surface; axillary leaf clusters absent. **Inflorescences** axillary, solitary flowers. **Pedicels** reflexed in fruit, 20–50 mm, glabrous. **Flowers:** sepals green, obscurely 1-veined, not keeled, broadly elliptic (herbaceous portion narrowly elliptic), 2.8–3.5 mm, to ca. 4 mm in fruit, apex obtuse to rounded, not pustulate, minutely ciliate basally; petals ovate, 5–6 mm, 1 1/2–2 times as long as sepals, apex rounded. **Capsules** tightly enclosed by calyx, ovoid, 4 mm, ca. equaling sepals. **Seeds** 15–20, dark brown, reniform, compressed, 0.8–0.9 mm, shiny, smooth.

Flowering late spring–summer. Boggy meadows, freshwater marshes; of conservation concern; 0–300 m; Calif.; c Mexico; Central America (Guatemala).

Arenaria paludicola is federally listed as endangered, and now is known only from a few sites in San Luis Obispo County; urban development and resultant habitat conversion have impacted it significantly. Historical collections of *A. paludicola* are known from other areas of the California coast and from Washington.

Arenaria paludicola occurs also in Mexico, where it grows (and is sometimes confused) with *A. bourgaei* Hemsley in lakes and wet areas above 2000 meters; the relationship of these species to each other and to *A. lanuginosa* requires study.

8. Arenaria pseudofrigida (Ostenfeld & Dahl) Schischkin & Knorring in V. L. Komarov et al., Fl. URSS 6: 537. 1936

Arenaria ciliata Linnaeus subsp. *pseudofrigida* Ostenfeld & Dahl, Nyt Mag. Naturvidensk. 55: 217. 1918; *A. ciliata* var. *pseudofrigida* (Ostenfeld & Dahl) B. Boivin

Plants perennial, forming cushions or mats. **Taproots** filiform to somewhat thickened. **Stems** 20–40+, procumbent, green, 3–14 cm; internodes terete to angular, 1/2–1+ times as long as leaves, shiny, pubescent in lines or throughout, hairs retrorsely curved. **Leaves** connate basally, with scarious sheath 0.2–0.3 mm, sessile; blade 1-veined (prominent proximally), oblanceolate to spatulate, 2–6 × 0.8–1.8 mm, succulent, margins thickened, minutely scarious-granular, ± shiny, ciliate in proximal 1/2, apex ± obtuse, sometimes pustulate, glabrous; axillary leaf clusters absent. **Inflorescences** terminal or axillary, open, leafy, 1–4-flowered cymes in distal leaf axils. **Pedicels** erect or nodding in fruit, 2–12 mm, retrorsely pubescent. **Flowers:** sepals green or distally purple, 3-veined (2 lateral veins 1/2 times as long as midvein), sometimes keeled, narrowly lanceolate (herbaceous portion narrowly lanceolate), 3–4 mm, to ca. 5 mm in fruit, apex ± acute, not pustulate, glabrous or sparsely stipitate-glandular; petals obovate, 5–8 mm, 1 2/3–2 times as long as sepals, apex acute to rounded. **Capsules** tightly enclosed by calyx, narrowly ellipsoid to cylindric, 3–5 mm, ca. equaling sepals. **Seeds** 15–20, black to reddish brown, reniform, compressed, 0.8–0.9 mm, shiny, low-tuberculate. $2n = 40$.

Flowering spring–summer. Open, dry to slightly moist sand or gravel; 0–50 m; Greenland; Europe.

We follow B. Jonsell (2001) in treating *Arenaria pseudofrigida* as one of several arctic species that formerly were included in *A. ciliata* Linnaeus, a species now considered restricted to alpine regions of central Europe.

9. Arenaria serpyllifolia Linnaeus, Sp. Pl. 1: 423. 1753 • Thymeleaf sandwort, sabline á feuilles de serpolet I W

Plants annual. **Taproots** filiform. **Stems** 1–100+, erect to ascending or sprawling, green, 3–40+ cm; internodes terete to ellipsoid, 2–8 times as long as leaves, dull, retrorsely pubescent throughout, sometimes also stipitate-glandular. **Leaves** often connate basally, with scarious or mostly herbaceous sheath 0.2–0.3 mm, petiolate (proximal leaves)

or usually sessile; petiole 1–4 mm; blade 3–5 veined, elliptic to broadly ovate or rarely orbiculate, 2–7 × 1–4 mm, herbaceous, margins flat, herbaceous, dull, ciliate especially proximally, apex acute to acuminate, pustulate, sparsely minutely pubescent or glabrous; axillary leaf clusters absent. **Inflorescences** terminal, open, leafy, 3–50+-flowered cymes. **Pedicels** erect or ascending in fruit, 1–12 mm, retrorsely pubescent. **Flowers:** sepals green, often prominently 3-veined, not keeled, lanceolate to ovate-lanceolate (herbaceous portion narrowly elliptic to broadly lanceolate), 2–3 mm, to 4 mm in fruit, apex narrowly acute to acuminate, ± minutely pustulate, stipitate-glandular; petals oblong, 0.6–2.7 mm, $^1/_5$–$^3/_4$ times as long as sepals, apex obtuse to rounded. **Capsules** loosely enclosed by calyx, ovoid to cylindric-ovoid, 3–3.5 mm, $^4/_5$–1$^1/_5$ times as long as sepals. **Seeds** 10–15, ashy black, reniform, plump, 0.4–0.6 mm, not shiny, with low-elongate, prominent tubercules.

Varieties 3+ (2 in the flora): introduced, North America; Eurasia, Africa; introduced in Australia.

Variation in *Arenaria serpyllifolia* in the broad sense is treated in various ways. The two varieties recognized here have been treated also as subspecies (e.g., A. O. Chater and G. Halliday 1993) or species (e.g., M. N. Abuhadra 2000; B. Jonsell 2001). Jonsell admitted that accepting them as species is questionable; while the morphological differences are slight (see esp. Abuhadra), the ploidy-level difference ($2n = 40$ in var. *serpyllifolia* vs. $2n = 20$ in var. *tenuior*) is important.

1. Capsules ovoid to ovoid-conic, broader at base; seeds 0.5–0.6 mm . . . .
. . . . 9a. *Arenaria serpyllifolia* var. *serpyllifolia*
1. Capsules cylindric to cylindric-ovoid, not significantly broader at base (walls nearly straight); seeds ca. 0.4 mm . . . .
. . . . 9b. *Arenaria serpyllifolia* var. *tenuior*

9a. Arenaria serpyllifolia Linnaeus var. **serpyllifolia** I W

Sepals 2–3 mm, to 4 mm in fruit. **Capsules** ovoid to ovoid-conic, broader at base. **Seeds** 0.5–0.6 mm. $2n = 40$, 44? (both Europe).

Flowering spring–summer. Roadsides, open, sandy or rocky, disturbed sites; 0–1800 m; introduced; St. Pierre and Miquelon; Alta., B.C., Man., N.B., N.S., Ont., P.E.I., Que., Sask.; Ala., Ark., Calif., Colo., Conn., Del., D.C., Fla., Ga., Idaho, Ill., Ind., Iowa, Kans., Ky., La., Maine, Md., Mass., Mich., Minn., Miss., Mo., Mont., Nebr., Nev., N.H., N.J., N.Mex., N.Y., N.C., Ohio, Okla., Pa., R.I., S.C., S.Dak., Tenn., Tex., Utah, Vt., Va., Wash., W.Va., Wis., Wyo.; Eurasia; n Africa; introduced in Australia.

9b. Arenaria serpyllifolia Linnaeus var. **tenuior** Mertens & W. D. J. Koch in J. C. Röhling et al., Deutschl. Fl. ed. 3, 3: 266. 1831 I W

Arenaria leptoclados (Reichenbach) Gussone; *A. serpyllifolia* subsp. *leptoclados* (Reichenbach) Nyman, *A. serpyllifolia* var. *leptoclados* Reichenbach

Sepals 2–3 mm, not enlarging in fruit. **Capsules** cylindric to cylindric-ovoid, not significantly broader at base (walls nearly straight). **Seeds** ca. 0.4 mm. $2n = 20$ (Europe).

Flowering spring–summer. Roadsides, open, sandy or rocky, disturbed sites; 0–400 m; introduced; Ark., D.C., Fla., Ga., Ky., La., Miss., Mo., N.C., Pa., S.C., Tenn., Va.; Europe; w Asia; n Africa; introduced in Australia.

Variety *tenuior* is to be expected elsewhere.

15. EREMOGONE Fenzl, Vers. Darstell. Alsin., 13, unnumbered plate. 1833 • Sandwort

[Greek *eremo-*, solitary or deserted, and *gone*, seed or offspring, allusion uncertain]

Ronald L. Hartman

Richard K. Rabeler

Frederick H. Utech

Plants perennial, often densely matted, usually with branched, woody base. **Taproots** slender or usually stout; rhizomes absent. **Stems** prostrate (nonflowering stems) or ascending to erect (flowering stems), simple or branched, terete. **Leaves** usually congested at or near base of flowering stems, usually connate, mostly sessile; blade 1-veined, needlelike or filiform to subulate or narrowly linear, succulent or not, apex blunt to usually apiculate or spinose. **Inflorescences** terminal, open to congested or umbellate cymes or sometimes flowers solitary; bracts paired (or clustered at summit of peduncle in some *E. congesta* varieties), reduced, foliaceous or scarious. **Pedicels** erect or flowers sessile. **Flowers:** perianth and androecium weakly perigynous; hypanthium shallowly cup-shaped; sepals 5, distinct or barely connate proximally, green (sometimes purplish in *E. capillaris* and *E. eastwoodiae*), linear-lanceolate to ovate, 1.8–12 mm, margins white, scarious, apex obtuse or rounded to acute, acuminate, or spinose, not hooded; petals 5, white, yellowish white, or occasionally pink or brownish, clawed or not, blade apex entire, erose, emarginate, or rarely 2-fid; nectaries usually 5, prominent at (or adjacent to in *E. eastwoodiae*) base of filaments opposite sepals, rarely absent; stamens 10, arising from hypanthium; filaments distinct; staminodes absent; styles 3, filiform, 2.5–3 mm, glabrous proximally; stigmas 3, subcapitate, smooth to papillate (50×). **Capsules** ovoid to urceolate, opening by 6 ascending to recurved teeth; carpophore present or usually absent. **Seeds** 1–10, dark reddish or greenish brown, tan, blackish purple, black, or gray, ovoid to pyriform or suborbicular, laterally compressed, smooth, rugulose, or tuberculate, marginal wing absent, appendage absent. $x = 11$.

Species ca. 89 (14 in the flora): north-temperate regions, esp. w North America, Eurasian mountains, Asia Minor.

Much of this treatment follows those by B. Maguire (1947, 1951) and M. F. Baad (1969) of *Arenaria* sect. *Eremogone*. The *Eremogone* "group" is morphologically distinctive (cespitose or matted, woody perennials with filiform to subulate leaves, stiffly erect to ascending flowering stems, and open to congested or umbellate cymes) and nearly all of the published chromosome counts are based on $x = 11$ (see also Baad), an uncommon base number in *Arenaria* in the narrow sense. The decision to recognize *Eremogone*, as used in W. A. Weber and R. C. Wittmann (1992) and R. D. Dorn (2001), is based largely on results of the molecular study of subfamily Alsinoideae by M. Nepokroeff et al. (2001). In that study, *Arenaria* (excluding *Minuartia* and *Moehringia*) was shown to be polyphyletic, with two distinct clades. One of the clades consists of seven species: five from *Arenaria* subg. *Eremogone* and two from the closely related subg. *Eremogoneastrum*; therefore the *Eremogone* group, as delimited by Maguire, is monophyletic.

While *Eremogone* is a distinct genus, there is considerable morphological variability among the species, with many, often geographically isolated, morphological types; see J. C. Hickman (1971) for a commentary on the situation, numerous examples of intermediate forms, and notes on their possible evolution. With this in mind, we have chosen not to adopt an infrageneric classification; J. McNeill's (1962) outline of *Arenaria* suggests that three species groups could be recognized within *Eremogone*.

Because of the aforementioned variability and intermediacy within and between taxa, the production of a satisfactory key has been a challenge and, as with those by B. Maguire (1947, 1951), likely will prove inadequate for at least a low percentage of specimens.

Reports of winged seeds in *Eremogone* possibly are derived from examination of immature material. The tubercles along the abaxial edge that have not yet expanded can appear as a lighter-colored band around a portion of the seed.

The morphology of the nectaries in *Eremogone* is very diverse, both among species and, in one case, among infraspecific taxa (*E. macradenia*). We feel that a morphogenetic study using scanning electron microscopy similar to the investigation of nectaries in *Schiedea* by W. L. Wagner and E. M. Harris (2000) may shed light on relationships within *Eremogone*.

Some suggestions for the identification of taxa of *Eremogone* follow.

In determining the shape of the sepals, several flowers should be evaluated and the membranous margins must be included, not just the central herbaceous portion. As this genus is in many ways a "problematic group," some effort beyond what may be considered "normal" may be needed for accurate identification. This is because of the frequent need to observe floral details at high magnification (20 or 30×), preferably with a dissecting microscope. Furthermore, on dried specimens, a flower or two may need to be "boiled up" or otherwise rehydrated so that good, clear observations can be made. This is a practice that too often is avoided, to the detriment of accurate observation and identification, and should be standard in the study of pressed material.

SELECTED REFERENCES Baad, M. F. 1969. Biosystematic Studies of the North American Species of *Arenaria*, Subgenus *Eremogone* (Caryophyllaceae). Ph.D. dissertation. University of Washington. Hickman, J. C. 1971. *Arenaria*, section *Eremogone* (Caryophyllaceae) in the Pacific Northwest: A key and discussion. Madroño 21: 201–207. Ikonnikov, S. S. 1973. Zametki o gvozdichnykh (Caryophyllaceae). I. O rode *Eremogone* Fenzl. Novosti Sist. Vyssh. Rast. 10: 136–142. Maguire, B. 1947. Studies in the Caryophyllaceae. III. A synopsis of the North American species of *Arenaria* sect. *Eremogone* Fenzl. Bull. Torrey Bot. Club 74: 38–56.

1. Inflorescences usually dense or ± open, capitate, umbellate cymes.
  2. Petals 1.5–2 times as long as sepals; sepals 3–6.5 mm, ovate to lanceolate . . . . .   4. *Eremogone congesta*
  2. Petals 0.8–1.1 times as long as sepals; sepals (5–)6–12 mm, lanceolate or usually linear-lanceolate.
    3. Stems glabrous, with 6–10 pairs of well-developed cauline leaves, leaves markedly overlapping . . . . 8. *Eremogone franklinii*
    3. Stems scabrid-puberulent, with 1–4 pairs of well-developed leaves or reduced distally, usually little overlapping . . . . 9. *Eremogone hookeri*
1. Inflorescences ± open, rarely compact cymes.
  4. Sepal apices obtuse to rounded or barely acute, sometimes abruptly mucronate.
    5. Basal leaves abundant, blade (2–)4–8 cm, straight to curved in one direction, flexuous, apex acute or acuminate to weakly spinose; valves of capsule at least sparsely stipitate-glandular . . . . 3. *Eremogone capillaris*
    5. Basal leaves abundant or sparse, blade 0.3–2(–4) cm, recurved in various directions, apex often stiff and spinose; capsules glabrous.
      6. Basal leaf blades ascending or often arcuate-spreading, rigid, herbaceous; plants glaucous or not.
        7. Sepals ± stipitate-glandular; plants glaucous; cauline leaves in 1–3 pairs; petals 1.5–3 times as long as sepals . . . . 2. *Eremogone aculeata*
        7. Sepals glabrous or nearly so; plants not glaucous; cauline leaves in 5–7 pairs; petals 1.3–1.5 times as long as sepals . . . . 1. *Eremogone aberrans* (in part)
      6. Basal leaf blades erect to spreading, flexuous or rigid, herbaceous or ± succulent; plants not glaucous.
        8. Sepals stipitate-glandular; basal leaf blades ± succulent, ± flexuous; Crater Lake and vicinity, sw Oregon . . . . 12. *Eremogone pumicola*
        8. Sepals glabrous or nearly so; basal leaf blades herbaceous, rigid.
          9. Sepals 1.8–3 mm in flower; nectaries neither grooved nor cupped, 0.3–0.4 mm; s California . . . . 14. *Eremogone ursina*
          9. Sepals 3–5(–6) mm in flower; nectaries with transverse groove or elongate cup, 0.6 mm; se Idaho, n Utah, Wyoming . . . . . 10. *Eremogone kingii* (in part)
  4. Sepal apices acute to acuminate, sometimes spinose.
    10. Sepal apices broadly acute, sometimes spinose.
      11. Stems usually 20–40(–100) cm; leaf blades usually 2–6 mm; basal leaves sparse or absent; nectaries rectangular, 2-lobed and 0.7–1.5 mm, or narrowly longitudinally rectangular, truncate, densely minutely pubescent with erect to spreading hairs, 0.7–0.8 mm, or laterally or abaxially rounded, 0.3–0.4 mm.
        12. Sepals 3–4.3 mm, in fruit to 5.5 cm; cymes open, branches spreading; nectaries as . . . lateral and abaxial rounding of filament bases, 0.3–0.4 mm . . . . 7. *Eremogone ferrisiae*
        12. Sepals 4.5–7.2 mm, in fruit to 8 mm; cymes ± compact, branches ascending or erect; nectaries rectangular, 2-lobed or truncate, 0.7–1.5 mm . . . . 11. *Eremogone macradenia*
      11. Stems usually 3–20 cm; leaf blades usually 0.3–3 cm; basal leaves abundant; nectaries rounded, 0.2–0.6 mm.

13. Cauline leaves in 5–7 pairs, reduced distally; petals yellowish white, 5.8–10 mm; capsules 7–10 mm; n Arizona . . . . 1. *Eremogone aberrans* (in part)
13. Cauline leaves in (1–)2–4+ pairs, reduced distally or not; petals white or rarely pink, (3–)4–7 mm; capsules 4.5–7 mm; Great Basin and to n and w
. . . . 10. *Eremogone kingii* (in part)

[10. Shifted to left margin.—Ed.]

10. Sepal apices narrowly acute to acuminate or spinose.
   14. Sepals glabrous throughout or essentially so.
      15. Petals 0.9–1.1 times as long as sepals; nectaries longitudinally rectangular, apically cleft or emarginate, 1–2 mm; Arizona, Colorado, New Mexico, Utah, Wyoming
. . . . 5. *Eremogone eastwoodiae*
      15. Petals 1.5–1.7 times as long as sepals; nectaries rounded at base of filaments, 0.3–0.4 mm; se Nevada . . . . 13. *Eremogone stenomeres*
   14. Sepals moderately to densely stipitate-glandular.
      16. Cauline leaves usually in 2–4 pairs, blade usually 0.3–3 cm; stems usually 2–20 cm; nectaries rounded lobe with transverse groove or elongate cup; Colorado Plateau and Great Basin . . . . 10. *Eremogone kingii* (in part)
      16. Cauline leaves in (4–)5+ pairs, blade usually 3–11 cm; stems usually 7–40 cm; nectaries rounded, not transversely grooved or cupped.
         17. Leaves flexuous; petals 0.9–13 times as long as sepals; Rocky Mountains proper and outlying mountains . . . . 6. *Eremogone fendleri*
         17. Leaves rigid; petals 1.5–1.7 times as long as sepals; se Nevada . . . .
. . . . 10. *Eremogone kingii* (in part)

1. **Eremogone aberrans** (M. E. Jones) Ikonnikov, Novosti Sist. Vyssh. Rast. 10: 139. 1973 • Mount Dellanbaugh sandwort [E]

Arenaria aberrans M. E. Jones, Contr. W. Bot. 16: 37. 1930

Plants tufted to mat forming, green, not glaucous, with woody base. **Stems** erect, (3–)10–23 cm, moderately to densely stipitate-glandular. **Leaves:** basal leaves abundant, persistent; cauline leaves in 5–7 pairs, reduced distally; basal blades spreading to arcuate-spreading, needlelike, 0.8–2 cm × 0.4–0.8 mm, ± rigid, not fleshy, herbaceous, apex spinose, glabrous, not glaucous. **Inflorescences:** (1–)3–6-flowered, open cymes. **Pedicels** 6–25 mm, stipitate-glandular. **Flowers:** sepals 1–3-veined, lateral veins less developed, narrowly elliptic to ovate, 3.5–4 mm, 4.8–5.2 mm in fruit, margins usually broadly winged, scarious, apex broadly acute to obtuse (at least in fruit), glabrous or nearly so; petals yellowish white, spatulate, 5.8–10 mm, 1.3–1.5 times as long as sepals, apex rounded; nectaries as lateral and abaxial rounding of base of filaments opposite sepals, 0.2–0.3 mm. **Capsules** 7–10 mm, glabrous. **Seeds** brownish black, suborbicular with hilar notch, 2–2.4 mm, tuberculate; tubercles rounded, elongate.

Flowering late spring–summer. Oak and yellow pine forests; 1500–2800 m; Ariz.

Eremogone aberrans is known only from northern Arizona and resembles a robust form of the more northerly occurring *E. aculeata*. In Arizona it is often confused with *E. fendleri*, which has sepals more or less glandular-pubescent whereas *E. aberrans* has sepals glabrous or with a few glandular hairs at their bases.

2. **Eremogone aculeata** (S. Watson) Ikonnikov, Novosti Sist. Vyssh. Rast. 10: 139. 1973 • Prickly sandwort [E]

Arenaria aculeata S. Watson, Botany (Fortieth Parallel), 40. 1871; *A. fendleri* A. Gray var. *aculeata* (S. Watson) S. L. Welsh; *A. pumicola* Coville & Leiberg var. *californica* Maguire

Plants strongly mat-forming, glaucous, often with woody base. **Stems** erect, 7–25(–30) cm, densely stipitate-glandular distally. **Leaves:** basal leaves persistent; cauline leaves in 1–3 pairs, abruptly reduced distal to lowest pair; basal blades ascending or often arcuate-spreading, needlelike, (0.5–)1–2.5(–3.5) cm × 0.5–1.5 mm, rigid, herbaceous, apex spinose, glabrous to puberulent, glaucous. **Inflorescences** 5–25+-flowered, open cymes. **Pedicels** 3–25 mm, stipitate-glandular. **Flowers:** sepals 1–3-veined, lateral veins less developed or all obscure, ovate, 3–4.5 mm, to 6 mm in fruit, margins broad, apex usually obtuse to rounded, abruptly acute, sparsely to densely stipitate-glandular; petals white, obovate to oblanceolate, 4.5–10 mm, 1.5–3 times as long as sepals, apex rounded; nectaries as lateral and abaxial rounding, with slight lateral expansion, at base of filaments opposite sepals, 0.3 mm. **Capsules** 5–9 mm, glabrous. **Seeds** yellowish tan to gray, ellipsoid-oblong,

1.8–2.5(–3.2) mm, tuberculate; tubercles rounded, elongate. **2***n* = 22.

Flowering summer–early fall. Rocky slopes, alluvium, volcanic areas; 1500–3400 m; Calif., Idaho, Mont., Nev., Oreg., Utah, Wash.

Some specimens from north-central California and southwestern Oregon have been named *Arenaria pumicola* var. *californica*. R. L. Hartman (1993) considered those plants to be robust forms of *Eremogone aculeata*, not deserving formal recognition. Based on work by M. F. Baad (1969), they warrant further study.

Reports of *Eremogone aculeata* from Arizona, New Mexico, and Wyoming are erroneous.

3. Eremogone capillaris (Poiret) Fenzl, Vers. Darstell. Alsin., 37. 1833 • Slender mountain sandwort

Arenaria capillaris Poiret in J. Lamarck et al., Encycl. 6: 380. 1804

Plants cespitose to somewhat matted, green or often glaucous, with woody base. **Stems** erect to ascending, (3–)5–20(–25) cm, entirely glabrous or glabrous proximally, stipitate-glandular distally. **Leaves:** basal leaves abundant, persistent; cauline leaves in 1–3 pairs, abruptly reduced distal to lowest pair; basal blades erect, straight or often curved in one direction, filiform, (2–)4–8 cm × 0.5–1 mm, flexuous, herbaceous, apex acute or acuminate to weakly spinose, glabrous, sometimes stipitate-glandular, often glaucous. **Inflorescences** 1–12(–18)-flowered, open or rarely subcongested cymes. **Pedicels** (2–)5–20 mm, glabrous or stipitate-glandular. **Flowers:** sepals often purplish, 1–3-veined, ovate to ovate-lanceolate, 3–4.7 mm, to 5 mm in fruit, margins broad, apex broadly rounded to barely acute, glabrous or stipitate-glandular; petals white, spatulate, 6–10 mm, 1.5–2.5 times as long as sepals, apex broadly rounded; nectaries as depressed transverse cup at base of filaments opposite sepals, 0.3–0.7 × 0.1–0.3 mm. **Capsules** 5–8 mm, valves at least sparsely stipitate-glandular or glabrous. **Seeds** brown to black, ellipsoid to ovoid with hilar notch, (1.2–)1.5–2(–2.5) mm, tuberculate; tubercles rounded, elongate.

Varieties 2 (2 in the flora): North America, Asia (Siberia).

B. Maguire (1947, 1951) and subsequent workers separated these varieties of *Eremogone capillaris* by sepal length: (3–)3.5–4.5(–4.8) mm in var. *americana* and (3.6–)5–6(–6.6) mm in var. *capillaris*. Material available to us did not exhibit that distinction.

1. Pedicels and sepals glabrous . . . .
. . . . 3a. *Eremogone capillaris* var. *capillaris*
1. Pedicels and sepals stipitate-glandular . . . .
. . . . 3b. *Eremogone capillaris* var. *americana*

3a. Eremogone capillaris (Poiret) Fenzl var. capillaris

Arenaria capillaris Poiret var. *nardifolia* (Ledebour) Regel; *A. formosa* Fischer ex Seringe

Stems 5–20 cm, glabrous. **Pedicels** glabrous. **Flowers:** sepals glabrous; ovary glabrous. **Capsules** glabrous. **2***n* = 22.

Flowering summer. Rock outcrops, dry, sandy hillsides, roadsides; 300–900 m; N.W.T., Yukon; Alaska; Asia (Mongolia, Siberia).

3b. Eremogone capillaris (Poiret) Fenzl var. americana (Maguire) R. L. Hartman & Rabeler, Sida 21: 239. 2004 • Fescue sandwort [E]

Arenaria capillaris Poiret subsp. *americana* Maguire, Bull. Torrey Bot. Club 74: 41. 1947; *A. capillaris* var. *americana* (Maguire) R. J. Davis; *Eremogone americana* (Maguire) Ikonnikov

Stems (3–)5–15(–25) cm, glabrous and glaucous proximally, stipitate-glandular distally. **Pedicels** stipitate-glandular. **Flowers:** sepals stipitate-glandular; ovary often with uniseriate, 3–8-celled hairs in proximal ¹⁄₂ or throughout. **Capsules** often with uniseriate, 3–8-celled hairs in distal ¹⁄₂ or throughout. **2***n* = 22.

Flowering late spring–summer. Alpine meadows, talus slopes, aspen woods; 1200–3000 m; Alta., B.C.; Idaho, Mont., Nev., Oreg., Wash.

M. F. Baad (1969) reported populations from Idaho, Nevada, Oregon, and Wyoming that he considered to be intermediate between var. *americana* and *Eremogone kingii* var. *glabrescens*, and one from central Washington that he considered to be intermediate between var. *americana* and *E. congesta* var. *prolifera*.

4. Eremogone congesta (Nuttall) Ikonnikov, Novosti Syst. Vyssh. Rast. 10: 139. 1973 • Ballhead sandwort [E]

Arenaria congesta Nuttall in J. Torrey and A. Gray, Fl. N. Amer. 1: 178. 1838

Plants tufted or sometimes matted, green, not glaucous, with woody base. **Stems** ± erect, 3–40(–50) cm, glabrous or often stipitate-glandular. **Leaves:** basal leaves persistent or not; cauline leaves in 3–5 pairs, similar, but reduced distally; basal blades erect-ascending to arcuate-spreading, subulate or needlelike to filiform, (0.8–)2–11(–14) cm × 0.4–2 mm, flexuous or rigid, herbaceous to ± fleshy, apex obtuse to sharply acute or spinose, glabrous, sometimes glaucous. **Inflorescences** 3–50+-flowered, congested and capitate or sometimes open, umbellate cymes. **Pedicels** 0.1–7 (–15) mm or ± absent, usually glabrous, rarely stipitate-glandular. **Flowers:** sepals 1–3-veined, sometimes obscurely so, ovate to lanceolate, 3–6.5 mm, not expanding in fruit, margins narrow, apex obtuse or acute to acuminate, rarely spinose, glabrous (or glandular in var. *prolifera*); petals white, oblong, 5–8(–10) mm, 1.5–2 times as long as sepals, apex entire to slightly emarginate; nectaries as lateral and abaxial mound with crescent-shaped groove at base of filaments opposite sepals, 0.3 × 0.15–0.2 mm. **Capsules** 3.5–6 mm, glabrous. **Seeds** reddish brown to black, broadly ellipsoid to ovoid, 1.4–3 mm, tuberculate; tubercles low, rounded, often elongate.

Varieties 9 (9 in the flora): w North America.

Eremogone congesta is highly polymorphic; it has been been divided into 11 varieties (nine recognized here), most of which are distinctive and locally distributed. M. F. Baad (1969) noted two patterns of variation of different origin within *E. congesta*, but he did not present a revised classification.

While most specimens of the four varieties with dense inflorescences do not exhibit evident pedicels, the occasional plant does bear one or more pedicels to 1–2 mm, sometimes in secondary inflorescences.

1. Pedicels usually 0.1–0.2 mm or ± absent; inflorescences dense, tight cymes.
  2. Basal leaf blades filiform, 3–14 cm.
    3. Inflorescences capitate, rounded cymes; sepal apices obtuse to rounded . . . .
    . . . . 4a. *Eremogone congesta* var. *congesta*
    3. Inflorescences capitate, pyramidal cymes; sepal apices narrowly acute to acuminate
    . . . . . 4b. *Eremogone congesta* var. *cephaloidea*
  2. Basal leaf blades needlelike, 1–3.5 cm.
    4. Sepals 5–6.5 mm, apex acute to acuminate; basal leaf blades herbaceous; California, Nevada . . . . 4g. *Eremogone congesta* var. *simulans* (in part)
    4. Sepals 3.5–4 mm, apex obtuse; basal leaf blades ± succulent; California, Oregon . . . . 4d. *Eremogone congesta* var. *crassula*
1. Pedicels 1–6(–15) mm; inflorescences somewhat to markedly open cymes or umbellate cymes.
  5. Sepal apices obtuse.
    6. Inflorescences proliferating, ± loose cymes; bracts often not closely enveloping sepals; Alberta, Saskatchewan, Colorado, Idaho, Montana, Utah, Wyoming . . . .
    . . . . 4e. *Eremogone congesta* var. *lithophila*
    6. Inflorescences umbels, bracts clustered at umbel base; California . . . .
    . . . . . 4i. *Eremogone congesta* var. *suffrutescens*

[5. Shifted to left margin.—Ed.]
5. Sepal apices acute to acuminate or spinose.
  7. Basal leaf blades (2–)3–8 cm, filiform; pedicels glabrous or sometimes stipitate-glandular (possible throughout range, true for Nevada populations); Idaho, Nevada, Oregon, Washington . . . . 4f. *Eremogone congesta* var. *prolifera*
  7. Basal leaf blades 0.5–2.5(–3.5) cm, needlelike or filiform; pedicels glabrous; California, Nevada, Utah.
    8. Inflorescences proliferating, ± loose cymes; sepals 3.5–4.5 mm . . . . 4h. *Eremogone congesta* var. *subcongesta*
    8. Inflorescences capitate and often rounded, to subcongested, proliferating or open cymes; sepals 4.5–6.5 mm.
      9. Sepals 4.5–5.5 mm, weakly to conspicuously 1–3-veined, apex spinose . . . . .
      . . . . 4c. *Eremogone congesta* var. *charlestonensis*
      9. Sepals usually 5.5–6.5 mm, conspicuously 3-veined, apex acute to acuminate
      . . . . 4g. *Eremogone congesta* var. *simulans* (in part)

4a. Eremogone congesta (Nuttall) Ikonnikov var. **congesta** 🇪

Stems 5–30 cm, glabrous. **Leaves:** basal blades erect to somewhat spreading, filiform, 3–8 cm × 0.5–1 mm, herbaceous. **Inflorescences** capitate, often rounded, dense cymes; bracts closely enveloping sepals. **Pedicels** ± absent, shorter than 0.2 mm, occasionally to 2 mm, glabrous. **Sepals** conspicuously 1 to weakly 3-veined, 3.5–5 mm, apex obtuse to rounded. **2n** = 22.

Flowering spring–summer. Open, gravelly or sandy soil of plains, mountain slopes, and ridges; 1200–3500 m; Calif., Colo., Idaho, Mont., Nev., Oreg., Utah, Wash., Wyo.

4b. Eremogone congesta (Nuttall) Ikonnikov var. **cephaloidea** (Rydberg) R. L. Hartman & Rabeler, Sida 21: 239. 2004 • Sharptip sandwort 🇪

Arenaria cephaloidea Rydberg, Bull. Torrey Bot. Club 39: 316. 1912; *A. congesta* Nuttall var. *cephaloidea* (Rydberg) Maguire

Stems 20–40(–50) cm, glabrous. **Leaves:** basal blades erect to somewhat spreading, filiform, 3–8(–14) cm × 0.4–0.7 mm, herbaceous. **Inflorescences** capitate, often pyramidal, dense cymes; bracts closely enveloping sepals. **Pedicels** ± absent, mostly shorter than 0.1 mm, glabrous. **Sepals** weakly to conspicuously 1–3-veined, 4–5 mm, apex acute to acuminate.

Flowering summer. Open, gravelly woods; 300–1000 m; Idaho, Oreg., Wash.

The pyramidal inflorescence of var. *cephaloidea* bears a fanciful resemblance to the grass *Dactylis glomerata* Linnaeus and some members of *Carex* Linnaeus sect. *Ovales* Kunth.

4c. Eremogone congesta (Nuttall) Ikonnikov var. **charlestonensis** (Maguire) R. L. Hartman & Rabeler, Sida 21: 239. 2004 • Charleston sandwort E

Arenaria congesta Nuttall var. *charlestonensis* Maguire, Bull. Torrey Bot. Club 73: 326. 1946

Stems 5–14 cm, glabrous. **Leaves:** basal blades straight to spreading, often arcuately so, needlelike, 0.7–2 cm × 0.5–0.8 mm, herbaceous. **Inflorescences** capitate, often rounded, open cymes; bracts often not closely enveloping sepals. **Pedicels** (1–)2–4 mm, glabrous. **Sepals** weakly to conspicuously 1–3-veined, 4.5–5.5 mm, apex spinose.

Flowering summer. Sandy ridges; 2200–3100 m; Calif., Nev.

Variety *charlestonensis* is known from the New York Mountains, Inyo County, California, and adjacent Nevada. It may be more closely allied to *Eremogone kingii* than to *E. congesta*.

4d. Eremogone congesta (Nuttall) Ikonnikov var. **crassula** (Maguire) R. L. Hartman & Rabeler, Sida 21: 239. 2004 E

Arenaria congesta Nuttall var. *crassula* Maguire, Bull. Torrey Bot. Club 74: 45. 1947

Stems 15–20 cm, glabrous. **Leaves:** basal blades curved inward, needlelike, (1.5–)2–3 cm × 1–2 mm, ± fleshy. **Inflorescences** capitate, rounded, dense cymes; bracts closely enveloping sepals. **Pedicels** ± absent, mostly shorter than 0.1 mm, glabrous. **Sepals** conspicuously 1- to weakly 3-veined, 3.5–4 mm, apex obtuse.

Flowering summer. Dry, upper montane ridges and rocky crevices; 1800–2500 m; Calif., Oreg.

4e. Eremogone congesta (Nuttall) Ikonnikov var. **lithophila** (Rydberg) Dorn, Vasc. Pl. Wyoming ed. 3, 376. 2001 • Loosehead sandwort E

Arenaria subcongesta Nuttall var. *lithophila* Rydberg, Mem. New York Bot. Gard. 1: 148. 1900; *A. lithophila* (Rydberg) Rydberg

Stems 10–30 cm, glabrous. **Leaves:** basal blades erect to somewhat spreading, filiform, 3–6 cm × 0.3–0.6 mm, herbaceous. **Inflorescences** proliferating, ± loose cymes, secondary axes to 2 cm; bracts often not closely enveloping sepals. **Pedicels** 1–5 mm, glabrous. **Sepals** weakly to conspicuously 1–3-veined, 3.5–5 mm, apex obtuse or rounded.

Flowering summer. Open, rocky, sagebrush slopes and hillsides; 800–3100 m; Alta., Sask.; Colo., Idaho, Mont., Utah, Wyo.

M. F. Baad (1969) suggested the possible relationship of var. *lithophila* to *Eremogone capillaris*, but he followed B. Maguire (1947) in retaining it in *E. congesta*, noting the existence of intermediates between var. *congesta* and var. *lithophila*.

4f. Eremogone congesta (Nuttall) Ikonnikov var. **prolifera** (Maguire) R. L. Hartman & Rabeler, Sida 21: 239. 2004 E

Arenaria congesta Nuttall var. *prolifera* Maguire, Bull. Torrey Bot. Club 74: 47. 1947; *A. congesta* var. *glandulifera* Maguire; *Eremogone congesta* var. *glandulifera* (Maguire) R. L. Hartman & Rabeler

Stems 15–25 cm, glabrous to stipitate-glandular. **Leaves:** basal blades mostly erect, filiform, (2–)3–8 cm × 0.3–0.7 mm, herbaceous. **Inflorescences** subcapitate, subcongested to proliferating cymes; bracts scattered within inflorescence. **Pedicels** 1–6(–15) mm, glabrous or sometimes stipitate-glandular. **Sepals** weakly to conspicuously 1–3-veined, (4.5–)5–6 mm, apex acute to acuminate or spinose, glabrous or glandular.

Flowering late spring–summer. Sagebrush plains and slopes; 500–900 m; Alta.; Idaho, Nev., Oreg., Wash.

We consider var. *glandulifera* to be a glandular extreme of var. *prolifera*. It is known from a single specimen (*Ownbey & Ownbey 2763*, NY) collected in Valley County, Idaho, in 1946. While most specimens of var. *prolifera* are glabrous, the pubescence on the stems and pedicels of a collection from Douglas County, Washington (*Hitchcock 17459*, WTU) closely approaches that seen on the type of var. *glandulifera*.

4g. Eremogone congesta (Nuttall) Ikonnikov var. **simulans** (Maguire) R. L. Hartman & Rabeler, Sida 21: 239. 2004 E

Arenaria congesta Nuttall var. *simulans* Maguire, Bull. Torrey Bot. Club 74: 48. 1947; *A. congesta* Nuttall var. *wheelerensis* Maguire; *Eremogone congesta* var. *wheelerensis* (Maguire) R. L. Hartman & Rabeler

Stems 10–20 cm, glabrous. **Leaves:** basal blades erect to spreading or widely so, needlelike, 2–3.5 cm × 0.3–0.6 mm, herbaceous. **Inflorescences** capitate, congested cymes; bracts scattered within inflorescence. **Pedicels** 1–5 mm, glabrous. **Sepals** conspicuously 3-veined, 5–6.5 mm, apex acute to acuminate.

Flowering spring–early summer. Open, rocky slopes; 1300–3700 m; Calif., Nev.

We consider var. *wheelerensis*, known from only a few sites in Elko, Lincoln, and White Pine counties, Nevada, to be an alpine extreme of the more widely distributed var. *simulans*. This is similar to the interpretation of M. F. Baad (1969), who considered that var. *wheelerensis* may be intermediate between var. *simulans* and var. *subcongesta*.

4h. Eremogone congesta (Nuttall) Ikonnikov var. **subcongesta** (S. Watson) R. L. Hartman & Rabeler, Sida 21: 239. 2004 E

Arenaria fendleri A. Gray var. *subcongesta* S. Watson, Botany (Fortieth Parallel), 40. 1871; *A. burkei* Howell; *A. congesta* Nuttall var. *subcongesta* (S. Watson) S. Watson; *A. subcongesta* (S. Watson) Rydberg

Stems 8–25 cm, glabrous. **Leaves:** basal blades erect to often arcuate-spreading, needlelike to filiform, 0.5–2.5 cm × 0.3–0.8 mm, herbaceous. **Inflorescences** subcapitate, proliferating, ± loose cymes, secondary axes to 2 cm; bracts scattered within inflorescence. **Pedicels** 1–5 mm, glabrous. **Sepals** conspicuously 3-veined, 3.5–4.5 mm, apex acute.

Flowering late spring–summer. Open, rocky, sagebrush slopes and flats, volcanic soils; 1300–2800 m; Calif., Nev., Utah.

M. F. Baad (1969) noted that var. *subcongesta* may be a xerophytic form of var. *simulans*. B. Maguire (1960) noted the existence of two historical records of var. *subcongesta* from Arizona, neither of which indicates a specific locality; we have not seen material from that state.

4i. Eremogone congesta (Nuttall) Ikonnikov var. **suffrutescens** (A. Gray) R. L. Hartman & Rabeler, Sida 21: 239. 2004 • Suffrutescent sandwort E

Brewerina suffrutescens A. Gray, Proc. Amer. Acad. Arts 8: 620. 1873; *Arenaria congesta* Nuttall var. *suffrutescens* (A. Gray) B. L. Robinson; *A. suffrutescens* (A. Gray) A. Heller

Stems 15–40 cm, glabrous or stipitate-glandular. **Leaves:** basal blades erect to somewhat spreading, needlelike, 1.5–8 cm × 0.5–1.5 mm, herbaceous. **Inflorescences** 1–3 umbellate cymes; bracts generally clustered at base of inflorescence. **Pedicels** 1–12 mm, glabrous. **Sepals** weakly to conspicuously 1–3-veined, 3–4 mm, apex obtuse.

Flowering late spring–early summer. Rocky mountain slopes and outcrops; 1200–3300 m; Calif.

5. Eremogone eastwoodiae (Rydberg) Ikonnikov, Novosti Syst. Vyssh. Rast. 10: 139. 1973 (as eastwoodii) • Eastwood's sandwort E

Arenaria eastwoodiae Rydberg, Bull. Torrey Bot. Club 31: 406. 1904; *A. fendleri* A. Gray var. *eastwoodiae* (Rydberg) S. L. Welsh

Plants densely matted, green, not glaucous, with woody base. **Stems** erect, (8–)10–25 cm, glabrous or stipitate-glandular. **Leaves:** basal leaves persistent; cauline leaves usually in 2–4 pairs, reduced distally; basal blades spreading to recurved, needlelike, 1–3(–3.5) cm × 0.5–0.7 mm, flexuous to rigid, herbaceous, apex spinose, glabrous to puberulent, not glaucous. **Inflorescences** (1–)3–17-flowered, ± open cymes. **Pedicels** 3–30 mm, glabrous or stipitate-glandular. **Flowers:** sepals green or purplish, 1–3-veined, lanceolate to ovate-lanceolate, (3.5–)4–6.5 mm, not enlarging in fruit, margins broad, apex narrowly acute to acuminate, glabrous or stipitate-glandular; petals yellowish white or sometimes brownish to reddish pink, broadly oblong-elliptic to oblanceolate, 4–6.5 mm, 0.9–1.1 times as long as sepals, apex rounded; nectaries narrowly longitudinally rectangular, apically cleft or emarginate, adjacent to filaments opposite sepals, 1–2 mm. **Capsules** 4–6 mm, glabrous. **Seeds** brown, ovoid to suborbicular with hilar notch, 1.2–1.7 mm, papillate, subechinate; tubercles conical.

Varieties 2 (2 in the flora): w North America.

The Hopi Indians may use *Eremogone eastwoodiae* as an emetic (B. Maguire 1960).

The nectaries in *Eremogone eastwoodiae* are different from those of most other species of the genus in North

America since they are a separate bilobed structure adjacent to, but not a direct enlargement of, the filament bases opposite the sepals.

1. Stems and pedicels glabrous . . . .
. . . . 5a. *Eremogone eastwoodiae* var. *eastwoodiae*
1. Stems and pedicels stipitate-glandular . . . .
. . . . 5b. *Eremogone eastwoodiae* var. *adenophora*

5a. Eremogone eastwoodiae (Rydberg) Ikonnikov var. **eastwoodiae** E

Stems glabrous. **Pedicels** glabrous. **2***n* = 22.

Flowering late spring–late summer. Dry, stony or sandy hills, mesas, and deserts; 1300–2200 m; Ariz., Colo., N.Mex., Utah, Wyo.

5b. Eremogone eastwoodiae (Rydberg) Ikonnikov var. **adenophora** (Kearney & Peebles) R. L. Hartman & Rabeler, Sida 21: 240. 2004 E

Arenaria eastwoodiae Rydberg var. *adenophora* Kearney & Peebles, J. Wash. Acad. Sci. 29: 475. 1939

Stems stipitate-glandular. **Pedicels** stipitate-glandular.

Flowering late spring–late summer. Dry, stony or sandy hills, mesas, and deserts; 1200–2100 m; Ariz., Colo., N.Mex., Utah.

Variety *adenophora* is largely restricted to western Colorado, northeastern Arizona, and northwestern New Mexico. The apparent allopatric distributions of the two varieties provide support for the recognition of var. *adenophora*. This is further justified in that the glandular variety abuts the range of the Rocky Mountain *Eremogone fendleri*. It is assumed that the presence of glandular stems in each of these taxa is of independent origin.

Plants of var. *adenophora* may be more likely than those of var. *eastwoodiae* to exhibit pink petals, and they often look smaller, with fewer flowers in a less-open inflorescence. Occasional specimens that we have examined suggest that this taxon may be polyphyletic.

6. Eremogone fendleri (A. Gray) Ikonnikov, Novosti Syst. Vyssh. Rast. 10: 139. 1973 • Fendler's sandwort E F

Arenaria fendleri A. Gray, Mem. Amer. Acad. Arts, n. s. 4: 13. 1849; *A. fendleri* subsp. *brevifolia* Maguire; *A. fendleri* var. *brevifolia* (Maguire) Maguire; *A. fendleri* var. *diffusa* Porter & Coulter; *A. fendleri* var. *porteri* Rydberg; *A. fendleri* var. *tweedyi* (Rydberg) Maguire; *A. tweedyi* Rydberg

Plants ± cespitose, bluish green, not glaucous, with woody base. **Stems** erect, (2–)10–30(–40) cm, stipitate-glandular. **Leaves:** basal leaves persistent; cauline leaves in (4–)5+ pairs, reduced or not; basal blades ascending or recurved, filiform, 1–10(–11) cm × 0.2–0.4 mm, flexuous, herbaceous, apex apiculate to spinose, glabrous to puberulent, not glaucous. **Inflorescences** (1–)3–35-flowered, ± open cymes. **Pedicels** 3–25 mm, stipitate-glandular. **Flowers:** sepals weakly to prominently 1–3-veined, linear-lanceolate, 4–7.5 mm, not enlarging in fruit, margins broad, apex acuminate, moderately to densely stipitate-glandular on herbaceous portion; petals white, oblong-elliptic to spatulate, 4–8 mm, 0.9–1.3 times as long as sepals, apex entire to somewhat erose; nectaries as lateral and abaxial rounding of base of filaments opposite sepals, 0.2 × 0.4 mm. **Capsules** 5–7 mm, glabrous. **Seeds** black, ovoid to pyriform with hilar notch, 1.5–1.9 mm, tuberculate; tubercles rounded, elongate to rounded-conic. **2***n* = 44.

Flowering spring–late summer. Sagebrush plains, pine forests, and mountain slopes to alpine zones; 1200–4300 m; Ariz., Colo., N.Mex., Tex., Utah, Wyo.

We agree with M. F. Baad (1969) in not formally recognizing varieties within *Eremogone fendleri*. B. Maguire (1947, 1951) recognized five varieties, defined chiefly on leaf and sepal characteristics. While some specimens can be "matched" to varieties, many appear intermediate between them, forming a continuum of variation.

B. Maguire (1947) noted that *Eremogone fendleri* is "probably to be found in the states of Sonora and Chihuahua, Mexico"; we have not seen any collections from that area.

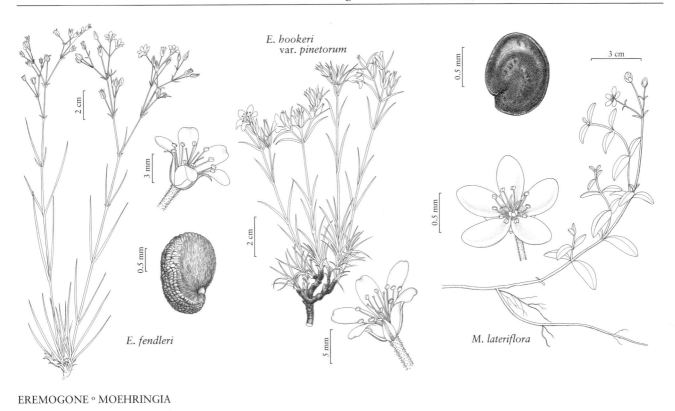

E. hookeri
var. *pinetorum*

E. fendleri

M. lateriflora

EREMOGONE ° MOEHRINGIA

7. Eremogone ferrisiae (Abrams) R. L. Hartman &
Rabeler, Sida 21: 754. 2004 • Ferris's sandwort E

Arenaria macradenia S. Watson
subsp. *ferrisiae* Abrams in L.
Abrams and R. S. Ferris, Ill. Fl.
Pacific States 2: 151. 1944;
Eremogone macradenia S. Watson
var. *ferrisiae* (Abrams) R. L.
Hartman & Rabeler

Plants tufted, green, not glaucous,
with woody base. **Stems** erect,
(10–)20–40(–100) cm, glabrous to stipitate-glandular.
Leaves: basal leaves sparse or absent; cauline leaves
usually in 5–7 pairs, not significantly reduced; basal
blades ascending, needlelike or narrowly linear, 2–6(–7)
cm × 0.5–1 mm, ± rigid, herbaceous to subsucculent,
apex blunt to spinose, usually glabrous, not glaucous.
Inflorescences 10–30(–80)-flowered, diffuse cymes;
branches spreading. **Pedicels** 15–55 mm, glabrous or
stipitate-glandular. **Flowers:** sepals 1–3-veined, ovate
to lanceolate or elliptic, 3–4.3 mm, to 5.5 mm in fruit,
margins narrow to broad, apex acute to acuminate,
glabrous to sparsely stipitate-glandular; petals white or
yellowish, oblanceolate to spatulate, 6–9 mm, 1–1.5

times as long as sepals, apex entire or erose; nectaries as
lateral and abaxial rounding of base of filaments opposite
sepals, 0.3–0.4 mm. **Capsules** 6–7 mm, glabrous. **Seeds**
reddish brown to blackish, suborbicular to pyriform or
ovoid, 1.3–3.2 mm, tuberculate; tubercles low, rounded
to conic.

Flowering spring–summer. Pine and oak woodlands,
granitic alluvium on foothills and mountain slopes;
1400–2900 m; Ariz., Calif., Nev., Utah.

We now believe that *Eremogone macradenia* (in the
sense of R. L. Hartman 1993) should be split into two
species, with *E. macradenia* var. *ferrisiae* (Abrams)
R. L. Hartman & Rabeler being elevated to species rank
(Hartman and R. K. Rabeler 2004), as here. This became
particularly obvious when comparing nectary
morphology of *E. macradenia* (rectangular, two-lobed
or truncate, 0.7–1.5 mm or narrowly longitudinally
rectangular, truncate, densely minutely pubescent with
erect to spreading hairs, 0.7–0.8 mm) with that of *E.
ferrisiae* (rounded, 0.3–0.4 mm). Furthermore, the
nectary types correlate well with sepal size and
inflorescence type, as indicated in the key. This
disposition agrees with the conclusions of M. F. Baad
(1969).

8. Eremogone franklinii (Douglas ex Hooker) R. L. Hartman & Rabeler, Sida 21: 240. 2004 • Franklin's sandwort [E]

Arenaria franklinii Douglas ex Hooker, Fl. Bor.-Amer. 1: 101, plate 35. 1831

Plants ± cespitose, bluish green, not glaucous, with woody base. **Stems** erect, 3–10(–15) cm, glabrous. **Leaves:** basal leaves persistent; cauline leaves in 6–10 pairs, closely overlapping, not reduced; basal blades arcuate-spreading, needlelike, (0.6–)1–2 cm × ca. 1 mm, ± rigid, herbaceous, apex spinose, glabrous, not glaucous. **Inflorescences** 3–45+-flowered, usually congested, subcapitate cymes. **Pedicels** ca. 0.1–3(–4) mm, glabrous. **Flowers:** sepals 1–3-veined, linear-lanceolate, (5–)8.5–12 mm, not enlarging in fruit, margins narrow, apex acuminate, glabrous; petals white, oblong-lanceolate to narrowly spatulate, 7–9 mm, 0.8–1.1 times as long as sepals, apex rounded to blunt; nectaries not apparent. **Capsules** 2.3–3.3 mm, glabrous. **Seeds** black, pyriform, 1.2–1.7 mm, tuberculate.

Varieties 2 (2 in the flora): w United States.

1. Inflorescences tightly congested cymes; pedicels 0.1–1(–3) mm; sepals 8.5–12 mm; petals 0.8–0.9 times as long as sepals ...... 8a. *Eremogone franklinii* var. *franklinii*
1. Inflorescences somewhat congested cymes; pedicels 1–4 mm; sepals 5–8 mm; petals 1–1.1 times as long as sepals ..... 8b. *Eremogone franklinii* var. *thompsonii*

8a. Eremogone franklinii (Douglas ex Hooker) R. L. Hartman & Rabeler var. **franklinii** [E]

Inflorescences tightly congested cymes. **Pedicels** 0.1–1(–3) mm. **Flowers:** sepals 8.5–12 mm; petals ca. 0.8–0.9 times as long as sepals.

Flowering summer. Sand and sage plains; 100–1600 m; Idaho, Nev., Oreg., Wash.

8b. Eremogone franklinii (Douglas ex Hooker) R. L. Hartman & Rabeler var. **thompsonii** (M. Peck) R. L. Hartman & Rabeler, Sida 21: 240. 2004 • Thompson's sandwort [E]

Arenaria franklinii Douglas ex Hooker var. *thompsonii* M. Peck, Torreya 32: 149. 1932 (as thompsoni)

Inflorescences somewhat congested cymes. **Pedicels** 1–4 mm. **Flowers:** sepals 5–8 mm; petals 7–9 mm, ca. 1–1.1 times as long as sepals.

Flowering summer. Sand dunes; 100–800 m; Oreg., Wash.

Variety *thompsonii*, at one time known only from the type (Gilliam County, Oregon), was documented in the 1980s in Benton County, Washington.

9. Eremogone hookeri (Nuttall) W. A. Weber, Brittonia 33: 326. 1981 • Hooker's sandwort [E] [F]

Arenaria hookeri Nuttall in J. Torrey and A. Gray, Fl. N. Amer. 1: 178. 1838

Plants densely or loosely matted, green, not glaucous, somewhat woody at base. **Stems** erect, 1–15(–20) cm, scabrid-puberulent. **Leaves:** basal leaves persistent; cauline leaves in 1–4 pairs, usually little overlapping, often larger than basal leaves; basal blades straight to arcuate-spreading, subulate to needlelike, 0.3–4 cm × 0.5–1.5 mm, flexible or rigid, herbaceous, apex spinose, glabrous, often glaucous. **Inflorescences** 3–30+-flowered, congested, capitate or subcapitate cymes. **Pedicels** 0.2–2 mm, scabrid-puberulent. **Flowers:** sepals 1–3-veined, often obscurely so, linear-lanceolate to lanceolate, (5–)6–10 mm, not enlarging in fruit, margins narrow, apex narrowly acute or acuminate, glabrous or pubescent; petals white, oblanceolate, 4.5–8.5 mm, ± equaling sepals, apex rounded to obtuse; nectaries as lateral and abaxial mounds with transverse groove at base of filaments opposite sepals, 0.2–0.3 mm. **Capsules** to 4 mm, glabrous. **Seeds** black, ellipsoid-oblong to pyriform with hilar notch, 1.8–2 mm, tuberculate; tubercles rounded, elongate.

Varieties 2 (2 in the flora): nc United States.

1. Basal leaf blades 0.3–1.5 cm, straight or recurved, rigid; sepals 5–8(–9) mm ..... 9a. *Eremogone hookeri* var. *hookeri*
1. Basal leaf blades 2–4 cm, straight, rigid or flexible; sepals (7–)8–10 mm .... 9b. *Eremogone hookeri* var. *pinetorum*

9a. Eremogone hookeri (Nuttall) W. A. Weber var. **hookeri** E

Arenaria hookeri subsp. *desertorum* (Maguire) W. A. Weber; *A. hookeri* Nuttall var. *desertorum* Maguire; *Eremogone hookeri* subsp. *desertorum* (Maguire) W. A. Weber

Basal leaf blades straight or recurved, 0.3–1.5 cm, rigid. **Sepals** 5–8(–9) mm. $2n = 44$.

Flowering spring–summer. Plains and exposed slopes and ridges; 1200–2900 m; Colo., Kans., Mont., Nebr., Nev., N.Mex., Okla., S.Dak., Utah, Wyo.

We include the densely pulvinate plants often known as var. *desertorum* within var. *hookeri*. R. L. Hartman (1971) analyzed leaf blade length and sepal length, the characters used by B. Maguire (1951) in recognizing three varieties of *Eremogone hookeri*, and found that variation is continuous between var. *hookeri* and var. *desertorum*.

9b. Eremogone hookeri (Nuttall) W. A. Weber var. **pinetorum** (A. Nelson) Dorn, Vasc. Pl. Wyoming ed. 3, 376. 2001 E F

Arenaria pinetorum A. Nelson, Bull. Torrey Bot. Club 26: 350. 1899; *A. hookeri* Nuttall subsp. *pinetorum* (A. Nelson) Maguire; *Eremogone hookeri* subsp. *pinetorum* (A. Nelson) W. A. Weber

Basal leaf blades straight, 2–4 cm, rigid or flexible. **Sepals** (7–)8–10 mm. $2n = 44, 66$.

Flowering spring–summer. Open pine and spruce woodlands; 1300–2400 m; Colo., Nebr., Okla., S.Dak., Wyo.

The two varieties recognized intergrade in some areas, but var. *pinetorum* is usually very distinctive and occurs in the eastern portion of the range of the species.

10. Eremogone kingii (S. Watson) Ikonnikov, Novosti Syst. Vyssh. Rast. 10: 140. 1973 • King's sandwort E

Stellaria kingii S. Watson, Botany (Fortieth Parallel), 39, plate 6, figs. 1–3. 1871; *Arenaria kingii* (S. Watson) M. E. Jones

Plants tufted or sometimes in compact cushion, green, not glaucous, woody or not at base. **Stems** erect, (1–)3–20(–25) cm, stipitate-glandular or glabrous proximally. **Leaves:** basal leaves abundant, persistent; cauline leaves in (1–)4+ pairs, reduced distally or not; basal blades erect or closely ascending to somewhat spreading, green to gray-green, filiform to needlelike or narrowly subulate, 0.3–3(–4) cm × 0.3–1.2 mm, flexuous or rigid, herbaceous, apex apiculate or stiff and spinose, glabrous to stipitate-glandular, not glaucous. **Inflorescences** (1–)3–13-flowered, ± open cymes. **Pedicels** 2–15 mm, glabrous to densely stipitate-glandular. **Flowers:** sepals 1–3-veined, lateral veins less developed, ovate or lanceolate, (2.5–)2.8–5(–6) mm, not expanding in fruit, margins broad, apex obtuse to broadly acute or acuminate, glabrous or stipitate-glandular on herbaceous portion; petals white or rarely pink, oblong to spatulate, (3–)4–7 mm, ca. 1.2–1.3 times as long as sepals, apex entire, erose, or 2-fid almost to base; nectaries as abaxial, rounded lobe with transverse groove or elongate cup at base of filaments opposite sepals, 0.6 × 0.3 mm. **Capsules** 4.5–7 mm, glabrous. **Seeds** black to brown, spheric or oblong to ovoid, 1.2–2.1 mm, low-tuberculate, sometimes papillate on abaxial ridge.

Varieties 2 (2 in the flora): w United States.

M. F. Baad (1969) considered *Eremogone kingii* to be monophyletic despite considerable morphological variation; J. C. Hickman (1971) thought otherwise, considering *E. kingii* to be a "genetic dumping ground for all the closely related taxa," but did not propose any new taxonomic alignment.

Eremogone kingii is extremely variable throughout its range with six infraspecific taxa recognized (under *Arenaria kingii*) by B. Maguire (1947, 1951). We have been unsuccessful in distinguishing more than two of those taxa. The others intergrade to such an extent that formal recognition is unwarranted. Most distinctive of these here-rejected taxa is var. *uintahensis*, said to have sepals (4.5–)5–6 mm, versus 3.6–4.5(–5) for the other taxa. Interestingly, the type specimen has sepals mostly 4.5 mm long. In the main portion of the range of var. *uintahensis*, the sepals are rounded to broadly obtuse, but they may also be acute. Furthermore, the sepals and pedicels are often glabrous, but the correlation of the above-mentioned characters varies over the range.

1. Petals white, apex 2-fid . . . . 10a. *Eremogone kingii* var. *kingii*
1. Petals white or rarely pink, apex entire or erose . . . . 10b. *Eremogone kingii* var. *glabrescens*

10a. Eremogone kingii (S. Watson) Ikonnikov var. **kingii** [E]

Petals white, apex deeply 2-fid into V-shaped notch.

Flowering late spring–summer. Pinyon-juniper woodlands, dry slopes; 1600–2900 m; Nev., Utah.

10b. Eremogone kingii (S. Watson) Ikonnikov var. **glabrescens** (S. Watson) Dorn, Vasc. Pl. Wyoming ed. 3, 376. 2001 [E]

Arenaria fendleri A. Gray var. *glabrescens* S. Watson, Botany (Fortieth Parallel), 40. 1871; *A. aculeata* S. Watson var. *uintahensis* (A. Nelson) M. Peck; *A. compacta* Coville; *A. kingii* subsp. *compacta* (Coville) Maguire; *A. kingii* S. Watson var. *glabrescens* (S. Watson) Maguire; *A. kingii* subsp. *plateauensis* Maguire; *A. kingii* var. *plateauensis* (Maguire) Reveal; *A. kingii* subsp. *rosea* Maguire; *A. kingii* subsp. *uintahensis* (A. Nelson) Maguire; *A. kingii* var. *uintahensis* (A. Nelson) C. L. Hitchcock; *A. uintahensis* A. Nelson; *Eremogone kingii* var. *plateauensis* (Maguire) R. L. Hartman & Rabeler; *E. kingii* var. *rosea* (Maguire) R. L. Hartman & Rabeler; *E. kingii* subsp. *uintahensis* (A. Nelson) W. A. Weber

Petals white (rarely pink), apex entire or erose.

Flowering late spring–summer. Dry hills, pinyon-juniper woods, sagebrush slopes; (1500–)2100–4000 m; Calif., Idaho, Nev., Oreg., Utah.

Arenaria kingii subsp. *compacta* intergrades completely with var. *glabrescens* in many areas of the Sierra Nevada and does not deserve recognition.

M. F. Baad (1969) noted populations from Idaho, Nevada, Oregon, and Wyoming that he considered intermediate between *Eremogone capillaris* var. *americana* and *E. kingii* var. *glabrescens*.

11. Eremogone macradenia (S. Watson) Ikonnikov, Novosti Syst. Vyssh. Rast. 10: 140. 1973 • Mohave sandwort [E]

Arenaria macradenia S. Watson, Proc. Amer. Acad. Arts 17: 367. 1882

Plants tufted, green, not glaucous, with woody base. **Stems** erect, (10–)20–40(–70) cm, glabrous to stipitate-glandular. **Leaves:** basal leaves sparse or absent; cauline leaves usually in 5–12+ pairs, not significantly reduced; basal blades ascending to arcuate-spreading or recurved, needlelike or narrowly linear, (0.7–)2–6(–7) cm × 0.8–2 mm, ± rigid, herbaceous to subsucculent, apex blunt to spinose, glabrous, not glaucous. **Inflorescences** 3–20(–30)-flowered, ± compact cymes; branches ascending to erect. **Pedicels** 3–45 mm, glabrous or stipitate-glandular. **Flowers:** sepals 1–3-veined, ovate to lanceolate or elliptic, 4.5–7.2 mm, to 8 mm in fruit, margins narrow to broad, apex acute to acuminate, glabrous to sparsely stipitate-glandular; petals white or yellowish, oblanceolate to spatulate, 6–11 mm, 1–2 times as long as sepals, apex entire or erose; nectaries thickened, molarlike, apically 2-lobed, 1–1.5 mm, or narrowly longitudinally rectangular, truncate, 0.7–0.8 mm, densely minutely pubescent with erect to spreading hairs. **Capsules** 6–8 mm, glabrous. **Seeds** greenish or reddish brown to blackish, suborbicular to pyriform or ovoid, 1.3–3.2 mm, tuberculate, sometimes echinate on abaxial ridge; tubercles low, rounded to conic.

Varieties 2 (2 in the flora): North America.

Eremogone macradenia is in the Center for Plant Conservation's National Collection of Endangered Plants.

1. Cauline leaves mostly in 5–8 pairs, blade ± ascending throughout, 0.8–1.2 mm wide . . . . 11a. *Eremogone macradenia* var. *macradenia*
1. Cauline leaves mostly in 6–12+ pairs, blade curved downward, especially proximal ones, 1.2–2 mm wide . . . . . 11b. *Eremogone macradenia* var. *arcuifolia*

11a. Eremogone macradenia (S. Watson) Ikonnikov var. **macradenia** [E]

Arenaria congesta Nuttall var. *parishiorum* (B. L. Robinson) B. L. Robinson; *A. macradenia* S. Watson var. *parishiorum* B. L. Robinson; *Eremogone parishiorum* (B. L. Robinson) Ikonnikov

Cauline leaves mostly in 5–8 pairs, blade ± ascending throughout, 0.8–1.2 mm wide. **Inflorescences:** branches erect to ascending. **Flowers:** sepals 4.5–7.2 mm, to 8 mm in fruit,

± glabrous; nectaries thickened, molarlike, apically 2-lobed, 1–1.5 mm.

Flowering spring–summer. Open woodlands, sagebrush flats, dry, rocky slopes, alluvial deposits, often on carbonates; 600–2200 m; Ariz., Calif., Nev., Utah.

Plants with petals equaling or barely exceeding the sepals and with cauline leaves fewer than five pairs have been recognized as *Arenaria macradenia* var. *parishiorum*; those features intergrade completely with var. *macradenia*. See M. F. Baad (1969) for his reasons for accepting var. *parishiorum* as a species.

B. Maguire (1947) included "Lower California" in his distribution statement for var. *macradenia*, implying Baja California; we have not seen any collections from that area.

11b. Eremogone macradenia (S. Watson) Ikonnikov var. **arcuifolia** (Maguire) R. L. Hartman & Rabeler, Sida 21: 240. 2004 [E]

Arenaria macradenia S. Watson var. *arcuifolia* Maguire, Bull. Torrey Bot. Club 74: 51. 1947; *A. kuschei* Eastwood; *A. macradenia* S. Watson var. *kuschei* (Eastwood) Maguire; *Eremogone macradenia* var. *kuschei* (Eastwood) R. L. Hartman & Rabeler

Cauline leaves in 6–12+ pairs; blade curved downward, especially proximal ones, 1.2–2 mm wide. **Inflorescences:** branches erect to ascending. **Flowers:** sepals 4.5–7 mm, to 7 mm in fruit, ± glabrous or sparsely to rarely densely stipitate-glandular; nectaries narrowly longitudinally rectangular, truncate, 0.7–0.8 mm, densely minutely pubescent with erect to spreading hairs.

Flowering late spring–early summer. Dry slopes and foothills, dry, yellow pine and oak forests; 600–2400 m; Calif.

While B. Maguire (1947, 1951) regarded var. *arcuifolia* as "a strongly marked variant," R. L. Hartman (1993) noted that it "intergrades with var. *macradenia* and might be considered the same as the latter."

The collection from "Forest Camp" described as *Arenaria macradenia* var. *kuschei* differs from var. *arcuifolia* in having densely stipitate-glandular pedicels and sepals; it does not deserve taxonomic recognition. Recently, populations of plants matching var. *kuschei* were discovered in the Liebre Mountains, northwestern Los Angeles County (T. S. Ross and S. Boyd 1996b). Four of the seven populations were mixed, some individuals having the stipitate-glandular pubescence of var. *kuschei*, and others, the glabrous inflorescences of var. *arcuifolia*. Furthermore, there was an east-to-west trend along the summit of Liebre Mountain from populations with a low frequency of glandular individuals to populations that were strictly glandular (Boyd, pers. comm.).

12. Eremogone pumicola (Coville & Leiberg) Ikonnikov, Novosti Syst. Vyssh. Rast. 10: 140. 1973 • Crater Lake sandwort [E]

Arenaria pumicola Coville & Leiberg, Proc. Biol. Soc. Wash. 11: 169. 1897

Plants tufted, green, not glaucous, slightly woody at base. **Stems** erect, 10–20 cm, stipitate-glandular. **Leaves:** basal leaves ± deciduous; cauline leaves (2–)3–5 of equal size distally; basal blades ascending to ± spreading, narrowly linear, 1.5–3.5 cm × 1–1.5 mm, ± flexuous, ± succulent, apex green, blunt or acute to apiculate, glabrous, not glaucous. **Inflorescences** 7–40+-flowered, ± open cymes. **Pedicels** 3–20 mm, stipitate-glandular. **Flowers:** sepals 1–3-veined, lateral pair obscure, ovate, 3–3.8 mm, 4–4.5 mm in fruit, margins broad, apex obtuse to rounded or barely acute, stipitate-glandular; petals white, narrowly spatulate, 6.6–7.5 mm, 2 times as long as sepals, apex rounded or emarginate; nectaries as lateral and abaxial rounding of base of filaments opposite sepals, 0.3 mm. **Capsules** 4.5–5.5 mm, glabrous. **Seeds** black to brown, oblong to pyriform, 1.8–2.4 mm, smooth to tuberculate.

Flowering summer. Areas with loose pumice; 1600–2800 m; Oreg.

Eremogone pumicola is restricted to Crater Lake and vicinity, southwestern Oregon.

13. Eremogone stenomeres (Eastwood) Ikonnikov, Novosti Syst. Vyssh. Rast. 11: 175. 1974 • Meadow Valley sandwort [C] [E]

Arenaria stenomeres Eastwood, Leafl. W. Bot. 4: 63. 1944

Plants densely cespitose, green, not glaucous, with woody base. **Stems** erect, 7–18 cm, glabrous. **Leaves:** basal leaves persistent; cauline leaves in 5 pairs, not reduced (stems leafy); basal blades spreading, needlelike, 1.5–2 cm × 0.5–1 mm, rigid, not fleshy, apex spinose, glabrous, glaucous. **Inflorescences** 5–13-flowered, open cymes. **Pedicels** 4–12 mm, glabrous to stipitate-glandular distally. **Flowers:** sepals 1-veined, narrowly to broadly lanceolate, 4.5–5 mm, 6.5–7.5 mm in fruit, margins narrow, apex acuminate to spinose, glabrous to stipitate-glandular distally; petals yellowish white, narrowly spatulate, 7.5–8.5 mm, 1.5–1.7 times as long as sepals, apex emarginate; nectaries as lateral and abaxial rounding of base of filaments opposite sepals, 0.3–0.4 mm. **Capsules** ca. 6 mm, glabrous. **Seeds** color not known, orbicular, ca. 1 mm, surface not known (mature seeds not observed).

Flowering spring. Limestone cliffs; of conservation concern; 900–1200 m; Nev.

Eremogone stenomeres is known only from a few sites in Clark and Lincoln counties. This species was recognized for its linear petals with notched apices. Rehydration of petals from representative specimens indicates that the lamina fold lengthwise upon drying, thus the linear appearance. When rehydrated, the petals were oblanceolate, with widths within the range of related taxa.

14. **Eremogone ursina** (B. L. Robinson) Ikonnikov, Novosti Syst. Vyssh. Rast. 10: 140. 1973 • Bear Valley sandwort [C] [E]

Arenaria ursina B. L. Robinson, Proc. Amer. Acad. Arts 29: 294. 1894

Plants tufted, green, not glaucous, with somewhat woody base. **Stems** ascending to erect, 10–18 cm, often glandular-hairy. **Leaves:** basal leaves persistent; cauline leaves in 3–5 pairs, reduced distally; basal blades erect to ± spreading, needlelike, 0.5–1.1(–1.6) cm × 0.5–0.7 mm, rigid, herbaceous, not

fleshy, apex blunt to apiculate, glabrous, ± glaucous. **Inflorescences** (1–)3–7-flowered, ± open cymes. **Pedicels** 0.3–2 mm, stipitate-glandular. **Flowers:** sepals 1–3-veined, lateral veins less developed, ovate, often broadly so, 1.8–3 mm, to 4.2 mm in fruit, margins broad, apex obtuse or rounded, glabrous; petals white, elliptic to oblanceolate, 2–4.5 mm, 1.4–1.6 times as long as sepals, apex rounded; nectaries as lateral and abaxial rounding of base of filaments opposite sepals, with terminal lateral groove, 0.3 mm. **Capsules** 4.5–6 mm, glabrous. **Seeds** blackish purple, suborbicular to broadly ellipsoid with hilar notch, 2.2–2.5(–3) mm, tuberculate; tubercles rounded, elongate.

Flowering late spring–summer. Pinyon-juniper woodlands on rocky (quartzite) soils; of conservation concern; 1900–2100 m; Calif.

Eremogone ursina is known from four counties in southern California, where it is threatened by development. It is relatively distinctive in appearance and markedly separated spatially from congeners.

16. MOEHRINGIA Linnaeus, Sp. Pl. 1: 359. 1753; Gen. Pl. ed. 5, 170. 1754 • Sandwort, moehringie, sabline [for P. H. G. Moehring, 1710–1791, Danzig naturalist]

Richard K. Rabeler

Ronald L. Hartman

Herbs, annual or perennial. **Taproots** slender, rhizomes slender or absent. **Stems** prostrate or ascending to erect, simple or branched, terete or angled. **Leaves** not connate, petiolate or sessile, not congested at or near base of flowering stem; blade 1–3(–7)-veined, lanceolate to elliptic or ovate to broadly ovate, not succulent, apex acute or obtuse. **Inflorescences** terminal or axillary, open cymes, or flowers solitary; bracts paired and foliaceous, or smaller and mostly scarious. **Pedicels** erect or recurved in fruit. **Flowers:** perianth and androecium weakly perigynous; hypanthium minute, disc-shaped; sepals (4–)5, distinct, green, ovate to obovate, 1.7–6 mm, herbaceous, margins white, scarious, apex obtuse or acute to acuminate, not hooded; petals (4–)5, white, not clawed, blade apex entire; nectaries as fleshy lobes at base of filaments opposite sepals, ca. 3 times width of filament, connate proximally into basal disc; stamens 10, occasionally 8, arising from nectariferous disc; filaments distinct; staminodes absent; styles 3, filiform, 1.5–1.8 mm, glabrous proximally; stigmas 3, linear along adaxial surface of styles, minutely papillate

(30×). **Capsules** broadly ovoid to subglobose, opening by 6 revolute teeth; carpophore absent. **Seeds** 2–6, reddish brown to blackish, ellipsoid to reniform, laterally compressed, shiny, smooth to minutely tuberculate, marginal wing absent, appendage white, ± elliptic, spongy. $x = 12$.

Species 25 (3 in the flora): north-temperate North America, Europe, Asia.

Members of *Moehringia* and *Petrocoptis* (a segregate from *Silene*, comprising four species in the Pyrenees) are the only Caryophyllaceae with strophioles (eliasomes), spongy seed appendages that attract ants. Foraging ants gather the seeds, eat only the strophiole, and "plant" the seeds in their nests.

We follow J. McNeill (1962) and V. Bittrich (1993) among others in recognizing *Moehringia*. The appendaged (strophiolate) seed and a chromosome base number of 12 are the chief characters distinguishing *Moehringia* from *Arenaria*. Although McNeill noted that this distinction is similar to features used to distinguish subgenera within *Arenaria*, he retained *Moehringia* and suggested that, among other evidence, cytological investigation of the North American species of *Moehringia* and members of *Arenaria* subg. *Leiosperma* McNeill (New World, especially Andean South America) would help support such action. Chromosome counts made since 1962 do show $x = 12$ in North American *Moehringia* and $x = (10)$ 11 for *Arenaria* subg. *Leiosperma*.

1. Plants annual (or short-lived perennial?); stems prostrate or ascending, pubescence in 2 lines; leaves petiolate, petioles 1–4 mm, blade ovate to broadly ovate . . . . 3. *Moehringia trinervia*
1. Plants perennial, with extensive rhizome systems; stems prostrate or ascending to erect, pubescence ± uniform; leaves sessile or subsessile, petioles 0.1–1 mm, blade mostly lanceolate to elliptic.
  2. Stem pubescence retrorse; sepals (herbaceous portion) oblong to elliptic, apex mostly rounded or obtuse, 1.7–2.8(–3) mm; petals ca. 2 times as long as sepals . . . . . . 2. *Moehringia lateriflora*
  2. Stem pubescence peglike, spreading; sepals (herbaceous portion) lanceolate, apex acute to acuminate, (2.8–)3–6 mm; petals ³/₄–1¹/₂ times as long as sepals . . . . 1. *Moehringia macrophylla*

1. **Moehringia macrophylla** (Hooker) Fenzl, Vers. Darstell. Alsin., 18, 38. 1833 • Bigleaf sandwort, moehringie ou sabline à grandes feuilles

Arenaria macrophylla Hooker, Fl. Bor.-Amer. 1: 102, plate 37. 1831

Plants perennial. **Rhizomes** forming extensive network. **Stems** ascending to erect, simple or branched, ± angled or grooved, 2–18 cm, hairs minute, spreading, peglike. **Leaves** sessile or subsessile; petiole 0.1–1 mm; blade 1–3-veined, lanceolate to elliptic, (8–)15–50(–70) × 2–9 mm, margins smooth to minutely granular, often ciliate in proximal ¹/₂, apex acute. **Inflorescences** 1–5-flowered; bracts 1–4 mm, margins broadly scarious. **Pedicels** ascending to erect, sometimes divergent in fruit, 2–15 mm. **Flowers:** sepals 5, midrib ± keeled, ovate, herbaceous portion lanceolate, (2.8–)3–6 mm, apex acute to acuminate; petals 5, 2–6 mm, ³/₄–1¹/₂ times as long as sepals. **Capsules** ovoid, 5 mm, ± equaling sepals. **Seeds** ovoid, 1.5–2.2 mm, tuberculate; tubercles minute, low, rounded. $2n = 48$.

Flowering late spring–summer. Moist, shaded slopes, rocky ridges, ultramafic outcrops and summits, shores; 200–3400 m.; B.C., Man., Nfld. and Labr. (Labr.), N.W.T., Ont., Que., Sask.; Calif., Colo., Conn., Idaho, Mass., Mich., Mont., N.Mex., Ore., Utah, Vt., Wash., Wis.; Asia.

2. **Moehringia lateriflora** (Linnaeus) Fenzl, Vers. Darstell. Alsin., 18, 38. 1833 • Grove or blunt-leaf sandwort, moehringie ou sabline latériflore [F] [W]

Arenaria lateriflora Linnaeus, Sp. Pl. 1: 423. 1753; *A. lateriflora* var. *angustifolia* H. St. John; *A. lateriflora* var. *taylorae* H. St. John; *A. lateriflora* var. *tenuicaulis* Blankinship

Plants perennial. **Rhizomes** forming extensive network. **Stems** ascending or decumbent, often branched, terete, 5–30 cm, uniformly retrorsely pubescent. **Leaves** sessile or subsessile; petiole 0.1–1 mm; blade 1–3-veined, broadly elliptic to oblong-elliptic or oblanceolate, 6–30(–35) × (2–)5–10(–17) mm, margins granular to minutely serrulate-ciliate, apex obtuse or rounded. **Inflorescences** 1–5-flowered; bracts 1–3 mm, margins scarious. **Pedicels** erect, 3–30 mm. **Flowers:** sepals 5, midrib not keeled, ovate or obovate, herbaceous portion oblong to elliptic, 1.7–2.8(–3) mm, margins narrow, apex mostly obtuse or rounded; petals 5, 3–6 mm, ca. 2 times as long as sepals. **Capsules** subglobose, 3–5 mm, 1¹/₂–2 times as long as sepals. **Seeds** reniform, 1 mm, smooth. $2n = 48$.

Flowering spring–early summer. Moist or dry woodlands, meadows, gravelly shores; 50–2700 m; St. Pierre and Miquelon; Alta., B.C., Man., N.B., Nfld. and Labr., N.S., Nunavut, Ont., P.E.I., Que., Sask., Yukon; Alaska, Colo., Conn., Idaho, Ill., Ind., Iowa, Maine, Md., Mass., Mich., Minn., Mo., Mont., Nebr., Nev., N.H., N.J., N.Mex., N.Y., N.Dak., Ohio, Oreg., Pa., R.I., S.Dak., Utah, Vt., Va., Wash., W.Va., Wis., Wyo.; Europe; Asia.

Four varieties of *Moehringia laterifolia* have been described based on variation in leaf width and pubescence; they have been little used, and the variation appears not to be correlated with geography.

3. **Moehringia trinervia** (Linnaeus) Clairville, Man. Herbor. Suisse, 150. 1811 • Three-nerved sandwort

I

Arenaria trinervia Linnaeus, Sp. Pl. 1: 423. 1753

Plants annual (or short-lived perennial?), not rhizomatous. **Stems** prostrate or ascending, branched, angled or grooved, 13–19 cm, retrorsely pubescent in 2 lines. **Leaves** petiolate; petiole 1–4 mm; blade 3(–7)-veined, ovate to broadly ovate, 5–18 × 4–8 mm, margins ciliate, apex acute. **Inflorescences** 3–10(–15)-flowered; bracts 1–15

mm, proximal foliaceous, distal much smaller, scarious. **Pedicels** ascending to erect, recurved in fruit, 4–15(–22) mm. **Flowers:** sepals 4 or 5, midrib keeled, ovate-lanceolate, 2–4 mm, margins broad, apex acuminate; petals 4 or 5, ca. 1.3 mm, ca. ¹/₂ as long as sepals. **Capsules** ovoid, 2–2.5 mm, shorter than sepals. **Seeds** reniform, 1 mm, smooth. $2n = 24$ (Europe).

Flowering early summer. Trails in forests; 400 m; introduced; Ohio; Eurasia.

Moehringia trinervia was collected near Cleveland in 1990 on property that formerly was a resort community (R. K. Rabeler and A. W. Cusick 1994). It closely resembles *Stellaria media*; the multiveined, ciliate leaves, retrorse pubescence in two lines on the stems, broadly scarious sepals, capsules shorter than the sepals, and smooth, appendaged seeds distinguish it from *S. media*.

17. **LEPYRODICLIS** Fenzl in S. L. Endlicher, Gen. Pl. 13: 966. 1840 • [Greek *lepyron*, rind or husk, and *diklis*, double-folding, alluding to two-valved capsule]   I

Richard K. Rabeler

Herbs, annual. **Taproots** slender. **Stems** ascending to decumbent or sprawling, simple or branched, terete or angled. **Leaves** connate proximally, sessile; blade 1-veined, lanceolate or rarely elliptic, not succulent, apex acute. **Inflorescences** axillary and terminal, compound cymes; bracts paired, foliaceous, smaller. **Pedicels** reflexed in fruit. **Flowers:** perianth and androecium weakly perigynous; hypanthium minimal; sepals 5, distinct, green, lanceolate to elliptic or ovate, 4–5 mm, herbaceous, margins white, scarious, apex somewhat acute, not hooded; petals 5, white or pink, not clawed, blade apex entire or emarginate; nectaries at base of filaments opposite sepals; stamens 10, arising from nectariferous disc at ovary base; filaments distinct nearly to base; staminodes absent; styles 2, clavate, 3.5–4 mm, glabrous proximally; stigmas

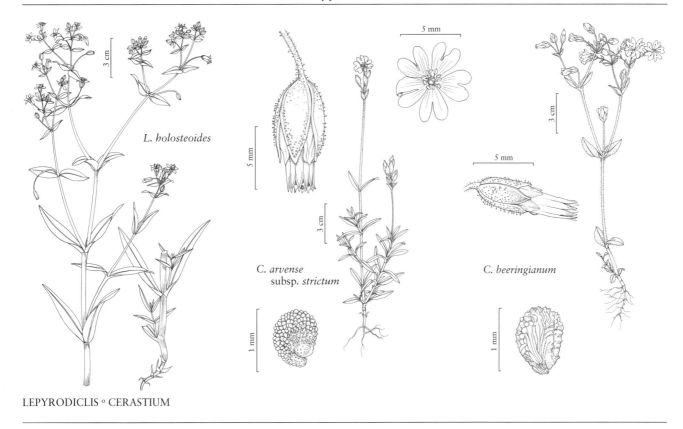

L. holosteoides

C. arvense
subsp. *strictum*

C. beeringianum

LEPYRODICLIS ○ CERASTIUM

2, subterminal, minutely roughened to papillate (50×). **Capsules** globose-ovoid, opening by 2 straight valves; carpophore absent. **Seeds** 1–2(–4), dark brown, reniform, laterally compressed, tuberculate, marginal wing absent, appendage absent. *x* = 17.

Species 3–4 (1 in the flora): introduced; c, sw Asia; introduced in Europe (Germany), Asia (Japan).

SELECTED REFERENCE   Rabeler, R. K. and R. R. Old. 1992. *Lepyrodiclis holosteoides* (Caryophyllaceae), "new" to North America. Madroño 39: 240–242.

1. Lepyrodiclis holosteoides (C. A. Meyer) Fenzl ex Fischer & C. A. Meyer, Enum. Pl. Nov. 1: 93, 110. 1841 [F][I][W]

Gouffeia holosteoides C. A. Meyer, Verz. Pfl. Casp. Meer., 217. 1831

Stems 40–100 cm, glabrous or papillate. **Leaf blades** 1–8 × 0.3–1 cm, margins glandular-dentate. **Flowers:** sepals 4–5 mm; petals 1–1.5 times as long as sepals. **Capsules** globose-ovoid, 3 mm, shorter than calyx. **Seeds** 1.8–2 mm. *2n* = 34, 68 (both Asia).

Flowering spring. Disturbed areas, pea and wheat fields; 500–800 m; introduced; Idaho, Wash.; c, sw Asia; introduced in Europe (Germany), Asia (Japan).

Although known in North America only since 1959, *Lepyrodiclis holosteoides* is a serious menace in pea and wheat plantings, where it grows quickly and often overtops the crops. It is listed as a noxious weed in Washington and Oregon; I am not aware of any documented occurrences in Oregon.

If *Lepyrodiclis stellarioides* Schrenk ex Fischer & C. A. Meyer is included as a variety of *L. holosteoides* (see Y. P. Kozhevnikov 1985), North American plants are var. *holosteoides*

18. CERASTIUM Linnaeus, Sp. Pl. 1: 437. 1753; Gen. Pl. ed. 5. 199. 1754 • Mouse-ear chickweed, céraiste [Greek, *ceras*, horn, alluding to shape of capsule]

John K. Morton

Herbs, annual, winter annual, or perennial. **Taproots** slender, perennial taxa often rhizomatous, rooting at nodes. **Stems** ascending to erect or decumbent, simple or branched, terete. **Leaves** basally connate, petiolate (basal in some species) or sessile (cauline); blade 1–5-veined, linear or elliptic to broadly ovate, not succulent (except in *C. bialynickii*, *C. regelii*, and *C. viride*), apex acute to obtuse. **Inflorescences** terminal, open or congested cymes, or flowers solitary, axillary (racemosely arranged in *C. axillare*); bracts paired, foliaceous or reduced, herbaceous or often with scarious margins. **Pedicels** erect, sometimes reflexed or hooked at apex in fruit, or flowers sometimes subsessile (*C. regelii*). **Flowers** bisexual, occasionally unisexual and pistillate; perianth and androecium hypogynous or weakly perigynous; hypanthium minimal; sepals (4–)5, distinct, green (red-tipped in *C. glomeratum* and *C. pumilum*, often violet-tipped in *C. alpinum*, purple in *C. bialynickii*, turning pale orange-brown in fruit in *C. texanum*), elliptic to ovate, 3–12 mm, herbaceous, margins translucent to purplish, scarious, apex acute, acuminate, or obtuse, not hooded; petals (4–)5 or sometimes absent, white (purple tinged in *C. pumilum* and *C. regelii*), clawed, blade apex 2-fid $^{1}/_{5}$–$^{1}/_{2}$ of length, notched, or emarginate; nectaries at base of filaments opposite sepals; stamens usually 10, sometimes 5 or 8, occasionally 4; filaments distinct, inserted at base of ovary; staminodes absent or 1–4 (via anther abortion), linear; styles (3–)5(–6), clavate to filiform, 0.5–2 mm, glabrous proximally; stigmas (3–)5(–6), subterminal to linear along adaxial surface of styles, roughened to papillate (30×). **Capsules** oblong or cylindric, usually ± curved, opening by 10, or occasionally 6 or 8, erect or spreading, convolute or revolute teeth, longer than sepals; carpophore absent. **Seeds** 15–150+, orange to brown, angular-obovate, often with abaxial groove, laterally compressed, papillate-tuberculate, marginal wing absent, appendage absent. x = [9?, 13, 15] 17, 18, 19.

Species ca. 100 (27 in the flora): worldwide, but mainly north-temperate region.

Two names that appear in many North American treatments, *Cerastium viscosum* Linnaeus and *C. vulgatum* Linnaeus, have been proposed for rejection (N. J. Turland and M. Wyse Jackson 1997) because they have been a long-standing source of confusion. American authors have frequently applied *C. viscosum* to *C. glomeratum* Thuillier but most of the possible lectotypes are referable to *C. fontanum*. Similarly *C. vulgatum* has been used for *C. fontanum* Baumgarten. However the possible lectotypes of *C. vulgatum* are mixed and most are referable to *C. arvense*, *C. fontanum*, and *C. glomeratum*.

I have not attempted to present an infrageneric classification for *Cerastium*. Several species groupings can be recognized based on capsule structure. Examples include species with the capsule teeth revolute (coiled outwards like a watch spring), e.g., *C. maximum* and *C. texanum*, whereas in most of our species the teeth are erect with their margins outwardly rolled (convolute). Also *C. cerastoides* is anomalous in having three styles, a straight ovoid-conic capsule, and deeply bilobed petals—characters that have led some authors to place it in the genus *Stellaria*. Accepting the most recent infrageneric classification (B. K. Schischkin 1936), even with subsequent modifications, does not appear to be warranted. Additional study is needed to determine relevant species relationships within the genus.

While the base chromosome number for *Cerastium* is often cited as x = 9, only a single count of $2n$ = 18 is known; see C. Favarger and M. Krähenbühl (1996) for a discussion of the diverse cytological conditions found in *Cerastium*.

SELECTED REFERENCES Böcher, T. W. 1977. *Cerastium alpinum* and *C. arcticum*, a mature polyploid complex. Bot. Not. 130: 303–309. Brysting, A. K. and R. Elven. 2000. The *Cerastium alpinum–C. arcticum* complex (Caryophyllaceae): Numerical analyses of morphological variation and a taxonomic revision of *C. arcticum* Lange s.l. Taxon 49: 189–216. Good, D. A. 1984. Revision of the Mexican and Central American species of *Cerastium*. Rhodora 86: 339–379. Hultén, E. 1956. The *Cerastium alpinum* complex. A case of world-wide introgressive hybridization. Svensk Bot. Tidskr. 50: 411–495.

1. Capsules straight, teeth becoming outwardly coiled.
    2. Petals less than 10 mm . . . . 24. *Cerastium texanum*
    2. Petals (15–)18–25 mm . . . . 18. *Cerastium maximum*
1. Capsules usually curved (rarely straight), teeth erect or sometimes spreading, never coiled, margins convolute.
    3. Plants annual, with all shoots producing flowers.
        4. Styles 3–4(–5); capsules with 6, 8 (10) teeth.
            5. Styles 4(–5); stamens 4(–5); capsules 1–1.5 times as long as sepals; cauline leaves ovate to oblong-ovate . . . . 12. *Cerastium diffusum*
            5. Styles 3(–4); stamens 10; capsules ca. 2 times as long as sepals; cauline leaves linear to linear-lanceolate . . . . 13. *Cerastium dubium*
        4. Styles 5; capsules with 10 teeth.
            6. Bracts, at least distalmost, with scarious margins.
                7. Stamens 10 . . . . 16. *Cerastium fontanum* (in part)
                7. Stamens 5.
                    8. Sepals and distal bracts with broad, scarious margins; petal veins not branched . . . . 22. *Cerastium semidecandrum*
                    8. Sepals and distal bracts with very narrow, scarious margins; petal veins branched . . . . 20. *Cerastium pumilum*
            6. Bracts all completely herbaceous.
                9. Sepals with long hairs exceeding sepal tips.
                    10. Pedicels shorter than capsules; cauline leaf blades broadly ovate or elliptic-ovate . . . . 17. *Cerastium glomeratum*
                    10. Pedicels longer than capsules; cauline leaf blades lanceolate or elliptic . . . . 8. *Cerastium brachypetalum*
                9. Sepals with hairs shorter than sepal tips.
                    11. Flowers racemosely arranged singly in axils of foliaceous bracts along length of stem . . . . 5. *Cerastium axillare*
                    11. Flowers in terminal dichotomous cymes or clusters.
                        12. Capsules narrowly conic, straight; stamens 5; pedicels shorter than sepals . . . . 11. *Cerastium dichotomum*
                        12. Capsules cylindric, curved; stamens 10; pedicels shorter than, equaling, or usually longer than sepals.
                            13. Pedicels equaling or shorter than capsules, often becoming deflexed proximally . . . . 9. *Cerastium brachypodum*
                            13. Pedicels longer than capsules, sharply deflexed distally.
                                14. Sepals ovate-lanceolate, apex broadly acute to obtuse, inner with broad, scarious margins (ca. as wide as herbaceous center); mid-stem leaf blades lanceolate to narrowly elliptic, 10–60 × 3–15 mm; capsules (9–)10–12(–13) mm . . . . 19. *Cerastium nutans*
                                14. Sepals narrowly lanceolate, apex sharply acute to acuminate, inner with narrow, scarious margins (narrower than herbaceous center); mid-stem leaf blades linear-lanceolate, 20–70 × 1.5–6 mm; capsules (5–)7–10(–11) mm . . . . 14. *Cerastium fastigiatum*

[3. Shifted to left margin.—Ed.]
3. Plants perennial, often with nonflowering shoots.
  15. Styles 3 (rarely to 6); capsules straight, ovoid-conic, oblong after dehiscence, teeth 6 (rarely to 12) . . . . 10. *Cerastium cerastoides*
  15. Styles 5; capsules usually curved, rarely straight, cylindric (rarely broadly conic), teeth 10.
    16. Small tufts of leaves in axils of mid and proximal stem leaves.
      17. Petals 7.5–9 mm; sepals 3.5–6(–7) mm; flowering stems 5–20(–30) cm; anthers 0.8–0.9 mm . . . . 4. *Cerastium arvense* (in part)
      17. Petals 10–15 mm; sepals 5–9 mm; flowering stems 15–45 cm; anthers 0.9–1.2 mm.
        18. Plants taprooted, tufted, with or without short, nonflowering, leafy shoots; pubescence on proximal stem spreading; e North America . . . . 26. *Cerastium velutinum*
        18. Plants with long-creeping rhizomes; pubescence on proximal stems deflexed.
          19. Capsules 2.5–4 mm wide; sepals 5–7 mm; seeds 0.6–1.2 mm; introduced lawn weed . . . . 4. *Cerastium arvense* (in part)
          19. Capsules 4–5 mm wide; sepals 6–9 mm; seeds 1–1.5 mm; Pacific coastal region . . . . 27. *Cerastium viride*
    16. Small tufts of leaves usually absent from axils of mid and proximal stem leaves (normal leafy shoots may be present).
      20. Leaf surfaces obscured by dense white tomentum. . . . . 25. *Cerastium tomentosum*
      20. Leaf surfaces clearly visible through pubescence or leaf blades glabrous.
        21. Long flexuous hairs present on stems and/or leaf blades . . . . 2. *Cerastium alpinum*
        21. Long flexuous hairs absent, pubescence of straight or deflexed, short or long hairs.
          22. Gland-tipped hairs absent, pubescence eglandular throughout.
            23. Inflorescences with 3+ flowers; stems 10–45 cm; leaves not marcescent; ubiquitous weed. . . . . 16. *Cerastium fontanum* (in part)
            23. Inflorescences with 1(–3) flowers; stems 3–7 cm; proximal leaves marcescent; w arctic native . . . . 1. *Cerastium aleuticum*
          22. Gland-tipped hairs present at least in inflorescences.
            24. Testa of seeds inflated, loose (rubs off when rolled between finger and thumb); Newfoundland . . . . 23. *Cerastium terrae-novae*
            24. Testa of seeds not inflated, closely and firmly enclosing seed; North America, rarely in Newfoundland.
              25. Plants pulvinate; stems 1–10 cm.
                26. Plants rarely flowering; leaf blades subglabrous or with few colorless cilia . . . . 21. *Cerastium regelii* (in part)
                26. Plants normally flowering; leaf blades densely hispid with long, fuscous hairs . . . . 7. *Cerastium bialynickii*
              25. Plants rhizomatous, mat-forming or tufted, not pulvinate; stems 5–50 cm.
                27. Petals ± equaling sepals.
                  28. Proximal bracts foliaceous. . . . . . . 6. *Cerastium beeringianum* (in part)
                  28. Proximal bracts herbaceous, lanceolate, reduced . . . . 16. *Cerastium fontanum* (in part)
                27. Petals 1.5–2 times as long as sepals.
                  29. Sepals 6–11 mm, narrowly lanceolate-triangular, apex acute or acuminate, scarious margins narrow.
                    30. Apices of mid and distal stem leaf blades acute; pubescence on stem and leaf blades long, dense, yellowish; inflorescences 2–10-flowered; w arctic . . . . 15. *Cerastium fischerianum*

30. Apices of mid and distal stem leaf blades usually round and obtuse; pubescence colorless to somewhat fuscous; inflorescences 1–3-flowered; e arctic . . . . 3. *Cerastium arcticum*
[29. Shifted to left margin.—Ed.]

29. Sepals 3–7 mm, lanceolate to broadly elliptic, apex obtuse to ± acute, scarious margins broad.

31. Sepals with ± acute apex; leaf blades not succulent, pubescent, those on flowering shoots with ± acute apex; petals usually equaling, rarely to 2 times as long as sepals; plants taprooted, forming clumps, rarely rhizomatous . . . . 6. *Cerastium beeringianum* (in part)

31. Sepals with obtuse apex; leaf blades succulent, subglabrous to ciliate with long, colorless hairs, rotund, obovate, broadly elliptic, or broadly lanceolate, apex obtuse; petals 1.5–2 times as long as sepals; plants strongly rhizomatous with nonflowering, creeping shoots 6–30 cm . . . . 21. *Cerastium regelii* (in part)

1. Cerastium aleuticum Hultén, Svensk Bot. Tidskr. 30: 520, figs. 3a,b. 1936 • Aleutian mouse-ear chickweed E

Cerastium beeringianum Chamisso & Schlechtendal var. *aleuticum* (Hultén) S. L. Welsh

Plants perennial, tufted, rhizomatous, eglandular. **Stems** branched, 3–7 cm, subglabrous proximally, softly pubescent distally, proximal internodes congested; small axillary tufts of leaves absent. **Leaves** proximal leaves marcescent, pseudopetiolate, spatulate, distal sessile; blade elliptic to lanceolate or oblanceolate, 4–12 × 2–5 mm, apex ± obtuse, hirsute with long, straight, eglandular hairs or subglabrous except for midrib and margins. **Inflorescences** 1–3-flowered, dense cymes; bracts foliaceous, margins not scarious, pubescent. **Pedicels** becoming curved at apex, slender, 2–12 mm, to 3 times as long as sepals, pubescence spreading, eglandular, fuscous hairs equaling or longer than pedicel diam. **Flowers:** sepals lanceolate to elliptic, concave, 4–5 mm, margins narrow, apex acute to obtuse, pubescent; petals 1–1.5 times as long as sepals, apex 2-fid; stamens 10; styles 5. **Capsules** cylindric, slightly curved, 8–11 mm, 1–2 times as long as sepals, teeth 10, ± erect, margins convolute. **Seeds** brown, 0.8–1 mm, shallowly and obtusely tuberculate; testa not inflated, tightly enclosing seed.

Flowering summer. Stony ground, screes, etc., mountain slopes; 200–700 m; Alaska.

Cerastium aleuticum is an eglandular relative of *C. beeringianum.* It is similar to *C. bialynickii* except in being eglandular and having less dense pubescence and narrower sepals. It is confined to the western arctic, mainly on the Aleutian, St. Lawrence, St. Paul, Popof, and Kodiak islands, but is not found on mainland Alaska.

2. Cerastium alpinum Linnaeus, Sp. Pl. 1: 438. 1753 • Alpine mouse-ear chickweed, céraiste alpin

Plants perennial, mat-forming, rhizomatous. **Stems** prostrate or ascending, tomentose (very rarely subglabrous), hairs white, translucent, long, soft, flexuous, some usually also short and glandular; flowering shoots ascending, 5–20 cm; small axillary tufts of leaves usually absent; nonflowering shoots ± prostrate, to 6 cm. **Leaves** marcescent or not, sessile; blade obovate or ovate to elliptic-oblanceolate, elliptic, or lanceolate, usually 10–18 × 5–7 mm, apex obtuse, pubescence as on stems. **Inflorescences** open, (1–)2–4-flowered cymes; bracts lanceolate, acute, margins narrow, scarious, glandular-pubescent. **Pedicels** straight but often becoming angled at base and curved at apex, slender, 5–30 mm, often elongating to 3 or 4 times as long as sepals, pubescence usually dense, hairs both long, flexuous, multicellular, and short, glandular, viscid. **Flowers:** sepals green, often violet tipped, narrowly elliptic-lanceolate, 7.5–10 mm, margins ± narrow, apex acute to obtuse, densely pubescent, hairs both long, eglandular and short, glandular; petals 1–2 times as long as sepals, apex shallowly 2-fid; stamens 10; styles 5. **Capsules** cylindric, slightly curved, 12–16 mm, to 2 times as long as sepals; teeth 10, erect, margins convolute. **Seeds** dark brown, 1–1.4 mm diam., acutely tuberculate; testa not inflated. $2n$ = 72, 108.

Subspecies 3+ (2 in the flora): Greenland, Canada, Europe (Iceland, Scandinavia).

The *Cerastium alpinum* group of species is a difficult complex of intergrading taxa. E. Hultén (1956) considered this complex to be the result of worldwide introgression among the various taxa. Members of this group in North America include *C. aleuticum, C. alpinum, C. arcticum, C. beeringianum, C. bialynickii, C. fischerianum, C. regelii,* and *C. terrae-novae.*

Cerastium alpinum itself is distinguished from all other members of the complex by its lanate pubescence, which consists of long, silvery, translucent, multicellular, flexuous, often tangled hairs; the more or less square base of the calyx; the convex margins of the sepals; and, in well-grown plants, the long, slender, divaricate pedicels.

In western North America, *Cerastium alpinum* is replaced by *C. beeringianum*, which has long, straight, strigose, somewhat fuscous hairs, usually smaller flowers, and smaller seeds. The two species intergrade in eastern Canada; intermediate specimens were named *C. alpinum* var. *strigosum* Hultén.

Cerastium arcticum differs from *C. alpinum*, with which it often grows, in its straight, somewhat fuscous hairs; calyx which is round at the base; long, narrowly lanceolate sepals; large, straight, broad capsules; and broad, obtuse cauline leaves. Like *C. alpinum*, it usually has large flowers with the petals much longer than the sepals.

Many infraspecific taxa have been named in *Cerastium alpinum* but in North America it is much less variable than elsewhere. Two forms can be recognized at either the varietal or subspecific level.

1. Leaf blades on flowering shoots narrowly elliptic or lance-elliptic to oblanceolate, apex ± acute, pubescence evenly distributed, not very dense; inflorescences usually (1–)2–4-flowered . . . . .
. . . . 2a. *Cerastium alpinum* subsp. *alpinum*
1. Leaf blades on flowering shoots lance-elliptic to ovate or obovate, apex obtuse, pubescence a tuft of longer, silvery, ± tangled hairs at apex; inflorescences usually 1-flowered . . . .
. . . . 2b. *Cerastium alpinum* subsp. *lanatum*

2a. Cerastium alpinum Linnaeus subsp. **alpinum**

Plants perennial, tufted, forming clumps or mats, taprooted, stoloniferous, rarely rhizomatous. **Flowering shoots** usually erect from decumbent base, 6–20 cm. **Leaves:** flowering shoots with blades narrowly elliptic or lance-elliptic to oblanceolate, apex ± acute, evenly, but not densely pubescent; those near base and on sterile shoots not marcescent, blade oblanceolate to obovate, apex round. **Inflorescences** (1–)2–4 flowered. **Pedicels** usually long, slender, divaricate in well-grown plants. **Petals** 1.5–2 times as long as sepals. $2n$ = 72, 108.

Flowering spring. Low-arctic tundra, cliff ledges, talus; 0–300 m; Greenland; Man., Nfld. and Labr., N.W.T., Nunavut, Ont., Que., Sask.; n Europe (Scandinavia).

2b. Cerastium alpinum Linnaeus subsp. **lanatum** (Lamarck) Cesati in C. Cattaneo, Not. Nat. Civ. Lombardia, 290. 1844 • Céraiste laineux

Cerastium lanatum Lamarck in J. Lamarck et al., Encycl. 1: 680. 1785; *C. alpinum* var. *lanatum* (Lamarck) Hegetschweiler; *C. alpinum* subsp. *squalidum* (Raymond) Hultén; *C. squalidum* Raymond

Plants perennial, tufted or mat-forming, taprooted or rhizomatous. **Flowering shoots** erect or decumbent, 6–10 cm. **Leaves:** flowering shoots with blade lance-elliptic to ovate or obovate, broad, apex obtuse, with tuft of silvery, ± tangled, long, woolly hairs, pubescent; those at base and on sterile shoots often marcescent, blade obovate, apex obtuse, densely pubescent. **Inflorescences** usually 1-flowered, sometimes with simple cyme. **Pedicels** straight, usually short, 5–20 mm. **Petals** 2 times as long as sepals. $2n$ = 72, 104 ± 2, 108.

Flowering spring. Northern arctic tundra; 0–400 m; Greenland; Nfld. and Labr., N.W.T., Nunavut, Ont., Que.; Europe (Iceland, Russia, Scandinavia, mountains).

3. Cerastium arcticum Lange in G. C. Oeder et al., Fl. Dan. 17(50): 7, plate 2863. 1880 • Arctic mouse-ear chickweed, céraiste arctique

Cerastium alpinum Linnaeus var. *procerum* Lange; *C. alpinum* var. *uniflorum* Durand; *C. arcticum* subsp. *hyperboreum* (Tolmatchew) Böcher; *C. arcticum* subsp. *procerum* (Lange) Böcher; *C. arcticum* var. *procerum* (Lange) Hultén; *C. arcticum* var. *vestitum* Hultén; *C. hyperboreum* Tolmatchew; *C. nigrescens* (H. C. Watson) Edmondston ex H. C. Watson subsp. *arcticum* (Lange) P. S. Lusby

Plants perennial, loosely mat-forming but not pulvinate, rhizomatous. **Stems** erect, from decumbent base, 5–30 cm, sparsely to densely pubescent, hairs spreading, straight, often glandular; small axillary tufts of leaves usually absent. **Leaves** subsessile; blade broadly elliptic to obovate or rotund, larger blades in mid-stem region, 5–22 × 2–8 mm, apex round and obtuse, rarely broadly acute, pubescence ciliate and strigose, hairs long and eglandular, mixed with shorter eglandular hairs, colorless to somewhat fuscous; proximal leaves not strongly marcescent, blade oblanceolate, often broadly so, spatulate. **Inflorescences** 1–3-flowered, lax cymes, with patent, multicellular, long, glandular hairs; proximal bracts broadly lanceolate, pubescence as leaves; distal bracts narrowly lanceolate, margins narrow, scarious, glandular-pubescent. **Pedicels** ± straight or somewhat curved at apex, 5–45 mm, 1–4 times as long as sepals (rarely longer), pubescence long and glandular. **Flowers:** sepals narrowly lanceolate, 8–11 mm, margins narrow, apex acuminate, hairs ascending, long, glandular and eglandular; petals 12–18 mm, 1–2 times as long as sepals, apex 2-fid; stamens 10; styles 5. **Capsules** broadly conic, straight or almost so, 14–18 mm, 1.5–2 times as long as sepals, teeth 10, ± erect, margins convolute. **Seeds** red-brown, 1.1–1.3 mm diam., tuberculate; tubercles acute; testa not inflated, closely enclosing seed. $2n = (54)\ 108$.

Flowering summer. Arctic tundra and solifluction areas, talus slopes, beaches, coastal grassland and seepage areas, exposed rocky headlands; 0–400 m; Greenland; Nfld. and Labr. (Labr.), N.W.T., Nunavut, Que.; n Europe (Franz Joseph Land, Novaya Zemlya, Svalbard).

Cerastium arcticum has been a much confused and misunderstood species, recently separated from *C. nigrescens* Edmonston ex H. C. Watson on seed characters. The latter has large seeds with loose, inflated testae, whereas *C. arcticum* has seeds that are smaller and have tight testae that cannot easily be removed (A. K. Brysting and R. Elven 2000). Much of the North American material previously placed in *C. arcticum* is now considered to belong to a distinct species, *C. bialynickii* Tolmatchew (see comments under that species). *Cerastium alpinum* differs from *C. arcticum* in having very long, flexuous, translucent hairs which often mat together. In *C. arcticum* and other members of the complex, the hairs are usually straight, yellowish, and not flexuous. *Cerastium arcticum* may on occasion have some pubescence like that of *C. alpinum*; such plants have been named *C. arcticum* var. *vestitum*. Whether they represent introgression of *C. alpinum* into *C. arcticum*, as suggested by E. Hultén (1956), is uncertain.

4. **Cerastium arvense** Linnaeus, Sp. Pl. 1: 438. 1753

• Field or prairie mouse-ear chickweed, céraiste des champs F W

Plants perennial, clumped and taprooted, or mat-forming and long-creeping rhizomatous. **Stems:** flowering shoots often decumbent proximally, 5–20(–30) cm, glandular-pubescent distally, pilose-subglabrous, deflexed or spreading proximally; non-flowering shoots present; small tufts of leaves present in axils of proximal leaves. **Leaves** not marcescent, sessile, ± spatulate proximally; blade linear-lanceolate to lanceolate or narrowly oblong, 4–30 × 0.5–6 mm, apex acute, rarely obtuse, subglabrous

to softly pubescent, sometimes glandular. **Inflorescences** lax, 1–20-flowered cymes, pubescence short, glandular; bracts lanceolate, margins narrow, scarious, glandular-pubescent. **Pedicels** curved just below calyx, 5–30 mm, 1–6 times as long as sepals, glandular-pubescent. **Flowers:** sepals narrowly lanceolate to lance-elliptic, 3.5–7 mm, margins narrow, softly pubescent; petals obovate, 7.5–12.5 mm, ca. 2 times as long as sepals, apex 2-fid; stamens 10; anthers 0.8–1.1 mm; styles 5. **Capsules** cylindric, curved, 7.5–11.5 × 2.5–4 mm, (1–)1.5–2 times as long as sepals; teeth 10, erect, margins convolute. **Seeds** brown, 0.6–1.2 mm diam., tuberculate; testa not inflated. $2n = 36, 72, (108,\ \text{Europe})$.

Subspecies 2+ (2 in the flora): worldwide.

The infraspecific taxonomy of *Cerastium arvense* is subject to many different interpretations. While many subspecies have been recognized, the "actual" number is uncertain because of worldwide distribution, wide range of variation, and conflicting taxonomies.

SELECTED REFERENCES Ugborogho, R. E. 1974. North American *Cerastium arvense* Linnaeus. IV. Phenotypic variation. Phyton (Horn) 32: 89–97. Ugborogho, R. E. 1977. North American *Cerastium arvense* Linnaeus: Taxonomy, reproductive system and evolution. Phyton (Horn) 35: 169–187.

1. Taproot absent, plant strongly rhizomatous with long-creeping shoots; flowering stems usually 25–30 cm, often purple pigmented proximally, pubescence eglandular (glandular hairs present in inflorescence), soft, short or subglabrous; sepals 5–7 mm; anthers 1–1.1 mm; petals usually turning brown when dried . . . .
. . . . 4a. *Cerastium arvense* subsp. *arvense*
1. Taprooted or shortly rhizomatous, forming clumps; flowering stems usually 5–20 cm, green or straw colored, glandular-pubescent; sepals 3.5–6(–7) mm; anthers 0.8–0.9 mm; petals usually remaining white when dried . . . .
. . . . 4b. *Cerastium arvense* subsp. *strictum*

4a. **Cerastium arvense** Linnaeus subsp. **arvense** I W

Plants straggling and creeping, strongly rhizomatous, not forming clumps, without taproot. **Stems:** flowering stems ascending, proximal ¹/₂ often purple tinged, usually 25–30 cm, softly pubescent to subglabrous, glandular hairs confined to inflorescences; nonflowering shoots well developed, producing obovate to oblanceolate, spatulate, overwintering leaves; small axillary tufts of leaves well developed, conspicuous. **Leaves** on flowering shoots often dimorphic; mid-stem leaves larger, blade lanceolate, 12–22 × 2.5–5 mm; proximal leaves oblong to linear, 5–15 × 0.5–2 mm.

Flowers: sepals 5–7 mm, with midrib; petals 10–12.5 mm, often turning brown when dried; anthers 1–1.1 mm. **Capsules** (8.5–)9.8(–11.5) × 3–4 mm, ca. 1.5(–2) times as long as sepals. **Seeds** 0.7–1.2 mm. **2*n*** = 72.

Flowering spring. Lawns, cemeteries, roadsides, riverbanks, old pastures; 0–1400 m; introduced; Nfld. and Labr. (Nfld.), Ont., Que.; Conn., Md., Mass., N.J., N.Y.; w Europe.

Subspecies *arvense* is probably more widespread in North America than present information suggests, but identification of herbarium specimens can be difficult and uncertain. However, in the field the two subspecies are readily distinguishable because of the larger size of subsp. *arvense* and its strongly rhizomatous habit. Hybrids with *C. tomentosum* (which have been called *C. ×maureri* Schulze, an invalid name) are readily formed when the two taxa grow together (J. K. Morton 1973).

4b. Cerastium arvense Linnaeus subsp. **strictum** Gaudin, Fl. Helv. 3: 245. 1828 • Céraiste dressé

F

Cerastium alsophilum Greene; *C. angustatum* Greene; *C. arvense* var. *angustifolium* Fenzl; *C. arvense* var. *fuegianum* Hooker f.; *C. arvense* var. *latifolium* Fenzl; *C. arvense* var. *ophiticola* Raymond; *C. arvense* var. *purpurascens* B. Boivin; *C. arvense* var. *sonnei* (Greene) Smiley; *C. arvense* var. *strictum* (Gaudin) W. D. J. Koch; *C. arvense* var. *viscidulum* Gremli; *C. campestre* Greene; *C. confertum* Greene; *C. effusum* Greene; *C. elongatum* Pursh; *C. fuegianum* (Hooker f.) A. Nelson; *C. graminifolium* Rydberg; *C. leibergii* Rydberg; *C. nitidum* Greene; *C. occidentale* Greene; *C. oreophilum* Greene; *C. patulum* Greene; *C. pensylvanicum* Hooker; *C. pubescens* Goldie; *C. scopulorum* Greene; *C. sonnei* Greene; *C. subulatum* Greene; *C. tenuifolium* Pursh; *C. thermale* Rydberg; *C. vestitum* Greene

Plants forming clumps or mats, rhizomatous, or tufted, taprooted; straggling, creeping shoots usually not well developed. **Stems:** flowering stems decumbent at base, usually green or straw colored, occasionally purple tinged (in some populations growing on serpentine rocks), 5–20(–30) cm, pubescent and often glandular distally, hairs patent or deflexed; nonflowering winter shoots, when present, elongating, with narrow, oblanceolate leaves; small axillary tufts of leaves always present. **Leaves** usually not strongly dimorphic; blade lanceolate or oblanceolate to linear, 2–25 × 1–5 mm. **Flowers:** sepals 3.5–6(–7) mm, midrib visible; petals 7.5–9 mm (–12 mm in western plants), usually remaining ± white when dried; anthers 0.8–0.9 mm. **Capsules** 7.5–11 × 2.5–4 mm, usually less than 1.5 times as long as sepals, rarely longer. **Seeds** 0.6–1.1 mm. **2*n*** = 36.

Flowering spring. Prairie grasslands, roadsides, arctic and alpine tundra, shores, dunes and rocky plains, rocky outcrops, alvars, sea cliffs and banks, favoring neutral to alkaline soils; 0–3800 m; Greenland; St. Pierre and Miquelon; Alta., B.C., Man., N.B., Nfld. and Labr., N.W.T., N.S., Nunavut, Ont., P.E.I., Que., Sask., Yukon; Alaska, Ariz., Calif., Colo., Conn., Idaho, Ind., Iowa, Maine, Md., Mass., Mich., Minn., Mo., Mont., Nebr., Nev., N.H., N.J., N.M., N.Y., N.Dak., Ohio, Oreg., Pa., S.Dak., Utah, Vt., Va., Wash., W.Va., Wis., Wyo.; Europe (Alps); South America (s to Tierra del Fuego).

Subspecies *strictum* is widely distributed and grows in a great diversity of habitats, making it difficult to circumscribe and distinguish, both from subsp. *arvense* and from forms of *Cerastium beeringianum*, *C. velutinum*, and *C. viride*. Forms of subsp. *strictum* growing at high elevations or latitudes often develop broader leaves and may be confused with *C. beeringianum* (S. J. Wagstaff and R. J. Taylor 1988). However, *C. arvense* always has small axillary tufts of leaves. In northern parts of the Ungava region of Labrador, subsp. *strictum* appears on occasion to intergrade with *C. alpinum*. The status of these plants is uncertain.

Subspecies *strictum* is a remarkably variable taxon. Plants from the western side of the continent often have larger petals and a ranker growth. Completely glabrous plants (var. *ophiticola*) occur on serpentine in southern Quebec. Plants with broader ovate-elliptic leaves and tomentose pubescence (similar to *Cerastium velutinum* var. *villosissimum* but smaller) occur in the same area. Populations on the limestone plains near Belleville, southern Ontario, are more robust, with broader, strongly marcescent leaves at the base and a woolly pubescence. Plants from river valleys in Idaho tend to be much larger, with long, very narrow leaves; these are the basis for the name *C. graminifolium*. However, all of these plants are completely interfertile and show no reduction in fertility when crossed. Most of this variation is under genic control but also is affected by environmental factors.

Subspecies *strictum* is not interfertile with subsp. *arvense* or with other similar taxa such as *Cerastium beeringianum*, *C. velutinum*, and *C. viride*. Differences in chromosome numbers present an effective barrier to interfertility. However, several sterile hybrids involving subsp. *strictum* and those species have been synthesized. Many workers, most recently R. E. Ugborogho (1977), have included *C. velutinum* and *C. viride* in *C. arvense* as varieties or subspecies. However specific status is more appropriate because of the strong sterility barriers between them and the presence of morphological characters that enable them to be distinguished, albeit with difficulty in some herbarium material. This difficulty arises from the remarkable degree of variation displayed by subsp. *strictum*.

5. **Cerastium axillare** Correll, Brittonia 18: 308. 1967 • Trans Pecos mouse-ear chickweed

Plants annual, viscid. **Stems** usually erect, simple or several-branched, from branched caudex, rarely bushy, 6–40 cm, glandular-pilose, hairs in mid-stem region equaling or longer than stem diam.; small axillary tufts of leaves absent **Leaves** sessile; blade 7–25 × 1–6 mm; basal rosette absent or poorly developed and withering when young; proximal with blade oblanceolate to spatulate; distal with blade linear-lanceolate or lanceolate to narrowly elliptic, apex usually acute, rarely obtuse, glandular-pilose. **Inflorescences** diffuse, elongate, usually with single dichotomy at or below mid stem, 3–18-flowered cymes, glandular-pubescent, flowers widely and racemosely spaced in axils of paired bracts along each branch; bracts herbaceous, lanceolate, glandular-pubescent. **Pedicels** sharply curved just below capsule, 2–10(–15) mm, shorter than to 2 times as long as capsule, with dense, patent, glandular pubescence. **Flowers:** sepals lanceolate, 3–5 mm, apex acute, glandular-hispid, hairs not extending beyond sepal tips, inner sepals with broad margins, outer sepals with very narrow margins; petals oblanceolate, 2–3 mm, shorter than sepals, apex 2-fid ca. ¼ of length; stamens 5; styles 5. **Capsules** narrowly cylindric, curved, 7–8.5 mm, ca. 2 times as long as sepals; teeth 10, erect, margins convolute. **Seeds** light brown, 0.4–0.7 mm diam., tuberculate; testa not inflated.

Flowering summer. Rocky canyons, woodland and mountain slopes; 1300–2800 m; N.Mex., Tex.; Mexico.

Cerastium axillare is similar to *C. brachypodum* in its short pedicels, but it is more viscid-pubescent, with a much more diffuse habit and solitary flowers widely spaced along the elongate, racemelike primary branches of the inflorescence. In addition, the leaves are usually acute, whereas in *C. brachypodum* they are usually obtuse. *Cerastium axillare* is confined in the United States to the trans-Pecos mountains of Texas and New Mexico.

6. **Cerastium beeringianum** Chamisso & Schlechtendal, Linnaea 1: 62. 1826 • Bering mouse-ear chickweed, céraiste du détroit de Bering E F

Cerastium alpinum Linnaeus var. *beeringianum* Regel; *C. alpinum* var. *capillare* (Fernald & Wiegand) B. Boivin; *C. beeringianum* var. *capillare* Fernald & Wiegand; *C. beeringianum* var. *glabratum* Hultén; *C. beeringianum* var. *grandiflorum* Hultén; *C. buffumiae* A. Nelson; *C. earlei* Rydberg; *C. fischerianum* Seringe ex de Candolle var. *beeringianum* (Chamisso & Schlechtendal) Hultén; *C. pilosum* Greene; *C. pulchellum* Rydberg; *C. scammaniae* Polunin; *C. variabile* Goodding; *C. vulgatum* Linnaeus var. *beeringianum* (Chamisso & Schlechtendal) Fenzl

Plants perennial, forming loose to dense mats or clumps, taprooted, with short, prostrate sterile shoots, rarely rhizomatous. **Stems:** flowering stems usually erect, 10–25 cm, glabrous to sparsely pubescent proximally, more densely so in mid and distal stem, hairs patent to slightly deflexed, multicellular, glandular and eglandular; internodes usually equaling or exceeding leaves; nonflowering shoots present; small axillary tufts of leaves absent. **Leaves:** sessile (flowering shoots) or petiolate (proximal stems and nonflowering shoots); flowering shoots with blade lanceolate, 5–20 × 2–5 mm, not succulent, apex ± acute, pubescent; proximal stem and nonflowering shoots with blade oblanceolate often spatulate, not succulent, apex obtuse, ± pubescent on both surfaces (rarely subglabrous), with straight, strigose hairs or cilia, hairs pale, fuscous, mostly eglandular. **Inflorescences** open, dichotomous, (1–)3–10-flowered cymes; bracts lanceolate, distal with scarious margins, proximal herbaceous and foliaceous, pubescence glandular and eglandular. **Pedicels** erect, angled at base of calyx in fruit, 8–55 mm, 1–5 times as long as capsule, densely pubescent with patent glandular and eglandular hairs. **Flowers:** sepals lanceolate to lance-elliptic, 3–7 mm, margins broad, apex ± acute, densely glandular-pubescent; petals broadly oblanceolate, 6–12 mm, ± equaling (rarely to 2 times as long as) sepals, apex deeply 2-fid; stamens 10; styles 5. **Capsules** cylindric, curved, 8–12 mm, ca. 2 times sepals, rarely shorter; teeth 10, erect, margins convolute. **Seeds** pale to dark brown, 0.7–1.1 mm, tuberculate; testa not inflated, tightly enclosing seed. $2n = 72$.

Flowering spring and early summer. Arctic and alpine tundra, meadows, open woodlands, rocky slopes, talus, cliff ledges, river and roadside gravel; 0–4000 m; Alta., B.C., Man., Nfld. and Labr., N.W.T., Nunavut, Ont., Que., Sask., Yukon; Alaska, Ariz., Calif., Colo., Idaho, Mont., Nev., N.Mex., Oreg., Utah, Wyo.

Cerastium beeringianum is distinguished from *C. alpinum* by the absence of the long, silvery, flexuous, translucent, glistening hairs of that species. *Cerastium beeringianum*'s pubescence consists of straight, strigose, multicellular, somewhat fuscous hairs of several lengths, many of those in the mid and distal stem and inflorescence being glandular and viscid. The nodes and the leaves, at least in the mid and distal stem, typically have long, strigose, eglandular, fuscous hairs; those on the adaxial surface of the leaf being appressed, and those on the nodes retrorse. However, plants from the many small, isolated populations on the mountains of western North America show a great deal of variation. Some of these populations tend to be subglabrous, lacking most of the long hairs normally found on this species. Others are small, delicate plants with slender divaricate pedicels and smaller capsules and seeds. Though names have been given to several of these variants, they frequently intergrade, and much of the variation is greatly influenced by the environment.

Cerastium beeringianum is self-compatible and often self-pollinates, but the flowers are also freely visited by insects, particularly Diptera and smaller Hymenoptera, resulting in varying degrees of outbreeding.

Cerastium beeringianum intergrades with *C. fischerianum*, and it may be appropriate to treat them as subspecies of a single species. Unfortunately, the name *C. fischerianum* has priority.

7. **Cerastium bialynickii**   Tolmatchew, Trudy Bot. Muz. 21: 81, fig. 1. 1927 · Bialynick's mouse-ear chickweed

Cerastium arcticum Lange var. *sordidum* Hultén

Plants perennial, compact, pulvinate, taprooted. **Stems** ascending, much-branched at base, 2–10 cm, densely hispid-pubescent, hairs patent, fuscous, multicellular; internodes short; small axillary tufts of leaves absent. **Leaves** dense below the inflorescence; blade broadly lanceolate, elliptic-lanceolate to ovate or obovate, 5–10 × 1.5–4 mm, thick and ± fleshy, apex broadly acute to obtuse, hispid, hairs fuscous, multicellular, long; proximal leaves marcescent, sometimes subglabrous. **Inflorescences** often 1-flowered and compact, sometimes 2–3-flowered cymes, hairs patent, dense, fuscous, eglandular and glandular; bracts lanceolate, with or without narrow, scarious margins, densely pubescent, often glandular. **Pedicels** straight or ± angled at base and/or apex, 5–12 mm, 1–2 times as long as sepals, densely pubescent with long, eglandular and short, glandular hairs. **Flowers:** sepals usually purplish, 5–6(–7) mm, herbaceous center narrowly lanceolate, margins broad, making apex obtuse, densely pubescent, hairs long, stiff, glandular

and eglandular; petals oblanceolate, 7–9 mm, 1–1.5 times as long as sepals, apex 2-fid; stamens 10; styles 5. **Capsules** cylindric, slightly curved, 9–12 mm, 1.5–2 times as long as sepals; teeth 10, erect, margins convolute. **Seeds** brown, 0.8–1 mm, tuberculate; testa not inflated, tightly enclosing seed. $2n = 108$.

Flowering summer. Tundra, rocky exposures, screes, nunataks in the high arctic; 0–1300 m; Greenland; N.W.T., Nunavut; Alaska; Eurasia (Russian Far East, arctic Siberia and associated islands, Spitzbergen).

Cerastium bialynickii previously was included in *C. arcticum* but is very different from that species, being a small, compact plant with a dense, hispid, fuscous pubescence, a small calyx with the broad, scarious margins making it obtuse, and small flowers, capsules, and seeds. *Cerastium bialynickii* resembles small compact plants of *C. beeringianum* but it differs in calyx shape and chromosome number.

8. **Cerastium brachypetalum**   Persoon, Syn. Pl. 1: 520. 1805 · Gray mouse-ear chickweed   [I]

Cerastium brachypetalum subsp. *tauricum* (Sprengel) Murbeck; *C. tauricum* Sprengel

Plants annual. **Stems** erect, simple or branched at base, 5.5–30 cm, shaggy, hairs silvery, spreading-ascending; small axillary tufts of leaves absent. **Leaves** not marcescent; blade 4–15(–20) × 1.5–5 (–7) mm, pubescence of long, glandular and/or eglandular hairs; basal ± crowded, blade oblanceolate, spatulate, apex obtuse; cauline sessile, blade lanceolate or elliptic, apex acute. **Inflorescences** lax, dichasiate, 3–30-flowered cymes; bracts herbaceous, lanceolate, densely pubescent, with long, ascending, glandular or eglandular hairs. **Pedicels** erect or ascending, bent distally in fruit, 6–15 mm, longer than capsule, shaggy, glandular or eglandular. **Flowers:** sepals lanceolate, 4–4.5 mm, foliaceous, with or without narrow margins, densely pubescent, hairs exceeding sepal tips; petals oblanceolate, 2–3 mm, ca. 0.5 times as long as sepals, apex 2-fid, sparsely ciliate proximally; stamens 10, with few long hairs near filament base; styles 5. **Capsules** cylindric, slightly curved near apex, 5–7 mm, ca. 1.5 times as long as sepals; teeth 10, erect, margins convolute. **Seeds** pale brown, 0.5 mm diam., acutely tuberculate; testa not inflated. $2n = 72$ (Europe), 88, 90.

Flowering spring. Dry, sandy places, roadsides, arable land, disturbed, open areas; 0–400 m; introduced; Ala., Ark., Fla., Ga., Idaho, Ill., Ind., Kans., Ky., Miss., Mo., N.Y., N.C., Ohio, Okla., Oreg., Pa., S.C., Tenn., Tex., Va., W.Va.; Eurasia.

The wholly herbaceous bracts of *Cerastium brachypetalum* distinguish it from *C. fontanum* subsp.

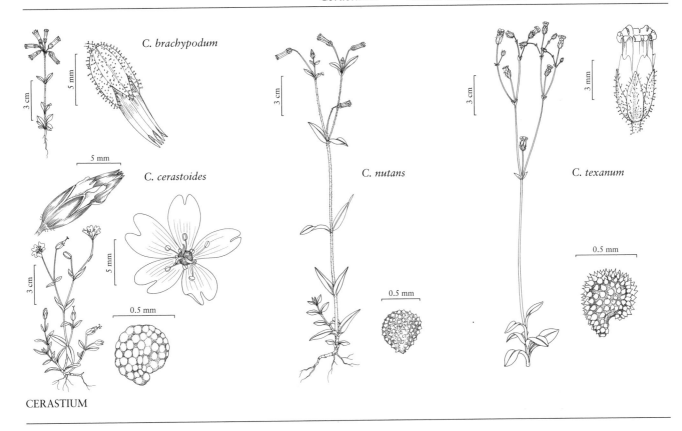

C. brachypodum

C. cerastoides

C. nutans

C. texanum

CERASTIUM

vulgare, C. *semidecandrum*, and C. *pumilum*; the ciliate petal and filament bases distinguish it from C. *diffusum* and C. *glomeratum*. *Cerastium brachypetalum* differs from all those species in the long, silvery hairs that give it a grayish appearance. In Europe C. *brachypetalum* is more variable and eight subspecies have been recognized, two of which—subsp. *brachypetalum* and subsp. *tauricum*—occur in North America. However, they differ only in the absence or presence of glandular hairs, an insufficient distinction for recognition at the subspecific level.

9. **Cerastium brachypodum** (Engelmann ex A. Gray) B. L. Robinson, Proc. Amer. Acad. Arts 29: 277. 1894

• Short-stalked mouse-ear chickweed   F

Cerastium nutans Rafinesque var. *brachypodum* Engelmann ex A. Gray, Manual ed. 5, 94. 1867; C. *adsurgens* Greene; C. *brachypodum* var. *compactum* B. L. Robinson

Plants annual, with filiform taproot. **Stems** erect, simple or several from branched caudex, 5–20 cm, glandular-pubescent, nodes without long, wooly hairs; small axillary tufts of leaves absent. **Leaves** sessile, not marcescent, usually confined to proximal ¹⁄₂ of plant; mid-stem leaves with blade lanceolate to narrowly elliptic or oblanceolate, 5–30 × 2–8 mm, apex ± acute, usually rather sparsely, softly pubescent; proximal leaves

sometimes forming loose rosette, blade ovate to obovate or spatulate, apex ± obtuse. **Inflorescences** compact to open, dichotomous, 3–30-flowered cymes, confined to distal portion of stem; bracts herbaceous, lanceolate, glandular-puberulent. **Pedicels** often becoming deflexed at base, 3–10(–12) mm, 0.5–1.25 times as long as sepals in flower, elongating to equaling capsule, glandular-pubescent. **Flowers:** sepals broadly lanceolate, 3–4.5 mm, apex subacute, glandular-puberulent to glabrate, hairs shorter than sepal tip, outer sepals herbaceous or with narrow margins, inner with broad margins; petals ovate-elliptic, 3–4 mm, equaling or shorter than sepals, apex 2-fid; stamens 10; styles 5. **Capsules** cylindric, curved, 6–12 mm, 2–2.25 times as long as sepals; teeth 10, erect, margins convolute. **Seeds** golden brown, 0.4–0.7 mm diam., tuberculate, especially around edge; testa not inflated. **2*n*** = 34.

Flowering spring and early summer. Grasslands, fields, meadows, open woods, roadsides, waste places, often in seasonally wet rocky or sandy ground; 100–2700 m; Alta., Man., Ont., Sask.; Ala., Ariz., Ark., Colo., Ga., Idaho, Ill., Ind., Iowa, Kans., La., Mich., Mo., Mont., Nebr., Nev., N.Mex., N.Dak., Okla., Oreg., S.C., S.Dak., Tenn., Tex., Utah, Va., Wis., Wyo.; Mexico.

Cerastium brachypodum differs from C. *nutans* in its smaller size; its short pedicels, which often become deflexed at the base instead of near the capsule; and its pubescence, which is usually sparser and lacks any long, lanate hairs. However, some specimens from the arid region of the southwest have a dense gray pubescence. Plants from the Arizona desert, which have particularly long and narrow capsules, are the basis for the erroneous report of *C. gracile* Dufour from that region.

Cerastium fastigiatum can be very similar to *C. brachypodum* but differs in its longer pedicels, narrowly acute leaves, glandular pubescence on the stems, and much more branched (fastigiate) habit.

10. **Cerastium cerastoides** (Linnaeus) Britton, Mem. Torrey Bot. Club 5: 150. 1894 • Starwort mouse-ear chickweed, céraiste à trois styles F

Stellaria cerastoides Linnaeus, Sp. Pl. 1: 422. 1753; *Arenaria trigyna* (Villars) Shinners; *Cerastium lapponicum* Crantz; *C. trigynum* Villars; *Dichodon cerastoides* (Linnaeus) Reichenbach; *Provencheria cerastoides* (Linnaeus) B. Boivin

Plants perennial, mat-forming, rhizomatous. **Stems** creeping, much-branched, rooting, glabrous except for line of small hairs down each internode; flowering shoots decumbent or ascending, 5–10 cm; nonflowering shoots prostrate, 5–15 cm; small axillary tufts of leaves usually absent. **Leaves** sessile, tending to be marcescent, somewhat succulent; blade elliptic-oblong or linear-lanceolate, 2–12 × 1–3 mm, apex obtuse, rarely acute, glabrous, sometimes ciliate at base. **Inflorescences** lax, 1–3-flowered terminal cymes; bracts lanceolate, 2–5 mm, glabrous or ciliate. **Pedicels** becoming curved, slender, 5–35 mm, equaling or exceeding sepals, glandular-puberulent. **Flowers:** sepals narrowly lanceolate, 4–5 mm, margins narrow, midrib present, apex obtuse, glandular-pubescent towards base; petals 5–8 mm, 1–1.5 times as long as sepals, apex deeply 2-fid; stamens 10; styles 3(–6). **Capsules** ovoid-conic, oblong after dehiscence, straight, 7–10 mm, 1.5–2 times as long as sepals; teeth 6(–12), erect to spreading, margins convolute. **Seeds** brown, 0.5 mm diam., shallowly rugose; testa not inflated. *2n* = 38.

Flowering summer. Wet, arctic areas, alpine rills, alpine and arctic snowbeds; 0–800 m; Greenland; Nfld. and Labr., Nunavut, Que.; Europe; amphi-Atlantic.

Cerastium cerastoides is an unusual member of the genus because it normally has only three styles and a straight, six-toothed capsule, rather than a curved capsule as in most of the other species. The blunt sepals help to distinguish this species from *C. arvense* subsp. *strictum*, with which it is most likely to be confused. The epithet of this species is often misspelled "cerastioides."

11. **Cerastium dichotomum** Linnaeus, Sp. Pl. 1: 438. 1753 • Forked mouse-ear chickweed I

Plants annual, with slender taproot. **Stems** erect, simple or several from branched caudex, 15–30 cm, densely

viscid-glandular; small axillary tufts of leaves usually absent. **Leaves** not marcescent, sessile; blade broadly

linear to linear lanceolate to oblong, 12–30(–50) × 3–10(–15) mm, apex usually acute, densely viscid-glandular. **Inflorescences** dense, 3–30-flowered cymes; bracts foliaceous, glandular-pubescent. **Pedicels** erect, 2–10 mm, shorter than sepals, with dense, spreading, glandular pubescence. **Flowers:** sepals lanceolate, 6–11 mm, margins narrow, apex acute, pubescence dense, stout, shorter than sepal tips, viscid-glandular; petals oblanceolate, 8–10 mm, ca. equaling sepals, apex shortly 2-fid; stamens 5; styles 5. **Capsules** narrowly conic, straight, 10–15 mm, ca. 2 times as long as sepals; teeth 10, erect, margins convolute. **Seeds** chestnut brown, ca. 1.3 mm, tuberculate; testa not inflated. *2n* = 38.

Flowering spring–early summer. Arable land, roadsides; 300–900 m; introduced; Calif., Oreg., Wash.; s Europe.

Cerastium dichotomum is very similar to *C. inflatum* Link from the Middle East [*C. dichotomum* subsp. *inflatum* (Link) Cullen] and is sometimes equated with it, but that species differs from *C. dichotomum* in having an inflated fruiting calyx. Reports of *C. siculum* Gussone in North America are referable to *C. dichotomum*.

Cerastium dichotomum is a rare weed of arable land and roadsides.

12. **Cerastium diffusum** Persoon, Syn. Pl. 1: 520. 1805 • Dark-green mouse-ear chickweed I

Cerastium atrovirens Babington; *C. tetrandrum* Curtis

Plants annual, with slender taproot. **Stems** decumbent or ascending, diffusely branched, 7.5–30 cm, densely covered and viscid with short, glandular hairs; small axillary tufts of leaves absent. **Leaves** not marcescent, sessile distally, spatulate to pseudopetiolate proximally; blade 5–10 × 2–4 mm, covered with short, glandular and eglandular hairs; proximal blades oblanceolate, apex obtuse; cauline blades ovate or oblong-ovate, apex acute. **Inflorescences** lax, 3–30-flowered cymes; bracts lanceolate to ovate, herbaceous, glandular-pubescent. **Pedicels** straight, ultimately erect in fruit, slender, 2–15 mm, much longer than capsule, glandular. **Flowers** 4(–5)-merous; sepals lanceolate, 4–7 mm, margins narrow distally, apex acute or acuminate, glandular-pubescent, hairs usually not projecting beyond apex; petals ca. 3 mm, ca. 0.75 times as long as sepals, apex 2-fid; stamens 4(–5); styles 4(–5). **Capsules** narrowly cylindric, nearly straight, 5–7.5 mm, 1–1.5 times as long as sepals; teeth

8 or 10, erect, margins convolute. **Seeds** reddish brown, 0.5–0.7 mm, bluntly tuberculate; testa not inflated. $2n$ = 72.

Flowering spring. Sandy places on coast, rarely inland in similar places and on railway ballast; 0–300 m; introduced; Calif., Ill.; Europe.

This species was abundant on the sandy shore at Fort Bragg, Mendocino County, California, in 1985 and should be looked for elsewhere. The entirely herbaceous bracts, short capsule, and the floral parts usually in fours identify this small weedy species.

Previous reports of this species (as *Cerastium tetrandrum*) by J. A. Steyermark (1963) from Missouri and M. L. Fernald (1950) from Virginia are referable to *C. pumilum* and *C. brachypetalum*, respectively.

13. **Cerastium dubium** (Bastard) Guépin, Fl. Maine et Loire ed. 2, 1: 267. 1838 • Anomalous mouse-ear chickweed [I]

Stellaria dubia Bastard, Suppl. Fl. Maine-et-Loire, 24. 1812; *Cerastium anomalum* Waldstein & Kitaibel; *Dichodon viscidum* (M. Bieberstein) Holub

Plants annual, taprooted. **Stems** erect, many-branched from base, 10–40 cm, minutely viscid-glandular; small axillary tufts of leaves usually absent. **Leaves** not marcescent, distal sessile, proximal spatulate; blade linear or linear-lanceolate to linear-oblong, 10–30 × 1–4 mm, apex obtuse to subacute, glabrous or sparsely and minutely viscid-glandular. **Inflorescences** lax, 3–21(–30)-flowered cymes; bracts narrowly lanceolate, glandular-pubescent. **Pedicels** erect, slender, 2–15 mm, 0.5–3 times as long as sepals, glandular-puberulent **Flowers:** sepals ovate-lanceolate, 5–6 mm, margins narrow, apex acute to obtuse, minutely viscid-glandular; petals oblanceolate, 5–8 mm, 1.5 times as long as sepals, apex 2-fid; stamens 10; styles 3(–4). **Capsules** oblong-ovoid, straight, 8–11 mm, ca. 2 times as long as sepals; teeth 6, occasionally 8, erect to spreading, margins convolute. **Seeds** pale brown, ovate, 0.6 mm diam., tuberculate; testa not inflated. $2n$ = 36, 38.

Flowering spring. Alien weed of cultivated land; 200–800 m; introduced; Ark., Idaho, Ill., Ind., Kans., Ky., Miss., Ohio, Oreg., Tenn., Va., Wash.; s Europe; Asia.

First collected in North America in 1966 in Washington, *Cerastium dubium* has now been gathered from many widely scattered sites, and appears to be spreading rapidly.

SELECTED REFERENCE Shildneck, P. and A. G. Jones. 1986. *Cerastium dubium* (Caryophyllaceae) new for the eastern half of North America (a comparison with sympatric *Cerastium* species, including cytological data). Castanea 51: 49–55.

14. **Cerastium fastigiatum** Greene, Pittonia 4: 303. 1901 • Fastigiate mouse-ear chickweed [E]

Plants annual, with slender taproot. **Stems** erect, branched from base, branches ascending (fastigiate), 10–50 cm, pubescent with stiff, gland-tipped, patent or slightly reflexed hairs with broadened base, pubescence shorter than diam. of stem, soft wooly hairs absent; small axillary tufts of leaves absent. **Leaves** sessile; distal and mid-stem blades linear-lanceolate, 20–70 × 1.5–6 mm, apex acute to acuminate; proximal leaves shortly connate basally, blade narrowly oblanceolate, tending to be spatulate, pubescence short, stiff, patent, glandular. **Inflorescences** very lax, making up at least ½ height of plant, 3–45-flowered cymes; bracts, linear-lanceolate, 2–22 mm, herbaceous, glandular-pubescent. **Pedicels** erect to spreading, bent distally, slender, 5–20 mm, 1–5 times as long as sepals, usually longer than capsules, glandular-pubescent. **Flowers:** sepals narrowly lanceolate, 4–5 mm, margins narrow (narrower than herbaceous center), apex sharply acute to acuminate, glandular-hispid, hairs shorter than sepal tips; petals oblanceolate, 4–5 mm, ± equaling sepals, apex 2-fid; stamens 10; styles 5. **Capsules** cylindric, curved, (5–)7–10(–11) mm, ca. 2 times as long as sepals; teeth 10, erect to slightly spreading, margins convolute. **Seeds** golden brown, 0.5–0.8 mm diam., coarsely tuberculate. $2n$ = 36.

Flowering summer. Sandy canyons and washes, open rocky and sandy places and dry pine woods in arid mountains; 1900–3000 m; Ariz., N.Mex.

Until recently *Cerastium fastigiatum* was included in *C. nutans* but it is readily separable by its bushy, ascending habit, shorter pubescence, long, narrow leaves, and smaller capsule. It can be very similar to forms of *C. brachypodum* but differs from that species in its longer pedicels, narrowly acute leaves, glandular pubescence on the stems, and much more branched (fastigiate) habit. From *C. nutans* var. *obtectum* it differs in its very narrow sepals, narrowly lanceolate leaves, and smaller capsule.

15. **Cerastium fischerianum** Seringe in A. P. de Candolle and A. L. P. P. de Candolle, Prodr. 1: 419. 1824 • Fischer's mouse-ear chickweed

Cerastium alpinum Linnaeus var. *fischerianum* (Seringe) Torrey & A. Gray; *C. unalaschkense* Takeda

Plants perennial, coarse, forming loose mats or clumps, ± rhizomatous. **Stems** erect to straggling, decumbent at base, sturdy, 10–50 cm, usually densely pubescent with hairs patent, yellowish, multicellular, glandular and eglandular, often subglabrous near base; nodes bearded with long, yellowish hairs; small axillary tufts of leaves usually absent. **Leaves** sessile, not marcescent; blade lanceolate, 7–50 × 3–15 mm, largest on mid and distal stem, apex usually acute, densely ciliate, hairs yellowish, multicellular, eglandular, of various lengths; proximal leaves with blade lanceolate to elliptic, oblanceolate, or oblong. **Inflorescences** lax, 2–10-flowered cymes, compact when young, branches elongating at maturity; proximal bracts foliaceous, herbaceous, pubescence as in leaves; distal bracts lanceolate, 5–10 mm, often with scarious tip, ciliate with long, fuscous hairs. **Pedicels** erect or becoming deflexed at base and curved at apex, 5–30(–60) mm, elongating in fruit, ca. 6 times as long as sepals, densely fuscous-pubescent with glandular and eglandular, patent hairs. **Flowers:** sepals narrowly lanceolate with round base, 6–10 mm, margins narrow, apex acute, strigose-ciliate, hairs long, fuscous; petals conspicuous, 10–14 mm, 1.5–2 times as long as sepals, apex deeply 2-fid; stamens 10; styles 5. **Capsules** slightly conic or cylindric, straight, 10–22 mm, ca. 2 times as long as sepals; teeth 10, erect, margins convolute. **Seeds** reddish brown, 1–1.5 mm diam., strongly papillate; testa not inflated, tightly enclosing seed. $2n = 66, 72$.

Flowering summer. Grassy areas, lake shores, riverbanks, gravel; 0–200 m; B.C.; Alaska; Asia.

Cerastium fischerianum is a distinctive species resembling a large and robust form of *C. fontanum*. The bush of yellow hairs underneath each node, together with the large flowers, sepals, and capsule, distinguish this species. It intergrades with *C. beeringianum*, and the intermediate plants often have exceptionally large petals; these have been named *C. beeringianum* var. *grandiflorum* Hultén.

16. **Cerastium fontanum** Baumgarten, Enum. Stirp. Transsilv. 1: 425. 1816 • Common mouse-ear chickweed, céraiste commun [I]

Plants perennial (rarely annual), tufted to mat-forming, often rhizomatous. **Stems:** flowering stems erect from decumbent base, branched proximally, 10–45 cm, softly pubescent, eglandular with straight hairs; nonflowering shoots, when present, produced proximally, decumbent, rooting at nodes, branched, 5–20 cm, often subglabrous with alternating lines of eglandular hairs; small axillary tufts of leaves usually absent. **Leaves** not marcescent; blade 10–25(–40) × 3–8(–12) mm, densely covered with patent to ascending, colorless, long, eglandular hairs; leaves of flowering shoots in distant pairs, sessile, blade elliptic to ovate-oblong, apex subacute; leaves of sterile shoots pseudopetiolate, often spatulate, blade oblanceolate,

apex obtuse. **Inflorescences** lax, 3–50-flowered cymes; bracts lanceolate, reduced, herbaceous, eglandular-pubescent, distal often with narrow, scarious margins. **Pedicels** somewhat curved distally, 2–10(–20) mm, longer than sepals, densely pubescent with patent, eglandular, rarely glandular hairs. **Flowers:** sepals ovate-lanceolate, 5–7 mm, margins narrow, apex acute, scarious, pubescent with eglandular, rarely glandular, hairs; petals oblanceolate, 1–1.5 times as long as sepals, apex deeply 2-fid; stamens 10, occasionally 5; styles 5. **Capsules** narrowly cylindric, curved, 9–17 mm, ca. 2 times sepals; teeth 10, erect, margins convolute. **Seeds** reddish brown, 0.4–1.2 mm, bluntly tuberculate; testa not inflated, tightly enclosing seed. $2n = 122–152$, usually 144.

Subspecies 4 (2 in the flora): introduced; Europe; introduced elsewhere.

Cerastium fontanum is a highly variable and complex species. It often has been reported as *C. vulgatum* Linnaeus, an ambiguous name; see discussion under the genus.

1. Inflorescences never glandular; petals 1.3–1.5 times as long as sepals; capsules 11–17 mm; seeds 0.9–1.2 mm . . . .
. . . . 16a. *Cerastium fontanum* subsp. *fontanum*
1. Inflorescences occasionally viscid-glandular; petals equaling sepals; capsules 9–13 mm; seeds 0.4–0.9 mm . . . .
. . . . 16b. *Cerastium fontanum* subsp. *vulgare*

16a. **Cerastium fontanum** Baumgarten subsp. **fontanum** • Mountain mouse-ear chickweed

Cerastium fontanum subsp. *scandicum* Gartner

Plants perennial. **Inflorescences** never glandular. **Petals** 1.3–1.5 times as long as sepals. **Capsules** 11–17 mm. **Seeds** 0.9–1.2 mm; tubercles coarse. $2n = 144$ (Europe).

Flowering summer. Grassy areas, scrub; 0–100 m; Greenland; Europe.

Subspecies *fontanum* is widespread in the mountains of northern and central Europe.

16b. **Cerastium fontanum** Baumgarten subsp. **vulgare** (Hartman) Greuter & Burdet, Willdenowia 12: 37. 1982 [I][W]

Cerastium vulgare Hartman, Handb. Skand. Fl., 182. 1820; *C. caespitosum* Gilibert; *C. fontanum* subsp. *triviale* (Link) Jalas; *C. triviale* Link; *C. vulgatum* Linnaeus var. *hirsutum* Fries

Plants perennial (very rarely annual). **Inflorescences** usually eglandular, occasionally viscid and glandular. **Petals** equaling sepals. **Capsules** 9–13 mm. **Seeds** 0.4–0.9 mm; tubercles small. $2n$ = ca. 122–152, usually 144.

Flowering throughout growing season. A common weed in grassy places: lawns, roadsides, pastures, open woodlands, wastelands; 0–3000 m; introduced; Greenland; St. Pierre and Miquelon; Alta., B.C., Man., N.B., Nfld. and Labr., N.W.T., N.S., Ont., P.E.I., Que., Sask., Yukon; Ala., Alaska, Ariz., Ark., Calif., Colo., Conn., Del., D.C., Fla., Ga., Idaho, Ill., Ind., Iowa, Kans., Ky., La., Maine, Md., Mass., Mich., Minn., Miss., Mo., Mont., Nebr., Nev., N.H., N.J., N.Mex., N.Y., N.C., N.Dak., Ohio, Okla., Oreg., Pa., R.I., S.C., S.Dak., Tenn., Tex., Utah, Vt., Va., Wash., W.Va., Wis., Wyo.; Europe; introduced worldwide.

Small annual forms of subsp. *vulgare* can be difficult to separate from *Cerastium pumilum*. The latter has smaller capsules, narrower and sharply acute sepals, and short, glandular hairs on the sepals, bracts, and inflorescence.

17. Cerastium glomeratum Thuillier, Fl. Env. Paris ed. 2, 226. 1799 • Sticky mouse-ear chickweed, céraiste aggloméré I W

Cerastium acutatum Suksdorf; *C. fulvum* Rafinesque

Plants annual, with slender taproots. **Stems** erect or ascending, branched, 5–45 cm, hairy, glandular at least distally, rarely eglandular; small axillary tufts of leaves absent. **Leaves** not marcescent, ± sessile; blade 5–20 (–30) × 2–8(–15) mm, apex apiculate, covered with spreading, white, long hairs; basal with blade oblanceolate or obovate, narrowed proximally, sometimes spatulate; cauline with blade broadly ovate or elliptic-ovate. **Inflorescences** 3–50-flowered, aggregated into dense, cymose clusters or in more-open dichasia; bracts: proximal herbaceous, distal lanceolate, apex acute, with long, mainly eglandular hairs. **Pedicels** erect to spreading, often arcuate distally, 0.1–5 mm, shorter than capsule, glandular-pubescent. **Flowers:** sepals green, rarely dark-red tipped, lanceolate, 4–5 mm, margins narrow, apex very acute, usually with glandular hairs as well as long white hairs usually extending beyond apex; petals oblanceolate, 3–5 mm, rarely absent, usually shorter than sepals, apex deeply 2-fid; stamens 10; styles 5. **Capsules** narrowly cylindric, curved, 7–10 mm; teeth 10, erect, margins convolute. **Seeds** pale brown, 0.5–0.6 mm, finely tuberculate; testa inflated or not. $2n$ = 72.

Flowering throughout growing season. Arable land, waste places, roadsides; 0–1800 m; introduced; B.C., Nfld. and Labr. (Nfld.), N.S., Ont., Que., Yukon; Ala., Alaska, Ariz., Ark., Calif., Conn., Del., D.C., Fla., Ga.,

Idaho, Ill., Ind., Kans., Ky., La., Md., Mass., Mich., Miss., Mo., Mont., Nev., N.J., N.Mex., N.Y., N.C., Ohio, Okla., Oreg., Pa., R.I., S.C., Tenn., Tex., Va., Wash., W.Va.; Europe; introduced and common in Mexico.

Cerastium glomeratum often has been reported as *C. viscosum* Linneaus, an ambiguous name; see discussion under the genus.

18. Cerastium maximum Linnaeus, Sp. Pl. 1: 439. 1753 • Great mouse-ear chickweed

Dichodon maximum (Linnaeus) Á. Löve & D. Löve

Plants perennial, subrhizomatous. **Stems** simple, or few together, erect or ascending, 20–70 cm, proximal internodes moderately pilose, becoming glandular distally; nonflowering, axillary branches usually present; small axillary tufts of leaves absent. **Leaves** sessile, not marcescent; blade narrowly lanceolate, with prominent midrib, 0.2–1 × 3–12 mm, apex acuminate, ± pubescent on both surfaces, short-ciliate. **Inflorescences** open or congested, usually 3–10-flowered cymes; bracts normally lanceolate, acuminate, herbaceous, pubescent. **Pedicels** erect, 2–25(–60) mm, usually ca. 2 times as long as sepals in fruit, glandular-pubescent. **Flowers** large, conspicuous, more than 2 cm diam.; sepals lanceolate, 8–11(–12) mm, outer sepal margins herbaceous, inner sepal margins narrow, membranous, apex acute, moderately to sparsely glandular-hairy; petals obovate, (15–)18–25 mm, at least 2 times as long as sepals, apex deeply 2-fid; stamens 10; styles 5. **Capsules** narrowly conic, straight, 15–22 mm, ca. 2 times as long as sepals; teeth 10, erect, short, becoming outwardly coiled. **Seeds** yellowish brown, round, 2–2.5 mm diam., finely rugose in concentric rings; testa not inflated. $2n$ = 38.

Flowering summer. Open woods, gravel bars, terraces by rivers; 0–1200 m; N.W.T., Yukon; Alaska; Asia.

This beautiful species is distinguished by its long, narrowly conic capsule with teeth that coil outward like a watch spring.

19. Cerastium nutans Rafinesque, Précis Découv. Somiol., 36. 1814 • Nodding mouse-ear chickweed, céraiste penché F W

Plants annual, slender, finely glandular-pubescent (often perennial and tomentose in var. *obtectum*), with slender taproot. **Stems** erect, simple or branched at or near base, sometimes with straggling, nonflowering basal shoots, 10–50 cm, softly pubescent, often with a few long, flexuous, woolly hairs at proximal nodes, glandular and somewhat viscid distally; small axillary tufts of leaves usually absent. **Leaves** marcescent or not, sessile; blade oblanceolate to spatulate in proximal leaves, becoming lanceolate to linear-lanceolate in distal leaves, occasionally elliptic, 10–60 × 3–15 mm, apex acuminate to acute, softly pubescent and glandular, sometimes tomentose. **Inflorescences** rather open, 3–21(–40)-flowered cymes, ultimately widely branched; bracts herbaceous, lanceolate, glandular-pubescent. **Pedicels** ascending, sharply deflexed at apex in fruit, 5–20(–35) mm, usually 1–3 times as long as sepals in flower, elongating to 5 times as long as sepals in fruit, longer than capsules, glandular-pubescent and viscid. **Flowers:** sepals ovate-lanceolate, 4–6 mm, outer sepals herbaceous or with narrow margins, inner with margins ca. as wide as herbaceous center, apex broadly acute to obtuse, glandular-puberulent, hairs shorter than sepal tips; petals oblanceolate, sometimes absent, 3–6(–8) mm, shorter to 1.5 times longer than sepals, apex 2-fid; stamens 10; styles 5. **Capsules** cylindric, curved, (9–)10–12(–13) **mm,** 2–3 times as long as sepals; teeth 10, erect, margins convolute. **Seeds** golden brown, 0.5–0.8 mm diam., shallowly tuberculate; testa not inflated. *2n* = 34, 36.

Varieties 2 (2 in the flora): North America, Mexico.

1. Leaves not marcescent, long, woolly pubescence confined to proximal nodes . . . .
. . . . 19a. *Cerastium nutans* var. *nutans*
1. Leaves tending to be marcescent, leaves and stem with long, woolly pubescence . . . .
. . . . 19b. *Cerastium nutans* var. *obtectum*

19a. Cerastium nutans Rafinesque var. **nutans**

E F W

Cerastium longipedunculatum Muhlenberg ex Britton; *C. nutans* var. *occidentale* B. Boivin

Stems erect, glandular and viscid distally, often with few long, woolly hairs at proximal nodes. **Leaves** not marcescent, sessile; blade oblanceolate to spatulate proximally, lanceolate to narrowly elliptic distally, 10–60 × 3–15 mm, apex mostly acute, softly pubescent and glandular. **Pedicels** usually 1–3 times as long as sepals in flower, elongating to 5 times as long as sepals in fruit. **Flowers:** sepals ovate-lanceolate, 4–6 mm; petals shorter than to slightly longer than sepals. **Capsules** (8–)10–12(–13) mm. *2n* = 36.

Flowering throughout growing season. Moist woods, stream banks, meadows, shores, rock ledges, boggy places, cultivated land; 0–2800 m; Alta., B.C., Man., N.W.T., Ont., Que., Sask., Yukon; Ala., Alaska, Ark., Colo., Conn., Del., D.C., Ga., Idaho, Ill., Ind., Iowa, Kans., Ky., Md., Mass., Mich., Minn., Miss., Mo., Mont., Nebr., N.J., N.Y., N.C., N.Dak., Ohio, Oreg., Pa., S.Dak., Tenn., Tex., Vt., Va., Wash., W.Va., Wis., Wyo.

19b. Cerastium nutans Rafinesque var. **obtectum** Kearney & Peebles, J. Wash. Acad. Sci. 29: 475. 1939

Cerastium sericeum S. Watson, Proc. Amer. Acad. Arts 20: 354. 1885, not Pourret 1788; *C. diehlii* M. E. Jones

Stems erect, stout, with straggling basal shoots, glandular and viscid distrally, woolly-pubescent proximally, sometimes densely so. **Leaves** often marcescent, sessile; midstem leaves with blades narrowly lanceolate-triangular, broadest at base (similar to *C. fastigiatum* but broader), 25–32 × 4–7 mm, apex mostly acute, woolly-pubescent, sometimes densely so. **Pedicels** 1.5–3 times as long as capsule. **Flowers:** sepals 5–6 mm; petals ca. 1.5 times as long as sepals. **Capsules** 12–13 × 3 mm. *2n* = 34.

Flowering summer. Sandy canyons and washes, ponderosa pine woodlands, among rocks in arid uplands; 2000–3000; Ariz., N.Mex.; Mexico.

As pointed out by Kearney and Peebles, var. *obtectum* tends to intergrade with var. *nutans*. Its status requires further study. It and *Cerastium fastigiatum* appear to replace the type variety in Arizona and New Mexico, but extreme plants of var. *obtectum* are infrequent and most of the material that I have examined is more or less intermediate with the type variety.

20. Cerastium pumilum Curtis, Fl. Londin. 2(6,69): plate 30. 1794 • Sticky mouse-ear chickweed   I

Cerastium glutinosum Fries; *C. pumilum* subsp. *glutinosum* (Fries) Jalas

Plants annual, with slender taproot. **Stems** erect or ascending, branching near base, 2–12 cm, covered with glandular and eglandular hairs; small axillary tufts of leaves usually absent. **Leaves** not marcescent, sessile; blade 5–15 × 3–6 mm, hairy; basal with blades oblanceolate, spatulate, petiolelike, apex obtuse; cauline with blades lanceolate, elliptic, or ovate, apex acute to obtuse. **Inflorescences** lax, 3–15-flowered (rarely more) cymes; bracts lanceolate: proximal usually foliaceous, distal smaller, usually with narrow, scarious margins and apex, glandular-pubescent. **Pedicels** erect, curved distally, 3–8(–10) mm, longer than capsule, glandular-pubescent. **Flowers:** sepals green, sometimes red tipped, oblong-lanceolate, 4–5 mm, margins narrow, apex acute, pubescent, hairs short, stiff, glandular, not projecting beyond scarious, glabrous apex; petals white or purple-tinged, with branching veins, oblanceolate, ca. 5 mm, ± equaling sepals, apex 2-fid for ca. 1/4 length; stamens 5;

styles 5. **Capsules** narrowly cylindric, slightly curved upward, 6–9 mm, ca. 2 times as long as sepals; teeth 10, erect, margins convolute. **Seeds** dark brown, deltoid, 0.6–0.7 mm, tuberculate; testa not inflated. $2n = 72$.

Flowering spring. Dry, sandy, gravelly places on roadsides and arable land; 0–900 m; introduced; B.C., N.B., N.S., Ont.; Ala., Ark., Ga., Ill., Ind., Kans., Ky., Maine, Md., Mich., Miss., Mo., Nebr., N.J., N.C., Ohio, Okla., Oreg., Pa., S.C., Tenn., Tex., Va., Wash., W.Va.; Eurasia.

North American material referred to here as *Cerastium pumilum* is very variable. At one extreme are plants resembling small annual forms of *C. fontanum*, with relatively short, broad capsules, petals slightly longer than the sepals, and sepals that are usually red at the tips. At the other extreme are plants with relatively long, narrow capsules resembling impoverished diffuse-inflorescenced *C. glomeratum*, with short petals and no red pigment. The latter are probably referable to *C. pumilum* subsp. *glutinosum*. B. Jonsell and T. Karlsson (2001+, vol. 2) treated *C. glutinosum* as a distinct species in Scandinavia, but the correlation of characters that they gave to distinguish *C. glutinosum* from *C. pumilum* does not occur in most North American material that I have examined. Hence, the recognition of a single species, possibly with two subspecies, as in *Flora Europaea* (T. G. Tutin et al. 1964–1980, vol. 1), appears to be more appropriate. The problem may arise from North American material having been introduced from several sources, whereas Scandinavian material may consist of two native genotypes that do not show the complete range of variation in the species.

Cerastium pumilum can look like a small annual form of *C. fontanum* but differs in its smaller capsules and the characteristic rather short, glandular hairs on the sepals, bracts, and inflorescence. It can be separated from *C. semidecandrum* by the much narrower scarious margins of the sepals and bracts and by the branching veins in the petals, which tend to be slightly longer and more conspicuous than in *C. semidecandrum*. Some forms of *C. glomeratum* have a very open inflorescence and may be confused with *C. pumilum*, but *C. glomeratum* has ten stamens, a narrower capsule, all the bracts herbaceous, and long, eglandular hairs (often mixed with glandular ones) on the bracts and sepals.

21. **Cerastium regelii** Ostenfeld, Skr. Vidensk.-Selsk. Christiana, Math.-Naturvidensk. Kl. 1909(8): 10, fig. 11. 1910 • Regel's mouse-ear chickweed, céraiste de Regel

Cerastium alpinum Linnaeus var. *caespitosum* Malmgren; *C. gorodkovianum* Schischkin; *C. jenisejense* Hultén; *C. regelii* subsp. *caespitosum* (Malmgren) Tolmatchew

Plants perennial, highly variable, ranging from small, pulvinate, 1–6(–8) cm, often not flowering, to rhizomatous. **Stems** erect or trailing, branched, pubescence

confined to alternating lines of short, crispate to more elongate hairs; flowering stems erect, 1–6(–8) cm in pulvinate plants, to 20 cm in creeping plants, pubescence spreading to retrorse; nonflowering shoots creeping, elongate, 1–30 cm, with terminal, fleshy, gemmaelike buds; small axillary tufts of leaves absent. **Leaves** sessile; blade rotund and obovate to ovate, elliptic, or broadly lanceolate, 3–9 × 2–6 mm, succulent, apex ± acute to obtuse; proximal leaves connate at base, blades becoming marcescent, broadly and shortly spatulate, margins and midrib ciliate, glabrous to sparsely pubescent, hairs long, colorless. **Inflorescences** erect, lax, 1–5(–11)-flowered cymes; bracts elliptic to lanceolate, with or without narrow, scarious margins, with long, strigose hairs. **Pedicels** erect, usually straight, 1–10 (–30) mm, shorter than sepals in pulvinate plants, longer in creeping plants, softly pubescent, hairs patent, of varying lengths, glandular and eglandular. **Flowers:** sepals often purple tinged, broadly elliptic, 4–6 mm, margins broad, apex round, obtuse, sparsely pubescent, hairs colorless, long, eglandular, and short, glandular; petals obovate, 4–8 mm, 1.5–2 times as long as sepals, apex deeply 2-fid; stamens 10; styles 5. **Capsules** cylindric, slightly curved, 6–12 mm, 1–2 times as long as sepals; teeth 10, erect to slightly spreading, margins convolute. **Seeds** dark brown, 1 mm diam., shallowly tuberculate; testa not inflated, tightly enclosing seed. $2n = 72$.

Flowering summer. High-arctic tundra to low-arctic taiga, often in wet solifluction areas or in moss; 0–900 m; Greenland; N.W.T., Nunavut, Que.; Alaska; nw Europe (Spitsbergen); Asia (Russian Far East, Siberia).

Until recently, *Cerastium gorodkovianum* (= *C. jenisejense*) was recognized as a species separate from *C. regelii*, the former being low arctic and the latter high arctic. A recent experimental study by O. M. Heide et al. (1990) elegantly demonstrated that the morphological differences are the result of the effects of day length and temperature on development. In the high arctic, *C. regelii* rarely flowers and is compact and pulvinate with small, broadly obovate, marcescent leaves that are fleshy, subglabrous, or with strigose cilia. In the low arctic and taiga zone, *C. regelii* becomes a slender, straggling plant with slender, erect inflorescences bearing a few large flowers. The creeping sterile shoots produce terminal bulbils for vegetative dispersal.

Apparent hybrids between *Cerastium regelii* and *C. beeringianum* were named *C. regelii* var. *hirsutum* Hultén. An apparent hybrid with *C. arvense* subsp. *strictum* also has been collected along the Nisling River in the Yukon.

SELECTED REFERENCE Heide, O. M., K. Pederson, and E. Dahl. 1990. Environmental control of flowering and morphology in the high arctic *Cerastium regelii*, and the taxonomic status of *C. jenisejense*. Nordic J. Bot. 10: 141–147.

22. **Cerastium semidecandrum** Linnaeus, Sp. Pl. 1: 438. 1753 • Five-stamen mouse-ear chickweed [I]

Plants annual, with slender taproot. **Stems** erect or ascending, branching at base, 1–20 cm, viscid, covered with short, dense, glandular and eglandular hairs; short axillary tufts of leaves absent. **Leaves** not marcescent, sessile but proximal leaves often spatulate; blade 5–18 × 2–5 mm, covered with short, white hairs; basal leaves with blade narrowly oblanceolate and ± spatulate, apex obtuse; cauline with blades ovate to elliptic-oblong, apex obtuse to acute. **Inflorescences** open, 3–30-flowered cymes; bracts lanceolate, with broad, scarious margins, glandular-pubescent. **Pedicels** curved at apex, often sharply angled at base, 3–8(–12) mm, 1–3 times as long as capsule, densely glandular-pubescent and viscid. **Flowers:** sepals narrowly lanceolate, 3–5 mm, margins broad, apex acute, glandular-pubescent; petals with unbranched veins, oblanceolate, 2–3 mm, shorter than sepals, apex notched; stamens 5; styles 5. **Capsules** cylindric, slightly curved, 4.5–6.5 mm, 1.5–2 times as long as sepals; teeth 10, erect, margins convolute. **Seeds** pale yellowish brown, 0.4–0.6 mm, finely tuberculate; testa not inflated. $2n = 36$.

Flowering spring. Common weed in dry, sandy, and gravelly places, roadsides and footpaths, parking lots, dunes; 0–300 m; introduced; B.C., N.S., Ont.; Ark., Conn., Fla., Ga., Idaho, Ill., Ind., Kans., La., Md., Mass., Mich., Mo., Nebr., N.J., N.Y., N.C., Ohio, Oreg., Pa., R.I., S.C., Wash., Wis.; Eurasia.

The very broad, scarious margins of the sepals and bracts distinguish this small, ephemeral species.

23. **Cerastium terrae-novae** Fernald & Wiegand, Rhodora 22: 176. 1921 • Newfoundland mouse-ear chickweed [E]

Cerastium beeringianum Chamisso & Schlechtendal subsp. *terrae-novae* (Fernald & Wiegand) Hultén

Plants perennial, tufted, rhizomatous. **Stems** loosely ascending to suberect, branched, very leafy, 10–15 cm, pubescence short, dense, glandular; small axillary tufts of leaves usually absent. **Leaves** tending to be marcescent, sessile but spatulate proximally; blade elliptic-oblong, 5–14 × 1.5–3.5 mm, apex obtuse, densely glandular-hirsute. **Inflorescences** lax, 1–3-flowered cymes; bracts ovate-lanceolate, margins very narrow, scarious, glandular-pubescent. **Pedicels** mostly erect, slender, 10–25 mm, 1–4 times as long as sepals, densely glandular-pubescent. **Flowers:** sepals ovate-oblong, 5.5–6.5 mm, elongating to 6–7 mm in fruit, margins broad,

apex obtuse, glandular-pubescent; petals narrowly oblanceolate, 7–10 mm, 1.5–2 times as long as sepals, apex 2-fid; stamens 10; styles 5. **Capsules** ovate-cylindric, ca. straight, short, broad, 9–13 mm, ca. 2 times as long as sepals; teeth 10, erect or partially spreading, margins convolute. **Seeds** brown, 1.3–1.7 mm diam., with prominent papillae around margins and rows of small, transverse ridges on sides; testa inflated, loose (rubs off when rolled between finger and thumb). $2n = 108$.

Flowering summer. Serpentine gravel, sands, rocky tablelands; 20–700 m; Nfld. and Labr. (Nfld.).

Cerastium terrae-novae is the only member of the European *C. nigrescens* group of species to occur in North America. The group is distinguished by large seeds with loose testae. *Cerastium terrae-novae* is distinguished by its narrowly elliptic leaves, usually purple-suffused stems and sepals, broad, straight capsule, short pubescence, and poorly developed inflorescence that often is reduced to a single flower.

24. **Cerastium texanum** Britton, Bull. Torrey Bot. Club 15: 97. 1888 • Chihuahuan mouse-ear chickweed [F]

Cerastium longepedunculatum Muhlenberg ex Britton var. *sordidum* (B. L. Robinson) Briquet; *C. sordidum* B. L. Robinson; *Stellaria montana* Rose

Plants annual, with slender taproot and branched caudex. **Stems** erect, sparingly branched proximally, slender, 15–35 cm, sparsely glandular-pilose; small axillary tufts of leaves absent. **Leaves** not marcescent; proximal blades broadly spatulate-petiolate, 8–55 × 3–16 mm, apex acute or obtuse, sometimes short-acuminate, softly pilose; cauline few, sessile, blade linear-lanceolate to narrowly oblanceolate, 7–30 mm, apex acute, pilose. **Inflorescences** very open and loose, 2–9(–25)-flowered cymes; bracts narrowly lanceolate, pilose. **Pedicels** straight, becoming sharply deflexed at base, slender, 5–20 mm, elongating in fruit, 1.5–4 times as long as sepals, glandular-pilose. **Flowers:** sepals green, turning pale orange-brown in fruit, lanceolate to ovate, 3–6 mm, margins narrow, apex acute, with short, glandular pubescence; petals oblanceolate, 5–8 mm, 1.5–2 times as long as sepals, apex 2-fid; stamens 5; styles 5. **Capsules** cylindric, straight, 5–12 mm, 1.5–2 times as long as sepals; teeth 10, becoming outwardly coiled. **Seeds** red-brown, 0.4–0.7 mm diam., tuberculate; tubercles ± pointed; testa not inflated. $2n = 36$.

Flowering spring. Canyons, sandy washes, oak woodlands, mountain pine forests; 1200–2800 m; Ariz., N.Mex.; Mexico.

Cerastium texanum is exceptionally variable in flower and capsule size. The extent to which this variation is due to environmental conditions or is genic in origin is not known. The broad, spatulate basal leaves and the straight, cylindric capsule with its outwardly coiled (revolute) teeth distinguish this species.

SELECTED REFERENCE Turner, B. L. 1995. *Cerastium texanum* does not occur in Texas. Phytologia 79: 356–363.

25. Cerastium tomentosum Linnaeus, Sp. Pl. 1: 440. 1753 • Dusty miller, snow-in-summer, céraiste tomenteux ⓘ

Plants perennial, mat-forming, rhizomatous. **Stems:** flowering stems ascending, branched, 15–40 cm; nonflowering stems prostrate proximally, rooting readily, pubescence dense, white-tomentose, eglandular; small axillary tufts of leaves often present. **Leaves** not marcescent, sessile; blade linear to linear-lanceolate or linear-oblong, 10–60 × 2–8 mm, apex ± obtuse, pubescence dense, whitish-tomentose, eglandular on both surfaces. **Inflorescences** lax, 3–13-flowered cymes; bracts lanceolate, margins scarious, pubescent. **Pedicels** ascending, straight, 10–40 mm, 2–7 times as long as sepals, white-tomentose. **Flowers** 12–20 mm diam.; sepals narrowly lanceolate-elliptic, 5–7 mm, margins narrow, often scarious, apex acute, white-tomentose; petals obtriangular, 10–18 mm, 2–2.5 times as long as sepals, apex 2-fid; stamens 10; styles 5. **Capsules** cylindric, slightly curved, 10–15 mm, 1.5–2 times as long as sepals; teeth 10, erect, margins convolute. **Seeds** brown, ca. 1.5 mm, round tubercles on margins, faces shallowly rugose; testa not inflated. $2n = 72$.

Flowering spring. A commonly grown rock-garden and wall plant, often escaping onto roadsides, riverbanks, old fields; 0–400 m; introduced; B.C., Man., N.B., Nfld. and Labr. (Nfld.), N.S., Ont., P.E.I., Que.; Maine, Mich., Mont., Nebr., N.Y., N.C., Ohio, Oreg., Pa., Wash., Wis., Wyo.; se Europe.

Cerastium tomentosum hybridizes readily with the introduced *C. arvense* subsp. *arvense* (J. K. Morton 1973).

North American reports of *Cerastium biebersteinii* de Candolle all appear to be referable to *C. tomentosum*. The two species are very similar, but *C. biebersteinii* has flat capsule teeth and is diploid ($2n = 36$); see M. K. Khalaf and C. A. Stace (2001).

26. Cerastium velutinum Rafinesque, Med. Repos., hexade 2, 5: 359. 1808 • Large field mouse-ear chickweed Ⓔ

Plants perennial, usually clumped, taproot present, sometimes with short rhizomes. **Stems:** flowering shoots ascending, often decumbent at base, branched, (15–)25–35(–40) cm, softly pubescent to subglabrous, hairs spreading, straight, concentrated in longitudinal lines

toward base, glandular and eglandular; nonflowering leafy shoots often present, straggling, 5–15 cm; small axillary tufts of leaves present. **Leaves** sessile; cauline with largest at mid stem, smaller distally and proximally, blade narrowly oblong, linear-lanceolate to lanceolate or ovate, 20–45 × 3–11 mm, apex usually obtuse, softly ciliate-pubescent on both surfaces, more so on margins and abaxial midrib, rarely subglabrous, sometimes densely villose; proximal leaf blades often oblanceolate, apex ± obtuse. **Inflorescences** lax, 5–20-flowered cymes, pubescent, glandular, not viscid; bracts lanceolate, proximal foliaceous, distal smaller, with scarious margins, glandular-pubescent. **Pedicels** curved immediately below capsule, slender, 10–24 mm, ca. 2 times as long as sepals, pubescence spreading, glandular. **Flowers:** sepals lanceolate-elliptic, 5–8 mm, margins narrow, apex acute, densely and softly glandular-pubescent; petals obovate, 10–15 mm, 2 times as long as sepals, apex deeply 2-fid; stamens 10; anthers 0.9–1.1 mm; styles 5. **Capsules** cylindric, slightly curved, slender, 10–14 × 3–4 mm, 2–2.3 times as long as sepals; teeth 10, erect, margins convolute. **Seeds** brown, 0.8–1.2 mm diam., tuberculate; testa not inflated, tightly enclosing seed. $2n = 72$.

Varieties 2 (2 in the flora): e North America.

Until recently *Cerastium velutinum* has been included in *C. arvense* as a variety or subspecies. However, morphological characters and chromosome number (diploid in the native forms of *C. arvense*, tetraploid in *C. velutinum*) distinguish the two taxa. They are not interfertile. Crosses between them can be produced with difficulty but they are completely sterile. None are known to occur naturally.

An apparent hybrid between *Cerastium velutinum* and *C. tometosum* has been collected on the south shore of Lake Erie. It was completely sterile.

1. Cauline leaves narrowly oblong to linear-lanceolate or lanceolate, not overlapping to obscure stem, not marcescent, softly pubescent to subglabrous . . . . 26a. *Cerstium velutinum* var. *velutinum*
1. Cauline leaves lanceolate to ovate, dense and over-lapping, obscuring stem on sterile shoots, often marcescent, densely villous . . . . 26a. *Cerstium velutinum* var. *villosissimum*

26a. Cerastium velutinum Rafinesque var. velutinum

Cerastium arvense Linnaeus var. *bracteatum* (Rafinesque) MacMillan; *C. arvense* var. *oblongifolium* (Torrey) Hollick & Britton; *C. arvense* subsp. *velutinum* (Rafinesque) Ugborogho; *C. arvense* var. *velutinum* (Rafinesque) Britton; *C. arvense* var. *villosum* Hollick & Britton; *C. arvense* var. *webbii* Jennings; *C. bracteatum* Rafinesque; *C. oblongifolium* Torrey

Decumbent sterile shoots absent, or when present, short, stems not obscured by overlapping leaves. **Leaf blades** linear-lanceolate to lanceolate or narrowly oblong, softly pubescent to subglabrous. $2n$ = 72.

Flowering spring. Limestone rocks, woodlands, serpentine barrens; 0–300 m; Ont.; Del., D.C., Ill., Ind., Iowa, Ky., Md., Mass., Mich., Minn., Mo., N.J., N.Y., Ohio, Pa., Tenn., Va., W.Va.

26b. Cerastium velutinum Rafinesque var. **villossissimum** (Pennell) J. K. Morton, Sida 21: 887. 2004 • Octoraro Creek chickweed E

Cerastium arvense Linnaeus var. *villosissimum* Pennell, Bartonia 12: 11. 1931

Decumbent sterile shoots well developed, 5–15(–20) cm. **Leaves** marcescent, overlapping and obscuring stem; blade broadly lanceolate to ovate, densely villous. $2n$ = 72.

Flowering spring. Serpentine outcrops and barrens; 100–300 m; Md., N.J., Pa.

Variety *villosissimum* grows in coarse grass in seepage areas on the serpentine slopes above Octoraro Creek at Lees Bridge, Chester County, Pennsylvania. Plants of var. *velutinum* with normal pubescence and leaf width grow nearby on the serpentine outcrops. The pubescence, leaf, and habit characters of var. *villosissimum* are controlled genetically (D. J. Gustafson et al. 2003) and segregation occurs in plants grown from seeds collected at Lees Bridge. These characters are displayed in varying degrees in plants growing on the serpentien barrens and outcrops in surrounding areas of Maryland, New Jersey, and Pennsylvania.

Variety *villosissimum* is in the Center for Plant Conservation's National Collection of Endangered Plants.

SELECTED REFERENCE Gustafson, D. J., G. Romano, R. E. Latham, and J. K. Morton. 2003. Amplified fragment length polymorphism analysis of genetic relationships among the serpentine barrens endemic *Cerastium velutinum* Rafinesque var. *villosissimum* Pennell (Caryophyllaceae) and closely related *Cerastium* species. J. Torrey Bot. Soc. 130: 218–223.

27. Cerastium viride A. Heller, Muhlenbergia 2: 281. 1907 • Western field mouse-ear chickweed E

Cerastium arvense Linnaeus subsp. *maximum* (Hollick & Britton) Ugborogho; *C. arvense* var. *maximum* Hollick & Britton

Plants perennial, mat-forming, strongly long-creeping rhizomatous. **Stems:** flowering stems ascending from elongate decumbent bases, branched, 15–45 cm, viscid, glandular in distal and mid-stem region, proximal portion with deflexed, long, soft, eglandular hairs, mostly in alternating longitudinal lines; nonflowering shoots horizontal, leafy. **Leaves** sometimes marcescent, sessile, highly variable; blade tending to be succulent, pubescent on both surfaces or ± glabrous abaxially except on midrib and margins; leaves of mid and distal stem largest, blade ovate-lan-ceolate to linear-lanceolate or narrowly oblong, 15–42 × 3.5–7 mm, apex acute; proximal leaves smaller, with tufts of small leaves in their axils, blade oblanceolate to linear-oblong, 10–30 × 2–7 mm, often spatulate, apex ± obtuse. **Inflorescences** lax, 1–12-flowered cymes; bracts glandular-pubescent, proximal bract often foliaceous, broadly lanceolate to ovate-lanceolate; distal bracts lance-elliptic to lanceolate, margins narrow, scarious. **Pedicels** straight or sometimes curving near apex in fruit, tending to be stout, 10–35 mm, ca. 2–4 times as long as capsules, rarely more, pubescence dense, glandular, viscid. **Flowers** large and showy; sepals lanceolate, 6–9 mm, margins broad, apex acute, densely glandular-pubescent, viscid; petals broadly oblanceolate, large, 10–15 mm, 2–3 times as long as sepals, apex 2-fid; stamens 10; anthers 0.9–1.2 mm; styles 5. **Capsules** broadly cylindric, ca. straight, 8–15 × 4–5 mm, 1.5–2 times as long as sepals; teeth 10, erect, margins convolute. **Seeds** reddish brown, 1–1.5 mm, tuberculate; testa not inflated, tightly enclosing seed. $2n$ = 72.

Flowering spring–early summer. Grassy slopes on coast, grassy and rocky slopes inland; 0–1000 m; Calif., Oreg.

Some forms of *Cerastium viride* can be difficult to distinguish from larger forms of *C. arvense* subsp. *strictum*, but the broader capsule and leaves usually are diagnostic. In rare cases, chromosome number or pollen size [38–70 (average 43) μ in *C. viride* versus 28–40 (average 34) μ in *C. arvense* subsp. *strictum*] may be needed to confirm identification. Some of the inland material from Catsop County, Oregon, is atypical in having more slender pedicels, longer and softer pubescence, and more acute sepals.

Cerastium viride often has been treated as a variety or subspecies of *C. arvense*. It differs from the native *C. arvense* subsp. *strictum* in being much larger in all its parts, and in chromosome number ($2n$ = 72 in *C. viride*, 36 in *C. arvense* subsp. *strictum*). The two taxa do not hybridize in the wild and attempts to cross them in cultivation have failed. The introduced *C. arvense* subsp. *arvense* may be confused with *C. viride*. Both are strongly rhizomatous and the ranges of measurements for key characters overlap. However, the two taxa have different appearances, *C. arvense* subsp. *arvense* being a more slender plant with narrow leaves that are never succulent. In contrast, *C. viride* tends to be larger in all its parts, with sturdier, decumbent stems, and usually succulent leaves.

Excluded species:
Cerastium clawsonii Correll, described from Texas, is a synonym of *Linum hudsonioides* Planchet (R. L. Hartman 1979).

19. MOENCHIA Ehrhart, Neues Mag. Aerzte 5: 203. 1783, name conserved • Upright chickweed [for Conrad Moench, 1744–1805, professor at Marburg, Germany] I

Richard K. Rabeler

Ronald L. Hartman

Herbs, annual. **Taproots** slender. **Stems** ascending to erect, simple or sometimes branched proximally, terete. **Leaves:** cauline leaves connate proximally, sessile or sometimes petiolate (basal leaves); blade 1-veined, linear to linear-oblanceolate, not succulent, apex acute. **Inflorescences** terminal, 1–3-flowered, spreading cymes or flowers solitary; bracts paired, foliaceous, those of axillary flowers with scarious margins. **Pedicels** erect. **Flowers:** perianth and androecium hypogynous; sepals 4, distinct, green, lanceolate, 3.8–7 mm, herbaceous, margins white or silvery, scarious, apex acute; petals 4 or rarely absent, white, claw absent, blade apex entire; nectaries at base of filaments opposite sepals; stamens 4, inserted at base of ovary; filaments distinct nearly to base; staminodes absent; styles 4, filiform, 0.7–1 mm, glabrous proximally; stigmas 4, linear along adaxial surface of styles, minutely papillate (50×). **Capsules** cylindric, opening by 8 revolute teeth, shorter than or equaling sepals; carpophore absent. **Seeds** 35–55, brown, reniform with deep abaxial groove, laterally compressed, papillate, marginal wing absent, appendage absent. *x* = 19.

Species 3 (1 in the flora): introduced; Europe (Mediterranean region); introduced in Africa (Republic of South Africa), Australia.

SELECTED REFERENCES Ketzner, D. M. 1996. *Crepis pulchra* (Asteraceae) and *Moenchia erecta* (Caryophyllaceae) in Illinois. Trans. Illinois State Acad. Sci. 89: 21–23. Rabeler, R. K. 1991. *Moenchia erecta* (Caryophyllaceae) in eastern North America. Castanea 56: 150–151.

1. **Moenchia erecta** (Linnaeus) P. Gaertner, B. Meyer & Scherbius, Oekon. Fl. Wetterau 1: 219. 1799   F I

Sagina erecta Linnaeus, Sp. Pl. 1: 128. 1753

Subspecies 2 (1 in the flora): introduced; Europe (w Mediterranean region); introduced in Africa (Republic of South Africa), Australia.

1a. **Moenchia erecta** (Linnaeus) P. Gaertner, B. Meyer & Scherbius subsp. **erecta**   F I

Stems 2.5–19 cm, glabrous, glaucous. **Leaf blades** 5–18 mm. **Pedicels** 1–7.6 cm. **Petals** ca. $^1/_2$–$^2/_3$ times as long as sepals. **Capsules** 5–6.5 mm, usually $^4/_5$–1 times as long as sepals. $2n$ = 38 (Europe).

Flowering spring. Lawns, disturbed areas; 0–600 m; introduced; B.C.; Calif., Ill., Oreg., S.C., Wash.; Europe.

Moenchia erecta is known in widely scattered areas of western North America; its occurrence in the east is limited. Historical records from the 1830s exist as specimens collected at Baltimore, Maryland, and Philadelphia, Pennsylvania; the species has not been recollected in either place since. The Illinois population, first documented in 1993, is scattered over an acre of a mowed field in a city park.

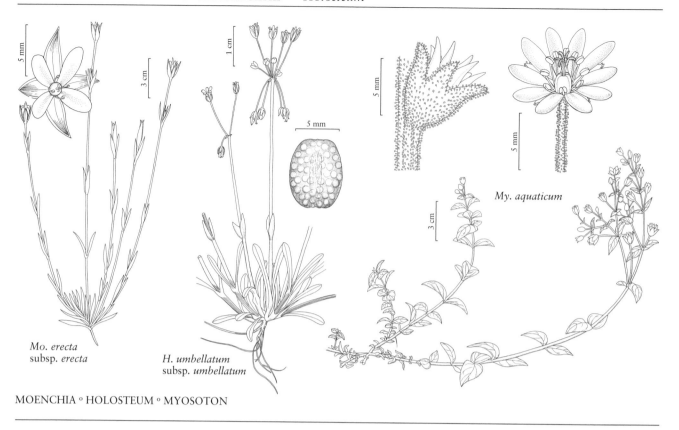

Mo. erecta
subsp. *erecta*

H. umbellatum
subsp. *umbellatum*

My. aquaticum

MOENCHIA ○ HOLOSTEUM ○ MYOSOTON

20. HOLOSTEUM Linnaeus, Sp. Pl. 1: 88. 1753; Gen. Pl. ed. 5, 39. 1754 • Jagged chickweed [Greek *holos*, whole or all, and *osteon*, bone, humorous allusion to frailty of the plant] ⊞

Richard K. Rabeler

Ronald L. Hartman

Herbs, annual or winter-annual. **Taproots** slender. **Stems** ascending to erect, simple or branched, terete. **Leaves** forming basal rosette, connate proximally into sheath, petiolate (proximal leaves) or sessile (cauline leaves); blade 1-veined, oblanceolate to spatulate (proximal leaves) or elliptic to ovate (cauline leaves), somewhat succulent, apex acute. **Inflorescences** terminal, umbellate cymes, sometimes not yet developed in young individuals, then with 2–4 bracts and often buds at base of pedicel; bracts clustered, foliaceous with scarious margins to entirely scarious. **Pedicels** reflexed after flowering, erect in fruit. **Flowers** bisexual or occasionally unisexual and pistillate; perianth and androecium hypogynous; sepals 5, distinct, green, lanceolate to ovate, 2.5–4.5 mm, herbaceous, margins white, scarious, apex acute to obtuse, not hooded; petals 5, white to pink, clawed, blade apex jagged but not 2-fid; nectaries not apparent; stamens 3–5, arising from base of ovary; filaments distinct nearly to base; staminodes absent; styles 3(–5), clavate to filiform, 0.5–1.5 mm, glabrous proximally; stigmas 3(–5), subterminal to linear along adaxial surface of styles, minutely papillate (50×). **Capsules** ovoid to cylindric, opening by 6 (rarely 8 or 10) revolute teeth; carpophore absent. **Seeds** 35–60, orange to brown, oblong, shield-shaped, dorsiventrally compressed, papillate, marginal wing absent, appendage absent; embryo central, straight. $x = 10$.

Species 3–4 (1 in the flora): introduced; Europe (e Mediterranean region), c, sw Asia, Africa (Mediterranean region, s to Ethiopia); introduced in South America (Argentina), w Europe, Africa (Republic of South Africa).

SELECTED REFERENCE Shinners, L. H. 1965. *Holosteum umbellatum* (Caryophyllaceae) in the United States: Population explosion and fractionated suicide. Sida 2: 119–128.

1. Holosteum umbellatum Linnaeus, Sp. Pl. 1: 88. 1753 F I W

Subspecies 5 (1 in the flora): introduced; Europe (e Mediterranean), c, sw Asia, Africa (Mediterranean region); introduced in South America (Argentina), w Europe, Africa (Republic of South Africa).

L. H. Shinners (1965) dem-onstrated that *Holosteum umbellatum* has been introduced in North America on several occasions. Collections from northeastern North America are mainly older ones from very localized populations, the first from near Lancaster, Pennsylvania, in 1856. Reports from the central United States show its occurrence there in several states in the 1940s, spreading rapidly along roadsides, railroads, and other calcareous sites. M. L. Fernald (1943f) suggested that *H. umbellatum* may have been spread as a contaminant in grass seed sown after highway construction in Virginia (see 34.2. *Petrorhagia prolifera* and 34.4. *P. dubia* for similar cases). The first collection from the western United States was made in 1926 and the species has since spread to various disturbed sites in the Pacific Northwest. Several plants in two recent collections from Oregon (e.g., *Joyal 463*, OSC) are infected with an ovary smut (*Microbotryum* sp.), the first evidence of such infection on *Holosteum* in North America known to us.

The early appearance and extremely brief life cycle of *Holosteum umbellatum* probably contribute to its being overlooked. It should be expected elsewhere in our range.

1a. Holosteum umbellatum Linnaeus subsp. **umbellatum** F I

Stems (1–)4–35 cm, glabrous to or often glandular-pubescent, especially proximally, often glaucous. **Leaves:** basal blades oblanceolate, 5–25 mm; cauline leaves in 1–4 pairs, blade 2–30 mm, margins ciliate. **Flowers:** sepals 2.5–4.5 mm; petals 3–5 mm. **Capsules** 4–8 mm. $2n = 20$, 40 (both Europe).

Flowering late winter–early spring. Lawns, roadsides, waste places; 50–1700 m; introduced; B.C., N.B., Ont.; Ala., Ark., Calif., Colo., Conn., Del., D.C., Ga., Idaho, Ill., Ind., Iowa, Kans., Ky., Md., Mass., Mich., Mo., Mont., Nebr., Nev., N.J., N.Y., N.C., Ohio, Okla., Oreg., Pa., R.I., S.C., S.Dak., Tenn., Utah, Va., Wash., W.Va., Wis., Wyo.; Europe (e Mediterranean region); c, sw Asia; Africa (Mediterranean region); introduced in South America (Argentina), w Europe, Africa (Republic of South Africa).

21. MYOSOTON Moench, Methodus, 225. 1794 • Giant or water chickweed, stellaire aquatique [Greek *myos*, mouse, and *otos*, ear, alluding to leaves] I

Richard K. Rabeler

Herbs, perennial. **Taproots** slender, rhizomes slender. **Stems** decumbent to ascending, simple or branched, terete to angular. **Leaves** sessile leaves clasping proximally, petiolate (proximal leaves) or sessile (most cauline leaves); blade 1-veined, ovate to broadly elliptic, not succulent, apex acute. **Inflorescences** axillary or terminal, open cymes; bracts paired, foliaceous, smaller. **Pedicels** ascending to erect, reflexed in fruit. **Flowers** bisexual, occasionally pistillate or reduced and sterile; perianth and androecium weakly perigynous; hypanthium minimal; sepals 5, distinct, green, ovate, 4–6(–9) mm, herbaceous, margins green or white, herbaceous or scarious, apex acute, not hooded; petals 5, white, claw absent, blade apex deeply 2-fid; nectaries at base of filaments opposite sepals; stamens 10, occasionally fewer, arising from nectariferous disc at ovary base; filaments distinct; staminodes sometimes present, linear; styles 5(–6), distinct, filiform, 0.7–1.8 mm, glabrous proximally; stigmas 5(–6), linear along adaxial surface of styles, papillate (30×). **Capsules** ovoid to globose-ovoid, opening by 5(–6), slightly recurved, briefly 2-fid valves; carpophore absent. **Seeds** 50–100, brown-blackish, reniform, laterally compressed, papillate, marginal wing absent, appendage absent. $x = 14$.

Species 1: introduced; Europe, temperate Asia.

The combination of 5(–6) styles and a capsule opening by the same number of briefly 2-fid valves characterizes *Myosoton* and is not known in *Stellaria*. Most chromosome counts of *M. aquaticum* are $2n = 28$, a very rare number in *Stellaria*.

1. **Myosoton aquaticum**   (Linnaeus) Moench, Methodus, 225. 1794   F I W

Cerastium aquaticum Linnaeus, Sp. Pl. 1: 439. 1753; *Alsine aquatica* (Linnaeus) Britton; *Stellaria aquatica* (Linnaeus) Scopoli

Stems 10–100 cm, minutely glandular-pilose distally. **Leaf blades** 2–3.5(–8.5) × 1–2(–4.4) cm. **Pedicels** 1–2(–3) cm, minutely glandular-pilose. **Flowers:** sepals 4–6 mm, to 9 mm in fruit; petals 4–7 mm, mostly exceeding sepals. **Capsules** 5–10 mm, usually slightly exceeding calyx. $2n = 20(?)$ (Asia), 28 (Europe, Asia), 29 (Europe).

Flowering spring–fall. Stream banks, low woods, marshes, meadows, occasionally cultivated areas; 100–700 m; introduced; B.C., Ont., Que.; Conn., Del., Ill., Ind., Iowa, Kans., Ky., Md., Mass., Mich., Minn., Mo., N.H., N.J., N.Y., N.C., Ohio, Pa., Tenn., Vt., Va., Wash., W.Va., Wis.; Europe; temperate Asia.

Reports of *Myosoton aquaticum* from Louisiana appear to be based on misidentified specimens of *Stellaria cuspidata* Willdenow ex Schlechtendal subsp. *prostrata* (Baldwin) J. K. Morton.

Although occurring over a wide area, *Myosoton aquaticum* is often noted as rare or occasional in particular states or provinces. Very few collections of this species from the flora area were made prior to 1900; two of the first gatherings were from port areas (Baltimore, Maryland, and as a ballast plant in Philadelphia, Pennsylvania, in 1877). Its presence outside the Japanese Pavillion at the Philadelphia Centennial Grounds in 1878 (*Scribner 50* and *51*, MO) suggests an escape from an intentional introduction.

22. **STELLARIA**   Linnaeus, Sp. Pl. 1: 421. 1753; Gen. Pl. ed. 5, 193. 1754 • Chickweed, stitchwort, starwort, stellaire [Latin *stella*, star, and *-aria*, pertaining to, alluding to shape of flower]

John K. Morton

Alsine Linnaeus

Plants annual, winter annual, or perennial. **Taproots** usually slender, perennial taxa often rhizomatous, rooting at nodes. **Stems** prostrate to ascending or erect, simple or branched, terete or 4-angled. **Leaves** sometimes connate basally into sheath, often sessile; blade 1-veined, linear or lanceolate to ovate or deltate, succulent (*S. crassifolia* [gemmae], *S. fontinalis*, *S. humifusa*, and *S. irrigua*) or not, apex acute or obtuse. **Inflorescences** terminal, open cymes, rarely axillary (*S. alsine*, *S. americana*) or umbellate (*S. umbellata*), or terminal or axillary solitary flowers; bracts paired (1 in *S. dicranoides*), foliaceous, scarious and reduced, or absent. **Pedicels** erect, sometimes reflexed in fruit, glabrous or pubescent, not glandular. **Flowers** usually bisexual (*S. dicranoides* unisexual); perianth and androecium hypogynous or weakly perigynous; hypanthium cup- or disc-shaped; sepals (4–)5, distinct, green, occasionally purple tinged (*S. irrigua*) or red proximally (*S. pallida*), lanceolate to ovate-triangular, 2–12 mm, herbaceous (rarely coriaceous), margins often white, scarious, apex acute, acuminate, or obtuse, not hooded; petals (1–)5 or absent, white (sometimes translucent in *S. borealis*), not clawed, blade apex 2-fid usually for ²/₃–⁴/₅ its length (*S. holostea* occasionally laciniate); nectaries at base of filaments opposite sepals usually present, disc sometimes prominent; stamens (1–)5 or 10 or absent, arising from nectariferous disc (prominent in *S. dicranoides* and *S. irrigua*) at ovary base; filaments distinct; staminodes absent; styles [2–]3(–5), capitate to clavate, 0.2–7 mm, glabrous proximally; stigmas [2–]3(–5), terminal or subterminal, papillate (30×). **Capsules** globose to conic, opening by 3 or 6, occasionally 4, 8, or 10 ascending to recurved valves; carpophore present or absent. **Seeds** (1–)3–20+, yellow-brown to dark brown, globose to ellipsoid, laterally compressed, rarely shiny, papillate or rugose, rarely smooth, marginal wing absent, appendage absent. $x = 10, 11, 12, 13, 15$.

Species ca. 120 (29 in the flora): worldwide, mainly north-temperate regions.

I have not attempted to present an infrageneric classification for *Stellaria*. Although several species complexes can be identified within the genus (e.g., *S. media*, *S. longipes*, *S. calycantha*), there are significant problems with accepting the most recent scheme (F. Pax and K. Hoffmann 1934c). *Stellaria jamesiana* is now placed in *Pseudostellaria* and *S. aquatica* in *Myosoton*. As well, *S. dicranoides* and *S. fontinalis* probably should be placed in other genera (see comments under those species). Additional study is warranted to determine both the generic circumscription of *Stellaria* and species relationships within the genus.

SELECTED REFERENCES Chinnappa, C. C. 1985. Studies in the *Stellaria longipes* complex (Caryophyllaceae): Interspecific hybridization. I. Triploid meiosis. Canad. J. Genet. Cytol. 27: 318–321. Chinnappa, C. C. and J. K. Morton. 1976. Studies on the *Stellaria longipes* Goldie complex—Variation in wild populations. Rhodora 78: 488–501. Chinnappa, C. C. and J. K. Morton. 1984. Studies on the *Stellaria longipes* Goldie complex (Caryophyllaceae): Biosystematics. Syst. Bot. 9: 60–73. Chinnappa, C. C. and J. K. Morton. 1991. Studies on the *Stellaria longipes* complex (Caryophyllaceae): Taxonomy. Rhodora 93: 129–135. Emery, R. J. N. and C. C. Chinnappa. 1994. Morphological variation among members of the *Stellaria longipes* complex: *S. longipes*, *S. longifolia*, and *S. porsildii* (Caryophyllaceae). Pl. Syst. Evol. 190: 69–78. Hultén, E. 1943. *Stellaria longipes* and its allies. Bot. Not. 1943: 251–270. Morton, J. K. and R. K. Rabeler. 1989. Biosystematic studies on the *Stellaria calycantha* (Caryophyllaceae) complex. I. Cytology and cytogeography. Canad. J. Bot. 67: 121–127. Porsild, A. E. 1963. *Stellaria longipes* Goldie and its allies in North America. Bull. Natl. Mus. Canada 186: 1–35. Rabeler, R. K. 1986. Revision of the *Stellaria calycantha* (Caryophyllaceae) Complex and Taxonomic Notes on the Genus. Ph.D. dissertation. Michigan State University.

1. Mid-stem and proximal leaves distinctly petiolate, broadly lanceolate to ovate or deltate.
  2. Leaf blade bases cordate to truncate (rarely abruptly rounded) . . . . 9. *Stellaria cuspidata*
  2. Leaf blade bases round or cuneate.
    3. Flowers large, usually 10 mm or more diam.; petals usually exceeding sepals.
      4. Sepals obtuse to acute, 3.5–6 mm . . . . 27. *Stellaria pubera* (in part)
      4. Sepals acuminate, (5–)7–10(–12) mm . . . . 6. *Stellaria corei*
    3. Petals equaling or shorter than sepals, or absent; flowers less than 10 mm diam.
      5. Flowers solitary in axils of mid and distal stem leaves . . . . 22. *Stellaria obtusa* (in part)
      5. Flowers several in foliaceous terminal cymes.
        6. Sepals 5–6.5 mm; stamens 8–10; seeds 1.1–1.7 mm; tubercles taller than broad, apex acute . . . . 20. *Stellaria neglecta*
        6. Sepals usually 3–5(–6) mm; stamens 1–5(–8); seeds 0.4–1.3 mm; tubercles usually broader than tall, apex usually obtuse.
          7. Stamens 3–5(–8); sepals 4.5–5(–6) mm; seeds 0.9–1.3 mm diam.; petals usually present; plants usually green . . . . 19. *Stellaria media*
          7. Stamens 1–3 or absent; sepals 3–4 mm; seeds mostly 0.5–0.9 mm diam.; petals usually absent; plants usually yellowish green . . . . 23. *Stellaria pallida*
1. Mid-stem and proximal leaves sessile to shortly petiolate; blades elliptic, ovate, lanceolate, linear, or linear-lanceolate.
  8. Inflorescences, or flowers when solitary, in axils of foliage leaves on mid or distal stem.
    9. Petals equaling or longer than sepals.

10. Sepals narrowly lanceolate-triangular, prominently 3-veined, margins straight
. . . . 7. *Stellaria crassifolia* (in part)
10. Sepals lanceolate to ovate-lanceolate, 1- or obscurely 3-veined, margins convex.
    11. Seeds prominently papillate, papillae taller than broad; sepals pubescent
. . . . 25. *Stellaria parva*
    11. Seeds smooth, rugose, or shallowly tuberculate, not papillate; sepals glabrous.
        12. Leaf blades succulent, elliptic to elliptic-lanceolate, 4–15 × 1–5 mm
. . . . 14. *Stellaria humifusa*
        12. Leaf blades not succulent, linear to linear-lanceolate, 27–35 × 2–3 mm . . . . 26. *Stellaria porsildii*
9. Petals shorter than sepals, or absent.
    13. Plants forming dense cushions, mid and distal stem internodes shorter than leaves.
        14. Stamens 5; petals present, deeply divided into 2 narrowly elliptic lobes
. . . . 15. *Stellaria irrigua*
        14. Stamens 10; petals absent . . . . 10. *Stellaria dicranoides*
    13. Plants creeping to straggling or ascending, sometimes forming mats, but not forming cushions, internodes equaling or longer than leaves.
        15. Leaf blades narrowly elliptic to lanceolate, oblanceolate, or linear.
            16. Flowers solitary in distal leaf axils . . . . 11. *Stellaria fontinalis*
            16. Flowers in axillary inflorescences in mid and distal leaf axils . . . . 2. *Stellaria alsine*
        15. Leaf blades broadly elliptic to ovate.
            17. Sepals obtuse at apex with ± obscure veins; styles shorter than 0.5 mm, curled; capsules globose to broadly ovoid . . . . 22. *Stellaria obtusa* (in part)
            17. Sepals acute to acuminate at apex with 3 prominent veins; styles ca. 1 mm, spreading to ascending; capsules ovoid to ovoid-ellipsoid . . . . . 8. *Stellaria crispa*
[8. Shifted to left margin.—Ed.]
8. Inflorescences with most flowers terminal, either several in bracteate inflorescence or solitary on long-ascending pedicels.
18. Bracts scarious or with scarious margins.
    19. Capsules ca. equaling or shorter than sepals.
        20. Plants annual . . . . 21. *Stellaria nitens*
        20. Plants perennial.
            21. Plants compact; stems ascending, 3–10(–20) cm; petals equaling or shorter than sepals . . . . 1. *Stellaria alaskana*
            21. Plants coarse; stems straggling with erect branches, 20–60 cm; petals 1.5–2 times as long as sepals . . . . 24. *Stellaria palustris*
18. Capsules longer than sepals.
    22. Inflorescences subumbellate; petals absent . . . . 29. *Stellaria umbellata*
    22. Inflorescences cymose or flowers solitary; petals present.
        23. Plants delicate, creeping, often forming mats; flowers solitary and axillary or in small, few-flowered, leafy cymes; leaf blades variable in shape, midrib obscure . . . . 7. *Stellaria crassifolia* (in part)
        23. Plants not with the above combination of characters; leaf blades with prominent midrib.
            24. Leaf blades linear-elliptic, broadest at middle or distally; angles of stems and/or margins of leaf blades minutely papillate-scabrid (30×) . . . . 17. *Stellaria longifolia*
            24. Leaf blades lanceolate to linear-lanceolate, broadest proximally; angles of stems and/or margins of leaf blades not papillate-scabrid (soft hairs of cilia may be present).
                25. Sepals narrowly triangular or lanceolate, margins straight, veins 3, forming prominent ridges; seeds coarsely rugose in concentric rings; plants coarse; stems ascending, straggling; inflorescences much-branched, many-flowered . . . . 12. *Stellaria graminea*

25. Sepals lanceolate to ovate-lanceolate, margins convex, veins 1–3, not forming ridges; seeds shallowly tuberculate to smooth; stems erect to straggling; flowers solitary or inflorescences few-flowered .... 18. *Stellaria longipes* (in part)

[18. Shifted to left margin.—Ed.]

18. Bracts or subtending leaf blades without scarious margins.

    26. Flowers 20–30 mm diam.; petals 8–14 mm, longer than sepals; leaf blades narrowly lanceolate, 4–8 cm, apex acuminate .... 13. *Stellaria holostea*

    26. Plants not having the above combination of characters.

       27. Leaves widest at or above middle.

          28. Plants glandular-puberulent; pedicels arcuate, pushing capsule into substrate; seeds 3–6, ca. 2.5 mm diam. .... 3. *Stellaria americana*

          28. Plants eglandular; other characters not as above.

             29. Sepals narrowly lanceolate-triangular; plants creeping, forming mats usually shorter than 5 cm .... 7. *Stellaria crassifolia* (in part)

             29. Sepals ovate to ovate-triangular; plants with diffusely branched or ascending flowering stems taller than 5 cm.

                30. Petals 4–8 mm, conspicuous, white; sepals 3.5–6 mm .... 27. *Stellaria pubera* (in part)

                30. Petals shorter than 3 mm, inconspicuous, white or translucent; sepals 2–3(–4) mm. .... 4. *Stellaria borealis* (in part)

       27. Leaves widest towards base.

          31. Plants straggling to scandent; stems to 60 cm; leaf blades ovate to ovate-lanceolate, 10–45 × 4–20 mm; inflorescences terminal leafy cymes .... 16. *Stellaria littoralis*

          31. Plants not having the above combination of characters.

             32. Sepals 2–3(–4) mm; open flowers less than 5 mm diam.; petals not exceeding sepals, inconspicuous or absent.

                33. Capsules broadly ovoid to globose, to 1.5 times as long as broad; styles less than 1 mm; sepals in open flowers less than 2.5 mm, veins obscure (rarely 1) .... 5. *Stellaria calycantha*

                33. Capsules ovoid, more than 1.5 times as long as broad; styles longer than 1 mm; sepals in open flowers more than 2.5 mm with 1–3 prominent veins .... 4. *Stellaria borealis* (in part)

             32. Sepals 3.5 mm or longer; open flowers 5 mm or more diam.; petals equaling or exceeding sepals, conspicuous.

                34. Sepals 4.5–6 mm; leaf blades ovate to broadly lanceolate, coriaceous, apex spinous; capsules equaling sepals .... 28. *Stellaria ruscifolia*

                34. Sepals 3.5–5 mm; leaf blades lanceolate to linear-lanceolate, not strongly coriaceous, apex acute to acuminate, not spinous; capsules 1.5–2 times as long as sepals .... 18. *Stellaria longipes* (in part)

1. **Stellaria alaskana** Hultén, Bot. Not. 1943: 264, fig. 5e, f. 1943 • Alaska starwort [E]

Plants perennial, forming compact clumps, from elongate rhizomes. **Stems** ascending, branched at base, square, 3–10(–20) cm, glabrous. **Leaves** clustered near base of each shoot, sessile; blade green, rarely glaucous, lanceolate (rarely narrowly so) to elliptic- or ovate-lanceolate, 0.8–2 cm × 1–7 mm, coriaceous, base round to cuneate, margins entire, apex acute to acuminate, glabrous. **Inflorescences** terminal, flowers usually solitary, rarely 2–3 on elongate pedicels; bracts narrowly lanceolate, 2–8 mm, scarious. **Pedicels** erect, 1–50 mm, glabrous. **Flowers** 10–20 mm diam.; sepals 5, 1–3-veined, narrowly lanceolate, triangular, (6.5–)7–10 mm, margins narrow, scarious, apex acuminate, glabrous; petals 5, equaling or shorter than sepals; stamens 10; styles 3, ascending. **Capsules** green to straw colored, narrowly conic, 6–8 mm, equaling sepals, opening by 6 valves; carpophore absent. **Seeds** light brown, broadly reniform, 0.8–1.2 mm diam., rugose.

Flowering summer. Rocky outcrops, talus slopes, gravelly moraines, marshy grasslands; 0–2300 m; Yukon; Alaska.

Stellaria alaskana is closely related to *S. longipes*; it differs in its exceptionally long, narrow, prominently veined sepals and larger flowers in which the petals are usually shorter than the sepals. Some specimens appear to intergrade with *S. longipes*. The single lowland record, from the Alaska Peninsula, is from one such intermediate population. Although it has the characteristic sepals of *S. alaskana*, it is a straggling plant with elongate stems and narrow, linear-lanceolate leaves.

2. Stellaria alsine Grimm, Nova Acta Phys.-Med. Acad. Caes. Leop.-Carol. Nat. Cur. 3(app.): 313. 1767

• Bog stitchwort or starwort, fausse alpine

Alsine uliginosa (Murray) Britton; *Stellaria uliginosa* Murray

Plants perennial, creeping, rhizomatous. **Stems** decumbent and ascending, branched, smoothly 4-angled, 10–40 cm, glabrous. **Leaves** sessile; blade narrowly elliptic, elliptic-lanceolate, or oblanceolate, 0.5–2(–3) cm × 2–10(–13) mm, base cuneate, margins thin with reticulate venation, entire, apex acute, glabrous, slightly ciliate basally. **Inflorescences** axillary with 1–5-flowered cymes in mid and distal axils of foliage leaves; bracts lanceolate, ca. 1 mm, scarious with green midrib. **Pedicels** 5–30 mm, glabrous. **Flowers** ca. 6 mm diam.; sepals 5, 3-veined, lanceolate-triangular, 2.5–3.5 mm, margins scarious, apex acute, glabrous; petals 5, 1.5–3 mm, shorter than (rarely equaling) sepals, blade apex 2-fid almost to base, with widely divergent lobes; stamens 10; styles 3, ca. 1 mm. **Capsules** green, ovoid, 2.5–3.5 mm, equaling sepals, apex broadly acute; carpophore absent. **Seeds** pale reddish brown, ± reniform, 0.3–0.4 mm diam., with small tubercles. $2n = 24$.

Flowering spring–early summer. Streamsides, flushes, wet tracks, ditches; 0–300 m; St. Pierre and Miquelon; B.C., N.B., Nfld. and Labr. (Nfld.), N.S., P.E.I., Que.; Del., D.C., Ga., La., Maine, Md., Mass., Minn., N.H., N.J., N.Y., N.C., Ohio, Pa., Tenn., Vt., Wash., W.Va.; Europe; introduced in South America (Chile).

Stellaria alsine is presumed to be native in eastern North America but has been introduced elsewhere in North America and Chile.

3. Stellaria americana (Porter ex B. L. Robinson) Standley, Contr. U. S. Natl. Herb. 22: 336. 1921

• American starwort [E]

Stellaria dichotoma Linnaeus var. *americana* Porter ex B. L. Robinson, Proc. Amer. Acad. Arts 29: 289. 1894; *Alsine americana* (Porter ex B. L. Robinson) Rydberg; *Arenaria stephaniana* (Willdenow ex Schlechtendal) Shinners var. *americana* (Porter ex B. L. Robinson) Shinners

Plants perennial, forming loose, prostrate mats, from rhizomatous rootstocks. **Stems** spreading, branched, very leafy, 4-angled, 10–20 cm, short glandular-puberulent on internodes. **Leaves** sessile; blade ovate to ovate-lanceolate, widest at or above middle, 8–30 × 2–13 mm, base round to cuneate, margins not scarious, apex usually obtuse, viscid. **Inflorescences** terminal, 1–5-flowered, very leafy cymes; bracts foliaceous. **Pedicels** ca. 10 mm in flower, elongating, recurved, and tortuous in fruit, glandular-pubescent, pushing capsule into substrate. **Flowers** 5–10 mm; sepals 5, obscurely veined, ovate-obtuse, 3–5 mm, margins narrow, scarious, glandular-pubescent; petals 5, 4–6 mm; stamens 5; styles 3, ascending, equaling petals. **Capsules** green to straw colored, broadly ovoid to globose, 5–6 mm, apex obtuse, tardily dehiscent with 3 valves; carpophore absent. **Seeds** 3–6, rusty brown, ± ovate, ca. 2.5 mm diam., finely tuberculate.

Flowering late Jul–Aug. Rocky slopes, talus; 1400–2800 m; Alta.; Mont.

Stellaria americana is remarkable for its tortuous fruiting pedicels, which push the opening capsule with its small number of large seeds into the substrate.

4. Stellaria borealis Bigelow, Fl. Boston. ed. 2, 182. 1824 • Boreal starwort [F]

Alsine borealis (Bigelow) Britton

Plants perennial, often matted, rhizomatous. **Stems** prostrate to ascending or erect, usually diffusely branched, sharply 4-angled, (5–)25–50 cm, glabrous to finely papillate, rarely pubescent. **Leaves** sessile; blade linear-lanceolate to ovate-lanceolate, rarely elliptic-lanceolate, 1–6 cm × 2–8 mm, base cuneate, margins eciliate or scabrid, sometimes ciliate towards base, apex acute. **Inflorescences** with flowers solitary, terminal and axillary, or terminal, often copious, very lax, leafy cymes; bracts foliaceous, lanceolate, reduced distally to ca. 2 mm, ± scarious. **Pedicels** erect or patent, usually reflexed at maturity, 10–40 mm, glabrous.

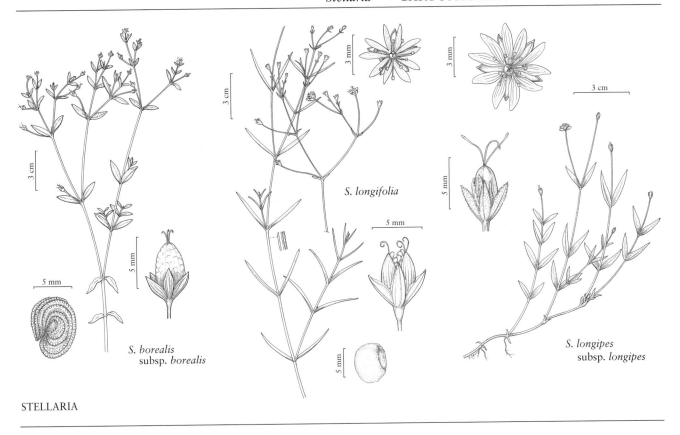

S. longifolia

S. borealis
subsp. *borealis*

S. longipes
subsp. *longipes*

STELLARIA

Flowers 3–5 mm; sepals 5, 1–3-veined, lanceolate to ovate, 2–5 mm, margins scarious, apex acute, glabrous; petals 5, rarely absent, white or translucent, 1–3 mm, usually shorter than sepals; stamens 5; styles 3, erect to spreading, 0.9–2 mm. **Capsules** greenish brown or straw colored, ovoid, 3–7 mm, more than 1–1.5 times as long as broad, exceeding sepals, apex acute, opening by 3 valves; carpophore very short or absent. **Seeds** 10–20, brown, obovate, 0.7–0.9 mm on longest axis, smooth or slightly rugose. $2n = 52$.

Subspecies 2 (2 in the flora): circumboreal.

Plants infected with an anther smut, *Microbotyrum stellariae* (Sowerby) G. Deml & Oberwinkler [*Ustilago violacea* (Persoon) Roussel, in the broad sense], exhibit flowers with enlarged, reddish anthers. This condition is known in both subspecies, especially in northern areas of the range, but is as yet unknown in *Stellaria calycantha*, a species previously united with *S. borealis* by some authors.

SELECTED REFERENCE Rabeler, R. K. 1993. The occurrence of anther smut, *Ustilago violacea* s.l., on *Stellaria borealis* (Caryophyllaceae) in North America. Contr. Univ. Michigan Herb. 19: 165–169.

1. Sepals ovate to ovate-triangular, with midvein extending to near apex, lateral veins visible only at base; capsules dark brown, opaque, less than 2 times as long as wide, usually 3–4.5(–5) mm; leaf blades elliptic-lanceolate to narrowly elliptic or linear-lanceolate, widest just below middle, usually 2–3 cm; circumboreal . . . . 4a. *Stellaria borealis* subsp. *borealis*
1. Sepals narrowly triangular, with 3 prominent, ridged veins extending to near apex; capsule straw colored, translucent, 2 times as long as wide, usually 5–7 mm; leaf blades narrowly lanceolate, widest near base, usually 3–6 cm; w North America . . . . 4b. *Stellaria borealis* subsp. *sitchana*

4a. **Stellaria borealis** Bigelow subsp. **borealis**

• Boreal starwort, stellaire boréale ⚐F⚐

Spergulastrum lanceolatum Michaux; *Stellaria borealis* var. *floribunda* Fernald; *S. borealis* var. *isophylla* Fernald; *S. calycantha* (Ledebour) Bongard var. *floribunda* (Fernald) Fernald; *S. calycantha* subsp. *interior* Hultén; *S. calycantha* var. *isophylla* (Fernald) Fernald; *S. calycantha* var. *latifolia* B. Boivin; *S. calycantha* var. *laurentiana* Fernald

Plants straggling to bushy and erect. **Stems** glabrous. **Leaf blades** elliptic-lanceolate (rarely elliptic) to narrowly elliptic or linear-lanceolate, usually 2–3 cm, rarely longer, widest just below (rarely at) middle. **Pedicels** not reflexed in fruit (except in small, erect plants from nw North America). **Flowers** usually 5–6 mm diam.; sepals with midvein extending to near apex, lateral veins visible only at base, ovate to ovate-triangular, 2–3.3, to 0.5 times as long as capsule. **Capsules** dark brown, opaque, 3–4.5 mm, somewhat less than 2 times as long as broad. **Seeds** smooth or indistinctly rugulose. $2n = 52$.

Flowering spring–summer. Open marshy woodlands and grasslands, river and roadside gravel, among boulders on talus slopes, edges of moist lowland and floodplain forests; 0–3500 m; Greenland; St. Pierre and Miquelon; Alta., B.C., Man., N.B., Nfld. and Labr., N.W.T., N.S., Nunavut, Ont., P.E.I., Que., Sask., Yukon; Alaska, Calif., Colo., Conn., Idaho, Maine, Mass., Mich., Minn., Mont., Nev., N.H., N.J., N.Y., Oreg., Pa., R.I., Utah, Vt., Wash., W.Va., Wis., Wyo.; Europe; Asia.

The circumboreal subsp. *borealis* is highly variable in habit, leaf shape, and inflorescence development. This variation is partly genetic but is also greatly influenced by the environment. The broad-leaved form named *Stellaria calycantha* var. *latifolia* is of occasional occurrence. A similar broad-leaved form occurs in subsp. *sitchana*. In both of these variants, the leaf blades are elliptic-lanceolate to elliptic, with a length-to-width ratio of 2.5–3 : 1. These variants retain their characters under cultivation. However, the occurrence of the same variational trend in both subspecies makes taxonomic recognition of this variation problematic.

Subspecies *borealis* hybridizes fairly frequently with *S. longifolia* to produce a sterile triploid ($2n = 39$). These hybrids are intermediate between the parents but have copious inflorescences with widely divaricate branches and numerous small, completely sterile flowers. Similar hybrids can be readily produced experimentally (C. C. Chinnappa 1985). European and North American workers (most recently B. Jonsell and T. Karlsson 2000+, vol. 2) often have called these hybrids *S. alpestris* Fries, but that name has caused a great deal of confusion and R. K. Rabeler (1986) concluded that it more correctly applies to subsp. *borealis*.

4b. Stellaria borealis Bigelow subsp. **sitchana** (Steudel) Piper & Beattie, Fl. N.W. Coast, 147. 1915 · Sitka starwort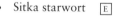

Stellaria sitchana Steudel, Nomencl. Bot. ed. 2, 2: 637. 1841, based on *S. brachypetala* Bongard, Mém. Acad. Imp. Sci. St.-Petersburg, Sér. 6, Sci. Math. 2: 126. 1832 (as brachipetala), not Bunge 1830; *Alsine bongardiana* (Fernald) Davidson & Moxley; *Stellaria borealis* subsp. *bongardiana* (Fernald) Piper & Beattie; *S. borealis* var. *bongardiana* (Fernald; *S. borealis* var. *sitchana* (Steudel) Fernald; *S. calycantha* (Ledebour) Bongard var. *bongardiana* (Fernald) Fernald; *S. calycantha* var. *sitchana* (Steudel) Fernald; *S. sitchana* var. *bongardiana* (Fernald) Hultén

Plants coarse, straggling. **Stems** usually finely papillate. **Leaf blades** narrowly lanceolate, usually 3–6 cm, widest near base. **Pedicels** usually reflexed in fruit. **Flowers** ca. 10 mm diam.; sepals with 3 prominent, ridged veins extending to near apex, narrowly triangular, 3.5–5 mm, longer than 0.5 times capsule length. **Capsules** straw colored, translucent, usually (3.6–)5–7 mm, 2 times as long as broad. **Seeds** usually rugulose. $2n = 52$.

Flowering May–Sep. River and stream gravel and banks, ditches, damp forests, forest openings; 0–2800 m; B.C.; Alaska, Calif., Idaho, Mont., Oreg., S.Dak., Utah, Wash.

Subspecies *sitchana* is sturdier than subsp. *borealis* and is readily distinguished by its leaf blades, which are narrowly lanceolate and widest at the base, and by its narrowly triangular, 3-veined sepals. It is a western taxon associated mainly with the slopes of the Coast Ranges and the Rocky Mountains. On the eastern side of its range and in the Aleutian Islands it tends to intergrade with subsp. *borealis*. In central California, a rare form has broad, elliptic leaves (length-to-width ratio 2.5–3 : 1) to 32 × 13 mm. It retains its characters in cultivation.

5. Stellaria calycantha (Ledebour) Bongard, Mém. Acad. Imp. Sci. St.-Pétersbourg, Sér. 6, Sci. Math. 2: 127. 1833 · Northern starwort

Arenaria calycantha Ledebour, Mém. Acad. Imp. Sci. St. Pétersbourg Hist. Acad. 5: 534. 1815; *Alsine calycantha* (Ledebour) Rydberg; *A. simcoei* Howell; *Stellaria borealis* Bigelow var. *simcoei* (Howell) Fernald; *S. calycantha* var. *simcoei* (Howell) Fernald; *S. simcoei* (Howell) C. L. Hitchcock

Plants perennial, forming clumps, from slender rhizomes. **Stems** erect or trailing, branched, square, weak, to 25 (–35) cm, glabrous or pilose, never papillate. **Leaves** sessile or subsessile; blade ovate to elliptic, rarely oblong, widest proximal to middle, 5–25 mm, thin, base round, margins entire, apex acute, glabrous or rarely ciliate. **Inflorescences** terminal, 1–5-flowered cymes; bracts foliaceous. **Pedicels** ascending, not reflexed, 5–25(–50) mm, glabrous. **Flowers** 3–5 mm; sepals 5, obscurely veined, ovate, 2–2.5 mm (rarely 3.5 in fruit), margins broad, scarious, apex broadly acute, glabrous; petals absent or 1–5, 1–1.5 mm, shorter than sepals, blade apex deeply lobed; stamens 5; styles 3, usually curved, 0.4–0.9 mm. **Capsules** green, semitransparent, globose to broadly ovoid, 3–5 mm, to 1.5 times as long as broad, longer than sepals, apex obtuse, opening by 3 valves; carpophore absent. **Seeds** brown, ovate, 0.5–0.9 mm, smooth or shallowly tuberculate. $2n = 26$.

Flowering summer. Alpine slopes, meadows, shores of lakes and streams; 0–3700 m; Alta., B.C., N.W.T., Yukon; Alaska, Calif., Colo., Idaho, Mont., Nev., N.Mex., Oreg., Utah, Wash., Wyo.; Asia (Russian Far East).

6. Stellaria corei Shinners, Sida 1: 103. 1962 • Tennessee chickweed E

Stellaria pubera Michaux subsp. *silvatica* Béguinot, Nuovo Giorn. Bot. Ital., n. s. 17: 385. 1910, not *S. silvatica* Jessen 1879; *Alsine tennesseensis* (C. Mohr) Small; *Stellaria pubera* var. *silvatica* (Béguinot) Weatherby; *S. silvatica* (Béguinot) Maguire; *S. tennesseensis* (C. Mohr) Strausbaugh & Core

Plants perennial, rhizomatous. **Stems** erect, branched, square, 10–40 cm, with alternating lines of soft, spreading, flexuous, mainly eglandular hairs. **Leaves** petiolate (proximal) or subsessile (distal); blade elliptic, broadly lanceolate to ovate, 1–5 cm × 5–16 mm, base cuneate, margins entire, apex acute, glabrous, sparsely pubescent adaxially on midrib. **Inflorescences** terminal, 3–7-flowered, cymes dichotomously branched; bracts foliaceous, lanceolate, 5–30 mm, soft, margins entire, distal ones ciliate on margins and adaxial vein. **Pedicels** erect, 5–45 mm, softly pubescent. **Flowers** 10–16 mm diam.; sepals 5, obscurely veined, narrowly triangular, (5–)7–10(–12) mm, margins narrow, membranous, apex acuminate, glabrous or with shortly ciliate margins; petals 5, equaling to slightly shorter than sepals; stamens 10; styles 3, ascending, ca. 2.5 mm. **Capsules** straw colored to pale brown, broadly ovoid, ca. 5 mm, shorter than sepals, apex obtuse, opening by 3 valves; carpophore absent. **Seeds** brown, broadly reniform, ca. 2 mm diam., coarsely sulcate-papillate. $2n = 60$.

Flowering spring. Rocky woods; 300–1000 m; Ala., Conn., Ind., Ky., Miss., N.C., Ohio, Pa., Tenn., Va., W.Va.

Stellaria corei has been introduced in Connecticut. It is very similar to *S. pubera* but differs in its long-acuminate sepals.

7. Stellaria crassifolia Ehrhart, Hannover. Mag. 8: 116. 1784 • Thick-leaved starwort

Stellaria crassifolia var. *eriocalycina* Schischkin; *S. crassifolia* var. *linearis* Fenzl; *S. gracilis* Richardson

Plants perennial, delicate, forming small to large tangled mats or straggling through grass, from slender rhizomes. **Stems** diffusely branched, 4-angled, 3–30 cm, glabrous. **Leaves** sessile or subsessile; blade with midrib obscure, broadly elliptic-lanceolate to linear-lanceolate, widest at or above middle, 0.2–0.8 (–1.5) cm × to 2 mm, ± succulent, base cuneate, margins entire, apex acute to acuminate, glabrous; leaf blades in terminal buds sometimes become fleshy and form gemmae. **Inflorescences** with flowers usually solitary, terminal and in axils of distal leaves forming open, diffuse cymes; bracts foliaceous, 1–10 mm; 1 or 2 pairs of bracteoles sometimes present, 1–3 mm, herbaceous or with narrow membranous margins. **Pedicels** erect or sharply angled at base, becoming sharply curved at apex, 3–40 mm, glabrous. **Flowers** 5–8 mm; sepals 5, 3-veined, narrowly triangular-lanceolate, 3–3.5(–4) mm, margins straight, narrow, scarious, apex acute, glabrous or rarely margins pubescent; petals 5, 2.5–5 mm, equaling to slightly longer than sepals; stamens 5 or 10; styles 3, ascending, curved at tip, ca. 2 mm. **Capsules** straw colored, conic to ellipsoid, 4–5 mm, longer than sepals, apex obtuse, opening by 6 valves; carpophore absent. **Seeds** reddish brown, reniform to round, 0.7–1 mm diam., rugose. $2n = 26$.

Flowering early summer. Marshes, streams, cold, wet, grassy places; 0–3000 m; Alta., B.C., Man., N.B., Nfld. and Labr., N.W.T., N.S., Nunavut, Ont., P.E.I., Que., Sask., Yukon; Alaska, Colo., Ill., Mich., Minn., N.Dak., S.Dak., Utah, Wyo.; Europe; Asia.

The sterile shoots of *Stellaria crassifolia* (described as forma *gemmificans* Norman) form fleshy terminal buds under suitable conditions of temperature and day length. These propagules survive under the snow and are readily dispersed in the spring runoff.

Leaf shape and size vary considerably. Leaves tend to be smaller and wider in exposed habitats, and longer and narrower in sheltered, more favorable habitats. Plants with pubescent margins to the sepals are referable to var. *eriocalycina* Schischk.

Stellaria crassifolia is often confused with *S. humifusa*, but the former is a much more slender, delicate species with long pedicels that are sharply angled below the capsule.

8. Stellaria crispa Chamisso & Schlechtendal, Linnaea 1: 51. 1826 • Crisp starwort E

Alsine crispa (Chamisso & Schlechtendal) Holzinger; *Stellaria borealis* Bigelow var. *crispa* (Chamisso & Schlechtendal) Fenzl ex Torrey & A. Gray

Plants perennial, forming small to large mats, from slender rhizomes. **Stems** trailing to ascending, branched, 4-angled, 10–60 cm, glabrous. **Leaves** subsessile; blade broadly elliptic to ovate, 0.4–2.6 cm × 2–15 mm, base round to cuneate, margins entire, apex acuminate, glabrous or with a few scattered cilia. **Inflorescences** with flowers solitary in leaf axils; bracts absent. **Pedicels** ascending, straight, mostly 5–30 mm, glabrous. **Flowers** 4–5 mm; sepals 5, prominently 3-veined, lanceolate, 2–4 mm, margins broadly scarious, apex acute to acuminate, glabrous; petals usually absent, rarely 1–5 and much shorter than sepals; stamens 10 or fewer; styles 3, spreading to ascending, curved but not curled, ca. 1 mm. **Capsules** straw colored or brownish, ovoid to ovoid-ellipsoid, 3.5–6 mm, equaling or slightly exceeding sepals, apex broadly acute, opening by 6 valves; carpophore absent. **Seeds** brown, broadly elliptic, 0.7–1 mm (longest axis), distinctly rugose. $2n = 26, 52$.

Flowering summer. Wet soil in woods, shaded streambanks and shores; 0–2300 m; Alta., B.C.; Alaska, Calif., Idaho, Mont., Oreg., Wash.

9. Stellaria cuspidata Willdenow ex Schlechtendal, Ges. Naturf. Freunde Berlin Mag. Neuesten Entdeck. Gesammten Naturk. 7: 196. 1816 • Mexican chickweed

Alsine cuspidata (Willdenow ex Schlechtendal) Wooton & Standley

Plants annual; taproot slender. **Stems** decumbent, much-branched, 4-sided, 15–70 cm, softly glandular-pubescent. **Leaves** petiolate (proximal) or sessile (distal), flaccid; blade ovate to deltate, 1–4.5 cm × 6–28 mm, base cordate, truncate, or rarely abruptly rounded, margins entire, apex acuminate, glabrous, rarely ciliate on margins. **Inflorescences** terminal, (3–)5–35-flowered cymes; bracts sessile, foliaceous, lanceolate to ovate, 3–30 mm, distally reduced. **Pedicels** ascending to spreading, sometimes deflexed in fruit, slender, 5–20(–30) mm, softly glandular. **Flowers** 3–8 mm diam.; sepals 5, with prominent midrib, lanceolate to ovate-lanceolate, 4–5 mm, to 8 mm in fruit, margins narrow, scarious, apex acuminate, blunt, pubescent on midrib, ± ciliate on margins; petals 4–5, 2–8 mm, shorter than to 2 times as long as sepals, blade apex deeply emarginate with 2 narrow lobes; stamens 3–8; styles 3, ascending, outwardly curved, 1.5–3 mm. **Capsules** green, transparent, ovoid, 4–6 mm, ± equaling sepals, apex obtuse, opening by 6 valves, recurved at tip; carpophore absent. **Seeds** reddish brown, round, 1–1.2 mm diam., covered with prominent, stalked glands. $2n = 26, 52$.

Subspecies 2 (2 in the flora): s United States, Mexico, Bermuda, South America.

Petal length, along with sepal and capsule size and shape, vary in the two subspecies. Although they appear to be distinct in the southern United States, in Mexico where their ranges overlap, plants of uncertain identification are frequently encountered.

1. Petals 5–8 mm, 1.5–2 times as long as sepals; sepals narrowly lanceolate, apex acuminate, 7–8 mm in fruit; montane . . . .
. . . . 9a. *Stellaria cuspidata* subsp. *cuspidata*
1. Petals 2–4 mm, equaling or shorter than sepals; sepals ovate-lanceolate, 4–5 mm in fruit; lowlands
. . . . 9b. *Stellaria cuspidata* subsp. *prostrata*

9a. Stellaria cuspidata Willdenow ex Schlechtendal subsp. **cuspidata**

Sepals narrowly lanceolate, enlarging to 7–8 mm in fruit; petals 5–8 mm, 1.5–2 times as long as sepals. $2n = 52$ (Peru).

Flowering Mar–Oct, after rains. Gulleys, among boulders in mountains; 2400–2700 m; N.Mex., Tex.; Mexico; South America.

9b. Stellaria cuspidata Willdenow ex Schlechtendal subsp. **prostrata** (Baldwin) J. K. Morton, Sida 21: 888. 2004

Stellaria prostrata Baldwin in S. Elliott, Sketch Bot. S. Carolina 1: 518. 1821; *Alsine baldwinii* Small; *A. prostrata* (Baldwin) A. Heller

Sepals ovate-lanceolate, enlarging to 4–5 mm in fruit; petals 2–4 mm, equaling or shorter than sepals. $2n = 26$.

Flowering Mar–Sep. Wooded flood plains of rivers, seasonal water courses at low elevations; 0–2000 m; Fla., La., N.Mex., Tex.; Mexico; Bermuda.

10. Stellaria dicranoides (Chamisso & Schlechtendal) Fenzl in C. F. von Ledebour, Fl. Ross. 1: 395. 1842

• Chamisso's starwort

Cherleria dicranoides Chamisso & Schlechtendal, Linnaea 1: 63. 1826; *Arenaria chamissonis* Maguire; *A. dicranoides* (Chamisso & Schlechtendal) Hultén

Plants perennial, dioecious, forming dense cushions to 10 cm or more diam., with branching caudex, arising from taproot. **Stems** branched, 4-angled, 1–4 cm, glabrous; branches erect or ascending, thickly clothed with marcescent leaves. **Leaves** sessile; basal blades oblanceolate to obovate or elliptic, 3–5 × 1–1.5 mm, succulent, base cuneate, margins entire, apex acute or abruptly acuminate to obtuse, glabrous; cauline shorter. **Inflorescences** solitary-flowered in axils of foliage leaves; bract 1, foliaceous, ca. 1 mm. **Pedicels** 1–5 mm, glabrous. **Flowers** unisexual, 3–4 mm diam.; sepals 5, 3-veined, keeled, 2.5–3 mm, margins narrow, apex acute, glabrous or with sparse, short, glandular pubescence; petals absent; stamens 10, shorter than sepals; styles 3, erect, becoming outwardly curved, ca. 1 mm; staminate flowers with brownish, peglike, conspicuous nectaries alternating with and attached to base of stamens; pistillate flowers with well-developed but nonfunctional stamens and nectaries. **Capsules** straw colored, broadly ovoid, ca. 3 × 2 mm, ca. equaling sepals, apex obtuse, opening by 3 valves, each of which splits into 2; carpophore absent. **Seeds** 1, brown, broadly reniform with thickened rim, ca. 1.1 mm diam., finely verrucate. *2n* = 26.

Flowering summer. Arctic screes, fellfields, gravelly tundra, rocky knolls on wide variety of rock types; 300–1700 m; Yukon; Alaska; Asia (Russian Far East).

Stellaria dicranoides is of uncertain generic position. Many workers have placed it in the genus *Arenaria*. The absence of petals deprives us of a key character separating *Stellaria* from *Arenaria*. The ovate capsule with its three valves, each tardily dehiscent into two, suggests *Arenaria* or *Minuartia*. However, the chromosome number of *2n* = 26 is more often associated with *Stellaria*. The single large seed, which fills the capsule, is unusual. In its floral structure, including its large nectaries and unisexual flowers, *S. dicranoides* closely resembles the European *M. (Cherleria) sedoides* (Linnaeus) Hiern. In fact, Chamisso, who first described this species, placed it in the genus *Cherleria*.

11. Stellaria fontinalis (Short & R. Peter) B. L. Robinson, Proc. Amer. Acad. Arts 29: 286. 1894

• Kentucky starwort [E]

Sagina fontinalis Short & R. Peter, Transylvania J. Med. Assoc. Sci. 7: 600. 1836; *Alsine fontinalis* (Short & R. Peter) Britton; *Arenaria fontinalis* (Short & R. Peter) Shinners; *Spergula fontinalis* (Short & R. Peter) Dietrich

Plants annual; taproot slender. **Stems** straggling to ascending, branched, square, 10–25 cm, glabrous. **Leaves** sessile; blade with obscure midrib, narrowly oblanceolate to linear-spatulate, 1–3 cm × 0.8–4 mm, somewhat fleshy, base cuneate, margins entire, apex ± acute, glabrous. **Inflorescences** with flowers solitary in distal leaf axils; bracts absent. **Pedicels** ascending or erect, 10–40 mm, glabrous. **Flowers** 2.5–4 mm diam.; sepals 4(–5), 3-veined, ovate-lanceolate, 2.5–3 mm, margins narrow, scarious, apex ± acute, glabrous; petals absent; stamens 4(–5), shorter than sepals; styles 3 or 4, ascending, ca. 0.5 mm. **Capsules** green or straw colored, ovoid, ca. 3 mm, ± equaling sepals, apex obtuse, opening to base into 3 or 4 valves; carpophore absent. **Seeds** dark red-brown, orbiculate-reniform, 0.6 × 0.8 mm, shiny, tuberculate; tubercles prominent, stalked and knoblike.

Flowering spring. Seasonally wet, rocky openings in wooded glades, on wet cliffs; 400–500 m; Ky., Tenn.

Stellaria fontinalis is a very rare and poorly known species of uncertain affinity. Its characters are closer to *Sagina* and *Minuartia* than *Stellaria*, in particular the absence of petals, the 4(–5)-merous flowers, and the distinctive sculpturing of the seeds.

12. Stellaria graminea Linnaeus, Sp. Pl. 1: 422. 1753

• Common or grass-leaved stitchwort or starwort, mouron des champs [I][W]

Alsine graminea (Linnaeus) Britton

Plants perennial, coarse, rhizomatous; rhizomes slender, elongate. **Stems** decumbent or ascending, straggling, diffusely branched, smoothly 4-angled, 20–90 cm, brittle, glabrous. **Leaves** sessile; blade linear-lanceolate to narrowly lanceolate, widest near base, 1.5–4 cm × 1–6 mm, base round, margins smooth, apex acute, often ciliate near base, otherwise glabrous, not glaucous. **Inflorescences** terminal, 5–many-flowered, open, conspicuously branched cymes; bracts narrowly lanceolate, 1–5 mm, wholly scarious, margins ciliate, apex acuminate. **Pedicels** divaricate, 10–30 mm, glabrous. **Flowers** 5–12 mm diam., rarely larger; sepals

5, distinctly 3-veined, narrowly lanceolate to triangular, 3–7 mm, margins narrow, straight, scarious, apex acute, glabrous; petals 5, 3–7 mm, equaling or longer than sepals; stamens 10, all, some, or none fully developed and fertile; styles 3, ascending, ca. 3 mm. **Capsules** green or straw colored, narrowly ovoid, 5–7 mm, longer than sepals, apex acute, opening by 3 valves, splitting into 6; carpophore absent. **Seeds** reddish brown, reniform-rotund, ca. 1 mm diam., rugose in concentric rings. $2n$ = 39, 52.

Flowering late spring–early summer. Rough grasslands, pastures, hayfields, roadsides; 0–1200 m; introduced; St. Pierre and Miquelon; B.C., N.B., Nfld. and Labr., N.S., Ont., P.E.I., Que.; Calif., Colo., Conn., D.C., Idaho, Ill., Ind., Iowa, Kans., Ky., Maine, Mass., Mich., Minn., Mo., Mont., Nev., N.H., N.J., N.Y., N.C., Ohio, Oreg., Pa., R.I., S.C., Tenn., Vt., Va., Wash., W.Va., Wis.; Europe.

In Europe, both diploid and tetraploid cytotypes of *Stellaria graminea* occur with occasional triploid hybrids. Only the tetraploid form has been found in North America, except for a triploid colony in Newfoundland. This species is often confused with *S. longifolia* but differs in its stems, which are very angular, glabrous, and not scabrid; the narrowly triangular leaves on the flowering stems; the smooth leaf margins; the stiff, triangular, prominently 3-veined sepals; and the larger, rugulose seeds.

The sterile overwintering shoots of *Stellaria graminea* have broader elliptic to elliptic-lanceolate leaf blades measuring 5–15 × 1.5–4 mm. They are broadest near the middle. This state of the plant has been named var. *latifolia* Petermann. Usually *S. graminea* has perfect flowers but occasionally plants that are entirely staminate-sterile are encountered. The flowers in these are partially fertile depending on the occurrence of cross-pollination.

13. **Stellaria holostea** Linnaeus, Sp. Pl. 1: 422. 1753

• Greater stitchwort, Easter-bell [I] [W]

Alsine holostea (Linnaeus) Britton

Plants perennial, scrambling to ascending, from slender, creeping rhizomes. **Stems** branched distally, 4-angled, 15–60 cm, glabrous or hispid-puberulent distally. **Leaves** sessile; blade narrowly lanceolate, widest near base, 4–8 cm × 2–10 mm, somewhat coriaceous, base round and clasping, margins and abaxial midrib very rough, apex narrowly and sharply acuminate, scabrid, otherwise glabrous, slightly glaucous. **Inflorescences** terminal, loose, 3–31-flowered cymes; bracts foliaceous, 5–50 mm, margins and abaxial midrib scabrid. **Pedicels** ascending, 1–60 mm, slender,

pubescent. **Flowers** 20–30 mm diam.; sepals 5, inconspicuously 3-veined, ovate-lanceolate, 6–8 mm, margins narrow, scarious, apex acute, glabrous; petals 5 (rarely absent), 8–14 mm, longer than sepals, blade apex 2-fid to middle; stamens 10, sometimes fewer by degeneration; styles 3, ascending, ca. 4 mm. **Capsules** green, subglobose, 5–6 mm, ± equaling sepals, apex obtuse, opening by 3 valves, tardily splitting into 6; carpophore absent. **Seeds** reddish brown, reniform, 2–3 mm diam., papillose. $2n$ = 26 (Europe).

Flowering spring. Woodlands, hedgerows; 0–500 m; introduced; Conn., Mass., N.J., N.Y., N.C., Ohio, Pa.; Eurasia.

Stellaria holostea is sometimes cultivated and occasionally naturalizes.

14. **Stellaria humifusa** Rottbøll, Skr. Kiøbenhavnske Selsk. Laerd. Elsk. 10: 447, plate 4, fig. 14. 1770

• Salt-marsh starwort

Alsine humifusa (Rottbøll) Britton; *Stellaria humifusa* var. *marginata* Fenzl; *S. humifusa* var. *oblongifolia* Fenzl; *S. humifusa* var. *suberecta* B. Boivin

Plants perennial, forming small to large mats or clumps, from slender rhizomes. **Stems** decumbent, freely branched, square, 2–20 cm, glabrous, rooting at proximal nodes. **Leaves** sessile; blade elliptic to elliptic-lanceolate, 0.4–1.5 cm × 1–5 mm, succulent, base cuneate to rounded, margins entire, apex acute to obtuse, glabrous or with few cilia along margins. **Inflorescences** with flowers solitary in axils of foliage leaves; bracts absent. **Pedicels** ascending, straight or nearly so, usually 5–10(–30) mm, glabrous. **Flowers** ca. 10 mm diam.; sepals 5, prominently 1–3-veined, lanceolate, 4–5 mm, margins convex, narrow, scarious, apex acute, glabrous; petals 5, 4–6 mm, equaling sepals; stamens 10; styles 3, ascending and outwardly curved, 1–1.5 mm. **Capsules** straw colored, ovoid, 4–5 mm, equaling sepals, apex obtuse, opening by 6 valves; carpophore absent. **Seeds** pale brown, broadly and obliquely reniform, 0.8–1 mm diam., smooth to slightly rugose. $2n$ = 26.

Flowering summer. Lake shores, beaches, marshes, salt marshes, mainly northern coastal; 0–100 m; Greenland; St. Pierre and Miquelon; B.C., Man., N.B., Nfld. and Labr., N.W.T., N.S., Nunavut, Ont., P.E.I., Que., Yukon; Alaska, Maine; arctic Europe; Asia (Russian Far East, Siberia).

Stellaria humifusa is often confused with *S. crassifolia*, but has thicker stems and fleshy leaves that wrinkle and tend to turn brownish when dried. Also, in *S. crassifolia* the long pedicels are very slender and sharply angled below the capsule.

SELECTED REFERENCE Boivin, B. 1954. Les variations du *Stellaria humifusa* Rottboell. Ann. A. C. F. A. S. 20: 97–98.

15. Stellaria irrigua Bunge, Mém. Acad. Imp. Sci. St.-Pétersbourg Divers Savans 2: 548. 1835 • Altai chickweed or starwort

Alsine polygonoides Greene ex Rydberg

Plants perennial, forming mats or low cushions, with elongate rhizomes. **Stems** ascending to spreading, striate, branched, 4-sided, 2–10 cm, glabrous, internodes usually shorter than leaves. **Leaves** sessile; blade with prominent midrib, elliptic or lanceolate to oblanceolate, 0.1–1 cm × 0.5–4 mm, fleshy, base cuneate to spatulate, margins transparent, entire, narrow, apex acute, glabrous, rarely with few cilia near base, with sessile, often purple glands impressed on abaxial surface; proximal leaves marcescent. **Inflorescences** with flowers solitary in distal leaf axils, often ± concealed among leaves; bracts, when present, paired near base of pedicel, with midrib, ovate-lanceolate, ca. 1 mm, scarious. **Pedicels** erect, becoming curved and deflexed distally, 3–15 mm, glabrous. **Flowers** ca. 6 mm diam.; sepals 5, purple tinged, outer prominently 3-veined, carinate, inner with midrib only, lanceolate, 3–4 mm, margins narrow, membranous, apex acute, glabrous; petals 5, ca. 2 mm, shorter than sepals, blade apex deeply 2-fid, lobes narrowly elliptic; stamens 5, inserted in prominent nectary disc; styles 3, outwardly ascending, curved, 0.75 mm. **Capsules** green, becoming straw colored, ovoid-obtuse, ca. 3 mm, ± equaling and enclosed in sepals, opening with 6 teeth. **Seeds** pale brown, reniform, 1–1.2 mm diam., not glossy, sides smooth to shallowly rugose, margins thickened with shallow, parallel, longitudinal ridges.

Flowering summer. Mountain rills and screes; 2500–4000 m; Colo.; Asia (Siberia: Altai Mountains).

Stellaria irrigua has a remarkable and perhaps uniquely disjunct distribution. It is known to occur only in Colorado and the Altai Mountains of Siberia. This raises the possibility that the two populations may not be conspecific, which requires further study.

SELECTED REFERENCE Weber, W. A. 1961. *Stellaria irrigua* Bunge in America. Univ. Colorado Stud., Ser. Biol. 7: 12–15.

16. Stellaria littoralis Torrey in War Department [U.S.], Pacif. Railr. Rep. 4(5): 69. 1857 • Beach starwort or chickweed [E]

Plants perennial, straggling to scandent, from elongate rhizomes. **Stems** ascending, often decumbent at base, branched, 4-sided, 10–60 cm, uniformly and softly pubescent. **Leaves** sessile; blade ovate to ovate-lanceolate, widest proximal to middle, 1–4.5 cm × 4–20 mm, base round, margins densely ciliate, apex shortly acuminate, pubescent on both surfaces. **Inflorescences** terminal, 5–many-flowered, leafy cymes; bracts foliaceous, 4–40 mm, margins ciliate, not scarious. **Pedicels** ascending to erect, straight, spreading to reflexed at base in fruit, 5–20 mm. **Flowers** 9–10 mm diam.; sepals (4–)5, 3-veined, lanceolate, 2.8–5 mm, margins narrow, scarious, apex acuminate, ciliate-pubescent mainly on margins and veins; petals 5, 4–6 mm, equaling or slightly longer than sepals, blade apex deeply 2-fid; stamens 10; styles 3, ascending, ca. 1.5 mm. **Capsules** green to straw colored, lanceoloid-ovoid, 5–6 mm, slightly longer than sepals, apex obtuse, opening by 3, tardily 6, ascending valves; carpophore absent. **Seeds** reddish brown, broadly and obliquely ovate, ± 1 mm diam., minutely rugose.

Flowering spring. Marshy fields, marshes, coastal bluffs; less than 100 m; Calif.

Stellaria littoralis is very similar to *S. dichotoma* Linnaeus from China, the Russian Far East, and Siberia. It may be conspecific with the latter and may have been introduced into the San Francisco area in the early days of exploration of the Pacific coast. A more detailed study is warranted.

17. Stellaria longifolia Muhlenberg ex Willdenow, Enum. Pl., 479. 1809 • Long-leaved starwort, stellaire à longues feuilles [F]

Alsine longifolia (Muhlenberg ex Willdenow) Britton; *Stellaria atrata* (J. W. Moore) B. Boivin; *S. atrata* var. *eciliata* B. Boivin; *S. diffusa* Willdenow ex Schlechtendal; *S. longifolia* var. *atrata* J. W. Moore; *S. longifolia* var. *eciliata* (B. Boivin) B. Boivin

Plants perennial, forming loose clumps, from elongate rhizomes. **Stems** erect or straggling, branched, square, 10–35 cm, glabrous but angles minutely papillate-scabrid. **Leaves** sessile; blade green to yellowish green, never glaucous, linear to very narrowly elliptic, widest at or beyond middle, 0.8–4 cm × 1–3 mm, not coriaceous, base attenuate, apex acuminate to acute, glabrous to sparingly ciliate at base, margins minutely papillate-scabrid; proximal leaves shorter and wider. **Inflorescences** terminal, widely divaricate, 2–many-flowered cymes; bracts lanceolate, 1–5 mm, scarious, apex acuminate. **Pedicels** straight or somewhat arcuate, commonly 3–30 mm, glabrous or scabrous. **Flowers** 5–9 mm diam.; sepals 5, obscurely 3-veined, ovate-elliptic, 2–4 mm, margins scarious, apex acute, glabrous; petals 5, 2–3.5 mm, ± equaling sepals; stamens 5–10; styles 3, ascending, ca. 1 mm. **Capsules** blackish purple or straw colored, ovoid-conic, 3–6 mm, much longer than sepals, opening by 6 valves; carpophore absent. **Seeds** brown, broadly reniform, 0.7–0.8 mm diam., slightly rugose. $2n = 26$.

Flowering late spring–summer. Wet meadows and woodlands, marshes, muskegs, grassy roadsides, usually in circumneutral to calcareous sites; 0–2800 m; St. Pierre and Miquelon; Alta., B.C., Man., N.B., Nfld. and Labr., N.W.T., N.S., Ont., Que., Sask., Yukon; Alaska, Ariz., Calif., Colo., Conn., Del., D.C., Idaho, Ill., Ind., Iowa, Ky., Maine, Md., Mass., Mich., Minn., Mo., Mont., Nebr., Nev., N.H., N.J., N.Y., N.Dak., Ohio, Pa., R.I., S.C., S.Dak., Tenn., Vt., Va., Wash., W.Va., Wis., Wyo.; Europe.

Stellaria longifolia often is confused with forms of *S. longipes* but differs in having leaves that are widest at or above the middle and in having the angles of the stem and/or the leaf margins minutely papillate-scabrid. The capsules can be either straw colored or black. Plants with black capsules have been named var. *atrata*.

Hybrids with *Stellaria borealis* subsp. *borealis* often occur; see note under that species.

18. Stellaria longipes Goldie, Edinburgh Philos. J. 6: 327. 1822 • Long-stalked starwort [F]

Alsine longipes (Goldie) Coville

Plants perennial, forming small to large clumps or mats, or diffuse, from slender rhizomes. **Stems** erect to straggling, branched or not, 4-angled, 3–32 cm, glabrous or softly pubescent, angles not minutely papillate-scabrid. **Leaves** sessile; blade green, frequently glaucous, 1–3-veined, midrib prominent, lanceolate to linear-lanceolate, widest at base, 0.4–2.6(–4) cm × 1–4 mm, strongly coriaceous or not, base round, margins entire, convex, glabrous or ciliate, apex acute to acuminate, not spinescent, shiny, smooth, glabrous or sparingly villous, base usually glabrous, rarely with few cilia. **Inflorescences** with flowers solitary, or terminal, 3–30-flowered (rarely more) cymes; bracts lanceolate, 2–10 mm, herbaceous with scarious margins, or scarious throughout, glabrous or ciliate. **Pedicels** ascending to erect, straight, 5–30 mm, glabrous or softly pubescent.

Flowers 5–10 mm diam.; sepals 5, 3-veined, midrib prominent, lanceolate to ovate-lanceolate, 3.5–5 mm, margins convex, narrow, scarious, sometimes ciliate, apex acute, glabrous or pubescent; petals 5, 3–8 mm, 1–1.5 times as long as sepals; stamens 5–10; styles 3(–6), ascending, curled at tip, ca. 1.5 mm. **Capsules** blackish purple or straw colored, ovoid to ovoid-lanceoloid, 4–6 mm, 1.5–2 times as long as sepals, apex broadly acute, opening by 6 valves; carpophore absent. **Seeds** brown, reniform to globose, 0.6–0.9 mm diam., shallowly tuberculate to smooth. $2n = 52$–104, (107).

Subspecies 2 (2 in the flora): circumpolar.

SELECTED REFERENCE Chinnappa, C. C. and J. K. Morton. 1974. The cytology of *Stellaria longipes*. Canad. J. Genet. Cytol. 16: 499–514.

1. Capsules purplish black; stems variable, commonly compact, erect; leaf blades very variable, from linear-lanceolate to ovate-triangular . . . . . . 18a. *Stellaria longipes* subsp. *longipes*
1. Capsules straw colored; stems elongate, straggling; leaf blades narrowly lanceolate . . . . 18b. *Stellaria longipes* subsp. *arenicola*

18a. Stellaria longipes Goldie subsp. **longipes**
  • Goldie's starwort [F]

Alsine palmeri Rydberg; *A. strictiflora* Rydberg; *A. validus* Goodding; *Stellaria arctica* Schischkin; *S. ciliatosepala* Trautvetter; *S. crassipes* Hultén; *S. dulcis* Gervais; *S. edwardsii* R. Brown; *S. edwardsii* var. *arctica* (Schischkin) Hultén; *S. edwardsii* var. *crassipes* (Hultén) B. Boivin; *S. hultenii* B. Boivin; *S. laeta* Richardson; *S. laeta* var. *altocaulis* (Hultén) B. Boivin; *S. laxmannii* Fischer; *S. longifolia* Muhlenberg ex Willdenow var. *laeta* (Richardson) S. Watson; *S. longipes* var. *altocaulis* (Hultén) C. L. Hitchcock; *S. longipes* var. *edwardsii* (R. Brown) A. Gray; *S. longipes* var. *laeta* (Richardson) S. Watson; *S. longipes* var. *minor* Hooker; *S. longipes* subsp. *monantha* (Hultén) W. A. Weber; *S. longipes* var. *monantha* (Hultén) S. L. Welsh; *S. longipes* subsp. *stricta* (Richardson) W. A. Weber; *S. longipes* var. *subvestita* (Greene) Polunin; *S. monantha* Hultén; *S. monantha* var. *altocaulis* Hultén; *S. monantha* subsp. *atlantica* Hultén; *S. monantha* var. *atlantica* (Hultén) B. Boivin; *S. palmeri* (Rydberg) Tidestrom; *S. stricta* Richardson; *S. subvestita* Greene; *S. valida* (Goodding) Coulter & A. Nelson

Plants usually forming clumps, cushions, or mats, but sometimes with diffuse, straggling stems. **Leaf blades** linear-lanceolate to ovate-triangular. **Capsules** purplish black. $2n = 52$–107, mostly 52, 78, or 104.

Flowering late spring–summer. Grassy places along streams, moist gravel or sand, arctic and alpine tundra, among rocks, on cliffs; 0–3800 m; Greenland; Alta., B.C., Man., N.B., Nfld. and Labr., N.W.T., Nunavut, Ont., Que., Sask., Yukon; Alaska, Ariz., Calif., Colo., Idaho, Mich., Minn., Mont., N.Mex., N.Y., N.Dak., Oreg., S.Dak., Utah, Wash., Wyo.; Europe; Asia.

Subspecies *longipes*, as shown by the above synonymy, is an exceptionally variable polyploid complex. Variation in this circumboreal, circumpolar taxon is under both genic and environmental control. Character states that are under genic control, such as pubescence, are scattered more or less at random throughout the range of distribution of the complex, and there is little discernible correlation among them, or with chromosome number or environment. All populations that have been studied are interfertile, and most show a high degree of phenotypic plasticity, being affected by such environmental conditions as temperature, day length, and shade. Most populations are self-compatible, although pollination by small insects, mainly flies, is the norm. Vegetative reproduction through fragmentation of the rhizome is also common. Hence, the complex consists of a large number of interfertile but more or less self-perpetuating, highly plastic biotypes.

A presumed hybrid between *Stellaria longipes* and *S. borealis* has been collected in the La Sal Mountains of Utah, although attempts to cross the two species experimentally were unsuccessful. Hybrids between *S. longipes* and *S. longifolia* occasionally occur and have been produced artificially (C. C. Chinnappa 1985); they are sterile triploids ($2n = 39$).

18b. Stellaria longipes Goldie subsp. **arenicola** (Raup) C. C. Chinnappa & J. K. Morton, Rhodora 93: 132. 1991 • Lake Athabasca starwort [E]

Stellaria arenicola Raup, J. Arnold Arbor. 17: 248, plate 196. 1936; *S. longipes* var. *arenicola* (Raup) B. Boivin

Plants forming loose clumps, with elongate, straggling stems. **Leaf blades** narrowly lanceolate. **Capsules** straw colored. $2n = 52$.

Flowering late spring–summer. Inland sand dunes; 200–300 m; Alta., Sask.

Confined to the extensive mobile sand dunes on the south side of Lake Athabasca, subsp. *arenicola* is largely self-pollinating but is interfertile with subsp. *longipes*, with which it intergrades in its natural habitat.

Very rarely, individual plants of subsp. *longipes* with straw-colored capsules are encountered in other localities. These are probably due to the presence of a recessive gene for capsule color that is of widespread but rare occurrence in these populations. Only on the Lake Athabasca sand dunes have selective pressures been sufficient for it to evolve into a distinct biotype.

19. Stellaria media (Linnaeus) Villars, Hist. Pl. Dauphiné 3: 615. 1789 • Common chickweed, mouron des oiseaux [I] [W]

Alsine media Linnaeus, Sp. Pl. 1: 272. 1753; *Stellaria apetala* Ucria ex Roemer; *S. media* var. *procera* Klatt & Richter

Plants annual or winter annual, green, with slender taproot. **Stems** decumbent or ascending, diffusely branched, 4-sided, 5–40 cm, with single line of hairs along each internode. **Leaves** petiolate (proximal) or ± sessile (distal); blade usually green, ovate to broadly elliptic, 0.5–4 cm × 2–20 mm, base round to cuneate, margins entire, apex acute or shortly acuminate, ± glabrous or ciliate at base. **Inflorescences** terminal, 5–many-flowered cymes; bracts ovate and shortly acuminate to lanceolate-acute, 1–40 mm, herbaceous. **Pedicels** ascending, usually straight, deflexed at base in fruit, 3–40 mm, usually with line of hairs. **Flowers** 2–5 mm diam.; sepals 5, with obscure midrib, ovate-lanceolate, 4.5–5(–6) mm, margins narrow, scarious, apex obtuse, usually glandular-hairy; petals absent or 5, 1–4 mm, shorter than to equaling sepals; stamens 3–5(–8); anthers red-violet; styles 3, outwardly curved, becoming curled, 0.5–1 mm. **Capsules** green to straw colored, ovoid-oblong, 3–5 mm, somewhat longer than sepals, apex obtuse, opening by 6 valves; carpophore absent. **Seeds** reddish brown, broadly reniform to round, 0.9–1.3 mm diam., with obtuse, round, or flat-topped (broader than tall) tubercles. $2n = 40, 42, 44$.

Flowering year-round where climatic conditions permit. Cultivated ground, waste places, open woodlands; 0–2500 m; introduced; Greenland; St. Pierre and Miquelon; Alta., B.C., Man., N.B., Nfld. and Labr., N.W.T., N.S., Ont., P.E.I., Que., Sask., Yukon; Ala., Alaska, Ariz., Ark., Calif., Colo., Conn., Del., D.C., Fla., Ga., Idaho, Ill., Ind., Iowa, Kans., Ky., La., Maine, Md., Mass., Mich., Minn., Miss., Mo., Mont., Nebr., Nev., N.H., N.J., N.Mex., N.Y., N.C., N.Dak., Ohio, Okla., Oreg., Pa., R.I., S.C., S.Dak., Tenn., Tex., Utah, Vt., Va., Wash., W.Va., Wis., Wyo.; Europe.

Stellaria media, now a cosmopolitan weed, is a very polymorphic species, varying in size, habit, pubescence, petal length, stamen number, and seed size and surface detail.

The *Stellaria media* complex consists of three very similar and closely related species, *S. media*, *S. neglecta*, and *S. pallida*. They can almost always be distinguished by the characters given in the key, but in a few doubtful cases a chromosome count is desirable for positive identification. The problem arises from the considerable phenotypic variation which is displayed by *S. media*, and to a lesser extent by *S. pallida*. There is no evidence for gene exchange between these species. *Stellaria pallida* is autogamous and sometimes cleistogamous; *S. media* is both autogamous and occasionally cross-pollinated by flies; *S. neglecta* is usually cross-pollinated by flies but is self-compatible.

SELECTED REFERENCES Turkington, R., N. C. Kenkel, and G. C. Franko. 1980. The biology of Canadian weeds. 42. *Stellaria media* (L.) Vill. Canad. J. Pl. Sci. 60: 981–982. Whitehead, F. H. and R. P. Sinha. 1967. Taxonomy and taximetrics of *Stellaria media* (Linnaeus) Vill., *S. neglecta* Weihe and *S. pallida* (Dumort.) Piré. New Phytol. 66: 769–784.

20. **Stellaria neglecta** Weihe in M. J. Bluff et al., Comp. Fl. German. 1: 560. 1825 • Greater chickweed [I]

Alsine neglecta (Weihe) Á. Löve & D. Löve; *Stellaria media* (Linnaeus) Villars subsp. *neglecta* (Weihe) Murbeck

Plants annual or winter annual, from slender taproot. **Stems** decumbent proximally, ascending distally, diffusely branched, 4-angled, to 80 cm, with single line of hairs along each internode. **Leaves** petiolate (proximal and those on sterile shoots) or sessile (distal and mid stem); blade ovate to broadly elliptic, 0.5–4 cm × 2–18 mm, base round to cuneate, margins entire with few cilia at base, thin, apex acute to short-acuminate, ± glabrous. **Inflorescences** terminal, 9–many-flowered cymes; bracts ovate to lanceolate, 3–25 mm, reduced distally, herbaceous, sparsely ciliate on margins and underside. **Pedicels** erect, often becoming deflexed, 5–40 mm, pubescence a single line of hairs. **Flowers** 5–7 mm diam.; sepals 5, veins obscure, lanceolate, 5–6.5 mm, margins narrow, membranous, apex acute, pubescent; petals 5 (rarely absent), 2–5 mm, shorter than or equaling sepals; stamens 8–10; styles 3, ascending, outwardly curved, 0.5–1 mm. **Capsules** green to straw colored, ovoid-oblong, 5–7 mm, ca. equaling sepals, apex obtuse, opening by 6 slightly recurved valves; caropohore absent. **Seeds** very dark brown when mature, round, 1.1–1.7 mm diam., tuberculate; tubercles conic, taller than broad, apex acute. $2n = 22$.

Flowering spring. Hedge banks, open woodlands, along streams, semishaded grassy places; 100–200 m; introduced; Ark., Calif., Ky., La., Md., N.C., Okla., Tenn.; Europe.

Formerly, *Stellaria neglecta* was rare in North America, but during the last ten to 15 years it has spread rapidly and become weedy. It is very like larger forms of *S. media* (see note under that species), but usually differs in having larger flowers, sepals, and seeds; having a larger number of stamens; and having seeds with acute conic tubercles. Flowers are self-compatible but usually are pollinated by flies.

21. **Stellaria nitens** Nuttall in J. Torrey and A. Gray, Fl. N. Amer. 1: 185. 1838 • Shining starwort [E]

Stellaria praecox A. Nelson

Plants annual, from threadlike taproots. **Stems** erect, sparingly branched below inflorescence, 4-sided, 3–25 cm, glabrous or sparsely hairy. **Leaves** sessile, crowded at base, shiny; blade oblanceolate to obovate and spatulate (proximal) or linear-lanceolate (distal), 0.5–1.5 cm × 0.5–2 mm, base round, apex acuminate, glabrous, often ciliate on margins. **Inflorescences** terminal, 3–21-flowered (rarely more) cymes; bracts linear-lanceolate, 1–12 mm, scarious distally, herbaceous proximally, often ciliate on margins. **Pedicels** ascending to erect, ± straight in fruit, 2–25 mm, glabrous. **Flowers** 2–3 mm diam.; sepals 5, with 3 prominent, ridged veins, very narrowly lanceolate, to acicular, 2.8–4.2 mm, margins wide, scarious, apex acuminate, glabrous; petals 5 or absent, 1–3 mm, shorter than sepals, blade apex 2-lobed; stamens 3–5; styles 3, spreading, becoming curled, ca. 0.3 mm. **Capsules** green or straw colored, narrowly ovoid, 2–3 mm, shorter than sepals, apex obtuse, opening by 3 valves, splitting into 6; carpophore absent. **Seeds** brown, round, 0.5–0.7 mm diam., minutely tuberculate. $2n = 20, 40$.

Flowering spring. Dry, open habitats: sand dunes, stream banks, rocky outcrops, open woodlands, beneath boulders, disturbed areas; 0–2000 m; B.C.; Ariz., Calif., Idaho, Mont., Nev., Oreg., Utah, Wash.

22. **Stellaria obtusa** Engelmann, Bot. Gaz. 7: 5. 1882 • Blunt-sepaled starwort, Rocky Mountain starwort [E]

Alsine obtusa (Engelmann) Rose; *A. viridula* Piper; *A. washingtoniana* (B. L. Robinson) A. Heller; *Stellaria viridula* (Piper) St. John; *S. washingtoniana* B. L. Robinson

Plants perennial, creeping, often matted but not forming cushions, rhizomatous. **Stems** prostrate, branched, 4-sided, 3–23 cm, internodes equaling or longer than leaves, glabrous, rarely pilose. **Leaves** sessile or short-petiolate; blade broadly ovate to elliptic, 0.2–1.2 cm × 0.9–7 mm, base round or cuneate, margins entire, apex acute, shiny, glabrous or ciliate near base. **Inflorescences** with flowers solitary, axillary; bracts absent. **Pedicels** spreading, 3–12 mm, glabrous. **Flowers** 1.5–2 mm diam.; sepals 4–5, veins obscure, midrib sometimes apparent, ± ovate, 1.5–3.5 mm, margins narrow, scarious, apex ± obtuse, glabrous; petals absent; stamens 10 or fewer; styles 3

(–4), curled, shorter than 0.5 mm. **Capsules** green to pale straw colored, translucent, globose to broadly ovoid, 2.3–3.5 mm, 1.9–2 times as long as sepals, apex obtuse, opening by 6 valves; carpophore absent. **Seeds** grayish black, broadly elliptic, 0.5–0.7 mm diam., finely reticulate. $2n$ = 26, 52, ca. 65, ca. 78.

Flowering late spring–summer. Moist areas in woods, shaded edges of creeks, talus slopes; 300–3400 m; Alta., B.C.; Calif., Colo., Idaho, Mont., Oreg., Utah, Wash., Wyo.

23. Stellaria pallida (Dumortier) Crépin, Man. Fl. Belgique ed. 2, 19. 1866 • Lesser chickweed [I]

Alsine pallida Dumortier, Fl. Belg., 109. 1827; *Stellaria boraeana* Jordan; *S. media* (Linnaeus) Villars subsp. *pallida* (Dumortier) Ascherson & Graebner

Plants annual, usually yellowish green, with slender taproot. **Stems** prostrate, much-branched, 4-sided, usually 10–20(–40) cm, glabrous, with single line of hairs along each internode. **Leaves** petiolate (proximal) or sessile (distal); blade ovate to elliptic, usually 0.3–1.5 cm × 1–7 mm, base round to cuneate, margins entire, apex shortly acuminate, glabrous or with few cilia on margins and abaxial midrib. **Inflorescences** terminal, 3–35-flowered cymes; bracts lanceolate, 2–10 mm, herbaceous, margins entire. **Pedicels** spreading, sometimes deflexed at base in fruit, 1–10 mm, pubescent. **Flowers** 2–3 mm diam.; sepals 4–5, veins obscure, midrib sometimes present, lanceolate, 3–4 mm, margins narrow, herbaceous, apex acute, pubescent; petals usually absent; stamens 1–3 or absent; anthers gray-violet; styles 3, ascending, becoming curled, 0.2–0.5 mm. **Capsules** pale straw colored, ovoid, 2–4 (–5) mm, equaling to slightly longer than sepals, apex obtuse, opening by 6 valves, outwardly curled at tip; carpophore absent. **Seeds** pale yellowish brown, reniform to round, 0.5–0.9 mm diam., tuberculate; tubercles prominent, broader than tall, apex obtuse. $2n$ = 22.

Flowering spring. Dunes, sandy waste places, rest areas on interstate highways; 0–1500 m; introduced; Ont.; Ariz., Ark., Calif., Colo., Fla., Ind., Kans., Ky., La., Mich., Mo., Nebr., N.C., Ohio, Okla., Pa., S.C., Tenn., Tex., Va., Wash., W.Va.; Mexico; Europe.

Stellaria pallida is automatically self-pollinated and often cleistogamous. It usually can be distinguished from apetalous forms of *S. media* by its smaller size, yellowish green color, its small sepals and small, pale seeds. Also the base and tip of the sepals occasionally are dark-red pigmented.

SELECTED REFERENCE Morton, J. K. 1972. On the occurrence of *Stellaria pallida* in North America. Bull. Torrey Bot. Club 99: 95–97.

24. Stellaria palustris Ehrhart ex Hoffmann, Deutschl. Fl. 1: 152. 1791 • Marsh stitchwort [I]

Alsine glauca (Withering) Britton; *Stellaria glauca* Withering

Plants perennial, with slender creeping rhizomes. **Stems** straggling, with erect branches, smoothly 4-angled, (20–)30–60 cm, glabrous. **Leaves** sessile; blade linear-lanceolate, 1.5–5 cm × 1–4 mm, base cuneate, margins smooth, apex acute, glabrous, usually glaucous. **Inflorescences** terminal, (1–)2–21-flowered cymes; bracts narrowly lanceolate, 2–7 mm, herbaceous or scarious with green midrib, not ciliate. **Pedicels** ascending, 30–100 mm, glabrous. **Flowers** 12–18 mm diam.; sepals 5, distinctly 3-veined, lanceolate, 6–8 mm, margins wide, scarious, apex acute, glabrous; petals 5, 7–10 mm, 1.5–2 times as long as sepals; stamens 10; styles 3, erect, 5–7 mm; stigmas club-shaped. **Capsules** green to straw colored, ovoid-oblong, 8–10 mm, ± equaling sepals, apex acute, opening by 6 valves; carpophore absent. **Seeds** dark reddish brown, round, 1.2–1.4 mm diam., tuberculate; tubercles shallow, round. $2n$ = 130–188 (Europe), ca. 198.

Flowering early summer. Hayfields and pastures subject to seasonal flooding; 0–20 m; introduced; Que.; Europe.

Stellaria palustris is found along the Saint Lawrence estuary.

SELECTED REFERENCE McNeill, J. and J. N. Findlay 1972. Introduced perennial species of *Stellaria* in Quebec. Naturaliste Canad. 99: 59–60.

25. Stellaria parva Pederson, Bot. Tidsskr. 57: 44, fig. 4. 1961 • Small starwort [I]

Plants annual or perennial, forming mats, rhizomatous. **Stems** creeping, much-branched, rooting at lower nodes, 4-angled, 5–30 cm, glabrous or sparsely pubescent on 2 sides. **Leaves** sessile to subsessile; blade elliptic to obovate, 0.3–1.5 cm × 1–6 mm, base cuneate to spatulate, margins entire, apex obtuse to ± acute, tip blunt, glabrous or pubescent at base and stem nodes. **Inflorescences** with flowers solitary, axillary in distal nodes; bracts absent. **Pedicels** spreading, becoming reflexed in fruit, 2–10 mm, glandular-pubescent. **Flowers** 3–4 mm diam.; sepals 4–5, herbaceous with midrib, ovate-lanceolate, 2.5–3 mm, margins convex, membranous, apex obtuse, glandular-pubescent; petals 4–5, ca. 3 mm, equaling sepals; stamens 4–5; styles 3–4, ascending. **Capsules** broadly ovoid, ca.

4.5 mm, longer than sepals, apex obtuse, opening by 6–8 valves; carpophore absent. **Seeds** ca. 20, brown, round, ca. 1 mm diam., prominently papillate; papillae obtuse, taller than broad. $2n = 34$.

Flowering spring–fall. Short grass in ditches and marshy areas; less than 100 m; introduced; La., Tex.; South America (Argentina, Brazil, Paraguay).

SELECTED REFERENCES Brown, L. E. and S. J. Marcus. 1998. Notes on the flora of Texas with additions and other significant records. Sida 18: 315–324. Landry, G. P., W. D. Reese, and C. M. Allen. 1988. *Stellaria parva* Pederson new to North America. Phytologia 64: 497.

26. **Stellaria porsildii** C. C. Chinnappa, Syst. Bot. 17: 29, fig. 1. 1992 · Porsild's starwort [C] [E]

Plants perennial, erect to straggling, rarely clumped, never compact and cushion-forming, from slender rhizomes. **Stems** erect, diffusely branched, rarely elongate and straggling, branched mainly at base, 4-sided, 9–20 cm, glabrous. **Leaves** sessile; blade green, never glaucous, linear to linear-lanceolate, widest at or near middle, 2.7–3.5 cm × 2–3 mm, not succulent, base cuneate, margins entire, apex gradually acuminate, acute, glabrous with few cilia at base. **Inflorescences** with flowers solitary, terminal or axillary in distal foliage leaves. **Pedicels** erect, 18–50 mm, glabrous. **Flowers** 7–10 mm diam.; sepals 5, midrib prominent, lateral veins obscure, ovate-lanceolate to lanceolate, 4–6 mm, margins narrow, membranous, apex acute, glabrous; petals 5, 4–6 mm, equaling or slightly longer than sepals, blade apex deeply divided into 2 oblanceolate lobes; stamens 10; styles 3, ascending, curled at tip, 2–3 mm. **Capsules** black, oblong, 6–8 mm, slightly longer than sepals, apex obtuse, opening by 6 valves; carpophore absent. **Seeds** dark brown, broadly ovate, 0.8–1 mm diam., shallowly tuberculate. $2n = 26$.

Flowering early summer. Willow thickets, open forests and woodlands on slopes of mountains; of conservation concern; 2400–3600 m; Ariz., N.Mex.

Stellaria porsildii is closely related to *S. longipes* and *S. longifolia* and tends to be intermediate between them, with somewhat larger, solitary flowers. Its leaves tend to be more like those of *S. longifolia*, but they lack the papillate-scabrid margins, and have a few long cilia at the base. It is postulated that the polyploid *S. longipes* complex arose through hybridization between *S. porsildii* and *S. longifolia*, both of which are diploid (C. C. Chinnappa 1992). The two species can be hybridized but the artificial hybrid is diploid.

Stellaria porsildii can be very difficult to distinguish from forms of *S. longipes*, and a confirmatory chromosome count is desirable, at least for records from new locations. The total absence of minute papillae on the stems and leaf margins distinguishes both species from *S. longifolia*. The presence of a few long cilia at the base of the leaves is a useful indication of *S. porsildii*, but such cilia often are present in *S. longipes*. Confirmatory characters for *S. porsildii* are the open, erect to straggling habit of the plant (never compact and cushion-forming), and the leaves, which are green (never glaucous), soft (not stiff or coriaceous), always narrowly linear-lanceolate, and tending to be widest near the center of the lamina (not lanceolate and widest at the base).

27. **Stellaria pubera** Michaux, Fl. Bor.-Amer. 1: 273. 1803 · Star chickweed [E] [F] [W]

Alsine pubera (Michaux) Britton; *A. pubera* var. *tennesseensis* C. Mohr

Plants perennial, with stems loosely tufted, rhizomatous. **Stems** erect, branched, 4-sided, 10–40 cm, with alternating lines of spreading, soft, flexuous, mainly eglandular hairs. **Leaves** usually sessile (distal), often short-petiolate (proximal); blade elliptic, obovate, or lanceolate, widest at or beyond middle, 1–10 cm × 5–35 mm, base cuneate, margins entire, apex acute, glabrous to sparsely pubescent adaxially, ciliate on margins and abaxial midrib. **Inflorescences** terminal, 3–70-flowered cymes; bracts elliptic to lanceolate, 7–65 mm, herbaceous. **Pedicels** erect in flower, often deflexed at base in fruit, 5–40 mm, softly pubescent. **Flowers** (8–) 10–12 mm diam.; sepals 5, with midrib, ovate, 3.5–6 mm, margins narrow, scarious, apex obtuse to acute, softly and often sparsely pubescent; petals 5, 4–8 mm, longer than sepals; stamens 10; styles 3, ascending, 2.5 mm. **Capsules** green to straw colored, broadly ovoid, 3.5–5.5 mm, ca. equaling sepals, apex obtuse, opening by 6 valves; carpophore absent. **Seeds** brown, obliquely reniform, 1.5–2 mm diam., coarsely sulcate-papillate. $2n = 30$.

Flowering spring. Rich deciduous woods, alluvial bottomlands; 100–1000 m; Ala., D.C., Fla., Ga., Ind., Ky., Md., Nebr., N.C., Ohio, Pa., S.C., Tenn., Va., W.Va.

Stellaria pubera has been introduced in Nebraska and possibly in Illinois. It is very similar to *S. corei* but is distinguished by its shorter, more ovate sepals.

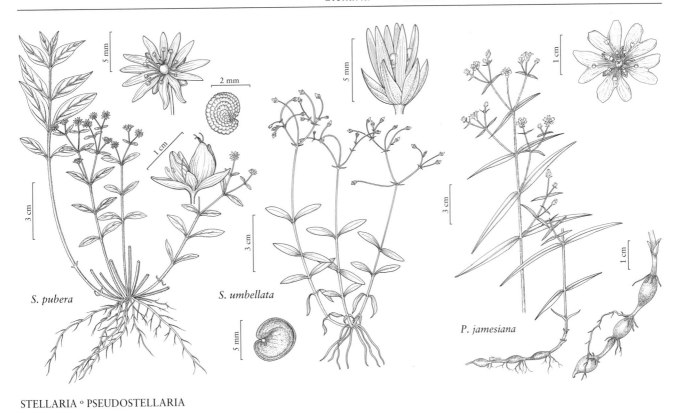

S. pubera

S. umbellata

P. jamesiana

STELLARIA ° PSEUDOSTELLARIA

28. Stellaria ruscifolia Pallas ex Schlectendal, Ges. Naturf. Freunde Berlin Mag. Neuesten Entdeck. Gesammten Naturk. 7: 194. 1816 • Prickly-leaved starwort

Stellaria ruscifolia subsp. *aleutica* Hultén

Plants perennial, forming small to moderate clumps, from elongate rhizomes. **Stems** erect, branched, 4-angled, 3–20 cm, glabrous. **Leaves** sessile; blade ovate to broadly lanceolate, widest below middle, 0.4–2 cm × 2–6 mm, coriaceous, base round, margins entire, apex acuminate, spinous, glabrous. **Inflorescences** with flowers solitary, subterminal in axils of foliage leaves, or terminal, 2–5-flowered cymes; bracts (when present) lanceolate, distally reduced, 3–14 mm, herbaceous, margins scarious, apex acuminate. **Pedicels** stiffly erect, 5–40 mm, glabrous. **Flowers** 10–13 mm diam.; sepals 5, 3-veined, lanceolate, 4.5–6 mm, margins narrow, scarious, apex acute, glabrous or sparsely pilose; petals 5, 5–7 mm, 1–1.5 times as long as sepals, blade apex with lobes oblanceolate; stamens 10, in 2 whorls; styles 3(–4), ascending and outwardly curved, 2 mm. **Capsules** olive green, ovoid, 4–6 mm, equaling and enclosed in sepals, opening by 6(–8) valves; carpophore absent. **Seeds** brown, reniform-rotund, 0.8–1.2 mm, rugose.

Flowering summer. Tundra, gravelly places; 0–1100 m; Alaska; Asia (Russian Far East).

North American material of *Stellaria ruscifolia* is variable but tends to be more compact and smaller than that from the Russian Far East. It is referable to subsp. *aleutica*. It appears to be a relative of *S. longipes*, and some forms of the latter with wider leaves (*S. crassipes*) are very similar but do not have the coriaceous, more or less prickly leaves.

29. Stellaria umbellata Turczaninow, Bull. Soc. Imp. Naturalistes Moscou 15: 173. 1842 • Umbellate starwort [F]

Alsine baicalensis Coville; *Stellaria gonomischa* B. Boivin; *S. weberi* B. Boivin

Plants perennial, forming small clumps or mats, rarely long-straggling, from slender rhizomes. **Stems** erect, branched at base, 4-angled, 5–20 cm (rarely to 40 cm when long and straggling), glabrous. **Leaves** spatulate-petiolate (proximal) or sessile (distal), bases clasping, connate around stem, ciliate; blade elliptic to lanceolate, 3–9 cm × 1–3 mm, somewhat succulent, base round to cuneate, margins entire, apex acute, glabrous. **Inflorescences** terminal, (1–)2–ca. 21-flowered, subumbellate, often with 1 or 2 axillary flowers below; bracts lanceolate, 1–7 mm, distal ones entire, scarious, proximal ones usually herbaceous. **Pedicels** sharply deflexed at base, often curved distally in fruit, 7–20 mm, glabrous. **Flowers** ca. 2 mm diam.; sepals 5, 3-veined, lanceolate, 2.5–3 mm, margins narrow, scarious, apex obtuse, glabrous; petals absent; stamens 5; styles 3, ascending, curled, ca. 0.25 mm. **Capsules** straw colored, conic, 3–4.5 mm, exceeding sepals, apex obtuse, opening by 6 valves; carpophore absent. **Seeds** brownish, round, 0.5–0.7 mm diam., shallowly rugose. $2n = 26$.

Flowering summer. Moist meadows, rocky summits, gravelly stream- and roadsides; 1000–2800 m; Alta., B.C., N.W.T., Yukon; Alaska, Ariz., Calif., Colo., Idaho, Mont., Nev., N.Mex., Oreg., Utah, Wash., Wyo.; Asia.

SELECTED REFERENCE Boivin, B. 1956. *Stellaria* sectio *Umbellatae* Schischkin (Caryophyllaceae). Svensk Bot. Tidskr. 50: 113–114.

23. PSEUDOSTELLARIA Pax in H. G. A. Engler and K. Prantl, Nat. Pflanzenfam. ed. 2, 16c: 318. 1934 • Sticky starwort [Greek *pseudo-*, false, and genus *Stellaria*, alluding to resemblance

Ronald L. Hartman

Richard K. Rabeler

Krascheninikovia Turczaninow ex Fenzl in S. L. Endlicher, Gen. Pl. 13: 968. 1840, not *Krascheninnikovia* Gueldenstaedt 1772

Herbs, perennial. **Taproots** absent, rhizomes usually with spherical or elongate, tuberous thickenings or vertical fleshy roots, rooting at nodes. **Stems** ascending to erect, simple or branched, terete or 4-angled. **Leaves** connate proximally into sheath, sessile; blade 1-veined, linear to lanceolate or elliptic, not succulent, apex acute. **Inflorescences** terminal, open cymes, or flowers solitary; bracts paired, smaller, herbaceous or scarious. **Pedicels** recurved to reflexed from base or abruptly bent downward near distal end in fruit. **Flowers:** perianth and androecium hypogynous; sepals 5, distinct, green, narrowly ovate to lanceolate, 3–7 mm, herbaceous, margins white, scarious, apex obtuse or acute to acuminate, not hooded; petals 5 [absent in cleistogamous flowers], white, not clawed, blade apex 2-fid to V-shaped notch $1/10$–$1/5$ of length; nectaries adnate to base of filaments opposite sepals, circular, thickened, 2–2$1/2$ times filament width; stamens 5 or 10, arising from base of ovary; filaments distinct; styles 3, capitate to clavate, 2–4.5 mm, glabrous proximally; stigmas 3, terminal or linear along adaxial surface of styles, minutely papillate (30×). **Capsules** ovoid, opening by 6, ± 2–3 times recoiled valves; carpophore absent. **Seeds** 1–3, red-brown or brown, circular to oblong or elliptic, plump or laterally compressed, tuberculate, marginal wing absent, appendage absent. $x = 8$.

Species 21 (3 in the flora): w United States, se Europe, e Asia.

We follow W. A. Weber and R. L. Hartman (1979) in recognizing *Pseudostellaria* in North America. The combination of a V-shaped petal sinus cut to only $1/10$–$1/5$ of the petal length, often tuberous rhizomes and thickened roots, and the more or less recoiled capsule valves segregates our species from *Stellaria*. The relationship between the North American and Asian species requires further investigation. While the seeds of *P. sierrae* and *P. oxyphylla* closely resemble those of *P. rupestris* (Turczaninow) Pax, the type species, the presence of five (rather than ten) stamens, absence of tubers, and absence of cleistogamous flowers complicate placing our species in the current infrageneric classification (M. Mizushina 1965).

1. Pedicels and sepals uniformly stipitate-glandular, often densely so; anthers 10, purple
    .... 1. *Pseudostellaria jamesiana*
1. Pedicels glabrous or with narrow internodal line of hairs; sepals glabrous or with margins
    ciliolate in proximal ¹/₂; anthers 5, yellow.
    2. Stems and pedicels each with narrow internodal line of hairs; leaf blades 6–12 cm;
       inflorescences with flowers paired (single by abortion), pedicellate, in distal 3–7 axils
       or in terminal cymes .... 2. *Pseudostellaria oxyphylla*
    2. Stems and pedicels glabrous; leaf blades 0.7–3.5 cm; inflorescences with flowers solitary,
       terminal, sometimes on 1–2 axillary branches .... 3. *Pseudostellaria sierrae*

1. Pseudostellaria jamesiana (Torrey) W. A. Weber &
R. L. Hartman, Phytologia 44: 314. 1979 [E] [F]

Stellaria jamesiana Torrey, Ann. Lyceum Nat. Hist. New York 2: 169. 1827; *Alsine glutinosa* A. Heller; *Arenaria jamesiana* (Torrey) Shinners

Rhizomes with spherical or elongate tuberous thickenings 0.5–2.5 cm. **Stems** 4-angled, 12–45(–60) cm, glabrous or stipitate-glandular throughout or at least in inflorescence, often densely so. **Leaf blades** linear to linear-lanceolate or broadly lanceolate, (1.5–)2–10(–15) × 0.2–1.5(–2) cm, margins flat to briefly revolute, ± smooth or granular to serrulate, glabrous or stipitate-glandular. **Inflorescences** open cymes, flowers often proliferating with age. **Pedicels** recurved to reflexed from base in fruit, uniformly stipitate-glandular. **Flowers:** sepals lanceolate to narrowly ovate, 3–5.5(–7) × 0.8–2 mm, stipitate-glandular, often densely so; petals 7–9.5 × 3–4 mm, apex notch 1–2 mm deep, lobes broadly rounded; anthers 10, purple; styles 3.5–4.5 mm; stigmas terminal, 0.1–0.2 mm. **Capsules** 4.5–5 mm. **Seeds** 1–3, reddish brown, broadly elliptic, ± plump, 2–3.4 mm; tubercles conic to elongate, rounded. $2n = 96$.

Flowering summer. Meadows, sagebrush-grasslands, dry understory of aspen and coniferous forests; 600–3400 m.; Ariz., Calif., Colo., Idaho, Mont., Nev., N.Mex., Oreg., Tex., Utah, Wash., Wyo.

2. Pseudostellaria oxyphylla (B. L. Robinson) R. L. Hartman & Rabeler, Sida 21: 176. 2004 • Robinson's starwort [C][E]

Stellaria oxyphylla B. L. Robinson, Bot. Gaz. 25: 165. 1898

Rhizomes without tuberous thickenings or fleshy storage roots. **Stems** 4-angled, 15–30 cm, pubescent in thin internodal line. **Leaf blades** lanceolate to lance-elliptic, 6–12 × 0.8–1.2 cm, margins often revolute, smooth or granular, sometimes papillate, sparsely ciliate

proximately, glabrous or margins and midrib (adaxial) ciliolate. **Inflorescences** paired flowers (single by abortion) in distal 3–7 axils or in terminal cymes. **Pedicels** recurved to reflexed from base in fruit, with thin internodal line of pubescence. **Flowers:** sepals narrowly ovate to lanceolate, 6.5–7 × 1.5–2.5 mm, essentially glabrous or margins ciliolate in proximal ¹/₂; petals 8–9 × 1.5–2 mm, apex notch 0.8–1 mm deep, lobes narrowly rounded; anthers 5, yellow; styles 3.5–4 mm; stigmas adaxial, linear, 2.5–3 mm. **Capsules** 4–4.5 mm. **Seeds** 1–2, reddish brown, circular, plump, 2.5–2.6 mm; tubercles broadly conic to elongate, rounded, each with 8–12+ stipitate glands smaller than ca. 0.015 mm (50×).

Flowering late spring, summer. Banks along perennial streams, often adjacent to coniferous forests; of conservation concern; 800–900(–1800?) m; Idaho.

Pseudostellaria oxyphylla is known from Kootenai and Shoshone counties in northern Idaho.

3. Pseudostellaria sierrae Rabeler & R. L. Hartman, Novon 12: 82, figs. 1, 2. 2002 • Sierra starwort [C][E]

Rhizomes with ± vertical, fleshy roots 5–15 cm × 0.3 mm, enlarging distally to 2–3 mm diam. **Stems** terete, obtusely angled or grooved when pressed, 9–27 cm, glabrous. **Leaf blades** narrowly lanceolate or elliptic, 0.7–3.5 × 0.1–0.5(–0.8) cm, margins flat, thinly scarious, smooth, glabrous. **Inflorescences** terminal, solitary flowers, sometimes on 1–2 axillary branches. **Pedicels** abruptly bent downward near distal end in fruit, glabrous. **Flowers:** sepals lanceolate to narrowly ovate, 4–6.5(–7) × 1–1.7 mm, glabrous; petals 5–8(–9) × 1–1.5 mm, apex notch 0.2–0.7 mm deep, lobes ± acute; anthers 5, yellow; styles 2–3.5 mm; stigmas terminal, 0.1–0.2 mm. **Capsules** 4–4.5 mm. **Seeds** 1(–2), light brown, circular to somewhat oblong, flattened laterally, 3–3.4 mm; tubercles broadly conic to elongate, rounded, each with 6–12 minute, red bumps that develop into 5–8 conic projections ca. 0.02 mm (50×).

Flowering summer. Dry understory of mixed oak or coniferous forests; of conservation concern; 1400–2000 m; Calif.

First collected in the 1880s, *Pseudostellaria sierrae* appears to be limited to the northern High Sierra region (J. C. Hickman 1993b).

24. MINUARTIA Linnaeus, Sp. Pl. 1: 89. 1753; Gen. Pl. ed. 5, 39. 1754 • Sandwort [for J. Minuart, 1693–1768, Spanish botanist and pharmacist]

Richard K. Rabeler

Ronald L. Hartman

Frederick H. Utech

Alsinanthe (Fenzl ex Endlicher) Reichenbach; *Alsinopsis* Small; *Lidia* Á. Löve & D. Löve; *Minuopsis* W. A. Weber; *Porsildia* Á. Löve & D. Löve; *Sabulina* Reichenbach; *Tryphane* (Fenzl) Reichenbach; *Wierzbickia* Reichenbach

Herbs, annual, winter annual, or perennial, sometimes mat-forming. **Taproots** filiform to stout and woody, perennial plants often with branched caudex or with rhizomes or trailing stems. **Stems** ascending to erect or prostrate, simple or branched, ± terete. **Leaves** mostly connate proximally, petiolate (*M. cumberlandensis, M. godfreyi* proximal leaves) or sessile; blade 1–3-veined, sometimes obscurely so, filiform-linear to subulate, lanceolate or oblanceolate, rarely to ovate, herbaceous to succulent, apex blunt, rounded, or obtuse to acute, acuminate, or spinescent. **Inflorescences** terminal, open or seldom congested cymes or flowers solitary and terminal or axillary, rarely absent; bracts paired, herbaceous or scarious, rarely absent (*M. pusilla, M. rossii*). **Pedicels** erect to arcuate-spreading, rarely reflexed (*M. drummondii*) in fruit. **Flowers:** perianth and androecium perigynous; hypanthium usually disc-, occasionally dish- or cup-shaped; sepals 5, distinct, green (herbaceous portion purple in *M. arctica, M. macrocarpa,* and *M. rossii*), linear, lanceolate, or oblong to elliptic, ovate, or broadly ovate, 1.5–6(–9) mm, margins herbaceous or silvery and scarious, apex rounded or obtuse to acute, acuminate, or spinescent, sometimes hooded; petals 5 or rarely absent, white, rarely pink (*M. biflora*) or lilac (*M. marcescens*), clawed (*M. glabra, M. groenlandica*) or not, blade apex entire, emarginate, or notched; nectaries 5, at base of filaments opposite sepals, sometimes prominent and 2-lobed; stamens 10 (8–10 in *M. godfreyi*), arising from hypanthium; filaments distinct; staminodes absent; styles 3 (to 4 in *M. cumberlandensis, M. godfreyi*), filiform, 0.6–2.5 mm, glabrous proximally; stigmas 3 (to 4 in *M. cumberlandensis, M. godfreyi*), linear along adaxial surface of styles, minutely papillate (30×). **Capsules** ovoid to broadly ellipsoid or rarely globose, opening by 3 incurved or erect to recurved valves; carpophore absent or sometimes present. **Seeds** 1–25, reddish brown to brown or black (or rarely yellowish or purplish brown), spherical or suborbiculate to reniform or obliquely triangular, plump or variously compressed, smooth, reticulate, tuberculate, muriculate-papillate, or rarely with long marginal papillae (*M. macrocarpa*), marginal wing absent (present in *M. douglasii*), appendage absent. x = 7, 9, 10, 12, 13, 14, 15, 23.

Species ca. 175 (33 in the flora): temperate and arctic Northern Hemisphere, n Africa, Asia Minor.

The nectaries in *Minuartia* flowers are often enlarged (to 0.5 mm) and variously lobed; they may not be apparent in fruiting material, possibly due to resorption by the developing flower following pollination. The hypanthium varies from disc- to cup-shaped and ranges in size from less than 1 mm to 3 to 4 mm in diameter (measured on the curve if cup-shaped). The cup-shaped hypanthium is best developed in fruiting material of *M. arctica*, *M. obtusiloba*, and relatives.

Minuartia is the second largest genus of Caryophyllaceae in our flora. It is the largest that here includes strictly native taxa. Of the eight genera with ten or more species, only *Eremogone* also is represented solely by native species.

J. McNeill (1962) outlined an infrageneric classification of *Minuartia* that included four subgenera and 12 sections within subg. *Minuartia*; our 33 species would be distributed among seven of those 12 sections. While we follow McNeill (1962, 1980b) in recognizing *Arenaria* and *Minuartia*, we have chosen not to adopt his hierarchy formally. Some of his groups do appear to represent natural assemblages; others do not. One of the latter includes most of the *Minuartia* species native to the southeastern United States. McNeill (1962) placed these species in sect. *Uninerviae* (Fenzl) Mattfeld; J. Mattfeld (1922) divided them among three series within that section; Á. Löve and D. Löve (1975) segregated two species as the genus *Porsildia*. There has been no thorough subsequent study of the entire group that could further resolve the question. Molecular investigation of *Minuartia* (M. Nepokroeff et al. 2001) suggested that it is not monophyletic. One or more of the various segregate genera that have been proposed, originally based on morphological features, with some resurrected by Löve and Löve on cytological grounds, and that are now in use (e.g., *Alsinanthe*, *Alsinopsis*, *Lidia*, *Minuopsis*, *Tryphane*; W. A. Weber and R. C. Wittmann 1992) may prove to be supported by DNA analysis as well.

Minuartia rossii, a densely pulvinate plant of moist arctic areas, may be difficult to identify in the key since flowers are often absent, especially in northern populations. Vegetative reproduction via easily detached axillary fascicles in the upper leaf axils is more common in these plants.

SELECTED REFERENCES Mattfeld, J. 1922. Geographisch-genetische Untersuchungen über die Gattung *Minuartia* (L.) Hiern. Repert. Spec. Nov. Regni Veg. Beih. 15. McCormick, J. F., J. R. Bozeman, and S. A. Spongberg. 1971. A taxonomic revision of granite outcrop species of *Minuartia* (*Arenaria*). Brittonia 23: 149–160. McNeill, J. and I. J. Bassett. 1974. Pollen morphology and the infrageneric classification of *Minuartia* (Caryophyllaceae). Canad. J. Bot. 52: 1225–1231.

1. Plants annual, not cespitose, rarely mat-forming.
    2. Sepal apices obtuse or merely acutish.
        3. Pedicels stipitate-glandular.
            4. Sepals 1–3-veined; seeds compressed, winged . . . . 10. *Minuartia douglasii* (in part)
            4. Sepals obscurely veined; seeds somewhat compressed, not winged.
                5. Leaf blades glabrous; seeds 0.7–0.8 mm, orbiculate, echinate; sc United States . . . . 11. *Minuartia drummondii*
                5. Leaf blades stipitate-glandular; seeds 1.4–1.7 mm, suborbiculate, low-tuberculate; California, Oregon . . . . 16. *Minuartia howellii* (in part)
        3. Pedicels glabrous.
            6. Sepals 1- or 3-veined, 2.5-2.8 mm; petals not clawed; California, Oregon . . . . 4. *Minuartia californica* (in part)
            6. Sepals obscurely veined, 1.5–4 mm; petals clawed or not; e United States.
                7. Petals broadly obovate, clawed; seeds 0.5–0.8 mm, obliquely triangular with adaxial groove . . . . 13. *Minuartia glabra*
                7. Petals oblanceolate to spatulate, not clawed; seeds 0.4–0.6 mm, suborbiculate, not grooved . . . . 32. *Minuartia uniflora*
    2. Sepal apices acuminate to distinctly acute.
        8. Sepals obscurely veined; leaves stipitate-glandular . . . . 16. *Minuartia howellii* (in part)
        8. Sepals 1–5-veined, often prominently so; leaves glabrous or seldom stipitate-glandular.

9. Pedicels glabrous.
   10. Leaves obscurely veined; petals 0.5–1 times as long as sepals or absent
       . . . . 25. *Minuartia pusilla*
   10. Leaves 1–3-veined; petals 1–2.2(–3) times as long as sepals.
      11. Leaves weakly 1-veined abaxially; sepals 1- or 3-veined, apex rounded
          to acute . . . . 4. *Minuartia californica* (in part)
      11. Leaves 1-veined abaxially; sepals prominently 3–5-veined, apex acute
          to acuminate.
         12. Sepals 3(–5)-veined; petals 1–1.5 times as long as sepals; capsules
             equaling or longer than sepals; California, Oregon . . . . . . 6. *Minuartia cismontana*
         12. Sepals (3–)5-veined; petals 1.5–2.2(–3) times as long as sepals;
             capsules usually shorter than sepals; se and c United States . . .
             . . . . 24. *Minuartia patula* (in part)
9. Pedicels stipitate-glandular, sometimes sparsely so.
   13. Petal apices entire, w United States.
      14. Seeds 1.3–2 mm, winged . . . . 10. *Minuartia douglasii* (in part)
      14. Seeds 0.4–0.6 mm, not winged . . . . 31. *Minuartia tenella*
   13. Petal apices emarginate or notched; se and c United States.
      15. Flowering stems arising from slender wintering stems; inflorescences
          3–5(–7)-flowered; petals oblong-spatulate . . . . 14. *Minuartia godfreyi*
      15. Wintering stems absent; inflorescences 5–50+-flowered; petals obovate.
         16. Sepals 3-veined, apex acute; leaf blades linear-lanceolate to
             oblanceolate, (0.6–)1.5–3.2 mm wide; seeds black, muriculate-
             papillate . . . . 21. *Minuartia muscorum*
         16. Sepals (3- or) 5-veined, apex acute or acuminate; leaf blades linear,
             0.5–1.5(–1.8) mm wide; seeds reddish brown to black, tuberculate
             . . . . 24. *Minuartia patula* (in part)
1. Plants perennial, often cespitose, pulvinate, or mat-forming.
   17. Sepal apices broadly obtuse or rounded.
      18. Stems pubescent, at least distally, often stipitate-glandular.
         19. Leaf blade apices acuminate-pungent; stems 10–30 cm; inflorescences 3–13-
             flowered . . . . 33. *Minuartia yukonensis*
         19. Leaf blade apices rounded to truncate; stems 1–12 cm; inflorescences 1–5-
             flowered.
            20. Leaf blades flat, obscurely 1- or 3-veined.
               21. Inflorescences with flowers solitary; capsules narrowly ellipsoid, 10–
                   18 mm; seeds ringed with elongate papillae . . . . 18. *Minuartia macrocarpa* (in part)
               21. Inflorescences 3–5-flowered; capsules broadly ellipsoid, 5.5 mm; seeds
                   smooth or obscurely sculptured . . . . 3. *Minuartia biflora* (in part)
            20. Leaf blades 3-angled or rounded 3-angled, obscurely or prominently 1- or
                3-veined.
               22. Axillary leaves absent; leaf blades obscurely 1-veined; seeds 1.2–1.6
                   mm, brown, minutely tuberculate . . . . 1. *Minuartia arctica*
               22. Axillary leaves present among vegetative leaves; leaf blades 1- or 3-
                   veined abaxially; seeds 0.6–1.2 mm, brown or reddish tan, smooth or
                   obscurely sculptured.
                  23. Capsules 6–10 mm; seeds 0.9–1.2 mm, brown . . . . 19. *Minuartia marcescens*
                  23. Capsules 3.5–6 mm; seeds 0.6–0.7 mm, reddish tan . . . . . 23. *Minuartia obtusiloba*
      18. Stems glabrous.
         24. Stems 8–28 cm; inflorescences 5–12+-flowered; sepals broadly ovate . . 5. *Minuartia caroliniana*
         24. Stems 1–10(–20) cm; inflorescences 1–5-flowered; sepals linear to ovate or
             obovate.
            25. Petals with apex emarginate or notched, petals 1.4–2.2 times as long as
                sepals; inflorescences 2–5(–7)-flowered cymes or flowers solitary.
               26. Pedicels stipitate-glandular . . . . 3. *Minuartia biflora* (in part)

26. Pedicels glabrous.
    27. Leaf blades (10–)20–30(–40) × 1–3 mm, linear-oblanceolate to linear-spatulate; inflorescences 1(–3)-flowered . . . . . 7. *Minuartia cumberlandensis*
    27. Leaf blades 4–12(–15) × 0.5 mm, ± linear; inflorescences 3–5-flowered . . . . 15. *Minuartia groenlandica*
25. Petals with apex entire, petals 0.8–2 times as long as sepals (sometimes absent); inflorescences with flowers solitary.
    28. Leaf blades 3-veined, often prominently so abaxially; capsules narrowly ellipsoid, 10–18 mm . . . . 18. *Minuartia macrocarpa* (in part)
    28. Leaf blades 1-veined; capsules spherical or ellipsoid, 1.5–4 mm.
        29. Stems 1–3 cm; sepals 1-veined. . . . . 27. *Minuartia rossii* (in part)
        29. Stems 3–15 cm; sepals 3-veined.
            30. Sepals, linear to lanceolate; petals rudimentary or absent . . . . 2. *Minuartia austromontana* (in part)
            30. Sepals, ovate to lanceolate; petals 0.8–1 times as long as sepals . . . . 12. *Minuartia elegans* (in part)

[17. Shifted to left margin.—Ed.]
17. Sepal apices acute to acuminate or spinescent.
    31. Petals 2.5–3 times as long as sepals, apex shallowly notched. . . . . 14. *Minuartia godfreyi* (in part)
    31. Petals with apex entire, if slightly emarginate, then petals 0.7–0.9 times as long as sepals, or absent.
        32. Pedicels stipitate-glandular.
            33. Stems cespitose to mat-forming; taproot not woody, filiform to somewhat thickened, rhizomes or trailing stems absent; seeds 0.4–0.5 mm. . . . . 28. *Minuartia rubella*
            33. Stems mat-forming; taproot woody, stout; rhizomes, stolons, or trailing stems present; seeds 1.5–2.7 mm.
                34. Plants densely stipitate-glandular throughout; w North America . . . . 22. *Minuartia nuttallii*
                34. Plants sparsely stipitate-glandular on pedicels and sepals; California.
                    35. Petals 0.7–0.9 times as long as sepals; stems and leaves green . . . . 9. *Minuartia decumbens*
                    35. Petals 1.4–2.2 times as long as sepals; stems and leaves glaucous or gray-green.
                        36. Stems and leaves glaucous, proximal leaves longer than internodes . . . . 26. *Minuartia rosei*
                        36. Stems and leaves gray-green, proximal leaves often shorter than internodes . . . . 29. *Minuartia stolonifera*
        32. Pedicels glabrous.
            37. Petals 1.1–2 times as long as sepals.
                38. Leaf blades 1–4 mm, 3-angled, 1-veined; sepals 1-veined; high arctic . . . . 27. *Minuartia rossii* (in part)
                38. Leaf blades 5–30 mm, flat to 3-angled, 1–3 veined; sepals 3-veined.
                    39. Leaf blades flat to navicular, apex blunt to pungent; inflorescences 5–30-flowered; Great Plains and east . . . . 20. *Minuartia michauxii*
                    39. Leaf blades navicular, apex rounded; inflorescences 1- or 2–5(–8)-flowered; Great Basin and Rocky Mountains . . . . . . 17. *Minuartia macrantha* (in part)
            37. Petals 0.5–1 times as long as sepals, rudimentary or absent.
                40. Inflorescences with flowers solitary, terminal.
                  41. Sepals linear to lanceolate; petals rudimentary or absent . . . . 2. *Minuartia austromontana* (in part)
                  41. Sepals ovate to lanceolate; petals 0.8–1 times as long as sepals . . . . 12. *Minuartia elegans* (in part)
                40. Inflorescences with flowers solitary and terminal or often 2–5(–15)-flowered.
                42. Sepals 3.5–5 mm, to 5.5 mm in fruit; seeds 0.7–1 mm . . . . 17. *Minuartia macrantha* (in part)
                42. Sepals (1.5–)2.5–3.2 mm, to 4 mm in fruit; seeds 0.4–0.6 mm.

[43. Shifted to left margin.—Ed.]

43. Inflorescences 7–15-flowered (rarely fewer); seeds dark brown to black, 0.5–0.6 mm

.... 8. *Minuartia dawsonensis*

43. Inflorescences 1–3(–5)-flowered; seeds brown or reddish brown, 0.4–0.6 mm........ 30. *Minuartia stricta*

1. Minuartia arctica (Steven ex Seringe) Graebner in P. F. A. Ascherson et al., Syn. Mitteleur. Fl. 5(1): 772. 1918 • Arctic stitchwort or sandwort

Arenaria arctica Steven ex Seringe in A. P. de Candolle and A. L. P. P. de Candolle, Prodr. 1: 404. 1824; *Alsine arctica* (Steven ex Seringe) Fenzl; *Lidia arctica* (Steven ex Seringe) Á. Löve & D. Löve

Plants perennial, mat-forming. **Taproots** stout, woody. **Stems** erect to ascending, green, 3–10 cm, retrorsely puberulent or stipitate-glandular, internodes of flowering stems 2–6 times as long as leaves. **Leaves** tightly overlapping (vegetative) or variably spaced (cauline), usually connate proximally, with tight, scarious to herbaceous sheath 1.2–1.5 mm; blade straight to outwardly curved, green, obscurely 1-veined, linear (proximal vegetative) or subulate (cauline), rounded 3-angled (abaxial surface thickened, rounded, adaxial surface flat to concave), 5–20 × 0.4–1 mm, flexuous, margins not thickened, herbaceous, often ciliate, apex often purple, rounded to truncate, shiny, glabrous (vegetative) or glabrous to stipitate-glandular (cauline); axillary leaves absent. **Inflorescences:** flowers solitary, terminal; bracts narrowly lanceolate to oblong, herbaceous. **Pedicels** 0.5–3 cm, usually densely stipitate-glandular. **Flowers:** hypanthium cup-shaped; sepals prominently 3-veined proximally, lanceolate to narrowly ovate (herbaceous portion often purple, ovate to oblong), 4–8 mm, enlarging slightly in fruit, apex often purple, rounded, hooded, stipitate-glandular; petals oblanceolate, 1.5–2 times as long as sepals, apex broadly rounded, entire. **Capsules** narrowly ellipsoid, 9–10 mm, longer than sepals. **Seeds** brown, suborbiculate with radicle prolonged into beak, compressed, 1.2–1.6 mm, minutely tuberculate (50×). $2n$ = 22 (Russia), 26 (Russia), 38 (Russia), 52, ca. 80.

Flowering summer. Dry ridges, rocky mountain slopes, heathlands, alpine snowbed slopes, stony tundra; 0–1000 m; N.W.T., Yukon; Alaska; Asia.

Minuartia arctica is an amphi-Beringian species that is known to intergrade with *M. obtusiloba*. Hybrids between *M. arctica* and *M. macrocarpa* are known as well.

2. Minuartia austromontana S. J. Wolf & Packer, Canad. J. Bot. 57: 1676, fig. 1. 1979 • Columbian stitchwort, Rocky Mountain sandwort E

Plants perennial, mat-forming. **Taproots** moderately stout, not woody. **Stems** spreading to erect, cespitose, green, 3–15 cm, glabrous, internodes of all stems 0.2–0.6 times as long as leaves. **Leaves** tightly overlapping, usually connate proximally, with ± loose, scarious to herbaceous sheath 0.2–0.8 mm; blade straight to outwardly curved, green, flat, prominently 1-veined abaxially, linear to subulate, 3–10 × 0.1–1 mm, flexuous, margins not thickened, scarious, smooth, apex green or purple, rounded, navicular, shiny, glabrous; axillary leaves present among cauline leaves. **Inflorescences:** flowers solitary, terminal; bracts linear to subulate, herbaceous. **Pedicels** 0.3–1.5 (–2) cm, glabrous. **Flowers:** hypanthium disc-shaped, sepals prominently 3-veined, linear to lanceolate (herbaceous portion linear to lanceolate), 2–3 mm, enlarging slightly in fruit, apex usually purple, acute or rounded, not hooded, glabrous; petals usually absent, if present, rudimentary, linear to oblong, shorter than sepals, apex entire. **Capsules** ellipsoid, 2–3 mm, equaling sepals. **Seeds** 0.6–1 mm, brown, suborbiculate with radicle prolonged into beak, somewhat compressed, minutely tuberculate (50×). $2n$ = 30.

Flowering summer. Dry, rocky, calcareous slopes and fell-fields in alpine areas; 1200–2800 m; Alta., B.C.; Idaho, Mont., Oreg., Utah, Wyo.

Minuartia austromontana is the Rocky Mountains member of the *M. rossii* complex (S. J. Wolf et al. 1979). Specimens from that region identified as *M. rossii* subsp. *columbiana* (Raup) Maguire are probably *M. austromontana*; contrary to B. Maguire's (1958) interpretation, the former is a synonym of *M. elegans*.

3. **Minuartia biflora** (Linnaeus) Schinz & Thellung, Bull. Herb. Boissier, sér. 2, 7: 404. 1907 • Mountain stitchwort or sandwort, minuartie à deux fleurs

Stellaria biflora Linnaeus, Sp. Pl. 1: 422. 1753; *Alsinopsis biflora* (Linnaeus) Rydberg; *Lidia biflora* (Linnaeus) Á. Löve & D. Löve

Plants perennial, mat-forming. **Taproots** stout, woody. **Stems** suberect to ascending, green, 2–10-cm, retrorsely pubescent in lines, internodes of flowering stems 2–7 times as long as leaves. **Leaves** tightly overlapping or not (vegetative and proximal cauline) or variably spaced (distal cauline), usually connate proximally, with tight, scarious to herbaceous sheath 0.5–1 mm; blade straight to outwardly curved, green, flat, obscurely 1-veined abaxially, oblong or spatulate to elliptic, 5–10 × 0.7–2 mm, flexuous, margins not thickened, scarious, rarely ciliate proximally, apex green or purple, rounded, flat to navicular, shiny, glabrous; axillary leaves mostly absent. **Inflorescences** 3–5-flowered, open cymes; bracts lanceolate, herbaceous. **Pedicels** 0.5–1 cm, usually densely stipitate-glandular. **Flowers:** hypanthium cup-shaped; sepals 3-veined prominently in fruit, oblong to narrowly lanceolate (herbaceous portion often purple, oblong to narrowly oblanceolate), 3.5–4.5 mm, not enlarging in fruit, apex rarely purple, rounded, hooded (at least inner sepals) or not, glabrous to stipitate-glandular proximally; petals white or often lilac, broadly oblanceolate, 1.4–1.7 times as long as sepals, apex truncate, often shallowly notched. **Capsules** broadly ellipsoid, 5.5 mm, longer than sepals. **Seeds** brown, suborbiculate with radicle prolonged into beak, slightly compressed, 0.7–0.8 mm, smooth or obscurely sculptured (50×). $2n = 26$.

Flowering spring–summer. Dry, calcareous, gravelly to rocky slopes, fell-fields, snow beds, heath in low arctic to alpine areas; 0–2500 m; Greenland; Alta., B.C., Nfld. and Labr. (Labr.), N.W.T., Nunavut, Que., Yukon; Alaska; circumpolar; Europe; Asia.

Specimens labeled *Arenaria sajanensis* Willdenow ex Schlechtendal from western North America, although sometimes referred to *M. biflora* (e.g., H. J. Scoggan 1978–1979, part 3), are likely to be *M. obtusiloba*.

4. **Minuartia californica** (A. Gray) Mattfeld, Bot. Jahrb. Syst. 57(Beibl. 126): 28. 1921 • California sandwort E F

Arenaria brevifolia Nuttall var. *californica* A. Gray, Proc. Calif. Acad. Sci. 3: 101. 1864; *Alsinopsis californica* (A. Gray) A. Heller; *Arenaria californica* (A. Gray) Brewer; *A. pusilla* S. Watson var. *diffusa* Maguire; *Minuartia pusilla* (S. Watson) Mattfeld var. *diffusa* (Maguire) McNeill

Plants annual. **Taproots** filiform. **Stems** widely spreading to erect, green, (1–)2–12 cm, glabrous, internodes of flowering stems 5–10 times as long as leaves. **Leaves** not overlapping, all evenly spaced, connate proximally, with loose, herbaceous or scarious sheath 0.5–0.7 mm; blade widely spreading, green, concave proximally, flat distally, weakly 1-veined abaxially, linear to awl-shaped or narrowly oblong, 2–5 × 0.2–1.5 mm, flexuous, margins not thickened, scarious proximally, smooth, apex green or purple, acute to often rounded, flat, dull to shiny, glabrous; axillary leaves absent. **Inflorescences** 5–7(–9)-flowered, open cymes, or rarely flowers solitary; bracts subulate, herbaceous, often scarious-margined proximally. **Pedicels** 0.2–1.5 cm, glabrous. **Flowers:** hypanthium disc-shaped; sepals 1–3-veined, midvein sometimes more prominent than lateral veins, broadly lanceolate to elliptic (herbaceous portion narrowly elliptic to oblong), 2.5–2.8 mm, not enlarging in fruit, apex green to purple, rounded to acute, slightly hooded, glabrous; petals elliptic to oblanceolate, 1.5–1.8 times as long as sepals, apex obtuse to rounded, entire. **Capsules** on stipe ca 0.1 mm, ± ovoid, 3–3.5 mm, longer than sepals. **Seeds** reddish brown, suborbiculate with radicle prolonged into beak, slightly compressed, 0.4–0.5 mm, minutely papillate. $2n = 26$.

Flowering spring–summer. Gravelly or sandy slopes, meadows, chaparral, vernal ponds, roadsides; 10–700 (–1500) m; Calif., Oreg.

5. **Minuartia caroliniana** (Walter) Mattfeld, Bot. Jahrb. Syst. 57(Beibl. 126): 28. 1921 • Pine-barren stitchwort or sandwort, long-root E

Arenaria caroliniana Walter, Fl. Carol., 141. 1788; *Alsinopsis caroliniana* (Walter) Small; *Minuopsis caroliniana* (Walter) W. A. Weber; *Sabulina caroliniana* (Walter) Small

Plants perennial. **Taproots** stout, woody; crown with radiating subterranean branches. **Stems** erect to ascending, green, 8–28 cm, glabrous, internodes of flowering stems 0.3–10 times as long as leaves. **Leaves**

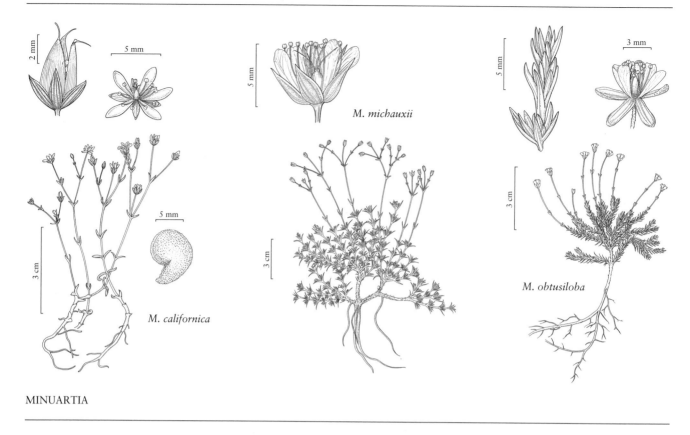

M. michauxii

M. californica

M. obtusiloba

MINUARTIA

variably spaced distally, overlapping (proximal ¹/₃), connate proximally, with tight, scarious to herbaceous sheath 0.2–1.5 mm; blade straight to slightly spreading, green, concave, 3-veined, lateral veins less prominent, lanceolate to subulate, 2–13 × 1–5 mm, rigid, margins rounded, scarious in proximal ¹/₂–²/₃, smooth, apex green, blunt to apiculate, navicular, shiny, glabrous; axillary leaves present. **Inflorescences** 5–12+-flowered, narrow cymes; bracts ovate to subulate, ± scarious. **Pedicels** 0.2–3 cm, densely stipitate-glandular. **Flowers:** hypanthium disc- to cup-shaped; sepals obscurely veined, ± broadly ovate (herbaceous portion ± broadly ovate), 2.5–3 mm, not enlarging in fruit, apex green, rounded, not hooded, stipitate-glandular proximally; petals spatulate, 2.5–3.2 times as long as sepals, apex broadly rounded, entire. **Capsules** on stipe ca 0.1 mm, ovoid, 4.7–5 mm, longer than sepals. **Seeds** brown, suborbiculate, without prolonged beak, not compressed, 0.6–0.65 mm, tuberculate; tubercles low, rounded.

Flowering spring–early summer. Oak or pine woodlands, dry, open, sandy areas; 0–100 m; Del., Fla., Ga., Md., N.J., N.Y., N.C., R.I., S.C., Va.

6. Minuartia cismontana Meinke & Zika, Madroño 39: 289, figs. 1, 3. 1992 • Cismontane minuartia E

Plants annual. **Taproots** filiform. **Stems** erect, green or reddish purple, (5–)8–20(–25) cm, glabrous, internodes of all stems 5–7 times as long as leaves. **Leaves** not overlapping, connate proximally, with loose, scarious sheath 0.1–0.3 mm; blade green or reddish purple, 3-veined proximally, midvein prominent abaxially, lateral veins 0.2–0.25 times as long as blade, straight to outwardly curved, flat, lance-attenuate to linear, 2–7(–9) × 0.5–1.2(–1.8) mm, flexuous, margins not thickened, scarious proximally, smooth, apex green or purple, rounded to acute, often mucronate, shiny, glabrous; axillary leaves occasionally present. **Inflorescences** 5–20-flowered, open cymes; bracts subulate, herbaceous, often scarious-margined proximally. **Pedicels** (0.7–)1–3(–3.5) cm, glabrous. **Flowers:** hypanthium disc-shaped; sepals strongly 3(–5)-veined, lance-linear to lanceolate (herbaceous portion narrowly lanceolate to lance-oblong), 3.2–5.5 mm, not enlarging in fruit, apex green to purple, acute, not hooded, glabrous; petals oblanceolate to oblong-elliptic, 1–1.5 times as long as sepals, apex obtuse to rounded, entire. **Capsules** on stipe about 0.2 mm, ± ovoid, 3.5–5.8 mm, equaling or longer than sepals. **Seeds** brown or reddish, asymmetically reniform with radicle prolonged into beak, not compressed, 0.7–1 mm, minutely papillate.

Flowering spring–summer. Dry woodlands, chaparral, often on serpentine, (100–)400–1700 m; Calif., Oreg.

Minuartia cismontana is closely related to *M. californica* and *M. pusilla*, and has been overlooked as the former for many years. Phenology and elevation appear to segregate the species in areas of California where the ranges of *M. cismontana* and *M. californica* overlap (R. J. Meinke and P. F. Zika 1992).

7. **Minuartia cumberlandensis**   (Wofford & Kral) McNeill, Rhodora 82: 498. 1980 • Cumberland stitchwort   C E

Arenaria cumberlandensis Wofford & Kral, Brittonia 31: 257, fig. 1. 1979

Plants perennial, cespitose from decumbent bases. **Taproots** filiform; basal offshoots present. **Stems** erect or ascending, green, (8–)10–15(–20) cm, glabrous, internodes of flowering stems 0.8–1.2 times as long as leaves. **Leaves** overlapping proximally, variably spaced distally, connate proximally, with ± loose, scarious sheath 0.1–0.2 mm; blade spreading or ascending to outwardly curved, green, flat, 1-veined, linear-oblanceolate to linear-spatulate, (10–)20–30(–40) × 1–3 mm, flexuous, margins not thickened, minutely scarious, smooth, apex green, obtuse to broadly acute, shiny, glabrous, axillary leaves absent. **Inflorescences** flowers solitary, terminal, or 1–3-flowered cymes; bracts narrowly lanceolate, herbaceous. **Pedicels** 12–30 cm, glabrous. **Flowers:** hypanthium dish-shaped; sepals very weakly 3-veined, broadly oblong (herbaceous portion broadly oblong), 2–3 mm, not enlarging in fruit, apex green, obtuse or rounded, not hooded, glabrous; petals oblong or obovate, 1.6–2 times as long as sepals, apex rounded to truncate, entire or slightly emarginate. **Capsules** broadly ovoid, (2–)3–3.5 mm, equaling or longer than sepals. **Seeds** reddish brown, asymmetrically reniform with radicle prolonged into beak, not compressed, 0.5–0.7 mm, reticulate. $2n = 20$.

Flowering summer. Shaded sand-rock ledges and bluffs; of conservation concern; 400–600 m; Ky., Tenn.

Minuartia cumberlandensis may be most closely related to *M. groenlandica* and *M. glabra*; R. Kral (1983) noted that it may be distinguished from either of those taxa by leaf size and shape, seed sculpture, phenology, and habitat preference (shaded sandstone versus sunny granitic flat-rocks).

Minuartia cumberlandensis is in the Center for Plant Conservation's National Collection of Endangered Plants.

8. **Minuartia dawsonensis**   (Britton) House, Amer. Midl. Naturalist 7: 132. 1921 • Rock stitchwort, minuartie de Dawson   E

Arenaria dawsonensis Britton, Bull. New York Bot. Gard. 2: 169. 1901; *Alsinopsis dawsonensis* (Britton) Rydberg; *Arenaria litorea* Fernald, *A. stricta* Michaux var. *dawsonensis* (Britton) Scoggan; *A. stricta* var. *litorea* (Fernald) B. Boivin; *Minuartia litorea* (Fernald) House; *Sabulina dawsonensis* (Britton) Rydberg

Plants perennial, sometimes mat-forming, green. **Taproots** filiform to somewhat thickened. **Stems** erect to ascending, green, 4–30 cm, glabrous, internodes of flowering stems 1–10 times as long as leaves. **Leaves** overlapping or crowded proximally, variably spaced distally, connate proximally, with tight, scarious to herbaceous sheath 0.2–0.5 mm; blade straight to slightly outwardly curved, green, flat, 1-veined, occasionally 3-veined abaxially, linear to subulate, 4–15 × 0.5–2 mm, flexuous, margins not thickened, scarious, smooth, apex green or purple, mostly rounded, slightly navicular, shiny, glabrous; axillary leaves present among proximal cauline leaves. **Inflorescences** 7–15-flowered (rarely fewer), open cymes; bracts subulate, herbaceous. **Pedicels** 0.3–2.5 cm, glabrous. **Flowers:** hypanthium disc-shaped; sepals prominently 3-veined, ovate to broadly lanceolate (herbaceous portion ovate to broadly lanceolate), 2.5–3.2 mm, to 4 mm in fruit, apex green to purple, acute to apiculate, not hooded, glabrous; petals lancolate to spatulate, 0.5–0.8 times as long as sepals, apex rounded, entire, or petals absent. **Capsules** on stipe ca. 0.2 mm, ovoid, 3.5–4.5 mm, longer than sepals. **Seeds** dark brown to black, suborbiculate with radicle prolonged into tiny beak, 0.5–0.6 mm, tuberculate; tubercles low, rounded, somewhat elongate. $2n = 30, 60$.

Flowering late spring–summer. Moist, calcareous ledges and gravelly areas (dry, open, and sometimes disturbed slopes, calcareous-gravel raised beach ridges, thin soil over limestone) in mesic forest openings and meadows in montane and subalpine areas and boreal plains, dry, open outcrops in oak or juniper savannas or prairies; 0–900 m; Alta., B.C., Man., Nfld. and Labr., N.W.T., Nunavut, Ont., Que., Sask., Yukon; Alaska, Mich., Minn., N.Dak., Wis.

Although sometimes included in *Minuartia michauxii* [e.g., H. J. Scoggan's (1978–1979, part 3) treatment of *Arenaria stricta*], *M. dawsonensis* is more closely related to the circumpolar *M. stricta*.

Minuartia litorea, known from Quebec and Ontario, may deserve recognition. An unpublished chromosome count suggests that it is a recent allopolyploid derived from *M. dawsonensis* and *M. rubella* (L. Brouillet, pers. comm.).

9. **Minuartia decumbens**   T. W. Nelson & J. P. Nelson, Brittonia 33: 162, fig. 1. 1981   • Lassicus stitchwort
C E

Plants perennial, mat-forming. **Taproots** stout, woody. **Stems** ascending to erect, green, 4–15 cm, trailing stems to 30 cm, glabrous, internodes of flowering stems ca. as long as leaves. **Leaves** overlapping proximally, all evenly spaced, connate proxi-mally, with tight, scarious sheath 0.5–0.7 mm; blade arcuate, green, flat, 3-veined, needlelike to subulate, 3–6(–9) × 0.7–2 mm, ± rigid, margins scarious proximally, apex green, blunt to ± acute, dull, glabrous; axillary leaves present among vegetative leaves. **Inflorescences** 5–20-flowered, open cymes; bracts narrowly lanceolate, herbaceous, thinly scarious-margined. **Pedicels** 0.5–2.5 cm, sparsely stipitate-glandular. **Flowers:** hypanthium dish-shaped; sepals (1- or) 3-veined, narrowly lanceolate (herbaceous portion narrowly lanceolate), 5–6 mm, not enlarging in fruit, apex often purple, acute to acuminate, not hooded, sparsely stipitate-glandular; petals broadly linear to oblong-elliptic, 0.7–0.9 times as long as sepals, apex rounded, entire to slightly emarginate. **Capsules** ellipsoid, 4–4.8 mm, shorter than sepals. **Seeds** purplish brown, elliptic-oblong, ± compressed, 1.8–2.2 mm, tuberculate; tubercles low, rounded.

Flowering spring–summer. Jeffrey pine woodlands, serpentine soils; of conservation concern; 1200–1500 m; Calif.

Minuartia decumbens, like *M. rosei* and *M. stolonifera*, is restricted to serpentine soils of northwestern California, specifically to Mule Ridge in Trinity County. The three species are most closely related to the polymorphic *M. nuttallii*.

10. **Minuartia douglasii**   (Fenzl ex Torrey & A. Gray) Mattfeld, Bot. Jahrb. Syst. 57(Beibl. 126): 27. 1921   • Douglas's stitchwort   E

Arenaria douglasii Fenzl ex Torrey & A. Gray, Fl. N. Amer. 1: 674. 1840; *Minuartia douglasii* var. *emarginata* (H. K. Sharsmith) McNeill

Plants annual. **Taproots** filiform to somewhat thickened. **Stems** erect to widely spreading, green or purple, 4–30 cm, stipitate-glandular distally, internodes of all stems 1–5 times as long as leaves. **Leaves** sometimes overlapping proximally, ± evenly spaced, connate proximally, with loose, scarious sheath 0.3–0.7 mm; blade straight to variously curved or coiled, green or purple, concave, 1–3 veined, linear, 5–40 × 0.2–0.4 mm, flexuous, margins not thickened, often scarious, sometimes ciliate or stipitate-glandular, apex green to purple, acute, navicular, dull, glabrous to stipitate-glandular; axillary leaves present proximally. **Inflorescences** 7–15+-flowered, open cymes; bracts subulate, scarious. **Pedicels** 0.2–4 cm, stipitate-glandular. **Flowers:** hypanthium disc-shaped; sepals 1(–3)-veined, midvein often more prominent than lateral veins, ± ovate (herbaceous portion lanceolate to oblong-lanceolate to narrowly ovate), 2.5–3.7 mm, not enlarging in fruit, apex often purple, obtuse to acute, not hooded, stipitate-glandular at least proximally; petals ovate, (1–)1.7–2.1 times as long as sepals, apex broadly rounded, entire or irregularly dentate. **Capsules** on stipe 0.3–0.5 mm, broadly ovoid, 4 mm, longer than sepals. **Seeds** reddish brown, winged, orbiculate with radicle not prolonged, compressed, 1.3–2 mm, tuberculate; tubercles low, elongate.

Flowering spring–early summer. Sandy and rocky slopes in chaparral, oak or pine woodlands; 100–1800 m; Ariz., Calif., Oreg.

The winglike margins on the seeds of *Minuartia douglasii* are unique among North American members of the genus; *M. howellii* and *M. tenella*, the other members of sect. *Greniera* (Gay) Mattfeld, do not share this feature.

Plants with petal apices usually emarginate rather than entire and obtuse and often shorter than in typical plants have been recognized by some as var. *emarginata*.

11. **Minuartia drummondii**   (Shinners) McNeill, Notes Roy. Bot. Gard. Edinburgh 24: 147. 1962   • Drummond's stitchwort   E

Arenaria drummondii Shinners, Field & Lab. 17: 89. 1949, based on *Stellaria nuttallii* Torrey & A. Gray, Fl. N. Amer. 1: 183. 1838, not *A. nuttallii* Pax 1893

Plants annual. **Taproots** filiform. **Stems** erect to ascending, green, 5–20 cm, stipitate-glandular, often densely so, internodes of all stems 1–3 times as long as leaves. **Leaves** overlapping proximally, perfoliate proximally, with ± loose, scarious to herbaceous sheath 0.5–1 mm; blade green, flat, 1-veined, oblanceolate to cuneate (proximal) to oblong-lanceolate to ovate (remaining cauline), 5–30(–35) × 2–4 mm, flexuous, margins not thickened, ± scarious, smooth, apex green to purple, obtuse to abruptly pointed, dull, glabrous; axillary leaves absent. **Inflorescences** 7–12-flowered, open cymes, or rarely solitary, terminal; bracts ± lanceolate, herbaceous, sometimes scarious-margined proximally. **Pedicels**

reflexed in fruit, 0.5–2.5 cm, stipitate-glandular. **Flowers:** hypanthium disc-shaped; sepals obscurely veined, ovate to broadly elliptic (herbaceous portion ovate to broadly elliptic), 3–6 mm, to 7 mm in fruit, apex green or purple, acute to acuminate, not hooded, stipitate-glandular; petals obovate to oblanceolate, 2–2.5 times as long as sepals, apex rounded, broadly notched. **Capsules** sessile, broadly ellipsoid, 6–7.5 mm, equaling or longer than sepals. **Seeds** dark brown to blackish, orbiculate with radicle prolonged into beak, only slightly compressed, 0.7–0.8 mm, echinate with rounded tubercles.

Flowering late winter–early summer. Open grassy woodlands, sandy soils; 0–500 m; Ark., La., Okla., Tex.

Minuartia drummondii is easily recognized by the proportionally large corollas (petals to three times as long as sepals) and pedicels reflexing in fruit.

12. **Minuartia elegans** (Chamisso & Schlechtendal) Schischkin in V. L. Komarov et al., Fl. URSS 6: 508. 1936 • Elegant stitchwort

Arenaria elegans Chamisso & Schlechtendal, Linnaea 1: 57. 1826; *Alsinanthe elegans* (Chamisso & Schlechtendal) Á. Löve & D. Löve; *Arenaria rossii* R. Brown ex Richardson subsp. *columbiana* (Raup) Maguire; *A. rossii* var. *columbiana* Raup; *A. rossii* subsp. *elegans* (Chamisso & Schlechtendal) Maguire; *A. rossii* var. *elegans* (Chamisso & Schlechtendal) S. L. Welsh; *Minuartia rossii* (R. Brown ex Richardson) Graebner subsp. *elegans* (Chamisso & Schlechtendal) Rebristaya; *M. rossii* var. *elegans* (Chamisso & Schlechtendal) Hultén

Plants perennial, loosely cespitose. **Taproots** filiform to slightly thickened. **Stems** erect to arcuate-ascending, green, commonly purplish, 3–8 cm, glabrous, internodes of all stems 0.2–1.5 times as long as leaves. **Leaves** tightly overlapping, usually connate proximally, with ± loose, scarious sheath 0.2–0.7 mm; blade ascending to variously curved, green, commonly purplish, flat, prominently 1-veined abaxially, linear to subulate, 3–10 × 1–2 mm, flexuous, margins not thickened, scarious, smooth, apex green to purple, rounded, navicular, shiny, glabrous; axillary leaves present among cauline leaves. **Inflorescences** solitary flowers, terminal; bracts linear to subulate, herbaceous. **Pedicels** 1–4 cm, glabrous. **Flowers:** hypanthium disc-shaped; sepals 3-veined, midrib prominent, lateral veins 1/4–1/2 times as long as sepals, ovate to lanceolate (herbaceous portion ovate to lanceolate), 2–4 mm, not enlarging in fruit, apex often purple, rounded to acute, not hooded, glabrous; petals oblong to obovate, 0.8–1 times as long as sepals, apex rounded, entire, rarely absent. **Capsules** on stipe ca. 0.2 mm, ellipsoid, 2–4 mm, equaling sepals. **Seeds** reddish brown, suborbiculate with radicle prolonged into rounded beak, somewhat compressed, 0.6–1 mm, tuberculate; tubercles low, rounded, elongate. **2***n* = 30, 60.

Flowering spring–summer. Rocky talus, montane ridges and meadows, moist tundra; 0–200 m; Alta., B.C., N.W.T., Yukon; Alaska; Asia (Russian Far East, e Siberia).

Minuartia elegans is a part of the *M. rossii* complex (S. J. Wolf et al. 1979), and is an amphi-Beringian species. The plants are tufted and are known in the flora area only from northwestern Canada and Alaska. Reports from the Pacific Northwest and southern Rocky Mountains likely are referable to *M. austromontana.*

13. **Minuartia glabra** (Michaux) Mattfeld, Bot. Jahrb. Syst. 57(Beibl. 126): 28. 1921 • Appalachian stitchwort E

Arenaria glabra Michaux, Fl. Bor.-Amer. 1: 274. 1803; *Alsinopsis glabra* (Michaux) Small; *Arenaria groenlandica* (Retzius) Sprengel var. *glabra* (Michaux) Fernald; *Minuartia groenlandica* (Retzius) Ostenfeld subsp. *glabra* (Michaux) Á. Löve & D. Löve; *Porsildia groenlandica* (Retzius) Á. Löve & D. Löve subsp. *glabra* (Michaux) Á. Löve & D. Löve; *Sabulina glabra* (Michaux) Small

Plants winter annual or annual. **Taproots** filiform. **Stems** erect or ascending, green, 9–20 cm, glabrous, internodes of all stems 2–7 times as long as leaves. **Leaves** overlapping proximally (basal rosette absent at flowering), connate proximally, with ± loose, scarious sheath 0.2–0.5 mm; blade ascending to spreading, green, 1-veined abaxially, flat, linear, 5–20(–30) × 0.5 mm, flexuous, margins not thickened, slightly scarious, smooth, apex green to purple, rounded to acute, dull, glabrous; axillary leaves absent. **Inflorescences** 8–15-flowered, open, leafy cymes; bracts linear to subulate, mostly herbaceous. **Pedicels** 0.5–2 cm, glabrous. **Flowers:** hypanthium disc-shaped; sepals obscurely veined, oblong-lanceolate, lanceolate, to elliptic (herbaceous portion oblong-lanceolate or lanceolate to elliptic), 1.5–4 mm, to 4.5 mm in fruit, apex green, ± rounded, not hooded, glabrous; petals clawed, broadly obovate, 1.5–2.2 times as long as sepals, apex rounded, shallowly notched. **Capsules** on stipe 0.1 mm or shorter, broadly ellipsoid, 3.5 mm, shorter than sepals. **Seeds** brown, obliquely triangular with adaxial groove, radicle prolonged into short beak, compressed, 0.5–0.8 mm, low-tuberculate. **2***n* = 20.

Flowering spring–summer. Siliceous rock outcrops in woods; 100–500 m; Ala., Conn., Ga., Ill., Ky., Maine, N.H., N.Y., N.C., Pa., R.I., S.C., Tenn., Va., W.Va.

Minuartia glabra, along with *M. groenlandica* (Retzius) Ostenfeld and *M. uniflora* (Walter) Mattfeld, comprise the so-called granite outcrop arenarias of the southeastern United States. These species have been studied extensively, both systematically (e.g., J. F. McCormick et al. 1971; R. E. Weaver 1970) and for pollination biology (R. Wyatt 1984).

At this time, we follow R. E. Weaver (1970) in maintaining *Minuartia glabra* separate from *M. groenlandica*. The species are very similar morpho-logically, including sharing clawed petals and obliquely triangular seeds, like those in most *Sagina* species but unique among North American *Minuartia* species; the annual versus perennial habit and, at least in the southeast, phenology and elevation can be used to distinguish these taxa. Further studies in northern populations may be warranted to resolve the question completely.

14. Minuartia godfreyi (Shinners) McNeill, Rhodora 82: 498. 1980 • Godfrey's stitchwort [C] [E]

Arenaria godfreyi Shinners, Sida 1: 51. 1962, based on *Stellaria paludicola* Fernald & B. G. Schubert, Rhodora 50: 197, plate 1104. 1948, not *A. paludicola* B. L. Robinson 1894

Plants short-lived perennial or winter annual. **Taproots** filiform. **Stems** erect or ascending, arising from mats of slender, prostrate or ascending, wintering stems, green, 10–40 cm, occasionally stipitate-glandular at nodes, internodes of all stems 1–3 times as long as leaves. **Leaves** sometimes overlapping proximally, connate proximally, with ± loose, herbaceous sheath 0.2–0.4 mm; blade widely spreading to erect, green, 1-veined abaxially, flat, narrowly elliptic (proximal petiolate blades) to linear (remaining cauline blades), 8–30 × 0.5–3 mm, flexuous, margins not thickened, scarious, occasionally stipitate-glandular, apex green, acute to acuminate, shiny, glabrous; axillary leaves absent. **Inflorescences** 3–5(–7)-flowered, open, leafy cymes; bracts linear to lanceolate, herbaceous. **Pedicels** 1–5 cm, sparsely glandular. **Flowers:** hypanthium disc-shaped; sepals 3-veined, lanceolate (herbaceous portion linear-lanceolate to lanceolate), 3.5–5 mm, not enlarging in fruit, apex often purple, acute, not hooded, stipitate-glandular; petals oblong-spatulate, 2.5–3 times as long as sepals, apex rounded, shallowly notched. **Capsules** on stipe ca. 0.2 mm, ovoid, 4 mm, shorter than sepals. **Seeds** black, suborbiculate to reniform, radicle obscure, laterally compressed, 0.6–0.8 mm, muriculate-papillate.

Flowering spring. Open mesic meadows, seeps, and stream banks; of conservation concern; 0–400 m; Ala., Ark., Fla., Ga., N.C., S.C., Tenn., Va.

Minuartia godfreyi is known from few, very widely scattered populations; it may be most closely related to *M. patula* and *M. muscorum*, sharing prominently ribbed sepals with both taxa and black, muriculate-papillate seeds with the latter (R. Kral 1983).

15. Minuartia groenlandica (Retzius) Ostenfeld, Meddel. Grønland 37: 226. 1920 • Greenland stitchwort, mountain sandwort, minuartie du Groenland

Stellaria groenlandica Retzius, Fl. Scand. Prodr. ed. 2, 107. 1795; *Alsinopsis groenlandica* (Retzius) Small; *Arenaria groenlandica* (Retzius) Sprengel; *Porsildia groenlandica* (Retzius) Á. Löve & D. Löve; *Sabulina groenlandica* (Retzius) Small

Plants perennial, mat-forming. **Taproots** filiform to slightly thickened. **Stems** ascending to erect, green, 3–10 cm, glabrous, internodes of all stems 2–4 times as long as leaves. **Leaves** overlapping proximally (basal rosette), perfoliate, connate proximally, with ± tight, herbaceous to scarious sheath 0.5–1 mm; blade erect to spreading, green, weakly 1-veined abaxially, flat, ± linear, 4–12(–15) × 0.5 mm, flexuous, margins not thickened, slightly scarious to herbaceous, smooth, apex green, rounded, dull, glabrous; axillary leaves absent. **Inflorescences** 3–5-flowered, open, leafy cymes or sometimes solitary, terminal; bracts linear to subulate, mostly herbaceous. **Pedicels** 0.2–1(–2) cm, glabrous. **Flowers:** hypanthium disc-shaped; sepals obscurely veined, elliptical-oblong to obovate (herbaceous portion elliptical-oblong to obovate), 2–4.5 mm, not enlarging in fruit, apex green, obtuse to rounded, not hooded, glabrous; petals clawed, broadly obovate, 2–2.2 times as long as sepals, apex rounded, shallowly notched. **Capsules** on stipe shorter than 0.1 mm, broadly ellipsoid, 5.5 mm, longer than sepals. **Seeds** brown, obliquely triangular with adaxial groove, radicle prolonged into short beak, compressed, 0.5–0.8 mm, obscurely tuberculate. $2n = 20$.

Flowering late spring–summer. Rocky and gravelly slopes, ledges in alpine areas, cracks in exposed bedrock; 0–1800 m; Greenland; St. Pierre and Miquelon; Nfld. and Labr., N.S., Nunavut, Ont., Que.; Maine, N.H., N.Y., N.C., S.C., Tenn., Vt., Va., W.Va.; South America (Brazil).

Minuartia groenlandica is morphologically very similar to *M. glabra* (Michaux) Mattfeld; the two are clearly separable by habit, phenology, and elevation at the southern end of the range of *M. groenlandica* (R. E. Weaver 1970).

E. Hultén (1964) confirmed the report of *Minuartia groenlandica* from a mountain in southern Brazil (Morro de Igreja, Santa Catarina). This remains the only report of *Minuartia* in South America.

16. **Minuartia howellii** (S. Watson) Mattfeld, Bot. Jahrb. Syst. 57(Beibl. 126): 27. 1921 • Howell's stitchwort E

Arenaria howellii S. Watson, Proc. Amer. Acad. Arts 20: 354. 1885; *Alsinopsis howellii* (S. Watson) A. Heller

Plants annual. **Taproots** moderately stout. **Stems** erect to spreading, green, becoming purple, 12–30 cm, mostly glabrous, internodes of all stems 2–7 times as long as leaves. **Leaves** sometimes overlapping proximally, often connate proximally, with loose, scarious to herbaceous sheath 0.2–0.5 mm; blade straight to recurved, green, becoming purple, concave, prominently 1(–3)-veined abaxially, linear-lanceolate, 5–15 × 1–2.5 mm, rigid, margins not thickened, slightly scarious, smooth, apex green to purple, obtuse to acute, often apiculate, flat to navicular, dull, densely stipitate-glandular; axillary leaves absent. **Inflorescences** 5–25+-flowered, open cymes; bracts lanceolate to ovate, herbaceous. **Pedicels** 0.3–4+ cm, stipitate-glandular. **Flowers:** hypanthium disc-shaped; sepals weakly veined proximally, ovate (herbaceous portion ovate to narrowly so), 1.9–3 mm, not enlarging in fruit, apex often purple, ± acute to acuminate, not hooded, stipitate-glandular proximally; petals oblanceolate, 1.8–2.3 times as long as sepals, apex rounded, entire. **Capsules** on stipe ca. 0.2 mm, broadly ovoid, 2.7–3.5 mm, usually longer than sepals. **Seeds** blackish brown, suborbiculate with radicle prolonged to rounded beak, somewhat compressed, 1.4–1.7 mm, tuberculate; tubercles low, rounded, often elongate.

Flowering spring–summer. Chaparral, Jeffrey pine-oak woodlands, serpentine; 500–1000 m; Calif., Oreg.

17. **Minuartia macrantha** (Rydberg) House, Amer. Midl. Naturalist 7: 132. 1921 • House's stitchwort, large-flower sandwort E

Alsinopsis macrantha Rydberg, Bull. Torrey Bot. Club 31: 407. 1904; *Alsinanthe macrantha* (Rydberg) W. A. Weber; *Arenaria filiorum* Maguire; *A. macrantha* (Rydberg) A. Nelson; *A. rubella* (Wahlenberg) Smith var. *filiorum* (Maguire) S. L. Welsh; *Minuartia filiorum* (Maguire) McNeill

Plants perennial, cespitose or mat-forming. **Taproots** occasionally filiform or often woody, somewhat thickened to moderately stout. **Stems** erect to procumbent, green, 2–15 cm, glabrous, internodes of all stems 0.3–1(–2) times as long as leaves. **Leaves** moderately to tightly overlapping (proximal cauline), variably spaced, progressively more so distally (distal cauline), connate proximally, with loose, scarious sheath 0.3–0.8 mm; blade straight to slightly outcurved, green, flat, to 3-angled distally, 1–3-veined, midvein more prominent than 2 lateral veins, subulate to linear, 5–10 × 0.5–1.2 mm, flexuous, margins not thickened, scarious, smooth, apex green, rounded, thickened and navicular, shiny, glabrous; axillary leaves present among proximal cauline leaves. **Inflorescences** solitary flowers, terminal, or 2–5(–8)-flowered, open cymes; bracts broadly subulate, herbaceous or scarious-margined proximally. **Pedicles** 0.2–1.5 cm, glabrous. **Flowers:** hypanthium disc-shaped; sepals strongly 3-veined, ovate to lanceolate (herbaceous portion lanceolate), 3.5–5 mm, to 5.5 mm in fruit, apex green or purple in part, sharply acute to acuminate, not hooded, glabrous; petals oblong to obovate, 0.7–1.8 times as long as sepals, apex rounded to blunt, entire. **Capsules** on stipe ca. 0.2 mm, broadly ovoid, 3–3.8 mm, shorter than sepals. **Seeds** black, suborbiculate with radicle prolonged to rounded beak, somewhat compressed, 0.7–1 mm, tuberculate; tubercles low, rounded.

Flowering spring–summer. Rocky, often limestone, areas, spruce-fir forests, alpine lake shores, tundra; 2100–3700 m; Ariz., Colo., Nev., N.Mex., Utah, Wyo.

B. Maguire (1958) segregated *Minuartia filiorum* (as *Arenaria filiorum*) from *M. macrantha* on the basis of habit (annual or at most a weak perennial), 3–7 flowers per inflorescence, and petals shorter than the sepals. Some populations may be distinguished using those features; the number of flowers per inflorescence is more variable than Maguire noted, and the seeds of the plants are identical with those of typical *M. macrantha*. We concur with W. A. Weber's herbarium annotations that *M. filiorum* and *M. macrantha* are conspecific.

18. **Minuartia macrocarpa** (Pursh) Ostenfeld, Meddel. Grønland 37: 226. 1920 • Long-pod stitchwort, large-fruited sandwort

Arenaria macrocarpa Pursh, Fl. Amer. Sept. 1: 318. 1813; *Alsinopsis macrocarpa* (Pursh) A. Heller; *Wierzbickia macrocarpa* (Pursh) Reichenbach

Plants perennial, mat-forming. **Taproots** stout, woody. **Stems** erect to ascending, green, 3–10 cm, glabrous or sometimes stipitate-glandular, internodes of flowering stems 1–5 times as long as leaves. **Leaves** tightly overlapping (vegetative), variably spaced (cauline), usually connate proximally, with tight, scarious to herbaceous sheath 1–1.5 mm; blade straight to outwardly curved, green, flat, 3-veined, often prominently so abaxially, linear to oblong or narrowly lanceolate, 4–14 × 0.5–2 mm, flexuous, margins thickened, ± coriaceous, ciliate, often densely so, apex green, rounded, navicular, shiny, glabrous or essentially so throughout or abaxially, sometimes pubescent adaxially, hairs resembling cilia; axillary leaves present among vegetative leaves. **Inflorescences** solitary flowers, terminal; bracts linear, herbaceous. **Pedicels** 0.4–1 cm, usually densely stipitate-glandular. **Flowers:** hypanthium cup-shaped; sepals prominently 3-veined proximally, lanceolate to oblong (herbaceous portion often purple, lanceolate to oblong), 4.5–6 mm, to 9 mm in fruit, apex often purple, rounded, hooded, stipitate-glandular; petals broadly obovate, 1.2–1.6 times as long as sepals, apex blunt or rounded, entire. **Capsules** narrowly ellipsoid, 10–18 mm, longer than sepals. **Seeds** red-brown to brown, orbiculate with radicle prominent and notch filled with papillae, somewhat compressed, 1–1.1 mm (excluding papillae), rounded-tuberculate, ringed with longitudinal, cylindrical, tan papillae 0.5–0.8 mm. $2n = 44$ (Russia), 46, 48 (Russia).

Flowering spring–summer. Rocky, montane ridges, sandy slopes, well-drained alpine tundra and heathlands; 0–2200 m; B.C., N.W.T., Yukon; Alaska; Asia (Japan, Russian Far East, Siberia).

An amphi-Beringian species, *Minuartia macrocarpa* is easily distinguished by having the largest capsules of any North American *Minuartia*.

19. **Minuartia marcescens** (Fernald) House, Amer. Midl. Naturalist 7: 132. 1921 • Serpentine stitchwort or sandwort, minuartie de la serpentine C E

Arenaria marcescens Fernald, Rhodora 21: 15. 1919; *A. laricifolia* Linnaeus var. *marcescens* (Fernald) B. Boivin

Plants perennial, mat-forming or more commonly straggly. **Taproots** stout, woody. **Stems** ascending, green, 4–6 cm, glabrous proximally, stipitate-glandular distally, internodes of flowering stems 6–8 times as long as leaves. **Leaves** tightly overlapping (vegetative), variably spaced (cauline), usually connate proximally, with tight, scarious to herbaceous sheath 0.5–1.5 mm; blade straight to outwardly curved, green, 3-angled, prominently 1-veined abaxially, subulate, 4–8 × 0.3–0.8 mm, flexuous, margins not thickened, herbaceous, smooth, apex green, rounded to truncate, sometimes apiculate, shiny, glabrous; axillary leaves present among

vegetative leaves. **Inflorescences** solitary flowers, terminal; bracts lance-subulate, herbaceous. **Pedicels** 0.5–1.5 cm, usually densely stipitate-glandular. **Flowers:** hypanthium cup-shaped; sepals 3-veined, ovate to broadly lanceolate (herbaceous portion oblong to narrowly ovate), 3–4 mm, not enlarging in fruit proximally, apex often purple, rounded, hooded or not, stipitate-glandular; petals white or rarely lilac, spatulate to spatulate-obovate, 1.5–2 times as long as sepals, apex rounded, entire. **Capsules** narrowly ellipsoid, 6–10 mm, longer than sepals. **Seeds** brown, suborbiculate with radicle prolonged into beak, somewhat compressed, 0.9–1.2 mm, smooth.

Flowering summer. Ultramafic ledges and barrens; of conservation concern; 200–1000 m; Nfld. and Labr. (Nfld.), Que.; Vt.

Marcescent leaves are a characteristic feature of this species.

20. **Minuartia michauxii** (Fenzl) Farwell, Rep. (Annual) Michigan Acad. Sci. 20: 177. 1919 • Michaux's stitchwort, rock sandwort E F

Alsine michauxii Fenzl, Vers. Darstell. Alsin., plate at 18. 1833, based on *Arenaria stricta* Michaux, Fl. Bor.-Amer. 1: 274. 1803, not *Alsine stricta* (Swartz) Wahlenberg 1812; *Arenaria stricta* subsp. *texana* (B. L. Robinson) Maguire; *A. stricta* var. *texana* (Britton) B. L. Robinson; *A. texana* Britton; *Minuartia michauxii* var. *texana* (Britton) Mattfeld; *Minuopsis michauxii* (Fenzl) W. A. Weber; *Sabulina stricta* (Michaux) Small

Plants perennial, cespitose, sometimes matted. **Taproots** thickened, woody; crown many-branched, thickened. **Stems** erect to ascending, green, 8–40 cm, glabrous, internodes of flowering stems 0.5–10+ times as long as leaves. **Leaves** tightly overlapping in proximal ¹/₃ of stem, variable spaced, usually connate proximally, with loose, scarious sheath 0.2–1 mm; blade erect to spreading, green, flat or convex to 3-angled, 1–3-veined, prominently so abaxially, filiform, linear to linear-lanceolate, 8–30 × 0.5–1.8 mm, rigid, margins not thickened, ± scarious, smooth, apex green, blunt to pungent, flat to navicular, shiny, glabrous; axillary leaves present among vegetative leaves. **Inflorescences** 5–30-flowered, lax to congested cymes; bracts narrowly lanceolate to subulate, herbaceous, margins scarious. **Pedicels** 0.3–6 cm, glabrous. **Flowers:** hypanthium dish-shaped; sepals 3-veined, ovate to lanceolate (herbaceous portion ovate to narrowly lanceolate), 3–6 mm, not enlarging in fruit, apex green, acute to mostly acuminate, not hooded, glabrous; petals oblong-obovate, 1.3–2 times as long as sepals, apex rounded, entire. **Capsules**

on stipe ca. 0.1–0.2 mm, ellipsoid, 3–4 mm, usually shorter than sepals. **Seeds** black, suborbiculate, compressed, 0.8–0.9 mm, tuberculate; tubercles elongate. **2n** = 30(?), 44.

Flowering spring–summer. Dry, calcareous gravel and ledges; 0–1000 m; Ont., Que.; Ala., Ark., Conn., Del., Ill., Ind., Iowa, Kans., Ky., Md., Mass., Mich., Mo., Nebr., N.H., N.J., N.Mex., N.Y., Ohio, Okla., Pa., R.I., S.Dak., Tex., Vt., Va., W.Va., Wis.

We concur with J. A. Steyermark (1963) and G. Yatskievych and J. Turner (1990) that the concept of *Arenaria texana* originally put forth by Britton requires further study. Plants labeled as *Minuartia michauxii* var. *texana* have slightly shorter leaves that are crowded into the proximal one-third rather than one-half of the somewhat shorter stem. The plants are often in the southern part of the range and may, as Steyermark noted, be associated with more open, xeric habitats.

21. Minuartia muscorum (Fassett) Rabeler, Sida 15: 95. 1992 • Dixie stitchwort E

Stellaria muscorum Fassett, Rhodora 39: 460. 1937; *Arenaria muriculata* Maguire; *A. patula* Michaux var. *robusta* (Steyermark) Maguire; *Minuartia muriculata* (Maguire) McNeill; *M. patula* (Michaux) Mattfeld var. *robusta* (Steyermark) McNeill

Plants annual. **Taproots** filiform. **Stems** erect, green, 10–55 cm, glabrous or weakly stipitate-glandular distally, internodes of all stems 0.5–2.5 times as long as leaves; wintering stems absent. **Leaves** not overlapping, connate proximally, with loose, scarious sheath 0.4–0.8 mm; blade straight to variously curved, green, flat, 1-veined, linear-lanceolate to oblanceolate, (5–)10–35(–50) × (0.6–)1.5–3.2 mm, flexuous, margins not thickened, herbaceous or thinly scarious, smooth, apex green, acute, flat, dull, glabrous; axillary leaves absent. **Inflorescences** 5–50+-flowered, open cymes; bracts lanceolate to subulate, herbaceous. **Pedicels** 0.6–5.5 cm, stipitate-glandular. **Flowers:** hypanthium shallowly disc-shaped; sepals prominently 3-veined, lanceolate (herbaceous portion narrowly lanceolate), 3–4 mm, to 5 mm in fruit, apex green, acute, not hooded, stipitate-glandular; petals obovate, 1.6–3 times as long as sepals, apex rounded, broadly notched. **Capsules** on stipe ca. 0.1 mm or shorter, ovoid to broadly so, 5.2–7 mm, longer than sepals. **Seeds** black, suborbiculate, radicle obscure, plump to slightly compressed, 0.6–0.8 mm, muriculate-papillate.

Flowering spring–summer. Prairies, meadows, roadsides; 200–500 m; Ala., Ark., La., Mo., Okla., Tenn., Tex.

Minuartia muscorum is closely related to *M. patula*, and is distinguished by the often longer and wider leaves, often longer distal stem internodes, consistently three-veined sepals, and shiny, black, muriculate-papillate seeds. B. Maguire (1951) treated this taxon as both a variety of *Arenaria patula* and a new species; see R. K. Rabeler (1992) for a review of the curious nomenclatural history.

22. Minuartia nuttallii (Pax) Briquet, Annuaire Conserv. Jard. Bot. Genève 13–14: 385. 1911 • Nuttall's sandwort E

Arenaria nuttallii Pax, Bot. Jahrb. Syst. 18: 30. 1893; *Minuopsis nuttallii* (Pax) W. A. Weber

Plants perennial, mat-forming. **Taproots** thickened, woody; crown, many-branched, woody; rhizomes and trailing stems to 60 cm. **Stems** ascending to erect, ± green, 2–20 cm, densely glandular-hairy throughout, internodes of flowering stems 0.2–2 times as long as leaves. **Leaves** tightly appressed to spreading, ± evenly spaced, connate proximally, with ± loose, scarious sheath 0.1–0.7 mm; blade straight to recurved, ± green, flat, prominently 1-veined abaxially, broadly lanceolate to linear, 5–20 × 0.5–1.5 mm, ± rigid, margins rounded, scarious in proximal 1/3–1/4, apex green to purple, acute to acuminate or spinescent, navicular with small mucro or spinescent, dull, stipitate-glandular; axillary leaves present proximally to throughout. **Inflorescences** (3–)6–30-flowered, open cymes; bracts lanceolate to subulate, usually scarious. **Pedicels** 0.2–2 cm, stipitate-glandular. **Flowers:** hypanthium disc-shaped; sepals 1–3-veined, narrowly lanceolate to lanceolate or ovate (herbaceous portion narrowly lanceolate to lanceolate or ovate), 3–6(–7) mm, not enlarging in fruit, apex often purple, acute to acuminate or spinescent, not hooded, stipitate-glandular; petals obovate, 0.5–1.6 times as long as sepals, apex rounded, entire. **Capsules** on stipe ca. 0.1–0.2 mm, ovoid, 5 mm, usually shorter than sepals. **Seeds** reddish brown to dark brown, oblong-elliptic with hilar notch on 1 end, 1.5–2.7 mm, tuberculate; tubercles low-rounded.

Varieties 4 (4 in the flora): w North America.

Minuartia nuttallii, M. decumbens, M. rosei, and *M. stolonifera* form a complex that, together with the eastern species *M. caroliniana* and *M. michauxii,* comprise sect. *Sclerophylla* Mattfeld. The four western species all have capsules that contain one to three(?) large (1.5–2.8 mm) seeds; unfortunately, these plants appear to be collected only rarely in fruit.

Minuartia nuttallii includes four varieties, which can, for the most part, be easily recognized. There is some overlap between var. *gracilis* and var. *fragilis* in western Nevada and southeastern Oregon, where some plants exhibit prominently arcuate-spreading leaves (as in var. *fragilis*) and weakly veined sepals (as in var. *gracilis*).

1. Leaf blade apices spinecent; sepal apices spinescent.
   2. Leaves prominently arcuate-spreading, blade 10–20 mm; sepals (1–)3-veined . . . . 22b. *Minuartia nuttallii* var. *fragilis*
   2. Leaves appressed to occasionally arcuate-spreading, blade 5–7 mm; sepals 1(–3)-veined . . . . 22c. *Minuartia nuttallii* var. *gracilis*
1. Leaf blade apices acute to acuminate, somewhat navicular; sepal apices acuminate.
   3. Sepals lanceolate to narrowly so; petals 0.5–1.2 times as long as sepals . . . . 22a. *Minuartia nuttallii* var. *nuttallii*
   3. Sepals ovate to lanceolate; petals 1.1–1.6 times as long as sepals . . . . 22d. *Minuartia nuttallii* var. *gregaria*

22a. Minuartia nuttallii (Pax) Briquet var. nuttallii E

Arenaria pungens Nuttall; *Minuopsis pungens* (Nuttall) Mattfeld

Leaves closely appressed to prominently arcuate; blade 7–11 mm, apex acute to acuminate, navicular, with small mucro, not spine-tipped. **Flowers:** sepals prominently 1(–3)-veined, lance-olate to narrowly so (herbaceous portion lanceolate to narrowly so), 3.5–5.5(–7) mm, apex acuminate; petals 0.5–1.2 times as long as sepals. $2n = 36$.

Flowering spring–summer. Sandy and rocky slopes and ridges, chaparral, open pine woodlands, alpine slopes; 600–3800 m; Alta., B.C.; Colo., Idaho, Mont., Oreg., Utah, Wash., Wyo.

22b. Minuartia nuttallii (Pax) Briquet var. fragilis (Maguire & A. H. Holmgren) Rabeler & R. L. Hartman, Sida 21: 753. 2004 • Brittle sandwort E

Arenaria nuttallii Pax subsp. *fragilis* Maguire & A. H. Holmgren, Madroño 8: 260. 1946; *A. nuttallii* var. *fragilis* (Maguire & A. H. Holmgren) C. L. Hitchcock; *Minuartia nuttallii* subsp. *fragilis* (Maguire & A. H. Holmgren) McNeill

Leaves prominently arcuate-spreading; blade 10–20 mm, apex spinescent. **Flowers:** sepals (1–)3-veined, lanceolate (herbaceous portion lanceolate), (3–)5–6 mm, apex spinescent; petals 0.8–1.2 times as long as sepals.

Flowering spring–summer. Metamorphic limestone talus; 1600–2400 m; Calif., Nev., Oreg.

Collections of var. *fragilis* often appear more yellowish than those of the other varieties.

22c. Minuartia nuttallii (Pax) Briquet var. gracilis (B. L. Robinson) Rabeler & R. L. Hartman, Sida 21: 753. 2004 • Brittle sandwort E

Arenaria nuttallii Pax var. *gracilis* B. L. Robinson, Proc. Amer. Acad. Arts 29: 304. 1894; *A. nuttallii* subsp. *gracilis* (B. L. Robinson) Maguire; *Minuartia nuttallii* subsp. *gracilis* (B. L. Robinson) McNeill

Leaves tightly appressed to occasionally arcuate-spreading; blade 5–7 mm, apex spinescent. **Flowers:** sepals 1(–3)-veined, lanceolate (herbaceous portion lanceolate to often narrowly so), 4–5.7 mm, apex spinescent; petals 0.7–1 times as long as sepals.

Flowering spring–summer. Loose talus, sandy flats, barren rock; 2600–3800 m; Calif., Nev., Oreg.

22d. Minuartia nuttallii (Pax) Briquet var. gregaria (A. Heller) Rabeler & R. L. Hartman, Sida 21: 754. 2004 • Brittle sandwort E

Arenaria gregaria A. Heller, Bull. S. Calif. Acad. Sci. 2: 67. 1903; *Alsinopsis gregaria* (A. Heller) A. Heller; *Arenaria nuttallii* Pax subsp. *gregaria* (A. Heller) Maguire; *A. nuttallii* var. *gregaria* (A. Heller) Jepson; *Minuartia nuttallii* subsp. *gregaria* (A. Heller) McNeill

Leaves tightly appressed to slightly arcuate-spreading; blade 5–10 mm, apex acute to acuminate, navicular, with small mucro, not spine-tipped. **Flowers:** sepals 1- or 3-veined, lateral veins less distinct, ovate to lanceolate (herbaceous portion ovate to lanceolate), 3–6 mm, apex acute to acuminate; petals 1.1–1.6 times as long as sepals.

Flowering spring–summer. Sandy and rocky slopes and ridges, scree, barren rock, serpentine chaparral, Jeffrey pine woodlands; 600–3200 m; Calif., Oreg.

Variety *gregaria* is the variety likely the most closely related to *M. decumbens*, *M. rosei*, and *M. stolonifera*; it is the only one found in northwestern California and adjacent southwestern Oregon.

23. **Minuartia obtusiloba** (Rydberg) House, Amer. Midl. Naturalist 7: 132. 1921 • Twin-flower sandwort, alpine stitchwort [F]

Alsinopsis obtusiloba Rydberg, Bull. Torrey Bot. Club 33: 140. 1906, based on *Arenaria obtusa* Torrey, Ann. Lyceum Nat. Hist. New York 2: 170. 1827, not Allioni 1785; *A. obtusiloba* (Rydberg) Fernald; *Lidia obtusiloba* (Rydberg) Á. Löve & D. Löve

Plants perennial, cespitose to mat-forming. **Taproots** stout, woody. **Stems** erect, green, 1–12 cm, trailing stems 2–20+ cm, stipitate-glandular, internodes of flowering stems 1–6 times as long as leaves. **Leaves** tightly overlapping (vegetative), variably spaced (cauline), usually connate proximally, with tight, scarious to herbaceous sheath 0.3–1.5 mm; blade straight to outwardly curved, green, 3-angled, 3-veined abaxially, midrib prominent, lateral veins weak in distal $^1/_3$, needlelike to subulate, 1–8 × 0.4–1 mm, flexuous, margins not thickened, herbaceous, sometimes finely ciliate, apex green, rounded to acute, often apiculate, somewhat navicular, shiny, glabrous; axillary leaves present among vegetative leaves. **Inflorescences** solitary flowers, terminal, or occasionally in 2–3-flowered, open cymes; bracts subulate, herbaceous. **Pedicels** 0.3–1.5 cm, stipitate-glandular. **Flowers:** hypanthium cup-shaped; sepals prominently 3-veined proximally, narrowly ovate to oblong (herbaceous portion lanceolate to oblong), 2.9–6.5 mm, not enlarging in fruit, apex often purple, narrowly rounded, hooded; petals ovate to spatulate, 1.2–2 times as long as sepals, apex rounded, entire. **Capsules** narrowly ellipsoid, 3.5–6 mm, equaling sepals. **Seeds** reddish tan, suborbiculate with radicle prolonged into beak, somewhat compressed, 0.6–0.7 mm, obscurely sculptured (50×). $2n = 26$, ca. 52, 78.

Flowering summer. Dwarf willow communities, fell-fields, snow beds in subalpine and alpine areas; 0–4000 m; Alta., B.C., N.W.T., Yukon; Alaska, Ariz., Calif., Colo., Idaho, Mont., Nev., N.Mex., Oreg., Utah, Wash., Wyo.; Asia (Russian Far East).

Minuartia obtusiloba, an amphi-Beringian species, sometimes forms hybrid swarms with *M. arctica*. Specimens labeled *Arenaria sajanensis* Willdenow ex Schlechtendal from western North America, sometimes referred to *M. biflora* (e.g., H. J. Scoggan 1978–1979, part 3), are likely to be *M. obtusiloba*.

24. **Minuartia patula** (Michaux) Mattfeld, Bot. Jahrb. Syst. 57(Beibl. 126): 28. 1921 • Pitcher's stitchwort [E] [F]

Arenaria patula Michaux, Fl. Bor.-Amer. 1: 273. 1803; *Alsinopsis patula* (Michaux) Small; *A. pitcheri* Nuttall; *Sabulina patula* (Michaux) Small ex Rydberg

Plants winter annual or annual. **Taproots** filiform. **Stems** erect to ascending, green, 5–30 cm, glabrous or sometimes stipitate-glandular distally or throughout, internodes of all stems 1–7 times as long as leaves; wintering stems absent. **Leaves** overlapping proximally, connate proximally, with loose, scarious to herbaceous sheath 0.1–0.5 mm; blade straight to variously curved, green, flat, prominently 1-veined abaxially, linear, 2–20 × 0.5–1.5(–1.8) mm, flexuous, margins not thickened, herbaceous, smooth, apex green or purple, blunt to acute, flat, ± shiny, glabrous to stipitate-glandular; axillary leaves absent. **Inflorescences** 5–30-flowered, open cymes; bracts subulate to ovate, herbaceous. **Pedicels** 0.3–3 cm, stipitate-glandular. **Flowers:** hypanthium shallowly disc-shaped; sepals prominently (3- or) 5-veined, narrowly to broadly lanceolate (herbaceous portion narrowly to broadly lanceolate), 4–5.5 mm, not enlarging in fruit, apex green or purple, narrowly acute to acuminate, not hooded, glabrous to sparsely stipitate-glandular; petals obovate, 1.5–2.2(–3) times as long as sepals, apex rounded, broadly notched. **Capsules** on stipe ca. 0.1 mm or shorter, narrowly ellipsoid, 3–4.2 mm, shorter than sepals. **Seeds** reddish brown to black, suborbiculate, radicle obscure, slightly compressed, 0.5–0.6 mm, tuberculate; tubercles low, rounded.

Flowering spring–early summer. Prairies, meadows, limestone barrens, and rocky outcrops in sandy, clayey, or gravelly soils; 0–500; Ala., Ark., Ga., Ill., Ind., Kans., Ky., La., Miss., Mo., Ohio, Okla., Pa., Tenn., Tex., Va., Wis.

Minuartia patula and the related *M. muscorum* have received little attention in comparison to the granite-outcrop minuartias, the *M. uniflora* complex. J. A. Steyermark (1941) studied these taxa and described three forms, based chiefly on pubescence variation. Plants entirely glabrous [forma *pitcheri* (Nuttall) Steyermark] and those with sepals and pedicels somewhat stipitate-glandular (forma *media* Steyermark) were segregated from densely stipitate-glandular plants (forma *patula*). We do not feel that such variations deserve formal taxonomic recognition. Forma *robusta*, as defined by Steyermark, is here referred to *M. muscorum*.

Most specimens of *Minuartia patula* have prominently five-veined sepals (seen especially easily in the glabrous forms); occasional plants from Georgia, Kentucky, and

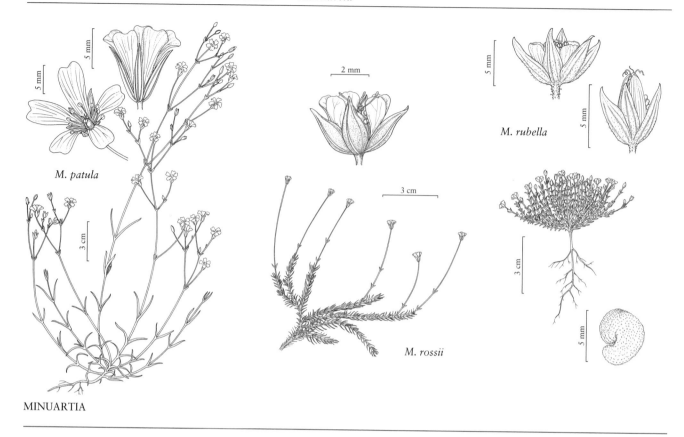

M. patula

M. rubella

M. rossii

MINUARTIA

Virginia have glabrous sepals with only three strong veins, resembling those of *M. muscorum*; in other features, including the seeds, they are clearly referable to *M. patula*. The status of the plants with three-veined sepals remains ambiguous; J. A. Steyermark (1941) included them in his forma *media* and B. Maguire (1951) included them (in our opinion incorrectly) in his var. *robusta*.

25. **Minuartia pusilla** (S. Watson) Mattfeld, Bot. Jahrb. Syst. 57(Beibl. 126): 28. 1921 · Annual sandwort, dwarf stitchwort [E]

Arenaria pusilla S. Watson, Proc. Amer. Acad. Arts 17: 367. 1882; *Alsinopsis pusilla* (S. Watson) Rydberg

Plants annual. **Taproots** threadlike. **Stems** spreading to erect, green, 1–5 cm, glabrous, internodes of all stems 2–10 times as long as leaves. **Leaves** not overlapping, irregularly spaced, connate proximally, with loose, mostly herbaceous sheath 0.2–0.3 mm; blade as-

cending to widely spreading, green, concave proximally, flat distally, obscurely 1-veined, awl-shaped to lanceolate, 1.5–5 × 0.2–1.5 mm, flexuous, margins not thickened, scarious throughout or proximally, smooth, apex green or purple, acute to obtuse, flat to navicular, often shiny, glabrous; axillary leaves absent. **Inflorescences** 2–9-flowered, open cymes; bracts subulate, herbaceous, margins scarious proximally. **Pedicels** 0.1–0.5 cm, glabrous. **Flowers:** hypanthium disc-shaped; sepals 1-veined, or weakly 3-veined in proximal $1/5$, ovate to lanceolate (herbaceous portion lanceolate to narrowly so), 1.5–3.5 mm, not enlarging in fruit, apex green or purple, acute to acuminate, not hooded, glabrous; petals narrowly lanceolate, 0.5–1 times as long as sepals, apex narrowly acute, entire, or absent. **Capsules** on stipe ca. 0.1 mm, ± ovoid, ca. 3 mm, equaling or longer than sepals. **Seeds** brown or reddish, asymmetically reniform with radicle prolonged into beak, not compressed, 0.5–0.6 mm, minutely papillate.

Flowering spring–summer. Plains, pine barrens, dry rock cliffs; 50–2400 m; B.C.; Calif., Idaho, Nev., Oreg., Utah, Wash.

26. Minuartia rosei (Maguire & Barneby) McNeill, Rhodora 82: 499. 1980 • Peanut stitchwort E

Arenaria rosei Maguire & Barneby, Leafl. W. Bot. 8: 56. 1956

Plants perennial, mat-forming. **Taproots** moderately stout, woody. **Stems** ascending to erect, green, 5–20 cm, glabrous except in inflorescence, glaucous, internodes of stems 0.5–4 times as long as leaves (proximal leaves longer than internodes); rhizomes and trailing stems 5–20 cm. **Leaves** loosely overlapping proximally, ± evenly spaced, connate proximally, with tight, scarious sheath 0.2–0.6 mm; blade straight or outwardly curved, green, shallowly concave (dorsiventrally flattened, curved into trough), 1-veined abaxially, needlelike, 4–15 × 0.5–1.2 mm, ± flexuous, margins not thickened, scarious in proximal ¹/₅, smooth, apex green to purple, acute to obtuse, sometimes apiculate, navicular, dull, glabrous, glaucous; axillary leaves well developed among proximal cauline leaves. **Inflorescences** 12–25-flowered, open cymes; bracts subulate, herbaceous, margins scarious proximally. **Pedicels** 0.4–3 cm, often stipitate-glandular. **Flowers:** hypanthium disc-shaped; sepals obscurely 1-veined, narrowly ovate to lanceolate (herbaceous portion narrowly ovate to lanceolate), 2.5–4 mm, not enlarging in fruit, apex green to purple, acute to acuminate, not hooded, glabrous or very sparsely stipitate-glandular; petals oblanceolate to narrowly oblong-elliptic, 1.4–2.2 times as long as sepals, apex rounded, entire. **Capsules** sessile, ovoid, 3.5–4.3 mm, longer than sepals. **Seeds** reddish brown to brown, oblong-elliptic, compressed, 2.3–2.8 mm, tuberculate; tubercles low, rounded.

Flowering spring–summer. Open, serpentine slopes with scattered oak and Jeffrey pine; 700–1400 m; Calif.

Minuartia rosei, like *M. decumbens* and *M. stolonifera*, is restricted to serpentine soils of northwestern California. The three species are most closely related to the polymorphic *M. nuttallii*.

27. Minuartia rossii (R. Brown ex Richardson) Graebner in P. F. A. Ascherson et al., Syn. Mitteleur. Fl. 5(1): 772. 1918 • Ross's sandwort F

Arenaria rossii R. Brown ex Richardson in J. Franklin et al., Narr. Journey Polar Sea, 738. 1823; *Alsinanthe rossii* (R. Brown ex Richardson) Á. Löve & D. Löve; *Alsinopsis rossii* (R. Brown ex Richardson) Rydberg; *Arenaria rossii* var. *apetala* Maguire; *Minuartia orthotrichoides* Schischkin; *M. rolfii* Nannfeldt; *M. rossii* var. *orthotrichoides* (Schischkin) Hultén

Plants perennial, densely pulvinate to loosely cespitose. **Taproots** stout, woody. **Stems** ascending to spreading, green or often purple, 1–3 cm, glabrous, internodes of flowering stems 0.2–1 times as long as leaves. **Leaves** overlapping, ± tightly (vegetative), ± evenly spaced proximally (cauline), connate-perfoliate proximally, with tight, herbaceous sheath 0.2–0.3 mm; blade upwardly curved, green or often purple, keeled, prominently 1-veined abaxially, subulate, 3-angled, 1–4 × 0.5–0.7 mm, flexuous, margins rounded, herbaceous, smooth, apex green to purple, rounded, navicular, shiny, glabrous; axillary leaves well developed. **Inflorescences** solitary flowers, axillary or terminal (rarely present); bracts absent. **Pedicels** 0.1–2 cm, glabrous. **Flowers:** hypanthium disc-shaped; sepals 1-veined, oblong-ovate (herbaceous portion usually purple, oblong-ovate), 1.5–2.5 mm, not enlarging in fruit, apex often purple, obtuse to acuminate, navicular, not hooded, glabrous; petals obovate to spatulate, 1.5–2 times as long as sepals, apex obtuse, entire. **Capsules** on stipe ca. 0.1–0.2 mm, spheric, 1.5–2.5 mm, equaling sepals. **Seeds** brown, suborbiculate, compression unknown, ca. 0.6 mm, obscurely reticulate. $2n$ = 58 (Russia), 60.

Flowering spring–summer. Wet, turfy, gravelly, or sandy calcareous barrens, high arctic, alpine tundra, heathlands; 0–500 m; Greenland; N.W.T., Nunavut, Yukon; Alaska; Europe (Spitzbergen); Asia (Russian Far East).

Minuartia rossii is the northernmost member of the *M. rossii* complex (S. J. Wolf et al. 1979; B. Maguire 1958), a pulvinate species of moist arctic areas. While specimens occasionally have many flowers, some specimens have few if any, instead reproducing via small axillary fascicles of leaves or short shoots in the upper leaf axils (see also Ö. Nilsson 2001).

28. Minuartia rubella (Wahlenberg) Hiern, J. Bot. 37: 320. 1899 • Beautiful or reddish sandwort, boreal stitchwort, minuartie rougeâtre [F]

Alsine rubella Wahlenberg, Fl. Lapp., 128, plate 6. 1812; *A. hirta* (Wormskjöld) Hartman var. *rubella* (Wahlenberg) Hartman; *Arenaria propinqua* Richardson; *A. rubella* (Wahlenberg) Smith; *A. verna* Linnaeus var. *propinqua* (Richardson) Fernald; *A. verna* var. *pubescens* (Chamisso & Schlechtendal) Fernald; *A. verna* var. *rubella* (Wahlenberg) S. Watson; *Tryphane rubella* (Wahlenberg) Reichenbach

Plants perennial, cespitose or mat-forming. **Taproots** filiform to somewhat thickened; rhizomes absent. **Stems** ascending to erect, green, 2–8(–18) cm, moderately to densely stipitate-glandular (very rarely glabrous), internodes of stems 1–10 times as long as leaves; trailing stems absent. **Leaves** overlapping, ± tightly, distally (cauline), concentrated proximally (cauline), connate proximally, with often loose, usually scarious sheath 0.2–0.7 mm; blade ± straight or outwardly curved, green, flat to 3-angled, prominently 3-veined abaxially, subulate, 1.5–10 × 0.3–1.3 mm, flexuous, margins not thickened, scarious, smooth, apex green or purple, acute to apiculate, often navicular, shiny, sparsely to densely ciliate, often stipitate-glandular; axillary leaves present among vegetative leaves. **Inflorescences** 3–7+-flowered, open cymes or rarely flower solitary, terminal; bracts broadly subulate to narrowly lanceolate, herbaceous, margins scarious. **Pedicels** 0.2–1.5 cm, densely stipitate-glandular. **Flowers:** hypanthium disc-shaped; sepals prominently 3-veined, ovate to lanceolate (herbaceous portion oblong to narrowly ovate), 2.5–3.2 mm, not enlarging in fruit, apex green to purple, acute to acuminate, not hooded, stipitate-glandular; petals elliptic, 0.8–1.3 times as long as sepals, apex rounded, entire. **Capsules** on stipe ca. 0.2 mm, ovoid, 4.5–5 mm, longer than sepals. **Seeds** reddish brown, suborbiculate with radicle prolonged into beak, somewhat compressed, 0.4–0.5 mm, tuberculate; tubercles low, elongate, rounded (to angled on edge) (50×). $2n = 24$.

Flowering summer. Arctic lowlands to rocky ridges and gravelly, montane, calcareous slopes in arctic and alpine tundra, heath and open woods, ± coastal gravelly limestone barrens in the Gulf of St. Lawrence area; 0–3800 m; Greenland; Alta., B.C., Man., Nfld. and Labr., N.W.T., Nunavut, Ont., Que., Sask., Yukon; Alaska, Ariz., Calif., Colo., Idaho, Maine, Mont., Nev., N.Mex., Oreg., S.Dak., Utah, Vt., Wash., Wyo.; arctic Eurasia.

Distinct among the arctic/alpine *Minuartia* species with its stiff, three-veined leaves, *M. rubella* is a circumpolar calciphile. We follow Ö. Nilsson (2001) in not recognizing infraspecific taxa that have been described based at least partly on pubescence. Variety *propinqua* has been applied to glabrous plants, which occur infrequently and sporadically throughout the range of the species. Where they do occur they are often intermixed with sparsely stipitate-glandular plants. This glabrous variety is rarely encountered in western North America.

29. Minuartia stolonifera T. W. Nelson & J. P. Nelson, Brittonia 43: 17, fig. 1. 1991 • Stolon or Scott Mountain sandwort [C] [E]

Plants perennial, mat-forming. **Taproots** moderately stout, woody. **Stems** ± erect, gray-green, 10–20 cm, glabrous or often stipitate-glandular, especially distally, internodes of stems 1–6 times as long as leaves (proximal leaves often shorter than internodes), 2–3 stolons radiating from crown, 6–20 cm. **Leaves** overlapping, loosely proximally, evenly spaced, connate proximally, with tight, scarious sheath 0.3–0.8 mm; blade ± straight to outwardly curved, gray-green, shallowly concave, 3-veined, often prominently so abaxially, needlelike, 5–11 × 0.5–0.9 mm, rigid, margins not thickened, scarious in proximal 1/2, stipitate-glandular, apex green to purple, acute to obtuse, navicular, dull, stipitate-glandular throughout; axillary leaves weakly developed among proximal cauline leaves. **Inflorescences** 7–25-flowered, open cymes; bracts lanceolate to subulate, herbaceous, margins scarious. **Pedicels** 0.3–1.5 cm, often stipitate-glandular. **Flowers:** hypanthium disc-shaped; sepals 1–3-veined (weakly in flower), ovate to lanceolate, (herbaceous portion narrowly lanceolate to linear-oblong), 3.5–4.8 mm, not enlarging in fruit, apex green to purple, narrowly acute to acuminate, not hooded, stipitate-glandular; petals broadly oblanceolate, 1.6–1.8 times as long as sepals, apex rounded, entire. **Capsules** sessile, ovoid, 3.5–5 mm, equaling sepals. **Seeds** reddish brown to brown, oblong-elliptic, 2–2.4 mm, tuberculate.

Flowering spring–summer. Jeffrey pine woodlands, serpentine soils; of conservation concern; 1200–1400 m; Calif.

Minuartia stolonifera, like *M. decumbens* and *M. rosei*, is restricted to serpentine soils of northwestern California, specifically to Scott Mountain in Siskiyou County. The three species are most closely related to the polymorphic *M. nuttallii*.

30. Minuartia stricta (Swartz) Hiern, J. Bot. 37: 320. 1899 • Bog stitchwort, rock sandwort, minuartie raide

Spergula stricta Swartz, Kongl. Vetensk. Acad. Nya Handl. 20: 229. 1799; *Alsinanthe stricta* (Swartz) Reichenbach; *Arenaria stricta* Michaux var. *uliginosa* (Schleicher ex Lamarck & de Candolle) B. Boivin; *A. uliginosa* Schleicher ex Lamarck & de Candolle

Plants perennial, cespitose to mat-forming. **Taproots** filiform to slightly thickened. **Stems** ascending to erect or procumbent, green, (0.8–)3–12 cm, glabrous, internodes of all stems 0.5–10 times as long as leaves. **Leaves** barely if at all overlapping, connate proximally, with tight, scarious to herbaceous sheath 0.2–0.5 mm; blade straight to slightly outward curved, green, flat, 1-veined, occasionally 3-veined abaxially, linear to linear-oblong or subulate, (2–)4–10 × (0.3–)0.5–1.5 mm, flexuous, margins not thickened, scarious, smooth, apex green or purple, mostly rounded, slightly navicular, shiny, glabrous; axillary leaves present among proximal cauline leaves. **Inflorescences** 1–3(–5)-flowered, open cymes or flowers solitary, terminal; bracts subulate, herbaceous. **Pedicels** 0.3–3 cm, glabrous. **Flowers:** hypanthium disc-shaped; sepals 1–3-veined, often prominently, becoming ribbed in fruit, broadly elliptic to ovate (herbaceous portion broadly elliptic to ovate), (1.5–)2.5–3.2 mm, to 4 mm in fruit, apex green to purple, acute to acuminate, not hooded, glabrous; petals lanceolate to spatulate or orbiculate, (0.6–)0.8–1 times as long as sepals, apex rounded, entire, or petals rudimentary or absent. **Capsules** on stipe ca. 0.1–0.2 mm, ovoid, 2.5–3.2 mm, shorter than or equaling sepals. **Seeds** brown or reddish brown, orbiculate with radicle prolonged into rounded bump, somewhat compressed, 0.4–0.6 mm, smooth or obscurely low-tuberculate (50×). **2*n*** = 22 (Europe), 26, 30 (Europe).

Flowering spring–summer. Moist, granitic gravels, sedge meadows, heath, alpine or arctic tundra; 100–3800 m; Greenland; B.C., Nfld. and Labr. (Labr.), N.W.T., Nunavut, Ont., Que., Yukon; Alaska, Calif., Colo.; Europe; Asia.

Minuartia stricta is circumpolar and sometimes has been included within *M. michauxii* (e.g., B. Maguire 1958; H. J. Scoggan 1978–1979, part 3).

Plants from alpine sites in California are, in spite of smaller stature and often smaller floral parts, *Minuartia stricta*. Colorado populations appear to vary widely in habit, but floral and fruit features match *M. stricta*.

31. Minuartia tenella (J. Gay) Mattfeld, Bot. Jahrb. Syst. 57(Beibl. 126): 29. 1921 • Slender stitchwort, slender sandwort [E]

Greniera tenella J. Gay, Ann. Sci. Nat., Bot., sér. 3, 4: 27. 1845, based on *Arenaria tenella* Nuttall in J. Torrey and A. Gray, Fl. N. Amer. 1: 179. 1838, not Kitaibel 1814; *Alsinopsis tenella* (J. Gay) A. Heller; *Arenaria macra* A. Nelson & J. F. Macbride; *A. stricta* Michaux; *A. stricta* subsp. *macra* (A. Nelson & J. F. Macbride) Maguire; *A. stricta* var. *puberulenta* (M. Peck) C. L. Hitchcock

Plants annual. **Taproots** filiform. **Stems** erect, green, 5–25 cm, stipitate-glandular distally or throughout, internodes of stems 2–5 times as long as leaves. **Leaves** overlapping proximally, often connate basally, with loose, scarious sheath 0.2–0.5 mm; blade straight to outwardly curved, green, flat to concave, prominently 1-veined abaxially, narrowly lanceolate to subulate, 5–17 × 0.5–1.5 mm, flexuous, margins not thickened, often scarious, sometimes ciliate or stipitate-glandular, apex purple, apiculate, navicular, shiny to dull, glabrous or stipitate-glandular; axillary leaves often present. **Inflorescences** 7–25+-flowered, open cymes; bracts subulate to lanceolate, scarious. **Pedicels** 0.2–1.5 cm, stipitate-glandular. **Flowers:** hypanthium disc-shaped; sepals prominently 3-veined, ovate to narrowly so (herbaceous portion narrowly ovate to lanceolate), 2.5–3 mm, not enlarging in fruit, apex green to purple, acute to acuminate, not hooded, densely stipitate-glandular; petals obovate, 1.5–2 times as long as sepals, apex rounded, entire. **Capsules** on stipe ca. 0.1 mm, ovoid, 3–4 mm, longer than sepals. **Seeds** brown, suborbiculate with radicle prolonged to rounded beak, somewhat compressed, 0.4–0.6 mm, tuberculate; tubercles low, rounded, elongate. **2*n*** = 24.

Flowering spring–summer. Coastal bluffs and forest openings; 0–700 m; B.C.; Oreg., Wash.

Although B. Maguire (1951, 1958) included *Minuartia tenella* within his concept of *Arenaria stricta* (*M. michauxii*), we see little more than a superficial resemblance between the taxa as we circumscribe them.

32. Minuartia uniflora (Walter) Mattfeld, Bot. Jahrb. Syst. 57(Beibl. 126): 28. 1921 • One-flower stitchwort E

Stellaria uniflora Walter, Fl. Carol., 141. 1788; *Alsine uniflora* (Walter) A. Heller; *Alsinopsis uniflora* (Walter) Small; *Arenaria alabamensis* (J. F. McCormick, Bozeman & Spongberg) R. E. Wyatt; *A. brevifolia* Nuttall; *Minuartia alabamensis* J. F. McCormick, Bozeman & Spongberg; *Sabulina uniflora* (Walter) Small

Plants annual. **Taproots** filiform. **Stems** erect to ascending, green, 7–20 cm, glabrous, internodes of stems 1–7 times as long as leaves. **Leaves** not overlapping, connate proximally, with tight, herbaceous or scarious sheath 0.1–0.3 mm; blade straight to outwardly curved, widely spreading, green, flat, 1-veined abaxially, especially proximal, narrowly lanceolate to oblong, commonly linear, 2–20 × 0.3–1.5 mm, flexuous, margins not thickened, scarious, smooth, apex green to purple, rounded to acute, dull, glabrous; axillary leaves poorly developed. **Inflorescences** 7–25+-flowered, open cymes; bracts subulate to ovate, herbaceous, margins scarious. **Pedicels** 0.5–5 cm, glabrous. **Flowers:** hypanthium disc-shaped; sepals obscurely veined, ovate to elliptic or lanceolate (herbaceous portion elliptic to lanceolate), 2–3.5 mm, not enlarging in fruit, apex green, obtuse to rounded, not hooded, glabrous; petals oblanceolate to spatulate, 1.5–2.5 times as long as sepals, apex rounded, entire to shallowly notched. **Capsules** on stipe shorter than 0.1 mm, pyramidal-ovoid, 3.5–4 mm, longer than sepals. **Seeds** yellowish brown, suborbiculate with radicle obscure, slightly compressed, 0.4–0.6 mm, tuberculate; tubercles low, rounded. $2n = 14$.

Flowering spring. Sandy or granitic outcrops; 70–200 m; Ala., Ga., N.C., S.C.

Minuartia alabamensis was originally described to accommodate much-reduced plants from Alabama (J. F. McCormick et al. 1971). Subsequent studies have shown them to be conspecific with *M. uniflora* (R. Wyatt 1984).

33. Minuartia yukonensis Hultén, Ark. Bot., n. s. 7: 52 1968 • Yukon stitchwort

Lidia yukonensis (Hultén) Á. Löve & D. Löve

Plants perennial, mat-forming. **Taproots** stout, woody. **Stems** erect to ascending, green, 10–30 cm, minutely retrorsely pubescent proximally, flowering stems stipitate-glandular, usually densely so, internodes of flowering stems 3–5 times as long as leaves. **Leaves** tightly overlapping (vegetative), variably spaced (cauline), usually connate proximally, with tight, scarious to herbaceous sheath 0.5–1.8 mm; blade straight to outwardly curved, green, flat, 3-veined abaxially, midvein more prominent than 2 lateral veins, filiform-linear, 10–18 × 0.8–1.3 mm, flexuous, margins not thickened, scarious, ciliate, often sparsely so, apex green, often with white callosity, acuminate-pungent, flat to slightly navicular, shiny, glabrous; axillary leaves present among vegetative leaves. **Inflorescences** 3–13-flowered, open cymes; bracts lanceolate, herbaceous. **Pedicels** 2–5 cm, densely stipitate-glandular. **Flowers:** hypanthium cup-shaped; sepals prominently 3-veined, lanceolate to oblong-lanceolate (herbaceous lanceolate to oblong-lanceolate), 6–8 mm, not enlarging in fruit, apex green to purplish, rounded, hooded, stipitate-glandular, especially proximally; petals oblanceolate, 1.3–1.5 times as long as sepals, apex rounded, entire. **Capsules** narrowly ellipsoid, 7–10 mm, longer than sepals. **Seeds** reniform, ca. 1 mm, tuberculate.

Flowering spring–summer. Dry, rocky slopes and meadows, scree slopes into alpine zone; 0–1000 m; B.C., N.W.T., Yukon; Alaska; Asia (Russian Far East).

Minuartia yukonensis is an amphi-Beringian species, reported from two sites in the Russian Far East. Some collections of it may be labeled as *Arenaria laricifolia* Linnaeus, a European species to which Alaskan material has been misattributed.

25. **WILHELMSIA** Reichenbach, Consp. Regn. Veg., 206. 1828 • Merckia [possibly for Christian Wilhelms, fl. 1819–1837, plant collector in the Caucasus]

Warren L. Wagner

Merckia Fischer ex Chamisso & Schlechtendal, Linnaea 1: 59. 1826, not *Merkia* Borkhausen 1792

Herbs, perennial; rhizomes somewhat fleshy, often with prominent nodal buds. **Taproots** usually not evident. **Stems** prostrate to decumbent, flowering stems erect or ascending, simple or branched, subterete, main stems often rooting at nodes. **Leaves** not connate proximally, sessile; blade 1-veined, elliptic-obovate to ovate-rhombic or ovate, slightly succulent, apex weakly acuminate. **Inflorescences** terminal, solitary flowers; bracts absent. **Pedicels** erect in fruit. **Flowers:** perianth and androecium weakly perigynous; hypanthium minute, cup-shaped; sepals 5, distinct, green, lanceolate to ovate or elliptic, usually navicular, 2–2.5 mm, herbaceous, margins white to pink or purple, scarious, apex obtuse to acute or short-acuminate, not hooded; petals 5, white, not or narrowly clawed, blade apex entire to weakly emarginate; nectaries as mounds around filaments opposite sepals, swollen on adaxial side; stamens 10, arising from narrow disc; filaments distinct; staminodes absent; ovary 3-locular when young (incompletely 3-celled by inward extension of margins of carpels in fruit); styles 3, narrowly clavate, 2.5–3 mm, glabrous proximally; stigmas 3, subterminal, minutely papillate (30×). **Capsules** subglobose, inflated, 6-lobed, eventually opening along false septa, sometimes irregularly, into 3, 2-toothed parts; carpophore absent. **Seeds** 8–16, dark red or brown, subpyriform, laterally compressed, smooth, marginal wing absent, appendage absent. $x = 15$.

Species 1: nw North America, ne Asia.

Wilhelmsia always has been treated as an isolated monospecific genus because of its inflated, partially septate ovary, except by American authors who have included it in *Arenaria*. The resemblance in habit between *Honckenya* and *Wilhelmsia*, previously presumed to represent convergence, has proven to reflect a close relationship based on recent molecular studies (M. Nepokroeff et al., unpubl.).

1. **Wilhelmsia physodes** (Fischer ex Seringe) McNeill, Taxon 9: 110. 1960 [F]

Arenaria physodes Fischer ex Seringe in A. P. de Candolle and A. L. P. P. de Candolle, Prodr. 1: 413. 1824; *Merckia physodes* (Fischer ex Seringe) Fischer ex Chamisso & Schlechtendal

Rhizomes spreading. **Stems** slender, 2–10(–20) cm, pubescent with spreading, purplish, jointed, weakly glandular-tipped hairs. **Leaf blades** 2–8 × 5–15 mm, sparsely pubescent with purplish, jointed hairs, becoming glabrate on faces, ciliate on margins and midrib. **Pedicels** 5–25 mm, pubescent. **Flowers:** sepals often purple tinged, oblong-ovate, 2–2.5 × 4.5–6 mm, apex acute to short-acuminate, sparsely pubescent; petals narrowly oblanceolate, abruptly constricted towards base, 5–6 mm; stamens moderately exserted, 2–3 mm. **Capsules** purplish, globose or slightly depressed, 7–10 mm diam., glabrous. **Seeds** 1.2–1.5 mm. $2n = 50$–60, 66, 70 (Siberia), 72 (Siberia), 100–110.

Flowering summer (late Jun–Aug). In sand or mud along streams and on gravel bars; 0–1500 m; N.W.T., Yukon; Alaska; Asia (Russian Far East, Siberia).

26. HONCKENYA Ehrhart, Neues Mag. Aerzte 5: 206. 1783 • Seaside sandwort, sea purslane [for Gerhard August Honckeny, 1724–1805, German botanist]

Warren L. Wagner

Adenarium Rafinesque; *Ammonalia* Desvaux; *Halianthus* Fries

Herbs, perennial, forming large mats or clumps by leafy rhizomes; rhizomes fleshy, often with prominent nodal buds and small membranous leaves. **Taproots** slender. **Stems** prostrate to decumbent, flowering stems ascending or weakly erect, simple or branched, terete or weakly 4-angled. **Leaves** not basally connate, sessile; blade 1-veined or obscurely so, usually elliptic to ovate, less commonly lanceolate to oblanceolate, obovate, or broadly elliptic, succulent, apex acute to acuminate or apiculate. **Inflorescences** terminal, open, leafy, 1–6-flowered cymes or

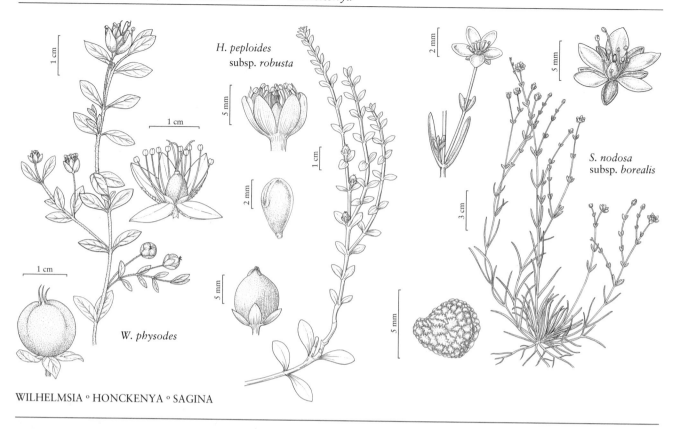

H. peploides
subsp. *robusta*

W. *physodes*

S. *nodosa*
subsp. *borealis*

WILHELMSIA ° HONCKENYA ° SAGINA

axillary and flowers solitary; bracts paired, foliaceous. **Pedicels** erect. **Flowers** functionally unisexual or, occasionally, staminate plants also with some bisexual; perianth and androecium subperigynous; hypanthium minimal; sepals 5(–6), distinct, green, narrowly ovate to elliptic, 3.5–7 mm, herbaceous, margins pale, scarious, apex obtuse or acute to apiculate, not hooded; petals absent or 5(–6), white, base clawed, blade apex emarginate; nectaries at base of filaments opposite sepals enlarged on both sides of filament, slightly reduced in pistillate flowers; stamens 10, fertile in staminate flowers, fewer or abortive in pistillate flowers, arising from rim of very brief hypanthium disc; filaments distinct; staminodes absent; styles (2–)3–5(–6), filiform, 1–2 mm, shorter and erect in staminate flowers, glabrous proximally; stigmas (2–)3–5(–6), linear along adaxial surface of styles, minutely papillate (30×). **Capsules** globose, inflated, opening by 3 spreading valves; carpophore absent. **Seeds** 3–15, reddish brown to dark reddish or yellowish brown, narrowly to broadly obovate, laterally compressed, smooth to minutely papillate, marginal wing absent, appendage absent. $x = 15$.

Species 1: temperate and arctic North America, n Eurasia.

The resemblance in habit between *Honckenya* and *Wilhelmsia* previously presumed to represent convergence has proven to indicate a close relationship based on recent molecular studies (M. Nepokroeff et al., unpubl.).

1. **Honckenya peploides** (Linnaeus) Ehrhart, Neues Mag. Aerzte 5: 206. 1783 [F]

Arenaria peploides Linnaeus, Sp. Pl. 1: 423. 1753

Plants maritime, glabrous; rhizomes spreading. **Stems** simple or branched, 5–25(–50) cm. **Leaf blades** usually long-elliptic to ovate, sometimes lanceolate to oblanceolate, obovate, or broadly elliptic, 4–46 × 0.5–20 mm, succulent, margins often crenulate, apex acute to acuminate or apiculate. **Pedicels** 2–10(–26) mm. **Flowers** strongly honey-scented; sepals ovate, 3.5–7 mm; petals spatulate to narrowly oblanceolate, abruptly constricted toward base, 2.5–6 mm in staminate flowers, 0.8–2 mm in pistillate flowers; filaments 2.5–5 mm in staminate flowers; anthers 0.5–0.7 mm, 0.5–0.8 mm in pistillate flowers. **Capsules** ovoid to subglobose, 5–10 × (4.5–) 5–12 mm, fleshy or chartaceous. **Seeds** 2–4 mm, rugulose. $2n = 66, 68, 70$.

Subspecies 4 (3 in the flora): coastal North America, Eurasia.

Honckenya peploides is polymorphic. A number of species and infraspecific taxa have been described from various parts of its geographical range. Recently, four subspecies of *H. peploides* have been recognized (A. Kurtto 2001b; V. V. Petrovsky 1971, 2000), as here; subsp. *peploides* occurs along European coasts. *Honckenya* is subdioecious (Petrovsky 1971, 2000; T. Tsukui and T. Sugawara 1992), and is pollinated largely by small bees, hover-flies, flies, and ants (Tsukui and Sugawara). The report of *Honckenya* (as *Ammodenia*) in Chile is an error based on G. Macloskie (1903–1914, vol. 1) (C. Marticorena, pers. comm.).

1. Plants stout; stems 3–6 mm diam., ascending to weakly erect; pedicels 2–3 mm . . . .
. . . . 1c. *Honckenya peploides* subsp. *robusta*
1. Plants slender; stems 0.5–3 mm diam., prostrate to decumbent; pedicels 3–26 mm.
  2. Stems 10–30(–50) cm × 1–3 mm, few; internodes of main stem 8–40(–60) mm; main stem leaves widely spaced, conspicuously larger than those of lateral branches, blade elliptic-obovate to broadly elliptic or narrowly elliptic or oblanceolate, (10–)20–46 × (0.5–)10–20 mm; pedicels (3–)8–16(–26) mm . . .
. . . . 1b. *Honckenya peploides* subsp. *major*
  2. Stems 5–15(–21) cm × 0.5–2.5 mm, numerous; internodes of main stem 5–30 mm; main stem leaves evenly spaced, similar in size to those of lateral branches, blade elliptic or ovate to obovate, 4–20(–30) × 3–9(–18) mm; pedicels 3–10(–15) mm . . . .
. . . . 1a. *Honckenya peploides* subsp. *diffusa*

1a. **Honckenya peploides** (Linnaeus) Ehrhart subsp. **diffusa** (Hornemann) Hultén ex V. V. Petrovsky in A. I. Tolmatchew, Fl. Arct. URSS 6: 71. 1971 (as Honkenya)

Arenaria peploides Linnaeus var. *diffusa* Hornemann, Fors. Oecon. Plantel. ed. 3, 1: 501. 1821; *A. diffusa* (Hornemann) Wormsköjd; *Halianthus peploides* (Linnaeus) Fries var. *diffusus* (Hornemann) Lange; *Honckenya diffusa* (Hornemann) Á. Löve & D. Löve; *H. peploides* var. *diffusa* (Hornemann) Ostenfeld

Plants with numerous stems, slender, succulent. **Stems** prostrate to decumbent, much-branched, 5–15(–21) cm × 0.5–2.5 mm, internodes of main stem 5–30 mm. **Leaves:** main stem leaves evenly spaced, of similar size on all stems; blade elliptic or ovate to obovate, 4–20 (–30) × 3–9(–18) mm, apex acute to apiculate. **Pedicels** 3–10(–15) mm. **Flowers:** sepals ovate, 3.5–6 mm, apex apiculate. **Capsules** 7–10 × 8–12 mm, chartaceous. **Seeds** reddish to yellowish brown, 2–4 mm, glossy. $2n = 66, 68, 70$ (Asia).

Flowering summer. Sea beaches, sandy flats, and dunes above high tide; ± 0–10 m; Greenland; Man., Nfld. and Labr. (Nfld.), N.W.T., Nunavut, Ont., Que., Yukon; Alaska; arctic Eurasia.

1b. **Honckenya peploides** (Linnaeus) Ehrhart subsp. **major** (Hooker) Hultén, Fl. Aleut. Isl., 171. 1937

Arenaria peploides Linnaeus var. *major* Hooker, Fl. Bor.-Amer. 1: 102. 1831; *A. peploides* subsp. *major* (Hooker) Calder & Roy L. Taylor; *A. peploides* var. *maxima* Fernald; *A. peploides* var. *oblongifolia* (Torrey & A. Gray) S. Watson; *Honckenya oblongifolia* Torrey & A. Gray; *H. peploides* var. *major* (Hooker) Abrams

Plants with few stems, slender, succulent. **Stems** prostrate to decumbent, not much-branched, 10–30(–50) cm × 1–3 mm, internodes of main stem 8–40(–60) mm. **Leaves:** main stem leaves widely spaced, conspicuously larger than those of lateral branches; blade elliptic-obovate to broadly elliptic or narrowly elliptic or oblanceolate, (10–)20–46 × (5–)10–20 mm, apex acute to apiculate. **Pedicels** (3–)8–16(–26) mm. **Flowers:** sepals ovate, 4–6 mm, apex apiculate. **Capsules** 5–8 × 5–10 mm, chartaceous. **Seeds** reddish to yellowish brown, 2–4 mm, glossy. $2n = 48, 64, 66, 68$.

Flowering late spring–summer. Sea beaches, sandy flats, and dunes above high tide; 0–10 m; B.C.; Alaska, Oreg., Wash.; Asia (Japan, Korea, Russia).

1c. Honckenya peploides (Linnaeus) Ehrhart subsp. **robusta** (Fernald) Hultén, Fl. Alaska Yukon 4: 677. 1944   E   F

Arenaria peploides Linnaeus var. *robusta* Fernald, Rhodora 11: 114. 1909; *Honckenya peploides* var. *robusta* (Fernald) House

Plants with few stems, stout, succulent-coriaceous. **Stems** ascending to weakly erect, simple or few-branched, 15–50 cm × 3–6 mm, internodes of main stem 10–55 mm. **Leaves:** main stem leaves evenly spaced, of similar size on all stems; blade lanceolate to ovate, 12–28 × 7–15 mm, apex acuminate. **Pedicels** 2–3 mm. **Flowers:** sepals ovate, 4–7 mm, apex obtuse to acute. **Capsules** 5–8 × 5–10 mm, fleshy. **Seeds** dark reddish brown, 2–4 mm, dull.

Flowering late spring–summer. Sea beaches, sandy flats, and dunes above high tide; 0–10 m; N.B., Nfld. and Labr.(Nfld.), N.S., P.E.I., Que.; Conn., Maine, Md., Mass., N.H., N.J., N.Y.

Subspecies *robusta* is possibly extirpated from Delaware and Virginia and now listed as rare/endangered in Maryland and New Jersey.

27. SAGINA Linnaeus, Sp. Pl. 1: 128. 1753; Gen. Pl. ed. 5, 62. 1754 • Pearlwort, sagine [Latin *sagina*, ancient name for *Spergula* once included in *Sagina*, a feasting, fatten, alluding to early use as forage]

Garrett E. Crow

Spergella Reichenbach

Herbs, annual, winter-annual, or perennial, often cespitose or matted. **Taproots** slender. **Stems** ascending, decumbent, or procumbent, simple or branched, terete to slightly angular. **Leaves:** basal and secondary rosettes present in perennial species, usually connate proximally, sometimes forming conspicuous, scarious cup, sessile, with or without axillary fascicles of leaves; blade 1-veined, linear to subulate, succulent or not, apex acute to mucronate or apiculate (or long-aristate in *S. subulata*). **Inflorescences** terminal or axillary cymes, or flowers solitary; bracts paired, foliaceous. **Pedicels** erect or spreading. **Flowers:** perianth and androecium hypogynous; sepals 4 or 5, distinct, green or sometimes purple (in *S. nivalis* and *S. decumbens*), lanceolate to elliptic or orbiculate, 1–5.5 mm, herbaceous, margins white or purple, scarious, apex obtuse or rounded to somewhat acute, frequently hood-shaped in bud; petals 4 or 5, sometimes absent (frequently absent or soon dropping in annual species), white, claw absent or minute, blade apex entire; nectaries at base of filaments opposite sepals; stamens 4, 5, 8, or 10, arising from base of ovary; filaments distinct; staminodes absent; styles 4 or 5, clavate to filiform, 0.5–1.5 mm, glabrous proximally; stigmas 4 or 5, subterminal to linear along adaxial surface of styles, minutely papillate (30×). **Capsules** globose to ovoid, opening by 4 or 5 valves, sutures running to base, but in some species dehiscing only ca. $^1/_4$–$^1/_2$ capsule length; carpophore absent. **Seeds** ca. 125, light tan to dark or reddish brown, obliquely triangular with abaxial groove, or reniform to nearly globose without abaxial groove (except *S. nodosa*, which is intermediate), laterally compressed or plump, smooth to tuberculate, marginal wing absent, appendage absent. $x = 6, 7, 11$.

Species 15–20 (10 in the flora): chiefly cold-temperate Northern Hemisphere, also on some tropical mountains.

SELECTED REFERENCES Crow, G. E. 1978. A taxonomic revision of *Sagina* (Caryophyllaceae) in North America. Rhodora 80: 1–91. Crow, G. E. 1979. The systematic significance of seed morphology in *Sagina* (Caryophyllaceae) utilizing SEM. Brittonia 31: 52–63.

1. Petals nearly 2 times sepals, 3–4.5 mm; distal cauline leaves and lateral branches usually with axillary fascicles of minute, succulent leaves, giving knotted appearance . . . . 1. *Sagina nodosa*
1. Petals shorter than, equaling, or barely longer than sepals, 0.8–3 mm, or absent; cauline leaves and procumbent stems sometimes with axillary fascicles resembling cauline leaves, not minute, succulent.
   2. Flowers 5-merous; leaf blades fleshy; seeds reniform or nearly globose, plump, abaxial groove absent.
      3. Calyx bases and distal portion of pedicels glabrous . . . . 9. *Sagina maxima* (in part)
      3. Calyx bases and distal portion of pedicels glandular-pubescent.
         4. Seeds reddish brown, smooth or slightly pebbled . . . . 9. *Sagina maxima* (in part)
         4. Seeds dark brown, usually distinctly tuberculate or strongly pebbled . . . . . . 10. *Sagina japonica*
   2. Flowers 4–5-merous; leaf blades not fleshy, or if slightly fleshy, then flowers predominantly 4-merous; seeds obliquely triangular, with abaxial groove.
      5. Plants annual; stems filiform; distal cauline leaf blades subulate, becoming shorter distally.
         6. Flowers 5-merous (rarely 4-merous, and then apetalous); leaf blade bases never ciliate; capsules longer than sepals . . . . 7. *Sagina decumbens*
         6. Flowers 4-merous, rarely 4- and 5-merous; leaf blade bases distinctly ciliate, especially of distal cauline leaves; capsules equaling or barely longer than sepals
         . . . . 8. *Sagina apetala*
      5. Plants perennial; stems not filiform; distal cauline leaf blades linear, linear-subulate, or, if subulate, then plants cespitose.
         7. Plants cespitose, forming low cushions; cauline leaf blades subulate; hyaline sepal margins purple.
            8. Flowers 4-merous, sometimes accompanied by 5-merous flowers; petals shorter than or equaling sepals, 1.5–2 mm; primary basal rosette of fleshy, subulate leaves present, secondary rosettes absent . . . . 5. *Sagina nivalis*
            8. Flowers 5-merous, sometimes accompanied by 4-merous flowers; petals longer than or rarely equaling sepals, 2.5–3 mm; primary basal rosette of leaves absent, secondary rosettes of linear leaves often present . . . . . . 6. *Sagina caespitosa*
         7. Plants with stems ascending, spreading, procumbent, or mat-forming (sometimes cespitose in alpine sites); leaf blades linear; hyaline sepal margins white.
            9. Flowers 4-merous, sometimes accompanied by 5-merous flowers; petals 0.8–1 mm, sometimes absent; sepals divergent at time of capsule dehiscence
            . . . . 3. *Sagina procumbens*
            9. Flowers 5-merous; petals (1–)1.5–2 mm; sepals appressed to loosely appressed at time of capsule dehiscence.
               10. Plants glabrous; cauline leaf blade apices apiculate; connate leaf bases not forming conspicuous cup. . . . . 2. *Sagina saginoides*
               10. Plants with leaves, stems, pedicels, and calyx bases glandular-pubescent; cauline leaf blade apices aristate, aristae equaling or exceeding leaf width; connate leaf bases forming conspicuous, scarious cup . . . . . . 4. *Sagina subulata*

1. Sagina nodosa (Linnaeus) Fenzl, Vers. Darstell. Alsin., 18. 1833 • Knotted pearlwort [F]

Spergula nodosa Linnaeus, Sp. Pl. 1: 440. 1753

Plants perennial, tufted, pubescent or glabrous. **Stems** ascending to loosely spreading or prostrate, simple or with few to many lateral branches bearing only tiny, succulent, subulate leaves, not filiform. **Leaves:** axillary fascicles of subulate, fleshy leaves among distal cauline leaves, giving knotted appearance; basal ascending, forming tuft, not in rosettes, blade linear, ca. 15–30 mm, herbaceous, apex apiculate to mucronate, glabrous or glandular; cauline conspicuously connate basally, cuplike, proximal with blade linear to subulate, apex apiculate to mucronate, glabrous or glandular-pubescent, axillary fascicles of leaves absent; distal with blade subulate, 1–1.5 mm, fleshy, apex mucronate, glabrous. **Pedicels** filiform, base glabrous or glandular-pubescent. **Flowers** terminal and axillary, 5-merous or 4- and 5-merous;

calyx base glabrous or glandular-pubescent; sepals elliptic, 2–3 mm, hyaline margins rarely purplish, apex frequently purplish, obtuse to rounded, remaining appressed after capsule dehiscence; petals elliptic to orbiculate, 3–4.5 mm, ca. 2 times sepal length; stamens 8 or 10. **Capsules** 3–4 mm, longer than sepals, dehiscing to base. **Seeds** dark brown, ovoid to reniform, laterally compressed, 0.5 mm, smooth or distinctly pebbled, abaxial groove present or absent.

Subspecies 2 (2 in the flora): North America, Europe.

1. Stems glandular-pubescent; pedicels distally glandular-pubescent; bases of calyces glandular-pubescent; basal leaf blades often glandular-pubescent (especially on margins and midrib) or glabrous . . . . 1a. *Sagina nodosa* subsp. *nodosa*
1. Stems glabrous or occasionally weakly pubescent at nodes; pedicels distally glabrous or occasionally weakly glandular-pubescent; bases of calyces glabrous or weakly glandular-pubescent; basal leaf blades glabrous . . . . 1b. *Sagina nodosa* subsp. *borealis*

1a. Sagina nodosa (Linnaeus) Fenzl subsp. **nodosa**

• Sagine noueuse I

Sagina nodosa var. *pubescens* (Besser) Koch

Stems glandular-pubescent. **Basal leaf blades** glandular-pubescent, especially on margins and midrib, or glabrous. **Pedicels** glandular-pubescent distally. **Flowers:** calyx base glandular-pubescent. $2n = 56$.

Flowering mid–late summer. Coastal, moist crevices of rocks along seashore and on sea cliffs, wet sand flats at river mouths; 0–300 m; introduced; St. Pierre and Miquelon; N.B., Nfld. and Labr. (Nfld.), N.S.; Maine, Mass., N.H.; Europe.

Subspecies *nodosa* is probably introduced in North America. Its localities tend to be correlated with coastal regions that had an early history of fishing by Europeans; it may have been introduced with the dumping of ballast. It was collected at least once from New Hampshire where it apparently has been extirpated.

Some variation occurs in the amount and distribution of pubescence on the leaf surface in subsp. *nodosa*. In plants with a lesser amount of pubescence, the glandular hairs are restricted chiefly to the leaf margins; the leaves may even be glabrous. This seems to be the case primarily when subsp. *nodosa* and the native subsp. *borealis* occur in the same vicinity, such as some populations along the Saint Lawrence Seaway, Grand Manan Islands, New Brunswick, and Machaias, Maine. More typically, the plants are pubescent and the trichomes are more frequent along the veins on the abaxial surface as well as the leaf margins.

1b. Sagina nodosa (Linnaeus) Fenzl subsp. **borealis**
G. E. Crow, Rhodora 80: 28, figs. 7, 8. 1978

• Sagine boréale F

Sagina nodosa var. *borealis* (G. E. Crow) Cronquist

Stems glabrous or weakly pubescent at nodes. **Basal leaf blades** glabrous. **Pedicels** glabrous or weakly pubescent distally. **Flowers:** calyx base glabrous or weakly pubescent. $2n = 56$.

Flowering mid–late summer. Chiefly shoreline, rock crevices, wet gravels and sands, tufts of moss along rocky coasts and shores of large lakes; 0–300 m; Greenland; Alta., Man., N.B., Nfld. and Labr. (Nfld.), N.W.T., N.S., Nunavut, Ont., P.E.I., Que., Sask.; Maine, Mich., Minn.; n Europe.

In populations along the shores of Lake Superior, plants sometimes occur with glandular hairs sparsely distributed on the stems, chiefly or solely at the bases of the nodes, as well as on the pedicels and calyces.

2. Sagina saginoides (Linnaeus) H. Karsten, Deut. Fl., 539. 1882 • Sagine des alpes

Spergula saginoides Linnaeus, Sp. Pl. 1: 441. 1753; *Sagina linnaei* C. Presl; *S. micrantha* (Bunge) Fernald; *S. saginoides* var. *hesperia* Fernald

Plants perennial, tufted or becoming cespitose in alpine habitats, glabrous. **Stems** ascending or sometimes pro-cumbent, few- to many-branched, not filiform. **Leaves:** axillary fascicles absent; basal frequently in primary and secondary rosettes 9–45 mm diam., blade linear, 10–20 mm, not succulent, apex apiculate, rarely aristate, glabrous; cauline not conspicuously connate basally, rarely forming inflated cup in cespitose, alpine plants, blade linear, sometimes linear-subulate in cespitose plants, 4–20(–25) mm, not fleshy, apex apiculate, glabrous. **Pedicels** frequently recurved during capsular development, erect in fruit, filiform, glabrous. **Flowers** axillary or terminal, 5-merous, very rarely some 4-merous; calyx base glabrous; sepals elliptic, 2–2.5 mm, hyaline margins white, rarely purple in alpine specimens, apex obtuse to rounded, remaining appressed following capsule dehiscence; petals elliptic, (1–)1.5–2 mm, shorter than or equaling sepals; stamens (5 or) 10. **Capsules** 2.5–3(–3.5) mm, 1.5–2 times sepals, dehiscing to base. **Seeds** brown, obliquely triangular with distinct abaxial groove, 0.3–0.4 mm, smooth to slightly pebbled. $2n = 22$.

Flowering mid–late summer. Montane sites, open or light shade, wet places on lake margins, along stream gravels and seepages in rock ledges and roadcuts, subalpine and alpine zones; 1000–4000 m; Greenland; Alta., B.C., Nfld. and Labr. (Nfld.), N.W.T., Nunavut, Que., Yukon; Alaska, Ariz., Calif., Colo., Idaho, Mont., Nev., N.Mex., Oreg., Utah, Wash., Wyo.; Mexico; Eurasia.

Some specimens from alpine habitats in Montana and Alberta are intermediate between *Sagina saginoides* and the typically arctic *S. nivalis*.

3. Sagina procumbens Linnaeus, Sp. Pl. 1: 128. 1753

• Matted pearlwort, sagine couchée [1] [W]

Sagina procumbens var. *compacta* Lange

Plants perennial, often mat-forming, glabrous. **Stems** ascending or, more frequently, procumbent, rooting at nodes, giving rise to secondary tufts or rosettes, few- to many-branched, slender. **Leaves:** axillary fascicles often present on procumbent stems; basal frequently in primary rosettes in younger plants; blade linear, 8–17 mm, herbaceous, apex apiculate to somewhat aristate, glabrous; cauline not conspicuously connate basally, never forming an inflated cup, blade linear, 4–15 mm proximally, becoming shorter toward apex, 2.5–6 mm distally, sometimes slightly fleshy, apex apiculate to aristate, rarely with minute glandular cilia. **Pedicels** frequently recurved during capsule development, filiform, glabrous. **Flowers** axillary or terminal, 4-merous, occasionally 4- and 5-merous; calyx base glabrous; sepals elliptic to orbiculate, 1.5–(–2.5) mm, hyaline margins white, never purple tinged, apex obtuse to rounded, appressed during capsular development, divergent following dehiscence; petals (1–)4(–5), orbiculate to elliptic, 0.8–1(–1.5) mm, shorter than or equaling sepals, or sometimes absent; stamens 4 (8). **Capsules** (1.5–)2–2.5(–3) mm, slightly exceeding sepals, dehiscing to base. **Seeds** brown, obliquely triangular with distinct abaxial groove, (0.3–)0.4(–0.5) mm, smooth to pebbled. $2n = 22$.

Flowering late spring–early fall. Weedy, wet or damp, gravelly or sandy soils along roadsides, sidewalk cracks, margins of paths or lawns, pond and lake margins, coastal rocks and sands, sea cliffs; 0–3500 m; introduced; Greenland; St. Pierre and Miquelon; B.C., N.B., Nfld. and Labr. (Nfld.), N.W.T., N.S., Ont., P.E.I., Que.; Alaska, Ark., Calif., Colo., Conn., Del., Idaho, Ill., Iowa, Maine, Md., Mass., Mich., Minn., Mont., N.H., N.J., N.Y., N.C., Ohio, Oreg., Pa., R.I., Utah, Vt., Wash., W.Va., Wis.; Europe; introduced in Mexico (Chiapas, México), Central America (Costa Rica, Guatemala), s South America (Bolivia, s Argentina), Asia (w Siberia), Antarctica (sub-Antarctic Islands).

4. Sagina subulata (Swartz) C. Presl, Fl. Sicul., 158. 1826 • Scottish moss [1]

Spergula subulata Swartz, Kongl. Vetensk. Acad. Nya Handl. 10: 45, plate 1, fig. 3a–c. 1789

Plants perennial, cespitose, frequently forming dense mats or tufted, glandular-pubescent or glabrous. **Stems** ascending or decumbent, few-branched, not filiform, densely glandular-pubescent, or less frequently glabrous; horizontal stems becoming slightly woody with extensive mat formation. **Leaves:** axillary fascicles absent; basal forming tufts, blade linear, curled inward, 3–12 mm, not fleshy, apex long-aristate, aristae equaling or exceeding leaf width, densely glandular-pubescent, or with glandular hairs restricted to margins, and then often minutely glandular-ciliate, rarely glabrous; cauline connate basally, forming conspicuous cup, blade linear-subulate, 3–10 mm, scarious. **Pedicels** filiform, densely to weakly glandular-pubescent. **Flowers** axillary or terminal, usually solitary, 5-merous, rarely 4- and 5-merous; calyx bases glandular-pubescent; sepals elliptic, 1.5–2 mm, hyaline margins white, apex obtuse to rounded, glandular-pubescent, remaining appressed following capsule dehiscence; petals elliptic, 1.5–2 mm, shorter than or equaling sepals; stamens (8 or) 10. **Capsules** 2–3(–3.5) mm, slightly longer than sepals, dehiscing to base. **Seeds** brown, obliquely triangular with abaxial groove, 0.4(–0.5) mm, smooth.

Flowering mid–late summer. Wet, gravelly sands of stream margins; 0–1800 m; introduced; Oreg.; Mexico (Baja California Sur); Europe.

Three specimens from the alpine zone of Steens Mountain, Harney County, are referable to *Sagina subulata*. Introduction of the species into that remote area is without explanation.

A strongly mat-forming cultivar of *Sagina subulata* is sometimes grown as a ground cover; plants flower profusely, but no subsequent capsule development typically occurs. The cultivar differs from the native European mat-forming plants by being glabrous except for the minutely glandular-ciliate leaf margins. Occasional waifs have been collected in the San Francisco area of California.

5. Sagina nivalis (Lindblom) Fries, Novit. Fl. Suec. Mant. 3: 31. 1842

Spergula saginoides Linnaeus var. *nivalis* Lindblom, Physiogr. Sällsk. Tidskr. 1: 328. 1838; *Sagina intermedia* Fenzl ex Ledebour; *Spergella intermedia* (Fenzl ex Ledebour) Á. Löve & D. Löve

Plants perennial, cespitose, forming low cushions, glabrous. **Stems** ascending or spreading, radiating from axils of basal rosette leaves, sometimes purple tinged, many-branched, slender. **Leaves:** axillary fascicles absent; basal in primary rosettes, secondary rosettes absent, blade subulate to linear, to 20(–30) mm, fleshy, apex apiculate, glabrous; cauline connate basally into shallow cup, blade often purplish, subulate to linear, 4–16 mm, becoming shorter toward stem apex, scarious, apex apiculate, glabrous. **Pedicels** filiform, glabrous. **Flowers** mostly terminal, 4-merous or 4- and 5-merous; calyx base glabrous; sepals frequently purplish, nearly orbiculate to elliptic, 1.5–2 mm, hyaline margins nearly always purple, sometimes only at apex, apex rounded, glabrous, remaining appressed following capsule dehiscence; petals narrowly elliptic, 1.5–2 mm, equaling to slightly shorter than sepals; stamens 8 or 10. **Capsules** 2–3 mm, usually shorter than sepals, dehiscing to base. **Seeds** brown, obliquely triangular with abaxial groove, 0.5 mm, lateral surfaces frequently with elongate ridges, abaxial surface appearing smooth to pebbled. $2n = 56, 88$.

Flowering mid–late summer. Sandy or gravelly beaches, coastal rocks, alluvial plains, fresh glacial moraines, low, swampy tundra, alpine areas; 0–2800 m; Greenland; Alta., B.C., Nfld. and Labr. (Labr.), N.W.T., Nunavut, Que., Yukon; Alaska, Mont.; arctic Eurasia.

6. Sagina caespitosa (J. Vahl) Lange in H. J. Rink, Grønland 2: 133. 1857 • Sagine cespiteuse

Arenaria caespitosa J. Vahl in G. C. Oeder et al., Fl. Dan. 13(39): 4. 1840; *Sagina nivalis* (Lindblom) Fries var. *caespitosa* (J. Vahl) B. Boivin; *Spergella caespitosa* (J. Vahl) Á. Löve & D. Löve

Plants perennial, cespitose, forming small mats or cushions, glandular-pubescent or glabrous. **Stems** ascending to spreading, frequently purple tinged, many-branched, not filiform. **Leaves:** axillary fascicles absent; basal not in primary rosettes, secondary rosettes usually present, blade linear to linear-subulate, 2–13 mm, not fleshy, apex apiculate to acute, glabrous; cauline connate basally, forming shallow, often purplish, scarious cup, blade with frequently conspicuous midvein, subulate, 3–9(–12) mm, becoming shorter toward apex, not fleshy, apex apiculate to acute, glabrous. **Pedicels** filiform, glandular-pubescent, rarely glabrous. **Flowers** terminal, 5-merous or 4- and 5-merous; calyx base glandular-pubescent or glabrous; sepals broadly ovate to lanceolate, 2–2.5 mm, hyaline margins usually purple tinged, at least at apex, apex obtuse to somewhat acute, glandular-pubescent or glabrous, remaining appressed after capsule dehiscence; petals elliptic to narrowly obovate, 2.5–3 mm, longer than or seldom equaling sepals; stamens 8 or 10. **Capsules** 3–3.5 mm, exceeding sepals, dehiscing to base. **Seeds** brown, obliquely triangular with abaxial groove, 0.5 mm, lateral surfaces frequently with elongate ridges, abaxial surface appearing smooth to pebbled.

Flowering mid–late summer. Wet sands and gravels of shorelines and stream margins, wet mossy places, dry rocky barrens, gravelly hillocks; 0–1300 m; Greenland; Man., Nfld. and Labr., Nunavut, Que.; arctic Europe.

7. Sagina decumbens (Elliott) Torrey & A. Gray, Fl. N. Amer. 1: 177. 1838 • Sagine décombante F

Spergella decumbens Elliott, Sketch Bot. S. Carolina 1: 523. 1821

Plants annual, glabrous or glandular-pubescent. **Stems** ascending or decumbent, frequently purple tinged, few- to many-branched, filiform. **Leaves:** axillary fascicles absent; basal rosette quickly deciduous; proximal cauline leaves connate basally, not appearing inflated, blade frequently purple tinged, linear, 4–23 mm, not fleshy, base never ciliate, margins conspicuously hyaline basally, apex apiculate, glabrous, distal blades becoming subulate and shorter toward apex, 1–5 mm, apex apiculate, glabrous. **Pedicels** filiform, glabrous or glandular-pubescent. **Flowers** terminal or often axillary, 5-merous, rarely 4-merous and then apetalous; calyx base glabrous or glandular-pubescent, often sparsely so; sepals ovate to orbiculate, (1–)1.5–2(–3) mm, hyaline margins conspicuous, margins or apex frequently purple, apex acute to rounded, glabrous or glandular-pubescent at calyx base, remaining appressed to capsule; petals elliptic, (1–)1.5–2(–2.3) mm, slightly longer than sepals at anthesis, equaling or shorter than sepals during capsule development; stamens 5 or 10, occasionally 8. **Capsules** 2–3(–3.5) mm, longer than sepals, dehiscing to 1/2 capsule length or less. **Seeds** light tan to light brown, obliquely triangular with abaxial groove, (0.2–)0.3–1.4 mm, smooth or pebbled to strongly tuberculate, protrusions sometimes borne on delicate ridges, ridges forming reticulate pattern (50–80×).

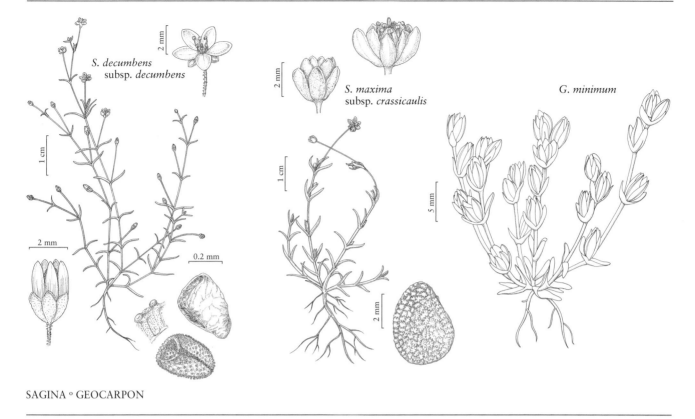

S. decumbens subsp. *decumbens*

S. maxima subsp. *crassicaulis*

G. minimum

SAGINA ° GEOCARPON

Subspecies 2 (2 in the flora): North America (including n Mexico).

1. Seeds light tan, smooth or pebbled to strongly tuberculate, protrusions borne on delicate ridges, ridges forming reticulate pattern (50–80×); chiefly se United States . . . . 7a. *Sagina decumbens* subsp. *decumbens*
1. Seeds light brown, usually smooth to slightly pebbled, never with reticulate ridge pattern; Pacific Coast region . . . . 7b. *Sagina decumbens* subsp. *occidentalis*

7a. Sagina decumbens (Elliott) Torrey & A. Gray subsp. **decumbens** • Sagine décombante E F

Sagina decumbens var. *smithii* (A. Gray) S. Watson

Leaves: basal rosette forming in winter annual plants, absent later. **Sepals** ovate, hyaline margins or apex frequently purple. **Seeds** light tan, with delicate reticulate ridge pattern (50–80×), smooth or pebbled to strongly tuberculate. $2n = 36$.

Flowering spring–early summer. Moist or dryish sandy places, field margins, open pine woods, paths, roadsides, sidewalk cracks, lawns; 0–500 m; Que.; Ala., Ark., Conn., Del., Fla., Ga., Ill., Ind., Kans., Ky., La., Md., Mass., Miss., Mo., N.J., N.Y., N.C., Ohio, Okla., Pa., S.C., Tenn., Tex., Vt., Va.

Specimens of subsp. *decumbens* from Alberta, New Brunswick, Saskatchewan, and Arizona represent historical collections that apparently did not persist.

Subspecies *decumbens* is extremely variable. A slender, nearly apetalous variation with a greater frequency of 4-merous flowers has previously been recognized as var. *smithii*; the range of variability is continuous and it seems best to consider the material as a single taxon.

7b. Sagina decumbens (Elliott) Torry & A. Gray subsp. **occidentalis** (S. Watson) G. E. Crow, Rhodora 80: 68. 1978

Sagina occidentalis S. Watson, Proc. Amer. Acad. Arts 10: 344. 1875

Leaves: basal rosette absent. **Sepals** ovate to orbiculate, body and hyaline margins frequently purple. **Seeds** light brown, seldom ridged, and then never forming reticulate pattern, smooth to slightly pebbled.

Flowering spring–early summer. Dryish hillsides, margins of vernal pools, streams, open spots in redwood and pine woods, roadsides, dwellings; 0–900 m; B.C.; Calif., Oreg., Wash.; Mexico (Baja California).

Except by geography, subsp. *occidentalis* is very difficult to distinguish from subsp. *decumbens*. In plants of subsp. *occidentalis* the sepals tend to be more orbiculate and the capsules, prior to dehiscence, tend to be more globose. Extremely variable, subsp. *decumbens* generally can be recognized on the basis of presence of tuberculate seeds (60% frequency) and 80% have a combination of tuberculate seeds and glandular-pubescent pedicels and calyx bases. But when seeds are smooth, seeing the reticulate ridge pattern requires high magnification, and while SEM readily clarifies the differences, its use is hardly practical. Subspecies *decumbens* has a greater tendency to possess purple sepal tips or sepal margins, and purplish coloration frequently at the nodes.

8. **Sagina apetala** Arduino, Animadv. Bot. Spec. Alt. 2: 22, fig. 1. 1764   ⓘ

Sagina apetala var. *barbata* Fenzl ex Ledebour

Plants annual, glandular-pubescent. **Stems** ascending to decumbent, much-branched, filiform, glabrous or sometimes glandular-pubescent. **Leaves:** axillary fascicles absent; basal sometimes in rosettelike whorl, withering early; cauline leaves connate basally, proximal blades linear, 4–8(–12) mm, not fleshy, apex aristate, with hyaline portion of leaf base long-ciliate, cilia occasionally occurring length of leaf (sometimes lacking cilia), distal blades linear to subulate, 1–3 mm toward apex. **Pedicels** filiform, glandular-pubescent. **Flowers** axillary, sometimes terminal, 4-merous, very rarely 4- and 5-merous; calyx glandular-pubescent; sepals ovoid to elliptic, sometimes lanceolate, 1.5–2 mm, hyaline margins white, apex somewhat acute, glandular-pubescent, divergent at capsule dehiscence; petals nearly always absent, minute if present; stamens 4(–5). **Capsules** 1.5–2(–2.5) mm, equaling or barely longer than sepals, dehiscing to base. **Seeds** brown, obliquely triangular with abaxial groove, 0.3–0.4 mm, smooth, pebbled, or frequently papillose. $2n = 12$.

Flowering spring–early summer. Open places, hard-packed soils around buildings, paths, roadsides, sidewalk cracks, grassy hillsides, streambanks; 0–1300 m; introduced; B.C.; Calif., Kans.; Europe.

Historical collections of *Sagina apetala* are known from Illinois, Louisiana, Maryland, New Jersey, Oregon, and Washington.

9. **Sagina maxima** A. Gray, Mem. Amer. Acad. Arts, n. s. 6: 382. 1858   Ⓕ

Plants annual or perennial, tufted, glabrous or glandular-pubescent. **Stems** spreading to decumbent or procumbent, much-branched, stout, rarely filiform, distal portion glandular-pubescent. **Leaves:** axillary fascicles absent; basal rosette or tuft of ascending leaves usually present; basal blades linear, 10–30 mm, succulent, apex apiculate, glabrous; cauline leaves conspicuously connate basally, forming shallow, scarious cup, blade linear, fleshy, apex apiculate, glabrous; proximal blades 6–15(–20) mm, distal blades rarely subulate, (2.5–)3.5–7(–9) mm. **Pedicels** slender to stout, glabrous or glandular-pubescent distally. **Flowers** axillary, 5-merous; calyx bases glabrous or glandular-pubescent; sepals ovate to orbiculate, (2–)2.5–3.5 mm, hyaline margins whitish, occasionally purple tinged on margins or apex, apex obtuse to rounded, glabrous or glandular-pubescent, remaining appressed following capsule dehiscence; petals elliptic to nearly orbiculate, (1.5–)2–2.5(–3) mm, shorter than sepals; stamens 10. **Capsules** (3–)3.5–4.5 mm, exceeding sepals, dehiscing ca. 1/4 length. **Seeds** reddish brown, reniform with abaxial groove absent, plump, 0.5 mm, smooth or slightly pebbled.

Subspecies 2 (2 in the flora): w North America, e Asia.

1. Calyx bases and distal portion of stems and pedicels glandular-pubescent . . . . 9a. *Sagina maxima* subsp. *maxima*
1. Calyx bases, stems, and distal portion of pedicels entirely glabrous . . . . 9b. *Sagina maxima* subsp. *crassicaulis*

9a. **Sagina maxima** A. Gray subsp. **maxima**

Sagina litoralis Hultén; *S. crassicaulis* S. Watson var. *litoralis* (Hultén) Hultén

Plants annual or perennial, glandular-pubescent. **Stems** spreading to decumbent, frequently glandular-pubescent distally, nodes green. **Leaves:** basal leaves often in tuft of ascending linear leaves, secondary fascicles or basal rosette rarely present; cauline leaf blades: proximal (6–)8–15(–20) mm, distal (2.5–)3.5–7(–9) mm, both glabrous or rarely minutely glandular-ciliate. **Pedicels** usually stout or sometimes slender, densely glandular-pubescent distally, less densely glandular-pubescent proximally, proximal 1/4 usually glabrous. **Flowers:** calyx base densely glandular-pubescent; sepals ovate to orbiculate, (2–)2.5–3.5 mm; petals elliptic to nearly orbiculate, 2–2.5(–3) mm, slightly shorter than sepals. **Capsules** (3–)3.5–4.5 mm. **Seeds** pebbled or, less frequently, smooth. $2n = 44$.

Flowering early–late summer. Coastal, rocky or sandy bluffs, rocky shores, gravelly beaches; 0 m; B.C.; Alaska, Wash.; e Asia.

In contrast to the eastern Asian members of subsp. *maxima*, specimens from the Aleutian Islands and the western coast of North America tend to have slightly larger flowers and smooth seeds. Additionally, pubescence is less dense and seldom occurs on the stems.

9b. Sagina maxima A. Gray subsp. **crassicaulis** (S. Watson) G. E. Crow, Rhodora 80: 79. 1978 [F]

Sagina crassicaulis S. Watson, Proc. Amer. Acad. Arts 18: 191. 1883

Plants perennial, glabrous or mostly so. **Stems** spreading, decumbent, or procumbent, glabrous, nodes frequently purple tinged. **Leaves:** basal leaves in rosette of broadly linear, fleshy leaves, or absent with primary or secondary tufts of ascending, linear basal leaves, these usually less fleshy than rosette leaves (rosettes rarely present in plants occurring north of Washington); cauline leaf blades: proximal 6–15 mm, distal 3–5 mm, glabrous. **Pedicels** slender to stout, glabrous. **Flowers:** calyx glabrous; sepals ovate to nearly orbiculate, (2–)2.5–3(–3.5) mm; petals elliptic to orbiculate, (1.5–)2–2.5(–3) mm, slightly shorter than sepals. **Capsules** (3–)3.5–4(–4.5) mm. **Seeds** smooth to slightly pebbled. $2n = 46, 66$.

Flowering spring–early autumn. Coastal, moist, sandy bluffs, crevices of rock cliffs, at or near high-tide mark, gravelly–sandy beaches; 0–10 m; B.C.; Alaska, Calif., Oreg., Wash.; Asia (Kamchatka).

Integradation occurs where the range of subsp. *crassicaulis* overlaps with that of subsp. *maxima*. Variation of pubescence in populations on Vancouver Island and the Queen Charlotte Islands ranges from completely glabrous specimens typical of subsp. *crassicaulis* to individuals with pedicels and calyx bases weakly pubescent, to others with densely pubescent pedicels. Subspecies *crassicaulis* is far more common than subsp. *maxima*.

10. Sagina japonica (Swartz) Ohwi, J. Jap. Bot. 13: 438. 1937 • Sagine du Japon [I]

Spergula japonica Swartz, Ges. Naturf. Freude Berlin Neue Schriften 3: 164, plate 1, fig. 2. 1801

Plants annual, glandular-pubescent. **Stems** ascending to spreading, much-branched, usually filiform, frequently glandular-pubescent distally. **Leaves:** axillary fascicles rarely present; basal frequently in tuft of ascending leaves, rosette rarely present, blade linear, 4–10 mm, succulent, apex apiculate, glabrous; cauline leaves conspicuously connate basally, forming shallow, scarious cup, blade linear, fleshy, apex apiculate, glabrous or rarely pubescent; proximal leaf blades 9–20 mm, becoming shorter distally, 4–7 mm. **Pedicels** slender, distal portion densely glandular-pubescent, becoming less densely glandular-pubescent proximally, proximal 1/4 usually glabrous. **Flowers** axillary, 5-merous; calyx glandular-pubescent basally; sepals elliptic to orbiculate, 2–2.5 mm, hyaline margins whitish, apex obtuse to rounded, glandular-pubescent, remaining appressed following capsule dehiscence; petals ovate to orbiculate, 1–2 mm, shorter than sepals; stamens 5 or 10. **Capsules** 2.5–3 mm, exceeding sepals, dehiscing ca. 1/4 length. **Seeds** dark brown, reniform to nearly globose, plump, abaxial groove absent, 0.4–0.5 mm, densely tuberculate or strongly pebbled (e North America).

Flowering early–late summer. Dryish sites, waste places; 200 m; introduced; B.C., Nfld. and Labr. (Nfld.), Ont., Que.; Conn., Ill., Mass., N.H., N.Y., Ohio, Oreg., Pa., R.I., Vt.; e Asia; introduced in Mexico (Veracruz).

Sagina japonica was recently introduced in widely scattered locations in northeastern North America, and can be found especially in gravelly roadsides, walkways, and driveways. The plants tend to be much less robust and without distinctly succulent leaves.

SELECTED REFERENCES Mitchell, R. S. and G. C. Tucker. 1991. *Sagina japonica* (Sw.) Ohwi (Caryophyllaceae), an overlooked adventive in the northeastern United States. Rhodora 93: 192–194. Rabeler, R. K. 1996. *Sagina japonica* (Caryophyllaceae) in the Great Lakes region. Michigan Bot. 35: 43–44. Rabeler, R. K. and J. W. Thieret. 1997. *Sagina* (Caryophyllaceae) range extensions in Canada: *S. japonica* new to Newfoundland, *S. procumbens* new to the Northwest Territories. Canad. Field-Naturalist 111: 309–310.

28. GEOCARPON Mackenzie, Torreya 14: 67. 1914 • [Greek *ge*, earth, and *carpos*, fruit]

C E

Mark A. Nienaber

Herbs, annual or winter annual. **Taproots** slender. **Stems** erect or spreading-ascending, simple or few-branched basally, terete. **Leaves** connate basally into sheath, sessile; blade 1-veined, narrowly oblong, not succulent, apex acute. **Inflorescences** axillary, flowers borne singly at alternate nodes; bracts paired at upper nodes, foliaceous. **Pedicels:** flowers sessile. **Flowers:** perianth and androecium perigynous; hypanthium green or reddish, cup-shaped, 10-veined, commisural veins branched ca. at base of sepals; sepals 5, connate proximally ca. 1 mm, green or reddish, 3-veined (lateral pair ⅛ length of midvein), triangular-ovate, margins white, scarious, apex acute; petals absent; nectaries unknown; stamens 5, arising between sepals; filaments distinct; staminodes 5, alternating with stamens, scalelike; styles 3, filiform, ca. 0.3 mm, without stiff hairs proximally; stigmas 3, linear along adaxial surface, minutely papillate (50×). **Capsules** ovoid, opening by 3 apical, spreading valves; carpophore absent. **Seeds** ca. 30(–60), yellowish green to brown, translucent, elliptic to ovoid, reniform, slightly laterally compressed, muriculate, marginal wing absent, appendage absent.

Species 1: sc United States.

When MacKenzie described *Geocarpon* he placed it in Aizoaceae, thinking that it was closest to *Cypselea* among the North American representatives of that family. He noted that *Geocarpon* differs markedly in the absence of stipules and styles and in the valvular rather than circumscissile dehiscence of the fruit. Other authors following this alignment included M. L. Fernald (1950), H. A. Gleason (1952, vol. 2), and P. Wilson (1934). After closer scrutiny of several morphological characteristics, E. J. Palmer and J. A. Steyermark (1950) moved *Geocarpon* to Caryophyllaceae. Features they noted in suggesting this change included exstipulate leaves, connate sepals, five stamens and, particularly, five staminodes inserted perigynously on the floral tube opposite the calyx lobes, the unilocular ovary with numerous amphitropous ovules on a free-central placenta, and three sessile, recurved stigmas alternating with three bifid tips of the ovary valves. Within Caryophyllaceae, Palmer and Steyermark erected a new monotypic tribe, Geocarpeae, between Alsineae (subtribe Sabulinae) and Sclerantheae. In addition to the morphological evidence for placing *Geocarpon* in Caryophyllaceae, there are also supporting chemical and anatomical data. The reddish pigment in *Geocarpon* is an anthocyanin; no betalain-type pigments have been found (A. L. Bogle et al. 1971). Furthermore, H.-D. Behnke (1982) found that the sieve-element plastids of *Geocarpon* are the type found in Caryophyllaceae (PIIIc′f), not like those found in members of Aizoaceae.

SELECTED REFERENCES Bogle, A. L., T. Swain, R. D. Thomas, and E. D. Kohn. 1971. *Geocarpon*: Aizoaceae or Caryophyllaceae? Taxon 20: 473–477. Palmer, E. J. and J. A. Steyermark. 1950. Notes on *Geocarpon minimum* Mackenzie. Bull. Torrey Bot. Club 77: 268–273. Steyermark, J. A., J. W. Voigt, and R. H. Mohlenbrock. 1959. Present biological status of *Geocarpon minimum* Mackenzie. Bull. Torr. Bot. Club 86: 228–235.

1. Geocarpon minimum Mackenzie, Torreya 14: 67. 1914 • Tiny tim [C][E][F]

Stems greenish brown or strongly suffused with red, 1–4 cm × 0.5 mm or less; proximal nodes 1–3 mm apart, internodes zigzag, distal nodes 5–10 mm apart. **Leaf blades** 3–4 mm, margins somewhat involute. **Inflorescence bracts** connate, strongly red or purple tinged, keeled, triangular, 2.5–3.5 mm, apex acute. **Flowers** funnelform-campanulate, 3–4 mm; sepal lobes ± erect to slightly spreading, subequal; ovary greenish, ca. 3 mm, apex narrow at anthesis, minutely glandular-toothed or retuse. **Capsules** opening along 3 wirelike valve margins. **Seeds** 0.3–0.5 mm.

Flowering late winter–early spring. Sandstone glades and alkali barrens; of conservation concern; 100–300 m; Ark., La., Mo., Tex.

Geocarpon minimum is known from fewer than 35 sites, about one-third of which have relatively large, vigorous populations. The species is listed as federally threatened. In Missouri, it is found only in shallow depressions in slightly tilted sandstone strata within sandstone glade plant communities. In Arkansas and Louisiana, it is found in saline-alkaline soils on the edge of highly localized, surficial concentrations of sodium and magnesium, locally known as "slick spots"; a similar "saline barren" hosts the recently discovered (2004) Texas population. These austere and nearly barren patches of mineral soil are scattered across savannalike formations classified as saline soil prairies.

Geocarpon minimum is in the Center for Plant Conservation's National Collection of Endangered Plants.

29. SCLERANTHUS Linnaeus, Sp. Pl. 1: 406. 1753 (as Schleranthus); Gen. Pl. ed. 5, 190. 1754 • Knawel, scléranthe [Greek *skleros*, hard, and *anthos*, flower, alluding to the indurate hypanthium] [I]

John W. Thieret

Richard K. Rabeler

Herbs, annual, biennial, or perennial. **Taproots** slender. **Stems** erect to prostrate, branched, terete. **Leaves** connate proximally, sessile; blade 1-veined, subulate to linear, not succulent, apex acute or obtuse. **Inflorescences** terminal or axillary, lax to dense cymes; bracts paired, foliaceous. **Flowers** sessile to subsessile; perianth and androecium perigynous; hypanthium urceolate, abruptly expanded distally; sepals 5, distinct, greenish, lanceolate to awl-shaped, 1.5–4 mm, herbaceous, margins whitish, scarious, apex acute to blunt or obtuse; petals absent; nectariferous disc at base of stamens; stamens 2–10, arising from hypanthium rim; filaments distinct; staminodes absent or 5–8, arising from hypanthium rim, filiform; styles 2, capitate, 0.8–1 mm, glabrous proximally; stigmas 2, terminal, minutely papillate (50×). **Utricles** ovoid, enclosed in persistent, indurate, shallowly or strongly furrowed, sepal-crowned hypanthium and falling with it, the whole constituting the indehiscent "fruit"; carpophore present. **Seeds** 1, yellowish, globose, not compressed, smooth, marginal wing absent, appendage absent. $x = 11$ [12].

Species ca. 10 (2 in the flora): introduced; temperate Europe (including Mediterranean region), Asia, Africa, Australia; widely naturalized elsewhere.

In spite of their small size, most *Scleranthus* flowers secrete nectar and are visited by insects, including small flies and ants. Pollination in species within the flora area varies from chiefly protandrous outcrossing (*S. perennis*) to chiefly autogamous (*S. annuus*). L. Svensson (1988) reported that these two species hybridize regularly in Europe, producing flowers with ten reduced, sterile stamens.

The common name "knawel" apparently refers to the glomerules of flowers (German *Knäuel*).

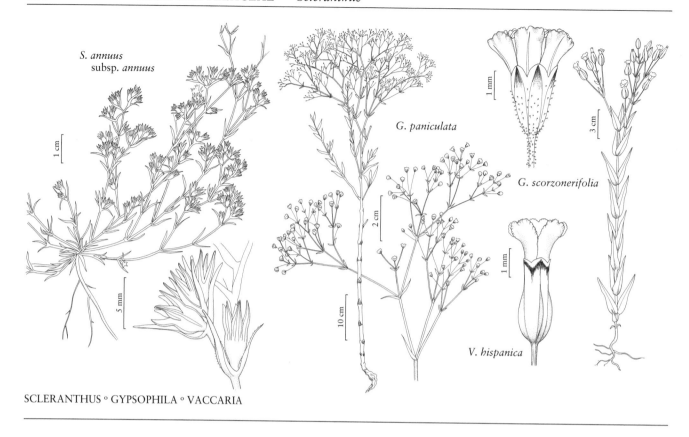

S. annuus
subsp. *annuus*

G. paniculata

G. scorzonerifolia

V. hispanica

SCLERANTHUS ° GYPSOPHILA ° VACCARIA

SELECTED REFERENCE   Smissen, R. D. and P. J. Garnock-Jones. 2002. Relationships, classification and evolution of *Scleranthus* (Caryophyllaceae) as inferred from analysis of morphological characters. Bot. J. Linn. Soc. 140: 15–29.

1. Sepals with acute apices, spreading to erect in fruit, not overlapping, margins 0.1 mm or less wide; flowers usually equaling or shorter than bracts . . . . 1. *Scleranthus annuus*
1. Sepals with blunt or rounded apices, usually bending together in fruit, overlapping, margins 0.3–0.5 mm wide; flowers usually longer than bracts . . . . 2. *Scleranthus perennis*

1. Scleranthus annuus Linnaeus, Sp. Pl. 1: 406. 1753
   • Annual knawel, scléranthe annuel   F I W

Subspecies 8 (1 in the flora): introduced; Europe, w Asia, n Africa.

1a. Scleranthus annuus Linnaeus subsp. **annuus**
   F I W

Plants annual or biennial. **Stems** erect to prostrate, 2–25 cm. **Leaf blades** 3–24 mm. **Flowers** usually equaling or exceeded by bracts; hypanthium becoming strongly 10-ribbed in fruit, 1.2–2 mm; sepals not overlapping, spreading to erect in fruit, 1.5–4 mm, margins 0.1 mm or less wide, apex acute; stamens ca. 1/2 length of sepals. "**Fruits**" 3.2–5 mm including sepals.   *2n* = 44 (Europe).

Flowering winter–fall. Sandy fields, roadsides, weedy areas, lawns; 0–1500 m; introduced; Alta., B.C., N.B., Nfld. and Labr. (Nfld.), N.S., Ont., P.E.I., Que., Sask.; Ala., Ark., Calif., Conn., Del., D.C., Fla., Ga., Idaho, Ill., Ind., Kans., Ky., La., Maine, Md., Mass., Mich., Minn., Miss., Mo., Mont., Nebr., N.H., N.J., N.Y., N.C., Ohio, Okla., Oreg., Pa., R.I., S.C., Tenn., Vt., Va., Wash., W.Va., Wis.; Europe; w Asia; n Africa; widely naturalized elsewhere, including Mexico, Central America (Costa Rica), South America (Ecuador), e Asia (South Korea), Africa (Kenya, Republic of South Africa), Pacific Islands (New Zealand).

2. **Scleranthus perennis** Linnaeus, Sp. Pl. 1: 406. 1753

  • Perennial knawel, scléranthe vivace   I

Subspecies 5 (1 in the flora): introduced; c Europe, w Asia.

In the flora area, *Scleranthus perennis* is probably best established in Wisconsin; R. A. Schlising and H. H. Iltis (1961) reported it from four counties there, the first collection made in 1928.

2a. **Scleranthus perennis** Linnaeus subsp. **perennis**   I

Plants perennial. **Stems** procumbent to erect, 3–25 cm, becoming somewhat woody basally. **Leaf blades** 3–13 mm. **Flowers** usually longer than bracts; hypanthium becoming shallowly furrowed in fruit, 1–1.3 mm; sepals overlapping, usually bending together in fruit, 2.5–4 mm, margins 0.3–0.5 mm wide, apex blunt or rounded; stamens ca. equaling sepals. "**Fruits**" 2.5–4.5 mm including sepals. $2n = 22$ (Europe).

Flowering summer–fall. Roadsides, sandy areas, shores; 0–300 m; introduced; Que.; Conn., Mass., Mich., Wis.; c Europe; w Asia.

43d. CARYOPHYLLACEAE Jussieu subfam. CARYOPHYLLOIDEAE Arnott in M. Napier, Encycl. Brit. ed. 7, 5: 99. 1832 (as Caryophylleae)

Richard K. Rabeler

Ronald L. Hartman

Caryophyllaceae subfam. Silenoideae Fenzl

Herbs, annual, biennial, or perennial; taprooted or rhizomatous, sometimes stoloniferous. **Stems** erect or ascending, seldom sprawling, decumbent, or prostrate, simple or branched. **Leaves** opposite, rarely whorled, connate proximally, petiolate (basal leaves) or often sessile, not stipulate; blade linear or subulate to ovate, not succulent or rarely so (*Silene*). **Inflorescences** terminal cymes, thyrses, fascicles, or capitula, or flowers solitary, axillary; bracts foliaceous, scarious, or absent; involucel bracteoles present or often absent. **Pedicels** present or rarely flowers sessile or subsessile. **Flowers** bisexual or seldom unisexual (the species then often dioecious), often conspicuous; perianth and androecium hypogynous; sepals 5, connate (^{1}/4–) ^{1}/2+ their lengths into cup or tube, (1–)5–40(–62) mm, apex not hooded or awned; petals absent or 5, often showy, white to pink or red, usually clawed, auricles absent or sometimes present, coronal appendages sometimes present, blade apex entire or emarginate to 2-fid, sometimes dentate to lacinate; stamens (5 or) 10 (absent in pistillate flowers), in 1 or 2 whorls, arising from base of ovary; staminodes absent or rarely 1–10; ovary 1-locular, sometimes 2-locular proximally (*Vaccaria*), or 3–5-locular (some *Silene*); styles 2–3(–5) (absent in staminate flowers), distinct; stigmas 2–3(–5) (absent in staminate flowers). **Fruits** capsules, opening by 4–6(–10) valves or teeth; carpophore usually present. **Seeds** 4–150(–500+), reddish to gray or often brown or black, usually reniform and laterally compressed to globose, sometimes oblong or shield-shaped and dorsiventrally compressed; embryo peripheral and curved, or central and straight. x = 7, 10, 12, [13?,] 14, 15, 17, [18].

Genera 20 or 26, species ca. 1500 (8 genera, 89 species in the flora): north-temperate regions, Europe (esp. Mediterranean region), Asia (esp. Mediterranean region e to c Asia), Africa (Mediterranean region, Republic of South Africa).

Caryophylloideae can be characterized by the presence of sepals connate into a cup or (usually) long tube, clawed petals (often with appendages and auricles), and a lack of stipules. The largest genera in the family [*Silene* (incl. *Lychnis*), about 700 species; *Dianthus*, about 320 species] are in the Caryophylloideae; together with *Gypsophila* (about 150 species), these three genera include about three-quarters of the species found in the family. Three tribes are often differentiated on calyx venation and number of styles, with two, Caryophylleae and Sileneae, incorporating nearly all of the genera.

Caryophylloideae share the caryophyllad type of embryogeny with Alsinoideae and, as postulated by V. Bittrich (1993), the two may form a monophyletic group. Results from preliminary molecular studies by M. Nepokroeff et al. (2002) and R. D. Smissen et al. (2002) reinforce that hypothesis, but the relationships among members of the two subfamilies remain unclear.

Most of the molecular work within the subfamily has focused on Sileneae and more specifically on trying to determine whether or not *Silene* is monophyletic.

SELECTED REFERENCES Maguire, B. 1950. Studies in the Caryophyllaceae. IV. A synopsis of the North American species of the subfamily Silenoideae. Rhodora 52: 233–245. Oxelman, B., M. Lidén, R. K. Rabeler, and M. Popp. 2000. A revised generic classification of the tribe Sileneae (Caryophyllaceae). Nordic J. Bot. 21: 743–748.

30. GYPSOPHILA Linnaeus, Sp. Pl. 1: 406. 1753; Gen. Pl. ed. 5, 191. 1754 • Baby's-breath, soupir-de-bébé [Greek *gypsos*, gypsum, and *philios*, loving, alluding to habitat of some species] ☐

James S. Pringle

Psammophiliella Ikonnikov

Plants annual or perennial. **Taproots** slender to stout, sometimes absent; perennials often with stout, branched caudices, some with adventitious roots from decumbent stems or elongating rhizomes. **Stems** erect, ± sprawling, or less often decumbent or prostrate, usually branched, terete. **Leaves** briefly connate proximally, sessile; blade 1- or 3–5-veined, linear to oblong or ovate, apex rounded or obtuse to acuminate. **Inflorescences** dichasial cymes or thyrses, diffuse (to subcapitate in *G. oldhamiana*); bracts paired, proximal bracts foliaceous, distal ones smaller, herbaceous with scarious margins; involucel bracteoles absent. **Pedicels** erect in fruit. **Flowers:** sepals connate proximally into cup, 1–5 mm, cup green and white, 5-veined, not winged, obconic to campanulate, terete to 5-angled, commissures between sepals veinless, broad, scarious; lobes green at least along midrib, usually ovate to elliptic, equaling or longer than cup, margins white, scarious, apex rounded to obtuse, sometimes mucronate; petals 5, white, pink, or rose-purple, claw poorly differentiated, auricles absent, coronal appendages absent; blade apex entire or shallowly emarginate to 2-fid, nectaries at filament bases; stamens 10, arising with petals from low nectariferous disc; filaments distinct nearly to base; staminodes absent; ovary 1-locular; styles 2(–3), clavate, 1.2–2.5 mm, glabrous proximally; stigmas 2(–3), subterminal, papillate (30×). **Capsules** globose or ellipsoid-ovoid, opening by 4(–6) slightly distally recurving valves; carpophore absent. **Seeds** 4–36, brown to black, reniform to snail-shell-shaped, laterally compressed, tuberculate, marginal wing absent, appendages absent; embryo peripheral, curved. *x* = 17, 12 (Eurasia), 18 (Eurasia); aneuploidy occasional.

Species ca. 150 (4 in the flora): introduced; temperate Eurasia, Africa, Pacific Islands, Australia; introduced in South America.

Gypsophila species are widely grown as ornamentals. In addition to those treated below, other European and Asiatic species have appeared sporadically in disturbed habitats in the flora area, sometimes remote from any site where likely to have been planted, but have not become established. *Gypsophila pilosa* Hudson [*G. porrigens* (Linnaeus) Boissier], which differs from *G. elegans* in its stems villous or hispid proximal to the inflorescence, slender pedicels that persist after the flowers and fruits have fallen, and consistently pink petals, has been found at waste-disposal sites in Maryland, New York, and Oregon. *Gypsophila repens* Linnaeus, a rhizomatous perennial species with prostrate to decumbent primary stems and more or less erect flowering branches to 3 dm, similar to *G. elegans* in floral characters, has been found escaped from cultivation in British Columbia and Maine. *Gypsophila oldhamiana* F. A. W. Miquel was found in a field in Alabama in 1969 [*Rebois 049* (AUA)]. It has pink petals and differs from other species described here in its densely corymboid to subcapitate inflorescences. Additional species are cultivated in the flora area.

All reports of *Gypsophila acutifolia* Steven ex Sprengel, *G. perfoliata* Linnaeus in the narrow sense, *G. stevenii* Fischer ex Schrank, *G. arrostii* Gussone, and *G. pacifica* Komarov (*G. perfoliata* var. *latifolia* Maximowicz) as naturalized species in the flora area appear to have been based on misidentified *G. scorzonerifolia.* The inflorescences of *G. acutifolia* are denser than those of *G. scorzonerifolia*, with the pedicels of the former being less than two times the calyx length, and those of the latter mostly being more than two times as long. True *G. perfoliata* and *G. pacifica*, neither of which is known in North America outside of cultivation, differ from *G. scorzonerifolia* in having glabrous pedicels and calyces; *G. perfoliata* differs also in having almost completely green sepals with the narrow white margins not sharply defined, and minutely rather than coarsely tuberculate seed coats.

All of the species described below, especially *Gypsophila scorzonerifolia*, can be expected to be found elsewhere in the flora area.

SELECTED REFERENCE   Barkoudah, Y. I. 1962. A revision of *Gypsophila, Bolanthus, Ankyropetalum* and *Phryna*. Wentia 9: 1–203.

1. Plants annual; stems diffusely branched throughout, 0.4–3(–4) dm; leaf blades linear, 0.2–2(–3) mm wide; petals usually pink . . . . 3. *Gypsophila muralis*
1. Plants annual or perennial; stems simple or few-branched proximal to inflorescence in annuals, or variously branched in perennials, 0.4–20 dm; leaf blades lanceolate to ovate, (1–)2–20(–35) mm wide; petals purplish pink or white.
  2. Plants annual; petals 6–15 mm . . . . 2. *Gypsophila elegans*
  2. Plants perennial; petals 1–6 mm.
    3. Leaf bases clasping; pedicels and calyces glandular-puberulent; petals at least tinged with purplish pink, 4–6 mm . . . . 4. *Gypsophila scorzonerifolia*
    3. Leaf bases not clasping; pedicels and calyces glabrous; petals usually white, or rarely light purplish pink, 1–4 mm . . . . 1. *Gypsophila paniculata*

1. Gypsophila paniculata Linnaeus, Sp. Pl. 1: 407. 1753 • Perennial or tall baby's-breath, gypsophile paniculée, oeillet d'amour F I W

Plants perennial. **Stems** erect or ± sprawling, diffusely much-branched at or near crown, 4–10 dm, glabrous or occasionally glandular-puberulent or scabrous near base. **Leaves** cauline, bases not clasping; blade linear-lanceolate to oblong-lanceolate, larger leaves 2–9 cm × 2–10 mm, apex acute to acuminate, glaucous. **Pedicels** 1–20 mm, glabrous. **Flowers:** calyx 1–3 mm, lobes glabrous, apex rounded to obtuse; petals white or rarely light purplish pink, 1–4 mm. **Capsules** globose. **Seed coats** coarsely tuberculate. $2n = 34, 68$ (both Europe).

Flowering summer–fall. Fields, roadsides, beaches, other open, sandy, disturbed sites; 0–2600 m; introduced; Alta., B.C., Man., N.B., N.W.T., Ont., Que., Sask.; Calif., Colo., Conn., Fla., Idaho, Ill., Ind., Iowa, Kans., Maine, Mass., Mich., Minn., Mont., Nebr., Nev., N.J., N.Y., N.Dak., Ohio, Okla., Oreg., Pa., S.Dak., Utah, Vt., Wash., Wis., Wyo.; Eurasia.

Gypsophila paniculata, which is widely cultivated as a garden ornamental and florists' crop, occurs rather sporadically in eastern North America, mostly in waste places, although it occasionally forms sizeable local populations in soils that are not strongly acidic. In parts of central and western North America, in contrast, it has become an abundant and widespread weed in hayfields and pastures, often being dispersed as a tumbleweed. Its roots may extend four meters into the soil. It has been designated a noxious weed in California, Washington, and Manitoba. Also, there is concern that the spread of *G. paniculata* in the dune-swale complexes around the upper Great Lakes presents a threat to some of the rare species that are largely restricted to these habitats (K. D. Herman 1996).

Botanical varieties of *Gypsophila paniculata* have been based on the distribution of pubescence on stems and leaves, but are doubtfully worthy of taxonomic recognition. Both glabrous-stemmed plants and plants with proximally glandular-puberulent stems are naturalized in North America. Cultivars have been selected for compact habit, supernumerary petals, petals to ca. 8 mm, and/or pink petals. Occasional naturalized plants have purple-tinged pedicels and calyces. A form with supernumerary petals has spread from cultivation in northern Michigan. The ploidy level(s) of populations in the flora area have not been investigated.

SELECTED REFERENCE   Darwent, A. L. 1975. The biology of Canadian weeds. 14. *Gypsophila paniculata*. Canad. J. Pl. Sci. 55: 1049–1058.

2. **Gypsophila elegans** M. Bieberstein, Fl. Taur.-Caucas. 1: 319. 1808 • Annual or showy baby's-breath, gypsophile élégante [I]

Plants annual. **Stems** erect, simple or few-branched proximal to inflorescence, 0.4–6 dm, glabrous. **Leaves** cauline, proximal leaves with clasping bases, gradually transitional to distal leaves with ± rounded bases; blade linear-lanceolate to narrowly oblong, 1.5–7 cm × (1–)3–16 mm, apex obtuse to acute in proximal leaves, acute in distal leaves, glaucous. **Pedicels** 10–35 mm, glabrous. **Flowers:** calyx 2.5–5 mm, lobes glabrous, apex obtuse or mucronate; petals white, occasionally with pinkish purple veins, or rarely pink, 6–15 mm. **Capsules** globose. **Seed coats** coarsely tuberculate. $2n = 34$ (Europe).

Flowering summer–early fall. Roadsides and other open, sandy or rocky, disturbed sites; 0–2100 m; introduced; Alta., N.W.T., Ont., Que., Sask., Yukon; Alaska, Calif., Colo., Conn., Ga., Ill., Iowa, Kans., Maine, Mass., Mich., Minn., N.H., N.J., N.Y., N.C., N.Dak., Ohio, Oreg., Pa., R.I., Tex., Utah, Vt., Va., W.Va., Wis.; Eurasia; introduced in the West Indies (Dominican Republic), Central America (Guatemala); widely cultivated elsewhere.

Gypsophila elegans is frequently included in mixtures of "wildflower" seeds used for roadside planting and other revegetation projects. A specimen specifically from such a mixture was seen from Louisiana, but such mixtures are used widely elsewhere and are believed to account for the presence of this species in Colorado and in at least one Utah locality. A report of this species from Labrador appears to have been based on garden plants.

If *Gypsophila elegans* is divided into two varieties, following Y. I. Barkoudah (1962), plants in the flora area are var. *elegans*. Cultivars are much used by florists and are frequently grown as garden ornamentals. These may have supernumerary petals, petals to 25 mm, and/or pink to maroon petals.

3. **Gypsophila muralis** Linnaeus, Sp. Pl. 1: 408. 1753 • Cushion or low baby's-breath, gypsophile de murs [I]

Psammophiliella muralis (Linnaeus) Ikonnikov

Plants annual. **Stems** erect, diffusely much-branched throughout, 0.4–3(–4) dm, proximally puberulent, distally glabrous, or occasionally glabrous throughout. **Leaves** cauline, bases not clasping; blade linear, (0.2–)0.3–3.2 cm × 0.2–2(–3) mm, not glaucous, apex acute to acuminate.

Pedicels 2–20 mm, glabrous. **Flowers:** calyx 2–4 mm, lobes glabrous, apex rounded to obtuse; petals pink or rarely white, 3.5–6(–10) mm. **Capsules** ellipsoid-ovoid. **Seed coats** minutely tuberculate. $2n = 30, 34$ (both Europe).

Flowering summer–fall. Roadsides, yards, cemeteries, other open, sandy or rocky, disturbed sites; 0–1000 m; introduced; Ont., Que.; Conn., D.C., Ind., Maine, Mass., Mich., N.H., N.J., N.Y., Ohio, Pa., R.I., S.Dak., Vt., Wis.; Europe.

Gypsophila muralis is well established in noncalcareous soils in the eastern part of its North American range. Populations in the Great Lakes region and westward are less likely to be long-persistent, e.g., Minnesota, where collections are known from 1910–1911. Its recent spread in Eau Claire, Wisconsin, has been associated with sites flooded in winter for skating rinks and with sites where snow and ice have been piled in winter (J. R. Rohrer 1998). This species has been confused with *Petrorhagia saxifraga* (Linnaeus) Link, but can readily be distinguished by its annual habit, lack of epicalyces, and snail-shell-shaped rather than pear-shaped seeds (R. K. Rabeler 1981). *Spergularia rubra* (Linnaeus) J. Presl & C. Presl is also similar in aspect; it differs in having distinct, narrowly triangular stipules and distinct sepals.

In recent years, cultivars of *Gypsophila muralis* have been selected for density of branching, flower size, supernumerary petals, and depth of and other variations in petal color.

4. **Gypsophila scorzonerifolia** Seringe in A. P. de Candolle and A. L. P. P. de Candolle, Prodr. 1: 352. 1824 • Glandular baby's-breath [F] [I]

Plants perennial. **Stems** ± erect, simple or few-branched proximal to inflorescence, 5–20 dm, proximally glabrous, distally glandular-puberulent. **Leaves** basal and cauline, bases clasping; blade oblong-lanceolate to narrowly ovate, larger leaves 2–15 cm × 7–22(–35) mm, glaucous, apex obtuse to acute. **Pedicels** 1–12 mm, glandular-puberulent. **Flowers:** calyx 2.5–4 mm, lobes glandular-puberulent, apex obtuse; petals white with pink tinge to light purplish pink (drying darker), 4–6 mm. **Capsules** globose. **Seed coats** coarsely tuberculate. $2n = 68$ (Europe, introduced population).

Flowering summer–fall. Beaches, roadsides, railroad grades, quarries, and other open, calcareous, sandy or rocky, disturbed sites; 0–1700 m; introduced; Alta., B.C., Ont.; Calif., Colo., Conn., Ill., Ind., Mich., Nev., N.Mex., N.Y., Ohio, Utah, Wis., Wyo.; Europe (se; introduced elsewhere in Europe).

There is a historic record of *Gypsophila scorzonerifolia* from Massachusetts collected in 1921.

Some Ohio specimens of *Gypsophila scorzonerifolia* have calyces rather sparsely glandular or appearing to lack glands at maturity, but all of the young flowers have obviously glandular calyces.

SELECTED REFERENCE Pringle, J. S. 1976. *Gypsophila scorzonerifolia* (Caryophyllaceae), a naturalized species in the Great Lakes region. Michigan Bot. 15: 215–219.

31. VACCARIA Wolf, Gen. Pl., 3. 1776 • [Latin *vacca*, cow, and *-aria*, pertaining to, alluding to alleged value for fodder] ☐

John W. Thieret

Richard K. Rabeler

Herbs, annual. **Taproots** stout. **Stems** simple proximally, branched distally, terete. **Leaves** opposite, somewhat clasping or connate proximally into sheath, petiolate (basal) or sessile (cauline); blade 1-veined, lanceolate to oblong- or ovate-lanceolate, apex acute to obtuse. **Inflorescences** terminal, lax to erect, open, often flat-topped cymes; bracts paired, foliaceous; involucel bracteoles absent. **Pedicels** erect. **Flowers:** sepals connate proximally into tube, 9–17 cm, tube whitish green, 5-veined, cylindric to ovoid, 5-angled or winged, especially in fruit, commissures between sepals absent; lobes green, 1-veined, obovate to broadly triangular, shorter than tube, margins green or reddish, scarious, apex acute or acuminate; petals 5, pink to purplish, clawed, auricles absent, coronal appendages absent, blade apex entire or sometimes briefly 2-fid; nectaries at filament bases; stamens 10, adnate to petals; filaments distinct; staminodes absent; ovary 1-locular or sometimes 2-locular proximally; styles 2(–3), filiform, 10–12 mm, glabrous proximally; stigmas 2(–3), linear along adaxial surface of styles, papillate (30×). **Capsules** oblong to subglobose, exocarp opening by 4(–6) slightly spreading teeth, endocarp opening irregularly; carpophore present. **Seeds** ca. 10, black, subglobose, laterally compressed, papillose, marginal wing absent, appendage absent; embryo peripheral, curved. $x = 15$.

Species 1 or 4 (1 in the flora): introduced; Eurasia; introduced in South America, Africa (Republic of South Africa), Australia.

1. **Vaccaria hispanica** (Miller) Rauschert, Feddes Repert. 73: 52. 1966 • Cowherb, cowcockle, saponaire des vaches Ⓕ Ⓘ Ⓦ

Saponaria hispanica Miller, Gard. Dict. ed. 8, Saponaria no. 4 (in errata). 1768; *S. vaccaria* Linnaeus; *Vaccaria pyramidata* Medikus; *V. segetalis* (Necker) Garcke ex Ascherson; *V. vaccaria* (Linnaeus) Britton; *V. vulgaris* Host

Plants glabrous, glaucous. **Stems** 20–100 cm. **Leaf blades** 2–10 cm, base cuneate to cordate. **Cymes** open, 16–50(–100)-flowered. **Pedicels** (5–)10–30(–55) mm. **Flowers:** calyx 9–17 mm, with 5 prominent, usually green, winged angles or ridges, each ridge with strong, cordlike marginal vein; petals with claw 8–14 mm, blade 3–8 mm. **Capsules** included in calyx tube. **Seeds** 2–2.5 mm wide. $2n = 30$.

Flowering spring–summer. Fields, waste places; 0–2400 m; introduced; Alta., B.C., Man., N.B., N.S., Ont., Que., Sask., Yukon; Ala., Alaska, Ariz., Ark., Calif., Colo., Conn., Del., D.C., Fla., Idaho, Ill., Ind., Iowa, Kans., Ky., La., Maine, Md., Mass., Mich., Minn., Miss., Mo., Mont., Nebr., Nev., N.H., N.J., N.Mex., N.Y., N.Dak., Ohio, Okla., Oreg., Pa., R.I., S.C., S.Dak., Tenn., Tex., Utah, Vt., Va., Wash., W.Va., Wis., Wyo.; Eurasia; widely naturalized elsewhere.

If the genus *Vaccaria* is treated as monotypic, *V. hispanica* then includes four subspecies, and our material is subsp. *hispanica*. *Vaccaria hispanica* still occasionally is included in *Saponaria* (e.g., F. Swink and G. S. Wilhelm 1994). Once a common weed of grain fields (like *Agrostemma githago*), it is now increasingly rare or has been extirpated in many localities; the distribution stated above may be the historical maximum, rather than current, North American distribution. The saponin-containing seeds of this species are poisonous upon ingestion.

32. SAPONARIA Linnaeus, Sp. Pl. 1: 408. 1753; Gen. Pl. ed. 5, 191. 1754 • Soapwort, saponaire [from Latin *saponis*, soap, and *-aria*, pertaining to, alluding to sap] [I]

John W. Thieret

Richard K. Rabeler

Spanizium Grisebach

Herbs, [annual, biennial, or] perennial. **Rhizomes** stout or slender. **Stems** erect to spreading, simple or branched, terete. **Leaves** connate proximally, petiolate or sessile; blade 3(–5)-veined, spatulate to elliptic or ovate, apex acute or rounded. **Inflorescences** terminal, dense to open, lax cymes; bracts paired, foliaceous; involucel bracteoles absent. **Pedicels** erect. **Flowers:** sepals connate proximally into tube, greenish, reddish, or purple, 7–25 mm, tube 15–25-veined, oblong-cylindric, terete, commissures between sepals absent; lobes green, reddish, or purple, 3–5-veined, triangular-attenuate, shorter than tube, margins white, scarious, apex acute or acuminate; petals 5 (doubled in some cultivars), pink to white, clawed, auricles absent, with 2 coronal scales, blade apex entire or emarginate; nectaries at filament bases; stamens 10, adnate with petals to carpophore; filaments briefly connate proximally; staminodes absent (present in some cultivars); ovary 1-locular; styles 2(–3), filiform, 12–15 mm, glabrous proximally; stigmas 2(–3), linear along adaxial surface of styles, papillate (30×). **Capsules** cylindric to ovoid, opening by 4(–6) ascending or recurving teeth; carpophore present. **Seeds** 15–75, dark brown, reniform, laterally compressed, papillose, marginal wing absent, appendage absent; embryo peripheral, curved. $x = 7$.

Species ca. 40 (2 in the flora): introduced; Europe, c, w Asia, Africa (Mediterranean region); *S. officinalis* widely naturalized elsewhere.

Saponaria pumilio (Linnaeus) Fenzl ex A. Braun [= *Silene pumilio* (Linnaeus) Wulfen], a species of the Alps and the Carpathians, was collected once from a ledge on Mount Washington, New Hampshire, in 1964; the collector, S. K. Harris (1965), suggested that it may have been an intentional planting. A cespitose plant, it differs from the two species below also in its one-flowered, rather than several-flowered, stems.

SELECTED REFERENCE Shults, V. A. 1989. Rod Myl'nyanka (*Saponaria* L. s.l.) vo Flore SSSR. Riga.

1. Stems erect, 30–90 cm; calyx 15–25 mm, glabrous or rarely with scattered trichomes; capsules ca. 15–20 mm .... 1. *Saponaria officinalis*
1. Stems trailing, procumbent, or ascending, 5–25 cm; calyx 7–12 mm, glandular-pubescent; capsule 6–8 mm... 2. *Saponaria ocymoides*

1. Saponaria officinalis Linnaeus, Sp. Pl. 1: 408. 1753
• Bouncing-bet, saponaire officinale [F] [I] [W]

Plants perennial, colonial. **Stems** erect, simple or branched distally, 30–90 cm. **Leaves:** petiole often absent or winged, 0.1–1.5 cm; blade strongly 3(–5)-veined, elliptic to oblanceolate or ovate, 3–11(–15) × 1.5–4.5 cm. **Cymes** dense to open. **Pedicels** 1–5 mm. **Flowers** sometimes double; calyx green or reddish, often cleft, 15–25 mm, glabrous or rarely with scattered trichomes; petals pink to white, often drying to dull purple, blade 8–15 mm. **Capsules** ca. 15–20 mm. **Seeds** 1.6–2 mm wide. $2n = 28$.

Flowering spring–fall. Waste places, streamsides, fields, roadsides; 0–2600 m; introduced; Alta., B.C., Man., N.B., Nfld. and Labr. (Nfld.), N.S., Ont., P.E.I., Que., Sask.; Ala., Ariz., Ark., Calif., Colo., Conn., Del., D.C., Fla., Ga., Idaho, Ill., Ind., Iowa, Kans., Ky., La., Maine, Md., Mass., Mich., Minn., Miss., Mo., Mont., Nebr., Nev., N.H., N.J., N.Mex., N.Y., N.C., N.Dak.,

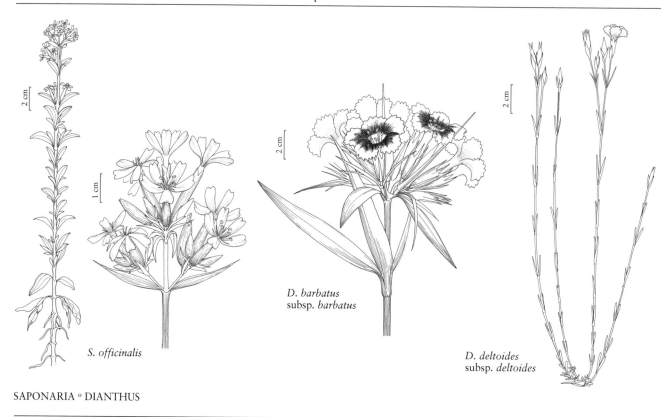

S. officinalis

D. barbatus
subsp. *barbatus*

D. deltoides
subsp. *deltoides*

SAPONARIA ○ DIANTHUS

Ohio, Okla., Oreg., Pa., R.I., S.C., S.Dak., Tenn., Tex., Utah, Vt., Va., Wash., W.Va., Wis., Wyo.; Eurasia; introduced in Mexico, South America, Asia (India), Africa (Egypt), Australia.

Saponaria officinalis, long cultivated for its showy flowers, is a widely naturalized, sometimes troublesome weed. It may persist for years about abandoned home sites. "Double"-flowered horticultural forms, which may lack functional stamens, also occur in the wild, where locally they may be as common as, or even more common than, "single"-flowered forms.

In former times, the leaves of this species were gathered and either soaked or boiled in water, the resulting liquid being used for washing as a liquid soap. Because of its saponin content, the species can be poisonous upon ingestion, in much the same manner as *Agrostemma githago*.

2. Saponaria ocymoides Linnaeus, Sp. Pl. 1: 409. 1753

• Rock soapwort   I

Plants perennial, with over-wintering leafy shoots. **Stems** trailing, procumbent, or ascending, much-branched, 5–25 cm. **Leaves:** petiole not winged, (0.1–)0.5–1(–3) cm; blade 1-veined, spatulate to ovate-lanceolate, 0.6–2.5 × 0.3–1.4 cm. **Cymes** spreading, lax. **Pedicels** 2–6 mm. **Flowers** sometimes double; calyx usually purple, not cleft, 7–12 mm, glandular-pubescent; petals red or pink to white, blade 8–15 mm. **Capsules** 6–8 mm. **Seeds** 1.6–2 mm wide. $2n = 28$ (Europe).

Flowering summer. Waste sites, rocky places, old gardens; 0–2200 m; introduced; Calif., Colo., Ind., Mass., Mich., N.Y., Oreg.; Europe.

Saponaria ocymoides is a long-cultivated rock-garden and wall plant that is only rarely persistent outside of gardens.

33. DIANTHUS Linnaeus, Sp. Pl. 1: 409. 1753; Gen. Pl. ed. 5, 191. 1754 • Pink, carnation, oeillet [Greek *dios*, divine, and *anthos*, flower, alluding to beauty or fragrance]

Richard K. Rabeler

Ronald L. Hartman

Herbs, perennial (*D. armeria* annual or biennial), sometimes mat-forming. **Taproots** stout, rhizomes (when present) slender or stout. **Stems** erect or ascending, simple or branched, terete or angled. **Leaves** connate proximally into sheath, petiolate (basal leaves) or sessile; blade 1-veined, linear or oblong to ovate, apex acute. **Inflorescences** terminal, open cymes, dense bracteate clusters or heads, or flowers solitary; bracts paired, herbaceous to scarious, or absent; involucel bracteoles 1–3 pairs, herbaceous or scarious. **Pedicels** erect in fruit. **Flowers:** sepals connate proximally into tube, 10–22 mm, tube green or reddish, 20–60-veined, ± cylindric, terete, commissures between sepals absent, lobes green or reddish, 3–8-veined, triangular to lanceolate, shorter than tube, margins white or reddish, mostly scarious, apex acute or obtuse; petals often pink or red, sometimes white or purple, sometimes spotted or with darker center, clawed, auricles absent, coronal appendages absent, blade apex dentate or fimbriate to ¹/₂ of length; nectaries at filament bases; stamens 10, adnate with petals to carpophore; filaments distinct; staminodes absent; ovary 1-locular; styles 2, filiform, 0.7–6 mm, glabrous proximally; stigmas 2, linear along adaxial surface of styles, papillate (30×). **Capsules** ovoid to cylindric, opening by 4 teeth; carpophore present. **Seeds** 40–100+, blackish brown, shield-shaped, dorsiventrally compressed, papillose-striate to papillate, marginal wing absent, appendage absent; embryo central, straight. $x = 15$.

Species ca. 320 (6 in the flora): n North America, Eurasia (Balkans to c Asia), Africa; introduced in North America (except *D. repens*), South America, Pacific Islands (Hawaii), possibly Australia.

Dianthus species have been popular garden subjects for years; there are now over 27,000 registered cultivar names (A. C. Leslie 1983 and 19 subsequent supplements). Although they are most popular in Great Britain, many species and cultivars are grown in North America. While some popular taxa (e.g., *D. caryophyllus* Linnaeus, clove pink, and the hybrids called *D.* 'allwoodii', Allwood's pink) do not appear to escape and/or persist after cultivation, others do so readily. Five of the six species treated here are introduced and readily persist; *D. repens* is our only native species.

In spite of the popularity of *Dianthus* in horticulture, the genus requires a thorough study using modern methods. It is the second largest genus in the family (surpassed only by *Silene*) but there is no recent monograph or comprehensive infrageneric classification. The genus is sometimes divided into two subgenera [*Dianthus* and *Carthusianastrum* F. Williams; e.g., F. A. Pax and K. Hoffmann (1934c) and T. G. Tutin and S. M. Walters (1993)], corresponding to the division indicated in couplet one of the key below. Others, including M. Kuzmina (2002, 2003), have considered this an artificial separation.

Besides the six species treated here, five others have been collected at least once outside of cultivation in North America and could be expected elsewhere in the flora area. On the basis of bracteole length, one species would key near *Dianthus deltoides*, and four near *D. plumarius*. *Dianthus chinensis* Linnaeus, the rainbow pink, similar to *D. deltoides* but with four involucral bracteoles, a glabrous calyx, and basal leaves absent when the plants flower, was noted as "persisting" in Lambton County, Ontario (C. K. Dodge 1915). *Dianthus sequieri* Villars, with bracteoles over one-half as long as the calyx, is otherwise similar to *D. plumarius* but with broader leaves [2–5 versus 1(–2) mm] and dentate petals. It was collected along a roadside at Fort Saskatchewan, Alberta, in 1948, 1949, and 1950 (*Turner 6416*, ALTA, DAO; *7019*, DAO; *7269*, ALTA). *Dianthus arenarius* Linnaeus, similar to *D. plumarius* but with green leaves shorter than 4 cm and petal blades with a greenish central spot and divided over one-half their length, was reported as a garden escape near Ottawa, Ontario, in 1958. *Dianthus sylvestris* Wulfen, similar to *D. plumarius* but with green leaves and entire or toothed, glabrous petals, was found in a ditch near Hancock, Michigan, in 1982 (*Rabeler 711*, MICH, MSC), quite possibly an escape from cultivation. *Dianthus gratianopolitanus* Villars, a popular rock-garden plant similar to *D. plumarius* but with glaucous leaves shorter than 4.5 cm, petal blades divided less than one-fourth their length, and narrowly scarious sepal-lobe margins, was collected along a roadside near Berrie, Ontario, in 1973 (*Reznicek 3640*, MICH, TRT).

1. Inflorescences 3–15(–20+)-flowered, dense heads or open cymes, seldom flowers solitary; flowers subsessile, pedicels 0.1–2(–3) mm; bracts herbaceous to scarious, equaling or longer than calyx; bracteoles $^2/_3$–1$^1/_4$ times as long as calyx [subg. *Carthusianastrum*].
   2. Inflorescences 3–6-flowered, open cymes, sometimes flowers solitary; calyx pubescent
     . . . . 1. *Dianthus armeria*
   2. Inflorescences 4–20+-flowered, dense heads; calyx glabrous, lobe margins may be ciliate.
     3. Leaf blades lanceolate to ovate; leaves connate proximally, forming sheath 1(–2) times as long as stem diam. . . . . 2. *Dianthus barbatus*
     3. Leaf blades linear to spatulate; leaves connate proximally, forming sheath 3–8 times as long as stem diam. 3. *Dianthus carthusianorum*
1. Inflorescences 1(–4)-flowered, open cymes or usually flowers solitary; flowers pedicellate, pedicels 4–25(–30) mm; bracts absent; bracteoles $^1/_4$–$^2/_3$(–$^3/_4$) as long as calyx [subg. *Dianthus*].
   4. Bracteoles ca. $^1/_2$–$^3/_4$ as long as calyx . . . . 6. *Dianthus repens*
   4. Bracteoles $^1/_4$–$^1/_2$ as long as calyx.
     5. Petals 4–9 mm, apex dentate; bracteoles $^1/_3$–$^1/_2$ as long as calyx . . . . 4. *Dianthus deltoides*
     5. Petals 8–15 mm, apex divided into narrow segments to $^1/_2$ as long as blade; bracteoles $^1/_4$–$^1/_3$ as long as calyx . . . . 5. *Dianthus plumarius*

1. **Dianthus armeria** Linnaeus, Sp. Pl. 1: 410. 1753

   • Deptford pink, oeillet arméria   [I] [W]

Subspecies 2 (1 in the flora): introduced; Europe, c, sw Asia; introduced in South America (Chile), Pacific Islands (Hawaii).

1a. Dianthus armeria Linnaeus subsp. **armeria**   [I] [W]

Plants annual or biennial. **Stems** erect, simple or branched, 4–88 cm, often pubescent. **Leaves:** sheath 1–4 mm, 1–2 times as long as stem diam.; blade linear to narrowly oblanceolate, 2–10.5 cm, green, margins basally ciliate. **Inflorescences** usually open 3–6-flowered cymes or flowers sometimes solitary; bracts linear, equaling or longer than calyx, herbaceous, apex acute; bracteoles 4, green, linear to lanceolate, $^3/_4$–1 times as long as calyx, herbaceous, apex acute. **Pedicels** 0.1–3 mm. **Flowers** subsessile; calyx 20–25-veined, 11–20 mm, pubescent, lobes acuminate, 2–7 mm; petals reddish with white dots (rarely all white), bearded, 3–8(–10) mm, apex dentate. **Capsules** 10–16 mm, slightly shorter than calyx. **Seeds** 1.1–1.4 mm. *2n* = 30 (Europe).

Flowering spring and summer. Roadsides, fields, shores, open woods, waste places; 0–2300 m; introduced; Alta., B.C., N.B., N.S., Ont., P.E.I., Que.; Ala., Ariz., Ark., Calif., Colo., Conn., Del., Fla., Ga., Idaho, Ill., Ind., Iowa, Kans., Ky., Maine, Md., Mass., Mich., Minn., Miss., Mo., Mont., Nebr., N.H., N.J., N.Mex., N.Y., N.C., Ohio, Okla., Oreg., Pa., R.I., S.C., S.Dak., Tenn., Tex., Vt., Va., Wash., W.Va., Wis., Wyo.; Europe; w Asia; introduced in South America (Chile), Pacific Islands (Hawaii).

Subspecies *armeria* is occasionally cultivated, as in Louisiana, where apparently it has not yet escaped. It was collected at Stephenville, Newfoundland, in 1972 (*Rouleau 11293*, CAN, MT), but apparently has not become established there.

2. Dianthus barbatus Linnaeus, Sp. Pl. 1: 409. 1753

• Sweet William, oeillet barbu [F] [I]

Subspecies 2 (1 in the flora): introduced; Europe, e, sw Asia; introduced in Mexico, Central America, South America, se Asia (Java).

2a. Dianthus barbatus Linnaeus subsp. **barbatus** [F] [I]

Plants cespitose, not matted. **Stems** erect, simple, 10–70 cm, glabrous. **Leaves:** sheath 3–6 mm, 1(–2) times as long as stem diam.; blade lanceolate to ovate, (1.5–)2.5–10 cm, green, margins finely ciliate. **Inflorescences** usually dense, 5–20-flowered heads, or rarely flowers solitary, axillary; bracts lanceolate, equaling or longer than calyx, herbaceous, apex acute; bracteoles 4 (or 6), green, ovate, ³/₄–1¹/₄ times as long as calyx, herbaceous, apex long-aristate in distal ¹/₃(–¹/₂). **Pedicels** 0.1–2 mm. **Flowers** subsessile; calyx 40-veined, (10–)12–19(–21) mm, glabrous (lobe margins may be ciliate), lobes ovate to short-tapered, 2–7 mm; petals purple, red, spotted with white, or pink or white with (or without) dark center, bearded, 4–10 mm, apex dentate. **Capsules** 10–13 mm, equaling calyx length. **Seeds** 2–2.5(–2.7) mm. *2n* = 30 (Europe).

Flowering spring–early summer. Waste places, roadsides, fallow fields, disturbed riverbanks, forest edges; 10–2500 m; introduced; Alta., B.C., Man., N.S., Ont., Que., Sask.; Ala., Calif., Conn., Ga., Idaho, Ill., Ind., Kans., Ky., La., Maine, Md., Mass., Mich., Minn., Mo., Mont., Nebr., N.H., N.Y., N.C., Ohio, Oreg., Pa., R.I., S.C., Tenn., Tex., Vt., Va., Wash., W.Va., Wis.; Europe; e, sw Asia; introduced in Mexico, Central America, South America, se Asia (Java).

Subspecies *barbatus* is cultivated widely and both persists at old homesites and sometimes escapes to nearby fields, etc.

3. Dianthus carthusianorum Linnaeus, Sp. Pl. 1: 409. 1753 • Clusterhead pink [I]

Plants cespitose, not matted. **Stems** erect, simple or branched proximally, 25–65 cm, glabrous. **Leaves:** sheath (5–)8–18(–25) mm, 3–8 times as long as stem diam.; blade linear to spatulate, 3–7(–13) cm, green, margins glabrous. **Inflorescences** dense, 4–15(–20+)-flowered heads; bracts lanceolate, equaling or longer than calyx, herbaceous to scarious, apex aristate; bracteoles 4 or 6, brown, oblong-obovate to oblanceolate, ²/₃–³/₄ times as long as calyx, scarious, apex aristate in distal ¹/₃–¹/₅. **Pedicels** 0.1–2 mm. **Flowers** subsessile; calyx 40–45-veined, 10–15 mm, glabrous, lobes ovate, 3–4 mm; petals deep pink to scarlet or purple, bearded, 10–15 mm, apex dentate. **Capsules** 8–10 mm, slightly shorter than calyx. **Seeds** 1.6–2 mm. *2n* = 30, 60 (both Europe).

Flowering summer. Disturbed areas, fields; 200–400 m; introduced; Mich., N.H., N.Y., Wis.; Europe.

Dianthus carthusianorum is an extremely variable species; at least 12 subspecies and 35 varieties have been described. A comprehensive study is warranted to determine how many of those taxa should be recognized or whether, as proposed by T. G. Tutin and S. M. Walters (1993), the variation in many of these characters is indeed continuous.

The New Hampshire collections of *Dianthus carthusianorum* were made between 1914 and 1926 in Coos County (R. K. Rabeler and R. E. Gereau 1984); all other observations and collections are post-1975. *Dianthus carthusianorum* is sometimes cultivated and may escape.

4. Dianthus deltoides Linnaeus, Sp. Pl. 1: 411. 1753

• Maiden pink, oeillet à delta [F] [I]

Subspecies 2 (1 in the flora): introduced; Europe; introduced in e Asia (China, Japan, Russian Far East).

4a. Dianthus deltoides Linnaeus subsp. **deltoides** [F] [I]

Plants cespitose, matted. **Stems** ascending, simple proximally, branched distally, (4.5–)10–40 cm, puberulent; sterile shoots matted, 3–6 cm. **Leaves:** sheath 0.5–1.5 mm, ca. equaling stem diam.; blade linear to linear-lanceolate (flowering stems) or oblanceolate (sterile stems), 0.6–2.6 cm, green, margins ciliate. **Inflorescences** open, 2–4-flowered cymes or often flowers solitary; bracts absent; bracteoles 2 (or 4), green, ovate, ca. ¹/₃–¹/₂ times as long as calyx, herbaceous with ± scarious margins, apex aristate. **Pedicels** 4–12(–30) mm. **Flowers:** calyx 25–30-veined, 10–17 mm, glabrous or minutely pubescent, lobes narrowly lanceolate to linear, 3–5 mm; petals light or dark pink to purple (rarely white), often with darker band at base of blade, bearded, 4–9 mm, apex dentate. **Capsules** 12–13 mm, somewhat shorter than calyx. **Seeds** 1.1 mm. *2n* = 30 (Europe).

Flowering early summer. Roadsides, meadows, shores, open woods, waste places; 40–2700 m; introduced; Alta., B.C., Man., N.B., N.S., Ont., P.E.I., Que., Sask.; Ark., Calif., Colo., Conn., Ill., Iowa, Maine,

Mass., Mich., Minn., Mo., Mont., N.H., N.J., N.Y., N.C., Ohio, Pa., R.I., Utah, Vt., Va., Wash., Wis., Wyo.; Europe; introduced in e Asia (China, Japan, Russian Far East).

Subspecies *deltoides* is often cultivated and occasionally escapes, perhaps not persisting in the northern part of its range.

5. **Dianthus plumarius** Linnaeus, Sp. Pl. 1: 411. 1753
  • Garden or cottage pink, oeillet mignardise   [I]

Subspecies 4(–8?) (1 in the flora): introduced; e Europe; introduced in Central America.

5a. **Dianthus plumarius** Linnaeus subsp. **plumarius**
  [I]

Plants cespitose, not matted. **Stems** erect, simple, 13–40 cm, glabrous, often glaucous. **Leaves:** sheath (1–)2–4 mm, 1–2 times as long as stem diam.; blade linear, 2–7.5 cm, margins glabrous, often glaucous. **Inflorescences** open, 2–4-flowered cymes, or sometimes flowers solitary; bracts absent; bracteoles 4, green, obovate, 1/4–1/3 times as long as calyx, herbaceous, apex abruptly acuminate or truncate. **Pedicels** 8–25 mm. **Flowers:** calyx 40–45-veined, 14–22 mm, glabrous, lobes ovate, 3–6 mm; petals white or pale pink, often with darker center, bearded, 8–15 mm, apex divided into narrow segments to 1/2 as long as blade.

Capsules 23–27 mm, slightly exceeding calyx. **Seeds** 2.4–3 mm. $2n$ = 30, 60, 90 (all Europe).

Flowering spring and summer. Roadsides, sandy fields; 0–400 m; introduced; N.S., Ont., Que., Yukon; Ala., Calif., Conn., Maine, Mass., Mich., Mo., N.H., N.Y., N.C., Pa., S.C., Vt., Va., Wis.; e Europe.

Subspecies *plumarius* is often cultivated and occasionally escapes, perhaps not persisting in the northern part of its range.

6. **Dianthus repens** Willdenow, Sp. Pl. 2: 681. 1799
  • Northern pink

Plants cespitose, matted. **Stems** erect, simple, 5–17(–25) cm, glabrous. **Leaves:** sheath 1–2 mm, 1–2 times as long as stem diam.; blade lance-linear to oblong or linear, 1–4.5 cm, green, margins glabrous. **Inflorescences** open, 2–4-flowered cymes, or usually flowers solitary; bracts absent; bracteoles 2 (or 4), green or reddish, lanceolate, ca. 1/2–3/4 times as long as calyx, herbaceous, apex acuminate. **Pedicels** 4–12 mm. **Flowers:** calyx (45–)50–60-veined, 10–14 mm, glabrous, lobes triangular, 2–4 mm; petals pink to pink-purple, bearded or not, 7–13 mm, apex dentate. **Capsules** 12–17 mm, exceeding calyx. **Seeds** 1.2–2 mm. $2n$ = 30, 60 (Asia).

Flowering summer. Rock outcrops, talus slopes; 10–1400 m; Yukon; Alaska; Asia (China, Russian Far East).

34. **PETRORHAGIA**   (Seringe) Link, Handbuch 2: 235. 1831 • [Greek *petra-*, rock, and *rhagas*, rent or chink, translation of Latin *saxifraga*, rockbreaking, alluding to prevalence in rock crevices]   [I]

Richard K. Rabeler

Ronald L. Hartman

Gypsophila Linnaeus sect. *Petrorhagia* Seringe in A. P. de Candolle and A. L. P. P. de Candolle, Prodr. 1: 354. 1824; *Kohlrauschia* Kunth; *Tunica* Ludwig

Herbs, annual or perennial with woody bases. **Taproots** slender to stout **Stems** erect or ascending, simple or branched proximally, terete or angular. **Leaves** connate proximally into sheath, sessile; blade 1- or 3-veined, linear to narrowly oblanceolate, apex acute. **Inflorescences** terminal, dense capitula or lax cymes, or flowers solitary; bracts paired, brown-scarious and often enclosing inflorescence; involucel bracteoles of 1–3 pairs [or absent], similar in size and texture. **Pedicels** erect. **Flowers** bisexual, occasionally unisexual and female; sepals connate proximally into tube, 4–15 mm; tube green or reddish and white or brown-scarious, 15-veined, cylindric, terete, commissures between sepals veinless, broad, scarious; lobes green, reddish, or brown, 3-veined, oblong, shorter than tube, margins white or brown, scarious, apex rounded; petals 5, pink or purplish to white, clawed (or not in *P. saxifraga*), auricles absent, coronal appendages absent, blade apex entire and obtuse to 2-fid to $^1/_{16}$ of length; nectaries at filament bases; stamens 10; filaments distinct; staminodes absent; ovary 1-locular; styles 2, filiform, 2–9 mm, glabrous proximally; stigmas 2, linear along adaxial surface of styles, papillate (30×). **Capsules** 4-lobed, oblong, shorter than sepals, opening by 4 slightly recurving or straight teeth; carpophore present. **Seeds** 8–15, blackish brown, shield- or helmet-shaped, dorsiventrally compressed, reticulate to papillate, marginal wing absent, appendage absent; embryo central, straight. $x = [13?, 14?], 15$.

Species 33 (4 in the flora): introduced; Europe (Mediterranean region), c, sw Asia, Africa (Mediterranean region); introduced in South America, Africa (Republic of South Africa), Pacific Islands (Hawaii), Australia.

Some authors, e.g., V. Bittrich (1993), prefer to split *Petrorhagia* as shown in couplet one of the key below, recognizing the five species in the genus with broad, brown-scarious bracts enclosing much of the inflorescence as the genus *Kohlrauschia* Kunth. While a dorsiventrally compressed seed with a straight, central embryo is common to all species, *Petrorhagia* is morphologically diverse, with five sections recognized by P. W. Ball and V. H. Heywood (1964), and in many ways morphologically intermediate between *Dianthus* and *Gypsophila*. If *Kohlrauschia* is recognized, the inflorescence is the only character not shared by at least a few other species of *Petrorhagia*.

SELECTED REFERENCES Ball, P. W. and V. H. Heywood. 1964. A revision of the genus *Petrorhagia*. Bull. Brit. Mus. (Nat. Hist.), Bot. 3: 121–172. Rabeler, R. K. 1985. *Petrorhagia* (Caryophyllaceae) in North America. Sida 11: 6–44. Thomas, S. M. and B. G. Murray. 1983. Chromosome studies in species and hybrids of *Petrorhagia* sect. *Kohlrauschia* (Caryophyllaceae). Pl. Syst. Evol. 141: 243–255.

1. Flowers in lax cymes or solitary; pedicels 5–20 mm; inflorescence bracts brown-scarious, linear to narrowly ovate, not enclosing flowers; involucel bracteoles scarious, narrowly ovate, less than $^1/_2$ length of sepals; sepals 4–6 mm [sect. *Petrorhagia*] . . . . 1. *Petrorhagia saxifraga*
1. Flowers in capitate inflorescences, rarely appearing solitary; pedicels 0.1–3 mm; inflorescence bracts and involucel bracteoles brown-scarious, broadly ovate, ± equaling sepals, enclosing flowers; sepals (5–)10–15 mm [sect. *Kohlrauschia* (Kunth) P. W. Ball & Heywood].
   2. Leaf sheaths ± as long as wide, usually 1–2 mm; petals with apex truncate or emarginate, dark-colored veins absent . . . . 2. *Petrorhagia prolifera*
   2. Leaf sheaths 1.5–3 times as long as wide, usually 3–9 mm; petals with apex obcordate or 2-fid, dark-colored veins 1–6.
      3. Seeds (1.3–)1.5–1.8 mm, shield-shaped, tuberculate; leaf sheaths (2–)3–4 mm; apices of inner inflorescence bracts obtuse or mucronate; dark-colored petal veins 1(–3) . . . . 3. *Petrorhagia nanteuilii*
      3. Seeds 1–1.4 mm, helmet-shaped, covered with conical papillae; leaf sheaths (3–)4–9 mm; apices of inner inflorescence bracts mucronate; dark-colored petal veins 3(–6) . . . . 4. *Petrorhagia dubia*

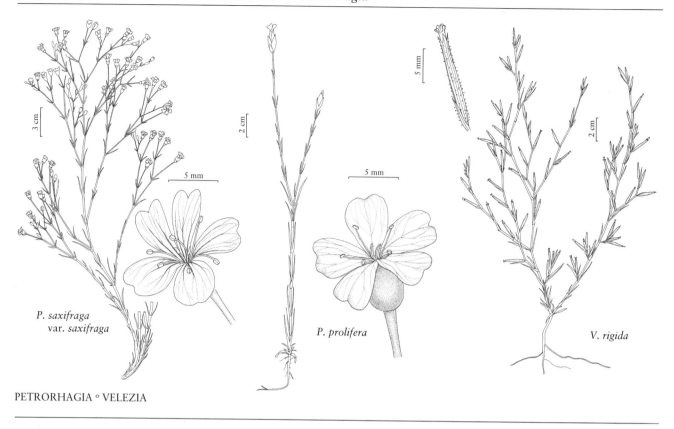

P. saxifraga
var. *saxifraga*

P. prolifera

V. rigida

PETRORHAGIA ° VELEZIA

1. Petrorhagia saxifraga (Linnaeus) Link, Handbuch 2: 235. 1831 • Tunic flower, coat flower [F] [I]

Dianthus saxifragus Linnaeus, Sp. Pl. 1: 413. 1753; *Tunica saxifraga* (Linnaeus) Scopoli

Varieties 2 (1 in the flora): introduced; c, s Europe, sw Asia; introduced in Europe (Great Britain, Sweden).

Petrorhagia saxifraga is often grown as a rock-garden or border plant, with several cultivars available in North America. Some North American records of *P. saxifraga* are from plants that either escaped from or persisted after cultivation. Most collections made since 1960 are from western Michigan, where *P. saxifraga* is known from naturalized populations in ten counties.

1a. Petrorhagia saxifraga (Linnaeus) Link var. **saxifraga** [F] [I]

Plants perennial with woody bases. **Stems** ascending to erect, much-branched, 5–40 cm; internodes glabrous distally, scabrous proximally. **Leaves:** sheath 1 mm or less, ± as long as wide; blade 1-veined, linear, 5–20 (–30) mm, margins basally ciliate. **Inflorescences** lax cymes or flowers solitary; involucel bracteoles enclosing to ¹/₂ of calyx, narrowly ovate, scarious, apex acute. **Pedicels**

5–20 mm. **Flowers:** sepals 4–6 mm; petals white to pink, primary veins 3, veins often dark pink near base of blade, apex obcordate. **Seeds** shield-shaped, 0.8–1.3 mm, tuberculate. $2n = 30, 60$ (both Europe).

Flowering summer–early autumn. Lawns, roadsides, sandy waste areas; 0–1100 m; introduced; B.C., Ont.; Idaho, Ill., Maine, Mass., Mich., Minn., N.Y., Ohio, Pa., S.Dak., Va., Wis.; c, s Europe; sw Asia; introduced in Europe (Great Britain, Sweden).

2. Petrorhagia prolifera (Linnaeus) P. W. Ball & Heywood, Bull. Brit. Mus. (Nat. Hist.), Bot. 3: 161. 1964 • Proliferous or childing pink [F] [I]

Dianthus prolifer Linnaeus, Sp. Pl. 1: 410. 1753; *Kohlrauschia prolifera* (Linnaeus) Kunth; *Tunica prolifera* (Linnaeus) Scopoli

Plants annual. **Stems** erect, simple or branched, (6–)20–30(–60) cm; internodes glabrous or midstem ones slightly scabrous. **Leaves:** sheath 1–2 mm, ± as long as wide; blade 3-veined, linear to linear-lanceolate, 10–30 mm, margins serrate-scabrous. **Inflorescences** capitate; inflorescence bracts and involucel bracteoles enclosing flowers, broadly ovate, brown-scarious, apex obtuse or of outer bracts mucronate. **Pedicels** 0.1–1.5 mm.

Flowers: sepals (7–)10–12 mm; petals pink to slightly purplish (rarely white), primary veins 1, veins not darkly colored near base of blade, apex truncate or emarginate. **Seeds** shield-shaped, 1.1–1.6(–1.8) mm, fine to coarsely reticulate. *2n* = 30 (Europe).

Flowering summer. Roadsides, ballast, fields; 0–1100 m; introduced; B.C.; Ala., Ark., Del., Ga., Idaho, Ky., Md., Mich., Mo., N.J., N.Y., N.C., Okla., Pa., Tenn., Va.; c, s Eurasia; introduced in Europe (Great Britain).

Historical records for *Petrorhagia prolifera* exist also for California (1902; *Congdon s.n.*, MIN), Ohio (last collected in 1896; *Stair s.n.*, OS), and South Carolina (1800s; *Durand s.n.*, NY).

Petrorhagia prolifera has been known in the northeastern United States since at least 1837, and its range has since expanded, with isolated populations occurring southwestward from New Jersey toward Arkansas and Oklahoma as well as western Michigan. Some introductions may have been as a contaminant in grass seed used for highway planting in Tennessee (B. E. Wofford et al. 1977). Literature reports of *P. prolifera* in Louisiana and West Virginia have not been confirmed.

3. **Petrorhagia nanteuilii** (Burnat) P. W. Ball & Heywood, Bull. Brit. Mus. (Nat. Hist.), Bot. 3: 164. 1964 [I]

Dianthus nanteuilii Burnat in E. Burnat et al., Fl. Alpes Marit. 1: 221. 1892

Plants annual. **Stems** erect, simple or branched, (20–)30(–52) cm; internodes glabrous or midstem and proximal ones minutely stipitate-glandular. **Leaves:** sheath (2–)3–4 mm, 1.5–2 times as long as wide; blade 3-veined, linear, 10–25 mm, margins scabrous. **Inflorescences** capitate; inflorescence bracts and involucel bracteoles enclosing flowers, broadly ovate, brown-scarious, apex of outer bracts mucronate, of inner bracts obtuse or mucronate. **Pedicels** 0.1–2 mm. **Flowers:** sepals (5–)10–12 mm; petals pink to slightly purplish, primary veins 3, at least center vein darkly colored near base of blade, apex obcordate or 2-fid. **Seeds** shield-shaped, (1.3–)1.5–1.8 mm, tuberculate. *2n* = 60 (Europe).

Flowering late spring–summer. Roadsides; 0–200 m.; introduced; B.C.; Calif.; w Europe; nw Africa; introduced in South America, Australia.

Petrorhagia nanteuilii is known from two counties in northern California (first collected in 1956) and a single site in British Columbia. It may be an alloploid derived through hybridization between *P. prolifera* and *P. dubia*; recent attempts at crossing the latter two have failed (S. M. Thomas and B. G. Murray 1983).

4. **Petrorhagia dubia** (Rafinesque) G. López & Romo, Anales Jard. Bot. Madrid 45: 363. 1988 [I]

Dianthus dubius Rafinesque, Caratt. Nuov. Gen., 75. 1810; *Kohlrauschia velutina* (Gussone) Reichenbach; *Petrorhagia velutina* (Gussone) P. W. Ball & Heywood

Plants annual. **Stems** erect, simple or branched, (9.5–)25–40(–91) cm; internodes glabrous or midstem ones densely stipitate-glandular. **Leaves:** sheath (3–)4–9 mm, 2–3 times as long as wide; blade 3-veined, linear to oblanceolate, 10–60 mm, margins scabrous. **Inflorescences** capitate; inflorescence bracts and involucel bracteoles enclosing flowers, broadly ovate, brown-scarious, apex mucronate. **Pedicels** 0.1–3 mm. **Flowers:** sepals (8–)12–15 mm; petals pink or purplish, primary veins 3, (3–)5–6 veins darkly colored near base of blade, apex 2-fid, sometimes obcordate. **Seeds** helmet-shaped, 1–1.4 mm, covered with conical papillae. *2n* = 30 (Europe).

Flowering spring–early summer. Roadsides, woodland savannas; 0–1800 m; introduced; Calif., La., Miss., Okla., Tex.; Europe (Mediterranean region); Africa; introduced in South America, Africa (Republic of South Africa), Australia.

All material of *Petrorhagia dubia* from California, where it appears to be have been introduced in the Sierra Nevada foothills in the 1920s, and one population from northeastern Texas, have stipitate-glandular internodes. The presence of glabrous internodes in the Louisiana, Mississippi, Oklahoma, and most Texas populations (where it was first seen along roadsides in 1967) led to early confusion with *P. prolifera*. This suggests that these populations were derived from seed that came from Italy or Sicily, the only area in the native range where plants with glabrous stems are known. Roadside planting of either Italian rye grass [*Lolium perenne* Linnaeus var. *italicum* (A. Braun) Parnell] or crimson clover (*Trifolium incarnatum* Linnaeus) in Texas is the likely source of *P. dubia* in that region.

35. VELEZIA Linnaeus, Sp. Pl. 1: 332. 1753; Gen. Pl. ed. 5, 155. 1754 • [For Cristóbal Velez, ca. 1710–1753, a friend of the botanist Pehr Loefling] I

Ronald L. Hartman

Richard K. Rabeler

Herbs, annual. **Taproots** slender. **Stems** simple or branched, terete. **Leaves** connate proximally by line of tissue or sheath, sessile; blade 3-veined, linear to subulate, apex acute. **Inflorescences** axillary, flowers generally solitary; bracts foliaceous; involucel bracteoles absent. **Pedicels** erect. **Flowers:** sepals 11–15 mm; tube green, 15-veined, narrowly cylindric, terete, commissures between sepals absent, lobes green or reddish, 3-veined proximally, lanceolate-acuminate, shorter than tube, margins silvery, scarious, apex acute; petals 5, pink or purple, clawed, auricles absent, corona of 6–8, linear to lanceolate segments, blade apex entire or notched; nectaries at filament bases; stamens 5; filaments distinct; staminodes absent; ovary 1-locular; styles 2, 9–10 mm, glabrous proximally; stigmas 2, linear along adaxial surface of styles, papillate (30×). **Capsules** narrowly cylindric, opening by 4 ascending teeth; carpophore present. **Seeds** 6–8, black, ovate-oblong, dorsiventrally compressed, finely papillate, marginal wing absent, appendage absent; embryo central, straight. $x = 14$.

Species 6 (1 in the flora): introduced; Europe (Mediterranean region), sw, c Asia.

1. Velezia rigida Linnaeus, Sp. Pl. 1: 332. 1753 F I

Stems green to purplish, with widely divergent or dichotomous branches, 7–40 cm, rigid, glandular-hairy at least distally. **Leaf blades** 0.5–2 cm, margins scarious near base, ciliate. **Pedicels** 1.5–3.5 mm. **Flowers:** calyx tube 10–14 × 0.8–1 mm, lobes 1–1.2 mm, base swollen, hardened; petals narrowly obovate, 11–16 mm, coronal appendages 0.4–0.6 mm. **Capsules** with rounded tip; carpophore 0.2–0.7 mm. **Seeds** in 1 row, with abrupt but round point, 1.3–1.8 mm. $2n = 28$ (Europe).

Flowering spring–early summer. Oak and pine woodlands, open ridges, gravelly streambeds, serpentine areas; 50–800 m; introduced; Calif.; Europe (Mediterranean region); sw, c Asia; Africa (Mediterranean region).

Velezia rigida is now known from at least eight counties in northern California; it was first collected in North America in 1896 (*Jepson s.n.*, JEPS).

36. SILENE Linnaeus, Sp. Pl. 1: 416. 1753; Gen. Pl. ed. 5, 193. 1754, name conserved • Campion, catchfly [Greek *seilenos*, probably derived from Silenus, the intoxicated foster father of the Greek god Bacchus, who was described as covered with foam; perhaps alluding to the viscid secretion covering many species]

John K. Morton

Anotites Greene; *Atocion* Adanson; *Coronaria* Guettard; *Gastrolychnis* (Fenzl) Reichenbach; *Lychnis* Linnaeus; *Melandrium* J. C. Röhling; *Physolychnis* (Bentham) Ruprecht; *Viscaria* Bernhardi; *Wahlbergella* Fries

Herbs, annual, biennial, or perennial, often decumbent at base or sometimes cespitose. **Taproots** slender or often stout, deep, branched caudex often present, some species stoloniferous or rhizomatous. **Stems** simple or branched, terete or sometimes angular. **Leaves** opposite or occasionally whorled, connate proximally, petiolate (basal leaves) or sessile (most cauline leaves); blade 1–5-veined, linear to obovate or spatulate, herbaceous, apex acute to obtuse. **Inflorescences** terminal or sometimes axillary, simple or branched, sometimes condensed cymes, frequently flowers few or solitary, frequently glandular-pubescent and viscid; bracts paired, herbaceous or scarious, or absent; involucel bracteoles absent. **Pedicels** erect, rarely flowers sessile or subsessile. **Flowers** bisexual, sometimes unisexual (rarely so on separate plants); sepals connate proximally into tube, (4–)10–28(–40) mm; tube green, whitish, and/or purplish, 10–30-veined, cylindric to campanulate, urceolate, or clavate, terete, frequently inflated, membranous or more rarely herbaceous, commissures between sepals 1-veined, herbaceous; lobes green or purplish, 1–5-veined, broadly triangular to lance-oblong or linear, usually shorter than tube, margins whitish, scarious, apex acute to obtuse; petals 5, white, pink, scarlet, dusky purple, or off-white tinged with purple, clawed, claw usually conspicuous, sometimes small, rarely absent, auricles 2, coronal appendages 2, variously shaped or dissected; limb usually exserted and conspicuous, oblanceolate to obovate, apex 2-lobed, sometimes dissected into 1–4 linear lobes or irregular teeth, or fimbriate, rarely entire; nectaries at filament bases; stamens 10, rarely fewer or absent, frequently dimorphic with longer opposite petals, arising with petals from carpophore; filaments distinct nearly to base; staminodes absent (rarely to 10 in pistillate flowers, arising with petals from carpophore, filiform); ovary 1- or 3–5-locular; styles 3 or 5, occasionally 4 (absent in staminate flowers), filiform, 1.5–20 mm, glabrous proximally; stigmas 3 or 5, occasionally 4, linear along adaxial surface of styles, papillate (30×). **Capsules** ovoid to globose, opening along sutures into 3–5 valves, frequently splitting into 6–10 equal teeth; carpophore usually present. **Seeds** ca. (5–)15–100(–500+), reddish to gray or black, reniform to globose, usually tuberculate or papillate, papillae around margins sometimes larger and inflated, marginal wing sometimes present, appendage absent; embryo peripheral, curved. *x* = (10) 12.

Species ca. 700 (70 in the flora): mainly Northern Hemisphere.

Silene includes several important weeds and some very beautiful horticultural plants. In addition to the species described in this account, several others have occurred in the flora area as chance introductions or garden escapes, but they have not become established and most have not been seen recently. They include *S. coeli-rosa* (Linnaeus) Godron, *S. cretica* Linnaeus, *S.* (*Lychnis*) *fulgens* (Fischer) E. H. L. Krause, *S. italica* Persoon, and *S. nutans* Linnaeus.

In this account, *Lychnis*, *Melandrium*, and *Viscaria* have been included in *Silene*, their previous recognition as distinct genera having resulted in a great deal of confusion in both nomenclature and taxonomy. I have not presented an infrageneric classification of *Silene* because existing systems either do not include those other genera (e.g., P. K. Chowdhuri 1957) or do not deal with most of our native North American taxa [e.g., W. Greuter (1995) and the molecular studies by Oxelman and coworkers (e.g., B. Oxelman et al. 1997, 2000]. The recent molecular study by J. G. Burleigh and T. P. Holtsford (2003) provides little support for existing morphologically based sectional classifications within *Silene* insofar as they relate to endemic North American taxa. However, it does indicate the distinctness of our arctic alpine species (*S. involucrata*—as *S. furcata*, and *S. acaulis*) that are circumpolar in their distribution.

SELECTED REFERENCES Bocquet, G. 1969. Revisio *Physolychnidum* (*Silene* Sect. *Physolychnis*).... Lehre. Burleigh, J. G. and T. P. Holtsford. 2003. Molecular systematics of the eastern North American *Silene* (Caryophyllaceae): Evidence from nuclear ITS and chloroplast *trn*L intron sequences. Rhodora 105: 76–90. Hitchcock, C. L. and B. Maguire. 1947. A Revision of the

North American Species of *Silene*. Seattle. [Univ. Wash. Publ. Biol. 13.] Kruckeberg, A. R. 1962. Intergeneric hybrids in the Lychnideae (Caryophyllaceae). Brittonia 14: 311–321. McNeill, J. 1978. *Silene alba* and *S. dioica* in North America and the generic delimitation of *Lychnis*, *Melandrium* and *Silene* (Caryophyllaceae). Canad. J. Bot. 56: 297–308. Oxelman, B. and M. Lidén. 1995. Generic boundaries in the tribe Sileneae (Caryophyllaceae) as inferred from nuclear rDNA sequences. Taxon 44: 525–542. Oxelman, B., M. Lidén, and D. Berglund. 1997. Chloroplast *rps*16 intron phylogeny of the tribe Sileneae (Caryophyllaceae). Pl. Syst. Evol. 206: 411–420. Oxelman, B., M. Lidén, R. K. Rabeler, and M. Popp. 2000. A revised generic classification of the tribe Sileneae (Caryophyllaceae). Nordic J. Bot. 20: 743–748. Williams, F. N. 1896b. A revision of the genus *Silene* Linn. J. Linn. Soc., Bot. 32: 1–196.

1. Stigmas 5 (rarely 4 in individual flowers); capsule with 5–10 teeth (rarely 4 or 8).
    2. Limbs of petals shorter than calyx, not differentiated from claw, ascending to spreading but rarely in horizontal plane; petals off-white, dingy pink, dusky purple, or purple tinged.
        3. Seeds winged.
            4. Calyces inflated in flower or fruit, thin and papery; petals dusky purple-red; seeds 1.5–2.5 mm, with broad wing; flowers nodding (erect in subsp. *porsildii* but seeds 2–2.5 mm) . . . . 64. *Silene uralensis*
            4. Calyces neither inflated nor very thin; petals white or pink; seeds 0.5–1.5 mm, wing often narrow; flowers erect.
                5. Seeds (0.5–)0.8–1(–1.3) mm, wing to ca. ¼ diam. of seed; calyces elliptic, 7–10(–12) mm; stem 2–10(–12) cm; flowers 1(–3) . . . . 22. *Silene hitchguirei*
                5. Seeds 1–1.5 mm, wing to ½ diam. of seed; calyces often campanulate or ovate, 8–20 mm; stems 10–45 cm; often with 1–3 flowers per peduncle
                . . . . 25. *Silene involucrata* (in part)
        3. Seeds not winged.
            6. Fruiting calyces 2–3 times longer than broad, broadly tubular to narrowly ellipsoid . . . . 18. *Silene drummondii*
            6. Fruiting calyces to 2 times as long as broad, campanulate, ovate, or elliptic.
                7. Fruiting calyces 10–12 mm, not or only slightly contracted at mouth
                . . . . 38. *Silene ostenfeldii*
                7. Fruiting calyces 13–17 mm, contracted at mouth to ½–⅔ diam.
                    8. Inflorescences 1(–3)-flowered; pedicels ½–3 times calyx, length of pubescence less than ½ pedicel diam.; veins of calyx prominent . . . . . . 26. *Silene kingii*
                    8. Inflorescences (1–)3-flowered; flowers sessile or pedicellate; pedicels usually much shorter than calyx, longest hairs ca. equaling pedicel or peduncle diam.; veins of calyx partially obscured by dense pubescence
                    . . . . 57. *Silene sorensenis*
    2. Limbs of petals clearly differentiated from claw, spreading horizontally, equaling or longer than calyx (shorter in *S. chalcedonica*, *S. flos-cuculi*, and *S. involucrata*); petals showy, scarlet, bright purple or pink, or white.
        9. Inflorescences dense (interrupted in *S. viscaria*) cymes of (3–)5–50 flowers; flowers sessile or pedicels shorter than calyx.
            10. Petals scarlet (rarely white or pink); coarse hispid herbs 50–100 cm; leaf blades broadly ovate or elliptic to lanceolate. 9. *Silene chalcedonica*
            10. Petals bright pink or purple (rarely white); glabrous or sparsely pubescent herbs 5–90 cm; leaf blades narrowly lanceolate.
                11. Flowers 14–22 mm diam.; limb of corolla obovate, ± entire; carpophore 3–5 mm; nodes of stem viscid . . . . 67. *Silene viscaria*
                11. Flowers 5–10 mm diam.; limb of corolla 2-lobed to middle; carpophore ca. 1 mm; nodes of stem not viscid. . . . . 61. *Silene suecica*
        9. Inflorescences open cymes, 1–3 to many-flowered; most pedicels longer than calyx.
            12. Open flowers 8–16 mm diam. . . . . 25. *Silene involucrata* (in part)
            12. Open flowers 20–35 mm diam.
                13. Petal limbs deeply divided with 4 linear segments . . . . 19. *Silene flos-cuculi*
                13. Petal limbs unlobed, emarginate or shallowly 2-lobed.
                    14. Leaf blades with dense, silky, grayish white tomentum; flowers bisexual
                    . . . . 13. *Silene coronaria*

14. Pubescence hirsute or sparingly to densely pubescent; flowers unisexual.
 15. Petals white; capsule teeth slightly reflexed or spreading . . . . 28. *Silene latifolia*
 15. Petals bright pink; capsule teeth revolute . . . . 16. *Silene dioica*
1. Stigmas 3 (rarely 4 or 5 in individual flowers, usually 4 in *S. scaposa*); capsules with 3 or 6 teeth (rarely 4, 5, 8, or 10).
 16. Petals scarlet.
  17. Limbs of petals unlobed or emarginate, shallowly dentate, or crenate, rarely shallowly 2-lobed.
   18. Stems and leaf blades glabrous . . . . 60. *Silene subciliata*
   18. Stems and leaf blades pubescent.
    19. Stems erect, 50–160 cm; inflorescences many-flowered; leaf blades ovate to lanceolate, apex acute, not acuminate . . . . 48. *Silene regia*
    19. Stems ascending, 10–20 cm; inflorescences with flowers usually solitary, terminal; leaf blades linear-lanceolate, apex sharply acuminate . . . . 44. *Silene plankii*
  17. Limbs of petals deeply lobed or laciniate.
   20. Petals deeply 4–6-lobed, small lateral teeth may be present . . . . 27. *Silene laciniata*
   20. Petals deeply 2-lobed, lobes often with lateral teeth.
    21. Leaf blades subrotund to broadly lanceolate; petal lobes ciliate . . . . . 50. *Silene rotundifolia*
    21. Leaf blades lanceolate, oblanceolate, or obovate to narrowly elliptic; petal lobes glabrous or nearly so (claw may be ciliate).
     22. Stems 4–10(–15) cm; inflorescences 1–3(–4)-flowered; leaves longest in mid stem, basal tufts absent; w United States . . . . 55. *Silene serpentinicola*
     22. Stems 20–80 cm; inflorescences (3–)7–11(–20)-flowered; leaves longest at base of stem, in basal tufts; e United States . . . . 66. *Silene virginica*
 16. Petals white, cream, pink, or dark red, off-white tinged with purple.
  23. Calyces with 20–30 parallel veins.
   24. Calyces glabrous, veins obscure . . . . 14. *Silene cserei* (in part)
   24. Calyces pubescent and usually glandular, veins prominent.
    25. Mature calyces 20–30 mm; seeds 1.3–1.8 mm broad . . . . 12. *Silene conoidea*
    25. Calyces 8–15 mm; seeds 0.6–1 mm broad.
     26. Petals ± equaling calyx, limb 1–3 mm, inconspicuous; corollas cream to dull orange abaxially, often purple tinged adaxially; calyx lobes 2–3 mm . . . . 11. *Silene coniflora*
     26. Petals much exceeding calyx, conspicuous, limb 3–6 mm; corollas pink (rarely white or dark red); calyx lobes ca. 5 mm . . . . 10. *Silene conica*
  23. Calyces with 10 or fewer veins or venation obscure.
   27. Inflorescences congested, capitate, or flowers clustered; flowers sessile or pedicels shorter than calyx.
    28. Plants glabrous . . . . 4. *Silene armeria*
    28. Plants pubescent.
     29. Calyx veins purple tinged or entirely purple; corollas rose pink, limb 2-lobed . . . . 49. *Silene repens* (in part)
     29. Calyx veins green; corollas greenish yellow to white, limb deeply laciniate into 6 or more linear lobes . . . . 40. *Silene parishii* (in part)
   27. Inflorescences open; flowers pedicellate, if pedicels short, then flowers in several separate whorls, or solitary.
    30. Leaves whorled at each node, 4 to many.
     31. Leaf blades lanceolate to ovate-lanceolate, 4 per node . . . . 59. *Silene stellata*
     31. Leaf blades linear to very narrowly elliptic-lanceolate, more than 4 per node . . . . 56. *Silene sibirica*
    30. Leaves opposite, 2 at each node (smaller axillary shoots sometimes present), or all or mostly proximal.
     32. Calyces herbaceous, conspicuous pale commissures between veins absent, veins obscure.
      33. Limbs of petals with several narrow, linear lobes . . . . 7. *Silene campanulata*
      33. Limbs of petals unlobed, sometimes emarginate, or 2-lobed.

34. Calyces glabrous.
   35. Petals bright pink, rarely white . . . . 1. *Silene acaulis*
   35. Petals white.
      36. Fruiting calyces ovoid, contracted at mouth to ca. ¹/₂ its diam.; stamens ca. 2 times calyx; filaments purple . . . . 14. *Silene csereii* (in part)
      36. Fruiting calyces clavate or campanulate, not contracted at mouth; stamens ca. 1–1¹/₃ times calyx; filaments usually white.
         37. Fruiting calyces broadly clavate, strongly contracted at base around carpophore; carpophores 5–6 mm; inflorescences leafy . . . . 33. *Silene nivea* (in part)
         37. Fruiting calyces campanulate, not contracted at base; carpophores 2–3 mm; inflorescences with leaves reduced, resembling bracts . . . . 68. *Silene vulgaris*
34. Calyces pubescent.
   38. Seeds winged . . . . 58. *Silene spaldingii*
   38. Seeds not winged.
      39. Mature calyces contracted towards base around carpophore; carpophores 4–7 mm.
         40. Corollas white; calyces green . . . . 33. *Silene nivea* (in part)
         40. Corollas pink (rarely white); calyces purple, purple veined, or purple tinged . . . . 49. *Silene repens* (in part)
      39. Mature calyces not contracted at base; carpophores less than 2 mm.
         41. Leaf blades oblanceolate to elliptic-lanceolate, rarely obovate or elliptic, broadest distally or at middle . . . . 31. *Silene menziesii*
         41. Leaf blades narrowly lanceolate to elliptic- or ovate-lanceolate, broadest proximally.
            42. Petals dark red (rarely white); calyces 6–9 × 3–4 mm . . . . 54. *Silene seelyi*
            42. Petals white; calyces 9–11 × 4–6 mm . . . . 69. *Silene williamsii*
[32. Shifted to left margin.—Ed.]
32. Calyces with pale commissures between prominent parallel veins.
   43. Plants annual or biennial, often with slender taproot.
      44. Mature calyces clavate, urceolate, or fusiform, contracted proximally around carpophore.
         45. Petals white, often pink tinged . . . . 34. *Silene noctiflora*
         45. Petals bright pink.
            46. Petals unlobed . . . . 46. *Silene pseudatocion*
            46. Petals 2-lobed . . . . 42. *Silene pendula*
      44. Mature calyces campanulate, elliptic, or ovoid, not contracted at base around carpophore.
         47. Stems subglabrous to puberulent, often glutinous on distal internodes . . . . . . 2. *Silene antirrhina*
         47. Stems coarsely hispid.
            48. Mature calyces (7–)10–15 mm; seeds grayish brown to black, ca. 1 mm diam.; stems (20–)50–100 cm . . . . 15. *Silene dichotoma*
            48. Mature calyces 7–10 mm; seeds dark reddish brown, ca. 0.5 mm diam.; stems 15–45 cm . . . . 20. *Silene gallica*
   43. Plants perennial, rhizomatous or with stout, often woody taproot and branched (or sometimes in *S. oregana*), woody caudex.
      49. Calyx lobes equaling or exceeding tube . . . . 3. *Silene aperta*
      49. Calyx lobes shorter than tube.

[50. Shifted to left margin.—Ed.]

50. Mature calyces clavate, funnelform, turbinate, urceolate, tubular, or fusiform, narrowed proximally around carpophore.
  51. Corolla limbs unlobed or nearly so . . . . 8. *Silene caroliniana*
  51. Corolla limbs lobed or laciniate.
    52. Calyx lobes 4–8 mm, lanceolate or narrowly triangular.
      53. Corollas coral pink, pink, or white, limb 7–25 mm; calyces umbilicate or turbinate in fruit.
        54. Corolla limbs laciniate into many narrow segments; calyces thinly pilose, funnelform in flower, umbilicate in fruit . . . . 45. *Silene polypetala*
        54. Corolla limbs 4-lobed, usually deeply so, rarely 2-lobed with 2 smaller lateral teeth; calyces canescent, eglandular or glandular, broadly tubular in flower, turbinate in fruit. . . . . 23. *Silene hookeri*
      53. Corollas pale greenish yellow or pale yellow to white, sometimes purple tinged, limb 5–8 mm; calyces clavate in fruit.
        55. Corolla limbs deeply laciniate into 6 or more linear lobes; calyces (20–)25–30 mm, densely puberulent and viscid-glandular, tubular in flower . . . . 40. *Silene parishii* (in part)
        55. Corolla limbs cleft ca. to middle into (2–)4–8 lanceolate to oblong lobes; calyces 16–20 mm, densely and coarsely pubescent and viscid-glandular, tubular to narrowly obconic in flower . . . . 70. *Silene wrightii*
    52. Calyx lobes 2–4 mm, lanceolate to ovate or triangular-acute, or 2–5 mm with broad, obtuse, membranous margins.
      56. Leaf blades ovate-acuminate, sessile; mature calyces turbinate . . . . 39. *Silene ovata*
      56. Leaf blades linear or spatulate to ovate-lanceolate or oblanceolate, at least proximal ones petiolate; mature calyces clavate to tubular or fusiform.
        57. Corolla limbs pink or rose red (rarely white), (7–)10–20 mm, with 4 (rarely 2 with 2 teeth) divergent, lanceolate lobes; mature calyces 15–38 mm . . . . 36. *Silene occidentalis*
        57. Corolla limbs white, greenish, pink, or dingy red, 3–8 mm, with 2–6(–10) linear lobes; mature calyces 8–17(–20) mm.
          58. Corolla limbs 2-lobed (rarely with small lateral teeth).
            59. Inflorescences open cymes with many elongate, ascending branches and pedicels longer than calyx . . . . 65. *Silene verecunda*
            59. Inflorescences pseudoracemose and nodding with flowers usually paired at nodes and pedicels ± equaling calyx, or erect with dense, many-flowered whorls and most flowers sessile or shortly pedicellate . . . . 53. *Silene scouleri* (in part)
          58. Corolla limbs 4–6-lobed.
            60. Corolla appendages 2, laciniate, apex rounded, 2–3 mm; claw and filaments ciliate; corolla lobes 4, linear . . . . 5. *Silene bernardina*
            60. Corolla appendages 4–6, linear, apex acute, 1–1.5 mm; claw and filaments glabrous; corolla lobes 4–6, often forked, producing to 10 linear segments . . . . 37. *Silene oregana*
50. Mature calyces campanulate, elliptic, or ovate, not contracted proximally around carpophore, or if so, broadly turbinate or obconic, at least ½ as broad as long.
  61. Styles exceeding 2 times calyx; stamens exserted from calyx.
    62. Leaves mostly in dense basal tufts. . . . . 29. *Silene lemmonii*
    62. Leaves mostly cauline . . . . 6. *Silene bridgesii*
  61. Styles and stamens ± equaling calyx and/or corolla.
    63. Sepals and pedicels without glandular hairs.
      64. Petals to 2 times calyx . . . . 17. *Silene douglasii* (in part)
      64. Petals slightly longer than calyx . . . . 32. *Silene nachlingerae*
    63. Glandular hairs present at least on pedicels and calyces.

[65. Shifted to left margin.—Ed.]

65. Leaves mostly basal, densely tufted, cauline leaves usually in 1–3 pairs and/or reduced.
    66. Corollas slightly longer than sepals, limbs 1–2 mm, unlobed, or apex notched
        .... 24. *Silene invisa*
    66. Corollas exceeding or sometimes equaling calyces; limbs 2–10 mm, 2(–4)-lobed.
        67. Plants small alpine species; stems 3–20(–30) cm; leaf blades fleshy, 0.5–5(–6) cm.
            68. Pubescence on calyces and pedicels with purple septa................. 62. *Silene suksdorfii*
            68. Pubescence on calyces and pedicels with colorless septa.
                69. Seeds 2–3 mm, margins thickened and winglike; proximal leaf blades
                    oblanceolate-spatulate, 2–7 mm wide .... 21. *Silene grayi*
                69. Seeds ca. 1.5 mm, margins with large inflated papillae; proximal leaf blades
                    linear-oblanceolate, 0.5–3 mm wide........................ 51. *Silene sargentii*
        67. Plants lowland and montane species; stems usually exceeding 30 cm; leaf blades not
            fleshy, 2–15 cm.
            70. Calyx lobes narrowly lanceolate with narrow membranous margins
                proximally .... 35. *Silene nuda*
            70. Calyx lobes ovate to broadly triangular with broad membranous margins.
                71. Claws of corollas exceeding calyx, ciliate proximally; veins to calyx lobes
                    conspicuously broadened distally and lance-shaped, commissural veins slen-
                    der, not forked distally .... 52. *Silene scaposa*
                71. Claws of corollas equaling calyx, glabrous; veins of calyx lobes not con-
                    spicuously broadened distally, commissural veins slender and forked
                    distally, connate with those to lobes....................... 41. *Silene parryi* (in part)
65. Basal leaves not densely tufted; cauline leaves in 3–12 pairs.
    72. Cymes sessile (proximal ones may be pedunculate); pedicels usually shorter than calyx
        and becoming deflexed at base of calyx.......................... 53. *Silene scouleri* (in part)
    72. Cymes pedunculate or flowers solitary; pedicels shorter than to exceeding calyx,
        ascending (may be partially deflexed in *S. thurberi* and *S. petersonii*).
        73. Petals bright pink, limb 5–15 mm; plants rhizomatous from thick rootstock; stems
            5–15 cm .... 43. *Silene petersonii*
        73. Petals white, pale pink, or yellowish, often tinged green or dusky purple, limb 3–8
            mm; plants not rhizomatous; stems erect, usually exceeding 15 cm.
            74. Pubescence on calyces and pedicels short, deflexed, grayish white, eglandular
                (glandular hairs often present in var. *rupinae*) .... 17. *Silene douglasii* (in part)
            74. Pubescence scabrid-puberulent or often viscid glandular-pubescent with
                septate hairs, glandular hairs present at least distally.
                75. Mature calyces ca. as broad as long; flowers nocturnal .... 30. *Silene marmorensis*
                75. Mature calyces ca. 2 times as long as broad, flowers diurnal.
                    76. Corolla limbs 5–7 mm, mature calyces 12–16 mm; stems softly viscid-
                        puberulent distally............................... 41. *Silene parryi* (in part)
                    76. Corolla limbs 2–4 mm, mature calyces 7–12 mm; stems scabrid-
                        puberulent.
                        77. Calyx lobes 1–2 mm, ovate, as long as broad, corolla appendages
                            entire .... 47. *Silene rectiramea*
                        77. Calyx lobes 3–4 mm, narrowly lanceolate, much longer than broad,
                            corolla appendages erose .... 63. *Silene thurberi*

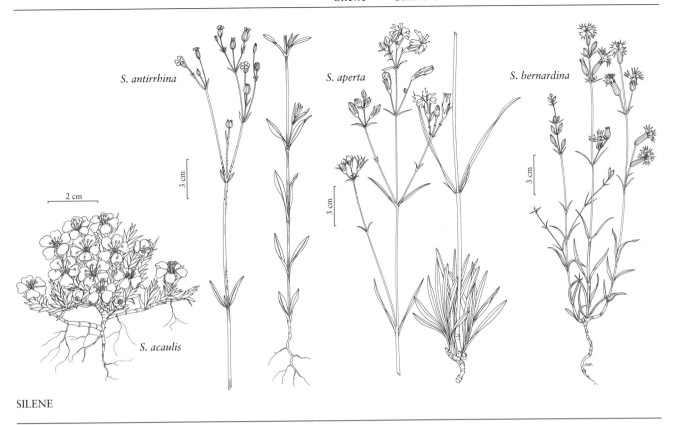

S. antirrhina

S. acaulis

S. aperta

S. bernardina

2 cm

3 cm

3 cm

3 cm

SILENE

1. Silene acaulis (Linnaeus) Jacquin, Enum. Stirp. Vindob., 78, 242. 1762 • Moss campion, silène acaule F

Cucubalus acaulis Linnaeus Sp. Pl. 1: 415. 1753; *Silene acaulis* subsp. *arctica* Á. Löve & D. Löve; *S. acaulis* subsp. *exscapa* (Allioni) de Candolle; *S. acaulis* subsp. *subacaulescens* (F. M. Williams) Hultén; *S. exscapa* Allioni; *Xamilensis acaulis* (Linnaeus) Tzvelev

Plants perennial, mat- or cushion-forming, subglabrous; taproot stout; caudex much-branched, becoming woody. **Flowering stems** erect, leafy proximally, 3–6(–15) cm, old leaves persistent at base. **Leaves** mostly basal, densely crowded and imbricate, sessile; blade 1(–3)-veined, linear-subulate to lanceolate, 0.4–1(–1.5) cm × 0.8–1.5 (–2) mm, margins cartilaginous, often ciliolate especially proximally, apex acute, glabrous to scabrous. **Inflorescences** solitary flowers. **Pedicels** 2–40 mm. **Flowers** bisexual or unisexual, all plants having both staminate and pistillate flowers, others having only pistillate flowers, subsessile or borne singly on peduncle; calyx 10-veined, lateral veins absent, tubular to campanulate, (5–)7–10 mm, herbaceous, margins often purple tinged, dentate, sometimes ciliate, ± scarious, glabrous, lobes lanceolate to ovate, 1–2 mm; petals bright pink, rarely white, limb unlobed to shallowly 2-fid, 2.5–3.5 mm, base tapered into claw, auricles and appendages poorly developed; stamens exserted in staminate flowers, not so or aborted in pistillate flowers; styles 3. **Capsules** 3-locular, cylindric, equaling or to 2 times calyx, opening by 6 recurved teeth; carpophore ca. 1 mm. **Seeds** light brown, reniform, 0.8–1(–1.2) mm broad, dull, shallowly rugose. $2n = 24$.

Flowering early summer. Arctic and alpine tundra, gravelly, often wet places, rocky ledges; 0–4200 m; Greenland; Alta., B.C., Nfld. and Labr., N.W.T., N.S., Nunavut, Ont., Que., Sask., Yukon; Alaska, Ariz., Colo., Idaho, Maine, Mont., Nev., N.H., N.Mex., N.Y., Oreg., Utah, Wash., Wyo.; Europe; Asia (Russian Far East).

Silene acaulis is a variable species, and most workers have recognized infraspecific taxa in North America: subsp. *acaulis* (subsp. *exscapa* and subsp. *arctica*), which is predominantly arctic; and subsp. *subacaulescens*, which extends down the Rocky Mountains from Alaska to Arizona and New Mexico. In subsp. *acaulis*, the leaves are flat and short and the flowers are subsessile and smaller in size. Subspecies *subacaulescens* is typically a larger, less-compact plant with longer, narrower leaves and larger, pedunculate flowers. However, in many populations, these two variants are poorly differentiated, and in others both occur together, connected by intermediates.

Silene acaulis is widely distributed in arctic and alpine Europe.

2. Silene antirrhina Linnaeus, Sp. Pl. 1: 419. 1753

· Sleepy catchfly, silène mufler F W

Silene antirrhina var. *confinis* Fernald; *S. antirrhina* var. *depauperata* Rydberg; *S. antirrhina* var. *divaricata* B. L. Robinson; *S. antirrhina* var. *laevigata* Engelmann & A. Gray; *S. antirrhina* var. *subglaber* Engelmann & A. Gray; *S. antirrhina* var. *vaccarifolia* Rydberg

Plants annual; taproot slender. **Stems** erect, simple or branched, slender, to 80 cm, subglabrous to retrorsely puberulent especially proximally, distal internodes frequently glutinous. **Leaves** 2 per node; blade with margins ciliate toward base, apex acute to obtuse; basal blades oblanceolate, spatulate; cauline narrowly oblanceolate to linear, 1–9 cm × 2–15 mm, scabrous or puberulent, rarely glabrous on both surfaces. **Inflorescences** cymose, open, branches usually ascending, several- to many-flowered, 1-flowered in depauperate specimens. **Flowers:** mature calyx prominently 10-veined, campanulate to ovate, 5–9 × 3–5 mm, margins dentate, glabrous, veins parallel, with pale commissures; lobes usually purple, triangular, acute, ca. 1 mm; petals white, often suffused with dark red, rarely wholly dark red, limb ovate, usually 2-lobed, ca. 2.5 mm, slightly longer than calyx, rarely petals absent, claw narrow, appendages 0.1–0.4 mm; stamens included; styles 3; stigmas included. **Capsules** equaling calyx, opening by 6 teeth; carpophore less than 1 mm. **Seeds** dull gray-black, reniform, 0.5–0.8 mm diam., finely papillate. $2n = 24$.

Flowering spring–late summer. Dry, sandy or gravelly places, roadsides, fields, waste places, open woods, often appearing after burning; 0–2300 m; Alta., B.C., Man., N.B., Ont., Que., Sask.; Ala., Ariz., Ark., Calif., Colo., Conn., Del., D.C., Fla., Ga., Idaho, Ill., Ind., Iowa, Kans., Ky., La., Maine, Md., Mass., Mich., Minn., Miss., Mo., Mont., Nebr., Nev., N.H., N.J., N.Mex., N.Y., N.C., N.Dak., Ohio, Okla., Oreg., Pa., R.I., S.C., S.Dak., Tenn., Tex., Utah, Vt., Va., Wash., W.Va., Wis., Wyo.; Mexico; South America; adventive in Europe.

The six varieties and forms of *Silene antirrhina* noted above were named on the basis of stature and flower color, but none appear to be worthy of recognition. The species is very plastic, being greatly affected by moisture, exposure, and nutrients.

3. Silene aperta Greene, Leafl. Bot. Observ. Crit. 1: 75. 1904 · Naked catchfly E F

Plants perennial, cespitose, puberulent throughout; caudex woody, branched, with clusters of leaves. **Stems** several, erect, not much-branched, slender, 15–60 cm, flowering above middle. **Leaves:** cauline in 2–4 pairs, gradually reduced distally, blade linear with broadened base, 1–8 cm × 1–2 mm, apex acute; basal leaves tending to wither by flowering time, blade with midrib present, linear-oblanceolate, 5–12 cm × 1–4 mm, apex acute. **Inflorescences** few-flowered, bracteate, narrow, flowers terminal and axillary; bracts linear, 2–10 mm. **Pedicels** ascending, straight, slender, very short in bud but equaling or exceeding flower at anthesis. **Flowers:** calyx 10-veined, campanulate, lobed to middle or below, 6–10 mm; lobes 6, recurved, 1–3-veined, lanceolate to linear-lanceolate, margins membranous, ciliate, apex acute; petals white to pale greenish, lobed, clawed, 12–20 mm including claw, ca. 2 times length of calyx; lobes 4, small, 1.5–2 mm, claw woolly towards base, appendages absent; stamens equaling petals; filaments pubescent at base; styles 3, shorter than to equaling stamens. **Capsules** ovoid, exceeding calyx, dehiscing with 6 spreading teeth; carpophore 1–2 mm. **Seeds** brown, broadly reniform, less than 1.75 mm, margins coarsely tuberculate to papillate, with concentric rings of tubercles on both faces. $2n = 48$.

Flowering Jun–Sep. Open, grassy areas in fir and pine forests; 1800–3000 m; Calif.

A deeply lobed calyx and grasslike leaves give *Silene aperta* a very distinct appearance. The species is found only in Tulare County.

4. Silene armeria Linnaeus, Sp. Pl. 1: 420. 1753

· Sweet-William catchfly, silène arméria I W

Atocion armeria (Linnaeus) Rafinesque

Plants annual, glabrous throughout, ± glaucous, some-times glutinous in distal parts; taproot slender. **Stems** simple, branched in inflorescence, (10–)20–40(–70) cm. **Leaves:** basal withering before flowering, blade lanceolate-spatulate, 2–5 cm; cauline sessile to amplexicaulous, blade lanceolate to ovate or elliptic, 1–6 cm × 5–25 mm, apex acute. **Inflorescences** cymose, bracteate; cyme capitate or with flowers clustered at end of slender branches; bracts lanceolate-acicular, 2–10 mm. **Pedicels** 0.1–0.5 cm. **Flowers:** calyx usually purple tinged, 10-

veined, elongate, clavate, lobed, constricted proximally into narrow tube, 13–17 × 2.5–4 mm, rather membranous; lobes ovate-triangular, ca. 1 mm, apex obtuse; petals pink (rarely white), unlobed, limb obovate, ca. 5 mm, base cuneate into claw 6–8 mm, auricles absent, appendages linear to lanceolate, 2–3 mm, apex acute; stamens slightly longer than petal claws; styles 3(–4), exserted. **Capsules** oblong, 7–10 mm, opening by 6 (or 8) spreading teeth; carpophore 7–8 mm, glabrous. **Seeds** dark brown, reniform-rotund, less than 1 mm diam., rugose. $2n = 24$ (Europe).

Flowering summer. Waste places, disturbed ground; 0–1200 m; introduced; B.C., N.B., N.S., Ont., Que.; Calif., Conn., Del., D.C., Fla., Ill., Ind., Ky., Maine, Md., Mass., Mich., Minn., Mo., N.H., N.J., N.Y., N.C., Ohio, Oreg., Pa., S.C., Utah, Vt., Va., Wash., W.Va., Wis.; Europe.

The long-tubular, clavate calyx enclosing the unusually long carpophore helps to distinguish *Silene armeria*. It is an occasional and adventive garden escape.

5. **Silene bernardina** S. Watson, Proc. Amer. Acad. Arts 24: 82. 1889 • Palmer's catchfly [F]

Silene bernardina subsp. *maguirei* Bocquet; *S. bernardina* var. *rigidula* (B. L. Robinson) Tiehm; *S. bernardina* var. *sierrae* (C. L. Hitchcock & Maguire) Bocquet; *S. occidentalis* S. Watson var. *nancta* Jepson; *S. shockleyi* S. Watson

Plants perennial, loosely cespitose; taproot stout; caudex branched, woody, bearing tufts of leaves. **Stems** not much-branched, slender, (15–)30–60 cm, sparsely pubescent proximally, viscid-glandular distally. **Leaves** mostly basal; blade linear-lanceolate to oblanceolate, 2–8 cm × 2–6(–15) mm (including petiole), base tapered into slender petiole, apex acute to obtuse, subglabrous to glandular-pubescent on both surfaces; cauline leaves to 4 pairs below inflorescence, narrower than basal leaves, blade usually linear but rarely elliptic-lanceolate. **Inflorescences** erect, with several short, ascending branches, few-flowered, open, bracteate, shortly pubescent and viscid-glandular; bracts narrowly lanceolate, 3–10 mm, rigid. **Flowers:** calyx prominently 10-veined, broadly tubular, umbilicate, moderately or not clavate, narrowed around carpophore, lobed, 12–15 × 4–6 mm, thin and papery, with short glandular-viscid pubescence, veins parallel, usually red pigmented, with pale commissures; lobes lanceolate, 2–4 mm, apex acute; petals white, pink, or dingy red, 1½–2 times calyx, claw equaling calyx, ciliate at base, limb obtriangular, 4–6 mm, deeply divided into 4 linear

lobes, appendages 2, conspicuous, laciniate, 2–3 mm, apex rounded; stamens slightly exserted; filaments ciliate at base; styles 3(–4), equaling or longer than stamens. **Capsules** 1-locular, narrowly ovoid, exceeding calyx, opening by 6 (or 8) ascending teeth; carpophore 3–6 mm. **Seeds** brown, reniform, 1.5–2 mm broad, shallowly tuberculate on both surfaces, papillate around margins. $2n = 48$.

Flowering summer. Dry, grassy or gravelly slopes, open woodlands; 1300–3600 m; Calif., Idaho, Nev., Oreg., Wash.; Mexico (Baja California).

Silene bernardina is the earliest valid name for this species. Watson had previously (1875) named it *S. montana*, and that name was taken up by C. L. Hitchcock and B. Maguire (1947), who cited *S. bernardina* as a subspecies of *S. montana*. Unfortunately, the epithet *montana* is pre-occupied in *Silene* by *S. montana* Arrondeau (1863), an unrelated European species. The situation was further complicated by Watson in 1877, when he used the name *Lychnis montana* for another unrelated species now transferred to *Silene* and called *S. hitchguirei*.

Silene bernardina varies in leaf width, pubescence, and flower color. The broader-leaved and more sparsely pubescent forms have been referred to subsp. *bernardina*, and the more-common, narrower-leaved, more-densely pubescent, and viscid forms have been referred to subsp. *maguirei*.

Some forms of *Silene bernardina* can be difficult to distinguish from *S. verecunda*, *S. sargentii*, and *S. oregana*. *Silene verecunda* differs in its smaller, clavate calyx and in its petals being only shortly two-lobed. *Silene sargentii* is a small, densely cespitose, high-alpine species with very narrow, linear leaves (1–2 mm wide), shortly two-lobed petals, and seeds with much larger papillae around the margins. In *S. oregana* the petals are larger (two times the calyx) and deeply divided into many very narrow segments; the claw and the filaments are glabrous; the leaves, particularly the basal ones, are broader; and the inflorescences are narrower, with the more numerous flowers arranged on short, ascending branches; also, the calyx lobes are ovate and obtuse instead of lanceolate and acute. The Idaho material tends to be intermediate with *S. oregana* but has open, dichotomously branched inflorescences, and the petals are nearer to those of *S. bernardina*. These plants from Valley County in the Payette National Forest need further study, preferably in the field. They may represent a distinct taxon.

6. Silene bridgesii Rohrbach, Index Seminum (Berlin), App. 2: 5. 1867 • Bridges's catchfly [E]

Silene engelmannii Rohrbach; *S. incompta* A. Gray; *S. longistylis* Engelmann ex S. Watson

Plants perennial; taproot stout; caudex much-branched, woody. **Flowering stems** erect, with 3–6 pairs of leaves below inflorescence, 30–80 cm, short-pubescent, glandular and somewhat viscid distally. **Leaves:** proximal petiolate, blade oblanceolate, 3–6(–8) cm × 5–15 mm (including petiole), base tapered into short petiole, apex acute to obtuse and apiculate, short-pubescent on both surfaces, pubescence rather sparse adaxially; cauline leaves sessile, blade elliptic-lanceolate, 2–6 cm × 5–15 mm. **Inflorescences** branched, several–many-flowered, open, bracteate, flowering portion to 15 cm and ca. ¹⁄₂ as broad, glandular and viscid; cymules usually 1–3-flowered; bracts narrowly lanceolate, shorter than pedicel; peduncle shorter than internodes. **Pedicels** divaricate, sharply bent distally, 5–15 mm. **Flowers** nodding; calyx prominently 10-veined, tubular to campanulate, umbilicate but narrowed at base, lobed, 9–11 × 3–5 mm in flower, in fruit ovate to turbinate, 5–8 mm broad, viscid-pubescent, veins parallel, green, papery between; lobes 5, narrowly lanceolate, obtuse, 2–3 mm, ciliate; corolla ± white, often greenish abaxially and pink tinged, 2 times calyx; petals 2-lobed, margins entire to erose, appendages linear, narrow, 1–2.5 mm; stamens often long-exserted; filaments pubescent at base; styles persistent, 3, long-exserted, filamentous, exceeding 2 times calyx. **Capsules** broadly ovoid, ca. equaling calyx, opening by 6 ascending, triangular teeth; carpophore 2–3 mm. **Seeds** reddish brown, reniform, 1.2–1.8 mm broad, coarsely papillate. *2n* = 48.

Flowering summer. Coniferous forest openings and mixed woodlands, dry slopes; 500–2800 m; Calif., Oreg.

Silene bridgesii is similar to *S. lemmonii* but usually can be distinguished by its larger size, broader and larger leaves, the near-absence of sterile basal shoots, and larger floral parts and fruits. Although *S. longistylis* has often been cited as a synonym of *S. lemmonii*, examination of the holotype (*Henderson s.n.*, GH) indicates that it is referable to *S. bridgesii*.

Silene bridgesii is found in the Sierra Nevada and southern Cascades.

7. Silene campanulata S. Watson, Proc. Amer. Acad. Arts 10: 341. 1875 [E] [F]

Plants perennial; taproot stout; caudex much-branched, woody, producing many erect-to-straggling, little-branched flowering shoots. **Stems** erect, 5–40 cm, softly pubescent to scabrous, eglandular or viscid-glandular, especially distally, very rarely glabrous, with several pairs of leaves equaling or shorter than internodes. **Leaves** sessile, or basal with short pseudopetiole; blade linear to lanceolate or broadly ovate, base round to cuneate, apex acute to shortly acuminate, puberulent on both surfaces, sometimes glandular. **Inflorescences** usually with single dichotomy, rarely double, open, bracteate, branches often elongate, flowers 1 per node; bracts foliaceous. **Pedicels** sharply reflexed at base, especially after anthesis, equaling calyx. **Flowers** nodding; calyx obscurely 10-veined, broadly campanulate, lobed, 6–8 mm, enlarging to 13–16 mm in fruit, herbaceous, usually with short, dense pubescence throughout, often glandular-viscid, veins green, rarely purplish tinged, conspicuous pale commissures absent; lobes ovate-triangular, ¹⁄₂ to equaling tube, herbaceous; petals creamy white, often greenish abaxially, rarely pink tinged to dusky pink (subsp. *campanulata*), clawed, to 2 times calyx, claw villose, limb deeply divided and fan-shaped with many narrow, linear lobes, lobes rapidly curling, margins deeply divided or erose, appendages 2–4, to 2 mm; stamens exserted; filaments hairy at base; styles 3, to 2 times calyx. **Capsules** ovoid, ca. equaling calyx and often splitting it, opening by 6 broadly triangular teeth; carpophore 1–2.5 mm. **Seeds** brown, reniform, 2–2.5 mm broad, coarsely and ± evenly papillate; papillae ca. as long as broad.

Subspecies 2 (2 in the flora): California, Oregon.

1. Petals dusky pink; leaf blades linear to lanceolate, usually less than 10 mm broad; stem pubescence scabrid, hairs 2 or 4 times as long as broad . . . . 7a. *Silene campanulata* subsp. *campanulata*
1. Petals creamy white; leaf blades lanceolate to ovate, usually more than 10 mm broad; stem pubescence not scabrid, hairs at least 5 times as long as broad . . . . 7b. *Silene campanulata* subsp. *glandulosa*

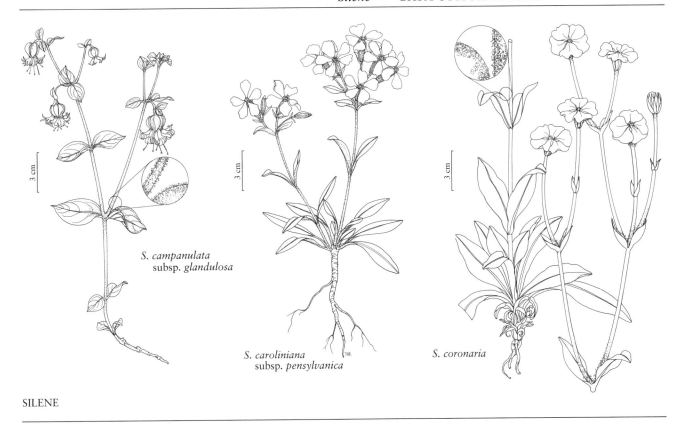

S. *campanulata*
subsp. *glandulosa*

S. *caroliniana*
subsp. *pensylvanica*

S. *coronaria*

SILENE

7a. Silene campanulata S. Watson subsp. **campanulata**
• Red Mountain catchfly [E]

Silene campanulata var. *angustifolia* F. N. Williams

Stems 4–20 cm, scabrous, pubescent, hairs 2–4 times as long as broad. **Leaf blades** narrowly linear to lanceolate, 1.5–5 cm × 2–10(–15) mm. **Petals** dusty pink.

Flowering summer. Serpentine outcrops; 1200–1500 m; Calif.

Subspecies *campanulata* intergrades freely with subsp. *glandulosa*, which is the common and more widely occurring form of this species. Subspecies *campanulata* is found in Colusa, Mendocino, Shasta, Tehama, and Trinity counties.

7b. Silene campanulata S. Watson subsp. **glandulosa**
C. L. Hitchcock & Maguire, Revis. N. Amer. Silene, 23, plate 5, fig. 34b. 1947 • Bell catchfly [E][F]

Silene campanulata subsp. *greenei* (S. Watson) C. L. Hitchcock & Maguire; *S. campanulata* var. *greenei* S. Watson; *S. campanulata* var. *latifolia* F. N. Williams; *S. campanulata* var. *orbiculata* B. L. Robinson; *S. campanulata* var. *petrophila* Jepson

Stems 15–50 cm, glandular-pubescent or eglandular, hairs long, 5 times as long as broad or longer, not scabrous. **Leaf blades** lanceolate to ovate, usually wider than 10 mm. **Petals** creamy white. $2n = 48$.

Flowering summer. Chaparral, open coniferous forest, rocky areas, coastal slopes; 100–2300 m; Calif., Oreg.

Subspecies *glandulosa* is found in the Cascades and Sierra Nevada. Eglandular plants have been treated by some workers as var. (or subsp.) *greenei*, a distinction that is of doubtful value.

8. Silene caroliniana Walter, Fl. Carol., 142 [as 241]. 1788 • Wild pink [E][F]

Plants perennial, cespitose; taproot stout; caudex much-branched, woody. **Stems** ascending, scarcely branched, 8–20(–30) cm, softly pubescent, stipitate-glandular or eglandular, rarely glabrate. **Leaves** mostly basal, petiolate, 3–12 cm (including petiole); cauline leaves in 2–4 pairs, those of mid and distal stem sessile, shorter and narrower; blade narrowly to broadly oblanceolate, base spatulate into winged petiole, apex acute to obtuse, glabrous, puberulent or pilose on both surfaces and frequently stipitate-glandular, at least petioles usually ciliate. **Inflorescences** (1–)3–15-flowered, open, bracteate; bracts foliaceous. **Pedicels** ascending or erect, 0.2–0.8(–1.5)

cm, densely pubescent and frequently stipitate-glandular. **Flowers:** calyx usually green, prominently 8–10-veined, in flower narrowly tubular, lobed, narrowed proximally around carpophore, 15–22 × to 5 mm, becoming broader and clavate in fruit, pilose or stipitate-glandular, veins parallel, with pale commissures, lobes round, 1–3 mm, margins usually purple tinged, broad, membranous; petals spreading, usually bright pink, rarely white, broadly to narrowly obovate, 2 times longer than calyx, base tapered into ciliate claw equaling or slightly longer than calyx, margins entire or shallowly lobed and crenulate, auricles absent, appendages oblong, unlobed, 1.5–2 mm; stamens equaling claw; filaments glabrous; styles 3(–4), ultimately slightly exceeding claw. **Capsules** ellipsoid to obovoid, 8–10 mm, equaling calyx, opening by 6 (or 8) recurved teeth; carpophore 5–8 mm. **Seeds** dark brown, reniform-rotund, 1.3–1.5 mm, coarsely and evenly papillate.

Subspecies 3 (3 in the flora): e United States.

Although the three subspecies of *Silene caroliniana* have overlapping ranges of distribution, subsp. *caroliniana* occurs predominently in the southeastern United States, subsp. *pensylvanica* in the northeast, and subsp. *wherryi* on the western side of the Appalachians. Intermediate plants are occasionally encountered. A hybrid between subsp. *wherryi* and *S. virginica* was reported by J. A. Steyermark (1963), and a hybrid swarm between subsp. *pensylvanica* and *S. virginica* by R. S. Mitchell and L. J. Uttal (1969).

1. Calyces with long, straight, nonglandular
   pubescence ...... 8c. *Silene caroliniana* subsp. *wherryi*
1. Calyces glandular-pubescent.
   2. Basal leaf blades pubescent on both surfaces,
      typically obovate to oblanceolate, apex obtuse;
      petioles broadly winged ....
      .... 8a. *Silene caroliniana* subsp. *caroliniana*
   2. Basal leaf blades ± glabrous on both surfaces
      (margins and veins pubescent abaxially),
      narrowly oblanceolate, apex acute; petioles
      with very narrow wing ....
      .... 8b. *Silene caroliniana* subsp. *pensylvanica*

8a. Silene caroliniana Walter subsp. **caroliniana**

• Carolina wild pink   [E]

Silene rubicunda A. Dietrich

Basal leaves: petiole broadly winged; blade obovate to broadly oblanceolate, usually 1.5–3 cm broad, apex obtuse, puberulent to subpilose and stipitate-glandular on both surfaces, rarely sub-glabrous. **Calyces** 15–18(–20) mm, glandular-pubescent. **Cap-**sules ca. 8(–10) mm. **2*n*** = 48.

Flowering spring. Open and often rocky, mainly deciduous woodlands; 0–1000 m; Ga., Md., N.J., N.C., Pa., S.C., Tenn., W.Va.

8b. Silene caroliniana Walter subsp. **pensylvanica** (Michaux) R. T. Clausen, Rhodora 41: 580. 1939

• Pennsylvania wild pink   [E] [F]

Silene pensylvanica Michaux, Fl. Bor.-Amer. 1: 272. 1803; *S. caroliniana* var. *pensylvanica* (Michaux) Fernald

Basal leaves: petiole very narrowly winged; blade narrowly ob-lanceolate, usually 0.5–1.5 cm broad, apex acute, glabrous or nearly so on both surfaces (margins and veins pubescent abaxially). **Calyces** 15–18 (–20) mm, glandular-pubescent. **Capsules** ca. 8(–10) mm. **2*n*** = 48.

Flowering spring. Open, often gravelly or rocky, mainly deciduous woodlands; 0–1200 m; Conn., Del., D.C., Md., Mass., N.H., N.J., N.Y., N.C., Ohio, Pa., R.I., S.C., Tenn., Va., W.Va.

8c. Silene caroliniana Walter subsp. **wherryi** (Small) R. T. Clausen, Rhodora 41: 582. 1939 • Wherry's pink   [E]

Silene wherryi Small, Torreya 26: 66, unnumbered plate. 1926; *S. caroliniana* var. *wherryi* (Small) Fernald

Basal leaves: petiole broadly winged; blade narrowly ob-lanceolate, usually 0.5–1.5 cm broad, apex acute, ± glabrous except for ciliate margins. **Calyces** 18–22 mm, with long, straight, nonglandular pubescence, hairs spreading to retrorse, white, soft. **Capsules** ca. 10 mm. **2*n*** = 48.

Flowering spring. Open, usually calcareous woods; 0–500 m; Ala., Kans., Ky., Mo., Ohio, Va.

9. **Silene chalcedonica** (Linnaeus) E. H. L. Krause in J. Sturm et al., Deutsch. Fl. ed. 2, 5: 96. 1901 • Maltese-cross, scarlet lychnis, lychnide de Chalcédoine, croix de Jérusalem 🔲

Lychnis chalcedonica Linnaeus, Sp. Pl. 1: 436. 1753; *Agrostemma chalcedonica* (Linnaeus) Doellinger

Plants perennial, coarse, rhizomatous; rhizome branched, stout. **Stems** erect, few-branched, 50–100 cm, hispid. **Leaves** rounded into tightly sessile base; blade lanceolate to ovate, 5–12 cm × 20–60 mm, apex acute, sparsely scabrous-pubescent on both surfaces, scabrous-ciliate on abaxial margins and midrib; basal leaf blades broadly spatulate. **Inflorescences** subcapitate between terminal pair of leaves, 10–50-flowered, congested, bracteate; bracts lanceolate, herbaceous, ciliate. **Flowers** sessile to subsessile, 10–16 mm diam.; calyx 10-veined, narrow and tubular in flower, clavate in fruit, 12–17 mm, margins dentate, lobes triangular-lanceolate, 2.5–2.5 mm, coarsely hirsute; petals scarlet, sometimes white or pink, clawed, claw equaling calyx, limb spreading, obovate, deeply 2-lobed, 6–11 mm, shorter than calyx, appendages tubular, 2–3 mm; stamens equaling calyx; stigmas 5, equaling calyx. **Capsules** ovoid, 8–10 mm, opening by 5 teeth; carpophore 4–6 mm. **Seeds** dark reddish brown, reniform-rotund, 0.7–1 mm diam., coarsely papillate; papillae ca. as high as wide. $2n = 24$ (Europe).

Flowering summer. Roadsides, waste places, open woodlands; 0–300 m; introduced; Alta., B.C., Man., N.B., N.S., Ont., P.E.I., Que., Sask.; Conn., Idaho, Ill., Ind., Iowa, Maine, Mass., Mich., Minn., N.H., N.J., N.Y., Pa., Vt., Wis.; Europe.

Silene chalcedonica is widely cultivated but rarely escapes and probably does not persist.

10. **Silene conica** Linnaeus, Sp. Pl. 1: 418. 1753 • Sand catchfly 🔲 🔲

Subspecies 3 (1 in the flora): introduced; Eurasia.

10a. **Silene conica** Linnaeus subsp. **conica** 🔲 🔲

Plants annual; taproot slender. **Stems** erect or decumbent, simple or branched at base, 15–50 cm, coarsely puberulent, stipitate-glandular distally. **Leaves:** proximal leaf pairs connate, blade 1–3-veined, linear to narrowly lanceolate or oblanceolate, 2–5 cm × 1.5–6 mm, apex acute; basal leaf blades tending to be oblanceolate and somewhat broader, coarsely puberulent on both surfaces.

Inflorescences 1–13-flowered, bracteate; bracts narrowly lanceolate-acuminate, to 20 mm. **Pedicels** ascending, 1–3 cm, stipitate-glandular, puberulent. **Flowers** 10–12 mm diam.; calyx prominently ca. 30-veined, umbilicate, narrowly conic in flower, conic-ovoid but scarcely inflated in fruit, lobed to $^1/_3$ of length with 5 narrow, acuminate lobes ca. 5 mm, 8–15 × 5–7 mm, puberulent, glandular or not, veins parallel; corolla conspicuous, pink, rarely white or dark red, clawed, much exceeding sepals, claw ca. equaling calyx, limb shallowly 2-lobed, oblong, 3–6 mm, appendages 2, 1–2 mm; stamens equaling calyx; stigmas 3, equaling calyx. **Capsules** ovoid, 8–10 mm, with narrow orifice, opening by 6 teeth; carpophore less than 1 mm. **Seeds** dark brown, reniform, 0.6–0.9 mm broad, tuberculate. $2n = 20$ (Europe).

Flowering summer. Dry, sandy or gravelly waste places, roadsides, pasture and arable land; 0–300 m; introduced; Del., Mass., Mich., N.J., N.Y., Ohio, Wash.; Eurasia.

Subspecies *conica*, a rare adventive weed, is very similar to *Silene conoidea*, from which it can be separated, with difficulty, by the smaller size of most of its parts. It is also similar to *S. coniflora*, which has a calyx with fewer veins and shorter teeth.

11. **Silene coniflora** Nees ex de Candolle in A. P. de Candolle and A. L. P. P. de Candolle, Prodr. 1: 371. 1824 • Multinerved catchfly 🔲

Silene multinervia S. Watson

Plants annual; taproot slender. **Stems** erect, simple or branched, 20–65 cm, glandular-pubescent. **Leaves:** proximal forming rosette, blade oblanceolate, spatulate, 3–8(–12) cm × 5–13(–25) mm, apex ± obtuse, sparingly pubescent and glandular; cauline reduced distally, blade lanceolate, 1–7 cm × 2–15 mm, apex ± acute, sparingly pubescent and glandular. **Inflorescences** dichasiate, open, bracteate; bracts leaflike, to 15 mm. **Pedicels** ascending, 1–3(–5) cm, densely glandular-pubescent, viscid. **Flowers:** calyx prominently 20–25-veined, ovate-conic, 8–12 mm, margins dentate, coarsely pubescent, glandular, lobes erect, narrowly lanceolate, 2–3 mm, margins membranous ca. $^1/_4$ length of calyx; corolla inconspicuous, cream, purple tinged adaxially, dull orange abaxially, equaling or slightly longer than calyx, limb ovate, 1–3 mm, apex notched, appendages absent, auricles round, small; stamens shorter than corolla; stigmas 3, shorter than corolla. **Capsules** tightly enclosed in calyx, ovoid, with narrow opening, opening by 6 triangular teeth ca. 1 mm; carpophore ca. 1 mm. **Seeds** dark brown to black, rotund, 0.6–1 mm broad, papillate. $2n = 20$ (Asia).

Flowering spring–early summer. Open places, oak parklands, especially after burning; 0–2000 m; introduced; Calif.; Mexico (Baja California); Asia.

Silene coniflora apparently was introduced into North America in the early days of European exploration and settlement of the Pacific coast. It occurs as a native species from the eastern shores of the Mediterranean to Pakistan and Afghanistan. The report by C. V. Piper (1906) of its occurrence in Washington is based on a specimen of *S. conica*. I have been unable to confirm the statement by M. E. Peck (1961) that *S. multinervia* is "sparingly introduced" along the coast of Oregon.

12. Silene conoidea Linnaeus, Sp. Pl. 1: 418. 1753 • Large sand catchfly I W

Plants annual; taproot slender. **Stems** erect, simple or with ascending branches, (20–)40–100 cm, coarsely puberulent, stipitate-glandular, viscid distally. **Leaves:** mid and proximal stem pairs connate, blade 1–several-veined, oblanceolate to narrowly lance-olate, (3–)5–12 cm × (3–)8–15 mm, apex acute, veins parallel; basal leaf blades oblanceolate and ± obtuse, sparsely to moderately puberulent on both surfaces, rarely subglabrous. **Inflorescences** several–many-flowered, open, bracteate; bracts resembling leaves but smaller. **Pedicels** ascending, straight, equaling or longer than calyx, densely stipitate-glandular, viscid. **Flowers:** calyx prominently 25–30-veined, lobed to ⅓ its length but splitting further in fruit, umbilicate, narrowly conic in flower, conic-ovoid and inflated in fruit, 20–30 × to 15 mm, margins dentate, puberulent and stipitate-glandular, lobes 5, lanceolate, narrow, acuminate, veins parallel; corolla deep pink, clawed, claw equaling or longer than calyx, limb slightly lobed or unlobed, broadly obovate, spatulate, 8–12 mm, appendages 2–4 mm, lobed or dentate; stamens equaling claw; stigmas 3, equaling claw. **Capsules** flask-shaped, 15–20 mm, opening by 6 recurved, lanceolate teeth; carpophore to 2 mm. **Seeds** brown, reniform, 1.2–1.8 mm broad, tuberculate. $2n = 20, 24$ (Europe, Asia).

Flowering early summer. Dry waste places, roadsides, arable land; 0–1000 m; introduced; Alta., B.C., Sask.; Calif., Idaho, Mo., Mont., Oreg., Tex., Wash.; Eurasia.

Similar to *Silene conica* but larger in all its parts, *S. conoidea* is a rare adventive weed with showy flowers and inflated fruiting calyces.

13. Silene coronaria (Linnaeus) Clairville, Man. Herbor. Suisse, 145. 1811 • Rose campion, dusty-miller, mullein pink, lychnide coronaire F I

Agrostemma coronaria Linnaeus, Sp. Pl. 1: 436. 1753; *Lychnis coronaria* (Linnaeus) Murray

Plants perennial, grayish white-tomentose, eglandular; taproot slender to stout; caudex branched, slightly woody. **Stems** several, erect, branched distally, stout, 40–100 cm. **Leaves:** basal blade oblanceolate, spatulate, 5–10 cm × 10–25(–30) mm, margins entire, apex acute, apiculate, with tuft of white hairs; cauline in 5–10 pairs, sessile, reduced distally, blade with both surfaces obscured by dense, silky, grayish-white tomentum. **Inflorescences** with 1–several dichotomies, several-flowered, open, bracteate; branches ascending, elongate; bracts leaflike, 10–20 mm. **Pedicels** straight, stout, to 10 cm. **Flowers** ca. 35 mm; calyx thickly 10-veined, obovate, ca. 15 × 10 mm in fruit, margins dentate with 5 narrowly lanceolate lobes ca. ¼ length of tube, tomentose; corolla rich magenta-pink, sometimes white, clawed, claw equaling calyx, limb spreading horizontally, broadly obovate, shallowly 2-lobed, appendages 2, narrow, 2–4 mm; stamens equaling claw; stigmas 5, equaling claw. **Capsules** equaling to tightly enclosed within calyx, obovate-elliptic, ca. 14 mm, opening by 5 spreading, lanceolate teeth; carpophore ca. 2 mm. **Seeds** grayish brown, reniform-rotund, plump, 1–1.5 mm, coarsely verrucate. $2n = 24$.

Flowering summer. Roadsides, fields, waste or rocky places; 0–300 m; introduced; B.C., N.B., N.S., Ont., Que.; Ala., Ark., Calif., Conn., Idaho, Ill., Ind., La., Md., Mass., Mich., Mo., N.H., N.J., N.Y., N.C., Ohio, Oreg., Pa., S.C., Utah, Vt., Va., Wash., W.Va., Wis.; Europe.

Silene coronaria is commonly cultivated and occasionally escapes.

14. Silene csereii Baumgarten, Enum. Stirp. Transsilv. 3: 345. 1816 (as cserei) • Biennial campion, silène bisannuel I W

Plants annual or biennial, gla-brous and somewhat glaucous; tap-root stout. **Stems** erect, sparingly branched below inflorescence, robust, to 65 cm. **Leaves:** basal few, usually withering by time of anthesis, blade spatulate; cauline numerous, 2 per node, blade 1-veined, ovate-lanceolate to oblan-ceolate, 3–7 cm × 7–30 mm, margins entire, apex acute. **Inflorescences** many-flowered, open, bracteate; primary branches racemose, elongate, with sessile or shortly pedunculate cymes of 1–6 flowers per node; bracts narrowly lanceolate, 3–12 mm, hyaline-margined, apex acute. **Pedicels** ascending, ± straight, 1–2 times calyx, broadening at calyx base. **Flowers:** calyx often obscurely ca. 20-veined, elliptic, abruptly contracted at base, opening constricted to ¹/₂ its diam., slightly inflated in flower, 7–10 × 3–4 mm, in fruit tightly enveloping capsule, ovoid, thin, enlarging to 9–13 × 5–7 mm, herbaceous, margins narrow, membranous, dentate with broadly triangular lobes to 1 mm, glabrous, veins obscure, usually purple tinged, without conspicuous, pale commissures, longitudinal, parallel, not obviously reticulate; petals white, clawed, claw equaling calyx, limb deeply 2-fid into 2 spatulate lobes, to 5 mm, appendages ca. 0.5 mm; stamens exserted, to 2 times length of calyx; filaments usually dark purple; stigmas 3, exserted, to 2 times length of calyx. **Capsules** ovoid, equaling calyx and sometimes splitting it, opening by 6 recurved, narrowly lanceolate teeth; carpophore ca. 1 mm. **Seeds** grayish brown, plump, broadly reniform, 0.6–1 mm, with concentric rings of papillae; papillae slightly longer than broad. *2n* = 24.

Flowering summer. Cultivated fields, roadsides, waste land; 0–1600 m; introduced; Alta., B.C., Man., Ont., Que., Sask.; Colo., Conn., Idaho, Ill., Ind., Iowa, Maine, Md., Mass., Mich., Minn., Mo., Mont., N.H., N.Y., N.C., N.Dak., Ohio, Pa., S.Dak., Vt., Wash., Wis., Wyo.; Europe.

Often confused with *Silene vulgaris*, *S. csereii* may be readily separated by the long, racemose primary branches of its inflorescence, the elliptic calyx that is constricted at both ends, tightly enclosing the capsule and lacking obvious reticulate venation, and the purple filaments.

15. **Silene dichotoma** Ehrhart, Beitr. Naturk. 7: 143. 1792 • Forked catchfly, silène fourchu [I][W]

Subspecies 3 (1 in the flora): introduced; Europe.

15a. **Silene dichotoma** Ehrhart subsp. **dichotoma** [I][W]

Plants annual, robust; taproot slender. **Stems** erect, branched distally, usually reddish, (20–)50–100 cm, coarsely hispid. **Leaves** 2 per node; proximal petiolate, blade lanceolate-spatulate, 3–10 cm × 6–30 mm (including petiole), base tapering into petiole as long as lamina; cauline sessile or nearly so, smaller distally, blade lanceolate-cuneate, 1.5–5 cm × 3–15 mm, apex acute, hispid on both surfaces. **Inflorescences** with many ascending, elongate branches, many-flowered, open, setose, glandular or not but not viscid; flowers 1 per node; bracts ranging from

resembling leaves to narrowly lanceolate, 5–10 mm, apex caudate-acuminate. **Flowers** subsessile; calyx prominently 10-veined, not inflated, setose, lobed, tubular in flower, ovoid and (7–)10–15 × 2.5–4 mm in fruit, veins parallel, with pale commissures; lobes 5, spreading to recurved, narrowly lanceolate, 2–3 mm, apex acuminate; petals white, rarely pink, clawed, claw equaling calyx, limb obovate but deeply 2-lobed, 5–9 mm, apex truncate, appendages 0.2 mm; stamens exceeding petals, reduced to staminodes in some flowers; filaments white; styles 3, white, ca. 2 times calyx. **Capsules** enclosed in calyx, ellipsoid, opening by 6 spreading, lanceolate teeth; carpophore stout, 1.5–4 mm, glabrous. **Seeds** grayish brown to black, broadly reniform with concave faces, ca. 1 mm diam., rugose. *2n* = 24.

Flowering summer. Cultivated land, roadsides, burnt clearings in forests, disturbed prairies; 0–2200 m; introduced; B.C., Nfld. and Labr. (Nfld.), Ont., Que., Sask.; Calif., Colo., Conn., Ga., Idaho, Ill., Ind., Iowa, Ky., Maine, Mass., Mich., Minn., Mo., Mont., Nebr., N.H., N.J., N.Y., N.C., Ohio, Oreg., Pa., Tenn., Tex., Vt., Va., Wash., W.Va., Wis., Wyo.; Europe.

Subspecies *dichotoma* is an occasional adventive weed.

16. **Silene dioica** (Linnaeus) Clairville, Man. Herbor. Suisse, 146. 1811 • Red campion, silène dioïque [I][W]

Lychnis dioica Linnaeus, Sp. Pl. 1: 437. 1753; *L. rubra* (Weigel) Patze, E. H. F. Meyer & Elkan; *Melandrium dioicum* (Linnaeus) Cosson & Germain; *M. dioicum* subsp. *rubrum* (Weigel) D. Löve; *M. rubrum* (Weigel) Garcke

Plants perennial; taproot slender. **Stems** ascending, decumbent at base, branched, shortly rhizomatous, to 80 cm, softly pubescent, ± glandular, at least distally, rarely subglabrous. **Leaves** sessile at mid and distal stem, petiolate to spatulate proximally; petiole equaling or longer than blade of basal leaves; blade ovate to elliptic, 3–13 cm × 10–50 mm (not including petiole), apex acute to acuminate, sparingly pubescent, densely so on abaxial midrib. **Inflorescences** dichasial cymes, several- to many-flowered, open, bracteate; bracts lanceolate, 4–20 × 2–7 mm, herbaceous, softly hairy throughout with long-septate hairs, not glandular, or with some glandular hairs. **Pedicels** ascending, 0.2–3 cm, usually shorter than calyx. **Flowers** unisexual, some plants having only staminate flowers, others having only pistillate flowers, 20–25 mm diam.; calyx 8–12-veined, campanulate, narrowly so in staminate flowers, broadly in pistillate, 10–15 × to 7 mm in flower, 11 mm broad in fruit, herbaceous, margins dentate, softly pubescent, lobes 5, erect, lanceolate, 2–3 mm; petals bright pink, clawed, claw equaling or longer than calyx, limb spreading horizontally, broadly obovate, unlobed or 2-lobed, to 12 × 12 mm, appendages 4, ca. 1 mm; stamens and stigmas equaling petal claw; styles 5. **Capsules** broadly ovoid to globose, equaling and often splitting calyx, opening by 5 (splitting into 10) revolute teeth; carpophore absent. **Seeds** dark brown to black, broadly reniform, plump, 1–1.6 mm, densely and evenly papillate. $2n = 24$.

Flowering summer. Woodlands, hedges, gardens, riverbanks, open waste places; 0–500 m; introduced; B.C., Man., N.B., Nfld. and Labr. (Nfld.), N.S., Ont., Que.; Conn., Ill., Iowa, Maine, Md., Mass., Mont., N.H., N.J., N.Y., Ohio, Oreg., Pa., R.I., S.Dak., Vt., Wash., Wis.; Europe.

Silene dioica is closely related to *S. latifolia* and completely interfertile with it. The two species hybridize wherever they grow in close proximity, and the offspring (*S. ×hampeana* Meusel & K. Werner) usually have pale pink flowers. *Silene dioica* and *S. latifolia* are difficult to separate in herbarium material unless flower color has been noted. The characters that distinguish *S. dioica* are the usually dense, long, and soft pubescence covering at least the distal portion of the plant; the broad, almost globose, thin, and brittle capsule with revolute teeth; and the softer, thinner, usually broader leaves. Occasionally, double-flowered plants are encountered as garden escapes.

17. Silene douglasii Hooker, Fl. Bor.-Amer. 1: 88. 1830
 • Douglas's catchfly [E]

Plants perennial; taproot stout; caudex branched, woody. **Stems** several–many, ascending from decumbent base, usually unbranched, slender, 10–40(–70) cm, with short, fine, dense, retrorse or curled grayish white hairs, rarely subglabrous, typically not glandular but occasionally somewhat glandular distally. **Leaves** 2 per node, finely retrorse; blade with no visible lateral veins, midrib distinct, oblanceolate, elliptic to linear, 2–10 cm × 1.5–13 mm, apex acute, puberulent to glabrous; basal leaves numerous, blade spatulate; cauline in 1–8 pairs, distal ones sessile. **Inflorescences** typically cymose, occasionally with reduced lateral cymes, 1- or 3-flowered, open, bracteate, grayish white retrorse-puberulent, typically not glandular, rarely with few stipitate glands; bracts narrowly lanceolate, 2–10 × 0.5–2 mm, herbaceous, puberulent. **Pedicels** ascending, straight, 0.5–4 cm.

Flowers: calyx green, sometimes suffused with purple, prominently 10-veined, tubular in flower, campanulate and ± inflated in fruit, occasionally somewhat constricted near base, 12–15 × 3–10 mm, papery, grayish white retrorse-puberulent and eglandular, often ciliate, rarely glabrous or with few stipitate glands, veins parallel, green, forked and connate between lobes, lobes 5, erect, ovate-triangular, 2–3 mm, margins membranous, apex blunt; corolla creamy white, often greenish and occasionally tinged with dark pink, clawed, to 2 times calyx, claw slightly longer than calyx, limb obovate-lanceolate, deeply 2-lobed, 4–11 mm, lobes oblong, rounded, margins entire to erose, appendages 1–2(–3) mm; stamens equaling corolla claw; styles 3–5, 1–1½ times corolla claw. **Capsules** ovoid-ellipsoid, 1⅓ times calyx, opening by 3–5 ascending to spreading teeth; carpophore 3–4 mm. **Seeds** rusty brown, broadly reniform, sides flat, 1.2–1.5 mm, margins coarsely papillate, verrucate-tuberculate.

Varieties 3 (3 in the flora): w North America.

Silene douglasii is usually readily recognized by its characteristic short, retrorse, grayish-white, eglandular pubescence on the calyx and pedicels. This, together with the usual absence of lateral teeth on the petals, appears to be the only reliable character separating it from *S. parryi*. However, intermediates between the two species occur, and it is probable that the occasional plants of *S. douglasii* with stipitate-glandular hairs in the inflorescence (e.g., var. *rupinae* and some plants that have been referred to var. *monantha*) have arisen through gene exchange with *S. parryi*, although it should be noted that A. R. Kruckeberg (1961) reported that such hybrids, when artificially produced, were sterile.

1. Leaf blades 3–13 mm wide, 0.3–0.7 mm thick, fleshy; calyces strongly inflated, (4–)7–10(–12) mm diam. at anthesis; petal limbs 5–11 mm wide, often with lateral tooth on each side; coastal Oregon . . . . 17b. *Silene douglasii* var. *oraria*
1. Leaf blades 1.5–9 mm wide, 0.1–0.5 mm thick, not fleshy; calyces not strongly inflated, (3–)5–8 (–10) mm diam. at anthesis; petal limbs 1–5 mm wide, usually without lateral teeth; widespread in w North America.
   2. Leaf blades 1.5–5 mm wide, typically over 15 times longer than wide, ± linear; pubescence on inflorecences and calyces often glandular; cliffs and ledges of Columbia River gorge . . . . 17c. *Silene douglasii* var. *rupinae*
   2. Leaf blades 2–9 mm wide, typically 5–10 times longer than wide, oblanceolate to elliptic to broadly linear; pubescence on inflorescences and calyces eglandular (very rarely with few stipitate glands); widespread in w North America . . . . 17a. *Silene douglasii* var. *douglasii*

17a. Silene douglasii Hooker var. **douglasii** [E]

Silene douglasii var. *brachycalyx* B. L. Robinson; *S. douglasii* var. *macrocalyx* B. L. Robinson; *S. douglasii* var. *monantha* (S. Watson) B. L. Robinson; *S. douglasii* var. *multicaulis* (Nuttall ex Torrey & A. Gray) B. L. Robinson; *S. douglasii* var. *villosa* C. L. Hitchcock & Maguire; *S. lyallii* S. Watson; *S. macrocalyx* (B. L. Robinson) Howell; *S. monantha* S. Watson; *S. multicaulis* Nuttall ex Torrey & A. Gray

Leaf blades oblanceolate or elliptic to broadly linear, 2–10 cm × 2–9 mm × 0.1–0.5 mm, 5–10 times as long as wide, not fleshy. **Inflorescences** eglandular, rarely with few stipitate glands. **Flowers:** calyx usually green, not strongly inflated, (3–)5–8(–10) mm diam. at anthesis, eglandular, rarely with few stipitate glands; petal limb 1–5 mm wide, usually without lateral teeth. *2n* = 48.

Flowering summer. Sagebrush, grassy or wooded mountain slopes, alpine meadows, rocky ledges, talus; 0–3200 m; B.C.; Calif., Idaho, Mont., Nev., Oreg., Utah, Wash., Wyo.

17b. Silene douglasii Hooker var. **oraria** (M. Peck) C. L. Hitchcock & Maguire, Revis. N. Amer. Silene, 40. 1947 [E]

Silene oraria M. Peck, Torreya 32: 148. 1932

Leaf blades oblanceolate, 2–5(–8) cm × 3–13 mm × 0.3–0.7 mm, fleshy. **Inflorescences** eglandular. **Flowers:** calyx strongly suffused with purple, strongly inflated, (4–)7–10(–12) mm diam. at anthesis, eglandular; petal limb 5–11 mm wide, often with lateral tooth on each side.

Flowering summer. Exposed coastal bluffs; 0–100 m; Oreg.

Variety *oraria* is in the Center for Plant Conservation's National Collection of Endangered Plants.

17c. Silene douglasii Hooker var. **rupinae** Kephart & Sturgeon, Madroño 40: 96, fig. 2. 1993 [E]

Leaf blades very narrow, linear, 1.5–5 mm wide, typically over 15 times longer than wide, not fleshy. **Inflorescences** often glandular-pubescent. **Flowers:** calyx green, not strongly inflated, (3–)5–8 (–10) mm diam. at anthesis, often glandular-pubescent; petal limb 1–5 mm wide, with lateral teeth on each side.

Flowering summer. Rocky crevices, cliff ledges; 0–100 m; Oreg., Wash.

Variety *rupinae* is found along the Columbia River gorge.

18. Silene drummondii Hooker, Fl. Bor.-Amer. 1: 89. 1830 • Drummond's catchfly, forked catchfly [E] [F]

Gastrolychnis drummondii (Hooker) Á. Löve & D. Löve; *Lychnis drummondii* (Hooker) S. Watson; *Melandrium drummondii* (Hooker) Porsild; *Wahlbergella drummondii* (Hooker) Rydberg

Plants perennial; taproot stout; caudex branched, somewhat fleshy. **Stems** erect, simple or several from base, retrorsely puberulent proximally, densely so and viscid distally, with stipitate glands. **Leaves:** blade with stiff, appressed pubescence on both surfaces; basal petiolate, blade lanceolate to elliptic or oblanceolate, (1.5–)3–10 cm × 4–12 mm (including petiole); cauline in 2–5 pairs, blade linear to linear-lanceolate, 3–9 cm × 2–7 mm. **Inflorescences** 1–20-flowered, bracteate, strongly viscid-glandular or less densely pubescent, longer hairs sometimes purple-septate; bracts narrowly lanceolate, thick, 3–15 mm, herbaceous, apex acuminate. **Pedicels** stiffly erect, 0.1–5 cm, varying in length within same inflorescence. **Flowers:** calyx 10-veined, broadly tubular to narrowly ellipsoid, not inflated, 12–18 × 4–8 mm in fruit, 2–3 times as long as broad, membranous between veins, margins dentate with 5 triangular, 1.2–2 mm lobes erect in flower and spreading in fruit, apex acuminate, veins green; petals off-white to dusky pink or dingy reddish purple, clawed, equaling or to 1½ times calyx, claw broadened distally, limb not differentiated from claw, narrower than claw, 1–3 mm; stamens included in calyx; styles (4–)5, included in calyx. **Capsules** 12–15 mm, equaling calyx (rarely to 1½ times calyx), opening by (4–)5 spreading teeth. **Seeds** dark brown, not winged, reniform to angular, 0.7–1 mm diam., margins finely papillate; papillae triangular, slender, longer than broad.

Subspecies 2 (2 in the flora): c, w North America.

Subspecies *drummondii* is characteristically a prairie taxon, while subsp. *striata* is associated with the Rocky Mountains. However, the two taxa frequently appear to intergrade; e.g., in the Cypress Hills of southern Alberta and Saskatchewan, and in the southern Rockies. Variety *kruckebergii* appears to be a luxuriant form with a more elongate capsule and calyx. *Silene invisa*, a Californian species, is similar to *S. drummondii*, some plants of which, from Nevada and Arizona, tend to be intermediate (see note under *S. invisa*).

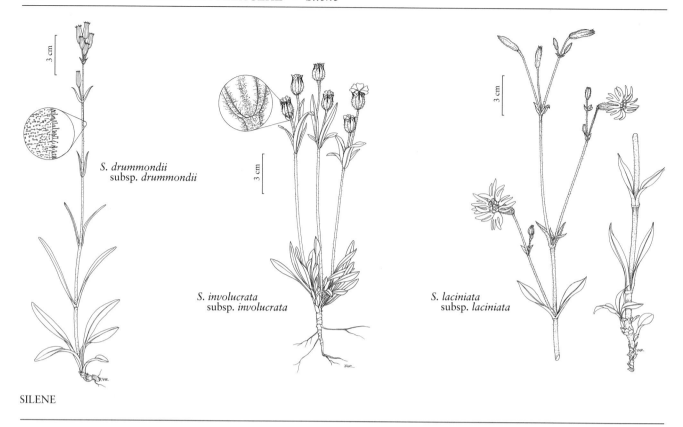

S. drummondii
subsp. *drummondii*

S. involucrata
subsp. *involucrata*

S. laciniata
subsp. *laciniata*

SILENE

1. Petals equaling calyx; fruiting calyces 12–15 × 4–6 mm, 2¼–3 times as long as broad; seeds ca. 0.7 mm diam.; inflorescences typically (1–)3–10 (–20)-flowered . . . . 18a. *Silene drummondii* subsp. *drummondii*

1. Petals 1¼–1½ times calyx and clearly exserted from it; fruiting calyces 13–18 × 6–8 mm, ca. 2 times as long as broad; seeds ca. 1 mm diam.; inflorescences typically 1–4(–8)-flowered . . . . 18b. *Silene drummondii* subsp. *striata*

18a. Silene drummondii Hooker subsp. drummondii

E F

Lychnis pudica B. Boivin; *Silene drummondii* var. *kruckebergii* Bocquet

Inflorescences 1–10(–20)-flowered, strongly viscid-glandular. **Fruiting calyces** 12–15 × 4–6 mm, 2¼–3 times as long as broad. **Petals** off-white to dusky pink, equaling calyx. **Seeds** ca. 0.7 mm diam. **2n = 48.**

Flowering late spring–summer. Dry, often sandy or gravelly places, prairies, dunes, bluffs, hillsides, sagebrush, open montane woodlands and forests; 500–3000 m; Alta., B.C., Man., Sask.; Ariz., Colo., Idaho, Minn., Mont., Nebr., Nev., N.Mex., N.Dak., S.Dak., Utah, Wyo.

18b. Silene drummondii Hooker subsp. striata (Rydberg) J. K. Morton, Sida 21: 887. 2004 E

Lychnis striata Rydberg, Bull. Torrey Bot. Club 31: 408. 1904; *L. drummondii* Hooker var. *heterochroma* B. Boivin; *L. drummondii* var. *striata* (Rydberg) Maguire; *Silene drummondii* var. *striata* (Rydberg) Bocquet; *Wahlbergella striata* (Rydberg) Rydberg

Inflorescences 1–4(–8)-flowered, usually not densely pubescent, longer hairs often purple-septate. **Fruiting calyces** 13–18 × 6–8 mm, ca. 2 times as long as broad. **Petals** clearly exserted from calyx, dingy reddish purple, 1¼–1½ times calyx. **Seeds** ca. 1 mm diam. **2n = 72.**

Flowering summer. Alpine meadows and rocky places, high prairies, open montane woodlands and forests; 1500–3800 m; Sask.; Ariz., Colo., Mont., Nev., N.Mex., Utah, Wyo.

19. Silene flos-cuculi (Linnaeus) Clairville, Man. Herbor. Suisse, 146. 1811 • Ragged robin, lychnide fleur-de-coucou I

Lychnis flos-cuculi Linnaeus, Sp. Pl. 1: 436. 1753; *Coronaria flos-cuculi* (Linnaeus) A. Braun

Subspecies 2 (1 in the flora): introduced; Europe.

19a. Silene flos-cuculi (Linnaeus) Clairville subsp. **flos-cuculi** [I]

Plants perennial; taproot slender; with branched caudex. **Stems** erect, nonflowering shoots decumbent, branched, 30–90 cm, with coarse deflexed hairs distally, subglabrous proximally. **Leaves:** blade glabrous except sometimes ciliate at base; those of proximal stem with blade oblanceolate, spatulate, 2–15 cm × 4–30 mm; those of mid and distal stem subsessile to sessile, connate at base, blade oblanceolate to narrowly lanceolate, 3–8 cm × 4–15 mm. **Inflorescences** cymose, open, bracteate; cyme branched, long-stalked, 3–30-flowered, compound; bracts narrowly lanceolate, 2–25 mm, apex acuminate. **Pedicels** 0.3–2 cm. **Flowers** 30–35 mm diam.; calyx prominently 10-veined, campanulate, 6–10 × 4–7 mm in fruit, submembranous, margins dentate with 5 ovate lobes ca. 3 mm, apex acuminate; petals rose pink (rarely white), clawed, claw equaling calyx, limb deeply divided into 4 spreading, linear lobes, ca. 1 cm, appendages 2, deeply forked, narrow, ca. 4 mm; stamens shortly exserted; stigmas 5, shortly exserted. **Capsule** broadly ovoid, equaling calyx, opening by 5 acute, revolute teeth; carpophore very short (less than 0.5 mm) or absent. **Seeds** dark brown to black, 0.5–0.7 mm, coarsely tuberculate; tubercules triangular, ca. as long as broad. $2n = 24$.

Flowering summer. Wet meadows, roadside ditches, river banks, lake shores; 0–800 m; introduced; N.B., Nfld. and Labr. (Nfld.), N.S., Ont., Que.; Conn., Maine, Md., Mass., Mont., N.H., N.J., N.Y., Ohio, Pa., R.I., Vt., Wash.; Europe.

20. Silene gallica Linnaeus, Sp. Pl. 1: 417. 1753, name conserved • Small-flowered catchfly, silène de France [I] [W]

Silene anglica Linnaeus; *S. quinquevulnera* Linnaeus

Plants annual; taproot slender. **Stems** erect, branched, rarely simple, 15–45 cm, with long, often crinkled hairs mixed with short pubescence, viscid-glandular distally. **Leaves** 2 per node, blade with coarse, ascending, scabrous pubescence on both surfaces; basal few, withering, blade oblanceolate to spatulate-petiolate, 0.5–5 cm × 3–15 mm; cauline blades oblanceolate to lanceolate, 1–7 cm × 1–15 mm, apex obtuse or shortly acuminate to acute. **Inflorescences** open, with racemose branches, internodes and bracts usually ca. equaling fruiting calyx, 1–5 mm,

longer proximally. **Flowers** 5–8 mm diam.; calyx prominently 10-veined, narrowly tubular-ovoid in flower, ovoid in fruit, constricted at mouth, 7–10 × 3–5 mm, membranous between veins, margins dentate, hispid, hairs ca. 2 mm, veins parallel, lobes lanceolate, 2–2.5 mm, apex greenish purple, acute; petals white or pink, often with dark spot or dark pink throughout, clawed, claw equaling calyx, limb elliptic to obovate, lobed or unlobed, to 6 mm, appendages 2, oblong to narrowly lanceolate, 1–1.5 mm; stamens equaling or shorter than calyx; stigmas 3, included in calyx. **Capsules** equaling calyx, opening with 6 recurved, narrowly triangular teeth; carpophore shorter than 1 mm, pubescent. **Seeds** dark reddish brown, reniform, angular with concave, radially ridged faces, broad outer edge transversely ridged and verrucose, ca. 0.5 mm broad. $2n = 24$.

Flowering spring–early summer. Dry, open places, sandy and gravelly ground, roadsides, waste land; 0–2000 m; introduced; B.C., N.B., N.S., Ont., P.E.I.; Ala., Alaska, Ariz., Calif., Fla., Idaho, La., Maine, Mass., Miss., Mo., N.H., N.Y., N.C., Oreg., Pa., R.I., S.C., Tex., Wash.; Europe; introduced worldwide.

21. Silene grayi S. Watson, Proc. Amer. Acad. Arts 14: 291. 1879 (as grayii) • Gray's catchfly [E]

Silene deflexa Eastwood

Plants perennial, with numerous, dense basal tufts of leaves; taproot stout; caudex much-branched, woody. **Stems** erect from sometimes decumbent base, little-branched, subscapose with 2–3 pairs of reduced leaves, 10–20 (–30) cm, finely retrorse-puberulent proximally, stipitate-glandular and viscid in inflorescence. **Leaves:** basal petiolate, blade oblanceolate to spatulate, (1.5–)2–5(–6) cm × 2–7 mm, thick and ± fleshy, apex broadly acute, puberulent on both surfaces; cauline blades linear-oblanceolate to linear-lanceolate, 0.5–2 cm × 1–3 mm. **Inflorescences** open, 1–3 (–5)-flowered, bracteate; bracts lanceolate, 2–7 mm, herbaceous. **Pedicels** erect and straight or slightly deflexed near apex, 5–20 mm, stipitate-glandular, hairs with colorless septa. **Flowers:** calyx prominently 10-veined, in flower broadly cylindric, 8–10 × 3–4 mm, in fruit becoming campanulate and somewhat contracted at base, 8–12 × 5–7 mm, membranous between veins, margins dentate, hairs with colorless septa, veins parallel, purplish, with pale commissures; lobes ovate, ca. 2 mm, shorter than tube, apex flushed with dark red, shortly apiculate with broad, scarious margins, glandular, puberulent; petals exserted, pink to dusky purple, clawed, claw equaling or slightly longer than calyx, limb 2-lobed, 3–5 mm, each lobe with lateral tooth, tooth usually small, rarely larger and equaling lobes, appendages 2, 0.7–1.5 mm; stamens equaling petals; stigmas 3(–4), equaling petals. **Capsules** slightly exceeding calyx, ovoid, opening with 6 (or 8) ascending to slightly recurved teeth; carpophore 2–3 mm. **Seeds** pale brown, reniform, 2–3 mm, sides with close radiating ridges, margins broadened and winglike. $2n = 48$.

Flowering summer. Loose talus, among boulders in mountains, chaparral, open coniferous forests; 1000–3100 m; Calif., Nev., Oreg.

Silene grayi is a small montane relative of *S. parryi*, but it differs in having small, fleshy leaves (ca. 2–4 cm), most of which are in basal tufts. The seeds are also larger and have a thickened wing. Some plants in the mountains of Washington and Oregon appear to intergrade and need further study.

A hybrid between *Silene grayi* and *S. campanulata* has been collected in the Siskiyou Mountains of northern California, an area where both species occur.

22. Silene hitchguirei Bocquet, Candollea 22: 29. 1967

* Mountain campion   E

Lychnis montana S. Watson, Proc. Amer. Acad. Arts 12: 247. 1877, not *Silene montana* Arrondeau 1863; *L. apetala* Linnaeus subsp. *montana* (S. Watson) Maguire; *L. apetala* var. *montana* (S. Watson) C. L .Hitchcock; *Silene uralensis* (Ruprecht) Bocquet subsp. *montana* (S. Watson) McNeill; *S. wahlbergella* Chowdhuri subsp. *montana* (S. Watson) Hultén; *Wahlbergella montana* (S. Watson) Rydberg

Plants perennial, with dense tuft of basal leaves; taproot stout. **Flowering stems** several, stiffly erect, 2–10(–12) cm, densely pubescent distally, hairs multicellular, with glandular tip; stem leaves in 1 or 2 pairs. **Leaves:** blade narrowly oblanceolate and long-spatulate, to 2.5 cm × 4 mm, somewhat fleshy, apex acute, ± glabrous except for ciliate margins. **Inflorescences** 1(–3) per flowering stem. **Flowers** erect; mature calyx veined, elliptic, not inflated or thin, 7–10(–12) mm, densely pubescent with purple-septate hairs, margins dentate with 1–1.5 mm lobes, apex acute, outwardly curved, veins not much-broadened distally, intermediate ones shorter than calyx, veins and calyx lobes dark purple; petals white or pink, to 1¼ times calyx, claw narrow, 10–12 mm, limb not differentiated from claw, obovate, emarginate, ca. 3 mm; stamens equaling calyx; styles 5, equaling calyx. **Capsules** equaling calyx, opening by 5 teeth, tardily splitting into 10, triangular, outwardly curved. **Seeds** brown, reniform to angular, (0.5–)0.7–1(–1.3) mm diam., wrinkled, wing narrow, less than ¼ diam. of seed. $2n = 24$.

Flowering summer. Alpine tundra; 3000–4300 m; Alta.; Colo., Mont., Utah, Wyo.

Silene hitchguirei is similar to *S. suksdorfii*, except that the latter species has larger seeds, an urceolate fruiting calyx with a contracted base, an inflorescence that is sometimes branched with up to three flowers, and short, erect stems that have three or four pairs of leaves. It is probably closely related to *S. involucrata* subsp. *tenella* and *S. ostenfeldii* but differs in its small size, its usually solitary flowers, and short petals. The wing on its small seeds is narrower than that of *S. involucrata* subsp. *tenella*, whereas in *S. ostenfeldii* the wing is completely absent.

23. Silene hookeri Nuttall in J. Torrey and A. Gray, Fl. N. Amer. 1: 193. 1838   E

Herbs, perennial; caudex much-branched, thick and woody. **Stems** solitary or numerous, decumbent and rooting at base, becoming erect, 5–14(–25) cm, with gray, soft, curly to retrorsely crispate pubescence, rarely glandular. **Leaves:** blade spatulate or narrowly lanceolate to oblanceolate, some-times broadly so, 4–7(–10) cm × 8–12(–20) mm, reduced toward base, apex acute, pubescent on both surfaces, especially on midrib; subterranean bractlike, papery. **Inflorescences** reduced to single, terminal flower or open, (1–)3–5(–9)-flowered cyme, bracteate; bracts leaflike, reduced distally to ca. 1 cm. **Pedicels** ascending, straight, 1–6 cm, with a short canescence. **Flowers:** calyx 10-veined, broadly tubular in flower, 12–25 × 5–8 mm, turbinate in fruit and swelling in middle to ca. 10 mm broad, canescent, rarely sparsely pubescent or glandular; lobes lanceolate, 4–7 mm, with narrow, membranous margins, apex acute; corolla coral pink or white, clawed, claw equaling calyx; limb 4-lobed, usually deeply so, rarely 2-lobed with smaller lateral teeth, lobes 7–22 mm, appendages 2, linear, 1.5–3.5 mm (absent in subsp. *bolanderi*); stamens slightly longer than corolla claw; stigmas 3, slightly longer than corolla claw. **Capsules** ovoid to oblong, equaling calyx, dehiscing by 6 teeth; carpophore 2–5 mm. **Seeds** dark brown to black, reniform, ca. 2 mm broad, with concentric rings of small papillae.

Subspecies 2 (2 in the flora): w North America.

1. Corolla lobes either 4 of unequal size or 2 with lateral teeth, lobes lanceolate to broadly oblong, 5–10 mm, corolla coral pink or white, appendages 2, 1.5–3.5 mm, linear . . . . 23a. *Silene hookeri* subsp. *hookeri*
1. Corolla lobes 4, equal, linear, 15–25 mm, corolla white with greenish center, appendages absent . . . . 23b. *Silene hookeri* subsp. *bolanderi*

23a. Silene hookeri Nuttall subsp. hookeri
* Hooker's Indian pink E

Silene hookeri subsp. *pulverulenta* (M. Peck) C. L. Hitchcock & Maguire; *S. ingramii* Tidestrom & Dayton; *S. pulverulenta* M. Peck

Plants completely eglandular or with glandular hairs on calyx and pedicels. **Petals** coral pink or white, limb with 4 unequal lobes or 2 lobes with lateral teeth, lobes lanceolate to broadly oblong, 5–10 mm, appendages 2, linear, 1.5–3.5 mm. $2n = 72$.

Flowering spring–early summer. Dry, sandy, gravelly, or rocky slopes, grassy areas, open woodlands, coniferous forests, serpentine areas; 100–1400 m; Calif., Oreg.

Although subsp. *hookeri* normally is eglandular, plants with stipitate-glandular hairs intermixed with the eglandular pubescence occur in several localities. They have been named subsp. *pulverulenta*.

23b. Silene hookeri Nuttall subsp. bolanderi
(A. Gray) Abrams in L. Abrams and R. S. Ferris, Ill. Fl. Pacific States 2: 162. 1944 • Bolander's Indian pink E

Silene bolanderi A. Gray, Proc. Amer. Acad. Arts 7: 330. 1868

Plants eglandular. **Petals** white with greenish center or coral pink, limb with 4 equal linear lobes 15–25 mm, appendages absent.

Flowering spring. Oak and douglas fir woodlands; 100–1000 m; Calif., Oreg.

24. Silene invisa C. L. Hitchcock & Maguire, Revis. N. Amer. Silene, 31, plate 4, fig. 25. 1947 • Short-petalled campion E

Plants perennial; taproot stout; caudex becoming branched, bearing tufts of leaves. **Stems** several, erect, unbranched proximal to inflorescence, 10–40 cm, puberulent. **Leaves** mostly basal, petiolate, blade oblanceolate or spatulate, 1.5–5 cm × 2–6 mm, apex acute, glabrous except for a few cilia on petiole; cauline leaves in 2–4 pairs, reduced distally, blade linear to narrowly lanceolate or oblanceolate, 2–7 cm × 2–6 mm. **Inflorescences** cymose, 1–3-flowered, open, bracteate; cyme 1, terminal, often with 1 flower at next node; bracts linear-lanceolate, 5–20 mm. **Pedicels** erect, from 0.5 cm, lengthening to 3 cm in fruit, gray and somewhat retrorse-puberulent with stipitate-glandular hairs. **Flowers:** calyx prominently 10-veined, veins parallel, green, with pale commissures, broadened in lobes, narrowly campanulate and 7–11 × 3–4 mm in flower, campanulate and 8–12 × 4–5 mm in fruit, tending to broaden proximally, glandular-puberulent, lobes 5, erect, lanceolate, 1–2 mm, apex blunt; petals cream to pink, often tinged with dusky purple, slightly longer than calyx, limb 1–2 mm, unlobed or apex notched; stamens equaling calyx; stigmas 3, equaling calyx. **Capsules** narrowly ovoid, 10–13 mm, slightly longer than calyx, opening with 6 outwardly curved teeth; carpophore shorter than 1 mm. **Seeds** brown, reniform, angular, 0.7–1 mm, margins with large, balloonlike papillae, sides rugose. $2n = 48$.

Flowering summer. Moist openings in coniferous forests on mountain slopes; 900–2900 m; Calif.

Silene invisa is found in the Cascades and Sierra Nevada. It is a rare species very similar to *S. drummondii*, from which it can usually be distinguished by its smaller size, glabrous leaves, and the large, inflated papillae of the seeds. Plants with intermediate characters occur in Nevada and Arizona.

25. Silene involucrata (Chamisso & Schlechtendal) Bocquet, Candollea 22: 22. 1967 • Arctic campion F

Lychnis apetala Linnaeus var. *involucrata* Chamisso & Schlechtendal, Linnaea 1: 43. 1826; *L. gillettii* B. Boivin

Plants perennial, sometimes with dense tufts of basal rosettes of leaves, subglabrous to pubescent and glandular; taproot slender or stout. **Flowering stems** several, erect, simple or branched, 10–45 cm, usually with 2–5

pairs of leaves. **Leaves** mainly basal, petiolate; blade narrowly oblanceolate, ± spatulate, 20–60 × 3–5(–10) mm (including petiole), glabrous to pubescent, especially on margins and abaxial veins, pubescence spreading, short, stiff, mainly eglandular. **Inflorescences** 1–3-flowered, open, bracteate, pubescent, usually densely so, hairs long, flexuous, purple-septate, mostly glandular; bracts narrowly lanceolate, 4–10 mm, usually pubescent. **Pedicels** usually several times longer than calyx. **Flowers** pedicellate, rarely sessile, erect, 8–16 mm diam.; calyx prominently 10-veined, not inflated or thin, campanulate or ovate, 8–20 mm, pubescent, especially on veins, rarely almost glabrous, hairs long and short purple-septate, ± glandular, veins heavily suffused with purple (rarely green), sinuses between veins pale, cream colored; petals white, often pink or purple tinged, claw equaling calyx, limb not differentiated from claw, emarginate to obovate, often 2-lobed, ca. ¹/₂ as long as calyx; stamens included in calyx, spreading horizontally; styles 5, included in calyx. **Capsules** equaling calyx, opening by 5 teeth, tardily splitting into 10. **Seeds** brown, winged, ± reniform to angular, 1–1.5 mm diam.; wing to ¹/₂ seed diam.

Subspecies 2 (2 in the flora): Greenland, nw Canada, Alaska, arctic Europe, e Asia (Russian Far East, Siberia).

Silene involucrata is a very variable circumpolar and arctic-alpine species complex. Many of the variants have been treated as species by earlier workers. Here, a single species with two subspecies is recognized.

1. Calyx campanulate to ovoid, 10–20 mm in fruit; flowering stems sturdy, usually less than 20 cm, internodes equaling or shorter than leaves . . . .
. . . . 25a. *Silene involucrata* subsp. *involucrata*
1. Calyx campanulate, 8–10(–12) mm in fruit; flowering stems slender, usually over 30 cm, internodes longer than leaves . . . .
. . . . 25b. *Silene involucrata* subsp. *tenella*

25a. Silene involucrata (Chamisso & Schlechtendal) Bocquet subsp. **involucrata** • Greater arctic campion, silène involucré

Gastrolychnis affinis (J. Vahl ex Fries) Tolmatchew & Kozhanchikov; *G. involucrata* (Chamisso & Schlechtendal) Á. Löve & D. Löve; *G. involucrata* subsp. *elatior* (Regel) Á. Löve & D. Löve; *Lychnis affinis* J. Vahl ex Fries; *L. apetala* Linnaeus var. *elatior* Regel; *L. brachycalyx* Raup; *L. furcata* (Rafinesque) Fernald; *L. furcata* subsp. *elatior* (Regel) Maguire; *L. triflora* R. Brown ex Sommerfelt var. *elatior* (Regel) B. Boivin; *Melandrium affine* (J. Vahl ex Fries) J. Vahl; *M. affine* var. *brachycalyx* (Raup) Hultén; *M. furcatum* (Rafinesque) Hadac; *Silene furcata* Rafinesque; *S. involucrata* subsp. *elatior* (Regel) Bocquet

Plants: taproot thick, with dense tufts of basal rosettes of leaves. **Flowering stems** simple or sparingly branched, sturdy, 10–20(–30) cm, usually with 2–3 pairs of lanceolate leaves, internodes equaling or shorter than leaves. **Pedicels** variable in length, ranging from much shorter than calyx to greatly exceeding it. **Flowers** 8–16 mm diam.; calyx campanulate to ovoid, 10–20 mm in fruit, 1–1¹/₂ times longer than broad. $2n$ = 24 (Siberia), 48.

Flowering summer. Arctic tundra, gravelly and grassy places; 0–2000 m; Greenland; B.C., Man., Nfld. and Labr. (Labr.), N.W.T., Nunavut, Ont., Que., Yukon; Alaska; Asia (Russian Far East, Siberia).

Subspecies *involucrata*, a circumpolar and panarctic taxon, is quite variable. In the western arctic there is a tendency towards larger, almost glabrous, leafier plants that have a markedly ovoid mature calyx to 20 mm. These plants often lack purple pigment and have been named subsp. *elatior*, but they probably do not warrant taxonomic recognition.

25b. Silene involucrata (Chamisso & Schlechtendal) Bocquet subsp. **tenella** (Tolmatchew) Bocquet, Candollea 22: 24. 1967 • Taylor's arctic campion

Melandrium affine (J. Vahl ex Fries) J. Vahl subsp. *tenellum* Tolmatchew, Trudy Bot. Muz. 24: 258, fig. 4. 1932 (as Melandryum); *Gastrolychnis angustiflora* Ruprecht subsp. *tenella* (Tolmatchew) Tolmatchew & Kozhanchikov; *G. involucrata* (Chamisso & Schlechtendal) Á. Löve & D. Löve subsp. *tenella* (Tolmatchew) Á. Löve & D. Löve; *Lychnis tayloriae* B. L. Robinson; *Melandrium angustiflorum* (Ruprecht) Walpers subsp. *tenellum* (Tolmatchew) Kozhanchikov; *M. taimyrensis* Tolmatchew; *M. tayloriae* (B. L. Robinson) Tolmatchew; *M. tayloriae* var. *glabrum* Hultén; *M. tenellum* (Chamisso & Schlechtendal) Tolmatchew; *S. tayloriae* (B. L. Robinson) Hultén; *Wahlbergella tayloriae* (B. L. Robinson) Rydberg

Plants short-lived; taproot slender, usually with fewer basal leaves. **Flowering stems** branched distally, slender, (20–)30–45 cm, usually with 3–5 pairs of leaves, internodes longer than leaves. **Pedicels** usually several times longer than calyx. **Flowers** 8–12 mm diam.; calyx campanulate, 8–10(–12) mm in fruit. $2n$ = 48.

Flowering summer. Gravelly and grassy places in the low arctic; 0–1000 m; Greenland; Man., N.W.T., Nunavut, Ont., Yukon; Alaska; Asia (Siberia).

Silene ostenfeldii often appears to be inseparable from subsp. *tenella* except for its wingless seeds. It may be that the two subspecies of *S. involucrata*, together with *S. ostenfeldii* and *S. hitchguirei*, are components of a single variable species with diploid, tetraploid, and hexaploid populations.

26. Silene kingii (S. Watson) Bocquet, Candollea 22: 29. 1967 • King's catchfly [C] [E]

Lychnis kingii S. Watson, Proc. Amer. Acad. Arts 12: 247. 1877; *Gastrolychnis kingii* (S. Watson) W. A. Weber; *Lychnis apetala* var. *kingii* (S. Watson) S. L. Welsh; *Melandrium kingii* (S. Watson) Tolmatchew; *Wahlbergella kingii* (S. Watson) Rydberg

Plants perennial, cespitose, compact; taproot thick; caudex much-branched, fleshy. **Stems** erect, simple proximal to inflorescence, 7–20 cm, with 2–4 pairs of leaves, short retrorse-pubescent proximally, stipitate-glandular and viscid distally, with purple-septate hairs. **Leaves:** basal pseudopetiolate, tufted, blade narrowly oblanceolate, 1.5–5 cm × 1.5–5 mm, base narrowed into long pseudopetiole, retrorse-puberulent on both surfaces; cauline sessile, connate at base, blade linear to linear-lanceolate or linear-oblanceolate, 1–4 cm × 1.5–4 mm, apex acute. **Inflorescences** 1(–3)-flowered. **Pedicels** erect, rarely somewhat curved or reflexed near tip, 0.5–2 cm, ½–3 times calyx, viscid with stipitate-glandular and purple-septate hairs, hairs less than ½ pedicel diam. **Flowers** 10–12 mm diam.; calyx prominently 10-veined, ellipsoid, somewhat inflated and contracted at mouth to ½–⅔ diam., 12–14 × 6–7 mm in flower, 14–17 × 7–8 mm in fruit, to 2 times as long as broad, papery, margins dentate, lobes 5, patent, ca. 2 mm, with broad, membranous margins, apex obtuse, pubescent, hairs short and stipitate-glandular and long purple-septate; corolla pink to purple, slightly longer than calyx, limb 2-lobed, 4–5 mm, with 2 short (ca. 0.5 mm) appendages; stamens equaling corolla; stigmas (4–)5, equaling corolla. **Capsules** equaling calyx, opening by 5 recurved teeth, which later split into 10; carpophore very short or absent. **Seeds** dark brown, reniform, not winged, 0.7–1 mm, papillate-tuberculate.

Flowering summer. Alpine slopes, tundra, stony ridges; of conservation concern; 3000–4000 m; Colo., Utah, Wyo.

Silene kingii is very similar to and probably a close relative of *S. uralensis* subsp. *uralensis*, from which it is distinguished by its nonwinged seeds and elliptic fruiting calyx. However, some material from the southern Rocky Mountains, growing with subsp. *uralensis*, is intermediate between the two species in having narrowly winged seeds. The status of *S. kingii* requires further study.

27. Silene laciniata Cavanilles, Icon. 6: 44, plate 564. 1801 • Mexican pink or campion [F]

Melandrium laciniatum (Cavanilles) Rohrbach

Plants perennial; taproot thick, fleshy; caudex branched. **Stems** 1–several, straggling to erect, sometimes decumbent at base, simple or much-branched distally, 20–120 cm, puberulent or scabrous-pubescent, sometimes retrorse, often glandular distally, becoming glabrate proximally. **Leaves** sessile or narrowed at base into pseudopetiole, blade pubescent on both surfaces; proximal with blade lanceolate and oblanceolate, broadest distally or distal often reduced, cauline longest near mid stem, blade linear to lanceolate or elliptic, 1.5–10 cm × 2–30 mm. **Inflorescences** from 1-flowered to simple dichotomy to compound and 3–5–many-flowered with elongate branches; bracts small, linear-lanceolate, or resembling leaves. **Pedicels** elongate, much exceeding calyx. **Flowers:** calyx green, 10-veined, broadly tubular, 12–25 × 4–8 mm in flower, narrower towards base, middle broadening to 6–13 mm in fruit, narrower at both ends, pubescent, glandular, lobes lanceolate, 2.5–5 mm, margins membranous or not, apex ± obtuse; corolla scarlet, clawed, claw equaling or longer than calyx, limb lobed, often deeply so, lobes 4–6, linear, lanceolate, or oblong, small lateral teeth may be present, 6–15 mm, appendages inconspicuous, 1–2 mm, dentate; stamens longer than corolla claw but shorter than lobes; stigmas 3, equaling corolla. **Capsules** oblong to ovoid or broadly tubular, equaling calyx, opening by 6 ascending teeth; carpophore 2–4 mm. **Seeds** reddish brown, broadly reniform, 1.7–2.3 mm, sides tuberculate, margins papillate.

Subspecies 3 (3 in the flora): sw United States, Mexico.

The large, beautiful scarlet flowers of *Silene laciniata* are bird- and butterfly-pollinated. The species occurs in three forms. Subspecies *laciniata* has narrow leaves, much-branched and many-flowered, open inflorescences, and oblong capsules. Subspecies *californica* has ovate to lanceolate leaves, few-flowered inflorescences, and ovoid capsules. These two entities both occur in California and Mexico. Although they are usually distinguishable, apparent intermediates occur. The third entity is subsp. *greggii*, which occurs in Arizona, southwestern Texas, and Mexico. It combines characters of the other two subspecies, having broader leaves, a branched and many-flowered open inflorescence, and a capsule of intermediate shape.

1. Inflorescences poorly developed, usually 1–3(–5)-flowered; stems straggling, sparsely branched, leafy throughout; distal leaf blades and bracts lanceolate to ovate-lanceolate; capsules ovoid; fruiting calyces obovate to turbinate, more than ¹/₂ as broad as long . . . .
. . . . 27b.   *Silene laciniata* subsp. *californica*
1. Inflorescences well-developed, (3–)5+-flowered; stems erect or ascending, sometimes with decumbent base, branched distally; distal leaves reduced, blade linear or lanceolate to elliptic; capsules oblong or broadly tubular; fruiting calyces less than ¹/₂ as broad as long.
    2. Distal leaf blades linear to lanceolate, proximal leaf blades lanceolate to oblanceolate; California, Mexico . . . .
    . . . . 27a.   *Silene laciniata* subsp. *laciniata*
    2. Distal leaf blades lanceolate to elliptic, proximal leaf blades oblanceolate, often broadly so; Arizona to sw Texas, s into Mexico
    . . . . 27c.   *Silene laciniata* subsp. *greggii*

27a. Silene laciniata Cavanilles subsp. **laciniata**   [F]

Lychnis pulchra Willdenow ex Schlechtendal & Chamisso; *Silene allamanii* Otth; *S. laciniata* var. *angustifolia* C. L. Hitchcock & Maguire; *S. laciniata* subsp. *brandegeei* C. L. Hitchcock & Maguire; *S. laciniata* var. *latifolia* C. L. Hitchcock & Maguire; *S. laciniata* subsp. *major* C. L. Hitchcock & Maguire; *S. mexicana* Otth; *S. pulchra* (Willdenow ex Schlechtendal & Chamisso) Torrey & A. Gray; *S. simulans* Greene

Stems ascending, decumbent at base, much-branched distally, glabrate proximally, glandular-pubescent distally. **Leaves:** proximal, blade oblanceolate to lanceolate, 1.5–10 cm × 2–10(–20) mm, with short, ± retrorse pubescence on both surfaces, sometimes glandular; distal sessile, blade linear to lanceolate, reduced into inflorescence, 3–10 cm × 2–10(–20) mm, apex acute. **Inflorescences** well developed, (3–)5–many-flowered, open, compound, long-branched cymes, bracteate; bracts resembling distal leaves, reduced. **Calyces** broadly tubular, 12–26 × 4–6 mm in flower, to 8 mm broad in fruit, less than ¹/₂ as broad as long, narrowed to base. **Capsules** oblong. $2n = 96$.

Flowering spring–summer. Chaparral, oak woodlands, sea slopes and cliffs; 0–1200 mm; Calif.; Mexico.

A very narrow-leaved form of subsp. *laciniata* from the cliffs of Santa Cruz Island off the coast of California has been named *Silene simulans*.

27b. Silene laciniata Cavanilles subsp. **californica** (Durand) J. K. Morton, Sida 21: 888. 2004
• California pink

Silene californica Durand, J. Acad. Nat. Sci. Philadelphia, n. s. 3: 83. 1855; *S. laciniata* var. *californica* (Durand) A. Gray

Stems decumbent at base, straggling to erect, leafy throughout, sparsely branched, scaly proximally below ground, with soft, short pubescence. **Leaves:** proximal blades oblanceolate, sometimes broadly so, narrowed into short pseudopetiole, 2–6 cm × 6–25 mm, shortly pubescent abaxially, subglabrous adaxially; distal sessile, blade lanceolate or ovate-lanceolate to broadly elliptic, 2–9 cm × 5–30 mm, apex acute to shortly acuminate. **Inflorescences** poorly developed, usually 1–3(–5)-flowered, open cymes, bracteate; bracts leaflike, lanceolate to ovate-lanceolate. **Calyces** broadly tubular, widened distally, 15–25 × 4–8 mm in flower, obovate to turbinate in fruit and to 13 mm broad, more than ¹/₂ as broad as long. **Capsules** ovoid. $2n = 48, 72$.

Flowering spring–summer. Dry, open woodlands, chaparral, rocky hillsides and cliffs; 0–2200 m; Calif., Mexico.

The recently described *Silene serpentinicola* is similar to subsp. *californica* but differs in its short, erect, more or less solitary flowering stems and the much larger appendages of the flowers.

27c. Silene laciniata Cavanilles subsp. **greggii** (A. Gray) C. L. Hitchcock & Maguire, Revis. N. Amer. Silene, 56. 1947 • Gregg's Mexican pink or campion

Silene greggii A. Gray, Smithsonian Contr. Knowl. 5(6): 17. 1853; *Melandrium laciniatum* (Cavanilles) Rohrbach var. *greggii* (A. Gray) Rohrbach; *Silene laciniata* var. *greggii* (A. Gray) S. Watson

Stems erect, much-branched distally. **Leaves:** proximal blades oblanceolate, often broadly so, 3–6(–8) cm × 15–30 mm, scabrid-pubescent on both surfaces; distal short-petiolate, blade lanceolate to elliptic, reduced, 1–4 cm × 3–10 mm, apex acute. **Inflorescences** well developed, open, 3–5-flowered cymes, bracteate; bracts lanceolate, 6–20 mm, apex acute. **Calyces** broadly tubular, 17–20 × 4–5 mm in flower, broadening to 6–9 mm in fruit, less than ¹/₂ as broad as long, narrowed to base and ± umbilicate. $2n = 48$.

Flowering in summer–early fall. Dry oak, pine, and juniper woodlands; 1600–3000 m; Ariz., N.Mex., Tex.; Mexico.

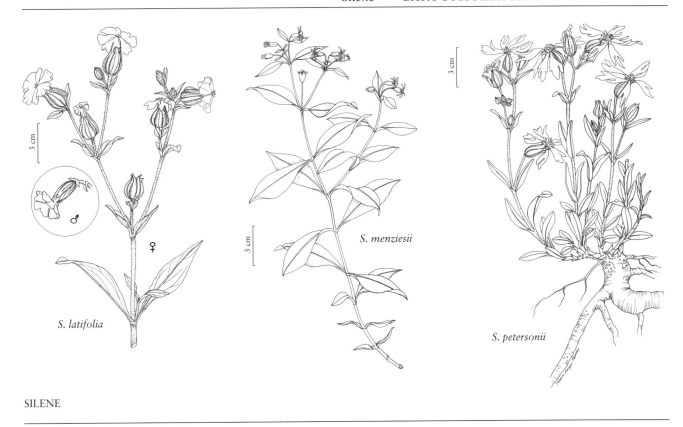

S. latifolia

♀

♂

S. menziesii

S. petersonii

SILENE

28. Silene latifolia Poiret, Voy. Barbarie 2: 165. 1789

• White campion or cockle, silène blanc [F] [I]

Lychnis alba Miller; *L. ×loveae* B. Boivin; *L. vespertina* Sibthorp; *Melandrium album* (Miller) Garcke; *M. dioicum* (Linnaeus) Cosson & Germain subsp. *alba* (Miller) D. Löve; *Silene alba* (Miller) E. H. L. Krause 1901, not Muhlenberg ex Rohrbach 1868; *S. latifolia* Poiret subsp. *alba* (Miller) Greuter & Burdet; *S. pratensis* (Rafinesque) Grenier & Godron

Plants annual or short-lived perennial; taproot woody. **Stems** erect or decumbent at base, branched, to 100 cm, finely hirsute, glandular-puberulent distally. **Leaves:** blade hirsute on both surfaces; basal usually withering by flowering time, petiolate, blade oblong-lanceolate to elliptic; cauline sessile, reduced into inflorescence, blade lanceolate to elliptic, 3–12 cm × 6–30 mm, apex acute. **Inflorescences** several–many-flowered (fewer in pistillate plants), open, dichasial cymes, bracteate; bracts much reduced, lanceolate, herbaceous. **Pedicels** 1–5 cm. **Flowers** unisexual, some plants having only staminate flowers, others having only pistillate flowers, fragrant, 25–35 mm diam.; in veined staminate plants subsessile to short-pedicellate, in pistillate plants pedicellate; calyx prominently 10-veined in staminate flowers, 20-veined in pistillate, tubular, becoming ovate in pistillate flowers, 10–20(–24) × 8–15 mm in fruit, margins dentate, hirsute and shortly glandular-pubescent, lobes to 6 mm, broadly ovate with apex obtuse, to lanceolate with apex acuminate; petals white, broadly obovate, ca. 2 times calyx, limb spreading, unlobed to 2-lobed; stamens equaling to slightly longer than calyx; stigmas (4–)5, slightly longer than calyx. **Capsules** ovate, ca. equaling calyx, opening by (4–)5, spreading to slightly reflexed, 2-fid teeth; carpophore 1–2 mm. **Seeds** dark gray-brown, reniform-rotund, plump, ca. 1.5 mm, coarsely tuberculate. $2n = 24$.

Flowering summer–fall. Arable land, roadsides, waste land; 0–2800 m; introduced; Greenland; St. Pierre and Miquelon; Alta., B.C., Man., N.B., Nfld. and Labr. (Nfld.), N.S., Ont., P.E.I., Que., Sask.; Alaska, Ariz., Ark., Calif., Colo., Conn., Del., D.C., Ga., Idaho, Ill., Ind., Iowa, Kans., Ky., Maine, Md., Mass., Mich., Minn., Mo., Mont., Nebr., N.H., N.J., N.Mex., N.Y., N.C., N.Dak., Ohio, Oreg., Pa., R.I., S.C., S.Dak., Tenn., Utah, Vt., Va., Wash., W.Va., Wis., Wyo.; Eurasia.

European botanists recognize several subspecies of *Silene latifolia*, at least two of which appear to occur in North America: subsp. *latifolia* [*S. alba* subsp. *divaricata* (Reichenbach) Walters], a commonly occurring form here, with acuminate calyx teeth and patent to recurved capsule teeth; and subsp. *alba* (Miller) Greuter & Burdet, less common in North America, with short, obtuse calyx teeth and erect teeth in the dehisced capsule. However, most of our material tends to be intermediate, making recognition of subspecies here of little value. Presumably there has been extensive gene exchange between populations of this outbreeding species since its introduction into North America.

The name *Silene latifolia* has been misapplied to *S. vulgaris* by some authors, which has been a cause of confusion.

29. Silene lemmonii S. Watson, Proc. Amer. Acad. Arts 10: 342. 1875 (as lemmoni) • Lemmon's catchfly E

Silene palmeri S. Watson

Plants perennial; taproot stout; caudex much-branched, woody, producing short, decumbent, leafy sterile shoots and erect flowering shoots. **Stems** 15–45 cm, pubescent and glandular-viscid distally, sparsely pubescent to ± glabrous proximally. **Leaves** mostly in dense basal tufts; basal blades oblanceolate to elliptic, 1–3.5 cm × 3–10 mm, narrowed to base, apex acute, scabrous-puberulent to subglabrous; cauline in 2–3 pairs, distal sessile, reduced, blade linear-lanceolate to lanceolate, 1.5–4 cm × 2–6 mm. **Inflorescences** cymose, (1–)3–5(–7)-flowered, open, bracteate, bracteolate, pubescent and viscid with stipitate glands; cyme open, slender-branched; bracts and bracteoles narrowly lanceolate, 2–15 mm, herbaceous. **Pedicels** divaricate, often curved near apex and/or at base, slender, 1/2–2 times longer than calyx. **Flowers:** calyx prominently 10-veined, campanulate, 6–10 × 2–4 mm in flower, broadening in fruit and becoming obconic with ± constricted base, ± as broad as long, pubescent and glandular, veins parallel, with pale commissures, lobes triangular, 1–2 mm, margins broad, membranous, apex acute; corolla yellowish white, sometimes tinged with pink, clawed, claw equaling or longer than calyx, limb deeply lobed, lobes 4, linear, 4–8 mm, appendages 2, narrow, ca. 1 mm; stamens exserted, equaling petals; styles 3, filamentous, much longer than petals and stamens, exceeding 2 times calyx. **Capsules** obovoid, equaling calyx and often splitting it, opening by 6 recurved teeth; carpophore 2–3 mm. **Seeds** rusty brown, often with gray bloom, broadly reniform, 1–1.8 mm, coarsely papillate. $2n = 48$.

Flowering spring–summer. Woodlands and forests, often in moist situations; 200–2800 m; Calif., Oreg.

Silene lemmonii has typical moth-pollinated flowers. It is closely related to *S. bridgesii* and appears to intergrade with it. However, the small size of *S. lemmonii* and the presence of a compact growth of short, leafy sterile shoots usually distinguish it from *S. bridgesii*.

30. Silene marmorensis Kruckeberg, Madroño 15: 174, figs. 1–3. 1960 • Marble Mountain or Somes Bar campion C E

Plants perennial; taproot long, stout; caudex branched, woody, producing several erect flowering shoots. **Stems** erect, simple proximal to inflorescence, 25–80 cm, puberulent, glandular distally. **Leaves** in 5–7 pairs proximal to inflorescence, sessile, blade lanceolate, narrowed to base, apex acute; proximal withering, becoming smaller in inflorescence, 2–5 cm × 3–10 mm, sparsely scabrous-pubescent on both surfaces. **Inflorescences** cymose, terminal, pedunculate, 1–3-flowered, open, bracteate, bracteolate, 10–25 cm, pubescence dense, hairs septate-glandular, septa colorless; cymes paired at each node; peduncle ascending, 1–3 cm; bracts and bracteoles leaflike, reduced distally to 2 mm. **Pedicels** not bent in fruit, 1/2 to equaling calyx. **Flowers** nocturnal; calyx prominently 10-veined, campanulate, 12–14 × 4–6 mm in flower, becoming obovate to obconic and to 10 mm broad in fruit, not contracted proximally around carpophore, margins dentate, glandular-pubescent, veins parallel, with pale commissures, lobes lanceolate-acuminate, 3–4 mm, membranous, margins narrow, apex blunt, veins green; corolla pale pink, greenish abaxially, clawed, claw equaling calyx, limb oblong, deeply 2-lobed, 4–6 mm, appendages 2, oblong, ca. 1 mm; stamens equaling petals; stigmas 3, equaling petals. **Capsules** obovoid, equaling calyx and often splitting it at maturity, opening by 5 teeth; carpophore 3–4 mm. **Seeds** black, reniform, 2–3 mm, tuberculate; tubercles conic, in concentric rows. $2n = 48$.

Flowering summer. Oak woodlands, coniferous forests; of conservation concern; 800–1000 m; Calif.

Silene marmorensis is closely related to *S. bridgesii* but has a narrower inflorescence, pedicels that are ascending instead of deflexed, and styles and stamens that are about equal to the petals. As in *S. bridgesii* and *S. lemmonii*, the flowers open at night and are probably moth-pollinated. The species is known only from Humboldt and Siskiyou counties.

31. Silene menziesii Hooker, Fl. Bor.-Amer. 1: 90, plate 30. 1830 • Menzies' catchfly [E] [F]

Anotites alsinoides Greene; *A. bakeri* Greene; *A. costata* Greene; *A. debilis* Greene; *A. diffusa* Greene; *A. discurrens* Greene; *A. dorrii* (Kellogg) Greene; *A. elliptica* Greene; *A. halophila* Greene; *A. jonesii* Greene; *A. latifolia* Greene; *A. macilenta* Greene; *A. menziesii* (Hooker) Greene; *A. nodosa* Greene; *A. picta* Greene; *A. tenerrima* Greene; *A. tereticaulis* Greene; *A. villosula* Greene; *A. viscosa* Greene; *Silene dorrii* Kellogg; *S. menziesii* subsp. *dorrii* (Kellogg) C. L. Hitchcock & Maguire; *S. menziesii* var. *viscosa* (Greene) C. L. Hitchcock & Maguire; *S. obovata* A. E. Porsild; *S. stellarioides* Nuttall ex Torrey & A. Gray

Plants perennial, with several–many decumbent, sometimes cespitose or matted shoots; taproot slender. **Stems** decumbent to erect, simple or branched, leafy throughout, 5–30(–70) cm, usually glandular-puberulent distally, proximal pubescence varying from short and sparse to multicellular, crinkled and deflexed, glandular or not. **Leaves** 2 per node, sessile or short-petiolate; blade usually oblanceolate to elliptic-lanceolate, rarely obovate or elliptic, 2–6(–10) cm × 3–20(–35) mm, broadest at or above middle, narrowed to base, margins ciliate with short, somewhat scabrid hairs, apex acute to acuminate, puberulent to pubescent. **Inflorescences** cymose, or flowers axillary or solitary and terminal; cyme loose, compound, leafy. **Pedicels** slender, 0.5–3 cm, glandular-puberulent. **Flowers** functionally unisexual, usually bisexual; calyx obscurely 10-veined, campanulate, 5–8 mm, ca. ¹/₂ as wide, herbaceous, margins dentate, lobes lanceolate, 1.5–3 mm, apex recurved, subacute to acuminate, puberulent to pilose and glandular, veins without conspicuous, pale commissures; corolla white, clawed, 1–1¹/₂ times calyx, claw shorter than calyx, limb oblong, 2-lobed, 1.5–3 mm, lobes oblong, apex obtuse, appendages 2, small, 0.1–0.4 mm; stamens in functionally staminate flowers equaling corolla, otherwise reduced and included in calyx; stigmas 3(–4), equaling corolla in functionally pistillate flowers, otherwise included in calyx, papillate along whole length. **Capsules** green, becoming black, ovoid-ellipsoid, slightly longer than calyx, opening by 3 erect teeth which often split into 6; carpophore ca. 1.5 mm. **Seeds** black, not winged, broadly reniform, 0.5–1 mm, glossy, obscurely reticulate. $2n = 24, 48$.

Flowering throughout summer. Common in open woodlands and forests, grasslands, gravelly places, river banks, mountains farther south; 200–3000 m; Alta., B.C., Man., N.W.T., Sask., Yukon; Alaska, Ariz., Calif., Colo., Idaho, Mont., Nebr., Nev., N.Mex., Oreg., Utah, Wash., Wyo.

Silene menziesii is quite variable in the extent to which the inflorescence is developed and in its pubescence. This, coupled with the functionally dioecious nature of the species, has spawned a plethora of names, none of which appear to warrant recognition. The similar *S. williamsii* from Alaska and the Yukon Territory can be separated by its narrower lanceolate leaves that are broadest near the base and dull, usually brown, tuberculate seeds. Also, its stigmas are papillate only near the top. *Silene seelyi* is also very similar to *S. menziesii* but has dark red flowers and leaves that are smaller (to 2 cm in length) and broadest below the middle.

32. Silene nachlingerae Tiehm, Brittonia 37: 344, fig. 1. 1985 • Nachlinger's or Jan's catchfly [C] [E]

Plants perennial; taproot stout; caudex usually branched, woody, producing tufts of basal leaves and 1–several flowering shoots. **Stems** subscapose, with (1–)2–3 (–4) pairs of leaves, 6–25 cm, much reduced above base, retrorsely puberulent, eglandular. **Leaves** 2 per node; basal blades oblanceolate, spatulate, 1.4–4 cm × 2–5 mm, base ciliate, apex acute, subglabrous to retrorse-puberulent; cauline shortly connate proximally, blade narrowly oblanceolate to linear-lanceolate, 0.5–3.5 cm × 1–2 mm, apex purple tinged, retrorse-puberulent, often sparsely so. **Inflorescences** 1–3(–4)-flowered, open. **Pedicels** erect, 1–4 times longer than fruiting calyx, retrorsely puberulent, without glandular hairs. **Flowers:** calyx prominently 10-veined, cylindric, becoming narrowly ovoid, not inflated, not contracted proximally around carpophore, 6–11.5 × 3 mm in flower, broadening to ca. 5 mm in fruit, membranous, uniformly puberulent, without glandular hairs, veins parallel, green, with pale commissures, lobes triangular, 1–1.5 mm, margins narrowly membranous, apex purple-tipped; corolla white, usually flushed with pink or purple towards apex, ligulate to ± oblanceolate, slightly longer than calyx, not clearly differentiated into claw and limb, unlobed to notched, auricles and appendages absent; stamens equaling calyx; styles 3, included in calyx. **Capsules** ellipsoid-ovoid, slightly longer than calyx, opening by 6 ascending teeth; carpophore ca. 1 mm. **Seeds** rust colored, reniform, 0.7–1 mm, margins papillate; papillae large, inflated.

Flowering summer. Alpine limestone ridges and slopes; of conservation concern; 2500–3000 m; Nev., Utah.

Silene nachlingerae is a small version of *S. invisa*, although the latter species has glabrous leaves. Both have the characteristic inflated (balloonlike) papillae around the outer edge of the seed.

33. Silene nivea (Nuttall) Muhlenberg ex Otth in A. P. de Candolle and A. L. P. P. de Candolle, Prodr. 1: 377. 1824 • Snowy campion E

Cucubalus niveus Nuttall, Gen. N. Amer. Pl. 1: 287. 1818; *Silene alba* Muhlenberg ex Rohrbach

Plants perennial, rhizomatous; rhizome elongate. **Stems** erect, simple to sparingly branched, leafy, 20–70 cm, glabrous to puberulent, especially distally. **Leaves** 2 per node, sessile or short-petiolate, largest near mid-stem region, reduced and withering proximally, blade elliptic-lanceolate, base cuneate or rounded, apex gradually acuminate and acute, glabrous to puberulent. **Inflorescences** cymose, (1–)3–5(–12)-flowered, open, leafy. **Pedicels** ¹/₂–2¹/₂ times calyx, apex often becoming deflexed, glabrous to hirsute. **Flowers:** calyx green, obscurely 10-veined, broadly tubular to campanulate, ± constricted at base around carpophore with broad umbilicate base, becoming broadly clavate in fruit, 14–17 × 5–9 mm, herbaceous, glabrous or hirsute, veins green, without pale commissures, lobes triangular, 2–3 mm; corolla white, clawed, claw equaling calyx, expanded distally into 2-lobed limb, limb oblong, 6–7 mm, appendages oblong, 1–1.6 mm, margins ± entire; stamens short-exserted; stigmas 3, short-exserted. **Capsules** globose, equaling calyx, opening by 3 broad teeth that sometimes split to form 6; carpophore 5–6 mm. **Seeds** dark brown to black, with grayish bloom, broadly reniform, not winged, 0.7–1 mm, sides with concentric crescents of low tubercles, larger and deeper on outer margins. 2*n* = 48.

Flowering late spring–summer. Alluvial woodlands; 0–400 m; D.C., Ill., Ind., Iowa, Maine, Md., Mass., Minn., Mo., Ohio, Pa., Va., W.Va., Wis.

The green, obscurely veined, umbilicate calyx with its broad base constricted around the carpophore is unique among the North American members of the genus. *Silene nivea* is occasionally weedy. It was introduced near Québec City (ca. 1969) but did not persist, and probably is not native also in Maine.

34. Silene noctiflora Linnaeus, Sp. Pl. 1: 419. 1753 • Night-flowering catchfly, silène noctiflore I W

Melandrium noctiflorum (Linnaeus) Fries

Plants annual, densely pubescent throughout, viscid-glandular, especially distally; taproot slender. **Stems** erect, simple proximal to inflorescence or with few basal branches, branched distally, to 75 cm. **Leaves** 2 per node, gradually reduced distally; basal blades oblanceolate, 6–12(–14) cm × 20–45 mm; cauline blades ascending, conspicuously veined, broadly elliptic to lanceolate, 1–11 cm × 3–40 mm, apex acute, shortly acuminate, densely pubescent on both surfaces. **Inflorescences** cymose, 3–15-flowered, bracteate; cyme open, flowers held on ascending branches; bracts leaflike, narrowly lanceolate, 1–5 cm, apex acuminate. **Pedicels** ascending, straight, ¹/₃–3 times longer than calyx. **Flowers** nocturnal, 20–25 mm diam.; calyx prominently 10-veined, ovate-elliptic, fusiform, narrowed to both ends and constricted around carpophore, 15–24(–40) × ca. 3 mm in flower, swelling to 10 mm diam. in fruit, thin and papery, margins dentate, with pale commissures; lobes erect, often recurved in fruit, linear-lanceolate, long, narrow, (3–)5–10(–15) mm, apex acuminate, short-pubescent, glandular, interspersed with long eglandular hairs, veins anastomosing; corolla white, often pink tinged, clawed, claw equaling calyx lobes, limb deeply 2-lobed, lobes usually narrow, appendages 0.5–1.5 mm broad, margins entire or erose; stamens shorter than petals; styles 3, shorter than petals. **Capsules** ovoid, constricted at mouth, equaling or slightly longer than calyx tube, opening by 6 recurved teeth; carpophore 1–3 mm. **Seeds** dark brown to black, with gray bloom, broadly reniform, 0.8–1 mm, strongly tuberculate. 2*n* = 24.

Flowering summer. Arable land, disturbed ground; 0–3000 m; introduced; Alta., B.C., Man., N.B., Nfld. and Labr. (Nfld.), N.S., Ont., P.E.I., Que., Sask., Yukon; Alaska, Ala., Calif., Colo., Conn., Del., D.C., Fla., Idaho, Ill., Ind., Iowa, Ky., La., Maine, Mass., Mich., Minn., Miss., Mo., Mont., Nebr., N.H., N.J., N.Mex., N.Y., N.C., N.Dak., Ohio, Oreg., Pa., R.I., S.Dak., Tenn., Utah., Vt., Va., Wash., W.Va., Wis., Wyo.; Europe.

Silene noctiflora is sometimes confused with *S. latifolia*, but they are very different species. *Silene noctiflora* differs in having perfect flowers with long, very narrow calyx teeth and an elliptic, fruiting calyx that is narrow at the mouth and constricted around the capsule base. It also has three styles and a capsule that dehisces by six teeth; *S. latifolia* has (four or) five styles and a capsule that dehisces by five bifid teeth. The flowers of *S. noctiflora*, as its name indicates, are nocturnal and moth-pollinated.

35. Silene nuda (S. Watson) C. L. Hitchcock & Maguire, Revis. N. Amer. Silene, 45. 1947 • Sticky catchfly E

Lychnis nuda S. Watson, Botany (Fortieth Parallel), 37. 1871; *Silene insectivora* L. F. Henderson; *S. nuda* subsp. *insectivora* (L. F. Henderson) C. L. Hitchcock & Maguire; *S. pectinata* S. Watson; *S. pectinata* var. *subnuda* B. L. Robinson

Plants perennial; taproot thick; caudex erect, branched, thick and woody, producing tufts of leaves. **Stems** erect, branched distally, with 2–4 pairs of reduced leaves, 15–50 cm; flowering shoots usually subscapose, coarsely pubescent with hairs colorless, septate, and long, viscid-glandular, especially distally. **Leaves** mostly basal; basal long-petiolate, blade oblanceolate to elliptic, 6–15 cm × 10–30 mm, narrowed to base, not fleshy; cauline few, sessile, reduced distally, blade lanceolate, 0.8–4 cm × 3–8 mm, not fleshy. **Inflorescences** thyrsate, subscapose, rarely simple, (3–)5–12(rarely more)-flowered, open, bracteate, bracteolate, densely pubescent, glandular, viscid; alternate branches often suppressed or developing unevenly; proximal nodes often with single flower; bracts and bracteoles resembling stem leaves but much reduced. **Pedicels** 1/4–2 times length of calyx. **Flowers:** calyx prominently 10-veined, veins parallel, those of lobes broadened distally, tubular in flower, 10–13 × 2.5–4 mm, campanulate-ovate in fruit, broadest near middle and contracted towards mouth, not contracted proximally, 12–18 × 5–8 mm, with pale commissures, lobes 5, erect, narrowly lanceolate, 4–6 mm, margins narrow, membranous proximally, apex blunt, with glandular hairs; petals 1 1/2–2 times longer than calyx tube; corolla pink, clawed, claw equaling calyx tube, limb obovate, deeply 2-lobed, 5–10 mm, appendages 2, linear, ca. 1.5 mm; stamens exserted, shorter than petals; styles 3–5, included in calyx, ± equaling calyx or corolla. **Capsules** conic to ellipsoid, equaling calyx lobes, opening by 6–10 recurved teeth; carpophore 1–2 mm. **Seeds** dark brown, reniform, 1–1.5 mm, prominently papillate; papillae larger around margins. $2n = 48$.

Flowering summer. Scrubby grasslands and openings in woodland and coniferous forests; 1100–2300 m; Calif., Nev., Oreg.

Silene nuda may be confused with the other scapose species, *S. scaposa*, but *S. nuda* has larger, more conspicuous petals that are one and one-half to two times as long as the calyx tube. It is found in the Sierra Nevada and southern Cascades. The Nevada populations tend to grow in drier situations and on saline flats.

36. Silene occidentalis S. Watson, Proc. Amer. Acad. Arts 10: 343. 1875 • Western catchfly E

Silene occidentalis subsp. *longistipitata* C. L. Hitchcock & Maguire

Plants perennial; taproot stout; caudex simple or branched, woody, bearing tufts of basal leaves. **Stems** erect, simple proximal to inflorescence, 30–60 cm, softly pubescent and stipitate-glandular. **Leaves** 2 per node; basal ± petiolate, petiole ciliate, blade oblanceolate, spatulate, 5–12 cm × 7–20 mm, apex acute, short-pubescent on both surfaces; cauline in 3(–4) pairs, reduced distally, blade oblanceolate to lanceolate. **Inflorescences** open, narrow, with ascending branches, 9–25-flowered, bracteate, pubescent and stipitate-glandular; bracts narrowly lanceolate, ciliate. **Pedicels** 1/2–3 times longer than calyx. **Flowers:** calyx prominently 10-veined, tubular in flower and fruit, umbilicate, somewhat constricted around carpophore, 15–38 × 3–6 mm, papery, sparsely pubescent and stipitate-glandular, veins parallel, green, with pale commissures, lobes 5, broadly ovate, 2–4 mm, scarious around green midrib; corolla pink or rose red (rarely white), clawed, claw equaling calyx, limb oblong, fanlike, deeply 4-lobed, lobes divergent, lanceolate (rarely with only 2 lobes, each with small lateral tooth), (7–)10–20 mm, appendages linear, 2–4 mm; stamens exserted, shorter than petals; stigmas 3, shorter than petals. **Capsules** narrowly ovate-elliptic, longer than calyx, opening by 6 recurved teeth; carpophore 4–18 mm. **Seeds** grayish brown, reniform, 1–1.5 mm, verrucate. $2n = 48$.

Flowering summer. Grassy openings in chaparral, coniferous forests, and woodlands; 700–2300 m; Calif.

The long, tubular calyx and the proportionally long carpophore of *Silene occidentalis* are remarkable. Plants with the longest calyx tubes (more than 30 mm) have been referred to subsp. *longistipitata* and appear to be confined to Butte County. Calyx length varies greatly, however, and it is doubtful whether it is a justifiable basis for taxonomic recognition. The deeply lobed pink petals, together with its habit, give *S. occidentalis* a superficial resemblance to the European *S. flos-cuculi*, which occurs as an introduction on both sides of the North American continent.

37. Silene oregana S. Watson, Proc. Amer. Acad. Arts 10: 343. 1875 • Oregon catchfly [E]

Silene filisecta M. Peck; *S. gormanii* Howell; *S. oregana* var. *filisecta* (M. Peck) M. Peck

Plants perennial; taproot stout; caudex simple or sparsely branched, woody. **Stems** usually simple proximal to inflorescence, 30–50(–70) cm, puberulent and shortly stipitate-glandular, especially distally. **Leaves** 2 per node, gradually reduced distally; basal petiolate, blade oblanceolate, spatulate, 5–9 cm × 7–15 mm (including petiole), apex acute to obtuse, usually glabrous adaxially, sparsely pubescent abaxially; cauline in 4–6 pairs, blade linear-lanceolate, 1–6(–8) cm × 2–6 mm, puberulent and shortly stipitate-glandular. **Inflorescences** thyrsate, 3–25-flowered, open, bracteate, pedunculate, stipitate-glandular, viscid; bracts narrowly lanceolate, 2–25 mm, apex acuminate. **Pedicels** ascending. **Flowers:** calyx prominently 10-veined, narrowly campanulate, umbilicate, somewhat clavate and constricted below middle around carpophore, 9–15 × 3–4 mm in flower, broadening to 7 mm in fruit, membranous, shortly stipitate-glandular, veins parallel, slender, tinged dark red, with pale commissures, lobes ovate-lanceolate, obtuse, 2–3 mm, margins scarious; corolla creamy white, sometimes pink tinged, clawed, claw equaling calyx, glabrous, broadening only slightly into limb 3–8 mm, limb with 4–6 linear lobes, some splitting to 10 linear segments, appendages 4–6, linear, 1–1.5 mm, apex acute; stamens ca. equaling petals; filaments glabrous; stigmas 3(–5), ca. equaling petals. **Capsules** ellipsoid, slightly longer than calyx, opening by 6 (or 8 or 10) very brittle teeth; carpophore 2–4 mm. **Seeds** brown, ± reniform, angular, glossy, shallowly tuberculate. $2n = 48$.

Flowering summer. Dry, grassy slopes, rocky areas, open woodlands and forests; 1500–2800 m; Calif., Idaho, Mont., Nev., Oreg., Utah, Wash., Wyo.

The creamy white laciniate petals are the best field (and herbarium) guide to distinguishing this species from *Silene parryi* and *S. scouleri,* both of which have 2–4-lobed petals that are usually dingy cream to greenish or purple tinged.

38. Silene ostenfeldii (A. E. Porsild) J. K. Morton, Sida 21: 888. 2004

Melandrium ostenfeldii A. E. Porsild, Sargentia 4: 37. 1943; *Gastrolychnis ostenfeldii* (A. E. Porsild) V. V. Petrovsky; *G. taimyrensis* (Tolmatchew) Czerepanov; *G. triflora* (R. Brown ex Sommerfelt) Tolmatchew & Kozhanchikov subsp. *dawsonii* (B. L. Robinson) Á. Löve & D. Löve; *Lychnis dawsonii* (B. L. Robinson) J. P. Anderson; *L. ostenfeldii* (A. E. Porsild) B. Boivin; *L. taimyrense* (Tolmatchew) Polunin; *L. triflora* R. Brown ex Sommerfelt subsp. *dawsonii* (B. L. Robinson) Maguire; *L. triflora* var. *dawsonii* B. L. Robinson; *Melandrium dawsonii* (B. L. Robinson) Hultén

Plants perennial, densely cespitose; taproot stout, fleshy; caudex tightly branched. **Stems** 1–many, erect, simple, slender, 10–30 cm, glandular-pubescent. **Leaves** connate basally, blade ciliate at base, pubescent on both surfaces; basal numerous, ± petiolate, blade linear-oblanceolate, 1–5 cm × 1–5 mm, somewhat fleshy, apex ± acute; cauline in 1–3 pairs, sessile, blade linear to linear-lanceolate, 1–4 cm × 1–3 mm, apex ± acute. **Inflorescences** cymose, terminal, (1–)3-flowered, rarely with 1–2 flowers at proximal nodes, bracteate, bracteolate; bracts and bracteoles leaflike, 2–10 mm. **Pedicels** usually shorter than calyx, rarely much longer, densely pubescent with purple-septate glandular hairs but not viscid. **Flowers:** calyx prominently 10-veined, elliptic to campanulate, not inflated, not or slightly contracted at mouth, 8–9 × 3–5 mm in flower, 10–12 × 5–6 mm in fruit, to 2 times as long as broad, papery, veins green or purple, densely pubescent, with purple-septate hairs, lobes spreading, lanceolate-triangular, ca. 2 mm, margins purple tinged, round, broad, membranous; corolla white to pink, clawed, ca. 1¼ times longer than calyx, claw equaling calyx, limb not differentiated from claw, obovate, emarginate to 2-lobed, shorter than calyx, appendages 2, ca. 1 mm; stamens equaling petals; styles 5, equaling petals. **Capsules** ovoid-ellipsoid, slightly longer than calyx, opening by 5 recurved teeth; carpophore shorter than 1 mm. **Seeds** brown, not winged, reniform, angular, less than 1 mm broad, finely papillate. $2n = 24, 48, 72$.

Flowering early summer. Gravelly tundra, rocky ledges, talus, river outwash, grassy areas; 0–1800 m; B.C., N.W.T., Nunavut, Yukon; Alaska; e Asia (Russian Far East).

Silene ostenfeldii, an amphi-Beringian species, is very similar to *S. involucrata* subsp. *tenella.* However, it lacks the wing on the seeds, and its mature calyx tends to be more elliptic than campanulate. It may be confused also with *S. sorensenis,* but that species has larger seeds and calyces and is a sturdier plant with a denser, longer, somewhat woolly pubescence.

39. Silene ovata Pursh, Fl. Amer. Sept. 1: 316. 1813

• Ovate-leaved campion or catchfly [C] [E]

Plants perennial, rhizomatous; rhizome creeping. **Stems** erect, usually simple, 30–150 cm, with short, dense, eglandular pubescence, sparsely so toward base. **Leaves** sessile, 2 per node; blade prominently 3–5-veined, ovate-acuminate, round at base, (4–)6–10(–13) cm × (20–)30–50 (–90) mm, appressed-pubescent on both surfaces. **Inflorescences** paniculate, narrow, many-flowered, open, bracteate, pedunculate, 10–50 × 3–5 cm, densely puberulent; bracts narrowly lanceolate, 3–15 mm, apex acuminate; peduncle ascending. **Pedicels** ascending, recurved near apex, ca. equaling calyx. **Flowers** nocturnal; calyx prominently 10-veined, tubular to narrowly campanulate and 6–9 × 3–4 mm in flower, turbinate and 10–12 × 4–5 mm in fruit, narrowed proximally around carpophore, veins parallel, green, broad, with pale commissures, puberulent, sometimes with few glands, lobes triangular-acute, 2–3 mm; corolla white, clawed, claw equaling calyx, broadened into limb, limb obtriangular, deeply lobed, 7–9 mm, lobes ca. 8, linear, appendages minute; stamens slightly longer than corolla; styles 3, ca. 2 times as long as corolla. **Capsules** narrowly ovoid, slightly longer than calyx, opening by 3 (splitting into 6) ascending teeth; carpophore 2–2.5 mm. **Seeds** dark brown, reniform, 0.8–1.5 mm, shallowly tuberculate. $2n = 48$.

Flowering late summer–fall. Rich woods; of conservation concern; 1000–1900 m; Ala., Ark., Ga., Ky., Miss., N.C., S.C., Tenn., Va.

Silene ovata is a very distinctive species with large, ovate, acuminate, sessile, paired leaves, and very narrowly lobed white petals. The flowers open at night and are moth-pollinated.

40. Silene parishii S. Watson, Proc. Amer. Acad. Arts 17: 366. 1882 • Parish's catchfly [E]

Silene parishii var. *latifolia* C. L. Hitchcock & Maguire; *S. parishii* var. *viscida* C. L. Hitchcock & Maguire

Plants perennial; taproot stout; caudex much-branched, woody. **Stems** many, decumbent to erect, usually simple, 10–40 cm, woody, pilose and glandular (rarely eglandular proximally). **Leaves** in 5–8 pairs, sessile but proximal ones narrowed into pseudopetiole, largest in mid-stem region, reduced proximally, scalelike at base, blade narrowly lanceolate to oblanceolate-elliptic or obovate, 2–6 cm × 3–20 mm, apex acute and acuminate, usually thick, leathery, densely puberulent and viscid-glandular or eglandular. **Inflorescences** cymose, open or congested, 3–15(–30)-flowered, sometimes compound, leafy. **Pedicels** to 1(–1.5) cm, shortly pilose, viscid-glandular, flowers sometimes sessile. **Flowers:** calyx prominently 10-veined, tubular, clavate in fruit, constricted proximally around carpophore, (20–)25–30 × 4–7 mm, papery, densely glandular-puberulent, viscid, veins parallel, green, with pale commissures, lobes narrowly lanceolate, acuminate, 5–8 mm, herbaceous; corolla pale greenish yellow to white, clawed, claw equaling calyx, ligulate, as broad as limb, limb 7–8 mm, deeply laciniate into 6 or more linear lobes, appendages oblong, laciniate, 2 mm; stamens equaling calyx; styles 3, exserted. **Capsules** equaling calyx, opening by 6 ascending teeth; carpophore ca. 3 mm. **Seeds** brown, reniform, 1.5–2 mm, margins papillate; papillae large, inflated. $2n = 48$.

Flowering spring and summer. Rocky ledges and slopes, stream banks, open coniferous woodlands; 1400–3400 m; Calif.

Silene parishii varies considerably in pubescence and leaf shape. C. L. Hitchcock and B. Maguire (1947) recognized three varieties on the basis of this variation. However, the characters vary independently and have only a weak geographical correlation. Hence recognition of the three varieties serves little useful purpose.

Silene parishii is confined to the mountains of southern California.

41. Silene parryi (S. Watson) C. L. Hitchcock & Maguire, Revis. N. Amer. Silene, 36. 1947 • Parry's catchfly [E]

Lychnis parryi S. Watson, Proc. Amer. Acad. Arts 12: 248. 1877; *Silene douglasii* Hooker var. *macounii* (S. Watson) B. L. Robinson; *S. macounii* S. Watson; *S. scouleri* Hooker var. *macounii* (S. Watson) B. Boivin; *S. tetonensis* E. E. Nelson; *S. tetragyna* Suksdorf; *Wahlbergella parryi* (S. Watson) Rydberg

Plants perennial; taproot thick; caudex branched, woody, with tufts of basal leaves. **Stems** erect, simple, (10–)20–60 cm, softly puberulent, viscid-glandular distally. **Leaves** mostly basal; basal petiolate, blade oblanceolate, spatulate, 3–8 cm × 2–14 mm, not fleshy, margins shortly ciliate, apex ± acute, glabrous to puberulent on both surfaces; cauline usually in 2–4 pairs, blade narrowly oblanceolate, lanceolate to linear-lanceolate, 0.2–0.8 cm × 10–80 mm, not fleshy, puberulent on both surfaces, at least distal ones glandular. **Inflorescences** (1–)3–7-flowered, open, bracteate; bracts linear-lanceolate, broadened at base, 2–10 mm. **Pedicels** ascending, usually longer than calyx, puberulent, viscid stipitate-glandular. **Flowers:** calyx prominently 10-veined, campanulate, inflated, ± umbilicate, not or only slightly constricted toward base, (10–)12–16 × 7–9 mm in fruit, glandular-pubescent, strongly viscid, veins parallel, purplish, with pale commissures, not much broadened distally, commissural veins slender, forked distally and fused to those of lobes, lobes ovate to broadly triangular with lanceolate midrib, 2–3 mm, margins purple tinged, broad, membranous; corolla white, often tinged green or purple, clawed, claw equaling calyx, glabrous, broadened distally, limb deeply 2-lobed, rarely 4-lobed, 5–7 mm, lobes with 2 prominent lateral teeth, appendages 2(–4), 1.5–2 mm; stamens equaling calyx; stigmas 3(–5), exserted. **Capsules** included in calyx, opening by 3(–5) teeth, each tardily splitting into 2; carpophore 2–3 mm. **Seeds** brown, not winged, broadly reniform and often flattened, 1.5–2.5 mm, rugose to shallowly tuberculate on sides, larger tubercules on margins. $2n$ = 48, 96.

Flowering summer. Mountains, gravelly ridges, rocky and grassy slopes, subalpine meadows, grassy openings in montane forests; 1500–3000 m; Alta., B.C.; Idaho, Mont., Oreg., Wash., Wyo.

Silene parryi is very similar to *S. douglasii*, but the latter is normally eglandular with a characteristic short, gray, retrorse pubescence. The two species may hybridize, accounting for the occurrence of populations of *S. douglasii* with some glandular pubescence in the inflorescence. *Silene parryi* is closely related also to *S. scouleri*, but the latter is normally readily distinguished by its pink flowers; taller stature; long, narrow, many-flowered inflorescences; and fusiform fruiting calyces that are constricted around the carpophore. However, some depauperate specimens of *S. scouleri* from montane habitats are difficult to place. Also, small plants of *S. parryi* from alpine habitats can easily be mistaken for *S. grayi*. The anthers of *S. parryi* are often smutted with *Microbotryum violaceum* (Persoon) G. Deml & Oberwinker [= *Ustilago violacea* (Persoon) Roussel], e.g., in the type collection of *S. tetonensis*.

42. **Silene pendula** Linnaeus, Sp. Pl. 1: 418. 1753
   • Drooping catchfly   I W

Plants annual, with several decumbent shoots; taproot slender. **Stems** procumbent to ascending, branched, leafy, 15–45 cm, lanuginose, often sparsely so, viscid distally. **Leaves** 2 per node; proximal with blade obovate, spatulate, apex obtuse; distal sessile, blade ovate to lanceolate, 2–5 cm × 2–20 mm, apex acute, sparsely pubescent adaxially, more densely so abaxially. **Inflorescences** pseudoracemose, lax, solitary flowers in axils of leafy bracts. **Pedicels** erect in flower, sharply deflexed at base in fruit, usually shorter than calyx, pilose and stipitate-glandular. **Flowers:** calyx prominently 10-veined, obovoid, especially in fruit, clavate, constricted around carpophore and narrowed at mouth, umbilicate, inflated, 13–18 mm, loose and papery, pubescence glandular and eglandular, sparsely lanuginose, veins parallel, green or purple, with pale commissures, lobes triangular, ca. 2 mm, apex obtuse; corolla bright pink, clawed, claw equaling calyx, limb obtriangular, 2-lobed, 7–11 mm, lobes divergent, ovate, appendages 2, shorter than 1 mm, apex acute; stamens slightly longer than petal claw; stigmas 3, equaling petals. **Capsules** included in calyx, ovoid-conic, opening by 6 teeth; carpophore 3–6 mm. **Seeds** dark brown, broadly reniform, 1.3–1.5 mm, with concentric crescents of shallow tubercles on both sides, margins with larger, deeper tubercles. $2n$ = 24 (Europe).

Flowering early summer. Roadsides; 0–2900 m; introduced; B.C.; Calif., Maine, N.J., N.Y., Oreg., Wyo.; Europe.

Silene pendula is an attractive, rarely escaping and persisting garden plant readily recognized by its beautiful pink flowers, straggling leafy stems, racemelike inflorescences with axillary flowers, and the obovoid, papery, strongly veined calyx that is constricted below the middle. It is occasionally used in seeding roadsides.

43. **Silene petersonii** Maguire, Madroño 6: 24. 1941
   • Peterson's campion or catchfly   C E F

Silene clokeyi C. L. Hitchcock & Maguire; *S. petersonii* var. *minor* C. L. Hitchcock & Maguire

Plants perennial, rhizomatous; caudex thick, with many rhizomatous, creeping, branched, slender subterranean shoots, terminating in tight tufts of leaves and erect flowering stems. **Stems** simple, 5–15 cm, pubescent and viscid, with stipitate glands. **Leaves:** basal with blade 1-veined, oblanceolate, broadly spatulate, 1–4 cm × 2–8 mm, apex obtuse to acute, glandular-puberulent throughout, rarely subglabrous adaxially; cauline in 3–6 pairs, sessile, reduced distally, blade lanceolate, narrowly elliptic, or oblanceolate, 2–4 cm × 2–8 mm, apex ± acute, glandular-puberulent throughout. **Inflorescences** usually with solitary terminal flower, sometimes cymose, to 8-flowered, open. **Pedicels** erect or angled near tip with flowers slightly nodding, 1–3 times longer than calyx, glandular-puberulent, often densely so. **Flowers:** calyx prominently 10-veined, campanulate, not contracted proximally around carpophore, 15–20 × 4–8 mm, papery, margins dentate, veins parallel, usually purple tinged, with pale commissures; lobes ovate, 3–5 mm, glandular-puberulent, midrib triangular, margins purple tinged, broad, membranous, apex obtuse; corolla bright pink, clawed, claw equaling calyx, broad and ligulate but abruptly contracted into limb, limb broadly cuneate, shallowly to deeply 2–4-lobed, 5–15 mm, lobes broad or narrow, appendages absent or to 2 mm, margins erose; stamens slightly longer than corolla claw; stigmas 3 (–5), slightly longer than corolla claw. **Capsules** equaling calyx, opening by 6 (or 8 or 10) lanceolate teeth; carpophore 1–2.5 mm. **Seeds** brown, broadly reniform, flattened, 2–2.5 mm, rugose, more coarsely so on margins. $2n = 96$.

Flowering summer. Calcareous gravel, clay, talus, and rocks on ridges, slopes, and barren ground; of conservation concern; 2000–3400 m; Nev., Utah.

This beautiful alpine species is variable with respect to density of pubescence, flower size, and petal structure. As this variation occurs both within and among populations, little useful purpose is served by giving names to it. The Nevada population, which is the basis for the name *Silene clokeyi*, is interfertile (A. R. Kruckeberg 1961) with populations in Utah (the basis for the name *S. petersonii*). Accordingly, a single species is recognized here without infraspecific taxa.

44. **Silene plankii** C. L. Hitchcock & Maguire, Revis. N. Amer. Silene, 56, plate 7, fig. 55. 1947 • Rio Grande fire pink C

Plants perennial, cespitose; taproot stout; caudex with many often subterranean branches, woody. **Stems** ascending, branched, wiry, leafy, slender, 10–20 cm, finely retrorse gray-puberulent. **Leaves** largest in mid-stem region; blade linear to narrowly lanceolate or oblance-olate, 1–4 cm × 1–5 mm, apex sharply acuminate, glandular-puberulent. **Inflorescences** with flowers usually solitary, terminal on branches. **Pedicels** shorter than calyx, glandular-puberulent. **Flowers:** calyx 10-veined, tubular, constricted around carpophore, umbilicate, 20–30 × 3–6 mm, papery, green, glandular-puberulent, lobes lanceolate, 2–4 mm, margins membranous, apex acute; corolla scarlet, clawed, claw equaling calyx, limb obconic, 2-lobed, 7–10 mm, margins entire or crenate, appendages ± lacerate, 1–1.5 mm; stamens exserted, ± equaling corolla lobes; styles 3, exserted, ± equaling corolla lobes. **Capsules** narrowly ellipsoid, equaling calyx, opening by 6 recurved, brittle teeth; carpophore ca. 5 mm. **Seeds** brown, reniform, 1.5 mm, rugose in concentric rings on sides, margins papillate. $2n = 48$.

Flowering summer–early autumn. Crevices in granite and quartzite cliffs; of conservation concern; 1300–2600 m; N.Mex., Tex.; Mexico.

Silene plankii is a close relative of *S. laciniata*, differing in its compact tufted growth, small and narrow leaves, and shallowly two-lobed petals. It is endemic to the Del Carmen Mountains on either side of the Rio Grande valley. Plants of *S. laciniata* with a habit and leaves similar to *S. plankii* but the deeply laciniate petals of *S. laciniata* occur on the cliffs of Santa Cruz Island off the coast of California.

45. **Silene polypetala** (Walter) Fernald & B. G. Schubert, Rhodora 50: 198. 1948 • Fringed campion C E

Cucubalus polypetalus Walter, Fl. Carol., 141. 1788; *Silene baldwinii* Nuttall; *S. fimbriata* Baldwin ex Elliott

Plants perennial, rhizomatous; taproot stout; caudex branched, woody. **Stems** decumbent to ascending, scarcely branched, weak, leafy, with several basal, procumbent, nonflowering shoots, 10–40 cm, softly pilose. **Leaves** 2 per node, blade 2–9 cm × 10–25 mm; proximal petiolate, petiole ciliate, blade broadly oblanceolate; distal sessile, blade obovate to elliptic or lanceolate, spatulate, margins ciliate, apex acute or obtuse, apiculate, glabrous to sparsely pilose. **Inflorescences** terminal, cymose, usually 3-flowered, open, bracteate; bracts foliaceous. **Pedicels** shorter than calyx, pilose, with colorless, large, stipitate glands. **Flowers:** calyx prominently 10-veined, funnelform, umbilicate in fruit, constricted around carpophore, 20–28 × 5–9 mm, broadening distally, papery, thinly pilose, veins parallel, green, with pale commissures, lobes lanceolate, 4–7 mm, apex acute; corolla showy, pink or white, large, clawed, claw equaling calyx, limb fan-shaped, 15–25 mm, laciniate into many narrow segments, appendages absent; stamens equaling calyx; stigmas 3(–4), equaling calyx. **Capsules** ovoid, included in calyx, opening by 6 (or 8) teeth; carpophore 7–10 mm. **Seeds** dark reddish brown, broadly reniform, ca. 1 mm, rugose on sides, margins papillate. $2n = 48$.

Flowering spring. Rich calcareous loam in deciduous forests and on wooded bluffs; of conservation concern; 0–100 m; Fla., Ga.

There is an old collection (*Mohr 4000*, MO) labelled "Alabama" without date. I have no other records of the occurrence of *Silene polypetala* in that state.

46. Silene pseudatocion Desfontaines, Fl. Atlant. 1: 353. 1798 (as pseudo-atocion) [I]

Plants annual; taproot slender. Stems straggling to erect, much-branched, elongate, 20–70 cm, sparsely retrorse-puberulent. Leaves 2 per node, sessile with spatulate, broad, ciliate base, blade oblanceolate, 1.3–5 cm × 4–15 mm, apex acuminate, glabrous or sparsely setose. Inflorescences cymose, open, compound, pedunculate; bracts leaflike, lanceolate, 3–15 mm, apex acuminate; peduncle ascending, elongate, viscid stipitate-glandular. Pedicels ascending, elongate, viscid stipitate-glandular. Flowers: calyx prominently 10-veined, clavate, with long, slender tube surrounding carpophore, 17–20 × 4–6 mm, veins parallel, green or purple with purple stipitate glands, with pale commissures, lobes lanceolate, 2–3 mm, margins ciliate, apex acute; corolla bright pink, clawed, claw slightly longer than calyx, limb obovate, unlobed, ca. 1 cm, appendages 2, oblong, 2 mm, entire; stamens included in tip of calyx tube; styles 3, exserted. Capsules ovoid, slightly longer than calyx, opening by 6 recurved teeth; carpophore 9–10 mm, pubescent. Seeds dull and very dark brown, almost globose, inrolled like a clenched fist, 1–1.3 mm, sides with radiating wrinkles, finely tuberculate abaxially. $2n = 24$ (Balearic Islands).

Flowering spring. Neglected gardens, roadsides, waste places; 0–500 m; introduced; Calif.; Europe (Balearic Islands); n Africa.

Silene pseudatocion is occasionally grown in gardens and rarely occurs as a weed in California. It is similar to another garden escape, *S. pendula*, but it differs in having a calyx tube with a very long, slender base and unlobed petals.

47. Silene rectiramea B. L. Robinson, Bot. Gaz. 28: 134. 1899 [C][E]

Plants perennial; taproot stout; caudex much-branched, woody. Stems several, erect from decumbent base, branched, 20–40 cm, sparsely scabrid-puberulent, stipitate-glandular distally. Leaves 2 per node; basal shortly petiolate, blade oblanceolate, apex acute, sparsely puberulent; cauline sessile, in 3–6 pairs, blade lanceolate to narrowly elliptic-lanceolate, 3–8 cm × 3–8 mm, apex acute, scabrid-puberulent. Inflorescences cymose, open, branches few, ascending, elongate, 5–10 cm, flowers 1–3 per branch, bracteate; bracts lanceolate, 3–5 mm, apex acute. Pedicels usually shorter than or to 1½ times calyx,

scabrous-puberulent, ± viscid. Flowers bisexual and unisexual, all flowers having bisexual flowers and staminate unisexual flowers, 6–8 mm diam.; calyx prominently 10-veined, narrowly campanulate in flower, campanulate in fruit, 7–10(–11) × 4–6 mm, only slightly contracted around carpophore, papery and membranous, sparsely scabrid with short glandular-setose hairs, veins parallel, green, with pale commissures, margins dentate, lobes broadly ovate (as broad as long), 1–2 mm, not ciliate; corolla cream or white, tinged purple, claw slightly exceeding calyx, narrowly triangular, auriculate, puberulent at base, limb oblong, 2-lobed, 2–4 mm, lobes entire, appendages 0.1–0.5 mm and broad, entire; stamens equaling petals; styles 3, equaling petals, reduced in functionally staminate flowers. Capsules ovoid-ellipsoid, equaling calyx, opening by 6 spreading teeth; carpophore ca. 2 mm. Seeds brown, reniform-triangular, ca. 1 mm; papillate mainly around margin.

Flowering April–May. Pinyon pine woodland, dry slopes; of conservation concern; 1800–2100 m; Ariz.

Silene rectiramea is found in the Grand Canyon in Arizona. It is very similar to forms of *S. verecunda* and to *S. thurberi*. It is distinguished by its smaller flowers and fruiting calyces that are not or only slightly constricted around the carpophore and tend to be more sparsely pubescent with broader, shorter lobes. Also the appendages smaller and entire.

48. Silene regia Sims, Bot. Mag. 41: plate 1724. 1815 • Royal catchfly [E]

Melandrium illinoense Rohrbach; *M. reginum* (Sims) A. Braun

Plants perennial; taproot stout, fleshy. Stems several, erect, simple proximal to inflorescence, 50–160 cm, glabrous to sparsely retrorse-puberulent proximally, more densely so and glandular distally. Leaves withered towards base at anthesis, in 10–20 pairs, sessile, gradually reduced distally; blade 3-veined, lanceolate to ovate, rounded into base, 4–12 cm × 20–50 mm, apex acute, not acuminate, glabrous to scabrous-puberulent adaxially, scabrous-puberulent abaxially. Inflorescences terminal, cymose, compound, many-flowered, bracteate, pedunculate, 15–25(–30) × 5–10 cm; bracts lanceolate, 3–30 mm; peduncle ascending. Pedicels 1–4 times longer than calyx, stipitate-glandular, puberulent. Flowers: calyx prominently 10-veined, tubular in flower, 18–25 × 3–4 mm, swelling in middle to 6–9 mm in fruit, umbilicate, constricted near base around carpophore, glandular-pubescent, especially on veins, veins green, lobes 2–4 mm, midrib lanceolate, margins often red tinged, broad, membranous; corolla scarlet, clawed, claw equaling calyx, limb oblong to elliptic, unlobed, rarely emarginate or shallowly 2-lobed, 10–20 × 5–6 mm, apex rounded, appendages 2, tubular, 2–4 mm; stamens exserted; styles 3(–5), exserted. **Capsules** ovoid-ellipsoid, narrowed at both ends, equaling calyx, opening by 6 (or 8 or 10) ascending teeth; carpophore 3–5 mm. **Seeds** dark reddish brown, reniform, 1.5–2 mm, with smooth and glossy, flat or concave sides, margins angled, shallowly rugose. $2n = 48$.

Flowering summer–fall. Dry prairies, rocky open woods, woodland edges and thickets; 100–400 m; Ala., Ark., Ga., Ill., Ind., Kans., Ky., Mo., Ohio, Tenn.

Silene regia is a very distinct, spectacular species with tall stems, a narrow, many-flowered inflorescence, and bright red petals. It is probably related to *S. laciniata* via *S. subciliata*.

Silene regia is in the Center for Plant Conservation's National Collection of Endangered Plants.

49. Silene repens Patrin ex Persoon, Syn. Pl. 1: 500. 1805 • Pink campion C

Silene purpurata Greene; *S. repens* subsp. *australis* C. L. Hitchcock & Maguire; *S. repens* var. *australis* (C. L. Hitchcock & Maguire) C. L. Hitchcock; *S. repens* var. *costata* (Williams) B. Boivin; *S. repens* subsp. *purpurata* (Greene) C. L. Hitchcock & Maguire

Plants perennial; taproot slender; caudex branched, woody, producing subterranean creeping stems and clumps of flowering and vegetative shoots. **Stems** erect to straggling, branched or simple, leafy, 7–35 cm, retrorse-puberulent. **Leaves** 2 per node, sessile, blade 1-veined, linear, lanceolate, or narrowly oblong, tapered to base, apex acute, puberulent on both surfaces. **Inflorescences** cymose, compound, ca. (2–)5–20-flowered, usually compact, bracteate; bracts narrowly lanceolate, 2–15 mm. **Pedicels** ascending, 0.1–1(–2) cm, shorter or longer than calyx, eglandular or with scattered glands, pubescence white, dense, short. **Flowers:** calyx obscurely 10-veined, tubular-campanulate, 10–15 × 3–5 mm in flower, becoming clavate and 5–6 mm broad in fruit, herbaceous, villous, veins purple tinged or entirely purple, without conspicuous pale commissures, lobes ovate, 1–2.5 mm, margins broad, scarious, apex obtuse; corolla rose pink, rarely white, clawed, claw equaling calyx, broadened distally, limb obconic, 2-lobed, 3–7 mm, appendages 0.7–1.2 mm; stamens equaling calyx; styles 3(–4), exserted. **Capsules** ovoid, equaling calyx, opening by 6 (or 8) spreading teeth; carpophore 4–7 mm. **Seeds** grayish brown, not winged, reniform, 0.8–1 mm, rugose on sides, margins shortly papillate. $2n = 24, 48$.

Flowering summer. Arctic and mountain areas, dry grassy slopes, open woods, sagebrush, rocky outcrops, talus, gravel flats; of conservation concern; 0–3200 mm; B.C., N.W.T., Yukon; Alaska, Idaho, Mont., Wash., Wyo.; Eurasia.

Three subspecies have been recognized within *Silene repens* on the basis of stature and the development of purple pigment in the calyx. Northern populations in the flora area have been referred to subsp. *purpurata* because of the unusually heavy pigment of the calyx, while the disjunct populations in the central Rocky Mountains have been recognized as subsp. *australis*. However, both of these forms occur among collections from Eurasia that have been referred to subsp. *repens*. When material from populations of subsp. *purpurata* was grown farther south, it took on the appearance of subsp. *australis*, suggesting that the differences are under environmental influence.

50. Silene rotundifolia Nuttall, Gen. N. Amer. Pl. 1: 288. 1818 • Round-leaved catchfly E

Melandrium rotundifolium (Nuttall) Rohrbach

Plants perennial; taproot stout, fleshy; caudex branched. **Stems** several, straggling to erect, freely branched, 2–7 cm, pilose and glandular, sparsely so proximally. **Leaves:** basal leaves withered at time of flowering, distal sessile, proximal petiolate, largest on mid to distal stem; blade subrotund to broadly ovate-lanceolate, 3–10 cm × 20–70 mm, base cuneate into petiole, apex short-acuminate, sparsely short-pilose. **Inflorescences** cymose, open, few-flowered, leafy, bracteate; bracts resembling distal leaves. **Pedicels** 1–3(–4) cm, viscid, with long septate-glandular hairs. **Flowers:** calyx indistinctly veined, tubular, broadened distally, constricted towards base around carpophore, ± umbilicate, 20–25 × 5–8 mm, herbaceous, glandular-pilose, lobes triangular, 3–4 mm, margins narrow, membranous, ciliate; corolla scarlet, clawed, claw equaling calyx, limb deeply 2-lobed, 10–15 mm, lobes lanceolate, sometimes with 2 smaller lateral teeth, ciliate, appendages saccate, 1–1.5 mm, with clear area abaxially; stamens shortly exserted; styles 3, shortly exserted. **Capsules** narrowly ellipsoid, not distending calyx, included within it, opening by 6 teeth; carpophore 6–8 mm. **Seeds** gray, broadly reniform, plump, ca. 1 mm, shallowly papillate. $2n = 48$.

Flowering late spring–summer. Woodlands, partially shaded cliffs and bluffs; 200–600 m; Ala., Ga., Ky., Ohio, Tenn., W.Va.

Silene rotundifolia is clearly related to *S. laciniata* but is a well-marked species of the deciduous forest region.

51. Silene sargentii S. Watson, Proc. Amer. Acad. Arts 14: 290. 1879 • Sargent's catchfly [E]

Lychnis californica S. Watson;
Silene lacustris Eastwood;
S. watsonii B. L. Robinson

Plants perennial; taproot stout; caudex much-branched, woody, producing many decumbent, leafy, short shoots and erect flowering shoots, often densely cespitose. **Stems** erect, decumbent at base, simple, branched in flowering region, slender, 10–20(–25) cm, sparsely pubescent and glandular. **Leaves** mostly basal; basal marcescent, densely tufted, long-petiolate, blade linear-oblanceolate, 1–3 cm × 0.5–3 mm, somewhat fleshy, setose-puberulent on both surfaces; cauline in 1–3 pairs proximal to inflorescence, reduced distally, blade linear, 1–4 cm × 0.5–2 mm. **Inflorescences** 1–3-flowered, with terminal flower, usually with 1 or 2 flowers at proximal nodes. **Pedicels** erect, sometimes slightly bent at apex, ¼–2 times calyx, viscid glandular-puberulent, septa of hairs colorless. **Flowers:** calyx prominently 10-veined, narrowly campanulate, ± umbilicate, not contracted proximally around carpophore, 9–18(–22) × 3–6 mm, papery, glandular-puberulent, viscid, hairs with colorless septa, veins parallel, purple, with pale commissures, lobes with midrib present, triangular, 2–3 mm, margins purple tinged, membranous, often broadened distally into round, crenulate lobe; corolla off-white, usually pink or purple tinged, to 1½ times longer than calyx, limb 2-lobed, 2–3 mm, sometimes with 2 small lateral teeth, appendages 2, ovate, 1–1.5 mm; stamens slightly longer than calyx; styles 3(–5), equaling petals. **Capsules** included in calyx, opening by 6 (or 8 or 10) ascending teeth; carpophore 1.5–3 mm, woolly. **Seeds** brown, reniform, ca. 1.5 mm, rugose on both surfaces, margins with large, inflated papillae. $2n = 48$.

Flowering summer. Alpine grassy, gravelly, or rocky slopes and ridges, openings in subalpine forests, sagebrush, and on juniper slopes; 2400–3800 m; Calif., Idaho, Nev., Wash.

Silene sargentii is very similar to *S. suksdorfii*, but that species has purple-septate hairs on the calyx and pedicels, whereas those on *S. sargentii* are colorless (see M. A. T. Showers 1987). Also, the cauline leaves of *S. suksdorfii* are narrowly oblanceolate rather than linear, and the basal leaves are marcescent. *Silene sargentii* may be confused with *S. bernardina* and can intergrade with it in Nevada. However, that species is usually larger with a longer, tubular calyx and petals that are deeply divided into 4–6 narrow lobes, unlike *S. sargentii*, which has 2-lobed petals.

52. Silene scaposa B. L. Robinson, Proc. Amer. Acad. Arts 28: 145. 1893 • Robinson's catchfly [E]

Silene scaposa var. *lobata*
C. L. Hitchcock & Maguire

Plants perennial, subscapose, cespitose; taproot stout; caudex branched, woody. **Stems** several, erect, simple, 15–50 cm, puberulent, viscid-glandular distally. **Leaves** mostly basal; basal marcescent, long-petiolate, densely tufted, blade 1-veined, narrowly oblanceolate, 2–10 (–20) cm × 2–12(–20) mm, not fleshy, base tapering to petiole, apex acute to obtuse, finely puberulent on both surfaces; cauline in 1–3 pairs, sessile, much reduced, blade linear-lanceolate, not fleshy. **Inflorescences** 1–5 (–7)-flowered, with terminal flower and lateral, open, pedunculate cymes often reduced to single flowers, bracteate; bracts narrowly lanceolate, 3–10(–20) mm. **Pedicels** erect, elongate, 0.5–4.5 cm, glandular-puberulent. **Flowers:** calyx prominently 10-veined, those to lobes lance-shaped broadened and thickened distally, commissural veins slender, not forked distally, campanulate, 10–12 × 3.5–5 mm in flower, enlarging to 15 × 10 mm in fruit, not contracted around carpophore, papery, margins dentate, glandular-pubescent, viscid, veins parallel, with pale commissures, lobes patent, ovate, 1.5–4 mm, rigid, margins broad, membranous; corolla off-white to dingy purple-red, clawed, claw exceeding calyx, ciliate proximally, broadened distally, limbs erect, 2–4-lobed, less than ½ length of calyx, lobes 2–5 mm, appendages 2–4, 0.5–1 mm; stamens slightly exserted; filaments lanate, expanded at base; styles 3–5, ± equaling calyx. **Capsules** slightly longer than calyx, opening by 3–5 teeth; carpophore 1.5–2.5 mm. **Seeds** brown, reniform, 1.2–2 mm, margins with large, inflated papillae, rugose on sides. $2n = 48$.

Flowering early summer. Subalpine grassy, gravelly, or rocky slopes, ponderosa pine forests, juniper scrub, sagebrush; 900–3000 m; Colo., Idaho, Nev., Oreg.

Silene scaposa is a very distinct species with its subscapose inflorescence, coronalike ring of short petals, and distended fruiting calyx in which the veins to the lobes are markedly broadened and lanceolate. Variation in lobing of the corolla has been the basis for recognizing two varieties: var. *scaposa* (var. *typica* C. L. Hitchcock & Maguire), which has two-lobed petals, and var. *lobata*, which has four-lobed petals. However, these differences appear to be of little significance.

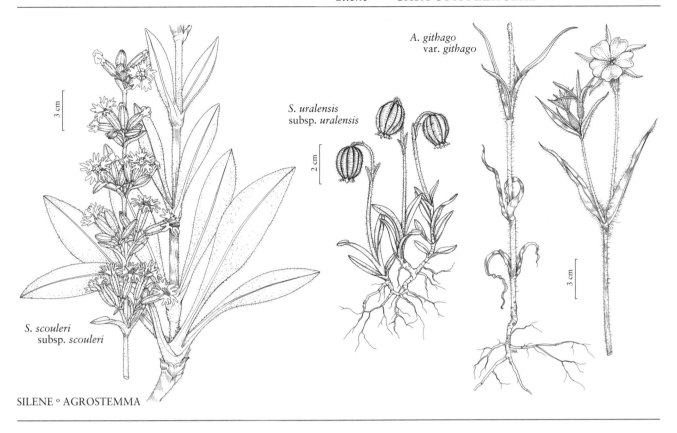

S. scouleri
subsp. *scouleri*

S. uralensis
subsp. *uralensis*

A. githago
var. *githago*

SILENE ∘ AGROSTEMMA

53. Silene scouleri Hooker, Fl. Bor.-Amer. 1: 88. 1830

F

Plants perennial; taproot stout; caudex branched, woody, crowns 1–several. **Stems** erect, simple proximal to inflorescence, slender or stout, 10–80 cm, puberulent. **Leaves** 2 per node; basal petiolate, blade oblanceolate, 6–25 cm × 4–30 mm, retrorsely puberulent on both surfaces; cauline in 1–12 pairs, usually sessile, blade well developed, lanceolate to ovate-lanceolate, oblanceolate, or rarely linear or linear-lanceolate. **Inflorescences** cymose, pseudo-racemose, or rarely paniculate, erect or nodding, with 1–12 flowering nodes, 2–20-flowered, open or dense, flowers paired or in many-flowered whorls, bracteate, cymes often sessile; bracts 3–60 mm. **Pedicels** becoming deflexed at base of calyx, ¼–2 times calyx, glandular-pubescent. **Flowers** shortly pedicellate or sessile; calyx prominently 10-veined, campanulate or tubular in flower, clavate, turbinate, or fusiform in fruit, constricted or not at base around carpophore in fruit, 8–20 × 3–8 mm, veins parallel, purplish or green, with pale commissures; lobes lanceolate, 2–5 mm, apex obtuse with broad membranous margin and tip; corolla white, greenish white, or pink, sometimes tinged pink or purple, clawed, claw longer than calyx, limb deeply 2–4-lobed, often with smaller lateral teeth, 2.5–8 mm, appendages 1–3

mm; stamens ± equaling corolla claw; styles 3–4, ± equaling corolla claw. **Capsules** ovoid to ellipsoid, equaling or slightly longer than calyx, opening by 6 or 8 teeth; carpophore 1.5–6 mm. **Seeds** brown or grayish brown, reniform, 1–1.5 mm, margins papillate, rugose on sides.

Subspecies 3 (3 in the flora): w North America, Mexico.

Silene scouleri is a very complex species that appears to be in the process of diverging into at least three different entities. Subspecies *scouleri* is a plant of the Pacific coast and lowlands. It has tall, stiffly erect stems, lanceolate to broadly lanceolate leaves, and a viscid inflorescence with many-flowered whorls of almost sessile flowers ranging in color from greenish white to rich pink. At the other extreme is subsp. *pringlei*, a plant of the mountains in Mexico extending northwards into Arizona and New Mexico. It has slender, somewhat nodding flowering stems with very narrow leaves. The flowers are usually paired at each node and secund on slender pedicels about equaling the calyx in length. The petals are off-white, sometimes tinged with dusky purple. Between the two extremes is subsp. *hallii*, a short, stocky plant of the Rocky Mountains and foothills with a few-flowered inflorescence. It has a larger, campanulate calyx, and some of the flowers usually become deflexed. Differentiation among these three forms is incomplete and plants indeterminate to subspecies are frequently encountered in areas away from the main distribution centers of the three subspecies. In northern Oregon and Idaho there appear to be populations connecting *S. scouleri* with *S. oregana*. They have some of the characteristics of *S. oregana* but not its laciniate petals. They may represent a more luxuriant form growing in taller vegetation, but their status needs further study.

1. Calyces campanulate, not or only slightly clavate in fruit, 13–18(–20) × (5–)6–8 mm; inflorescences with (1–)3–6(–8) flowering nodes; plants 10–40 cm; pedicels stout . . . . . 53b. *Silene scouleri* subsp. *hallii*
1. Calyces tubular to narrowly clavate in flower, clavate, turbinate, or fusiform in fruit, (8–)10–16 × 3.5–7 mm; inflorescences with 3–12 flowering nodes; plants 20–80 cm; pedicels slender.
   2. Inflorescences erect, flowers in dense pseudowhorls of usually sessile, (2–)5–20-flowered cymes, both sessile and pedicellate flowers in each cyme; pedicels erect . . . . . .
   . . . . 53a. *Silene scouleri* subsp. *scouleri*
   2. Inflorescences nodding, flowers usually paired at each node, all pedicellate; pedicels ± equaling calyx, often sharply deflexed at base of calyx . . . . 53c. *Silene scouleri* subsp. *pringlei*

53a. Silene scouleri Hooker subsp. **scouleri**
• Scouler's catchfly [E] [F]

Silene grandis Eastwood; *S. pacifica* Eastwood; *S. scouleri* subsp. *grandis* (Eastwood) C. L. Hitchcock & Maguire; *S. scouleri* var. *pacifica* (Eastwood) C. L. Hitchcock

Stems ± unbranched, stout, 20–80 cm, densely puberulent, glandular-puberulent and viscid distally. **Leaves:** basal petiolate, blade lanceolate to ovate-lanceolate, 6–17 cm × 6–30 mm (including petiole), base cuneate into petiole, apex acute, appressed-puberulent on both surfaces; cauline in (3–)5–8(–12) pairs, distal leaves sessile, reduced, blade lanceolate, proximal oblanceolate. **Inflorescences** erect, cymose, elongate with 4–12 nodes, each with dense pseudowhorl of (2–)5–20 flowers, densely glandular-puberulent, viscid; both sessile and pedicillate flowers in each cyme, rarely with short, erect peduncle shorter than to equaling calyx; bracts lanceolate, 5–60 mm. **Pedicels** slender, shorter than to equaling calyx in flower, some to 1¹/₂ times calyx in fruit, densely glandular-pubescent, viscid, some flowers may be sessile. **Flowers:** calyx tubular to narrowly campanulate and somewhat clavate, (8–)10–15 × 3.5–5 mm in flower, broadening to 5–7 mm and clavate in fruit, constricted at base around carpophore, glandular-pubescent with purple-septate hairs, viscid, veins purplish, lobes lanceolate, 2–4 mm, membranous, margins suffused with purple, apex broadly obtuse and membranous; corolla greenish white

tinged with pink to dull pink, claw slightly longer than calyx, broadened distally, ciliate at base, limb ± obovate, 2–4-lobed, 3–5 mm, often with small lateral teeth, appendages 2, oblong, 1–3 mm, margins erose; stamens equaling petals; styles equaling petals. **Capsules** ovoid, equaling calyx; carpophore 3–6 mm. **Seeds** gray-brown, reniform, 1–1.5 mm. **2***n* = 48.

Flowering summer. Coastal bluffs, rocky and grassy slopes, dry prairie, woodlands; 0–3800 m; B.C.; Calif., Oreg., Wash.

Although subsp. *scouleri* is primarily a taxon of the Pacific coast and foothills, plants approaching this subspecies occur at higher elevations in Idaho, Wyoming, and the interior of Washington and Oregon. However, the characters of plants of this subspecies growing in inland habitats often tend to be less extreme than those in plants from near the coast.

53b. Silene scouleri Hooker subsp. **hallii** (S. Watson) C. L. Hitchcock & Maguire, Revis. N. Amer. Silene, 26. 1947 • Hall's catchfly [E]

Silene hallii S. Watson, Proc. Amer. Acad. Arts 21: 446. 1886; *Lychnis elata* S. Watson; *Silene concolor* Greene; *S. scouleri* var. *concolor* (Greene) C. L. Hitchcock & Maguire

Stems stout, 10–40 cm, glandular-puberulent. **Leaves:** basal broadly petiolate, petiole ciliate, blade oblanceolate, 7–15 cm × 7–18(–30) mm (including petiole), base cuneate, apex acute, puberulent on both surfaces; cauline in (1–)2–3(–6) pairs, sessile, blade lanceolate to elliptic-oblanceolate, 1.5–13 cm × 3–15 (–30) mm, base cuneate. **Inflorescences** erect, cymose, sturdy, with (1–)3–6(–8) flowering nodes, 3–10(–20)-flowered; cymes usually sessile; bracts lanceolate, usually 3–30 mm. **Pedicels** ascending in flower, often spreading to reflexed proximal to calyx in fruit, stout, ¹/₄–1(–1¹/₂) times length of calyx, ± equaling calyx in fruit, viscid glandular-pubescent. **Flowers:** calyx campanulate, not or only slightly clavate in fruit, 13–18(–20) × (5–)6–8 mm, veins green or purple tinged, often heavily so, those to lobes lanceolate, much-broadened distally, viscid glandular-pubescent, lobes 3–5 mm, margins membranous, apex acute with obtuse, broad, ciliate margins; corolla white or greenish white, pink tinged to purple tipped, claw lanceolate, broadened distally, ciliate, limb 2-lobed, 2.5–8 mm, sometimes with lateral teeth, appendages 1–3 mm; stamens equaling petals; styles equaling petals. **Capsules** ellipsoid, equaling calyx; carpophore 1.5–3.5 mm. **Seeds** grayish brown, reniform, ca. 1 mm. **2***n* = 48, 96.

Flowering summer. Grasslands, woodlands, open forests, alpine meadows; 800–3800 m; Alta., B.C.; Colo., Idaho, Mont., N.Mex., Oreg., Wash., Wyo.

The main center of distribution of subsp. *hallii* is Colorado, but plants referable to or approaching this subspecies occur along the Rocky Mountains from New Mexico to southern British Columbia and Alberta.

53c. Silene scouleri Hooker subsp. **pringlei**
(S. Watson) C. L. Hitchcock & Maguire, Revis. N. Amer. Silene, 26. 1947 • Pringle's catchfly

Silene pringlei S. Watson, Proc. Amer. Acad. Arts 23: 269. 1888; *S. scouleri* var. *eglandulosa* C. L. Hitchcock & Maguire; *S. scouleri* var. *grisea* C. L. Hitchcock & Maguire; *S. scouleri* var. *leptophylla* C. L. Hitchcock & Maguire

Stems slender, 20–70 cm, densely puberulent with deflexed, gray, short, eglandular hairs intermixed with stipitate-glandular hairs. **Leaves:** blade finely puberulent on both surfaces, often sparsely so; basal tufted, blade narrowly oblanceolate, 8–12(–25) cm × 4–20 mm (including petiole), base cuneate into petiole, apex acute; cauline in 3–8 pairs, ± connate basally, reduced distally, ± sessile, blade narrowly lanceolate, 2–10(–17) cm × 3–10 mm, ciliate at base. **Inflorescences** tending to be nodding with secund flowers, usually with 3–7 flowering nodes, slender, elongate, 4–30(–60) cm, retrorsely gray-puberulent, glandular or not, not strongly viscid, distal nodes often with cymes sessile and reduced to 2 pedicellate flowers, cymes of proximal nodes sometimes pedunculate on peduncles to 10 cm, with 2–5 pedicellate flowers; peduncle ascending; bracts leaflike, 4–20 mm. **Pedicels** ascending, frequently sharply deflexed at base of calyx, slender, ± equaling calyx. **Flowers:** calyx tubular to narrowly campanulate in flower, ± umbilicate, 11–14 (–16) × 3–5(–6) mm, becoming turbinate or fusiform in fruit, puberulent, glandular with short-stipitate glands, veins almost always green, those of sinus slender, shorter than tube, those of lobes lanceolate, broadened distally, lobes 3–4 mm, margins membranous, apex obtuse with broad, membranous, ciliate tip; corolla white, often suffused with purple, sometimes greenish, clawed, claw lanceolate, broadened distally, ciliate, limb deeply 2–4-lobed, 5–8 mm, lobes narrow, usually with smaller lateral teeth, appendages 1–2.5 mm; stamens equaling claw; styles shortly exserted. **Capsules** ovoid-ellipsoid, slightly longer than calyx; carpophore 2–5 mm. **Seeds** brown, ± reniform, ca. 1 mm; papillae inflated, large. *2n* = 60.

Flowering late summer. Subalpine meadows, open woodlands and scrub, rocky hillsides and canyons; 2000–3300 m; Ariz., N.Mex., Utah; Mexico.

Most of the collections of subsp. *pringlei* from Arizona tend to have larger calyces that suggest an affinity with Colorado material of subsp. *hallii*. Although subsp. *pringlei* is primarily a plant of high elevations in the arid regions of the southwest, its influence appears to extend as far north along the Rocky Mountains as the Canada–United States border, with many plants there showing some of the characteristics of this taxon.

54. Silene seelyi C. V. Morton & J. W. Thompson,
Torreya 33: 70. 1933 • Seely's catchfly or silene
C E

Anotites seelyi (C. V. Morton & J. W. Thompson) W. A. Weber

Plants perennial; taproot slender; caudex with much-branched crown, finely pubescent throughout with mainly glandular hairs. **Stems** numerous, decumbent to ascending, branched, tufted, leafy, slender, 5–30 cm. **Leaves** 2 per node, sessile or nearly so, blade reticulate-veined, lanceolate to ovate-lanceolate, thin, 0.8–2 cm × 3–8 mm, broadest proximally, base rounded, apex acute. **Inflorescences** cymose, open, compound, leafy, single flowers borne terminally and in axils of distal leaves; bracteoles, when present, 2. **Pedicels** straight, slender, 1/2–3 times longer than calyx. **Flowers:** calyx green, obscurely 10-veined, narrowly campanulate, in fruit 6–9 × 3–4 mm, herbaceous, pubescent, veins without conspicuous pale commissures; lobes triangular, 2–3 mm; corolla dark red, sometimes white, clawed, claw slightly longer than calyx, limb deeply 2-lobed, 2–3 mm, appendages 2, very small; stamens exserted; styles 3, exserted. **Capsules** ellipsoid, included in calyx, opening by 6 teeth; carpophore ca. 1.5 mm. **Seeds** brown, not winged, broadly reniform, flattened, ca. 0.8 mm, reticulate. *2n* = 24.

Flowering summer. Crevices and ledges on granite and basalt cliffs; of conservation concern; 800–1800 m; Wash.

Silene seelyi is confined to the Wenatchee Mountains. It closely related to *S. menziesii*, but is distinguished by its usually dark red flowers and broadly lanceolate leaves.

Silene seelyi is in the Center for Plant Conservation's National Collection of Endangered Plants.

55. **Silene serpentinicola** T. W. Nelson & J. P. Nelson, Madroño 51: 384, fig. 1. 2004 • Serpentine Indian pink C E

Plants perennial, rhizomatous; taproot stout, rhizomes thin, branching. **Flowering shoots** 4–10(–15) cm, softly pubescent. **Leaves:** cauline in 4–8 pairs, crowded; blade gray-green, oblanceolate to obovate, spatulate, 2.5–4.5 cm × 5–15 mm, longest near middle of stem, sparsely pubescent on both surfaces, reduced and bractlike on subterranean base. **Inflorescences** terminal, 1–3(–4)-flowered cymes, densely glandular-pubescent; bracts leaflike, (0.5–)0.7–1.1 cm. **Pedicels** ascending and straight, (5–)7–10 mm, glandular-pubescent. **Flowers** ca. 30 mm diam.; calyx purple tinged, distinctly 10-veined, tubular, inflated and expanding in fruit, 13–17 mm, densely glandular-pubescent, lobes lanceolate; corolla scarlet, clawed, limb carmine red, turning purple on drying, ± equally deeply 2-lobed, each lobe with lateral tooth, ca. 11 mm, glabrous, claw narrowly obtriangular, equaling calyx, appendages 2, prominent, petaloid, linear, truncate, 2.5–4.5 mm; stamens long-exserted; stigmas 3, long-exserted. **Capsules** ovoid to oblong, equaling calyx, (8–)12–15 mm; carpophore 0.5–1 mm. **Seeds** dark brown, reniform, 1.8–2 mm diam., strongly papillate. $2n = 72$.

Flowering early summer. Grassy, gravelly, or rocky openings in chaparral, woodlands, and coniferous forest on serpentine; of conservation concern; 100–800 m; Calif.

Silene serpentinicola is a recently described endemic of the serpentines of the Smith River basin of northwestern Del Norte County and probably occurs on the same rock system across the border in Oregon. It differs from *S. hookeri* in flower color, and from both *S. hookeri* and *S. laciniata* subsp. *californica* in its erect, more or less solitary flowering stems and large, clearly visible petaloid appendages in the flowers.

56. **Silene sibirica** (Linnaeus) Persoon, Syn. Pl. 1: 497. 1805 • Siberian catchfly I

Cucubalus sibiricus Linnaeus, Syst. Nat. ed. 10, 2: 1031. 1759

Plants perennial, producing several decumbent, short, woody, underground shoots bearing perennating buds and terminated by erect flowering shoots; taproot stout, woody; caudex crown branched. **Stems** simple but with short axillary leafy shoots, leafy, 40–60 cm, glossy, ± glabrous, with 10–20 nodes proximal to inflorescence. **Leaves:** basal withered at time of flowering; cauline whorled, appearing fasciculate (more than 4) at each node, connate at base, reduced distally, tapered at both ends, blade linear to very narrowly elliptic-lanceolate, 0.5–8 cm × 1–12 mm, base broadened, margins scabrous, apex acute, subglabrous to minutely puberulent on both surfaces. **Inflorescences** thyrsate, verticillate, open, with many short, ascending branches bearing large numbers of small, unisexual flowers, bracteate; bracts connate basally, triangular, 2–10 mm, margins finely ciliate, apex acuminate. **Pedicels** ascending, straight, fasciculate, wiry, ± equaling calyx, glabrous or sparsely and minutely puberulent. **Flowers** mostly unisexual; calyx pale green, indistinctly 10-veined, elliptic, constricted near base, umbilicate, ca. 5 × 2 mm, lobes with thickened midrib, ovate-obtuse, 0.2 mm, margins crenate, membranous; corolla greenish white, clawed, claw and limb not differentiated, narrowly oblong, unlobed, less than 2 times length of calyx, appendages absent; stamens exserted, longer than corolla; styles 3, exserted, longer than corolla. **Capsules** green, ovoid-conic, longer than calyx, opening by 6 teeth; carpophore shorter than 1 mm. **Seeds** brown, winged, reniform, ca. 1 mm, faces with shallow radiating ridges.

Flowering summer. Arable land, pastures, roadsides, sand dunes; 300–500 m; introduced; Sask.; Europe (Ukraine); Asia (Mongolia).

Silene sibirica is a very distinct species with masses of small flowers on stiffly erect stems, and narrow, fascicled leaves. It was abundant and persisted for a number of years in the 1950s near Duck Lake and Bladworth in central Saskatchewan.

57. **Silene sorensenis** (B. Boivin) Bocquet, Candollea 22: 21. 1967 • Three-flowered campion E

Lychnis sorensenis B. Boivin, Canad. Field-Naturalist 65: 6. 1951; *Agrostemma triflorum* (R. Brown ex Sommerfelt) G. Don; *Lychnis affinis* J. Vahl ex Fries var. *triflora* (R. Brown ex Sommerfelt) Hart; *L. triflora* R. Brown ex Sommerfelt, not *Silene triflora* Bornmüller; *Melandrium triflorum* (R. Brown ex Sommerfelt) J. Vahl; *Wahlbergella triflora* (R. Brown ex Sommerfelt) Fries

Plants perennial, cespitose; taproot long, stout; caudex usually branched. **Stems** simple below flowering region, stout, 5–30 cm, pubescent, viscid-glandular, densely so distally, ciliate at nodes, hairs with purple septa. **Leaves:** basal petiolate, tufted, petiole to length of blade, broad, blade oblanceolate, 1–8 cm × 2–8 mm, fleshy, base blunt, tapering into petiole, margins ciliate, apex ± acute, glabrous (rarely pubescent) on both surfaces; cauline in 1–2 pairs, sessile, connate proximally, blade narrowly oblong-lanceolate, 5–30 × 1.5–5 mm, apex purple-tipped,

± acute, pubescence as in basal leaves. **Inflorescences** cymose, single, terminal, congested, 1–3-flowered, bracteate, rarely with 1 or 2 flowers in axil of mid-stem leaves (occasionally branched with 2 or 3 erect, elongate branches), densely woolly with purple septate hairs of various lengths, longest equaling pedicel diam.; bracts leaflike, lanceolate, 4–10 mm. **Pedicels** stout, usually much shorter than calyx, rarely to 2 times as long, or flowers sessile. **Flowers:** calyx broadly 10-veined, ovate-campanulate, ca. 10 × 6 mm in flower, enlarging to 15 × 10 mm in fruit, base round, narrowed to ca. ¹/₂ its diam. at mouth, margins dentate, teeth purple, ovate-obtuse, ca. 2 mm, pubescence densely glandular, viscid, partially obscuring the broad veins; corolla white to dingy pink, clawed, claw equaling calyx, limb obovate, 2-lobed, 3–5 mm, appendages 2, oblong, ca. 1 mm, margins crenulate; stamens equaling petals; styles 5, equaling petals. **Capsules** included in calyx, dehiscing by 5 teeth, often splitting into 10; carpophore 1–1.5 mm. **Seeds** brown, not winged, triangular-reniform, ca. 1 mm, spinose-papillate. $2n = 72$.

Flowering summer. Arctic tundra in gravel and clay; 0–300 m; Greenland; N.W.T., Nunavut.

Silene sorensenis usually is readily separable from most other arctic silenes by the dense purplish pubescence that tends to obscure its calyx venation, the nonwinged seeds, and the congested flowers. Specimens of *S. taimyrensis* in the western arctic can resemble *S. sorensenis* but are distinguishable by their smaller seeds and calyx, more-slender stems, and hairs that are shorter than the diameter of the pedicel. Apparent hybrids with *S. involucrata* are occasionally encountered. A. Nygren (1951) considered *S. sorensenis* to be of amphidiploid origin involving *S. uralensis* and *S. involucrata*.

58. **Silene spaldingii** S. Watson, Proc. Amer. Acad. Arts 10: 344. 1875 • Spalding's catchfly or campion

C E

Plants perennial, viscid; taproot stout; caudex branched, woody, producing several to many shoots. **Stems** erect, branched, leafy, 20–60 cm, villose-tomentose, viscid-glandular. **Leaves** 2 per node, connate proximally, sessile, largest in mid stem; blade lanceolate, 3–7 cm × 5–15 mm, apex acute, glandular-tomentose throughout. **Inflorescences** open, leafy cymes, bracteate, viscid and glandular-tomentose, branches ascending, mostly floriferous, flowers terminal and at distal nodes; bracts leaflike, 5–30 mm. **Pedicels** shorter than calyx. **Flowers:** calyx obscurely 10-veined, tubular-campanulate, 10–15 × 4–5 mm in flower,

becoming clavate and 15–20 × 6–8 mm in fruit, narrowed toward base around carpophore, herbaceous, viscid-pubescent, veins more distinct at base, without conspicuous pale commissures, lobes narrowly lanceolate, 3–6 mm, margins very narrow, membranous, apex blunt; corolla greenish white, clawed, claw equaling calyx, widened distally, limb emarginate, 2 × 4 mm, appendages 4(–6), ca. 0.5 mm; stamens equaling petals; styles 3, equaling petals. **Capsules** ellipsoid, slightly longer than calyx, opening by 6 teeth; carpophore 1.5–2.5 mm. **Seeds** yellowish brown, winged, reniform, ca. 2 mm, rugose; wing broad, wrinkled. $2n = 48$.

Flowering summer. Mixed prairie and ponderosa pine forests in swales and on dry hillsides; of conservation concern; 800–1100 m; B.C.; Idaho, Mont., Oreg., Wash.

59. **Silene stellata** (Linnaeus) W. T. Aiton in W. Aiton and W. T. Aiton, Hortus Kew. 3: 84. 1811 • Starry campion, widow's frill E

Cucubalus stellatus Linnaeus, Sp. Pl. 1: 414. 1753; *Silene scabrella* (Nieuwland) G. N. Jones; *S. stellata* var. *scabrella* (Nieuwland) E. J. Palmer & Steyermark

Plants perennial; taproot thick; caudex branched. **Stems** several, simple proximal to inflorescence, 30–80 cm, puberulent, becoming subglabrous near base. **Leaves** withering proximally, in whorls of 4, ± sessile to short-petiolate, largest in mid-stem region; blade lanceolate to ovate-lanceolate, 3–10 cm × 4–40 mm, apex acuminate, puberulent on both surfaces, sparsely so adaxially. **Inflorescences** paniculate, open, bracteate, bracteolate, branches elongate, puberulent; bracts and bracteoles linear-lanceolate, 2–15 mm. **Pedicels** straight, often with 1 or 2 pairs of bracteoles, slender, ¹/₂–3 times calyx, glabrous or scabrous-puberulous. **Flowers:** calyx obscurely 10-veined, broadly campanulate, becoming obtriangular in fruit, 7–11 × 6–10 mm, herbaceous, margins dentate, very narrow, membranous, sparsely puberulent, lobes broadly triangular, 2–3 mm; corolla white, ca. 2 times longer than calyx, limb obtriangular, narrowed into claw, divided ca. ¹/₂ its length into 4–12 lobes, appendages absent; stamens equaling petals; styles 3, longer than petals. **Capsules** globose, opening by 3 broadly triangular teeth; carpophore 2–3 mm. **Seeds** dark brown, reniform, ca. 1 mm, papillate. $2n = (34), 48$.

Flowering summer. Rich deciduous woods, river flats, tall-grass prairies; 0–1300 m; Ala., Ark., Conn., Del., D.C., Ga., Ill., Ind., Iowa, Kans., Ky., La., Md., Mich.,

Minn., Miss., Mo., Nebr., N.J., N.Y., N.C., Ohio, Okla., Pa., R.I., S.C., S.Dak., Tenn., Tex., Vt., Va., W.Va., Wis.

Silene stellata is a very distinct species with its broadly lanceolate leaves in groups of four at each node, and its brilliant white, multilobed petals. Two varieties are recognized by some workers: var. *stellata*, with glabrous pedicels; and var. *scabrella*, with scabrous pedicels. The former tends to have longer, more slender pedicels and be more common towards the northeast, whereas the latter tends to be more western. The correlation of characters and distribution is poor, however, and intermediate plants are often encountered.

Silene stellata was collected near the Grand River, Cambridge, Ontario, in 1941, but was probably introduced there and has not been seen since.

60. **Silene subciliata** B. L. Robinson, Proc. Amer. Acad. Arts 29: 327. 1894 • Prairie-fire pink [E]

Plants perennial; taproot thick; caudex branched. **Stems** erect, scarcely branched, 20–100 cm, glabrous. **Leaves:** blade linear to narrowly oblanceolate, 3–16 cm × 3–12 mm, ± fleshy, base tapering into short petiole, apex acute, glabrous except for few cilia at base. **Inflorescences** cymose, terminal and axillary, elongate, 1–3-flowered, open, bracteate, bracteolate, pedunculate; peduncle slender, 2–10 cm; bracts and bracteoles much-reduced, linear-lanceolate, often ciliate. **Pedicels** slender, 2–10 cm. **Flowers:** calyx tubular, 17–22 × 4–5 mm in flower, broadening to 7 mm in fruit but contracted around carpophore, glabrous, lobes lanceolate, 3–4 mm, margins membranous and ciliate near broad, obtuse apex; corolla scarlet, 2–2¹/₂ times longer than calyx, limb lanceolate, narrowed into claw, 4–5 mm wide, margins entire or shallowly dentate, appendages linear, 4–5 mm; stamens exserted; styles 3, exserted. **Capsules** clavate, equaling calyx, opening by 6 teeth; carpophore 3–4 mm. **Seeds** brown, broadly reniform, 2–2.3 mm, rugose. *2n* = 48.

Flowering late summer–fall. Sandy soil, open woodlands, river banks; 10–200 m; La., Tex.

Silene subciliata, a very rare species, is closely related to *S. laciniata* and *S. virginica* but is readily distinguished by its subglabrous stems and leaves and by having unlobed petals.

Silene subciliata is in the Center for Plant Conservation's National Collection of Endangered Plants.

61. **Silene suecica** (Loddiges) Greuter & Burdet, Willdenowia 12: 190. 1982 • Alpine pink, lychnide alpine

Lychnis suecica Loddiges, Bot. Cab. 9: plate 881. 1824; *L. alpina* Linnaeus 1753, not *Silene alpina* Gray 1821; *Steris alpina* (Linnaeus) M. Sourkova; *Viscaria alpina* (Linnaeus) G. Don; *V. alpina* subsp. *americana* (Fernald) Böcher

Plants perennial, cespitose, glabrous to sparsely pubescent, nonviscid; taproot stout. **Stems** erect, simple, 5–35 cm, glabrous or very sparsely short-pubescent. **Leaves:** basal crowded, blade narrowly oblanceolate, 1–5 cm × 1–5 mm, tapered into broad ciliate base, apex acute; cauline in 2–5 pairs, sessile, connate proximally, blade narrowly lanceolate, 1–4 cm × 2–7 mm, margins ciliate, apex acute. **Inflorescences** cymose, congested, 6–30-flowered, bracteate, pedunculate, often with smaller pedunculate branches in distal nodes; bracts purple, lanceolate, 2–20 mm; peduncle glabrous to sparsely puberulent. **Pedicels** glabrous to sparsely puberulent. **Flowers** sessile or short-petiolate, 5–10 mm diam.; calyx purple, faintly 10-veined, campanulate, 4–6 × 3–5 mm, base attenuate into pedicel, lobes ovate, 1–1.5 mm, margins broad, membranous, apex obtuse; corolla bright pink (rarely white), limb spreading, 2-lobed to middle, 3.5–7 mm, cuneate into claw, ca. 1¹/₄–1¹/₂ times calyx, appendages absent; stamens ca. equaling petals; stigmas 5, ca. equaling petals. **Capsules** ovoid, equaling to slightly longer than calyx, opening by 5 recurved teeth; carpophore ca. 1 mm. **Seeds** dark brown, reniform, 0.5–0.8 mm, verrucate with crescent-shaped pattern. *2n* = 24.

Flowering summer. Tundra, rocky barrens, gulleys and river outwashes, grassy slopes, sea cliffs; 0–1100 m; Greenland; Nfld. and Labr., Nunavut, Que.; Europe (Iceland).

North American material of this arctic-alpine species has been regarded as distinct at the varietal and subspecific levels (M. L. Fernald 1940b; T. W. Böcher 1963) because it tends to be larger. However, the distinction is arbitrary, and some European material is as large as that from North America. A recent electrophoretic study (K. B. Haraldsen and J. Wesenberg 1993) of allozymes in populations from both continents provides no support for subdivision of this species.

62. Silene suksdorfii B. L. Robinson, Bot. Gaz. 16: 44, plate 6, figs. 9–11. 1891 • Suksdorf's catchfly [E]

Plants perennial, cespitose, with decumbent subterranean shoots; taproot stout; caudex branched, woody. **Stems** numerous, erect, simple, 3–15 cm, pubescent, viscid-glandular distally. **Leaves** mostly basal, densely tufted; basal numerous, pseudopetiolate, blade narrowly oblanceolate, tapering into base, 0.5–3 cm × 1.5–4 mm, ± fleshy, apex acute, puberulent; cauline in 1–3 pairs, ± sessile, reduced, blade narrowly oblanceolate to linear-lanceolate, 0.7–2 cm × 1–3 mm, apex acute, puberulent. **Inflorescences:** flowers terminal, solitary, or in single dichotomy, bracteate; bracts leaflike, 3–15 mm. **Pedicels** erect, ca. equaling calyx, viscid glandular-pubescent, hairs with purple septa. **Flowers:** calyx prominently 10-veined, campanulate, not contracted proximally around carpophore, 10–15 × 5–7 mm, papery, veins parallel, purplish, with pale commissures, with purple-septate glandular hairs (rarely septa not purple), lobes ovate, ca. 2 mm, margins broad, membranous, apex obtuse; corolla off-white or tinged with dusky purple, clawed, claw equaling calyx, broadened distally, limb 2-lobed, 3–5 mm, appendages ca. 1 mm; stamens equaling calyx; styles 3(–4), equaling calyx. **Capsules** equaling calyx, opening by 6 (or 8) teeth; carpophore 2.5–3.5 mm. **Seeds** brown, broadly winged, reniform, 1–2 mm, rugose-tessellate. 2*n* = 48.

Flowering summer. Alpine ridges, gravel slopes, talus; 1600–3000 m; Calif., Oreg., Wash.

Silene suksdorfii appears to be closely related to *S. parryi* but differs in its broadly winged seeds, smaller size, cespitose habit, and the prominent purple-septate hairs of the calyx, although the latter occasionally are present in *S. parryi*. It is very similar to, and in Idaho appears to intergrade with, another alpine species, *S. sargentii*, which has linear leaves and lacks the purple septa in the hairs and the broad wing on the seeds. It is similar also to *S. hitchguirei*; see discussion under that species.

63. Silene thurberi S. Watson, Proc. Amer. Acad. Arts 10: 343. 1875 • Thurber's catchfly

Silene plicata S. Watson

Plants perennial; taproot stout; caudex branched, woody. **Stems** several, erect, freely branched, leafy, elongate, 30–80 cm, scabrid-puberulent and glandular-viscid. **Leaves** 2 per node; basal long-petiolate, blade oblanceolate, (2–)5–18 cm × 5–30 mm, base narrowed into petiole, apex ± acute, glandular-puberulent on both surfaces; cauline not greatly reduced in distal stem, blade 5–15 cm × 10–20 mm, viscid glandular-pubescent, sparsely so in shade forms, proximal petiolate and blade oblanceolate, distal sessile and blade lanceolate, apex acute. **Inflorescences** open, elongate, dichotomously branched, branches ascending, bracteate; bracts leaflike, reduced distally. **Pedicels** ascending, may be partially deflexed, rather slender, longer than calyx, viscid glandular-pubescent with septate hairs, septa colorless. **Flowers:** calyx prominently 10-veined, tubular in flower, 8–12 × 2–4 mm, swelling to campanulate and 5–7 mm broad in fruit, not contracted proximally around carpophore, viscid glandular-pubescent, especially on veins, veins parallel, green, with pale commissures, those to lobes broadened distally, lobes erect, narrowly lanceolate, 3–4 mm, rigid, setose-scabrous, glandular-viscid, apex recurved; corolla greenish white, claw equaling calyx, limb 2-lobed, ca. 3 mm, appendages ca. 0.5 mm, margins erose; stamens equaling petals; styles 3, equaling petals. **Capsules** slightly exserted from calyx, narrowly ovoid, opening by 6 teeth; carpophore 1–2 mm. **Seeds** almost black, ellipsoid-reniform, ca. 1 mm, coarsely papillate; papillae inflated. 2*n* = 48.

Flowering late summer–fall. Open rocky places and canyons; 1500–2000 m; Ariz., N.Mex.; Mexico.

Silene thurberi is a rare, coarse, scabrous, and viscid herb with small, inconspicuous flowers on elongate pseudoracemes or open dichasia. The veins on the small calyces are usually green and conspicuously broaden into the recurved teeth. It is more common in Mexico.

Silene rectiramea is very similar but differs in its somewhat smaller calyx; short, ovate calyx lobes; and entire corolla appendages.

64. Silene uralensis (Ruprecht) Bocquet, Candollea 22: 25. 1967 • Nodding campion [F]

Gastrolychnis uralensis Ruprecht, Verbr. Pfl. Ural, 30. 1850

Plants perennial, cespitose or not; taproot stout. **Stems** erect, branched or simple, 5–40 cm, glabrous or pubescent. **Leaves:** basal few or numerous, petiolate, blade narrowly oblanceolate, spatulate, 1–13 cm × 1–7 mm, margins ciliate, glabrous or pubescent; cauline in 1–5 pairs, sessile, much reduced distally, blade linear to lanceolate, 0.1–4 cm × 1–5 mm. **Pedicels** slender or stout. **Flowers** erect or nodding; calyx prominently 10-veined, ovate-elliptic to broadly campanulate, inflated, 11–18 × 6–13 mm, thin and papery, veins purple or brown, lobes ovate to triangular, 2–3 mm; corolla dingy pink, purple, or red, claw shorter than to equaling calyx, limb not differentiated from claw, ovate or obovate, unlobed to

2-lobed, 1–4 mm; stamens shorter than petals; styles 5, shorter than petals. **Capsules** equaling to slightly longer than calyx, opening by 10 recurved teeth; carpophore 1–2 mm. **Seeds** brown or sooty brown, broadly winged, round to ± angular, 1.5–2.5 mm diam. including wing.

Subspecies 3 (3 in the flora): Greenland, Canada, nw United States including Alaska, e Asia (Russian Far East, Siberia).

Silene uralensis is a very variable species complex. Recently, A. Kurtto (2001) has resurrected *S. wahlbergella* as a species distinct from *S. uralensis*, differing in its shorter petals and larger capsule. It appears to be confined to Scandanavia and adjoining arctic Russia. North American reports of *S. wahlbergella* (and its synonym *Lychnis apetala*) are all referable to *S. uralensis*.

1. Inflorescences usually simple, pedicels slender with single nodding (deflexed) flower (fruiting pedicels erect); seeds 1.5–2(–2.5) mm diam. (incl. broad wing) . . . . 64a. *Silene uralensis* subsp. *uralensis*
1. At least some of the inflorescences branched with 2 to several flowers, pedicels erect to angled but not deflexed except at tip in flower; seeds 2–2.5 mm diam.
   2. Stems usually glabrous to sparsely, rarely pubescent, (15–)20–40 cm, slender; inflorescences branched with (1–)2–10 flowers; corolla only slightly exceeding calyx . . . . 64b. *Silene uralensis* subsp. *ogilviensis*
   2. Stems densely pubescent with purple-septate hairs, 10–35 cm, stout; at least some inflorescences forked with 1–3(–4) flowers; corolla usually ca. 1½ times calyx . . . . 64c. *Silene uralensis* subsp. *porsildii*

64a. Silene uralensis (Ruprecht) Bocquet subsp. **uralensis** • Silène de l'Oural F

Gastrolychnis apetala (Linnaeus) Tolmatchew & Kozhanchikov subsp. *arctica* (Fries) Á. Löve & D. Löve; *Lychnis apetala* Linnaeus subsp. *arctica* Hultén; *L. apetala* Linnaeus var. *arctica* (Fries) Macoun & Holm; *L. apetala* subsp. *attenuata* (Farr) Maguire; *L. apetala* var. *attenuata* (Farr) C. L. Hitchcock; *L. apetala* var. *glabra* Regel; *L. apetala* var. *nutans* B. Boivin; *L. attenuata* Farr; *L. nesophila* Holm; *L. soczavianum* (Schischkin) J. P. Anderson; *Melandrium apetalum* (Linnaeus) Fenzl subsp. *arcticum* (Fries) Hultén; *M. apetalum* subsp. *attenuatum* (Farr) H. Hara; *M. apetalum* var. *glabrum* (Regel) Hultén; *Silene attenuata* (Farr) Bocquet; *S. uralensis* subsp. *arctica* (Fries) Bocquet; *S. uralensis* subsp. *attenuata* (Farr) McNeill; *S. uralensis* var. *mollis* (Chamisso & Schlechtendal) Bocquet; *S. wahlbergella*

subsp. *arctica* (Fries) Hultén; *S. wahlbergella* subsp. *attenuata* (Farr) Hultén; *Wahlbergella attenuata* (Farr) Rydberg

Plants cespitose; taproot stout. **Stems** simple (very rarely 1-branched), 5–30 cm, pubescent, often densely so distally, with long purple-septate glandular and eglandular hairs, rarely subglabrous. **Leaves:** basal numerous, blade 1–5 cm × 2–5 mm (including petiole), glabrous or softly pubescent, eglandular; cauline in 1–3 pairs, blade linear to narrowly lanceolate, 0.5–2.5 cm × 2–4 mm. **Inflorescences** simple, slender, with single terminal flower, occasionally branched with 2 (rarely 3) flowers, densely glandular-pubescent, slightly viscid. **Pedicels** slender. **Flowers:** nodding, becoming erect in fruit; calyx veined, ovate-elliptic, 11–17 × 6–10 mm, veins purple with long purple-septate hairs, lobes ovate to triangular, ca. 2.5 mm, margins purple tinged, broad, membranous; corolla dingy pink to purple, 1–1¼ times length of calyx, claw shorter than calyx, limb ovate, emarginate to 2-lobed, ca. 1–3 mm, appendages 2, lacerate, ca. 0.3 mm. **Capsules** equaling to slightly longer than calyx; carpophore 1–2 mm. **Seeds** brown, ± round to angular, 1.5–2(–2.5) mm diam.; wing flat, ca. as broad as body. $2n = 24$.

Flowering summer. Arctic and alpine tundra, gravelly and grassy places; 0–4000 m; Greenland; Alta., B.C., Man., Nfld. and Labr. (Labr.), N.W.T., Nunavut, Ont., Que., Yukon; Alaska, Colo., Mont., Utah, Wyo.; arctic Asia.

Subspecies *uralensis* is an arctic-alpine taxon. The populations that extend through the Rocky Mountains from Alaska to Utah are often referred to subsp. *attenuata*. They tend to have a less-inflated calyx, slightly longer purple petals, flowers that are angled at less than 45° rather than nodding, and less well-developed cauline leaves. These differences are minor, however, and populations of subsp. *attenuata* often contain plants referable to subsp. *uralensis*, while plants resembling subsp. *attenuata* are scattered across the range of subsp. *uralensis*. Some collections from the southern Rocky Mountains (Colorado and Utah) appear to intergrade with *S. kingii* in having a narrow wing to the seeds. Hybrids with *S. involucrata* occasionally occur in nature, and A. Nygren (1951) reported a synthesized hybrid that was triploid ($2n = 36$) and sterile (see discussion under 57. *S. sorensenis*).

64b. Silene uralensis (Ruprecht) Bocquet subsp. **ogilviensis** (A. E. Porsild) D. F. Brunton, Canad. J. Bot. 59: 1362. 1981 E

Melandrium apetalum (Linnaeus) Fenzl subsp. *ogilviense* A. E. Porsild, Publ. Bot. (Ottawa) 4: 23. 1974; *Gastrolychnis soczaviana* (Schischkin) Tolmatchew &

Kozhanchikov subsp. *ogilviensis* (A. E. Porsild) Á. Löve & D. Löve

Plants not cespitose; taproot producing 1–several shoots. **Stems** erect, branched, slender, (15–)20–40 cm, usually glabrous, rarely pubescent. **Leaves:** basal few, petiolate, petiole broadened at base and clasping stem, to equaling blade, blade narrowly lanceolate, spatulate, 2.5–15 cm × 1–6 mm, apex acute, glabrous to sparsely puberulent; cauline in 2–5 pairs, sessile, connate proximally, blade linear to linear-lanceolate, 1–12 cm × 1–5 mm, apex acute. **Inflorescences** branched, (1–)2–10-flowered; bracts and bracteoles very narrowly lanceolate, 5–15 mm. **Pedicels** ascending, deflexed at tip in flower, erect in fruit, slender, elongate, 1–7 cm, glabrous and sparsely pubescent. **Flowers** nodding; calyx veined, ovate-elliptic, 11–17 × 6–10 mm, glabrous to sparsely pubescent, veins purple, lobes spreading, ovate-triangular, 2–3 mm, apex short-acuminate; corolla dingy purple, slightly longer than calyx, claw shorter than calyx, limb ovate, 2-lobed, ca. 2 mm. **Capsules** 1¼ times length of calyx, teeth recurved; carpophore ca. 1 mm. **Seeds** brown, broadly winged, round, flat, 2 mm diam. $2n = 48$.

Flowering summer. Damp calcareous tundra, river flats, heath, talus; 0–800 m; Man., N.W.T., Nunavut, Ont., Que., Yukon; Alaska.

Subspecies *ogilviensis* is rare in scattered localities across the low arctic. Its appearance is distinct, with its taller, more-slender, branched, and usually glabrous stems. However, it appears to intergrade with subsp. *uralensis*, and further study is required to determine its status.

64c. Silene uralensis (Ruprecht) Bocquet subsp. **porsildii** Bocquet, Candollea 22: 27. 1967 • Large-seeded nodding campion

Melandrium macrospermum A. E. Porsild, Rhodora 41: 225, plate 552, figs. 1–3. 1939; *Gastrolychnis macrosperma* (A. E. Porsild) Tolmatchew & Kozhanchikov; *G. soczaviana* (Schischkin) Tolmatchew & Kozhanchikov; *Lychnis apetala* Linnaeus var. *macrosperma* (A. E. Porsild) B. Boivin; *L. macrosperma* (A. E. Porsild) J. P. Anderson; *Melandrium soczavianum* Schischkin; *Silene macrosperma* (A. E. Porsild) Hultén; *S. soczaviana* (Schischkin) Bocquet

Plants cespitose; taproot stout. **Stems** erect, simple or branched, stout, 10–35 cm, densely pubescent with long, purple-septate, eglandular and glandular hairs. **Leaves:** basal numerous, blade 1.5–6 cm × 3–7 mm, somewhat

fleshy, subglabrous; cauline in 1–3 pairs, blade narrowly lanceolate, 1.5–4 cm × 1–5 mm. **Inflorescences** sometimes forked, 1–3(–4)-flowered. **Pedicels** erect or angled in flower and erect in fruit, stout, usually elongate, 1–10 cm, densely glandular-pubescent with long purple-septate hairs, ± viscid. **Flowers** erect; calyx veined, ovate-elliptic to broadly campanulate, 11–14(–18) × 8–13 mm, veins brown or purple, densely glandular-pubescent with purple-septate hairs, lobes ovate-triangular, 2–3 mm, margins broad, membranous; corolla dingy purple-red, 1⅓–1½ times length of calyx, claw equaling calyx, limb obovate, 2-lobed, 2–4 mm. **Capsules** not contracted at mouth, equaling calyx; carpophore 1–2 mm. **Seeds** sooty brown, winged, round, flat, 2–2.5 mm diam.; wing inflated, broad, strongly rugose. $2n = 48$.

Flowering summer. Tundra, gravel slopes, talus, cliffs; 0–2500 m; Yukon; Alaska; Asia (Russian Far East, Siberia).

The usually branched stout stems, erect flowers, broad capsule, and very large seeds with broad rugose wing distinguish subsp. *porsildii*.

65. Silene verecunda S. Watson, Proc. Amer. Acad. Arts 10: 344. 1875 • San Francisco campion

Silene andersonii Clokey; *S. behrii* (Rohrbach) F. N. Williams; *S. luisana* S. Watson; *S. occidentalis* S. Watson var. *nancta* Jepson; *S. platyota* S. Watson; *S. verecunda* subsp. *andersonii* (Clokey) C. L. Hitchcock & Maguire; *S. verecunda* var. *eglandulosa* C. L. Hitchcock & Maguire;

S. verecunda subsp. *platyota* (S. Watson) C. L. Hitchcock & Maguire; *S. verecunda* var. *platyota* (S. Watson) Jepson

Plants perennial; taproot stout; caudex branched, woody. **Stems** usually several–many, rarely 1, erect, leafy, 10–55 cm, base often decumbent with marcescent leaf bases, scabrous-puberulent to pubescent, usually ± viscid-glandular distally, rarely densely so. **Leaves** 2 per node; basal petiolate, blade linear-lanceolate to lanceolate, 3–10 cm × 2–13 mm (including petiole), apex acute, glabrous to scabrous-puberulent or softly pubescent, petiole often ciliate; cauline sessile or petiolate, connate proximally, reduced distally, blade linear to narrowly lanceolate, 2–10 cm × 2–8 mm. **Inflorescences** cymose with elongate ascending branches, open, flowers (1–)3 to many, bracteate, bracteolate; bracts and bracteoles lanceolate, 3–20 mm, apex acuminate. **Pedicels** ¼–3 times longer than calyx, scabrous-puberulous to pubescent and glandular, ± viscid. **Flowers:** calyx prominently 10-veined, tubular in flower, ± clavate and contracted around carpophore in fruit, 10–14 × 4–6 mm, margins dentate, veins parallel, green (rarely purplish), with pale commissures, lobes ascending, lanceolate, 2–

3 mm, margins usually with obtuse, membranous border, apex spreading, shortly glandular-pubescent, usually viscid; corolla off-white (greenish) to dusky pink, clawed, claw equaling calyx, limb obovate, 2-lobed, 3–7 mm, shorter than claw, rarely with small lateral teeth, appendages 2, usually lacerate, 1–2 mm; stamens ca. equaling petals; styles 3(–4), often much longer than petals. **Capsules** ovoid-ellipsoid, slightly longer than calyx, opening by 6 (or 8) recurved teeth; carpophore 2–5 mm. **Seeds** dark reddish brown to black, reniform, ca. 1.5 mm, papillate, with larger papillae around margins. $2n = 48$.

Flowering summer. Meadows, chaparral, sagebush, open woodlands, dry pine forests, alpine ridges, dry canyons; 0–3400 m; Ariz., Calif., Nev., Oreg., Utah; Mexico.

Silene verecunda is an exceptionally variable species, very difficult to circumscribe and tending to intergrade with *S. bernardina*, *S. oregana*, and *S. grayi*. It differs from the first two of those species mainly in having two-lobed petals. *Silene grayi* is a much smaller cespitose alpine plant with very large seeds. Hitchcock and Maguire divided *S. verecunda* into subsp. *verecunda*, subsp. *platyota*, and subsp. *andersonii*. Of these, subsp. *andersonii* is the most distinct, with a scabrous-puberulent indumentum, very narrow, stiff leaves, and rigid stems that are decumbent at the base, with marcescent leaf bases. The claw of the petals also is often more uniformly ciliate. Subspecies *verecunda* has a very different appearance, its mature calyx being shorter, broader, and markedly clavate. It is a short, stocky, viscid-glandular plant of exposed coastal habitats and may simply be a local ecotype. Subspecies *platyota* encompasses the remainder of the variation in the complex. Most of this variation consists of plants with fewer flowering stems; softer pubescence; broader, flat leaves; and thinner, more papery calyces. All these forms of *S. verecunda* appear to intergrade freely and, based on current information, any separation would be arbitrary. The species is in need of an experimental study to determine the nature of variation and its taxonomic value.

66. Silene virginica Linnaeus, Sp. Pl. 1: 419. 1753

· Fire pink   [E]

Melandrium virginicum (Linnaeus) A. Braun; *Silene catesbaei* Walter; *S. coccinea* Moench; *S. virginica* var. *hallensis* Pickens & M. C. W. Pickens; *S. virginica* var. *robusta* Strausbaugh & Core

Plants perennial; taproot slender; caudex decumbent, branched, producing tufts of leaves and erect flowering shoots. **Stems** simple proximal to inflo-

rescence, 20–80 cm, glandular-pubescent, often subglabrous near base. **Leaves:** basal numerous, tufted, petiolate, petiole ciliate, blade oblanceolate, 3–10 cm × 8–18 mm, base spatulate, apex acute to obtuse, glabrous on both surfaces, rarely puberulent; cauline in 2–4 pairs, broadly petiolate to sessile, reduced distally, blade oblanceolate to narrowly elliptic or lanceolate, 1–10 (–30) cm × 4–16(–30) mm, margins ciliate, apex acute, shortly acuminate, glabrous. **Inflorescences** open, with ascending, often elongate branches, (3–)7–11(–20)-flowered, bracteate, glandular-pubescent, often densely so, viscid; bracts leaflike, lanceolate, 4–40 mm. **Pedicels** erect in flower, sharply deflexed at base in fruit, $\frac{1}{2}$–1 times length of calyx. **Flowers:** calyx green to purple, 10-veined, tubular to narrowly obconic in flower, 16–22 × 5–6 mm, clavate and swelling to 7–12 mm in fruit, glandular-pubescent, lobes lanceolate to oblong, 3–4 mm, margins usually narrow, membranous, apex acute or obtuse; corolla scarlet, 2 times longer than calyx, clawed, claw ciliate, gradually widening into limb, longer than calyx, limb obtriangular to oblong, deeply 2-lobed with 2 small lateral teeth, 10–14 mm, glabrous or nearly so, appendages 2, tubular, 3 mm; stamens exserted, shorter than petals; styles 3(–4), equaling stamens. **Capsules** ovoid, equaling calyx, opening by 3 (or 4) teeth that sometimes split into 6 (or 8); carpophore 2–3(–4) mm. **Seeds** ash gray, reniform, 1.5 mm, with large inflated papillae. $2n = 48$.

Flowering spring. Deciduous woodlands, bluffs, moist wooded slopes; 200–1300 m; Ala., Ark., Del., Ga., Ill., Ind., Iowa, Kans., Ky., La., Mich., Miss., Mo., N.Y., N.C., Ohio, Okla., Pa., S.C., Tenn., Va., W.Va.

Silene virginica is related to the scarlet-flowered species from the southwest, *S. laciniata* and *S. subciliata*. It makes a beautiful garden plant in semishaded locations. J. A. Steyermark (1963) recorded the occurrence of a hybrid between *S. virginica* and *S. caroliniana* subsp. *wherryi* in Shannon County, Missouri. Reports of the occurrence of *S. virginica* in Ontario are based on a collection (CAN, K) made in 1873 from "islands in the Detroit River" in "Canada West."

67. Silene viscaria (Linnaeus) Jessen, Deut. Excurs.-Fl., 280. 1879 · Sticky catchfly   [I]

Lychnis viscaria Linnaeus, Sp. Pl. 1: 436. 1753; *L. viscosa* Scopoli; *Steris viscaria* (Linnaeus) Rafinesque; *Viscaria viscosa* (Scopoli) Ascherson; *V. vulgaris* Bernhardi

Subspecies 2 (1 in the flora): introduced; Europe.

67a. Silene viscaria (Linnaeus) Jessen subsp. **viscaria**

Plants perennial; taproot stout, branched; caudex compact, multicapitate. **Stems** erect, simple proximal to inflorescence, 30–90 cm, glabrous, viscid, especially at nodes. **Leaves:** basal numerous, tufted, petiolate, blade oblanceolate to linear-oblanceolate, 4–15 cm × 2–10 mm (including petiole), apex acute, glabrous, ciliate at base; cauline in 2–4 pairs, connate proximally, sessile, reduced distally, blade linear-lanceolate, 2–10 cm × 2–8 mm, apex acute, glabrous, ciliate at base. **Inflorescences** paniculate, consisting of short, interrupted cymes, narrow, bracteate, bracteolate; cyme 3–10-flowered, proximal ones pedunculate; peduncle to 3 cm; bracts leaflike, to 3 cm; bracteoles broadly lanceolate, 3–10 mm, margins membranous, apex acuminate. **Pedicels** shorter than calyx, glabrous but viscid. **Flowers** 14–22 mm diam.; calyx purple, obscurely veined, narrowly campanulate in flower, 6–10 × 3–4 mm, clavate in fruit, 10–15 × ca. 5 mm, papery, puberulent, lobes ovate, ca. 1.5 cm, margins narrow, membranous, apex obtuse; corolla purple or dark pink (rarely white), ca. 1½–2 times calyx, clawed, claw equaling calyx, limb spreading, obovate, entire or slightly notched, 5–9 mm, appendages 2, oblong, 3 mm; stamens shortly exserted; stigmas 5, shortly exserted. **Capsules** ovoid, equaling calyx, opening by 5 teeth; carpophore 3–5 mm. **Seeds** dark brown, reniform, 0.3–0.5 mm, finely tuberculate. $2n = 24$ (Europe).

Flowering early summer. Roadsides, waste ground, fields; 0–500 m; introduced; Conn., Maine, Mass., N.H., N.Y., Ohio; Europe.

Subspecies *viscaria* is an attractive garden plant similar to *S. suecica* but somewhat larger and sturdier with viscid pedicels. It occasionally escapes but does not persist.

68. Silene vulgaris (Moench) Garcke, Fl. N. Mitt.-Deutschland ed. 9, 46. 1869 • Bladder campion, silène enflé I W

Behen vulgaris Moench, Methodus, 709. 1794; *Silene cucubalus* Wibel; *S. inflata* Smith; *S. latifolia* (Miller) Britten & Rendle var. *pubescens* (de Candolle) Farwell

Plants short-lived perennial, glabrous, rarely pubescent, glaucous; taproot stout; caudex woody. **Stems** several–many, erect, branched and decumbent at base, rarely simple, 20–80 cm. **Leaves** mainly cauline, 2 per node, sessile, almost clasping, reduced proximal to inflorescence, blade broadly oblong to oblanceolate or lanceolate, rarely ± linear, 2–8 cm × 5–30 mm, base round, apex acute to acuminate. **Inflorescences** open dichasial cyme, 5–40-flowered, bracteate; bracts much-reduced, lanceolate. **Pedicels** 0.5–3 cm. **Flowers** bisexual and unisexual, some plants having bisexual flowers, others having pistillate unisexual flowers, 15–20 mm diam.; calyx pale green, rarely purplish, campanulate, not contracted at mouth or base, inflated, 9–12 mm in flower, 12–18 × 7–11 mm in fruit, herbaceous, papery, venation obscure, reticulate, without conspicuous pale commissures, margins dentate, lobes broadly triangular, 2–3 mm, glabrous; petals white, ca. 2 times as long as calyx; limb obovate, emarginate to 2-lobed; stamens exserted by 2–4 mm; styles 3, cream to greenish, at most slightly pink tinged, 2 times longer than calyx. **Capsules** ovoid to globose, equaling calyx, opening by 6 teeth; carpophore 2–3 mm. **Seeds** black or nearly so, globose-reniform, 1–1.5 mm, finely tuberculate. $2n = 24$.

Flowering summer–fall. Roadsides, waste ground, gravel pits and shores, arable land; 0–2000 m; introduced; Alta., B.C., Man., N.B., Nfld. and Labr. (Nfld.), N.S., Ont., P.E.I., Que., Sask., Yukon; Alaska, Ariz., Ark., Calif., Colo., Conn., Del., D.C., Ga., Idaho, Ill., Ind., Iowa, Kans., Ky., Maine, Md., Mass., Mich., Minn., Mo., Mont., Nebr., N.H., N.J., N.Y., N.C., N.Dak., Ohio, Oreg., Pa., R.I., S.C., S.Dak., Vt., Va., Wash., W.Va., Wis., Wyo.; Europe.

Silene vulgaris is less variable in North America than in its native Europe, where five subspecies are recognized on the basis of capsule size, petal color, leaf shape, and habit. All North American material appears to belong to subsp. *vulgaris*, although a few collections from sandy habitats tend to have unusually narrow leaves. Similar plants from Europe have been named var. *litoralis* (Ruprecht) Jalas and subsp. *angustifolia* Hayek.

69. Silene williamsii Britton, Bull. New York Bot. Gard. 2: 168. 1901 • Williams's catchfly E

Silene menziesii Hooker subsp. *williamsii* (Britton) Hultén; *S. menziesii* var. *williamsii* (Britton) B. Boivin

Plants perennial; taproot slender; rootstock much-branched. **Stems** several–many, decumbent to erect, much-branched and sometimes matted, leafy, 5–30 cm, pubescent and glandular, at least distally. **Leaves** 2 per node, leafy above, sessile; blade narrowly lanceolate to elliptic-lanceolate, broadest proximally, narrowed to base, 1–5 cm × 3–8 mm, apex acute, short-pubescent and ± glandular on both surfaces. **Inflorescences** cymose, loose, leafy, compound, or flowers terminal, axillary in distal nodes. **Pedicels** 0.5–1(–3) cm, glandular-pubescent. **Flowers** unisexual, all plants having both staminate and pistillate flowers; calyx obscurely 10-veined, ovate-campanulate, 9–11 × 4–6 mm, herbaceous, papery, pubescence rather dense, glandular, ± obscuring veins, veins slender, without conspicuous pale commissures, lobes lanceolate, ca. 2 mm, apex acute to acuminate; corolla white, clawed, claw shorter than calyx, limb oblong, 2-lobed, 1.5–3 mm, lobes lanceolate, apex acute, appendages linear, 0.3–0.5 mm; stamens equaling corolla; stigmas 3, slender, equaling corolla, papillate only at tip. **Capsules** straw colored, ovoid-ellipsoid, slightly longer than calyx, opening by 6 teeth; carpophore ca. 1 mm. **Seeds** dull brown, not winged, angular-reniform, ca. 1 mm, tuberculate. *2n* = 24.

Flowering summer. Heaths, disturbed ground, river gravel and bluffs, roadsides; 100–700 m; Yukon; Alaska.

Silene williamsii is similar to *S. menziesii* but is readily separated by its leaves, which are broadest below the middle, its dull tuberculate seeds, and its stigmas, which are papillate only at the tip. It is also monoecious instead of functionally dioecious.

70. Silene wrightii A. Gray, Smithsonian Contr. Knowl. 5(6): 17. 1853 · Wright's catchfly E

Plants perennial, viscid; taproot stout; caudex branched, woody. **Stems** several, simple or branched, spreading to ascending, leafy, 10–30 cm, densely pubes-cent, glandular. **Leaves** 2 per node, mostly cauline, blade 1.5–6 cm × 3–14 mm, apex sharply acuminate, pubescent and viscid on both surfaces; distal sessile, blade elliptic-lanceolate; proximal short-petiolate, blade oblanceolate. **Inflorescences** leafy, flowers terminal and axillary.

Pedicels straight, rather slender, ¹/₅ times to equaling calyx. **Flowers:** calyx prominently 10-veined, tubular to narrowly obconic in flower, 16–20 × 4–5 mm, clavate and broadening to 7 mm in fruit, narrowed proximally around carpophore, coarsely glandular-pubescent and viscid, veins parallel, green, with pale commissures, lobes narrowly lanceolate, 5–7 mm, margins narrow, membranous, apex acuminate; corolla white to pale yellow, sometimes purple tinged, clawed, ca. 2 times calyx, claw longer than calyx, broadened into obtriangular limb, limb 5–8 mm, cleft ca. to middle into (2–)4–8 lanceolate to oblong lobes, appendages 2, very short; stamens exserted, shorter than petals; styles 3, exserted, slender, shorter than petals. **Capsules** narrowly ovoid, equaling calyx, opening by 3 teeth that tardily split into 6; carpophore 3–6 mm. **Seeds** brown, broadly reniform, flattened, ca. 1.5 mm, sides rugose, margins papillate; papillae conic, acute. *2n* = 96.

Flowering summer. Cliff crevices in mountains; 1800–2800 m; N.Mex.

Silene wrightii is an uncommon, distinct species with large, pale yellowish flowers and tubular to narrowly funnelform calyces with long, narrow, lanceolate lobes. The leaves are mainly cauline, with the largest in the mid-stem region. The stems are few-branched and arise in tufts from the very woody caudex.

37. AGROSTEMMA Linnaeus, Sp. Pl. 1: 435. 1753; Gen. Pl. ed. 5, 198. 1754 · [Greek *agros*, field, and *stemma*, crown or wreath, alluding to the flowers' use in garlands] I

John W. Thieret

Herbs, annual. **Taproots** stout. **Stems** simple or branched, terete. **Leaves** connate proximally into sheath, sessile; blade 1-veined or obscurely 3-veined, linear to narrowly lanceolate, apex acute. **Inflorescences** terminal, lax cymes or of solitary, mostly axillary flowers; bracts, when present, paired, foliaceous; involucel bracteoles absent. **Pedicels** erect. **Flowers** bisexual or rarely unisexual and pistillate; sepals connate proximally into tube, 25–62 mm; tube green, 10-veined, cylindric to ovoid, terete, commissures between sepals 1-veined, herbaceous; lobes green, 1-veined, linear-lanceolate, [shorter or] longer than tube, often equaling or longer than petals, margins green, herbaceous, apex acute; petals 5, purplish red or white, clawed, auricles absent, coronal appendages absent, blade apex obtuse, entire or briefly emarginate; nectaries near bases of filaments opposite sepals; stamens 10, 5 adnate to petals, 5 at base of gynoecium; filaments distinct; staminodes absent; styles (4–)5, clavate, 10–12 mm, with dense, stiffly ascending hairs proximally; stigmas (4–)5, subterminal, papillate (30×). **Capsules** ovoid, opening by (4–)5 ascending teeth; carpophore absent. **Seeds** ca. 30–60, black, reniform, laterally compressed, tuberculate, marginal wing absent, appendage absent; embryo peripheral, curved. $x = 12$.

Species 2 (1 in the flora): introduced; Eurasia; introduced worldwide.

Agrostemma brachyloba (Fenzl) Hammer (*A. gracilis* Boissier), a species native to Greece and Asia Minor, has been reported as a garden waif in Boulder, Colorado, where it persisted for at least four years (W. A. Weber et al. 1979). It differs from *A. githago* in that its calyx lobes are shorter (rather than longer) than the calyx tube, and its petals, spotted on the limb, are longer (rather than shorter) than the calyx lobes.

SELECTED REFERENCES Firbank, L. G. 1988. Biological flora of the British Isles. 165. *Agrostemma githago* L. (*Lychnis githago* (L.) Scop.). J. Ecol. 76: 1232–1246. Hammer, K., P. Hanelt, and H. Knupffer. 1982. Vorarbeiten zur monographischen Darstellung von Wildpflanzensortimenten: *Agrostemma* L. (Studies towards a monographic treatment of wild plant collections: *Agrostemma* L.) Kulturpflanze 30: 45–96.

1. **Agrostemma githago** Linnaeus, Sp. Pl. 1: 435. 1753
   • Common corncockle, nielle [F] [I] [W]

Lychnis githago (Linnaeus) Scopoli

Varieties 3 (1 in the flora): introduced; Eurasia; introduced in South America, s Africa, Pacific Islands (New Zealand), Australia.

Formerly a common weed of grain fields, *Agrostemma githago* is becoming increasingly scarce, both in North America and in its native environs in Europe (R. Svensson and M. Wigren 1986). Mechanical screening of grain, which removes contaminants, and modern herbicides have more or less eliminated the plant from grain fields in the flora. The saponin-containing seeds, occurring as contaminants in grain, are poisonous to livestock, birds, and humans. This species is sometimes cultivated in flower gardens.

1a. **Agrostemma githago** Linnaeus var. **githago** [F] [I]

Plants 30–100 cm, spreading-pilose, villous. **Leaf blades** 4–15 cm. **Inflorescences** of solitary, axillary flowers. **Pedicels** 5–20 cm. **Flowers:** calyx 2.5–6.2 cm; tube 1.2–1.7 cm, spreading-pilose; lobes linear to linear-lanceolate, 1.2–4.5 cm; petals purplish red to pink (rarely white), unspotted, 1.5–4 cm, equaling or shorter than calyx lobes. **Capsules** 16–24 mm. **Seeds** 3–4 mm wide. $2n = 48$ (Europe).

Flowering spring–summer. Fields, roadsides, waste places; 0–1800 m; introduced; Man., N.B., Ont., P.E.I., Que., Sask.; Ala., Alaska, Ark., Calif., Conn., Del., D.C., Fla., Ga., Idaho, Ill., Ind., Iowa, Kans., Ky., La., Maine, Md., Mass., Mich., Minn., Miss., Mo., Mont., Nebr., N.H., N.J., N.Mex., N.Y., N.C., N.Dak., Ohio, Okla., Oreg., Pa., R.I., S.C., S.Dak., Tenn., Tex., Vt., Va., Wash., W.Va., Wis., Wyo.; Eurasia; widely introduced worldwide.

Historical collections are known from British Columbia (1915) and Newfoundland (1890).

44. POLYGONACEAE Jussieu

• Buckwheat Family

Craig C. Freeman

James L. Reveal

Trees, shrubs, vines, or herbs, perennial, biennial, or annual, homophyllous (heterophyllous in some species of *Polygonum*), polycarpic (rarely monocarpic in *Eriogonum*); roots fibrous or a solid or, rarely, chambered taproot, rarely tuberous. **Stems** prostrate to erect, sometimes scandent or scapose, solid or hollow, rarely with recurved spines (some species of *Persicaria*), glabrous or pubescent, sometimes glandular; nodes swollen or not; tendrils absent (except in *Antigonon* and *Brunnichia*); branches free (adnate to stems distal to nodes and appearing to arise internodally in *Polygonella*); caudex stems (subfam. Eriogonoideae) tightly compact to spreading and at or just below the soil surface or spreading to erect and above the soil surface, woody; aerial flowering stems prostrate or decumbent to erect, arising at nodes of caudex branches, at distal nodes of aerial branches, or directly from the root, slender to stout and solid or slightly to distinctly fistulose, rarely disarticulating in ringlike segments (*Eriogonum*). **Leaves** deciduous (persistent in *Coccoloba* and sometimes more than 1 year in *Antigonon*, *Eriogonum*, *Chorizanthe*, and *Polygonella*), basal or basal and cauline, rosulate, mostly alternate, infrequently opposite or whorled; stipule (ocrea) absent (subfam. Eriogonoideae, possibly vestigial in some perennial species of *Chorizanthe*) or present (subfam. Polygonoideae), persistent or deciduous, cylindric to funnelform, sometimes 2-lobed (*Polygonum*), chartaceous, membranous, coriaceous or partially to entirely foliaceous; petiole present or absent, rarely articulate basally (*Fagopyrum*, *Polygonella*, *Polygonum*), rarely with extrafloral nectaries (*Fallopia*, *Muehlenbeckia*); blade simple, margins entire, occasionally crenulate, crisped, undulate, or lobed, rarely awn-tipped (*Goodmania*). **Inflorescences** terminal or terminal and axillary, cymose and dichotomously or trichotomously branched, or racemose, umbellate, or capitate (subfam. Eriogonoideae; or spikelike, racemelike, paniclelike, cymelike, or, rarely, capitate (subfam. Polygonoideae), comprising simple or branched clusters of compound inflorescences; bracts absent (subfam. Polygonoideae), or 2–10+, usually connate proximally or to ½ their length, rarely perfoliate, foliaceous or scalelike, margins entire, sometimes awn-tipped (subfam. Eriogonoideae, rarely absent in *Eriogonum*), glabrous or pubescent; peduncle present or absent; clusters of flowers subtended by involucral bracts or enclosed in typically nonmembranous tubular involucres (subfam. Eriogonoideae) or subtended by connate bracteoles forming a persistent membranous tube (ocreola) (subfam. Polygonoideae). **Pedicels** present or absent, rarely accrescent (*Brunnichia*), articulate to flowers. **Flowers** usually bisexual, sometimes bisexual and unisexual on same or different plants, rarely unisexual only, 1–many, often with stipelike base distal to articulation; perianth persistent, often accrescent in fruit, often greenish, white, pink, yellow, red, or purple, rarely winged or keeled (*Fallopia* and some species of *Polygonum*), campanulate to urceolate, sometimes membranous, indurate (*Brunnichia* and *Emex*), or fleshy (*Coccoloba*, *Muehlenbeckia*, and some species of *Persicaria*) in fruit, rarely developing raised tubercles proximally (*Rumex*), glabrous or pubescent, sometimes glandular or glandular-punctate; tepals 2–6, distinct or connate proximally and forming tube, usually in

2 whorls, petaloid or sepaloid, dimorphic or monomorphic, rarely coriaceous (*Lastarriaea*), entire, emarginate, or lobed to laciniate apically, rarely awn-tipped (*Lastarriaea*); nectary a disk at base of ovary or glands associated with bases of filaments; stamens (1–)6–9, staminode rarely present; filaments distinct, or connate basally and sometimes forming staminal tube, free or adnate to perianth tube, glabrous or pubescent proximally; anthers dehiscing by longitudinal slits; pistil 1, (rudimentary pistil sometimes present in staminate flowers of monoecious or dioecious taxa), (2–)3(–4)-carpellate, homostylous (heterostylous in some species of *Fagopyrum* and *Persicaria*); ovary 1-locular (sometimes with vestigial partitions proximally); ovule 1, orthotropous or, rarely, anatropous, placentation basal or free-central; styles 1–3, erect to spreading or recurved, distinct or connate proximally; stigmas 1 per style, peltate, capitate, fimbriate, or penicillate. **Fruits** achenes, included or exserted, yellowish, brown, red, or black, homocarpic (sometimes heterocarpic in *Polygonum*), winged or unwinged, 2-gonous, 3-gonous, discoid, biconvex, lenticular, rarely 4-gonous or spheroidal, glabrous or pubescent. **Seeds** 1; endosperm usually abundant, mealy, development nuclear; embryo straight or curved, rarely folded.

Genera 48, species ca. 1200 (35 genera, 442 species in the flora): widespread, well represented in the north-temperate zone.

Monophyly of Polygonaceae is well supported by molecular studies (M. W. Chase et al. 1993; P. Cuénoud et al. 2002; A. S. Lamb Frye and K. A. Kron 2003), which place it sister to Plumbaginaceae. Wood anatomy also supports this relationship (S. Carlquist 2003). Polygonaceae has been divided variously into subfamilies (S. A. Graham and C. E. Wood Jr. 1965; K. Haraldson 1978; J. Brandbyge 1993); two often are recognized based on morphological evidence: Eriogonoideae and Polygonoideae. Recent studies using the chloroplast gene *rbc*L suggest Eriogonoideae are monophyletic and Polygonoideae are paraphyletic (Lamb Frye and Kron). Generic limits have been much debated, particularly the circumscription of *Polygonum* in the broad sense. Morphological, cytological, palynological, and anatomical data, although useful in resolving some questions about the circumscriptions and relationships of genera, have not provided a consensus regarding relationships within the family. Pollen types (J. W. Nowicke and J. J. Skvarla 1977) and chromosome numbers (Brandbyge) are diverse. Multiple base chromosome numbers have been documented in some genera, and polyploidy is common. The classification used here generally follows J. L. Reveal (1978) for Eriogonoideae and Haraldson for Polygonoideae.

A characteristic feature of the family is the ocrea, a nodal sheath variously interpreted as an outgrowth of the sheathing base of the petiole, as connate stipules, or as an expanded axillary stipule (S. A. Graham and C. E. Wood Jr. 1965). It is absent in subfamily Eriogonoideae except in some perennial, South American members of *Chorizanthe*, where it is rudimentary (J. L. Reveal 1978).

The flowers of Polygonaceae have been studied extensively. Most researchers concluded that six tepals is the primitive condition in the family (R. A. Laubengayer 1937), but A. S. Lamb Frye and K. A. Kron (2003) concluded that five tepals is the primitive condition, and that taxa with six or four tepals evolved multiple times within the family. The same molecular study also suggested parallel evolution of growth forms and woodiness. Floral nectaries are useful generic characters in the Polygonoideae (L.-P. Ronse Decraene and J. R. Akeroyd 1988) but their utility in floristic treatments is limited because of their small size. In some genera of Polygonaceae, the outer tepals are connate and form a slender, stipelike hypanthium base above the articulation with the true pedicel. In the Polygonoideae treatments, pedicel refers to the true pedicel plus the stipelike hypanthium above the articulation; in Eriogonoideae, stipe refers only to the stipelike hypanthium base, and pedicel is applied only to the true pedicel.

Members of the family are of relatively minor economic importance (A. N. Steward 1930; S. A. Graham and C. E. Wood Jr. 1965). *Fagopyrum* has been cultivated for millennia, *Coccoloba* produce edible fruits, and the petioles of *Rheum* and leaves of some species of *Rumex* are edible. Many species have been used as food or medicine or for ceremonies by various tribes of Native Americans. Outside the flora area, the family includes some tropical trees harvested for timber (*Coccoloba* and *Triplaris*). Some species of *Fallopia*, *Persicaria*, *Polygonum*, and *Rumex* are cosmopolitan weeds, and *Emex* is a pernicious weed in parts of the world. Of the 35 genera in the flora, four are entirely non-native: *Emex*, *Fagopyrum*, *Muehlenbeckia*, and *Rheum*. *Antigonon* and *Coccoloba* are popular ornamentals in the southern part of the flora region. Many species of *Erigonoum* are planted in rock gardens. *Persicaria* species are important wetland plants because waterfowl consume their fruits.

Mechanisms for achene dispersal vary greatly. Hooked or awned involucres occur in many genera of subfam. Eriogonoideae. The outer tepals of the perianth expand into wings (*Fallopia*) or spines (*Emex*), or the achene has wings (*Eriogonum alatum*, *Oxyria*, *Rheum*, and *Rumex*), which aid in dispersal by wind or water. The achenes are forcibly tossed from the plant in *Persicaria virginiana*, and persistent, hooked styles aid in their transport by animals. In *Coccoloba*, the fleshy tube of the tepals aids in dispersal by birds.

SELECTED REFERENCES Brandbyge, J. 1993. Polygonaceae. In: K. Kubitzki et al., eds. 1990+. The Families and Genera of Vascular Plants. 4+ vols. Berlin etc. Vol. 2, pp. 531–544. Carlquist, S. 2003. Wood anatomy of Polygonaceae: Analysis of a family with exceptional wood diversity. Bot. J. Linn. Soc. 141: 25–51. Graham, S. A. and C. E. Wood Jr. 1965. The genera of Polygonaceae in the southeastern United States. J. Arnold Arbor. 46: 91–113. Jaretzky, R. 1928. Histologische und karyologische Studien an Polygonaceen. Jahrb. Wiss. Bot. 69: 357–490. Lamb Frye, A. S. and K. A. Kron. 2003. *rbc*L phylogeny and character evolution in Polygonaceae. Syst. Bot. 28: 326–332. Laubengayer, R. A. 1937. Studies in the anatomy and morphology of the polygonaceous flower. Amer. J. Bot. 24: 329–343. Roberty, G. E. and S. Vautier. 1964. Les genres de Polygonacées. Boissiera 10: 7–128.

1. Ocreae absent; nodes not swollen; flowers usually enclosed in involucres or subtended by involucral bracts, these rarely absent (in *Gilmania*) . . . . . . 44a. Polygonaceae subfam. Eriogonoideae, p. 218
1. Ocreae present, persistent or deciduous; nodes usually swollen; flowers not enclosed in involucres or associated with involucral bracts . . . . 44b. Polygonaceae subfam. Polygonoideae, p. 479

44a. POLYGONACEAE Jussieu subfam. ERIOGONOIDEAE Arnott in M. Napier, Encycl. Brit. ed. 7, 5: 126. 1832 · Wild Buckwheat

James L. Reveal

Shrubs, subshrubs, or herbs, sometimes nearly arborescent (*Eriogonum*), perennial, biennial, or annual, homophyllous, polycarpic (rarely monocarpic in *Eriogonum*); taproot solid or, rarely, chambered (*Eriogonum*), slender to stout. **Stems** prostrate or decumbent to spreading or erect, sometimes scapose, rarely absent (*Eriogonum*), without recurved spines, glabrous or pubescent, sometimes glandular; nodes not swollen; tendrils absent; caudex stems tightly compact to spreading and at or just below the soil surface or spreading to erect and above the soil surface, woody; aerial flowering stems decumbent to spreading or erect, arising at nodes of caudex branches, at distal nodes of aerial branches, or directly from the root, slender to stout and solid or slightly to distinctly fistulose, rarely disarticulating in ringlike segments (*Eriogonum*).

Leaves deciduous (persistent in some shrubby and matted *Eriogonum* species), basal or basal and cauline, rarely only cauline, rosulate, alternate, or infrequently opposite (*Goodmania*) or in whorls of 3 (*Gilmania*); stipules absent (possibly vestigial in some perennial species of *Chorizanthe*); petiole present, sometimes indistinct, not articulate or with extrafloral nectaries; blade simple, rarely lobed (*Pterostegia*), rarely awn-tipped (*Goodmania*). **Inflorescences** terminal or terminal and axillary, cymose and dichotomously or trichotomously branched, or racemose, simple or compound umbellate, or capitate; bracts usually connate proximally, leaflike or scalelike, entire apically, sometimes awn-tipped, glabrous or pubescent. **Peduncles** absent or erect to deflexed relative to inflorescence branch, sometimes reflexed, straight or curved. **Involucral structures** tubular (involucre) or consisting of a series of individual bractlike lobes (involucral bracts) arranged in whorls or spirals, rarely absent (*Gilmania*), awns present or absent; involucre cylindric, prismatic, turbinate, campanulate, urceolate, or funnelform with 3–8(–36) usually erect teeth or 4–12 spreading to reflexed lobes (teeth and lobes are distal portions of proximally connate involucral bracts); involucral bracts in 1–3 whorls, rarely in spirals (*Johanneshowellia*), free or connate only at base, linear to oblanceolate or ovate. **Flowers** (1–)2–30(–100) per involucral structure, occasionally with stipelike base distal to articulations (*Eriogonum*); perianth accrescent in fruit, mostly white to red, yellow, light green, greenish white, maroon, or purple, urceolate to campanulate, occasionally glandular or pustulose abaxially, nearly always minutely glandular along midvein adaxially, glabrous or pubescent; tepals (5–)6, in 2 whorls of 3, connate proximally, typically not forming tube (except *Chorizanthe*, *Lastarriaea*, *Mucronea*, *Pterostegia*), petaloid or, rarely, coriaceous (*Lastarriaea*), monomorphic or dimorphic, entire, emarginate, or lobed to laciniate apically, rarely awn-tipped (*Lastarriaea*) or apiculate (*Eriogonum*); nectary a disk at base of ovary; stamens 3, 6, or 9 (variously 3–9 in *Chorizanthe*, *Mucronea*); staminodes absent; filaments usually distinct, occasionally forming staminal tube (*Chorizanthe*); pistils 3-carpellate, homostylous; ovary 1-locular; ovule 1, orthotropous, placentation basal; styles 3, distinct; stigmas capitate. **Achenes** brown to black or maroon, homocarpic, winged or unwinged, 3-gonous, less often lenticular or globose-lenticular to globose. **Seeds:** embryo straight or curved.

Genera 20, species ca. 325 (19 genera, 281 species in the flora): mainly temperate regions of w North America (Alaska to Mexico), uncommon in South America (Argentina and Chile) and e North America (West Virginia s to c Florida, e to Missouri, Oklahoma, and Texas).

Detailed habitat, elevation, and distribution data for the eriogonoid genera are maintained by the author and available on the Web at: "Eriogonoideae (Polygonaceae) of North America north of Mexico" (http://www.life.umd.edu/emeritus/reveal/pbio/eriog/key.html).

SELECTED REFERENCES Reveal, J. L. 1978. Distribution and phylogeny of the Eriogonoideae (Polygonaceae). Great Basin Naturalist Mem. 2: 169–190. Reveal, J. L. 1989. The eriogonoid flora of California (Polygonaceae: Eriogonoideae). Phytologia 66: 295–416. Reveal, J. L. 1989b. Notes on selected genera related to *Eriogonum* (Polygonaceae: Eriogonoideae). Phytologia 66: 236–245. Reveal, J. L. 2004b. Nomenclatural summary of Polygonaceae subfamily Eriogonoideae. Harvard Pap. Bot. 9: 143–230. Reveal, J. L. and C. E. Hardham. 1989. Notes on selected genera related to *Chorizanthe* (Polygonaceae: Eriogonoideae). Phytologia 66: 199–220.

1. Plants annual with involucre forming distinct tube, this cylindric or prismatic and awn-tipped typically with at least some awns uncinate or curved, rarely turbinate or urceolate to campanulate and then with 5 divergent, straight awns, or reduced to 1–4, often awn-tipped involucral bracts; flowers mostly 1–2(–6) per involucre.
    2. Involucres reduced to 1 highly modified, 2-winged bract; leaf blades entire or variously lobed; stems sprawling and spreading (Pterostegieae) . . . . 19. *Pterostegia*, p. 477
    2. Involucres tubular or reduced to 3(–4) bracts; leaf blades entire; stems erect to prostrate.
        3. Involucres reduced to 3(–4) bracts; tepals mucronate or awn-tipped; California.
            4. Perianths yellowish, densely tomentose abaxially; tepals mucronate apically; stamens 6 or 9; inflorescence branches tomentose (Hollisteriineae) . . . . . 11. *Hollisteria*, p. 444

4. Perianths light green to greenish white, thinly pubescent abaxially; tepals acute or awn-tipped apically; stamens 3; inflorescence branches thinly pubescent (Chorizanthineae).. 18.   *Lastarriaea*, p. 476

3. Involucres tubular; tepals not mucronate or awn-tipped (Chorizanthineae); widespread.

    5. Inflorescence bracts opposite, mostly 2, sometimes numerous and whorled; flowers 1(–2) per involucre; w North America................... 12.   *Chorizanthe*, p. 445

    5. Inflorescence bracts alternate and positioned on one side of branch or perfoliate around branch, 3-lobed or parted; flowers (1–)2–6 per involucre; widespread.

        6. Involucres 5-toothed, each terminated by divergent awn; flowers 4(–6) per involucre; perianth pubescent abaxially; wc California .... 17.   *Aristocapsa*, p. 475

        6. Involucres 2–4(6)-lobed or -toothed, or, if 5-awned apically, involucre with additional basal awns; flowers (1–)2–3 per involucre, perianth glabrous by densely papillate or pubescent abaxially; California.

            7. Involucres not awned basally; California.

                8. Perianths pubescent abaxially; flowers bisexual ....   13.   *Mucronea*, p. 470

                8. Perianths glabrous, densely papillate abaxially; flowers bisexual and unisexual, with proximal 1 pistillate and distal 1 bisexual .... 14.   *Systenotheca*, p. 471

            7. Involucres awned basally; sw North America.

                9. Basal awns 3, on saccate lobes; terminal awns 5, straight, involucres 3-angled; flowers 2 per involucre; sw North America..... .... 15.   *Centrostegia*, p. 473

                9. Basal awns 6, on nonsaccate basal lobes; terminal awns 6, uncinate; involucres 6-angled; flowers 3 per involucre; sw California .... 16.   *Dodecahema*, p. 474

1. Plants perennial or biennial, or, if annual, with involucre forming distinct tube, this turbinate to campanulate or hemispheric and awnless or tipped with erect, straight awns or reduced to awnless involucral bracts, absent entirely in *Gilmania*; flowers mostly (2–)6–200 per involucre (Eriogoneae, Eriogonineae).

  10. Plants perennial, if annual or biennial involucre tubular and awnless.

    11. Involucres tubular, the lobes rarely connate proximally; plants annual, biennial, or perennial; widespread... 1.   *Eriogonum*, p. 221

    11. Involucres reduced to series of 2–5 obscure, awnless bracts; plants perennial shrubs; ec California... 2.   *Dedeckera*, p. 430

  10. Plants annual; involucres tubular and awn-tipped or series of free or basally connate, awnless bracts, rarely absent (*Gilmania*).

    12. Involucres not awned or absent.

      13. Involucral bracts in 2 whorls of 3; sw Wyoming to ne Arizona and nw New Mexico.. 3.   *Stenogonum*, p. 431

      13. Involucral bracts absent or in 1 whorl or tight spiral; California, Nevada, sw Utah.

        14. Stamens 3; s California, sw Arizona......................... 9.   *Nemacaulis*, p. 441

        14. Stamens 9; se California, s Nevada, sw Utah.

            15. Perianths thinly pubescent abaxially; involucral bracts absent; Death Valley, California .... 5.   *Gilmania*, p. 433

            15. Perianths glabrous, smooth or minutely pustulose; involucral bracts (3–)4(–7); se California, s Nevada, sw Utah .... 10.   *Johanneshowellia*, p. 443

    12. Involucres awn-tipped.

      16. Involucres not tubular; se California, ec Nevada .... 4.   *Goodmania*, p. 433

      16. Involucres tubular; widespread.

1. ERIOGONUM Michaux, Fl. Bor.-Amer. 1: 246, plate 24. 1803 • Wild buckwheat

[Greek *erion*, wool, and *gony*, knee, alluding to the hairy nodes of the species first described, *E. tomentosum*]

James L. Reveal

Shrubs, subshrubs, or herbs, sometimes nearly arborescent, perennial, biennial, or annual, polycarpic or, rarely, monocarpic (subg. *Pterogonum*), synoecious (sometimes polygamodioecious in subg. *Micrantha* and *Oligogonum*, rarely dioecious in subg. *Oligogonum*); taproot slender to stout, solid, or rarely chambered (subg. *Pterogonum*). **Stems** prostrate or decumbent to erect, infrequently absent, glabrous or pubescent, sometimes glandular; caudex stems absent or woody, tightly compact to spreading and at or just below surface, or spreading to erect and above surface; aerial flowering stems arising at nodes of caudex branches, at distal nodes of aerial branches, or directly from the root, prostrate or decumbent to erect, slender to stout, solid or slightly to distinctly hollow and fistulose, rarely disarticulating into ringlike segments (subg. *Clastomyelon*). **Leaves** usually persistent through anthesis, occasionally persistent through growing season or longer, sometimes marcescent or quickly deciduous, basal and sometimes sheathing up stems, cauline, or basal and cauline, alternate, opposite, or whorled, 1 per node or fasciculate; petiole usually present, sometimes obscure; blade linear to orbiculate, entire apically. **Inflorescences** terminal or terminal and axillary, cymose and dichotomously or trichotomously branched, or racemose, simple or compound-umbellate, subcapitate, or capitate, occasionally distally uniparous due to suppression of secondary branches; branches mostly dichotomous except for initial trichotomous node, not brittle or disarticulating into segments, round and smooth, rarely grooved, angled or ridged, variously lanate, tomentose, floccose, sericeous, hispid, pilose-pubescent, or puberulent, occasionally glandular, rarely scabrellous; bracts 2–13 or more at proximal nodes, usually 3 distally, connate proximally, leaflike, semileaflike, or scalelike, not awn-tipped, glabrous or variously pubescent or glandular. **Peduncles** absent or erect to deflexed. **Involucres** 1–8 or more per cluster, smooth or ribbed, tubular, cylindric or narrowly turbinate to broadly campanulate or hemispheric; teeth 5–10, sometimes lobelike, not awned. **Flowers** bisexual or, infrequently, unisexual, (2–)6–100 per involucre at any single time during full anthesis, sometimes with stipelike base; perianth usually white to red or variously yellow, broadly campanulate when open, cylindric to urceolate when closed, glabrous or pubescent or glandular abaxially; tepals 6, connate proximally to ¹/₂ their length, monomorphic or dimorphic, usually entire apically, rarely emarginate; stamens 9; filaments adnate basally, glabrous or pubescent; anthers usually red to cream or yellow, oblong to ellipsoid or oval. **Achenes** included to exserted, various shades of brown, black, or occasionally yellow, rarely winged or ridged (subg. *Pterogonum*), lenticular or 3-gonous, glabrous or pubescent. **Seeds:** embryo curved or straight. $x = 10$.

Species ca. 250 (224 in the flora): North America (including n Mexico).

Eriogonum is the basal group of subfam. Eriogonoideae. Like all of its related genera, *Eriogonum* is a highly derived tetraploid taxon that has undergone rapid evolution in arid regions of western North America. The circumscription of the genera in the subfamily is now being studied molecularly and cladistically. The approach taken here is to divide the group into numerous genera, acknowledging that the resulting *Eriogonum* remains paraphyletic and that all genera of Eriogoneae are imbedded within *Eriogonum* as presently circumscribed. Resolution may well come with the reduction of the subfamily to two genera, *Eriogonum* and *Pterostegia* (including *Harfordia* Greene, a genus of Baja California, Mexico), or, at the other extreme, reducing *Eriogonum* to just two species. What the future will hold is difficult to ascertain at this time.

As presently circumscribed, *Eriogonum* is one of the larger genera in the flora area, being exceeded in numbers of species only by *Carex* (ca. 480), *Astragalus* (ca. 350), and *Penstemon* (ca. 250). As a native North American genus, *Eriogonum* (ca. 250) is second only to *Penstemon*. Ecologically, species of *Eriogonum* occur from the seashore to the highest mountains in the United States. They are among the last plants seen atop the Sierra Nevada and on the "outskirts" of Badwater in Death Valley. About one-third of the species are uncommon to rare in their distribution. The United States Department of the Interior currently lists some as endangered or threatened species. Some species tend to be weedy, and some of the annual species are aggressively so.

Species of *Eriogonum* have long been regarded as among the most difficult in North America to distinguish. Regional treatments should be consulted before attempting to use this review, especially for plants found outside California or the Intermountain West. Geographic distribution is a useful character, and such information is given fully in keys and discussion here to aid with identification. In addition to regional keys noted below, keys exist for Texas (J. L. Reveal 1970b), the Pacific Northwest (J. L. Reveal 1973), and the Great Plains (R. Kaul 1986). In each instance the nomenclature should be compared with that presented here. To aid in the identification of species belonging to the largest subgenus, *Eucycla*, regional keys are given here, thereby avoiding a long and complex key to the more than 100 species.

In collecting specimens of *Eriogonum*, try to obtain leaves (especially for annuals), fruits (especially those belonging to subg. *Pterogonum*), and ample flowers (rarely difficult to accomplish). Field observations on flower color, pubescence, and overall size and habit are useful. Some species (especially those of subg. *Oligogonum*) are dioecious, with the mature staminate and pistillate plants occasionally markedly different in aspect. It is not uncommon for several annual species to grow intermixed in disturbed places, so care must be taken to prevent mixed collections. Finally, as in all cases, collectors should try to sample the range of variation rather than concentrate on extremes.

Eriogonum has a long history of aboriginal use. Today, several members of the genus are in cultivation, especially in the rock or alpine garden (G. Nicholls 2002).

Members of *Eriogonum* are hosts for a number of butterfly species, including such endangered ones as the El Segundo dotted-blue (*Euphilotes battoides allyni*), Smith's dotted blue (*Euphilotes enoptes smithi*), and Lange's metalmark (*Apodemia mormo langei*). Species of the genus *Euphilotes* spend their entire life on particular species complexes. Other butterfly species found in association with *Eriogonum* and relatives (see P. A. Opler and A. B. Wright 1999) include the western green hairstreak (*Callophrys affinis*), desert green or Comstock's hairstreak (*C. comstocki*), bramble hairstreak (*C. dumetorum*), Lembert's hairstreak (*C. lemberti*), Sheridan's green hairstreak (*C. sheridani*), green hairstreak (*C. viridis*), varied blue (*Chalceria heteronea*), Rocky Mountain dotted-blue (*Euphilotes ancilla*), Bauer's dotted-blue (*E. baueri*), Bernardino dotted-blue (*E. bernardino*), Ellis's dotted-blue (*E. ellisi*), Pacific dotted-blue (*E. enoptes*), intermediate dotted-blue (*E. intermedia*), Mojave dotted-blue (*E. mojave*), pallid dotted-blue (*E. pallescens*), Rita dotted-blue (*E. rita*), Spalding's dotted-blue (*E. spaldingi*), Gorgon copper (*Gaeides gorgon*), gayas or Edward's blue (*Hemiargus ceranus gyas*), blue copper (*Lycaena heteronea*), small blue (*Philotiella speciosa*), Boisduval's blue (*Plebeius icarioides*), acmon blue (*P. acmon*), lupine blue (*P. lupini*), veined blue (*P. neurona*), California hairstreak (*Satyrium californica*), nut-brown hairstreak (*S. saepium*), Avalon scrub-hairstreak (*Strymon avalona*), and gray hairstreak (*S. melinus*). Flowering plants of *Eriogonum* are infrequently visited by the sooty hairstreak (*Satyrium fulginosum*), the flowers being a source of nectar for adults. According to Opler, several additional species and subspecies of these butterflies remain to be described.

SELECTED REFERENCES Reveal, J. L. 1973b. *Eriogonum* (Polygonaceae) of Utah. Phytologia 25: 169–217. Reveal, J. L. 1976. *Eriogonum* (Polygonaceae) of Arizona and New Mexico. Phytologia 34: 409–484. Reveal, J. L. 1985. An annotated key to *Eriogonum* (Polygonaceae) of Nevada. Great Basin Naturalist 45: 493–519. Stokes, S. G. 1936. The Genus *Eriogonum*, a Preliminary Study Based on Geographical Distribution. San Francisco.

Key to the Subgenera of *Eriogonum*

1. Flowers attenuate at base, sometimes only weakly so, stipelike base not at all winged; inflorescence bracts 2–10 or sometimes more.
    2. Tall, erect perennials; perianths white-tomentose . . . . 1d. *Eriogonum* subg. *Eriogonum*, p. 329
    2. Low, spreading, cespitose to shrubby, perianths glabrous or, if pubescent, not white-tomentose . . . . 1e. *Eriogonum* subg. *Oligogonum*, p. 331
1. Flowers not attenuate at base, stipelike base absent or, if present, then slightly winged; inflorescence bracts usually 3.
    3. Plants annual or, if perennial, then involucres usually distinctly pedunculate; leaves usually basal, usually tomentose at least abaxially, or if pilose or strigose then plants of desert regions.
        4. Involucres smooth, not ribbed or angled, usually on a peduncle or, if sessile, then involucres not appressed to inflorescence branches . . . . 1g. *Eriogonum* subg. *Ganysma*, p. 380
        4. Involucres angled to strongly ribbed, strongly appressed to inflorescence branches and sessile or terminal on bracteated branchlets, rarely pedunculate . . . .
        . . . . 1h. *Eriogonum* subg. *Oregonium*, p. 413
    3. Plants perennial herbs, subshrubs, or shrubs or, if annual or biennial, then stems usually 1, leafy throughout, and plants polygamodioecious; involucres (except 1 at fork of node in some) typically sessile, or if pedunculate throughout then leaves neither roundish and plants from central and southern Texas, nor perianths yellow-hispid and leaves hirsute.
        5. Plants herbaceous perennials, subshrubs, or shrubs, sometimes nearly arborescent, without jointed stems, without winged or distinctly ridged achenes . . . .
        . . . . 1a. *Eriogonum* subg. *Eucycla*, p. 224
        5. Plants annuals or biennials or, if perennials, then stems jointed basally, or achenes winged or distinctly ridged.
            6. Plants biennial or annual . . . . 1b. *Eriogonum* subg. *Micrantha*, p. 325
            6. Plants perennial.
                7. Involucres sessile in axils of inflorescence bracts, becoming ruptured at maturity by numerous bractlets and pubescent flowers; stems jointed basally . . . . 1c. *Eriogonum* subg. *Clastomyelon*, p. 327
                7. Involucres pedunculate, not ruptured by bractlets and (usually glabrous) flowers; stems not jointed basally . . . . 1f. *Eriogonum* subg. *Pterogonum*, p. 376

1a. ERIOGONUM Michaux subg. EUCYCLA (Nuttall) Kuntze in T. E. von Post and O. Kuntze, Lex. Gen. Phan., 204. 1903

Eucycla Nuttall, Proc. Acad. Nat. Sci. Philadelphia 4: 16. 1848

Shrubs, subshrubs, or herbs, sometimes nearly arborescent, perennial, glabrous or pubescent, rarely glandular; taproot woody. **Stems** matted to spreading, decumbent, or erect, infrequently absent, with or without persistent leaf bases, glabrous or variously pubescent or glandular; caudex stems woody, tightly compact to spreading and at or just below surface, or spreading to erect and above surface; aerial flowering stems arising at nodes of caudex branches, at distal nodes of aerial branches, or directly from root, prostrate or decumbent to erect, slender to stout and usually solid, infrequently slightly to distinctly fistulose and hollow. **Leaves** basal, sometimes in rosettes, sometimes sheathing up stems, cauline, or basal and cauline, 1 per node or fasciculate on flowering stems, at tips of dwarf branches, or on exposed woody caudices, usually persistent, occasionally persistent through growing season or longer, rarely quickly deciduous; blade glabrous or floccose to tomentose, occasionally also glandular. **Inflorescences** cymose, cymose-umbellate, umbellate, virgate, or racemose, mostly spreading and open to diffuse, sometimes dense, congested, or compact, sometimes reduced and in subumbellate, subcapitate, or capitate heads or reduced to a single terminal involucre; branches open to diffuse, spreading to erect, usually dichotomously branched except for initial trichotomous node, round and smooth, rarely grooved, angled, or ridged, tomentose to floccose or glabrous, occasionally lanate or glandular, rarely scabrellous; bracts usually 3, connate basally, usually scalelike, sometimes semileaflike or leaflike. **Peduncles** absent or erect, usually stout. **Involucres** 1–8 or more per cluster, narrowly turbinate to campanulate; teeth (3–)5–6(–10), erect or occasionally spreading, apex acute to obtuse or rounded. **Flowers** bisexual or, infrequently, unisexual, not attenuate at base, without stipelike base (except for slightly winged stipelike bases in *E. saxatile* and *E. crocatum*); perianth various shades of white, yellow, pink, or red, glabrous, glandular, or variously pubescent abaxially, usually glabrous adaxially except for minute glands and a few scattered hairs; tepals connate only basally or in proximal 1/2, monomorphic or dimorphic; stamens usually exserted, sometimes included; filaments variously pubescent but usually pilose proximally, infrequently glabrous. **Achenes** light to dark brown, not winged (but nearly so in *E. saxatile*), 3-gonous, glabrous, occasionally with minutely papillate beak. **Seeds:** embryo curved.

Species 107 (101 in the flora): w North America (including nw Mexico), mainly arid regions.

Species of subg. *Eucycla* not accounted for in this treatment are: *Eriogonum encelioides* Reveal & C. A. Hanson, *E. fastigiatum* Parry, *E. molle* Greene, *E. orcuttianum* S. Watson, *E. pondii* Greene, and *E. zapatoense* Moran. All are native to Baja California, Mexico.

The keys to species for subg. *Eucycla* are broken into geographic regions to aid in identification. They are: California (key 1, p. 225); British Columbia, Idaho, Oregon, and Washington (key 2, p. 228); Nevada and Utah (key 3, p. 231); Arizona and New Mexico (key 4, p. 237); Alberta, Colorado, Montana, and Wyoming (key 5, p. 239); and Manitoba, Saskatchewan, Great Plains states, and Texas (key 6, p. 242).

Key 1—California

1. Flowers with winged stipelike bases.
   2. Perianths white to rose or yellowish; inflorescences cymose, 10–25 × 5–12 cm; achenes 3.5–4 mm; s Coast Ranges, Transverse Ranges, s Sierra Nevada, desert ranges to east . . . . 100. *Eriogonum saxatile*
   2. Perianths bright yellow; inflorescences cymose-umbellate, 0.5–3 × 3–8 cm; achenes 2.5–3 mm; Santa Monica Mountains, Ventura County . . . . 101. *Eriogonum crocatum*
1. Flowers without stipelike bases.
   3. Plants forming cespitose or pulvinate mats; inflorescences capitate, or if racemose, subumbellate, cymose, or cymose-umbellate, then plants 0.1–1 dm.
      4. Inflorescences cymose, racemose, subumbellate, cymose-umbellate, or virgate, not capitate.
         5. Perianths glandular-hairy, whitish to reddish; flowering stems glandular-puberulent; Kern and Tulare counties . . . . 54. *Eriogonum breedlovei*
         5. Perianths glabrous, white to pink or rose, or yellowish with reddish spot; flowering stems not glandular; Inyo and Tulare counties.
            6. Flowers (1.5–)2–2.5 mm; tepals monomorphic; involucres 0.8–1.3 mm; s Sierra Nevada, Olancha Peak . . . . 72. *Eriogonum wrightii* (in part)
            6. Flowers 3–4 mm; tepals dimorphic, outer globose; involucres 1.5–2 mm; Cottonwood, Last Chance, and Panamint mountains . . . . 99. *Eriogonum gilmanii* (in part)
      4. Inflorescences capitate, rarely umbellate.
         7. Tepals dimorphic, outer often 2 times as wide as inner; scapes floccose to tomentose or lanate or nearly glabrous; common in e and s California . . . . 98. *Eriogonum ovalifolium*
         7. Tepals monomorphic, or if somewhat dimorphic, flowers, involucres, and scapes glandular-pubescent; ec and ne California.
            8. Perianths yellow or ochroleucous to reddish yellow, not white.
               9. Involucres glandular or glandular-hairy; scapes glandular; Sierra Nevada from Placer County to Fresno County and in desert ranges of Inyo and Mono counties, usually above 2500 m . . . . 53. *Eriogonum rosense*
               9. Involucres glabrous or floccose, occasionally glandular; scapes glabrous or glandular; Sierra Nevada and Great Basin; often below 1900 m.
                  10. Leaf blades oblanceolate to elliptic or ovate to obovate, 1–2 × 0.5–1(–1.5) cm; scapes 0.5–1.5(–2) dm, glandular at least proximal to inflorescence; involucres floccose abaxially; clay soils, Lassen County . . . . 62. *Eriogonum ochrocephalum*
                  10. Leaf blades oblanceolate to spatulate, elliptic or ovate, 0.4–0.8 (–1) × 0.15–0.4(–0.6) cm; scapes 0.2–0.8 dm, glabrous; involucres glabrous or sparsely floccose; volcanic soils, Lassen and Modoc counties . . . . 47. *Eriogonum prociduum*
            8. Perianths white to rose or pink, not yellow or ochroleucous (except sometimes yellow in *E. shockleyi*).
               11. Involucres glandular or pubescent to glandular-puberulent, membranous; c Sierra Nevada and White Mountains, Inyo and Mono counties, 3000–3900 m . . . . 59. *Eriogonum gracilipes*
               11. Involucres glabrous or tomentose, rigid; s Sierra Nevada, Transverse Ranges, and desert ranges to the east from Mono County s to Los Angeles and San Bernardino counties, typically below 3000 m.

12. Perianths and achenes pubescent; Last Chance Range, Inyo and
Mono counties .... 68. *Eriogonum shockleyi*
12. Perianths and achenes glabrous; s Sierra Nevada, Transverse
Ranges, and desert ranges to the east from Mono County s to Los
Angeles and San Bernardino counties .... 73. *Eriogonum kennedyi*
[3. Shifted to left margin.—Ed.]
3. Plants shrubs to subshrubs or spreading to erect perennials, not cespitose or matted; inflorescences branched or, if capitate, plants 1–5 dm.
13. Involucres 2–10 or more per node or in heads.
14. Tepals distinctly dimorphic, those of outer whorl to 2 times as wide as those of inner whorl.
15. Inflorescences cymose-umbellate or capitate; outer tepals globose; Inyo County
.... 99. *Eriogonum gilmanii* (in part)
15. Inflorescences umbellate-cymose to cymose; outer tepals not globose; Plumas
County n to Siskiyou and Modoc counties .... 96. *Eriogonum strictum*
14. Tepals essentially monomorphic, those of outer whorl ca. as wide as those of inner whorl.
16. Plants herbaceous or woody at base.
17. Leaf blades lanceolate to lance-ovate, 4–15(–25) cm, villous and green on both surfaces; Sierra Nevada and Great Basin region, Tulare and Mono counties n .... 94. *Eriogonum elatum*
17. Leaf blades oblanceolate to oblong-ovate, ovate, or elliptic, usually 2–6 cm, tomentose abaxially, tomentose or floccose to subglabrous, or glabrous adaxially; widespread.
18. Involucres (4–)5–7 × 3–8(–10) mm; flowering stems and inflorescence branches glabrous, rarely floccose; inflorescences subcapitate to cymose; leaf blades densely tomentose abaxially, greenish and glabrate adaxially; plants forming large mats; Channel Islands .... 93. *Eriogonum grande*
18. Involucres (2.5–)3–5(–7) × (1.5–)2–4 mm; flowering stems and inflorescence branches glabrous, floccose, or tomentose; inflorescences capitate to umbellate or cymose; leaf blades variously lanate to tomentose on both surfaces or tomentose to floccose or glabrous adaxially; plants erect, occasionally forming large mats; mainland.
19. Involucres and flowering stems glabrous or, if tomentose, plants not along the immediate coast; leaf blades variously lanate to tomentose on both surfaces or tomentose, floccose, or glabrous adaxially; plants mostly erect, not forming dense mats; widespread
.... 91. *Eriogonum nudum* (in part)
19. Involucres and flowering stems tomentose to floccose, rarely glabrous; inflorescences capitate, umbellate, or cymose; leaf blades whitish-lanate to tawny-tomentose on both surfaces, or tomentose to floccose or glabrous and green adaxially; plants forming dense mats; cliffs and bluffs along immediate coast from San Luis Obispo County n .... 92. *Eriogonum latifolium*
16. Plants usually shrubs.
20. Leaf blades narrowly linear or nearly so to oblanceolate, shorter than 2 cm and leaves usually strongly fasciculate; s coastal and desert ranges ....
.... 80. *Eriogonum fasciculatum*
20. Leaf blades linear-oblong to orbiculate or, if linear, longer than 2 cm and leaves not fasciculate; insular or coastal mesas and foothills.

21. Perianths glabrous; coastal and near-coast areas, Monterey County s to Los Angeles County . . . . 82. *Eriogonum parvifolium*
21. Perianths villous; coastal Santa Barbara s to Los Angeles County and insular.
    22. Leaf blades linear to narrowly oblong, 2–4(–5) cm, white-tomentose abaxially, cinereous to glabrate adaxially, margins revolute; Santa Cruz, Santa Rosa, and Anacapa islands . . . . 84. *Eriogonum arborescens*
    22. Leaf blades ovate or oblong-ovate to ovate or lanceolate to narrowly oblong, 1.5–10 cm, white-tomentulose or tomentose on both surfaces, margins not revolute; coastal regions, Santa Barbara s to Los Angeles County and on Santa Rosa, Santa Catalina, San Clemente, and Santa Barbara islands.
        23. Leaf blades ovate, 1.5–3 × 1–2.5(–3) cm, white-tomentulose; inflorescences congested, capitate; involucres narrowly turbinate; coastal regions, Santa Barbara s to Los Angeles County, and on Santa Rosa islands . . 81. *Eriogonum cinereum*
        23. Leaf blades oblong-ovate to ovate or lanceolate to narrowly oblong, 2–7(–10) × 1–5 cm, white-tomentose; inflorescences often open, cymose; involucres campanulate; Santa Catalina, San Clemente, and Santa Barbara islands . . . . 83. *Eriogonum giganteum*

[13. Shifted to left margin. —Ed.]
13. Involucres 1 per node.
    24. Involucres disposed in elongate racemes at tips of inflorescence branches.
        25. Involucres usually 6–7 mm; plants perennial herbs, (6–)12–18 dm; s Coast Ranges and Transverse Ranges south . . . . 71. *Eriogonum elongatum*
        25. Involucres (0.8–)1–6 mm; plants shrubs, subshrubs, or herbs, (1–)1.5–5 dm; widespread.
            26. Plants shrubs; Great Basin region, Mono and n Inyo counties . . . . . . 5. *Eriogonum nummulare*
            26. Plants subshrubs or herbs; widespread.
                27. Plants suffrutescent and much-branched basally; leaf blades oblanceolate to elliptic, 0.1–1 cm wide, margins often revolute.
                    28. Involucres 0.7–4 mm; flowers 2–4 mm; perianth usually white to pink; widespread . . . . 72. *Eriogonum wrightii* (in part)
                    28. Involucres (4–)5–6 mm; flowers usually 4–5 mm; perianth ochroleucous to rose; Santa Lucia Mountains, Monterey County . . 74. *Eriogonum butterworthianum*
                27. Plants not suffrutescent basally; leaf blades rounded to broadly ovate, (0.3–)1–3.5(–4) cm wide, margins not revolute.
                    29. Leaf blades round, (0.3–)0.5–1.5 × (0.3–)1–1.5 cm; petioles 1–2 cm; involucres campanulate; Inyo Mountains and Panamint and Coso ranges, Inyo County . . . . 77. *Eriogonum mensicola*
                    29. Leaf blades elliptic, oblong, ovate, or obovate, 1.5–4.5(–5) × 1–3.5 (–4) cm; petioles 1–7 cm; involucres turbinate or turbinate-campanulate; Inyo and White mountains, Mono and Inyo counties s to New York Mountains, San Bernardino County.
                        30. Leaf blades elliptic to oblong, (2–)2.5–4.5(–5) × 1.5–3.5(–4) cm; petioles (2–)3–7 cm; involucres turbinate-campanulate, 3–4 × 3–4 mm; White and Inyo mountains, Mono and Inyo counties . . . . 75. *Eriogonum rupinum*
                        30. Leaf blades elliptic to ovate or obovate, 1.5–4 × 1–2.5 cm; petioles 1–5 cm; involucres turbinate, 3–5 × 2–4 mm; White and Inyo mountains south to New York mountains . . . . 76. *Eriogonum panamintense*

[24. Shifted to left margin.—Ed.]
24. Involucres disposed at forks of branching system.
    31. Plants herbaceous perennials.
        32. Leaf blades round-ovate, 0.3–0.5(–1) × 0.3–0.5(–1) cm; Amador County... 90. *Eriogonum apricum*
        32. Leaf blades oblanceolate to elliptic or ovate, 1–6 × (0.3–)1–4 cm; widespread
                . . . . 91. *Eriogonum nudum* (in part)
    31. Plants subshrubs or shrubs.
        33. Perianths pubescent.
            34. Flowers (2.5–)3–4(–4.5) mm; perianth greenish yellow to yellow; involucres
                1.5–2.5(–3) mm, 4-toothed; leaf blades (0.3–)0.5–1.5(–3.5) cm; Imperial and
                San Diego counties. . . . . 85. *Eriogonum deserticola*
            34. Flowers 3–6(–7) mm; perianth white; involucres 2–5 mm, 5–8-toothed; leaf
                blades (0.7–)1.2–5 cm; Del Norte and Plumas counties.
                35. Involucres 3.5–5 mm; Del Norte County . . . . 86. *Eriogonum pendulum*
                35. Involucres 2–3 mm; Plumas County. . . . . 87. *Eriogonum spectabile*
        33. Perianths glabrous.
            36. Inflorescences usually compact, terminally cymose; involucres tomentose, floc-
                cose, or glabrous.
                37. Plants erect to spreading subshrubs in montane places, shrubs at lower
                    elevations, 0.2–15 dm; leaf blades (0.3–)1–3.5 cm, margins not revolute,
                    or if so, plants distinctly shrubby; e and s California . . . . 1. *Eriogonum microthecum*
                37. Plants spreading or matted subshrubs of desert mountains, 0.5–1.5 dm;
                    leaf blades 0.4–0.6 cm, margins revolute; New York Mountains, San Ber-
                    nardino County . . . . 15. *Eriogonum thornei*
            36. Inflorescences divaricatedly branched panicles or cymes; involucres glabrous.
                38. Outer tepals obovate to orbiculate; inflorescence branches green, dichoto-
                    mous, ascending; mountain slopes, s Coast Ranges, Transverse Ranges, s
                    Sierra Nevada, and desert ranges from Los Angeles and San Bernardino
                    counties n to Monterey, San Benito, and Mono counties . . . . . . 88. *Eriogonum heermannii*
                38. Outer tepals obovate; inflorescence branches grayish, mostly horizontal,
                    tiered; desert valleys, Kern County s to e San Diego and eastward . . . . .
                    . . . . 89. *Eriogonum plumatella*

Key 2—British Columbia, Idaho, Oregon, and Washington
1. Plants subshrubs or shrubs.
    2. Perianths villous; involucres 3.5–5 mm; peduncles (1–)3–10 cm; Josephine County,
      Oregon . . . . 86. *Eriogonum pendulum*
    2. Perianths glabrous; involucres usually shorter than 3.5 mm; peduncles absent or shorter
      than 1.5 cm; widespread.
        3. Leaves cauline; widespread . . . . 1. *Eriogonum microthecum*
        3. Leaves basal; Deschutes, Harney, Klamath, Lake, and Malheur counties, Oregon.
            4. Plants (3–)3.5–9 dm; perianths greenish white to pale yellow; Malheur County,
              Oregon. . . . . 63. *Eriogonum novonudum* (in part)
            4. Plants 0.3–1 dm; perianths yellow; Deschutes, Harney, Klamath, and Lake
              counties, Oregon . . . . 48. *Eriogonum cusickii* (in part)
1. Plants herbaceous, sometimes cespitose perennials.
    5. Plants erect to spreading, decumbent, or prostrate, not forming compact, dense, cespitose
      mats; inflorescences branched.
        6. Perianths usually hairy abaxially, sometimes sparsely so near base.
            7. Leaf blades lanceolate to lanceolate-ovate, 4–15(–25) cm, loosely villous and
              green on both surfaces; e of Cascade Ranges in e Oregon and Washington e to
              Elmore and Washington counties, Idaho, and Siskiyou Mountains of s Oregon
              e into desert ranges of Lake County, Oregon . . . . 94. *Eriogonum elatum*

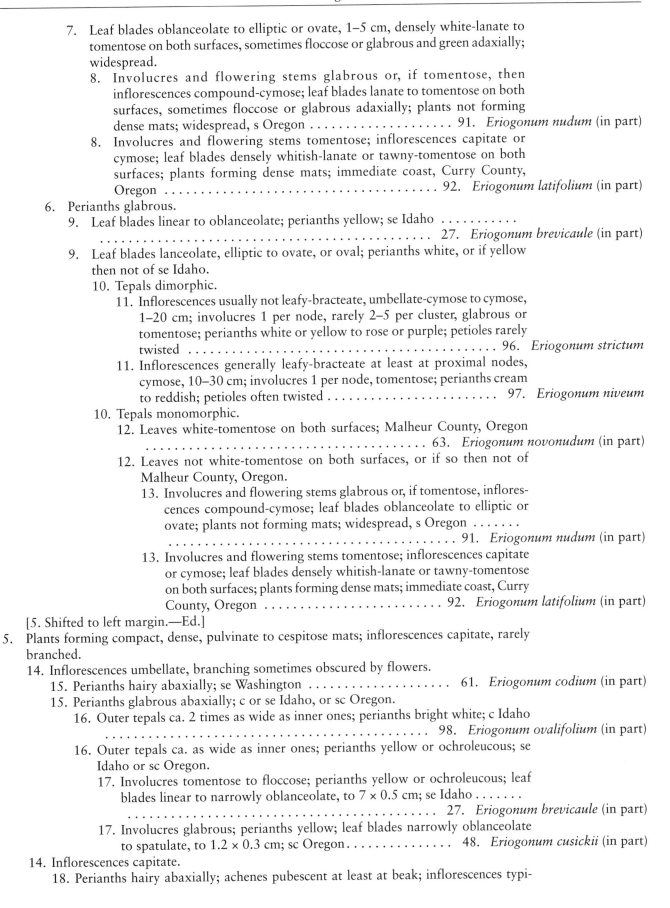

7. Leaf blades oblanceolate to elliptic or ovate, 1–5 cm, densely white-lanate to tomentose on both surfaces, sometimes floccose or glabrous and green adaxially; widespread.

    8. Involucres and flowering stems glabrous or, if tomentose, then inflorescences compound-cymose; leaf blades lanate to tomentose on both surfaces, sometimes floccose or glabrous adaxially; plants not forming dense mats; widespread, s Oregon . . . . 91. *Eriogonum nudum* (in part)

    8. Involucres and flowering stems tomentose; inflorescences capitate or cymose; leaf blades densely whitish-lanate or tawny-tomentose on both surfaces; plants forming dense mats; immediate coast, Curry County, Oregon . . . . 92. *Eriogonum latifolium* (in part)

6. Perianths glabrous.

    9. Leaf blades linear to oblanceolate; perianths yellow; se Idaho . . . . 27. *Eriogonum brevicaule* (in part)

    9. Leaf blades lanceolate, elliptic to ovate, or oval; perianths white, or if yellow then not of se Idaho.

      10. Tepals dimorphic.

        11. Inflorescences usually not leafy-bracteate, umbellate-cymose to cymose, 1–20 cm; involucres 1 per node, rarely 2–5 per cluster, glabrous or tomentose; perianths white or yellow to rose or purple; petioles rarely twisted . . . . 96. *Eriogonum strictum*

        11. Inflorescences generally leafy-bracteate at least at proximal nodes, cymose, 10–30 cm; involucres 1 per node, tomentose; perianths cream to reddish; petioles often twisted . . . . 97. *Eriogonum niveum*

      10. Tepals monomorphic.

        12. Leaves white-tomentose on both surfaces; Malheur County, Oregon . . . . 63. *Eriogonum novonudum* (in part)

        12. Leaves not white-tomentose on both surfaces, or if so then not of Malheur County, Oregon.

          13. Involucres and flowering stems glabrous or, if tomentose, inflorescences compound-cymose; leaf blades oblanceolate to elliptic or ovate; plants not forming mats; widespread, s Oregon . . . . 91. *Eriogonum nudum* (in part)

          13. Involucres and flowering stems tomentose; inflorescences capitate or cymose; leaf blades densely whitish-lanate or tawny-tomentose on both surfaces; plants forming dense mats; immediate coast, Curry County, Oregon . . . . 92. *Eriogonum latifolium* (in part)

[5. Shifted to left margin.—Ed.]

5. Plants forming compact, dense, pulvinate to cespitose mats; inflorescences capitate, rarely branched.

  14. Inflorescences umbellate, branching sometimes obscured by flowers.

    15. Perianths hairy abaxially; se Washington . . . . 61. *Eriogonum codium* (in part)

    15. Perianths glabrous abaxially; c or se Idaho, or sc Oregon.

      16. Outer tepals ca. 2 times as wide as inner ones; perianths bright white; c Idaho . . . . 98. *Eriogonum ovalifolium* (in part)

      16. Outer tepals ca. as wide as inner ones; perianths yellow or ochroleucous; se Idaho or sc Oregon.

        17. Involucres tomentose to floccose; perianths yellow or ochroleucous; leaf blades linear to narrowly oblanceolate, to 7 × 0.5 cm; se Idaho . . . . 27. *Eriogonum brevicaule* (in part)

        17. Involucres glabrous; perianths yellow; leaf blades narrowly oblanceolate to spatulate, to 1.2 × 0.3 cm; sc Oregon. . . . . 48. *Eriogonum cusickii* (in part)

  14. Inflorescences capitate.

    18. Perianths hairy abaxially; achenes pubescent at least at beak; inflorescences typi-

cally branched but occasionally seemingly or actually capitate.
  19. Perianths lemon yellow; se Washington .... 61. *Eriogonum codium* (in part)
  19. Perianths white to rose or yellow; s Idaho .... 68. *Eriogonum shockleyi*
[18. Shifted to left margin.—Ed.]
18. Flowers glabrous or glandular, not hairy; achenes glabrous, except on beaks; inflorescences
  usually capitate, rarely umbellate to cymose.
  20. Involucres membranous.
    21. Perianths cream; se Idaho .... 50. *Eriogonum mancum* (in part)
    21. Perianths pale yellow to yellow; c and w Idaho, or e Oregon.
      22. Perianths pale yellow; flowering stems glandular-hairy; leaf blades white-
        tomentose and glandular on both surfaces; Wallowa County, Oregon, and Adams
        and Idaho counties, Idaho .... 55. *Eriogonum scopulorum*
      22. Perianths yellow; flowering stems glandular or thinly floccose to tomentose;
        leaf blades tomentose on both surfaces or less so and grayish to greenish
        adaxially, not glandular; Blaine, Butte, Camas, Custer, Lemhi, and Owyhee
        counties, Idaho, or Malheur County, Oregon.
        23. Flowering stems glandular or thinly floccose; involucres tomentose or glan-
          dular; Blaine, Butte, Camas, Custer, Lemhi, and Owyhee counties, Idaho
          .... 49. *Eriogonum crosbyae* (in part)
        23. Flowering stems floccose to tomentose; involucres glabrous except for
          margins of teeth; Malheur County, Oregon .... 56. *Eriogonum chrysops*
  20. Involucres rigid.
    24. Tepals dimorphic, those of outer whorl generally 2 times as wide as those of inner
      whorl or, if dimorphic but less than 2 times as wide, plants of subalpine or alpine
      habitats, or leaf blades elliptic to round and perianths white to cream or rose ..
      .... 98. *Eriogonum ovalifolium* (in part)
    24. Tepals monomorphic, those of outer whorl ca. as wide as those of inner whorl or, if
      slightly dimorphic, plants not of subalpine or alpine habitats, or leaf blades oblan-
      ceolate to narrowly elliptic and perianths usually yellow.
      25. Inflorescences of 1 involucre atop peduncle; inflorescence bracts absent;
        Beaverhead Mountains, Lemhi County, Idaho .... 51. *Eriogonum soliceps*
      25. Inflorescences of 2–8 involucres atop scape; inflorescence bracts subtending
        cluster of involucres; widespread.
        26. Perianths distinctly pustulose along midribs and base abaxially.
          27. Perianths yellow; Blaine, Butte, Camas, Custer, Lemhi, and Owyhee
            counties, Idaho .... 49. *Eriogonum crosbyae* (in part)
          27. Perianths white to cream; Butte, Clark, Custer, and Lemhi counties,
            Idaho.. 50. *Eriogonum mancum* (in part)
        26. Perianths not pustulose abaxially.
          28. Involucres turbinate, 2.5–4.5(–6) × 1.5–2.5(–4) mm; leaf blades
            (0.2–)1–10(–12).
            29. Flowering stems to 1.4 dm, tomentose; se Idaho ....
              .... 27. *Eriogonum brevicaule* (in part)
            29. Flowering stems (0.6–)1–4(–5) dm, glabrous or rarely slightly floc-
              cose; sw Idaho, se Oregon.................... 62. *Eriogonum ochrocephalum*
          28. Involucres campanulate, 2–4 × 2–4 mm; leaf blades 0.5–1.5 cm.
            30. Involucres (2–)3–4 × (2.5–)3–4 mm, glabrous or floccose only on
              teeth; flowers 2–3.5(–4) mm, perianth bright yellow; Harney, Lake,
              and Baker counties, Oregon .... 47. *Eriogonum prociduum*
            30. Involucres 2–3.5 × 2–3.5 mm, floccose throughout; flowers 1.5–3
              mm, perianth yellow; Harney and Lake counties, Oregon.....
              .... 49. *Eriogonum crosbyae* (in part)

Key 3—Nevada and Utah

1. Plants subshrubs or shrubs.
   2. Perianths pubescent.
      3. Leaves not fasciculate, blade lanceolate to elliptic or oblanceolate, tomentose or nearly so; Thousand Lake Mountain, Wayne County, Utah .... 16. *Eriogonum corymbosum* (in part)
      3. Leaves usually fasciculate, blade usually oblanceolate, canescent on both surfaces or densely tomentose abaxially and canescent adaxially; s Nevada and sw Utah . . . . 80. *Eriogonum fasciculatum*
   2. Perianths glabrous.
      4. Involucres usually racemosely arranged along tips of inflorescence branches.
         5. Flowering stems and inflorescence branches angled and ribbed or scabrous, or if round and smooth then rigid or stout, glabrous and with short, spinose lateral branches; s Nevada and Washington County, Utah . . . . 88. *Eriogonum heermannii* (in part)
         5. Flowering stems and inflorescence branches not angled and ribbed or scabrellous or with short, spinose lateral branches, tomentose to floccose or, if glabrous, slender and flexible; widespread.
            6. Leaves fasciculate, blade 0.2–1.5 × 0.25–0.7(–0.9) cm, oblanceolate to elliptic; wc and s Nevada, and Washington County, Utah . . . . 72. *Eriogonum wrightii*
            6. Leaves not fasciculate, blade 1.5–3.5 × 0.2–0.8(–1.2) cm, linear-lanceolate or linear-oblanceolate to narrowly oblong; Nevada, s and e Utah.
               7. Leaf blades usually linear-lanceolate to narrowly oblong or narrowly elliptic; inflorescence branches tomentose to floccose or glabrate, occasionally glabrous; e Utah . . . . 4. *Eriogonum leptocladon*
               7. Leaf blades broadly oblanceolate to broadly elliptic; inflorescence branches usually tomentose; Nevada and s Utah . . . . 5. *Eriogonum nummulare*
      4. Involucres dichotomously arranged even at tips of inflorescence branches.
         8. Margins of leaf blades revolute or at least inrolled.
            9. Leaf blades (0.5–)2–6 cm; inflorescences densely branched, usually glabrous; involucres narrowly turbinate, glabrous; San Juan County, Utah . . . . 7. *Eriogonum leptophyllum*
            9. Leaf blades 0.5–1.8(–2.5) cm; inflorescences usually sparsely branched, tomentose to floccose; involucres turbinate to turbinate-campanulate, tomentose to floccose or rarely glabrous; widespread.
               10. Tepals monomorphic; widespread . . . . 1. *Eriogonum microthecum* (in part)
               10. Tepals dimorphic; e Utah.
                  11. Plants 1–2.5 dm; leaf blades oblanceolate, usually thinly tomentose and grayish, rarely glabrous and green adaxially; involucres 2.5–4.5 mm wide; San Juan County, Utah . . . . 8. *Eriogonum clavellatum*
                  11. Plants 0.2–0.8 dm; leaf blades linear-oblanceolate to narrowly elliptic, white- to reddish- or tannish-tomentose adaxially; involucres 1.5–3 mm wide; Carbon, Emery, and Grand counties s to Garfield, San Juan, Sevier, and Wayne counties, Utah . . . . 10. *Eriogonum bicolor*
         8. Margins of leaf blades usually plane.
            12. Inflorescences flat-topped and in tiers, branches zigzag; leaf blades tomentose on both surfaces; plants erect; s Clark County, Nevada . . . . 89. *Eriogonum plumatella*
            12. Inflorescences usually open, not flat-topped, branches not zigzag; leaf blades variable but typically not tomentose on both surfaces; plants spreading to rounded, typically wider than tall; widespread.
               13. Leaf apices sharply acute; leaf blades usually narrowly elliptic to elliptic, 0.2–1.2 cm wide . . . . 1. *Eriogonum microthecum* (in part)

[13. Shifted to left margin.—Ed.]
13. Leaf apices acute to rounded; leaf blades lanceolate to oblanceolate or elliptic to oval, cordate, or nearly orbiculate, (0.2–)0.3–3(–3.5) cm wide.
  14. Flowering stems and inflorescence branches tomentose to floccose, rarely subglabrous.
    15. Leaf blades linear-lanceolate to lanceolate, 3.5–7(–9) × 0.3–0.6(–0.8) cm, densely white-tomentose abaxially, less so and greenish adaxially; leaf margins rarely crenulate; Duchesne County, Utah. . . . . 18. *Eriogonum hylophilum*
    15. Leaf blades lanceolate to oblanceolate or elliptic to nearly orbiculate, rarely cordate, 1–5 × (0.3–)0.5–3(–3.5) cm, densely silvery to white-, tannish-, or brownish-tomentose on both surfaces or less so or glabrous and green abaxially; leaf margins occasionally crenulate; not of Duchesne County, Utah.
      16. Leaf blades densely white-, tannish-, or brownish-tomentose on both surfaces or less so or nearly glabrous and green abaxially, (0.5–)1–3(–4.5) cm; Clark County, Nevada, and s and e Utah. . . . . 16. *Eriogonum corymbosum* (in part)
      16. Leaf blades densely silvery-tomentose on both surfaces or rarely less so or subglabrous and greenish abaxially, 3–5 cm; Wellington area, Carbon and Emery counties, Utah . . . . 17. *Eriogonum lancifolium*
  14. Flowering stems and inflorescence branches glabrous.
    17. Leaf blades green and glabrous, rarely thinly floccose adaxially; flowers 3–4 mm, perianth yellow; San Rafael Desert, Emery and Wayne counties, Utah . . . . . . 20. *Eriogonum smithii*
    17. Leaf blades tomentose to floccose at least on one surface; flowers 1.5–4 mm; perianth usually white, sometimes yellow, cream, or pink; widespread.
      18. Involucres campanulate; leaf blades floccose adaxially, thinly floccose or glabrous abaxially; Nevada. . . . . 88. *Eriogonum heermannii* (in part)
      18. Involucres turbinate; leaf blades variable but always with one surface tomentose; Utah.
        19. Perianths usually yellow to pale yellow, occasionally cream; leaf blades lanceolate to oblanceolate or elliptic, usually densely tomentose on both surfaces, sometimes less so and greenish adaxially; Washington County, Utah . . . . 16. *Eriogonum corymbosum* (in part)
        19. Perianths white; leaf blades lanceolate or broadly elliptic, usually densely tomentose only abaxially; Millard or Uintah counties, Utah.
          20. Plants spreading to somewhat sprawling subshrubs, (1.5–)2–4 dm; leaves commonly basal or sheathing up proximal ¼ of flowering stems, blade broadly elliptic, 1–2(–2.5) cm; Millard County, Utah . . . . 6. *Eriogonum ammophilum*
          20. Plants erect shrubs, 3–5 dm; leaves cauline, sheathing up proximal ½ of flowering stems, blade usually lanceolate, 3–6 cm; Uintah County, Utah . . . . 23. *Eriogonum lonchophyllum* (in part)
1. Plants herbs.
  21. Plants cespitose, matted, or pulvinate; inflorescences usually capitate.
    22. Perianths hairy.
      23. Achenes tomentose; widespread . . . . 68. *Eriogonum shockleyi*
      23. Achenes glabrous; c and e Nevada, c, ne, and w Utah.
        24. Flowers 2–2.5(–3) mm; perianth pale yellow to yellow; Emery and Garfield counties, Utah . . . . 67. *Eriogonum aretioides*
        24. Flowers 3–4.5 mm; perianth white to rose; e Nevada and ne and w Utah.
          25. Flowering stems spreading to prostrate, (0.5–)1.5–5(–8) cm; leaf blades narrowly elliptic, 0.3–1 × 0.1–0.2(–0.3) cm; e Nevada and w Utah . . . . 65. *Eriogonum villiflorum*
          25. Flowering stems erect, 0.1–1 cm; leaf blades oblanceolate to elliptic, 0.3–0.4 × 0.07–0.1 cm; ne Utah . . . . 66. *Eriogonum tumulosum*

[22. Shifted to left margin.—Ed.]
22. Perianths glabrous, sparsely pilose, or glandular.
  26. Tepals strongly dimorphic or, if only slightly so, leaf blades oval to rotund . . . .
    . . . . 98. *Eriogonum ovalifolium* (in part)
  26. Tepals monomorphic or, if slightly dimorphic, leaf blades not oval to rotund.
    27. Involucre tubes membranous.
      28. Perianths greenish yellow to pale yellow, rarely yellow.
        29. Involucres 2.5–3.5 mm, glabrous or sparsely pubescent with white hairs
          abaxially; flowers (2.5–)3–3.5 mm; perianths glabrous or sparsely glandu-
          lar, usually greenish yellow or pale yellow; dry limestone and granitic mon-
          tane slopes, Elko County and n White Pine County, Nevada . . . . 57. *Eriogonum kingii*
        29. Involucres 2–2.5 mm, sparsely to densely tomentose with scattered glands
          along ribs; flowers (2–)2.5–3 mm; perianths sparsely glandular, yellow;
          moist, crusted, sandy, alkaline flats, Ruby Valley, Elko County, Nevada
          . . . . 58. *Eriogonum argophyllum*
      28. Perianths white to rose.
        30. Leaf blades oblanceolate to elliptic, (0.3–)1–1.5(–2) × (0.2–)0.3–0.6 cm,
          glandular and densely white-tomentose abaxially; involucres 5–7 per clus-
          ter, sparsely glandular to glandular-puberulent or pubescent; White Moun-
          tains, Esmeralda County, Nevada . . . . 59. *Eriogonum gracilipes*
        30. Leaf blades oblanceolate to spatulate, 0.2–0.5(–1) × (0.1–)0.2–0.4 cm, glan-
          dular and densely white- or greenish-tomentose abaxially; involucres 2–4
          per cluster, sparsely glandular-puberulent; Snake Range, White Pine County,
          Nevada . . . . 60. *Eriogonum holmgrenii*
    27. Involucre tubes rigid.
      31. Leaf blades linear to narrowly oblanceolate or narrowly elliptic, (1.5–)3–9
        (–12) × 0.1–0.5(–0.7) cm; n Utah.
        32. Leaf blades white- to gray-tomentose abaxially, less so and grayish or occa-
          sionally greenish to thinly floccose and bright green adaxially; perianths usu-
          ally yellow, occasionally ochroleucous; widespread . . . . 27. *Eriogonum brevicaule* (in part)
        32. Leaf blades densely reddish- to brownish-lanate abaxially, less so to to-
          mentose adaxially; perianths ochroleucous, rarely yellow; Cache and n
          Morgan counties, Utah . . . . 30. *Eriogonum loganum*
      31. Leaf blades lanceolate or oblanceolate to ovate, 0.2–3 × 0.05–1 cm; Nevada or
        c and n Utah.
        33. Scapes glandular or glandular-hairy.
          34. Involucres glandular and sparsely hairy; scapes densely glandular-hairy
            throughout, 0.1–0.9(–1.1) dm; wc Nevada . . . . 53. *Eriogonum rosense*
          34. Involucres sparsely to densely floccose; scapes glandular or glandular
            only proximal to inflorescences, (0.5–)1–4.5(–5) dm; nw Nevada . . .
            . . . . 62. *Eriogonum ochrocephalum*
        33. Scapes glabrous or tomentose to floccose.
          35. Scapes glabrous.
            36. Leaf margins crenulate or, if plane, leaf blades 1.5–6(–7) cm, scapes
              8–25(–30) cm, and plants of s Utah.
              37. Perianths usually greenish white to creamy white or pale yel-
                lowish white, rarely pale yellow or yellow; n Utah . . . .
                . . . . 27. *Eriogonum brevicaule* (in part)
              37. Perianths white; s Utah . . . . 42. *Eriogonum panguicense*
            36. Leaf margins plane.
              38. Perianths white to pink or rose; Esmeralda County, Nevada
              . . . . 73. *Eriogonum kennedyi*
              38. Perianths yellow; Humboldt and Washoe counties, Nevada.

39. Flowers pustulose; leaf blades obovate, (1–)1.5–2(–2.2) cm; Humboldt and Washoe counties, Nevada . . . . 49. *Eriogonum crosbyae* (in part)
39. Perianths not pustulose; leaf blades oblanceolate to spatulate or elliptic to ovate, 0.3–1(–1.5) cm; n Washoe and Humboldt counties, Nevada . . . . 47. *Eriogonum prociduum*

[35. Shifted to left margin.—Ed.]

35. Scapes tomentose to floccose.
   40. Leaf margins crenulate; Wasatch Mountains, Utah, above 3000 m . . . . 27. *Eriogonum brevicaule* (in part)
   40. Leaf margins not crenulate; Nevada or, if in Utah, below 2800 m.
      41. Perianths yellowish white or whitish to cream, glandular; Silver Peak Range, Esmeralda County, Nevada . . . . 45. *Eriogonum tiehmii*
      41. Perianths cream or yellow, glabrous; not of Esmeralda County, Nevada.
         42. Perianths yellow.
            43. Leaf blades narrowly oblanceolate to oblanceolate, (0.8–)1–4(–4.5) cm, densely grayish-tomentose abaxially, tomentose to floccose and grayish to green adaxially; Elko and White Pine counties, Nevada, and Box Elder, Rich, and Tooele counties, Utah . . . . 27. *Eriogonum brevicaule* (in part)
            43. Leaf blades oblanceolate to elliptic or rarely ovate, 0.4–2(–2.5) cm, densely greenish- to white-tomentose on both surfaces, or densely greenish-white-tomentose abaxially, thinly tomentose and whitish green adaxially; n Nevada and nw Utah.
               44. Perianths not pustulose; e Elko County, Nevada, and Box Elder, Juab, and Tooele counties, Utah . . . . 43. *Eriogonum desertorum*
               44. Perianths pustulose; Washoe County s to Mineral County, e to Humboldt and w Elko counties, Nevada . . . . 49. *Eriogonum crosbyae* (in part)
         42. Perianths cream to white, pink, or rose.
            45. Flowering stems 0.02–0.15 dm; leaf blades (0.2–)0.25–0.45(–0.7) × (0.07–)0.1–0.25 cm; San Francisco Mountains, Beaver County, Utah . . . . 52. *Eriogonum soredium*
            45. Flowering stems 0.1–2 dm; leaf blades (0.5–)0.8–2(–2.3) × (0.2–)0.3–1.2 (–1.8) cm; not in San Francisco Mountains, Beaver County, Utah.
               46. Leaves sheathing flowering stem (1.5–)2–4 cm; Churchill Narrows, Lyon County, Nevada . . . . 46. *Eriogonum diatomaceum*
               46. Leaves not sheathing flowering stem; not of Churchill County, Nevada.
                  47. Perianths not pustulose; leaf blades elliptic to obovate or suborbiculate, (0.7–)0.9–1.3(–2) × (0.3–)0.5–0.9(–1.1) cm; Elko, Eureka, Humboldt, Lander, and Pershing counties, Nevada . . . . 44. *Eriogonum anemophilum*
                  47. Perianths pustulose; leaf blades oblanceolate to spatulate or ovate, 0.5–1.5(–1.8) × (0.2–)0.3–0.5(–1) cm; House Range, Millard County, Utah . . . . 50. *Eriogonum mancum*

[21. Shifted to left margin.—Ed.]

21. Plants herbaceous; inflorescences branched.
   48. Flowers with stipelike bases; leaf blades obovate to rounded and lanate to tomentose; Esmeralda and Nye counties, Nevada . . . . 100. *Eriogonum saxatile*
   48. Flowers without stipelike bases; leaf blades not obovate to rounded and densely lanate to tomentose; widespread.
      49. Perianths usually hairy abaxially.
         50. Plants matted; leaf blades 0.3–0.8 × 0.1–0.2 cm, margins revolute; to be expected in Daggett County, Utah . . . . 34. *Eriogonum acaule*
         50. Plants not matted; leaf blades 1–15(–25) × 0.3–6 cm, margins plane; Nevada.

51. Leaf blades oblanceolate to elliptic, 1–4 × 0.3–2 cm, densely tomentose abaxially, tomentose to floccose or glabrous adaxially; w Nevada . . . . 91. *Eriogonum nudum* (in part)
51. Leaf blades lanceolate to lance-ovate, 4–15(–25) × 1.5–6 cm, loosely villous and green on both surfaces; nw and n Nevada . . . . 94. *Eriogonum elatum*

[49. Shifted to left margin.—Ed.]

49. Perianths glabrous.
  52. Inflorescences with involucres racemosely arranged.
    53. Perianths pale yellow to yellow; flowering stems and inflorescence branches usually glabrous, rarely tomentose, occasionally fistulose; s Utah . . . . 79. *Eriogonum zionis*
    53. Perianths white to rose; flowering stems and inflorescence branches usually tomentose, rarely glabrous, not fistulose; widespread.
      54. Inflorescences with 5–20 or more racemosely arranged involucres along most of length of inflorescence branches; plants 3–8(–10) dm; e Nevada and Utah . . . . 78. *Eriogonum racemosum*
      54. Inflorescences with 3–5 racemosely arranged involucres at tips of distalmost inflorescence branches; plants 1–5 dm; c and sw Nevada.
        55. Leaf blades round, (0.3–)0.5–1.5 cm; involucres campanulate, 2–3(–4) mm; Sheep Range, Clark County, Nevada . . . . 77. *Eriogonum mensicola*
        55. Leaf blades elliptic to oblong, ovate, or obovate, 1.5–4.5(–5) cm; involucres turbinate to turbinate-campanulate, 3–5 mm; Clark, Esmeralda, Lander, Lincoln, Mineral, and Nye counties, Nevada.
          56. Plants 3–5 dm; leaf blades elliptic to oblong, (2–)2.5–4.5(–5) × 1.5–3.5(–4) cm, thinly tomentose abaxially, less so to floccose and greenish adaxially; Esmeralda, Lander, Lincoln, Mineral, and Nye counties, Nevada . . . . 75. *Eriogonum rupinum*
          56. Plants 1.5–3 dm; leaf blades elliptic to ovate or obovate, 1.5–4 × 1–2.5 cm, white-tomentose abaxially, slightly less so and often greenish adaxially; w Esmeralda, sw Nye, and n Clark counties, Nevada . . . . 76. *Eriogonum panamintense*
  52. Involucres not racemosely arranged.
    57. Flowering stems and inflorescence branches usually tomentose to floccose.
      58. Leaf blades linear or narrowly oblanceolate to narrowly elliptic or narrowly lanceolate, 0.1–0.5(–0.7) cm wide; n Utah.
        59. Peduncles absent; plants 1–3.5 dm, tomentose; n Utah . . . 27. *Eriogonum brevicaule* (in part)
        59. Peduncles 0.2–1 cm; plants 0.5–0.8(–1) dm, thinly floccose or glabrous; e Emery and Grand counties, Utah . . . . 33. *Eriogonum contortum*
      58. Leaf blades lanceolate to elliptic or spatulate, 0.3–1.3 cm wide; c Utah.
        60. Perianths bright yellow; leaf blades elliptic, (0.8–)1–1.3 cm wide; Millard County, Utah . . . . 32. *Eriogonum natum*
        60. Perianths ochroleucous or pale yellow; leaf blades lanceolate to narrowly elliptic or narrowly spatulate, 0.3–1 cm wide; Beaver, Iron, Juab, Millard, Piute, Sanpete and Sevier counties, Utah . . . . 35. *Eriogonum spathulatum* (in part)
    57. Flowering stems and inflorescence branches usually glabrous.
      61. Tepals dimorphic; plants forming loose mats; involucres 4–6 mm; inflorescences umbellate-cymose to cymose; n Nevada . . . . 96. *Eriogonum strictum*
      61. Tepals usually monomorphic; plants not forming mats, plants of Utah with subcapitate to umbellate-cymose inflorescences; inflorescences usually cymose; widespread.
        62. Inflorescences elongate, rather narrow; plants 2–3(–5) dm; involucres usually 3–5 mm; wc Nevada . . . . 91. *Eriogonum nudum* (in part)
        62. Inflorescences spreading or, if narrow, plants of ne Utah; plants 0.4–4.5 (–6) dm; involucres usually 1.5–3.5 mm; Utah.

[63. Shifted to left margin.—Ed.]

63. Leaf blades linear, 1.5–6 × 0.05–0.1 cm; Sevier County, Utah . . . . 37. *Eriogonum mitophyllum*
63. Leaf blades linear to oval, (0.5–)1–10(–12) × 0.1–0.7 cm; widespread.
  64. Leaf blades linear to narrowly lanceolate, narrowly elliptic, or oblanceolate; n and ne Utah.
    65. Leaf blades 3–10(–12) cm, linear to narrowly oblanceolate; n and ne Utah . . . . 27. *Eriogonum brevicaule* (in part)
    65. Leaf blades 1–3(–4) cm, linear to narrowly elliptic or lanceolate; ne Utah.
      66. Leaf blades lanceolate, margins plane; inflorescences sparsely branched; perianths ochroleucous to pale yellow, rarely yellow; Uintah County, Utah . . . . 29. *Eriogonum ephedroides*
      66. Leaf blades linear to narrowly elliptic, margins revolute or nearly so; inflorescences densely branched; perianths yellow; Duchesne, Emery, Grand, and Uintah counties.
        67. Plants 1–3.5 dm; peduncles 0.1-0.2 cm; Duchesne and Uintah counties, Utah . . . . 28. *Eriogonum viridulum*
        67. Plants 0.5–0.8(–1) dm; peduncles 0.2–1 cm; Emery and Grand counties, Utah . . . . 33. *Eriogonum contortum* (in part)
  64. Leaf blades elliptic, spatulate, oblong, or ovate to round, or if oblanceolate, then plants of sw Utah.
    68. Leaf blades tomentose on both surfaces; Beaver and Millard counties, Utah.
      69. Perianths ochroleucous to pale yellow, rarely yellow; flowers 3.5–4 mm; peduncles 0.2–1 cm; volcanic sands . . . . 38. *Eriogonum artificis*
      69. Perianths white to ochroleucous; flowers 2.5–3 mm; peduncles absent; limestone gravel and outcrops.
        70. Leaf blades lanceolate to narrowly elliptic or narrowly spatulate . . . . 35. *Eriogonum spathulatum* (in part)
        70. Leaf blades ovate to round . . . . 39. *Eriogonum eremicum*
    68. Leaf blades tomentose abaxially, less so to floccose or glabrous adaxially; widespread.
      71. Leaf margins crenulate; leaves sheathing up stem; inflorescences subcapitate to cymose; plants spreading.
        72. Perianths greenish white to creamy white or pale yellowish white, rarely pale yellow or yellow; Wasatch Mountains, Box Elder and Weber counties, Utah . . . . 27. *Eriogonum brevicaule* (in part)
        72. Perianths white; Bull Mountain, Garfield County, Utah . . . . 41. *Eriogonum cronquistii*
      71. Leaf margins not crenulate; leaves usually basal; inflorescences cymose; plants usually erect.
        73. Flowering stems and inflorescence branches bright green; Kane and Washington counties, Utah . . . . 22. *Eriogonum thompsoniae*
        73. Flowering stems and inflorescence branches usually gray to reddish or, if dark green, plants of wc Utah.
          74. Leaf blades oblanceolate to narrowly lanceolate, 0.2–0.7 cm wide; San Juan and Uintah counties, Utah . . . . 23. *Eriogonum lonchophyllum* (in part)
          74. Leaf blades elliptic to spatulate or broadly elliptic, 0.5–1(–1.5) cm wide; Carbon, Duchesne, Emery, Garfield, Grand, Piute, Sevier, and Uintah counties, Utah.
            75. Peduncles (0.5–)1–3(–5) cm; Garfield, Piute, and Sevier counties, Utah . . . . 36. *Eriogonum ostlundii*
            75. Peduncles absent; Carbon, Duchesne, Emery, Garfield, Grand, and Uintah counties, Utah . . . . 40. *Eriogonum batemanii*

Key 4—Arizona and New Mexico

1. Plants herbaceous perennials, sometimes cespitose and matted.
  2. Perianths pubescent.
    3. Achenes glabrous; flowering stems and inflorescence branches glabrous; plants erect, (2–)3–6 dm; se New Mexico . . . . 70. *Eriogonum havardii*
    3. Achenes villous-tomentose; flowering stems and inflorescence branches tomentose or floccose; plants erect, (0.5–)1–3.5 dm, or cespitose mats; n Arizona and n New Mexico.
      4. Involucres (2–)4–6 mm, deeply 5–10-toothed; perianths white to rose or yellow; n Arizona and nw New Mexico . . . . 68. *Eriogonum shockleyi*
      4 Involucres (2–)3–4 mm, 5-toothed; perianths yellow; ne Arizona and n New Mexico. . . . . 69. *Eriogonum lachnogynum*
  2. Perianths glabrous.
    5. Tepals dimorphic; inflorescences usually capitate; n Arizona and nw New Mexico . . . . 98. *Eriogonum ovalifolium*
    5. Tepals usually monomorphic; inflorescences not capitate; widespread.
      6. Inflorescences virgate or racemose; involucres racemose along proximal inflorescence branches; plants usually erect, 3–8(–10) dm.
        7. Perianths white to pinkish; flowering stems and involucres usually tomentose to floccose; n Arizona and n New Mexico . . . . 78. *Eriogonum racemosum*
        7. Perianths pale yellow to yellow or bright red; flowering stems and involucres usually glabrous, rarely tomentose; nw Arizona . . . . 79. *Eriogonum zionis*
      6. Inflorescences broadly cymose; involucres not racemose along proximal inflorescence branches; plants spreading to erect, 1–4(–6) dm.
        8. Flowering stems and inflorescence branches tomentose; leaf blades elliptic to ovate or obovate, 1.5–4 × 1–2.5 cm; perianths white to whitish brown; Mohave County, Arizona . . . . 76. *Eriogonum panamintense*
        8. Flowering stems and inflorescence branches glabrous; leaf blades usually linear to ovate or cordate to truncate, rarely reniform, (1–)3–8(–10) × 0.2–2.5(–3) cm; perianths white or yellow; widespread.
          9. Leaf blades usually cordate to truncate, rarely reniform, (1–)1.5–2.5 cm, glabrous except for fine hairs on margins and veins; flowers 1–2 mm; perianth yellow; Eddy County, New Mexico . . . . 26. *Eriogonum gypsophilum*
          9. Leaf blades linear to ovate, 1.5–8(–10) cm, tomentose to floccose on both surfaces, sometimes glabrous adaxially; flowers 2–3.5(–4) mm; perianth white or pale yellow to yellow; nw Arizona and n New Mexico.
            10. Involucres 1 per node, turbinate, 2–3(–3.5) × 1–1.5(–2) mm; leaf margins plane or slightly revolute; Arizona . . . . 22. *Eriogonum thompsoniae*
            10. Involucres 1 per node or 2–5 per cluster, turbinate to turbinate-campanulate, 2.5–4 × (1.3–)1.5–3.5(–4) mm; leaf margins plane or occasionally crenulate; New Mexico. . . . 23. *Eriogonum lonchophyllum* (in part)

1. Plants shrubs or subshrubs.
  11. Perianths pubescent.
    12. Perianths white to pinkish; leaf blades usually oblanceolate, canescent adaxially; widespread, Arizona . . . . 80. *Eriogonum fasciculatum*
    12. Perianths greenish yellow to yellow; leaf blades oblong-ovate to round-oblong or orbiculate, tomentose on both surfaces; Yuma County, Arizona . . . . 85. *Eriogonum deserticola*
  11. Perianths glabrous.
    13. Flowering stems and inflorescence branches angled and ridged or scabrous or, if round and smooth, glabrous, dark green, and with spinose lateral branches. . . . . 88. *Eriogonum heermannii* (in part)
    13. Flowering stems smooth, glabrous, floccose, or tomentose, grayish to greenish.
      14. Involucres racemosely disposed at tips of inflorescence branches.

15. Leaves fasciculate, blade 0.2–1.5 × 0.2–0.7(–0.9) cm, oblanceolate to elliptic; widespread . . . . 72. *Eriogonum wrightii*
15. Leaves not fasciculate, blade usually 1.5–3.5 × 0.2–0.8(–1.2) cm, linear-lanceolate or linear-oblanceolate to narrowly oblong or broadly elliptic; n Arizona and nw New Mexico.
    16. Leaf blades linear-lanceolate to linear-oblanceolate to narrowly oblong or lanceolate to narrowly elliptic; inflorescence branches tomentose to floccose or glabrous; ne Arizona and San Juan County, New Mexico . . . . 4. *Eriogonum leptocladon*
    16. Leaf blades broadly oblanceolate to broadly elliptic, rarely orbiculate; inflorescence branches tomentose to thinly floccose; nw Arizona
    . . . . 5. *Eriogonum nummulare*
[14. Shifted to left margin.—Ed.]
14. Involucres dichotomously arranged at tips of inflorescence branches.
  17. Leaf margins revolute or at least inrolled or thickened.
    18. Leaf blades (0.5–)2–6 cm; inflorescences densely branched, thinly pubescent or glabrous; involucres narrowly turbinate, glabrous; n Arizona, nw New Mexico . . .
    . . . . 7. *Eriogonum leptophyllum*
    18. Leaf blades 0.2–1.8(–2.5) cm; inflorescences sparsely branched, usually tomentose to floccose; involucres turbinate to campanulate, tomentose to floccose or, rarely, glabrous; widespread.
      19. Plants sprawling to decumbent subshrubs; inflorescences cymose-umbellate or capitate and reduced to single involucre; Coconino, se Yavapai, and nw Maricopa counties, Arizona . . . . 11. *Eriogonum ripleyi*
      19. Plants erect to spreading subshrubs; inflorescences umbellate, umbellate-cymose, or cymose; widespread.
        20. Tepals monomorphic . . . . 1. *Eriogonum microthecum* (in part)
        20. Tepals dimorphic.
          21. Leaf blades oblanceolate, 1–1.5(–2) cm, thinly tomentose and grayish adaxially, rarely glabrous; involucres turbinate-campanulate, 2.5–4.5 mm wide; nw San Juan County, New Mexico . . . . 8. *Eriogonum clavellatum*
          21. Leaf blades linear to narrowly elliptic, 0.3–1 cm, floccose or glabrous and green adaxially; involucres narrowly turbinate to turbinate, 1–2 mm wide, or campanulate, 2.5–3.5 mm wide; Arizona.
            22. Leaf blades linear, glabrous and green adaxially; involucres turbinate, 1.5–3(–3.5) × 1–2 mm; Verdi River Valley, c Yavapai County, Arizona . . . . 14. *Eriogonum ericifolium*
            22. Leaf blades linear-oblanceolate to narrowly elliptic, floccose or glabrous and greenish adaxially; involucres narrowly turbinate and 1.5–3 × 1–1.5 mm or campanulate and 3–4.5 × 2.5–3.5 mm; not of Verdi River Valley, Arizona.
              23. Flowers 1.5–2(–2.5) mm; involucres narrowly turbinate, 1–1.5 mm wide; plants 0.8–1.2(–1.5) dm; Apache, Coconino, ne Mohave, Navajo, and n Yavapai counties, Arizona . . . . . .
              . . . . 12. *Eriogonum pulchrum*
              23. Flowers 3.5–4.5(–5) mm; involucres campanulate, 2.5–3.5 mm wide; plants 1–4(–5) dm; Cochise and Pima counties, Arizona
              . . . . 13. *Eriogonum terrenatum*
  17. Leaf margins not revolute or inrolled.
    24. Inflorescences flat-topped and tiered, branches zigzag; leaf blades tomentose on both surfaces; plants erect, taller than wide; Mohave and w Yavapai counties, Arizona . . . . 89. *Eriogonum plumatella*
    24. Inflorescences open, not flat-topped, branches not zigzag; leaf blades typically not tomentose on both surfaces; plants spreading to rounded, wider than tall; widespread.

[25. Shifted to left margin.—Ed.]

25. Leaf apices sharply acute; leaf blades usually narrowly elliptic, 0.1–0.2 cm wide; nw and nc
Arizona . . . . 1. *Eriogonum microthecum* (in part)
25. Leaf apices acute to rounded; leaf blades lanceolate to oblanceolate, elliptic, or cordate,
(0.2–)0.3–3 cm wide; widespread.
  26. Flowering stems and inflorescence branches glabrous.
    27. Leaf blades bright green and glabrous; Arizona . . . . 21. *Eriogonum mortonianum*
    27. Leaf blades densely white-tomentose abaxially, sparsely tomentose to thinly floc-
cose or glabrous and green adaxially; New Mexico . . . . 23. *Eriogonum lonchophyllum* (in part)
  26. Flowering stems and inflorescence branches usually tomentose to floccose.
    28. Tepals dimorphic.
      29. Leaf blades cordate, 1–2(–2.5) cm wide . . . . 19. *Eriogonum jonesii*
      29. Leaf blades linear, oblanceolate to spatulate or elliptic, 0.1–0.8 cm wide . . .
. . . . 88. *Eriogonum heermannii* (in part)
    28. Tepals monomorphic.
      30. Leaf blades oblanceolate to oblong or obovate, (0.2–)0.3–0.7 cm wide, densely
white-tomentose abaxially, white-floccose to glabrate or green and glabrous
adaxially, margins plane; inflorescences usually dense; n New Mexico w to e
San Juan County . . . . 2. *Eriogonum effusum*
      30. Leaf blades usually lanceolate to oblanceolate or elliptic to nearly orbiculate,
rarely cordate, (0.3–)0.5–3 cm wide, densely white-, tannish-, or brownish-
tomentose on both surfaces or less so to nearly glabrous and green abaxially,
margins crisped, occasionally crenulate; inflorescences usually open; n Arizona
and nw New Mexico . . . . 16. *Eriogonum corymbosum*

Key 5—Alberta, Colorado, Montana, and Wyoming
1. Plants shrubs or subshrubs.
  2. Leaf blades oblanceolate to lanceolate or elliptic, cordate to nearly orbiculate, apices
rounded or nearly so.
    3. Flowering stems and inflorescence glabrous, rarely tomentose . . . .
. . . . 23. *Eriogonum lonchophyllum* (in part)
    3. Flowering stems and inflorescence tomentose to floccose, glabrate, or subglabrous.
      4. Inflorescences 10–30(–40) × 10–40 cm, floccose to glabrate or subglabrous; e
Colorado . . . . 2. *Eriogonum effusum* (in part)
      4. Inflorescences (1–)3–20 × 2–25(–30) cm, tomentose to floccose; w Colorado
. . . . 16. *Eriogonum corymbosum* (in part)
  2. Leaf blades linear to narrowly elliptic or narrowly lanceolate, apices sharply acute or
nearly so.
    5. Leaf margins plane or merely rolled.
      6. Leaf blades (0.5–)1–3 cm; widespread.
        7. Leaf blades usually elliptic, (0.5–)1–2(–2.5) × (0.1–)0.2–0.6(–0.8) cm,
densely to sparsely white-tomentose abaxially, less so to sparsely white-
floccose adaxially; nw Colorado, sw Wyoming and sw Montana . . . . .
. . . . 1. *Eriogonum microthecum* (in part)
        7. Leaf blades oblanceolate to oblong or obovate, (1–)1.5–3 × (0.2–)0.3–0.7
cm, densely white-tomentose abaxially, white-floccose to glabrate or green
and glabrous adaxially; e Colorado, w Nebraska and se Wyoming to se
Montana and sw South Dakota . . . . 2. *Eriogonum effusum* (in part)
      6. Leaf blades 3–6 cm; Colorado, sw Wyoming.
        8. Flowering stems and inflorescence branches tomentose to floccose . . . .
. . . . 16. *Eriogonum corymbosum* (in part)
        8. Flowering stems and inflorescence branches glabrous; Colorado . . . . .
. . . . 23. *Eriogonum lonchophyllum* (in part)

5. Leaf margins usually revolute.
   9. Leaf blades 1.5–4(–6) cm.
      10. Flowering stems and inflorescence branches tomentose to floccose;
         Montezuma County, Colorado . . . . 4. *Eriogonum leptocladon*
      10. Flowering stems and inflorescence branches glabrous or thinly pubescent;
         widespread.
         11. Leaf blades linear to linear-oblanceolate, (0.03–)0.1–0.3 cm wide,
            greenish adaxially; inflorescences green and compactly branched;
            Montezuma County, Colorado . . . . 7. *Eriogonum leptophyllum*
         11. Leaf blades narrowly lanceolate or oblanceolate to elliptic, 0.2–2 cm
            wide, grayish; inflorescences mostly grayish and openly branched;
            widespread . . . . 23. *Eriogonum lonchophyllum* (in part)
   9. Leaf blades 0.5–1.8(–2.5) cm.
      12. Tepals monomorphic.
         13. Leaf blades usually narrowly elliptic, 0.5–1.8(–2.5) cm, mostly white-
            floccose adaxially, margins slightly revolute; tepals connate proximal
            $^{1}/_{4}$; widespread . . . . 1. *Eriogonum microthecum* (in part)
         13. Leaf blades oblanceolate, 0.5–1.2(–1.5) cm, subglabrous or glabrous
            and green adaxially, margins tightly revolute; tepals connate proximal
            $^{1}/_{2}$; Delta and Montrose counties, Colorado . . . . 9. *Eriogonum pelinophilum*
      12. Tepals dimorphic.
         14. Flowers (2.5–)3–3.5 mm; leaf blades oblanceolate, thinly grayish-
            pubescent or glabrous and greenish adaxially, margins tightly revo-
            lute; plants 1–2.5 dm; Montezuma County, Colorado . . . . . . 8. *Eriogonum clavellatum*
         14. Flowers 2.5–4(–4.5) mm; leaf blades linear-oblanceolate to narrowly
            elliptic, reddish- or tannish-tomentose adaxially, margins revolute;
            plants 0.2–0.8 dm; Mesa County, Colorado . . . . 10. *Eriogonum bicolor*
1. Plants herbs.
   15. Perianths pubescent.
      16. Achenes villous to tomentose.
         17. Plants matted, 0.3–0.7 dm; w Colorado . . . . 68. *Eriogonum shockleyi*
         17. Plants erect, not matted, 1–3.5 dm; e Colorado . . . . 69. *Eriogonum lachnogynum*
      16. Achenes glabrous.
         18. Perianths yellow; e Montana, ne Wyoming . . . . 27. *Eriogonum brevicaule* (in part)
         18. Perianths white to rose; widespread.
            19. Perianths pustulose; involucres campanulate, not rigid; Big Horn County,
            Wyoming, and Carbon County, Montana . . . . 50. *Eriogonum mancum* (in part)
            19. Perianths not pustulose; involucres turbinate or campanulate, rigid;
            widespread.
            20. Involucres narrowly turbinate; leaf blades 1–4 cm; Great Plains . . .
              . . . . 64. *Eriogonum pauciflorum* (in part)
            20. Involucres campanulate; leaf blades 0.3–0.4 cm; nw Colorado . . . .
              . . . . 66. *Eriogonum tumulosum*
   15. Perianths glabrous.
      21. Tepals dimorphic.
         22. Inflorescences compoundly branched . . . . 96. *Eriogonum strictum*
         22. Inflorescences capitate or umbellate . . . . 98. *Eriogonum ovalifolium*
      21. Tepals usually monomorphic.
         23. Inflorescences virgate or racemose, branches bearing 5–20(–30) racemosely ar-
            ranged involucres; leaf blades (1.5–)2–6(–10) cm, elliptic to ovate or oval to
            nearly rotund; Colorado . . . . 78. *Eriogonum racemosum*
         23. Inflorescences umbellate to cymose or capitate or if racemose, plants forming
            dense, compact mats; leaf blades 0.5–1.5 cm or, if longer, linear to oblanceolate,
            3–10 cm, or spatulate to elliptic, (0.3–)1.5–4 cm; widespread.

[24. Shifted to left margin.—Ed.]
24. Plants erect to spreading, not forming mats.
    25. Inflorescences capitate or umbellate-cymose; c Colorado . . . . 31. *Eriogonum brandegeei*
    25. Inflorescences cymose; n or w Colorado northward.
        26. Leaf blades densely white-tomentose on both surfaces, elliptic; nw Colorado
        . . . . 40. *Eriogonum batemanii*
        26. Leaf blades tomentose abaxially, thinly floccose and greenish or glabrous and green
            adaxially, linear to elliptic, lanceolate, or spatulate; n Colorado north.
            27. Leaf blades linear to oblanceolate or spatulate, (1–)1.5–10 × 0.1–0.9(–1.2) cm,
                margins not revolute; widespread . . . . 27. *Eriogonum brevicaule* (in part)
            27. Leaf blades usually linear to lanceolate, rarely narrowly elliptic, (0.5–)1–3(–4)
                × 0.1–0.2(–0.5) cm, margins often revolute; nw Colorado.
                28. Peduncles absent; to be expected in Moffat County, Colorado . . . .
                    . . . . 28. *Eriogonum viridulum*
                28. Peduncles erect, 0.2–1.5 cm; Colorado.
                    29. Leaf blades linear to narrowly lanceolate, margins revolute; perianths
                        yellow; flowering stems and inflorescence branches floccose to nearly
                        glabrous; Garfield and Mesa counties, Colorado . . . . 33. *Eriogonum contortum*
                    29. Leaf blades lanceolate, margins plane; perianths ochroleucous or pale
                        yellow or rarely yellow; flowering stems and inflorescence branches
                        glabrous; Rio Blanco County, Colorado . . . . 29. *Eriogonum ephedroides*
24. Plants forming mats.
    30. Perianths usually yellow.
        31. Scapes and involucres glandular; w Montana . . . . 49. *Eriogonum crosbyae*
        31. Scapes and involucres not glandular; sw Wyoming, nw Colorado . . . .
        . . . . 27. *Eriogonum brevicaule* (in part)
    30. Perianths white to ochroleucous or rose.
        32. Perianths pustulose; leaf blades 0.4–1.5 cm; involucres tomentose; Montana, ne
            Wyoming.
            33. Involucres 2–5 per cluster, not rigid; flowering stem scapelike, inflorescence
                bracts 3–5; peduncles absent . . . . 50. *Eriogonum mancum* (in part)
            33. Involucres 1 per node, rigid; flowering stems and inflorescence bracts absent;
                peduncles present . . . . 51. *Eriogonum soliceps*
        32. Perianths not pustulose; leaf blades 1–6 cm; involucres usually glabrous except for
            hairs on teeth; widespread.
            34. Peduncles 0.1–0.4 cm; inflorescences cymose-umbellate; flowering stems
                glabrous or floccose to tomentose; leaf blades linear to linear-oblanceolate,
                (2–)3–6 × 0.1–0.3 cm; nc Colorado and se Wyoming . . . . 25. *Eriogonum exilifolium*
            34. Peduncles absent; inflorescences capitate or subcapitate to cymose-umbellate;
                flowering stems glabrous or tomentose; leaf blades linear-oblanceolate to oblan-
                ceolate to lanceolate or narrowly spatulate to elliptic, 1–4(–5) × 0.1–1 cm;
                Great Plains or Rocky Mountains.
                35. Flowering stems glabrous; Rocky Mountains, Colorado, above 2800 m
                    . . . . 24. *Eriogonum coloradense*
                35. Flowering stems tomentose; Great Plains, Montana to Colorado, below
                  1800 m . . . . 64. *Eriogonum pauciflorum* (in part)

Key 6—Manitoba, Saskatchewan, Great Plains States, and Texas
1. Plants shrubs or subshrubs.
    2. Tepals strongly dimorphic; w Texas . . . . 95. *Eriogonum suffruticosum*
    2. Tepals monomorphic; Great Plains, including Texas.
        3. Leaf blades linear to linear-oblanceolate, 0.1–0.3 cm wide; wc Kansas . . . .
          . . . . 3. *Eriogonum helichrysoides*
        3. Leaf blades oblanceolate to oblong, elliptic or obovate, 0.2–2.5 cm wide;
           widespread.
            4. Leaf blades (1–)1.5–2.5 cm wide; Deaf Smith County, Texas . . . . . 16. *Eriogonum corymbosum*
            4. Leaf blades 0.2–0.7 cm wide; widespread.
                5. Inflorescences cymose, with involucres not racemosely arranged; leaves
                   not fasciculate; n Great Plains . . . . 2. *Eriogonum effusum*
                5. Inflorescences virgate or somewhat cymose, with involucres racemosely
                   arranged; leaves often fasciculate; w Texas . . . . 72. *Eriogonum wrightii*
1. Plants herbs.
    6. Perianths glabrous; n Great Plains.
        7. Perianths usually yellow, rarely white or cream. . . . . 27. *Eriogonum brevicaule*
        7. Perianths whitish brown to rose . . . . 64. *Eriogonum pauciflorum* (in part)
    6. Perianths pubescent; Great Plains or Texas.
        8. Perianths whitish brown to rose, sparsely pubescent; plants loosely matted; n Great
           Plains . . . . 64. *Eriogonum pauciflorum* (in part)
        8. Perianths yellow, densely white-pubescent; plants not matted; s Great Plains or
           Texas.
            9. Achenes villous to tomentose; sw Kansas, w Oklahoma, Texas . . . . . 69. *Eriogonum lachnogynum*
            9. Achenes glabrous; w Texas . . . . 70. *Eriogonum havardii*

1. **Eriogonum microthecum** Nuttall, Proc. Acad. Nat.
    Sci. Philadelphia 4: 15. 1848 (as microtheca) [E] [F]

Subshrubs or shrubs, erect to
spreading, not scapose, 0.2–1.5 ×
(0.6–)1–13(–16) dm, white- to
tannish-tomentose, floccose, or
glabrous. **Stems** spreading to
erect, typically without persistent
leaf bases, up to ¹⁄₂ height of plant;
caudex stems absent or spreading;
aerial flowering stems erect to
spreading, slender, solid, not fistulose, 0.05–1.5 dm,
lanate, tomentose, floccose, subglabrous, or glabrous.
Leaves cauline, 1 per node or fasciculate; petiole 0.1–
0.5 cm, tomentose to floccose or glabrous; blade usually
elliptic, sometimes linear to obovate, 0.3–3.5 ×
(0.07–)0.1–1.2 cm, tomentose abaxially, less so or
glabrous adaxially, margins occasionally revolute.
Inflorescences cymose, compact, often flat-topped, 0.5–
6(–12) × 1–10(–13) cm; branches dichotomous, whitish-
lanate to brownish- or reddish-tomentose to floccose or
glabrate, infrequently green or gray and subglabrous or
glabrous; bracts 3, scalelike, linear to triangular, 1–5 mm.
Peduncles absent or mostly erect, slender, 0.3–1.5 cm,
tomentose to floccose. **Involucres** 1 per node, turbinate,
(1.5–)2–3.5(–4) × 1.3–2.5(–3) mm, tomentose, floccose,
subglabrous, or glabrous; teeth 5, erect, (0.3–)0.5–1
(–1.7) mm. **Flowers** 1.5–3(–4) mm; perianth yellow or
white to pink, orange, rose, red, or occasionally cream,
glabrous; tepals connate proximal ¹⁄₅–²⁄₅, essentially
monomorphic, oblong to obovate; stamens usually
exserted, 2.5–4 mm; filaments sparsely to densely
puberulent proximally. **Achenes** brown, 1.5–3 mm,
glabrous.

Varieties 13 (13 in the flora): w United States.

Eriogonum microthecum is used as browse by deer
and to a lesser degree by cattle and sheep. Some forms
are now in cultivation. The species is reportedly used
by the Piute of Nevada in the treatment of tuberculosis,
lameness, rheumatism, and bladder trouble (P. Train et
al. 1941). S. A. Weber and P. D. Seaman (1985) stated
that A. F. Whiting found the plants being used as a tea
by the Havasupai in northern Arizona. Members of *E.
microthecum* are food plants for subspecies of the rare
pallid blue butterfly (*Euphilotes pallescens*). Also found
on this species is the cythera metalmark (*Apodemia
mormo cythera*). Some authors have referred *E. effusum*
to this species, even though the ranges of the two species
do not overlap and intermediates are unknown.

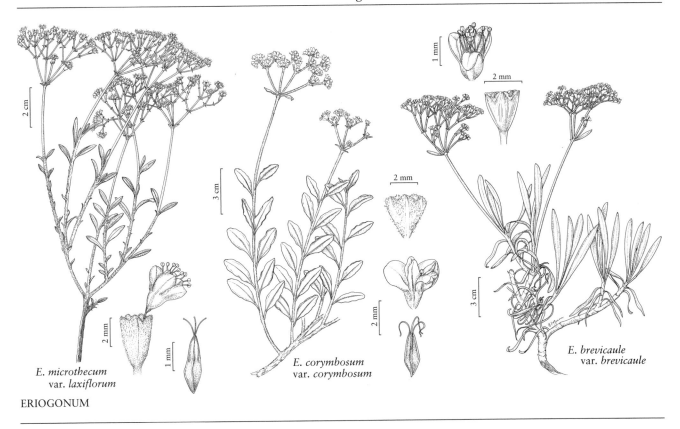

E. *microthecum*
var. *laxiflorum*

E. *corymbosum*
var. *corymbosum*

E. *brevicaule*
var. *brevicaule*

ERIOGONUM

1. Perianths yellow.
  2. Flowering stems and inflorescence branches usually glabrous; e Oregon and wc Idaho .... 1k. *Eriogonum microthecum* var. *microthecum*
  2. Flowering stems and inflorescence branches tomentose to floccose; se Oregon and sw Idaho s to e California and w Nevada.
    3. Leaf blades (0.2–)0.3–0.6(–0.8) cm wide; flowers (1.5–)2–2.5(–3) mm; involucres 2–2.5 mm; achenes 1.5–2 mm; se Oregon and sw Idaho s to e California and w Nevada .... 1l. *Eriogonum microthecum* var. *ambiguum*
    3. Leaf blades 0.5–1.2 cm wide; flowers 2.5–3 mm; involucres 2.5–4 mm; achenes 2.5–3 mm; ne California, nw Nevada .... 1m. *Eriogonum microthecum* var. *schoolcraftii*
1. Perianths various shades of white, cream, orange, pink, or red, not yellow.
  4. Tomentum whitish (see also var. *alpinum* of the Sierra Nevada, California); flowering stems and inflorescence branches infrequently glabrous.
    5. Leaf margins not revolute; flowering inflorescence branches floccose or glabrous; northern phase of species........... ... 1a. *Eriogonum microthecum* var. *laxiflorum*

  5. Leaf margins revolute or nearly so; flowering inflorescence branches lanate to tomentose, or if subglabrous or glabrous, then southern phase of species........ .... 1b. *Eriogonum microthecum* var. *simpsonii*
[4. Shifted to left margin.—Ed.]
4. Tomentum brownish or reddish (may be white in var. *alpinum*), or flowering stems and inflorescence branches essentially glabrous.
  6. Plants shrubs, 3–6 dm.
    7. Flowering stems and inflorescence branches tomentose when young, becoming floccose at maturity; flowers 1.5–2(–2.5) mm; achenes 1.8–2 mm; Death Valley region, California........ 1c. *Eriogonum microthecum* var. *panamintense*
    7. Flowering stems and inflorescence branches lanate to tomentose at maturity; flowers 2–2.5(–3) mm; achenes 2.5–3 mm; Transverse Ranges, California .... 1d. *Eriogonum microthecum* var. *corymbosoides*
  6. Plants subshrubs, 0.2–1.5(–2) dm.
    8. Leaf blades elliptic or ovate, margins not revolute; flowers (1.5–)2–3.5(–4) mm.
      9. Leaf blades 0.5–1 × (0.2–)0.3–0.5 (–0.6) cm; involucres (2–)2.5–3 mm; flowers (2.5–)3–3.5(–4) mm; San Gabriel Mountains, California .... 1e. *Eriogonum microthecum* var. *johnstonii*

9. Leaf blades 0.3–0.7(–0.8) × 0.1–0.4 cm;
involucres 2.5–3.5 mm; flowers (1.5–)
2–3.5 mm; desert ranges of se
California, c Nevada, and Utah
. . . . 1f. *Eriogonum microthecum*
var. *lapidicola*

[8. Shifted to left margin—Ed.]

8. Leaf blades linear or linear-oblanceolate to narrowly elliptic, margins often revolute; flowers 1.5–
2.5 mm.
10. Involucres 3–4 mm; perianths cream; San Bernardino Mountains, California . . . .
. . . . 1g. *Eriogonum microthecum*
var. *lacus-ursi*
10. Involucres (1.5–)2–3 mm; perianths white,
pink, red, or rose; Sierra Nevada of California, or e Nevada and w Utah.
11. Flowering stems glabrous; leaf blades
sparsely floccose or glabrous adaxially;
desert ranges, se Nevada . . . . . 1j. *Eriogonum
microthecum* var. *arceuthinum*
11. Flowering stems white- to brownish-
floccose to subglabrous, or reddish-
tomentose to floccose; leaf blades
floccose to subglabrous adaxially;
California or wc Utah.
12. Tomentum white to brownish; Sierra
Nevada, California . . . .
. . . . 1h. *Eriogonum microthecum*
var. *alpinum*
12. Tomentum reddish; desert ranges,
wc Utah . . . .
. . . . 1i. *Eriogonum microthecum*
var. *phoeniceum*

1a. **Eriogonum microthecum** Nuttall var. **laxiflorum**
Hooker, Hooker's J. Bot. Kew Gard. Misc. 5: 264.
1853 (as microtheca) • Great
Basin wild buckwheat E F

Eriogonum confertiflorum
Bentham; *E. microthecum* subsp.
confertiflorum (Bentham) S. Stokes;
E. microthecum subsp. *laxiflorum*
(Hooker) S. Stokes

Subshrubs, (1–)2–4(–5) × 3–8 dm.
Stems: caudex absent; aerial
flowering stems 0.2–0.6(–0.8) dm, floccose or tomentose,
rarely glabrous. **Leaf blade** usually elliptic, (0.5–)1–2
(–2.5) × (0.1–)0.2–0.6(–0.8) cm, densely to sparsely
whitish-tomentose abaxially, less so to sparsely whitish-
floccose adaxially, margins not revolute. **Inflorescences**
(1–)2–4(–8) cm; branches floccose or glabrous.
Involucres 2–3(–3.5) mm, subglabrous, glabrous, or
merely floccose between angled ridges. **Flowers** 2–3 mm;
perianth white to pink or rose. **Achenes** 2–3 mm.

Flowering Jun–Oct. Sandy to gravelly flats and slopes,
mixed grassland, saltbush, blackbrush, and sagebrush
communities, pinyon-juniper and montane conifer
woodlands; (400–)1500–3200 m; Ariz., Calif., Colo.,
Idaho, Mont., Nev., Oreg., Utah, Wash., Wyo.

Variety *laxiflorum* is the common expression of the
species in the northern part of the species' range. It
occurs in northern Arizona, eastern California, western
Colorado, central and southern Idaho, southwestern
Montana, Nevada, eastern Oregon, northern and
western Utah, eastern Washington, and southwestern
Wyoming. It overlaps morphologically with var.
simpsonii in northern Arizona. The variety is the primary
host plant for the rare Mattoni blue butterfly (*Euphilotes
rita mattoni*).

1b. **Eriogonum microthecum** Nuttall var. **simpsonii**
(Bentham) Reveal, Taxon 32: 293. 1983
• Simpson's wild buckwheat E

Eriogonum simpsonii Bentham in
A. P. de Candolle and A. L. P. P. de
Candolle, Prodr. 14: 18. 1856; *E.
microthecum* var. *foliosum* (Torrey
& A. Gray) Reveal; *E. microthecum*
var. *macdougalii* (Gandoger)
S. Stokes

Shrubs, (1–)2–15 × 4–16 dm.
Stems: caudex absent; aerial
flowering stems 0.2–0.7 dm, densely lanate to tomentose,
sometimes floccose, rarely glabrous. **Leaves:** blade
narrowly elliptic, 0.5–1.8(–2.5) × 0.1–0.2 cm, densely
whitish-tomentose abaxially, mostly whitish-floccose
adaxially, margins revolute or nearly so. **Inflorescences**
(1.5–)2–4(–6) cm; branches tomentose to floccose, rarely
subglabrous or glabrous. **Involucres** 2–3 mm, tomentose
to floccose or subglabrous. **Flowers** 2–3 mm; perianth
white to pink or rose. **Achenes** 2–3 mm.

Flowering Jun–Oct. Clayey to gravelly or occasionally
sandy washes, flats and slopes, mixed grassland,
saltbush, blackbrush, and sagebrush communities,
pinyon-juniper and montane conifer woodlands; 1400–
2300 m; Ariz., Calif., Colo., Idaho, Mont., Nev., N.Mex.,
Utah, Wyo.

Variety *simpsonii* is the common expression of the
species in the southern part of the species' range. It occurs
in northern Arizona, southeastern California, western
and south-central Colorado, central and southern
Nevada, northern New Mexico, Utah, and southwestern
Wyoming.

Populations most similar to the type occur mainly on
the Colorado Plateau in northern New Mexico, southern
Colorado, and in northeastern Arizona and southwestern
Utah; these are low shrubs usually less than 3 dm tall
with thinly tomentose to floccose leaf blades, flowering
stems, and inflorescence branches. As such, these plants

resemble, and are sometimes confused with, small forms of *Eriogonum leptophyllum*. In east-central and northeastern Utah, and in northwestern Colorado, there are similar plants but, unlike the more southern phase with its tightly revolute, linear leaf blades, here the leaf blades are somewhat broader and tend to be glabrous. These plants are outwardly similar to *E. microthecum* var. *phoeniceum*. Similar depauperate individuals are found elsewhere, most notably in east-central Nevada where the plants have even broader leaf blades and instead of being dark green the plants are light green and whitish. Such plants are found on clayey substrates and will either key out here or, with some difficulty, to var. *lapidicola*. The variety is disjunct in the Salmon River Valley of Lemhi County, Idaho (*Reveal & Welsh 4485 & 4486*, BRY, NY, UTC) and in Beaverhead County, Montana (*Shelly & King 1150*, MONTU). These populations are found in close association with known Native American villages and may well have been introduced in pre-Columbian times as var. *simpsonii* is known to have been used in certain rites associated with "witchcraft."

West of the Colorado Plateau, in northwestern New Mexico and across the northern tier of counties in Arizona, plants are tall, open shrubs. This expression is common throughout the southern portion of the Great Basin and on the desert ranges of eastern California. At its extreme, the expression is markedly different from that represented by the type of the variety, but there is a complete intergradation from one to the other, and no attempt is made here to parse them into taxonomic entities.

1c. Eriogonum microthecum Nuttall var. **panamintense** S. Stokes, Eriogonum, 74. 1936 • Panamint wild buckwheat [E]

Eriogonum effusum Nuttall var. *limbatum* S. Stokes

Shrubs, 3–6 × 5–12(–15) dm. **Stems:** caudex absent; aerial flowering stems 0.5–1(–1.5) dm, tomentose when young, becoming floccose. **Leaves:** blade usually broadly elliptic, 0.6–1.8 × 0.3–0.8 cm, brown-tomentose abaxially, floccose to subglabrous adaxially, margins not revolute. **Inflorescences** 1–4 cm; branches tomentose when young, becoming floccose. **Involucres** 2–2.5 mm, subglabrous or glabrous. **Flowers** 1.5–2(–2.5) mm; perianth white to reddish. **Achenes** 1.8–2 mm.

Flowering Jul–Oct. Gravelly slopes, sagebrush communities, pinyon-juniper woodlands; 1900–2800 m; Calif.

Variety *panamintense* is a localized expression known only from the Inyo Mountains and the Panamint Range of Inyo County.

1d. Eriogonum microthecum Nuttall var. **corymbosoides** Reveal, Sci. Bull. Brigham Young Univ., Biol. Ser. 13(1): 25, fig. 13. 1971 • San Bernardino wild buckwheat [E]

Shrubs, 3–6 × 6–12(–15) dm. **Stems:** caudex absent; aerial flowering stems 0.5–1.3(–1.5) dm, tannish- to reddish-brown-lanate to tomentose or rarely floccose. **Leaves:** blade elliptic to obovate, (0.8–)1–2(–2.5) × (0.4–)0.6–1 cm, densely whitish-brown-tomentose abaxially, floccose to subglabrous adaxially, margins not revolute. **Inflorescences** 1–4 cm; branches lanate to tomentose. **Involucres** 2–3 mm, floccose to subglabrous. **Flowers** 2–2.5(–3) mm; perianth white to reddish. **Achenes** 2.5–3 mm.

Flowering Jul–Sep. Gravelly to rocky granitic slopes, chaparral communities, oak and montane conifer woodlands; (1500–)1800–2900 m; Calif.

Variety *corymbosoides* is found infrequently at the eastern end of the San Bernardino Mountains, San Bernardino County; Little Rock Canyon of the San Gabriel Mountains, Los Angeles County; and the Mt. Pinos area of southwestern Kern County.

1e. Eriogonum microthecum Nuttall var. **johnstonii** Reveal, Sci. Bull. Brigham Young Univ., Biol. Ser. 13(1): 28, fig. 15. 1971 • Johnston's wild buckwheat [E]

Subshrubs, 0.6–1.3 × 2–5 dm. **Stems:** caudex spreading; aerial flowering stems 0.3–0.6 dm, floccose to subglabrous. **Leaves:** blade elliptic to ovate, 0.5–1 × (0.2–)0.3–0.5(–0.6) cm, densely whitish-brown-tomentose abaxially, floccose to subglabrous adaxially, margins not revolute. **Inflorescences** 0.5–3 cm; branches floccose to subglabrous. **Involucres** (2–)2.5–3 mm, usually glabrous. **Flowers** (2.5–)3–3.5(–4) mm; perianth white to reddish. **Achenes** 2.5–3 mm.

Flowering Jul–Sep. Granitic slopes, montane conifer woodlands; 2600–2900 m; Calif.

Variety *johnstonii* is rare and known only from the upper reaches of the San Gabriel Mountains near the San Bernardino–Los Angeles county line.

1f. Eriogonum microthecum Nuttall var. **lapidicola**
Reveal, Sci. Bull. Brigham Young Univ., Biol. Ser. 13(1):
28, fig. 17. 1971 • Pahute Mesa wild buckwheat
Ⓔ

Subshrubs, 0.5–1.5 × 1–3 dm.
Stems: caudex absent or
sometimes spreading; aerial
flowering stems 0.2–0.6 dm,
tomentose to floccose. Leaves:
blade elliptic, 0.3–0.7(–0.8) × 0.1–
0.4 cm, densely reddish-brown-
tomentose abaxially, tomentose to
floccose adaxially, margins not
revolute. **Inflorescences** 2–6 cm; branches tomentose to
floccose. **Involucres** 2.5–3.5 mm, floccose to
subglabrous. **Flowers** (1.5–)2–3.5 mm; perianth whitish
red to pink, rose, or orange. **Achenes** 2.5–3 mm.

Flowering Jul–Sep. Gravelly to rocky volcanic or
sandstone outcrops, sagebrush communities, pinyon-
juniper woodlands; 2600–3100 m; Calif., Nev., Utah.

Variety *lapidicola* is found infrequently in the southern
mountains of the Great Basin from Inyo County,
California, eastward across central Nevada (Clark,
Esmeralda, Lincoln, Nye, and White Pine counties) to
western Utah (Juab and Washington counties). It
approaches the more depauperate forms of var. *simpsonii*
in habit and is difficult to separate from var. *phoeniceum*
in western Utah. An expression possibly worthy of
formal recognition with sparse, whitish tomentum and
narrow, greenish adaxial leaf surfaces from White River
Valley in central Nye County, Nevada (e.g., *Harrison &
Harrison 13211*, BRY; *Holmgren & Holmgren 8219*,
NY; *Reveal 8331*, NY; *B. T. Welsh & K. H. Thorne 402*
and *567*, BRY), tentatively here assigned to var.
lapidicola, is somewhat similar to var. *simpsonii*.

1g. Eriogonum microthecum Nuttall var. **lacus-ursi**
Reveal & A. Sanders, Phytologia 86: 137. 2004
• Bear Lake wild buckwheat   Ⓔ

Subshrubs, 1.5–2 × 4–6 dm.
Stems: caudex absent; aerial
flowering stems 0.4–0.8 dm,
sparsely white-floccose or
glabrous. **Leaves:** blade narrowly
elliptic, 0.7–1.5 × (0.07–)0.1–0.3
cm, densely white-tomentose
abaxially, subglabrous or glabrous
adaxially, margins usually
revolute. **Inflorescences** 1–3 cm; branches sparsely
floccose or glabrous. **Involucres** 3–4 mm, glabrous or
thinly floccose near base of teeth. **Flowers** 2–2.5 mm;
perianth cream. **Achenes** 2–2.5 mm.

Flowering Jul–Aug. Clayey outcrops, montane conifer
woodlands; 2000–2100 m; Calif.

Variety *lacus-ursi* is known only from Bear Valley in
the San Bernardino Mountains, San Bernardino County,
where it is restricted to a few hundred plants in a single
area subject to development.

1h. Eriogonum microthecum Nuttall var. **alpinum**
Reveal, Sci. Bull. Brigham Young Univ., Biol. Ser. 13(1):
31, fig. 19. 1971 • Sonora Pass wild buckwheat
Ⓔ

Subshrubs, 0.4–1 × 1–3 dm.
Stems: caudex spreading; aerial
flowering stems 0.1–0.3 dm,
white- or brownish-floccose to
subglabrous. **Leaves:** blade linear-
oblanceolate to narrowly elliptic,
0.3–0.7(–0.9) × 0.1–0.3 cm,
densely white-tomentose abaxially, floccose to subglabrous
adaxially, margins usually revolute. **Inflorescences** 0.5–
2(–3) cm; branches floccose to subglabrous. **Involucres**
(1.5–)2–2.5 mm, floccose or glabrous. **Flowers** 1.5–2.5
mm; perianth white to pink or rose. **Achenes** 1.5–2 mm.

Flowering Jul–Sep. Sandy to gravelly granitic or
volcanic slopes, subalpine and alpine conifer woodlands;
2500–3300 m; Calif.

Variety *alpinum* is known only from the central Sierra
Nevada and the Sweetwater Range in Alpine, Mono,
and Tuolumne counties. It appears to be a high-elevation
expression that evolved from var. *laxiflorum*.

1i. Eriogonum microthecum Nuttall var. **phoeniceum**
(L. M. Schulz) Reveal, Harvard Pap. Bot. 9: 183. 2004
• Scarlet wild buckwheat   Ⓔ

Eriogonum phoeniceum L. M.
Shultz, Harvard Pap. Bot. 3: 49,
fig. 1. 1998

Subshrubs, 0.2–0.4 × 0.6–1.2 dm.
Stems: caudex absent or
sometimes spreading; aerial
flowering stems 0.05–0.1 dm,
reddish-tomentose to floccose.
Leaves: blade linear, 0.4–0.6 ×
0.08–0.12 cm, densely white-tomentose abaxially,
sparsely floccose and green adaxially, margins revolute.
Inflorescences 0.5–1 cm; branches reddish-floccose to
glabrate. **Involucres** 2–3 mm, floccose to subglabrous.
Flowers 2–2.5 mm; perianth white to pink, rose, or red.
Achenes 1.5–2 mm.

Flowering Jul–Sep. Tuffaceous ash outcrops,
sagebrush communities, pinyon-juniper woodlands;
1600–2100 m; Utah.

Variety *phoeniceum* is known only from widely scattered populations in Juab and Millard counties of western Utah. It closely approaches var. *lapidicola*, and a distinction between the two is somewhat arbitrary, the leaf blades of var. *phoeniceum* being tightly revolute and a brighter green on the adaxial surface. Specimens of var. *phoeniceum* also approach specimens of the variable var. *simpsonii* found mainly in Emery County in eastern Utah.

1j. Eriogonum microthecum Nuttall var. **arceuthinum** Reveal, Phytologia 86: 135. 2004 • Juniper Mountain wild buckwheat E

Subshrubs, 0.5–0.7 × 1–1.5 dm. **Stems:** caudex spreading; aerial flowering stems 0.1–0.3 dm, glabrous. **Leaves:** blade linear, 0.5–0.8 × 0.05–0.1 cm, densely white-tomentose adaxially, sparsely floccose or glabrous and green adaxially, margins revolute. **Inflorescences** 0.5–2.5 cm; branches sparsely floccose or glabrous. **Involucres** 2–3 mm, glabrous. **Flowers** 1.5–2 mm; perianth white. **Achenes** 1.5–2 mm.

Flowering Jul–Sep. Volcanic slopes and cliffs, oak and pinyon-juniper woodlands; 1900–2100 m; Nev.

Variety *arceuthinum* was gathered originally by Carl Purpus in the "Juniper Mountains" in 1898. It is uncertain whether his collection came from southwestern Iron County, Utah, or from southeastern Lincoln County, Nevada. A single, sterile collection (*Shultz & Shultz 7109*, BRY, GH) with dubious label data supposedly was found in Nevada, and a second collection from that state, made by A. Jerry Tiehm and sent to L. M. Schulz, apparently has been lost.

1k. Eriogonum microthecum Nuttall var. **microthecum** • Slender wild buckwheat E

Eriogonum microthecum var. *idahoense* (Rydberg) S. Stokes

Shrubs, (2.5–)3–5 × 3–7 dm. **Stems:** caudex absent; aerial flowering stems 0.3–0.7 dm, floccose or glabrous. **Leaves:** blade oblanceolate to elliptic, (0.8–)1–2(–2.7) × (0.3–)0.4–0.9 (–1.2) cm, densely white-tomentose abaxially, floccose or glabrous adaxially, margins not revolute. **Inflorescences** 3–10 cm; branches floccose or glabrous. **Involucres** 2.5–3 mm, floccose or glabrous. **Flowers** 2–2.5 mm; perianth yellow. **Achenes** 2–2.5 mm.

Flowering Jul–Sep. Gravelly to rocky slopes, saltbush and sagebrush communities, juniper and conifer woodlands; 600–1100 m; Idaho, Oreg.

Variety *microthecum* is found in the foothills and low mountains north of the Intermountain Region in Baker, Crook, Grant, Malheur, Sherman, Wasco, and Wheeler counties in Oregon. Scattered populations also occur in Washington County, Idaho. The variety is disjunct onto the Pueblo Mountains of Harney County, Oregon.

1l. Eriogonum microthecum Nuttall var. **ambiguum** (M. E. Jones) Reveal in P. A. Munz, Suppl. Calif. Fl., 61. 1968 • Yellow-flowered wild buckwheat E

Eriogonum aureum M. E. Jones var. *ambiguum* M. E. Jones, Proc. Calif. Acad. Sci., ser. 2, 5: 719. 1895

Subshrubs or shrubs, 0.5–5 × 1–8 dm. **Stems:** caudex absent; aerial flowering stems 0.2–1 dm, mostly floccose. **Leaves:** blade 0.8–2.5 × (0.2–)0.3–0.6(–0.8) cm, densely white- or reddish-brown-tomentose abaxially, floccose adaxially, margins rarely revolute. **Inflorescences** 1–5(–12) cm; branches tomentose to floccose. **Involucres** 2–2.5 mm, tomentose to floccose. **Flowers** (1.5–)2–2.5(–3) mm; perianth yellow. **Achenes** 1.5–2 mm.

Flowering Jul–Sep. Sandy to rocky soil, sagebrush and saltbush communities, pinyon-juniper and conifer woodlands; (1100–)1900–3300 m; Calif., Idaho, Nev., Oreg.

Variety *ambiguum* occurs in eastern California, northwestern Nevada, and southeastern Oregon. In portions of its range, var. *ambiguum* appears to be little more than a yellow-flowered phase of var. *laxiflorum*. However, where the two occur together, intermediates are rarely found, and limited garden studies indicate that flower color is genetically fixed. Nonflowering specimens from southern Owyhee County, Idaho, have been tentatively assigned to this taxon.

1m. Eriogonum microthecum Nuttall var. **schoolcraftii** Reveal, Phytologia 86: 138. 2004 • Schoolcraft's wild buckwheat E

Shrubs, 3–8 × (4–)5–10(–12) dm, floccose. **Stems:** caudex absent; aerial flowering stems 0.5–1.5 dm, floccose. **Leaves:** blade 1.5–3.5 × 0.5–1.2 cm, densely reddish-brown-tomentose abaxially, floccose adaxially, margins not revolute. **Inflorescences** 1–8 cm; branches floccose. **Involucres** 2.5–4 mm, floccose. **Flowers** 2.5–3 mm; perianth yellow. **Achenes** 2.5–3 mm.

Flowering Jul–Sep. Sandy to rocky soil, sagebrush communities, pinyon-juniper woodlands; 1400–2200 m; Calif., Nev.

Variety *schoolcraftii* is known only from low desert hills west of Doyle, Lassen County, California, and on Seven Lake Mountain, Washoe County, Nevada. It has the potential of being an attractive garden shrub.

2. Eriogonum effusum Nuttall, Proc. Acad. Nat. Sci. Philadelphia 4: 15. 1848 • Spreading wild buckwheat E

Eriogonum microthecum Nuttall var. *effusum* (Nuttall) Torrey & A. Gray

Shrubs, spreading, not scapose, (1.5–)2–5(–7) × 5–15 dm, grayish- to reddish-brown-tomentose to floccose and gray or, rarely, thinly floccose and greenish. **Stems** spreading to erect, typically without persistent leaf bases, up to 1/2 height of plant; caudex stems absent; aerial flowering stems erect to spreading, slender, solid, not fistulose, 0.3–0.8 dm, floccose or glabrous. **Leaves** cauline, 1 per node; petiole 0.2–0.7 cm, tomentose to floccose; blade oblanceolate to oblong or obovate, (1–)1.5–3 × (0.2–)0.3–0.7 cm, densely white-tomentose abaxially, white-floccose to glabrate or green and glabrous adaxially, margins plane. **Inflorescences** cymose, 10–30(–40) × 10–40 cm; branches dichotomous, white-floccose to glabrate or subglabrous; bracts 3, scalelike, triangular, 0.5–2(–5) mm. **Peduncles** absent or mostly erect, slender, 0.3–2.5 cm, floccose. **Involucres** 1 per node, turbinate, 1.5–2.5(–3) × 1–2 mm, tomentose to floccose; teeth 5, erect, 0.3–0.6 mm. **Flowers** 2–4 mm; perianth yellow, glabrous; tepals connate proximal 1/4, essentially monomorphic, elliptic to obovate; stamens mostly exserted, 2–4.5 mm; filaments sparsely pilose proximally. **Achenes** brown, 2–2.5 mm, glabrous.

Flowering Jun–Sep. Sandy to rocky slopes and flats, mixed grassland and sagebrush communities, juniper and montane conifer woodlands; 1200–2500 m; Colo., Mont., Nebr., N.Mex., S.Dak., Wyo.

Eriogonum effusum is rather common on the northern Great Plains and along the eastern slope of the Rocky Mountains in central and eastern Colorado, southeastern Montana, western Nebraska, northern New Mexico, southwestern South Dakota, and southeastern Wyoming. Some specimens from Chaffee County, Colorado, are thinly floccose and greenish (*Atwood & Welsh 29689,* BRY) and thus similar to *E. leptocladon.* A collection from Pyramid Lake, Washoe County, Nevada (*Frandsen & Brown 182,* NESH) is clearly mislabeled. A roadside collection of *E. effusum* gathered near Little America,

Sweetwater County, Wyoming, in 1961 (*G. Mason 4025,* ASU) was an introduction that has not persisted.

The spreading wild buckwheat is occasionally merged with *Eriogonum microthecum* even though the two are morphologically distinct and their ranges do not overlap. Plants in New Mexico are sometimes difficult to distinguish from the related *E. leptocladon* var. *ramosissimum.* The species is the food plant for the Rita dotted-blue butterfly (*Euphilotes rita*). A hybrid between *Eriogonum effusum* and *E. pauciflorum* has been named *E.* ×*nebraskense* Rydberg [*E. multiceps* Nees subsp. *nebraskense* (Rydberg) S. Stokes; *E. pauciflorum* Pursh var. *nebraskense* (Rydberg) Reveal]. The hybrid is known from Weld County, Colorado; Cheyenne and Kimball counties, Nebraska; and Converse and Platte counties, Wyoming.

3. Eriogonum helichrysoides (Gandoger) Prain in B. D. Jackson et al., Index Kew., suppl. 4: 82. 1913 • Strawflower wild buckwheat E

Eriogonum microthecum Nuttall subsp. *helichrysoides* Gandoger, Bull. Soc. Roy. Bot. Belgique 42: 192. 1906; *E. effusum* Nuttall subsp. *helichrysoides* (Gandoger) S. Stokes; *E. effusum* var. *rosmarnioides* Bentham

Shrubs, spreading, not scapose, 1.5–3(–4) × 3–8 dm, floccose and dark green. **Stems** spreading, often without persistent leaf bases, up to 1/3 height of plant; caudex stems absent; aerial flowering stems erect to slightly spreading, slender, solid, not fistulose, 0.02–0.04 dm, floccose. **Leaves** cauline, 1 per node; petiole 0.2–0.7 cm, floccose; blade linear to linear-oblanceolate, (2–)3–6 × 0.1–0.3 cm, densely tomentose abaxially, floccose and green adaxially, margins revolute. **Inflorescences** cymose, compact, 0.5–1.3(–1.6) dm; branches dichotomous, floccose to nearly glabrous; bracts 3, scalelike, triangular, 0.5–2 mm. **Peduncles** absent or erect, 0.5–2.5 cm, floccose. **Involucres** 1 per node, turbinate, 2.5–3 × 1–2 mm, floccose or glabrous; teeth 5, erect, 0.3–0.6 mm. **Flowers** 2–2.5(–3) mm; perianth white, becoming rose, glabrous; tepals connate proximal 1/4, essentially monomorphic, elliptic to obovate; stamens exserted, 2.5–4.5 mm; filaments sparsely pilose. **Achenes** brown, 2–2.5 mm, glabrous.

Flowering Jul–Sep. Clay slopes or chalky limestone outcrops, mixed grassland communities; 900–1100 m; Kans.

Eriogonum helichrysoides is limited to west-central Kansas (Ellis, Gove, Lane, Logan, Scott, and Trego counties). The plants' dark green color and narrow leaf blades readily segregate the species from *E. effusum.*

4. **Eriogonum leptocladon** Torrey & A. Gray in War Department [U.S.], Pacif. Railr. Rep. 2(1): 129. 1857 E

Eriogonum effusum Nuttall subsp. *leptocladon* (Torrey & A. Gray) S. Stokes

Shrubs, erect to spreading, not scapose, (2–)3–10(–12) × 5–15 (–20) dm, white-tomentose to floccose, or green and nearly glabrous. **Stems** spreading, often without persistent leaf bases, up to ¹/₂ height of plant; caudex stems absent or spreading in moving sand; aerial flowering stems spreading, slender, solid, not fistulose, 0.3–1 dm, white-tomentose, floccose to glabrate or glabrous. **Leaves** cauline, 1 per node; petiole 0.2–0.5 cm, tomentose to floccose; blade linear-lanceolate to linear-oblanceolate to narrowly oblong, or lanceolate to narrowly elliptic, 1.5–4 × 0.2–0.8(–1.2) cm, densely white-tomentose abaxially, less so and greenish adaxially, margins plane, infrequently revolute. **Inflorescences** cymose, open, 5–40 × 10–50 cm; branches dichtomous proximally, often with involucres racemosely arranged proximally, tomentose to floccose or glabrate, occasionally glabrous; bracts 3, scalelike, linear to triangular, 1–3(–6) mm. **Peduncles** absent. **Involucres** 1 per node, turbinate to turbinate-campanulate, 1.5–3 × 1–2 mm, tomentose to floccose or glabrous; teeth 5, erect, 0.4–0.7 mm. **Flowers** (2–)2.5–3.5 mm; perianth white or pale yellow to yellow, glabrous; tepals connate proximal ¹/₄–¹/₃, essentially monomorphic, oblong to broadly obovate; stamens slightly exserted, 2–4 mm; filaments sparsely pilose proximally. **Achenes** light brown, 2.5–3.5 mm, glabrous except for minutely papillate beak.

Varieties 3 (3 in the flora): w United States.

Eriogonum leptocladon is localized on moving sands on the Colorado Plateau. It is not unusual, therefore, for portions of the shrub to be buried in sand, especially the main, woody trunk from which the flowering stems arise. In the Four Corners area three species share portions of overlapping ranges, and herbarium material from there can be difficult to separate. In northwestern New Mexico, var. *ramosissimum* approaches *E. effusum*, and in eastern San Juan County the two can be distinguished only by the open, spreading inflorescence branching pattern of var. *ramosissimum* as compared to the densely branched, compact inflorescence of *E.*

effusum. In the field, the former is confined to sandy areas whereas the latter is typically on heavier, usually clayey soils. In south-central Utah, var. *ramosissimum* approaches *E. nummulare*. There, both are found on moving sands, and only the degree of stoutness of the branches can be used to differentiate them when cauline leaves are absent. For the most part, the inflorescence branches of *E. nummulare* are more rigid and stouter than those of var. *ramosissimum*, which tend to be flexible and slender. When leaves are present, those of the former are broader and more apically rounded compared to those of the latter.

1. Perianths pale yellow to yellow . . . .
. . . . 4c. *Eriogonum leptocladon* var. *leptocladon*
1. Perianths white.
  2. Flowering stems and inflorescence branches tomentose, rarely floccose . . . .
     . . . . 4a. *Eriogonum leptocladon* var. *ramosissimum*
  2. Flowering stems and inflorescence branches glabrous . . . . 4b. *Eriogonum leptocladon* var. *papiliunculi*

4a. **Eriogonum leptocladon** Torrey & A. Gray var. **ramosissimum** (Eastwood) Reveal, Proc. Utah Acad. Sci. 42: 289. 1966   • San Juan wild buckwheat   E

Eriogonum ramosissimum Eastwood, Proc. Calif. Acad. Sci., ser. 2, 6: 322. 1896; *E. effusum* Nuttall subsp. *pallidum* (Small) S. Stokes

Plants 2–8 × 5–15 dm. **Aerial flowering stems** white-tomentose to floccose. **Leaves:** blade linear-lanceolate to narrowly oblong, 1.5–3.5 × (0.2–)0.3–1 cm, margins plane or infrequently revolute. **Inflorescence branches** tomentose, rarely floccose. **Perianths** white.

Flowering Jun–Oct. White or infrequently red blow sand on flats, washes and slopes, mixed grassland, saltbush, ephedra, and sagebrush communities, pinyon-juniper woodlands; 1200–2100 m; Ariz., Colo., N.Mex., Utah.

Variety *ramosissimum* is the more common of the expressions in the species. It occurs in Emery, Garfield, San Juan, and Wayne counties in Utah, and in Montezuma County, Colorado. It is common in Apache, western Coconino, and Navajo counties, Arizona, and in San Juan County, New Mexico.

4b. Eriogonum leptocladon Torrey & A. Gray var. **papiliunculi** Reveal, Brittonia 26: 92, fig. 2. 1974 • Butterfly wild buckwheat E

Plants (3–)4–10 × 5–15 dm. **Aerial flowering stems** glabrous. **Leaves:** blade lanceolate to narrowly elliptic, 2.5–3.5(–4) × (0.4–)0.5–0.8(–1.2) cm, margins plane. **Inflorescence branches** glabrous. **Perianths** white. 2*n* = 40.

Flowering Jun–Oct. White or red wind-blown sands on flats, washes and slopes, mixed grassland and sagebrush communities, pinyon-juniper woodlands; 1200–2200 m; Ariz., Utah.

Variety *papiliunculi* occurs from Emery, Garfield, Kane, and San Juan counties, Utah, southward to Apache, Coconino, and Navajo counties, Arizona, and just enters San Juan County, New Mexico.

4c. Eriogonum leptocladon Torrey & A. Gray var. **leptocladon** • Sand wild buckwheat E

Eriogonum effusum Nuttall var. *shandsii* S. Stokes

Plants 4–12 × 5–20 dm. **Aerial flowering stems** floccose to glabrate. **Leaves:** blade linear-lanceolate to linear-oblanceolate, 1.5–4 × 0.2–0.4 cm, margins plane. **Inflorescence branches** thinly floccose to glabrate, rarely white-tomentose. **Perianths** pale yellow to yellow.

Flowering Jun–Oct. White or red blow sand on flats, washes and slopes, mixed grassland, saltbush, ephedra, and sagebrush communities, pinyon-juniper woodlands; 1100–1900 m; Ariz., Utah.

Variety *leptocladon* is much more common in Utah's Emery, Garfield, Grand, and Wayne counties than is either of the other two varieties. It also occurs in San Juan County, but there it is restricted to portions of Canyonlands National Park, and is just inside the eastern edge of Sevier County. It is encountered also, albeit rarely, in Navajo County, Arizona. Variety *leptocladon* is worthy of cultivation, especially those expressions with densely white-tomentose inflorescence branches such as found in the Dolores Triangle area north of Ryan Creek in Grand County (e.g., *Atwood & Welsh 23213*, BRY).

5. Eriogonum nummulare M. E. Jones, Contr. W. Bot. 11: 13. 1903 • Money or Kearney wild buckwheat E

Eriogonum dudleyanum S. Stokes; *E. kearneyi* Tidestrom; *E. kearneyi* var. *monoense* (S. Stokes) Reveal; *E. kearneyi* subsp. *monoense* (S. Stokes) Munz ex Reveal; *E. nodosum* Small var. *kearneyi* (Tidestrom) S. Stokes; *E. nodosum* subsp. *monoense* S. Stokes

Shrubs, spreading or rarely sprawling, not scapose, (1.5–)3–8(–10) × 3–12(–15) dm, tomentose or rarely floccose to glabrate, grayish. **Stems** decumbent to spreading, often without persistent leaf bases, up to ¹/₃ height or more of plant; caudex stems absent or spreading in moving sand; aerial flowering stems spreading, slender to stout, solid, not fistulose, 0.5–1 dm, tomentose to floccose. **Leaves** cauline, 1 per node; petiole 0.2–1 cm, tomentose to floccose; blade broadly oblanceolate to elliptic or, rarely, orbiculate, (0.5–)1–3 × 0.4–1.2(–1.5) cm, densely white-tomentose abaxially, less so and greenish adaxially, margins plane. **Inflorescences** cymose, open, 5–50 × 5–80 cm; branches dichotomous proximally, often with involucres racemosely arranged distally, tomentose to thinly floccose, rarely glabrate; bracts 3, scalelike, linear to triangular, 0.5–6 mm. **Peduncles** absent or erect, 0.5–3 cm, tomentose. **Involucres** 1 per node, turbinate to turbinate-campanulate, 2–3 × 1.5–2.7 mm, tomentose or nearly so; teeth 5, erect, 0.1–0.4 mm. **Flowers** 1.5–3 mm; perianth white, glabrous; tepals connate proximal ¹/₄, essentially monomorphic, obovate; stamens slightly to long-exserted, 2–5 mm; filaments subglabrous or sparsely puberulent proximally. **Achenes** light brown, 2–3 mm, glabrous. 2*n* = 80.

Flowering Jul–Oct. Sandy to occasionally gravelly washes, flats, and slopes, saltbush and sagebrush communities, pinyon-juniper woodlands; 800–2600 m; Ariz., Calif., Nev., Utah.

Eriogonum nummulare occurs primarily in the Intermountain West from Inyo, Lassen, and Mono counties in California eastward across central and southern Nevada to western and southern Utah. It is found also in northern Mohave and northwestern Coconino counties in Arizona.

This species is the food plant of the pallid dotted-blue butterfly (*Euphilotes pallescens pallescens*) and the Sand Mountain blue butterfly (*Euphilotes pallescens arenamontana*) is an unprotected subspecies restricted to Sand Mountain in Churchill County, Nevada, where its host plant, *Eriogonum nummulare*, currently is subject to continued destruction due to unrestricted off-road vehicle activities.

6. **Eriogonum ammophilum** Reveal, Phytologia 23: 163. 1972 • Ibex wild buckwheat C E

Eriogonum nummulare M. E. Jones var. *ammophilum* (Reveal) S. L. Welsh

Subshrubs, spreading to somewhat sprawling, not scapose, (1.5–)2–4 × 2–5 dm, glabrous, grayish. **Stems** decumbent to spreading, without persistent leaf bases, up to ¼ height of plant; caudex stems absent or slightly spreading in moving sand; aerial flowering stems spreading, slender, solid, not fistulose, 0.5–1.5 dm, glabrous. **Leaves** basal or sheathing up proximal ¼ of flowering stems, 1 per node; petiole 0.1–0.5(–1) cm, tomentose to floccose; blade broadly elliptic, 1–2(–2.5) × 0.8–1.7 cm, densely white-tomentose abaxially, less so to subglabrous and green adaxially, margins plane. **Inflorescences** cymose, open, 5–20 × 5–20 cm; branches dichotomous, glabrous; bracts 3, scalelike, triangular, 1–2.5 mm. **Peduncles** absent or erect, restricted to proximal nodes, (0.2–)0.5–1(–1.5) cm, glabrous. **Involucres** 1 per node, turbinate, (2.5–)3–3.5 × 2–2.5 mm, glabrous; teeth 5, erect, 0.4–0.8 mm. **Flowers** 2–3 mm; perianth white, glabrous; tepals connate proximal ¼, essentially monomorphic, narrowly obovate; stamens slightly exserted, 2.5–3.5 mm; filaments puberulent proximally. **Achenes** light brown, 3–3.5 mm, glabrous except for distinctly papillate beak.

Flowering Jun–Sep. Sandy flats and washes, mixed grassland and sagebrush communities, pinyon-juniper woodlands; of conservation concern; 1600–1900 m; Utah.

Eriogonum ammophilum is known only from a few, scattered locations in Millard County. It is considered to be a "sensitive" species by the Bureau of Land Management.

7. **Eriogonum leptophyllum** (Torrey) Wooton & Standley, Contr. U.S. Natl. Herb. 16: 118. 1913 • Slender-leaf wild buckwheat E

Eriogonum effusum Nuttall var. *leptophyllum* Torrey in L. Sitgreaves, Rep. Exped. Zuni Colorado Rivers, 168. 1853

Shrubs or subshrubs, rounded to spreading, not scapose, (0.5–)2–8(–13) × (1–)3–15(–18) dm, thinly pubescent or glabrous and green, yellowish green or infrequently grayish, occasionally papillate. **Stems** spreading, without persistent leaf bases, up to ⅓ height of plant; caudex stems absent or compact; aerial flowering stems spreading, slender, solid, not fistulose, (0.05–)0.1–0.8 dm, thinly pubescent or glabrous. **Leaves** cauline, 1 per node or fasciculate; petiole 0.05–0.1 cm, tomentose to floccose or glabrous; blade linear to linear-oblanceolate, (0.5–)2–6 × (0.03–)0.1–0.3 cm, densely to thinly white-tomentose abaxially, thinly so or glabrous and green adaxially, margins tightly revolute. **Inflorescences** cymose, usually compact, (0.1–)2–12(–15) × (1–)4–15 (–30) cm; branches dichotomous, thinly pubescent or glabrous; bracts 3, scalelike, triangular, (0.5–)1–4 mm. **Peduncles** absent or erect, 0.05–0.2 cm, glabrous. **Involucres** 1 per node, narrowly turbinate, 2–4(–4.5) × 1–2 mm, glabrous; teeth 5, erect, 0.3–0.7 mm. **Flowers** 2.5–4 mm; perianth white, glabrous; tepals connate proximal ¼, essentially monomorphic, oblong to narrowly obovate; stamens long-exserted, (2–)3–6 mm; filaments subglabrous or sparsely puberulent proximally. **Achenes** brown, (2.5–)3.5–4 mm, glabrous. *2n* = 40.

Flowering Jul–Nov. Clayey flats, slopes, and outcrops, mixed grassland and sagebrush communities, pinyon-juniper woodlands; 1500–2300 m; Ariz., Colo., N.Mex., Utah.

Eriogonum leptophyllum is found mainly on southern Colorado Plateau in San Juan County, Utah, and adjacent Montezuma County, Colorado, southward into Apache County, Arizona, and in northwestern New Mexico (Bernalillo, Cibola, McKinley, Rio Arriba, Sandoval, San Juan, Santa Fe, and Taos counties). The species is found rarely in eastern Coconino and Navajo counties, Arizona, in the north, and in northern Gila County, Arizona, to the south. It is disjunct to the Bitter Spring Creek area of Capitol Reef National Park in Garfield County, Utah (*R. Fleming 199*, SJNM). The species, when dwarfed as it sometimes is on wind-swept ridges, resembles *E. microthecum* var. *simpsonii*, and has a form and aspect much more typical of that species than of the large, mature plants of *E. leptophyllum*. Mature plants on the southern edge of the range (as in McKinley County) tend to be grayish rather than the more common yellowish green seen elsewhere.

This species is considered a "life medicine" by the Navajo (Diné) people (C. Arnold, pers. comm.), being used in a variety of ways, including as an analgesic, a gynecological aid, a snake-bite remedy (D. E. Moerman 1986), and in casting spells (Arnold Clifford, pers. comm.). P. A. Vestal (1952) listed similar uses of this species by the Ramah Navajo of northwestern New Mexico, including an infusion of roots for stomach trouble, a decoction of the whole plant for snake bite, and for postpartum pain. The species is cultivated occasionally as a horticultural novelty.

There are two anomalous populations of particular interest. These occurred in the Broomfield area of San Juan County, New Mexico. The specimens are of low, spreading herbs to 0.8 dm with linear-oblong leaf blades 1–2 cm long but only 1–2 mm wide. The inflorescences are cymose but typically with one branch suppressed. A peduncle is present in some, this being up to 3.5 mm and erect; it is always at the basal node of the inflorescence. The involucres are turbinate and long (4–6 mm). A mature achene has not been observed. The plants flowered in late May and early June. Efforts to find such plants again have been unsuccessful. Searches in the late summer and early fall, when they ought to be in fruit, have found only plants that clearly can be assigned to *Eriogonum leptophyllum*. Generally, the two anomalous collections, both made by J. Mark Porter in the 1980s, resemble that species. Until such odd plants can be found again, and studied in detail, the significance of those populations cannot be ascertained.

8. Eriogonum clavellatum Small, Bull. Torrey Bot. Club 25: 48. 1898 • Comb Wash wild buckwheat [C] [E]

Subshrubs, spreading, not scapose, 1–2.5 × 3–8 dm, thinly floccose or glabrous, greenish to grayish. **Stems** spreading, without persistent leaf bases, up to ¹⁄₃ height of plant; caudex stems compact; aerial flowering stems spreading to erect, slender, solid, not fistulose, 0.06–0.2(–2.5) dm, thinly floccose or glabrous. **Leaves** cauline, 1 per node or fasciculate; petiole 0.05–0.1 cm, tomentose to floccose, rarely glabrous; blade oblanceolate, 1–1.5(–2) × 0.08–0.2 cm, densely white-tomentose abaxially, thinly tomentose and grayish or rarely glabrous and green adaxially, margins tightly revolute. **Inflorescences** umbellate to cymose, compact, 0.5–2 × 1–2 cm; branches dichotomous, thinly floccose or glabrous; bracts 3, scalelike, triangular, 1.5–2.5(–3) mm. **Peduncles** erect, 0.15–0.8 cm, glabrous. **Involucres** 1 per node, turbinate-campanulate, (3–)3.5–4.5(–5) × 2.5–4.5 mm, glabrous; teeth 5, erect, 0.6–0.9(–1.1) mm. **Flowers** (2.5–)3–3.5 mm; perianth white, glabrous; tepals connate proximal ¹⁄₄–¹⁄₃, dimorphic, those of outer whorl broadly obovate to nearly fan-shaped, 2–2.5 mm wide, those of inner whorl slightly shorter and oblanceolate to spatulate, 0.9–1.5 mm wide; stamens long-exserted, 3–6 mm; filaments sparsely pilose proximally. **Achenes** light brown, 3–3.5 mm, glabrous.

Flowering May–Jul. Sandy to heavy clay washes and slopes, saltbush communities; of conservation concern; 1300–1800 m; Colo., N.Mex., Utah.

Eriogonum clavellatum is relatively rare, known only from a few sites in the Four Corners area of San Juan County, Utah, Montezuma County, Colorado, and San

Juan County, New Mexico. It has yet to be found in Arizona. The Comb Wash wild buckwheat is considered a "sensitive" species by the Bureau of Land Management throughout its range.

9. Eriogonum pelinophilum Reveal, Great Basin Naturalist 33: 120. 1973 • Clay-loving wild buckwheat [C] [E]

Subshrubs, spreading, not scapose, 0.5–1(–1.2) × 0.8–3(–4) dm, floccose or glabrous, grayish. **Stems** spreading, without persistent leaf bases, up to ¹⁄₃ height of plant; caudex stems absent or compact; aerial flowering stems spreading to erect, slender, solid, not fistulose, 0.05–0.1 dm, thinly floccose or glabrous. **Leaves** cauline, 1 per node; petiole 0.05–0.1 cm, floccose; blade oblanceolate, 0.5–1.2(–1.5) × 0.08–0.2(–0.3) cm, densely white-tomentose abaxially, subglabrous or glabrous and green adaxially, margins tightly revolute. **Inflorescences** cymose, compact, 0.1–2 × 1–3 cm; branches dichotomous, thinly floccose or glabrous; bracts 3, scalelike, triangular, 0.5–1 mm. **Peduncles** absent or erect, 0.1–0.5 cm, floccose or glabrous. **Involucres** 1 per node, narrowly turbinate, (2–)2.5–3.5 × 1–1.5 mm, floccose or glabrous; teeth 5, erect, 0.3–0.4 mm. **Flowers** (2.5–)3–3.5 mm; perianth cream, glabrous; tepals connate proximal ¹⁄₂, essentially monomorphic, oblong; stamens slightly exserted, 2.5–4 mm; filaments sparsely pilose proximally. **Achenes** light brown, 3–3.5 mm, glabrous. $2n = 40$.

Flowering May–Jul. Heavy clay flats and slopes, saltbush communities; of conservation concern; 1600–1900 m; Colo.

Eriogonum pelinophilum is a federally listed endangered species with designated critical habitat. It is known only from Mancos Shale hills in Delta and Montrose counties. Much of the former habitat in the Montrose, Colorado, area has been destroyed since the species was listed in 1984. The type locality in Delta County was largely destroyed in 2001 by off-road vehicle activities in the designated critical habitat. A small population is preserved at the Fairview Natural Area east of Montrose.

Eriogonum pelinophilum is similar to *E. clavellatum* although the two are well-separated geographically. It is a smaller plant than *E. clavellatum* in habit. The flowers of *E. clavellatum* lack the pronounced, rounded, greenish-red to brownish-red base of the perianth seen in *E. pelinophilum*, and the tepals are distinctly dimorphic in *E. clavellatum* whereas they are essentially monomorphic in *E. pelinophilum*. Ants actively pollinate the flowers, being involved with both self- and cross-pollination. Some 50 additional visitors were found associated with the flowers, but none was confirmed as a pollinator (W. R. Bowlin et al. 1993).

The species is in the Center for Plant Conservation's National Collection of Endangered Plants.

SELECTED REFERENCE   Bowlin, W. R., V. J. Tepedino, and T. L. Griswold. 1993. The reproductive biology of *Eriogonum pelinophilum* (Polygonaceae). In: R. C. Sivinski and K. Lightfoot, eds. 1993. Southwestern Rare and Endangered Plants: Proceedings of the...Conference. Santa Fe. Pp. 296–300.

10. Eriogonum bicolor M. E. Jones, Zoë 4: 281. 1893
    • Pretty wild buckwheat   [E]

Eriogonum microthecum Nuttall subsp. *bicolor* (M. E. Jones) S. Stokes

Subshrubs, compact, not scapose, 0.2–0.8 × 0.5–2(–3) dm, reddish- or tannish-tomentose, reddish. **Stems** erect to slightly spreading, often with persistent leaf bases, up to ¼ height of plant; caudex stems compact; aerial flowering stems spreading to somewhat erect, slender, solid, not fistulose, 0.05–0.2 dm, tomentose. **Leaves** cauline, 1 per node or fasciculate; petiole 0.1–0.2 cm, tomentose to floccose; blade linear-oblanceolate to narrowly elliptic, 0.5–1.7(–2) × 0.1–0.2 (–0.3) cm, densely white-tomentose abaxially, slightly less so and often reddish or tannish adaxially, margins revolute. **Inflorescences** umbellate to cymose, compact, rarely reduced to single involucre, 0.5–1 × 0.5–1.5 cm; branches dichotomous, tomentose; bracts 3, scalelike, linear, 0.7–1.3 mm. **Peduncles** erect, 0.1–0.3(–0.4) cm, tomentose to floccose. **Involucres** 1 per node, turbinate-campanulate, 2–4 × 1.5–3 mm, tomentose to floccose or glabrous; teeth 5, erect, 0.4–0.7 mm. **Flowers** 2.5–4(–4.5) mm; perianth white, glabrous; tepals connate proximal ¼, dimorphic, those of outer whorl broadly obovate to nearly orbicular, (2–)2.5–3 mm wide, those of inner whorl oblanceolate to narrowly elliptic, 1–1.5 mm wide; stamens long-exserted, 3–5 mm; filaments sparsely pilose proximally or glabrous. **Achenes** light brown, 3–3.5 mm, glabrous.

Flowering Apr–Jul. Silty, sandy, or heavy clay washes, flats, and slopes, saltbush and blackbrush communities, juniper or pinyon-juniper woodlands; 1300–2000 (–2300) m; Colo., Utah.

Eriogonum bicolor is common from Castle Valley and the San Rafael Swells of Carbon and Emery counties, Utah, eastward to Grand Valley of Grand County, Utah, and Mesa County, Colorado. It occurs to the south in Garfield, San Juan, Sevier, and Wayne counties of Utah, especially in Canyonlands National Park and the Grand Staircase-Escalante National Monument. The species is worthy of cultivation, although being a slow-growing perennial of rather specialized soils it is a challenge. Seed germination in garden plots is reasonably successful; transplanting is difficult.

11. Eriogonum ripleyi J. T. Howell, Leafl. W. Bot. 4: 5. 1944 • Ripley's wild buckwheat   [C] [E]

Subshrubs, sprawling to decumbent, occasionally scapose, 0.5–2.5(–4) × 0.5–5(–7) dm, grayish-tomentose. **Stems** decumbent to slightly spreading or erect, often with persistent leaf bases, up to ¼ height of plant; caudex stems absent or compact; aerial flowering stems spreading to erect, slender, solid, not fistulose, 0.05–2.5(–3.5) dm, tomentose. **Leaves** cauline, 1 per node or fasciculate; petiole 0.05–0.15 cm, villous; blade narrowly oblanceolate, 0.2–0.6 × 0.05–0.1 cm, densely white-tomentose abaxially, thinly floccose to villous adaxially, margins revolute. **Inflorescences** cymose-umbellate or capitate and reduced to single involucre; branches usually dichotomous, sometimes with secondaries suppressed, occasionally absent, tomentose; bracts absent. **Peduncles** erect, 0.1–1.5 cm, thinly floccose. **Involucres** 1 per node, campanulate, 3–3.5 × 3–3.5 mm, thinly floccose or villous to subglabrous; teeth 3–5, erect, 0.7–1 mm. **Flowers** 3.5–4.5 mm; perianth white, glabrous; tepals connate proximal ⅕, dimorphic, those of outer whorl nearly orbiculate, 3–3.5 mm wide, those of inner whorl broadly obovate, 2–2.5 mm wide; stamens long-exserted, 4–5 mm; filaments villous to densely pilose proximally. **Achenes** light brown, 2–2.5 mm, glabrous.

Flowering Mar–Jun. Sandy clay flats and slopes on edge of sandstone outcrops, oak-juniper woodlands; of conservation concern; 1000–1900 m; Ariz.

Eriogonum ripleyi is known only from two areas in Arizona, one near Frazier's Well in Coconino County and a second in the Verdi Valley area of southeastern Yavapai and extreme northwestern Maricopa counties. The species is worthy of cultivation. Ripley's wild buckwheat is considered a "sensitive" species in Arizona.

12. Eriogonum pulchrum Eastwood, Proc. Calif. Acad. Sci., ser. 4, 20: 139. 1931 • Meteor Crater wild buckwheat [E]

Eriogonum ericifolium subsp. *pulchrum* (Eastwood) L. M. Shultz; *E. ericifolium* Torrey & A. Gray var. *pulchrum* (Eastwood) Reveal; *E. mearnsii* Parry var. *pulchrum* (Eastwood) Kearney & Peebles; *E. microthecum* Nuttall subsp. *pulchrum* (Eastwood) S. Stokes

Subshrubs, spreading and matted, not scapose, 0.8–1.2(–1.5) × 1–2.5(–3) dm, grayish- or reddish-tomentose to floccose, mostly reddish. **Stems** slightly spreading or erect, often with persistent leaf bases, up to ¼ height of plant; caudex stems absent or compact; aerial flowering stems spreading to erect, slender, solid, not fistulose, 0.1–0.5(–0.6) dm, slightly tomentose to floccose. **Leaves** cauline, 1 per node or fasciculate; petiole 0.1–0.2 cm, tomentose; blade oblanceolate to narrowly elliptic, 0.5–0.8(–1) × 0.1–0.2 cm, densely white-tomentose adaxially, floccose and greenish adaxially, margins slightly revolute or merely thickened. **Inflorescences** cymose, usually compact, 0.5–1.5 × 0.1–2(–25) cm; branches dichotomous, sparsely tomentose to floccose; bracts 3, scalelike, linear, 1–1.5 mm. **Peduncles** absent. **Involucres** 1 per node, narrowly turbinate, 1.5–3 × 1–1.5 mm, floccose; teeth 5, erect, 0.3–0.5 mm. **Flowers** 1.5–2(–2.5) mm; perianth white to rose, glabrous; tepals connate proximal ⅕, dimorphic, those of outer whorl nearly orbiculate, 2–2.5 mm wide, apex slightly notched, those of inner whorl oblanceolate, 1.5–2 mm wide; stamens slightly exserted, 2–3 mm; filaments pilose proximally. **Achenes** light brown, 2–2.5 mm, glabrous except for papillate beak.

Flowering Aug–Nov. Gravelly to rocky volcanic soil and outcrops, blackbrush and sagebrush communities, juniper or pinyon-juniper woodlands; (1000–)1600–2100 m; Ariz.

Eriogonum pulchrum occurs as a series of scattered populations north of the Mogollon Rim from northeastern Mohave County, eastward into Apache, Coconino, and Navajo counties; it is found also in northern Yavapai County. It is clearly related to *E. ericifolium* but is morphologically more distinct from that species than *E. thornei* is from *E. ericifolium*. The recognition of *E. pulchrum* at species rank is a logical extension of the treatment given to Thorne's wild buckwheat by L. M. Shultz (1998). The species is occasionally seen in cultivation.

13. Eriogonum terrenatum Reveal, Phytologia 86: 144. 2004 • San Pedro River wild buckwheat [E]

Shrubs, sprawling to erect, not scapose, 1–4(–5) × (1–)2–6(–9) dm, thinly tomentose or floccose to subglabrous, greenish. **Stems** spreading or erect, without persistent leaf bases, up to ½ height of plant; caudex stems absent; aerial flowering stems erect or nearly so, slender, solid, not fistulose, 0.05–0.3 dm, thinly tomentose. **Leaves** cauline, fasciculate; petiole 0.05–0.1 cm, glabrous; blade linear-oblanceolate or linear-elliptic, 0.3–0.8(–1) × (0.05–)0.1–0.2 cm, densely white-tomentose abaxially, thinly floccose or glabrous and greenish adaxially, margins rolled. **Inflorescences** cymose, compact, 1–3 × 1–3 cm; branches dichotomous, thinly tomentose; bracts 3, scalelike, triangular, (0.5–)1–2 mm. **Peduncles** absent or erect, 0.1–0.8(–1) cm, thinly tomentose. **Involucres** 1 per node, campanulate, 3–4.5 × 2.5–3.5 mm wide, thinly tomentose to subglabrous; teeth 5, erect, 0.5–1.2 mm. **Flowers** 3.5–4.5(–5) mm; perianth white, glabrous; tepals connate proximal ⅓, dimorphic, those of outer whorl broadly cordate, 2–3 mm wide, apex rounded, those of inner whorl oblanceolate, 1–2 mm wide; stamens exserted, 3.5–4.5 mm; filaments pilose proximally. **Achenes** light brown, 4–4.5 mm, glabrous.

Flowering Aug–Nov. Clayey slopes and flat, creosote bush communities; 1000–1200 m; Ariz.

Eriogonum terrenatum is confined to two geographically separate areas. Near Vail, Pima County, it is restricted to clayey outcrops of the Pantano Formation, whereas near Fairbanks, Cochise County, it is confined to the eroded, clay slopes and flats of the Saint David Formation.

14. Eriogonum ericifolium Torrey & A. Gray, Proc. Amer. Acad. Arts 8: 170. 1870 (as ericaefolium) • Heather-leaf wild buckwheat [E]

Eriogonum mearnsii Parry; *E. microthecum* Nuttall subsp. *ericifolium* (Torrey & A. Gray) S. Stokes; *E. microthecum* subsp. *mearnsii* (Parry) S. Stokes

Subshrubs, spreading and matted, not scapose, 0.5–1.5 × 0.8–2.5(–3) dm, mostly floccose or glabrous, greenish. **Stems** spreading, with persistent leaf bases, up to ¼ height of plant; caudex stems absent or compact; aerial flowering stems spreading to erect, slender, solid, not fistulose,

0.03–0.1 dm, floccose to thinly tomentose. **Leaves** cauline, 1 per node or fasciculate; petiole 0.1–0.15(–0.2) cm, tomentose; blade linear, 0.6–1 × (0.08–)0.1–0.2 cm, densely white-tomentose adaxially, glabrous and green adaxially, margins revolute. **Inflorescences** umbellate-cymose to cymose, compact, 0.5–1 × 0.5–2 cm; branches usually dichtomous proximally, otherwise with secondaries suppressed, floccose or glabrous; bracts 3, scalelike, linear, 0.5–1.5 mm. **Peduncles** absent. **Involucres** 1 per node, turbinate, 1.5–3(–3.5) × 1–2 mm, floccose; teeth 5, erect, 0.4–0.6 mm. **Flowers** 2–3(–3.5) mm; perianth white to pink or rose, glabrous; tepals connate proximal ¼, dimorphic, those of outer whorl obovate, 1.7–2 mm wide, those of inner whorl oblanceolate, 0.8–1 mm wide; stamens slightly exserted, 2–3 mm; filaments pilose proximally. **Achenes** light brown, 2–2.5 mm, glabrous except for papillate beak.

Flowering Aug–Nov. Gravelly or rocky slopes of lacustrine silt, mixed grasslands, chaparral and oak-woodlands; 900–1100 m; Ariz.

Eriogonum ericifolium is rare and localized, known only from the Verde River Valley of Yavapai County. The species is distinctive, yet some plants appear to approach *E. microthecum* var. *simpsonii*. The heather-leaf wild buckwheat is considered a "sensitive" species in Arizona and a candidate for federal protection.

15. **Eriogonum thornei** (Reveal & Henrickson) L. M. Shultz, Harvard Pap. Bot. 3: 51. 1998 • Thorne's wild buckwheat [E]

Eriogonum ericifolium Torrey & A. Gray var. *thornei* Reveal & Henrickson, Madroño 23: 205, fig. 1. 1975; *E. ericifolium* subsp. *thornei* (Reveal & Henrickson) Thorne

Subshrubs, spreading and matted, not scapose, 0.4–0.8 × 0.4–1 (–2.5) dm. **Stems** spreading, with persistent leaf bases, up to ¼ height of plant; caudex stems compact; aerial flowering stems spreading, slender, solid, not fistulose, 0.1–0.2 dm, floccose to slightly tomentose. **Leaves** cauline, 1 per node or fasciculate; petiole 0.01–0.02 cm, tomentose; blade linear, 0.4–0.6 × 0.05–0.1 cm, densely white-tomentose adaxially, finely villous and green adaxially, margins revolute. **Inflorescences** umbellate-cymose, compact, 0.5–1 × 0.5–1 cm; branches usually dichotomous, otherwise with secondaries suppressed, glabrous or nearly so; bracts 3, scalelike, linear, 0.5–1 mm. **Peduncles** absent. **Involucres** 1 per node, turbinate, 1.5–2 × 1–1.5 mm, floccose; teeth 5, erect, 0.4–0.6 mm. **Flowers** 1.5–2 mm; perianth white, glabrous; tepals connate proximal ¼, dimorphic, those of outer whorl obovate, 1–1.5 mm wide, those of inner

whorl oblanceolate, 0.8–1 mm wide, connate proximally; stamens slightly exserted, 2–3 mm; filaments pilose proximally. **Achenes** light brown, 2–2.5 mm, glabrous except for papillate beak.

Flowering May–Jul. Copper-rich quartzite gravel on ridges, pinyon woodlands; 1800 m; Calif.

Eriogonum thornei is known from a single canyon area in the New York Mountains, San Bernardino County. Shultz elevated Thorne's wild buckwheat to species status when she proposed *E. phoeniceum*, here considered a variety of *E. microthecum*. *Eriogonum thornei* and *E. ericifolium* are weakly differentiated, with minimal morphologic differences, and yet they occur on different substrates and are well-removed biogeographically from each other. Thorne's wild buckwheat is a protected plant in California.

16. **Eriogonum corymbosum** Bentham in A. P. de Candolle and A. L. P. P. de Candolle, Prodr. 14: 17. 1856 [E] [F]

Eriogonum effusum Nuttall subsp. *corymbosum* (Bentham) S. Stokes

Shrubs or subshrubs, spreading, rounded, occasionally erect, rarely somewhat matted, not scapose, (0.5–)1.5–8(–15) × (2–)3–15(–23) dm, grayish- to reddish-brown-tomentose to floccose or glabrous, grayish or greenish. **Stems** spreading or erect, often with persistent leaf bases, up to ¾ or more height of plant; caudex stems absent or somewhat matted; aerial flowering stems erect or nearly so, slender or occasionally stout, solid, not fistulose, (0.1–)1–2 dm, tomentose to floccose, occasionally glabrous. **Leaves** cauline, 1 per node; petiole 0.1–1.5 cm, tomentose to floccose; blade lanceolate to oblanceolate or elliptic to nearly orbiculate, rarely cordate, (0.5–)1–3(–4.5) × (0.3–)0.5–3(–3.5) cm, densely white-, tannish- or brownish-tomentose on both surfaces or less so to nearly glabrous and green adaxially, margins occasionally crenulate. **Inflorescences** cymose, rarely capitate or umbellate, diffuse to rather open, (1–)3–20 × 2–25(–30) cm; branches dichotomous, tomentose, floccose, or rarely glabrous; bracts 3, scalelike, usually triangular, and 1–3(–6) mm, or leaflike, 10–25 mm, and similar to leaf blades. **Peduncles** absent. **Involucres** 1 per node, turbinate, (1–)1.5–3.5 × 1–2(–2.5) mm; teeth 5, erect, 0.3–1 mm. **Flowers** (1.5–)2–3.5 mm; perianth white to cream, pink, or pale yellow to yellow, glabrous or rarely sparsely pilose; tepals connate proximal ¼–⅓, essentially monomorphic, oblanceolate to spatulate; stamens included to slightly exserted, 1–4 (–5) mm; filaments typically pilose proximally. **Achenes** brown, 2–2.5(–3) mm, glabrous except for occasional papillate beak.

Varieties 8 (8 in the flora): w United States.

Eriogonum corymbosum is a difficult complex of overlapping expressions, some of which are maintained here as taxonomically significant. Although perianth color is used to group the varieties, this feature is not consistent even in single populations. Therefore, population trends in perianth color must be noted in the field. Most of the varieties are then distinguished on the basis of leaf characters, and again, considerable variation can be seen in some populations. Still, the combination of flower color, leaf features, and geographic distribution should prove useful in distinguishing the varieties.

S. L. Welsh et al. (2003) alluded to hybrid combinations involving *Eriogonum corymbosum* and other species. Aside from the instances involving *E. brevicaule*, discussed below, none has been confirmed. Most of the putative hybrids are misidentified specimens of *E. lonchophyllum* or collections of var. *corymbosum* in which the leaf-margins are not decidedly crisped, a feature usually seen only in fully mature plants.

Eriogonum corymbosum was widely used by Native Americans. P. A. Vestal (1940) reported that the Hopi pressed boiled stalks into cakes that, when dried, were eaten with salt. J. W. Fewkes (1896) indicated that boiled leaves were mixed with cornmeal and water, and then baked into a kind of bread. S. A. Weber and P. D. Seaman (1985) indicated that A. F. Whiting was aware of a decoction of leaves (probably from var. *glutinosum*) being used for headaches. Variety *glutinosum* also was used primarily to treat tuberculosis, or at least as a cough medicine (D. E. Moerman 1986).

Some of the expressions of *Eriogonum corymbosum* are attracting the interest of gardeners, a few are coming into cultivation, and several selections are now being developed. The plants are slow growing but can be transplanted with some degree of success. Members of the varieties are food plants for Ellis's dotted-blue butterfly (*Euphilotes ellisi*).

1. Perianths usually pale yellow to yellow, occasionally white; s Utah, n Arizona, se Nevada.
  2. Inflorescence branches glabrous or nearly so; sw Utah . . . . 16h. *Eriogonum corymbosum* var. *aureum*
  2. Inflorescence branches tomentose to floccose.
    3. Leaf blades greenish-floccose adaxially; Colorado Plateau, n Arizona, s Utah . . . . 16f. *Eriogonum corymbosum* var. *glutinosum*
    3. Leaf blades silvery-floccose adaxially; Mojave Desert, se Nevada, sc Utah . . . . 16g. *Eriogonum corymbosum* var. *nilesii*

1. Perianths white to cream or pink; sw Wyoming s through e Utah and w Colorado to nw New Mexico and Arizona.
  4. Leaf blades elliptic-oblong to nearly orbiculate, 1–3(–3.5) × 1–2.5(–3.5) cm.
    5. Leaf blades elliptic-oblong to oblong or ovate, rarely cordate, densely white- to tan-lanate or less so to brownish-floccose adaxially; plants brownish-white-tomentose; flowers 2–2.5(–3) mm . . . . 16d. *Eriogonum corymbosum* var. *velutinum*
    5. Leaf blades usually broadly ovate to nearly orbiculate, densely grayish-tomentose abaxially, sometimes tomentose to floccose on both surfaces, or more often sub-glabrous or glabrous and green adaxially; plants greenish-tomentose; flowers 2.5–3 mm . . . . 16e. *Eriogonum corymbosum* var. *orbiculatum*
  4. Leaf blades lanceolate to oblanceolate or elliptic, (0.5–)1–3(–4.5) × (0.3–)0.5–1.5 cm.
    6. Plants sprawling subshrubs, 0.5–1.5 dm, occasionally shrubs, 2–4.5 dm; leaf blades 0.5–2 × 0.3–0.6(–0.8) cm; subalpine in bristlecone pine communities . . . . 16c. *Eriogonum corymbosum* var. *heilii*
    6. Plants erect subshrubs or shrubs, 3–10 dm; leaf blades 1–3(–4.5) × (0.3–)0.5–1.5 cm; desert scrub to montane forest communities.
      7. Leaves cauline ½ or more length of flowering stems; n Arizona, w Colorado, e and s Utah, sw Wyoming . . . . 16a. *Eriogonum corymbosum* var. *corymbosum*
      7. Leaves cauline ¼ or less length of flowering stems; c and e Utah . . . . 16b. *Eriogonum corymbosum* var. *revealianum*

16a. Eriogonum corymbosum Bentham var. **corymbosum** • Crisp-leaf wild buckwheat

E F

Eriogonum corymbosum var. *divaricatum* Torrey & A. Gray; *E. corymbosum* var. *erectum* Reveal & Brotherson; *E. divergens* Small; *E. effusum* Nuttall subsp. *divaricatum* (Torrey & A. Gray) S. Stokes; *E. effusum* subsp. *durum* S. Stokes; *E. effusum* subsp. *salinum* (A. Nelson) S. Stokes

Subshrubs or shrubs, 3–10 × 3–10 dm. **Leaves** cauline ½ or more length of flowering stem; petiole 0.2–0.6 (–0.8) cm; blade lanceolate to oblanceolate or elliptic, 1–3(–4.5) × (0.3–)0.5–1.5 cm, usually densely tomentose on both surfaces, sometimes less so and greenish adaxially. **Inflorescences** 3–10 cm; branches tomentose to floccose. **Involucres** 1.5–3.5 × 1–2 mm. **Flowers** 2–3.5 mm; perianth white to cream, glabrous. $2n = 40$.

Flowering Jul–Oct. Sandy to gravelly or clayey flats, washes, slopes, outcrops, and cliffs, saltbush, blackbrush, and sagebrush communities, pinyon-juniper and montane conifer woodlands; 1200–2700 m; Ariz., Colo., Utah, Wyo.

Variety *corymbosum* is the common white-flowered expression of the species. It occurs in northern Arizona, western Colorado, eastern and southern Utah, and southwestern Wyoming. Montane plants in the northern part of the range in Utah and Wyoming with long, narrow, erect leaves have been called var. *erectum*; desert plants in east-central Utah with small crenulate leaves have been called var. *divaricatum*. Hybrids between the shrubby var. *corymbosum* and the herbaceous *E. brevicaule* have been named *E.* ×*duchesnense* Reveal (as species, including *E. corymbosum* Bentham var. *albogilvum* Reveal). The hybrid is known from Rio Blanco County, Colorado (*Goodrich 21999*, BRY), Uintah County (*Neese & Sinclair 15056*, BRY), Utah (*Reveal & Reveal 725*, BRY, UTC), and Wasatch County (*Goodrich 16099*, BRY) in northeastern Utah, and from Sweetwater County, Wyoming (*Porter & Porter 10510*, BRY, RM; *Reveal & Reveal 2935*, BRY, UTC). Given the limited distribution of *E. corymbosum* in Wyoming (Sweetwater County), it is considered a "species of concern" in that state.

16b. Eriogonum corymbosum Bentham var. **revealianum** (S. L. Welsh) Reveal, Great Basin Naturalist 35: 362. 1976 • Bicknell wild buckwheat [E]

Eriogonum revealianum S. L. Welsh, Great Basin Naturalist 30: 17, fig. 2. 1970

Subshrubs, 3–6 × 2–7 dm. **Leaves** cauline ¹/₄ or less length of flowering stem; petiole 0.5–1(–1.5) cm; blade lanceolate to oblanceolate, 2–4.5 × 0.3–1 cm, densely tomentose abaxially, less so and greenish adaxially. **Inflorescences** 8–20 cm; branches floccose. **Involucres** 1.5–3.5 × 1–2 mm. **Flowers** 2–3.5 mm; perianth white to cream, glabrous.

Flowering Jul–Sep. Gravelly to rocky clay slopes, sagebrush communities, pinyon-juniper and conifer woodlands; 2100–2800 m; Utah.

Variety *revealianum* is known only from Utah (mainly Garfield, northern Kane, Piute and Sanpete counties), where it is encountered only infrequently. Elsewhere it is found in widely scattered disjunct populations on Bromley Ridge in the La Sal Mountains in northern San Juan County, and near the abandoned townsite of Rainbow in southeastern Uintah County. It is only marginally distinct from var. *corymbosum*. Some populations of *E. lonchophyllum* closely resemble var. *revealianum*, but that species always has glabrous inflorescences where such confusion might occur.

16c. Eriogonum corymbosum Bentham var. **heilii** Reveal, Phytologia 86: 125. 2004 • Heil's wild buckwheat [E]

Subshrubs, 0.5–1.5 × 2–4 dm or 2–4.5 × 4–9 dm. **Leaves** cauline ¹/₄ length of flowering stem or ³/₄ length of nonscapose flowering stem; petiole 0.1–0.3 cm; blade lanceolate to elliptic or oblanceolate, 0.5–2 × 0.3–0.6(–0.8) cm, densely tomentose abaxially, tomentose to floccose and greenish adaxially. **Inflorescences** capitate, 1–1.5 cm wide, or umbellate or cymose, 2–4(–8) × 3–6(–12) cm, tomentose to floccose; branches tomentose to floccose. **Involucres** (2–)2.5–3.5 × 1.5–2.5 mm. **Flowers** 2–3 mm; perianth white to pink, glabrous or, rarely, sparsely pilose.

Flowering Jul–Aug. Steep rocky slopes, subalpine conifer woodlands; 2500–2800 m; Utah.

Variety *heilii* is known only from the upper Deep Creek drainage on the eastern slope of Thousand Lake Mountain in Capitol Reef National Park, Wayne County, where it occurs with Rocky Mountain bristlecone pine (*Pinus aristata* Engelmann).

16d. Eriogonum corymbosum Bentham var. **velutinum** Reveal, Great Basin Naturalist 27: 224, fig. 14. 1968 • Velvety wild buckwheat [E]

Shrubs, 5–10 × 8–15(–20) dm. **Leaves** cauline ¹/₂ or more length of flowering stem; petiole 0.5–1.5 cm; blade elliptic-oblong to oblong or ovate, rarely cordate, 1.5–3(–3.5) × (1–)1.5–2.5(–3.5) cm, densely white-tomentose abaxially, densely white- to tan-lanate or less so to brownish-floccose adaxially. **Inflorescences** rather open, 3–10 cm; branches densely tomentose. **Involucres** 2–3.5 × 1.5–2.5 mm. **Flowers** 2–2.5(–3) mm; perianth white to cream, glabrous.

Flowering Aug–Oct. Sandy to gravelly or clayey flats, washes, and slopes, mixed grassland, saltbush, and sagebrush communities, pinyon-juniper woodlands; 1200–2300 m; Ariz., Colo., N.Mex., Texas, Utah.

Variety *velutinum* is the principal expression of the species in the southeastern portion of the species' range in northeastern Arizona (Apache County), western Colorado (Mesa, Montezuma, Montrose, and San Miguel counties), northwestern New Mexico (Quay, Rio Arriba, Sandoval, San Juan, Santa Fe, and Socorro counties), and in southeastern Utah (Garfield, Grand, Kane, San Juan, and Wayne counties). The Texas location remains to be confirmed by modern collections. This variety is not always distinct from var. *orbiculatum* in southwestern Colorado and adjacent southeastern Utah. Plants with cordate leaf blades from Gypsum Valley in San Miguel County, Colorado, assigned here to var. *orbiculatum*, are particularly difficult to place. Variety *velutinum* is currently being introduced into cultivation.

16e. Eriogonum corymbosum Bentham var. **orbiculatum** (S. Stokes) Reveal & Brotherson, Great Basin Naturalist 27: 221. 1968 • Orbicular-leaf wild buckwheat   E

Eriogonum effusum Nuttall subsp. *orbiculatum* S. Stokes, Eriogonum, 79. 1936

Shrubs, (3–)5–15 × 5–15 dm. **Leaves** cauline ½ or more length of flowering stem; petiole 0.5–1.5 cm; blade broadly ovate to nearly orbiculate, 1–3(–3.5) × 1–3(–3.5) cm, floccose to tomentose on both surfaces or floccose adaxially. **Inflorescences** compact and rather flat-topped, 3–10 cm; branches densely tomentose. **Involucres** 2–3.5 × 1.5–2.5 mm. **Flowers** 2.5–3 mm; perianth white, glabrous.

Flowering Aug–Oct. Sandy to gravelly flats, washes, slopes, mixed grassland, saltbush, blackbrush, and sagebrush communities, pinyon-juniper woodlands; 900–1600(–2300) m; Ariz., Colo., Utah.

Variety *orbiculatum* is encountered in northeastern Arizona (Coconino and Navajo counties), western Colorado (Mesa, Montezuma, San Juan, and San Miguel counties), and southeastern Utah (Emery, Garfield, Grand, Kane, San Juan, and Wayne counties). It is worthy of the serious horticultural efforts now being made to develop cultivars for the arid garden, with selection development emphasizing large, round, fully flowering shrubs capable of surviving a wide range of temperatures. A mature plant in full flower is a remarkable sight. The hemispheric shrubs wholly and densely blanketed by bright white flowers and set against the deep red sands and sandy-clay hills in Monument Valley are seen in numerous photographs and several John Ford movies. An expression from Gypsum Valley, San Miguel County, Colorado, may well reprecent an undescribed variety allied to var. *orbiculatum*.

16f. Eriogonum corymbosum Bentham var. **glutinosum** (M. E. Jones) M. E. Jones, Contr. W. Bot. 11: 14. 1903 • Sticky wild buckwheat   E

Eriogonum aureum M. E. Jones var. *glutinosum* M. E. Jones, Proc. Calif. Acad. Sci., ser. 2, 5: 719. 1895; *E. microthecum* Nuttall var. *crispum* (L. O. Williams) S. Stokes

Subshrubs or shrubs, 2–10 × 3–10 dm. **Leaves** cauline ½ or more length of flowering stem; petiole 0.5–1 cm; blade lanceolate to oblanceolate or elliptic, 1–4 × 0.5–1.5 cm, usually densely tomentose on both surfaces, sometimes floccose and greenish adaxially. **Inflorescences** 3–10 cm; branches tomentose to floccose. **Involucres** 1–2 × 1–1.5(–2) mm. **Flowers** 1.5–2.5 mm; perianth usually pale yellow to yellow, sometimes white, glabrous. $2n = 40$.

Flowering Jul–Oct. Sandy to gravelly or clayey flats, washes, slopes, outcrops, and cliffs, saltbush, blackbrush, sagebrush communities, pinyon-juniper woodlands; (800–)1200–2600 m; Ariz., Utah.

Variety *glutinosum* is the common yellow-flowered expression of the species. It occurs in Arizona (Coconino, Mohave, Navajo, and Yavapai counties) and Utah (Beaver, Emery, Garfield, Iron, Kane, Washington, and Wayne counties). Yellow-flowered plants in Garfield and Wayne counties of Utah might be better assigned to var. *corymbosum*.

16g. Eriogonum corymbosum Bentham var. **nilesii** Reveal, Phytologia 86: 128. 2004 • Niles's wild buckwheat   E

Shrubs, 3–12 × 4–23 dm. **Leaves** cauline ½ or more length of flowering stem; petiole 0.3–1.5 cm; blade elliptic to oblong, 0.8–2.5(–3) × 0.4–0.8 cm, white-lanate to densely white-tomentose abaxially, silvery-floccose adaxially. **Inflorescences** 2–20 cm; branches floccose. **Involucres** 1.5–2 × 1–1.5 mm. **Flowers** 2–3 mm; perianth yellow to pale yellow or, rarely, white, glabrous.

Flowering Aug–Nov. Sandy to gravelly or gypsum flats and washes, saltbush communities; 200–900 m; Nev., Utah.

Variety *nilesii* is a plant of the Mojave Desert, known for certain only from the Las Vegas and Muddy Mountains region of Clark County, Nevada. A collection from the flood plain of the Paria River in Kane County, Utah, is tentatively assigned to this variety.

16h. Eriogonum corymbosum Bentham var. **aureum** (M. E. Jones) Reveal, Taxon 32: 293. 1983

* Golden wild buckwheat [E]

Eriogonum aureum M. E. Jones, Proc. Calif. Acad. Sci., ser. 2, 5: 718. 1895; *E. microthecum* Nuttall subsp. *aureum* (M. E. Jones) S. Stokes

Subshrubs or shrubs, 2–6(–8) × 3–8(–12) dm. **Leaves** cauline ¹/₂ or more length of flowering stem; petiole 0.5–0.8 cm; blade usually elliptic, 2–4 × 1–1.5 cm, densely tomentose abaxially, usually glabrous and green adaxially. **Inflorescences** 3–10 cm; branches glabrous or nearly so. **Involucres** 1–2 × 1–1.5(–2) mm. **Flowers** 1.5–2.5 mm; perianth yellow to pale yellow, rarely nearly white, glabrous.

Flowering Aug–Oct. Gravelly slopes, creosote and blackbrush communities, juniper woodlands; 1000–1100 m; Utah.

Variety *aureum* is known only from Shivwits Hill near Castle Cliffs in Washington County. This rare expression is probably the result of past introgression with the no longer sympatric herbaceous perennial *Eriogonum thompsoniae.*

17. Eriogonum lancifolium Reveal & Brotherson, Great Basin Naturalist 27: 187, fig. 1. 1968

* Lance-leaf wild buckwheat [C][E]

Eriogonum corymbosum Bentham var. *davidsei* Reveal

Shrubs, spreading, not scapose, 3.5–8(–10) × 5–10(–13) dm, grayish-tomentose, grayish. **Stems** spreading, usually without persistent leaf bases, sheathing up to ¹/₂ or more height of plant; caudex stems absent; aerial flowering stems erect to spreading, slender, solid, not fistulose, 0.5–1.2 dm, tomentose. **Leaves** cauline, 1 per node; petiole 0.3–0.6 cm, tomentose; blade lanceolate to elliptic, 3–5 × 0.5–2 cm, densely silvery-tomentose on both surfaces or rarely less so to subglabrous and greenish abaxially, margins plane or slightly revolute, sometimes crenulate. **Inflorescences** cymose, often compact, (3–)6–14 × 3–10 dm; branches dichotomous, tomentose to subglabrous; bracts 3, scalelike, linear, and 1.5–3 mm, or leaflike, 10–15 mm, and similar to leaf blades. **Peduncles** absent. **Involucres** 1 per node, turbinate, 2.5–3 × 1.5–2 mm; teeth 5, erect, 0.4–1 mm. **Flowers** (2–)2.5–3.5 mm; perianth white to cream, glabrous; tepals connate proximal ¹/₄, essentially monomorphic, spatulate; stamens exserted, 1.5–4.5 mm;

filaments glabrous or slightly pilose proximally. **Achenes** brown, 2–2.5 mm, glabrous. **2***n* = 40.

Flowering Jul–Oct. Heavy clay flats and slopes, mixed grassland and saltbush communities, juniper woodlands; of conservation concern; 1500–1800 m; Utah.

Eriogonum lancifolium is known only from Mancos Shale hills and flats east and south of Wellington, in Carbon and Emery counties. The largest population is found on the flats south of Mounds Reef. S. L. Welsh et al. (2003) reduced this species to synonymy under *E. corymbosum* var. *corymbosum.*

18. Eriogonum hylophilum Reveal & Brotherson, Great Basin Naturalist 27: 190, fig. 2. 1968 • Gate Canyon wild buckwheat [C][E]

Eriogonum corymbosum Bentham var. *hylophilum* (Reveal & Brotherson) S. L. Welsh

Subshrubs, spreading, not scapose, 2.5–4 × 3–5 dm, tomentose, greenish. **Stems** spreading, often with persistent leaf bases, up to ¹/₄ or more height of plant; caudex stems somewhat matted; aerial flowering stems erect or nearly so, somewhat stout, solid, not fistulose, 0.5–1.5 dm, tomentose. **Leaves** cauline and sheathing proximal ¹/₃ of stem, 1 per node; petiole 0.5–1(–1.8) cm, tomentose; blade linear-lanceolate to lanceolate, 3.5–7(–9) × 0.3–0.6(–0.8) cm, densely white-tomentose abaxially, less so and greenish adaxially, margins plane and slightly revolute, rarely crenulate. **Inflorescences** cymose, open, 3–8 × 3–10 cm; branches dichotomous, tomentose; bracts 3, scalelike, triangular, and 2–3 mm, or leaflike, 1–2 cm, and similar to leaf blades. **Peduncles** absent. **Involucres** 1 per node or 2–3 per cluster, turbinate, 3.5–4 × 2.5–3 mm; teeth 5–6, erect, 0.5–1 mm. **Flowers** (3–)3.5–4(–4.5) mm; perianth white, glabrous; tepals connate proximal ¹/₄, slightly dimorphic, those of outer whorl spatulate, 1-3–1.7(–2) mm wide, those of inner whorl oblanceolate, 0.6–0.9(–1.2) mm wide; stamens included, 2–3(–3.5) mm; filaments pilose proximally. **Achenes** brown, 2.5–3 mm, glabrous except for slightly papillate beak.

Flowering Jul–Sep. Gravelly slopes, sagebrush communities, pinyon-juniper or montane conifer woodlands; of conservation concern; 2200–2600 m; Utah.

Eriogonum hylophilum is known from a few, scattered locations in the Gate Canyon area and just west along the rim of the Badland Cliffs in Duchesne County. The species is closely related to *E. corymbosum* and approaches the minor segregate of the latter known as var. *erectum.* It is possible that the Gate Canyon wild buckwheat has an evolutionary history involving that expression of *E. corymbosum* and the white- to cream-colored flower expression of *E. brevicaule* var. *laxifolium*. If so, *E. hylophilum* is a stable and persistent taxon of hybrid origin.

19. Eriogonum jonesii S. Watson, Proc. Amer. Acad. Arts 21: 454. 1886 • Jones's wild buckwheat Ⓒ Ⓔ

Eriogonum lanosum Eastwood

Subshrubs, spreading, not scapose, 2–5 × 2.5–5(–6) dm, white- to brownish-tomentose, reddish- to brownish-white. **Stems** spreading, without persistent leaf bases, up to ¼ height of plant; caudex stems absent; aerial flowering stems erect or nearly so, somewhat stout, solid, not fistulose, 1–2(–2.5) dm, floccose. **Leaves** cauline on proximal 1–7(–9) cm of stem, 1 per node; petiole 1.5–4(–5) cm, tomentose; blade cordate, (1.5–)2–3.5 × 1–2(–2.5) cm, densely white-tomentose abaxially, floccose and greenish or brownish white adaxially, margins plane or crenulate. **Inflorescences** cymose, usually open, 3–15 × 5–20 cm; branches dichotomous, tomentose to floccose; bracts 3, scalelike, triangular, 1–4 mm. **Peduncles** absent. **Involucres** 1 per node, turbinate, 1.5–2(–2.5) × 1–1.5 (–1.8) mm; teeth 5–6, erect, 0.3–0.5 mm. **Flowers** 2–3 mm; perianth brownish white, glabrous; tepals connate proximal ¼, dimorphic, those of outer whorl obovate, 1.8–2 mm wide, those of inner whorl lanceolate to narrowly elliptic, 1–1.2 mm wide; stamens slightly exserted, 2.5–4 mm; filaments glabrous or sparsely pilose proximally. **Achenes** brown, 2–2.5 mm, glabrous.

Flowering Aug–Nov. Rocky limestone, sandstone or pumice washes, flats, and outcrops, saltbush, blackbrush, and sagebrush communities, pinyon-juniper woodlands; of conservation concern; 1200–2100 m; Ariz.

Eriogonum jonesii is found primarily in Coconino County, with scattered populations just entering Mohave and Navajo counties. Jones's wild buckwheat would make an interesting addition to the garden. Although the plants are sometimes "leggy," the white to brownish tomentum is attractive, as are the brownish-white flowers.

20. Eriogonum smithii Reveal, Great Basin Naturalist 27: 202, fig. 6. 1968 • Smith's wild buckwheat Ⓒ Ⓔ

Eriogonum corymbosum Bentham var. *smithii* (Reveal) S. L. Welsh

Shrubs, erect to spreading, not scapose, (3–)4–10 × 5–20 dm, glabrous, bright green. **Stems** spreading, without persistent leaf bases, up to ⅓ height of plant; caudex stems absent; aerial flowering stems erect to spreading, slender, solid, not fistulose, 0.2–2 dm, glabrous. **Leaves** cauline on proximal ⅔ of stem, 1 per node; petiole 0.3–0.5 cm, floccose or glabrous; blade narrowly elliptic, 2.5–4.5 × (0.3–)0.6–1 cm, glabrous and green on both surfaces, rarely thinly floccose adaxially, margins occasionally slightly revolute. **Inflorescences** cymose, open to compact, 2–25 × 3–35 cm; branches often with involucres racemosely arranged at tips of inflorescence, glabrous; bracts 3, scalelike, linear to triangular, 1.5–4.5 mm. **Peduncles** absent. **Involucres** 1 per node, turbinate, (2.5–)3–3.5 × 2–2.5 mm, glabrous; teeth 5, erect, 0.3–0.5 mm. **Flowers** 3–4 mm; perianth yellow, glabrous; tepals connate proximal ¼, slightly dimorphic, those of outer whorl obovate, 1.5–2 mm wide, those of inner whorl oblanceolate, 1–1.5 mm wide; stamens exserted, 2–5(–7) mm; filaments pilose proximally. **Achenes** brown, 3–3.5 mm, glabrous except for minutely papillate beak. **2*n*** = 40.

Flowering Jul–Oct. Deep, moving, red blow sand, mixed grassland and scrub oak communities; of conservation concern; 1400–1900 m; Utah.

Eriogonum smithii is restricted to the San Rafael Desert of Emery and Wayne counties. Plants from Little Gilson Butte (*J. L. Anderson 86-228*, BRY) have leaf blades thinly floccose on the abaxial surface but are otherwise similar to the shrubs found at Flat Top to the east. Smith's wild buckwheat occurs in selenium-rich sands, which may limit its ability to be cultivated. The bright green, brilliantly yellow-flowered plants would make an elegant addition to the garden.

21. Eriogonum mortonianum Reveal, Brittonia 26: 90, fig. 1. 1974 • Fredonia wild buckwheat Ⓒ Ⓔ

Shrubs, erect to spreading, not scapose, (3–)4–10 × 5–20 dm, essentially glabrous and bright green. **Stems** spreading, without persistent leaf bases, up to ⅓ height of plant; caudex stems absent; aerial flowering stems erect to spreading, slender, solid, not fistulose, 0.2–2 dm, glabrous. **Leaves** cauline on proximal ⅔ of stem, 1 per node; petiole

(0.2–)0.3–0.8(–1) cm, floccose or glabrous; blade elliptic, 1.5–4(–4.5) × (0.3–)0.6–1(–1.3) cm, glabrous and green on both surfaces, margins plane. **Inflorescences** cymose, open, 15–25 × 13–30 cm; branches often with involucres racemosely arranged at tips of inflorescence branches, glabrous; bracts 3, scalelike, triangular, 1–2 mm. **Peduncles** absent. **Involucres** 1 per node, turbinate, 2–2.5 × 1.2–1.8(–2) mm, glabrous; teeth 5, erect, 0.2–0.3 mm. **Flowers** (2–)2.5–3 mm; perianth pale yellow to yellow or whitish, glabrous; tepals connate proximal 1/4, dimorphic, those of outer whorl obovate, 1–1.3 mm wide, those of inner whorl oblanceolate, 0.8–1 mm wide; stamens included to slightly exserted, 2–2.5 mm; filaments pilose proximally. **Achenes** brown, 3–3.5 mm, glabrous. **2*n*** = 40.

Flowering Jul–Sep. Red, gypsophilous clay flats and outcrops, mixed grassland, saltbush, and blackbrush communities; of conservation concern; 1400–1600 m; Ariz.

Eriogonum mortonianum is restricted to a series of low, gypsophilous hills southwest of Fredonia, Mohave County.

22. Eriogonum thompsoniae S. Watson, Amer. Naturalist 7: 302. 1873 (as thompsonae) [E]

Eriogonum corymbosum Bentham var. *thompsoniae* (S. Watson) S. L. Welsh

Herbs, spreading, not scapose, 2–5(–6) × 2–5 dm, glabrous, green. **Stems** spreading, without persistent leaf bases, up to 1/5 height of plant; caudex stems spreading, glabrous; aerial flowering stems erect, slender, solid, not fistulose, 1.2–3 dm, glabrous, tomentose among leaves. **Leaves** basal or sheathing up stem less than 2 cm, 1 per node; petiole (0.1–)3–7(–10) cm, tomentose to floccose or glabrous; blade linear or oblong to oblanceolate or elliptic, 2–8 (–10) × 0.2–2.5(–3) cm, thinly to densely white-tomentose abaxially, glabrous and green adaxially, margins plane or slightly revolute. **Inflorescences** cymose, open, 10–30 × 10–25 cm; branches dichotomous, glabrous; bracts 3, scalelike, linear, 2–5 (–7) mm. **Peduncles** absent. **Involucres** 1 per node, turbinate, 2–3(–3.5) × 1–1.5(–2) mm, glabrous; teeth 5, erect, 0.4–0.5 mm. **Flowers** 3–3.5 mm; perianth yellow or white, glabrous; tepals connate only at base, monomorphic, oblong to obovate; stamens slightly exserted, 3–3.5(–4) mm; filaments sparsely pilose proximally. **Achenes** light brown to brown, 2.5–3 mm, glabrous except for slightly papillate beak.

Varieties 3 (3 in the flora): w United States.

1. Leaf blades linear, 3–8(–10) × 0.2–0.4(–0.6) cm . . . . 22c. *Eriogonum thompsoniae* var. *atwoodii*
1. Leaf blades oblong to oblanceolate or elliptic; 2–5 × 0.8–2.5(–3) cm.
   2. Perianths yellow . . . . 22a. *Eriogonum thompsoniae* var. *thompsoniae*
   2. Perianths white . . . . . 22b. *Eriogonum thompsoniae* var. *albiflorum*

22a. Eriogonum thompsoniae S. Watson var. **thompsoniae** • Thompson's wild buckwheat [E]

Plants 2–4 dm. **Leaf blades** oblong to oblanceolate or elliptic, (2–)3–5 × 0.8–2.5(–3) cm, tomentose abaxially, margins plane. **Involucres** (2.5–)3–3.5 mm. **Perianths** yellow.

Flowering Jul–Nov. Sandy to clayey flats and slopes, saltbush, blackbrush, and sagebrush communities, pinyon-juniper woodlands; 1000–1700 m; Ariz., Utah.

Variety *thompsoniae* is found infrequently in Kane and Washington counties, Utah, and in Mohave County, Arizona.

22b. Eriogonum thompsoniae S. Watson var. **albiflorum** Reveal, Madroño 19: 299. 1969 (as thompsonae) • Virgin wild buckwheat [E]

Eriogonum corymbosum Bentham var. *albiflorum* (Reveal) S. L. Welsh; *E. corymbosum* var. *matthewsiae* Reveal; *E. thompsoniae* var. *matthewsiae* (Reveal) Reveal

Plants 2–5(–6) dm. **Leaf blade** oblong to oblanceolate or elliptic, 2–5 × 0.8–2.5 cm, tomentose abaxially, margins plane. **Involucres** 2–3 mm. **Perianths** white.

Flowering Jun–Oct. Sandy to gravelly, often volcanic flats and slopes, saltbush and blackbrush communities, pinyon-juniper and oak woodlands; 900–1400 m; Ariz., Utah.

Variety *albiflorum* is known only from scattered populations in Washington County, Utah, and in adjacent portions of Mohave County, Arizona.

22c. Eriogonum thompsoniae S. Watson var.
atwoodii Reveal, Great Basin Naturalist 34: 245.
1974 (as thompsonae) • Atwood's wild buckwheat
E

Eriogonum corymbosum Bentham
var. *atwoodii* (Reveal) S. L. Welsh

Plants 3–4 dm. **Leaf blade** linear,
3–8(–10) × 0.2–0.4(–0.6) cm,
thinly to moderately tomentose
abaxially, margins often slightly
revolute. **Involucres** 2–3 mm.
Perianths white or yellow.

Flowering Jul–Oct. Red,
gypsophilous clay flats and slopes, saltbush, blackbrush,
and sagebrush communities; 1400–1500 m; Ariz.

Variety *atwoodii* is known only from two locations,
one near Fredonia, and a second along the Honeymoon
Trail, both in Mohave County. It appears to be an
established, albeit rare, variety that probably is the result
of past hybridization involving *E. thompsoniae* var.
albiflorum and *E. mortonianum*. The Bureau of Land
Management in Arizona considers Atwood's wild
buckwheat a taxon of special concern.

23. Eriogonum lonchophyllum Torrey & A. Gray,
Proc. Amer. Acad. Arts 8: 173. 1870   E

Eriogonum ainsliei Wooton &
Standley; *E. corymbosum* Bentham
var. *humivagans* (Reveal) S. L.
Welsh; *E. effusum* Nuttall subsp.
ainsliei (Wooton & Standley)
S. Stokes; *E. effusum* subsp.
fendlerianum (Bentham) S. Stokes;
E. effusum var. *nudicaule* Torrey; *E.
effusum* subsp. *salicinum* (Greene)
S. Stokes; *E. fendlerianum* (Bentham) Small; *E. humivagans*
Reveal; *E. intermontanum* Reveal; *E. lonchophyllum* var.
fendlerianum (Bentham) Reveal; *E. lonchophyllum* var.
humivagans (Reveal) Reveal; *E. lonchophyllum* var.
intermontanum (Reveal) Reveal; *E. lonchophyllum* var.
nudicaule (Torrey) Reveal; *E. lonchophyllum* var. *saurinum*
(Reveal) S. L. Welsh; *E. nudicaule* (Torrey) Small; *E.
nudicaule* subsp. *scoparium* (Small) S. Stokes; *E. nudicaule*
subsp. *tristichum* (Small) S. Stokes; *E. salicinum* Greene; *E.
saurinum* Reveal; *E. scoparium* Small; *E. tristichum* Small

Shrubs, subshrubs, or herbs, spreading to erect, not
scapose, (1–)1.5–5 × 2–5(–8) dm, glabrous or rarely
floccose to tomentose, grayish. **Stems** spreading or erect,
usually without persistent leaf bases, up to ½ or more
height of plant; caudex stems absent or matted to
spreading; aerial flowering stems spreading to erect,
slender, solid, not fistulose, 0.3–3 dm, glabrous or, rarely,
floccose to tomentose, tomentose among leaves. **Leaves**
basal or cauline on proximal ½ of stem, 1 per node;

petiole 0.5–2 cm, tomentose to floccose or glabrous;
blade narrowly lanceolate or oblanceolate to elliptic, 1.5–
7(–9) × 0.2–2 cm, velvety- to densely white-tomentose
abaxially, sparsely tomentose to thinly floccose or
glabrous and green adaxially, margins plane or
occasionally crenulate. **Inflorescences** cymose, dense to
more commonly open, 2–25 × 2–20 cm; branches
dichotomous, glabrous or, rarely, floccose; bracts 3,
scalelike, usually triangular, and 1–3 mm, or occasionally
leaflike, 8–30 mm, and otherwise similar to leaf blades.
Peduncles absent or erect, 0.1–0.8 cm, glabrous.
Involucres 1 per node or 2–5 per cluster, turbinate to
turbinate-campanulate, 2.5–4 × (1.3–)1.5–3.5(–4) mm,
glabrous; teeth 5, erect, 0.4–0.9 mm. **Flowers** 2–3.5
(–4) mm; perianth white, glabrous; tepals connate
proximal ¼–⅓, monomorphic, oblanceolate, elliptic to
oblong or obovate; stamens exserted, 2–4 mm; filaments
pilose proximally. **Achenes** light brown to brown, 2–3
mm, glabrous except (typically) for slightly papillate
beak.

Flowering Jun–Oct. Heavy gumbo clay soil or (at
higher elevations) sandy-loam to gravelly or rocky soil
and outcrops, mixed grassland, saltbush, blackbrush,
and sagebrush communities, pinyon-juniper and
montane conifer woodlands; 1400–2900 m; Colo.,
N.Mex., Utah.

Eriogonum lonchophyllum is widespread and
occasionally rather common in sagebrush-dominated
communities in the Rocky Mountains of Colorado,
northern New Mexico and eastern Utah. The
distribution is fragmented, many of the populations are
markedly distinct, and several have been provided with
names. The continued recognition of *E. coloradense* is
probably dubious given what is already reduced here to
synonymy under *E. lonchophyllum*. *Eriogonum
lonchophyllum* typically has glabrous flowering stems
and inflorescence branches, yet floccose to tomentose
individuals are known from Eagle (*Reveal & Davidse
861*, UTC) and Garfield (*Goodrich & D. Nelson 24582*,
BRY) counties in Colorado. Plants representing the
typical expression of the species differ from the majority
of populations assigned here in having broad leaves and
a sprawling habit. Such plants grow in deep shade under
tall conifers and are known currently only from the type
locality. Variation in the size, shape, and distribution of
leaves has been used to distinguish varieties of *E.
lonchophyllum*. At their extremes, the differences are
remarkable. For example, plants with linear, basal leaves
have been named var. *nudicaule*. Such plants are found
in the New Mexico counties of Rio Arriba, Sandoval,
and Santa Fe. Another herbaceous phase also with basal,
but broader leaves and known from the Book Cliffs and
Tavaputs Plateau regions of northeastern Emery,
southern Uintah, and northern Grand counties, Utah,
and in Rio Blanco County, Colorado, has been named
var. *intermontanum*. The low, spreading subshrubby

phase found along the foothills of the Front Ranges in south-central Colorado (El Paso, Fremont, Jefferson, Las Animas, Otero, and Pueblo counties) and adjacent north-central New Mexico (Colfax and Taos counties) is known as var. *fendlerianum*. These plants have velvety, tomentose leaf blades. Another shrubby phase is found in and around the Dinosaur National Park in Daggett and Uintah counties, Utah, and in northwestern Rio Blanco County, Colorado. This is var. *saurinum*, characterized by having sheathing leaf blades along the flowering stems that vary from narrowly oblanceolate to narrowly elliptic; it is occasionally confused with *E. corymbosum* in that area. At the north end of Mesa Verde National Park and on Sleeping Ute Mountain of Montezuma County, Colorado, is a large shrub that outwardly resembles *E. corymbosum* in size and shape. It is similar to, but larger than, the more eastern var. *fendlerianum*. The most common phase of the species (as represented by var. *humivagans*, including *E. salicinum*, *E. scoparium*, and *E. tristichum*) is found mainly on fine, clayey to gravelly, alluvial soils in the Rocky Mountains and western desert ranges of Colorado (Archuleta, Delta, Dolores, Eagle, Garfield, Gunnison, La Plata, Mesa, Moffat, Montezuma, Montrose, Ouray, Rio Blanco, Saguache, and San Miguel counties), northern New Mexico (Rio Arriba, Sandoval, and San Juan counties), and San Juan County, Utah. The majority of the plants have long, narrow leaf blades (2–5 mm wide), but populations with broader leaves are common in southwestern Colorado. It is this latter expression of *E. lonchophyllum* that can be difficult to distinguish from *E. corymbosum* in portions of eastern Utah and western Colorado. A single collection (*Rammel s.n.*, 1872, MIN, US) was supposedly gathered in Texas. Until this is confirmed, the species is not considered to be part of the Texas flora.

24. **Eriogonum coloradense** Small, Bull. Torrey Bot. Club 33: 53. 1906 • Colorado wild buckwheat C E

Eriogonum multiceps Nees var. *coloradense* (Small) S. Stokes

Herbs, matted, scapose, (0.1–)0.3–0.6 × 0.5–5 dm, glabrous, grayish. **Stems** spreading, often with persistent leaf bases, up to $^1/_5$ height of plant; caudex stems matted, glabrous; aerial flowering stems erect, slender, solid, not fistulose, 0.3–1(–1.2) dm, glabrous. **Leaves** basal, 1 per node but congested; petiole (0.2–)0.5–3 cm, tomentose to floccose or glabrous; blade oblanceolate to lanceolate or narrowly spatulate, 1–4(–5) × 0.3–0.6 (–0.9) cm, densely white-tomentose abaxially, thinly floccose or glabrous and green adaxially, margins plane or thickened. **Inflorescences** capitate, 0.7–1.5 cm wide;

branches absent; bracts 3, semileaflike, usually triangular, 1.5–3(–4) mm. **Peduncles** absent. **Involucres** 3–4 per cluster, turbinate-campanulate, 2–3.5 × 1.5–2.5(–3) mm, glabrous except for cottony tomentum on teeth, rarely thinly floccose; teeth 5, erect, 0.3–0.7(–1) mm. **Flowers** 2.5–3.5(–4) mm; perianth white to rose, glabrous; tepals connate proximal $^1/_4$, monomorphic, oblong to ovate; stamens exserted, 3.5–4 mm; filaments sparsely pilose proximally. **Achenes** light brown to brown, 2.5–3.5 mm, glabrous.

Flowering Jul–Sep. Sandy to gravelly flats and slopes, high-elevation sagebrush and marginal meadow grassland communities, montane or subalpine conifer woodlands; of conservation concern; 2800–3900 m; Colo.

Eriogonum coloradense is a high-elevation species closely allied to *E. lonchophyllum*. It is restricted to the backbone of the central Colorado Rocky Mountains in Chaffee, Gunnison, Park, Pitkin, and Saguache counties. The Colorado wild buckwheat is listed as a "sensitive" species for Colorado. It is occasionally seen in cultivation.

25. **Eriogonum exilifolium** Reveal, Great Basin Naturalist 27: 114, fig. 1. 1967 • Drop-leaf wild buckwheat E

Herbs, matted, not scapose, 0.3–1 × 1–2 dm, glabrous, green. **Stems** spreading, usually with persistent leaf bases, up to $^1/_5$ height of plant; caudex stems matted; aerial flowering stems mostly erect, slender, solid, not fistulose, 0.03–0.1 dm, glabrous or floccose or sparsely tomentose. **Leaves** basal, 1 per node but congested; petiole 0.5–1 cm, tomentose to floccose or glabrous; blade linear to linear-oblanceolate, (2–)3–6 × 0.1–0.3 cm, densely white-tomentose abaxially, less so or glabrous and green adaxially, margins revolute. **Inflorescences** cymose-umbellate, 1.5–3 × 1.5–3 cm; branches dichotomous, glabrous; bracts 3, semileaflike, usually triangular, 3–5 mm. **Peduncles** stout, 0.1–0.4 cm, glabrous. **Involucres** 1 per node, turbinate-campanulate, 2.5–3.5(–4.5) × 2–3 mm, glabrous except for cottony tomentum between teeth; teeth 5, erect, 0.3–0.5 mm. **Flowers** 2–3.5 mm; perianth white, glabrous; tepals connate proximal $^1/_4$–$^1/_3$, monomorphic, oblanceolate to elliptic; stamens exserted, 3–4 mm; filaments sparsely pilose proximally. **Achenes** brown, 2.5–3.5 mm, glabrous.

Flowering Jun–Sep. Clay hills and flats or granitic sandy slopes, mixed grassland and sagebrush communities; 2200–2600 m; Colo., Wyo.

Eriogonum exilifolium is known only from Grand, Jackson, and Larimer counties, Colorado, and from

Carbon and Albany counties, Wyoming. The drop-leaf wild buckwheat is a "species of concern" in Wyoming.

26. Eriogonum gypsophilum Wooton & Standley, Contr. U.S. Natl. Herb. 16: 118, plate 49. 1913   • Seven River Hills or gypsum wild buckwheat C E

Herbs, erect, not scapose, 1.2–2 × 1–2 dm, glabrous, green. **Stems** spreading, without persistent leaf bases, up to ¹/₈ height of plant; caudex stems matted; aerial flowering stems erect, slender, solid, not fistulose, 0.8–1 dm, glabrous, densely tomentose among leaves. **Leaves** basal, 1 per node; petiole 3–5 cm, finely strigose; blade cordate to truncate or rarely reniform, (1–)1.5–2.5 × 1.5–2.5(–3) cm, glabrous except for fine hairs on margins and veins, margins plane. **Inflorescences** cymose, (4–)10–20 × 5–15(–20) cm; branches dichotomous, glabrous; bracts 3, scalelike, triangular, (2–)3–5 mm. **Peduncles** slender, erect, (0.5–)1–3 cm. **Involucres** 1 per node, campanulate, 1–1.5 × 2–2.5 mm, glabrous; teeth 5, erect to spreading, 1–1.3 mm. **Flowers** 1–2 mm; perianth yellow, glabrous except for fine white hairs along midrib; tepals connate proximal ¹/₃, slightly dimorphic, those of outer whorl lanceolate, 1.3–1.7 mm wide, those of inner whorl narrowly lanceolate, 0.7–0.9 mm wide; stamens exserted, 1.8–2.2 mm; filaments glabrous or nearly so. **Achenes** light brown, 1.5–2 mm, glabrous. $2n = 40$.

Flowering May–Sep. Eroded gypsum clay hills and fans, creosote bush communities; of conservation concern; 900–1100 m; N.Mex.

Eriogonum gypsophilum is a federally listed threatened species. It is known only from three locations (Seven River Hills, south of Black River Village, and in the Ben Slaughter Draw/Hay Hollow drainage) in Eddy County. Its relationship to the rest of the species in subg. *Eucycla* is obscure. It may be more closely related to *E. orcuttianum* S. Watson, a large shrub of Baja California, Mexico, than to anything in the Rocky Mountain flora.

27. Eriogonum brevicaule Nuttall, Proc. Acad. Nat. Sci. Philadelphia 4: 15. 1848 (as brevicaulis)   E F

Eriogonum campanulatum Nuttall subsp. *brevicaule* (Nuttall) S. Stokes

Herbs, matted, cespitose, pulvinate, erect or spreading, sometimes scapose, (0.3–)1–5 × 1–5 (–8) dm, tomentose to floccose or glabrous, grayish or greenish to green. **Stems** matted to spreading, occasionally with persistent leaf bases, up to ¹/₄ or more height of plant; caudex stems

matted or spreading; aerial flowering stems spreading to erect or nearly so, slender, rarely stout, solid, not fistulose, (0.4–)0.5–2(–2.5) dm, glabrous, floccose, or sparsely to densely tomentose to lanate. **Leaves** basal or more commonly sheathing 1–7(–15) cm up stem, 1 per node; petiole 0.2–2(–4) cm, tomentose to floccose; blade linear, oblanceolate, or spatulate to elliptic, (0.2–)1–10(–12) × 0.1–0.9(–1.2) cm, densely tomentose abaxially, less so to floccose adaxially, margins plane or revolute, sometimes crenulate. **Inflorescences** cymose, subumbellate, umbellate, or capitate, (1–)3–10(–25) × (0.7–)1–10(–15) cm; branches dichotomous, sometimes absent, tomentose to floccose or glabrous; bracts 3, triangular, scalelike, 1–3(–5) mm. **Peduncles** absent or erect, 0.3–3 cm, tomentose to floccose or glabrous. **Involucres** 1 per node or 3–7(–9) per cluster, turbinate to turbinate-campanulate, 1.5–4(–5) × (1–)1.5–3(–3.5) mm, tomentose to floccose or glabrous; teeth 5, erect to spreading, 0.3–1 mm. **Flowers** (1–)2–4 mm; perianth various shades of white to cream or yellow, glabrous or pubescent; tepals connate proximal ¹/₄–¹/₃, monomorphic, lanceolate, oblong to obovate or ovate to oval; stamens exserted, 2–4 mm; filaments pilose basally. **Achenes** light brown to brown, 2–3 mm, glabrous except for roughened to papillate beak.

Varieties 8 (8 in the flora): w United States.

Eriogonum brevicaule is highly variable, and the variation has yet to be fully resolved taxonomically. The expressions recognized here will encompass the vast majority of populations. The extreme variation previously under the name var. *laxifolium* is now reduced with the recognition of var. *bannockense* (low-elevation or northern phase), var. *nanum*, and var. *caelitum* (high-elevation, southern phases).

Essentially all of the following species (28–63 below) belong to the *Eriogonum brevicaule* complex. *Eriogonum desertorum, E. loganum, E. spathulatum, E. ostlundii,* and *E. artificis* are allied to the complex associated with var. *laxifolium,* while *E. natum* is related to var. *cottamii. Eriogonum viridulum* and *E. ephedroides* are allied to *E. brevicaule* var. *brevicaule* as are *E. contortum* and *E. acaule. Eriogonum brandegeei* is also related, but exactly how is less certain. Allied to this complex of species are on the one hand those related to *E. batemanii,* and on the other all of the matted perennials belonging to the *E. ochrocephalum* complex. Essentially all of these species form relatively small populations on discrete edaphic sites and are well isolated one from the other. Unfortunately, a clear separation of *E. brevicaule* from *E. desertorum, E. loganum,* and *E. spathulatum* is not always possible.

1. Flowering stems and inflorescence branches glabrous.
    2. Plants not pulvinate or cespitose, (0.8–)1.5–5 dm; inflorescences cymose, open, divided 3 times or more; widespread . . . . 27a. *Eriogonum brevicaule* var. *brevicaule*
    2. Plants pulvinate and cespitose, 0.3–1.5(–1.8) dm; inflorescences capitate or umbellate to cymose and divided 1–2 times; Utah . . . . 27g. *Eriogonum brevicaule* var. *nanum* (in part)
1. Flowering stems and inflorescence branches floccose to tomentose or lanate.
    3. Perianths pubescent; se Montana, ne Wyoming . . . . 27b. *Eriogonum brevicaule* var. *canum*
    3. Perianths glabrous; sc Idaho, ne Nevada, Utah, sw Wyoming, not se Montana or ne Wyoming.
        4. Inflorescences divided (2–)3–5 times; perianths usually yellow, rarely ochroleucous.
            5. Flowers (1–)1.5–2.5 mm; sw Wyoming . . . . 27c. *Eriogonum brevicaule* var. *micranthum*
            5. Flowers (2.5–)3–4 mm; Utah . . . . 27d. *Eriogonum brevicaule* var. *cottamii*
        4. Inflorescences capitate or umbellate to cymose and divided 1–2 times; perianths ochroleucous or yellow.
            6. Leaf blades linear to narrowly oblanceolate; inflorescences capitate or divided.
                7. Leaf blades (1.5–)3–9(–12) × 0.1–0.5(–0.7) cm, tomentose abaxially, less so and grayish or occasionally greenish adaxially, margins usually revolute, occasionally plane; perianths ochroleucous or yellow; n Utah, se Idaho, below 2800 m . . . . 27e. *Eriogonum brevicaule* var. *laxifolium* (in part)
                7. Leaf blades 0.2–4.5(–5) × (0.2–)0.3–0.6(–0.7) cm, tomentose abaxially, thinly floccose and bright green adaxially, margins plane or slightly thickened; perianths yellow; c Utah, above 2700 m . . . . 27f. *Eriogonum brevicaule* var. *caelitum*
            6. Leaf blades linear, oblanceolate, or narrowly oblanceolate to elliptic; inflorescences capitate.

[8. Shifted to left margin.—Ed.]
8. Leaf blades and flowering stems bright green under the thinly floccose pubescence; ne Nevada and nw Utah . . . . 27e. *Eriogonum brevicaule* var. *laxifolium* (in part)
8. Leaf blades and flowering stems usually grayish to dull greenish under tomentum, rarely thinly floccose; se Idaho and sw Wyoming s to n Utah and ne Nevada.
    9. Leaf margins crenulate; leaf blades narrowly oblanceolate to narrowly elliptic, (0.3–)0.5–1.5(–2) × 0.2–0.5(–7) cm, densely white-tomentose abaxially, floccose and greenish adaxially . . . 27g. *Eriogonum brevicaule* var. *nanum* (in part)
    9. Leaf margins usually plane (rarely crenulate in Wyoming); leaf blades narrowly oblanceolate to oblanceolate, (0.8–)1–4(–4.5) × (0.3–)0.4–0.8 cm, densely tomentose abaxially, tomentose to floccose and grayish to greenish adaxially . . . . 27h. *Eriogonum brevicaule* var. *bannockense*

27a. Eriogonum brevicaule Nuttall var. **brevicaule**

* Short-stem wild buckwheat E F

Eriogonum brevicaule subsp. *grangerense* (M. E. Jones) S. Stokes; *E. brevicaule* var. *wasatchense* (M. E. Jones) Reveal; *E. campanulatum* Nuttall; *E. grangerense* M. E. Jones; *E. wasatchense* M. E. Jones

Plants erect, (0.8–)1.5–4(–5) × 2–5(–8) dm. **Aerial flowering stems** erect, 1–2.5 dm, glabrous. **Leaves:** blade linear to oblanceolate, 3–10 × 0.1–0.7 cm, tomentose abaxially, less so to thinly floccose and green adaxially, margins plane. **Inflorescences** cymose, usually divided 3 or more times, (3–)5–25 cm; branches glabrous. **Involucres** 1 per node, turbinate to turbinate-campanulate, 2–4 × (1–)1.5–2.5(–3) mm, glabrous. **Flowers** (2–)2.5–3 mm; perianth yellow, rarely white or cream, glabrous.

Flowering Jun–Sep. Sandy to clayey flats, washes and slopes, mixed grassland and sagebrush communities, juniper or montane conifer woodlands; 1400–3000 m; Colo., Idaho, Nebr., S.Dak., Utah, Wyo.

Variety *brevicaule* is widely scattered in northwestern and north-central Colorado, southeastern Idaho, western Nebraska, southwestern South Dakota, northern Utah, and Wyoming. Plants with white or cream-colored perianths occur in southwestern Wyoming and have been called var. *grangerense*. Hybrids involving var. *brevicaule* and *E. corymbosum* var. *corymbosum* are known from southwestern Wyoming.

27b. Eriogonum brevicaule Nuttall var. **canum**
(S. Stokes) Dorn, Vasc. Pl. Wyoming, 299. 1988
· Parasol wild buckwheat   [E]

Eriogonum multiceps Nees subsp. *canum* S. Stokes, Eriogonum, 94. 1936; *E. lagopus* Rydberg; *E. pauciflorum* Pursh var. *canum* (S. Stokes) Reveal

Plants erect, 1.5–3 × 1.5–3.5 dm. **Aerial flowering stems** erect, 0.5–1 dm, tomentose. **Leaves:** blade narrowly oblanceolate to oblanceolate or spatulate, (1–)1.5–3.5(–4) × 0.2–0.7 (–0.9) cm, margins plane. **Inflorescences** subumbellate to cymose, usually divided 3 or more times, 5–10 cm; branches tomentose. **Involucres** 1 per node, turbinate, (2–)3–3.5(–5) × 2–2.5(–3) mm, tomentose. **Flowers** (1–)1.5–2.5(–3) mm; perianth yellow, sparsely pubescent.

Flowering Jul–Sep. Clayey flats and slopes, mixed grassland and sagebrush communities, juniper woodlands; 1100–1800 m; Mont., Wyo.

Variety *canum* is rare throughout its widely scattered range in Carbon, Park, and Yellowstone counties, Montana, and in Big Horn and Sheridan counties, Wyoming.

27c. Eriogonum brevicaule Nuttall var. **micranthum**
(Nuttall) Reveal, Taxon 16: 410. 1967 · Red Desert wild buckwheat   [E]

Eriogonum micranthum Nuttall, Proc. Acad. Nat. Sci. Philadelphia 4: 15. 1848; *E. brevicaule* subsp. *orendense* (A. Nelson) S. Stokes; *E. orendense* A. Nelson

Plants erect, (0.8–)1.5–5 × 2–5 dm. **Aerial flowering stems** erect, 0.5–1 dm, densely tomentose to lanate. **Leaves:** blade oblanceolate to narrowly spatulate, 1.5–5(–6) × (0.2–)0.3–0.8 (–1.2) cm, margins plane. **Inflorescences** cymose, usually divide 3 or more times, 3–6 cm; branches tomentose. **Involucres** 1 per node, turbinate-campanulate, 2.5–3 × 2–2.5 mm, tomentose. **Flowers** (1–)1.5–2.5 mm; perianth yellow or, rarely, ochroleucous, glabrous.

Flowering Jun–Sep. Sandy to gravelly flats and slopes, mixed grassland and sagebrush communities; (1700–)1800–2400(–2600) m; Wyo.

Variety *micranthum* is restricted to the Red Desert region of southwestern Wyoming in Carbon, Fremont, Lincoln, Natrona, Sublette, Sweetwater, and Uinta counties.

27d. Eriogonum brevicaule Nuttall var. **cottamii**
(S. Stokes) Reveal, Great Basin Naturalist 32: 113. 1972 · Cottam's wild buckwheat   [E]

Eriogonum tenellum Torrey subsp. *cottamii* S. Stokes, Eriogonum, 70. 1936 (as cottami)

Plants erect, 1–3.5 × 2–5 dm. **Aerial flowering stems** erect, 0.5–2 dm, tomentose. **Leaves:** blade narrowly oblanceolate to narrowly elliptic; leaf blades (2–)3–4(–4.5) × 0.3–0.5(–0.7) cm, margins plane. **Inflorescences** cymose, usually divided 3 or more times, 2–7 cm; branches tomentose. **Involucres** usually 1 per node, turbinate, 2–3(–3.5) × 1.5–2(–2.5) mm, tomentose. **Flowers** (2.5–)3–4 mm; perianth yellow, glabrous.

Flowering Jun–Sep. Gravelly limestone slopes and clay hills, sagebrush communities, oak and pinyon-juniper woodlands; 1300–2100 m; Utah.

Variety *cottamii* is infrequent in the desert ranges west of the Wasatch Range in eastern portions of Juab, Millard, and Tooele counties, the southwestern corner of Salt Lake County, and in western Utah County. The attractive tomentum of Cottam's wild buckwheat changes from a gray to tannish- or reddish-brown as the plant matures, and the species is worthy of horticultural consideration. The narrow leaf blades of var. *cottamii* readily distinguish it from the more western *E. natum*.

27e. Eriogonum brevicaule Nuttall var. **laxifolium**
(Torrey & A. Gray) Reveal, Great Basin Naturalist 27: 220. 1968 · Loose-leaf wild buckwheat   [E]

Eriogonum kingii Torrey & A. Gray var. *laxifolium* Torrey & A. Gray, Proc. Amer. Acad. Arts 8: 165. 1870; *E. brevicaule* var. *huberi* S. L. Welsh; *E. brevicaule* var. *promiscuum* S. L. Welsh; *E. brevicaule* var. *pumilum* S. Stokes ex M. E. Jones; *E. chrysocephalum* A. Gray; *E. medium* Rydberg

Plants mostly erect, 1–3.5 × 1–5 dm. **Aerial flowering stems** erect or nearly so, scapelike or dichotomous, (0.7–)1–2.5 dm, tomentose to floccose. **Leaves:** blade linear to narrowly oblanceolate, (1.5–)3–9(–12) × 0.1–0.5(–0.7) cm, tomentose abaxially, less so and grayish or occasionally greenish adaxially, margins plane or more often revolute. **Inflorescences** umbellate or cymose divided 1–3 times and 2–5(–8) cm, or capitate and 1–2 cm; branches tomentose to floccose. **Involucres** 1 per node or more commonly 3–7 per cluster, turbinate to turbinate-campanulate, (2–)2.5–4 × 1.5–3 mm, usually tomentose, occasionally thinly floccose or glabrous. **Flowers** (2–)3–3.5(–4) mm; perianth yellow or ochroleucous, glabrous.

Flowering Jun–Sep. Sandy to clayey flats, washes and slopes, mixed grassland, sagebrush, and mountain mahogany communities, juniper, oak and montane conifer woodlands; 1400–2800 m; Idaho, Nev., Utah.

Variety *laxifolium* is found in Idaho (Bear River County) and in Utah (Carbon, Duchesne, Emery, Juab, Millard, Salt Lake, Sanpete, Sevier, Tooele, Uintah, Utah, Wasatch, and Weber counties), where it occurs mainly on the western foothills of the Wasatch Range. Plants often have a reddish tomentum rather than the grayish tomentum of the consistently capitate expressions of vars. *bannockense*, *nanum*, and *caelitum*. Plants from the Tavaputs Plateau with exceedingly long, narrow leaf blades have been segregated as var. *huberi*, but plants with even longer leaves occur elsewhere (as in the Soldier Summit area of Utah County, Utah), and the difference is not taxonomically significant.

27f. Eriogonum brevicaule Nuttall var. **caelitum** Reveal, Phytologia 86: 123. 2004 • Wasatch Plateau wild buckwheat [E]

Plants cespitose, 0.6–1.8(–2) × 1–2.5 dm. **Aerial flowering stems** erect or nearly so, scapelike, 0.5–1.5 dm, densely floccose to tomentose. **Leaves:** blade linear to linear-oblance-olate, 0.2–4.5 (–5) × (0.2–)0.3–0.6(–0.7) cm, tomentose abaxially, thinly floccose and bright green adaxially, margins plane or slightly thickened. **Inflorescences** capitate, 1–2 cm; branches absent. **Involucres** 3–7 per cluster, turbinate to turbinate-campanulate, 2–3.5(–4) × 2–3(–3.5) mm, floccose to tomentose. **Flowers** 2.5–3.5(–4) mm; perianth yellow, glabrous.

Flowering Jul–Sep. Gravelly to rocky, mostly limestone soil, high-elevation sagebrush communities, subalpine conifer woodlands; 2700–3700 m; Utah.

Variety *caelitum* is restricted to the limestone Flagstaff Formation on the high mesa tops of the Wasatch Mountains in Sanpete and northern Sevier counties.

27g. Eriogonum brevicaule Nuttall var. **nanum** (Reveal) S. L. Welsh, Great Basin Naturalist 44: 531. 1984 • Dwarf wild buckwheat [E]

Eriogonum nanum Reveal, Phytologia. 25: 194. 1973; *E. grayi* Reveal

Plants pulvinate and cespitose, 0.3–1.5(–1.8) × 1–2.5 dm. **Aerial flowering stems** erect or nearly so, usually scapelike, (0.2–)0.4–1.5 dm, glabrous. **Leaves:** blade narrowly oblanceolate to narrowly elliptic, 0.3–2 × 0.2–0.5(–0.7) cm, densely white-tomentose abaxially, floccose and greenish adaxially, margins crenulate. **Inflorescences** capitate or open and umbellate, 1–5 cm; branches absent or typically glabrous. **Involucres** 3–7(–9) per cluster, turbinate, 1.5–3 × 1–2.5 mm, glabrous or thinly floccose. **Flowers** 2–3 mm; perianth greenish- to creamy- or pale yellowish-white, rarely pale yellow or yellow, glabrous.

Flowering Jul–Sep. Gravelly to rocky, mostly limestone slopes and ridge tops, high-elevation sagebrush communities, subalpine conifer woodlands; 2400–2600 m; Utah.

Variety *nanum* is restricted to the Wasatch Mountains in Box Elder and Weber counties.

27h. Eriogonum brevicaule Nuttall var. **bannockense** (S. Stokes) Reveal, Harvard Pap. Bot. 9: 160. 2004 • Bannock wild buckwheat [E]

Eriogonum chrysocephalum A. Gray subsp. *bannockense* S. Stokes, Leafl. W. Bot. 3: 200. 1943; *E. bannockense* (S. Stokes) R. J. Davis

Plants pulvinate and cespitose, (0.3–)0.5–1.2(–1.5) × 1–2.5 dm. **Aerial flowering stems** erect, scapelike, 0.3–1.5 dm, tomentose. **Leaves:** blade narrowly oblance-olate to oblanceolate, (0.8–)1–4(–4.5) × (0.3–)0.4–0.8 cm, densely tomentose abaxially, tomentose to floccose and grayish to greenish adaxially, margins plane (rarely crenulate in Wyoming). **Inflorescences** capitate, 1–2 cm; branches absent. **Involucres** 3–6 per cluster, turbinate, 2.5–4 × 2–3 mm, floccose to tomentose. **Flowers** 2–3 mm; perianth yellow or, rarely, ochrocephalous, glabrous.

Flowering Jun–Sep. Sandy or shaley to gravelly flats and slopes, mixed grassland, sagebrush, and mountain mahogany communities, oak-juniper, pinyon-juniper, or montane conifer woodlands; 1800–2800 m; Idaho, Nev., Utah, Wyo.

Variety *bannockense* is found in widely scattered sites in southeastern Idaho (Bannock, Bear Lake, Bonneville, Cassia, and Franklin counties), northeastern Nevada (Elko County), northwestern Utah (Box Elder and northern Tooele counties) and southwestern Wyoming (Fremont, Lincoln, Sublette, and Teton counties). In habit it is similar to the more western *E. desertorum*, and a sharp, morphologic distinction is not always obvious between the two. Geographically the two approach each other in Box Elder County, Utah, and in Elko County, Nevada.

28. Eriogonum viridulum Reveal, Proc. Utah Acad. Sci. 42: 287. 1966 · Clay hill wild buckwheat E

Eriogonum brevicaule Nuttall var. *viridulum* (Reveal) S. L. Welsh

Herbs, spreading, not scapose, 1–3.5 × 1–4 dm, glabrous, bright green. **Stems** spreading, without persistent leaf bases, up to ¹/₅ height of plant; caudex stems absent; aerial flowering stems erect, slender, solid, not fistulose, 0.5–1.2 dm, glabrous, tomentose among leaves. **Leaves** sheathing 1–4(–6) cm, 1 per node; petiole 0.1–0.2 cm, tomentose or glabrous; blade linear or rarely narrowly elliptic, 1–3(–4) × 0.1–0.2(–0.5) cm, densely white-tomentose abaxially, glabrous and green adaxially, margins usually tightly revolute. **Inflorescences** cymose, 3–15 × 2–10 cm; branches dichotomous, glabrous; bracts 3, scalelike, narrowly triangular, 1–3 mm. **Peduncles** absent or erect, slender, 0.05–0.2 cm. **Involucres** 1 per node, turbinate, 2–3 × 1.5–2 mm, glabrous; teeth 5, erect, 0.5–0.8 mm. **Flowers** 1.5–2 mm; perianth yellow, glabrous; tepals connate proximal ¹/₄–¹/₃, slightly dimorphic, those of outer whorl ovate, 1–1.2 wide, those of inner whorl oblong, 0.7–0.8 mm wide; stamens exserted, 1.5–2 mm; filaments pilose proximally. **Achenes** brown, 1.5–2 mm, glabrous.

Flowering Jul–Oct. Sandy or silty flats or clay slopes and hills, saltbush or sagebrush communities, pinyon-juniper woodlands; 1400–2000(–2200) m; Utah.

Eriogonum viridulum has a distinctive upright inflorescence composed of numerous, bright green and glabrous branches bearing essentially sessile involucres. These features and the linear to narrowly elliptic leaf blades that are typically tightly revolute combine to distinguish the species from the related *E. brevicaule* var. *brevicaule*. The clay hill wild buckwheat is known only from Duchesne and Uintah counties but to be expected in adjacent Moffat County, Colorado.

29. Eriogonum ephedroides Reveal, Madroño 19: 295, fig. 4. 1969 · Ephedra wild buckwheat E

Eriogonum brevicaule Nuttall var. *ephedroides* (Reveal) S. L. Welsh

Herbs, erect, not scapose, 2–3.5 × 2–3(–4) dm, glabrous, grayish green. **Stems** spreading, without persistent leaf bases, up to ¹/₈ height of plant; caudex stems absent; aerial flowering stems erect, slender, solid, not fistulose, 1–2 dm, glabrous, tomentose among leaves. **Leaves** basal, 1 per node; petiole 0.5–1 cm, glabrous; blade lanceolate, 1.5–2.5 × 0.2–0.3 cm, densely white-tomentose abaxially, mostly glabrous and green adaxially, margins plane. **Inflorescences** narrowly cymose, 15–25 × 5–15 cm; branches dichotomous, upper secondaries suppressed, glabrous; bracts 3, linear to triangular, scalelike, 1–4(–7) mm. **Peduncles** erect, 0.5–1.5 cm at proximal nodes, 0.1–0.5 cm at distal nodes. **Involucres** 1 per node, turbinate, 2–2.5 × 1–1.5 mm, glabrous; teeth 5, erect, 0.4–0.5 mm. **Flowers** 2–2.5 mm; perianth ochroleucous to pale yellow or, rarely, yellow, glabrous; tepals connate proximal ¹/₄, monomorphic, lanceolate; stamens exserted, 1.5–2 mm; filaments pilose proximally. **Achenes** brown, 2–2.5 mm, glabrous.

Flowering Jun–Oct. Shale and clay flats and slopes, saltbush and sagebrush communities, pinyon-juniper woodlands; 1500–2100 m; Colo., Utah.

Eriogonum ephedroides is infrequently encountered in eastern Uintah County, Utah, and western Rio Blanco County, Colorado. The narrow, upright flowering stems and inflorescence branches and the strictly basal leaves quickly set the species apart from *E. viridulum*. This is a species of potential horticultural value, being an elegant addition to the garden because of its habit.

30. Eriogonum loganum A. Nelson, Bot. Gaz. 54: 149. 1912 · Cache Valley wild buckwheat C E

Eriogonum brevicaule Nuttall var. *loganum* (A. Nelson) S. L. Welsh; *E. chrysocephalum* A. Gray subsp. *loganum* (A. Nelson) S. Stokes

Herbs, cespitose or spreading, scapose, (0.5–)1.5–3 × (0.5–)2–4(–5) dm, tomentose, reddish, greenish, or grayish. **Stems** spreading, occasionally with persistent leaf bases, up to ¹/₄ height of plant; caudex stems matted; aerial flowering stems slightly spreading to erect, slender, solid, not fistulose, (0.06–)1–2.5 dm, tomentose. **Leaves** sheathing up stem 1–4 cm, 1 per node; petiole (0.5–)1–2.5(–3) cm, lanate; blade oblanceolate to elliptic, (0.5–)1–5(–7) × (0.2–)0.3–1 cm, densely lanate abaxially, less so to tomentose adaxially, margins plane. **Inflorescences** capitate, 1 cm; branches absent; bracts 3, scalelike, triangular, 1.5–3 mm. **Peduncles** absent. **Involucres** 3–5 per cluster, turbinate, (2–)3.5–4.5 × (1.5–)2.5–3 mm, tomentose; teeth 5, erect, 0.5–1 mm. **Flowers** (2–)2.5–3.5(–4) mm; perianth ochroleucous or yellow, glabrous; tepals connate proximal ¹/₄, monomorphic, oblong; stamens exserted, (2.5–)3–5 mm; filaments glabrous or sparsely pilose proximally. **Achenes** brown, (2.5–)3–3.5 mm, glabrous.

Flowering May–Aug. Sandy to gravelly slopes, sagebrush communities, juniper woodlands, gravelly or limestone slopes, high-elevation sagebrush communities, subalpine conifer woodlands; of conservation concern; 1400–3100 m; Utah.

Eriogonum loganum is known from a few widely scattered, disjunct populations in Cache and Morgan counties. The type locality is on the southern edge of the Utah State University campus, but construction of three parking lots has resulted in its near total destruction. Fragmentary populations remain at the mouth of Logan Canyon, but more robust populations are known from Cart Hollow (south of Logan) and Smithfield Canyon (north of Logan). There is also a series of high-elevation populations found along the backbone of the Bear River Range, with both the Cart Hollow and Smithfield Canyon populations more similar to the material found around Tony Grove Lake than to that on the bench in Logan. The species is clearly related to *E. brevicaule*, especially to the ochroleucous forms of var. *laxifolium* found in Rich County, Utah.

31. **Eriogonum brandegeei** Rydberg, Fl. Rocky Mts., 220, 1061. 1917 (as brandegei) • Brandegee's wild buckwheat [C] [E]

Eriogonum spathulatum A. Gray var. *brandegeei* (Rydberg) S. Stokes

Herbs, spreading, sometimes scapose, 1–2.5 × (0.5–)1–2 dm, tomentose to floccose, grayish. **Stems** spreading, with persistent leaf bases, up to ¹/₄ height of plant; caudex stems matted, floccose or glabrous; aerial flowering stems erect, slender, solid, not fistulose, 1–2.5 dm, tomentose to floccose. **Leaves** strictly basal, 1 per node; petiole 1–3(–3.5) cm, lanate; blade oblanceolate to elliptic, 1.5–4(–5) × 0.4–1.2(–1.6) cm, densely white-tomentose adaxially, less so and greenish adaxially, margins plane. **Inflorescences** capitate or umbellate-cymose, 0.1–1.5 cm wide, tomentose to floccose; branches dichotomous, sometimes absent, tomentose to floccose; bracts 3,

scalelike, triangular, and 2–5 mm, or leaflike, lanceolate, and 1–3 × 0.2–0.7 cm. **Peduncles** absent. **Involucres** 4–8 per cluster, turbinate, 3.5–5 × 3–4 mm, floccose or glabrous; teeth 5, erect, 0.4–0.8 mm. **Flowers** (2.5–)3–3.5 mm; perianth ochroleucous, glabrous; tepals connate proximal ¹/₄, monomorphic, oblanceolate to oblong; stamens exserted, 3–3.5 mm; filaments pilose basally. **Achenes** light brown, 3–3.5 mm, glabrous.

Flowering Jul–Oct. Clay slopes and washes, sagebrush communities, juniper woodlands; of conservation concern; 1800–2600 m; Colo.

Eriogonum brandegeei is a rare and localized species known from nine occurrences along the Arkansas River in Chaffee and Fremont counties. It is regarded as a "sensitive" species in Colorado, by both the state and the U.S. Forest Service. The Droney Gulch Natural Area has been established, in part, to protect the species. Brandegee's wild buckwheat is related to *E. brevicaule* var. *laxifolium*, but well isolated from that taxon. An undated E. L. Greene specimen (ISC) supposedly was gathered in northern New Mexico, but that is most unlikely. A G. W. Letterman collection (MO) purportedly from Colorado Springs certainly is mislabeled.

32. **Eriogonum natum** Reveal, Great Basin Naturalist 35: 363. 1976 • Mark's wild buckwheat [C] [E]

Eriogonum spathulatum A. Gray var. *natum* (Reveal) S. L. Welsh

Herbs, spreading, not scapose, 1–3.5 × 1–4 dm, tomentose, greenish. **Stems** spreading, without persistent leaf bases, up to ¹/₄ height of plant; caudex stems absent; aerial flowering stems slightly erect, slender, solid, not fistulose, 1–2(–2.5) dm, tomentose. **Leaves** sheathing up stem 0.5–2(–3) cm, 1 per node; petiole (1–)2–3 cm, tomentose; blade elliptic, 2–2.5(–3) × (0.8–)1–1.3 cm, densely white-tomentose abaxially, less so and greenish-tomentose adaxially, margins plane. **Inflorescences** cymose-umbellate, 3–10(–15) × 3–5(–8) cm; branches dichotomous, tomentose; bracts 3, scalelike, triangular, and 1–2 mm, or leaflike, lanceolate, and 5–10(–12) × (1.5–)2–4(–5) mm. **Peduncles** absent. **Involucres** 1 per node, rarely 2 per cluster, turbinate-campanulate, 2.5–4 × 2–3 mm, tomentose; teeth 5, erect, 0.5–0.8 mm. **Flowers** 2–2.5(–3) mm; perianth bright yellow, glabrous; tepals connate proximal ¹/₄–¹/₃, monomorphic, oblong to obovate; stamens exserted, 2.5–4 mm; filaments pilose proximally. **Achenes** brown, 2–3 mm, glabrous.

Flowering Jul–Oct. Alkaline clay flats, saltbush communities; of conservation concern; 1400–1600 m; Utah.

Eriogonum natum is known only from Millard County, where it occurs near Sevier Lake. It is probably a recently evolved species, as its entire habitat was submerged under pluvial Lake Bonneville until some 10,000 years ago. Now isolated and morphologically distinct, it probably originated from *E. brevicaule* var. *cottamii*, a plant of limestone outcrops in the hills to the north and east. The species is worthy of introduction into the garden. An unnumbered M. E. Jones collection of *E. natum* dated 6 June 1913 (POM, UTC) supposedly was collected at Moab, Grand County. There is a question mark above the label on the POM sheet, and whoever added it may have been aware that the sheet was mislabeled.

33. **Eriogonum contortum** Small ex Rydberg, Fl. Colorado, 104, 107. 1906 • Grand Valley wild buckwheat E

Eriogonum effusum Nuttall subsp. *contortum* (Small ex Rydberg) S. Stokes

Herbs, spreading, not scapose, 0.5–0.8(–1) × 0.8–3.5 dm, thinly floccose or glabrous, greenish. **Stems** spreading, occasionally with persistent leaf bases, up to ⅓ height of plant; caudex stems matted; aerial flowering stems slightly spreading to erect, slender, solid, not fistulose, 0.2–0.6 dm, thinly floccose or glabrous, tomentose among leaves. **Leaves** sheathing up stem 1–4 cm, 1 per node; petiole 0.1–0.2 cm, glabrous or sparsely floccose; blade linear to narrowly lanceolate, (0.5–)1–2(–2.5) × 0.15–0.2(–0.25) cm, densely white-tomentose abaxially, floccose and greenish adaxially, margins revolute. **Inflorescences** cymose, 1–3 × 0.8–2 cm; branches dichotomous, thinly floccose or glabrous; bracts 3, scalelike, triangular, (0.5–)1–2 mm. **Peduncles** erect, slender, 0.2–1 cm, floccose or glabrous. **Involucres** 1 per node, turbinate to turbinate-campanulate, 1.5–2 (–2.5) × 1–2 mm, glabrous; teeth 5, erect, 0.3–0.5 mm. **Flowers** 1.5–2.5 mm; perianth yellow, glabrous; tepals connate proximal ½, monomorphic to slightly dimorphic, those of outer whorl slightly broader than those of inner whorl, oblong to obovate; stamens exserted, 1.5–2 mm; filaments pilose proximally. **Achenes** brown, 2–2.5 mm, glabrous.

Flowering May–Aug. Heavy shale or clay flats and slopes, saltbush communities; 1300–1700 m; Colo., Utah.

Eriogonum contortum is localized on Mancos Shale in Grand Valley of eastern Grand County, Utah, and western Mesa County, Colorado. Disjunct populations occur just outside Grand Valley in Garfield County, Colorado, and in Emery County, Utah. In Colorado, the Bureau of Land Management considers it a "sensitive" species. The Badger Wash Natural Area has been established to protect this species along with *Astragalus musiniensis* M. E. Jones (Ferron milkvetch) and *Cryptantha elata* Payson (tall cryptanth). Grand Valley wild buckwheat occasionally is seen in rock gardens. It is related to *E. brevicaule* var. *brevicaule* but shares many characters in common with *E. pelinophilum*, most notably a flower that in fruit has its proximal half decidedly rounded due to its close adherence to the globose base of the achene.

34. **Eriogonum acaule** Nuttall, Proc. Acad. Nat. Sci. Philadelphia 4: 13. 1848 • Single-stem wild buckwheat E

Eriogonum caespitosum Nuttall subsp. *acaule* (Nuttall) S. Stokes; *E. caespitosum* var. *acaule* (Nuttall) R. J. Davis

Herbs, matted, not scapose, 0.15–0.3 × 1–2 dm, tomentose, greenish. **Stems** matted, with persistent leaf bases, up to ⅛ height of plant; caudex stems matted; aerial flowering stems erect, slender, solid, not fistulose, 0.05–0.2 dm, tomentose. **Leaves** sheathing entire stem, 1 per node or fasciculate; petiole 0.05–0.15 cm, glabrous; blade linear to oblanceolate, 0.3–0.8 × 0.1–0.2 cm, densely white-tomentose on both surfaces, margins revolute. **Inflorescences** cymose or racemose, condensed, 0.2–0.5(–0.8) × 0.3–0.8 cm; branches dichotomous or secondaries suppressed, tomentose; bracts absent. **Peduncles** absent. **Involucres** 1 per node, campanulate, 1.5–3 × 1.5–3 mm, tomentose; teeth 4–5, erect to slightly spreading, 1–2 mm. **Flowers** 2–2.5 mm; perianth yellow, white-tomentose abaxially; tepals connate proximal ¼, monomorphic, lanceolate; stamens exserted, 2.5–3 mm; filaments pilose proximally. **Achenes** light brown, 2–2.5 mm, sparsely pubescent.

Flowering May–Jul. Gravelly or clayey flats and slopes, saltbush or sagebrush communities; 2000–2500 m; Colo., Wyo.

Eriogonum acaule is the ultimate reduction in the *E. brevicaule* complex. The low, matted habit of this rather elegant species makes it attractive to rock garden enthusiasts. In Colorado, the Bureau of Land Management considers it a "sensitive" species; it is known only from Moffat County. It is encountered more frequently in Wyoming, where it is found in Albany, Carbon, Fremont, Lincoln, Natrona, Sublette, Sweetwater, Teton, and Uinta counties. The species is to be expected just south of the Daggett County line in northeastern Utah.

35. Eriogonum spathulatum A. Gray, Proc. Amer. Acad. Arts 10: 76. 1874 • Spoon-leaf wild buckwheat E

Eriogonum nudicaule (Torrey) Small subsp. *ochroflorum* S. Stokes; *E. spathulatum* var. *kayeae* S. L. Welsh

Herbs, spreading, not scapose, 1.5–4 × 1–3 dm, tomentose or, rarely, glabrous, grayish. **Stems** spreading, without persistent leaf bases, up to ¼ height of plant; caudex stems absent; aerial flowering stems erect, slender, solid, not fistulose, 1–2 dm, tomentose or rarely glabrous. **Leaves** sheathing up stem 1–8 cm, 1 per node; petiole 0.5–1.5 cm, tomentose; blade lanceolate to narrowly elliptic or narrowly spatulate, 1–4(–6) × 0.3–1 cm, tomentose on both surfaces, margins plane, rarely crenulate or revolute. **Inflorescences** cymose, 3–10 × 3–10 cm; branches dichotomous, tomentose or rarely glabrous; bracts 3, narrowly triangular, scalelike, 1–3 mm. **Peduncles** absent. **Involucres** (1–)3–6 per cluster, turbinate-campanulate, 2–3(–3.5) × 2–2.5(–3) mm, tomentose, rarely glabrous; teeth 5, erect, 0.5–0.8 mm. **Flowers** 2.5–3 mm; perianth ochroleucous or pale yellow, glabrous; tepals connate proximal ¼, monomorphic, oblong; stamens exserted, 2.5–4 mm; filaments pilose proximally. **Achenes** brown, 3–3.5 mm, glabrous.

Flowering Jul–Oct. Clay flats, limestone slopes, or gypsum outcrops, saltbush, sagebrush, and mountain mahogany communities, pinyon-juniper and conifer woodlands; 1400–2200(–2600) m; Utah.

Eriogonum spathulatum is fairly common in west-central Utah. The typical expression, tomentose plants with ochroleucous flowers, occurs mainly in Iron, southern Juab, Millard, Piute, Sanpete, and Sevier counties. Plants with pale yellow to yellow perianths are seen infrequently in a few populations; these are always intermixed with plants bearing non-yellow perianths. Mixed populations of plants with either tomentose or glabrous flowering stems and inflorescence branches occur on gravelly limestone ridges in Beaver County; those with glabrous flowering stems and inflorescence branches have been separated recently as var. *kayeae*. In aspect, these approach *Eriogonum artificis*, a plant of sandy volcanic soils.

36. Eriogonum ostlundii M. E. Jones, Contr. W. Bot. 11: 12. 1903 (as ostlundi) • Elsinore wild buckwheat E

Eriogonum batemanii M. E. Jones var. *ostlundii* (M. E. Jones) S. L. Welsh; *E. spathuliforme* Rydberg; *E. spathulatum* A. Gray subsp. *spathuliforme* (Rydberg) S. Stokes; *E. tenellum* Torrey subsp. *ostlundii* (M. E. Jones) S. Stokes

Herbs, spreading, not scapose, 2–3.5(–4.5) × 1–2.5 dm, glabrous, grayish. **Stems** spreading, without persistent leaf bases, up to ¼ height of plant; caudex stems absent; aerial flowering stems erect, slender, solid, not fistulose, 0.8–2 dm, glabrous. **Leaves** basal, 1 per node; petiole 1–2.5 (–3) cm, tomentose; blade elliptic to spatulate or broadly elliptic, 0.6–1.5(–2) × 0.5–1(–1.5) cm, densely white-tomentose abaxially, floccose and greenish adaxially, margins plane. **Inflorescences** cymose, 5–25(–30) × 5–15 cm; branches dichotomous, glabrous; bracts 3, scalelike, triangular, 0.8–2 mm. **Peduncles** erect, slender, (0.5–)1–3(–5) cm, glabrous. **Involucres** 1 per node, turbinate to turbinate-campanulate, 2–3 × 1.8–2(–2.2) mm, glabrous; teeth 5, erect, 0.5–0.8 mm. **Flowers** 1.5–2.5 mm; perianth white, glabrous; tepals connate proximal ⅕, monomorphic, obovate; stamens exserted, 2–4 mm; filaments pilose proximally. **Achenes** light brown, 2.5–3 mm, glabrous.

Flowering May–Sep. Gravelly or clayey flats and slopes, sagebrush communities, pinyon-juniper and montane conifer woodlands; 1600–2100 m; Utah.

Eriogonum ostlundii is relatively common in Piute and Sevier counties, and just enters Garfield County. It is similar to *E. spathulatum* and could be confused with infrequent, glabrous individuals of that species that occur in Beaver County on the western deserts of Utah. The United States Forest Service considers *E. ostlundii* to be a "sensitive" species.

37. Eriogonum mitophyllum Reveal, Phytologia 86: 140. 2004 • Lost Creek wild buckwheat E

Herbs, erect, not scapose, (1.5–)2–3.5 × 0.5–1(–1.5) dm, glabrous, light gray to green. **Stems** spreading, without persistent leaf bases, up to ⅕ height of plant; caudex stems compact; aerial flowering stems erect, slender, solid, not fistulose, 1–2 dm, glabrous. **Leaves** basal, 1 per node; petiole 0.05–0.1 cm, glabrous; blade linear, 1.5–6 × 0.05–0.1 cm, sparsely floccose and green abaxially, glabrous and green adaxially, margins revolute.

Inflorescences narrowly cymose, 0.5–1.5 × 0.2–0.8 dm; branches dichotomous, glabrous; bracts 3, scalelike, elongate-triangular, 1–4.5 mm. **Peduncles** absent. **Involucres** 1 per node, turbinate, 2.5–4.5 × 1.5–2.5(–3) mm, glabrous; teeth 5, erect, 0.5–0.8 mm. **Flowers** (2–)2.5–4 mm; perianth pale or greenish yellow, rarely white, glabrous; tepals connate proximal $^1/_3$–$^1/_2$, monomorphic, oblong; stamens exserted, 2.5–4 mm; filaments glabrous or minutely pubescent. **Achenes** light brown, 2.5–3.5(–4) mm, glabrous.

Flowering Jul–Sep. Clay flats and slopes, saltbush communities, juniper woodlands; 1700–1800 m; Utah.

Eriogonum mitophyllum is known only from the Arapien Shale badlands above Lost Creek northeast of Sigurd, Sevier County. It is allied to *E. ostlundii*, but differs in its linear leaf blades, narrow, strict inflorescences with di- and trichotomous, U-shaped branches, and longer, pale yellow flowers. The leaf blades are threadlike and the narrowest of any species in the genus.

38. Eriogonum artificis Reveal, Phytologia 86: 121. 2004 • Kaye's wild buckwheat E

Herbs, spreading, not scapose, (2–)3–4.5 × 1–2(–3) dm, glabrous. **Stems** spreading, without persistent leaf bases, up to $^1/_3$ height of plant; caudex stems absent; aerial flowering stems erect, slender, solid, not fistulose, 1–2 dm, pubescent among leaves. **Leaves** basal or sheathing up stem 1–3 cm, 1 per node; petiole 2.5–4 cm, tomentose; blade narrowly elliptic, 2–5 × (0.8–)1–1.5(–1.7) cm, grayish-tomentose on both surfaces, margins plane. **Inflorescences** cymose, 10–25 × 5–15 cm; branches dichotomous, glabrous; bracts 3, scalelike, triangular, 2–3 mm. **Peduncles** erect, slender, 0.2–1 cm, glabrous. **Involucres** 1 per node, turbinate, 3.5–4 × 2–2.5 mm; teeth 5(–6), erect, 0.5–0.8 mm. **Flowers** 3.5–4 mm; perianth ochroleucous to pale yellow or, rarely, yellow, glabrous; tepals connate proximal $^1/_5$, monomorphic, oblong; stamens exserted, 5–6 mm; filaments pilose proximally. **Achenes** light brown, 3.5–4 mm, glabrous.

Flowering Aug–Sep. Sandy to somewhat gravelly, volcanic slopes, mixed grassland and sagebrush communities, juniper woodlands; 1800–1900 m; Utah.

Eriogonum artificis is known from a single location near Frisco in Beaver County. Care must be taken not to confuse Kaye's wild buckwheat with glabrous forms of *E. spathulatum* that occur nearby. The flowers of *E. artificis* are usually pale yellow, but occasionally plants are found with cream-colored and distinctly yellow flowers.

39. Eriogonum eremicum Reveal, Phytologia 23: 165. 1972 • Limestone wild buckwheat C E

Eriogonum batemanii M. E. Jones var. *eremicum* (Reveal) S. L. Welsh

Herbs, spreading, not scapose, 2.5–4.5 × 1–2.5 dm, glabrous, grayish. **Stems** spreading, without persistent leaf bases, up to $^1/_4$ height of plant; caudex stems absent; aerial flowering stems erect, slender, solid, not fistulose, 0.5–2 dm, glabrous, pubescent among leaves. **Leaves** basal, 1 per node; petiole 1–2.5 cm, tomentose; blade ovate to round, 1.2–2(–2.5) × 1–1.7(–2) cm, tomentose on both surfaces, margins plane. **Inflorescences** cymose, 1.2–2.5 × 1–2 dm; branches dichotomous, glabrous; bracts 3, scalelike, triangular, 1–3 mm. **Peduncles** absent. **Involucres** (1–)2–5 per cluster, turbinate, 2.5–4(–4.5) × 2–2.5 mm, glabrous; teeth 5, erect, 0.4–0.8 mm. **Flowers** 2.5–3 mm; perianth white, glabrous; tepals connate proximal $^1/_5$, monomorphic, obovate; stamens exserted, 2.5–3.5 mm; filaments pilose proximally. **Achenes** light brown to brown, 2.5–3 mm, glabrous.

Flowering Jun–Sep. Gravelly dolomite flats, saltbush communities, pinyon-juniper woodlands; of conservation concern; 1600–2100 m; Utah.

Eriogonum eremicum is the dolomitic counterpart to the silty-clay species *E. batemanii* of eastern Utah and western Colorado. The limestone wild buckwheat is known from a small area in southwestern Millard County and adjacent northern Beaver County of western Utah.

40. Eriogonum batemanii M. E. Jones, Contr. W. Bot. 11: 11. 1903 (as batemani) • Bateman's wild buckwheat E

Herbs, spreading, not scapose, (1–)1.5–3.5(–5) × 1–2.5 dm, glabrous, grayish. **Stems** spreading, usually without persistent leaf bases, up to $^1/_4$ height of plant; caudex stems absent; aerial flowering stems erect, slender, solid, not fistulose, 1–2 dm, glabrous. **Leaves** basal, 1 per node; petiole 0.8–2.5(–4) cm, tomentose; blade elliptic, (1–)1.5–3(–5) × 0.6–1(–1.2) cm, densely white-tomentose abaxially, less so and white adaxially, margins plane. **Inflorescences** cymose, 4–15 × 5–15 cm; branches dichotomous, glabrous; bracts 3, scalelike, triangular, 1–2 mm. **Peduncles** absent. **Involucres** (1–)2–5 per cluster, turbinate, 2–4 × 1.5–2.5(–3.5) mm, glabrous; teeth 5, erect, 0.5–0.8 mm. **Flowers** 1.5–3 mm; perianth white, glabrous; tepals connate proximally, monomorphic, obovate; stamens exserted, 1.5–3.5 mm; filaments sparsely pilose proximally. **Achenes** light brown to brown, 2.5–3 mm, glabrous. $2n = 40$.

Flowering Jun–Sep. Shaley, silty, gravelly, or clayey flats, washes, and slopes, saltbush and sagebrush communities, juniper woodlands; 1400–2600 m; Colo., Utah.

Eriogonum batemanii is fairly common throughout its range in northwestern Colorado (Moffat and Rio Blanco counties) and east-central Utah (Carbon, Duchesne, Emery, Garfield, Grand, and Uintah counties). In Duchesne County it hybridizes with *E. shockleyi*, producing an unnamed but distinctive hybrid.

41. Eriogonum cronquistii Reveal, Madroño 19: 289, fig. 1. 1969 • Cronquist's wild buckwheat C E

Eriogonum corymbosum Bentham var. *cronquistii* (Reveal) S. L. Welsh

Herbs, spreading, not scapose, (0.7–)1–1.5(–4) × 3–8 dm, glabrous, grayish. **Stems** spreading, usually with persistent leaf bases, up to 1/4 height of plant; caudex stems spreading; aerial flowering stems spreading to erect, slender, solid, not fistulose, 0.5–1 dm, glabrous, tomentose among leaves. **Leaves** sheathing up stems 1–4 cm, 1 per node; petiole 0.3–0.8 cm, tomentose; blade elliptic, 0.5–2 × 0.4–1 cm, densely white-tomentose abaxially, sparsely so and greenish adaxially, margins crenulate. **Inflorescences** subcapitate to cymose, 1–7 × 1–5 cm; branches dichotomous, glabrous; bracts 3, scalelike, triangular, 1 mm. **Peduncles** absent. **Involucres** 1 per node or 2–4 per cluster, turbinate, (2.5–)3–3.5 × 2–2.5 mm, glabrous; teeth 5, erect, 0.6–1 mm. **Flowers** (1.5–)2–3 mm; perianth white, glabrous; tepals connate proximal 1/3, monomorphic, oblanceolate; stamens exserted, 2.5–4 mm; filaments pilose proximally. **Achenes** light brown, 2–2.5 mm, glabrous. $2n = 40$.

Flowering Jul–Sep. Granitic talus slopes and outcrops, sagebrush communities, pinyon-juniper woodlands; of conservation concern; 2500–2800 m; Utah.

Eriogonum cronquistii is known only from the western slope and summit of Bull Mountain, Henry Mountains, Garfield County. The Bureau of Land Management considers it a "sensitive" species.

42. Eriogonum panguicense (M. E. Jones) Reveal, Proc. Utah Acad. Sci. 42: 291. 1966 E

Eriogonum pauciflorum Pursh var. *panguicense* M. E. Jones, Contr. W. Bot. 11: 9. 1903

Herbs, matted, pulvinate or cespitose, scapose, 0.3–3 × 0.5–2 dm, glabrous, grayish. **Stems** spreading, usually with persistent leaf bases, up to 1/5 height of plant; caudex stems matted; aerial flowering stems scapelike, erect or nearly so, slender, solid, not fistulose, 0.2–2.5(–3) dm, glabrous, tomentose among leaves. **Leaves** basal, fasciculate in terminal tufts; petiole 0.1–0.8 cm, tomentose or glabrous; blade linear-oblanceolate to elliptic, 0.5–4(–5) × 0.2–1 cm, densely white-tomentose abaxially, floccose to subglabrous and greenish adaxially, margins plane or crenulate. **Inflorescences** capitate, 0.7–1.5 cm; branches absent; bracts 3, scalelike, triangular, 1–2 mm. **Peduncles** absent. **Involucres** 3–7 per cluster, turbinate, (2–)2.5–3 × 1.5–2.5 mm, glabrous; teeth 5, erect, 0.3–0.5 mm. **Flowers** 2–3 mm; perianth white, glabrous; tepals connate proximal 1/4–1/3, monomorphic, obovate; stamens exserted, 2–3 mm; filaments pilose proximally. **Achenes** light brown, 3–4 mm, glabrous.

Varieties 2 (2 in the flora): Utah.

Eriogonum panguicense is an isolated species that may be related to the *E. batemanii* complex, or perhaps to *E. lonchophyllum*. It is a distinct taxon and worthy of horticultural consideration as a rock garden plant. The large (0.5–0.6 mm), oblong, deep purplish-red anthers against the whitish tepals are striking.

1. Leaf blades (1–)1.5–4(–5) cm; scapes 0.8–2.5(–3) dm; flowers 2–2.5 mm . . . 42a. *Eriogonum panguicense* var. *panguicense*
1. Leaf blades 0.5–1.5 cm; scapes 0.2–0.7(–1) dm; flowers (2–)2.5–3 mm . . . . 42b. *Eriogonum panguicense* var. *alpestre*

42a. Eriogonum panguicense (M. E. Jones) Reveal var. **panguicense** • Panguitch wild buckwheat E

Plants 0.8–3 × 0.5–1.3(–1.5) dm. **Leaves:** petiole 0.2–0.8 cm, base indistinct, tomentose abaxially; blade linear-oblanceolate to elliptic, (1–)1.5–4(–5) × 0.2–1 cm, margins usually plane. **Scapes** 0.8–2.5(–3) dm. **Flowers** 2–2.5 mm. **Achenes** 3–3.5 mm. $2n = 40$.

Flowering Jun–Sep. Clayey slopes, saltbush and sagebrush communities, pinyon-juniper and montane conifer woodlands; 1700–2400(–3100) m; Utah.

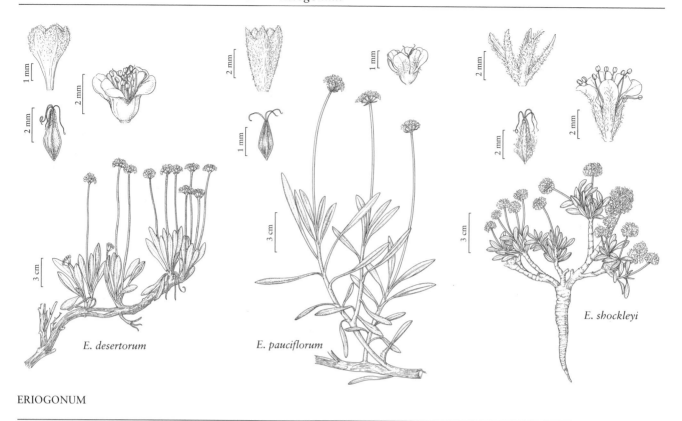

E. desertorum

E. pauciflorum

E. shockleyi

ERIOGONUM

Variety *panguicense* is widespread and infrequent to locally common in Garfield, Iron, Kane, Millard, Sanpete, Sevier and Washington counties. It would be an attractive addition to the rock garden.

42b. Eriogonum panguicense (M. E. Jones) Reveal var. **alpestre** (S. Stokes) Reveal, Proc. Utah Acad. Sci. 42: 292. 1966 • Cedar Breaks wild buckwheat E

Eriogonum chrysocephalum A. Gray subsp. *alpestre* S. Stokes, Eriogonum, 93. 1936

Plants 0.3–0.8 × 1–2 dm. **Leaves:** petiole 0.1–0.4 cm, base distinct, glabrous abaxially; blade elliptic, 0.5–1.5 × 0.2–0.45(–0.6) cm, margins often crenulate. **Scapes** 0.2–0.7(–1) dm. **Flowers** (2–)2.5–3 mm. **Achenes** 3.5–4 mm.

Flowering Jul–Sep. Clayey slopes, high-elevation sagebrush communities, subalpine conifer woodlands; 2900–3400 m; Utah.

Variety *alpestre* is the high-elevation expression of the species and is only marginally distinct. It is restricted essentially to the Cedar Breaks National Monument area of Iron County, where it is scattered but fairly common along the rim of the breaks. It is this component of the species that is now in cultivation.

43. Eriogonum desertorum (Maguire) R. J. Davis, Fl. Idaho, 246. 1952 • Great Basin Desert wild buckwheat E F

Eriogonum chrysocephalum A. Gray subsp. *desertorum* Maguire, Leafl. W. Bot. 3: 11. 1941; *E. brevicaule* Nuttall var. *desertorum* (Maguire) S. L. Welsh; *E. lewisii* Reveal

Herbs, matted, scapose, 0.5–1.2 × 0.7–4 dm, tomentose or floccose, grayish. **Stems** spreading, with persistent leaf bases, up to $^1/_5$ height of plant; caudex stems matted; aerial flowering stems scapelike, erect, slender, solid, not fistulose, (0.2–)0.4–1 dm, tomentose or floccose. **Leaves** basal, fasciculate in terminal tufts; petiole 0.3–1.5 cm, tomentose to floccose; blade oblanceolate to elliptic or, rarely, ovate, 0.4–2 (–2.5) × 0.2–1 cm, densely greenish- or grayish-white-tomentose on both surfaces or densely greenish-white tomentose abaxially, margins plane. **Inflorescences** capitate, 0.7–1.5 cm; branches absent; bracts 3, scalelike, triangular, 1.5–2 mm. **Peduncles** absent. **Involucres** 4–7(–9) per cluster, turbinate to turbinate-campanulate, 2–3.5 × 2–3.5 mm, weakly rigid, floccose at least on teeth; teeth 5–8, erect. **Flowers** 2–3.5 mm; perianth yellow, glabrous; tepals connate proximal $^1/_4$–$^1/_3$, monomorphic, lanceolate or oblong; stamens exserted,

2–4 mm; filaments glabrous or sparsely pilose proximally. **Achenes** brown, 2–3.5 mm, glabrous or sometimes with minute bristles on beak.

Flowering May–Aug. Gravelly or silty to clayey flats, slopes, and ridges, often on limestone soils, mixed grassland, saltbush, and sagebrush communities, pinyon-juniper woodlands; 1500–3000 m; Nev., Utah.

Eriogonum desertorum is a low- to mid-elevation species restricted to central and eastern Elko County, Nevada, and northwestern Box Elder County, Utah. The phase represented by the type is from the valley bottoms and lower foothills, although it extends onto the eastern slope of the East Humboldt Mountains to ca. 2600 m elevation. Such plants tend to have leaf blades that are grayish-tomentose on both surfaces. At higher elevations on isolated desert ranges (Jarbidge, Independent, and Kinsley mountains, where they occur as low as 1950 m) are plants that are smaller in all aspects and tend to have elliptic (rather than oblanceolate to narrowly elliptic) leaf blades; these have been named *E. lewisii*. Similar plants are in the Grouse Creek Mountains of Utah. As noted above, *E. brevicaule* var. *bannockense* occurs in eastern Elko County, where it is widespread and more common than *E. desertorum*. It is almost always at low elevations in the valley bottoms but can occur on some of the higher, isolated peaks. The leaf blades of var. *bannockense* are distinctly narrower and mostly longer.

44. Eriogonum anemophilum Greene, Pittonia 3: 199. 1897 • Wind-loving wild buckwheat C E

Herbs, matted, scapose, 0.5–1 × 0.5–3 dm, thinly tomentose to floccose, grayish. **Stems** spreading, with persistent leaf bases, up to ¹/₅ height of plant; caudex stems matted; aerial flowering stems scapelike, erect, slender, solid, not fistulose, 0.6–1(–1.5) dm, thinly tomentose to floccose. **Leaves** basal, fasciculate in terminal tufts; petiole 0.5–2.5(–3.5) cm, tomentose; blade elliptic to obovate or suborbiculate, (0.7–)0.9–1.5(–2) × (0.3–)0.5–0.9(–1.1) cm, densely white-tomentose abaxially, less so and greenish-tomentose adaxially, margins plane. **Inflorescences** capitate, 1–1.5 cm; branches absent; bracts 3, triangular, scalelike, 1.5–2 mm. **Peduncles** absent. **Involucres** 3–5 per cluster, turbinate to turbinate-campanulate, 2–3.5(–4) × 2–3(–3.5) mm, weakly rigid, floccose at least on teeth; teeth 5–7, erect, 0.4–1 mm. **Flowers** 2–3(–3.5) mm; perianth creamy white, glabrous; tepals connate proximal ¹/₅, monomorphic, oblanceolate; stamens exserted, 2–3 mm; filaments sparsely pubescent proximally. **Achenes** brown, 2.5–3.5 mm, glabrous, occasionally minutely papillate on beak.

Flowering May–Aug. Volcanic tuffaceous or gravelly to rocky (often limestone) slopes, saltbush and sagebrush communities, pinyon-juniper woodlands; of conservation concern; 1400–2600 m; Nev.

Eriogonum anemophilum, in a strict sense, is known only from the limestone ridges and slopes of Star Peak in the West Humboldt Range of Pershing County. Also included within the present circumscription are plants from tuffaceous hills in Reese River Valley in Lander County, and from scattered sites on low hills in northeastern Eureka and southwestern Elko counties. A third set of populations occurs elsewhere in Pershing County (Jersey Valley and the Tobin and Trinity ranges) and just over the line in Humboldt County. Each group differs slightly from the others, the plants of the mountain ranges tending to be more robust, with obovate to suborbiculate leaf blades, compared to the plants of lower elevations, which have narrower leaf blades. Plants from Eureka and Elko counties have a grayish leaf tomentum, while those from the valleys of Humboldt, Pershing, and Lander counties have a greenish or tawny leaf tomentum. What, if any, taxonomic recognition these groups merit has not been determined.

45. Eriogonum tiehmii Reveal, Great Basin Naturalist 45: 277. 1985 • Tiehm's wild buckwheat C E

Herbs, matted, scapose, 1–1.5 × 0.5–3 dm, floccose, grayish. **Stems** spreading, with persistent leaf bases, up to ¹/₅ height of plant; caudex stems matted; aerial flowering stems scapelike, erect, slender, solid, not fistulose, (0.6–)1–1.3(–1.5) dm, floccose. **Leaves** basal, fasciculate in terminal tufts; petiole 0.5–1.6(–2) cm, tomentose; blade elliptic to oblong, (0.8–)1–2.5(–3) × 0.5–0.8(–1) cm, densely white- or grayish-tomentose on both surfaces, margins plane. **Inflorescences** capitate, 0.9–1.4 cm wide; branches absent; bracts 3, scalelike, triangular, 1–1.5 mm. **Peduncles** absent. **Involucres** 4–8 per cluster, turbinate-campanulate, 4–5 × 3–4 mm, rigid, floccose; teeth 5–6, erect to slightly spreading, 1.5–2 mm. **Flowers** 2.5–3.5 (–4) mm; perianth yellowish white or whitish to cream, sparsely glandular abaxially; tepals connate proximal ¹/₄, monomorphic, oblong; stamens exserted, 3–4(–4.5) mm; filaments pilose proximally. **Achenes** light brown, 3–4 mm, glabrous.

Flowering May–Jun. Rocky clay slopes and washes, saltbush communities; of conservation concern; 1800–1900 m; Nev.

Eriogonum tiehmii is known only from the Silver Peak Range of Esmeralda County. It is considered a "sensitive" species in Nevada.

46. Eriogonum diatomaceum Reveal, J. Reynolds & Picciani, Novon 12: 87, fig. 1. 2002   •   Churchill Narrows wild buckwheat   [C] [E]

Herbs, matted or slightly spreading, scapose, 0.5–2 × 0.5–2.5 dm, tomentose, grayish. **Stems** spreading, with persistent leaf bases, up to ¼ height of plant; caudex stems matted or slightly spreading; aerial flowering stems scapelike, erect, slender, solid, not fistulose, (0.4–)0.5–1.5(–2) dm, tomentose. **Leaves** sheathing up flowering stem (1.5–)2–4 cm, 1 per node, fasciculate in terminal tufts on stemless caudex branches; petiole (0.3–)0.5–1.5(–1.8) cm, tomentose; blade elliptic, (0.5–)0.8–2(–2.3) × (0.3–)0.5–1.2(–1.8) cm, densely grayish-tomentose on both surfaces, margins plane. **Inflorescences** capitate, 1–1.5 cm wide; branches absent; bracts 3–8, elongate-triangular to triangular, scalelike, 1–3 mm, tomentose. **Peduncles** absent. **Involucres** 5–10 per cluster, turbinate, (2.5–)3–4.5 × 2–3 mm, rigid, tomentose; teeth 5–7, erect, 0.6–1 mm. **Flowers** (1.5–)2.5–3 mm; perianth creamy white, glabrous; tepals connate proximal ⅓–½, monomorphic, oblong-ovate; stamens exserted, 3–3.5 mm; filaments sparsely pilose proximally. **Achenes** light brown, 2–2.5 mm, glabrous.

Flowering Jun–Sep. White, chalky slopes, saltbush communities; of conservation concern; 1300–1400 m; Nev.

Eriogonum diatomaceum is known only from a few scattered populations in the Churchill Narrows area south of Fort Churchill State Park in Lyon County, occurring on silty diatomaceous deposits of the Coal Valley Formation. It is considered a "sensitive" species in Nevada, and is also a candidate for federal endangered listing, U.S. Fish and Wildlife Service.

47. Eriogonum prociduum Reveal, Aliso 7: 417. 1972   [E]

Herbs, matted, scapose, 0.3–1 × 1–3 dm, glabrous, greenish. **Stems** spreading, with persistent leaf bases, up to ⅕ height of plant; caudex stems matted; aerial flowering stems scapelike, weakly erect, slender, solid, not fistulose, 0.2–0.8(–0.9) dm, glabrous. **Leaves** basal, fasciculate in terminal tufts; petiole 0.2–1.5(–2) cm, tomentose; blade oblanceolate to spatulate or elliptic to ovate, 0.3–1.5 × 0.15–0.8 cm, densely white-tomentose on both surfaces or floccose and greenish adaxially, margins plane. **Inflorescences** capitate, 0.8–1.2 cm wide; branches absent; bracts 3–6, triangular, scalelike, 1–3(–4) mm.

Peduncles absent. **Involucres** 4–7 per cluster, campanulate, 2–4 × (2.5–)3–4 mm, weakly rigid, glabrous or sparsely floccose; teeth 5–6, erect, 0.8–1.4 mm. **Flowers** 2–4 mm; perianth bright yellow, glabrous; tepals connate proximal ¼–⅓, monomorphic, oblong to oblong-obovate; stamens exserted, 2.5–3.5 mm; filaments pilose proximally. **Achenes** light brown, 2–3.5 mm, glabrous.

Varieties 2 (2 in the flora): w United States.

1. Leaf blades 0.3–1(–1.4) × 0.15–0.4(–0.6) cm, tomentose on both surfaces; petioles 0.2–0.5 cm; ne California, nw Nevada, and sc Oregon . . . . **47a.** *Eriogonum prociduum* var. *prociduum*
1. Leaf blades 0.7–1.5 × 0.3–0.8 cm, floccose and greenish adaxially; petioles 1–1.5(–2) cm; se Oregon, sw Idaho, and nc Nevada . . . . **47b.** *Eriogonum prociduum* var. *mystrium*

47a. Eriogonum prociduum Reveal var. **prociduum**   •   Prostrate wild buckwheat   [E]

Leaves: petiole 0.2–0.5 cm; blade 0.3–1(–1.4) × 0.15–0.4(–0.6) cm, tomentose on both surfaces. **Scapes** 0.2–0.6(–0.8) dm. **Involucres** 2–3 mm. **Flowers** 2–3 mm. **Achenes** 2–3 mm. $2n = 40$.

Flowering May–Jul. Volcanic slopes, mixed grassland and sagebrush communities, pinyon-juniper woodlands; 1400–2700; Calif., Nev., Oreg.

Variety *prociduum* is an attractive, matted perennial that is cultivated in rock gardens. In the wild, it is known from northern Lassen and Modoc counties in California, northern Washoe County, Nevada, and south-central Lake County, Oregon. Also, a series of small, disjunct populations occurs in southern Baker County, Oregon. The name *E. chrysops* was misapplied to these plants by J. L. Reveal (1968b). The taxon is of "special concern" to the Bureau of Land Management in California, and is considered "sensitive" in Nevada.

47b. Eriogonum prociduum Reveal var. **mystrium** Reveal, Phytologia 86: 142. 2004   •   Pueblo Mountains wild buckwheat   [E]

Leaves: petiole 1–1.5(–2) cm; blade 0.7–1.5 × 0.3–0.8 cm, tomentose abaxially, floccose and greenish adaxially. **Scapes** 0.4–0.8(–0.9) dm. **Involucres** 3–4 mm. **Flowers** 2.5–3.5(–4) mm. **Achenes** 2.5–3.5 mm.

Flowering May–Jul. Sandy to gravelly slopes, sagebrush communities; 1400–2400; Idaho, Nev., Oreg.

Variety *mystrium* is known from the Owyhee Mountains (South Mountain) of Owyhee County, Idaho, the Santa Rosa Range (Auto Hill) of Humboldt County, Nevada, and the Oregon Canyon, Pueblo, Steens, and Trout Creek mountains of Harney County, Oregon.

48. Eriogonum cusickii M. E. Jones, Contr. W. Bot. 11: 10. 1903 • Cusick's wild buckwheat C E

Eriogonum chrysocephalum A. Gray subsp. *cusickii* (M. E. Jones) S. Stokes

Herbs, matted, usually not scapose, 0.3–1 × 0.5–3(–5) dm, glabrous, greenish. **Stems** spreading, with persistent leaf bases, up to $^1/_5$ height of plant; caudex stems matted; aerial flowering stems rarely scapelike, erect, slender, solid, not fistulose, 0.2–0.9 dm, glabrous, tomentose among leaves. **Leaves** basal, fasciculate in terminal tufts; petiole 0.2–0.5 cm, tomentose; blade narrowly oblanceolate to spatulate, 0.5–1(–1.2) × 0.15–0.25(–0.3) cm, densely white- or grayish-tomentose on both surfaces, margins plane. **Inflorescences** umbellate-cymose to cymose, rarely capitate, 3–5 × 3–5 cm; branches dichotomous, glabrous, rarely absent; bracts 3, triangular, scalelike, 0.5–1 mm. **Peduncles** absent. **Involucres** 1 per node, rarely 2–5, turbinate-campanulate to campanulate, 2–3 × 2–3 mm, rigid, glabrous; teeth 6–8, erect, 0.8–1.4 mm. **Flowers** 2–2.5(–3) mm; perianth yellow, glabrous; tepals connate proximally, monomorphic, oblong; stamens exserted, 2–3 mm; filaments pilose proximally. **Achenes** light brown, 2.5–3 mm, glabrous.

Flowering Jun–Jul. Sandy, volcanic flats, mixed grassland and sagebrush communities, montane conifer woodlands; of conservation concern; 1300–1500 m; Oreg.

Eriogonum cusickii is fairly common in scattered portions of Deschutes, Harney, Klamath, and Lake counties. It is gradually being introduced into cultivation.

Eriogonum prociduum is distinct from *E. cusickii*, but some populations of the latter in Lake County, Oregon, south of Christmas Valley and northwest of Wagontire, have capitate inflorescences even in fruit. These plants differ from typical *E. prociduum* by consistently having small, rather narrow leaf blades (3–6 × 1–4 mm) that sheath up the flowering stems.

49. Eriogonum crosbyae Reveal, Brittonia 33: 442, fig. 1. 1981 • Crosby's wild buckwheat E

Eriogonum capistratum Reveal; *E. capistratum* var. *muhlickii* Reveal; *E. capistratum* var. *welshii* Reveal; *E. meledonum* Reveal; *E. ochrocephalum* S. Watson var. *alexanderae* Reveal; *E. verrucosum* Reveal

Herbs, matted, scapose, 0.05–1.5(–2) × (0.1–)1–3 dm, floccose to tomentose or glabrous, sometimes glandular, greenish or grayish. **Stems** matted, with persistent leaf bases, up to $^1/_5$ height of plant; caudex stems matted; aerial flowering stems scapelike, weakly erect to erect, slender, solid, not fistulose, 0.02–1.5(–1.8) dm, floccose to tomentose or glabrous, occasionally also or only sparsely to densely glandular. **Leaves** basal, fasciculate in terminal tufts; petiole 0.2–3(–3.5) cm, tomentose, infrequently glandular; blade oblanceolate to spatulate or elliptic to obovate or ovate, (0.5–)1–2(–3) × 0.2–1 (–1.5) cm, densely white- or grayish-tomentose on both surfaces, sometimes less and greenish white adaxially, margins plane. **Inflorescences** capitate, 0.7–1.5 cm; branches absent; bracts 3, narrowly triangular to triangular, scalelike, 1–3 mm. **Peduncles** absent. **Involucres** (3–)5–8 per cluster, turbinate to campanulate, (1.5–)2–5(–5.5) × 2–4(–4.5) mm, rigid or membranous, tomentose to floccose, occasionally glabrous except for floccose teeth, rarely sparsely pilose and glandular; teeth 5–7, erect to spreading or reflexed, 0.5–1.5 mm. **Flowers** 1.5–3.5(–4) mm, glabrous or occasionally minutely glandular, pustulose in some; perianth yellow to pale yellow or, rarely, cream; tepals connate proximal $^1/_4$–$^1/_3$, monomorphic, oblong to oblong-obovate; stamens exserted, 1.5–4 mm; filaments glabrous or sparsely pilose proximally. **Achenes** light brown, 2–4 mm, glabrous or sometimes with minute bristles on beak. $2n = 40$.

Flowering May–Aug. White tuffaceous shale volcanic outcrops, metamorphic rock outcrops, or basaltic or granitic sandy flats, washes, slopes, and ridges, saltbush and sagebrush or high-elevation sagebrush to alpine tundra communities, juniper or montane conifer woodlands; (1200–)1400–3100; Idaho, Mont., Nev., Oreg.

Eriogonum crosbyae, as now defined, is widely scattered in the valley bottoms and foothills, and atop several mountain ranges of central Idaho (Blaine, Butte, Camas, Custer, and Lemhi counties) and in western Montana (Deer Lodge and Ravalli counties). It is disjunct to southwestern Idaho (Owyhee Mountains, Owyhee County), southeastern Oregon (Guano and Coleman valleys, Harney County, and Fish Fin Rim, Lake County), and in northwestern Nevada (Washoe and Humboldt counties south through Douglas, Lyon, and Pershing counties to Mineral County). It also occus in the Marys River Peak area of Elko County, Nevada.

The species may be subdivided into four phases, for which names are available. The vast majority of populations have bright yellow flowers with pustulose bases and midveins. The pustulose condition may also be observed in *Eriogonum chrysops* (on pale greenish-yellow flowers) and in *E. mancum* (with cream-colored flowers). Populations with pale yellow flowers here assigned to *E. crosbyae* occur in the mountains near Mackay, Idaho, and cream-colored flowers are found near Challis and around Salmon, both well outside the known ranges of *E. chrysops* and *E. mancum*. Final resolution of the taxonomy of this group awaits further study.

50. **Eriogonum mancum** Rydberg, Fl. Rocky Mts., 220, 1061. 1917 • Imperfect wild buckwheat [E]

Eriogonum chrysocephalum A. Gray subsp. *mancum* (Rydberg) S. Stokes

Herbs, matted, scapose, 0.5–1(–1.3) × 1–2 dm, tomentose, grayish. **Stems** spreading, with persistent leaf bases, up to ¹/₅ height of plant; caudex stems matted; aerial flowering stems scapelike, weakly erect to erect, slender, solid, not fistulose, 0.1–0.9(–1.2) dm, tomentose. **Leaves** basal, fasciculate in terminal tufts; petiole 0.5–1 cm; blade oblanceolate to spatulate or ovate, 0.5–1.5(–1.8) × (0.2–)0.3–1 cm, densely grayish-tomentose on both surfaces, margins plane. **Inflorescences** capitate, 0.8–1.2 cm; branches absent; bracts 3–5, scalelike, triangular, 1–3 mm. **Peduncles** absent. **Involucres** 2–5 per cluster, campanulate, 2–3(–3.5) × 2–3.5(–4) mm, membranous, tomentose; teeth 5, erect, 0.4–0.8 mm. **Flowers** 2–3 mm; perianth cream to pink or rose, glabrous or sparsely pilose, pustulose; tepals connate proximal ¹/₄–¹/₃, monomorphic, oblong; stamens exserted, 2–2.5 mm; filaments pilose proximally. **Achenes** light brown, (2–)2.5–3 mm, glabrous except for minutely papillate beak.

Flowering May–Jul. Gravelly to clayey flats and slopes, mixed grassland, saltbush, sagebrush, and mountain mahogany communities, juniper woodlands; 1100–2000 m; Idaho, Mont., Utah, Wyo.

Eriogonum mancum is widely scattered in Butte, Clark, Custer, and Lemhi counties of southeastern Idaho, in Beaverhead, Broadwater, Carbon, Deer Lodge, Gallatin, Granite, Jefferson, Lewis and Clark, Madison, Powell, Ravalli, and Silver Bow counties in western Montana, and in Big Horn County of northwestern Wyoming. It is disjunct on limestone outcrops in the House Range of Millard County, Utah, no doubt a Pleistocene relic.

Plants from the rim of Little Mountain west of Kane, Big Horn County, Wyoming (*B. E. Nelson 5407*, BRY, MARY, NY, RM, UTC), and those just to the north in southern Carbon County, Montana (*Lesica 2632, 5379, 5405, 5439,* and *5453,* MONTU), have tepals with a few long, pilose hairs along the midrib adaxially. These populations are well isolated from others of the species, but the taxonomic significance of this expression has yet to be determined.

51. **Eriogonum soliceps** Reveal & Björk, Brittonia 56: 296, fig. 1. 2004 • Railroad Canyon wild buckwheat [C][E]

Herbs, matted, not scapose, (0.02–)0.04–0.06(–0.08) × (0.5–)1–3(–4.5) dm, tomentose, grayish. **Stems** spreading, with persistent leaf bases, up to ¹/₈ height of plant; caudex stems matted; aerial flowering stems absent. **Leaves** basal, fasciculate in terminal tufts; petiole 0.1–0.3 cm, tomentose; blade narrowly spatulate, 0.4–0.8 × (0.15–)0.2–0.4 cm, densely grayish-white-tomentose on both surfaces, margins plane. **Inflorescences** capitate, 0.5–0.9 cm; branches absent; bracts absent. **Peduncles** erect, (1.5–)3–5(–7) cm, thinly tomentose, arising directly from caudices. **Involucres** 1 per node, turbinate to turbinate-campanulate, 2–3 × 1.5–2(–2.5) mm, rigid, thinly tomentose; teeth 5, erect, 0.4–0.6 mm. **Flowers** 2–3.5(–4) mm; perianth white, glabrous, pustulose; tepals connate proximal ¹/₅, slightly dimorphic, those of outer whorl oblong, those of inner whorl oblong-oblanceolate; stamens slightly exserted, 2–3 mm; filaments pilose proximally. **Achenes** light brown, 2.5–3.5 mm, glabrous except for slightly papillose beak.

Flowering Jul–Aug. Gravelly soil, sagebrush communities; of conservation concern; 2500–2600 m; Idaho, Mont.

Eriogonum soliceps occurs in the Beaverhead Mountains that border Lemhi County, Idaho, and Beaverhead County, Montana. Elsewhere in Beaverhead County the species is found in the Tendoy Mountains, and in the Cedar Pass and Bannack areas. Otherwise the Railroad Canyon wild buckwheat is known from Chalk Bluffs northeast of Wisdow in Deer Lodge County, and from an unknown location in the "mountains south of Virginia City" in Madison County, Montana. It may be quickly distinguished from *E. mancum* by its reduced inflorescence composed of a single, upright peduncle bearing a solitary involucre.

52. **Eriogonum soredium** Reveal, Great Basin Naturalist 41: 229, fig. 1. 1981 • Frisco wild buckwheat [C][E]

Herbs, matted, scapose, 0.01–0.1 × 1–3 dm, densely white-tomentose, whitish. **Stems** matted, with persistent leaf bases, up to ¹/₅ height of plant; caudex stems matted; aerial flowering stems scapelike, erect, slender, solid, not fistulose, 0.02–0.15 dm, tomentose. **Leaves** basal, fasciculate and sheathing nearly entire length of stem;

petiole 0.05–0.2(–0.3) cm, tomentose; blade narrowly elliptic to narrowly oblong, 0.2–0.7 × (0.07–)0.1–0.25 cm, densely white-tomentose, margins plane. **Inflorescences** capitate, 0.5–1 cm; branches absent; bracts 6–8, scalelike, lanceolate to narrowly triangular, 1.3–1.6 mm. **Peduncles** absent. **Involucres** 4–6 per cluster, turbinate, 2–2.5 × 1.3–1.5 mm, rigid; teeth (4–)5, erect, 0.5–0.6 mm. **Flowers** (1.5–)2–2.5 mm; perianth white, glabrous; tepals connate proximal ¼, slightly dimorphic, oblanceolate to ovate or obovate, those of outer whorl 1.5–2 mm wide, those of inner whorl 1–1.5 mm wide; stamens included, 2–2.5 mm; filaments sparsely pubescent proximally. **Achenes** light brown, 2–2.5 mm, glabrous.

Flowering Jun–Sep. Gravelly to rocky limestone slopes, mixed saltbush and sagebrush communities, pinyon-juniper woodlands; of conservation concern; 2000–2300 m; Utah.

Eriogonum soredium is a rare and localized species known only from the San Francisco Mountains of Beaver County. It is considered a "sensitive" species in Utah, and occasionally is seen in cultivation.

53. Eriogonum rosense A. Nelson & P. B. Kennedy, Proc. Biol. Soc. Wash. 19: 36. 1906 (as rosensis)   E

Herbs, matted, scapose, 0.2–1 (–1.3) × 0.5–5 dm, glandular-hairy, greenish. **Stems** matted, with persistent leaf bases, up to ⅕ height of plant; caudex stems matted; aerial flowering stems scapelike, weakly erect to erect, slender, solid, not fistulose, 0.1–0.9(–1.1) dm, densely glandular-hairy. **Leaves** basal, fasciculate in terminal tufts; petiole 0.4–2 cm, tomentose, sometimes also glandular; blade oblanceolate or broadly elliptic to oval, 0.4–2.5 × (0.15–)0.25–1.6 cm, densely white-tomentose and glandular on both surfaces or densely white-tomentose abaxially and greenish-tomentose adaxially, margins plane. **Inflorescences** capitate, 0.6–1.5 cm; branches absent; bracts 3, scalelike, triangular, 1–3.5 mm. **Peduncles** absent. **Involucres** 3–6 per cluster, turbinate to campanulate, (2.5–)3–5(–6) × 2.5–4(–6) mm, rigid, glandular and sparsely hairy; teeth 5–8, erect, 0.7–1.4 mm. **Flowers** 2–4 mm; perianth bright yellow to reddish yellow, occasionally cream, glabrous or glandular; tepals connate proximally, monomorphic, obovate or oblong; stamens exserted, 3–5.5 mm; filaments glabrous or sparsely pilose proximally. **Achenes** light brown, 1.5–3.5 mm, glabrous.

Varieties 2 (2 in the flora): California and Nevada.

1. Leaf blades oblanceolate to elliptic, (0.15–)0.25–0.6(–1) cm wide; involucres (2.5–)3–4(–4.5) × 2.5–3.5 mm; flowers 2–3 mm; tepals obovate; achenes 1.5–2.5 mm; Sierra Nevada of California and adjacent desert ranges of wc Nevada . . . .
. . . . 53a. *Eriogonum rosense* var. *rosense*
1. Leaf blades broadly elliptic to oval, 0.5–1.6 cm wide; involucres (2.5–)3–5(–6) × 3–5(–6) mm; flowers (2.5–)3–4 mm; tepals oblong; achenes (2.5–)3–3.5 mm; desert ranges, Nevada . . . . . .
. . . . 53b. *Eriogonum rosense* var. *beatleyae*

53a. Eriogonum rosense A. Nelson & P. B. Kennedy var. **rosense** • Mt. Rose wild buckwheat   E

Eriogonum ochrocephalum S. Watson var. *agnellum* Jepson; *E. ochrocephalum* subsp. *agnellum* (Jepson) S. Stokes

Plants 0.2–1 × 0.5–1.5(–2) dm. **Leaf blades** oblanceolate to elliptic, 0.4–1.5(–1.7) × (0.15–) 0.25–0.6(–1) cm, densely white-tomentose and glandular on both surfaces. **Involucres** turbinate to turbinate-campanulate, (2.5–)3–4(–4.5) × 2.5–3.5 mm; teeth (5–)6–8. **Flowers** 2–3 mm; perianth bright yellow to reddish yellow; tepals obovate. **Achenes** 1.5–2.5 mm. $2n = 40$.

Flowering Jul–Sep. Granitic or volcanic slopes and ridges, sagebrush communities, montane to alpine conifer woodlands; (2300–)2500–4000 m; Calif., Nev.

Variety *rosense* is common in the Sierra Nevada (Alpine, Amador, El Dorado, Fresno, Inyo, Madera, Mono, Nevada, Placer, Shasta, Sierra, and Tuolumne counties) and higher desert ranges to the east in Carson City, Douglas, Esmeralda, Storey, and Washoe counties, Nevada. It is an excellent rock-garden plant.

53b. Eriogonum rosense A. Nelson & P. B. Kennedy var. **beatleyae** (Reveal) Reveal, Harvard Pap. Bot. 9: 194. 2004 • Beatley's wild buckwheat   E

Eriogonum beatleyae Reveal, Aliso 7: 415. 1972

Plants 0.4–0.8(–1.3) × 1–5 dm. **Leaf blades** broadly elliptic to oval, (0.7–)1–2.5 × 0.5–1.6 cm, densely white-tomentose abaxially, greenish-tomentose adaxially. **Involucres** campan-ulate, (2.5–)3–5(–6) × 3–5(–6) mm; teeth 5. **Flowers** (2.5–)3–4 mm; perianth yellow or, infrequently, cream; tepals oblong. **Achenes** (2.5–)3–3.5 mm.

Flowering May–Aug. Volcanic tuffaceous soils, sagebrush communities, pinyon-juniper woodlands; 1700–2600(–2800) m; Nev.

Variety *beatleyae* has yellow flowers in Eureka, Mineral, and Nye counties; in Churchill and adjacent Lander counties it has yellow or cream-colored flowers. Unlike var. *rosense*, var. *beatleyae* is commonly encountered at lower elevations in the foothills of Great Basin ranges and occurs at higher elevations only in the Toiyabe Mountains.

54. Eriogonum breedlovei (J. T. Howell) Reveal, Leafl. W. Bot. 10: 335. 1966 E

Eriogonum ochrocephalum S. Watson var. *breedlovei* J. T. Howell, Leafl. W. Bot. 10: 14. 1963

Herbs, matted, occasionally scapose, 0.2–1 × 0.8–1.5(–2) dm, glandular-puberulent and slightly pilose, greenish. **Stems** matted, with persistent leaf bases, up to ¹/₅ height of plant; caudex stems matted; aerial flowering stems rarely scapelike, erect to prostrate, slender, solid, not fistulose, 0.15–0.6(–0.8) dm, thinly to densely glandular-puberulent and slightly pilose. **Leaves** basal, fasciculate in terminal tufts; petiole 0.2–0.6(–1) cm, tomentose; blade broadly elliptic, 0.2–0.8 (–1) × 0.2–0.4(–0.6) cm, densely white-tomentose abaxially, less so and olive green adaxially, margins plane. **Inflorescences** cymose-umbellate or cymose, 0.5–3 × 1–2.5 cm, rarely capitate, 1–1.3 cm; branches dichotomous, occasionally absent, pilose and glandular; bracts 3, scalelike, triangular, 1–1.2 mm. **Peduncles** erect to spreading, slender, 0.05–0.5(–1) cm, glandular or nearly glabrous, rarely absent. **Involucres** 1 per node, rarely 2–4, turbinate-campanulate, (2.5–)3.5–4 × 3.5–4 mm, rigid, glandular-puberulent or nearly glabrous; teeth 7–9, erect or slightly spreading, 1–1.5 mm. **Flowers** 2.5–3.5(–4) mm; perianth whitish to reddish, finely glandular-hairy abaxially; tepals connate proximally, slightly dimorphic, those of outer whorl obovate and obtuse, 1.5–2.3 mm wide, apex truncate or emarginate, those of inner whorl cuneate or spatulate, 0.7–1.3 mm wide, apex rounded; stamens exserted, 3–5.5 mm; filaments sparsely pubescent proximally. **Achenes** light brown, 2.5–3 mm, apex sparsely glandular.

Varieties 2 (2 in the flora): California.

Eriogonum breedlovei is geographically isolated from the rest of the *E. ochrocephalum* complex, and the only member nearby is *E. rosense*, found at higher elevations of the Sierra Nevada. Both varieties of *E. breedlovei* are rare.

1. Scapes erect; inflorescences cymose-umbellate, rarely capitate..........................
.... 54a. *Eriogonum breedlovei* var. *breedlovei*
1. Scapes suberect to prostrate; inflorescences cymose .... 54b. *Eriogonum breedlovei* var. *shevockii*

54a. Eriogonum breedlovei (J. T. Howell) Reveal var. **breedlovei** · Piute wild buckwheat E

Scapes erect, densely glandular-puberulent, slightly pilose. **Inflorescences** cymose-umbellate, rarely capitate. **Peduncles** 0.05–0.2 cm.

Flowering Jun–Sep. Quartz outcrops, conifer woodlands; 2300–2500 m; Calif.

Variety *breedlovei* is restricted to Piute Mountain and the Owens Peak areas of Kern County. It is occasionally grown in rock gardens.

54b. Eriogonum breedlovei (J. T. Howell) Reveal var. **shevockii** J. T. Howell, Mentzelia 1: 21. 1976 · Shevock's wild buckwheat E

Scapes suberect to prostrate, thinly glandular-puberulent, slightly pilose. **Inflorescences** cymose. **Peduncles** 0.3–0.5(–1) cm.

Flowering Jun–Oct. Granitic outcrops, conifer woodlands; 1800–2300 m; Calif.

Variety *shevockii* is restricted to The Needles, Baker Point, and Little Kern River Gorge areas of Kern and Tulare counties.

55. Eriogonum scopulorum Reveal, Phytologia 23: 170. 1972 · Cliff wild buckwheat E

Herbs, matted, scapose, 0.4–1 × 0.5–1.5(–2) dm, glandular-hairy, grayish. **Stems** matted, with persistent leaf bases, up to ¹/₅ height of plant; caudex stems matted; aerial flowering stems scapelike, erect, slender, solid, not fistulose, (0.3–)0.5–0.7 dm, glandular-hairy. **Leaves** basal, fasciculate in terminal tufts; petiole 0.2–0.4 cm, tomentose, glandular; blade oblanceolate to elliptic, (0.5–)0.6–0.9(–1) × 0.25–0.6 cm, densely white-tomentose and glandular on both surfaces or greenish adaxially, margins plane. **Inflorescences** capitate, 0.5–1.5 cm; branches absent; bracts 5–6, scalelike, triangular, 1.3–1.6 mm. **Peduncles** absent. **Involucres** 5–7 per cluster, turbinate-campanulate, 2.5–3 × (1.5–)2.5–3 mm, membranous, glabrous or glandular and sparsely floccose; teeth 5–6, erect, 1–1.5 mm. **Flowers** 2.5–3 (–3.5) mm; perianth pale yellow, minutely glandular; tepals connate proximal ¹/₅, monomorphic, oblanceolate; stamens exserted, 3–4 mm; filaments pilose proximally. **Achenes** brown, 2–2.5 mm, glabrous.

Flowering Jul–Aug. Rocky granitic outcrops, sagebrush communities, montane conifer woodlands; 2400–2900 m; Idaho, Oreg.

Eriogonum scopulorum is known only from a few scattered locations in the Wallowa Mountains of Wallowa County, Oregon, and in Adams and Idaho counties, Idaho. While rare, and of concern to the U.S. Forest Service, it is not considered a "sensitive species," given its remote location and lack of any immediate threat. The cliff wild buckwheat is worthy of cultivation as a rock garden plant.

56. Eriogonum chrysops Rydberg, Fl. Rocky Mts., 220, 1061. 1917 • One-eyed wild buckwheat C E

Eriogonum ochrocephalum S. Watson subsp. *chrysops* (Rydberg) S. Stokes; *E. ovalifolium* Nuttall var. *chrysops* (Rydberg) M. Peck

Herbs, matted, scapose, 0.2–1 × 0.5–2 dm, floccose to tomentose, greenish. **Stems** matted, with persistent leaf bases, up to ¹/₅ height of plant; caudex stems matted; aerial flowering stems scapelike, erect , slender, solid, not fistulose, 0.2–1 dm, floccose to tomentose. **Leaves** basal, fasciculate in terminal tufts; petiole 0.2–0.5(–0.8) cm, tomentose; blade oblanceolate to spatulate, (0.5–)0.7–1 × 0.2–0.4(–0.5) cm, densely white- or grayish-tomentose on both surfaces, margins plane. **Inflorescences** capitate, 0.5–1.5 cm wide; branches absent; bracts 3–5, scalelike, triangular, 1–2 mm. **Peduncles** absent. **Involucres** 3–5 per cluster, turbinate-campanulate to campanulate, 2.5–3(–3.5) × 2.5–3 mm, membranous, glabrous, sparsely floccose on teeth; teeth 5, erect, 0.6–1.2 mm. **Flowers** (2–)2.5–3 mm; perianth yellow, sparsely glandular, infrequently glabrous; tepals connate proximal ¹/₃, monomorphic, oblong to narrowly obovate; stamens exserted, 2–2.5(–3) mm; filaments pilose proximally. **Achenes** light brown, 2.5–3 mm, glabrous except for minutely bristly beak.

Flowering May–Jul. Gravelly basaltic or rhyolitic slopes and outcrops, sagebrush communities; of conservation concern; 1200–1400 m; Oreg.

Eriogonum chrysops is known from five scattered sites in the Skull Creek area of Malheur County. It is no longer considered a candidate for threatened status under the provisions of the Endangered Species Act. However, it still is considered a "threatened" species by the state of Oregon.

57. Eriogonum kingii Torrey & A. Gray, Proc. Amer. Acad. Arts 8: 165. 1870 • Ruby Mountains wild buckwheat E

Herbs, matted, scapose, 0.2–1.5 × 0.5–4 dm, tomentose or floccose to subglabrous, greenish or grayish. **Stems** matted, with persistent leaf bases, up to ¹/₅ height of plant; caudex stems matted; aerial flowering stems scapelike, erect, slender, solid, not fistulose, 0.2–1(–1.5) dm, tomentose or floccose to subglabrous or even glabrous. **Leaves** basal, fasciculate in terminal tufts; petiole (0.2–)0.4–2 (–3) cm, tomentose; blade oblanceolate to spatulate or elliptic, 0.5–1.5(–2) × (0.2–)0.3–0.8(–1) cm, densely white- or greenish-tomentose on both surfaces, margins plane. **Inflorescences** capitate, 0.8–1.7(–2) cm; branches absent; bracts 3–5, linear to narrowly scalelike, triangular, 1.5–3 mm. **Peduncles** absent. **Involucres** 3–5 per cluster, turbinate-campanulate to campanulate, 2.5–3.5 × 2.5–4 mm, membranous, sparsely pubescent or rarely sparsely glandular abaxially; teeth 5–6, erect to spreading, 0.7–1.8(–2) mm. **Flowers** (2.5–)3–3.5 mm; perianth greenish yellow or pale yellow, rarely yellow, glabrous or sparsely glandular abaxially; tepals connate proximal ¹/₄–¹/₂, monomorphic, spatulate to obovate; stamens exserted, 2–3.5 mm; filaments glabrous or sparsely pilose proximally. **Achenes** light brown, 2.5–3 mm, glabrous except for slightly papillate beak in some. $2n = 40$.

Flowering Jun–Sep. Limestone or granitic gravelly slopes and ridges, mixed grassland and sagebrush communities, pinyon-juniper and subalpine conifer woodlands; 1600–3300 m; Nev.

Eriogonum kingii is known only from Elko and northern White Pine counties. It is occasionally cultivated in rock gardens.

58. Eriogonum argophyllum Reveal, Phytologia 23: 168. 1972 • Sulphur Hot Springs wild buckwheat C E

Herbs, matted, scapose, 0.5–0.9 × 1–2 dm, floccose, grayish. **Stems** matted, with persistent leaf bases, up to ¹/₅ height of plant; caudex stems matted; aerial flowering stems scapelike, erect, slender, solid, not fistulose, (0.4–) 0.5–0.7 dm, floccose. **Leaves** basal, fasciculate in terminal tufts; petiole 0.05–0.1(–0.15) cm, tomentose; blade oblanceolate to elliptic, 0.4–0.8(–1) × 0.2–0.4(–0.5) cm, densely white-tomentose on both surfaces, margins plane. **Inflorescences** capitate, 0.5–1 cm; branches absent; bracts 5–6, lanceolate, scalelike, 2–2.5 mm. **Peduncles** absent. **Involucres** 5–7 per cluster, turbinate-campanulate, 2–2.5 × 2–2.5 mm, membranous, tomentose and sparsely glandular; teeth 6–7, erect to spreading, 1–1.5 mm. **Flowers** (2–)2.5–3 mm; perianth yellow, sparsely glandular; tepals connate proximal 1/4–1/2, monomorphic, oblong; stamens exserted, 3–3.5 mm; filaments pubescent proximally. **Achenes** light brown, 2.8–3 mm, glabrous except for minutely bristly beak.

Flowering Jun–Sep. Moist, crusted, sandy, alkaline flats near warm springs, saltgrass and saltbush communities; of conservation concern; 1800–1900 m; Nev.

Eriogonum argophyllum is known only from the Sulphur Hot Springs area in Ruby Valley, Elko County, where plants grow along runoff channels associated with the hot springs. In 2003 the species was removed as a candidate for federal protection on the basis of "additional pop[ulation]s, individuals, habitat," although it is still known only from the type location, where protection is provided by an interested private landowner. The species is considered to be "critically endangered" by the state of Nevada, and is in the Center for Plant Conservation's National Collection of Endangered Plants.

59. Eriogonum gracilipes S. Watson, Proc. Amer. Acad. Arts 24: 85. 1889   • White Mountains wild buckwheat E

Eriogonum kennedyi Porter ex S. Watson subsp. *gracilipes* (S. Watson) S. Stokes; *E. ochrocephalum* S. Watson var. *gracilipes* (S. Watson) J. T. Howell

Herbs, matted, scapose, 0.5–1 × 0.5–2 dm, glandular-hairy, greenish. **Stems** matted, with persistent leaf bases, up to 1/5 height of plant; caudex stems matted; aerial flowering stems scapelike, erect, slender, solid, not fistulose, (0.2–)0.3–1(–1.2) dm, glandular-hairy. **Leaves** basal, fasciculate in terminal tufts; petiole (0.5–)0.8–1.7(–2) cm, thinly tomentose, glandular; blade oblanceolate to elliptic, (0.3–)1–1.5(–2) × (0.2–)0.3–0.6 cm, densely white-tomentose and glandular abaxially, less so and greenish adaxially, margins plane. **Inflorescences** capitate, 1–2 cm; branches absent; bracts 3–5, lanceolate, scalelike, 1.5–3 mm. **Peduncles** absent. **Involucres** 5–7 per cluster, turbinate to campanulate, 2–4 × 2–3 mm, membranous, sparsely glandular to glandular-puberulent or pubescent; teeth 5, erect to spreading, 1–1.5 mm. **Flowers** 2–3 mm; perianth white to rose, glandular; tepals connate proximal 1/4, monomorphic, obovate;

stamens exserted, 2–3.5(–4) mm; filaments pilose proximally. **Achenes** light brown, 2–2.3(–2.5) mm, glabrous.

Flowering Jul–Sep. Granitic, sandstone or limestone sandy to gravelly outcrops, slopes, and ridge tops, high-elevation sagebrush communities, subalpine and alpine conifer woodlands; 2900–3900 m; Calif., Nev.

Eriogonum gracilipes is generally found with Great Basin bristlecone pine (*Pinus longaeva* D. K. Bailey) and foxtail pine (*Pinus balfouriana* S. Watson) in the higher elevations of the White Mountains and on the eastern edge of the Sierra Nevada immediately to the west. The White Mountains wild buckwheat is restricted to Mono and Inyo counties, California, and just enters Esmeralda County in Nevada. It is occasionally seen in cultivation.

60. Eriogonum holmgrenii Reveal, Leafl. W. Bot. 10: 184. 1965   • Snake Range wild buckwheat C E

Herbs, matted, scapose, 0.1–0.5 × 0.5–3 dm, glandular-hairy, greenish. **Stems** matted, with persistent leaf bases, up to 1/5 height of plant; caudex stems matted; aerial flowering stems scapelike, erect, slender, solid, not fistulose, 0.03–0.3 dm, glandular-hairy. **Leaves** basal, fasciculate in terminal tufts; petiole (0.1–)0.2–0.6(–1) cm, floccose and glandular; blade oblanceolate to spatulate, 0.2–0.5(–1) × (0.1–)0.2–0.4 cm, glandular and densely white-tomentose abaxially, less so to floccose and greenish adaxially, margins plane. **Inflorescences** capitate, 0.6–0.9(–1) cm; branches absent; bracts 4–6, scalelike, narrowly lanceolate, 0.8–1.4(–1.8) mm. **Peduncles** absent. **Involucres** 2–4 per cluster, campanulate, (1.5–)2–3 × 2–3 mm, membranous, sparsely glandular-puberulent; teeth 5, erect to spreading, 1–1.3 mm. **Flowers** 2.5–3(–3.5) mm; perianth white to rose, sparsely glandular; tepals connate proximal 1/4, monomorphic, spatulate to obovate; stamens exserted, 2.5–3(–3.5) mm; filaments pilose proximally. **Achenes** light brown, 2.5–3 mm, glabrous except for slightly roughened beak.

Flowering Jul–Sep. Quartzite or limestone outcrops, slopes, and ridge tops, high-elevation sagebrush communities, alpine conifer woodlands; of conservation concern; 3200–3600 m; Nev.

Eriogonum holmgrenii is closely related to *E. gracilipes*, and like that species it occurs with Great Basin bristlecone pine (*Pinus longaeva* D. K. Bailey). It is restricted to the Snake Range in White Pine County, and is regarded as a "sensitive" species by the U.S. Forest Service and the state of Nevada. The species is occasionally seen in cultivation.

61. Eriogonum codium Reveal, Caplow & K. A. Beck, Rhodora 97: 350, fig. 1. 1997 • Umtanum Desert wild buckwheat [C] [E]

Herbs, matted, not scapose, 0.3–1 × (1–)2–7(–9) dm, tomentose, greenish. **Stems** matted, with persistent leaf bases, up to ¼ height of plant; caudex stems matted; aerial flowering stems erect, slender, solid, not fistulose, 0.2–0.9 dm, floccose. **Leaves** basal, fasciculate in terminal tufts; petiole 0.2–0.8(–1) cm, tomentose; blade oblanceolate to elliptic, (0.5–)0.6–1.2 × 0.3–0.6 cm, densely white-tomentose on both surfaces, margins plane. **Inflorescences** cymose-umbellate or cymose, 1–2 × 1–4 cm, tomentose; branches dichotomous, floccose; bracts triangular, scalelike, 1–2.5 mm. **Peduncles** erect, slender, (0.15–)0.2–0.5(–0.7) cm, floccose. **Involucres** 1 per node, turbinate-campanulate, 2.5–4 × 2–2.5 mm, membranous, tomentose to floccose; teeth 5, erect, 0.8–1.2(–1.5) mm. **Flowers** 2–3 mm; perianth lemon yellow, thinly tomentose; tepals connate proximal ½, monomorphic, broadly oblong to oblong; stamens slightly exserted, 2.5–3.5 mm; filaments sparsely pilose proximally. **Achenes** light brown, 2.5–3 mm, sparsely tomentose.

Flowering May–Aug. Gravelly volcanic soils, mixed grassland, saltbush and sagebrush communities; of conservation concern; 300–400 m; Wash.

Eriogonum codium is a potentially endangered species known from a single site on volcanic bluffs overlooking the Columbia River in Hanford Research National Monument in Benton County. It is worthy of cultivation as a rock-garden plant, although little or no sexual reproduction is known in the natural population. The Umtanum Desert wild buckwheat is a candidate for federal listing and is considered an endangered species by the state of Washington. Much of the population was destroyed in a man-caused fire in 1997. The species is in the Center for Plant Conservation's National Collection of Endangered Plants.

62. Eriogonum ochrocephalum S. Watson in W. H. Brewer et al., Bot. California 2: 480. 1880 [E]

Herbs, matted, scapose, 0.5–4.5 (–5) × 0.5–3 dm, glabrous or glandular, grayish. **Stems** matted or spreading, occasionally with persistent leaf bases, up to ⅕ height of plant; caudex stems matted or absent; aerial flowering stems scapelike, erect, slender, solid, not fistulose, (0.5–)1.5–4.5(–5) dm, glabrous, rarely slightly floccose, or

glandular below inflorescence. **Leaves** basal, fasciculate in terminal tufts, sometimes 1 per node and sheathing up stem 1–4 cm; petiole 1–4(–5) cm, tomentose; blade oblanceolate or lanceolate to elliptic, oblong, or ovate to obovate, (0.6–)1–3.5(–4) × (0.3–)0.5–1.5 cm, densely tomentose abaxially, only slightly less so adaxially, margins plane. **Inflorescences** capitate, 1.5–2.5 cm; branches absent; bracts 3, scalelike, narrowly triangular, 1–2.5 mm. **Peduncles** absent or erect, slender, 0.1–0.5 (–0.6) cm, glabrous. **Involucres** 5–8 per cluster, turbinate to campanulate, 3.5–5.5(–6) × (1.5–)2–4.5(–5) mm, rigid, floccose at least on teeth, occasionally glandular, rarely glabrous; teeth 6–8, erect, 0.5–1 mm. **Flowers** 1.5–3 mm; perianth yellow, glabrous or weakly glandular abaxially; tepals connate proximally, monomorphic, broadly oblong; stamens exserted, 2.5–3.5(–4) mm; filaments pilose proximally. **Achenes** light brown, (1.5–)2–3 mm, glabrous.

Varieties 2 (2 in the flora): w United States.

Variety *ochrocephalum* is occasionally seen in rock gardens, and var. *calcareum* is worthy of cultivation.

1. Scapes glandular at least proximal to inflorescence; ne California, nw Nevada .... 62a. *Eriogonum ochrocephalum* var. *ochrocephalum*
1. Scapes glabrous or, rarely, slightly floccose; sw Idaho, se Oregon, wc Nevada .... 62b. *Eriogonum ochrocephalum* var. *calcareum*

62a. Eriogonum ochrocephalum S. Watson var. **ochrocephalum** • White-woolly wild buckwheat [E]

Leaf blades oblanceolate to elliptic or ovate to obovate, 1–2 × (0.3–)0.5–1(–1.5) cm. **Scapes** 0.5–1.5 (–2) dm, glandular at least proximal to inflorescence. **Involucres** turbinate-campanulate to campanulate, 3–5(–6) × 3–4.5(–5) mm, sparsely to densely floccose, occasionally sparsely glandular. **Flowers** 2–2.5(–3) mm; perianth glabrous or rarely weakly glandular. **Achenes** 1.5–2.5 mm.

Flowering May–Jun. Volcanic or gumbo clay flats and slopes, saltbush and sagebrush communities, juniper woodlands; 1300–1700(–2500) m; Calif., Nev.

Variety *ochrocephalum* occurs in Lassen and Modoc counties, California, and in Carson City, Douglas, Lyon, Pershing, Storey, and Washoe counties, Nevada.

62b. Eriogonum ochrocephalum S. Watson var. **calcareum** (S. Stokes) M. Peck, Leafl. W. Bot. 4: 178. 1945 • Harper wild buckwheat E

Eriogonum ochrocephalum subsp. *calcareum* S. Stokes, Eriogonum, 92. 1936; *E. ochrocephalum* var. *sceptrum* Reveal

Leaf blades lanceolate to elliptic or oblong, (0.6–)1–3.5(–4) × 0.5–1.5 cm. **Scapes** (0.6–)1–4(–5) dm, glabrous or, rarely, slightly floccose. **Involucres** turbinate, 3–5.5(–6) × (1.5–)2.5–3(–4) mm, sparsely floccose or glabrous. **Flowers** 1.5–3 mm; perianth glabrous. **Achenes** 1.5–3 mm.

Flowering May–Jun. Volcanic, diatomaceous or gumbo flats, washes, and slopes, saltbush and sagebrush communities, juniper woodlands; 600–1800 m; Idaho, Nev., Oreg.

Variety *calcareum* is known from southern Baker and Malheur counties, Oregon, and from Elmore, Owyhee, Payette, Twin Falls, and southern Washington counties, Idaho. It is considered "sensitive" in Idaho.

63. Eriogonum novonudum M. Peck, Leafl. W. Bot. 4: 178. 1945 • False naked wild buckwheat E

Herbs, erect, not scapose, (3–)3.5–9 × 0.5–1 dm, glabrous, grayish. **Stems** spreading, without persistent leaf bases, up to 1/5 height of plant; caudex stems absent; aerial flowering stems erect, slender, solid, not fistulose, 1.5–3.5(–4) dm, glabrous. **Leaves** basal, 1 per node; petiole (1–)2–4(–6) cm, tomentose; blade spatulate to narrowly obovate, 1–3(–4) × 0.3–0.8(–1.3) cm, densely white-tomentose on both surfaces, margins plane. **Inflorescences** elongate cymose-umbellate, 8–45(–50) × 5–10 cm; branches dichotomous, glabrous; bracts 3–5, scalelike, triangular, 1–2 mm. **Peduncles** absent. **Involucres** 2–5 per cluster, turbinate, 3–5 × 2–3 mm, rigid, glabrous; teeth 6–8, erect, 0.5–1 mm. **Flowers** (1.5–)2.5–3.5 mm; perianth dull greenish white to pale yellow, glabrous; tepals connate proximally, monomorphic, lanceolate to oblong; stamens exserted, 3–5 mm; filaments pilose proximally. **Achenes** light brown, 2.5–3 mm, glabrous.

Flowering May–Sep. Sandy clay slopes and washes, saltbush and sagebrush communities, juniper woodlands; 700–1200 m; Oreg.

Eriogonum novonudum is known only from a few scattered populations primarily in the Leslie Gulch area

of Malheur County. This tall, openly branched species is easily distinguished from the related *E. ochrocephalum* var. *calcareum*, in which the inflorescence is reduced to a capitate cluster of involucres.

64. Eriogonum pauciflorum Pursh, Fl. Amer. Sept. 2: 735. 1813 • Few-flower wild buckwheat E F

Eriogonum depauperatum Small; *E. gnaphalodes* Bentham; *E. multiceps* Nees; *E. pauciflorum* var. *gnaphalodes* (Bentham) Reveal

Herbs, loosely matted, usually scapose, 0.5–2 × 0.5–3 dm, tomentose, grayish. **Stems** spreading, usually with persistent leaf bases, up to 1/4 height of plant; caudex stems matted; aerial flowering stems scapelike, erect or nearly so, slender, solid, not fistulose, 0.3–2 dm, tomentose. **Leaves** fasciculate in terminal tufts, sometimes 1 per node and sheathing up stem 1–5 cm; petiole 1–5 cm, tomentose to lanate; blade linear-oblanceolate to oblanceolate or narrowly spatulate to elliptic, 1–4 × 0.1–1 cm, grayish- or whitish-tomentose abaxially, less so to subglabrous (or rarely glabrous) adaxially or white-lanate on both surfaces, margins plane. **Inflorescences** capitate, subcapitate, or cymose-umbellate, 1–5 × 1–2 cm; branches absent or dichotomous; bracts 2–6, linear to lanceolate, scalelike to semileaflike, 1.5–20 × 1–5 mm. **Peduncles** absent. **Involucres** 1 per node or 2–5(–7) per cluster, narrowly turbinate, (3.5–)4–5 × (1.5–)2–3 mm, rigid, floccose to tomentose; teeth 5, erect, 0.5–0.8 mm. **Flowers** 2–2.5 mm; perianth whitish brown to rose, pubescent, rarely glabrous; tepals connate proximal 1/3, monomorphic, oblong; stamens exserted, 2.5–3 mm; filaments pilose proximally. **Achenes** light brown to brown, 2–2.5 mm, glabrous.

Flowering May–Sep. Clay to gravelly flats, washes, and slopes, grassland and sagebrush communities, juniper woodlands; 600–1800 m; Man., Sask.; Colo., Mont., Nebr., N.Dak., S.Dak., Wyo.

Eriogonum pauciflorum is the common, matted wild buckwheat on the Great Plains. The species occurs in southern Manitoba and Saskatchewan, northeastern Colorado, Montana, western Nebraska, North Dakota, Nebraska, South Dakota, and Wyoming. Plants from southeastern Wyoming, northeastern Colorado and western Nebraska have somewhat broader leaf blades (spatulate to elliptic rather than linear-oblanceolate to oblanceolate) that are densely tomentose to lanate on both surfaces (rather than loosely tomentose abaxially and less so or glabrous adaxially). Such plants have been distinguished as var. *gnaphalodes*. This expression would be worthy of a place in the rock garden. A hybrid between var. *pauciflorum* and *E. effusum* has been named

E. ×nebraskense Rydberg (see 2. *E. effusum*). The hybrid is known from Weld County, Colorado, Cheyenne, Dawes, Kimball, and Sioux counties, Nebraska, and Converse and Platte counties, Wyoming.

65. Eriogonum villiflorum A. Gray, Proc. Amer. Acad. Arts 8: 630. 1873 • Gray's wild buckwheat E

Herbs, matted, not scapose, 0.01–0.05 × 0.1–0.6 dm, villous to silky-tomentose, grayish. **Stems** matted, with persistent leaf bases, up to ¹/₈ height of plant; caudex stems matted; aerial flowering stems spreading to prostrate, slender, solid, not fistulose, (0.05–)0.15–0.5(–0.8) dm, villous to silky-tomentose. **Leaves** basal, fasciculate in terminal tufts; petiole 0.05–0.1(–1.5) cm, silky-tomentose; blade narrowly elliptic, 0.3–1 × 0.1–0.2(–0.3) cm, villous to silky-tomentose, margins plane. **Inflorescences** subcapitate or cymose-umbellate, 0.3–0.9 × 0.5–1.2 (–2) cm; branches dichotomous, villous to silky-tomentose; bracts 4–8, semileaflike, linear-lanceolate to lanceolate, 2–6(–8) × 0.3–0.7(–1) mm. **Peduncles** absent or erect, slender, 0.05–0.15 cm, villous to silky-tomentose. **Involucres** 1 per node, campanulate, (3.5–)4–5 × (3.5–)4–5 mm, rigid, tomentose; teeth 6–8(–13), spreading, 2–3 mm. **Flowers** 3–4.5 mm; perianth white to rose, pilose; tepals connate proximal ¹/₄, monomorphic, oblong to spatulate; stamens exserted, 3–4 mm; filaments glabrous proximally. **Achenes** light brown, 2.5–3 mm, glabrous.

Flowering Apr–Jul. Gravelly to clayey flats and slopes, sagebrush communities, pinyon-juniper woodlands; 1500–2400 m; Nev., Utah.

Eriogonum villiflorum is probably more common than now documented, the plants being inconspicuous and easily overlooked. It is known from Eureka, Lincoln, northeastern Nye, and White Pine counties in eastern Nevada, and from Beaver, Iron, Juab, Kane, Millard, Sanpete, and Tooele counties in western Utah. The Kane County type location is questionable, as it is well outside the known range of the species today. The sprawling flowering stems with large clusters of involucres and flowers are distinctive. The species is occasionally encountered in cultivation as a rock-garden plant.

66. Eriogonum tumulosum (Barneby) Reveal, Phytologia 23: 173. 1972 • Woodside wild buckwheat E

Eriogonum villiflorum A. Gray var. *tumulosum* Barneby, Leafl. W. Bot. 5: 153. 1949

Herbs, matted, scapose, 0.01–0.1 × 1–4 dm, villous to silky-tomentose, greenish. **Stems** matted, with persistent leaf bases, up to ¹/₅ height of plant; caudex stems matted; aerial flowering stems scapelike, erect, slender, solid, not fistulose, 0.01–0.1 dm, villous to silky-tomentose. **Leaves** basal, fasciculate in terminal tufts, or cauline, 1 per node, sheathing nearly entire length of stem; petiole 0.04–0.07 cm, silky-tomentose; blade oblanceolate to elliptic, 0.3–0.4 × 0.07–0.1 cm, margins plane or slightly thickened. **Inflorescences** capitate, 0.5–0.8(–1) cm; branches absent; bracts 4–5, semileaflike, linear-lanceolate to lanceolate, 2.5–3.5 × 0.3–0.4 mm. **Peduncles** absent. **Involucres** 1 per node, campanulate, 2–4 × (4–)5–8 mm, rigid, tomentose; teeth 7–10, spreading, 1.6–2.2 mm. **Flowers** 3–4 mm; perianth white to rose, pilose; tepals connate proximal ¹/₄, monomorphic, oblong; stamens exserted, 3–4 mm; filaments glabrous. **Achenes** light brown, 2–2.5 mm, glabrous.

Flowering May–Jul. Gravelly to clayey flats and slopes, saltbush and sagebrush communities, pinyon and/or juniper woodlands; 1500–2300 m; Colo., Utah.

Eriogonum tumulosum is infrequent in Duchesne, Emery, eastern Sevier, and Uintah counties, Utah, and in western Moffat County, Colorado. It is often confused with *Parthenium ligulatum* (M. E. Jones) Barneby (Asteraceae); they grow together and both form dense hummock-like mats. Woodside wild buckwheat is seen infrequently in rock gardens.

67. Eriogonum aretioides Barneby, Leafl. W. Bot. 5: 154. 1949 • Widtsoe wild buckwheat C E

Herbs, matted, seemingly scapose, 0.01–0.05 × (0.5–)0.7–1.5 dm, pilose, greenish. **Stems** matted, with persistent leaf bases, up to ¹/₅ height of plant; caudex stems matted; aerial flowering stems absent. **Leaves** basal, fasciculate in terminal tufts, or cauline, 1 per node, sheathing entire length of stem; petiole absent; blade oblanceolate, 0.1–0.4(–0.6) × 0.09–0.12 cm, pilose on both surfaces, margins revolute. **Inflorescences** capitate, 0.4–0.8 cm; branches absent; bracts absent. **Peduncles** absent. **Involucres** 1 per node, campanulate, 2.5–3.5 × 3–4 mm, rigid, villous;

teeth 4, erect to spreading, 1–1.5 mm. **Flowers** 2–2.5 (–3) mm; perianth pale yellow to yellow, pilose; tepals connate proximal ¹/₃, monomorphic, narrowly ovate; stamens exserted, 1.5–2 mm; filaments glabrous. **Achenes** light brown, 1.8–2.3 mm, glabrous.

Flowering May–Jul.   Limestone gravelly soils, sagebrush communities, pinyon-juniper and subalpine conifer woodlands; of conservation concern; 2200–2600 m; Utah.

Eriogonum aretioides is known from a few scattered locations in Garfield and Emery counties. It is restricted largely to the Red Canyon Natural Research Area, where the U.S. Forest Service protects this and other rare species. A sterile collection from the southern loop of the Dirty Devil Loop Road in Emery County (*Heil 1899*, BRY, SJNM) almost certainly is this species, but that remains to be confirmed. Widtsoe wild buckwheat is one of the more elegant members of the genus and a classy addition in any rock garden.

68. Eriogonum shockleyi S. Watson, Proc. Amer. Acad. Arts 18: 194. 1883 • Shockley's wild buckwheat
E F

Eriogonum pulvinatum Small; *E. shockleyi* subsp. *candidum* (M. E. Jones) S. Stokes; *E. shockleyi* subsp. *longilobum* (M. E. Jones) S. Stokes; *E. shockleyi* var. *longilobum* (M. E. Jones) Reveal; *E. shockleyi* S. Watson var. *packardiae* Reveal; *E. villiflorum* A. Gray var. *candidum* M. E. Jones

Herbs, matted, scapose, 0.3–0.5(–0.7) × (0.5–)1–4(–20) dm, floccose to tomentose, greenish or grayish. **Stems** matted, sometimes only seemingly so, with persistent leaf bases, up to ¹/₅ height of plant; caudex stems matted; aerial flowering stems absent or scapelike, erect or nearly so, slender, solid, not fistulose, (0.05–)0.1–0.3 dm, floccose to tomentose. **Leaves** basal, fasciculate in terminal tufts; petiole 0.2–0.5 cm, tomentose to floccose; blade oblanceolate to elliptic or spatulate, (0.2–)0.3–0.8(–1.2) × 0.2–0.4(–0.6) cm, tomentose to floccose, margins plane or slightly thickened. **Inflorescences** capitate, 0.8–2 cm; branches absent; bracts 3–5, scalelike, linear to linear-lanceolate, 1.5–4 × 0.6–1 mm. **Peduncles** absent. **Involucres** 2–4(–6) per cluster, campanulate, (2–)2.5–5(–6) × 3–6(–7) mm, rigid, tomentose; teeth 5–10, erect to spreading, (0.5–)1–3 mm. **Flowers** 2.5–4 mm; perianth white to rose or yellow, densely pilose; tepals connate proximally, monomorphic, oblong to obovate; stamens exserted, 2.5–5 mm; filaments subglabrous or sparsely pilose proximally. **Achenes** light brown to brown, 2.5–3 mm, tomentose.

Flowering May–Aug.   Gravelly or clayey (rarely sandy) flats, washes, and slopes, saltbush, blackbrush,

and sagebrush communities, pinyon-juniper woodlands; (800–)1200–2600 m; Ariz., Calif., Colo., Idaho, Nev., N.Mex., Utah.

Eriogonum shockleyi is widely distributed in northern Arizona, east-central California, western Colorado, southern Idaho, northwestern New Mexico, Nevada, and Utah. On the Colorado Plateau, it has oblanceolate to spatulate leaf blades usually 0.3–1.2 × 0.3–0.6 cm, scapes 1–3 cm, involucres with long (2–3.5 mm) often spreading teeth, and flowers 3–4 mm. These plants have been distinguished as var. *longilobum*. The typical Great Basin expression has elliptic leaf blades 0.3–0.6 × 0.3–0.5 cm, scapes 0.5–2 cm, involucres with short (0.5–2 mm) erect teeth, and flowers 2.5–4 mm. Low, compact, hummock-like plants in southwestern Idaho with elliptic leaf blades 0.1–0.3(–0.35) × 0.1–0.15 cm, flowering stems absent or up to 0.5 cm, and involucres 2.5–3 mm with teeth 0.8–1 mm have been termed var. *packardiae*. These morphologic differences have been shown to be genetically insignificant, however (J. F. Smith and T. A. Bateman 2002). Plants on moving sand dunes at the southeast end of Baking Power Flat in Lincoln County, Nevada, can be up to 2 m across. Further studies may indicate that the various geographic expressions deserve taxonomic recognition.

The Great Basin expression is the food plant of the Bernardino dotted-blue butterfly (*Euphilotes bernardino*). Members of the species are occasionally found in cultivation.

SELECTED REFERENCES Moseley, R. K. and J. L. Reveal. 1995. The Taxonomy and Preliminary Conservation Status of *Eriogonum shockleyi* S. Wats. in Idaho. Boise. [U.S. Bur. Land Managem., Idaho, Techn. Bull. 96-4.] Smith, J. F. and T. A. Bateman. 2002. Genetic differentiation of rare and common varieties of *Eriogonum shockleyi* (Polygonaceae) in Idaho using ISSR variability. W. N. Amer. Naturalist 62: 316–326.

69. Eriogonum lachnogynum Torrey ex Bentham in A. P. de Candolle and A. L. P. P. de Candolle, Prodr. 14: 8. 1856 E F

Herbs, erect or matted, sometimes scapose, 1–3.5 × 1–2 dm or 0.05–0.2 × 0.5–3.5 dm, floccose or silky-tomentose, grayish. **Stems** spreading or matted, occasionally with persistent leaf bases, up to ¹/₄ or more height of plant; caudex stems matted; aerial flowering stems erect or nearly so, slender, solid, not fistulose, 1–2 dm and floccose, or (0.1–)0.02–0.5(–0.65) dm and silky-tomentose. **Leaves** basal, 1 per node or fasciculate in terminal tufts; petiole 0.1–2.5(–3) cm, tomentose to floccose; blade lanceolate or oblanceolate to narrowly elliptic, 1–2.5(–3) × 0.3–0.5 (–0.8) cm, or 0.4–1.2 × 0.1–0.3(–0.35) cm, densely white- or silvery-tomentose on both surfaces, margins plane.

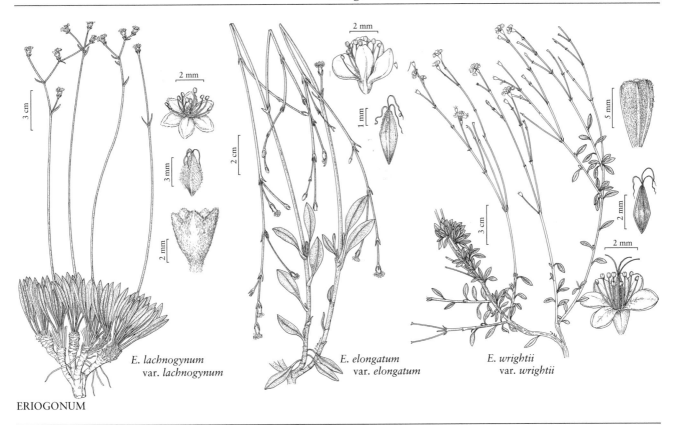

E. lachnogynum
var. lachnogynum

E. elongatum
var. elongatum

E. wrightii
var. wrightii

ERIOGONUM

Inflorescences capitate, subcapitate, umbellate-cymose, or cymose, 2–20 × 2–20 cm, floccose or, if matted, 0.3–1 × 0.3–1 cm; branches open and divided 1–3 times, 2–20 cm, floccose or if matted, (0–)0.3–0.85 cm, silky-tomentose; bracts 3, scalelike, triangular, (0.5–)1–3 mm. **Peduncles** absent or erect, slender, 2–15 mm, floccose. **Involucres** 1 per node or 2–5 per cluster, broadly campanulate, 1.5–4 × (2–)3–6(–8) mm, tomentose; teeth 5, erect to spreading, 1–1.5 mm. **Flowers** 2.5–5(–6) mm; perianth yellow, densely white-pubescent; tepals connate proximal ⅓, monomorphic, lanceolate to broadly lanceolate; stamens exserted, 3.5–5 mm; filaments glabrous. **Achenes** brown to dark brown, 3–4 mm, villous to tomentose.

Varieties 3 (3 in the flora): w United States.

The Navajo or Diné people consider *Eriogonum lachnogynum* to be a "life medicine," making a concoction of shredded roots for the treatment of internal, and some external, ailments. It is often given for diarrhea.

1. Plants erect; inflorescences subcapitate, umbellate-cymose, or cymose; flowering stems 1–2 dm at full anthesis; leaf blades 1–2.5(–3) × 0.3–0.5(–0.8) cm; widespread, not in Arizona . . . . 69a. *Eriogonum lachnogynum* var. *lachnogynum*

1. Plants cespitose mats; inflorescences capitate; flowering stems 0.01–0.65 dm at full anthesis; leaf blades 0.4–1.2 × 0.1–0.35 cm; Arizona, New Mexico.
2. Flowering stems 0.1–0.5(–0.65) dm, longer than leaves . . . . 69b. *Eriogonum lachnogynum* var. *sarahiae*
2. Flowering stems (0.01–)0.02–0.05(–0.12) dm, shorter than or just exceeding leaves . . . . 69c. *Eriogonum lachnogynum* var. *colobum*

69a. **Eriogonum lachnogynum** Torrey ex Bentham var. **lachnogynum** • Woolly-cup wild buckwheat E F

Eriogonum lachnogynum subsp. *tetraneuris* (Small) S. Stokes; *E. tetraneuris* Small

Plants erect, 1–3.5 × 1–2 dm. **Aerial flowering stems** 1–2 dm, floccose. **Leaves:** petiole 0.5–2.5(–3) cm; blade lanceolate or oblanceolate to narrowly elliptic, 1–2.5(–3) × 0.3–0.5(–0.8) cm. **Inflorescences** subcapitate, umbellate-cymose, or cymose, open and divided 1–3 times; branches 2–20 × 2–20 cm. **Involucres** 2–5 per cluster, (2–)3–4 mm. **Flowers** 2.5–5(–6) mm.

Flowering Jul–Oct. Sandy to gravelly (often calcareous) or shaley to clayey or gypsum flats and washes, grassland, creosote bush, or mesquite communities, pinyon and/or juniper woodlands; 900–2400 m; Colo., Kans., N.Mex., Okla., Tex.

Variety *lachnogynum* is widely scattered throughout its range, being mostly infrequent to locally common in eastern Colorado, southwestern Kansas, eastern New Mexico, western Oklahoma, and northern Texas.

69b. Eriogonum lachnogynum Torrey ex Bentham var. **sarahiae** (N. D. Atwood & A. Clifford) Reveal, Harvard Pap. Bot. 9: 178. 2004 • Sarah's wild buckwheat E

Eriogonum sarahiae N. D. Atwood & A. Clifford in S. L. Welsh et al., Utah Fl. ed. 3, 840. 2003

Plants cespitose, hummock-forming and matlike, 0.5–1.5 × 0.5–3 dm. **Aerial flowering stems** 0.1–0.5(–0.65) dm, silky-tomentose. **Leaves:** petiole 0.3–0.6 cm; blade narrowly elliptic, 0.4–1.2 × 0.15–0.35 cm. **Inflorescences** capitate, 0.5–1.5 cm; branches absent. **Involucres** 2–5 per cluster, 2–3.5 mm. **Flowers** 2.5–5 mm.

Flowering May–Jul. Rocky limestone and mesa tops, pinyon-juniper woodlands; 1800–2300 m; Ariz., N.Mex.

Variety *sarahiae* is known from caprock formed by the Owl Rock Member of the Chinle Formation in the Red Valley area of Apache County, Arizona, and McKinley County, New Mexico, where it occurs on windswept, limestone mesa tops on the Navajo Reservation. It occurs also in Petrified Forest National Park to the west. The scapes and capitate inflorescences extend well beyond the leaves. The variety is worthy of cultivation.

69c. Eriogonum lachnogynum Torrey ex Bentham var. **colobum** Reveal & A. Clifford, Phytologia 86: 169. 2004 • Clipped wild buckwheat E

Plants cespitose, flattened and matlike, 0.05–0.2 × 0.5–3.5 dm. **Aerial flowering stems** (0.01–)0.02–0.05(–0.12) dm, silky-tomentose. **Leaves:** petiole 0.1–0.3 cm; blade narrowly elliptic to oblanceolate, 0.4–0.8(–1.2) × 0.1–0.3(–0.35) cm. **Inflorescences** capitate, 0.3–0.85 cm; branches absent. **Involucres** 1 per node, 2–3.5 mm. **Flowers** 3–4.5 mm.

Flowering May–Jul. Rocky limestone flats and slopes, pinyon-juniper woodlands; 2100–2300 m; N.Mex.

Variety *colobum* occurs on a low, windswept, limestone ridgetop north and east of Thoreau, McKinley County. It also occurs to the northeast on gravelly soil

on the western edge of the Rio Grande Gorge in Taos County. The scapes and inflorescences are buried among the leaves, the flowers spotting the tops of the mats with specks of yellow. The variety is worthy of cultivation.

70. Eriogonum havardii S. Watson, Proc. Amer. Acad. Arts 18: 194. 1883 (as havardi) • Havard's wild buckwheat E

Eriogonum leucophyllum Wooton & Standley

Herbs, erect, not scapose, (2–)3–6 × 1.5–4 dm, glabrous, grayish. **Stems** spreading, without persistent leaf bases, up to 1/5 height of plant; caudex stems absent; aerial flowering stems erect, slender, solid, not fistulose, 0.5–2.5 dm, glabrous, tomentose among leaves. **Leaves** basal, 1 per node; petiole 0.5–1.5(–2.5) cm, tomentose; blade oblanceolate to elliptic, 1–3(–5) × 0.2–1(–1.3) cm, densely white- or silvery-tomentose on both surfaces, margins plane. **Inflorescences** cymose, open, divided 3–10 times, 10–40 × 10–30 cm; branches dichotomous, glabrous; bracts 3, scalelike, triangular, 1–2 mm. **Peduncles** erect, slender, 0.5–6 cm. **Involucres** 1 per node, campanulate, 1.5–2.5 × (1.5–)2–3 mm, tomentose; teeth 5, erect, 0.5–0.8 mm. **Flowers** 2.5–3 mm; perianth yellow, densely white-pubescent; tepals connate proximal 1/3, monomorphic, lanceolate; stamens exserted, 2.5–3.5 mm; filaments glabrous. **Achenes** brown, 2–2.5 mm, glabrous. $2n = 40$.

Flowering May–Sep. Gravelly (often calcareous) or shaley to clayey or gypsum flats and outcrops, mixed grassland, creosote, and mesquite communities, juniper woodlands; (400–)800–1800(–2100) m; N.Mex., Tex.

Eriogonum havardii is infrequent in southeastern New Mexico (Chaves, Eddy, Lea, Lincoln, Otero, and Socorro counties) and western Texas (Brewster, Culberson, El Paso, Hudspeth, Pecos, Presidio, Terrell, Val Verde, and Winkler counties). The leaf blades are demarcated by depressed lines along their entire length, appearing as a series of folds. The bright yellow of the flowers is seen only by looking at the adaxial surface of the tepals, the abaxial surface being wholly obscured by dense white hairs. The species is worthy of cultivation.

71. Eriogonum elongatum Bentham, Boy. Voy. Sulphur, 45. 1844 F

Varieties 3 (1 in the flora): California; nw Mexico.

All three varieties of *Eriogonum elongatum* (var. *areorivum* Reveal, var. *elongatum*, and var. *vollmeri* (Wiggins) Reveal) occur in Baja California, Mexico, where they typically flower during the fall and winter.

71a. Eriogonum elongatum Bentham var. **elongatum**
• Long-stem wild buckwheat [F]

Herbs, erect, not scapose, 6–12 (–18) × 5–15(–20) dm, tomentose to floccose, grayish. **Stems** spreading to erect, without persistent leaf bases, up to ¹/₅ height of plant; caudex stems absent; aerial flowering stems erect, slender, solid, not fistulose, 1–4 dm, tomentose to floccose. **Leaves** cauline, fasciculate or 1 per node; petiole 0.5–2 cm, tomentose to floccose; blade narrowly oblong to narrowly ovate, 1–3 × 0.5–2 cm, densely tomentose on both surfaces, margins plane, sometimes crisped. **Inflorescences** virgate, with involucres racemosely disposed at tips, 15–120 × 10–50 cm; branches dichotomous, tomentose to floccose; bracts 3, scalelike, triangular, 1–5 mm. **Peduncles** absent. **Involucres** 1 per node, narrowly turbinate, (4–)6–7 × 2–4 mm, tomentose to floccose; teeth 5, erect, 0.3–0.6 mm. **Flowers** 2.5–3 mm; perianth white, glabrous; tepals connate proximal ¹/₄, monomorphic, oblong to obovate; stamens exserted, 2–4 mm; filaments glabrous or sparsely villous proximally. **Achenes** dark brown, 2–3 mm, glabrous. $2n = 34$.

Flowering (Jan–)Jul–Nov(–Dec). Sandy to loamy flats and slopes, grassland, coastal sage, and chaparral communities, oak and conifer woodlands; 60–1900 m; Calif.; Mexico (Baja California).

Variety *elongatum* occurs in the Coast and Transverse Ranges of southeastern California from Monterey and San Benito counties southward to Mexico. The elongate, upright flowering stems and inflorescence branches often persist for several years with new stems produced annually. The long, "leggy" habit is not particularly attractive in the garden even though the variety is occasionally grown as an ornamental. The long-stem wild buckwheat is heavily visited by honey bees in some areas and is the host plant of the Gorgon copper butterfly (*Gaeides gorgon*). J. B. Romero (1954) reported that the variety was used by Native Americans to treat high blood pressure, and E. W. Voegelin (1938) mentioned that the stems were roasted and the resulting juice was used as a chewing gum. Care should be taken not to confuse fragmentary specimens of the annual *Eriogonum roseum* with the perennial *E. elongatum*.

72. Eriogonum wrightii Torrey ex Bentham in A. P. de Candolle and A. L. P. P. de Candolle, Prodr. 14: 15. 1856 • Bastard-sage [F]

Eriogonum trachygonum Torrey ex Bentham subsp. *wrightii* (Torrey ex Bentham) S. Stokes

Shrubs, subshrubs, or herbs, rarely scapose, (1–)1.5–10 × 1–15(–18) dm or, if matted, 0.1–2.5(–3) × 0.5–3(–5) dm, , lanate to thinly tomentose, or glabrous, grayish to greenish or reddish. **Stems** spreading to erect, with or without persistent leaf bases, up to ¹/₂ or more height of plant; caudex stems absent or spreading, occasionally matted; aerial flowering stems erect to spreading, stout to slender, solid, not fistulose, (0.1–)0.5–4(–6) dm, tomentose, floccose, or glabrous. **Leaves** basal and fasciculate in terminal tufts, or cauline and fasciculate, occasionally 1 per node; petiole 0.02–0.5(–1) cm, tomentose to floccose; blade oblanceolate to broadly elliptic, 0.1–3 × 0.1–1 cm, tomentose to floccose, sometimes subglabrous or glabrous and green adaxially, margins plane, sometimes revolute. **Inflorescences** virgate or cymose with involucres disposed at tips racemosely arranged involucres, rarely capitate, (1–)5–20 × (1–)10–40 cm; branches dichotomous, tomentose, floccose, or glabrous; bracts 3, triangular, scalelike, 0.5–3.5 mm. **Peduncles** absent. **Involucres** 1 per node, turbinate to narrowly campanulate, (0.7–)1–4 × 1–2.5 mm, tomentose, floccose, or glabrous; teeth 5, erect, 0.3–1 mm. **Flowers** 1–4 mm; perianth white to pink or rose, glabrous; tepals connate proximal ¹/₄, monomorphic, obovate; stamens exserted, 1.5–4 mm; filaments glabrous or sparsely pilose proximally. **Achenes** light brown to brown, (1–)1.5–3 mm, glabrous.

Varieties 9 (6 in the flora): w North America, including nw Mexico.

Eriogonum wrightii is subdivided into several varieties most of which are distinct, although a few have rather indistinct boundaries. Some of the variation has yet to be fully resolved, especially in the var. *nodosum* complex where, at least in Mexico, one additional expression remains to be named. A clear distinction between var. *subscaposum* and *E. kennedyi* is not possible in southern California (see discussion below). Nearly all of the varieties of *E. wrightii* are in cultivation, although the most elegant (var. *olanchense*) has yet to be so honored.

Few ethnobotanical uses of bastard-sage are reported in the literature. L. C. Wyman and S. K. Harris (1951) noted that the Kayenta Navajo use it (var. *wrightii*) as an emetic, while M. L. Zigmond (1981) stated that the Kawaiisu used the pounded seeds (probably of var. *subscaposum*) in a beverage or as a dry meal. Members of the species are food plants for the rare Rita dotted-blue butterfly (*Euphilotes rita*), the Pacific dotted-blue (*E. enoptes*), the veined blue (*Plebeius neurona*), and the Mormon metalmark (*Apodemia mormo mormo*).

1. Plants loosely to compactly matted or herbs.
   2. Plants 0.5–2.5(–3) dm; leaf blades 0.5–1(–1.2) cm; involucres 1.5–4 mm; e and s California, wc Nevada . . . . 72e. *Eriogonum wrightii* var. *subscaposum*
   2. Plants 0.1–0.3(–0.6) dm; leaf blades 0.1–0.25 cm; involucres 0.8–1.7(–2) mm; se California . . . . 72f. *Eriogonum wrightii* var. *olanchense*
1. Plants shrubs or subshrubs.
   3. Flowering stems and branches grayish, lanate to densely tomentose; sw Arizona, se California . . . . 72d. *Eriogonum wrightii* var. *nodosum*
   3. Flowering stems and inflorescence branches whitish, reddish, or greenish, tomentose to floccose; widespread.
      4. Petiole bases forming distinct ring around stem; leaf blades 0.2–0.6(–1) × 0.1–0.3 (–0.4) cm; s California . . . . 72c. *Eriogonum wrightii* var. *membranaceum*
      4. Petiole bases not forming distinct ring around stem; leaf blades 0.5–3 × 0.2–1 cm; sw United States, Mexico.
         5. Leaf blades 0.5–1.5 × 0.2–0.5(–0.7) cm; involucres 2–2.5 mm; flowers 2.5–3.5 mm; se California to w Texas . . . . 72a. *Eriogonum wrightii* var. *wrightii*
         5. Leaf blades 1.5–3 × 0.5–1 cm; involucres 3–4 mm; flowers 3–4 mm; c and nw California . . . . 72b. *Eriogonum wrightii* var. *trachygonum*

72a. Eriogonum wrightii Torrey ex Bentham var. **wrightii** • Wright's bastard-sage ⬚F

Eriogonum trachygonum Torrey ex Bentham subsp. *glomerulum* S. Stokes; *E. trachygonum* subsp. *helianthemifolium* (Bentham) S. Stokes; *E. wrightii* subsp. *glomerulum* (S. Stokes) S. Stokes; *E. wrightii* subsp. *helianthemifolium* (Bentham) S. Stokes

Subshrubs or shrubs, (1–)1.5–5 (–7.5) × 1–12(–18) dm, mostly tomentose. **Leaf blade** oblanceolate to elliptic, 0.5–1.5 × 0.2–0.5(–0.7) cm, margins usually plane. **Inflorescences** virgate or cymose; branches slender. **Involucres** 2–2.5 mm. **Flowers** 2.5–3.5 mm; perianth white to pink or rose. **Achenes** 2.5–3 mm.

Flowering Jul–Oct. Gravelly to rocky (often calcareous) slopes, mixed grassland, saltbush, blackbrush, creosote bush, and mesquite communities, oak or pinyon and/or juniper woodlands; (300–)900–2200 m; Ariz., Calif., Nev., N.Mex., Tex., Utah; Mexico (Chuhuahua, Coahuila, Durango, San Luis Potosi, Sonora, Zacatecas).

Variety *wrightii* is remarkably homogeneous throughout its extensive range in Arizona, southeastern California, southern Nevada, New Mexico, western Texas, and southeastern Utah. It is occasionally found in cultivation.

D. E. Moerman (1986) indicated that this variety was used by the Navajo-Keyenta as an emetic.

72b. Eriogonum wrightii Torrey ex Bentham var. **trachygonum** (Torrey ex Bentham) Jepson, Fl. W. Calif., 154. 1901 • Rough-node bastard-sage ⬚E

Eriogonum trachygonum Torrey ex Bentham in A. P. de Candolle and A. L. P. P. de Candolle, Prodr. 14: 15. 1856; *E. wrightii* subsp. *trachygonum* (Torrey ex Bentham) S. Stokes

Subshrubs, 1.5–4 × 1–5 dm, mostly densely tomentose. **Leaf blades** elliptic, 1.5–3 × 0.5–1 cm, margins usually plane. **Inflorescences** virgate or cymose; branches stout. **Involucres** 3–4 mm. **Flowers** 3–4 mm; perianth white to pink or rose. **Achenes** 2.5–3 mm.

Flowering Jul–Oct. Gravelly to rocky flats and slopes, mixed grassland and chaparral communities, oak and conifer woodlands; 40–800 m; Calif.

Variety *trachygonum* is the common expression of the species in the Coast Ranges of California (Alameda, Butte, Lake, Merced, Monterey, San Benito, Santa Clara, Shasta, Solano, Stanislaus, Tehama, and Yolo counties). It is found also in scattered locations along the foothills of the Sierra Nevada (mainly Kern, Merced, Nevada, Stanislaus, and Tuolumne counties). The distinction between var. *trachygonum* and var. *subscaposum* is imprecise. The difficulty arises in the foothills of the Sierra Nevada, where var. *trachygonum* is reduced in stature and resembles the larger, mid-elevation expressions of var. *subscaposum*. The problem is less pronounced in the Coast Ranges, but there one is uncertain whether the high-elevation plants are small forms of var. *trachygonum* or disjunct populations of var. *subscaposum*.

72c. Eriogonum wrightii Torrey ex Bentham var. **membranaceum** S. Stokes ex Jepson, Fl. Calif. 1: 416. 1914 • Ringed-stem bastard-sage

Eriogonum trachygonum Torrey ex Bentham subsp. *membranaceum* (S. Stokes ex Jepson) S. Stokes; *E. wrightii* subsp. *membranaceum* (S. Stokes ex Jepson) S. Stokes

Subshrubs, 2–4(–5) × 3–6 dm, mostly thinly tomentose. **Leaves:** petiole bases forming distinct ring around stem; blade elliptic, 0.2–0.6(–1) × 0.1–0.3(–0.4) cm. **Inflorescences** virgate or cymose; branches usually slender. **Involucres** 2–3 mm. **Flowers** 3–4 mm; perianth white to pink or rose. **Achenes** 2.5–3 mm.

Flowering Jul–Oct. Gravelly to rocky soils, chaparral communities, oak and conifer woodlands; 300–2200 m; Calif.; Mexico (Baja California).

Variety *membranaceum* is a basically Mexican taxon that barely enters the flora area (southwesternmost Kern, western Imperial, Los Angeles, western Riverside, southeastern San Bernardino, San Diego, and southern Ventura counties). The ring-like petiole bases that surround the node are distinctive, being seen otherwise only in some of the perennial species of *Chorizanthe* in Chile. Ringed-stem bastard-sage is occasionally planted as an ornamental; it is particularly attractive to honey bees. The variety is related to both var. *dentatum* (S. Stokes) Reveal and var. *taxifolium* (Greene) Parish of Baja California, Mexico.

72d. Eriogonum wrightii Torrey ex Bentham var. **nodosum** (Small) Reveal, Leafl. W. Bot. 10: 334. 1966 • Knot-stem bastard-sage

Eriogonum nodosum Small, Bull. Torrey Bot. Club 25: 49. 1898; *E. pringlei* J. M. Coulter & E. M. Fischer; *E. trachygonum* Torrey ex Bentham subsp. *pringlei* (J. M. Coulter & E. M. Fischer) S. Stokes; *E. wrightii* subsp. *pringlei* (J. M. Coulter & E. M. Fischer) S. Stokes; *E. wrightii* var. *pringlei* (J. M. Coulter & E. M. Fischer) Reveal

Shrubs, 3–10 × (3–)5–15 dm, densely tomentose to lanate. **Leaf blades** oblanceolate to narrowly elliptic, 0.8–1.2 × 0.25–0.5(–0.9) cm. **Inflorescences** open; branches slender or stout. **Involucres** 0.7–2.5 mm. **Flowers** 1–4 mm; perianth white to pink or rose. **Achenes** 1–3 mm.

Flowering Aug–Feb. Gravelly to rocky soils, mixed grassland, saltbush, creosote bush, and mesquite communities, scattered juniper woodlands; 100–1600 m; Ariz., Calif.; Mexico (Baja California, Sonora).

In the strict sense, var. *pringlei* differs from var. *nodosum* by its smaller (1–1.5 versus 3–4 mm) flowers, involucres (0.7–2 versus 1.5–2.5 mm), and achenes (1–1.5 versus 2–3 mm); also, the inflorescence branches of var. *pringlei* are more numerous and much more slender than those of var. *nodosum*. In the flora area, var. *pringlei* seems to be restricted to southwestern Arizona (La Paz, Maricopa, Pima, and Pinal counties) and is rare. It occurs also in northwestern Sonora, Mexico, where it is locally common. Variety *nodosum* is infrequent in California (Imperial, Riverside, San Bernardino, and San Diego counties) and northern Baja California, Mexico; it appears to be the common expression of the species in Yuma County, Arizona, rather than var. *pringlei*. Until adequate fieldwork is done on this complex, the uncertainty about var. *nodosum* (s. lat.) must remain. The varietal epithet *nodosum* has priority over *pringlei* by virtue of the autonym established by the publication of *E. nodosum* var. *jaegeri* Munz & I. M. Johnston.

72e. Eriogonum wrightii Torrey ex Bentham var. **subscaposum** S. Watson in W. H. Brewer et al., Bot. California 2: 29. 1880 (as subscaposa) • Short-stemmed bastard-sage E

Eriogonum kennedyi Porter ex S. Watson subsp. *pinorum* S. Stokes; *E. trachygonum* Torrey ex Bentham subsp. *subscaposum* (S. Watson) S. Stokes; *E. wrightii* var. *curvatum* (Small) Munz; *E. wrightii* subsp. *subscaposum* (S. Watson) S. Stokes

Herbs, loosely matted, 0.5–2.5 (–3) × 1–3(–5) dm, tomentose or glabrous. **Leaves:** blade oblanceolate to elliptic, 0.5–1(–1.2) × 0.2–0.4(–0.5) cm. **Inflorescences** virgate or cymose; branches dichotomous, mostly slender. **Involucres** 1.5–4 mm. **Flowers** 2–3 mm; perianth white to pink. **Achenes** 2–2.5 mm. $2n = 34$.

Flowering Jun–Sep. Gravelly to rocky, often volcanic or granitic flats, washes, slopes, and outcrops, sagebrush and chaparral communities, oak, pinyon-juniper, and montane conifer woodlands; 200–3400 m; Calif., Nev.

Variety *subscaposum* occurs mainly in the Transverse Ranges and the Sierra Nevada of California, with some populations in the desert ranges just to the east in west-central Nevada. The distinction between var. *subscaposum* and var. *trachygonum* is not always sharp, and some plants along the foothills of the Sierra Nevada may well be the latter. On the Transverse Ranges (and especially Mt. Pinos), var. *subscaposum* appears to merge with *E. kennedyi* var. *kennedyi*.

72f. Eriogonum wrightii Torrey ex Bentham var. **olanchense** (J. T. Howell) Reveal in P. A. Munz, Suppl. Calif. Fl., 63. 1968 • Olancha Peak bastard-sage E

Eriogonum kennedyi Porter ex S. Watson var. *olanchense* J. T. Howell, Leafl. W. Bot. 6: 151. 1951

Herbs, compactly matted, 0.1–0.3(–0.6) × 0.5–3 dm, mostly thinly tomentose. **Leaves:** blade elliptic, 0.1–0.25 × 0.06–0.12 cm. **Inflorescences** capitate or nearly so; branches absent or slender. **Involucres** (0.8–)1–1.7 (–2) mm. **Flowers** (1.5–)2–2.5 mm; perianth white to pink. **Achenes** 1.5–2 mm.

Flowering Jul–Aug. Gravelly to rocky granitic talus slopes; 3500–3600 m; Calif.

Variety *olanchense* is known only from Olancha Peak, Tulare County. It is worthy of cultivation as a rock-garden plant. In Baja California, var. *oresbium* Reveal is similar; that taxon is restricted to the Sierra Juárez and Sierra San Pedro Martír, but already has made its appearance as a cultivated rock-garden plant.

73. Eriogonum kennedyi Porter ex S. Watson, Proc. Amer. Acad. Arts 12: 263. 1877 E

Herbs, scapose, matted, 0.4–1.5 × 1–4 dm, glabrous or sparsely floccose to tomentose, grayish or reddish. **Stems** matted, occasionally with persistent leaf bases, up to 1/5 height of plant; caudex stems matted; aerial flowering stems scapelike, erect or nearly so, slender, solid, not fistulose, (0.5–) 1.5–4.5(–5) dm, glabrous or sparsely floccose to tomentose. **Leaves** basal, fasciculate in terminal tufts, sometimes 1 per node and sheathing up stem 1–2 cm; petiole 0.05–1 cm, tomentose; blade oblanceolate, elliptic, or oblong, 0.2–1(–1.2) × 0.05–0.4 cm, grayish-, brownish-, white-, or reddish-white-tomentose, usually on both surfaces, margins plane or revolute. **Inflorescences** capitate, 0.4–1 cm; branches absent; bracts 4–7, scalelike, triangular, 0.5–2.5(–3) mm. **Peduncles** absent. **Involucres** 3–7 per cluster, turbinate to turbinate-campanulate, 1.5–4 × 1–3.5 mm, rigid, glabrous or tomentose; teeth 5, erect to spreading, 0.4–1.5 mm.

Flowers 1.5–4 mm, glabrous; perianth white to pink or rose; tepals connate proximal 1/4, monomorphic, oblanceolate to elliptic or obovate to nearly oval; stamens exserted, 1.5–4 mm; filaments glabrous or sparsely pubescent proximally. **Achenes** light brown to brown, 1.8–4 mm, glabrous.

Varieties 5 (5 in the flora): w United States.

The capitate inflorescence and basal leaves of *Eriogonum kennedyi* distinguish it from the closely related *E. wrightii*. At higher elevations in the Transverse Ranges of California, the distinction between *E. kennedyi* var. *kennedyi* and *E. wrightii* var. *subscaposum* is not always clear, especially near the summit of Mt. Pinos. At lower elevations in the San Bernardino Mountains, in more arid habitats, the distinction between *E. kennedyi* var. *austromontanum* and *E. wrightii* var. *subscaposum* also is blurred. For the most part this intergradation seems to be associated with a reduction in the length of the inflorescence in depauperate expressions of var. *subscaposum* and not a result of hybridization. The extreme reduction seen in *E. wrightii* var. *olanchense* exemplifies the general trend in the California expressions of that species toward depauperate forms. It is felt that *E. kennedyi* itself is an established segregate that owes its origin to *E. wrightii*. Members of the species are food plants for Bauer's dotted-blue butterfly (*Euphilotes baueri*) and the Ord Mountain metalmark butterfly (*Apodemia mormo dialeuca*).

1. Leaf tomentum white; plants forming dense mats; desert ranges, ec California, wc Nevada . . . . 73e. *Eriogonum kennedyi* var. *purpusii*
1. Leaf tomentum grayish or reddish- to brownish-white; plants usually forming loose, open mats; southern Sierra Nevada and Transverse Ranges of s California.
  2. Leaf blades oblong, 0.3–0.5 × 0.1–0.4 cm . . . . 73d. *Eriogonum kennedyi* var. *pinicola*
  2. Leaf blades oblanceolate to elliptic, 0.2–1.2 × 0.05–0.2 cm.
    3. Scapes 0.5–2(–3) cm . . . . 73c. *Eriogonum kennedyi* var. *alpigenum*
    3. Scapes 4–15 cm.
      4. Leaf blades 0.2–0.4(–0.5) × 0.05–0.15 (–0.2) cm; scapes glabrous; involucres 1.5–2.5 mm . . . . . 73a. *Eriogonum kennedyi* var. *kennedyi*
      4. Leaf blades (0.4–)0.6–1(–1.2) × 0.1–0.2 mm; scapes sparsely tomentose to floccose; involucres 2.5–4 mm . . . . 73b. *Eriogonum kennedyi* var. *austromontanum*

73a. Eriogonum kennedyi Porter ex S. Watson var. **kennedyi** • Kennedy's wild buckwheat [E]

Herbs, loosely matted, 0.4–1.2 × 1–3 dm. **Leaf blades** oblanceolate to elliptic, 0.2–0.4(–0.5) × 0.05–0.15(–0.2) cm, grayish- to brownish-white-tomentose, margins occasionally revolute. **Scapes** 4–12 cm, glabrous. **Involucres** 1.5–2.5 mm, glabrous or sparsely tomentose. **Flowers** 1.5–2.5 mm. **Achenes** 2–2.5 mm.

Flowering Apr–Jul. Gravelly to rocky flats and slopes, sagebrush and montane conifer woodlands; 1700–2700 m; Calif.

Variety *kennedyi* is known from the Mt. Pinos-Lockwood Valley area of Ventura County and the eastern San Bernardino Mountains of San Bernardino County. It is not always clearly distinct morphologically from var. *austromontanum* in the latter location. However, Kennedy's wild buckwheat flowers earlier than the San Bernardino Mountains wild buckwheat, and the two can be distinguished on that basis.

73b. Eriogonum kennedyi Porter ex S. Watson var. **austromontanum** Munz & I. M. Johnston, Bull. Torrey Bot. Club 51: 295. 1924 • Southern mountain wild buckwheat [E]

Eriogonum kennedyi subsp. *austromontanum* (Munz & I. M. Johnston) S. Stokes

Herbs, loosely matted, 0.8–1.5 × 1.5–3.5 dm. **Leaf blades** oblanceolate to elliptic, (0.4–)0.6–1(–1.2) × 0.1–0.2 cm, grayish-white-tomentose, margins not revolute. **Scapes** 8–15 cm, sparsely tomentose to floccose. **Involucres** 2.5–4 mm, tomentose. **Flowers** 2–3 mm. **Achenes** 3.5–4 mm.

Flowering Jun–Aug. Gravelly flats and slopes, sagebrush and montane conifer woodlands; 2000–2200 m; Calif.

Variety *austromontanum* is rare but occasionally locally common and known only from the Bear Valley area of the San Bernardino Mountains in San Bernardino County. Its rarity is due to local habitat destruction around Bear and Baldwin lakes, and to people gathering the plants for dried miniature displays for model railroads, doll houses, and Christmas decorations. It

appears on the U.S. Fish and Wildlife Service list of threatened species. The variety is sometimes confused with *E. wrightii* var. *subscaposum*, which has a distinctly branched inflorescence. Both of these taxa can grow in mixed populations with var. *kennedyi*.

73c. Eriogonum kennedyi Porter ex S. Watson var. **alpigenum** (Munz & I. M. Johnston) Munz & I. M. Johnston in P. A. Munz, Man. S. Calif. Bot., 597. 1935 • San Gorgonio wild buckwheat [E]

Eriogonum kennedyi forma *alpigenum* Munz & I. M. Johnston, Bull. Torrey Bot. Club 51: 296. 1924; *E. kennedyi* subsp. *alpigenum* (Munz & I. M. Johnston) Munz

Herbs, loosely matted, 0.2–0.5 × 1–4 dm. **Leaf blades** oblanceolate to elliptic, 0.2–0.4 × 0.07–0.15 cm, reddish- or brownish-white-tomentose, margins often revolute. **Scapes** 0.5–3 cm, tomentose. **Involucres** 1.5–2 mm, glabrous or tomentose. **Flowers** (1.5–)2–2.5 mm. **Achenes** 1.8–2 mm.

Flowering Jul–Aug. Gravelly to rocky slopes and ridges, sagebrush communities, alpine conifer woodlands; 2500–3500 m; Calif.

Variety *alpigenum* is rare and localized, restricted to the higher mountains of the San Gabriel and San Bernardino mountains of Los Angeles and San Bernardino counties. In the Mt. Williamson and Mt. Baden-Powell area of the San Gabriel Mountains, it occurs from 2500–2800 m, whereas on Mt. San Gorgonio, it is found from 3350–3500 m.

73d. Eriogonum kennedyi Porter ex S. Watson var. **pinicola** Reveal in P. A. Munz, Suppl. Calif. Fl., 68. 1968 • Sweet Ridge wild buckwheat [E]

Herbs, 0.5–1.3 × 1–3 dm. **Leaf blade** oblong, 0.3–0.5 × 0.1–0.4 cm, grayish- or reddish-white-tomentose, margins not revolute. **Scapes** 5–13 cm, glabrous. **Involucres** 2.5–3.5 mm, sparsely tomentose. **Flowers** 2–4 mm. **Achenes** 2.5–3 mm.

Flowering May–Jun. Gravelly to rocky ridges, chaparral communities, montane conifer woodlands; 1700–1800 m; Calif.

Variety *pinicola* is known from two small populations in the mountains east of Tehachapi, Kern County.

73e. Eriogonum kennedyi Porter ex S. Watson var. **purpusii** (Brandegee) Reveal in P. A. Munz, Suppl. Calif. Fl., 67. 1968 • Purpus wild buckwheat E

Eriogonum purpusii Brandegee, Bot. Gaz. 27: 457. 1899 (as purpusi); *E. kennedyi* subsp. *purpusii* (Brandegee) Munz

Herbs, densely matted, 0.4–1.2 × 1–3 dm. **Leaf blades** elliptic, (0.25–)0.3–0.6 × 0.15–0.35 cm, white-tomentose on both surfaces, margins not revolute. **Scapes** 0.4–1 cm, glabrous or rarely floccose. **Involucres** 1.5–2 mm, glabrous or sparsely tomentose. **Flowers** 2–2.5 mm. **Achenes** 2.5–3 mm.

Flowering May–Jul. Sandy flats and slopes, mixed sagebrush and grassland communities, pinyon, juniper, and Jeffrey pine woodlands; 1500–2500 m; Calif., Nev.

Variety *purpusii* differs from other varieties of the species in its distribution disjunct to the north and east, and its more arid, desert-like habitat. It is common in Inyo and southern Mono counties of California, with scattered populations known near Inyokern in Kern County. In Nevada, it is known only from the Orchard Springs area of Esmeralda County. The bright white tomentum of the leaves is distinctive, and the low mats are sometimes used for dried decorations. Purpus wild buckwheat is an excellent rock-garden plant.

74. Eriogonum butterworthianum J. T. Howell, Leafl. W. Bot. 9: 153. 1961 • Butterworth's wild buckwheat C E

Subshrubs, rarely scapose, 1–3 × 1–4 dm, tomentose, grayish to reddish. **Stems** slightly spreading to erect, without persistent leaf bases, up to ¼ height of plant; caudex stems absent; aerial flowering stems erect, slender, solid, not fistulose, 0.1–0.3 dm, tomentose. **Leaves** sheathing up stem 0.5–1.5 cm, 1 per node; petiole 0.1–0.3 cm, tomentose; blade linear to narrowly elliptic, (0.5–)1–2 × 0.1–0.4 cm, grayish- to reddish-brown-tomentose, margins revolute. **Inflorescences** cymose-umbellate to cymose, rarely capitate, often with involucres racemosely disposed at tips, usually 1–2 × 1–2 cm; branches dichotomous, rarely scapelike, tomentose; bracts 3, scalelike, narrowly triangular, (1–)2–4(–5) mm. **Peduncles** absent. **Involucres** 1 per node, turbinate to turbinate-campanulate, (4–)5–6 × 3–4 mm, tomentose; teeth 5, erect, 1–2 mm. **Flowers** (3–)4–5 mm; perianth ochroleucous to rose, glabrous; tepals connate proximal ¼, monomorphic, obovate; stamens exserted, 5–6 mm; filaments pubescent proximally. **Achenes** brown, 3–3.5 mm, glabrous.

Flowering Jun–Sep. Sandstone outcrops, chaparral communities, oak and conifer woodlands; of conservation concern; 600–700 m; Calif.

Eriogonum butterworthianum is a rare and localized species that grows in the cracks of Vaqueros sandstone near The Indians in the Santa Lucia Range of Monterey County. It is considered a "sensitive species" by the U.S. Forest Service. Its relationship to other species in subg. *Eucycla* is uncertain. Butterworth's wild buckwheat would make an elegant addition to the garden. Given its rarity, great care must be taken even in collecting seeds.

75. Eriogonum rupinum Reveal, Aliso 7: 226. 1970 • Wyman Creek wild buckwheat E

Herbs, erect, 3–5 × 0.5–1 dm, tomentose, brownish. **Stems** slightly spreading to erect, without persistent leaf bases, up to ⅕ height of plant; caudex stems absent; aerial flowering stems erect, slender, solid, not fistulose, 1.5–2.5 dm, tomentose. **Leaves** basal, 1 per node; petiole (2–)3–7 cm, tomentose; blade elliptic to oblong, (2–)2.5–4.5(–5) × 1.5–3.5(–4) cm, thinly tomentose abaxially, less so to floccose and greenish adaxially, margins plane. **Inflorescences** cymose, with involucres racemosely disposed at tips, 15–25 × 5–15 cm; branches dichotomous, tomentose; bracts 3, scalelike, triangular, 2–4 mm. **Peduncles** absent. **Involucres** 1 per node, turbinate-campanulate, 3–4 × 3–4 mm, tomentose; teeth 5, erect, 0.1–0.3 mm. **Flowers** 2.5–4.5(–5) mm; perianth creamy white or cream, glabrous; tepals connate proximal ¼, monomorphic, broadly oblanceolate; stamens exserted, 2.5–5 mm; filaments pilose proximally. **Achenes** light brown, 2–4 mm, glabrous. $2n = 40$.

Flowering Jul–Sep. Gravelly flats, slopes, and ridges, saltbush and sagebrush communities, pinyon-juniper, oak, and montane conifer woodlands; 1700–3100 (–3500) m; Calif., Nev.

Eriogonum rupinum is the most widespread of the three robust species related to *E. racemosum*. The Wyman Creek wild buckwheat occurs from the White and Inyo mountains of Mono and Inyo counties, California, eastward in scattered mountain ranges across the central Great Basin of Nevada (Esmeralda, Lander, Lincoln, Mineral, and Nye counties). The bulk of the populations are concentrated in Nye County, Nevada. The Lincoln County location, based on an 1898 collection (*Purpus 6263*, DS, SD), requires confirmation.

76. Eriogonum panamintense C. V. Morton, J. Wash. Acad. Sci. 25: 308. 1935 • Panamint Mountain wild buckwheat E

Eriogonum racemosum Nuttall var. *desertorum* S. Stokes; *E. reliquum* S. Stokes

Herbs, erect, 1.5–3 × 1–4 dm, tomentose, brownish. **Stems** slightly spreading to erect, without persistent leaf bases, up to ¹/₅ height of plant; caudex stems absent; aerial flowering stems erect, slender, solid, not fistulose, 0.5–1.5 dm, tomentose. **Leaves** basal, 1 per node; petiole 1–5 cm, tomentose; blade elliptic to ovate or obovate, 1.5–4 × 1–2.5 cm, white-tomentose abaxially, less so and often greenish adaxially, margins plane. **Inflorescences** cymose with involucres racemosely disposed at tips, (10–)12–20 × 1–5 cm; branches dichotomous, tomentose; bracts 3, scalelike, narrowly triangular, and 1–4 mm, or leaflike, elliptic to oval, and 5–15 × 5–15 mm. **Peduncles** absent. **Involucres** 1 per node, turbinate, 3–5 × 2–4 mm, tomentose; teeth 5, erect, 0.4–0.8 mm. **Flowers** 3.5–5 mm; perianth white to whitish brown, glabrous; tepals connate proximal ¹/₄, monomorphic, oblanceolate; stamens exserted, 3–7 mm; filaments pilose proximally. **Achenes** light brown, 2.5–3 mm, glabrous.

Flowering May–Oct. Rocky to gravelly slopes and ridges, sagebrush and mountain mahogany communities, pinyon-juniper and montane conifer woodlands; (1600–)1800–2900 m; Ariz., Calif., Nev.

Eriogonum panamintense is widely scattered in the desert ranges from the southern tip of the White Mountains in Inyo County to the New York Mountains of San Bernardino County, California, eastward across Nevada in the Magruder and Silver Peak (Esmeralda County), Grapevine (Nye County), and Spring (Charleston) mountains (Clark County) of Nevada, on the Hualapai Mountains south of Kingman, and on Mt. Bangs south of Littlefield, Mohave County, Arizona. The Panamint Mountain wild buckwheat typically grows at elevations in California higher than does *E. rupinum*, but in Nevada their elevation ranges overlap, as in the Magruder Mountains and near Westgard Pass in the White Mountains.

77. Eriogonum mensicola S. Stokes, Leafl. W. Bot. 3: 16. 1941 • Scale-bract wild buckwheat E

Eriogonum panamintense C. V. Morton var. *mensicola* (S. Stokes) Reveal

Herbs, erect, (1–)1.5–3 × 1–2 dm, tomentose, grayish. **Stems** slightly spreading to erect, without persistent leaf bases, up to ¹/₅ height of plant; caudex stems absent; aerial flowering stems erect, slender, solid, not fistulose, 0.5–1(–1.3) dm, tomentose. **Leaves** basal, 1 per node; petiole 1–2 cm, tomentose; blade round, (0.3–)0.5–1.5 × (0.3–)1–1.5 cm, tomentose abaxially, thinly floccose and green adaxially, or white-tomentose on both surfaces, margins plane. **Inflorescences** cymose with involucres racemosely disposed at tips, 10–15(–20) × 0.6–3(–5) cm; branches dichotomous, tomentose; bracts 3, scalelike, linear to narrowly triangular, and 1–3 mm, or leaflike, ovate to orbiculate, and 5–12 × 5–12 mm. **Peduncles** absent. **Involucres** 1 per node, campanulate, 2–3(–4) × 2–4 mm, tomentose; teeth 5, erect, 0.4–0.8 mm. **Flowers** 3–4 mm, perianth white to whitish brown, glabrous; tepals connate proximal ¹/₄, monomorphic, oblanceolate; stamens exserted, 3–6 mm; filaments pilose proximally. **Achenes** light brown, 2.5–3 mm, glabrous.

Flowering Jul–Oct. Rocky to gravelly flats and slopes, sagebrush and mountain mahogany communities, pinyon-juniper and montane conifer woodlands; 1800–2700 m; Calif., Nev.

Eriogonum mensicola is infrequent in the Inyo Mountains and the Panamint and Coso ranges of Inyo County, California, with a disjunct population on the Sheep Range in Clark County, Nevada.

78. Eriogonum racemosum Nuttall, Proc. Acad. Nat. Sci. Philadelphia 4: 14. 1848 • Red-root wild buckwheat E F

Eriogonum racemosum var. *obtusum* (Bentham) S. Stokes; *E. racemosum* var. *orthocladon* (Torrey) S. Stokes

Herbs, erect to slightly spreading, 3–8(–10) × 0.5–1.5 dm, tomentose to floccose or rarely glabrous, grayish. **Stems** spreading to erect, without persistent leaf bases, up to ¹/₆ height of plant; caudex stems absent; aerial flowering stems erect to slightly spreading, slender to stout, solid, not fistulose, (1–)1.5–2.5(–3) dm, tomentose to floccose, rarely glabrous. **Leaves** basal, 1 per node; petiole (2–)3–10(–15) cm, tomentose to floccose; blade elliptic to ovate or oval to nearly rotund, (1.5–)2–6

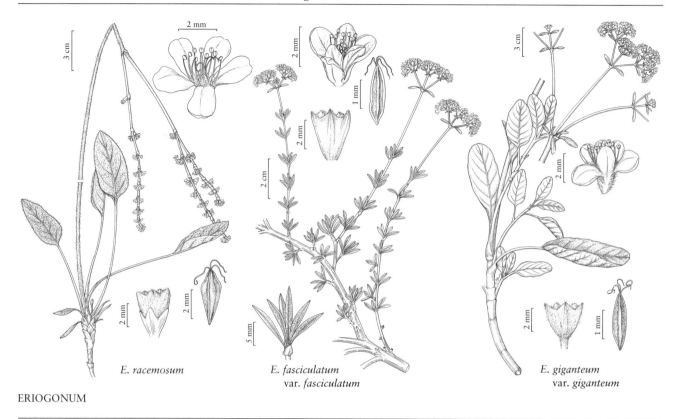

E. *racemosum*

E. *fasciculatum*
var. *fasciculatum*

E. *giganteum*
var. *giganteum*

ERIOGONUM

(–10) × 1–4(–5) cm, lanate to thinly tomentose abaxially, floccose or glabrous and green adaxially, margins plane. **Inflorescences** virgate or racemose with involucres racemosely disposed throughout or at tips, 15–50 × 05–20 cm, tomentose, rarely glabrous; branches dichotomous, upper secondaries suppressed and bearing 5–20(–30) racemosely arranged involucres; bracts 3, scalelike, triangular, and (1–)2.5–7 mm, or leaflike, linear-oblanceolate or oblanceolate to elliptic, and 10–40 × 5–20(–25) mm. **Peduncles** absent or erect, stout, 0.3–4 cm, tomentose to floccose. **Involucres** 1 per node, turbinate to turbinate-campanulate, (2–)3–5 × (2–)2.5–4 mm, tomentose to floccose; teeth 5, erect, (0.1–)0.2–0.5 mm. **Flowers** (2–)2.5–5 mm; perianth white to pinkish, glabrous; tepals connate proximal ¹/₄, monomorphic, oblong; stamens exserted, 2–5 mm; filaments pilose proximally. **Achenes** light brown, 3–4 mm, glabrous. $2n = 36$.

Flowering Jun–Oct. Sandy to gravelly flats and slopes, mixed grass, sagebrush, and mountain mahogany communities, scrub oak, pinyon, juniper, and conifer woodlands; 1400–2900(–3500) m; Ariz., Colo., Nev., N.Mex., Utah.

Eriogonum racemosum is highly variable in stature, the presence of leaflike bracts in the inflorescences, the size and shape of the leaves, and the length of the mature flowers. None of these features is geographically correlated and thus no taxonomic distinctions are attempted. The species is occasionally cultivated. The tomentose, nonfistulose flowering stems easily distinguish it from *Eriogonum zionis*. Individuals of *E. racemosum* with glabrous flowering stems are known (*Reveal & Holmgren 1893*, BRY, US, UTC; *Goodrich 17355*, BRY; *Neese & White 9237*, BRY), but are rare and clearly aberrant expressions.

The Navajo or Diné people use the roots of *Eriogonum racemosum* as a "life medicine," primarily in the treatment of internal problems, notably poisoning and diarrhea (C. Arnold, pers. comm.; P. A. Vestal 1952). They also use it as an analgesic and orthopedic aid (D. E. Moerman 1986; L. C. Wyman and S. K. Harris 1951); there are reports of its use for venereal disease. Leaves and stems were eaten raw by the Ramah Navajo in northwestern New Mexico (Wyman and Harris).

Eriogonum racemosum is the food plant for the Spalding dotted-blue butterfly (*Euphilotes spaldingi*) and is occasionally visited by the desert green or Comstock's hairstreak (*Callophrys comstocki*).

79. Eriogonum zionis J. T. Howell, Leafl. W. Bot. 2: 253. 1940 [E]

Herbs, erect to spreading, 3–8 (–10) × 0.5–1.5 dm, glabrous and glaucous, rarely tomentose, greenish. **Stems** spreading to erect, without persistent leaf bases, up to ⅛ height of plant; caudex stems absent; aerial flowering stems erect to spreading, slender to stout, solid or hollow, often fistulose, 1–2.5(–3) dm, glabrous, rarely tomentose. **Leaves** basal, 1 per node; petiole 3–6(–10) cm, tomentose to floccose; blade elliptic or oblong-ovate to ovate, 2–4.5(–6) × 1.5–2.5(–5) cm, lanate to densely tomentose abaxially, thinly floccose or glabrous and green adaxially, margins plane. **Inflorescences** narrowly virgate or racemose with involucres racemosely disposed throughout or at tips, 20–50(–60) × 5–25 cm, glabrous, rarely tomentose; branches dichotomous, upper secondaries suppressed and bearing (8–)10–15(–20) racemosely arranged involucres; bracts 3, scalelike, triangular, and 2–7 mm, or leaflike, elliptic, and 10–30 × 4–15 mm. **Peduncles** absent. **Involucres** 1 per node, turbinate to turbinate-campanulate, 1.5–3 × 1.5–2.5 mm, tomentose; teeth 5, erect, 0.2–0.4 mm. **Flowers** 2–6(–7) mm; perianth white to yellow or bright red, glabrous; tepals connate proximal ¼ their length, monomorphic, oblong; stamens exserted, 2–5 mm; filaments pilose proximally. **Achenes** light brown, (3–)4–5(–6) mm, glabrous.

Varieties 2 (2 in the flora): w United States.

1. Perianths white to yellow . . . .
. . . . 79a. *Eriogonum zionis* var. *zionis*
1. Perianths bright red . . . .
. . . . 79b. *Eriogonum zionis* var. *coccineum*

79a. Eriogonum zionis J. T. Howell var. **zionis**
 • Zion wild buckwheat [E]

Eriogonum racemosum Nuttall var. *nobilis* S. L. Welsh & N. D. Atwood

Plants erect to spreading, 3–8 (–10) dm. **Aerial flowering stems** usually fistulose, glabrous or rarely tomentose. **Leaf blades** oblong-ovate to ovate, 2–4.5 × 1.5–2.5 cm. **Inflorescences** 20–45 (–60) cm. **Involucres** 1.5–3 mm.

Flowers 2–3.5(–4) mm; perianth white to yellow.

Flowering Jul–Oct. Deep sandy soil, sagebrush communities, scrub oak, juniper, pinyon, and montane conifer woodlands; 1300–2100 m; Ariz., Utah.

Variety *zionis* is found in scattered populations in Kane, San Juan, Washington, and Wayne counties, Utah, and in northern Coconino and northeastern Mohave counties, Arizona. It is rare or only locally infrequent. Occasionally, plants with tomentose flowering stems occur with glabrous individuals. The Zion wild buckwheat, with a chromosome number $2n = 40$ (J. L. Reveal, unpubl.), is distinct from the related *E. racemosum*, with $2n = 36$, and no intermediates are known.

79b. Eriogonum zionis J. T. Howell var. **coccineum** J. T. Howell, Leafl. W. Bot. 3: 205. 1943 • Point Sublime wild buckwheat [E]

Eriogonum racemosum Nuttall var. *coccineum* (J. T. Howell) S. L. Welsh

Plants erect, 3–6 dm. **Aerial flowering stems** fistulose, glabrous or tomentose. **Leaf blades** elliptic to ovate, 3–4(–6) × 1.5–2(–5) cm. **Inflorescences** 20–45 cm. **Involucres** 2–3 mm. **Flowers** 4–6(–7) mm; perianth bright red.

Flowering Jul–Sep. Loamy to gravelly flats and rim edges, sagebrush communities, juniper, pinyon, and montane conifer woodlands; 1300–2300 m; Ariz.

Variety *coccineum* is rare and localized, known from scattered populations on the edge of the Grand Canyon (1950–2300 m) in Coconino County, and in the Hack Canyon area (1300–1500 m) of eastern Mohave County.

80. Eriogonum fasciculatum Bentham, Trans. Linn. Soc. London 17: 411. 1836 [F]

Shrubs or subshrubs, compact to spreading or rounded and more or less erect, occasionally decumbent, infrequently scapose, (1–)2–15 × 2–25(–30) dm, tomentose to canescent, floccose, or glabrous. **Stems** sprawling or spreading to erect, often with persistent leaf bases, up to ½ or more height of plant; caudex stems absent or matted to spreading; aerial flowering stems erect to spreading, slender, solid, not fistulose, 0.3–2.5(–3) dm, tomentose, canescent, or glabrous. **Leaves** cauline, 1 per node or fasciculate; petiole 0.1–0.3 cm, canescent; blade linear to linear-oblanceolate or oblanceolate, 0.6–1.5(–1.8) × 0.05–0.4(–0.6) cm, white-tomentose or canescent to subglabrous abaxially, tomentose or canescent and grayish, subglabrous, or glabrous and green adaxially, margins often revolute. **Inflorescences** cymose, infrequently cymose-umbellate or capitate, compact to open, occasionally flat-topped, 0.2–20 × 0.2–15 cm;

branches dichotomous, infrequently absent, tomentose to canescent or glabrous; bracts usually 3, scalelike, triangular, and 1–3 mm, or leaflike, linear to oblanceolate, and 3–10 × 1–3 mm. **Peduncles** absent. **Involucres** (1–)3–8 per cluster, turbinate to campanulate, 2–4 × 1.5–3 mm, canescent, pubescent, glabrous, or subglabrous; teeth 5, erect, 0.3–1.2 mm. **Flowers** 2.5–3 mm; perianth white to pinkish, glabrous or pubescent; tepals connate proximal ¼, monomorphic, usually elliptic to obovate; stamens exserted, 2.5–5 mm; filaments subglabrous or pubescent proximally. **Achenes** light brown to brown, 1.8–2.5 mm, glabrous.

Varieties 5 (4 in the flora): w North America, including nw Mexico.

Eriogonum fasciculatum is a complex, polyploid series of variants that are generally distinct but often difficult to distinguish morphologically. Variety *emphereium* Reveal is confined to central Baja California, Mexico. The introduction of *Eriogonum fasciculatum* as a decorative roadside plant by the California Department of Transportation is resulting in hybrid populations involving *E. cinereum*. The aggressively weedy and (for Arizona) exotic variety *foliosum* is rapidly invading the native habitat of var. *polifolium*. Members of *E. fasciculatum* are food plants for several butterflies, notably the Bernardino dotted-blue (*Euphilotes bernardino*), lupine blue (*Plebeius lupini*), Mormon metalmark (*Apodemia mormo*), and Behr's metalmark (*A. virgulti*). Probably the butterfly most commonly seen with the species is the nut-brown hairstreak (*Satyrium saepium*), which frequents plants in full flower. *Eriogonum fasciculatum* is also the most important native source of honey in California.

This widespread species was used extensively by Native Americans for a variety of ailments. Its application for pain and headaches (D. P. Barrows 1900; K. Hedges 1986; E. W. Voegelin 1938) was rather common, as was its general use for diarrhea (Hedges; Voegelin). M. L. Zigmond (1981) reported that the Kawaiisu lined their acorn granaries with leaves of var. *proliferum* to keep out rain—a daunting challenge given the size of the leaves! L. Hinton (1975) reported the use of a decoction of dried flowers and roots to maintain a healthy heart, and M. C. Stevenson (1915) indicated that a powder derived from the roots was used by the Zuñi to treat wounds, whereas a root decoction was taken for colds and hoarseness. B. R. Bocek (1984) reported that the Costanoan Indians of California used a decoction of the plant to treat unspecified urinary problems. F. H. Elmore (1943) reported the use of a decoction of var. *proliferum* by the Navajo (Diné) people as an anti-witchcraft medicine.

1. Leaves thinly white-tomentose abaxially, glabrous adaxially; perianths and involucres mostly glabrous; plants usually decumbent; primarily coastal . . . .
. . . . 80c. *Eriogonum fasciculatum* var. *fasciculatum*
1. Leaves, perianths, and involucres pubescent or, if subglabrous, then plants of desert regions; plants erect to rounded; generally inland.
  2. Leaves light yellow-green, subglabrous adaxially; involucres and perianths glabrous or subglabrous . . . . . 80b. *Eriogonum fasciculatum* var. *flavoviride*
  2. Leaves dark green or grayish, usually pubescent adaxially; involucres and perianths pubescent.
    3. Inflorescences capitate to cymose-umbellate, rarely cymose; leaves canescent on both surfaces or densely tomentose abaxially and canescent adaxially; leaf margins plane or infrequently revolute; mainly deserts . . . 80a. *Eriogonum fasciculatum* var. *polifolium*
    3. Inflorescences open and mostly cymose; leaves densely white-tomentose abaxially, less so to green and floccose adaxially; leaf margins usually tightly revolute; mainly inner coastal foothills . . . .
. . . . 80d. *Eriogonum fasciculatum* var. *foliolosum*

80a. Eriogonum fasciculatum Bentham var. **polifolium** (Bentham) Torrey & A. Gray, Proc. Amer. Acad. Arts 8: 169. 1870 • Mojave Desert California buckwheat

Eriogonum polifolium Bentham in A. P. de Candolle and A. L. P. P. de Candolle, Prodr. 14: 12. 1856; *E. fasciculatum* var. *revolutum* (Goodding) S. Stokes

Shrubs or subshrubs, compact to spreading, 2–5(–8) × 2–20(–30) dm, tomentose to canescent and grayish. **Aerial flowering stems** thinly tomentose to canescent, rarely glabrous. **Leaf blade** usually oblanceolate, 0.6–1.8 × (0.1–)0.2–0.6 cm, canescent on both surfaces or densely grayish-tomentose abaxially and canescent adaxially, margins plane or infrequently revolute. **Inflorescences** capitate to cymose-umbellate, rarely cymose; branches tomentose to canescent, rarely glabrous. **Involucres** turbinate-campanulate to campanulate, 2.5–3.5 × 2–3 mm, canescent. **Perianths** pubescent. $2n = 40$.

Flowering year-round. Sandy to gravelly flats and slopes, saltbush, blackbrush, and creosote bush communities, pinyon-juniper or juniper woodlands; (60–)300–2500 m; Ariz., Calif., Nev., Utah; Mexico (Baja California, Sonora).

Variety *polifolium* is a widespread, common to abundant, or occasionally dominant shrub of the Mojave and Sonoran deserts in Arizona, southern California, southern Nevada, and southwestern Utah. This is the common tetraploid expression of the species. It is occasionally planted as an ornamental in the more arid regions of the American Southwest.

Plants were used by several groups of Native Americans as a medicinal plant to treat a variety of symptoms (D. E. Moerman 1986). It was used also in the practice of witchcraft by the Navajo, in a potion against evil spells.

80b. Eriogonum fasciculatum Bentham var. **flavoviride** Munz & I. M. Johnston, Bull. Torrey Bot. Club 49: 350. 1923 • Sonoran Desert California buckwheat

Eriogonum fasciculatum subsp. *flavoviride* (Munz & I. M. Johnston) S. Stokes

Shrubs or subshrubs, rounded and somewhat compact, 2–5 × 3–6 (–10) dm, thinly floccose or glabrous, yellowish green. **Aerial flowering stems** usually glabrous. **Leaf blades** linear or linear-oblanceolate, 0.6–1 × 0.05–0.2 cm, thinly tomentose to subglabrous and light green abaxially, subglabrous or glabrous and green adaxially, margins tightly revolute. **Inflorescences** mostly capitate; branches glabrous. **Involucres** turbinate-campanulate, 2–3 × 2–3 mm, glabrous or subglabrous. **Perianths** glabrous or infrequently thinly pubescent. $2n = 40$.

Flowering year-round. Sandy to gravelly flats and slopes, saltbush and creosote bush communities, pinyon-juniper woodlands; 50–1300 m; Calif.; Mexico (Baja California, Sonora).

Variety *flavoviride* is a widespread, infrequent to common, warm-desert shrub found on the Mojave and Sonoran deserts in southern San Bernardino, Riverside and San Diego counties. The yellowish-green hue of the flowering stems, inflorescence branches, and leaves readily distinguish it from var. *polifolium*, with which it occasionally occurs (especially in Mexico), although this feature is not always obvious on herbarium specimens. Variety *flavoviride* is much more attractive in the garden than its more frequently planted Mojave Desert counterpart. Reports (e.g., R. S. Felger 2000) of var. *fasciculatum* along the coast in extreme northwestern Sonora are based on specimens of var. *flavoviride*.

80c. Eriogonum fasciculatum Bentham var. **fasciculatum** • Coastal California buckwheat

F

Eriogonum fasciculatum subsp. *aspalathoides* (Gandoger) S. Stokes

Shrubs or subshrubs, spreading and often decumbent, 1–5 × 5–30 dm, mostly glabrous. grayish. **Aerial flowering stems** glabrous, usually grayish or reddish. **Leaf blades** blade linear to linear-oblanceolate, 0.6–1(–1.2) × 0.05–0.2(–0.4) cm, thinly white-tomentose abaxially, glabrous and green adaxially, margins tightly revolute. **Inflorescences** usually capitate, occasionally cymose; branches mostly glabrous. **Involucres** narrowly turbinate, 3–4 × 1.5–2 mm, glabrous or nearly so. **Perianths** glabrous or with only a few hairs proximally. $2n = 40$.

Flowering year-round. Sandy mesa tops and slopes in coast scrub and chaparral communities; 0–300 m; Calif.; Mexico (Baja California).

Variety *fasciculatum* is the tetraploid coastal expression of the species, consisting mainly of low, spreading plants of the coastal bluffs and mesas near the ocean and on the offshore islands. It occurs along the immediate coast from San Luis Obispo County southward, but is found inland in Los Angeles, Orange, and San Diego counties, where plants typically are larger and more shrub-like. The tetraploid var. *fasciculatum* and the octoploid var. *foliolosum* are not always distinct morphologically. Variety *fasciculatum* hybridizes with *E. molle* Greene on Cedros Island in Mexico. The decumbent coastal expressions are occasionally cultivated as cover plants in rock gardens.

80d. Eriogonum fasciculatum Bentham var. **foliolosum** (Nuttall) S. Stokes ex Abrams, Bull. New York Bot. Gard. 6: 351. 1910 • Coastal California buckwheat

Eriogonum rosmarinifolium Nuttall var. *foliolosum* Nuttall, Proc. Acad. Nat. Sci. Philadelphia 4: 16. 1848; *E. fasciculatum* subsp. *foliolosum* (Nuttall) S. Stokes; *E. fasciculatum* var. *obtusiflorum* S. Stokes

Shrubs, rounded to erect, 6–15 × (8–)10–25 dm. **Aerial flowering stems** thinly tomentose or glabrous. **Leaf blades** linear-oblanceolate to oblanceolate, 0.6–1.2 × 0.1–0.4 cm, densely white-tomentose abaxially, less so to green and floccose adaxially, margins plane usually tightly revolute. **Inflorescences** open and mostly cymose; branches thinly tomentose or glabrous.

Involucres turbinate, 3–4 × (1.5–)2–2.5 mm, pubescent. **Perianths** pubescent at least proximally. $2n = 80$.

Flowering year-round. Sandy to gravelly flats and slopes, mixed grassland and chaparral communities, oak and conifer woodlands; 60–1300(–1600) m; Calif.; Mexico (Baja California).

Variety *foliolosum* is widespread and common to abundant, often being a dominant shrub in the chaparral in the Coast Ranges of California (Kern, Los Angeles, Monterey, Orange, Riverside, San Benito, San Bernardino, San Diego, San Joaquin, San Luis Obispo, San Mateo, Santa Barbara, Santa Clara, Santa Cruz, Stanislaus, and Ventura counties). It is an octoploid and may be the product of an ancient hybridization involving the coastal var. *fasciculatum* and desert var. *polifolium*.

Variety *foliolosum* is being widely planted by the California Department of Transportation along roadsides, where it has hybridized with *E. cinereum*. As a very unfortunate result, the distribution of var. *foliolosum* has now expanded into northern California (Alameda, Marin, San Francisco, Trinity, and likely other countries) and even into Oregon (Jackson County). In southern Arizona, var. *foliolosum* has been introduced as a roadside plant in Maricopa County, and has been found (*Bierner 90-45*, ARIZ, TEX) escaped along a roadside in Graham County. Since this is a potentially aggressive weedy shrub, efforts should be made to curtail its introduction into areas outside its native range.

81. **Eriogonum cinereum** Bentham, Bot. Voy. Sulphur, 45. 1844 · Coastal wild buckwheat [C][E]

Shrubs, round, 6–15 × 10–20 (–25) dm, tomentulose, grayish. **Stems** spreading, occasionally with persistent leaf bases, up to ³/₄ or more height of plant; caudex stems absent; aerial flowering stems erect to spreading, slender, solid, not fistulose, 1–4 dm, tomentulose. **Leaves** cauline, 1 per node; petiole 0.1–0.5(–1) cm, tomentulose; blade ovate, 1.5–3 × 1–2.5(–3) cm, white-tomentulose abaxially, less so and greenish adaxially, margins plane. **Inflorescences** capitate, congested, 1–2.5 × 1–2.5 cm; branches dichotomous, tomentulose; bracts usually 3, scalelike, triangular, and 1–3 mm, or leaflike, ovate, and 7–20 × 5–18 mm. **Peduncles** absent. **Involucres** 3–10 per cluster, narrowly turbinate, 3–5 × 2–3 mm, tomentulose; teeth 5, erect, 0.2–0.5 mm. **Flowers** 2.5–3 mm; perianth white to pinkish, densely white-villous; tepals connate proximally, monomorphic, spatulate to narrowly obovate; stamens exserted, 2.5–3.5 mm; filaments subglabrous. **Achenes** brown, 2–2.5 mm, glabrous. $2n = 80$.

Flowering year-round. Sandy beaches, coastal bluffs, mesas, canyon slopes, coast scrub and chaparral communities; of conservation concern; 0–400 m; Calif.

Eriogonum cinereum occurs naturally along the Pacific Coast in Los Angeles, Santa Barbara, and Ventura counties, and also on Santa Rosa Island. It is infrequently planted along highways (and is now found in San Diego County) and hybridizes with *E. fasciculatum* var. *foliolosum*. The species is the food plant of the Bernardino dotted-blue butterfly (*Euphilotes bernardino*).

82. **Eriogonum parvifolium** Smith in A. Rees, Cycl. 13(2): Eriogonum no. 2. 1809 · Seacliff wild buckwheat [E]

Eriogonum parvifolium var. *crassifolium* Bentham; *E. parvifolium* subsp. *lucidum* J. T. Howell ex S. Stokes; *E. parvifolium* var. *lucidum* (J. T. Howell ex S. Stokes) Reveal; *E. parvifolium* subsp. *paynei* C. B. Wolf ex Munz; *E. parvifolium* var. *paynei* (C. B. Wolf ex Munz) Reveal

Shrubs, matted to spreading or rounded, 3–10 × 5–20 (–25) dm, thinly tomentose or glabrous, greenish. **Stems** spreading, sometimes matted, often with persistent leaf bases, up to ¹/₂ or more height of plant; caudex stems absent or matted; aerial flowering stems prostrate, spreading, or erect, slender, solid, not fistulose, 0.2–1 dm, thinly tomentose or glabrous. **Leaves** cauline, fasciculate, infrequently 1 per node; petiole 0.1–0.7 cm, floccose; blade lanceolate to round, 0.5–3 × 0.3–0.8 (–1.2) cm, lanate to tomentose abaxially, mostly glabrous and olive green to green adaxially. **Inflorescences** capitate to cymose, 20–30 × 2–10 cm; branches dichotomous, thinly tomentose or glabrous; bracts 3, scalelike, triangular, and 1–2 mm, or leaflike, usually elliptic, and 5–20 × 2–10 mm. **Peduncles** absent. **Involucres** 2–7 per cluster, turbinate-campanulate, (2.5–)3–4 × 2–3.5 mm, floccose to glabrate; teeth 5, erect, 0.5–0.9 mm. **Flowers** 2.5–3 mm; perianth white to pinkish or greenish yellow, glabrous; tepals connate proximally, monomorphic, obovate; stamens exserted, 2.5–3.5 mm; filaments pilose proximally. **Achenes** brown, 2.5–3 mm, glabrous. $2n = 40$.

Flowering year-round. Sandy beaches, dunes, and bluffs or sandy to gravelly inland slopes and flats, coastal grassland and chaparral communities, oak and pine woodlands; 0–300 m; Calif.

The native range of *Eriogonum parvifolium* is restricted to coastal and near-coastal areas (Los Angeles, Monterey, Orange, San Diego, San Luis Obispo, Santa Barbara, and Ventura counties). The coastal expression (var. *parvifolium*) has thickened leaf blades (0.5–1.5 ×

0.3–0.8 cm) and simple or dichotomous inflorescences of compact clusters of involucres containing white to rose flowers. Highly compact and dense mat-forming plants on rocky bluffs immediately next to the ocean were named var. *crassifolium*; those with yellow flowers were named var. *lucidum*. The inland form with thin leaf blades (1.5–3 × 0.3–0.8 cm) and highly-branched, cymose, white-flowered inflorescences is perhaps worthy of continued recognition as var. *paynei*, although there is no sharp distinction between the extremes. Several expressions of the seacliff wild buckwheat are in cultivation, and unfortunately the California Department of Transportation is using the species in roadside plantings, with the result that it is now established in Santa Clara County. Every effort should be made to halt its introduction beyond its native range.

The species is the food plant for two federally endangered butterflies, the El Segundo dotted-blue (*Euphilotes battoides allyni*), near Los Angeles, and Smith's dotted-blue *(Euphilotes enoptes smithi)*, near Monterey.

83. Eriogonum giganteum S. Watson, Proc. Amer. Acad. Arts 20: 371. 1885 • St. Catherine's-lace

[C] [E] [F]

Shrubs, round to erect, 3–20(–35) × (3–)5–20(–35) dm, tomentose to floccose or glabrate, grayish to reddish. **Stems** spreading to erect, occasionally with persistent leaf bases, up to ³/₄ or more height of plant, often with a distinct main trunk up to 1 dm thick; caudex stems absent; aerial flowering stems erect to spreading, slender to stout, solid, not fistulose, 1–4 dm, tomentose to glabrate. **Leaves** cauline, 1 per node; petiole 0.5–4 cm, tomentose; blade oblong-ovate to ovate or lanceolate to narrowly oblong, 2.5–7(–10) × 1–5 cm, white-tomentose abaxially, sometimes slightly less so or cinereous to somewhat glabrate and greenish adaxially, margins plane, sometimes crisped. **Inflorescences** cymose, open or compact, 5–30 × (2–)5–50(–80) cm; branches dichotomous, tomentose to floccose or glabrate; bracts 3, scalelike, broadly triangular, and 1–2 mm, or leaflike, oblanceolate to elliptic, and 5–30 mm. **Peduncles** absent or erect, slender, 0.1–0.5 cm, tomentose. **Involucres** 1 per node, campanulate, 3–5 × 2.5–4 mm, tomentose; teeth 5, erect, 0.3–0.8 mm. **Flowers** 2–4 mm; perianth white to rose, white-villous abaxially; tepals connate proximal ¹/₄,

monomorphic, obovate; stamens exserted, 2–4 mm; filaments pilose proximally. **Achenes** brown, 2–3.5 mm, glabrous.

Varieties 3 (3 in the flora): California.

The California Department of Transportation is planting members of this insular species, especially var. *giganteum*, on the mainland, where it readily hybridizes with both *E. fasciculatum* and *E. cinereum*. Every effort should be made to remove the introduced *E. giganteum* from coastal California.

The Avalon scrub-hairstreak butterfly (*Strymon avalona*), which is endemic to Santa Catalina Island, uses *Eriogonum giganteum* as a food plant.

1. Leaf blades oblong-lanceolate to lanceolate; San Clemente Island . . . . 83c. *Eriogonum giganteum* var. *formosum*
1. Leaf blades oblong-ovate to ovate; not San Clemente Island.
    2. Leaf blades 3–7(–10) × 2–5 cm; inflorescences open; Santa Catalina Island . . . . 83a. *Eriogonum giganteum* var. *giganteum*
    2. Leaf blades 2.5–3.5(–6) × 1.5–2(–4) cm; inflorescences compact to rarely open; Santa Barbara and Sutil islands . . . . 83b. *Eriogonum giganteum* var. *compactum*

83a. Eriogonum giganteum S. Watson var. **giganteum** • Santa Catalina Island's St. Catherine's-lace

[C] [E] [F]

Plants (3–)5–35 dm. **Aerial flowering stems** 1–3 dm, tomentose to glabrate. **Leaves:** petiole 1–3 cm; blade oblong-ovate to ovate, 3–7(–10) × 2–5 cm. **Inflorescences** open, 10–50 (–80) cm wide. **Involucres** 3–4 mm. **Flowers** 2–2.5 mm. **Achenes** 2–2.5 mm. $2n = 40$.

Flowering year-round. Rocky cliff faces, slopes, and ridges, coastal scrub communities, oak woodlands; of conservation concern; 0–500 m; Calif.

Variety *giganteum* occurs naturally only on Santa Catalina Island, Los Angeles County. Mainland introductions occur from Santa Clara County to Los Angeles County; the variety has also been introduced onto Santa Cruz Island, where it threatens the genetic purity of *E. arborescens*. Garden hybrids between var. *giganteum* and *E. arborescens* have been named *E.* ×*blissianum* H. Mason.

83b. Eriogonum giganteum S. Watson var. **compactum** Dunkle, Bull. S. Calif. Acad. Sci. 41: 130. 1943 • Santa Barbara Island's St. Catherine's-lace [C] [E]

Eriogonum giganteum subsp. *compactum* (Dunkle) Munz

Plants 4–6(–10) dm. **Aerial flowering stems** (1.5–)2–3 dm, tomentose. **Leaves:** petiole 0.5–1 cm; blade oblong-ovate, 2.5–3.5 (–6) × 1.5–2(–4) cm, densely white-tomentose on both surfaces or sometimes slightly less so adaxially. **Inflorescences** compact to rarely open, 2–15 cm wide. **Involucres** (3–)3.5–4(–5) mm. **Flowers** 2–2.5 mm. **Achenes** 2–2.5 mm.

Flowering May–Oct. Rocky outcrops and cliffs, coastal scrub communities; of conservation concern; 0–100 m; Calif.

Variety *compactum* is a rare and localized plant on Santa Barbara and Sutil islands, Santa Barbara County. It is in cultivation, but only in botanical gardens.

83c. Eriogonum giganteum S. Watson var. **formosum** K. Brandegee, Erythea 5: 79. 1897 • San Clemente Island's St. Catherine's-lace [C] [E]

Eriogonum giganteum subsp. *formosum* (K. Brandegee) P. H. Raven

Plants 3–15(–25) dm. **Aerial flowering stems** 2–4 dm, tomentose to glabrate. **Leaves:** petiole (1–)2–4 cm; blade lanceolate to narrowly oblong, 5–8 × 1–2 cm. **Inflorescences** open, 10–50 cm wide. **Involucres** 4–5 mm. **Flowers** 3–4 mm. **Achenes** 3–3.5 mm.

Flowering May–Sep. Rocky cliffs, slopes, and ridges, coastal scrub communities, oak woodlands; of conservation concern; 0–300 m; Calif.

Variety *formosum* is known only from San Clemente Island, Los Angeles County. Extensive sheep grazing and military operations harmed much of the population prior to the 1960s. Reduction of the sheep population and greater concern for the local environment have greatly improved the ecology of the island and prospects for the long-term survival of this species. The variety is cultivated, and established populations occasionally are found on the mainland of southern California.

84. Eriogonum arborescens Greene, Bull. Calif. Acad. Sci. 1: 11. 1884 • Santa Cruz Island wild buckwheat [C] [E] [F]

Shrubs, erect to spreading, 6–15 (–20) × 5–30 dm, glabrous and cinereous, reddish. **Stems** spreading to erect, occasionally with persistent leaf bases, up to ³/₄ or more height of plant; caudex stems absent; aerial flowering stems erect, stout, solid, not fistulose, 0.5–1 (–1.5) dm, glabrous. **Leaves** cauline, 1 per node; petiole 0.1–0.5 cm, tomentose; blade linear to narrowly oblong, 2–4(–5) × 0.1–0.4(–0.6) cm, white-tomentose abaxially, cinereous to glabrate and greenish adaxially, margins often revolute. **Inflorescences** cymose, 5–10 × 5–15(–20) cm; branches dichotomous, glabrous; bracts 1–3, leaflike, lanceolate to oblong, 0.5–2 × 0.3–1.5 cm. **Peduncles** absent or erect, stout, 0.1–0.5 cm, cinereous. **Involucres** 1 per node, campanulate, 2–3 × 2.5–4 mm, cinereous; teeth 5–7, erect, 0.5–1.5 mm. **Flowers** 2–3.5(–4) mm; perianth white to pinkish, villous abaxially; tepals connate proximally, monomorphic, oblanceolate to narrowly obovate; stamens exserted, 3–5 mm; filaments glabrous. **Achenes** brown, 2.5–3.5 mm, glabrous. $2n = 40$.

Flowering Apr–Oct. Rocky slopes and canyon walls, coastal scrub communities; of conservation concern; 10–600 m; Calif.

Eriogonum arborescens is local and occasionally rare on Santa Cruz, Santa Rosa, and Anacapa islands in Santa Cruz and Ventura counties. The species is cultivated, and populations have become naturalized on the mainland from San Mateo County south to San Diego, in large part because of planting along highways. Every effort should be made to remove these naturalized populations from the mainland.

85. Eriogonum deserticola S. Watson, Proc. Amer. Acad. Arts 26: 125. 1891 • Colorado Desert wild buckwheat

Shrubs, erect or spreading, not scapose, 6–15(–18) × 10–20(–35) dm, white-tomentose to thinly floccose, greenish or grayish. **Stems** mostly spreading, without persistent leaf bases, up to ¹/₂ height or more of plant; caudex stems absent or spreading in moving sand; aerial flowering stems spreading, slender, solid, not fistulose, 1.5–4 dm, white-tomentose to thinly floccose, occasionally glabrate and light green. **Leaves** cauline, 1 per node, rarely fasciculate, quickly deciduous; petiole (0.3–)0.5–2 cm, tomentose to floccose; blade oblong-ovate to round-oblong or orbiculate, (0.3–)0.5–1.5(–3.5) × (0.5–)0.7–1.7(–2) cm, white-tomentose on both surfaces, margins plane, sometimes crisped. **Inflorescences** cymose, open, 15–90 × 20–100 cm; branches dichotomous, white-tomentose to thinly floccose, becoming floccose or glabrate and green; bracts 3, triangular, 0.5–1.5 mm. **Peduncles** absent or erect, slender, 0.1–0.5 cm, floccose to glabrate. **Involucres** 1 per node, turbinate-campanulate, 1.5–2.5(–3) × 1.3–1.6(–2) mm, thinly tomentose; teeth 4, erect, 0.3–0.5 mm. **Flowers** (2.5–)3–4(–4.5) mm; perianth greenish yellow to yellow, silky-villous abaxially; tepals connate proximally, monomorphic, oblong to oblong-obovate; stamens exserted, 2.5–4 mm; filaments pilose proximally. **Achenes** brown, 3–4(–4.5) mm, glabrous except for scabrellous beak.

Flowering Jul–Jan. Deep moving sand dunes and sandy flats, desert scrub communities; -60–200 m; Ariz., Calif.; Mexico (Baja California, Sonora).

Eriogonum deserticola is found primarily in the moving sand dunes in the southern Salton Sea basin of extreme eastern San Diego and Imperial counties, California, with small populations in Yuma County, Arizona, and Baja California and Sonora, Mexico, where the species can be locally abundant. The larger plants are continually being buried and then uncovered by moving sand, revealing slender, grotesquely twisted trunks. R. S. Felger (2000) reported seeing some of the "exposed lateral roots" being up to six or more meters in length. It appears that the leaves are frequently stripped away by wind-blown sand.

86. **Eriogonum pendulum** S. Watson, Proc. Amer. Acad. Arts 23: 265. 1888 • Waldo wild buckwheat Ⓔ

Eriogonum pendulum var. *confertum* S. Stokes

Shrubs, erect or slightly spreading, not scapose, 2–5 × 2–8 dm, densely tomentose, tannish. **Stems** mostly erect, without persistent leaf bases, up to 1/2 height or more of plant; caudex stems absent; aerial flowering stems erect, stout, solid, not fistulose, 2–4 dm, tomentose. **Leaves** 1 per node and sheathing up stems or fasciculate at tips of basal branches; petiole 0.1–0.5(–1) cm, tomentose; blade oblanceolate to narrowly oblong, (1.5–)2–4(–5) × 1–2.5(–3) cm, densely white-tomentose abaxially, floccose adaxially, margins plane. **Inflorescences** cymose, 15–30(–40) × 15–40 cm; branches dichotomous, tomentose; bracts 3, semileaflike, narrowly oblanceolate, 3–7 × 1–3 mm. **Peduncles** absent or erect, slender, (1–)3–10 cm, tomentose. **Involucres** 1 per node, turbinate-campanulate, 3.5–5 × 2.5–4 mm, tomentose; teeth 5–8, erect, 0.4–0.8 mm. **Flowers** 3–6(–7) mm; perianth white, densely villous abaxially; tepals connate proximal 1/3, monomorphic, narrowly oblong; stamens exserted, 3–7 mm; filaments pilose proximally. **Achenes** brown, 3–5 mm, villous.

Flowering Jul–Sep. Sandy to gravelly flats and slopes, sagebrush communities, oak and montane conifer woodlands; 200–800 m; Calif., Oreg.

Eriogonum pendulum is a rare to locally uncommon shrub in the O'Brien-Waldo area of Josephine County, Oregon, and more common in the Siskiyou Mountains of Del Norte County, California, as far south as Gasquet.

87. **Eriogonum spectabile** B. L. Corbin, Reveal & R. Barron, Madroño 47: 134, fig. 1. 2001 • Barron's wild buckwheat Ⓔ

Shrubs, spreading, not scapose, 1–1.5 × 1.7–2.5 dm, densely tomentose, grayish. **Stems** spreading, without persistent leaf bases, up to 1/3 height of plant; caudex stems absent; aerial flowering stems erect, slender, solid, not fistulose, (0.15–)0.6.–1.3(–1.7) dm, tomentose. **Leaves** fasciculate at tips of branches; petiole (0.2–)0.6–0.9 (–1.3) cm, tomentose; blade narrowly elliptic, (0.7–)1.2–1.7(–2.2) × (0.2–)0.4–0.7(–0.9) cm, tomentose on both surfaces, margins plane to revolute. **Inflorescences** umbellate, 2–10 × 2–9 cm; branches umbellate, tomentose; bracts 2–4(–6), semileaflike, narrowly oblong to narrowly elliptic, 2–5 × 1–2 mm. **Peduncles** erect, slender, (1.5–)2–8.5(–9.5) cm, tomentose to glabrate. **Involucres** 1 per node, broadly campanulate, 2–3 × (2–)3–4(–5) mm, tomentose and densely glandular abaxially; teeth 5–7, erect, 1–1.2 mm. **Flowers** 4–6 mm; perianth white, densely villous and glandular abaxially; tepals connate proximal 1/4, monomorphic, obovate; stamens included to exserted, 2.5–6.5 mm; filaments sparsely pilose proximally. **Achenes** light brown, 3–4 mm, villous.

Flowering Jul–Sep. Gravelly soil, manzanita scrub and conifer woodlands; 2000 m; Calif.

Eriogonum spectabile is known only from a population of approximately 250 individuals north of Chester in the Lassen National Forest of Plumas County. The U.S. Forest Service considers it to be a "sensitive species," and it is a candidate for possible state and federal protection.

88. Eriogonum heermannii Durand & Hilgard in War Department [U.S.], Pacif. Railr. Rep. 5(3): 14. 1857 (as heermanni) E F

Eriogonum geniculatum Durand & Hilgard, J. Acad. Nat. Sci. Philadelphia, n. s. 3: 45. 1855, not Nuttall 1848

Shrubs and subshrubs, spreading to rounded and occasionally erect, not scapose, (0.5–)1–20 × 2–25 dm, glabrous or occasionally floccose, sometimes scabrellous, greenish, infrequently grayish. **Stems** spreading or erect, without persistent leaf bases, up to ¹/₂ height of plant; caudex stems absent; aerial flowering stems erect or nearly so, slender to stout, solid, not fistulose, 0.02–0.5 dm, thinly tomentose or glabrous. **Leaves** cauline, 1 per node, quickly deciduous; petiole 0.1–1.5 cm, floccose or glabrous; blade linear, oblanceolate or spatulate or elliptic, or oblong, (0.4–)1–2(–4) × 0.1–0.8 cm, tomentose to floccose or glabrous abaxially, floccose to thinly floccose or glabrous adaxially, margins plane. **Inflorescences** cymose or racemose, 1–25(–30) × 1–30 (–35) cm; branches dichotomous, sometimes with secondaries suppressed, smooth or angled to ridged and grooved, glabrous or occasionally floccose or scabrous; bracts 3, scalelike, 0.3–2 mm. **Peduncles** absent. **Involucres** 1 per node, narrowly turbinate or campanulate, 0.7–3 × 0.7–4 mm, glabrous, infrequently floccose; teeth 5, erect, 0.3–0.7 mm. **Flowers** (1.5–)2–4 mm; perianth white, yellowish white, pink, or reddish, glabrous; tepals connate proximal ¹/₄, dimorphic, those of outer whorl obovate to orbiculate, those of inner whorl narrowly lanceolate to oblong; stamens exserted, 2–5 mm; filaments pilose proximally. **Achenes** light brown to brown, 2–5 mm, glabrous.

Varieties 8 (8 in the flora): w United States.

As *Eriogonum heermannii* is here circumscribed, the number of varieties is dramatically decreased from past presentations, with *E. apachense* reduced to synonymy under a now greatly expanded var. *argense*. Also included in that variety is the more stoutly branched var. *subracemosum.* The southern var. *heermannii* of basically desert ranges and the more northern var. *occidentale* of the Coast Ranges in California are maintained, but their separation is more traditional than certain. The fragile and bulky nature of many dried, often poorly prepared specimens, and the tendency for leaves to fall away have made varietal identification within *E. heermannii* difficult.

Eriogonum heermannii varieties are food plants for Ellis's dotted-blue butterfly (*Euphilotes ellisi*) and the Mormon metalmark (*Apodemia mormo mormo*).

1. Flowering stems and inflorescence branches round or angled, scabrellous or papillate-scabrous; plants often densely branched.
  2. Flowering stems and inflorescence branches round, usually distinctly scabrellous or infrequently papillate-scabrous; Arizona, se California, Nevada . . . . . . 88g. *Eriogonum heermannii* var. *argense*
  2. Flowering stems and inflorescence branches sharply ridged and deeply grooved, minutely scabrellous; nw Arizona, se California, s Nevada, sw Utah . . . . . 88h. *Eriogonum heermannii* var. *sulcatum*
1. Flowering stems and inflorescence branches round, not angled, usually smooth, glabrous or thinly tomentose; plants sparsely branched.
  3. Involucres not racemosely arranged on inflorescence branches, or only last 2–3 so disposed; inflorescences diffusely branched, glabrous; Great Basin or northeast edge of Mojave Desert.
    4. Inflorescence branches mostly slender, smooth, not ridged, not spinose; inflorescences 3–15(–23) × 5–20 cm; subshrubs or shrubs, 3–7 × 5–12(–15) dm; Great Basin, ec California and Nevada . . . . 88a. *Eriogonum heermannii* var. *humilius*
    4. Inflorescence branches stoutish, faintly ridged and grooved, spinose; inflorescences 3–7(–10) × 3–10(–12) cm; shrubs, 1–3 × 1.5–5(–8) dm; Mojave Desert, sw Utah and nw Arizona . . . . . . 88b. *Eriogonum heermannii* var. *subspinosum*
  3. Involucres racemosely arranged along inflorescence branches or at least at tips of branches; inflorescences openly branched, glabrous or floccose to thinly tomentose; Mojave and Sonoran deserts.
    5. Inflorescence branches thinly tomentose to floccose; s California, s Nevada, nw Arizona . . . . 88f. *Eriogonum heermannii* var. *floccosum*
    5. Inflorescence branches glabrous; s Nevada, California.
      6. Involucres racemosely arranged; inflorescence branches whiplike; s Nevada . . . . 88c. *Eriogonum heermannii* var. *clokeyi*
      6. Involucres racemosely arranged only distally; inflorescence branches not whiplike; California.
        7. Leaf blades 0.5–1.5 cm, glabrous abaxially; inflorescence branches stout; sc California . . . . 88d. *Eriogonum heermannii* var. *heermannii*
        7. Leaf blades 1.5–3(–4) cm, mostly tomentose to floccose abaxially (at least in early anthesis); inflorescence branches slender; sw California . . . . 88e. *Eriogonum heermannii* var. *occidentale*

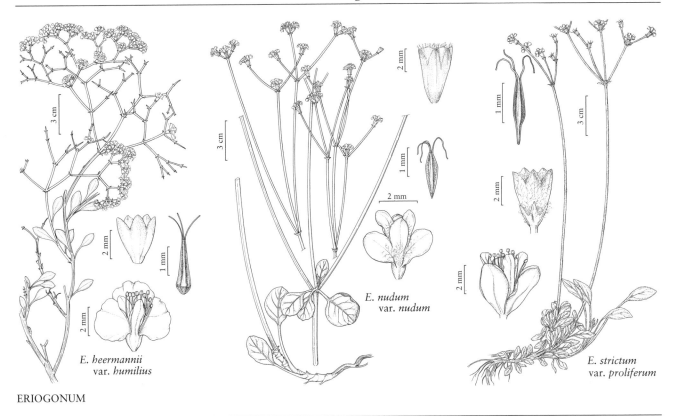

E. heermannii
var. *humilius*

E. nudum
var. *nudum*

E. strictum
var. *proliferum*

ERIOGONUM

88a. Eriogonum heermannii Durand & Hilgard var. **humilius** (S. Stokes) Reveal, Aliso 7: 226. 1970

• Heermann's Great Basin wild buckwheat   E F

Eriogonum heermannii subsp. *humilius* S. Stokes, Eriogonum, 90. 1936 (as heermanni)

Shrubs or subshrubs, spreading to rounded, usually sparsely branched, 3–7 × 5–12(–15) dm. **Leaf blades** oblanceolate to spatulate, 0.8–1.5 × 0.4–0.8 cm, floccose or glabrous abaxially. **Inflorescences** 3–15(–23) × 5–20 cm; branches usually diffuse with dichotomously arranged involucres, slender, round, smooth, glabrous, not spinose. **Involucres** campanulate, 1–1.5 × 1.5–3 mm. **Flowers** 2.5–3 mm; perianth white.

Flowering Jun–Oct. Limestone or volcanic flats, washes, and slopes, saltbush and sagebrush communities, pinyon-juniper and montane conifer woodlands; 1100–2500 m; Calif., Nev.

Variety *humilius* is the common expression of the species in the northern Great Basin, extending from Mono and Inyo counties, California, northeast across Nevada (Carson City, Churchill, Douglas, Elko, Esmeralda, Eureka, Humboldt, Lander, Lyon, Mineral, Nye, Pershing, Storey, and Washoe counties). Plants are usually scattered and the variety is rarely a local dominant; it often grows in volcanic soils. Variety

humilius is worthy of cultivation; the large shrubs, when in full flower, can be spectacular.

88b. Eriogonum heermannii Durand & Hilgard var. **subspinosum** Reveal, Phytologia 86: 134. 2004

• Tabeau Peak wild buckwheat   E

Shrubs, rounded, usually sparsely branched, 1–3 × 1.5–5 (–8) dm. **Leaf blades** lanceolate to narrowly elliptic or spatulate, 0.5–1.5 × 0.2–0.8 cm, thinly tomentose to floccose abaxially. **Inflorescences** 3–7(–10) × 3–10(–12) cm; branches usually diffuse, with dichotomously arranged involucres, stoutish, round, faintly grooved, glabrous, spinose. **Involucres** campanulate, 2–3 × 3–4 mm. **Flowers** 2.5–4 mm; perianth yellowish white.

Flowering Mar–May. Limestone flats, washes, and slopes, saltbush and blackbrush communities; 1000–2500 m; Ariz., Utah.

Variety *subspinosum* is known only from the southern end of the Beaver Dam Mountains in southern Washington County, Utah, and the eastern end of the Virgin Mountains and Virgin Narrows of Mohave County, Arizona. It approaches in habit the stoutly branched phase of var. *argense* known as var. *subracemosum*, but it lacks the papillate-scabrous condition. It also resembles var. *sulcatum*, but has only faintly grooved, rather than distinctly furrowed or deeply angled, inflorescence branches. It differs also in having the more open inflorescences of var. *humilius*. Both var. *subspinosum* and the *subracemosum* phase of var. *argense* have short, spinose lateral branches.

88c. Eriogonum heermannii Durand & Hilgard var. **clokeyi** Reveal, Phytologia 34: 437. 1976 • Clokey's wild buckwheat E

Shrubs, spreading to erect, sparsely branched, 4–8 × 6–10 dm. **Leaf blades** oblanceolate to spatulate, 0.5–1.5 × 0.2–0.6 cm, mostly glabrous. **Inflorescences** (5–)10–20 × 5–8 cm; branches open, with numerous racemosely arranged involucres, slender and whiplike, round, smooth, glabrous, not spinose. **Involucres** campanulate, 1–1.5 × 1.5–3 mm. **Flowers** 2.5–3 mm; perianth white.

Flowering Jun–Sep. Limestone gravelly to rocky flats, slopes, and washes, saltbush, blackbrush, or sagebrush communities, pinyon-juniper and montane conifer woodlands; 1200–1900 m; Nev.

Variety *clokeyi* is restricted to scattered locations mainly in the Spring (Charleston) Mountains and Sheep Range of Clark County, with outlying populations in the limestone mountains around Mercury and just entering Lincoln County in the Hiko Range.

88d. Eriogonum heermannii Durand & Hilgard var. **heermannii** • Heermann's wild buckwheat E

Shrubs, rounded, usually sparsely branched, 5–15 × 6–15(–20) dm. **Leaf blades** oblanceolate or spatulate to obovate, 0.5–1.5 × 0.2–0.6 cm, glabrous to slightly tomentose abaxially. **Inflorescences** (5–)10–20 × (5–)8–25 cm; branches rather diffuse, with racemosely arranged involucres at tips, stout, round, smooth, glabrous, not spinose. **Involucres** campanulate, 2–2.5 × 2–2.5 mm. **Flowers** (2.5–)3–4 mm; perianth yellowish white.

Flowering May–Nov. Sandy to gravelly washes, and slopes, mixed grassland and chaparral communities, oak and conifer woodlands; 900–1700 m; Calif.

Variety *heermannii* is infrequently encountered in widely scattered locations in the Transverse Ranges of Los Angeles and Ventura counties. It is disjunct in the Caliente Mountains of San Luis Obispo County and apparently was first encountered somewhere along Poso Creek in Kern County, although it has not been collected there since the type was collected in 1853.

88e. Eriogonum heermannii Durand & Hilgard var. **occidentale** S. Stokes, Leafl. W. Bot. 1: 30. 1932 (as heermanni) • Heermann's Coast Range wild buckwheat E

Eriogonum heermannii subsp. *occidentale* (S. Stokes) S. Stokes

Shrubs, rounded, usually sparsely branched, 10–20 × 10–25 dm. **Leaf blades** narrowly lanceolate, 1.5–3(–4) × 0.5–0.8 cm, tomentose to floccose abaxially (in early anthesis), occasionally glabrous with age. **Inflorescences** 10–30 × 10–35 cm; branches rather diffuse, with racemosely arranged involucres at tips, slender, round, smooth, glabrous, not spinose. **Involucres** campanulate, 2.5–3 × 2.5–3.5 mm. **Flowers** 3–4 mm; perianth white to pink.

Flowering Jul–Oct. Clay or shale slopes, mixed grassland communities, oak and conifer woodlands; (100–)400–1000 m; Calif.

Variety *occidentale* is basically restricted to scattered populations in San Benito County, although it is also found just inside Monterey County near Lonoak. A collection reported to be from King City (*Rose 678*, CAS) is suspect as to location.

88f. Eriogonum heermannii Durand & Hilgard var. **floccosum** Munz, Man. S. Calif. Bot., 121, 597. 1935 • Heermann's woolly wild buckwheat E

Eriogonum heermannii subsp. *floccosum* (Munz) Munz

Shrubs, rounded, usually sparsely branched, 3–6 × 4–8 dm. **Leaf blades** oblanceolate, (0.5–)0.8–1.5 × 0.2–0.5 cm, tomentose to floccose abaxially. **Inflorescences** rather dense 5–12 × 5–20 cm; branches rather diffuse, with racemosely arranged involucres at tips, slender, round, smooth, floccose to thinly tomentose, not spinose. **Involucres** campanulate, 1–1.5(–2) × 1–1.5 mm. **Flowers** 2–3 mm; perianth yellowish white.

Flowering May–Oct. Calcareous gravelly washes and slopes in saltbush and creosote bush communities, and occasionally in juniper woodlands; 900–2300 m; Ariz., Calif., Nev.

Variety *floccosum* is found in scattered desert ranges of eastern San Bernardino County and in the Kingston Mountains of Inyo County, California, in the McCullough Mountains of southern Clark County, Nevada, and just over the border in Mohave County, Arizona. It is disjunct in the Juniper Mesa area of west-central Yavapai County, Arizona, northwest of Prescott.

88g. Eriogonum heermannii Durand & Hilgard var. **argense** (M. E. Jones) Munz, Aliso 4: 89. 1958

• Heermann's rough wild buckwheat E

Eriogonum sulcatum S. Watson var. *argense* M. E. Jones, Contr. W. Bot. 11: 15. 1903; *E. apachense* Reveal; *E. heermannii* subsp. *argense* (M. E. Jones) Munz; *E. heermannii* var. *subracemosum* (S. Stokes) Reveal; *E. howellii* S. Stokes; *E. howellii* var. *subracemosum* S. Stokes

Shrubs, spreading to rounded, sparsely to densely branched, 1–10 × 2–10(–12) dm. **Leaf blades** linear to linear-lanceolate or oblanceolate to elliptic or oblong, 0.5–1.2(–1.5) × 0.1–0.6 cm, tomentose or glabrous abaxially. **Inflorescences** (1–)3–20 × (1–)3–25 cm; branches diffuse with racemosely arranged involucres at tips, slender to stout, round, smooth, glabrous, scabrellous, or infrequently papillate-scabrous, infrequently somewhat spinose. **Involucres** narrowly turbinate to campanulate, 0.9–1.8 × 0.7–1.3 mm. **Flowers** 1.5–4 mm; perianth white to yellowish white or reddish.

Flowering Apr–Nov. Limestone (rarely volcanic) cliffs, outcrops, and washes, or on gypsophilous outcrops, saltbush, blackbrush, sagebrush, and mountain mahogany communities, pinyon-juniper and montane conifer woodlands; 800–2800 m; Ariz., Calif., Nev.

Variety *argense* is widespread and occasionally locally common. It is found in the Mohave Desert from Inyo County south through San Bernardino County to northern Riverside County, California, then eastward across Nevada in northern Clark, Esmeralda, Lincoln, Nye, and White Pine counties, into Coconino, Mohave, Navajo counties, Arizona. It just enters the Sonoran Desert in Yavapai County, Arizona, and is disjunct farther to the east in Graham and Gila counties. A population near Vail (Gila County) sampled by M. E. Jones in 1903 (DS, POM, US) has not been rediscovered, although the variety is known from Pueblo Canyon in the Sierra Ancha (*Wagner 320*, DUKE).

The scabrellous condition of the inflorescence branches of this variety can be obscure, occasionally leading to its misidentification as var. *heermannii*. The flowering stems and inflorescence branches are always rounded, not angled or ribbed as in var. *sulcatum*, but the latter condition is sometimes difficult to see, resulting in plants of that variety being misidentified as var. *argense*. Specimens from above the Little Colorado River, and along the Colorado River near the mouth of the Little Colorado, with stout, often pointed branches have been distinguished as var. *subracemosum*. They are low, spreading, and usually sparsely branched, unlike

the typical expressions of var. *argense*. Based on herbarium specimens, it appears that var. *subracemosum* merges completely with classical var. *argense* on the cliff faces above the Colorado River on and to the east of the Kaibab Plateau. The disjunct *Eriogonum apachense* of gypsophilous outcrops northwest of Bylas in Graham County, Arizona, differs little from plants found on limestone ledges and slopes in the White Hills of Yavapai County. By and large, the involucres of typical *E. apachense* are narrowly turbinate, the leaf blades remain tomentose on the abaxial surface, and the inflorescence branches are decidedly more slender. The reduction of *E. apachense* seems justified, resulting in a var. *argense* that is highly variable but nonetheless easy to distinguish.

88h. Eriogonum heermannii Durand & Hilgard var. **sulcatum** (S. Watson) Munz & Reveal in P. A. Munz, Suppl. Calif. Fl., 62. 1968 • Heermann's grooved wild buckwheat E

Eriogonum sulcatum S. Watson, Proc. Amer. Acad. Arts 14: 296. 1879; *E. heermannii* subsp. *sulcatum* (S. Watson) S. Stokes

Subshrubs, spreading, densely branched, (0.5–)1–8 × 2–8 dm. **Leaf blades** linear-lanceolate to elliptic or spatulate, 0.4–1.2(–1.5) × 0.2–0.8 cm, thinly tomentose abaxially. **Inflorescences** 1–5(–8) × 3–10 cm; branches diffuse, with dichotomously arranged involucres, slender, sharply ridged and deeply grooved, minutely scabrellous, not spinose. **Involucres** campanulate, 0.7–1.5(–2) × 0.7–1.5(–2) mm. **Flowers** 1.5–2.5 mm; perianth yellowish white.

Flowering Apr–Oct. Limestone cliffs and outcrops, saltbush, blackbrush, sagebrush, and mountain mahogany communities, scrub oak and pinyon-juniper woodlands; 700–2700 m; Ariz., Calif., Nev., Utah.

Variety *sulcatum* is locally common and occurs typically on limestone outcrops and cliff-faces. It is found in desert ranges of southeastern Inyo and northeastern San Bernardino counties, California, in Clark and southern Lincoln and Nye counties, Nevada, and in Washington County, Utah. In Arizona the plants occur in northern Mohave County and into northwestern Coconino County. The variety would make an ideal addition to the rock garden as a subshrub if one were willing to allow sufficient time for the plants to mature. This slow-growing plant forms dense, dark green clumps of tangled inflorescence branches and is attractive even without flowers.

89. Eriogonum plumatella Durand & Hilgard, J. Acad. Nat. Sci. Philadelphia, n. s. 3: 45. 1855 • Yucca wild buckwheat [E]

Eriogonum plumatella var. *jaegeri* (Munz & I. M. Johnston) S. Stokes ex Munz

Shrubs, erect, not scapose, 3–10 (–12) × 3–6(–8) dm, tomentose or glabrous, grayish or greenish. **Stems** erect, without persistent leaf bases, up to ¹/₂ or more height of plant; caudex stems absent; aerial flowering stems erect, slender, solid, not fistulose, 0.5–2 dm, tomentose or glabrous. **Leaves** basal and cauline, 1 per node; petiole 0.1–0.5 cm, mostly tomentose; blade oblanceolate, 0.6–1.5 × 0.2–0.3 cm, tomentose on both surfaces, occasionally slightly less so adaxially, margins plane. **Inflorescences** cymose, flat-topped and in tiers, 15–40 × 15–40 cm; branches dichotomous, divaricately arranged in a zig-zag pattern, tomentose or glabrous; bracts 3, scalelike, 1–7 mm. **Peduncles** absent. **Involucres** 1 per node, turbinate, 2–2.5 × 1.5–2 mm, glabrous; teeth 5, erect, 0.4–0.6 mm. **Flowers** 2–2.5 mm; perianth white to pale yellow, glabrous; tepals connate proximal ¹/₄ dimorphic, those of outer whorl obovate, those of inner whorl oblong; stamens exserted, 2–4 mm; filaments pilose proximally. **Achenes** light brown to brown, 2.5–3 mm, glabrous.

Flowering Apr–Oct. Sandy flats, washes, and slopes, mixed grassland, saltbush, blackbrush, and creosote bush communities, oak and pinyon-juniper woodlands; 400–1700 m; Ariz., Calif., Nev.

Eriogonum plumatella is encountered infrequently, primarily on the Mojave Desert, where it occurs in two forms, one with pubescent (var. *plumatella*) and one with glabrous (var. *jaegeri*) inflorescence branches. As both forms occur occasionally in the same population, the varieties are not considered to be taxonomically significant. *Eriogonum plumatella* occurs from Kern County, California, south through Los Angeles (including the foothills of the San Gabriel Mountains), San Bernardino, and Riverside counties to eastern San Diego County. To the east the species is found in Clark County, Nevada, and Mohave and western Yavapai counties, Arizona. A single specimen (*Palmer s.n.*, GH) reportedly was gathered in Utah. Repeated efforts to find the plant in Washington County, Utah, have been unsuccessful, and the record is discounted here.

This species is the food plant for the rare pallid dotted-blue butterfly (*Euphilotes pallescens pallescens*). According to M. L. Zigmond (1981), the Kawaiisu pounded the seeds into a powder and made a mush.

Given the small size of the achenes, and the paucity of large populations, the task of gathering sufficient seed must have been daunting. In taste, these seeds differ little from other, more common members of wild buckwheat, so it is likely that achenes of several species were gathered and processed into a watery meal.

90. Eriogonum apricum J. T. Howell, Leafl. W. Bot. 7: 237. 1955 [C][E]

Herbs, erect to spreading or prostrate, not scapose, 0.08–0.2 × 0.1–0.25 dm, glabrous, usually grayish. **Stems** spreading, with persistent leaf bases, up to ¹/₅ height of plant; caudex stems matted; aerial flowering stems erect to slightly spreading or prostrate, slender, solid, not fistulose, 0.4–0.8 dm, glabrous. **Leaves** basal, 1 per node; petiole 0.3–1(–2.5) cm, tomentose; blade round-ovate, 0.3–0.5(–1) × 0.3–0.5(–1) cm, densely white-tomentose abaxially, floccose or glabrous and green adaxially, margins plane. **Inflorescences** cymose, 0.5–1.5 × 1–2 cm; branches dichotomous, glabrous; bracts 3, scalelike, 1–2 mm. **Peduncles** absent. **Involucres** 1 per node, campanulate, 2–2.5 × 2–2.5 mm, glabrous; teeth 5, erect, 0.5–0.8 mm. **Flowers** 2–3 mm; perianth white, glabrous; tepals connate proximal ¹/₄, monomorphic, oblong; stamens exserted, 2.5–3 mm; filaments sparsely pubescent proximally. **Achenes** light brown, 2.5–3 mm, glabrous.

Eriogonum apricum appears to be more akin to the Pacific Coast *E. nudum* complex than to the *E. brevicaule* complex of the Rocky Mountain West. The plants are sometimes infected by *Coniothrium eriogoni* Earle, a rust fungus that also attacks *E. nudum*.

Varieties 2 (2 in the flora): California.

1. Flowering stems erect to slightly spreading . . . .
. . . . 90a. *Eriogonum apricum* var. *apricum*
1. Flowering stems prostrate . . . .
. . . . 90b. *Eriogonum apricum* var. *prostratum*

90a. Eriogonum apricum J. T. Howell var. **apricum** • Ione wild buckwheat [C][E]

Aerial flowering stems erect to slightly spreading. 2*n* = 40.

Flowering Jun–Oct. Clay hills and barren outcrops, manzanita communities, conifer woodlands; of conservation concern; 80–200 m; Calif.

Variety *apricum* is federally listed as endangered. It is known only from a 10-acre site south of Ione in Amador County.

90b. Eriogonum apricum J. T. Howell var. **prostratum** Myatt, Madroño 20: 320, fig. 1. 1970 • Irish Hill wild buckwheat C E

Aerial flowering stems prostrate. Flowering Jun–Sep. Clay hills and outcrops, manzanita communities, conifer woodlands; of conservation concern; 90–200 m; Calif.

Variety *prostratum* is federally listed as endangered. It is known from three sites that together constitute less than one acre north of Ione in Amador County.

91. Eriogonum nudum Douglas ex Bentham, Trans. Linn. Soc. London 17: 413. 1836 F

Eriogonum latifolium Smith subsp. *nudum* (Douglas ex Bentham) S. Stokes

Herbs, mostly erect, infrequently scapose, (0.5–)1–15(–20) × 0.5–3 dm, glabrous or floccose to tomentose, usually greenish, occasionally grayish. **Stems** spreading to erect, with or without persistent leaf bases, up to ¼ height of plant; caudex stems absent; aerial flowering stems erect to spreading, slender to stout, solid or hollow, occasionally fistulose, 0.3–4(–10) dm, glabrous or lanate to tomentose or floccose. **Leaves** basal or sheathing up stem 0.5–4 dm; petiole 1–10 cm, glabrous or tomentose; blade oblanceolate to elliptic or ovate, 1–6 × (0.3–)1–4 cm, densely white-lanate or tomentose abaxially, tomentose to floccose or subglabrous to glabrous adaxially, margins plane or undulate-crisped. **Inflorescences** cymose, rarely umbellate or capitate, 2–100(–150) × 2–40(–80) cm; branches usually dichotomous, glabrous or tomentose to floccose or sparsely pubescent; bracts 3, scalelike, 0.5–3(–5) mm. **Peduncles** absent. **Involucres** 1 per node or 2–10 per cluster, turbinate to turbinate-campanulate, (2.5–)3–5(–7) × (1.5–)2–4 mm, glabrous, tomentose, or sparsely pubescent; teeth 5–8, 0.2–0.6 mm. **Flowers** (1.5–)2–4 mm; perianth white or yellow, sometimes pink or rose, glabrous or pubescent; tepals connate proximal ¼, monomorphic, oblong to obovate; stamens exserted, 2–5 mm; filaments pilose proximally. **Achenes** light brown to brown, 1.5–3.5 mm, glabrous.

Varieties 13 (13 in the flora): w United States, nw Mexico.

Several local groups of Native Americans in California used members of this species in a variety of ways. S. A. Barrett and E. W. Gifford (1933) and S. M. Schenck and E. W. Gifford (1952) reported the consumption of raw young stems that are rather moist and tasty, although there is a sour aftertaste. The Kawaiisu used the hollow stems (probably var. *westonii*, rather than var. *pauciflorum*) as drinking tubes and as pipes (M. L. Zigmond 1981). Zigmond reported also that the roots of var. *pauciflorum* are used as an infusion for coughs.

Members of *Eriogonum nudum* are food plants for the Bauer's dotted-blue butterfly (*Euphilotes baueri*), the Pacific dotted-blue (*E. enoptes*), the gorgon copper (*Gaeides gorgon*), and the Mormon metalmark (*Apodemia mormo*).

1. Involucres and inflorescence branches tomentose to floccose; leaves usually basal.
  2. Perianths white; flowering stems lanate to tomentose; leaf blades 2–3.5 cm; s Sierra Nevada . . . . 91k. *Eriogonum nudum* var. *regirivum* (in part)
  2. Perianths yellow or white to rose; flowering stems tomentose to floccose; leaf blades 2–4 cm; s Oregon, n California, wc Nevada . . . . . . 91m. *Eriogonum nudum* var. *oblongifolium*
1. Involucres and inflorescence branches glabrous or, if pubescent, leaves sheathing up stems.
  3. Leaves sheathing up stems, margins often strongly undulate-crisped.
    4. Flowering stems tomentose.
      5. Perianths pubescent; flowers 1.5–2 mm; involucres 3–4 mm; Sierra Nevada, California . . . . 91k. *Eriogonum nudum* var. *regirivum* (in part)
      5. Perianths glabrous; flowers 3–4 mm; involucres 4–6 mm; Coast Ranges, California . . . . 91l. *Eriogonum nudum* var. *decurrens*
    4. Flowering stems glabrous.
      6. Leaf blades densely lanate abaxially, tomentose adaxially; involucres 5–10 per cluster; Sierra Nevada, California . . . . . 91h. *Eriogonum nudum* var. *murinum*
      6. Leaf blades tomentose abaxially, less so to floccose, glabrous, or nearly so adaxially; involucres 1 per node or 2–5 per cluster; widespread, Coast Ranges, California.
        7. Flowering stems occasionally fistulose; involucres (2–)3–5 per cluster; perianths white to pink, rarely yellowish . . . . 91i. *Eriogonum nudum* var. *auriculatum*
        7. Flowering stems strongly fistulose; involucres 1 per node; perianths pale yellowish white to yellow or white . . . 91j. *Eriogonum nudum* var. *indictum*

[3. Shifted to left margin.—Ed.]
3. Leaves basal, margins plane or slightly undulate-crisped.
  8. Involucres 1(–2) per cluster . . . .
    . . . . 91g. *Eriogonum nudum* var. *westonii* (in part)
    9. Flowering stems fistulose; perianths yellow, infrequently white; c California . . . .
    . 91g. *Eriogonum nudum* var. *westonii* (in part)
    9. Flowering stems not fistulose or, if so, plants of sw California; perianths white, rarely yellow.
      10. Perianths pubescent or, infrequently, glabrous; s California . . . .
      . . . . 91e. *Eriogonum nudum* var. *pauciflorum* (in part)
      10. Flowers glabrous; sw Oregon, n California, Nevada.
        11. Leaf blades 1–5 cm; sw Oregon, n California (including w slope of Sierra Nevada), Washington . .
        . . . . 91a. *Eriogonum nudum* var. *nudum* (in part)
        11. Leaf blades 1–2 cm; Sierra Nevada of California and adjacent desert ranges of wc Nevada . . . .
        . . . . 91c. *Eriogonum nudum* var. *deductum*
  8. Involucres 2–10 per cluster.
    12. Perianths pubescent, often yellow.
      13. Flowering stems not fistulose; leaf blades floccose or glabrous adaxially, margins plane; plants of nonarid regions . . . . 91f. *Eriogonum nudum* var. *pubiflorum*
      13. Flowering stems slightly to distinctly fistulose; leaf blades tomentose to floccose adaxially, margins undulate-crisped; plants of arid regions . . .
      . . . . 91g. *Eriogonum nudum* var. *westonii* (in part)
    12. Perianths usually glabrous abaxially, white, rarely yellow.
      14. Inflorescences capitate or nearly so; alpine, Sierra Nevada . . . .
      . . . . 91d. *Eriogonum nudum* var. *scapigerum*
      14. Inflorescences cymose or, if capitate, not alpine.
        15. Involucres 1(–2) per cluster, 5–7 mm; mountains of s California
        . . . . 91e. *Eriogonum nudum* var. *pauciflorum* (in part)
        15. Involucres 2–10 per cluster, 3–5 mm; mountains and foothills of c and n California or coastal bluffs.

[16. Shifted to left margin—Ed.]
16. Inflorescences cymose and branched 2 or more times; involucres 2–5 per cluster; mountains and foothills of c and n California, Oregon, and Washington . . . . 91a. *Eriogonum nudum* var. *nudum* (in part)
16. Inflorescences capitate or cymose and branched 1–2 times; involucres 5–10 per cluster; coastal bluffs, sw Oregon, ne California . . . .
  . . . . 91b. *Eriogonum nudum* var. *paralinum*

91a. Eriogonum nudum Douglas ex Bentham var. **nudum** • Naked wild buckwheat   E   F

Eriogonum latifolium Smith var. *parvulum* S. Stokes

Plants 3–10 dm. **Aerial flowering stems** not fistulose, (1.5–)2–4 dm, glabrous. **Leaves** basal; blade 1–5 × 0.5–2 cm, tomentose abaxially, thinly floccose or glabrous adaxially, margins plane. **Inflorescences** cymose, 20–50 cm; branches glabrous. **Involucres** 2–5 per cluster, 3–5 mm, glabrous or sparsely pubescent. **Flowers** 2–3 mm; perianth white, rarely yellow, glabrous. $2n = 40$.

Flowering Jun–Sep. Sandy to gravelly flats, slopes, banks, and washes, mixed grassland and chaparral communities, oak and conifer woodlands; 10–2100 (–2400) m; Calif., Oreg., Wash.

Variety *nudum* is the low-elevation tetraploid expression of the species, found mainly in the Coast Ranges and interior valleys from southern Washington through Oregon to California. It is replaced by var. *deductum* at higher elevations in the Sierra Nevada. Yellow-flowered populations occur rarely in the Siskiyou Mountains of California and Oregon.

91b. Eriogonum nudum Douglas ex Bentham var. **paralinum** Reveal, Phytologia 66: 258. 1989 • Port Orford wild buckwheat   E

Plants 0.5–2 dm. **Aerial flowering stems** not fistulose, 0.3–1.2 dm, glabrous. **Leaves** basal; blade 1–2 × 0.5–1.5 cm, densely tomentose abaxially, glabrous adaxially, margins plane. **Inflorescences** capitate or cymose and branched 1–2 times, 5–20 × 10–20 cm; branches glabrous. **Involucres** 5–10 per cluster, 3.5–5 mm, glabrous. **Flowers** 3–3.5 mm; perianth white, glabrous.

Flowering Jun–Oct. Sandy to gravelly flats, mesas, or coastal bluffs, mixed grassland and manzanita communities, oak and scattered conifer woodlands; 0–80(–600) m; Calif., Oreg.

Variety *paralinum* occurs along the immediate coast in Curry County, Oregon, and Del Norte County, California. This variety bridges the morphologic gap between *Eriogonum nudum* and *E. latifolium*. S. G. Stokes (1936) combined the two species, and her action can be justified. That taxonomy, however, would entail recognition of two subspecies, one each for what are maintained here as species, and little would be gained by such action.

91c. Eriogonum nudum Douglas ex Bentham var. **deductum** (Greene) Jepson, Fl. Calif. 1: 420. 1914 • Reduced wild buckwheat [E]

Eriogonum deductum Greene, Pittonia 5: 71. 1902

Plants 2–3(–5) dm. **Aerial flowering stems** not fistulose, 0.5–1.5 dm, glabrous. **Leaves** basal; blade 1–2 × 0.3–0.8(–1) cm, tomentose abaxially, floccose or glabrous adaxially, margins plane. **Inflorescences** cymose, 5–15 × 5–15(–20) cm; branches glabrous. **Involucres** 1(–2) per cluster, 2.5–3.5 mm, glabrous or sparsely pubescent. **Flowers** 2–3 mm; perianth white, glabrous. $2n = 40$.

Flowering Jun–Sep. Granitic flats and slopes, manzanita, buckbrush, and sagebrush communities, scrub oak and conifer woodlands; (1100–)1500–3000 (–3200) m; Calif., Nev.

Variety *deductum* is the common mid-elevation tetraploid expression of the species, found primarily in the Sierra Nevada, including some desert ranges just to the east. It is not always readily distinguishable from var. *nudum*. At higher elevations, var. *deductum* is replaced by var. *scapigerum*.

91d. Eriogonum nudum Douglas ex Bentham var. **scapigerum** (Eastwood) Jepson, Fl. Calif. 1: 420. 1914 • Sierran crest wild buckwheat [E]

Eriogonum scapigerum Eastwood, Proc. Calif. Acad. Sci., ser. 3, 3: 286. 1902; *E. latifolium* Smith var. *scapigerum* (Eastwood) S. Stokes

Plants 1–2 dm. **Aerial flowering stems** not fistulose, 1–1.8 dm, glabrous. **Leaves** basal; blade 1–2 × 0.3–0.8(–1) cm, tomentose abaxially, mostly glabrous adaxially, margins plane. **Inflorescences** capitate or nearly so, 2–5 × 2–5 cm; branches glabrous. **Involucres** 3–6 per cluster, 2–3 mm, glabrous or sparsely pubescent. **Flowers** 2–3 mm; perianth white, usually glabrous, rarely slightly pubescent.

Flowering Jul–Sep. Granitic sandy flats, slopes, ridges, talus slopes, and alpine fell-fields, mountain meadow and high-elevation sagebrush communities, alpine conifer woodlands; 2800–3800 m; Calif.

Variety *scapigerum* is the alpine expression of the species and occurs in the Sierra Nevada (Fresno, Inyo, Madera, Mariposa, Mono, Tulare, and Tuolumne counties). It is not always distinct from var. *deductum*. Reduction of both var. *deductum* and var. *scapigerum* into an already variable var. *nudum* could be justified, but limited garden studies suggest that the extremes are genetically fixed. Specimens from the southern Sierra Nevada tend to have slightly pubescent flowers, suggesting that these may be from populations of depauperate var. *pauciflorum*.

91e. Eriogonum nudum Douglas ex Bentham var. **pauciflorum** S. Watson, Proc. Amer. Acad. Arts 12: 264. 1877 • Little-flower wild buckwheat [E]

Eriogonum latifolium Smith subsp. *pauciflorum* (S. Watson) S. Stokes

Plants 3–8 dm. **Aerial flowering stems** rarely slightly fistulose, 1.5–5 dm, glabrous. **Leaves** basal; blade 1.5–3 × 0.8–1.8 cm, densely tomentose abaxially, glabrous adaxially, margins plane. **Inflorescences** cymose, 20–50 × 10–30 cm; branches glabrous. **Involucres** 1(–2) per cluster, 5–7 mm, glabrous or sparsely pubescent. **Flowers** 2–2.5 mm; perianth white or, rarely, yellow, glabrous or pubescent. $2n = 40$.

Flowering Jun–Oct. Sandy, often granitic flats and slopes, grassland, saltbush, chaparral, and sagebrush communities, oak and montane conifer woodlands; 1100–2800 m; Calif.; Mexico (Baja California).

Variety *pauciflorum* is the southern montane, tetraploid expression of the species. It occurs in the Transverse Ranges, southward in the Peninsular Ranges of southern California (Kern, Los Angeles, Orange, Riverside, San Bernardino, San Diego, Santa Barbara, and Ventura counties), and south to the mountains of northern Baja California. Care must be taken when identifying specimens from the San Jacinto and Santa Rosa mountains in Riverside County to ascertain the age of the plants, as the annual *Eriogonum molestum* can closely resemble the perennial var. *pauciflorum*. The yellow-flowered expression of var. *pauciflorum* is worthy of cultivation.

91f. Eriogonum nudum Douglas ex Bentham var. **pubiflorum** Bentham in A. P. de Candolle and A. L. P. P. de Candolle, Prodr. 14: 13. 1856 • Frémont's wild buckwheat [E]

Plants 3–8(–10) dm. **Aerial flowering stems** not fistulose, 1–3 dm, glabrous. **Leaves** basal; blade 1–4 × 0.5–2 cm, tomentose abaxially, floccose or glabrous adaxially, margins plane. **Inflorescences** cymose, 20–70 × 20–40 cm; branches glabrous. **Involucres** 2–7 per cluster, 3–5 mm, glabrous or sparsely pubescent. **Flowers** 2.5–3 mm; perianth white or, infrequently, yellow, pubescent. 2*n* = 40.

Flowering Jun–Oct. Sandy to gravelly, often volcanic flats and slopes, mixed grassland, manzanita, buckbrush, and sagebrush communities, scrub oak and montane conifer woodlands; 50–2200 m; Calif., Nev., Oreg.

Variety *pubiflorum* is rather common throughout its extensive range in California, northwestern Nevada, and southern Oregon. It merges with var. *nudum* in scattered locations in northern California. In the Coast Ranges of central California, there is not always a sharp distinction between var. *pubiflorum* and var. *auriculatum*.

91g. Eriogonum nudum Douglas ex Bentham var. **westonii** (S. Stokes) J. T. Howell, Mentzelia 1: 20. 1976 • Weston's wild buckwheat [E]

Eriogonum latifolium Smith subsp. *westonii* S. Stokes, Eriogonum, 64. 1936 (as westoni); *E. gramineum* S. Stokes; *E. latifolium* subsp. *saxicola* (A. Heller) S. Stokes; *E. nudum* var. *gramineum* (S. Stokes) Reveal; *E. nudum* subsp. *saxicola* (A. Heller) Munz; *E. saxicola* A. Heller

Plants 3–6 dm. **Aerial flowering stems** fistulose, 1–3 dm, glabrous. **Leaves** basal; blade 1–2 × 0.3–0.8 dm, tomentose abaxially, tomentose to floccose adaxially, margins undulate-crisped. **Inflorescences** often cymose, 10–30 × 5–20 cm; branches glabrous. **Involucres** 1(–2) per cluster, 3–5 mm, glabrous. **Flowers** 2.5–3 mm; perianth yellow or, infrequently, white, pubescent.

Flowering May–Sep. Sandy, gravelly, or clayey flats, slopes or rocky outcrops, mixed grassland, saltbush, and sagebrush communities, oak, pinyon-juniper, and conifer woodlands; 700–1900 m; Calif.

Variety *westonii* is widespread and occasionally common along the foothills of the Inner Coast and Transverse ranges of Kern and Santa Barbara counties,

and onto the foothills of the Sierra Nevada in Tulare County, where the plants tend to have multiple, slender to slightly fistulose flowering stems arising from a large, woody caudex. This phase gives way to a different expression along the eastern slope of the Sierra Nevada and adjacent desert ranges in Inyo and Mono counties, where plants tend to have one or a few prominently inflated flowering stems arising from a much more compact caudex. The name var. *gramineum* is available for this latter phase.

91h. Eriogonum nudum Douglas ex Bentham var. **murinum** Reveal, Aliso 7: 228. 1970 • Mouse wild buckwheat [E]

Plants 3–6 dm. **Aerial flowering stems** not fistulose, 2–3 dm, glabrous. **Leaves** sheathing; blade 1.5–3.5 × 1–2 cm, lanate abaxially, tomentose adaxially, margins plane. **Inflorescences** cymose, 15–30 × 10–20 cm; branches glabrous. **Involucres** 5–10 per cluster, (4–)5–6 mm, glabrous. **Flowers** 3–4 mm; perianth white, pubescent.

Flowering May–Oct. Sandy flats and slopes, mixed grassland, manzanita, and buckbrush communities, oak and montane conifer woodlands; 400–700 m; Calif.

Variety *murinum* is a rare and localized taxon in the foothills of the Sierra Nevada in Tulare County.

91i. Eriogonum nudum Douglas ex Bentham var. **auriculatum** (Bentham) J. P. Tracy ex Jepson, Fl. Calif. 1: 420. 1914 • Ear-shaped wild buckwheat [E]

Eriogonum auriculatum Bentham, Trans. Linn. Soc. London 17: 412. 1836; *E. latifolium* Smith var. *alternans* S. Stokes; *E. latifolium* subsp. *auriculatum* (Bentham) S. Stokes

Plants 5–15(–20) dm. **Aerial flowering stems** occasionally fistulose, 2–5(–10) dm, glabrous. **Leaves** sheathing; blade 3–7 × 2–4 cm, tomentose abaxially, subglabrous or glabrous adaxially, margins strongly undulate-crisped. **Inflorescences** cymose, 30–100(–150) × 10–80 cm; branches glabrous. **Involucres** (2–)3–5 per cluster, 3–4 mm, glabrous or sparsely pubescent. **Flowers** 2.5–3 mm; perianth white to pink, rarely yellowish, glabrous or sparsely pubescent. 2*n* = 80.

Flowering May–Oct. Sandy to gravelly or clayey flats and slopes, mixed grassland and chaparral communities, oak and conifer woodlands; 0–1200 m; Calif.

Variety *auriculatum* is a highly variable taxon that occurs along the coast and in the adjacent mountains (Alameda, Contra Costa, Humboldt, Lake, Marin, Monterey, Napa, San Benito, San Francisco, San Luis Obispo, San Mateo, Santa Clara, Santa Cruz, Solano, Sonoma, and western Stanislaus counties).

A related and as yet undescribed variety, known only from the Antioch sand dunes area of Contra Costa County, California, is the host for the federally endangered Lange's metalmark butterfly (*Apodemia mormo langei*).

91j. Eriogonum nudum Douglas ex Bentham var. **indictum** (Jepson) Reveal in P. A. Munz, Suppl. Calif. Fl., 69. 1968 • Protruding wild buckwheat E

Eriogonum indictum Jepson, Fl. Calif. 1: 421. 1914; *E. latifolium* Smith var. *indictum* (Jepson) S. Stokes

Plants 5–8(–10) dm. **Aerial flowering stems** strongly fistulose, 2.5–3.5 dm, glabrous. **Leaves** sheathing; blade 1–6 × 1–2(–3) cm, densely tomentose abaxially, less so to floccose adaxially, margins strongly undulate-crisped. **Inflorescences** cymose, 25–50 × 15–25 cm; branches glabrous. **Involucres** 1 per node, 4–5 mm, glabrous. **Flowers** 2.5–3 mm; perianth pale yellowish white to yellow or white, glabrous. *2n* = 80.

Flowering May–Oct. Clayey slopes, grassland communities, oak and conifer woodlands; 100–1100 m; Calif.

Variety *indictum* is found on the inner Coast Ranges of central California. For the most part, its range is east of that of var. *auriculatum*, occurring in Fresno, western Kern, Kings, Merced, eastern Monterey, San Benito, and eastern San Luis Obispo counties.

91k. Eriogonum nudum Douglas ex Bentham var. **regirivum** Reveal & J. C. Stebbins, Phytologia 66: 247. 1989 • Kings River wild buckwheat E

Plants 5–10 dm. **Aerial flowering stems** not fistulose, 1–3(–4) dm, lanate to tomentose. **Leaves** usually basal, rarely sheathing; blade 2–3.5 × 1.5–2.5 cm, lanate abaxially, tomentose adaxially, margins plane. **Inflorescences** cymose, 30–100(–140) × 20–40 cm; branches lanate to tomentose. **Involucres** 1 per node, 3–4 mm, tomentose. **Flowers** 1.5–2 mm; perianth white, pubescent. *2n* = 40.

Flowering Aug–Nov. Limestone slopes, oak and pine woodlands; 200–600 m; Calif.

Variety *regirivum* is a rare and localized taxon presently known only from the Pine Flat Reservoir area and near Dunlap in Fresno County. Most of the historic range of the taxon is now under Pine Flat Reservoir.

91l. Eriogonum nudum Douglas ex Bentham var. **decurrens** (S. Stokes) M. L. Bowerman, Fl. Pl. Ferns Mt. Diablo, 140. 1944 • Ben Lomand wild buckwheat E

Eriogonum latifolium Smith subsp. *decurrens* S. Stokes, Eriogonum, 64. 1936

Plants 5–12(–15) dm. **Aerial flowering stems** occasionally slightly fistulose, 3–6 dm, tomentose. **Leaves** sheathing; blade 1–3 × 1–1.5 cm, tomentose abaxially, thinly floccose or glabrous adaxially, margins undulate-crisped. **Inflorescences** cymose, 50–100 × 30–60 cm; branches tomentose. **Involucres** 1–2 per cluster, 4–6 mm, tomentose. **Flowers** 3–4 mm; perianth white, glabrous.

Flowering Jul–Oct. Deep sandy flats and slopes, chaparral communities, oak and pine woodlands; 90–200 m; Calif.

Variety *decurrens* is restricted to the Ben Lomand sand hills area of Santa Cruz County, where it is uncommon.

91m. Eriogonum nudum Douglas ex Bentham var. **oblongifolium** S. Watson, Proc. Amer. Acad. Arts 12: 264. 1877 • Harford's wild buckwheat E

Eriogonum affine Bentham in A. P. de Candolle and A. L. P. P. de Candolle, Prodr. 14: 13. 1856; *E. capitatum* A. Heller; *E. harfordii* Small; *E. latifolium* Smith var. *affine* (Bentham) S. Stokes; *E. latifolium* var. *harfordii* (Small) S. Stokes; *E. latifolium* subsp. *sulphureum* (Greene) S. Stokes; *E. nudum* var. *sulphureum* (Greene) Jepson; *E. sulphureum* Greene

Plants 5–10(–18) dm. **Flowering stems** not fistulose, 2–5(–8) dm, tomentose to floccose. **Leaves** basal; blade 2–4 × 1.5–2 cm, tomentose abaxially, thinly floccose adaxially, margins plane. **Inflorescences** cymose, 20–50(–100) × 10–50 cm; branches tomentose to floccose. **Involucres** 3–6 per cluster, 3–5 mm, tomentose. **Flowers** 3–4 mm; perianth white to rose or yellow, pubescent. *2n* = 40.

Flowering May–Oct. Sandy to gravelly slopes and flats, grassland and chaparral communities, oak and montane conifer woodlands; 20–1900 m; Calif., Nev., Oreg.

Variety *oblongifolium* is the widespread and rather common tetraploid phase of the species, being the only expression with pubescent flowering stems. It occurs from southern Oregon south to northern California and western Nevada. The plants vary in flower color. This variety and var. *pubiflorum* occasionally grow together. In such instances, differentiation between the two based on the presence or absence of tomentum can be arbitrary.

92. **Eriogonum latifolium** Smith in A. Rees, Cycl. 13(2): Eriogonum no. 3. 1809 • Seaside wild buckwheat [E]

Subshrubs or herbs, often scapose, much-branched and matted, 2–7 × 5–20 dm, usually tomentose to floccose, rarely glabrous. **Stems** spreading to erect, with persistent leaf bases, up to ¼ height of plant; caudex stems matted; aerial flowering stems often scapelike, erect to spreading or decumbent, usually stout, solid, not fistulose, 2–6 dm, usually tomentose to floccose, rarely glabrous. **Leaves** cauline; petiole 2–6(–10) cm, tomentose; blade oblong to ovate, (1.5–)2.5–5 × 1.5–4 cm, white-lanate to tawny-tomentose on both surfaces, or tomentose to floccose or glabrous and green adaxially, margins plane, occasionally crisped. **Inflorescences** capitate to umbellate or cymose, 3–40 × 2–20 cm; branches usually tomentose to floccose, rarely glabrous; bracts usually 3, leaflike, oblong to ovate, and 5–20 × 5–15 mm proximally, scalelike, triangular, and 2–5 mm distally. **Peduncles** absent. **Involucres** (3–)5–20 per cluster, turbinate, 3.5–5(–6) × 2–4 mm, tomentose or glabrous; teeth 5–6, erect, 0.3–0.6 mm. **Flowers** 3–3.5 mm; perianth white to pink or rose, glabrous; tepals connate proximal ¼, monomorphic, obovate; stamens exserted, 3–6 mm; filaments pilose proximally. **Achenes** brown, 3.5–4 mm, glabrous. $2n = 40$.

Flowering year-round. Sandy coastal flats, slopes, bluffs, and mesas, coastal scrub and grassland communities; 0–80(–200) m; Calif., Oreg.

Eriogonum latifolium is found along the immediate coast of southwest Oregon (Curry County) and western California (Alameda, Contra Costa, Del Norte, Humboldt, Marin, Mendocino, Monterey, San Francisco, San Luis Obispo, San Mateo, Santa Cruz, and Sonoma counties). The species is rather variable as to size and aspect, these depending to a considerable degree on exposure to on-shore winds. The flowering stems are rarely glabrous, but plants with this expression are always intermixed with plants having tomentose to floccose stems. The brilliantly white-lanate, spreading shrubs become rather globose in shape under cultivation, and as a result make an attractive addition to the garden, especially as the flowers wither through various shades of pink to rose. The species should be used much more than at present in places where cool summer temperatures, good moisture, and sandy soils are available.

A decoction consisting of the roots, leaves, and stems of *Eriogonum latifolium* was taken by various Native American people along the California coast for colds and coughs (B. R. Bocek 1984; D. E. Moerman 1986). V. K. Chestnut (1902) reported that the native people of Mendocino County, California, used a decoction of the roots for stomach pain, "female complaints," and sore eyes. The species is the food plant for the bramble hairstreak butterfly (*Callophrys viridis*), Mormon metalmark (*Apodemia mormo*), western square-dotted blue (*Euphilotes comstocki comstocki*), and the federally endangered Smith's dotted-blue (*Euphilotes enoptes smithi*).

93. **Eriogonum grande** Greene, Pittonia 1: 38. 1887

Eriogonum latifolium Smith subsp. *grande* (Greene) S. Stokes; *E. nudum* Douglas ex Bentham var. *grande* (Greene) Jepson

Subshrubs or herbs, much-branched and matted, not scapose, (1–)5–15 × 2–8 dm, glabrous, rarely floccose. **Stems** spreading to erect, with persistent leaf bases, up to ¼ height of plant; caudex stems matted; aerial flowering stems erect to spreading, usually stout, solid or hollow, occasionally fistulose, (0.8–)2–6 dm, glabrous, rarely floccose. **Leaves** basal or sheathing up stems 0.5–3 dm; petiole 5–20 cm, floccose; blade oblong to oblong-ovate, (1.5–)3–10 × (1–)2–6 cm, densely white-tomentose abaxially, tomentose to floccose to subglabrous and greenish adaxially, margins plane, occasionally crisped. **Inflorescences** subcapitate to cymose, (5–)20–100 × 4–50 cm; branches usually dichotomous, glabrous, rarely floccose; bracts 3–8 or more, leaflike, oblong to oblong-ovate, and 5–20 × 5–10 mm proximally, scalelike, triangular, and 2–6 mm distally. **Peduncles** absent. **Involucres** 1–3 per cluster, turbinate-campanulate to campanulate, (4–)5–7 × 3–6 mm, floccose to subglabrous; teeth 5–8, erect, 0.3–0.6 mm. **Flowers** 2.5–3.5 mm; perianth white to pink, rose, or red, glabrous; tepals connate proximal ¼, monomorphic, oblong-obovate; stamens exserted, 3–6 mm; filaments pilose proximally. **Achenes** brown, 2.5–3 mm, glabrous.

Varieties 4 (3 in the flora): California; nw Mexico.

In the flora area, *Eriogonum grande* is the insular phase of the *E. nudum* complex; var. *testudinum* Reveal is found on the mainland in Baja California, Mexico.

1. Perianths pink to red or rose; plants 2–5 dm; involucres 5–7 mm; San Miguel, Santa Cruz, and Santa Rosa islands, California . . . .
. . . . 93c. *Eriogonum grande* var. *rubescens*
1. Perianths white; plants 1–15 dm; involucres 4–6 mm; various islands of California, but not Santa Rosa Island.
  2. Plants 5–15 dm; involucres 5–6 mm; Santa Cruz, San Miguel, Santa Catalina, Anacarpa, and San Clemente islands, California . . . . .
  . . . . 93a. *Eriogonum grande* var. *grande*
  2. Plants 1–2(–2.5) dm; involucres 4–5 mm; San Nicolas Island, California . . . .
  . . . . 93b. *Eriogonum grande* var. *timorum*

93a. Eriogonum grande Greene var. **grande**
  • Pacific Island wild buckwheat   E

Subshrubs, 5–15 dm, glabrous. **Aerial flowering stems** often fistulose, 2–6 dm. **Leaf blades** (2–)3–10 2–6 cm, tomentose abaxially, subglabrous adaxially. **Inflorescences** cymose, 20–100 × (5–)10–50 cm. **Involucres** turbinate-campanulate, 5–6 × 3–5 mm. **Flowers** 2.5–3 mm; perianth white.

Flowering Mar–Oct. Rocky cliffs and bluffs, coastal grassland and scrub communities; 0–300 m; Calif.

Variety *grande* is found on Santa Cruz, San Miguel, Santa Catalina, Anacapa, and San Clemente islands off the coast of southern California, being fairly common on each island. The more robust plants are seen inland, away from the immediate effects of ocean winds.

93b. Eriogonum grande Greene var. **timorum** Reveal, Aliso 7: 229. 1970 • St. Nicholas wild buckwheat E

Eriogonum grande subsp. *timorum* (Reveal) Munz

Subshrubs or herbs, 1–2(–2.5) dm, glabrous. **Aerial flowering stems** not fistulose, 0.8–1.2(–1.5) dm. **Leaf blades** (1.5–)2–3.5 (–4.5) × 1–2(–2.5) cm, tomentose abaxially, tomentose to densely floccose adaxially. **Inflorescences** subcapitate to cymose, 5–8 × 4–8 cm. **Involucres** campanulate, 4–5 × 3–6 mm. **Flowers** 2.5–3(–3.5) mm; perianth white.

Flowering Apr–Oct. Rocky cliffs and bluffs, coastal grassland and scrub communities; 20–60 m; Calif.

Variety *timorum* is rare and restricted to San Nicolas Island, Ventura County. It is in cultivation but has not become naturalized on the mainland.

93c. Eriogonum grande Greene var. **rubescens** (Greene) Munz, Man. S. Calif. Bot., 597. 1935
  • Red-flowered pacific island wild buckwheat   E

Eriogonum rubescens Greene, Pittonia 1: 39. 1887; *E. grande* var. *dunklei* Reveal; *E. latifolium* Smith subsp. *rubescens* (Greene) S. Stokes; *E. latifolium* var. *rubescens* (Greene) Munz

Subshrubs or herbs, 2–5 dm, occasionally floccose. **Aerial flowering stems** occasionally fistulose, 2–6 dm. **Leaf blades** 2–5(–6) × 1–3(–4.5) cm, tomentose abaxially, subglabrous adaxially. **Inflorescences** subcapitate to cymose, 5–15(–30) × 5–20 cm. **Involucres** campanulate, 5–7 × 4–6 mm. **Flowers** 2.5–3 mm; perianth pink to red or rose.

Flowering Apr–Sep. Cliffs and bluffs, coastal grassland and scrub communities; 10–200 m; Calif.

Variety *rubescens* occurs naturally on San Miguel, Santa Cruz, and Santa Rosa islands off the coast of southern California. This expression is highly attractive and frequently planted. It has yet to be found naturalized on the mainland.

94. Eriogonum elatum Douglas ex Bentham, Trans. Linn. Soc. London 17: 413. 1836   E

Herbs, erect, not scapose, 4–8 (–15) × 1–4 dm, glabrous or villous. **Stems** erect, without persistent leaf bases, up to $^1/_5$ height of plant; caudex stems absent; aerial flowering stems erect, slender to stout, solid or hollow, infrequently fistulose, 1.5–4(–8) dm, glabrous, tomentose, or villous. **Leaves** basal; petiole 5–25 cm, villous; blade lanceolate to lance-ovate, 4–15(–25) × 1.5–6 cm, loosely villous and green on both surfaces or infrequently thinly tomentose abaxially and glabrate adaxially, margins plane. **Inflorescences** cymose, 15–50 × 10–30 cm; branches glabrous or villous; bracts 3, semileaflike, linear, and 5–30 × 2–5 mm proximally, scalelike, triangular, and 1–4 mm distally. **Peduncles** absent or slender, erect, 0.5–4 cm, glabrous or tomentose. **Involucres** 1 per node or 2–5 per cluster, turbinate, (2.5–)3–4 × 2.5–3 mm, glabrous or slightly tomentose; teeth 5, erect, 0.4–0.9 mm. **Flowers** 2.5–4 mm, glabrous; perianth white; tepals connate proximal $^1/_4$, monomorphic, obovate; stamens exserted, 2.5–4 mm; filaments pilose proximally. **Achenes** light brown, 3.5–4 mm, glabrous.

Varieties 2 (2 in the flora): w United States.

Eriogonum elatum is widely distributed but rather scattered throughout its range. The plants are occasionally seen in the garden. Inflorescence branches were chewed or made into an infusion and taken as a physic by some Native American people (D. E. Moerman 1986).

1. Flowering stems and inflorescence branches glabrous . . . . 94a. *Eriogonum elatum* var. *elatum*
1. Flowering stems and inflorescence branches villous . . . . 94b. *Eriogonum elatum* var. *villosum*

94a. Eriogonum elatum Douglas ex Bentham var. **elatum** • Tall wild buckwheat ⊑

Aerial flowering stems glabrous. **Inflorescence branches** glabrous. $2n = 40$.

Flowering May–Oct. Sandy to gravelly slopes and flats, mixed grassland and sagebrush communities, pinyon, juniper, and conifer woodlands; 60–3000 m; Calif., Idaho, Nev., Oreg., Wash.

Variety *elatum* is found mainly along the eastern edge of the Cascade Ranges in Washington south into northern Oregon, and skips to the Siskiyou/Trinity mountains of southwestern Oregon and northwestern California. To the east in the desert ranges, it is found in south-central Oregon and northeastern California. The third major area of concentration is in the Sierra Nevada of California and east into some of the desert ranges in west-central Nevada. It is found as an isolated disjunct at Lloyd Meadows in Tulare County, and on the Kern Plateau in Kern County, California. An 1876 collection labeled as being from Auburn, Placer County, California (*Ames s.n.*, WIS), is discounted, as that location is doubtful.

94b. Eriogonum elatum Douglas ex Bentham var. **villosum** Jepson, Fl. Calif. 1: 421. 1914 • Tall woolly wild buckwheat ⊑

Eriogonum elatum var. *incurvum* Jepson; *E. elatum* subsp. *villosum* (Jepson) Munz ex Reveal

Aerial flowering stems villous. **Inflorescence branches** villous.

Flowering Jun–Sep. Sandy to gravelly slopes and flats, sagebrush communities, pinyon and/or juniper and montane conifer woodlands; 500–2000(–2500) m; Calif., Nev., Oreg.

Variety *villosum* is found primarily in scattered places in the southern portion of the species' range, namely in the Siskiyou/Trinity mountains of southern Oregon (Jackson, Josephine, and Klamath counties) and northern California (Siskiyou and Trinity counties), and in the Sierra Nevada of California (Nevada, Placer, Plumas, and Shasta counties) and Nevada (Carson City, Douglas, and Washoe counties). The taxonomic significance of this variety is questionable, as pubescence may be the result of minor genetic differences.

95. Eriogonum suffruticosum S. Watson, Proc. Amer. Acad. Arts 20: 370. 1885 • Bushy wild buckwheat ⊑ ⊑

Subshrubs, spreading, not scapose, 1–2 × 2–4(–5) dm, thinly tomentose. **Stems** spreading, without persistent leaf bases, up to ¼ height of plant; caudex stems absent; aerial flowering stems erect, slender, solid, not fistulose, 0.1–0.3 dm, thinly tomentose. **Leaves** cauline, 1 per node or fasciculate; petiole 0.1–0.2 cm; blade narrowly elliptic, 0.5–0.8 × 0.1–0.25 cm, densely white-tomentose abaxially, silky-tomentose adaxially, margins revolute. **Inflorescences** cymose, 0.5–1 × 0.5–1.5 cm; branches dichotomous, thinly tomentose; bracts 3, scalelike, narrowly elliptic, 1–3 mm. **Peduncles** slender, erect, 0.3–0.5 cm, thinly tomentose. **Involucres** 1 per node, campanulate, 2–3 × 2.5–3.5(–4) mm, thinly tomentose; lobes 6, reflexed, 1–1.5 mm. **Flowers** 3.5–6 mm; perianth white to yellowish white, with large reddish to maroon spots, glabrous; tepals connate proximally, dimorphic, those of the outer whorl fan-shaped, 2–4 × 2–4 mm, those of inner whorl oblanceolate, 3.5–6 × 1–2.5 mm; stamens slightly exserted, 2–3 mm; filaments pilose proximally. **Achenes** light brown, 3–3.5 mm, glabrous.

Flowering Apr–May. Rocky limestone slopes and outcrops, creosote bush communities; of conservation concern; 900–1400 m; Tex.

Eriogonum suffruticosum is a rare, localized species, known only from 13 scattered locations in Brewster and Presidio counties of western Texas. Its relationship to other species in the genus is obscure.

96. Eriogonum strictum Bentham, Trans. Linn. Soc.
London 17: 414. 1836 E F

Herbs, erect, loosely to densely matted, not scapose, 1–5 × 1–10 dm, tomentose or glabrous. **Stems** spreading to erect, with or without persistent leaf bases, up to ¼ height of plant; caudex stems absent or spreading to matted; aerial flowering stems erect to spreading, slender, solid, not fistulose, 1–3 dm, tomentose or glabrous. **Leaves** basal, 1 per node; petiole rarely twisted or curled, 1–6 cm, mostly tomentose; blade elliptic to ovate, 0.5–2.5(–4) × (0.3–)0.5–1.5 cm, lanate, tomentose to floccose on both surfaces, sometimes sparsely tomentose to floccose and greenish or floccose to subglabrous or glabrous adaxially, margins plane. **Inflorescences** umbellate-cymose to cymose, 1–20 × 3–25 cm; branches dichotomous, tomentose to floccose or less often glabrous; bracts 3, scalelike, triangular, 1–3 mm. **Peduncles** absent. **Involucres** 1 per node, rarely 2–5 per cluster, narrowly turbinate to turbinate-campanulate, 4–6 × 1.5–5 mm, tomentose or glabrous; teeth 5, erect, 0.5–1.3 mm. **Flowers** 3–5(–6) mm; perianth yellow or white to rose or purple, glabrous; tepals connate proximally, dimorphic, those of outer whorl elliptic to nearly orbiculate, 2–3 × 2–3 mm, those of inner whorl oblanceolate to oblong, 3–4 × 1–2 mm; stamens included to slightly exserted, 2–5 mm; filaments pilose proximally. **Achenes** light brown to brown, 3–3.5 mm, glabrous.

Varieties 4 (4 in the flora): w United States.

Eriogonum strictum, E. niveum, and *E. ovalifolium* form a complex of closely related species differing in leaf, inflorescence branching, and flower features. Variety *proliferum* appears to be the basal entity of the complex, approaching both *E. niveum* and *E. ovalifolium* var. *pansum* in its pubescence and branching pattern. Also, specimens of var. *proliferum* are sometimes difficult to differentiate from *E. nudum* var. *oblongifolium.* Careful observation, though, will permit well-made collections to be easily distributed among the individual species. An alternative taxonomy is to reduce all of the taxa to *E. ovalifolium* and recognize a series of subspecies and varieties. It is possible that additional study will show that *E. strictum* is sufficiently distinct from its tomentose to floccose counterparts to justify recognition of *E. proliferum.* In that case, both var. *anserinum* and var. *greenei* would be assigned to the latter species. Or, one could follow C. L. Hitchcock et al. (1955–1969, vol. 2) and recognize subsp. *strictum* as distinct from subsp. *proliferum,* with the latter consisting of varieties *proliferum, anserinum,* and *greenei.*

Members of the *Eriogonum strictum* are food plants for the Bauer's dotted-blue butterfly (*Euphilotes baueri*).

1. Inflorescence branches glabrous. . . . .
. . . . 96d. *Eriogonum strictum* var. *strictum*
1. Inflorescence branches tomentose to floccose.
2. Perianths yellow . . . . 96c. *Eriogonum strictum* var. *anserinum*
2. Perianths white to rose or purple.
3. Leaf blades grayish-tomentose to floccose on both surfaces, or greenish-tomentose to floccose adaxially . . . 96a. *Eriogonum strictum* var. *proliferum*
3. Leaf blades densely whitish-lanate to tomentose on both surfaces . . . . 96b. *Eriogonum strictum* var. *greenei*

96a. Eriogonum strictum Bentham var. **proliferum** (Torrey & A. Gray) C. L. Hitchcock in C. L. Hitchcock et al., Vasc. Pl. Pacif. N.W. 2: 132. 1964 • Proliferous wild buckwheat E F

Eriogonum proliferum Torrey & A. Gray, Proc. Amer. Acad. Arts 8: 164. 1870; *E. fulvum* S. Stokes; *E. strictum* var. *argenteum* S. Stokes; *E. strictum* subsp. *bellum* (S. Stokes) S. Stokes; *E. strictum* subsp. *proliferum* (Torrey & A. Gray) S. Stokes

Plants loose mats, 2–4 dm wide. **Leaf blades** broadly elliptic to ovate, 1–3 cm, grayish-tomentose to floccose on both surfaces, sometimes greenish-tomentose to floccose and greenish adaxially. **Inflorescences** 5–15 cm; branches tomentose to floccose. **Involucres** 4–6 × 2–4 mm, tomentose. **Flowers** 3–5 mm; perianth white to rose or purple.

Flowering Jun–Sep. Sandy to gravelly flats and slopes, sagebrush communities, montane conifer woodlands; (100–)400–2700 m; Calif., Idaho, Mont., Nev., Oreg., Wash.

Variety *proliferum* is widespread and often rather common throughout its range. The largest concentration is found in a gentle arc from northeastern Washington to southern Idaho and western Montana. The variety is widely distributed also in central and eastern Oregon, northern California, and Nevada. In portions of central Idaho and western Montana, some individuals clearly approach *Eriogonum ovalifolium* var. *pansum.*

96b. Eriogonum strictum Bentham var. **greenei** (A. Gray) Reveal, Phytologia 40: 467. 1978

• Greene's wild buckwheat [E]

Eriogonum greenei A. Gray, Proc. Amer. Acad. Arts 12: 83. 1876; *E. niveum* Douglas ex Bentham var. *greenei* (A. Gray) S. Stokes

Plants dense mats, 2–10 dm wide. **Leaf blades** ovate, 0.5–1 cm, densely white-lanate to tomentose on both surfaces. **Inflorescences** 3–5 cm; branches tomentose. **Involucres** 4–6 × 2–3 mm, tomentose. **Flowers** 4–5 mm; perianth white.

Flowering Jun–Sep. Rocky, typically serpentine flats, slopes, and ridges, sagebrush communities, montane conifer woodlands; 1500–2400 m; Calif.

Variety *greenei* is restricted to the North Coast Ranges of northwestern California (Humboldt, Mendocino, Siskiyou, Tehama, and Trinity counties). In the herbarium, it is occasionally difficult to distinguish from nonserpentine populations of var. *proliferum*. The variety is worthy of cultivation.

96c. Eriogonum strictum Bentham var. **anserinum** (Greene) S. Stokes in R. J. Davis, Fl. Idaho, 249. 1952 • Goose Lake wild buckwheat [E]

Eriogonum anserinum Greene, Pittonia 4: 320. 1901; *E. ovalifolium* Nuttall subsp. *flavissimum* (Gandoger) S. Stokes; *E. proliferum* Torrey & A. Gray subsp. *anserinum* (Greene) Munz; *E. strictum* subsp. *anserinum* (Greene) S. Stokes; *E. strictum* var. *flavissimum* (Gandoger) C. L. Hitchcock

Plants loose mats, 2–4 dm wide. **Leaf blades** usually ovate, 0.5–2 cm, grayish-tomentose or floccose on both surfaces. **Inflorescences** 1–3(–5) cm; branches tomentose to floccose. **Involucres** 4–5.5 × 4–5 mm, tomentose. **Flowers** 3–4.5 mm; perianth yellow.

Flowering May–Aug. Sandy to gravelly flats and slopes, sagebrush communities, conifer woodlands; (100–)400–2600 m; Calif., Idaho, Nev., Oreg., Wash.

Variety *anserinum* is the yellow-flowered phase of the species; it and var. *proliferum* are only occasionally found together. This taxon is widely scattered in most of its range in northeastern California, southwestern Idaho, northern Nevada, eastern Oregon, and eastern Washington. It is common mainly from south-central Oregon south into northwestern Nevada and eastern California. The plants are attractive and are occasionally seen in cultivation.

96d. Eriogonum strictum Bentham var. **strictum**

• Strict wild buckwheat [E]

Eriogonum strictum var. *glabrum* C. L. Hitchcock

Plants loose mats, 1–4 dm wide. **Leaf blades** elliptic to ovate, 1–4 cm, densely tomentose abaxially, floccose to subglabrous or glabrous adaxially. **Inflorescences** (2–)5–20 cm; branches glabrous. **Involucres** 4–5 × 1.5–3 mm, glabrous or nearly so. **Flowers** 3–6 mm; perianth white.

Flowering May–Aug. Sandy to gravelly flats and slopes, mixed grassland and sagebrush communities, conifer woodlands; 100–1400(–1800) m; Idaho, Oreg., Wash.

Variety *strictum* is infrequent and widely scattered throughout its range in west-central Idaho (Adams, Nez Perce, and Washington counties), northeastern Oregon (Douglas, Morrow, Umatilla, Union, and Wallowa counties), and southeastern Washington (Benton, Columbia, Douglas, Grant, Kittitas, and Yakima counties). Its greatest concentration is in the Blue Mountains of northeastern Oregon and extreme southeastern Washington.

97. Eriogonum niveum Douglas ex Bentham, Trans. Linn. Soc. London 17: 414. 1836 • Snow wild buckwheat [E]

Eriogonum niveum subsp. *decumbens* (Bentham) S. Stokes; *E. niveum* subsp. *dichotomum* (Douglas ex Bentham) S. Stokes

Herbs, erect to decumbent or prostrate, loosely matted, not scapose, 2–6 × 1–4(–6) dm, tomentose to floccose. **Stems** spreading, with or without persistent leaf bases, up to ¹/₄ height of plant; caudex stems absent or spreading to matted; aerial flowering stems erect to spreading or decument to prostrate, slender, solid, not fistulose, (0.5–)1–3 dm, tomentose to floccose. **Leaves** 1 per node, sheathing up stems (1–)3–10(–12) cm; petiole often twisted or curled, 1–10 cm, usually tomentose; blade lanceolate or broadly lanceolate to elliptic or narrowly ovate, (1–)1.5–6(–8) × 0.6–1.5 cm, densely lanate on both surfaces or densely white-tomentose abaxially and grayish-tomentose to floccose adaxially, margins plane, rarely brownish. **Inflorescences** cymose, 10–30 × 50–250 cm; branches dichotomous, tomentose to floccose; bracts 3, semileaflike to leaflike, lanceolate to narrowly elliptic, and 5–30 × 3–20 mm proximally, scalelike, triangular, and 1–4(–5) mm distally. **Peduncles** absent. **Involucres**

1 per node, turbinate, 3–5 × 2–3 mm, tomentose to floccose; teeth 5, erect, 0.5–1.5 mm. **Flowers** 3–6 mm; perianth cream to reddish, glabrous; tepals connate proximally, dimorphic, those of outer whorl oblong to oval, 3–6 × 2–4 mm, those of inner whorl oblanceolate to oblong, 4–6 × 1–2 mm; stamens included to slightly exserted, 2–5 mm; filaments pilose proximally. **Achenes** light brown to brown, (3.5–)4–4.5(–5) mm, glabrous.

Flowering Jun–Oct. Sandy to gravelly flats, slopes, bluffs, and rocky, often volcanic outcrops, mixed grassland and sagebrush communities, montane conifer woodlands; 30–1100(–1300) m; B.C.; Idaho, Oreg., Wash.

Eriogonum niveum is a highly variable species with a multitude of minor expressions that do not appear to have any biogeographic or taxonomic significance. The species is found mainly on the grassy plains east of the Cascade Range in southern British Columbia, west-central Idaho, northeastern Oregon, and eastern Washington. Some populations closely approach *E. strictum* var. *proliferum*, but the densely lanate leaves and semileaflike to leaflike bracts nearly always distinguish *E. niveum* from that taxon where their ranges overlap. It may well prove that *E. niveum* would be better treated as a subspecies of *E. strictum*, but the nomenclatural combination is not available and it is not suggested here. The plants do well in cultivation.

N. J. Turner et al. (1980) reported that the snow wild buckwheat was used by the Okanagan-Colville people for colds and as a wash for cuts.

98. Eriogonum ovalifolium Nuttall, J. Acad. Nat. Sci. Philadelphia 7: 50, plate 8, fig. 1. 1834 E F

Eucycla ovalifolia (Nuttall) Nuttall

Herbs, forming pulvinate to cespitose maps, usually scapose, 0.2–3 × 0.5–5 dm, floccose to tomentose or lanate. **Stems** decumbent to spreading, with persistent leaf bases, up to ¹/₅ height of plant; caudex stems matted to spreading; aerial flowering stems scapelike, spreading to erect, infrequently decumbent to ascending, slender, solid, not fistulose, 0.03–4 dm, floccose to tomentose or lanate or nearly glabrous. **Leaves** basal, 1 per node; petiole not twisted or curled, 0.1–10 cm, mostly tomentose; blade oblanceolate to elliptic or spatulate to rounded, 0.2–6 × (0.1–)0.2–1.5 cm, lanate to tomentose or floccose, sometimes less so adaxially, margins plane, occasionally brownish. **Inflorescences** capitate or rarely umbellate, 0.7–5(–7) × 1.5–5 cm; bracts 3, scalelike, linear to triangular, 0.8–4 mm. **Peduncles** absent. **Involucres** 1 per node or (2–)3–15 per cluster, turbinate to turbinate-campanulate, (2–)3.5–5(–8) × 2–4 mm, tomentose to floccose; teeth 5, erect, 0.1–1 mm. **Flowers** (2.5–)3–6 (–7) mm; perianth yellow or white to cream, rose, red, or purple, glabrous; tepals connate proximally, dimorphic, those of outer whorl usually oval to orbiculate, 2–4 × 2–4 mm, those of inner whorl oblanceolate to elliptic, 3–7 × 0.8–1.5 mm; stamens mostly included, 1–3 mm; filaments pilose proximally. **Achenes** light brown to brown, 2–3 mm, glabrous.

Varieties 11 (11 in the flora): w North America.

Eriogonum ovalifolium is a highly diverse and widespread complex of generally distinct but sometimes intergrading varieties. Several varieties are in cultivation and make worthy additions, especially to the rock garden. The dimorphic nature of the tepals is obvious only in fully mature flowers. In some populations of the more depauperate varieties, such as var. *nivale* and var. *depressum*, the tepals may not be as distinctly dimorphic. Nonetheless, the overall aspect of the species is unmistakable.

There are several reports of traditional use of *Eriogonum ovalifolium* by Native Americans. P. Train et al. (1941) indicated that a decoction of the roots is used in Nevada for colds. R. V. Chamberlin (1911) reported that the Gosiute Indians in southwestern Utah used it in a poultice or wash to treat venereal diseases.

Members of the species are food plants for Bauer's dotted-blue butterfly (*Euphilotes baueri*).

1. Leaf blades 0.2–1.2(–2) cm; scapes usually 0.3–5(–9) cm, rarely longer; involucres usually 2–4.5 (rarely 5–8) mm.
  2. Perianths yellow; c Nevada and ec California ..... 98k. *Eriogonum ovalifolium* var. *caelestinum*
  2. Perianths white, sometimes rose, purple, or red; widespread.
    3. Leaf margins brownish.
      4. Leaf blades densely lanate, margins brownish ..... 98g. *Eriogonum ovalifolium* var. *eximium* (in part)
      4. Leaf blades tomentose, margins not brownish ..... 98h. *Eriogonum ovalifolium* var. *williamsiae* (in part)
    3. Leaf margins not brownish or, if so, plants of high elevation, central Sierra Nevada.
      5. Involucres 5–8 mm.
        6. Leaf blades white-lanate; San Bernardino Mountains, s California 98d. .... *Eriogonum ovalifolium* var. *vineum* (in part)
        6. Leaf blades tomentose to floccose; s Sierra Nevada, California .... 98e. *Eriogonum ovalifolium* var. *monarchense*

5. Involucres 3–4.5 mm.
    7. Leaf blades greenish and thinly tomentose at least adaxially, usually elliptic, infrequently oblong to spatulate; scapes often suberect to decumbent, usually thinly floccose; Rocky Mountains of Alberta, Idaho, Montana, and Wyoming, desert ranges of ne Nevada and e Oregon .... 98i. *Eriogonum ovalifolium* var. *depressum*
    7. Leaves densely lanate to white-tomentose on both surfaces or only slightly less so adaxially, not at all greenish, usually round; scapes usually erect, lanate or tomentose; Sierra-Cascade cordillera, desert ranges of Great Basin .... 98j. *Eriogonum ovalifolium* var. *nivale*
1. Leaf blades usually 1–6 cm, occasionally shorter; scapes (1–)5–30(–40) cm; involucres (3.5–)4–6.5 (–8) mm.
    8. Scapes usually 1–5(–7.5) cm; leaf margins brownish; rare, wc Nevada, ec California.
        9. Leaf blades densely lanate, brownish margins .... 98g. *Eriogonum ovalifolium* var. *eximium* (in part)
        9. Leaf blades white-tomentose, margins not brownish .... 98h. *Eriogonum ovalifolium* var. *williamsiae* (in part)
    8. Scapes usually (4–)7–30(–40) cm, rarely shorter; leaf margins only rarely brownish; widespread.
      10. Inflorescences umbellate, branches 1–3 cm; c Idaho and Montana .... 98f. *Eriogonum ovalifolium* var. *pansum*
      10. Inflorescences capitate, branches absent; widespread.
        11. Flowers 5–7 mm; involucres 5–7 mm; s California .... 98d. *Eriogonum ovalifolium* var. *vineum* (in part)
        11. Flowers 4–5 mm; involucres 4–6.5 mm; widespread.
          12. Perianths yellow .... 98b. *Eriogonum ovalifolium* var. *ovalifolium*
          12. Perianths white to rose or purple.
            [13. Shifted to left margin.—Ed.]
13. Leaf blades spatulate, oblong, or obovate to oval; scapes (4–)5–20 cm; widespread .... 98a. *Eriogonum ovalifolium* var. *purpureum*
13. Leaf blades oblanceolate to narrowly elliptic; scapes (15–)20–30(–40) cm; Montana and Wyoming .... 98c. *Eriogonum ovalifolium* var. *ochroleucum*

98a. Eriogonum ovalifolium Nuttall var. **purpureum** (Nuttall) Durand, Trans. Amer. Philos. Soc., n. s. 11: 175. 1860 • Purple cushion wild buckwheat [E]

Eucycla purpurea Nuttall, Proc. Acad. Nat. Sci. Philadelphia 4: 17. 1848; *Eriogonum davisianum* S. Stokes; *E. orthocaulon* Small; *E. ovalifolium* var. *celsum* A. Nelson; *E. ovalifolium* var. *orthocaulon* (Small) C. L. Hitchcock; *E. ovalifolium* subsp. *purpureum* (Nuttall) A. Nelson ex S. Stokes

Plants 2.5–4 dm wide. **Leaf blades** spatulate, oblong, or obovate to oval, 0.5–2 cm, tomentose to floccose, margins not brownish. **Scapes** erect, (4–)5–20 cm, tomentose. **Inflorescences** capitate, 1.5–3.5 cm wide; branches absent. **Involucres** 3–15 per cluster, 4–5 mm. **Flowers** 4–5 mm; perianth white to rose or purple. $2n = 40$.

Flowering Apr–Aug. Sandy to gravelly flats, washes, slopes, and ridges, mixed grassland, saltbush, and sagebrush communities, pinyon and/or juniper and montane conifer woodlands; 700–3100 m; Alta., B.C.; Ariz., Calif., Colo., Idaho, Mont., Nev., N.Mex., Oreg., Utah, Wash., Wyo.

Variety *purpureum* is the most widespread and common expression of the species, being found in southern British Columbia and southwestern Alberta, and in northern Arizona, eastern California, western Colorado, Idaho, western Montana, Nevada, northwestern New Mexico, eastern Oregon, Utah, southeastern Washington, and Wyoming. It approaches var. *depressum* both geographically and morphologically in the Yellowstone National Park area, and a clear distinction is not always possible. The name var. *ovalifolium* was long misapplied to what is here termed var. *purpureum*.

98b. Eriogonum ovalifolium Nuttall var. **ovalifolium** • Cushion wild buckwheat [E] [F]

Eriogonum ovalifolium var. *multiscapum* Gandoger; *E. ovalifolium* var. *nevadense* Gandoger

Plants 2.5–4 dm wide. **Leaf blades** usually elliptic to spatulate or oblong, (1–)3–6 cm, tomentose to floccose, margins rarely brownish. **Scapes** erect, (4–)5–20 cm, thinly tomentose. **Inflorescences** capitate. 1–3.5 cm wide; branches absent. **Involucres** 3–15 per cluster, 4–5(–6.5) mm. **Flowers** 4–5 mm; perianth yellow. $2n = 40$.

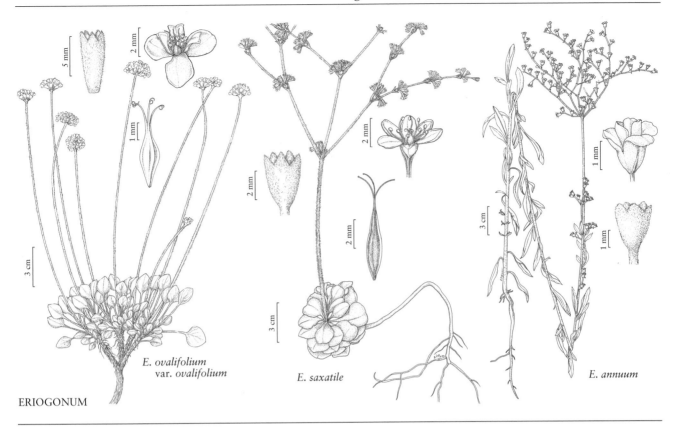

E. *ovalifolium*
var. *ovalifolium*

E. *saxatile*

E. *annuum*

ERIOGONUM

Flowering Apr–Aug. Sandy to gravelly flats, washes, slopes, and ridges, mixed grassland, saltbush, and sagebrush communities, pinyon and/or juniper and montane conifer woodlands; 600–2600 m; Calif., Colo., Idaho, Mont., Nev., Oreg., Utah, Wash., Wyo.

Variety *ovalifolium* is found in eastern California, northwestern Colorado, Idaho, Montana, Nevada, eastern Oregon, Utah, eastern Washington, and Wyoming. It is less widespread than var. *purpureum* and generally tends to flower earlier than that variety. The two sometimes occur together but do not seem to intergrade, although in some cases the only distinguishing feature is flower color. It is important to note that the yellowish hue of var. *ovalifolium* will fade in some herbarium material, making identification of older or less well-preserved material difficult. Brownish leaf margins occur on specimens of var. *ovalifolium* found in northeastern California (Lassen, Plumas, and Siskiyou counties).

98c. Eriogonum ovalifolium Nuttall var. **ochroleucum** (Small ex Rydberg) M. Peck, Man. Higher Pl. Oregon, 256. 1941 • Long-stemmed cushion wild buckwheat E

Eriogonum ochroleucum Small ex Rydberg, Mem. New York Bot. Gard. 1: 123. 1900; *E. ovalifolium* var. *macropodum* (Gandoger) Reveal; *E. ovalifolium* subsp. *ochroleucum* (Small ex Rydberg) S. Stokes

Plants 2.5–4 dm wide. **Leaf blades** usually oblanceolate to narrowly elliptic, 0.5–3 cm, tomentose, margins not brownish. **Scapes** erect, (15–)20–30(–40) cm, thinly tomentose. **Inflorescences** capitate, 1–2.5 cm wide; branches absent. **Involucres** 5–12 per cluster, 4–5 mm. **Flowers** 4–5 mm; perianth white.

Flowering Jun–Aug. Sandy to gravelly flats and slopes, mixed grassland and sagebrush communities, juniper and montane conifer woodlands; 900–2600 (–3100) m; Mont., Wyo.

Variety *ochroleucrum* is restricted to the western edge of the Great Plains as they interface with the eastern slope of the northern Rocky Mountains. The long scape atop a short-leafed mat is distinctive in the field. High-elevation populations occur in Park County, Wyoming; these can resemble var. *depressum*. In western Montana, great care must be taken to distinguish between early-flowering specimens of this variety and var. *pansum*. The involucres of the latter are distinctly longer and narrower, although the tight cluster of involucres often obscures the short branches of the umbellate inflorescence typical of var. *pansum*.

98d. Eriogonum ovalifolium Nuttall var. **vineum** (Small) A. Nelson, Bot. Gaz. 52: 262. 1911 • Cushionbury wild buckwheat [C] [E]

Eriogonum vineum Small, Bull. Torrey Bot. Club 25: 45. 1898; *E. ovalifolium* subsp. *vineum* (Small) S. Stokes

Plants 1.5–2.5 dm wide. **Leaf blades** usually round, 0.7–1.2 cm, densely white-lanate, margins not brownish. **Scapes** erect or nearly so, 3–6 cm, floccose to nearly glabrous. **Inflorescences** capitate, 1.5–3.5 cm wide; branches absent. **Involucres** 2–4(–5) per cluster, 5–7 mm. **Flowers** 5–7 mm; perianth white to cream.

Flowering May–Jun. Limestone talus slopes, creosote bush and saltbush communities, conifer woodlands; of conservation concern; 1500–2100 m; Calif.

Variety *vineum* is a rare and localized taxon restricted to the Cushionbury Grade area of San Bernardino County. It has been designated federally as endangered (see M. C. Neel et al. 2001; Neel and N. C. Ellstrand 2003) and is in the Center for Plant Conservation's National Collection of Endangered Plants. The plants are highly attractive and worthy of cultivation.

98e. Eriogonum ovalifolium Nuttall var. **monarchense** D. A. York, Madroño 49: 16, fig. 1. 2002 • Monarch wild buckwheat [E]

Plants 1.5–3 dm wide. **Leaf blades** elliptic to round, 0.3–1.2 cm, tomentose to densely floccose, margins not brownish. **Scapes** decumbent to ascending, 2–6(–9) cm, tomentose to floccose. **Inflorescences** capitate, 1.5–4 cm wide; branches absent. **Involucres** 4–6 per cluster, 5–8 mm. **Flowers** 4–6 mm; perianth white to cream.

Flowering Jun–Aug. Limestone outcrops, pinyon woodlands; 1800 m; Calif.

Variety *monarchense* is known from a single population of fewer than 30 individuals in the Kings River Canyon area of Fresno County.

98f. Eriogonum ovalifolium Nuttall var. **pansum** Reveal, Phytologia 66: 259. 1989 • Branched cushion wild buckwheat [E]

Plants 2–5 dm wide. **Leaf blades** usually elliptic, (0.6–)1–2(–2.5) cm, densely tomentose, margins not brownish. **Scapes** erect, (5–)7–20 cm, thinly tomentose. **Inflorescences** umbellate, 1–5(–7) × (1.5–)2–5 cm, thinly tomentose; branches absent. **Involucres** 1 per node, 4–5 mm. **Flowers** 2.5–5(–7) mm; perianth white.

Flowering Jun–Aug. Sandy to gravelly flats and slopes, sagebrush communities, conifer woodlands; 900–2200 m; Idaho, Mont.

Variety *pansum* occurs in two disjunct series of scattered populations. One is in central Idaho (Blaine, Boise, Custer, Elmore, Lemhi, and Valley counties); the second is in western Montana (Lewis and Clark, Lincoln, Missoula, Powell, Sanders, and Silver Bow counties). This variety bridges the morphologic gap between *E. ovalifolium* and *E. strictum* var. *proliferum*. The inflorescence of var. *pansum* is umbellate with branches up to 3 cm long, but never compoundly branched as in *E. strictum*. The umbellate condition becomes obvious only in late anthesis or during early fruit-set, so immature plants of var. *pansum* might be mistaken for var. *purpureum*, large specimens of var. *depressum*, or short-scaped plants of var. *ochroleucum*. Still, careful observation of such specimens will show an early branching condition, although this can be obscured by bracts subtending the inflorescence or an abundance of early flowers in the numerous involucres. The branched cushion wild buckwheat is certainly worth consideration as an ornamental.

98g. Eriogonum ovalifolium Nuttall var. **eximium** (Tidestrom) J. T. Howell, Mentzelia 1: 19. 1976 • Slide Mountain cushion wild buckwheat [E]

Eriogonum eximium Tidestrom, Proc. Biol. Soc. Wash. 36: 181. 1923; *E. ovalifolium* subsp. *eximium* (Tidestrom) S. Stokes

Plants 1–3 dm wide. **Leaf blades** elliptic or spatulate to oval, (0.3–)0.5–1.5(–2) cm, usually densely lanate, margins brownish. **Scapes** erect, 1–5(–7.5) cm, tomentose. **Inflorescences** capitate, 1–2.5 cm wide; branches absent. **Involucres** 3–10 per cluster, 4–5(–6.5) mm. **Flowers** 4–5 mm; perianth white.

Flowering Jun–Sep. Granitic sandy or gravelly to rocky or even talus slopes, sagebrush communities, montane conifer woodlands; 1700–2500 m; Calif., Nev.

Variety *eximium* is known only from the Carson Range in Carson City, Douglas, and Washoe counties, Nevada, and from the Job's Peak area of Alpine County, California. The Slide Mountain wild buckwheat morphologically approaches higher-elevation populations of var. *nivale* in the Carson Range, and the edaphically restricted var. *williamsiae* from Steamboat Springs. In addition, there is a series of populations in Siskiyou County, California, which approaches var. *eximium*. At this time, a definite disposition of those northern California populations has not been made. On infrequent occasions, the distinct brown leaf-blade margins are absent from individuals within an otherwise normal population. The variety is an ideal rock-garden plant, although it is slow to grow to the size seen in the field.

98h. Eriogonum ovalifolium Nuttall var. **williamsiae** Reveal, Brittonia 33: 446. 1981 • Steamboat Springs wild buckwheat [E]

Plants 1–5 dm wide. **Leaf blades** broadly elliptic to oval or reniform, 0.3–0.7 cm usually white-tomentose, margins brownish. **Scapes** erect, (1–)5–15(–18) cm, tomentose. **Inflorescences** capitate, 1–2 cm wide; branches absent. **Involucres** 3–10 per cluster, 4–5 mm. **Flowers** 4–5 mm; perianth white.

Flowering May–Jul. Heavy clay soil, sagebrush communities; 1400 m; Nev.

Variety *williamsiae* is listed federally as endangered and is known only from Steamboat Springs in southern Washoe County. The plants are restricted to an outcrop of sinter, a substrate derived from hot spring deposits. The population occupies ca. 150 hectares and is divided into three subpopulations.

J. K. Archibald et al. (2001) found that the "Steamboat buckwheat has high genetic variability, with levels of variation similar to that typical of a widespread species rather than a narrow endemic." They also found var. *williamsiae* to be allozymically most similar to var. *ovalifolium*, but conceded that the two are morphologically distinct. Evidence of male sterility in var. *williamsiae* suggested that the taxon might be "a hybrid or undergoing cytoplasmic introgression." The facts that these plants are capable of fragmentation and that they have low seed-set are not surprising, as these traits are found in other entities in subfam. Eriogonoideae (e.g., see discussion under *Dedeckera*). Given the long-lived nature of an individual plant, and the dense, crowded condition of the restricted habitat where the variety is found, the low rate of seed-set is obviously adequate for its long-term survival.

Variety *williamsiae* is in the Center for Plant Conservation's National Collection of Endangered Plants.

98i. Eriogonum ovalifolium Nuttall var. **depressum** Blankinship, Sci. Stud. Montana Coll. Agric., Bot. 1: 49. 1905 • Dwarf cushion wild buckwheat [E]

Eriogonum depressum (Blankinship) Rydberg

Plants 1–2.5 dm wide. **Leaf blades** elliptic or infrequently oblong to spatulate, 0.4–0.8 cm, tomentose to floccose, margins not brownish. **Scapes** often suberect to decumbent, 1–4(–8) cm, thinly floccose. **Inflorescences** capitate, 1–1.5(–2) cm wide; branches absent. **Involucres** 2–4 per cluster, 3–3.5 mm. **Flowers** 4–5 mm; perianth white to rose.

Flowering Jun–Aug. Sandy to gravelly flats, slopes, ridges, talus slopes, mixed grasslands, mountain meadows, sagebrush, alpine fell-field communities, montane to alpine conifer woodlands; 900–3500 m; Alta.; Idaho, Mont., Nev., Oreg., Wyo.

Variety *depressum* occurs mostly at higher elevations in widely scattered mountain ranges in southwestern Alberta and in Idaho, western Montana, eastern Oregon, and Wyoming. In Nevada the plants are isolated in the Independence Mountains in Elko County (*Maguire & Holmgren 22456*, NY, UTC). The variety is rather easy to distinguish from the related var. *nivale* except in the Steens Mountains of southeastern Oregon. Plants from that area are here assigned to var. *depressum* but with some caution. At Crater-of-the-Moon National Monument in Idaho, var. *depressum* is a local dominant, the densely white-tomentose plants distinctively spotting the black lava cinder cones and flats. Only there is the variety encountered at relatively low elevations. The dwarf cushion wild buckwheat is a highly attractive expression of the species and is frequently seen in cultivation.

98j. Eriogonum ovalifolium Nuttall var. **nivale** (Canby ex Coville) M. E. Jones, Contr. W. Bot. 11: 8. 1903 • Sierran cushion wild buckwheat [E]

Eriogonum nivale Canby ex Coville, Contr. U.S. Natl. Herb. 4: 187. 1893; *E. rhodanthum* A. Nelson & P. B. Kennedy

Plants 0.5–3 dm wide. **Leaf blades** usually round, 0.2–0.8 cm, densely lanate to white-tomentose, margins infrequently brownish. **Scapes** mostly erect, 0.3–

5(–13) cm, mostly lanate or tomentose. **Inflorescences** capitate, 1–2.5 cm wide; branches absent. **Involucres** 2–4 per cluster, 3–4.5 mm. **Flowers** 4–5 mm; perianth white to rose or red. $2n = 40$.

Flowering Jun–Sep. Sandy to gravelly often granitic slopes, ridges, and alpine fell-fields, mixed grassland, mountain meadow, and sagebrush communities, conifer woodlands; 1700–4200 m; B.C.; Calif., Nev., Oreg., Utah, Wash.

Variety *nivale* is the common high-elevation expression of the species in desert ranges of the Great Basin and in the Sierra-Cascade cordillera. In northwestern Washington, some plants of var. *nivale* have scapes to 13 cm (especially in Chellam County). They are well removed from var. *purpureum*, and have the dense, almost brilliant white tomentum of var. *nivale*. Plants in the Slide Mountain-Mt. Rose area of west-central Nevada occasionally have the brown leaf margins more commonly seen in the lower-elevation var. *eximium*. A few populations having off-yellowish flowers are known from the White Mountains of Inyo and Mono counties, California. There as well, the tepals are distinctly marked by a large red midrib. This is similar to the condition found in *Eriogonum gilmanii*, but in the White Mountains plants the tepals are not globose as in that species nor is the pubescence similar. The variety is ideal for the rock garden and is widely available for cultivation.

98k. Eriogonum ovalifolium Nuttall var. **caelestinum** Reveal, Great Basin Naturalist 32: 115. 1972

• Heavenly wild buckwheat [E]

Plants 0.5–1(–4) dm wide. **Leaf blades** usually elliptic, 0.2–0.5 cm, thinly tomentose, margins not brownish. **Scapes** erect, 1–6 cm, thinly floccose. **Inflorescences** capitate, 1–2 cm wide. **Involucres** 1–2 per cluster, 2–2.5 mm. **Flowers:** perianth yellow. $2n = 40$.

Flowering Jul–Aug. Granitic sand on alpine ridges and slopes, high-elevation sagebrush communities; 3000–3600 m; Calif., Nev.

Variety *caelestinum* is a high-elevation expression most closely related to var. *ovalifolium*. In both the Toiyabe and Toquima mountains of Lander and Nye counties, Nevada, and at Tioga Crest in Mono County, California, there is a pronounced elevational difference between the two, with var. *caelestinum* several thousand feet higher than var. *ovalifolium*. Heavenly wild buckwheat is occasionally seen in cultivation.

99. Eriogonum gilmanii S. Stokes, Leafl. W. Bot. 3: 16. 1941 (as gilmani) • Gilman's wild buckwheat [E]

Herbs, cespitose or pulvinate, occasionally scapose, 0.1–0.5 × 1–3 dm, tomentose. **Stems** spreading, with persistent leaf bases, up to ⅕ height of plant; caudex stems matted; aerial flowering stems scapelike, spreading to erect, slender, solid, not fistulose, 0.1–0.2 dm, tomentose. **Leaves** basal, 1 per node; petiole 0.2–0.5 cm, tomentose; blade elliptic, 0.2–0.4 × 0.1–0.2 cm, tomentose, margins plane. **Inflorescences** cymose-umbellate or capitate, (0.4–)0.8–1.5 × 0.5–1.5 cm; branches umbellate, tomentose; bracts 3, scalelike, triangular to linear-lanceolate, 1–5 mm. **Peduncles** spreading, 0.1–0.3(–0.5) cm, tomentose. **Involucres** 1 per node, turbinate, 1.5–2 × 1–1.5 mm, tomentose; teeth 5, erect, 0.4–0.7 mm. **Flowers** 3.5–5 mm; perianth yellowish, with reddish spot, glabrous; tepals connate proximally, dimorphic, those of outer whorl orbiculate and globose, 3–4 × 3–4 mm, those of inner whorl oblanceolate, 3.5–5 × 1–1.5 mm; stamens exserted, 3–4 mm; filaments pubescent proximally. **Achenes** brown, 2.5–3 mm, glabrous.

Flowering May–Sep. Limestone gravelly to rocky slopes and ridges, sagebrush communities, pinyon-juniper woodlands; 1500–2000 m; Calif.

Eriogonum gilmanii is a rare, localized species. It is known from the Cottonwood, Last Chance, and Panamint mountains of Inyo County.

100. Eriogonum saxatile S. Watson, Proc. Amer. Acad. Arts 12: 267. 1877 • Hoary wild buckwheat [E] [F]

Eriogonum saxatile subsp. *multicaule* S. Stokes

Herbs, loosely to densely matted, not scapose, (1–)2–4 × 0.5–2 dm, densely white- or grayish-lanate to tomentose or floccose. **Stems** spreading, often with persistent leaf bases, up to ¼ height of plant; caudex stems matted, decumbent to spreading; aerial flowering stems spreading to erect, slender, solid, not fistulose, 0.5–1.5 dm, lanate to tomentose or floccose. **Leaves** basal or sheathing up stems 4 cm, 1 per node or fasciculate at tips of caudex branches; proximal leaves: petiole 1–3(–4) cm, tomentose, blade obovate to rounded, 1–2(–2.5) × 1–2 cm, lanate to tomentose; distal leaves sessile, blade elliptic to rounded, 0.3–1 × 0.3–1 cm, lanate to tomentose. **Inflorescences** cymose, 10–25 × 5–12 cm; branches dichotomous, lanate to tomentose or floccose; bracts 3–4, scalelike, triangular, 1.5–7 mm. **Peduncles** absent. **Involucres** 1 per node, turbinate, 3–4 × 2–3 mm, tomentose to floccose; teeth 5–6, erect, 0.8–1.5 mm. **Flowers** (3–)5–7 mm, including elongate, sharply triangular, slightly winged, stipelike base; perianth white to rose or yellowish, glabrous; tepals connate proximally, dimorphic, those of outer whorl oblanceolate to lanceolate, 3–5 × 1.5–2 mm, those of inner whorl obovate, 4–6 × 2–3 mm; stamens included to slightly exserted, 2.5–5 mm; filaments pilose proximally. **Achenes** 3-gonous, nearly winged, 3.5–4 mm, glabrous. $2n = 40$.

Flowering May–Oct. Decomposed granitic or volcanic flats, slopes, and ridges, chaparral, saltbush, and sagebrush communities, pinyon-juniper and montane conifer woodlands; (300–)800–3400(–3500) m; Calif., Nev.

Eriogonum saxatile is found mainly in arid mountains of California (Fresno, Inyo, Kern, Los Angeles, Mono, Monterey, Riverside, San Benito, San Bernardino, Santa Barbara, Tulare, and Ventura counties) and Nevada (Esmeralda and western Nye counties). The plants vary considerably as to robustness, degree of branching, and sprawl of the caudex. The size and position of the leaves also vary, as does the density of tomentum on the blades. Flower color varies from white to rose or yellowish, but the deep yellow of *E. crocatum* is never seen in *E. saxatile*. The species is frequently cultivated and is an excellent plant for the rock garden.

101. **Eriogonum crocatum** Davidson, Bull. S. Calif. Acad. Sci. 23: 17, plate E. 1924 • Saffron wild buckwheat [C][E]

Eriogonum saxatile S. Watson var. *crocatum* (Davidson) Munz

Herbs, erect to spreading, not scapose, 3–5 × 5–10 dm, lanate to tomentose. **Stems** spreading, usually without persistent leaf bases, up to ¼ height of plant; caudex stems matted to spreading; aerial flowering stems spreading to erect, slender, solid, not fistulose, 0.2–1 dm, lanate to tomentose. **Leaves** cauline, 1 per node; petiole 0.3–1.2 cm tomentose; blade broadly ovate, 1–3(–3.5) × 0.8–2.5(–3) cm, lanate to tomentose. **Inflorescences** cymose-umbellate, 0.5–3 × 3–8 cm; branches umbellate, lanate to tomentose; bracts 3, scalelike, triangular, 1–2.5 mm. **Peduncles** absent. **Involucres** 1 per node, broadly campanulate, 3–4 × 3–5 mm, tomentose; teeth 5–6, erect, 0.5–1 mm. **Flowers** 5–6 mm, including elongate, rounded, slightly winged, stipelike base; perianth bright yellow, glabrous; tepals connate proximally, dimorphic, those of outer whorl narrowly oblong, 3–4 × 0.7–0.9 mm, those of inner whorl oblong to spatulate, 3.5–5 × 1–1.5 mm; stamens exserted, 3–5 mm; filaments pilose proximally. **Achenes** brown, 3-gonous, 2.5–3 mm. $2n = 40$.

Flowering Apr–Jul. Sandstone slopes, oak woodlands; of conservation concern; 60–200 m; Calif.

Eriogonum crocatum is known only from the Conejo Grade area of the Santa Monica Mountains in Ventura County. A population in the Malibu Hills of Los Angeles County sampled by M. E. Jones (26 Apr 1926, DS, POM) has not been rediscovered. The species is rare, and the state of California considers it worthy of protection. Federal action has been long delayed. The species is widely cultivated and is one of the more popular wild buckwheats in the garden. Great care should be taken to prevent its escape from cultivation.

1b. ERIOGONUM Michaux subg. MICRANTHA (Bentham) Reveal in J. E. Gunckel, Curr. Topics Pl. Sci., 237. 1969

Eriogonum sect. *Micrantha* Bentham, Trans. Linn. Soc. London 17: 413. 1836

Herbs, biennial or annual, polygamodioecious, floccose to tomentose; taproot somewhat woody. **Stems** erect, without persistent leaf bases, pubescent; caudex stems absent; aerial flowering stems erect, slender to stout, solid, not disarticulating in ringlike segments proximally, arising directly from the root. **Leaves** persistent or marcescent, in basal rosettes or fasciculate and scattered along stems, 1 per node; blade usually tomentose. **Inflorescences** terminal, cymose, compact to open; branches dichotomous, round, smooth, tomentose to floccose; bracts 3, connate basally, scalelike. **Peduncles** absent or erect, slender. **Involucres** 1 per node, not appressed to inflorescence branches, turbinate to campanulate; teeth 5–6, erect, apex acute. **Flowers** not attenuate at base, without stipelike base; perianth white to cream, rose, or reddish brown, glabrous abaxially, densely pubescent and minutely glandular adaxially; tepals connate proximally ¼–⅓ their length, dimorphic; stamens included or slightly exserted; filaments pilose proximally. **Achenes** brown, not winged, 3-gonous, glabrous. **Seeds:** embryo curved.

Species 2 (2 in the flora): c North America, including n Mexico.

1. Outer tepals obovate, 0.9–1.5 mm wide; involucres 2.5–4 mm, tomentose to floccose abaxially, glabrous adaxially; plants grayish . . . . 102. *Eriogonum annuum*
1. Outer tepals oblong-cordate, 1.5–2 mm wide; involucres 2–2.5 mm, usually tomentose on both surfaces; plants reddish . . . . 103. *Eriogonum multiflorum*

102. Eriogonum annuum Nuttall, Trans. Amer. Philos. Soc., n. s. 5: 164. 1835 • Annual wild buckwheat Ⓕ

Eriogonum annuum subsp. *cymosum* (Bentham) S. Stokes; *E. annuum* subsp. *hitchcockii* (Gandoger) S. Stokes

Herbs, 5–20 × 5–10 dm, grayish. **Aerial flowering stems** slender, 4–10(–15) dm, floccose to densely tomentose. **Leaves:** petiole (rosette) 0.3–1.2 cm, or petiole (cauline) 0.2–0.5 cm, tomentose to floccose; blade oblanceolate to oblong, 1–7 × 0.3–1.5 cm, densely tomentose abaxially, floccose adaxially not thickened and auriculate-subclasping proximally; margins entire or slightly revolute. **Inflorescences** 3–10 × 2–7 cm; bracts triangular, 1–4 mm. **Peduncles** 0.1–0.5 cm, tomentose to floccose. **Involucres** turbinate to campanulate, 2.5–4 × 2–3 mm, tomentose to floccose abaxially, glabrous adaxially; teeth 5–6, 0.4–1 mm. **Flowers** 1–2.5 mm; perianth white or cream to rose; tepals: those of outer whorl obovate, 1–2 × 0.9–1.5 mm, those of inner whorl narrowly ovate to oblong, 1.5–4 × 1.2–1.8 mm; stamens usually included, 1–2 mm. **Achenes** 1.5–2 mm. $2n = 40$.

Flowering Apr–Nov. Sandy flats, slopes, dunes, and banks, mixed grassland, oak and conifer woodlands; (0–)100–1900(–2300) m; Colo., Kans., Mont. Nebr., N.Mex., N.Dak., Okla., S.Dak., Tex., Wyo.; Mexico (Chihuahua).

Eriogonum annuum is widespread and common to locally abundant or even weedy on the Great Plains of the central United States and extreme north-central Mexico. It was collected in Sherburne County, Minnesota, in 1982, but that population did not persist. The species was recently found as an introduction at Sandy Hook in Monmouth County, New Jersey (*Snyder & McArthur s.n.*, NY), but its fate there remains to be determined. Unfortunately, this weedy species recently has been introduced into northern Arizona as a roadside wild flower. C. L. Perez et al. (1998) have demonstrated that the seed bank can be rich in seeds of this species, but germination rates are low.

The Lakota people traditionally used the annual wild buckwheat as an aid in the treatment of sore mouths in children, seemingly in association with teething (D. J. Rogers 1980). Leaves were used to stain buffalo and deer hides by the Kiowa (P. A. Vestal and R. E. Schultes 1939). Vestal (1952) stated that the species was considered a "life medicine" by the Navajo (Diné) people; it was used also for protection against witches. It is likely that *E. annuum* was obtained by the Navajo through trade, but it might have been grown locally in historic times where the species recently has been reintroduced.

103. Eriogonum multiflorum Bentham, Trans. Linn. Soc. London 17: 413. 1836 • Many-flowered wild buckwheat Ⓔ

Herbs, 5–20 × 3–6 dm, reddish. **Aerial flowering stems** slender or stout, 4–15 dm, floccose to tomentose. **Leaves:** petiole 0.3–1 cm (rosette), or 0.2–0.5 cm (cauline), tomentose to floccose; blade narrowly lanceolate to narrowly elliptic or oblong to narrowly ovate, 1.5–5(–8) × 0.3–2.2(–2.5) cm, densely tomentose abaxially, tomentose or floccose to glabrate adaxially, sometimes thickened and auriculate-subclasping proximally; margins entire or crisped, rarely revolute. **Inflorescences** 50–200 × 20–100 cm; bracts triangular, 1–3 mm. **Peduncles** 0.1–0.4 cm, tomentose to floccose. **Involucres** turbinate to turbinate-campanulate, 2–2.5 × 1.5–2(–2.5) mm, tomentose to floccose or occasionally glabrous abaxially, tomentose adaxially; teeth 5, 0.3–0.8 mm. **Flowers** 1.5–2.5 mm; perianth white to reddish brown; tepals: those of outer whorl oblong-cordate, 1.5–2 × 1.5–2 mm, those of inner whorl lanceolate to oblong, 1.5–2.5 × 0.2–0.5 mm; stamens slightly exserted, 1.5–3 mm. **Achenes** 1.5–2 mm.

Varieties 2 (2 in the flora): c United States.

1. Leaf blades narrowly lanceolate to narrowly elliptic, 2–5(–8) × 0.3–0.9(–1.5) cm, not thickened, attenuate to rarely rounded proximally, not auriculate-subclasping . . . .
. . . . 103a. *Eriogonum multiflorum* var. *multiflorum*
1. Leaf blades oblong to narrowly ovate, 1.5–4(–4.5) × 1–2.2(–2.5) cm, occasionally thickened, auriculate-subclasping proximally . . . .
. . . . 103b. *Eriogonum multiflorum* var. *riograndis*

103a. Eriogonum multiflorum Bentham var. **multiflorum** • Many-flowered wild buckwheat

E

Leaf blades narrowly lanceolate to narrowly elliptic, 2–5(–8) × 0.3–0.9(–1.5) cm, tomentose to floccose adaxially, not thickened, attenuate to rarely rounded proximally.

Flowering Aug–Nov. Sandy to gravelly flats and slopes, mixed grassland and mesquite communities, oak and conifer woodlands; 10–500(–800) m; Ark., La., Okla., Tex.

Variety *multiflorum* is widespread and infrequent to common in scattered and disjunct populations in southern Oklahoma, central and east-central Texas, central Arkansas, and north-central Louisiana.

103b. Eriogonum multiflorum Bentham var. **riograndis** (G. L. Nesom) Reveal, Harvard Pap. Bot. 9: 184. 2004 • Rio Grande wild buckwheat

E

Eriogonum riograndis G. L. Nesom, Sida 20: 32, fig. 1. 2003

Leaf blades oblong to narrowly ovate, 1.5–4(–4.5) × 1–2.2(–2.5) cm, floccose to glabrate adaxially, occasionally thickened, auriculate-subclasping proximally.

Flowering Aug–Nov. Sandy to gravelly flats and slopes, mixed grassland and mesquite communities, oak and conifer woodlands; 0–300 m; Tex.

Variety *riograndis* is mostly infrequent and widely scattered in central southern Texas (Aransas, Brooks, Calhoun, Cameron, Duval, Hidalgo, Jim Hogg, Jim Wells, Karnes, Kenedy, Kleberg, Nueces, Refugio, San Patricio, Webb, Willacy, Wilson, and Zapata counties), becoming common only near the coast.

Variety *riograndis* is found just to the west and south of var. *multiflorum*, and a series of overlapping (and intergrading) populations occurs along their shared boundary. Plants immediately along the coast of Texas often have thickened, almost crassulate leaf blades.

1c. ERIOGONUM Michaux subg. **CLASTOMYELON** Coville & C. V. Morton, J. Wash. Acad. Sci. 26: 304. 1936 E

Herbs, erect, herbaceous, polycarpic perennials, glabrous; taproot woody. **Stems** erect, without persistent leaf bases, glabrous; caudex absent; aerial flowering stems erect, stout, solid or partially hollow, slightly fistulose, jointed basally and disarticulating into ringlike segments, arising directly from the root. **Leaves** persistent or marcescent, basal, 1 per node; blade pilose. **Inflorescences** terminal, subspicate, compact; branches racemose distally, disarticulating into ringlike segments, round and smooth, sparsely pilose or glabrous; bracts 3–5, connate basally, scalelike or semileaflike. **Peduncles** absent. **Involucres** 1 per node, not appressed to inflorescence branches, campanulate; teeth 5, erect to spreading, apex acute. **Flowers** not attenuate at base, without stipelike base; perianth yellow to reddish yellow, hispid abaxially, glabrous adaxially; tepals connate proximally $^1/_4$–$^1/_3$ their length, dimorphic; stamens included; filaments glabrous. **Achenes** light brown, not winged, 3-gonous, glabrous except for pilose beak. **Seeds:** embryo curved.

Species 1: California.

The infrageneric relationships of subg. *Clastomyelon* are obscure. Preliminary molecular evidence suggests that it is allied to subg. *Eucycla*.

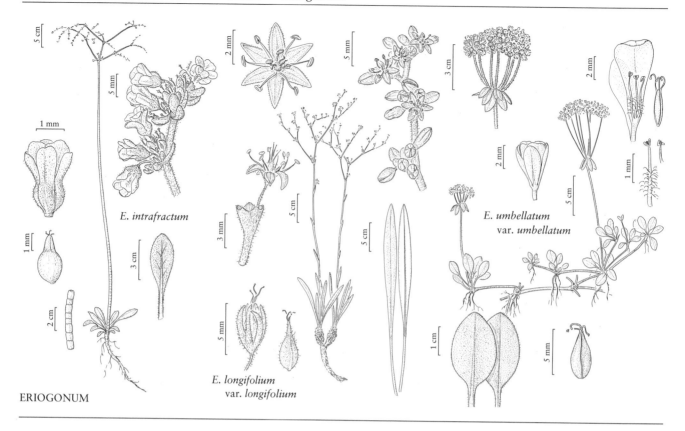

E. intrafractum

E. longifolium
var. longifolium

E. umbellatum
var. umbellatum

ERIOGONUM

104. Eriogonum intrafractum Coville & C. V. Morton, J. Wash. Acad. Sci. 26: 305. 1936 • Napkinring wild buckwheat C E F

Herbs, 6–15 × 0.3–0.9 dm, grayish. **Aerial flowering stems** 5–12 dm, disarticulating into ringlike segments 3–10(–16) mm. **Leaves:** petiole 3–8 cm, pilose; blade oblong-ovate, 2.5–7 × 0.7–2(–3) cm, pilose on both surfaces, margins entire. **Inflorescences** 1–3 × 2–6 cm; bracts scalelike distally, triangular, and 1–3 mm, semileaflike proximally, oblong, and 3–6 mm. **Involucres** 2.5–3.5 × 3–5 mm, pilose; teeth 0.5–2 mm. **Flowers** 1.5–3 mm; tepals: those of outer whorl oblanceolate, 0.9–1.2 mm wide, those of inner whorl broadly oblanceolate to fan-shaped, 1–1.5 mm wide; stamens 0.8–1.5 mm. **Achenes** 2–2.5 mm. $2n = 40$.

Flowering May–Oct. Limestone washes, slopes, and cliffs, saltbush and creosote bush communities, juniper woodlands; of conservation concern; (600–)800–1600 m; Calif.

Rare and restricted to the Cottonwood, Funeral, Grapevine, and Panamint ranges of Death Valley National Park, Inyo County, *Eriogonum intrafractum* is one of the more remarkable species of the genus. The tall, simple, glabrous stem is branched only near the top, and then with only two to five inflorescence branches. The small leaves are restricted to the base of the plant. The lower part of the stem, when dried, fragments into napkinring-like segments. The rings have no evident value to the species, as they are not capable of vegetative reproduction. Some parenchyma cells at the fracture points are large enough to be seen with the naked eye, and it is likely that they act as compression points. Much as the intervertebral disks in the vertebral column provide a degree of flexibility, so too may these large parenchyma cells allow the top-heavy flowering stem to move in the wind without breaking. The flowers are numerous, each involucre so filled with more than a hundred flowers that it is torn apart by late anthesis.

1d. ERIOGONUM Michaux subg. ERIOGONUM [E]

Herbs, erect, polycarpic perennials, tomentose or glabrous; taproot woody. **Stems** erect, tomentose to floccose or glabrous; caudex stems absent; aerial flowering stems erect, slender to stout, solid, not hollow, fistulose, or disarticulating in ringlike segments proximally, arising directly from the root. **Leaves** usually persistent, basal or cauline, 1 per node or fasciculate; blade tomentose abaxially. **Inflorescences** terminal, cymose, open; branches mostly dichotomous, not disarticulating into ringlike segments, round and smooth or finely striated or grooved, tomentose; bracts 3(–5), connate basally, scalelike or leaflike. **Peduncles** absent or erect and stout. **Involucres** 1 per node, not appressed to the inflorescence branches, turbinate to campanulate; teeth 5, erect, apex acute. **Flowers** attenuate at base, with stipelike base 0.5–7 mm; perianth cream to light tan or yellow, densely pubescent abaxially, glabrous adaxially; tepals connate proximally ¼–⅓ their length, monomorphic or dimorphic; stamens exserted; filaments glabrous or densely pilose. **Achenes** brown, not winged, 3-gonous, glabrous or pubescent. **Seeds:** embryo straight.

Species 2 (2 in the flora): United States.

1. Tepals monomorphic; perianths densely white- or silvery-tomentose abaxially; filaments glabrous; Arkansas, Kansas, New Mexico, Oklahoma, and Texas e to Alabama, Florida, Kentucky, and Tennessee . . . . 105. *Eriogonum longifolium*
1. Tepals dimorphic; perianths densely tannish- to rusty-lanate abaxially; filaments pilose; se Alabama, Florida, Georgia, and South Carolina . . . . 106. *Eriogonum tomentosum*

105. Eriogonum longifolium Nuttall, Trans. Amer. Philos. Soc., n. s. 5: 164. 1835 [E] [F] [W]

Herbs, 3–20 × 1.5–8 dm, tomentose or nearly glabrous. **Aerial flowering stems** erect, 2–17 dm, occasionally finely striated or grooved. **Leaves:** petiole 5–20 cm; blade lanceolate or oblanceolate to oblong, 0.5–20 × 0.3–2.5(–3) cm, tomentose abaxially, less so to floccose or glabrous adaxially. **Inflorescences** 5–50(–80) × 5–50 cm; bracts 3, scalelike, usually triangular, 1–5 mm. **Peduncles** 0.3–3 cm, thinly tomentose or nearly glabrous. **Involucres** turbinate to campanulate, 3–7 × 1.5–6 mm; teeth 0.2–0.8 mm. **Flowers** 5–15 mm, including (0.5–)1–4(–7) mm stipelike base; perianth yellow, densely white- to silvery-tomentose abaxially; tepals monomorphic, lanceolate to narrowly elliptic; stamens 1.7–2.5 mm; filaments glabrous. **Achenes** brown, 4–6 mm, densely tomentose.

Varieties 3 (3 in the flora): e North America.

1. Flowers 8–15 mm, including 2–4(–7) mm stipelike base; involucres 6–7 mm; c Florida . . . .
. . . . . 105c. *Eriogonum longifolium* var. *gnaphalifolium*
1. Flowers 5–11 mm, including 0.5–2.5 mm stipelike base; involucres 3–6 mm; not of c Florida.

2. Involucres 4–6 mm; flowers 5–11 mm, including (0.5–)1–2.5 mm stipelike base; s Arkansas, s Kansas, wc Louisiana, sw Missouri, se New Mexico, Oklahoma, Texas . . . . 105a. *Eriogonum longifolium* var. *longifolium*
2. Involucres 3–3.5 mm; flowers 5–7 mm, including 0.5–1(–1.2) mm stipelike base; nw Alabama, c Kentucky, sc Tennessee . . . . 105b. *Eriogonum longifolium* var. *harperi*

105a. Eriogonum longifolium Nuttall var. **longifolium**
· Long-leaf wild buckwheat [E] [F]

Eriogonum longifolium subsp. *diffusum* S. Stokes; *E. longifolium* var. *lindheimeri* Gandoger; *E. longifolium* var. *plantagineum* Engelmann & A. Gray; *E. vespinum* Shinners

Plants 3–20 dm, tomentose to nearly glabrous. **Leaves:** basal leaf blade 0.5–20 × 0.3–3 cm, tomentose abaxially, less so to floccose or nearly glabrous adaxially. **Peduncles** 0.3–3 cm. **Involucres** turbinate to campanulate, 4–6 × 2.5–6 mm. **Flowers** 5–11 mm, including 0.5–2.5 mm stipelike base. **Achenes** 4–6 mm.

Flowering May–Oct. Sandy to gravelly, often calcareous flats, slopes, and outcrops, mixed grassland, creosote bush, and mesquite communities, oak and conifer woodlands; 60–1100(–1300) m; Ark., Kans., La., Mo., N.Mex., Okla., Tex.

Variety *longifolium* is widespread and common in the south-central United States. Its roots were used as food by the Kiowa (P. A. Vestal and R. E. Schultes 1939), and an infusion of the roots was taken by the Comanche for stomach trouble (G. G. Carlson and V. H. Jones 1940).

105b. Eriogonum longifolium Nuttall var. **harperi** (Goodman) Reveal, Sida 3: 197. 1968 · Harper's wild buckwheat E

Eriogonum harperi Goodman, Bull. Torrey Bot. Club 74: 329, figs. 1–3. 1947

Plants 10–18 dm, thinly tomentose. **Leaves:** basal leaf blade 10–15 × 1.5–2.5 cm, tomentose abaxially, sparsely floccose or glabrous adaxially. **Peduncles** 0.3–0.8 cm. **Involucres** turbinate, 3–3.5 × 1.5–2.5 mm. **Flowers** 5–7 mm, including 0.5–1(–1.2) mm stipelike base. **Achenes** 4–4.5 mm. $2n = 40$.

Flowering Jun–Sep. Sandy to gravelly, often calcareous flats, bluffs, outcrops, and slopes, oak and conifer woodlands; 100–300 m; Ala., Ky., Tenn.

Variety *harperi* is known only from scattered populations in Alabama (Colbert and Franklin counties), Kentucky (Christian County), and Tennessee (DeKalb, Putnam, Smith, and Wilson counties). The Kentucky record is based on a report indicating that the plant had been extirpated there.

105c. Eriogonum longifolium Nuttall var. **gnaphalifolium** Gandoger, Bull. Soc. Roy. Bot. Belgique 42: 190. 1906 · Scrub wild buckwheat E

Eriogonum floridanum Small

Plants 5–13 dm, thinly tomentose. **Leaves:** basal leaf blade (3–)8–20 × 0.3–1 cm, tomentose abaxially, glabrous adaxially. **Peduncles** 1–3 cm. **Involucres** turbinate, 6–7 × 3–5 mm. **Flowers** 8–15 mm, including 2–4(–7) mm stipelike base. **Achenes** 5–6 mm.

Flowering Mar–Sep. Sandy flats and slopes, scrub oak and conifer woodlands; 10–50 m; Fla.

Variety *gnaphalifolium* is federally listed as threatened and is known only from DeSoto, Highlands, Lake, Marion, Orange, Polk, Putnam, Sumter, and Volusia counties in central Florida.

106. Eriogonum tomentosum Michaux, Fl. Bor.-Amer. 1: 246, plate 24. 1803 · Sandhill wild buckwheat E

Herbs, 5–12 × 2–5 dm, white- to rufous-tomentose. **Aerial flowering stems** erect, 1.5–7 dm, rounded and smooth. **Leaves:** petiole 2–7 cm; blade oblong to elliptic or spatulate, 4–15 × 1.5–4 cm, densely tomentose abaxially, subglabrous or glabrous adaxially. **Inflorescences** (3–)5–30 × 3–20 cm; bracts 3–5, leaflike, ovate to elliptic, 1–6 × 0.5–2.5 cm proximally, semileaflike, usually elliptic, 4–30 × 1–20 mm distally. **Peduncles** absent. **Involucres** turbinate-campanulate to campanulate, 3–4.5(–5) × 3–5 mm; teeth 0.1–0.5 mm. **Flowers** 8–13 mm, including 2.5–3.5(–4) mm stipelike base; perianth cream to light tan, densely tannish- to rusty-lanate abaxially; tepals dimorphic, those of outer whorl broadly lanceolate, 3–5 × 1.7–2.2 mm, those of inner whorl obovate to nearly orbiculate, 3–6 × 3–4 mm; stamens 3.5–5 mm; filaments pilose. **Achenes** brown, 4.5–5 mm, glabrous.

Flowering May–Oct. Sandy soil, mixed grassland communities, pine and oak woodlands; 0–100(–200) m; Ala., Fla., Ga., S.C.

Eriogonum tomentosum is widespread in a series of somewhat disjunct populations mainly on the coastal plain. It supposedly is found in North Carolina, but no specimens have been seen. An unnumbered J. K. Small collection from Miami, Florida (Nov. 1904, NY), surely is mislabeled. A Cuthbert collection (30 Jul. 1876, MICH) from Augusta may be from a garden, as that location is well out of the known range of the species. Mark Catesby made the first collections (BM-Sloane, OXF) of the species in the 1720s. T. Walter (1788) ascribed his specimen (BM-Walter) to a European species of Saxifragaceae.

1e. ERIOGONUM Michaux subg. OLIGOGONUM Nuttall, J. Acad. Nat. Sci. Philadelphia, n. s. 1: 166. 1848 (as Olygogonum)

Herbs, shrubs, or subshrubs, cespitose, matted or spreading to erect, polycarpic perennials, synoecious or polygamodioecious, rarely dioecious, lanate to tomentose, floccose, or glabrous occasionally sericeous; taproot woody. **Stems** prostrate to erect, with persistent leaf bases, glabrous, pubescent, or glandular; caudex woody, tightly compact to spreading or erect and at or just below surface, or spreading to erect and above surface; aerial flowering stems prostrate or decumbent to erect, slender to stout, usually solid or hollow, sometimes fistulose, not disarticulating in ringlike segments proximally, arising at nodes of caudex branches, at distal nodes of aerial branches or, rarely, directly from the root. **Leaves** persistent through the growing season or longer, tufted at tips of caudices or whorled at base of stems in rosettes; blade lanate, tomentose, floccose, villous, sericeous, or glabrous. **Inflorescences** simple- to compound-umbellate, usually open and spreading, occasionally reduced and subcapitate to capitate; branches absent or round and smooth, tomentose to floccose or glabrous; bracts usually 2–10 sometimes more, scalelike, semileaflike, or leaflike. **Peduncles** absent (or sometimes technically erect and slender, being that portion of "aerial flowering stem" above a whorl of leaflike bracts about midlength; not treated as a distinct structure here). **Involucres** 1 per node, rarely 2–3 per cluster, not appressed to inflorescence branches, turbinate to campanulate; teeth 5–10 or more, erect or lobelike and spreading to reflexed. **Flowers** bisexual or unisexual, attenuate at base, stipelike base 0.1–3 mm; perianth various shades of white, cream, yellow, pink, or reddish, glabrous or pubescent abaxially, glabrous adaxially; tepals connate proximally 1/4–1/3 their length, monomorphic or dimorphic; stamens usually exserted, occasionally included; filaments pilose proximally, very rarely glabrous. **Achenes** light to dark brown, not winged, 3-gonous, glabrous or pubescent only on beak. **Seeds**: embryo straight or curved.

Species 35 (34 in the flora): w North America (including n Mexico).

Only *Eriogonum turneri* Reveal is not treated here; it is related to *E. jamesii* var. *undulatum*, and is known only from gypsum outcrops in Nuevo León, Mexico.

1. Involucral teeth lobelike, at least half as long as tube, usually reflexed or spreading.
   2. Perianths pubescent abaxially.
      3. Flowering stems (actually scapes) without subtending bracts on stem or below involucre; e California to se Oregon, s Idaho, sw Montana, w Wyoming, Nevada, Utah, and nw Arizona . . . . 123. *Eriogonum caespitosum*
      3. Flowering stems with whorl of subtending bracts at base of umbel, near middle of stem, or near middle of branches.
         4. Involucres usually more than 1 per, subtended by (2–)3–several leafy bracts at base of umbel; inflorescences compound-umbellate or umbellate, occasionally with whorl of bracts midlength on central branch of an umbel; flowering stem usually without whorl of bracts near middle.
            5. Flowers (5–)6–9 mm; perianth bright or pale yellow or ochroleucous to cream; leaf blades lanate to tomentose, floccose, or glabrous on both surfaces; plants 0.5–4 dm; ne California to e Washington, s Idaho, and n Nevada . . . . 120. *Eriogonum sphaerocephalum*
            5. Flowers 4–5 mm; perianth bright yellow; leaf blades densely tomentose on both surfaces; plants 2.5–5 dm; n California . . . . 117. *Eriogonum tripodum*
         4. Involucre 1 per inflorescence, not immediately subtended by leafy bracts; inflorescences capitate; flowering stems with whorl of bracts near middle.

6. Leaf blades tomentose or lanate adaxially; e Washington to n and ec California, and nw and ne Nevada . . . . 121. *Eriogonum douglasii*

6. Leaf blades thinly tomentose to nearly glabrous and green adaxially; widespead.

   7. Leaf blades usually linear-oblanceolate or narrowly oblanceolate to narrowly spatulate; n California and nw Nevada to sw Idaho, Oregon, and c Washington . . . . 120. *Eriogonum sphaerocephalum* (in part)

   7. Leaf blades oblanceolate to elliptic; s Sierra Navada in Tulare County, California . . . . 122. *Eriogonum twisselmannii*

2. Perianths glabrous abaxially.

  8. Inflorescences capitate, without subtending bracts immediately below involucre; flowering stems with a whorl of leafy bracts ca. midlength; perianths yellow.

    9. Leaf blades 0.5–1.5(–2) cm, densely tomentose on both surfaces, or subglabrous and pale green adaxially; plants erect or rounded herbs or subshrubs, occasionally matted; Sierra Nevada from Nevada County to Frenso County . . . . 116. *Eriogonum prattenianum* (in part)

    9. Leaf blades (0.3–)0.5–0.8 cm, sparsely floccose to glabrate and bright green or olive green adaxially; plants matted herbs; Scott Mountain and Mt. Eddy region, Siskiyou and Trinity counties, California . . . . 110. *Eriogonum siskiyouense*

  8. Inflorescence umbellate or compound-umbellate, subtended by leafy bracts immediately below umbel; flowering stems occasionally with an additional whorl of bracts ca. midlength; perianths various shades of white, yellow, or rose.

   10. Flowering stems usually with whorl of leafy bracts ca. midlength; leaf blades usually linear to oblanceolate; s British Columbia and Washington to w Montana s to ne California, Idaho, n Nevada, Wyoming, c Utah, and nw Colorado . . . . 108. *Eriogonum heracleoides*

   10. Flowering stems without a whorl of leafy bracts ca. midlength (present in *E. umbellatum* var. *argus*); leaf blades variable; equally widespread, extending to s California and n Arizona.

    11. Stipelike bases of flowers 0.1–0.4 mm.

     12. Flowers 5–7 mm; achenes 4.5–6 mm; flowering stems prostrate to decumbent or weakly erect; inflorescences subcapitate to umbellate, infrequently 2-umbellate; n California, sw Oregon, and wc Nevada . . . . 138. *Eriogonum lobbii* (in part)

     12. Flowers 7–9 mm; achenes 6–8 mm; flowering stems erect or nearly so; inflorescences 2-umbellate; wc Nevada . . . . 139. *Eriogonum robustum* (in part)

    11. Stipelike bases of flowers 0.7–2 mm.

     13. Leaf blades 0.3–3(–4) cm; flowering stems not fistulose; w North America . . . . 107. *Eriogonum umbellatum*

     13. Leaf blades (2–)7–25 cm; flowering stems occasionally fistulose; nw Idaho n to California, Oregon, Washington . . . . 109. *Eriogonum compositum*

1. Involucral teeth not lobelike, much shorter than tube, erect or nearly so.

  14. Perianths pubescent abaxially.

   15. Leaf blades pilose or hirtellous to glabrescent abaxially.

    16. Leaf blades pilose, 1–3(–3.5) cm; perianths cream to pale yellow; flowers 3–6 mm; ec California, wc Nevada . . . . 135. *Eriogonum latens*

    16. Leaf blades hirtellous to glabrescent, 0.5–2(–2.5) cm; perianths bright yellow; flowers 3–3.5(–4) mm; Del Norte and Siskiyou counties, California . . . 136. *Eriogonum hirtellum*

   15. Leaf blades tomentose abaxially.

    17. Flowering stems with whorl of 6–10(–12) bracts ca. midlength; leaf blades silky-villous to sericeous adaxially; se Washington to ne Oregon and sw Idaho . . . . 118. *Eriogonum thymoides*

    17. Flowering stems without whorl of bracts ca. midlength; leaf blades not silky-villous to sericeous adaxially; widespread.

18. Perianths white to rose; inflorescences subtended by 2 bracts; flowering stems weakly erect to ascending or prostrate; s British Columbia and Washington to n California to w Montana . . . . 137. *Eriogonum pyrolifolium*
18. Perianths white to cream or yellow; inflorescences subtended by 3–10 or more bracts; flowerings stem weakly erect or erect to ascending; widespread.
    19. Perianths white to cream; e Arizona, Colorado, sw Kansas, New Mexico, w Texas . . . . 128. *Eriogonum jamesii*
    19. Perianths pale to bright yellow; n Arizona and Colorado n to Alaska and Canada.
        20. Inflorescences capitate or subcapitate to umbellate or, if compound-umbellate, plants of e Utah, Colorado, and nw New Mexico; plants matted, often cespitose, perennials usually shorter than 3 dm; Colorado Plateau, Rocky Mountains, and n Great Plains from n Arizona and New Mexico to Alaska and Canada.
           21. Achenes glabrous; perianths pale yellow; sw Alberta and nw Montana . . . . 134. *Eriogonum androsaceum*
           21. Achenes sparsely pubescent on beak; perianths yellow; widespread.
                22. Inflorescences compound-umbellate or, if umbellate or capitate, not of distribution of *E. flavum*; Colorado Plateau, s Rocky Mountains and w edge of Great Plains from s Wyoming to n Arizona and n New Mexico . . . . 129. *Eriogonum arcuatum*
                22. Inflorescences subcapitate or umbellate; n Great Plains and n Rocky Mountains, Wyoming and Nebraska north to Canada and Alaska, west to e Oregon and Washington . . . . 133. *Eriogonum flavum*
        20. Inflorescences compound-umbellate; plants not matted, taller than 3 dm or, if 1.5–3 dm, then plants of n Texas; se New Mexico, n Texas, nw Virginia, and e West Virginia.
           23. Leaf blades 4–8 cm wide; nw Virginia, e West Virginia . . . . 132. *Eriogonum allenii*
           23. Leaf blades 1–3.5 cm wide; se New Mexico, n Texas.
                24. Plants 4–5 dm; leaf blades (2.5–)3–6.5 cm; se New Mexico . . . . 130. *Eriogonum wootonii*
                24. Plants 1.5–3.5(–4) dm; leaf blades (2–)4–12(–15) cm; n Texas . . . . 131. *Eriogonum correllii*

[14. Shifted to left margin.—Ed.]
14. Perianths usually glabrous abaxially.
  25. Inflorescences not immediately subtended by bracts; flowering stems seemingly with whorl of bracts near middle (portion below whorl technically is a flowering stem, whereas portion above whorl but below solitary involucre technically is a peduncle); California.
    26. Perianths whitish to pinkish or rose-red; leaf blades silky-tomentose and silvery abaxially, 0.1–0.3(–0.4) cm wide; Red Mountain, Mendocino County, California . . . . 119. *Eriogonum kelloggii*
    26. Perianths yellow; leaf blades tomentose abaxially, 0.2–2(–3) cm wide; not of Mendocino County, California.
      27. Flowering stems (0.3–)0.4–0.6 dm; leaf blades 1–2(–3) cm wide; Mt. Eddy and Scott Mountain, Siskiyou and Trinity counties, California . . . . 140. *Eriogonum alpinum*
      27. Flowering stems 0.5–3 dm; leaf blades 0.2–0.7(–0.8) cm wide; not of Mt. Eddy and Scott Mountain, Siskiyou and Trinity counties, California.

28. Perianths sparsely pubescent abaxially; involucral teeth 5–8, 0.5–1.5 mm;
North Coast Ranges, Trinity and Tehama counties, California . . . .
. . . . 112. *Eriogonum libertini* (in part)
28. Perianths glabrous; involucral teeth 8–10, 1–3 mm; Sierra Nevada, Neva-
da County to Frenso County, California . . . . 116. *Eriogonum prattenianum* (in part)
[25. Shifted to left margin.—Ed.]
25. Inflorescences immediately subtended by whorl of bracts; flowering stems without a whorl
of bracts near middle; California, w Nevada, s Oregon, disjunct in Washington.
29. Perianths chalky white, white to rose, or creamy yellow to pale yellowish; stipelike
bases of flowers 0.1–0.4 mm.
30. Perianths chalky white; flowers 2.5–3.5 mm; involucres (2–)2.5–3.5(–4) mm; Fresno
and Tulare counties, California . . . . 124. *Eriogonum polypodum*
30. Perianths white to rose or creamy yellow to pale yellowish; flowers 5–9 mm; n
California, wc Nevada.
31. Flowers 5–7 mm; achenes 4.5–6 mm; flowering stems prostrate to decumbent
or weakly erect; inflorescences subcapitate to umbellate, infrequently 2-umbel-
late; n California, sw Oregon, and wc Nevada . . . . 138. *Eriogonum lobbii* (in part)
31. Flowers 7–9 mm; achenes 6–8 mm; flowering stems erect or nearly so; inflorescences 2-umbellate; wc Nevada . . . . 139. *Eriogonum robustum* (in part)
29. Perianths yellow to brownish or sulphur yellow, ochroleucous, or cream and sometimes
suffused with blush of pinkish red to maroon, rarely white; stipelike bases of flowers
0.3–1.5 mm.
32. Plants dioecious, with morphologically different pistillate and staminate plants in
fruit.
33. Leaf blades glabrate and bright green to olive green adaxially; ne California, wc
Nevada, s Oregon, disjunct to Yakima County, Washington . . . . . . 125. *Eriogonum marifolium*
33. Leaf blades lanate to tomentose on both surfaces, usually not green adaxially;
Sierra Nevada or Siskiyou and Trinity mountains, California, Oregon, and
Nevada.
34. Pistillate inflorescences capitate, umbellate after fertilization; leaf blades
oblong to oblong-ovate or spatulate, 0.5–1.5 × 0.3–0.7 cm, tomentose;
petioles (0.3–)0.5–1 cm; Sierra Nevada of ec California and wc Nevada
. . . . 126. *Eriogonum incanum*
34. Pistillate inflorescences usually persistently capitate, rarely umbellate after
fertilization; leaf blades elliptic to ovate, (0.5–)1–2 × 0.5–1.5 cm, lanate;
petioles 0.7–3 cm; Siskiyou and Trinity mountains, nw California and sw
Oregon . . . . 127. *Eriogonum diclinum*
32. Plants synoecious, without morphologically different plants in fruit.
35. Perianths white, cream, ochroleucous, pale yellow, or rarely yellow; involucres
villous.
36. Flowering stems 0.4–4 dm; inflorescences compound-umbellate; perianths
cream, pale yellow, or rarely yellow; Sierra Nevada, Placer County to Shasta
County, or n Coast Ranges, Shasta and Trinity counties to Siskiyou County,
California . . . . 114. *Eriogonum ursinum*
36 Flowering stems 0.2–0.6(–1) dm; inflorescences subcapitate; perianths white
to ochroleucous; n Coast Ranges, Sonoma and Lake counties, California
. . . . 115. *Eriogonum nervulosum*
35. Perianths sulphur yellow; involucres tomentose.
37. Inflorescences not immediately subtended by bracts; flowering stems seemingly with a whorl of bracts near middle (portion below whorl is flowering
stem, portion above whorl but below solitary involucre is peduncle); flowers 5–8 mm; n Coast Ranges in Trinity and Tehama counties, California
. . . . 112. *Eriogonum libertini* (in part)

[37. Shifted to left margin.—Ed.]

37. Inflorescences subtended by a whorl of bracts immediately below umbel or head of several involucres; flowering stems not with a whorl of bracts near middle; flowers 3–6 mm; Del Norte, Siskiyou, Trinity, and Tehama counties, California, n to Curry and Josephine counties, Oregon.

    38. Leaf blade margins not revolute, blade oblong to obovate, (0.4–)0.8–1.3 cm wide; plants matted; n Coast Ranges, Del Norte, Siskiyou, Trinity, and Tehama counties, California, to Curry and Josephine counties, Oregon . . . . 111. *Eriogonum ternatum*

    38. Leaf blade margins revolute, blade narrowly elliptic to narrowly oblong, 0.3–0.6(–0.8) cm wide; plants subshrubs; Trinity and Siskiyou counties, California . . . . . . 113. *Eriogonum congdonii*

107. Eriogonum umbellatum Torrey, Ann. Lyceum Nat. Hist. New York 2: 241. 1827 • Sulphur flower E F

Herbs, subshrubs, or shrubs, cespitose, matted or spreading, sometimes erect, often polygamo-dioecious, (0.2–)1–12(–20) × (0.5–)1–12(–20) dm, glabrous or tomentose. **Stems:** caudex spreading; aerial flowering stems spreading to erect or nearly so, slender, solid, not fistulose, arising at nodes of caudex branches and at distal nodes of short, nonflowering aerial branches, (0.1–)0.5–3(–4) dm, without a whorl of bracts at midlength. **Leaves** in loose to compact basal rosettes; petiole 0.1–3(–4) cm, mostly tomentose to floccose or glabrous; blade oblong-ovate or oblanceolate to elliptic to oval, 0.3–3(–4) × 0.1–2.5 cm, densely lanate to tomentose or floccose abaxially, tomentose to floccose or glabrous adaxially, occasionally glabrous on both surfaces, margins entire, plane or rarely wavy. **Inflorescences** umbellate or compound-umbellate, rarely subcapitate or capitate, 3–25 × 2–18 cm; branches tomentose to floccose or glabrous, rarely with whorl of bracts ca. midlength; bracts 3–several, semileaflike at proximal node, 0.3–2.5 × 0.2–1.8 cm, usually scalelike distally, 1–5 × 0.5–3 mm. **Involucres** 1 per node, turbinate to campanulate, 1–6 × (1–)1.5–10 mm, tomentose to thinly floccose or glabrous; teeth 6–12, lobelike, reflexed, 1–4(–6) mm. **Flowers** 2–10(–12) mm, including (0.7–)1.3–2 mm stipelike base; perianth various shades of white, yellow, or red, glabrous; tepals monomorphic, usually spatulate to obovate; stamens exserted, 2–8 mm; filaments pilose proximally. **Achenes** light brown to brown, 2–7 mm, glabrous except for sparsely pubescent beak.

Varieties 41 (41 in the flora): w North America.

Eriogonum umbellatum is a widespread and exceedingly variable species rivaling *Astragalus lentiginosus* Douglas ex Hooker in complexity. Only the variety *majus* is sometimes recognized at the species rank (as *E. subalpinum*).

In the following key and descriptions, reference is made to "glabrous" leaf surfaces. This is a function of both age and power of observation. High-power magnification may show some exceedingly fine hairs that are not readily observable to the naked eye. Furthermore, new leaves that ultimately will be "glabrous" will have some fine-tomentose pubescence that becomes less obvious (or even wholly inconspicuous) as the blade expands and matures. Here, the term "glabrous" is used to refer to leaves that are not obviously hairy and are typically bright green on both surfaces when the plant is at full anthesis.

There are several reported uses of sulphur flower in the enthobotany literature, mostly without an indication of the variety. J. B. Romero (1954) indicated that in California, where most of the variants are found, an infusion of the flowers is used for ptomaine poisoning, and M. L. Zigmond (1981) stated that the Kawaiisu used mashed flowers as a salve for gonorrheal sores. E. V. A. Murphey (1959), who worked with P. Train et al. (1941), reported that in Nevada members of the species (most likely var. *nevadense*) were used in the treatment of colds and stomachaches; J. H. Steward (1933) reported the same uses among the Owens Valley Piute in California. Train and his group noted that poultices of leaves and sometimes roots were used for lameness or rheumatism. In Oregon, leaves of var. *ellipticum* were used in a poultice to soothe pain, especially that resulting from burns (F. V. Coville 1897; L. Spier 1930). Most of the reports associated with the Navajo or Diné people probably relate to the use of var. *subaridum*. L. C. Wyman and S. K. Harris (1951) found the species used as a disinfectant or an emetic. The Cheyenne people employed a mixture of powdered stems and flowers to halt lengthy menses (G. B. Grinnell 1923; J. A. Hart 1981). According to A. Johnston (1987), the Blackfoot made a tea from boiled leaves.

The cythera metalmark butterfly (*Apodemia mormo cythera*) is found in association with a few varieties of sulphur flower (G. F. Pratt and G. R. Ballmer 1991). More commonly seen with the species are the Rocky Mountain dotted-blue (*Euphilotes ancilla*) and lupine blue (*Plebeius lupini*).

1. Inflorescences compound-umbellate, or with at least some branches seemingly with whorl of bracts about midlength.
    2. Inflorescence branches with a whorl of bracts at midlength (see also *E. umbellatum* var. *polyanthum*).
        3. Perianths bright yellow; flowers 7–10 mm; Sierra Nevada, c California . . . . 107nn. *Eriogonum umbellatum* var. *torreyanum*
        3. Perianths cream or whitish; flowers 4–7 mm; Warner Mountains, se Oregon and ne California . . . . 107oo. *Eriogonum umbellatum* var. *glaberrimum*
    2. Inflorescence branches without a whorl of bracts at midlength, bracts restricted to base of inflorescence or involucres.
        4. Flowers 7–10(–12) mm; shrubs 5–15(–20) × 5–20 dm; inflorescences branched 2–4 times; Humboldt and Trinity counties, California . . . . 107kk. *Eriogonum umbellatum* var. *speciosum*
        4. Flowers 3–8 mm; matted herbs or subshrubs, or, if shrubby, inflorescences branched 4 or more times, or not of Humboldt and Trinity counties, California.
            5. Perianths cream, whitish, or pale yellow to greenish yellow, becoming reddish brown to rose or pink.
                6. Perianths yellow, becoming reddish brown to rose or pink, with large reddish spot on each midrib; plants spreading to somewhat prostrate mats; se California, s Nevada . . . . 107gg. *Eriogonum umbellatum* var. *versicolor* (in part)
                6. Perianths cream, whitish, or pale yellow to greenish yellow, without large reddish spot on midrib; plants subshrubs or shrubs; nc Arizona, se California, s Nevada, s Utah . . . . 107hh. *Eriogonum umbellatum* var. *juniporinum*
            5. Perianths bright yellow, not becoming reddish brown to rose or pink.
                7. Leaf blades thinly floccose, glabrous, or densely lanate to tomentose on both surfaces at anthesis (see also *E. umbellatum* var. *munzii*, s California).
                    8. Leaf blades densely lanate on both surfaces; wc California . . . . . 107bb. *Eriogonum umbellatum* var. *bahiiforme*
                    8. Leaf blades lanate to tomentose abaxially, tomentose to densely floccose adaxially, sometimes thinly floccose or glabrous on both surfaces; nw or ec California, Intermountain West, or Pacific Northwest.

[9. Shifted to left margin.—Ed.]
9. Leaf blades lanate to tomentose abaxially, thinly tomentose to floccose and greenish adaxially; nw California.
    10. Leaf blades 0.3–0.7 cm wide; involucral lobes 1–3 mm; plants of serpentine soils; Glenn, Lake, Mendocino, Sonoma, Tehama, and Trinity counties, California . . . . 107cc. *Eriogonum umbellatum* var. *smallianum* (in part)
    10. Leaf blades (0.5–)0.8–1.8(–2) cm wide; involucral lobes (3–)4–6 mm; plants of nonserpentine soils; Siskiyou County, California . . . . 107ii. *Eriogonum umbellatum* var. *lautum* (in part)
9. Leaf blades usually thinly floccose or glabrous on both surfaces; not of nw California.
    11. Leaf blades thinly floccose on both surfaces, or glabrous and green adaxially (rarely glabrous on both surfaces in s Utah, or tomentose abaxially in se Utah); e and s California to sw Colorado and ne Arizona . . . . 107dd. *Eriogonum umbellatum* var. *subaridum*
    11. Leaf blades glabrous on both surfaces; e Idaho, ne Oregon, se Washington, and ec California.
        12. Herbs; leaf blades broadly elliptic, 1–1.5 cm wide; e Idaho, ne Oregon, and se Washington . . 107x. *Eriogonum umbellatum* var. *devestivum*
        12. Subshrubs or shrubs; leaf blades oblanceolate to narrowly elliptic, 0.3–1 cm wide; s Mono, Inyo, and ne Tulare counties, California . . . . 107ee. *Eriogonum umbellatum* var. *chlorothamnus*

[7. Shifted to left margin.—Ed.]
7. Leaf blades thinly to densely tomentose or lanate abaxially, less so to floccose, nearly glabrous, or glabrous adaxially (rarely both surfaces tomentose in *E. umbellatum* var. *munzii*, s California).
    13. Plants densely branched, shrubs; w foothills, Sierra Nevada, n California.
        14. Leaf blades white-tomentose abaxially; inflorescences branched 1–2(–3) times, branches thinly floccose or glabrous, central branch sometimes seemingly with a whorl of bracts ca. midlength; Butte, Plumas, and Sierra counties . . . . 107jj. *Eriogonum umbellatum* var. *polyanthum*
        14. Leaf blades rusty-lanate to tomentose abaxially; inflorescences branched 3–4 time, branches tomentose to floccose, central branch without a whorl of bracts ca. midlength; Butte County . . . . 107ll. *Eriogonum umbellatum* var. *ahartii*

[13. Shifted to left margin.—Ed.]

13. Plants usually matted herbs or more openly and sparsely branched subshrubs or shrubs; mountains, Sierra Nevada, c and n California, wc Nevada, or Siskiyou and Trinity mountains, nw California (but not on the western foothills), sw Oregon, n Arizona, Colorado, Idaho, Montana, or Washington.

    15. Flowering stems mostly sparsely floccose or glabrous.

        16. Leaf blades oblong-ovate to elliptic or oval, 1–1.5(–2) cm wide; plants compact mats; n Arizona . . . . 107mm. *Eriogonum umbellatum* var. *cognatum*

        16. Leaf blades usually narrowly elliptic, 0.3–1(–1.3) cm wide; plants subshrubs or spreading mats; California, Nevada, Oregon.

            17. Leaf margins plane; flowering stems without a single leaflike bract ca. midlength; Sierra Nevada, e California, wc Nevada . . . . 107y. *Eriogonum umbellatum* var. *furcosum*

            17. Leaf margins often finely wavy; flowering stems often with a single leaflike bract ca. midlength; Siskiyou and Trinity mountains, nw California, sw Oregon . . . . 107z. *Eriogonum umbellatum* var. *argus*

    15. Flowering stems usually tomentose or floccose, rarely nearly glabrous at maturity.

        18. Leaf blades usually broadly elliptic to oval.

            19. Leaf blades gray abaxially; c Colorado . . . . 107v. *Eriogonum umbellatum* var. *ramulosum*

            19. Leaf blades white abaxially; n California . . . . 107ii. *Eriogonum umbellatum* var. *lautum* (in part)

        18. Leaf blades elliptic.

            20. Leaf blades 1–1.5 cm wide, usually glabrous adaxially; Idaho, w Montana, Oregon, and e Washington . . . . 107w. *Eriogonum umbellatum* var. *ellipticum*

            20. Leaf blades 0.3–1 cm wide, usually floccose adaxially; California.

                21. Leaf blades 0.3–0.7 cm wide; nw California . . . . 107cc. *Eriogonum umbellatum* var. *smallianum* (in part)

                21. Leaf blades 0.5–1 cm wide; s California . . . . 107ff. *Eriogonum umbellatum* var. *munzii*

1. Inflorescences umbellate, not compound-umbellate or with any branches seemingly with whorl of bracts about midlength.

    22. Perianths usually whitish or cream to red, occasionally yellow.

        23. Leaf blades densely lanate on both surfaces; perianths lemon yellow to yellowish red; San Gabriel and San Bernardino mountains, e Los Angeles and w San Bernardino counties, California . . . . 107t. *Eriogonum umbellatum* var. *minus*

        23. Leaf blades tomentose to floccose or glabrous adaxially, sometimes glabrous on both surfaces at full anthesis; perianths not lemon yellow or yellowish red; not of e Los Angeles or w San Bernardino counties, California.

            24. Leaf blades glabrous on both surfaces; s Idaho and sw Montana to w Wyoming, ne Nevada, and n Utah . . . . . 107r. *Eriogonum umbellatum* var. *desereticum*

            24. Leaf blades densely floccose to tomentose or lanate at least abaxially at full anthesis; widespread.

[25. Shifted to left margin.—Ed.]

25. Leaf blades densely lanate abaxially, glabrous and olive green to bright green adaxially; plants compact mats; perianths cream; Rocky Mountains to Cascade Range . . . . 107u. *Eriogonum umbellatum* var. *majus*

25. Leaf blades densely floccose to tomentose abaxially, less so or glabrous and green adaxially; plants spreading mats, rarely compact; perianths pale yellow to cream or whitish, rarely greenish white or yellow, becoming reddish brown to rose or pink, with large reddish spots on each midrib; Rocky Mountains to Sierra Nevada.

    26. Perianths pale yellow to cream or whitish, rarely greenish white; leaf blades usually greenish adaxially; se Oregon and e California to w Montana, s Idaho, w Wyoming, c Nevada, and n Utah . . . . 107q. *Eriogonum umbellatum* var. *dichrocephalum*

    26. Perianths yellow, becoming reddish brown to rose or pink, with large reddish spot on each midrib; leaf blades usually reddish adaxially; se California, s Nevada . . . . 107gg. *Eriogonum umbellatum* var. *versicolor* (in part)

22. Perianths bright yellow (pale yellow in some individuals of var. *vernum* in Nevada).
    27. Umbels with branches usually longer than 2.5 cm.
        28. Leaf blades glabrous on both surfaces at full anthesis, occasionally with some individual blades thinly floccose abaxially . . . . . 107d. *Eriogonum umbellatum* var. *aureum* (in part)
        28. Leaf blades at least thinly tomentose or lanate abaxially.
            29. Leaf blades thinly tomentose to thinly floccose on both surfaces, or glabrous adaxially.
                30. Plants matted herbs.
                    31. Leaf blades broadly elliptic to ovate, (0.8–)2–3(–3.5) × (0.7–)1–2(–2.5) cm; s Idaho, ne Nevada, wc Wyoming . . . . 107c. *Eriogonum umbellatum* var. *stragulum*
                    31. Leaf blades narrowly elliptic, 0.7–2.5 × 0.3–1 cm; nw Arizona . . . . 107s. *Eriogonum umbellatum* var. *mohavense*
                30. Plants subshrubs or shrubs.
                    32. Flowers 4–7 mm; California, n Nevada, se Oregon . . . . 107n. *Eriogonum umbellatum* var. *nevadense* (in part)
                    32. Flowers (5–)6–10 mm; n Nye County, Nevada . . . . 107o. *Eriogonum umbellatum* var. *vernum*
            29. Leaf blades densely white-lanate or tomentose abaxially, less so to floccose or glabrous adaxially.
                33. Subshrubs 3–5 dm; nonserpentine soils; n California and sc Oregon . . 107k. *Eriogonum umbellatum* var. *dumosum*
                33. Herbs (0.7–)1–4.5(–5) dm; Rocky Mountains or, if in California and Oregon, often on serpentine soils.
                    34. Plants usually compact mats; nonserpentine soils; Rocky Mountains, sw Montana and e Idaho to Colorado, s and w Wyoming, and n Utah . . . . 107a. *Eriogonum umbellatum* var. *umbellatum* (in part)

[34. Shifted to left margin.—Ed.]
34. Plants spreading to prostrate mats; often on serpentine soils; n California and sw Oregon.
    35. Flowering stems 1–2.5(–4) dm; leaf blades 0.5–2(–3.5) cm; flowers 6–8(–9) mm; 400–1700(–2100) m . . . . 107l. *Eriogonum umbellatum* var. *goodmanii*
    35. Flowering stems 0.5–1.5 dm; leaf blades 0.5–1(–1.5) cm; flowers 3–6 mm; 1700–2800 m . . . . 107m. *Eriogonum umbellatum* var. *humistratum*
[27. Shifted to left margin.—Ed.]
27. Umbels with branches usually shorter than 2.5 cm (immature specimens should be keyed through the opposing couplet as well).
    36. Plants prostrate, mostly in montane to subalpine or alpine communities.
        37. Leaf blades glabrous on both surfaces at full anthesis . . . . 107e. *Eriogonum umbellatum* var. *porteri*
        37. Leaf blades tomentose at least abaxially at full anthesis.
            38. Leaf blades narrowly elliptic, 0.3–0.6(–1) × 0.2–0.4(–0.6) cm, tomentose abaxially, slightly less so and greenish adaxially; s Sierra Nevada and White Mountains, Mono, Inyo, and Tulare counties, California . . . . 107p. *Eriogonum umbellatum* var. *covillei*
            38. Leaf blades broadly elliptic, 0.5–1.5(–2.5) × 0.5–1.2(–1.5) cm, tomentose or glabrous and olive green adaxially; n Cascade Range, Benton, Clackamas, Hood River, and Wasco counties, Oregon, and Kittitas and Yakima counties, Washington . . . . 107h. *Eriogonum umbellatum* var. *haussknechtii*
    36. Plants erect to slightly spreading, not prostrate, usually not in subalpine or alpine communities.
        39. Leaf blades glabrous on both surfaces, sometimes with marginal hairs at full anthesis.
            40. Leaf blades without marginal hairs; se Oregon, Idaho, and Nevada to w Montana, Wyoming, Colorado, and Utah . . . . 107d. *Eriogonum umbellatum* var. *aureum* (in part)
            40. Leaf blades with marginal hairs; Cascade Range, Washington . . . . 107f. *Eriogonum umbellatum* var. *hypoleium*

39. Leaf blades lanate to tomentose or floccose at least abaxially at full anthesis.
    [41. Shifted to left margin.—Ed.]

41. Leaf blades thinly tomentose abaxially, less so to floccose or glabrous and green adaxially; Sierra Nevada and Great Basin ranges . . . . 107n. *Eriogonum umbellatum* var. *nevadense* (in part)

41. Leaf blades densely lanate to tomentose abaxially; widespread, generally not in Great Basin ranges.

   42. Mature leaf blades lanate to tomentose on both surfaces; Yellowstone National Park, Wyoming . . . . . . 107b. *Eriogonum umbellatum* var. *cladophorum*

   42. Mature leaf blades densely lanate or tomentose abaxially, less so to floccose or glabrous and greenish adaxially; w United States.

      43. Larger leaf blades usually 0.3–1.5(–2) cm, usually elliptic to oblong; plants subshrubs or herbs.

         44. Leaf blades usually elliptic; non-serpentine soils; ne California, se Oregon, sw Idaho, n Nevada . . . . 107j. *Eriogonum umbellatum* var. *modocense*

         44. Leaf blades elliptic to oblong; serpentine outcrops; nw California . . . . 107aa. *Eriogonum umbellatum* var. *nelsoniorum*

      43. Larger leaf blades 1–3(–3.5) cm, elliptic to ovate; plants subshrubs, or compact or spreading mats.

         45. Plants subshrubs; n Cascade Range, Washington. . . . . 107g. *Eriogonum umbellatum* var. *sandbergii*

         45. Plants compact or spreading mats; Rocky Mountains, Sierra Nevada, Transverse, and Argus ranges, California, Colorado, Idaho, Montana, Utah, and Wyoming.

           46. Leaf blades elliptic to ovate, densely white-lanate abaxially; Rocky Mountains, sw Montana and e Idaho to s Colorado, w and s Wyoming, and s Utah . . . . . 107a. *Eriogonum umbellatum* var. *umbellatum* (in part)

           46. Leaf blades elliptic, densely grayish-lanate abaxially; s Sierra Nevada, Transverse Ranges, and Argus Range, California . . . . . . 107i. *Eriogonum umbellatum* var. *canifolium*

107a. Eriogonum umbellatum Torrey var. **umbellatum** • Common sulphur flower [E] [F]

Herbs, spreading mats, 1–3.5 × 2–6 dm. **Aerial flowering stems** erect, 1–2.5(–3) dm, tomentose to floccose, without one or more leaf-like bracts ca. midlength. **Leaves** in loose rosettes; blade usually elliptic to ovate, 1–2.5(–3) × 0.5–1.5(–1.8) cm, white- to gray-lanate abaxially, less so to floccose or glabrous and green adaxially, margins plane. **Inflorescences** umbellate; branches 0.3–2.5(–8) cm, without a whorl of bracts about midlength; involucral tubes 2–3 mm, lobes 1.5–3 mm. **Flowers** 4–7(–8) mm; perianth bright yellow. $2n = 80$.

Flowering Jun–Sep. Sandy to gravelly flats and slopes, mixed grassland and sagebrush communities, scrub oak and montane conifer woodlands; (1000–)1200–2700 (–3100) m; Colo., Idaho, Mont., Utah, Wyo.

Variety *umbellatum* is widespread and rather common. Overlying a portion of the range of var. *umbellatum* is the glabrous-leaved var. *aureum*, but the latter occurs over a greater area and mainly much farther to the west. Still, clear distinction between the two is not always possible. Variety *umbellatum* is found in cultivation, especially in European gardens.

107b. Eriogonum umbellatum Torrey var. **cladophorum** Gandoger, Bull. Soc. Roy. Bot. Belgique 42: 198. 1906 • Yellowstone sulphur flower [E]

Eriogonum rydbergii Greene

Herbs, compact mats, 1–2 × 2–5 dm. **Aerial flowering stems** erect, 1–2 dm, tomentose to densely floccose, without one or more leaflike bracts ca. midlength. **Leaves** in tight rosettes; blade usually elliptic to ovate, 1–2 × 0.5–1.5 cm, lanate to densely tomentose on both surfaces, margins plane. **Inflorescences** umbellate; branches 0.3–2.5 cm, without a whorl of bracts about midlength; involucral tubes 2–3 mm, lobes 1.5–3 mm. **Flowers** 4–7 mm; perianth bright yellow.

Flowering Jun–Sep. Sandy to gravelly flats, sagebrush communities, montane conifer woodlands; 2000–2300 m; Wyo.

Variety *cladophorum* is a rare and localized expression known from three locations (Upper Geyser Basin, Old Faithful, and Madison Junction) within Yellowstone National Park, Teton County.

107c. Eriogonum umbellatum Torrey var. **stragulum**
Reveal, Phytologia 86: 156. 2004 • Spreading
sulphur flower [E]

Herbs, spreading mats, 1–3.5(–4)
× 2.5–10(–12) dm. **Aerial flowering stems** erect, 1–3 dm, thinly
floccose, without one or more
leaflike bracts ca. midlength.
Leaves in loose rosettes; blade
broadly elliptic to ovate, (0.8–)2–
3(–3.5) × (0.7–)1–2(–2.5) cm,
thinly tomentose to sparsely
floccose abaxially, thinly floccose or glabrous and green
adaxially, margins plane. **Inflorescences** umbellate;
branches (1–)2.5–5(–8) cm, without a whorl of bracts
ca. midlength; involucral tubes 2–3 mm, lobes 3–5 mm.
Flowers 4–7(–8) mm; perianth bright yellow. *2n* = 80.

Flowering May–Sep. Sandy to gravelly or occasionally
rocky flats and slopes, mixed grassland and sagebrush
communities, juniper and montane conifer woodlands;
1400–2500 m; Idaho, Nev., Wyo.

Variety *stragulum* is the common expression of the
species across southern Idaho (Bannock, Boise, Camas,
Custer, Elmore, Gooding, Twin Falls, and Valley
counties), mainly in the foothills and mountains adjacent
to the Snake River Plains. It occurs also just to the east
in Teton County, Wyoming. A collection from extreme
northeastern Elko County, Nevada (*Morefield & Price
5566*, NESH, NY, RENO) with much shorter leaves and
inflorescence branches is tentatively included. The
variety forms large, spreading mats with largish leaves
in rather loose rosettes. It is often found growing
intermixed with *E. heracleoides*. Spreading sulphur
flower is worthy of cultivation.

107d. Eriogonum umbellatum Torrey var. **aureum**
(Gandoger) Reveal, Taxon 17: 532. 1968
• Golden sulphur flower [E]

Eriogonum glaberrimum Gandoger
var. *aureum* Gandoger, Bull. Soc.
Roy. Bot. Belgique 42: 195. 1906;
E. neglectum Greene; *E.
umbellatum* var. *glabrum* S. Stokes;
E. umbellatum var. *intectum* A.
Nelson; *E. umbelliferum* Small

Herbs, often rather compact mats,
1–3.5 × 2–6 dm. **Aerial flowering
stems** erect, 1–2(–3) dm, thinly floccose or glabrous,
without one or more leaflike bracts ca. midlength. **Leaves**
in loose rosettes; blade usually elliptic, 1–2 × 0.5–1.5
cm, glabrous on both surfaces (except in central Idaho,
where sparsely floccose even at full maturity), margins
plane. **Inflorescences** umbellate; branches 0.3–2(–2.5)
cm, without a whorl of bracts ca. midlength; involucral

tubes 1.8–3 mm, lobes 1–3 mm. **Flowers** 4–7 mm;
perianth bright yellow.

Flowering Jun–Sep. Sandy to gravelly or occasionally
rocky flats and slopes, mixed grassland and sagebrush
communities, pinyon and/or juniper and montane conifer
woodlands; (1400–)1600–3200(–3400) m; Colo., Idaho,
Mont., Nev., Oreg., Utah, Wyo.

Variety *aureum* is closely related to var. *umbellatum*,
differing mainly in having glabrous leaf blades at full
anthesis. However, var. *aureum* is much more widely
distributed. At higher elevations, it is replaced by the
weakly differentiated var. *porteri*. It is occasionally found
in cultivation, especially in European gardens.

107e. Eriogonum umbellatum Torrey var. **porteri**
(Small) S. Stokes, Eriogonum, 109. 1936
• Porter's sulphur flower [E] [F]

Eriogonum porteri Small, Bull.
Torrey Bot. Club 25: 41. 1898

Herbs, prostrate, cespitose mats,
0.2–0.6 × 1–5 dm. **Aerial flowering stems** erect or nearly so, 0.1–
0.5 dm, floccose to nearly
glabrous, without one or more
leaflike bracts ca. midlength.
Leaves in tight rosettes; blade
usually elliptic to spatulate, 0.4–1.1 × 0.3–0.8 cm,
glabrous on both surfaces, margins plane. **Inflorescences**
umbellate and subcapitate or capitate; branches shorter
than 0.5 cm, without a whorl of bracts ca. midlength;
involucral tubes 2–3 mm, lobes 2–3 mm. **Flowers** 3–6
mm; perianth bright yellow.

Flowering Jul–Sep. Rocky slopes and ridges, high-
elevation sagebrush and meadow communities,
subalpine to alpine conifer woodlands; (2400–)2700–
3700 m; Colo., Nev., Utah.

Variety *porteri* is a high-elevation counterpart of var.
aureum that may not be worthy of taxonomic
recognition. At their extremes, the two are markedly
distinct, and based on limited observations of cultivated
plants, each maintains its basic habit when grown under
uniform conditions. In Nevada (Elko, Lander, and Nye
counties), var. *porteri* grows with a similarly reduced
alpine expression of the cream-colored var. *desereticum*.
Elsewhere it is widely scattered, mainly in the Uinta and
Wasatch mountains of Utah (Beaver, Duchesne, Iron,
Piute, Salt Lake, Sevier, Summit, Uintah, Utah, and
Wasatch counties), as well as in the Colorado Rocky
Mountains (Chaffee, Delta, Gunnison, La Plata, Moffat,
and Pitkin counties). A collection supposedly gathered
near the Goodman Ranch in Uinta County, Wyoming
(*Payson 4920*, DS, RM, S, UTC), is almost certainly
mislabeled. The variety is occasionally seen in
cultivation, mainly in Europe.

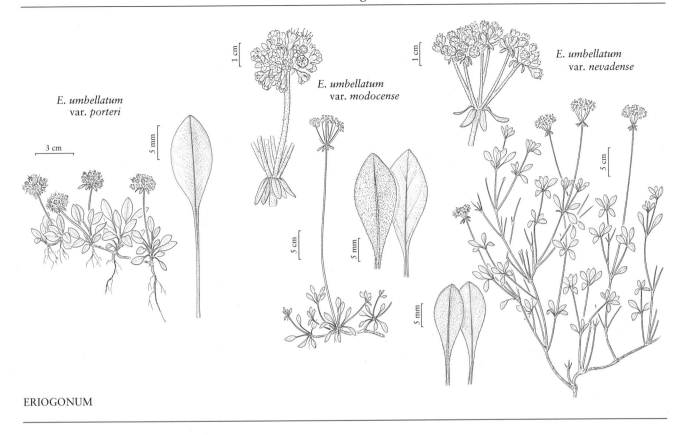

E. umbellatum
var. *porteri*

E. umbellatum
var. *modocense*

E. umbellatum
var. *nevadense*

ERIOGONUM

107f. Eriogonum umbellatum Torrey var. **hypoleium** (Piper) C. L. Hitchcock in C. L. Hitchcock et al., Vasc. Pl. Pacif. N.W. 2: 135. 1964 • Kittitas sulphur flower E

Eriogonum umbellatum subsp. *hypoleium* Piper, Contr. U.S. Natl. Herb. 11: 238. 1906

Herbs, mostly prostrate, compact mats, 1–2.5 × 2–5 dm. **Aerial flowering stems** erect, 0.7–1.5(–2) dm, thinly floccose or glabrous, without one or more leaflike bracts ca. midlength. **Leaves** in loose rosettes; blade usually elliptic, (0.5–)1–2 × (0.4–)0.5–1.5 cm, glabrous on both surfaces except for fringed, pubescent margins and veins, margins plane, pubescent. **Inflorescences** umbellate; branches 1–2(–3) cm, without a whorl of bracts ca. midlength; involucral tubes 2–4 mm, lobes 2–4 mm. **Flowers** 4–6 mm; perianth bright yellow.

Flowering Jun–Sep. Gravelly to rocky slopes and ridges, high-elevation sagebrush communities, montane to subalpine conifer woodlands; (900–)1200–2100 m; Wash.

Variety *hypoleium* is restricted to Chelan and Kittitas counties, Washington, extending from the Mt. Stuart Range south to the Bald Mountain area west of Ellensburg. It is doubtfully distinct from var. *aureum,* although geographically well isolated

107g. Eriogonum umbellatum Torrey var. **sandbergii** Reveal, Phytologia 86: 154. 2004 • Sandberg's sulphur flower E

Subshrubs, sprawling, 1–4 × 2–6 dm. **Aerial flowering stems** erect, 1–2.5(–3) dm, thinly tomentose to floccose, without one or more leaflike bracts ca. midlength. **Leaves** in loose rosettes; blade elliptic to ovate, 1–2.5(–3) × 0.5–1.5(–2) cm, densely lanate abaxially, glabrous or nearly so and bright green adaxially, margins plane. **Inflorescences** umbellate; branches 0.3–2.5 cm, thinly tomentose to floccose, without a whorl of bracts ca. midlength; involucral tubes 2–3 mm, lobes 1.5–4 mm. **Flowers** 4–7 mm; perianth bright yellow.

Flowering May–Aug. Sandy to gravelly slopes and ravines, sagebrush communities, montane conifer woodlands; 300–1200 m; Wash.

Variety *sandbergii* is restricted to a series of scattered populations in the foothills and low mountains of the north Cascade Ranges of north-central Oregon (Hood River and Wasco counties) and central Washington (Chelan, Kittitas, Okanogan, and Yakima counties). It appears to be most closely related to the more southern, matted var. *modocense.*

107h. Eriogonum umbellatum Torrey var. **haussknechtii** (Dammer) M. E. Jones, Contr. W. Bot. 11: 6. 1903 (as hausknechtii)

• Haussknecht's sulphur flower   E

Eriogonum haussknechtii Dammer, Gartenflora 40: 493, fig. 92. 1891 (as *hausknechtii*); *E. umbellatum* subsp. *haussknechtii* (Dammer) S. Stokes

Herbs, typically prostrate, sprawling mats, 0.5–1.5 × 1–4 dm. **Aerial flowering stems** spreading to erect, 0.3–0.6(–1.5) dm, thinly tomentose, without one or more leaflike bracts ca. midlength. **Leaves** typically in tight rosettes; blade usually broadly elliptic, 0.5–1.5(–2.5) × 0.5–1.2(–1.5) cm, tannish-tomentose abaxially, thinly tomentose or glabrous and olive green adaxially, margins plane. **Inflorescences** compact-umbellate; branches 0.1–1.5 (–2) cm, thinly tomentose, without a whorl of bracts ca. midlength; involucral tubes 1.5–3.5 mm, lobes 1–4 mm. **Flowers** 2–6 mm; perianth bright yellow.

Flowering Jun–Sep. Volcanic, sandy to gravelly slopes and ridges, mixed grassland and sagebrush communities, montane to subalpine conifer woodlands; 1000–2500 (–3100) m; Oreg., Wash.

Variety *haussknechtii*, as here circumscribed, is a high-elevation taxon found mainly on volcanic peaks in north-central Oregon (Benton, Clackamas, Hood River, and Wasco counties) and south-central Washington (Kittitas and Yakima counties). It is common on Mt. Hood and Mt. Adams. It typically grows with *E. marifolium*, and mixed collections often are found in herbaria; the two taxa have in common a distinctive olive green color of the adaxial leaf surfaces. Haussknecht's sulphur flower is not always clearly distinct from var. *modocense*. It is occasionally seen in cultivation, especially in European gardens.

107i. Eriogonum umbellatum Torrey var. **canifolium** Reveal, Phytologia 86: 147. 2004

• Sherman Pass sulphur flower   E

Herbs, spreading mats, 1–2.5 × 3–10 dm. **Aerial flowering stems** erect, slender, (0.5–)1–1.8(–2) dm, tomentose, without one or more leaflike bracts ca. midlength. **Leaves** in rather tight rosettes; blade elliptic, 0.4–2 × 0.3–0.7 (–0.9) cm, densely grayish-lanate on both surfaces or tomentose and grayish to greenish adaxially, rarely thinly floccose or glabrous on individual leaf blades, margins plane. **Inflorescences** umbellate; branches 1–2.5(–4) cm, tomentose to floccose, without a whorl of bracts ca. midlength; involucral tubes 2–3 mm, lobes 2–3.5 mm. **Flowers** (4–)5–7(–8) mm; perianths bright yellow.

Flowering Jun–Sep. Sandy granitic slopes, montane conifer woodlands; 2400–2600 m; Calif.

Variety *canifolium* is found infrequently in the southern Sierra Nevada of Inyo and Tulare counties, and in the Argus Mountains to the east. It also occurs in the Transverse Ranges of southwestern Kern and western Los Angeles counties. The densely lanate leaf surfaces and the low, matted habit are distinctive features, especially when this variety is compared with the more commonly encountered var. *nevadense* of the Sierra Nevada. As the plants mature, the amount of tomentum on the adaxial surface of the leaf blades thins, and individual leaves can become floccose or even glabrous.

107j. Eriogonum umbellatum Torrey var. **modocense** (Greene) S. Stokes, Eriogonum, 110. 1936

• Modoc sulphur flower   E   F

Eriogonum modocense Greene, Pittonia 5: 68. 1902

Herbs, mats, 1–3.5(–4) × 1–3(–5) dm. **Aerial flowering stems** erect, 0.7–2.5 dm, usually tomentose, without one or more leaflike bracts ca. midlength. **Leaves** in somewhat compact rosettes; blade usually elliptic, (0.3–)1–1.5(–2) × 0.1–1(–1.2) cm, densely tomentose abaxially, mostly glabrous and green adaxially, margins plane. **Inflorescences** umbellate; branches 3–10(–15) cm, usually tomentose, without a whorl of bracts ca. midlength; involucral tubes 2–3 mm, lobes 2–3 mm. **Flowers** 4–8 mm; perianth bright yellow.

Flowering Jun–Sep. Sandy to gravelly flats and slopes, mixed grassland and sagebrush communities, pinyon and/or juniper and montane conifer woodlands; (200–) 600–2300(–2500) m; Calif., Idaho, Nev., Oreg.

Variety *modocense* is the common form of the species encountered mainly east of the Cascade Range from central Oregon (Crook, Deschutes, Douglas, Grant, Harney, Jackson, Jefferson, Josephine, Klamath, Lake, and Malheur counties) to northern California (Butte, Lassen, Modoc, Shasta, and Siskiyou counties). It is found less frequently in northernmost Nevada (Humboldt and Washoe counties) and southwestern Idaho (Camas, Gem, Gooding, Owyhee, and Twin Falls counties). In the northern Sierra Nevada, var. *modocense* merges with var. *nevadense*. In California and Nevada, plants now assigned here generally have gone under the name *Eriogonum umbellatum* var. *polyanthum* in the pre-1989 literature.

107k. Eriogonum umbellatum Torrey var. **dumosum** (Greene) Reveal, Harvard Pap. Bot. 9: 202. 2004

• American Valley sulphur flower [E]

Eriogonum dumosum Greene, Pittonia 3: 199. 1897; *E. umbellatum* subsp. *dumosum* (Greene) S. Stokes

Shrubs, round to erect, 3–5 × 3–10 dm. **Aerial flowering stems** erect, 0.8–2(–2.5) dm, tomentose to floccose, without one or more leaflike bracts ca. midlength. **Leaves** in loose rosettes; blade elliptic to ovate, 1–2.5 (–3) × 0.5–1.2 cm, densely tomentose abaxially, thinly floccose or glabrous adaxially, margins plane. **Inflorescences** umbellate; branches 2–6(–10) cm, tomentose to floccose, without a whorl of bracts ca. midlength; involucral tubes 2–3(–5) mm, lobes 2.5–4 mm. **Flowers** 5–9 mm; perianth bright yellow.

Flowering Jun–Sep. Sandy to gravelly flats and slopes, mixed grassland communities, oak and conifer woodlands; (300–)600–1200 m; Calif., Oreg.

Variety *dumosum* has been called var. *polyanthum* in the recent California literature. It is known only from widely scattered locations, where it is localized and typically infrequent, in Amador, Placer, Plumas, Shasta, and Siskiyou counties in California, and just inside Oregon (in Jackson County). It approaches var. *nevadense* in size but differs in having much more densely tomentose leaf blades. It can be confused also with var. *modocense*, which is much smaller and nearly always to the east or north of var. *dumosum*. Unlike the related var. *ahartii*, which is always on serpentine soils, var. *dumosum* is only occasionally found on that substrate. The American Valley sulphur flower is a largish shrub that is ideal for the garden and may well prove popular if successfully cultivated.

107l. Eriogonum umbellatum Torrey var. **goodmanii** Reveal, Phytologia 66: 259. 1989

• Goodman's sulphur flower [E]

Herbs, spreading to prostrate mats, 1–4.5(–5) × 4–7 dm. **Aerial flowering stems** erect, 1–2.5(–4) dm, floccose, without one or more leaflike bracts ca. midlength. **Leaves** in mostly loose rosettes; blade elliptic to ovate, 0.5–2(–3.5) × 0.5–1(–1.5) cm, densely lanate abaxially, tomentose to rarely floccose adaxially, margins plane. **Inflorescences** umbellate; branches (2.5–)3–5(–8) cm, floccose, without a whorl of bracts ca. midlength; involucral tubes 2–3 mm, lobes 2.5–4 mm. **Flowers** 6–8(–9) mm; perianth bright yellow.

Flowering May–Sep. Sandy to gravelly serpentine flats and slopes, mixed grassland communities, oak and montane conifer woodlands; 400–1700(–2100) m; Calif., Oreg.

Variety *goodmanii* is an elegant taxon of potential horticultural value. It is encountered rather infrequently in its restricted range, being common only in the Waldo area of Josephine County, Oregon, although it is found also in Benton, Deschutes, and Douglas counties, Oregon, and Del Norte, Humboldt, and Siskiyou counties, California. Variety *goodmanii* merges with var. *modocense* on the eastern edge of its range, particularly in Siskiyou and Deschutes counties.

107m. Eriogonum umbellatum Torrey var. **humistratum** Reveal, Phytologia 66: 260. 1989

• Scott Mountain sulphur flower [E]

Herbs, prostrate mats, 0.7–2 × 1–3 dm. **Aerial flowering stems** erect or nearly so, 0.5–1.5 dm, floccose, without one or more leaflike bracts ca. midlength. **Leaves** in rather compact rosettes; blade broadly elliptic, 0.5–1(–1.5) × 0.5–1 cm, densely white-tomentose on both surfaces or slightly less so adaxially. **Inflorescences** umbellate; branches (2–)2.5–5 cm, floccose, without a whorl of bracts ca. midlength; involucral tubes 1.5–2 mm, lobes 1.5–3 mm. **Flowers** 3–6 mm; perianth bright yellow.

Flowering Jun–Sep. Gravelly serpentine slopes and ridges, montane conifer woodlands; 1700–2800 m; Calif.

Variety *humistratum* is restricted to exposed serpentine in four areas in Siskiyou and Trinity counties of northern California: White Mountain, Mt. Eddy-Scott Mountain, the Marble Mountains, and Mt. Shasta.

107n. Eriogonum umbellatum Torrey var. **nevadense** Gandoger, Bull. Soc. Roy. Bot. Belgique 42: 198. 1906 • Nevada sulphur flower [E] [F]

Subshrubs, mostly spreading, 1–5 × 2–6 dm. **Aerial flowering stems** erect, 1–3 dm, thinly tomentose to nearly glabrous, without one or more leaflike bracts ca. midlength. **Leaves** in rather open rosettes; blade usually elliptic, 1–2(–2.5) × 0.5–1.5 cm, thinly tomentose to floccose abaxially, less so to thinly floccose or glabrous and green adaxially, margins plane. **Inflorescences** umbellate; branches 0.3–3(–5) cm, thinly tomentose to subglabrous, without a whorl of bracts ca. midlength; involucral tubes 2–3.5 mm, lobes 1.5–3.5 mm. **Flowers** 4–7 mm; perianth bright yellow. $2n = 80$.

Flowering Jun–Sep. Sandy to gravelly flats and slopes, mixed grassland and sagebrush communities, pinyon-juniper and montane to subalpine conifer woodlands; (1000–)1500–3000(–3400) m; Calif., Nev., Oreg.

Variety *nevadense* is widespread and often common in California, Nevada, and southeastern Oregon. It is the common expression of the species at middle and lower elevations on the eastern slope of the Sierra Nevada and the Great Basin portion of the Intermountain West. In the northern Sierra Nevada of California, and in the Granite Range of Washoe County, Nevada, it merges with var. *modocense*. At higher elevations in the Sierra Nevada, it grades into var. *covillei*. The Nevada sulphur flower occasionally is found in cultivation.

A single collection from the Charley Johnson Canyon, Canyon Mountains, eastern Millard County, Utah (*Goodrich 15832*, BRY), may well be var. *nevadense*, but it needs to be confirmed.

107o. Eriogonum umbellatum Torrey var. vernum
Reveal, Great Basin Naturalist 28: 157. 1968 · Spring-flowering sulphur flower [E]

Shrubs, dome-shaped, 3–6(–9) × 3–9(–13) dm. **Aerial flowering stems** erect, 0.5–1.5 dm, floccose or glabrous, without one or more leaflike bracts ca. midlength. **Leaves** in rather open rosettes; blade elliptic, 0.5–2.5 × 0.3–1 cm, thinly floccose abaxially, less so to thinly floccose or glabrous and green adaxially, with some blades glabrous on both surfaces, margins plane. **Inflorescences** umbellate; branches 3–8 cm, mostly glabrous, without a whorl of bracts ca. midlength; involucral tubes 1.5–2.5 mm, lobes 2–3 mm. **Flowers** (5–)6–9(–10) mm; perianth pale to bright yellow.

Flowering May–Jul. Sandy to gravelly, often volcanic flats and slopes, saltbush and sagebrush communities; 1400–2000(–2200) m; Nev.

Variety *vernum* is known from several scattered populations in northern Nye County. It flowers mainly in May and early June, with some fruit-bearing flowers persisting into early July. Flower color can vary from bright to pale yellow in a single population. The taxon clearly is related to var. *nevadense*, which in central Nevada occurs at higher elevations (mainly in pinyon-juniper communities). The shrubs certainly are worthy of cultivation.

107p. Eriogonum umbellatum Torrey var. covillei
(Small) Munz & Reveal in P. A. Munz, Suppl. Calif. Fl., 43. 1968 · Coville's sulphur flower [E]

Eriogonum covillei Small, Bull. Torrey Bot. Club 25: 42. 1898; *E. umbellatum* subsp. *covillei* (Small) Munz

Herbs, typically prostrate mats, 0.5–1 × 1–5 dm. **Aerial flowering stems** spreading to erect, 0.3–0.9 dm, thinly tomentose to nearly glabrous, without one or more leaflike bracts ca. midlength. **Leaves** in tight rosettes; blade usually narrowly elliptic, 0.3–0.6(–1) × 0.2–0.4 (–0.6) cm, white-tomentose abaxially, slightly less so and greenish adaxially, margins plane. **Inflorescences** compact-umbellate; branches 0.2–1 cm, thinly tomentose to floccose, without a whorl of bracts ca. midlength; involucral tubes 1.5–2.5 mm, lobes 1–3 mm. **Flowers** 2–4(–5) mm; perianth bright yellow.

Flowering Jul–Sep. Gravelly to rocky or talus slopes and ridges, high-elevation sagebrush communities, alpine conifer woodlands; 3000–3600 m; Calif.

Variety *covillei* is restricted to the backbone of the Sierra Nevada in Inyo and Tulare counties, and the White Mountains in Mono County; it is rare throughout its range. Clearly it is an alpine derivative of var. *nevadense*, and some specimens can be difficult to place.

107q. Eriogonum umbellatum Torrey var. dichrocephalum
Gandoger, Bull. Soc. Roy. Bot. Belgique 42: 199. 1906 · Bicolor sulphur flower [E]

Eriogonum latum Small ex Rydberg; *E. umbellatum* subsp. *aridum* (Greene) S. Stokes; *E. umbellatum* var. *aridum* (Greene) C. L. Hitchcock

Herbs, spreading (rarely compact) mats, 1–3(–3.5) × 5–10 dm. **Aerial flowering stems** erect (rarely spreading), (0.5–)1–2.5 dm, thinly tomentose to floccose or rarely glabrous, without one or more leaflike bracts ca. midlength. **Leaves** in mostly loose rosettes; blade elliptic to broadly elliptic, 1–2(–2.5) × 0.5–1.5(–2) cm, tomentose to floccose abaxially, thinly tomentose or glabrous and green adaxially, margins plane. **Inflorescences** umbellate; branches 0.5–4 cm, thinly tomentose, without a whorl of bracts ca. midlength; involucral tubes 2–3 mm, lobes 1–2.5 mm. **Flowers** 4–8 mm; perianth pale yellow to cream or whitish, rarely greenish white.

Flowering Jun–Sep. Sandy to gravelly flats, slopes, and ridges, mixed grassland and sagebrush communities, pinyon and/or juniper and montane to subalpine conifer woodlands; (1200–)1400–3100(–3400) m; Calif., Idaho, Mont., Nev., Oreg., Utah, Wyo.

Variety *dichrocephalum* is widespread and occasionally locally common from southeastern Oregon, southern Idaho, western, and western Wyoming south to central-eastern California, central Nevada, and northern Utah.

In western Montana and Wyoming var. *dichrocephalum* is not always clearly distinct from var. *majus*, the key difference being that the tomentum of the former tends to be more whitish, while that of the latter is rusty. Furthermore, the adaxial surface of the leaf blades in the latter tends to be olive green, a color not seen in var. *dichrocephalum*. Finally, the leaves of var. *majus* tend to be longer and narrower, and the entire plant forms a flat, dense mat, a condition not usually seen in the more loosely arranged mats of var. *dichrocephalum*.

107r. Eriogonum umbellatum Torrey var. **deureticum** Reveal, Great Basin Naturalist 35: 365. 1976

  • Deseret sulphur flower E

Herbs, spreading mats, 1–3.5(–4) × 3–6 dm. **Aerial flowering stems** erect, 1–3 dm, thinly floccose or glabrous, without one or more leaflike bracts ca. midlength. **Leaves** in loose rosettes; blade usually elliptic, 1–2(–2.5) × 0.5–1.5(–2) cm, glabrous on both surfaces at full anthesis, margins plane. **Inflorescences** umbellate; branches 2–4.5(–5.5) cm, glabrous, without a whorl of bracts ca. midlength; involucral tubes 2–3 mm, lobes 1–2.5 mm. **Flowers** 4–8 mm; perianth pale yellow to cream.

Flowering Jun–Sep. Sandy to gravelly slopes and ridges, mixed grassland and sagebrush communities, oak, aspen, and montane to subalpine conifer woodlands; (1500–)1900–3300 m; Idaho, Mont., Nev., Utah, Wyo.

Variety *deureticum* is widely scattered in southern Idaho (Bear Lake, Blaine, Custer, and Owyhee counties), southwestern Montana (Park County), and southwestern Wyoming (Carbon, Teton, and Uinta counties) south into northeastern Nevada (Elko County) and northern Utah (Box Elder, Carbon, Daggett, Duchesne, Emery, Juab, Millard, Piute, Rich, Salt Lake, Sanpete, Summit, Tooele, Utah, and Wasatch counties). It is common only in Utah and southeastern Idaho. Variety *deureticum* is related to var. *dichrocephalum*, but the two rarely occur together. Late in the season, high-elevation plants can have attractive, bright red leaves. Such plants approach var. *porteri* in aspect, especially in the Jarbidge, Ruby, and East Humboldt mountains of Nevada. The Deseret sulphur flower is worthy of cultivation.

107s. Eriogonum umbellatum Torrey var. **mohavense** Reveal, Phytologia 86: 149. 2004 • Mohave sulphur flower E

Herbs, spreading mats, 0.5–2 × 1–3 dm. **Aerial flowering stems** spreading to erect, (0.3–)0.5–1.5(–2) dm, floccose, without one or more leaflike bracts ca. midlength. **Leaves** mostly in tight rosettes; blade narrowly elliptic, 0.7–2.5 × 0.3–1 cm, thinly floccose on both surfaces or glabrous adaxially, margins plane. **Inflorescences** umbellate; branches 2.5–8 cm, thinly floccose or glabrous, without a whorl of bracts ca. midlength; involucral tubes 2–3 mm, lobes 1.5–3 mm. **Flowers** 3–7 mm; perianth bright yellow.

Flowering May–Jun. Sandy to gravelly flats and slopes, sagebrush communities, oak, pinyon-juniper, and montane conifer woodlands; 1200–1600 m; Ariz.

Variety *mohavense* is known only from the Black Rock and Wolf Hole mountains area on the Arizona Strip of Mohave County, Arizona. The rays or branchlets of the inflorescences are rather long (2.5–8 cm). The taxon is related to the much more widely distributed, late-season-flowering var. *subaridum* found to the north and west.

107t. Eriogonum umbellatum Torrey var. **minus** I. M. Johnston, Bull. S. Calif. Acad. Sci. 17: 64. 1918 • Old Baldy sulphur flower E F

Eriogonum umbellatum subsp. *minus* (I. M. Johnston) Munz

Herbs, dense, prostrate mats, 0.3–1 × 0.5–2 dm. **Aerial flowering stems** spreading, 0.2–0.8(–1.5) cm, tomentose, without one or more leaflike bracts ca. midlength. **Leaves** in tight rosettes; blade usually round-ovate, 0.3–0.8(–1) × 0.3–0.8 cm, densely lanate on both surfaces, margins plane. **Inflorescences** usually compact-umbellate; branches 0.5–2.5 cm, tomentose, without a whorl of bracts ca. midlength; involucral tubes 1.5–2 mm, lobes 1.5–2 mm. **Flowers** (2.5–)4–6 mm; perianth lemon yellow to yellowish red, becoming red to rose-red.

Flowering Jul–Sep. Gravelly to rocky or talus slopes and ridges, sagebrush communities, montane to subalpine conifer woodlands; (1800–)2400–3100 m; Calif.

Variety *minus* is a rare and localized taxon in the San Bernardino and San Gabriel mountains of Los Angeles and San Bernardino counties. It is one of the more distinctive expressions of the species and is attractive in

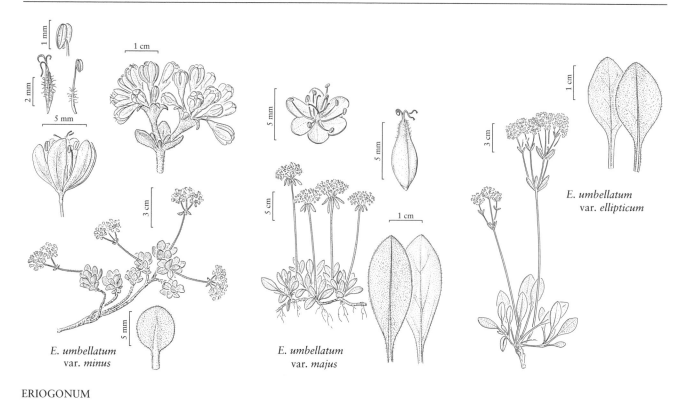

E. *umbellatum*
var. *minus*

E. *umbellatum*
var. *majus*

E. *umbellatum*
var. *ellipticum*

ERIOGONUM

the garden. A Parish collection (Aug 1915, DS) supposedly from the San Jacinto Mountains of Riverside County is presumed to be mislabeled. This variety is closely related to var. *bahiiforme*.

107u. Eriogonum umbellatum Torrey var. **majus** Hooker, Hooker's J. Bot. Kew Gard. Misc. 5: 264. 1853 • Subalpine sulphur flower [E] [F]

Eriogonum subalpinum Greene; *E. umbellatum* subsp. *majus* (Hooker) Piper; *E. umbellatum* subsp. *subalpinum* (Greene) S. Stokes

Herbs, prostrate to spreading, compact mats, 1–4.5(–5) × 2–8 (–10) dm. **Aerial flowering stems** erect, 1.5–3 dm, usually floccose, without one or more leaflike bracts ca. midlength. **Leaves** in loose rosettes; blade oblanceolate to elliptic, (0.3–)0.5–2(–4) × 0.3–1(–1.5) cm, densely whitish-, greenish-, or reddish-lanate abaxially, glabrous and olive green to bright green adaxially, margins plane. **Inflorescences** umbellate; branches (2–)3–9 cm, usually floccose, without a whorl of bracts ca. midlength; involucral tubes 2–3.5 mm, lobes 1–4 mm. **Flowers** 3–7(–9.5) mm; perianth cream. *2n* = 76.

Flowering Jun–Sep. Sandy to gravelly flats, slopes and ridges, mixed grassland and sagebrush communities, oak, aspen, and montane to subalpine conifer woodlands, mountain meadows, or in alpine tundra; (800–)1200–2800(–3500) m; Alta., B.C.; Colo., Idaho, Mont., Utah, Wash., Wyo.

Variety *majus* is widespread and common in the Rocky Mountains. These plants are often locally common in Idaho and northern Utah, but they are rather rare in the Cascade Range of Washington. The high-elevation plants in Washington are often markedly different from similarly situated Rocky Mountain plants, having smaller leaves and flowers, and tighter, more compact, umbellate inflorescences.

Variety *majus* is distinct from and often grows with var. *umbellatum* in Colorado, leading many local taxonomists to distinguish the two at species rank. In Wyoming and Montana, however, var. *majus* occasionally is difficult to differentiate from var. *dichrocephalum*. Variety *majus* often occurs with *Eriogonum heracleoides*, and mixed collections occasionally are encountered. Care must be taken in the herbarium to differentiate the narrow-leaved *E. heracleoides* var. *leucophaeum* from the broader-leaved *E. umbellatum* var. *majus*, although the two do not grow together.

107v. Eriogonum umbellatum Torrey var. **ramulosum** Reveal, Phytologia 86: 153. 2004 • Buffalo Bill's sulphur flower E

Herbs, often rather compact mats, 1–3.5 × 2–4 dm. **Aerial flowering stems** erect, 1–3 dm, floccose, without one or more leaflike bracts ca. midlength. **Leaves** in loose rosettes; blade usually broadly elliptic to oval, 1–2.5 × 0.5–1.5 cm, densely gray-tomentose abaxially, floccose and green adaxially, margins plane. **Inflorescences** compound-umbellate, branched 1–3 times; branches floccose, without a whorl of bracts ca. midlength; involucral tubes 2–3 mm, lobes 1.5–3 mm. **Flowers** 4–7 mm; perianth bright yellow.

Flowering Jun–Sep. Sandy to gravelly slopes, sagebrush communities, montane conifer woodlands; 1600–2700 m; Colo.

Variety *ramulosum* is encountered rarely, mainly along the Front Range of the Colorado Rocky Mountains. It is known only from Jefferson and Larimer counties. It is related to var. *umbellatum*, consistently differing in having a branched inflorescence.

107w. Eriogonum umbellatum Torrey var. **ellipticum** (Nuttall) Reveal, Taxon 32: 294. 1983 • Starry sulphur flower E F

Eriogonum ellipticum Nuttall, Proc. Acad. Nat. Sci. Philadelphia 4: 14. 1848; *E. croceum* Small; *E. stellatum* Bentham; *E. umbellatum* var. *chrysanthum* Gandoger; *E. umbellatum* var. *croceum* (Small ex Rydberg) S. Stokes; *E. umbellatum* subsp. *stellatum* (Bentham) S. Stokes; *E. umbellatum* var. *stellatum* (Bentham) M. E. Jones

Herbs, often rather compact mats, 1–3.5 × 2–5 dm. **Aerial flowering stems** erect, 0.7–2(–3) dm, thinly floccose or glabrous, without one or more leaflike bracts ca. midlength. **Leaves** in loose rosettes; blade usually elliptic, 1.5–3(–4) × 1–1.5 cm, tomentose abaxially, usually glabrous adaxially, margins plane. **Inflorescences** compound-umbellate, branched 2–4 times; branches thinly floccose or glabrous, without a whorl of bracts ca. midlength; involucral tubes 2–4 mm, lobes 2–4 mm. **Flowers** 6–8 mm; perianth bright yellow.

Flowering Jun–Sep. Sandy to gravelly flats and slopes, mixed grassland and sagebrush communities, montane conifer woodlands; 700–2400 m; Idaho, Mont., Oreg., Wash.

Variety *ellipticum* is widely scattered but locally common in the mountains of the Pacific Northwest. It has long been known as var. *stellatum*, the name being altered to var. *ellipticum* only for technical nomenclatural reasons. This is the northern phase of the species, with compound inflorescences. Considerable variation in plant size is retained within the circumscription adopted here. Plants from northeastern Oregon and adjacent west-central Idaho are large and showy, and it is this phase (called *Eriogonum croceum* or *E. umbellatum* var. *chrysanthum*) that occasionally is seen in cultivation.

107x. Eriogonum umbellatum Torrey var. **devestivum** Reveal, Great Basin Naturalist 32: 115. 1972 • Emperor's sulphur flower E

Herbs, spreading mats, 1–3.5 × 2–6 dm. **Aerial flowering stems** erect, 1.5–2.5 dm, thinly floccose or glabrous, without one or more leaflike bracts ca. midlength. **Leaves** in loose rosettes; blade broadly elliptic, 1.5–3(–4) × 1–1.5 cm, glabrous on both surfaces at full anthesis, margins plane. **Inflorescences** compound-umbellate, branched 2–4 times; branches thinly floccose or glabrous, without a whorl of bracts ca. midlength; involucral tubes 2–3.5 mm, lobes 1–2.5 mm. **Flowers** 4–7 mm; perianth bright yellow.

Flowering Jun–Sep. Sandy to gravelly flats and slopes, sagebrush communities, montane conifer woodlands; 800–1800 m; Idaho, Oreg., Wash.

Variety *devestivum* is a glabrous-leaved expression obviously related to var. *ellipticum*. It is infrequently encountered in Asotin and Columbia counties, Washington; Baker, Grant, and Union counties, Oregon; and Ada, Adams, Blaine, Lemhi, Valley, and Washington counties, Idaho. The plants are bright and showy and would be attractive in the garden.

107y. Eriogonum umbellatum Torrey var. **furcosum** Reveal, Great Basin Naturalist 45: 278. 1985 • Sierra Nevada sulphur flower E

Subshrubs, spreading to rounded, 3–6 × 3–8 dm. **Aerial flowering stems** erect, 0.5–2 dm, sparsely floccose or glabrous, without one or more leaflike bracts ca. midlength. **Leaves** in loose rosettes; blade elliptic to oblong, (0.7–)1–2.5(–3) × 0.3–0.8(–1.3) cm, white-tomentose abaxially, floccose to mostly glabrous and green adaxially, margins plane. **Inflorescences** compound-umbellate, branched

2–4 times; branches without a whorl of bracts ca. midlength; involucral tubes 2–3(–4.5) mm, lobes 1.5–3(–4) mm. **Flowers** (5–)6–8 mm; perianth bright yellow.

Flowering Jun–Sep. Sandy to gravelly flats and slopes, sagebrush communities, oak and montane conifer woodlands; 1200–3000 m; Calif., Nev.

Variety *furcosum* is the common, compound-umbellate form of the species encountered in the Sierra Nevada of California (Alpine, Amador, Calaveras, El Dorado, Fresno, Inyo, Kern, Madera, Mariposa, Mono, Nevada, Placer, Sierra, Tulare, and Tuolumne counties) and in the Mt. Rose/Slide Mountain area of southern Washoe County, Nevada.

107z. Eriogonum umbellatum Torrey var. **argus** Reveal, Phytologia 66: 261. 1989 • One-eyed sulphur flower E

Herbs, usually spreading mats, 1–3 × 5–15 dm. **Aerial flowering stems** erect, (0.8–)1–2 dm, thinly tomentose to floccose or glabrous, often with single leaflike bract ca. midlength. **Leaves** in loose rosettes; blade oblanceolate to elliptic, (0.7–)1–2(–2.5) × 0.4–1 cm, mostly thinly tomentose abaxially, floccose or glabrous adaxially, margins often finely wavy. **Inflorescences** umbellate or compound-umbellate, branched 2–4 times; branches usually floccose or glabrous, without a whorl of bracts ca. midlength; involucral tubes 2–3 mm, lobes 2–3(–4) mm. **Flowers** 3–8 mm; perianth bright yellow.

Flowering Jun–Sep. Gravelly to rocky serpentine slopes and ridges, oak and montane conifer woodlands; (900–)1500–2200(–2500) m; Calif., Oreg.

Variety *argus* occurs in the Siskiyou/Trinity mountains of Josephine and Jackson counties, Oregon, and in Del Norte, Glenn, Humboldt, Plumas, Siskiyou, and Trinity counties, California. It is the serpentine counterpart to var. *furcosum* of volcanic and granitic soils in the Sierra Nevada, but is more elegant and much more attractive. The inflorescences are nearly always compound, and the flowering stem frequently has a single leaflike bract about midlength. Based on the latter feature, the one-eyed sulphur flower is probably more closely related to var. *bahiiforme* than it is to var. *furcosum*.

107aa. Eriogonum umbellatum Torrey var. **nelsoniorum** Reveal, Phytologia 86: 151. 2004 • Nelson's sulphur flower E

Subshrubs or herbs, spreading, 1–2.5 × 2–7 dm. **Aerial flowering stems** erect, 1–2 dm, thinly floccose, without one or more leaflike bracts ca. midlength. **Leaves** in rather loose rosettes; blade elliptic to oblong, (0.5–)1–1.5(–2) × 0.4–0.8 cm, densely white-tomentose to lanate abaxially, white-tomentose to floccose or rarely glabrate and greenish adaxially, margins plane. **Inflorescences** umbellate, 1–3 cm; branches 1–2 cm, thinly floccose, without a whorl of bracts ca. midlength; involucral tubes 3–4 mm, lobes 2–4 mm. **Flowers** (5–)6–7 mm; perianth bright yellow.

Flowering Jul–Sep. Sandy to gravelly serpentine slopes, oak and montane conifer woodlands; 1500–2300 m; Calif.

Variety *nelsoniorum* is somewhat similar morphologically to var. *argus* but probably more closely related to var. *bahiiforme*. It differs from both in consistently having a simple, rather than compound-umbellate, inflorescence. The large bracts that subtend the umbel are long (1–1.5 cm) and leaflike, like those of var. *bahiiforme*, but the distribution of tomentum of the leaf blades is like that of var. *argus*. All three occur on serpentine soils. However, var. *nelsoniorum* is found only in Humboldt and Trinity counties, or basically north of var. *bahiiforme* and southwest of var. *argus*. A series of small populations in the Scott Bar Mountains of Siskiyou County may prove to be var. *nelsoniorum*.

107bb. Eriogonum umbellatum Torrey var. **bahiiforme** (Torrey & A. Gray) Jepson, Fl. Calif. 1: 425. 1914 (as bahiaeforme) • Santa Clara sulphur flower E

Eriogonum polyanthum Bentham var. *bahiiforme* Torrey & A. Gray, Proc. Amer. Acad. Arts 8: 159. 1870 (as bahiaeforme)

Herbs, spreading mats, 0.8–2 (–2.5) × 3–6 dm. **Aerial flowering stems** erect, usually 0.5–1.5 dm, tomentose, without one or more leaflike bracts ca. midlength. **Leaves** in rather compact rosettes; blade elliptic to oval, 0.5–1.5(–1.7) × 0.3–0.7 cm, densely white- to gray-lanate on both surfaces, margins plane. **Inflorescences** compound-umbellate, branched 2–5 times; branches tomentose, without a whorl of bracts ca. midlength; involucral tubes 2–3 mm, lobes 2–3.5(–4) mm. **Flowers** 5–8 mm; perianth bright yellow.

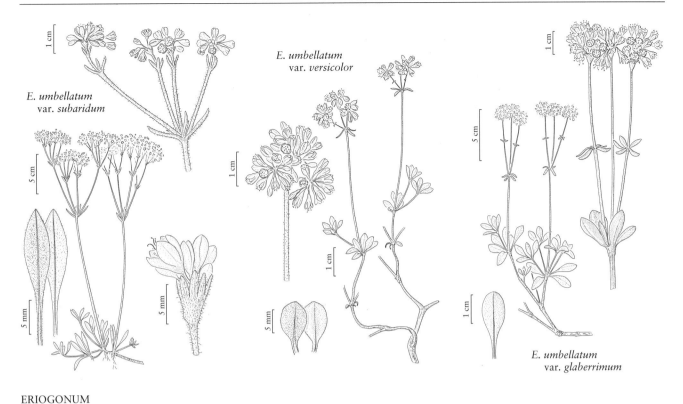

E. umbellatum
var. *subaridum*

E. umbellatum
var. *versicolor*

E. umbellatum
var. *glaberrimum*

ERIOGONUM

Flowering Jul–Sep. Sandy to gravelly, mostly serpentine flats and slopes, oak and montane conifer woodlands; 700–2000 m; Calif.

Variety *bahiiforme* occurs in widely scattered locations in the Central Coast Ranges (Colusa, Contra Costa, Monterey, Napa, San Benito, Santa Clara, and Sonoma counties) with a disjunct population in the San Gabriel Mountains of Los Angeles County. The more depauperate relative, var. *minus*, occurs in the mountains to the east.

107cc. Eriogonum umbellatum Torrey var.
smallianum (A. Heller) S. Stokes, Eriogonum,
113. 1936 • Small's sulphur flower E

Eriogonum smallianum A. Heller,
Bull. S. Calif. Acad. Sci. 2: 68.
1903

Herbs, spreading mats, 0.8–2
(–2.5) × 3–5 dm. **Aerial flowering
stems** erect, 0.5–1.5 dm, tomentose, without one or more leaflike
bracts ca. midlength. **Leaves** in rather compact rosettes; blade elliptic, 0.5–1.5 × 0.3–0.7 cm, white-lanate to tomentose abaxially, thinly tomentose to floccose and greenish adaxially, margins plane. **Inflorescences** compound-umbellate, branched 1–3 times; branches tomentose, without a whorl of bracts ca. midlength; involucral tubes

2–3 mm, lobes 1–3 mm. **Flowers** 4–6 mm; perianth bright yellow.

Flowering Jul–Sep. Sandy to gravelly, mostly serpentine flats and slopes, oak and montane conifer woodlands; 700–2000 m; Calif.

Variety *smallianum* occurs in the North Coast Ranges (Glenn, Lake, Mendocino, Sonoma, Tehama, and Trinity counties) and is the counterpart to var. *bahiiforme*. It too is widely scattered in small, localized populations.

107dd. Eriogonum umbellatum Torrey var.
subaridum S. Stokes, Leafl. W. Bot. 2: 53. 1937
• Ferris's sulphur flower E F

Eriogonum biumbellatum Rydberg;
E. umbellatum subsp. *ferrissii* (A.
Nelson) S. Stokes; *E. umbellatum*
subsp. *subaridum* (S. Stokes) Munz

Subshrubs or shrubs, spreading to
erect, sometimes round, 2–7 × 3–
9(–12) dm. **Aerial flowering stems**
erect, (0.5–)1–3 dm, floccose or
glabrous, without one or more
leaflike bracts ca. midlength. **Leaves** in rather loose rosettes; blade elliptic, 1–3 × 0.5–2 cm, thinly floccose on both surfaces or glabrous and green adaxially (glabrous on both surfaces in s Utah, rarely tomentose abaxially in se Utah), margins plane. **Inflorescences** compound-umbellate, branched 2–5 times; branches

floccose or glabrous, without a whorl of bracts ca. midlength; involucral tubes 2–3(–3.5) mm, lobes 1–3 mm. **Flowers** 3–7 mm; perianth bright yellow.

Flowering Jun–Oct. Sandy to gravelly flats and slopes, mixed grassland, saltbush, and sagebrush communities, oak, pinyon-juniper, and montane conifer woodlands; 1200–3100 m; Ariz., Calif., Colo., Nev., Utah.

Variety *subaridum* is widespread and often common throughout its extensive range from southeastern California, southern Nevada, and northern Arizona to Utah and southwestern Colorado.

Variety *subaridum* is more closely related to var. *nevadense* than it is to var. *ellipticum*. Some populations in southwestern Utah and southwestern Colorado are distinctly shrubby, and around Zion National Park some may even have leaves glabrous on both surfaces by late fruit (e.g., *Neese & Neese 9633*, BRY; *Thorne & Welsh 15194*, BRY; *Welsh et al. 27166*, BRY). Such plants never have leaf blades glabrous at anthesis, as is the case with var. *chlorothamnus*. A few populations in southeastern Utah (e.g., *Welsh & Thorne 25091*, BRY) have densely tomentose abaxial leaf blades and approach var. *munzii*. The taxonomic significance of those differences remains to be determined.

107ee. Eriogonum umbellatum Torrey var. **chlorothamnus** Reveal in P. A. Munz, Suppl. Calif. Fl., 44. 1968 • Sherwin Grade sulphur flower E

Subshrubs or shrubs, erect to rounded, (4–)5–10(–12) × 5–12 dm. **Aerial flowering stems** erect, 1–2.5 dm, glabrous, without one or more leaflike bracts ca. midlength. **Leaves** in rather loose rosettes; blade oblanceolate to narrowly elliptic, 0.5–2 × 0.3–1 cm, glabrous on both surfaces, margins plane. **Inflorescences** compound-umbellate, branched 2–5 times; branches glabrous, without a whorl of bracts ca. midlength; involucral tubes 2–2.5 mm, lobes 1–1.5 mm. **Flowers** 3–6 mm; perianth bright yellow.

Flowering Jul–Sep. Sandy granitic flats and slopes, sagebrush communities, conifer woodlands; 1600–2600(–2900) m; Calif.

Variety *chlorothamnus* is closely related to var. *subaridum* and is known only from scattered populations on the eastern slope of the Sierra Nevada in Inyo, southern Mono, and northeastern Tulare counties. Frequently it grows with var. *nevadense*, and mixed collections may be encountered in herbaria. The plants do well in cultivation and certainly are worthy of a place in the garden.

107ff. Eriogonum umbellatum Torrey var. **munzii** Reveal, Aliso 7: 218. 1970 • Munz's sulphur flower E

Eriogonum umbellatum subsp. *munzii* (Reveal) Thorne ex Munz

Herbs, spreading mats, 1–3.5 × 3–6 dm. **Aerial flowering stems** erect, 1–2 dm, mostly tomentose, without one or more leaflike bracts ca. midlength. **Leaves** in loose rosettes; blade elliptic, 1–2 × 0.5–1 cm, densely white-tomentose abaxially, floccose and greenish adaxially, rarely tomentose on both surfaces, margins plane. **Inflorescences** compound-umbellate, branched 2–4 times; branches mostly tomentose, without a whorl of bracts ca. midlength; involucral tubes 2–3 mm, lobes 1–2 mm. **Flowers** 3–8 mm; perianth bright yellow.

Flowering Jun–Sep. Sandy to gravelly flats and slopes, sagebrush communities, oak and montane conifer woodlands; 1500–2600(–2900) m; Calif.

Variety *munzii* occurs mainly in the Transverse Ranges, basically along the edge of the Mojave Desert from southern Santa Barbara and northern Ventura counties eastward across southern Kern County to northwestern San Bernardino County, then south into portions of northern Los Angeles County (as far west as the Liebre Mountains). It is disjunct in the San Jacinto Mountains, Riverside County (where the leaf blades are tomentose on both surfaces) and the Laguna Mountains, San Diego County. It is the southwestern element related to var. *subaridum*, and the distinction between the two varieties is not always sharp. Variety *munzii* occasionally is seen in cultivation.

107gg. Eriogonum umbellatum Torrey var. **versicolor** S. Stokes, Leafl. W. Bot. 3: 17. 1941 • Panamint sulphur flower E F

Herbs, spreading to somewhat prostrate mats, 1–3 × 1–4 dm. **Aerial flowering stems** usually erect, 0.5–1.5 dm, floccose, without one or more leaflike bracts ca. midlength. **Leaves** mostly in loose rosettes; blade usually elliptic, 0.5–1.5 × 0.3–1 cm, thinly tomentose on both surfaces, sometimes nearly or quite glabrous and green or becoming reddish adaxially, margins plane. **Inflorescences** umbellate or compound-umbellate; branches 0.5–1 dm, without a whorl of bracts ca. midlength, floccose; involucral tubes 2–3 mm, lobes 1–2 mm. **Flowers** 3–6 mm; perianth yellow, becoming reddish brown to rose or pink, with large reddish spot on each midrib.

Flowering Jun–Sep. Gravelly to rocky flats, slopes, and ridges, sagebrush communities, montane conifer woodlands; 1900–3300 m; Calif., Nev.

Variety *versicolor* occurs infrequently in scattered mountain ranges across southern Nevada (Clark, Eureka, Lincoln, Lyon, Nye, and White Pine counties) and into California (eastern Inyo and Mono counties), where it may be somewhat more common. It occasionally is difficult to distinguish, as it approaches both var. *dichrocephalum* and var. *juniporinum* in nearly all features except flower color, a trait that is distinctive and consistent in the field. Throughout its range, plants with both simple and branched inflorescences may be observed, the branched expression being the more common one. Variety *versicolor* is found infrequently in cultivation and ought to be more commonly used.

107hh. Eriogonum umbellatum Torrey var. **juniporinum** Reveal, Great Basin Naturalist 45: 279. 1985 • Juniper sulphur flower [E]

Subshrubs or shrubs, mostly spreading to erect, 4–8 × 5–10 dm. **Aerial flowering stems** erect, 1–2.5 dm, floccose or glabrous, without one or more leaflike bracts ca. midlength. **Leaves** in loose rosettes; blade elliptic, (0.7–)1–2 × (0.3–)0.5–1(–1.2) cm, floccose or glabrous on both surfaces, margins plane. **Inflorescences** compound-umbellate, branched 2–5 times; branches floccose or glabrous, without a whorl of bracts ca. midlength; involucral tubes (2.5–)3–3.5 mm, lobes 1–2.5 mm. **Flowers** (4–)5–6 mm; perianth cream, whitish, or pale yellow to greenish yellow, without large reddish spot on midrib.

Flowering Jun–Oct. Sandy to gravelly flats and slopes, saltbush and sagebrush communities, pinyon-juniper and occasionally montane conifer woodlands; 1300–2300 (–2500) m; Ariz., Calif., Nev., Utah.

Variety *juniporinum* is widespread and infrequent in widely scattered and disjunct populations in isolated desert mountain ranges from southern Utah (Navajo Mountains, San Juan County and Beaverdam Mountains, Washington County) and north-central Arizona (Kaibab Plateau, Coconino County) westward across southern Nevada (Spring Range of Clark County, Mt. Irish and Mormon Range in Lincoln County, and Schell Creek and Snake ranges in White Pine County) to California (southeastern Inyo and northeastern San Bernardino counties, mainly Clark Mountain, Kingston,

New York, and Providence ranges, and in the Mid Hills). A collection supposedly from the San Bernardino Mountains of California (*Meebold 20381*, M) surely is mislabeled.

Variety *juniporinum* is most closely related to var. *subaridum*, but immature plants (or specimens without habit information) may be confused with var. *versicolor*. These plants most frequently are seen in San Bernardino County, California, and in White Pine County, Nevada. They would make an attractive addition to the garden.

107ii. Eriogonum umbellatum Torrey var. **lautum** Reveal, Phytologia 86: 148. 2004 • Scott Valley sulphur flower [E]

Herbs, spreading mats, 1–3 × 3–10 dm. **Aerial flowering stems** erect, 1–2 dm, tomentose, without one or more leaflike bracts ca. midlength. **Leaves** in rather loose rosettes; blade broadly elliptic, (1–)1.5–4 × (0.5–)0.8–1.8(–2) cm, densely white-tomentose to lanate abaxially, white-tomentose and greenish, rarely some floccose or even glabrous and green adaxially, margins plane. **Inflorescences** compound-umbellate, 3–10(–12) cm; branches tomentose, without a whorl of bracts ca. midlength; involucral tubes (3–)3.5–5 mm, lobes (3–)4–6 mm. **Flowers** 4–7 mm; perianth bright yellow.

Flowering Jul–Sep. Sandy to gravelly flats, oak and conifer woodlands; 800–900 m; Calif.

Variety *lautum* is restricted to the Scott Valley area of Siskiyou County. It is by far one of the most attractive expressions of the species that could be introduced into the garden.

107jj. Eriogonum umbellatum Torrey var. **polyanthum** (Bentham) M. E. Jones, Contr. W. Bot. 11: 5. 1903 • American River sulphur flower [E]

Eriogonum polyanthum Bentham in A. P. de Candolle and A. L. P. P. de Candolle, Prodr. 14: 12. 1856; *E. umbellatum* Torrey subsp. *polyanthum* (Bentham) S. Stokes

Shrubs, round, rather open, 4–10 × 5–10 dm. **Aerial flowering stems** erect, 1–2 dm, floccose, without one or more leaflike bracts ca. midlength. **Leaves** in rather open, terminal rosettes; blade oblanceolate to narrowly elliptic, 1–3 × 0.3–1(–1.3) cm, densely white-tomentose abaxially, thinly floccose or glabrous and light green adaxially, margins plane. **Inflorescences** umbellate or compound-umbellate, branched 1–2(–3) times; branches thinly floccose or glabrous, occasionally central branch seemingly with a whorl of bracts ca. midlength; involucral tubes 2.5–4 mm, floccose, lobes 2–3.5 mm. **Flowers** 4–7 mm; perianth bright yellow.

Flowering Jun–Sep. Serpentine flats and slopes, oak and montane conifer woodlands; 800–1500 m; Calif.

The inflorescences of var. *polyanthum* are commonly compound-umbellate, but plants with reduced yet bracteated inflorescences do occur. Those with a reduced inflorescence technically consist of a long (6–10 cm), central, bractless peduncle and two lateral branches (3–4 cm), with each of the latter bearing a peduncle (3–6 cm). Such branches seem to have a whorl of leaflike bracts, but actually the bracts are positioned between the branch and involucres (technically at the base of the peduncle) and thus are like other members of the genus.

The name var. *polyanthum* has been misapplied in California to plants here attributed to var. *modocense* and var. *dumosum*.

107kk. Eriogonum umbellatum Torrey var. **speciosum** (Drew) S. Stokes, Eriogonum, 112. 1936 • Beautiful sulphur flower [E]

Eriogonum speciosum Drew, Bull. Torrey Bot. Club 16: 152. 1889

Shrubs, spreading to rounded, 5–15(–20) × 5–20 dm. **Aerial flowering stems** erect, 1.5–3 dm, mostly tomentose, without one or more leaflike bracts ca. midlength. **Leaves** in loose rosettes; blade elliptic to ovate, 1–3 × 0.5–2.5 cm, densely gray-lanate to tomentose abaxially, floccose or glabrous and green adaxially, margins plane. **Inflorescences** compound-umbellate, branched 2–4 times; branches tomentose, without a whorl of bracts ca. midlength; involucral tubes 4–6 mm, lobes 3–5 mm. **Flowers** 7–10(–12) mm; perianth bright yellow.

Flowering Jun–Sep. Serpentine flats and slopes, oak and conifer woodlands; 100–800 m; Calif.

Variety *speciosum* is known only from a few scattered locations in Humboldt and Trinity counties. These large and showy shrubs are worthy of cultivation. In early-flowering material, the flowers can be shorter than noted above.

The name var. *speciosum* has been widely misapplied to var. *dumosum*, var. *polyanthum*, and especially var. *ahartii*.

107ll. Eriogonum umbellatum Torrey var. **ahartii** Reveal, Phytologia 86: 146. 2004 • Ahart's sulphur flower [E]

Shrubs, spreading to rounded, 3–8 × 5–13 dm. **Aerial flowering stems** erect, 1–2 dm, mostly tomentose or at least densely floccose, without one or more leaflike bracts ca. midlength. **Leaves** in loose rosettes; blade elliptic to broadly elliptic or ovate, 1–2.5(–3) × 0.7–1.5 cm, densely rusty-lanate to tomentose abaxially, floccose or glabrous and olive green adaxially, margins plane. **Inflorescences** compound-umbellate, branched 3–4 times; branches tomentose to floccose, without a whorl of bracts ca. midlength; involucral tubes 2.5–4 mm, lobes 2–3 mm. **Flowers** 5–8 mm; perianth bright yellow.

Flowering Jun–Sep. Serpentine slopes, oak and conifer woodlands; 400–1000(–2000) m; Calif.

Variety *ahartii* is restricted to the Paradise and Lumpkin Ridge areas of Butte County. These large shrubs are among the more elegant of the sulphur flowers and are worthy of widespread cultivation.

107mm. Eriogonum umbellatum Torrey var. **cognatum** (Greene) Reveal, SouthW. Naturalist 13: 357. 1968 • Flagstaff sulphur flower [E]

Eriogonum cognatum Greene, Pittonia 3: 201. 1897; *E. umbellatum* subsp. *cognatum* (Greene) S. Stokes

Herbs, compact mats, 1–2(–2.5) × 1–3 dm. **Aerial flowering stems** erect, 1–1.5 dm, glabrous, without one or more leaflike bracts ca. midlength. **Leaves** in loose rosettes; blade oblong-obovate to elliptic or oval, (1–)1.5–2(–2.5) × 1–1.5(–2) cm, densely white-tomentose abaxially, thinly floccose or glabrous and dark green adaxially, margins plane. **Inflorescences** compound-umbellate, branched 2–4 times; branches glabrous, without a whorl of bracts ca. midlength; involucral tubes 3–5 mm, lobes 1–2.5 mm. **Flowers** 4–7 mm; perianth bright yellow.

Flowering Jul–Sep. Sandy flats and slopes, sagebrush communities, oak and montane conifer woodlands; 1600–2200 m; Ariz.

Variety *cognatum* is localized and rather infrequent in south-central Coconino County and also barely enters Gila and Yavapai counties. It is seen most frequently in the greater Flagstaff area.

107nn. Eriogonum umbellatum Torrey var. **torreyanum** (A. Gray) M. E. Jones, Contr. W. Bot. 11: 5. 1903 • Donner Pass sulphur flower E

Eriogonum torreyanum A. Gray, Proc. Amer. Acad. Arts 8: 158. 1870

Herbs, compact mats, 1–3.5 × 4–8 dm. **Aerial flowering stems** erect, 1–2 dm, glabrous, without one or more leaflike bracts ca. midlength. **Leaves** in loose rosettes; blade elliptic to broadly elliptic, 1–3(–4) × 1–2 cm, glabrous and dark green on both surfaces, margins plane. **Inflorescences** umbellate; branches 3–10 cm, glabrous, with a whorl of bracts ca. midlength; involucral tubes 5–7 mm, lobes 2–5 mm. **Flowers** 7–10 mm; perianth bright yellow. $2n = 40$.

Flowering Jul–Sep. Sandy to gravelly granitic slopes, buckbrush and manzanita communities, montane conifer woodlands; (1800–)2100–2400 m; Calif.

Variety *torreyanum* is found only in Nevada, Placer, and Sierra counties in the Sierra Nevada. Many of its historic sites were destroyed by road construction and development, especially to the west and north of Lake Tahoe, and the number of extant populations is limited. In fact, most collections of these plants were gathered prior to 1900. This variety and var. *glaberrimum* are distinctive within *E. umbellatum*, as both have glandular rather than hairy bractlets subtending the pedicel. Variety *torreyanum* might reasonably be recognized at the species level were it not for its occasional introgression with var. *nevadense* (T. M. Kan 1992). It rarely is seen in cultivation.

107oo. Eriogonum umbellatum Torrey var. **glaberrimum** (Gandoger) Reveal, Taxon 17: 532. 1968 • Warner Mountains sulphur flower E F

Eriogonum glaberrimum Gandoger, Bull. Soc. Roy. Bot. Belgique 42: 195. 1906

Herbs, compact mats, 1–2.5(–3) × 3–8 dm. **Aerial flowering stems** erect, 1–2 dm, glabrous, without one or more leaflike bracts ca. midlength. **Leaves** in loose rosettes; blade narrowly elliptic to elliptic, 1–2 × 0.3–1 cm, glabrous and green on both surfaces, margins plane. **Inflorescences** umbellate; branches 3–8 cm, glabrous, with a whorl of bracts ca. midlength; involucral tubes 4–5 mm, lobes 1–3.5(–4) mm. **Flowers** 4–7 mm; perianth cream or whitish.

Flowering Jul–Sep. Sandy to gravelly slopes, sagebrush communities, aspen and montane conifer woodlands; 1600–2300 m; Calif., Oreg.

Variety *glaberrimum* is localized and known with certainty from only five locations in the Warner Mountains of Lake County, Oregon, and Modoc County, California. A collection supposedly gathered somewhere in Klamath County, Oregon (*Austin & Bruce 1758*, DS) in 1897 is discounted. This expression of the species is worthy of introduction into cultivation.

108. Eriogonum heracleoides Nuttall, J. Acad. Nat. Sci. Philadelphia 7: 49, plate 7. 1834 E F

Herbs, spreading mats, infrequently polygamodioecious, 1–6 × 2–10, tomentose to floccose. **Stems:** caudex spread-ing; aerial flowering stems erect, slender, solid, not fistulose, arising at nodes of caudex branches and at distal nodes of short, nonflowering aerial branches, (0.5–)1–3 (–4) dm, often with a whorl of (2–)5–10, linear to oblanceolate, leaflike bracts ca. midlength, 0.5–4(–5) × 0.2–1 (1.5) cm, mostly tomentose to floccose. **Leaves** in loose rosettes; petiole 0.5–3 cm, usually floccose; blade usually linear to oblanceolate, (1.5–)2–5 × 0.2–1(–1.5) cm, densely white-lanate to tomentose on both surfaces or only abaxially, thinly floccose or glabrous and green adaxially, margins entire, plane. **Inflorescences** simple or compound-umbellate, rarely reduced and compact, 1–10 × 1–10 cm; branches tomentose to floccose; bracts 3–10 or more, leaflike at proximal node, oblanceolate to linear, 0.3–1 × 0.2–0.4 cm, usually scalelike distally, 1–5 × 0.5–2.5 mm. **Involucres** 1 per node, turbinate to campanulate, 3–4.5 × 2.5–5(–6) mm, tomentose, rarely glabrous; teeth 6–12 or more, lobelike, reflexed, 1.5–5 mm. **Flowers** 4–9 mm, including 1.5–3 mm stipelike base; perianth white to cream or ochroleucous, glabrous; tepals monomorphic, spatulate to oblong-ovate; stamens exserted, 4–8 mm; filaments pilose proximally. **Achenes** light to dark brown, (2–)3.5–5 mm, glabrous except for sparsely pubescent beak.

Varieties 2 (2 in the flora): w North America.

According to N. J. Turner et al. (1980), plants of the narrow-leaved phase (var. *angustifolum*) were used for colds, tuberculosis, and other lung ailments, and to treat infected cuts and sores (a decoction of roots and stems). V. F. Ray (1933) reported that a decoction of roots of such plants was taken for diarrhea. E. V. Steedman (1930) said that they were taken for stomachaches, used in steambaths to treat aching joints and muscles, and had a role in a purifying ceremony held in sweatlodges. Steedman also indicated that a strong decoction was taken by the Thompson Indians to treat syphilis.

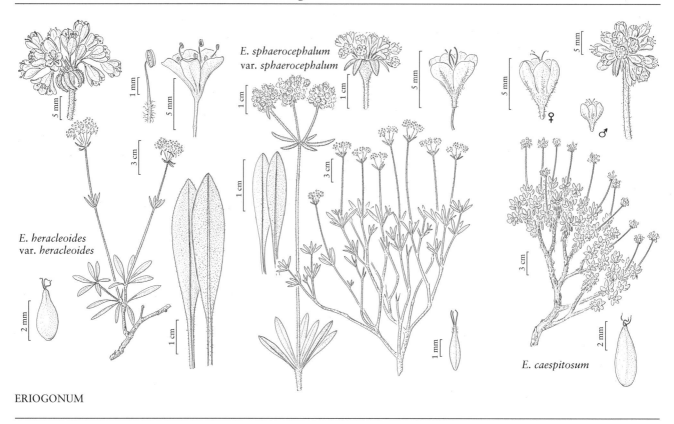

E. heracleoides
var. *heracleoides*

E. sphaerocephalum
var. *sphaerocephalum*

E. caespitosum

ERIOGONUM

1. Flowering stems with a distinct whorl of leaflike
   bracts ca. midlength . . . .
   . . . . . 108a. *Eriogonum heracleoides* var. *heracleoides*
1. Flowering stems bractless, or with 1 leaflike bract
   ca. midlength . . . .
   . . . . . 108b. *Eriogonum heracleoides* var. *leucophaeum*

108a. Eriogonum heracleoides Nuttall var.
   heracleoides • Parsnip-flower wild buckwheat
   E F

Eriogonum angustifolium Nuttall;
E. heracleoides subsp.
angustifolium (Nuttall) Piper;
E. heracleoides var. *angustifolium*
(Nuttall) Torrey & A. Gray;
E. heracleoides var. *minus* Bentham

Aerial flowering stems (0.5–)
1–3(–4) dm, with a distinct whorl
of leaflike bracts ca. midlength.
Leaf blades usually linear to linear-oblanceolate or
oblanceolate, 2–5 × 0.2–1(–1.5) cm, densely lanate to
tomentose abaxially, floccose or glabrous and green

adaxially, rarely densely lanate on both surfaces.
Involucres turbinate to campanulate; tubes 2–4.5 mm;
teeth 1.5–5 mm. **Flowers** 4–8 mm; perianth white to
cream or ochroleucous.

Flowering May–Sep. Sandy to gravelly flats, slopes,
and ridges, mixed grassland and sagebrush communities,
oak, aspen, and montane to subalpine conifer
woodlands; (300–)600–3100(–3500) m; B.C.; Calif.,
Colo., Idaho, Mont., Nev., Oreg., Utah, Wash., Wyo.

Variety *heracleoides* is widespread and usually
common. It is highly variable. Narrow-leaved
populations of southern British Columbia, northern
Washington, Idaho, and northwestern Montana
occasionally are segregated as var. *angustifolium*, but
plants in southeastern Oregon and northeastern Nevada
can have narrower leaves during years of limited
precipitation, and the same condition is found in
scattered populations elsewhere. Plants found farther
to the east consistently have broader leaves. This
expression of the species is most frequently seen in
cultivation.

108b. Eriogonum heracleoides Nuttall var. **leucophaeum** Reveal, Phytologia 25: 200. 1973

• Bractless parsnip-flower wild buckwheat E

Eriogonum caespitosum Nuttall subsp. *ramosum* (Piper) S. Stokes

Aerial flowering stems 1–1.5(–2) dm, bractless or with 1 leaflike bract ca. midlength. **Leaf blades** usually linear-oblanceolate, 1.5–3 × 0.2–0.5 cm, densely lanate to tomentose on both surfaces. **Involucres** usually campanulate; tubes 3–4.5 mm; teeth 2.5–3.5 mm. **Flowers** (5–)6–9 mm; perianth cream or ochroleucous.

Flowering May–Jul. Sandy to gravelly flats and slopes, mixed grassland and sagebrush communities; 200–1400(–1900) m; Idaho, Wash.

Variety *leucophaeum* comprises a series of populations scattered in eastern Washington (Adams, Chelan, Douglas, Kittitas, Lincoln, Okanogan, Spokane, Stevens, and Whitman counties) and adjacent Idaho (Gooding and Kootenai counties); a few outlying ones in Elmore and Valley counties are questionably assigned to this taxon.

109. Eriogonum compositum Douglas ex Bentham, Edwards's Bot. Reg. 21: plate 1774. 1835 E

Herbs, erect, infrequently polygamodioecious, 2–4(–7) × 2–5 dm; floccose or glabrous. **Stems:** caudex spreading; aerial flowering stems erect, slender or stout, hollow, often slightly fistulose, arising at nodes of caudex branches and at distal nodes of short, nonflowering aerial branches, 1–5 dm, floccose or glabrous. **Leaves** basal, occasionally in rosettes; petiole 4–10(–15) cm, tomentose; blade lanceolate or ovate to deltoid, (2–)7–25 × (0.7–)1–8 cm, densely white-lanate to tomentose abaxially, less so to glabrate and greenish adaxially, margins entire, plane. **Inflorescences** umbellate or compound-umbellate, 3–20 × 3–20 cm; branches floccose or glabrous; bracts 3–several, leaflike or semileaflike at proximal nodes, linear to linear-lanceolate, 1–3(–6) cm, scalelike distally, usually 1–5 × 0.5–3 mm. **Involucres** 1 per node, turbinate-campanulate to campanulate, 6–10 × 4–10 mm, sparsely to densely lanate, weakly glandular-puberulent, or glabrous; teeth (5–)7–10, usually not lobelike, erect to weakly reflexed, 2–4 mm. **Flowers** 5–6 mm, including 0.7–1.5 mm

stipelike base; perianth pale to bright yellow, occasionally ochroleucous, glabrous; tepals monomorphic, oblong to oblong-ovate; stamens slightly exserted, 4–8 mm; filaments pilose proximally. **Achenes** light brown, 5–6 mm, glabrous except for sparsely pubescent beak.

Varieties 3 (3 in the flora): w United States.

Eriogonum compositum is one of the more attractive members of the genus and does well in cultivation. The three varieties are only weakly differentiated, and both var. *lancifolium* and var. *leianthum* merge with var. *compositum*. The butterfly *Euphilotes enoptes* is a pollinator of this species.

The Okanagan-Colville Indians of British Columbia and Washington used plants of *Eriogonum compositum* as a cold remedy, an antidiarrheal, and a wash for infected cuts. Not surprisingly, children used the hollow, often slightly inflated stems as a toy (N. J. Turner et al. 1980).

1. Leaf blades lanceolate; Chelan, Kittitas, Okanogan, and Yakima counties, Washington . . . . 109c. *Eriogonum compositum* var. *lancifolium*
1. Leaf blades ovate to deltoid; widespread.
  2. Involucres sparsely to densely lanate; n California, wc Idaho, Oregon, and Washington . . . . 109a. *Eriogonum compositum* var. *compositum*
  2. Involucres glabrous or weakly glandular-puberulent; wc Idaho, ne Oregon, and e Washington . . . . 109b. *Eriogonum compositum* var. *leianthum*

109a. Eriogonum compositum Douglas ex Bentham var. **compositum** • Arrow-leaf wild buckwheat E

Eriogonum compositum var. *citrinum* S. Stokes; *E. compositum* var. *pilicaule* H. St. John & F. A. Warren

Plants 4–7 dm. **Aerial flowering stems** stout, often fistulose, 2–5 dm, usually glabrous, rarely floccose. **Leaf blades** ovate to deltoid, (2–)7–25 × 2–8 cm. **Involucres** sparsely to densely lanate. **Flowers** 5–6 mm; perianth pale to bright yellow. $2n = 40$.

Flowering Apr–Jul. Sandy to gravelly flats and slopes, mixed grassland and sagebrush communities, oak and montane conifer woodlands; 30–2500 m; Calif., Idaho, Oreg., Wash.

Variety *compositum* is widespread and common from central-northern Washington and west-central Idaho south through Oregon to northern California.

109b. Eriogonum compositum Douglas ex Bentham var. **leianthum** Hooker, Hooker's J. Bot. Kew Gard. Misc. 5: 264. 1853 • Smooth arrow-leaf wild buckwheat E

Eriogonum compositum var. *simplex* (S. Watson ex Piper) H. St. John & F. A. Warren

Plants 2–4(–5) dm. **Aerial flowering stems** usually slender, infrequently fistulose, 1–3 dm, usually glabrous, rarely floccose. **Leaf blades** ovate to deltoid, 3–10(–15) × 2–4(–4.5) cm. **Involucres** glabrous or more commonly weakly glandular-puberulent. **Flowers** 5–6 mm; perianth usually bright yellow.

Flowering May–Jul. Sandy to gravelly slopes, mixed grassland and sagebrush communities, oak and montane conifer woodlands; 30–1600(–2000) m; Idaho, Oreg., Wash.

Variety *leianthum* is mostly occasional to locally common in eastern Washington, northwestern and west-central Idaho, and northeastern Oregon.

109c. Eriogonum compositum Douglas ex Bentham var. **lancifolium** H. St. John & F. A. Warren, Res. Stud. State Coll. Wash. 1: 88. 1929 • Wenatchee wild buckwheat E

Eriogonum compositum subsp. *lancifolium* (H. St. John & F. A. Warren) S. Stokes

Plants 2–4 dm. **Aerial flowering stems** slender, not fistulose, 1–3 dm, usually glabrous rarely floccose. **Leaf blades** lanceolate, 2–7 × (0.7–)1–2(–2.5) cm. **Involucres** glabrous. **Flowers** 5–6 mm; perianth bright yellow or ochroleucous.

Flowering May–Jul. Sandy to gravelly slopes, mixed grassland and sagebrush communities, montane conifer woodlands; 250–1900(–2500) m; Wash.

Variety *lancifolium* is local and usually uncommon in the mountains of Chelan, Kittitas, Okanogan, and Yakima counties in eastern Washington.

110. Eriogonum siskiyouense Small, Bull. Torrey Bot. Club 25: 44. 1898 (as siskiyouensis) • Siskiyou wild buckwheat E

Eriogonum ursinum S. Watson var. *siskiyouense* (Small) S. Stokes

Herbs, spreading, matted, 0.5–2 × 1–5 dm, glabrate or glabrous. **Stems:** caudex spreading; aerial flowering stems erect, slender, solid, not fistulose, arising at nodes of caudex branches and at distal nodes of short, non-flowering aerial branches, 0.5–1.5(–2) dm, usually glabrous, with a whorl of 2–4 leaflike bracts ca. midlength, similar to leaf blade, 0.3–0.5 × 0.1–0.2 cm. **Leaves** in dense compact basal rosettes; petiole 0.2–0.6 cm, glabrate or glabrous; blade spatulate to round, (0.3–)0.5–0.8 × (0.2–)0.3–0.5(–0.7) cm, densely white to thinly tomentose abaxially, sparsely floccose to glabrate and green or olive green adaxially, rarely glabrous on both surfaces, margins entire, plane. **Inflorescences** capitate, rarely umbellate, 0.8–1.5 cm wide or 1–2 × 1–3 cm; branches usually glabrate; bracts absent immediately below involucre. **Involucres** 1 per node, turbinate-campanulate to campanulate, (3–)3.5–4 × 4–6 mm, arachnoid-tomentose; teeth 6–10, lobelike, reflexed, (2–)2.5–3.5 mm. **Flowers** (4–)4.5–6 mm, including 0.6–1 mm stipelike base; perianth sulphur yellow, glabrous; tepals monomorphic, oblong; stamens exserted, 3.5–5 mm; filaments pilose proximally. **Achenes** light brown, 4.5–5 mm, glabrous.

Flowering Jul–Sep. Gravelly serpentine slopes and outcrops, manzanita communities, montane conifer woodlands; 1600–2800 m; Calif.

Eriogonum siskiyouense is restricted to the ridge system that extends from the Scott Mountain area to the Mt. Eddy region of Siskiyou and Trinity counties. The vast majority of individuals in a population have a single involucre atop each flowering stem, but at lower elevations and in somewhat more protected sites the inflorescence may be umbellate. The Siskiyou wild buckwheat does well in cultivation if its soil requirements are fulfilled.

111. Eriogonum ternatum Howell, Fl. N.W. Amer., 570. 1902 • Ternate wild buckwheat E

Eriogonum umbellatum Torrey var. *ternatum* (Howell) S. Stokes; *E. ursinum* S. Watson var. *confine* S. Stokes

Herbs, spreading, matted, synoecious, 1–2.5(–3) × 3–13 dm, tomentose. **Stems:** caudex spreading; aerial flowering stems erect, slender, solid, not fistulose, arising at nodes of caudex branches and at distal nodes of short, nonflowering aerial branches, 1–2.5(–3) dm, tomentose. **Leaves** in loose basal rosettes; petiole 0.5–1 cm, tomentose; blade oblong to obovate, (0.7–)1–1.5 × (0.4–)0.8–1.3 cm, usually densely brown-tomentose abaxially, tomentose to thinly tomentose, floccose or glabrate adaxially, margins entire, plane. **Inflorescences** umbellate, rarely with an additional bracteate branch and 1 peduncle with 1 involucre, 0.8–3 × 1–3.5 cm; branches thinly tomentose; bracts 4–6(–8) at first node, lanceolate, 0.5–1.5 × 0.2–0.9 cm, scalelike distally, 1–3 × 0.5–1.5 mm. **Involucres** 1 per node, turbinate, 5–8 × 3–5 mm, tomentose; teeth 5–8, erect, 0.8–2 mm. **Flowers** 3–5 mm, including 0.3–0.6 mm stipelike base; perianth sulphur yellow, glabrous; tepals monomorphic, spatulate to obovate; stamens exserted, 3–4 mm; filaments pilose proximally, infrequently glabrous. **Achenes** light brown, 3.5–5 mm, glabrous except for sparsely pubescent beak.

Flowering Jun–Aug. Sandy to gravelly often serpentine slopes and outcrops, manzanita and sagebrush communities, oak and montane conifer woodlands; 400–1700(–2200) m; Calif., Oreg.

Eriogonum ternatum is infrequently encountered in the Coast Ranges of southwestern Oregon (Curry and Josephine counties) and adjacent northwestern California (Del Norte and western Siskiyou counties). A collection from "Dodkin's Lake" in the Mt. Eddy area of Siskiyou County (*Bracelin 455*, IND) remains to be confirmed, as this is well to the east of other populations. Some workers have confused this species with *E. umbellatum* (especially var. *humistratum*). No doubt the poor representation of this species in herbaria is due at least in part to collectors having assumed that it is some variant of the much more common *E. umbellatum*.

112. Eriogonum libertini Reveal, Madroño 28: 163, fig. 1. 1981 • Dubakella Mountain wild buckwheat [E]

Herbs, spreading, matted, synoecious, 0.5–2 × 1–4 dm, thinly floccose. **Stems:** caudex spreading; aerial flowering stems erect, slender, solid, not fistulose, arising at nodes of caudex branches and at distal nodes of short, nonflowering aerial branches, 0.5–1.5(–1.8) dm, thinly floccose, with a whorl of 3 bracts ca. midlength, similar to leaf blades, 0.5–0.7 × 0.2–0.4 cm. **Leaves** in compact basal rosettes; petiole 0.2–0.6 cm, tomentose; blade oblong to elliptic, 0.5–1.5(–2) × 0.3–0.5(–0.8) cm, densely white-tomentose abaxially, thinly tomentose and greenish adaxially, margins entire, plane. **Inflorescences** capitate, 1–1.5 cm wide; branches thinly floccose; bracts 3, leaflike,

narrowly lanceolate to narrowly elliptic, 0.5–0.7 × 0.2–0.5 cm. **Involucres** 1 per node, turbinate-campanulate to campanulate, 4–8 × 5–8 mm, tomentose; teeth 5–8, erect, 0.5–1.5 mm. **Flowers** 5–8 mm, including 1–1.5 mm stipelike base; perianth sulphur yellow, sparsely pubescent abaxially; tepals monomorphic, spatulate to oblong; stamens mostly included, 4–6 mm; filaments pilose proximally. **Achenes** light brown, 4–5 mm, glabrous except for sparsely pubescent beak.

Flowering Jun–Aug. Sandy to gravelly serpentine slopes and outcrops, manzanita communities, oak and montane conifer woodlands; 1100–1600 m; Calif.

Eriogonum libertini is encountered infrequently in the Dubakella and Tedoc mountains area of Tehama and Trinity counties in northwestern California. It is closely related to *E. ternatum*, found to the north.

113. Eriogonum congdonii (S. Stokes) Reveal, Aliso 7: 220. 1970 • Congdon's wild buckwheat [E]

Eriogonum ursinum S. Watson var. *congdonii* S. Stokes, Eriogonum, 114. 1936 (as congdoni); *E. ternatum* Howell var. *congdonii* (S. Stokes) J. T. Howell

Subshrubs, spreading to erect, synoecious, 1.5–5 × 1–3 dm, floccose or glabrous. **Stems:** caudex spreading to somewhat erect; aerial flowering stems erect, slender, solid, not fistulose, arising at nodes of caudex branches and at distal nodes of short, nonflowering aerial branches, 1–2(–2.5) dm, thinly floccose or glabrous. **Leaves** in rather compact basal rosettes; petiole 0.2–0.8 cm; blade narrowly elliptic to narrowly oblong, 0.5–2 × 0.3–0.6 (–0.8) cm, densely white-tomentose abaxially, thinly floccose or glabrous and olive green adaxially, margins entire, revolute. **Inflorescences** umbellate, 0.3–3 × 1–3.5 cm; branches thinly floccose or glabrous; bracts 3–4, semileaflike, lanceolate, 0.1–0.5 × 0.1–0.3 mm. **Involucres** 1 per node, turbinate, 5–6 × 3–4 mm, tomentose; teeth 6–8, erect, 0.5–2 mm. **Flowers** 4–6 mm, including 0.4–0.6 mm stipelike base; perianth sulphur yellow, glabrous; tepals monomorphic, obovate; stamens exserted, 4–5 mm; filaments pilose proximally. **Achenes** light brown, 4–5.5 mm, glabrous except for sparsely pubescent beak.

Flowering Jul–Sep. Gravelly serpentine slopes and outcrops, manzanita communities, montane conifer woodlands; (1000–)1500–2300 m; Calif.

Eriogonum congdonii is an elegant subshrub restricted to the mountains of central-northern California (southern Siskiyou and northern Trinity counties). It is worthy of serious consideration as a garden introduction even though it typically occurs on nearly raw, greenish-black serpentine outcrops. In that setting, the olive green leaf blades, the bright green and nearly glabrous flowering stems and inflorescence branches, and the dense clusters of rich sulphur-yellow flowers make a beautiful contrast.

114. Eriogonum ursinum S. Watson, Proc. Amer. Acad. Arts 10: 347. 1875 [E]

Herbs, spreading, matted, synoecious, 0.5–4 × (1.5–)2–8 (–12), thinly tomen-tose to floccose or glabrate. **Stems:** caudex spreading; aerial flowering stems erect, slender, solid, not fistulose, arising at nodes of caudex branches and at distal nodes of short, non-flowering aerial branches, 0.4–4 dm, thinly tomentose to floccose or glabrate. **Leaves** in rather compact basal rosettes; petiole 0.1–1.5 cm, tomentose; blade elliptic, 0.7–2 (–2.2) × 0.4–1(–1.2) cm, or ovate, 0.8–1.4(–2.5) × 0.5–1.2(–2) cm, densely white- or rufous-tomentose abaxially, thinly tomentose, sparsely floccose, or glabrous and greenish adaxially, margins slightly wavy. **Inflorescences** compound-umbellate, 1–3(–6) × 1–5 cm, thinly tomentose to glabrate; bracts: proximal 3–8, lanceolate to narrowly elliptic, 0.5–2 × 0.25–1 cm, distal 3, semileaflike, midway along branch, linear to narrowly lanceolate, 0.7–1 × 0.1–0.25 cm, or ± scalelike below involucre, 1–5 mm. **Involucres** 1 per node, turbinate to turbinate-campanulate, 3.5–8 × 2.5–4 mm, villous; teeth 5–8, erect, 0.5–2 mm. **Flowers** 5–7 mm at anthesis, 6.5–9 mm in fruit, including 0.5–0.8 mm stipelike base, with perianth cream or rarely yellow, suffused with blush of pinkish red to maroon, glabrous, or flowers 4–6 mm, including 1–1.3 mm stipelike base, with perianth pale yellow or rarely yellow, not suffused with blush of color glabrous; tepals slightly dimorphic or monomorphic, broadly ovate or obovate; stamens exserted, 3–5 mm; filaments pilose proximally. **Achenes** light brown, (5–)5.5–8 mm or 3–3.5 mm, glabrous except for sparsely pubescent beak.

Varieties 2 (2 in the flora): California.

1. Flowers 5–9 mm, perianth cream, rarely yellow; achenes (5–)5.5–8 mm; North Coast Ranges, Siskiyou and Trinity counties . . . .
. . . . 114a. *Eriogonum ursinum* var. *erubescens*
1. Flowers 4–6 mm, perianth pale yellow or rarely yellow; achenes 3–3.5 mm; n Sierra Nevada, Lassen and Shasta counties to Nevada, Placer, Plumas, and Sierra counties . . . .
. . . . 114b. *Eriogonum ursinum* var. *ursinum*

114a. Eriogonum ursinum S. Watson var. **erubescens** Reveal & J. Knorr, Phytologia 86: 161. 2004
• Blushing wild buckwheat [E]

Plants 1.5–2.5 × (1.5–)2–8(–12) dm. **Aerial flowering stems** 0.4–2 dm, thinly tomentose to floccose. **Leaves:** petiole 0.5–1.5 cm; blade elliptic, 0.7–2(–2.2) × 0.4–1(–1.2) cm, densely white-tomentose abaxially, sparsely floccose or glabrous and green adaxially. **Inflorescences** thinly tomentose; proximal bracts 4–6, lanceolate to narrowly elliptic, 1–2 × 0.25–0.6 cm, distal bracts semileaflike, midway along branch, linear to narrowly lanceolate, 0.7–1 × 0.1–0.25 cm. **Involucres** turbinate to turbinate-campanulate, 5–8 × 3–4 mm. **Flowers** 5–7 mm at anthesis, 6.5–9 mm in fruit; perianth cream, rarely yellow, becoming suffused with blush of pinkish red to maroon. **Achenes** (5–)5.5–8 mm.

Flowering Jun–Sep. Gravelly metavolcanic soils in montane chaparral, conifer and mountain mahogany communities; 1600–1900 m; Calif.

Variety *erubescens* is localized and rare, known only from the eastern Scott Bar Mountains west of Yreka in Siskiyou County, and Trinity Mountain, Trinity County. Technically, the three bracts along the "branchlet" terminate a branch, and a peduncle extends from the whorl of bracts to the base of the involucre.

114b. Eriogonum ursinum S. Watson var. **ursinum**
• Bear Valley wild buckwheat [E]

Eriogonum ovatum Greene

Plants 0.5–4 × 3–6 dm. **Aerial flowering stems** 0.4–4 dm, thinly tomentose to glabrate. **Leaves:** petiole 0.1–0.5(–0.8) cm; blade ovate, 0.8–1.4(–2.5) × 0.5–1.2 (–2) cm, densely white- or rufous-tomentose abaxially, thinly to-mentose or glabrous and greenish adaxially. **Inflorescences** thinly tomentose to glabrate; proximal bracts 3–8, lanceolate, 0.5–1.5(–2) × 0.3–1 cm, distal bracts scalelike, not midway along branch, 1–5 mm. **Involucres** turbinate, 3.5–4.5(–5) × 2.5–4 mm. **Flowers** 4–6 mm at anthesis, 5–6 mm in fruit; perianth pale yellow, rarely yellow, not suffused with blush of color. **Achenes** 3–3.5 mm.

Flowering May–Sep. Sandy to gravelly, mostly volcanic flats and slopes, oak and montane conifer woodlands; (500–)900–2500(–2800) m; Calif.

Variety *ursinum* is rather common in the northern Sierra Nevada (Lassen, Nevada, Placer, Plumas, Shasta, and Sierra counties), with a series of disjunct populations on the Trinity-Tehama-Shasta county lines. A low-elevation population (*Hutchinson et al. 2693*, JEPS) is known from near Pulga along the Feather River in Butte County.

Bear Valley wild buckwheat forms large, colorful mats on the forest floor, with rather compact but compound umbels of pale yellow flowers. The plants do well in the garden. A specimen (*T. J. Howell s.n.*, NY) supposedly gathered somewhere in Oregon is discounted as to location.

115. Eriogonum nervulosum (S. Stokes) Reveal,
Phytologia 40: 467. 1978 • Snow Mountain wild buckwheat C E

Eriogonum ursinum S. Watson var. *nervulosum* S. Stokes, Eriogonum, 114. 1936

Herbs, spreading, matted, synoecious, 0.2–1 × 1–3(–5) dm, tomentose. **Stems:** caudex spreading; aerial flowering stems erect or nearly so, slender, solid, not fistulose, arising at nodes of caudex branches and at distal nodes of short, nonflowering aerial branches, 0.2–0.6(–1) dm, tomentose. **Leaves** in compact basal rosettes; petiole 0.5–1 cm; blade broadly ovate, 0.4–0.8(–1) × 0.5–1 cm, densely white- to brownish-tomentose abaxially, thinly floccose or glabrous and green adaxially, margins plane to slightly revolute. **Inflorescences** subcapitate, 0.5–1.5 × 1–2 cm; branches tomentose; bracts 3–6, leaflike, lanceolate, 0.3–0.6 × 0.1–0.3 mm. **Involucres** 1 per node, turbinate, 3–4 × 2–3 mm, villous; teeth 6–8, erect, 0.4–0.8 mm. **Flowers** 3.5–5.5 mm, including 0.5–0.8 mm stipelike base; perianth white to ochroleucous, becoming pinkish rose or deep red, glabrous; tepals monomorphic, obovate; stamens exserted, 4–5 mm; filaments pilose proximally. **Achenes** light brown, 4.5–5 mm, glabrous.

Flowering May–Oct. Serpentine slopes and outcrops, mixed grassland communities, oak and conifer woodlands; of conservation concern; 300–2100 m; Calif.

Eriogonum nervulosum is found infrequently in small, widely scattered populations in the southern half of the North Coast Ranges. There are six known locations, and no doubt more remain to be found. Except for a Sonoma County site, the plants are found on ridges that separate Colusa and Lake counties, from Snow Mountain in the north to southwest of Clear Lake in the south. The species is found also in the Confusion Canyon area south and west of the lake. It is an ideal rock-garden introduction, although it grows only on grayish or reddish-brown serpentine soils in the wild.

116. Eriogonum prattenianum Durand, J. Acad. Nat.
Sci. Philadelphia, n. s. 3: 100. 1855 E

Herbs or subshrubs, spreading and matted or rounded to erect, 1–5 × 3–5 dm, glabrous. **Stems:** caudex spreading to erect; aerial flowering stems erect, slender, solid, not fistulose, arising at nodes of caudex branches and at distal nodes of short, nonflowering aerial branches, (0.5–) 1–3 dm, glabrous, with a whorl of 3–6, leaflike bracts ca. midlength, similar to leaf blades, 0.7–1 × 0.2–0.4 cm. **Leaves** in compact basal rosettes; petiole 0.1–0.5 cm, tomentose; blade elliptic to ovate, 0.5–1.5(–2) × 0.2–0.7 cm, densely white-tomentose on both surfaces or sometimes subglabrous and pale green adaxially, margins entire, plane. **Inflorescences** capitate, 0.8–2 cm wide; branches absent; bracts absent immediately below involucre. **Involucres** 1 per node, campanulate, 3–4 × 3–8 mm, tomentose; teeth 8–10, occasionally lobelike, erect to reflexed, 1–3 mm. **Flowers** 3–8 mm, including 1–1.5 mm stipelike base; perianth yellow, glabrous; tepals monomorphic, spatulate to obovate; stamens exserted, 3–4 mm; filaments pilose proximally. **Achenes** light to rusty brown, 4–5 mm, glabrous.

Varieties 2 (2 in the flora): California.

1. Leaf blades tomentose on both surfaces; n and c Sierra Nevada, Nevada County to Mariposa County . . . .
. . . . 116a. *Eriogonum prattenianum* var. *prattenianum*
1. Leaf blades tomentose abaxially, subglabrous adaxially; s Sierra Nevada, Madera and Fresno counties . . . 116b. *Eriogonum prattenianum* var. *avium*

116a. Eriogonum prattenianum Durand var.
prattenianum • Nevada City wild buckwheat
E

Eriogonum umbellatum Torrey subsp. *serratum* S. Stokes

Herbs, mats, to erect **subshrubs**, 3–5 dm. **Leaf blades** tomentose on both surfaces. **Involucres** 5–8 mm wide; teeth erect to slightly reflexed, 1–2 mm. **Flowers** usually 5–8 mm.

Flowering May–Jul. Volcanic or granitic flats and slopes, manzanita and buckbrush communities, oak and montane conifer woodlands; 800–2200 m; Calif.

Variety *prattenianum* is widely scattered in disjunct populations along the western slope of the northern and central Sierra Nevada (Calaveras, El Dorado, Madera,

Mariposa, Nevada, Placer, and Tuolumne counties) from Nevada City in the north to Chowchilla Mountain in the south.

116b. Eriogonum prattenianum Durand var. avium Reveal & Shevock, Phytologia 66: 249. 1989 • Kettle Dome wild buckwheat E

Subshrubs, rounded, 1–2 dm. **Leaf blades** tomentose abaxially, subglabrous adaxially. **Involucres** 3–5 mm wide; teeth lobelike, reflexed, 2–3 mm. **Flowers** usually 3–6 mm.

Flowering Jul–Aug. Granitic slopes and outcrops, manzanita communities, montane conifer woodlands; 2500–2900 m; Calif.

Variety *avium* is restricted to a few scattered populations on the western slope of the southern Sierra Nevada in Madera and Fresno counties.

117. Eriogonum tripodum Greene, Pittonia 1: 39. 1887 • Tripod wild buckwheat E

Subshrubs, spreading, 2.5–5 × 3–6 dm, glabrate. **Stems:** caudex spreading to somewhat erect; aerial flowering stems erect, slender, solid, not fistulose, arising at nodes of caudex branches and at distal nodes of short, nonflowering aerial branches, 2–3 dm, glabrate. **Leaves** in compact basal rosettes; petiole 0.2–0.4 cm, tomentose; blade narrowly oblanceolate, 1.5–2(–2.5) × 0.5–1 cm, densely white-tomentose on both surfaces, margins entire, plane. **Inflorescences** umbellate, 0.4–1 × 0.4–1 cm; branches glabrate; bracts 4–8, leaflike, occasionally midway along branch, lanceolate, 0.5–1.5 × 0.3–0.7 cm. **Involucres** 1 per node, campanulate, 3–4 × 3–4.5 mm, tomentose; teeth 6–10, lobelike, spreading, 1–1.5 mm. **Flowers** 4–5 mm, including 1–1.5 mm stipelike base; perianth bright yellow, villous abaxially; tepals monomorphic, obovate; stamens included to slightly exserted, 3–4 mm; filaments pilose proximally. **Achenes** light brown, 2.5–3 mm, glabrous except for sparsely pubescent beak.

Flowering May–Jul. Serpentine flats, slopes, and outcrops, mixed grassland communities, oak and conifer woodlands; (100–)300–800(–1500) m; Calif.

Eriogonum tripodum occurs in widely scattered locations on the inner coast ranges (Colusa, Lake, and Tehama counties) and along the western foothills of the Sierra Nevada (Amador, El Dorado, Mariposa, Placer, and Tuolumne counties). These subshrubs would make an impressive addition to the garden.

118. Eriogonum thymoides Bentham in A. P. de Candolle and A. L. P. P. de Candolle, Prodr. 14: 9. 1856 • Thyme-leaf wild buckwheat E

Eriogonum sphaerocephalum Douglas ex Bentham subsp. *minimum* (Small) S. Stokes; *E. thymoides* subsp. *congestum* S. Stokes

Subshrubs, spreading, polygamodioecious, 0.5–2(–3) × 1–3(–4) dm, tomentose to sericeous. **Stems:** caudex spreading; aerial flowering stems erect, slender, solid, not fistulose, arising at nodes of caudex branches and at distal nodes of short, nonflowering aerial branches, (0.1–)0.3–0.8(–1.2) dm, tomentose to sericeous, with a whorl of 6–10(–12), leaflike bracts ca. midlength, these similar to leaf blades, 0.3–0.8(–1) × 0.1–0.2 cm. **Leaves** in compact basal rosettes, fasciculate, and sheathing up stems; petiole 0.05–0.2 cm; blade linear to narrowly spatulate, (0.2–)0.3–1(–1.5) × 0.1–0.2(–0.3) cm, densely white-tomentose abaxially, silky-villous or sericeous adaxially, margins entire, inrolled to tightly revolute. **Inflorescences** capitate, 0.8–2 cm wide; branches absent; bracts absent immediately below involucre. **Involucres** 1 per node, turbinate, 3–5 × 2.5–4 mm, villous to sericeous; teeth 6–8, erect, 0.5–1 mm. **Flowers** 4–10 mm, including 0.5–1 mm stipelike base; perianth white to pale yellow or yellow, becoming pink to rose, villous abaxially; tepals monomorphic, obovate; stamens included to slightly exserted, 2–4 mm; filaments pilose proximally. **Achenes** light brown, 2–2.5 mm, glabrous except for densely pubescent beak.

Flowering Apr–Jul. Sandy to gravelly, often volcanic flats, slopes, and outcrops, mixed grassland and sagebrush communities, montane conifer woodlands; (200–)600–1700 m; Idaho, Oreg., Wash.

Eriogonum thymoides is an exquisite species concentrated in three regions of the Pacific Northwest. The first is along the eastern edge of the Cascade Range from near Wenatchee, Washington (Adams, Benton, Chelan, Douglas, Franklin, Grant, Kittitas, Klickitat, Lincoln, and Yakima counties), to near the Dalles in extreme north-central Oregon (Union County). The second is from Baker and northern Malheur counties, Oregon, to Adams, Canyon, and Washington counties, Idaho. A third series of populations is in the Mount Bennett Hills area of Gooding County, Idaho, and just over the borders in Blaine, Camas, Elmore, and Lincoln counties. Staminate plants tend to have yellow flowers that quickly fade after pollen release. Pistillate plants tend to have white to pale yellow flowers that persist and greatly elongate as the achene matures.

119. Eriogonum kelloggii A. Gray, Proc. Amer. Acad. Arts 8: 293. 1870 (as kellogii) • Red Mountain wild buckwheat [C] [E]

Herbs, spreading, loosely cespitose, probably polygamodioecious, 0.5–1 × 1–5 dm, tomentose to sericeous. **Stems:** caudex spreading; aerial flowering stems weakly erect, slender, solid, not fistulose, arising at nodes of caudex branches and at distal nodes of short, nonflowering aerial branches, (0.4–)0.5–0.8 dm, tomentose to sericeous, with a whorl of 2–4, leaflike bracts ca. midlength, these similar to leaf blades, 0.3–0.8 × 0.1–0.25 cm. **Leaves** in tight basal rosettes; petiole 0.05–0.2 cm, silky-tomentose; blade oblanceolate to narrowly spatulate, 0.4–1 × 0.1–0.3(–0.4) cm, silky-tomentose and silvery abaxially, less so and greenish adaxially, margins entire, plane. **Inflorescences** capitate, 0.5–1.5 cm wide; branches absent; bracts absent immediately below involucre. **Involucres** 1 per node, turbinate, (4–)5–6 × 2.5–4 mm; teeth 6–8, erect, 0.5–1 mm. **Flowers** 5–7 mm, including 0.5–0.8(–1) mm stipelike base; perianth whitish to pinkish or rose-red, glabrous; tepals monomorphic, obovate; stamens exserted, 4–5 mm; filaments pilose proximally. **Achenes** light brown, 4–5 mm, glabrous except for sparsely pubescent beak.

Flowering May–Aug. Heavy clayey soils, montane conifer woodlands; of conservation concern; 1000–1200 m; Calif.

Eriogonum kelloggii is known from only five occurrences on Red Mountain, Mendocino County. It is listed as endangered by the state of California, and the Bureau of Land Management considers it a species of "special" status. It is also a candidate for federal endangered status, U.S. Fish and Wildlife Service.

A long sought, undescribed *Eriogonum* from Black Butte in Glenn County remains to be rediscovered. J. T. Howell (*19120*, CAS) made a single specimen of it from a single mat that he saw in 1943. The leaf blades are small (3–4 × 1.5–2 mm) and held on a short petiole (1–2 mm). Based on the vegetative features, that plant appears to be related to *E. kelloggii*.

120. Eriogonum sphaerocephalum Douglas ex Bentham, Trans. Linn. Soc. London 17: 407. 1836 [E] [F]

Subshrubs, spreading to erect, infrequently polygamodioecious, 0.5–4 × 3–5(–6) dm, floccose or glabrous. **Stems:** caudex spreading; aerial flowering stems erect, slender, solid, not fistulose, arising at nodes of caudex branches and at distal nodes of short, nonflowering aerial branches, (0.3–)0.5–1 dm, thinly floccose or glabrous, with a whorl of 4–8 leaflike bracts ca. midlength or proximally, linear-oblanceolate to oblanceolate, 1–2 × 0.1–0.4 cm. **Leaves** in loose to congested basal rosettes; petiole 0.05–0.3 cm, floccose or glabrous; blade linear-oblanceolate, narrowly oblanceolate, or narrowly spatulate, 1–3(–4) × (0.1–)0.3–0.6(–1) cm, lanate or tomentose to floccose or glabrous, margins entire, plane or revolute. **Inflorescences** capitate, compound-umbellate, or umbellate, 1–2 cm wide or 1–5 × 1–5 cm; branches thinly floccose or glabrous; bracts absent immediately below involucre. **Involucres** 1 per node, turbinate-campanulate to campanulate, 3–4 × 2.5–4.5 mm; teeth 6–10, lobelike, slightly reflexed, 1.5–3(–5) mm. **Flowers** (5–)6–9 mm, including 1–2 mm stipelike base; perianth pale to bright yellow or cream to ochroleucous, villous abaxially; tepals monomorphic, obovate to oblong-ovate; stamens exserted, 4–6 mm; filaments pilose proximally. **Achenes** light brown, 3–4 mm, glabrous except for sparsely pubescent beak.

Varieties 4 (4 in the flora): w United States.

The Piute of Nevada have used a decoction of the root of *Eriogonum sphaerocephalum* (either var. *sphaerocephalum* or var. *halimoides*) to treat colds and diarrhea (P. Train et al. 1941).

1. Perianths bright yellow; inflorescences umbellate or compound-umbellate; ne California, sw Idaho, n and wc Nevada, e Oregon, and e Washington . . . . 120a. *Eriogonum sphaerocephalum* var. *sphaerocephalum*
1. Perianths pale yellow to cream or ochroleucous; inflorescences umbellate, compound-umbellate, or capitate; ne California, sw or wc Idaho, c Oregon, and c Washington.
   2. Leaf blades usually thinly floccose or glabrous on both surfaces; inflorescences umbellate or compound-umbellate; sw Idaho . . . . 120b. *Eriogonum sphaerocephalum* var. *fasciculifolium*
   2. Leaf blades tomentose on both surfaces, or floccose adaxially; inflorescence capitate or umbellate; ne California to nw Nevada, c Oregon to wc Idaho, and c Washington.

[3. Shifted to left margin.—Ed.]

3. Leaf blades usually narrowly oblanceolate to narrowly spatulate, margins plane; ne California to
nw Nevada, c Oregon to wc Idaho, and c Washington . . . . 120c. *Eriogonum sphaerocephalum*
var. *halimioides*

3. Leaf blades linear-oblanceolate, margins usually
revolute; nc Oregon, sc Washington . . . .
. . . . 120d. *Eriogonum sphaerocephalum* var. *sublineare*

120a. Eriogonum sphaerocephalum Douglas ex Bentham var. **sphaerocephalum** · Rock wild buckwheat E F

Leaf blades usually oblanceolate, mostly lanate to tomentose on both surfaces, margins usually plane. **Inflorescences** umbellate or compound-umbellate. **Perianths** bright yellow.

Flowering May–Jul. Sandy to gravelly flats and slopes, mixed grassland and sagebrush communities, juniper and montane conifer woodlands; (100–)300–2000(–2300) m; Calif., Idaho, Nev., Oreg., Wash.

Variety *sphaerocephalum* is common and widespread in eastern Washington, eastern Oregon, and southwestern Idaho, less so in northern and central-western Nevada, and infrequent in California. A collection supposedly obtained in 1883 from the "Flathead region" of Montana (*Ayres s.n.*, NY) is discounted as to location.

120b. Eriogonum sphaerocephalum Douglas ex Bentham var. **fasciculifolium** (A. Nelson) S. Stokes, Eriogonum, 104. 1936 · Weiser wild buckwheat E

Eriogonum fasciculifolium A. Nelson, Bot. Gaz. 54: 407. 1912

Leaf blades usually narrowly oblanceolate to narrowly spatulate, thinly floccose or glabrous on both surfaces or thinly floccose abaxially and eventually glabrous adaxially, margins usually plane. **Inflores-cences** umbellate or compound-umbellate. **Perianths** pale yellow to cream or ochroleucous.

Flowering May–Jul. Sandy to gravelly, often volcanic flats and slopes, sagebrush communities, montane conifer woodlands; 600–2200 m; Idaho.

Variety *fasciculifolium* is known only from southwestern Idaho in two disjunct areas (Adams and Washington counties in the west, and Blaine and Gem counties to the east). This variety is the most attractive expression of the species and would do well in the garden.

120c. Eriogonum sphaerocephalum Douglas ex Bentham var. **halimioides** (Gandoger) S. Stokes, Eriogonum, 104. 1936 · Halimium wild buckwheat E

Eriogonum halimioides Gandoger, Bull. Soc. Roy. Bot. Belgique 42: 197. 1906; *E. fruticulosum* S. Stokes

Leaf blades mostly narrowly oblanceolate to narrowly spatulate, mostly tomentose on both surfaces or floccose adaxially, margins usually plane. **Inflorescences** capitate or umbellate. **Perianths** pale yellow to cream or ochroleucous. $2n = 40$.

Flowering May–Jul. Sandy to gravelly, occasionally volcanic flats and slopes, sagebrush communities, juniper and montane conifer woodlands; 300–2300 m; Calif., Idaho, Nev., Oreg., Wash.

Variety *halimioides* is common and widespread in three areas of concentration. The northernmost is east of the Cascade Range in central Washington (Douglas, Kittitas, Klickitat, and Yakima counties). The middle series of populations occurs from central Oregon (Gilliam, Jefferson, Union, Wallowa, and Wasco counties) east into Idaho (Blaine, Elmore, Gem, Gooding, and Washington counties). The southernmost series is in central-southern Oregon (Baker, Grant, Harney, Jackson, Klamath, Lake, Malheur, and Wheeler counties), northeastern California (Lassen, Modoc, Shasta, and Siskiyou counties), and northwestern Nevada (Humboldt and Washoe counties).

Variety *halimioides* is highly variable, and a clear distinction between it and some populations assigned here to *E. douglasii* var. *douglasii* is not always possible. Of particular concern are those plants of var. *halimioides* in northeastern Oregon and adjacent southeastern Washington with capitate rather than umbellate inflorescences. Much of what has passed for *E. douglasii* (especially its *sublineare* phase) in that area actually may be var. *halimioides*.

120d. Eriogonum sphaerocephalum Douglas ex
Bentham var. **sublineare** (S. Stokes) Reveal,
Harvard Pap. Bot. 9: 197. 2004 • Scabland wild
buckwheat E

Eriogonum caespitosum Nuttall var.
sublineare S. Stokes, Leafl. W. Bot.
2: 72. 1938; *E. douglasii* var.
sublineare (S. Stokes) Reveal;
E. douglasii var. *tenue* (Small)
C. L. Hitchcock

Leaf blades linear-oblanceolate,
mostly tomentose on both surfaces, margins usually revolute.
Inflorescences capitate or umbellate. **Perianths** pale yellow to ochroleucous.

Flowering May–Jul. Gravelly, volcanic scablands,
mixed grassland and sagebrush communities; 100–1300
m; Oreg., Wash.

Variety *sublineare* is found in south-central Washington and adjacent north-central Oregon. It is frequently
confused with *Eriogonum douglasii*.

121. Eriogonum douglasii Bentham in A. P. de
Candolle and A. L. P. P. de Candolle, Prodr. 14: 9.
1856 E

Eriogonum caespitosum Nuttall
subsp. *douglasii* (Bentham)
S. Stokes

Herbs, matted, occasionally
polygamodioecious, 0.4–1.5 ×
0.5–6 dm, thinly tomentose to
glabrate. **Stems:** caudex spreading; aerial flowering stems erect
or nearly so, slender, solid, not
fistulose, arising at nodes of caudex branches and at
distal nodes of short, nonflowering aerial branches, 0.4–
1.2 dm, with a whorl of 4–8 leaflike bracts ca. midlength,
similar to leaf blades, 0.3–1.5 × 0.1–0.3 cm. **Leaves** in
basal rosettes; petiole 0.05–0.5(–1) cm, tomentose; blade
oblanceolate or elliptic to spatulate, 0.4–1.5(–1.9) × 0.1–
0.5 cm, lanate on both surfaces, or tomentose abaxially
and slightly less so and greenish adaxially, margins entire,
plane. **Inflorescences** capitate, 0.8–1.5 cm wide; branches
absent; bracts absent immediately below involucre.
Involucres 1 per node, turbinate to turbinate-campanulate, 2.5–3.5 × 2–2.5 mm; teeth 6–14, lobelike,
strongly reflexed, 1.5–4(–6) mm. **Flowers** 4–9 mm,
including 0.7–2 mm stipelike base; perianth yellow,
cream, or ochroleucous to rose-red, sparsely to densely
villous abaxially; tepals monomorphic, obovate; stamens
exserted, 4–6 mm; filaments pilose proximally. **Achenes**
light brown, 3–4.5 mm, glabrous except for pubescent
beak.

Varieties 3 (3 in the flora): w United States.

1. Leaf blades tomentose abaxially, less so adaxially,
   narrowly oblanceolate to spatulate; flowers 5–9
   mm; e Washington, ne Oregon . . . .
   . . . . 121c. *Eriogonum douglasii* var. *douglasii*
1. Leaf blades usually densely lanate on both
   surfaces, elliptic to spatulate; flowers 4–8 mm; s
   Oregon, ne California, n Nevada.
   2. Flowers 4–5(–6) mm; s Oregon, ne California,
      nw Nevada . . . . 121a. *Eriogonum douglasii*
      var. *meridionale*
   2. Flowers 5–8 mm; ne Nevada . . . .
      . . . . 121b. *Eriogonum douglasii* var. *elkoense*

121a. Eriogonum douglasii Bentham var. **meridionale**
Reveal, Phytologia 86: 130. 2004 • Southern
wild buckwheat E

Leaf blades broadly elliptic to
spatulate, densely grayish-lanate
on both surfaces. **Flowers** 4–5
(–6) mm; perianth yellow or
ochroleucous to rose-red, densely
villous.

Flowering Apr–Jun. Sandy to
gravelly or rocky flats and slopes,
manzanita and sagebrush com-
munities, oak and montane conifer woodlands; 1300–
2500(–2900) m; Calif., Nev., Oreg.

Variety *meridionale* is found infrequently in five
widely scattered areas. The northernmost is along the
Jackson County, Oregon, and Siskiyou County,
California, border. In the latter county, the variety is
found also in the Marble Mountains area. It is seen
occasionally in and near the Warner Mountains of Lake
County, Oregon, and Modoc County, California. The
plants are much more common along the eastern slope
of the northern Sierra Nevada in Lassen County, Nevada,
Plumas and Sierra counties, California, and in the Reno
area of Washoe County, Nevada. The last and most
restricted population is in the Intermountain Region near
Pyramid Lake in Washoe County. Variety *meridionale*
occasionally is seen in cultivation and is an ideal addition
to any rock garden.

121b. Eriogonum douglasii Bentham var. **elkoense**
Reveal, Phytologia 86: 129. 2004 • Sunflower
Flat wild buckwheat E

Leaf blades elliptic, densely lanate
on both surfaces, sometimes
slightly less so and faintly greenish
adaxially. **Flowers** 5–8 mm;
perianth cream or ochroleucous,
sparsely villous.

Flowering May–Jul. Sandy to
gravelly flats and slopes, mixed
grassland and sagebrush communities; 1900–2100 m; Nev.

Variety *elkoense* is known only from the Sunflower Flat area northeast of Wild Horse State Park in northwestern Elko County.

121c. Eriogonum douglasii Bentham var. **douglasii**

• Douglas's wild buckwheat E

Leaf blades narrowly oblanceolate to spatulate, tomentose abaxially, less so adaxially. **Flowers** 5–9 mm; perianth yellow to ochroleucous, sparsely to densely villous abaxially.

Flowering May–Jul. Sandy to gravelly or rocky flats and slopes, mixed grassland and sagebrush communities, oak and montane conifer woodlands; 100–1700(–1900) m; Oreg., Wash.

Variety *douglasii* is widespread in scattered, disjunct populations in southeastern Washington (Columbia, Douglas, Ferry, Kittitas, Klickitat, and Yakima counties) and northeastern Oregon (Baker, Gilliam, Grant, Jefferson, Malheur, Sherman, Union, Wallowa, and Wasco counties).

122. Eriogonum twisselmannii (J. T. Howell) Reveal in P. A. Munz, Suppl. Calif. Fl. 40. 1968

• Twisselmann's wild buckwheat C E

Eriogonum douglasii Bentham var. *twisselmannii* J. T. Howell, Leafl. W. Bot. 10: 13. 1963

Herbs, matted, occasionally polygamodioecious, (0.2–)0.5–1.5 × 3–4 dm, tomentose. **Stems:** caudex spreading; aerial flowering stems erect or nearly so, slender, solid, not fistulose, arising at nodes of caudex branches and at distal nodes of short, nonflowering aerial branches, (0.2–)0.5–1.2 dm, tomentose, with a whorl of 4–8 leaflike bracts ca. midlength, similar to leaf blades, 0.4–0.8 × 0.1–0.3 cm. **Leaves** in loose to congested basal rosettes; petiole 0.2–0.4 cm, thinly tomentose; blade oblanceolate to elliptic, 0.5–1 × (0.2–)0.4–0.6 cm, white- or grayish-tomentose abaxially, thinly tomentose to nearly glabrous and dull green adaxially. **Inflorescences** capitate, 0.7–1.5 cm wide; branches absent; bracts absent immediately below involucre. **Involucres** 1 per node, campanulate, (2.5–)4–5 × 4–6 mm; teeth 6–9, lobelike, strongly reflexed, (2–)3–5 mm. **Flowers** (4–)5–6 mm, including 1–1.3 mm stipelike base; perianth pale yellow, villous abaxially; tepals slightly dimorphic, those of outer whorl obtuse, 3–5 × 3–3.5 mm, those of inner whorl broadly oblanceolate, 5–6 × 2.5–3 mm; stamens exserted, 4–5 mm; filaments pilose proximally. **Achenes** light brown, 5–5.5 mm, glabrous except for sparsely pubescent beak.

Flowering Jun–Sep. Rocky outcrops, montane conifer woodlands; of conservation concern; 2300–2600 m; Calif.

Eriogonum twisselmannii is a rare species of the southern Sierra Nevada and is known only from Slate Mountain and The Needles in Tulare County. It is worthy of being on a "watch" list of rare plants in California, but because of its remote location, it is not threatened.

123. Eriogonum caespitosum Nuttall, J. Acad. Nat. Sci. Philadelphia 7: 50, plate 8, fig. 2. 1834

• Matted wild buckwheat E F

Eriogonum sphaerocephalum Bentham var. *sericoleucum* (Greene ex Tidestrom) S. Stokes

Herbs, matted, polygamodioecious, (0.1–)0.3–1 × 1–5(–12) dm. **Stems:** caudex spreading; aerial flowering stems scapelike, ascending to weakly erect, slender, solid, not fistulose, arising at nodes of caudex branches and at distal nodes of short, nonflowering aerial branches, (0.1–)0.3–0.8(–1) dm, mostly floccose or glabrous, without a whorl of bracts. **Leaves** in compact basal rosettes; petiole 0.05–0.4(–0.7) cm, usually tomentose; blade elliptic to obovate or nearly oval, 0.2–1(–1.5) × 0.15–0.4(–0.5) cm, grayish- to whitish-tomentose on both surfaces or less so and greenish adaxially, margins entire, plane. **Inflorescences** capitate, 0.5–2 cm wide; branches absent; bracts absent immediately below involucre. **Involucres** 1 per node, campanulate, 2–3.5 × 2–4 mm; teeth 6–9, lobelike, strongly reflexed, 2–3.5 mm. **Flowers** 2.5–10 mm, including 0.5–1 mm stipelike base; perianth yellow to reddish or rose, densely pilose to villous abaxially; stamens exserted, 3–4 mm; filaments pilose proximally; staminate flowers shorter, 2.5–4 mm, tepals monomorphic, oblong-oblanceolate; pistillate flowers 2.5–10 mm, tepals slightly dimorphic. **Achenes** light brown to brown, (3.5–)4–5 mm, glabrous or sparsely pubescent on beak.

Flowering Apr–Jul. Sandy to gravelly flats and slopes, mixed grassland, saltbush, sagebrush, and mountain mahogany communities, oak, pinyon and/or juniper and montane conifer woodlands; (1300–)1500–3000(–3700) m; Ariz., Calif., Idaho, Mont., Nev., Oreg., Utah, Wyo.

Eriogonum caespitosum is widespread and usually common. It is variable throughout its range but no taxonomic subunits have been noted. The functionally staminate plants can be morphologically different from the functionally pistillate ones, and that may cause some confusion in the field, especially when the latter are in fruit. The plants are widely cultivated and worthy of consideration for the rock or sand garden. Plants from along the eastern edge of the Sierra Nevada in southern Mono County, California, form large, dense mats and would be most attractive in the garden. The scapelike flowering stem is technically a peduncle, being a further reduction of the inflorescence from that seen in *E. douglasii*. A specimen supposedly gathered in Nebraska (*Abbott s.n.*, CAS) is discounted as to location.

124. Eriogonum polypodum Small, Bull. Torrey Bot. Club 25: 46. 1898 • Fox-tail wild buckwheat [E]

Eriogonum umbellatum Torrey var. *polypodum* (Small) S. Stokes; *E. ursinum* S. Watson var. *venosum* S. Stokes ex Smiley

Herbs, matted, polygamo-dioecious, 0.5–1.5 × 1–4(–5) dm, mostly tomentose. **Stems:** caudex spreading; aerial flowering stems erect, slender, solid, not fistulose, arising at nodes of caudex branches and at distal nodes of short, nonflowering aerial branches, 0.5–1.5 dm, mostly tomentose. **Leaves** in congested basal rosettes; petiole 0.1–0.6(–1) cm, usually tomentose; blade ovate, 0.2–1 × 0.1–0.6(–0.8) cm, densely tomentose on both surfaces but adaxial surface soon glabrous and green, margins entire, plane. **Inflorescences** capitate, 0.5–2 cm wide, those with pistillate flowers becoming umbellate in fruit, 1.5–3 × 2–3 cm; branches tomentose to floccose; bracts 4–6 proximally, usually leaflike, 0.2–0.8 cm, absent immediately below involucre. **Involucres** 1 per node, turbinate, (2–)2.5–3.5(–4) × 2–3 mm, tomentose; teeth 5–7, spreading, 1–2 mm. **Flowers** 2.5–3.5 mm, including 0.2–0.4 mm stipelike base; perianth chalky white, glabrous; tepals monomorphic, oblong to narrowly obovate; stamens exserted, 1–2 mm; filaments pilose proximally. **Achenes** light brown, 3–3.5 mm, glabrous.

Flowering Jul–Sep. Sandy to gravelly granitic flats, slopes, and outcrops, high-elevation sagebrush communities, subalpine and alpine conifer woodlands; (2400–)2800–3500(–3700) m; Calif.

Eriogonum polypodum is found in the southern Sierra Nevada in Fresno and Tulare counties, with the bulk of its scattered populations west of Mt. Whitney. The chalky white flowers are distinctive, especially when observed in the field, and unique in subg. *Oligogonum*.

125. Eriogonum marifolium Torrey & A. Gray, Proc. Amer. Acad. Arts 8: 161. 1870 • Marum-leaf wild buckwheat [E] [F]

Eriogonum cupulatum S. Stokes; *E. marifolium* var. *apertum* S. Stokes

Herbs, matted, dioecious, (0.1–) 0.5–5 × 2–8 dm, floccose or glabrous. **Stems:** caudex spreading; aerial flowering stems erect, slender, solid, not fistulose, arising at nodes of caudex branches and at distal nodes of short, non-flowering aerial branches, (0.1–)0.5–4 dm, floccose or glabrous. **Leaves** in loose to compact basal rosettes; petiole 0.3–2 cm, often tomentose; blade ovate to oval, 0.3–3 × 0.3–1 cm, densely tannish- to brownish-lanate abaxially, glabrate and green to olive green adaxially, margins entire, plane. **Inflorescences** capitate, 0.5–2 cm wide, mature pistillate plants open and elongate-umbellate, 1–5 × 1–7 cm; branches usually glabrous; bracts 5–10, leaflike, 0.2–0.8 cm, often absent immediately below involucre. **Involucres** 1 per node but occasionally appearing congested, turbinate and 2–3 × 1.5–2 mm or campanulate and 4–5 × 5–7 mm, sparsely tomentose or glabrous; teeth 5–6, erect, 0.4–0.7 mm. **Flowers** stipelike base 0.5–1 mm; perianth yellow, glabrous; staminate flowers 1.5–3 mm, tepals ovate; pistillate flowers 4–7 mm, tepals oblanceolate, often becoming reddish in fruit; stamens exserted, 2–3 mm; filaments pilose proximally. **Achenes** light brown to brown, 3.5–5 mm, glabrous except for sparsely pubescent beak. $2n = 32$.

Varieties 2 (2 in the flora): w United States.

Eriogonum marifolium is dioecious, with the two sexes having such markedly different morphology that collectors often consider each to represent a different species. In Oregon, male plants of *E. marifolium* and *E. umbellatum* var. *haussknechtii* can be difficult to distinguish. In El Dorado and Placer counties, California, herbarium specimens of *E. marifolium* and *E. incanum* can be difficult to separate, as their distribution ranges overlap there.

1. Involucres turbinate, 2–3 × 1.5–2 mm; ne California, wc Nevada, s Oregon, disjunct to Yakima County, Washington . . . . 125a. *Eriogonum marifolium* var. *marifolium*
1. Involucres campanulate, 4–5 × 5–7 mm; nc California . . . 125b. *Eriogonum marifolium* var. *cupulatum*

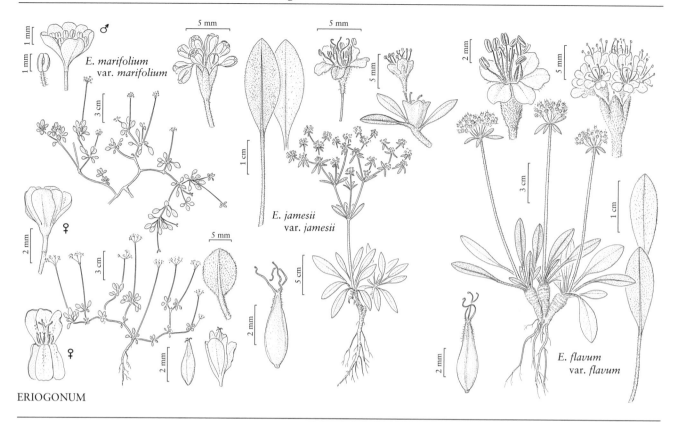

E. *marifolium* var. *marifolium*

E. *jamesii* var. *jamesii*

E. *flavum* var. *flavum*

ERIOGONUM

125a. Eriogonum marifolium Torrey & A. Gray var. **marifolium** • Marum-leaf wild buckwheat E F

Eriogonum marifolium var. *apertum* S. Stokes

Leaf blades 0.3–1.5 cm. **Involucres** turbinate, 2–3 × 1.5–2 mm.

Flowering Jun–Aug. Sandy, volcanic or occasionally granitic flats and slopes, mixed grassland, manzanita, and sagebrush communities, montane conifer woodlands; (900–)1100–3100(–3300) m; Calif., Nev., Oreg., Wash.

Variety *marifolium* occurs in widely scattered locations, often on volcanic peaks, in Washington (Yakima County), Oregon (Crook, Deschutes, Douglas, Hood River, Jackson, Jefferson, Klamath, Lane, Linn, and Marion counties), and north-central California (to Shasta County). The variety is more common in the northern Sierra Nevada (Alpine, Amador, Butte, Calveras, El Dorado, Lassen, Mono, Nevada, Place, Plumas, Sierra, and Tuohumne counties) and west-central Nevada (Carson City, Douglas, and Washoe counties). An isolated population is found on the Pine Forest Range of Humboldt County, Nevada

125b. Eriogonum marifolium Torrey & A. Gray var. **cupulatum** (S. Stokes) Reveal, Harvard Pap. Bot. 9: 182. 2004 • McCloud wild buckwheat E

Eriogonum cupulatum S. Stokes, Leafl. W. Bot. 1: 34. 1933

Leaf blades 1–3 cm. **Involucres** campanulate, 4–5 × 5–7 mm.

Flowering Jun–Aug. Sandy flats, manzanita communities, conifer woodlands; 1100–1200 m; Calif.

Variety *cupulatum* occurs in Southern Siskiyou County.

126. Eriogonum incanum Torrey & A. Gray, Proc. Amer. Acad. Arts 8: 161. 1870 • Frosted wild buckwheat E

Eriogonum marifolium Torrey & A. Gray var. *incanum* (Torrey & A. Gray) M. E. Jones; *E. ursinum* S. Watson var. *rosulatum* (Small) S. Stokes

Herbs, matted, dioecious, 0.5–3 × 1–4 dm wide, tomentose to floccose. **Stems:** caudex spreading; aerial flowering stems erect or nearly so, slender, solid, not fistulose, arising at nodes of caudex branches and at distal nodes of short, nonflowering aerial branches, 0.1–2(–2.5) dm, tomentose to floccose. **Leaves** in loose to congested basal rosettes; petiole (0.3–)0.5–1 cm, tomentose; blade oblong to oblong-ovate or spatulate, 0.5–1.5 × 0.3–0.7 cm, densely white- or grayish-tomentose on both surfaces, sometimes greenish adaxially, margins entire, plane. **Inflorescences** capitate and 0.5–2 cm wide, mature pistillate plants open and umbellate, 1–3 × 1–4 cm; branches tomentose to floccose; bracts 2–6, leaflike, 0.1–0.5 × 0.1–0.3 cm, often absent immediately below involucre. **Involucres** 1 per node but occasionally appearing congested, turbinate-campanulate, 2.5–3 × 2–2.5 mm, tomentose; teeth 5–8, erect, 0.5–1 mm. **Flowers** stipelike base 0.5–1 mm; perianth yellow, glabrous; staminate flowers 2–3 mm, tepals ovate; pistillate flowers 4–6 mm, tepals oblanceolate, often becoming reddish in fruit; stamens exserted, 2–3 mm; filaments pilose proximally. **Achenes** light brown to brown, 3–3.5 mm, glabrous except for sparsely pubescent beak.

Flowering Jun–Sep. Sandy to gravelly or rocky granitic or occasionally volcanic flats, slopes, and outcrops, mixed grassland, manzanita, and sagebrush communities, montane conifer woodlands; (1900–)2100–4000 m; Calif., Nev.

Eriogonum incanum is common throughout the central and southern Sierra Nevada of California (Alpine, El Dorado, Fresno, Inyo, Madera, Mariposa, Nevada, Tulare, and Tuolumne counties) and extreme west-central Nevada (Carson City, Douglas, and Washoe counties). It is a food plant for the green hairstreak butterfly (*Callophrys lemberti*), the Pacific dotted-blue (*Euphilotes enoptes*), and the gorgon copper (*Gaeides gorgon*).

127. Eriogonum diclinum Reveal, Aliso 7: 218. 1970

• Jaynes Canyon wild buckwheat E

Herbs, matted, dioecious, (0.5–)1–2 × 3–8 dm, sparsely tomentose. **Stems:** caudex spreading; aerial flowering stems erect, slender, solid, not fistulose, arising at nodes of caudex branches and at distal nodes of short, nonflowering aerial branches, (0.5–)1–2 dm, sparsely tomentose. **Leaves** in loose basal rosettes; petiole 0.7–3 cm, lanate to tomentose; blade elliptic to ovate, (0.5–)1–2 × 0.5–1.5 cm, densely grayish-lanate on both surfaces, margins entire, plane. **Inflorescences:** staminate plants capitate, rarely umbellate, 1–2 × 1–2 cm; pistillate plants capitate, rarely umbellate, 1–4 × 1–4 cm; branches tomentose to floccose; bracts 4–6, semileaflike to leaflike, 3–5 × 1–3 mm. **Involucres** 1 per node or 2–3 per cluster, turbinate to turbinate-campanulate, tomentose; staminate plants 2.5–3 × 3–4 mm, teeth 5–6, erect, 1.5–2 mm; pistillate plants 3–4 × 3–4 mm, teeth 4–5, erect, 2.5–3.5 mm. **Flowers:** perianth yellow to brownish yellow, glabrous; staminate flowers 2–3 mm, including 0.5–0.7 mm stipelike base; pistillate flowers (3.5–)5–8 mm, including 0.5–0.8 mm stipelike base; tepals monomorphic, oblanceolate to obovate; stamens exserted, 1–2 mm; filaments pilose proximally. **Achenes** light brown to brown, 3–4 mm, glabrous except for sparsely pubescent beak.

Flowering Jun–Sep. Gravelly serpentine flats and slopes, sagebrush communities, montane conifer woodlands; 1800–2400 m; Calif., Oreg.

Eriogonum diclinum is found infrequently in the Siskiyou and Trinity mountains of northwest California (Siskiyou and Trinity counties) and southwest Oregon (Jackson and Josephine counties). The species approaches *E. incanum* in aspect, and reports of that species in Oregon were based on staminate specimens of *E. diclinum*.

128. Eriogonum jamesii Bentham in A. P. de Candolle and A. L. P. P. de Candolle, Prodr. 14: 7. 1856

• Antelope sage F

Herbs or subshrubs, compact or spreading, matted, 0.5–2.5 × 3–15 dm, tomentose to floccose. **Stems:** caudex absent or spreading; aerial flowering stems erect, slender, solid, not fistulose, usually arising directly from a taproot, 0.5–1.5 dm, tomentose to floccose. **Leaves** basal, typically not in rosettes; petiole 0.5–6 cm, tomentose to floccose; blade usually narrowly elliptic, 1–3(–3.5) × (0.3–)0.5–1(–1.2) cm, densely tomentose abaxially, thinly tomentose, floccose or glabrous and grayish to greenish adaxially, margins entire, plane or undulate and crisped. **Inflorescences** umbellate or compound-umbellate, 10–30 × 10–25 cm; branches tomentose to floccose; bracts 3–9, semileaflike at proximal node, 0.5–2 × 0.2–1 cm, often scalelike distally. **Involucres** 1 per node, turbinate, 1.5–7 × 2–5 mm, tomentose to floccose; teeth 5–8, erect, 0.1–0.5 mm. **Flowers** 3–8 mm, including 0.7–2 mm stipelike base; perianth white to cream, densely pubescent abaxially; tepals dimorphic, those of outer whorl lanceolate to elliptic, 2–5 × 1–3 mm, those of inner whorl lanceolate to fan-shaped, 1.5–6 × 2–4 mm; stamens exserted, 2–4 mm; filaments pilose proximally. **Achenes** light brown to brown, 4–5 mm, glabrous except for sparsely pubescent beak.

Varieties 3 (3 in the flora): w North America, including Mexico.

Eriogonum jamesii is a nectar source for the rare Spalding dotted-blue butterfly (*Euphilotes spaldingi*).

Eriogonum jamesii and *E. arcuatum* (see below) are considered "life medicines" and used ceremonially by Native Americans (C. Arnold, pers. comm.; A. B. Reagan 1929; P. A. Vestal 1952).

1. Leaf margins undulate, frequently crisped; flowers 3–5(–6) mm; se Arizona, s New Mexico, sw Texas
   . . . . 128c. *Eriogonum jamesii* var. *undulatum*
1. Leaf margins plane, not crisped; flowers 4–8 mm; e Arizona, c and s Colorado, wc Kansas, New Mexico, w Oklahoma, n and w Texas.
   2. Inflorescences compound-umbellate; e Arizona, c and s Colorado, New Mexico, w Oklahoma, n and w Texas . . . .
      . . . . 128a. *Eriogonum jamesii* var. *jamesii*
   2. Inflorescences simple-umbellate; wc Kansas
      . . . . 128b. *Eriogonum jamesii* var. *simplex*

128a. Eriogonum jamesii Bentham var. **jamesii**
   • James's antelope sage   E F

Herbs, spreading, loose mats, 3–8 dm wide. **Aerial flowering stems** tomentose to floccose. **Leaf blades** 1–3 × 0.5–1 cm, densely whitish- to grayish-tomentose abaxially, less so to thinly tomentose and greenish adaxially, margins plane. **Inflorescences** compound-umbellate, branched 2–5 times; bracts usually semileaflike, those of proximal node 0.5–2 × 0.3–1 cm. **Involucres** 4–7 × 2–5 mm. **Flowers** 4–8 mm. $2n = 40$.

Flowering Jun–Oct. Sandy to gravelly or infrequently rocky flats and slopes, mixed grassland, saltbush, blackbrush, creosote bush, mesquite, and sagebrush communities, oak, pinyon and/or juniper, and montane conifer woodlands; (1000–)1300–2900(–3100) m; Ariz., Colo., N.Mex., Okla., Tex.

Variety *jamesii* is widespread, and although generally scattered and infrequent, can be locally common in places. It occurs from southern Colorado south into eastern Arizona, New Mexico, western Oklahoma, and northern and western Texas. It merges with var. *undulatum* in southeastern Arizona, southern New Mexico, and especially western Texas.

128b. Eriogonum jamesii Bentham var. **simplex**
   Gandoger, Bull. Soc. Roy. Bot. Belgique 42: 190. 1906   • Kansas antelope sage   E

Herbs, compact, tight mats, 3–6 dm wide. **Aerial flowering stems** tomentose to floccose. **Leaf blades** 1–3(–3.5) × (0.3–)0.5–1 (–1.2) cm, densely whitish- to grayish-tomentose abaxially, less so to thinly tomentose and greenish adaxially, margins plane. **Inflorescences** umbellate; bracts usually semileaflike, those of proximal node 0.5–2 × 0.3–1 cm. **Involucres** 4–6 × 2–5 mm. **Flowers** 4–7(–8) mm.

Flowering Jun–Aug. Clayey or chalk flats, slopes, and outcrops, mixed grassland communities; 800–1000 m; Kans.

Variety *simplex* is known only from a few scattered populations in Logan County and one site just inside Scott County.

128c. Eriogonum jamesii Bentham var. **undulatum** (Bentham) S. Stokes ex M. E. Jones, Contr. W. Bot. 11: 8. 1903   • Wavy-margined antelope sage

Eriogonum undulatum Bentham in A. P. de Candolle and A. L. P. P. de Candolle, Prodr. 14: 7. 1856; *E. jamesii* subsp. *undulatum* (Bentham) S. Stokes

Subshrubs, loosely matted, 5–15 dm wide. **Aerial flowering stems** tomentose. **Leaf blades** 1–2 × 0.3– 0.8(–1) cm, densely whitish- to grayish-tomentose abaxially, thinly tomentose to floccose or glabrous and greenish adaxially, margins undulate, frequently crisped. **Inflorescences** compound-umbellate, branched 2–4 times; bracts usually semileaflike, those of proximal node 0.5–1.5 × 0.2–0.6 cm. **Involucres** 1.5– 4 × 2–3 mm. **Flowers** 3–5(–6) mm. $2n = 40$.

Flowering Jul–Oct. Sandy to gravelly or rocky, often limestone flats and slopes, mixed grassland, saltbush, creosote bush, and mesquite communities, oak and conifer woodlands; 1600–2900 m; Ariz., N.Mex., Tex.; Mexico (Chihuahua, Coahuila, Durango, Nuevo León, San Luis Potosí, Tamaulipas, Zacatecas).

Variety *undulatum*, the southern expression of the species, is common in northern Mexico from Hidalgo and Zacatecas northward; there, the plants (called "yerba chuchaca") are found at elevations up to 3700 m on the higher volcanic peaks. In the southwestern United States, the variety is common only in the Santa Rita Range in southeastern Arizona (Cochise, Gila, and Santa Cruz counties) and the Chisos Mountains of Brewster County, Texas, although it does occur in Hidalgo County, New

Mexico. In the flora area, the plants are smallish subshrubs compared to those seen farther to the south. Variety *undulatum* is worthy of cultivation. The related Mexican endemic *Eriogonum turneri* Reveal is a low, compact expression known only from gypsum outcrops in Nuevo León.

129. Eriogonum arcuatum Greene, Pittonia 4: 319. 1901 E

Eriogonum jamesii Bentham var. *arcuatum* (Greene) S. Stokes

Herbs, spreading to often compact mats, 0.2–2.5 × 2–10 dm, floccose. **Stems:** caudex absent or spreading; aerial flowering stems erect or nearly so, slender, solid, not fistulose, usually arising directly from a taproot, 0.2–2 dm, floccose. **Leaves** basal, typically not in rosettes; petiole 0.5–2 cm, tomentose to floccose; blade oblanceolate or oblong to elliptic, 0.5–3 × 0.5–1.5 cm, densely tomentose abaxially, mostly floccose and grayish or greenish adaxially, margins entire, plane. **Inflorescences** umbellate or compound-umbellate, occasionally capitate or even reduced to single involucre, 0.3–2 × 0.3–2 dm; branches floccose; bracts 3–8, semileaflike at proximal node, 0.3–2 × 0.2–1 cm, often scalelike distally. **Involucres** 1 per node, usually campanulate, 3–7 × 3–8 mm, floccose; teeth 5–8, erect, 0.1–0.5 mm. **Flowers** 4–8 mm, including 0.7–2 mm stipelike base; perianth yellow, densely pubescent abaxially; tepals dimorphic, those of outer whorl lanceolate to elliptic, 1.5–5 × 1–3 mm, those of inner whorl lanceolate to fan-shaped, 1.5–7 × 2–4 mm; stamens exserted, 2–4 mm; filaments pilose proximally. **Achenes** light brown to brown, 4–5 mm, glabrous except for sparsely pubescent beak.

Varieties 3 (3 in the flora): w United States.

Eriogonum arcuatum has long been included under *E. jamesii*, primarily as the var. *flavescens*. It has also gone under the name *E. bakeri*, published by Greene about six weeks after he published *E. arcuatum*; the Fort Collins, Colorado, entomologist and botanist Charles Fuller Baker collected the types of both.

1. Involucres (3–)4–8 mm wide; inflorescences umbellate or compound-umbellate, rarely capitate; flowers (4–)5–8 mm; below 3000 m; n Arizona, c and w Colorado, nw New Mexico, e Utah, se Wyoming . . . . . . 129a. *Eriogonum arcuatum* var. *arcuatum*
1. Involucres 3–5 mm wide; inflorescences capitate, rarely umbellate or reduced to single involucre; flowers 4–5(–6) mm or 6–7 mm; usually above 3000 m in c Colorado, or 1600–1900 m in sw Utah.

2. Flowers 4–5(–6) mm; Zion National Park, Utah . . . . 129b. *Eriogonum arcuatum* var. *rupicola*
2. Flowers 6–7 mm; Rocky Mountains, Colorado . . . . 129c. *Eriogonum arcuatum* var. *xanthum*

129a. Eriogonum arcuatum Greene var. **arcuatum**
• Baker's wild buckwheat E

Eriogonum bakeri Greene; *E. flavum* Nuttall var. *tectum* (A. Nelson) S. Stokes; *E. jamesii* Bentham subsp. *bakeri* (Greene) S. Stokes; *E. jamesii* subsp. *flavescens* (S. Watson) S. Stokes; *E. jamesii* var. *flavescens* S. Watson; *E. jamesii* var. *higginsii* S. L. Welsh

Plants spreading, compact mats, (1–)2–3(–6) dm wide. **Aerial flowering stems** (0.3–)0.5–2 dm. **Leaf blades** oblanceolate to elliptic, (0.5–)1–3 × 0.5–1.5 cm. **Inflorescences** umbellate or compound-umbellate, branched 1–3 times, rarely capitate; bracts (0.3–)0.5–2 × (0.3–)0.4–1 cm. **Involucres** 3–7 × (3–)4–8 mm. **Flowers** (4–)5–8 mm.

Flowering Jun–Oct. Sandy, clayey, gravelly, or rocky flats, slopes and outcrops, ledges and cliffs, mixed grassland, saltbush, creosote bush, blackbrush, and sagebrush communities, pinyon, juniper, or conifer woodlands; 1000–3000 m; Ariz., Colo., N.Mex., Utah, Wyo.

Variety *arcuatum* is widely distributed and occasionally common from southeastern Wyoming south in central and western Colorado, eastern Utah, northern Arizona, and northwest New Mexico. Some specimens of var. *arcuatum* gathered in southeastern Wyoming can be difficult to distinguish from *Eriogonum flavum*. Plants with capitate inflorescences, sometimes recognized as var. *higginsii*, occur throughout the range of the variety. If such distinction is to be made, the earliest available epithet is *tectum*.

129b. Eriogonum arcuatum Greene var. **rupicola** (Reveal) Reveal, Harvard Pap. Bot. 9: 157. 2004
• Slickrock wild buckwheat E

Eriogonum jamesii Bentham var. *rupicola* Reveal, Phytologia. 25: 202. 1973

Plants spreading, compact mats, 2–6(–8) dm wide. **Aerial flowering stems** 0.3–1 dm. **Leaf blades** elliptic, 0.5–1.2 × 0.5–0.8 cm. **Inflorescences** capitate, rarely umbellate; bracts 0.3–1 × 0.3–0.5 cm. **Involucres** 3–5 × 3–4 mm. **Flowers** 4–5(–6) mm.

Flowering Jun–Oct. Sandstone outcrops, ledges, and cliffs, blackbrush and sagebrush communities, oak and pinyon-juniper woodlands; 1600–1900 m; Utah.

Variety *rupicola* is restricted to Zion National Park in Kane and Washington counties. It is marginally distinct from var. *arcuatum*.

129c. Eriogonum arcuatum Greene var. xanthum (Small) Reveal, Harvard Pap. Bot. 9: 157. 2004 · Ivy League wild buckwheat E

Eriogonum xanthum Small, Bull. Torrey Bot. Club. 33: 51. 1906; *E. chloranthum* Greene; *E. flavum* Nuttall subsp. *chloranthum* (Greene) S. Stokes; *E. flavum* var. *xanthum* (Small) S. Stokes; *E. jamesii* Bentham var. *xanthum* (Small) Reveal

Plants compact mats, 3–10 dm wide. **Aerial flowering stems** 0.2–0.8 dm. **Leaf blades** oblanceolate to oblong, 1–2.5 × 0.5–1 cm. **Inflorescences** capitate, often reduced to single involucre; bracts 0.3–1 × 0.2–0.7 cm. **Involucres** 4–6 × 3–5 mm. **Flowers** 6–7 mm.

Flowering Jul–Sep. Sandy, gravelly to rocky or talus slopes, high-elevation grassland, sagebrush, and tundra communities, alpine conifer woodlands; (2600–)3000–4000(–4200) m; Colo.

Variety *xanthum* is the subalpine to alpine expression of the species and is distributed along the backbone of the Colorado Rocky Mountains on many of the highest peaks (Chaffee, Clear Creek, Custer, El Paso, Fremont, Grand, Hinsdale, Huerfano, Lake, Larimer, Park, Saguache, Summit, and Teller counties). The plants are rather exquisite in the field and do well in cultivation.

130. Eriogonum wootonii (Reveal) Reveal, Harvard Pap. Bot. 9: 206. 2003 · Wooton's wild buckwheat E

Eriogonum jamesii Bentham var. *wootonii* Reveal, Phytologia 25: 201. 1973

Herbs, erect, 4–5 × 3–5 dm, tomentose to floccose. **Stems:** caudex absent; aerial flowering stems erect, stout, solid, not fistulose, arising directly from a taproot, 1–2.5 dm, tomentose to floccose. **Leaves** basal, not in rosettes; petiole 2–5 cm, tomentose to floccose; blade elliptic, (2.5–)3–6.5 × 1.5–3 cm, densely whitish- to tannish-tomentose abaxially, thinly tomentose to floccose and greenish adaxially, margins entire, plane. **Inflorescences** compound-umbellate, 5–15 × 5–20 cm; branches tomentose to floccose; bracts 3–6, leaflike at proximal node, 2–4(–5) × 0.7–1.8 cm, often scalelike distally. **Involucres** 1 per node, campanulate, 3–6 × 3–7 mm, thinly floccose; teeth

5–7, erect, 0.3–0.6 mm. **Flowers** 4–7(–9) mm, including 1–1.5 mm stipelike base; perianth bright yellow, densely pubescent abaxially; tepals dimorphic, those of outer whorl lanceolate to elliptic, 3–6 × 1.5–2.5, those of inner whorl elliptic, 5–9 × 2–3.5 mm; stamens exserted, 3–5 mm; filaments pilose proximally. **Achenes** light brown to brown, 4–8 mm, glabrous except for sparsely pubescent beak.

Flowering Jul–Sep. Sandy to gravelly slopes, sagebrush communities, montane conifer woodlands; 2000–2800 m; N.Mex.

Eriogonum wootonii is known only from the San Francisco Mountains of Lincoln and Otero counties in southern New Mexico. It is related to, but significantly disjunct from, *E. arcuatum*, and is part of a three-species complex that includes *E. correllii* of northern Texas and *E. allenii* of Virginia and West Virginia, and that skips across the central United States, the result of fragmentation associated with the southward push of Pleistocene glaciation.

131. Eriogonum correllii Reveal, Sida 3: 198, fig. 1. 1968 · Correll's wild buckwheat E

Herbs, erect, 1.5–3.5(–4) × 0.5–1 dm, tomentose. **Stems:** caudex absent; aerial flowering stems erect, stout, solid, not fistulose, arising directly from a taproot, (0.5–)1–2 dm, tomentose. **Leaves** basal, not in rosettes; petiole 4–10 cm, tomentose to floccose; blade lanceolate to elliptic, (2–)4–12(–15) × 1–3.5 cm, densely white-tomentose abaxially, floccose or glabrous and green adaxially, margins entire, plane. **Inflorescences** compound-umbellate, 5–15(–20) × 3–10(–12) cm; branches tomentose; bracts 3–6, leaflike at proximal node, 1–4 × 0.5–1.5(–1.8) cm, often scalelike distally. **Involucres** 1 per node, turbinate to campanulate, 3–5 × 2–4 mm, tomentose; teeth 5–7, erect or slightly spreading, 0.3–0.6 mm. **Flowers** (2.5–)3–7 mm, including 1–1.5(–2) mm stipelike base; perianth bright yellow, densely pubescent; tepals dimorphic, those of outer whorl lanceolate to elliptic, 2.5–5 × 1.5–2.5 mm, those of inner whorl elliptic to fan-shaped, 4–7 × 2–3.5 mm; stamens exserted, 3–5 mm; filaments pilose proximally. **Achenes** light brown to brown, 3.5–6 mm, glabrous except for sparsely pubescent beak. **2***n* = 40.

Flowering Jul–Oct. On clayey flats and mounds, mesquite communities, oak and juniper woodlands; 400–1000 m; Tex.

Eriogonum correllii is known from a few scattered locations in northern Texas (Armstrong, Briscoe, Floyd, Foard, Hardeman, Knox, and Oldham counties). Published references to this species (R. B. Kaul 1986) being in Oklahoma were based on specimens of *E. jamesii* var. *jamesii*.

132. Eriogonum allenii S. Watson in A. Gray et al., Manual ed. 6, 734. 1890 (as alleni) • Shale barren wild buckwheat E

Herbs, erect, 3–5(–7) × 0.5–1 dm, tomentose. Stems: caudex absent; aerial flowering stems erect, stout, solid, not fistulose, arising directly from a taproot, 2–4 dm, tomentose. Leaves basal, not in rosettes; petiole 5–15 cm, tomentose to floccose; blade oblong to ovate, (5–)10–15 × 4–8 cm, densely brownish-white-tomentose abaxially, floccose or glabrous and green adaxially, margins entire, plane. Inflorescences compound-umbellate, 10–40(–50) × 8–30(–50) cm; branches tomentose; bracts 3–6, leaflike at proximal node, 1–10(–12) × 0.5–4(–5) cm, often scalelike distally. Involucres 1 per node, campanulate, 3–5(–7) × (4–)5–8 mm, tomentose; teeth 5–7, erect or slightly spreading, 0.8–1.5 mm. Flowers 3–7 mm, including 1–1.5 mm stipelike base; perianth bright yellow, densely pubescent abaxially; tepals dimorphic, those of outer whorl broadly lanceolate to elliptic, 3–6 × 1.5–3 mm, those of inner whorl elliptic to fan-shaped, 4–7 × 2.5–4 mm; stamens exserted, 3–5(–7) mm; filaments pilose proximally. Achenes light brown to brown, 4–6 mm, glabrous except for sparsely pubescent beak.

Flowering Jun–Oct. Rocky shale slopes, oak and pine woodlands; 400–800 m; Va., W.Va.

Eriogonum allenii is restricted to the Appalachian shale barrens of Virginia (Alleghany, Augusta, Bath, Botetourt, Craig, Frederick, Highland, Montgomery, and Shenandoah counties) and West Virginia (Greenbrier, Monroe, and Pendleton counties), where it is local and infrequent, and found in four geographic areas of concentration. The plants are protected at The Nature Conservancy's Slaty Mountain site in Monroe County, West Virginia, and in Douthat State Park in Bath County, Virginia. The species occasionally is cultivated.

133. Eriogonum flavum Nuttall, Cat. Pl. Upper Louisiana, no. 34. 1813 E F

Eriogonum flavum var. *polyphyllum* (Small ex Rydberg) M. E. Jones; *E. polyphyllum* Small ex Rydberg

Herbs, matted, 0.2–4 × 1–10 dm, tomentose to floccose. Stems: caudex absent or spreading; aerial flowering stems erect or nearly so, slender, solid, not fistulose, arising at nodes of caudex branches and at distal nodes of short, nonflowering aerial branches, (0.1–)0.5–2(–3) dm, tomentose to floccose. Leaves basal,

occasionally in rosettes; petiole 0.5–4 cm, tomentose to floccose; blade linear-oblanceolate to oblong or elliptic, 1–7(–9) × (0.3–)0.5–1.5 cm, densely whitish- or grayish-tomentose abaxially, mostly tomentose to floccose or glabrous and green adaxially, margins entire, plane. Inflorescences subcapitate or umbellate, 0.5–3(–5) × 0.3–2.5(–3) dm; branches tomentose to floccose; bracts 4–6, leaflike to semileaflike at proximal node, 0.5–2 × 0.2–0.5 cm, sometimes absent immediately below involucre. Involucres 1 per node, turbinate to campanulate, 3–9 × 2–5 mm, tomentose to floccose; teeth 5–8, erect, 0.2–1 mm. Flowers 3–7 mm, including 0.2–1.5 mm stipelike base; perianth pale to bright yellow, densely pubescent abaxially; tepals monomorphic, oblong; stamens exserted, 3–6 mm; filaments pilose proximally. Achenes light brown to brown, 3–5 mm, glabrous except for sparsely pubescent beak.

Varieties 3 (3 in the flora): w North America.

Eriogonum flavum approaches *E. arcuatum* morphologically in southeastern Wyoming, and a clear distinction is not always possible when the plants are immature or the herbarium material is poor.

Variety *polyphyllum*, as traditionally circumscribed, is an alpine phase of both var. *flavum* and var. *piperi*; it is encountered infrequently in scattered range in the high mountains of southwestern Montana and northwestern Wyoming.

1. Stipelike bases of flowers 1–1.5 mm; usually w of Continental Divide; sw Alberta and British Columbia, n Idaho, w Montana, ne Oregon, e Washington, nw Wyoming . . . . 133c. *Eriogonum flavum* var. *piperi*
1. Stipelike bases of flowers 0.2–1 mm; usually e of Continental Divide; ec Alaska, s Alberta, sw Manitoba, s Saskatchewan, wc Yukon, Montana, w Nebraska, North Dakota, South Dakota, e Wyoming, and Colorado.
   2. Leaf blades (0.3–)1–1.5 cm wide; inflorescence branches 4–20 cm; s Alberta, sw Manitoba, s Saskatchewan, Montana, w Nebraska, North Dakota, South Dakota, e Wyoming, and Colorado . . . . 133a. *Eriogonum flavum* var. *flavum*
   2. Leaf blades 0.4–1 cm wide; inflorescence branches 0.5–2 cm; ec Alaska, wc Yukon . . . . 133b. *Eriogonum flavum* var. *aquilinum*

133a. Eriogonum flavum Nuttall var. **flavum**
  • Alpine golden wild buckwheat E F

Eriogonum flavum subsp.
crassifolium (Bentham) S. Stokes; *E. flavum* var. *linguifolium* Gandoger; *E. flavum* var. *muticum* S. Stokes

Plants tight, compact mats, 1–3 (–5) dm wide. **Aerial flowering stems** mostly erect, 0.1–2 dm. **Leaf blades** usually elliptic, 1–5 × 0.3–1.5 cm, densely tomentose abaxially, tomentose to floccose and greenish adaxially. **Inflorescences** umbellate; branches 4–20 cm. **Involucres** turbinate to campanulate, 3–8 mm. **Flowers** 3–7 mm, including 0.3–1 mm stipelike base; perianth bright yellow. $2n = 80$.

Flowering Jun–Sep. Clayey or sandy to gravelly flats and slopes, mixed grassland and sagebrush communities, montane conifer woodlands, high-elevation sagebrush communities, subalpine or alpine conifer woodlands; 500–3200 m; Alta., Man., Sask.; Colo., Mont., Nebr., N.Dak., S.Dak., Wyo.

Variety *flavum* is widespread and rather common on the short-grass prairies of the Great Plains from Alberta, southwestern Manitoba, and Saskatchewan south through North Dakota, South Dakota, Nebraska, and eastern Colorado, westward across the plains and into the mountains of Montana and Wyoming. It occasionally is cultivated and, while slow-growing, it will, given time, form nice mats, with a fair profusion of inflorescences bearing bright yellow flowers.

133b. Eriogonum flavum Nuttall var. **aquilinum**
  Reveal, Ark. Bot., n. s. 7: 46. 1968 • Yukon wild buckwheat E

Plants tight, compact mats, 3–10 dm wide. **Aerial flowering stems** mostly erect, 0.2–0.8 dm. **Leaf blades** narrowly elliptic, 1–2(–3.5) × 0.4–1 cm, densely tomentose abaxially, floccose and greenish adaxially. **Inflorescences** umbellate; branches 0.5–2 cm. **Involucres** usually turbinate-campanulate, (4–)5–7(–8) mm. **Flowers** 3–5(–6) mm, including 0.2–0.3 mm stipelike base; perianth bright yellow.

Flowering Jul–Aug. Exposed bluffs, mixed grassland communities, willow thickets; 300–500 m; Yukon; Alaska.

Variety *aquilinum* is rare and isolated, known only from the Eagle area of extreme east-central Alaska (including the Kathul Mountains, Southeast Fairbanks County) and from the Aishihik Lake area in the Yukon Territory. It is closely related to and marginally distinct from var. *flavum*, its isolation the result of Pleistocene glaciation.

133c. Eriogonum flavum Nuttall var. **piperi** (Greene) M. E. Jones, Contr. W. Bot. 11: 7. 1903 • Piper's wild buckwheat E

Eriogonum piperi Greene, Pittonia 3: 263. 1898; *E. flavum* subsp. *piperi* (Greene) S. Stokes

Plants loose to compact mats, 1–6 dm wide. **Aerial flowering stems** mostly erect, (0.5–)1–2(–3) dm. **Leaf blades** linear-oblanceolate to oblong or narrowly elliptic, 2–7 (–9) × 0.3–0.8(–1) cm, densely tomentose abaxially, tomentose to floccose or glabrous and greenish adaxially. **Inflorescences** subcapitate to umbellate; branches 0.2–1 cm. **Involucres** usually turbinate, 4–9 mm. **Flowers** (4–)5–7 mm, including 1–1.5 mm stipelike base; perianth pale to bright yellow. $2n = 76, 80$.

Flowering Jun–Sep. Sandy to gravelly flats and slopes in mixed grassland and sagebrush communities, and in montane to subalpine conifer woodlands; (700–)1200–2800(–3200) m; Alta., B.C.; Idaho, Mont., Oreg., Wash., Wyo.

Variety *piperi* is the common and widespread phase of the species, found mainly west of the Continental Divide in southern Alberta, southern British Columbia, eastern Washington, northern Idaho, and western Montana south into northeastern Oregon and northwestern Wyoming. It is only slightly variable, the major exception being depauperate individuals at high elevations in harsh exposures; these have been recognized by some as var. *polyphyllum*. The length of the stipelike base shortens from west to east, but only rarely are individuals in Montana troublesome to place either here or in var. *flavum*. The plants do well in cultivation and are now widely available.

134. Eriogonum androsaceum Bentham in A. P. de Candolle and A. L. P. P. de Candolle, Prodr. 14: 9. 1856 • Rock-jasmine wild buckwheat E

Eriogonum flavum Nuttall subsp. *androsaceum* (Bentham) S. Stokes

Herbs, matted, 0.5–3 dm wide, tomentose to floccose. **Stems:** caudex spreading; aerial flowering stems ascending to erect, slender, solid, not fistulose, arising at nodes of caudex branches and at distal nodes of short, non-flowering aerial branches, (0.1–)0.3–0.7(–1) dm, tomentose to floccose or subglabrous. **Leaves** basal, occasionally in rosettes; petiole 0.3–1.5 cm, tomentose; blade narrowly elliptic, (0.5–)1–2 × 0.2–0.5 cm, densely white-lanate or grayish-tomentose abaxially, floccose and green adaxially, margins entire, usually slightly revolute. **Inflorescences** subcapitate or umbellate, 0.5–1.5 × 0.3–2 cm; branches tomentose to floccose; bracts 5–7, semileaflike at proximal node, 0.4–1 × 0.1–0.3 cm, often absent immediately below involucre. **Involucres** 1 per node, narrowly turbinate to turbinate-campanulate, 3–5 × 3–4.5 mm, tomentose to floccose; teeth 5–8, erect, 0.2–0.5 mm. **Flowers** (3.5–)4–5(–6.5) mm, including 0.1–0.2 mm stipelike base; perianth pale yellow, sparsely pubescent abaxially; tepals monomorphic, narrowly oblong; stamens exserted, 4–5 mm; filaments pilose proximally. **Achenes** light brown, (3–)4–6 mm, glabrous.

Flowering Jul–Aug. Sandy to gravelly or rocky to talus slopes, ridges, and outcrops, mixed grassland, sagebrush, or alpine meadow communities, montane conifer woodlands; 1700–2700 m; Alta.; Mont.

Eriogonum androsaceum is common in the high northern Rocky Mountains in southern Alberta and northwestern Montana (Glacier, Lincoln, Park, Pondera, and Teton counties). It is clearly related to *E. flavum* but is sufficiently distinct to merit recognition as a species. It is seen occasionally in cultivation but deserves more horticultural attention.

A decoction of the rock-jasmine wild buckwheat was used in sweatbaths for rheumatism, and for internal pain by the Thompson Indians of British Columbia (E. V. Steedman 1930). Interestingly, Steedman indicated also that a strong decoction of the plants was used for syphilis.

135. **Eriogonum latens** Jepson, Fl. Calif. 1: 427. 1914 • Inyo wild buckwheat [E]

Eriogonum elatum Douglas ex Bentham subsp. *glabrescens* S. Stokes; *E. monticola* S. Stokes

Herbs, erect, (1.5–)2.5–4.5(–5) × 1–2 dm, glabrous. **Stems:** caudex absent or nearly so; aerial flowering stems erect, slender, solid, not fistulose, arising at nodes of caudex branches and at distal nodes of short, nonflowering aerial branches, 1–4 dm, essentially glabrous. **Leaves** basal, occasionally in rosettes; petiole 1–4(–7) cm, pilose; blade elliptic-obovate to round-ovate, 1–3(–3.5) × 0.8–2.5 cm, pilose, margins entire, plane. **Inflorescences** capitate, 2–3.5(–4) cm wide; branches absent; bracts 5–8, essentially scalelike, 1–8 × 1–3 mm. **Involucres** 2–5 or more per cluster, campanulate, 6–8 × 6–8 mm, pilose; teeth 5–8, erect to slightly spreading, 1–2 mm. **Flowers** 3–6 mm, including 0.1–0.2 mm stipelike base; perianth cream to pale yellow, sparsely pubescent abaxially; tepals monomorphic, obovate to spatulate; stamens exserted, 3–7 mm; filaments pilose proximally. **Achenes** light brown, 3–5 mm, glabrous.

Flowering Jun–Aug. Sandy to gravelly granitic slopes and ridges, sagebrush communities, montane to subalpine conifer woodlands; 2600–3400 m; Calif., Nev.

Eriogonum latens is locally uncommon to rare in the White Mountains of western Nevada (Esmeralda County) and along the eastern slope of the Sierra Nevada in California (Inyo and Mono counties). An isolated population occurs on Waucoba Mountain in the Inyo Mountains. A collection supposedly from Long Valley in Mono County (*Noldeke s.n.*, Jul 1938, CAS) remains to be confirmed as to location.

136. **Eriogonum hirtellum** J. T. Howell & Bacigalupi, Leafl. W. Bot. 9: 174. 1961 • Klamath Mountain wild buckwheat [C][E]

Herbs, spreading, 1–3.5 × 2–4 (–6) dm, hirtellous to glabrescent. **Stems:** caudex spreading; aerial flowering stems erect, slender, solid, not fistulose, arising at nodes of caudex branches and at distal nodes of short, nonflowering aerial branches, 0.8–2.5(–3) dm, essentially glabrous. **Leaves** basal, occasionally in rosettes; petiole 0.3–4 cm, hirtellous; blade broadly oblanceolate to elliptic or ovate, 0.5–2(–2.5) × 0.3–0.8(–1.2) cm, hirtellous to glabrescent, margins plane, entire. **Inflorescences** subcapitate, (1.5–)2–5 cm wide; branches subglabrous; bracts 5, semileaflike, 0.2–0.5 × 0.1–0.2 cm, absent immediately below involucre. **Involucres** 1 per node (often appearing clustered), narrowly turbinate, 5–6 × 2–3 mm, hirtellous; teeth 5–6, erect, 0.8–1 mm. **Flowers** 3–3.5(–4) mm, including 0.1–0.2 mm stipelike base; perianth bright yellow, sparsely white-pilose abaxially; tepals dimorphic, those of outer whorl spatulate, 2.5–3 mm wide, those of inner whorl oblanceolate, 1 mm wide; stamens exserted, 3–5 mm; filaments pilose proximally. **Achenes** light brown, 3–3.5 mm, glabrous except for pubescent beak.

Flowering Jul–Sep. Serpentine slopes and outcrops, oak and conifer woodlands; of conservation concern; (1100–)1300–1700(–2000) m; Calif.

Eriogonum hirtellum is known only from a few scattered locations in the Klamath Mountains of Siskiyou County and just over the boundary in Del Norte County. It is considered to be "sensitive" by the Bureau of Land Management.

137. Eriogonum pyrolifolium Hooker, Hooker's J. Bot.
Kew Gard. Misc. 5: 395, plate 10. 1853 (as
pyrolaefolium) [E]

Herbs, compact or spreading, matted, 0.3–2 × 0.5–3 dm, floccose or glabrous. **Stems:** caudex absent or spreading; aerial flowering stems prostrate to weakly erect, slender, solid, not fistulose, usually arising directly from a taproot, 0.3–1.5(–1.8) dm, villous to floccose or glabrous. **Leaves** basal, mostly in loose rosettes; petiole 1–4 cm, tomentose to floccose; blade ovate to round, 1–2.5(–4) × 0.8–2 cm, grayish- to tannish-lanate or tomentose abaxially, glabrous and green adaxially, or glabrous on both surfaces, margins entire, plane. **Inflorescences** capitate or umbellate, 1–3(–5) × 1–4 cm; branches floccose or glabrous; bracts 2, leaflike, (0.4–)1–2 × 0.1–0.2(–0.4) cm, sometimes absent immediately below involucre. **Involucres** 1 per node (occasionally appearing clustered), campanulate, 4–6 × (3–)4–8 mm, pilose or glabrous; teeth 4–5, erect, 0.6–1 mm. **Flowers** 4–6 mm, including 0.1–0.2 mm stipelike base; perianth white to rose, pilose to villous with intermixed glandular hairs abaxially; tepals monomorphic, obovate; stamens exserted, 4–8 mm; filaments pilose proximally. **Achenes** light brown to brown, 4–5 mm, glabrous except for pilose beak.

Varieties 2 (2 in the flora): w North America.

The taxonomic merit of the two varieties of *Eriogonum pyrolifolium* is dubious. Plants with both hairy and nonhairy leaves are found growing together, and yet there are many populations in which only a single expression is found.

1. Leaf blades densely lanate to tomentose abaxially, mostly glabrous adaxially . . . .
. . . . . . 137a. *Eriogonum pyrolifolium* var. *coryphaeum*
1. Leaf blades glabrous on both surfaces . . . . . .
. . . . . 137b. *Eriogonum pyrolifolium* var. *pyrolifolium*

137a. Eriogonum pyrolifolium Hooker var.
coryphaeum Torrey & A. Gray, Proc. Amer. Acad.
Arts 8: 162. 1870 (as pyrolaefolium) • Hairy
Shasta wild buckwheat [E]

Eriogonum pyrolifolium var. *bellingeranum* M. Peck

Aerial flowering stems prostrate to weakly erect, 0.4–1.5(–1.8) dm. **Leaf blades** densely lanate to tomentose abaxially, mostly glabrous and green adaxially. **Perianths** usually copiously pilose to villous abaxially with glandular hairs obscure. *2n* = 40.

Flowering Jun–Sep. Sandy to gravelly, usually nonvolcanic flats, slopes, and ridges, mixed grassland, sagebrush, and mountain meadow communities, oak, montane, and subalpine conifer woodlands; 1600–3100 m; B.C.; Calif., Idaho, Mont., Oreg., Wash.

Variety *coryphaeum* is by far the more common of the two varieties, being found in the mountains from southern British Columbia south through Washington and Oregon into northwestern California, eastward into central Idaho and central-western Montana.

137b. Eriogonum pyrolifolium Hooker var.
pyrolifolium • Shasta wild buckwheat [E]

Aerial flowering stems mostly ascending, 0.3–1 dm. **Leaf blades** glabrous on both surfaces. **Perianths** sparsely pilose, with glandular hairs evident.

Flowering Jul–Sep. Sandy to gravelly, usually volcanic flats, slopes, and ridges, mixed grassland, sagebrush, and mountain meadow communities, oak, montane, and subalpine conifer woodlands; (800–)1600–3300 m; Calif., Idaho, Wash.

Variety *pyrolifolium* is widely scattered; its range overlaps to a considerable degree with that of var. *coryphaeum*. In the Cascade Range, var. *pyrolifolium* occurs only in Kittitas County, Washington, and on Little Mt. Hoffman, Mt. Lassen, and Mt. Shasta in Siskiyou and Shasta counties, California. Only in Blaine, Boise, Custer, Elmore, Lemhi, and Valley counties of central Idaho are the plants relatively common, and then they nearly always occur with var. *coryphaeum*.

138. Eriogonum lobbii Torrey & A. Gray, Proc. Amer.
Acad. Arts 8: 162. 1870 • Lobb's wild buckwheat
[E]

Herbs, compact or sprawling, matted, 0.3–3 × 1–2.5 dm, tomentose to floccose. **Stems:** caudex absent or nearly so; aerial flowering stems prostrate to decumbent or weakly erect, slender or stout, solid, not fistulose, usually arising directly from a taproot, 0.5–1.5(–2) dm, tomentose to floccose. **Leaves** basal, in well-defined rosettes; petiole (0.8–)1–3.5(–5) cm, tomentose to floccose; blade ovate to obovate or round-oval, 1–4(–5) × 1–4(–5) cm, densely white- to grayish- or reddish-tomentose abaxially, less so to floccose or glabrous and greenish adaxially, margins entire, plane. **Inflorescences** subcapitate to umbellate or infrequently 2-umbellate, 1–4 × 1–4 cm; branches tomentose to floccose; bracts 3–5 at proximal node, leaflike, 0.6–1.5(–2.5) × 0.1–0.5(–0.8) cm, sometimes absent immediately below involucre. **Involucres** 1 per node, turbinate-campanulate to campanulate, 5–10(–12) × 5–10(–12) mm, thinly tomentose to lanate; teeth 6–10, usually lobelike, mostly reflexed, 2–6 mm. **Flowers** 5–7 mm, including 0.1–0.4 mm stipelike base; perianth white to rose, glabrous; tepals monomorphic, oblong-obovate; stamens exserted, 5–7 mm; filaments pilose proximally. **Achenes** light brown to brown, 4.5–6 mm, glabrous. *2n* = 40.

Flowering Jun–Aug. Gravelly to rocky or talus slopes, mixed grassland, buckbrush, manzanita, and sagebrush communities, montane, subalpine, or alpine conifer woodlands; (1000–)1600–3800 m; Calif., Nev., Oreg.

Eriogonum lobbii is rather infrequent throughout its range. It is found in three widely scattered areas of concentration: the high mountains of northwestern California and southwestern Oregon; the southern portion of the North Coast Range of California; and the Sierra Nevada of eastern California and west-central Nevada. In the first two areas, the plants frequently are associated with serpentine soils; elsewhere the species is found almost exclusively on granitic soils and infrequently on volcanic ones. There are some morphologic differences between the Sierran plants and those of the coastal mountains, but no taxonomic separation is suggested. The species is a food plant for the intermediate dotted-blue butterfly (*Euphilotes intermedia*).

139. Eriogonum robustum Greene, Bull. Calif. Acad. Sci. 1: 126. 1885 • Altered andesite wild buckwheat [C][E]

Eriogonum lobbii Torrey & A. Gray var. *robustum* (Greene) M. E. Jones

Herbs, erect, matted, 0.5–3 × 1–2 dm, tomentose to floccose. **Stems:** caudex absent; aerial flowering stems erect or nearly so, stout, solid, not fistulose, usually arising directly from a taproot, 0.5–1.2 (–1.6) dm, tomentose to floccose. **Leaves** basal, in well-defined rosettes; petiole 1–4(–5.5) cm, tomentose to floccose; blade ovate to obovate, 2.5–4(–5) × 1.6–2.5(–3.5) cm, densely white- to grayish- or reddish-tomentose abaxially, less so to floccose or glabrous and greenish adaxially, margins entire, plane. **Inflorescences** 2-umbellate, 5–10 × 5–10 cm; branches tomentose to floccose; bracts 3–5 at proximal node, leaflike, 1.5–2.5(–3.5) × 0.3–0,8(–1) cm, sometimes absent immediately below involucre. **Involucres** 1 per node, campanulate, 8–11(–13) × 8–12 mm, thinly tomentose to lanate; teeth 6–10, usually lobelike, mostly reflexed, 2–6 mm. **Flowers** 7–9 mm, including 0.1–0.4

mm stipelike base; perianth creamy yellow to pale yellowish, glabrous; tepals monomorphic, oblong-obovate; stamens exserted, 7–9 mm; filaments pilose proximally. **Achenes** light brown to brown, 6–8 mm, glabrous.

Flowering Jun–Aug. Heavy clayey slopes, montane and subalpine conifer woodlands; of conservation concern; 1300–2000(–2500) m; Nev.

Eriogonum robustum is restricted primarily to altered andesite soils in west-central Nevada, essentially at the confluence of Carson City, Lyon, Storey, and Washoe counties. There, due to the unusual soil, the plants typically occur in areas without sagebrush but among conifer species usually found at higher elevations. The species is cultivated infrequently and is a food plant for the intermediate dotted-blue butterfly (*Euphilotes intermedia*).

SELECTED REFERENCE Kuyper, K. F., U. Yandell, and R. S. Nowak. 1997. On the taxonomic status of *Eriogonum robustum* (Polygonaceae), a rare endemic in western Nevada. Great Basin Naturalist 57: 1–10.

140. Eriogonum alpinum Engelmann, Bot. Gaz. 7: 6. 1882 • Trinity wild buckwheat [C][E]

Herbs, spreading, compact, 0.2–0.6 × (0.3–)0.4–1 dm, tomentose. **Stems:** caudex absent; aerial flowering stems erect or nearly so, slender, solid, not fistulose, arising directly from a taproot, (0.3–)0.4–0.6 dm, floccose, with a whorl of 3–5 leaflike bracts ca. or slightly below midlength, 0.4–0.8(–1) × 0.15–0.4(–0.7) cm. **Leaves** basal, in well-defined rosettes; petiole 1–2(–3) cm, tomentose; blade oval to round-oval, 1–2(–3) × 1–2(–3) cm, densely whitish- or grayish-tomentose on both surfaces, margins entire, plane. **Inflorescences** capitate, 1–2(–2.5) cm wide; branches absent; bracts absent immediately below involucre. **Involucres** 1 per node, broadly campanulate, 3–5(–6) × 4–7(–10) mm, tomentose; teeth 6–12, erect, 0.5–0.9 mm. **Flowers** (3–)4–8 mm, including 0.5–0.8 mm stipelike base; perianth bright yellow, glabrous; tepals monomorphic, oblong-obovate; stamens included to exserted, 3–6 mm; filaments pilose proximally. **Achenes** light brown, 4–5 mm, glabrous.

Flowering Jul–Sep. Serpentine slopes and ridges, montane conifer woodlands; of conservation concern; 2000–2800 m; Calif.

Eriogonum alpinum is a rare and localized species found only on the slopes of Scott Mountain (where it is known only from the type collection), Cory Peak, and Mt. Eddy on the border of Siskiyou and Trinity counties. It is considered to be a "sensitive" species by the U.S. Forest Service, but is listed as endangered by the state of California.

1f. ERIOGONUM Michaux subg. PTEROGONUM (H. Gross) Reveal, Sida 3: 82. 1967

Pterogonum H. Gross, Bot. Jahrb. Syst. 49: 239. 1913

Herbs, erect, polycarpic or monocarpic perennials, strigose, silky-pubescent, or glandular-puberulent, occasionally glabrous [tomentose]; taproot woody occasionally chambered. **Stems** erect, with persistent leaf bases, silky-pubescent to glandular-puberulent or glabrous; caudex absent or compact, spreading; aerial flowering stems erect, slender to stout, solid or hollow, sometimes fistulose, not disarticulating in ringlike segments proximally, usually arising directly from root. **Leaves** persistent or marcescent, basal or basal and cauline, not fasciculate; blade glabrous or strigose to hispid or stipitate-glandular. **Inflorescences** cymose, open, occasionally with 1 branch of cyme suppressed; bracts 3, connate basally, semileaflike to scalelike. **Peduncles** ascending to erect, slender. **Involucres** 1 per node, not appressed to inflorescence branches, turbinate to campanulate; teeth 5, erect. **Flowers** not attenuate at base, without stipelike base; perianth various shades of white, yellow, red, or purple, glabrous on both surfaces or occasionally pubescent abaxially; tepals connate proximally $1/4$–$1/3$ their length, monomorphic; stamens exserted; filaments pilose proximally or glabrous. **Achenes** various shades of brown, 3-winged or -ridged, 3-gonous, glabrous or pilose [strigose]. **Seeds:** embryo straight.

Species 11 (6 in the flora): sw United States, Mexico.

W. J. Hess and J. L. Reveal (1976) subdivided subg. *Pterogonum* into four sections: sect. *Pterogonum* (only *Eriogonum atrorubens* in the flora area), sect. *Peregrina* W. J. Hess & Reveal (*E. greggii*), sect. *Astra* W. J. Hess & Reveal (*E. hemipterum, E. hieracifolium,* and *E. nealleyi*), and sect. *Alata* Bentham (*E. alatum*). The rest of the species occur in northern Mexico and all belong to sect. *Pterogonum*: *E. ciliatum* Torrey ex Bentham, *E. clivosum* W. J. Hess & Reveal, *E. fimbriatum* W. J. Hess & Reveal, *E. henricksonii* Reveal, and *E. viscanum* W. J. Hess & Reveal.

SELECTED REFERENCE Hess, W. J. and J. L. Reveal. 1976. A revision of *Eriogonum* (Polygonaceae) subgenus *Pterogonum*. Great Basin Naturalist 36: 281–333.

1. Leaves basal or, if cauline, whorled; stems glandular-puberulent or, if glabrous, stems fistulose.
    2. Perianths glabrous, purple to red or maroon; cauline leaves absent; stems glabrous, often fistulose; Hidalgo County, New Mexico . . . . 141. *Eriogonum atrorubens*
    2. Perianths strigose, yellowish white; cauline leaves whorled; stems glandular-puberulent, not fistulose; Hidalgo County, Texas . . . . 142. *Eriogonum greggii*
1. Leaves basal; stems strigose to glandular-pubescent or, if glabrous, stems not fistulose.
    3. Plants monocarpic; taproot often chambered; perianths yellow to yellowish green, rarely maroon in anthesis; achenes winged entire length; e Utah, se Wyoming and sw Nebraska to n Arizona, New Mexico, and n and w Texas . . . . 146. *Eriogonum alatum*
    3. Plants polycarpic; taproot not chambered; perianths yellow, white to greenish white, reddish purple, or maroon; achenes ribbed or winged only along distal $1/2$; ec Arizona, wc and s New Mexico, w Texas.
        4. Inflorescence branches glabrous; perianths white to greenish white, glabrous or sparsely strigose; Coke, Howard, Irion, and Sterling counties, Texas . . . . . 144. *Eriogonum nealleyi*
        4. Inflorescence branches strigose; perianths yellow to reddish purple or maroon, strigose or pilose; ec Arizona, wc and s New Mexico to sw Texas.
            5. Perianths reddish purple or maroon, rarely yellow; Brewster County, Texas . . . . 143. *Eriogonum hemipterum*
            5. Perianth yellow in anthesis, reddish in fruit; ec Arizona, wc and s New Mexico to w Texas . . . . 145. *Eriogonum hieracifolium*

141. Eriogonum atrorubens Engelmann in F. A. Wislizenus, Mem. Tour N. Mexico, 108. 1848

Pterogonum atrorubens (Engelmann) H. Gross

Varieties 5 (1 in the flora): w North America, including n Mexico.

All five varieties occur in northern Mexico (Chihuahua, Coahuila, Durango, Nuevo León, Sonora, and Zacatecas) in oak and conifer woodlands, commonly above 2200 m elevation. In addition to var. *atrorubens* (including var. *psuedociliatum* Reveal), they are var. *auritulum* W. J. Hess & Reveal, var. *intonsum* Reveal, var. *nemorosum* W. J. Hess & Reveal, and var. *rupestre* (S. Stokes) W. J. Hess & Reveal.

141a. Eriogonum atrorubens Engelmann var. **atrorubens** • Red wild buckwheat

Herbs, polycarpic, 5–10(–12) dm, glabrous, glaucous; taproot not chambered. **Stems:** caudex absent; aerial flowering stems usually 1, usually fistulose, 1–4 dm, glabrous, glaucous. **Leaves** basal; petiole 3–8(–10) cm, sparsely strigose; blade oblanceolate or oblong to elliptic, 4–8 × 1–3 cm, strigose on both surfaces, often slightly less so adaxially. **Inflorescences** 4–8 dm; branches glabrous, glaucous; bracts semileaflike to scalelike, 1–4(–10) × 1–3 mm. **Peduncles** erect, straight or curving upward, (1–)2–6(–12) cm, glabrous. **Involucres** turbinate to turbinate-campanulate, 1.5–4(–4.5) × 1–2.5(–3) mm, glabrous; teeth 0.5–1.5 mm. **Flowers** (1.5–)2–2.5 mm in anthesis, 3–6 mm in fruit; perianth purple to red or maroon, glabrous; tepals broadly spatulate to obovate; stamens 2–2.5 mm; filaments glabrous. **Achenes** light greenish brown to brown, (2–)3–5 mm, slightly 3-winged distally, nearly beakless, glabrous. *2n* = 40.

Flowering Jun–Oct. Sandy to loamy flats and slopes, mixed grassland communities, oak and montane conifer woodlands; 1800–2000 m; N.Mex.; Mexico (Chihuahua, Durango, Sonora, Zacatecas).

Variety *atrorubens* enters the flora area in the San Luis Mountains of southern Hidalgo County, New Mexico, literally along the border with Mexico. The root is used medicinally in Mexico as a treatment for toothache.

142. Eriogonum greggii Torrey & A. Gray, Proc. Amer. Acad. Arts 8: 187. 1870 • Gregg's wild buckwheat [C]

Eriogonum ciliatum Torrey ex Bentham var. *foliosum* Torrey

Herbs, polycarpic, 1–4 dm, glandular-puberulent; taproot not chambered. **Stems:** caudex absent or compact; aerial flowering stems usually 1, erect or nearly so, not fistulose, 0.5–1.5 dm, glandular-puberulent. **Leaves** basal and cauline; basal petiolate, petiole often winged, 0.5–2 cm, hispid, blade elliptic to broadly spatulate, 2–6(–10) × 0.5–2(–2.5) cm, hispid and uniformly stipitate-glandular on both surfaces or glabrous except for ciliate midvein and margin; cauline sessile, blades in whorls of 3–10 per node, oblanceolate, obovate, or spatulate, 0.5–4 × 0.2–1.5(–2) cm, similar to basal blade except more hispid and less glandular adaxially than abaxially, or glabrous except for ciliate midvein and margins. **Inflorescences** 1–3 × 1–3 dm; branches glandular-puberulent; bracts semileaflike to scalelike, 3–15 × 1–3(–5) mm. **Peduncles** ascending to erect, straight or slightly curving upward, 1–6(–7) cm, glandular and hispid. **Involucres** campanulate, 1.5–3 × 3–5 mm, slightly to densely strigose and glandular; teeth 1–1.5 mm. **Flowers** 1.5–2.5 mm in anthesis, 2.5–3.5 mm in fruit; perianth yellowish white with prominent reddish brown to brown midribs and bases, becoming more reddish in fruit, strigose abaxially; tepals lanceolate to oblong; stamens 1.5–2 mm; filaments glabrous. **Achenes** reddish brown, 3–4 mm, 3-ridged distally, short-beaked, glabrous. *2n* = 32.

Flowering Jul–May. Calcareous flats and slopes, mixed grassland, creosote bush, and saltbush communities; of conservation concern; 50 m; Tex.; Mexico (Coahuila, Nuevo León).

Eriogonum greggii just enters the flora area in Hildago and Starr counties, where it is rare. In Mexico, the plants occur in Coahuila and Nuevo León, nearly always below 1850 m. There, Gregg's wild buckwheat occasionally is found in montane conifer woodlands.

143. Eriogonum hemipterum (Torrey & A. Gray) Torrey ex S. Stokes, Eriogonum, 21. 1936

Eriogonum hieracifolium Bentham var. *hemipterum* Torrey & A. Gray, Proc. Amer. Acad. Arts 8: 154. 1870

Varieties 2 (1 in the flora): sw United States; Mexico (Chihuahua, Coahuila).

143a. Eriogonum hemipterum (Torrey & A. Gray)
Torrey ex S. Stokes var. **hemipterum** · Chisos
Mountains wild buckwheat

Herbs, polycarpic, 2–6 dm,
strigose; taproot not chambered.
Stems: caudex compact; aerial
flowering stems usually 1, not
fistulose, 1–4 dm, strigose. Leaves
basal and cauline; petiole 0.5–3
cm, strigose; basal blade oblanceolate to elliptic or spatulate, 2–
8 × 0.5–1.5 cm, thinly strigose and
greenish on both surfaces; cauline blade sessile,
oblanceolate, 1–5 × 0.2–1 cm, similar to basal blade.
Inflorescences 1–3 dm; branches strigose; bracts scalelike,
2.5–8 × 1–3 mm. Peduncles erect, straight or slightly
curving upward, 0.5–7 cm, strigose. Involucres
turbinate-campanulate to campanulate, 2–4 × 1.5–4 mm,
strigose; teeth 0.5–1 mm. Flowers 1.5–2 mm in anthesis,
2.5–3 mm in fruit; perianth reddish purple or maroon,
rarely yellow, becoming slightly less reddish in fruit,
strigose abaxially; tepals narrowly spatulate to obovate;
stamens 2–3 mm; filaments pilose proximally. Achenes
greenish brown to reddish brown, 3.5–5 mm, 3-winged
along distal ¹/₃, nearly beakless, strigose. *2n* = 40.

Flowering Jun–Oct. Calcareous slopes, mixed
grassland communities, oak, juniper, and montane
conifer woodlands; 1400–2200 m; Tex.; Mexico
(Coahuila).

Variety *hemipterum* is known in the flora area only
from the Chisos Mountains of southern Brewster County.
There the variety is infrequent to locally common.
Variety *griseum* I. M. Johnston of Mexico differs in
having leaf blades that are densely grayish-tomentose
abaxially and strigose adaxially. It occurs in
northwestern Coahuila and northeastern Chihuahua.

144. Eriogonum nealleyi J. M. Coulter, Contr. U.S.
Natl. Herb. 1: 48. 1890 · Irion County wild
buckwheat [C] [E]

Herbs, polycarpic, 5–12 dm,
glabrous; taproot not chambered.
Stems: caudex compact; aerial
flowering stems usually 1, not
fistulose, 2–4.5(–5) dm, glabrous.
Leaves basal and cauline; basal
petiolate, petiole 1–2.5 cm,
strigose, blade oblanceolate to
spatulate, 4–8 × 0.5–1.6 cm,
strigose and grayish on both surfaces; cauline blade
absent or sessile, oblanceolate, (0.5–)1–4 × (0.2–)0.3–
0.8 cm, similar to basal blade. Inflorescences 2–7 dm;
branches glabrous; bracts scalelike, 0.7–2.5 × 0.5–2 mm.
Peduncles erect, straight or slightly curving upward, 1–
8 cm, glabrous. Involucres turbinate-campanulate to

campanulate, 2–3 × 2–4 mm, glabrous; teeth 0.5–1 mm.
Flowers 1.5–2 mm at anthesis, 2.5–3 mm in fruit;
perianth white to greenish white at anthesis, reddish in
fruit, glabrous or sparsely strigose abaxially; tepals
narrowly elliptic to oblong; stamens 2–3 mm; filaments
pilose proximally. Achenes greenish brown to reddish
brown, 4–6 mm, 3-winged along distal ¹/₃, nearly
beakless, strigose. *2n* = 40.

Flowering Jun–Sep. Calcareous flats and slopes,
mixed grassland, saltbush, and creosote bush
communities, oak woodlands; of conservation concern;
500–800 m; Tex.

Historically, *Eriogonum nealleyi* occurred in Coke,
Howard, Irion, and Sterling counties of southwestern
Texas. Recent collections are from Irion and Sterling
counties.

145. Eriogonum hieracifolium Bentham in A. P. de
Candolle and A. L. P. P. de Candolle, Prodr. 14: 6.
1856 · Hawkweed wild buckwheat [F]

Eriogonum leucophyllum Wooton
& Standley subsp. *pannosum*
(Wooton & Standley) S. Stokes;
Pterogonum hieracifolium
(Bentham) H. Gross

Herbs, polycarpic, 4–7 dm,
strigose; taproot not chambered.
Stems: caudex compact; aerial
flowering stems usually 2–5, not
fistulose, 3.5–5.5 dm, strigose. Leaves basal and cauline;
basal petiolate, petiole 0.5–5 cm, strigose, blade
oblanceolate to spatulate, 3–15 × 0.5–2(–2.5) cm,
sparsely to densely strigose on both surfaces; cauline
sessile, blade oblanceolate, 0.5–5(–6) × 0.3–1 cm, similar
to basal blade. Inflorescences 0.4–1.5(–1.8) dm;
branches strigose; bracts scalelike, 2–8 × 1–3 mm.
Peduncles erect, straight, 0.5–3(–3.5) cm, strigose.
Involucres turbinate-campanulate to campanulate, 2.5–
4 × 2.5–5 mm, hirsute to strigose; teeth 0.5–1.5 mm.
Flowers 1.5–2.5 mm in anthesis, 3–5 mm in fruit;
perianth yellow in anthesis, reddish in fruit, pilose
abaxially; tepals narrowly ovate; stamens 2–3 mm;
filaments pilose proximally. Achenes yellowish green to
light brown, 4.5–6 mm, strigose, 3-winged along distal
¹/₂, nearly beakless. *2n* = 40.

Flowering Jun–Oct. Sandy to gravelly, often
calcareous flats and slopes, grassland, saltbush, creosote
bush, and mesquite communities, oak, juniper, and
montane conifer woodlands; (900–)1300–2600 m; Ariz.,
N.Mex., Tex.; Mexico (Chihuahua).

Eriogonum hieracifolium is widely scattered from
east-central Arizona (Apache, Gila, Graham, and Navajo
counties) and adjacent west-central and southern New
Mexico (Bernalillo, Catron, Dona Ana, Eddy, Grant,
Lincoln, Otero, and Sierra counties) into Texas (Brewster,

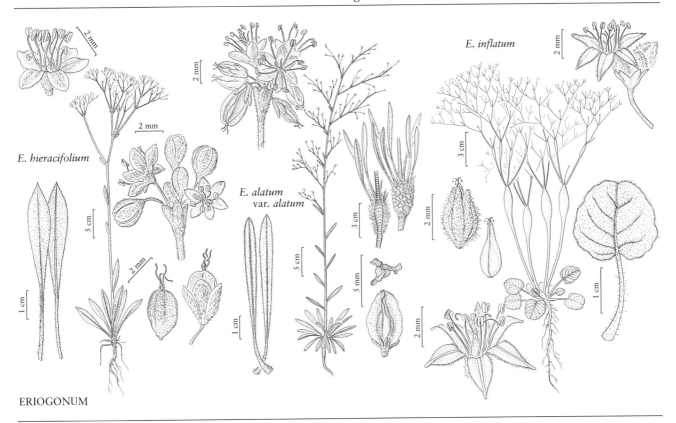

ERIOGONUM

Culberson, Jeff Davis, Pecos, and Presidio counties). The species occurs in northeastern Chihuahua, Mexico, as well. Similar plants found near La Linda in Coahuila appear to represent a different taxon, as their vegetative structures are tomentose rather than strigose.

146. Eriogonum alatum Torrey in L. Sitgreaves, Rep. Exped. Zuni Colorado Rivers, 168, plate 8. 1853 F

Pterogonum alatum (Torrey) H. Gross

Herbs, monocarpic, 5–20(–25) dm, strigose or glabrous; taproot often chambered. **Stems:** caudex absent; aerial flowering stems usually 1, not fistulose, 2–13 dm, strigose or glabrous. **Leaves** basal and sometimes cauline; basal petiolate, petiole 2–6 cm, stigose to woolly or glabrous, blade linear-lanceolate or lanceolate to oblanceolate to spatulate, (3–)5–20 × 0.3–2 cm, strigose, becoming glabrous and green on both surfaces except for margins and midvein; cauline sessile, blade linear-oblanceolate to lanceolate, 1–9 × 0.3–0.8(–1) cm, similar to basal blade. **Inflorescences** 2–10 dm; branches strigose or glabrous; bracts semileaflike proximally, linear to linear-lanceolate, 2–9 × 1–3 mm, scalelike distally, triangular, 0.8–5 × 0.5–2 mm. **Peduncles** erect, straight or curving upward, 0.5–3.5 cm, strigose or glabrous. **Involucres** turbinate to campanulate, 2–4(–4.5) × 2–4(–4.5) mm, strigose or glabrous; teeth 1–1.8 mm. **Flowers** 1.5–2.5 mm in anthesis, 3–6 mm in fruit; perianth yellow to yellowish green, rarely maroon in anthesis, often reddish or maroon in fruit, glabrous; tepals lanceolate; stamens 1.5–3 mm; filaments glabrous. **Achenes** yellowish green to reddish brown, 5–9 mm, glabrous, 3-winged entire length, beakless.

Varieties 2 (2 in the flora): w North America, including Mexico.

The mature reddish roots of *Eriogonum alatum* can be distinctively chambered. The Navajo (Diné) people consider the species to be a "life medicine" (L. C. Wyman and S. K. Harris 1951), using a mixture of shredded roots and water primarily to treat internal ailments. The species is used also as a ceremonial medicine (P. A. Vestal 1952). The Zuni use it as an emetic for stomachaches (S. Camazine and R. A. Bye 1980).

1. Flowering stems and inflorescence branches usually strigose; peduncles strigose or nearly so; involucres strigose, rarely glabrous; plants 5–13 (–17) dm; n Arizona, Colorado, w Kansas, sw Nebraska, New Mexico, w Texas, e Utah, se Wyoming . . . . 146a. *Eriogonum alatum* var. *alatum*
1. Flowering stems and inflorescence branches glabrous or nearly so; peduncles glabrous, occasionally slightly strigose; involucres glabrous; plants 10–20(–25) dm; ne New Mexico, w Oklahoma, n Texas . . . . 146b. *Eriogonum alatum* var. *glabriusculum*

146a. Eriogonum alatum Torrey var. **alatum**

• Winged wild buckwheat F

Eriogonum alatum subsp. *mogollense* (S. Stokes ex M. E. Jones) S. Stokes; *E. alatum* var. *mogollense* S. Stokes ex M. E. Jones; *E. alatum* subsp. *triste* (S. Watson) S. Stokes; *E. triste* S. Watson

Plants 5–13(–17) dm. **Aerial flowering stems** strigose. **Leaves** basal, infrequently cauline; basal: petiole 2–6 cm, strigose to woolly or glabrous, blade lanceolate to oblanceolate, (3–)5–15 × 0.3–2 cm, strigose or glabrous except for margins and midveins; cauline: blade 1–6 cm. **Inflorescences** 2–10 dm; branches thinly strigose, infrequently glabrous at maturity. **Peduncles** strigose or nearly so. **Involucres** strigose or rarely glabrous. **Perianths** yellow to yellowish green. **Achenes** 5–8 × 3–6 mm. $2n = 40$.

Flowering Jun–Oct. Sandy to gravelly flats and slopes, mixed grassland, saltbush, and sagebrush communities, oak, pinyon and/or juniper, and montane conifer woodlands; (500–)1300–3100(–4000) m; Ariz., Colo., Kans., Nebr., N.Mex., Tex., Utah, Wyo.; Mexico (Chihuahua).

Variety *alatum* is widespread and often common from southeastern Wyoming and adjacent southwestern Nebraska southward through eastern Utah, Colorado, and western Kansas to northern Arizona, New Mexico, and northern and western Texas to northern Chihuahua, Mexico. The variety rarely is cultivated, as an individual plant may go up to seven (or more) years before flowering, after which it dies.

Inclusion here of var. *mogollense*, characterized by spatulate leaves and woolly petiole bases, is now suggested, as both features fail to hold in some populations near Flagstaff, Arizona, the center of its distribution.

146b. Eriogonum alatum Torrey var. **glabriusculum** Torrey in War Department [U.S.], Pacif. Railr. Rep. 4(5): 131. 1857 • Canadian River wild buckwheat E

Plants 10–20(–25) dm. **Aerial flowering stems** glabrous or nearly so. **Leaves** basal and cauline; basal: petiole 2–5, strigose, blade linear-lanceolate to lanceolate, 8–20 × 0.5–1.5 cm, slightly strigose or glabrous adaxially, glabrous abaxially except for strigose margins and midveins; cauline: blade 1–9 cm. **Inflorescences** 2–6.5 dm; branches glabrous. **Peduncles** glabrous or occasionally slightly strigose. **Involucres** glabrous. **Perianths** yellow to yellowish green or maroon. **Achenes** 5.5–9 × 3–5.5 mm. $2n = 40$.

Flowering Jul–Oct. Sandy to gravelly flats and gentle slopes, mixed grassland, saltbush, and mesquite communities, oak woodlands; 300–1400 m; N.Mex., Okla., Tex.

Variety *glabriusculum* is a distinctive taxon of the southern Great Plains, often being the tallest plants on the low, rolling hills. It is geographically isolated from var. *alatum*, being found near Ruth in Curry County, New Mexico, in western Oklahoma, and in northern Texas.

1g. Eriogonum Michaux subg. **Ganysma** (S. Watson) Greene, Fl. Francisc., 151. 1891

Eriogonum sect. *Ganysma* S. Watson, Proc. Amer. Acad. Arts 12: 259. 1877

Herbs, spreading to erect or prostrate polycarpic perennials and annuals, glabrous, hispid, tomentose, floccose, hirsute, villous, viscid, or sometimes glandular; taproot usually not woody. **Stems** spreading to erect, usually without persistent leaf bases, tomentose, floccose, hirsute, villous, viscid, short-hispid, or glabrous, sometimes glandular; caudex absent or compact, spreading; aerial flowering stems prostrate to erect, slender to stout, solid or hollow, sometimes fistulose, not disarticulating in ringlike segments proximally, usually arising directly from root. **Leaves** persistent or marcescent, basal, sometimes sheathing, or basal and cauline, not fasciculate; blade variously lanate, tomentose, floccose, hirsute, villous, pilose, hispid, or glabrous, sometimes glandular. **Inflorescences** cymose, rarely cymose-paniculate, paniculate or racemose, open or diffuse; bracts 3(–8), semileaflike or more often scalelike. **Peduncles** absent or erect, ascending, spreading, horizontal, cernuous, recurved, or deflexed, filiform to capillary to slender. **Involucres** 1 per node, not appressed to the inflorescence branches, turbinate to campanulate or hemispheric; teeth 4–5(–8), erect to spreading, very rarely lobelike and reflexed. **Flowers** not attenuate at base, without stipelike base; perianth various shades of white, yellow, pink, or red, glabrous, hispid, pilose, hirsute, puberulent, villous, pustulose, or glandular abaxially, usually glabrous (rarely tomentose) adaxially; tepals connate proximally $^{1}/_{5}$–$^{1}/_{2}$ their length, monomorphic or dimorphic; stamens included or exserted; filaments glabrous or pilose proximally. **Achenes** various shades of brown, or black, not winged, lenticular to 3-gonous, glabrous [sparsely pilose]. **Seeds:** embryo curved.

Species 62 (52 in the flora): w North America, mainly in arid regions.

The species of subg. *Ganysma* not found in the flora area are: *Eriogonum angelense* Moran, *E. austrinum* (S. Stokes) Reveal, *E. galioides* I. M. Johnston, *E. intricatum* Bentham, *E. moranii* Reveal, *E. pilosum* S. Stokes, *E. preclarum* Reveal, *E. repens* (S. Stokes) Reveal, and *E. scalare* S. Watson. All occur only in Baja California, Mexico.

The existence of an undescribed species restricted to San Bernardino County, California, has been known for some two decades. It was briefly in cultivation. It is related to *Eriogonum austrinum* and *E. moranii* of east-central Baja California, Mexico. Until this species can be studied in the field and adequate material obtained, the California plants cannot be named formally.

Various species of subg. *Ganysma* are food plants for the small dotted-blue butterfly (*Philotiella speciosa*).

1. Plants perennials, sometimes flowering first year.
  2. Leaf blades pilose or hirsute on one or both surfaces.
    3. Perianths densely hirsute; leaf blades short-hirsute on both surfaces or at least adaxially; Arizona, s and ec California, nw and sw Colorado, s Nevada, nw New Mexico, s and e Utah . . . . 147. *Eriogonum inflatum* (in part)
    3. Perianths glabrous; leaf blades sparsely pilose on both surfaces; nc and nw Arizona . . . . 160. *Eriogonum arizonicum*
  2. Leaf blades densely tomentose on one or both surfaces.
    4. Peduncles absent or horizontal, 0.1–2 mm; leaves basal, blade densely tomentose abaxially, tomentose to floccose or glabrous adaxially; se California . . . . *Eriogonum* cf. *moranii* (see discussion under subg. *Ganysma*, page 381)
    4. Peduncles erect to spreading, 6–60 mm; leaves basal or basal and sheathing up stems, blade densely tomentose on both surfaces; se Colorado, New Mexico, w Oklahoma, w and c Texas . . . . 176. *Eriogonum tenellum*
1. Plants annuals, rarely perennials.
  5. Leaves basal and cauline, not basal only or basal and sheathing up stem.
    6. Involucres glandular-puberulent or puberulent.
      7. Tepals monomorphic, oblong to elliptic, not inflated.
        8. Involucres not densely tomentose abaxially, 1.8–2 mm; flowers 2–2.5 mm; s California . . . . 197. *Eriogonum gracillimum*
        8. Involucres densely tomentose abaxially, 2.7–3 mm; flowers 1.5–1.7 mm; sc California . . . . 198. *Eriogonum gossypinum*
      7. Tepals dimorphic, outer ones elliptic, obovate, spatulate, or roundish, sometimes inflated apically or basally, inner ones narrowly lanceolate to spatulate.
        9. Outer tepals elliptic to obovate, not obviously inflated or, if so, only proximally; inner tepals narrowly spatulate; stamens exserted; s California . . . . 194. *Eriogonum angulosum*
        9. Outer tepals elliptic to roundish or obovate to spatulate, obviously inflated proximally or distally; inner tepals narrowly spatulate or lanceolate; stamens included; widespread w United States.
          10. Outer tepals inflated from base to middle; peduncles and involucres glandular-puberulent; Arizona, ec and s California, sw Idaho, Nevada, nw New Mexico, e Oregon, w Utah, se Washington . . . . . 195. *Eriogonum maculatum*
          10. Outer tepals inflated above middle; peduncles and involucres glandular; s California . . . . 196. *Eriogonum viridescens*

6. Involucres glabrous, hispid, or villous, not puberulent or glandular-puberulent.
   11. Perianths hirsutulous with hooked hairs; California (*E. inerme* introduced and disjunct in Boise County, Idaho).
      12. Involucres 2-flowered; achene beaks exserted; flowers 0.8–1.1 mm ... . 192. *Eriogonum hirtiflorum*
      12. Involucres 4–6-flowered; achene beaks included; flowers 1.2–1.8 mm . . . . 193. *Eriogonum inerme*
   11. Perianths glabrous or, if hairy, without hooked hairs; widespread.
      13. Basal leaves essentially glabrous on both surfaces; e Montana, sc North Dakota and w South Dakota or ne New Mexico, or if glabrous or thinly floccose then perianths densely short-villous and plants of California.
         14. Peduncles capillary, 0.5–2 cm; perianths densely short-villous; sw California . . . . 156. *Eriogonum ordii* (in part)
         14. Peduncles slender, absent or confined to first node, 0.3–1.5 cm; perianths sparsely hispid or glabrous; e Montana to n New Mexico.
            15. Perianths sparsely hispid, (1.2–)1.5–2.5 mm; involucres turbinate, 1–1.5(–2) × 0.8–1.5 mm; e Montana, sc North Dakota, w South Dakota . . . . 157. *Eriogonum visheri*
            15. Perianths glabrous, 1.2–2 mm; involucres turbinate-campanulate, 1–1.3 × 1–1.2 mm; ne (Colfax County) New Mexico . . . . 158. *Eriogonum aliquantum*
      13. Basal leaves villous to hoary-tomentose or hispid on both surfaces; Arizona, Nevada, New Mexico (not Colfax County), Utah, Texas, Idaho, Oregon, and California.
         16. Basal leaf blades oblong to obovate, villous to hoary-tomentose on both surfaces; tepals dimorphic, those of outer whorl orbiculate-cordate, not basally 2-saccate; Arizona, New Mexico, w Texas . . . . 189. *Eriogonum abertianum*
         16. Basal leaf blades linear-lanceolate to linear-oblanceolate and densely lanate abaxially, or linear and hispid on both surfaces; tepals dimorphic, those of outer whorl oblong-ovate and 2-saccate proximally, or monomorphic; widespread.
            17. Tepals dimorphic, those of outer whorl 2-saccate proximally; involucres 5-toothed; n and e Arizona, se Nevada, New Mexico, sw Utah . . . . 190. *Eriogonum pharnaceoides*
            17. Tepals monomorphic, not 2-saccate proximally; involucres 4-toothed; e California, c Idaho, w Nevada, s Oregon . . . . 191. *Eriogonum spergulinum*
5. Leaves basal and sometimes sheathing up stem, not basal and cauline.
   18. Leaf blades variously glabrous or floccose, villous, pilose, strigose, hirsute, or hispid on one or both surfaces.
      19. Perianths glabrous.
         20. Peduncles 0.1–0.5 at proximal nodes; involucres sessile distally; nw Nevada . . . . 165. *Eriogonum lemmonii*
         20. Peduncles (0.05–)0.2–3(–4) cm at least on proximal nodes; involucres usually pedunculate distally or, if sessile distally, not of nw Nevada; widespread.
            21. Involucres narrowly turbinate to turbinate.
               22. Peduncles 0.2–1.5 cm, slender to filiform, spreading to reflexed; leaf blades obovate to round-obovate; involucres 0.8–1.8 × 0.5–1.2 mm; perianths white to pinkish; flowering stems not fistulose; ec California, c and ne Nevada, c Utah . . . . 162. *Eriogonum esmeraldense*
               22. Peduncles (0.05–)0.1–0.25 cm, slender, deflexed; leaf blades subcordate; involucres 1.2–1.6 × 1–1.4 mm; perianths greenish white to reddish; flowering stems usually slightly fistulose; sc Nevada . . . . 163. *Eriogonum concinnum*
            21. Involucres campanulate.
               23. Perianths pale yellow to yellow or yellowish red; involucres 2.5–3 × 2.5–3 mm; nw Nevada . . . . 164. *Eriogonum rubricaule*

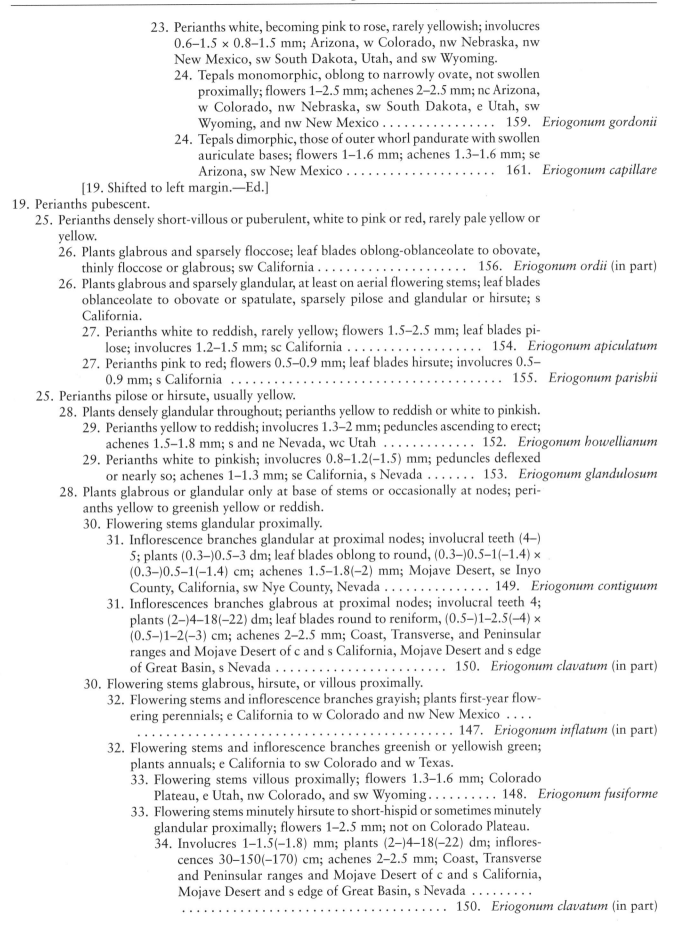

23. Perianths white, becoming pink to rose, rarely yellowish; involucres 0.6–1.5 × 0.8–1.5 mm; Arizona, w Colorado, nw Nebraska, nw New Mexico, sw South Dakota, Utah, and sw Wyoming.

    24. Tepals monomorphic, oblong to narrowly ovate, not swollen proximally; flowers 1–2.5 mm; achenes 2–2.5 mm; nc Arizona, w Colorado, nw Nebraska, sw South Dakota, e Utah, sw Wyoming, and nw New Mexico . . . . 159. *Eriogonum gordonii*

    24. Tepals dimorphic, those of outer whorl pandurate with swollen auriculate bases; flowers 1–1.6 mm; achenes 1.3–1.6 mm; se Arizona, sw New Mexico . . . . 161. *Eriogonum capillare*

[19. Shifted to left margin.—Ed.]

19. Perianths pubescent.

  25. Perianths densely short-villous or puberulent, white to pink or red, rarely pale yellow or yellow.

    26. Plants glabrous and sparsely floccose; leaf blades oblong-oblanceolate to obovate, thinly floccose or glabrous; sw California . . . . 156. *Eriogonum ordii* (in part)

    26. Plants glabrous and sparsely glandular, at least on aerial flowering stems; leaf blades oblanceolate to obovate or spatulate, sparsely pilose and glandular or hirsute; s California.

      27. Perianths white to reddish, rarely yellow; flowers 1.5–2.5 mm; leaf blades pilose; involucres 1.2–1.5 mm; sc California . . . . 154. *Eriogonum apiculatum*

      27. Perianths pink to red; flowers 0.5–0.9 mm; leaf blades hirsute; involucres 0.5–0.9 mm; s California . . . . 155. *Eriogonum parishii*

  25. Perianths pilose or hirsute, usually yellow.

    28. Plants densely glandular throughout; perianths yellow to reddish or white to pinkish.

      29. Perianths yellow to reddish; involucres 1.3–2 mm; peduncles ascending to erect; achenes 1.5–1.8 mm; s and ne Nevada, wc Utah . . . . 152. *Eriogonum howellianum*

      29. Perianths white to pinkish; involucres 0.8–1.2(–1.5) mm; peduncles deflexed or nearly so; achenes 1–1.3 mm; se California, s Nevada . . . . 153. *Eriogonum glandulosum*

    28. Plants glabrous or glandular only at base of stems or occasionally at nodes; perianths yellow to greenish yellow or reddish.

      30. Flowering stems glandular proximally.

        31. Inflorescence branches glandular at proximal nodes; involucral teeth (4–)5; plants (0.3–)0.5–3 dm; leaf blades oblong to round, (0.3–)0.5–1(–1.4) × (0.3–)0.5–1(–1.4) cm; achenes 1.5–1.8(–2) mm; Mojave Desert, se Inyo County, California, sw Nye County, Nevada . . . . 149. *Eriogonum contiguum*

        31. Inflorescences branches glabrous at proximal nodes; involucral teeth 4; plants (2–)4–18(–22) dm; leaf blades round to reniform, (0.5–)1–2.5(–4) × (0.5–)1–2(–3) cm; achenes 2–2.5 mm; Coast, Transverse, and Peninsular ranges and Mojave Desert of c and s California, Mojave Desert and s edge of Great Basin, s Nevada . . . . 150. *Eriogonum clavatum* (in part)

      30. Flowering stems glabrous, hirsute, or villous proximally.

        32. Flowering stems and inflorescence branches grayish; plants first-year flowering perennials; e California to w Colorado and nw New Mexico . . . . 147. *Eriogonum inflatum* (in part)

        32. Flowering stems and inflorescence branches greenish or yellowish green; plants annuals; e California to sw Colorado and w Texas.

          33. Flowering stems villous proximally; flowers 1.3–1.6 mm; Colorado Plateau, e Utah, nw Colorado, and sw Wyoming. . . . . 148. *Eriogonum fusiforme*

          33. Flowering stems minutely hirsute to short-hispid or sometimes minutely glandular proximally; flowers 1–2.5 mm; not on Colorado Plateau.

            34. Involucres 1–1.5(–1.8) mm; plants (2–)4–18(–22) dm; inflorescences 30–150(–170) cm; achenes 2–2.5 mm; Coast, Transverse and Peninsular ranges and Mojave Desert of c and s California, Mojave Desert and s edge of Great Basin, s Nevada . . . . 150. *Eriogonum clavatum* (in part)

34. Involucres 0.7–1 mm; plants 1–4.5(–6) dm; inflorescences 5–30
cm; achenes 1–1.5 mm; s Mojave, Sonoran, and nw Chihuahuan
deserts, Arizona, s California, s Nevada, s New Mexico, sw Utah
. . . . 151. *Eriogonum trichopes*

[18. Shifted to left margin.—Ed.]
18. Leaf blades densely tomentose to floccose-tomentose or floccose on one or both surfaces.
35. Outer tepals usually oblong, obovate, or orbiculate, cordate proximally.
36. Peduncles absent or erect, 0.1–5 mm.
37. Flowering stems 0.05–0.3 dm; plants 1–4 dm; inflorescences spreading and
flat-topped; se Inyo County, California, nw Clark and sw Nye counties,
Nevada . . . . 170. *Eriogonum bifurcatum*
37. Flowering stems (0.5–)1–2 dm; plants (0.5–)3–6(–10) dm ; inflorescences narrowly erect and strict; nw Arizona, s Nevada, sw Utah, and s California
. . . . 171. *Eriogonum exaltatum*
36. Peduncles absent or deflexed, 0.1–15 mm.
38. Flowering stems and inflorescence branches glandular; s and w edge of Great
Basin, Mojave and nw Sonoran deserts, nw Arizona, se and ec California, wc
and s Nevada, sw Utah. . . . . 169. *Eriogonum brachypodum*
38. Flowering stems and inflorescence branches glabrous; Great Basin, Mojave and
Sonoran deserts, and Colorado Plateau; Arizona, e and s California, w Colorado, se Idaho, Nevada, w New Mexico, se Oregon, sw Wyoming, and Utah.
39. Involucres broadly campanulate to hemispheric; peduncles absent; perianths yellow to reddish yellow; Intermountain Region and Colorado Plateau, n Arizona, ec California, w Colorado, se Idaho, Nevada, nw New
Mexico, se Oregon, sw Wyoming, and Utah . . . . 168. *Eriogonum hookeri*
39. Involucres narrowly turbinate or narrowly campanulate; peduncles absent
or 0.1–15 mm; tepals as long or longer than wide; perianths white to pink;
Arizona, s California, s Nevada, sw New Mexico, wc and s Utah.
40. Involucres 1.5–2.5 mm; inflorescences variously branched, not of many
horizontal tiers; Arizona, se California, s Nevada, sw New Mexico, wc
and s Utah. . . . . 166. *Eriogonum deflexum* (in part)
40. Involucres 1–1.5 mm; inflorescences of many horizontal tiers of
branches; e Inyo County, California, and sw Nye County, Nevada
. . . . 167. *Eriogonum rixfordii*
35. Outer tepals oblong to oblanceolate or ovate, truncate to obtuse proximally.
41. Perianths pustulose at least proximally.
42. Flowering stems and inflorescence branches scabrellous; e Utah, nw and sw
Colorado, nw New Mexico . . . . 174. *Eriogonum scabrellum*
42. Flowering stems and inflorescence branches glabrous, except for proximal floccose stems in some; ne California, sw Idaho, nw Nevada se Oregon.
43. Peduncles curving, ascending, 1–5 cm; achenes 2–2.5 mm; ne California
nw Nevada, se Oregon. . . . . 187. *Eriogonum collinum*
43. Peduncles straight, erect, 0.1–0.5 cm; achenes 1.6–2 mm; sw Idaho, nc
Nevada, se Oregon . . . . 188. *Eriogonum salicornioides*
41. Perianths not pustulose.
44. Tepals monomorphic.
45. Involucres 1–2 mm, 5-toothed; Death Valley region, California.
46. Flowering stems and branches glabrous; peduncles erect; flowers
1.5–1.8 mm. . . . . 172. *Eriogonum hoffmannii*
46. Flowering stems and branches glandular; peduncles deflexed; flowers
2–2.5 mm . . . . 175. *Eriogonum eremicola*
45. Involucres 0.3–1.2 mm, 4–5-toothed; n Arizona, sw Colorado, e Nevada,
nw New Mexico, s Utah.
47. Flowering stems and branches minutely viscid; perianths pale yellow
to yellow, glabrous; flowers 1.3–2 mm; se Nevada, nw Arizona . . .
. . . . 186. *Eriogonum viscidulum*

47. Flowering stems and branches glabrous or floccose to glabrescent at base of stems, not viscid; perianths white to rose, sometimes yellowish to red, glabrous or puberulent to sparsely hirsute; flowers 0.5–1.5(–2) mm; n Arizona, sw Colorado, nw New Mexico, s Utah.

 48. Perianths yellowish to red, becoming pink to rose; flowers 0.5–1.5 mm; involucral teeth 4; inflorescences diffuse; ne Arizona, sw Colorado, nw New Mexico, se Utah . . . . 184. *Eriogonum wetherillii*

 48. Perianths white to rose, rarely yellowish; flowers 0.6–1.6(–2) mm; involucral teeth 5; inflorescences open to somewhat diffuse; n Arizona, nw New Mexico, s Utah . . . . 185. *Eriogonum subreniforme*

[44. Shifted to left margin.—Ed.]

44. Tepals dimorphic or, if monomorphic, then pandurate to flatbellate or ovate, rarely oblong and perianth glandular-pubescent.

49. Perianths glandular.

 50. Perianths white to red, glandular-puberulent, with a tuft of long white hairs adaxially; outer tepals broadly pandurate or flabellate; Arizona, s California, sw New Mexico . . . . 180. *Eriogonum thurberi*

 50. Perianths yellow or, if becoming whitish, outer tepals saccate-dilated proximally, glandular or short-hispidulous, without a tuft of long white hairs adaxially; Arizona, s California, Nevada, se Oregon, w Utah, and Mexico.

  51. Outer tepals cordate, becoming saccate-dilated proximally; perianths yellow in early anthesis, becoming white to rose; involucres 0.6–1.2 mm, glabrous; Arizona, se California, s Nevada, sw Utah . . . . 181. *Eriogonum thomasii*

  51. Outer tepals oblong-elliptic to obovate or ovate, not saccate-dilated proximally; perianths yellow; involucres 1–2 mm, glabrous or glandular; w Arizona, s and e California, w and s Nevada, se Oregon, w Utah.

   52. Involucres glandular-puberulent; nw Arizona, se California, wc and s Nevada, se Oregon, and w Utah . . . . 182. *Eriogonum pusillum*

   52. Involucres glabrous; w Arizona, s California and s Nevada s to Mexico . . . . 183. *Eriogonum reniforme*

49. Perianths glabrous.

 53. Involucres narrowly turbinate; sw Idaho, c Nevada, se Oregon . . . . 173. *Eriogonum watsonii*

 53. Involucres turbinate to campanulate; widespread w United States and Canada.

  54. Peduncles deflexed; outer tepals oblong; ec California, Nevada . . . . 166. *Eriogonum deflexum* (in part)

  54. Peduncles cernuous, ascending, or erect; widespread w United States and Canada.

   55. Outer tepals oblong to oval; peduncles cernuous, usually curved, glandular or infrequently glabrous; nw Arizona, ec California, se Oregon, w and ec Utah, and Nevada . . . . 179. *Eriogonum nutans*

   55. Outer tepals pandurate or flabellate; peduncles cernuous, ascending or erect, infrequently absent, glabrous; widespread w United States and Canada.

    56. Outer tepals pandurate; peduncles absent or cernuous to ascending, (1–)2–25 mm; involucres 1–1.5 mm wide; widespread, sw Canada to w Nebraska, ec California, s Nevada, n Arizona, and n New Mexico . . . . 177. *Eriogonum cernuum*

    56. Outer tepals flabellate; peduncles erect, 3–15 mm; involucres 1.5–2.5 mm wide; se Arizona, New Mexico, w Texas . . . . 178. *Eriogonum rotundifolium*

147. Eriogonum inflatum Torrey & Frémont in J. C. Frémont, Rep. Exped. Rocky Mts., 317. 1845 • Desert trumpet, Indian pipeweed, bottle stopper F

Eriogonum glaucum Small; *E. inflatum* var. *deflatum* I. M. Johnston; *E. trichopes* Torrey subsp. *glaucum* (Small) S. Stokes

Herbs, erect, perennial, occasionally flowering first year, 1–10(–15) dm, glabrous, usually glaucous, grayish. **Stems:** caudex compact; aerial flowering stems erect, solid or hollow and fistulose, (0.2–)2–5 dm, glabrous, usually glaucous, occasionally hirsute proximally. **Leaves** basal; petiole 2–6 cm, hirsute; blade oblong-ovate to oblong or rounded to reniform, (0.5–)1–2.5(–3) × (0.5–)1–2 (–2.5) cm, short-hirsute and grayish or greenish on both surfaces, sometimes less so or glabrous and green adaxially, margins occasionally undulate. **Inflorescences** cymose, open, spreading to erect, 5–70 × 5–50 cm; branches occasionally fistulose, glabrous, usually glaucous; bracts 3, scalelike, 1–2.5(–5) × 1–2.5 mm. **Peduncles** erect, straight, filiform to capillary, 0.5–2 (–3.5) cm, glabrous. **Involucres** turbinate, 1–1.5 × 1–1.8 mm, glabrous; teeth 5, erect, 0.4–0.6 mm. **Flowers** (1–)2–3(–4) mm; perianth yellow with greenish or reddish midribs, densely hirsute with coarse curved hairs; tepals monomorphic, narrowly ovoid to ovate; stamens exserted, 1.3–2.5 mm; filaments glabrous or sparsely pubescent proximally. **Achenes** light brown to brown, lenticular to 3-gonous, 2–2.5 mm, glabrous. $2n = 32$.

Flowering year-round. Sandy to gravelly washes, flats, and slopes, mixed grassland, saltbush, creosote bush, mesquite, and sagebrush communities, pinyon and/or juniper woodlands; -30–1800(–2000) m; Ariz., Calif., Colo., Nev., N.Mex., Utah; Mexico (Baja California, Sonora).

The cause of the fistulose stem and inflorescence branches in *Eriogonum inflatum* was imaginatively attributed by A. M. Stone and C. T. Mason (1979) to the larvae of gall insects. This fallacy continues to appear in the literature. Greenhouse studies have shown that stems of this and some other species of the genus inflate without the presence of any insects. Other researchers have shown that the inflation involves a build-up of CO_2 within the stems, which take over as the primary photosynthetic body as leaves wilt or eventually dry up and fall away from the plant (C. D. Osmond et al. 1987). Not all individuals of *E. inflatum* will have fistulose stems and branches, as this feature is partly a function of available moisture: the drier the conditions, the less pronounced the inflation. Stems produced in the summer tend to be inflated less frequently than those produced in the spring.

The "annual" phase of *Eriogonum inflatum* is distinct from its truly annual relatives. Its flowering stems and inflorescence branches are distinctly grayish, whereas those of the true annuals are green or yellowish green.

As circumscribed here, *Eriogonum inflatum* occurs in Arizona, southern and east-central California, western Colorado, northwestern New Mexico, central and southern Nevada, and southern and eastern Utah.

Some Native Americans occasionally ate newly emerged stems of *Eriogonum inflatum* (S. A. Weber and P. D. Seaman 1985; M. L. Zigmond 1981). The hollow stems were used as drinking tubes (Weber and Seaman) and pipes (E. W. Gifford 1936). This wild buckwheat is a food plant for the desert metalmark butterfly (*Apodemia mormo deserti*).

148. Eriogonum fusiforme Small, Bull. Torrey Bot. Club 33: 56. 1906 • Grand Valley desert trumpet E

Eriogonum inflatum Torrey & Frémont var. *fusiforme* (Small) Reveal

Herbs, spreading, annual, (0.3–) 0.5–4 dm, essentially glabrous, green or occasionally yellow-green. **Stems:** caudex absent; aerial flowering stems erect, usually hollow and fistulose, 0.5–1.5 dm, glabrous, villous proximally. **Leaves** basal; petiole 1–3 cm, hirsute; blade round, 0.5–3 × 0.5–2.5 cm, short-hirsute and greenish on both surfaces, margins plane. **Inflorescences** cymose, open, spreading, 5–30 × 5–30 cm; branches fistulose, glabrous; bracts 3, scalelike, 1–2 × 1–1.5 mm. **Peduncles** erect, straight, filiform to capillary, 1–2 cm, glabrous. **Involucres** turbinate, 1–1.2 × 0.7–1 mm, glabrous; teeth (4–)5, erect, 0.4–0.6 mm. **Flowers** 1.3–1.6 mm; perianth yellow with greenish to reddish midribs, densely hirsute with coarse curved hairs; tepals monomorphic, ovate; stamens exserted, 1.3–1.8 mm; filaments sparsely pubescent proximally. **Achenes** light brown, lenticular to 3-gonous, 1.3–1.8(–2) mm, glabrous.

Flowering Apr–Jul. Heavy clay, sometimes gravelly flats and slopes, saltbush and greasewood communities, pinyon and/or juniper woodlands; (900–)1100–2000 m; Colo., Utah, Wyo.

Eriogonum fusiforme is widespread and common in southwestern Wyoming (Sweetwater County), eastern Utah (Carbon, Duchesne, Emery, Garfield, Grand, Kane, San Juan, Uintah, and Wayne counties), and adjacent western Colorado (Delta, Garfield, Mesa, Montrose, and Rio Blanco counties). In a "good" year, millions of individuals carpet the heavy clay flats and slopes (typically Mancos Shale), especially in Utah and Colorado. The plants can be so abundant and closely arranged that it can be difficult to walk through the tangle of stems and branches. This species is absolutely distinct from *E. inflatum* and no intermediates have ever been observed.

149. Eriogonum contiguum (Reveal) Reveal, Phytologia 23: 175. 1972 • Annual desert trumpet C E

Eriogonum inflatum Torrey & Frémont var. *contiguum* Reveal, Aliso 7: 221. 1970

Herbs, spreading to erect, annual, (0.3–)0.5–3 dm, glabrous and sparsely glandular, green or reddish green. **Stems:** caudex absent; aerial flowering stems erect, solid, not fistulose, 0.1–0.5 dm, glabrous, glandular proximally. **Leaves** basal; petiole 0.4–1.5 cm, hirsute; blade oblong to round, (0.3–)0.5–1(–1.4) × (0.3–)0.5–1(–1.4) cm, hirsute and greenish to yellowish on both surfaces, margins plane. **Inflorescences** cymose, dense, spreading, (2–)5–18 × 2–15 cm; branches not fistulose, glabrous except glandular near proximal nodes; bracts 3, scalelike, 0.5–1.5 × 0.3–1 mm. **Peduncles** mostly erect, straight or slightly curved, capillary, 0.3–1.2(–2) cm, glabrous, glandular proximally. **Involucres** turbinate, 1–1.3 × 0.6–1 mm, glabrous; teeth (4–)5, erect, 0.2–0.4 mm. **Flowers** 1–2.5 mm; perianth golden yellow to reddish, densely short-hirsute with coarse curved hairs; tepals monomorphic, lanceolate; stamens exserted, 1–2.5 mm; filaments sparsely pubescent proximally. **Achenes** light brown, lenticular to 3-gonous, 1.5–1.8 (–2) mm, glabrous.

Flowering Apr–Jun. Sandy to gravelly flats and slopes, saltbush, creosote bush, and mesquite communities; of conservation concern; -20–900 m; Calif., Nev.

Eriogonum contiguum is confined to the greater Death Valley region in Inyo County, California, and southern Nye County, Nevada.

150. Eriogonum clavatum Small, Bull. Torrey Bot. Club 25: 50. 1898 • Hoover's desert trumpet

Eriogonum trichopes Torrey var. *hooveri* Reveal

Herbs, erect, annual, (2–)4–18 (–22) dm, glabrous, bright green. **Stems:** caudex absent; aerial flowering stems erect, hollow and fistulose, 0.5–3.5(–5) dm, glabrous, minutely glandular proximally, infrequently also short-hispid. **Leaves** basal; petiole 1–6 cm, hirsute; blade round to reniform, (0.5–)1–2.5(–4) × (0.5–)1–2(–3) cm, short-hirsute on both surfaces and greenish, margins plane.

Inflorescences cymose, open, spreading to erect, 30–150(–170) × 10–80 cm; branches usually fistulose, glabrous; bracts 3, scalelike, 1–4 × 1–2 mm. **Peduncles** mostly spreading, straight, slightly bent distally, capillary, 1–4 cm, glabrous. **Involucres** turbinate, 1–1.5(–1.8) × 0.6–0.9(–1.2) mm, glabrous; teeth 4, erect, 0.3–0.5 mm. **Flowers** 1.5–2.5 mm; perianth yellow with greenish to reddish midribs, densely hirsute with coarse curved hairs; tepals monomorphic, narrowly ovate; stamens exserted, 0.9–1.5 mm; filaments glabrous or sparsely pubescent proximally. **Achenes** light brown to brown, 3-gonous, 2–2.5 mm, glabrous. $2n = 32$.

Flowering year-round. Clayey flats and slopes, mixed grassland, chaparral, saltbush, and sagebrush communities, oak and montane conifer woodlands; 400–1100 m; Calif., Nev.; Mexico (Baja California).

Eriogonum clavatum is the tall, slender member of the *E. inflatum* complex found on the Inner Coast, Peninsular, and Transverse ranges of southern California. It is rather common from Monterey and San Benito counties southward to Ventura, northwestern Los Angeles, and extreme southwestern Kern counties, and then disjunct in the Peninsular Ranges of Riverside and San Diego counties, where it just enters north-central Baja California, Mexico. It extends eastward onto desert ranges, where it rarely exceeds 6 dm in height, from southern Mono County and Inyo County south through western San Bernardino County to northwestern Riverside County. From Inyo County, the species occurs across southern Nye and northern Clark counties into western Lincoln County, Nevada.

151. Eriogonum trichopes Torrey in W. H. Emory, Not. Milit. Reconn., 150. 1848 • Little desert trumpet

Eriogonum trichopes subsp. *minus* (Bentham) S. Stokes

Herbs, spreading to somewhat erect, annual, 1–4.5(–6) dm, glabrous and often glaucous, yellow-green. **Stems:** caudex absent; aerial flowering stems erect, occasionally hollow and fistulose, 0.5–2(–3) dm, glabrous, minutely hirsute or short-hispid proximally. **Leaves** basal; petiole 1–6 cm, hirsute; blade broadly oblong, (0.5–)1–2.5(–4) × (0.5–)1–2(–3) cm, short-hirsute on both surfaces and greenish, margins wavy. **Inflorescences** cymose, open to dense, usually spreading, 5–30 × 5–50 cm; branches usually not fistulose, glabrous; bracts 3, scalelike, 1–4 × 1–2 mm. **Peduncles** mostly erect, straight, capillary, 0.5–1.5 cm, glabrous. **Involucres** turbinate, 0.7–1 × 0.6–0.9 mm, glabrous; teeth 4(–5), erect, 0.3–0.4 mm. **Flowers** 1–2 mm; perianth yellow to greenish yellow with greenish to reddish midribs, densely hirsute with coarse curved hairs; tepals monomorphic, narrowly

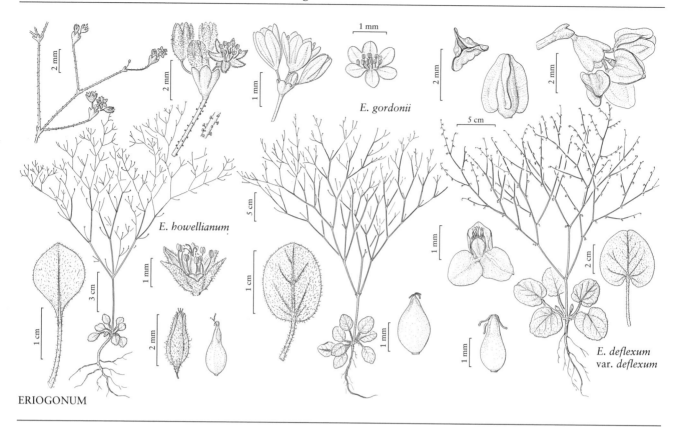

E. gordonii

E. howellianum

E. deflexum
var. deflexum

ERIOGONUM

ovate; stamens exserted, 0.9–1.5 mm; filaments sparsely pubescent proximally. **Achenes** light brown to brown, lenticular to 3-gonous, 1–1.5 mm, glabrous.

Flowering year-round. Clayey, sandy to gravelly flats, washes, and slopes, mixed grassland, saltbush, creosote bush, blackbrush, and mesquite communities, pinyon and/or juniper woodlands; -60–1500(–1900) m; Ariz., Calif., Nev., N.Mex., Utah; Mexico (Baja California, Chihuahua, Sonora).

Eriogonum trichopes is common on the Sonoran Desert and the southern portion of the Mojave Desert from southern California (Imperial, Inyo, Riverside, San Bernardino, and San Diego counties), eastward across southern Nevada (Clark, Lincoln, and Nye counties), southwestern Utah (Washington County), and Arizona (Coconino, Cochise, Gila, Graham, La Paz, Maricopa, Mohave, Pima, Pinal, Santa Cruz, Yavapai, and Yuma counties) to southern New Mexico (Cibola, Dona Ana, Grant, Hidalgo, Luna, Otero, and Valencia counties). It is also found to the south in northern Mexico (Baja California Norte, Sonora, and Chihuahua). Two related species occur in Baja California, the perennial *E. scalare* and the annual *E. intricatum*. Unlike *E. trichopes*, which occasionally has inflated stems, neither of those Mexican species exhibits that trait.

152. Eriogonum howellianum Reveal, Phytologia 25: 204. 1973 • Howell's wild buckwheat [E] [F]

Herbs, erect to spreading, annual, 0.5–3 dm, glandular, greenish or reddish green. **Stems:** caudex absent; aerial flowering stems erect, solid, not fistulose, 0.3–1 dm, glandular. **Leaves** basal; petiole 0.5–4 cm, glabrous or sparsely pilose; blade broadly elliptic to oval, 0.7–2.5 × 0.7–2.5 cm, pilose-hirsutulous, rarely slightly glandular and green on both surfaces, margins entire. **Inflorescences** cymose, open, spreading, 5–25 × 5–35 cm; branches not fistulose, glandular; bracts 3, scalelike, 1–2 × 0.5–1.5(–2) mm. **Peduncles** absent or ascending to erect, straight or curved, slender, 0.2–0.5 cm at proximal nodes, 0.01–0.1 cm distally, proximal ¹/₂ sparsely glandular. **Involucres** turbinate-campanulate, 1.3–2 × 1–2 mm, glabrous; teeth (4–)5, erect, 0.5–0.8 mm. **Flowers** 1–1.5(–2) mm; perianth yellow with reddish midribs to entirely reddish, densely pilose; tepals monomorphic, lanceolate; stamens exserted, 1–1.5 mm; filaments glabrous. **Achenes** dull brown, 3-gonous, 1.5–1.8 mm, glabrous.

Flowering Jun–Sep. Sandy to gravelly, often volcanic slopes, saltbush, greasewood, and sagebrush communities, pinyon-juniper woodlands; (700–)1200–2100 m; Nev., Utah.

Eriogonum howellianum is encountered infrequently in widely scattered locations in southern Nevada (Clark, Lincoln, Nye, and White Pine counties) and west-central Utah (Juab, Millard, and Tooele counties). A disjunct population occurs on Pilot Peak in extreme eastern Elko County, Nevada. This species is rarely common. The name *E. glandulosum* was long misapplied to these plants.

153. Eriogonum glandulosum (Nuttall) Nuttall ex Bentham in A. P. de Candolle and A. L. P. P. de Candolle, Prodr. 14: 21. 1856 • Glandular wild buckwheat C E

Oxytheca glandulosa Nuttall, Proc. Acad. Nat. Sci. Philadelphia 4: 19. 1848; *Eriogonum carneum* (J. T. Howell) Reveal; *E. glandulosum* var. *carneum* J. T. Howell; *E. trichopes* Torrey subsp. *glandulosum* (Nuttall) S. Stokes

Herbs, spreading, annual, (0.5–)1–2.5 dm, glandular, greenish or reddish green. **Stems:** caudex absent; aerial flowering stems erect, solid, not fistulose, 0.3–0.7 dm, glandular. **Leaves** basal, often sheathing up stem 1–2 cm; petiole 0.3–2 cm, glandular; blade broadly elliptic to oval, 0.5–1.5 × 0.5–1.5 cm, pilose-hirsutulous, slightly glandular and greenish on both surfaces, margins entire. **Inflorescences** cymose, diffuse, spreading and usually flat-topped, 5–20 × 5–30 cm; branches not fistulose, glandular; bracts 3, scalelike, 0.8–1.2 × 0.5–1 mm. **Peduncles** deflexed or nearly so, straight, slender, 0.2–0.5 cm at proximal nodes, 0.01–0.1 cm distally, sparsely glandular nearly entire length. **Involucres** narrowly turbinate, 0.8–1.2(–1.5) × 0.6–1(–1.3) mm, glabrous; teeth 5, erect, 0.3–0.5 mm. **Flowers** 1–1.8 mm; perianth white to pinkish with dark reddish brown to red midribs, densely pilose; tepals monomorphic, narrowly lanceolate; stamens exserted, 1–1.5 mm; filaments glabrous. **Achenes** black, 3-gonous, 1–1.3 mm, glabrous.

Flowering May–Nov. Sandy to gravelly, often calcareous slopes, saltbush and creosote bush communities; of conservation concern; 900–1600 m; Calif., Nev.

Eriogonum glandulosum is a localized species that varies from rare to common in southeastern Inyo and extreme northeastern San Bernardino counties, California, and in southern Nye, northern Clark, and southwestern Lincoln counties, Nevada. The name *E.*

glandulosum may not be the earliest for the taxon. The type of *E. cordatum* Torrey & Frémont (1845, as "cordalum") was collected by Frémont most likely near the California-Nevada border, where *E. trichopes*, *E. contiguum*, and *E. glandulosum* occur. Because the description is not decisive, and because J. Torrey and A. Gray (1870) reported the type as missing, it is impossible to place the name unequivocally. Nonetheless, *E. cordatum* was reported to be glandular ("glaneous") with the leaves "pubescent above, hairy underneath." The glandular condition probably eliminates *E. trichopes*, while those leaf features probably exclude *E. contiguum*.

154. Eriogonum apiculatum S. Watson, Proc. Amer. Acad. Arts 17: 378. 1882 E

Herbs, erect to spreading, annual, 2–9 dm, glabrous and sparsely glandular especially (or at least) near nodes, grayish to greenish. **Stems:** caudex absent; aerial flowering stems erect, solid, not fistulose, 0.5–1.5 dm, glabrous, glandular distally or sparsely so throughout. **Leaves** basal; petiole 1–4 cm, pilose, slightly winged; blade oblanceolate to obovate, (0.5–)1–4 × 0.5–1.5 cm, sparsely pilose, glandular and greenish on both surfaces, margins entire. **Inflorescences** cymose, open, 30–80 × 10–50 cm; branches not fistulose, glabrous, sparsely glandular at nodes; bracts 3, scalelike, 1–2 × 1–2 mm. **Peduncles** deflexed, straight, filiform, (0.1–)0.2–0.35 cm, sparsely glandular. **Involucres** turbinate, 1.2–1.5 × 1–1.3 mm, glabrous; teeth 4, erect or nearly so, 0.3–0.7 mm. **Flowers** 1.5–2.5 mm; perianth white with reddish brown midribs, becoming reddish in fruit, rarely yellow, puberulent; tepals monomorphic, oblong-obovate, sometimes with apiculate tip 1–2 mm; stamens exserted, 1.5–2 mm; filaments glabrous. **Achenes** light brown to brown, lenticular, 1.3–1.7(–2.2) mm, glabrous. *2n* = 40.

Flowering May–Nov. Sandy granitic flats and slopes, chaparral communities, oak and conifer woodlands; (200–)700–2700 m; Calif.

Eriogonum apiculatum is restricted to the San Jacinto, Santa Rosa, Palomar, and Cuyamaca mountains of Riverside and San Diego counties. Two collections are discounted as to location: *Hall 1025* (MIN), purportedly from the San Bernardino Mountains, certainly was labeled erroneously by the Parish brothers, who redistributed the sheet; and *Jaeger s.n.* (19 May 1940, DS, IDS, UTC, WTU) from "Falcon Flat" in the Little San Bernardino Mountains appears to be another of the small but significant list of Jaeger specimens with highly dubious label data. No such place is known in Joshua Tree National Park. The phase of the species named var. *subvirgatum*, with nearly sessile involucres, occurs intermixed with the typical expression.

A yellow-flowered population was found at the eastern end of the San Bernardino Mountains, San Bernardino County. Recent attempts to relocate those plants have been unsuccessful, and they remain undescribed.

155. Eriogonum parishii S. Watson, Proc. Amer. Acad. Arts 17: 379. 1882 • Parish's wild buckwheat

Herbs, spreading, annual, 1–4(–5) dm, glabrous, greenish to reddish brown. **Stems:** caudex absent; aerial flowering stems erect, solid, not fistulose, 0.3–1 dm, glabrous, glandular distally. **Leaves** basal; petiole 0.5–2.5 cm, hirsute, slightly winged; blade spatulate, 2–6 × 0.5–2 cm, hirsute and greenish on both surfaces, margins plane, often ciliate. **Inflorescences** cymose, usually diffuse, 10–35 × 10–50 cm; branches not fistulose, glabrous, glandular at nodes; bracts 3, scalelike, 1–2 × 1–2 mm. **Peduncles** spreading, straight, capillary, 0.4–1.5(–2.5) cm, glabrous or sparsely glandular at least proximally. **Involucres** turbinate, 0.5–0.9 × 0.5–0.7 mm, glabrous; teeth 4, erect or nearly so, 0.3–0.5 mm. **Flowers** 0.5–0.9 mm; perianth pink to red with red midribs, puberulent, becoming white with pink to red midribs; tepals slightly dimorphic, those of outer whorl ovate and 0.5–0.7 mm wide, those of inner whorl oblong and 0.4–0.5 mm wide; stamens included, 0.5–0.6 mm; filaments glabrous. **Achenes** dark brown to blackish, 3-gonous, 1–1.3 mm, glabrous. $2n = 40$.

Flowering Jun–Oct. Granitic sandy flats and slopes, mixed grassland, chaparral, and sagebrush communities, oak and montane conifer woodlands; (1000–)1300–3200 m; Calif.; Mexico (Baja California).

Eriogonum parishii occurs from southern Mono County in the White Mountains southward in the Sierra Nevada of Inyo and Tulare counties to scattered desert ranges in San Bernardino (San Bernardino Mountains), Riverside (San Jacinto and Santa Rosa mountains), and San Diego (Laguna Mountains) counties. In Mexico, the species occurs as far south as the Sierra San Pedro Mártir in Baja California Norte. A disjunct population from Crown King, Yavapai County, Arizona (*Beaty s.n.*, 6 Sep 1951, CAS) requires confirmation. Possibly it is a recent introduction.

156. Eriogonum ordii S. Watson, Proc. Amer. Acad. Arts 21: 468. 1886 • Fort Mohave wild buckwheat [E]

Eriogonum tenuissimum Eastwood

Herbs, erect, annual, (0.5–)1–7 dm, glabrous and sparsely floccose, greenish. **Stems:** caudex absent; aerial flowering stems erect, solid, not fistulose, (0.3–) 0.7–3 dm, thinly floccose or glabrous, floccose proximally. **Leaves** basal, occasionally cauline; basal: petiole 2–6(–10) cm, floccose, blade oblong-oblanceolate to obovate, (1.5–)2–8 × (0.8–)1–3 cm, thinly floccose or glabrous and green on both surfaces, margins entire; cauline: petiole 0.5–3 cm, thinly floccose, blade elliptic to obovate, 0.7–3 × 0.2–2 cm, similar to basal blade. **Inflorescences** paniculate, open to diffuse, (5–)10–50 × 5–50 cm; branches not fistulose, glabrous except for floccose nodes and proximal branches; bracts 3, scalelike, 0.5–3 × 0.3–1 mm. **Peduncles** erect, straight, capillary, 0.5–2 cm, glabrous or thinly floccose. **Involucres** narrowly turbinate to turbinate, 1–1.5(–1.8) × 0.6–1.2 mm, glabrous; teeth 4, erect, 0.2–0.5 mm. **Flowers** 1–2.5(–3) mm; perianth white with greenish or reddish midribs to pale yellow with greenish midribs, becoming pink to reddish, densely short-villous; tepals monomorphic, oblong to narrowly ovate; stamens exserted, 1–1.5 mm; filaments glabrous. **Achenes** dark brown to black, 3-gonous, 1.8–2 mm, glabrous.

Flowering Mar–Jul. Gravelly to clayey flats and slopes, mixed grassland communities, oak and conifer woodlands; 200–1400 m; Calif.

Eriogonum ordii is infrequently encountered (rarely locally common) along the inner Coast Ranges from Monterey and San Benito counties south through Fresno, Merced, and San Luis Obispo counties to Ventura County, then eastward across northern Los Angeles County to the hills east of Bakersfield in Kern County. This distribution is based on confirmed modern collections. The type (*J. G. Lemmon 4189*, ASU, BM, DS, G, GH, ISC, K, P, UC, US) supposedly was collected near Fort Mohave, Mohave County, Arizona, in 1884. Two T. Brandegee specimens reportedly were gathered on the boundary of San Diego and Imperial counties, one at Split Mountain (Apr 1905, UC) and the second along San Felipe Creek (4 Apr 1901, UC). Brandegee's label data often are dubious, and these disjunct sites are discounted.

157. Eriogonum visheri A. Nelson, Bot. Gaz. 56: 64. 1913 • Visher's wild buckwheat E

Herbs, erect to spreading, annual, (1–)1.5–3.5(–4) dm, sparsely villous, grayish. **Stems:** caudex absent; aerial flowering stems erect, solid, not fistulose, 0.3–0.8(–1) dm, sparsely villous. **Leaves** basal and cauline; basal: petiole 1–3 cm, villous to pilose, blade elliptic to rotund, (0.8–)1–2.5 × (0.6–)1–2.5 cm, glabrous and green on both surfaces, margins entire, villous; cauline: petiole 0.5–1.5 cm, sparsely villous, blade elliptic, 0.5–1.5 × 0.5–1 cm, similar to basal blade. **Inflorescences** cymose, open, 5–35 × 5–35 cm; branches sparsely villous; bracts 3, scalelike, 1–2.5(–3) × 1–2.5 mm. **Peduncles** absent except in fork of proximal node, erect, straight, slender, 0.3–1(–1.5) cm, sparsely villous. **Involucres** turbinate, 1–1.5(–2) × 0.8–1.5 mm, glabrous; teeth 5, erect, 0.3–0.6(–0.8) mm. **Flowers** (1.2–)1.5–2.5 mm; perianth yellowish with darker yellow to greenish yellow or reddish brown midribs, sparsely hispid; tepals monomorphic, oblance-olate to oblong; stamens exserted, 1.2–1.7(–2) mm; filaments glabrous. **Achenes** shiny dark brown, 3-gonous, (2–)2.5–3 mm, glabrous.

Flowering Jun–Aug. Loamy to clayey flats and outcrops, mixed grassland and saltbush communities; 500–900 m; Mont., N.Dak., S.Dak.

Eriogonum visheri is rare to infrequent throughout its range. It appears to be concentrated in two areas, one primarily in the "badlands" of western South Dakota and a second from the Cheyenne River northward to just north of the Cannonball River area in south-central North Dakota. It occurs also in Carter County, Montana. In South Dakota, the species is protected on the Buffalo Gap National Grassland, and is considered "sensitive" by the U.S. Forest Service. It is in the Center for Plant Conservation's National Collection of Endangered Plants.

158. Eriogonum aliquantum Reveal, Phytologia 34: 460. 1976 • Cimarron wild buckwheat E

Herbs, erect to spreading, annual, 1.5–3.5 dm, sparsely villous, grayish. **Stems:** caudex absent; aerial flowering stems erect, solid, not fistulose, 0.3–0.7 dm, sparsely villous. **Leaves** basal and cauline; basal: petiole 1.5–2.5 cm, sparsely villous, blade elliptic to broadly elliptic, 1.5–2 × 1–1.5 cm, glabrous and green on both surfaces, margins entire, villous; cauline: petiole 0.5–1 cm, sparsely villous, blade elliptic, 0.5–1.5 × 0.2–1 cm, similar to basal blade.

Inflorescences cymose, open, 15–25 × 15–25 cm; branches not fistulose, sparsely villous; bracts 3, scalelike, 1–4 × 1–3 mm. **Peduncles** absent or rarely 1 at proximal node, erect, straight, slender, 0.3–0.7 cm, villous. **Involucres** turbinate-campanulate, 1–1.3 × 1–1.2 mm; teeth 5, erect, 0.4–0.6 mm. **Flowers** 1.2–2 mm, glabrous; perianth yellowish with darker yellowish brown to reddish brown midribs, glabrous; tepals monomorphic, oblanceolate; stamens exserted, 1.3–1.5 mm; filaments glabrous. **Achenes** shiny light brown to brown, 3-gonous, 1.7–2.3 mm, glabrous.

Flowering Jun–Aug. Clayey flats, saltbush and sagebrush communities, montane conifer woodlands; 1900–2100 m; N.Mex.

Eriogonum aliquantum is known only from the Cimarron, Vermejo, and Canadian river basins in Colfax County. Morphologically it is similar to *E. visheri* from the northern Great Plains, and a clear distinction is not always possible. Although considered a candidate for federal protection since the late 1970s, *E. aliquantum* remains essentially forgotten.

159. Eriogonum gordonii Bentham in A. P. de Candolle and A. L. P. P. de Candolle, Prodr. 14: 20. 1856 (as gordoni) • Gordon's wild buckwheat E F

Herbs, erect to spreading, annual, (0.5–)1–5(–7) dm, glabrous or sparsely hispid, grayish. **Stems:** caudex absent; aerial flowering stems erect, solid, not fistulose, 0.5–1.5(–3) dm, glabrous or sparsely hispid. **Leaves** basal; petiole 1–5 cm, glabrous or sparsely hirsute; blade obovate to round or reniform, 1–5 × 1–5 cm, sparsely villous to hirsute and green on both surfaces, becoming glabrous with age, margins plane. **Inflorescences** cymose, open to dense, 5–40(–50) × 5–50 cm; branches not fistulose, glabrous or sparsely hispid; bracts 3, scalelike, 0.5–2(–3) × 0.5–2.5 mm. **Peduncles** erect, straight, slender, 0.5–2.5(–4) cm, glabrous. **Involucres** campanulate, 0.6–1.5 × 0.8–1.5 mm, glabrous; teeth 5, erect, 0.2–0.4 mm. **Flowers** 1–2.5 mm; perianth white with greenish or reddish midribs, rarely yellowish, becoming pink to rose, glabrous; tepals monomorphic, oblong to narrowly ovate; stamens included to exserted, 1–1.8 mm; filaments glabrous. **Achenes** shiny light brown to brown, 3-gonous, 2–2.5 mm, glabrous.

Flowering May–Oct. Sandy to clayey flats and slopes, saltbush, greasewood, and sagebrush communities, pinyon and/or juniper or montane conifer woodlands; 900–2200 m; Ariz., Colo., Nebr., N.Mex., S.Dak., Utah, Wyo.

Eriogonum gordonii is common to even locally abundant throughout a range that extends in an arc from the Colorado Plateau of extreme northern Coconino and Apache counties, Arizona, and San Juan County, New Mexico, northward through the desert regions of eastern Utah and western Colorado into southwestern Wyoming, then northeastward mainly in short-grass prairie through east-central Wyoming and extreme northwestern Nebraska to South Dakota. A series of disjunct populations occurs in Fremont and Las Animas counties in Colorado. The species can be weedy in appearance but usually is absent from areas of significant disturbance, unlike other annual wild buckwheats.

160. **Eriogonum arizonicum** S. Stokes ex M. E. Jones, Contr. W. Bot. 11: 16. 1903 · Arizona wild buckwheat C E

Herbs, erect to slightly spreading, perennial, 2.5–5(–6) × 3–8 dm, glabrous, glaucous. **Stems:** caudex compact to spreading; aerial flowering stems erect, solid, not fistulose, 1–3 dm, glabrous or thinly floccose. **Leaves** basal, sheathing up stems 0.5–15 cm; petiole (0.5–)1–2.5(–4) cm, densely pilose; blade ovate to rounded, (0.5–)1–2(–2.3) × (0.5–)1–2(–2.5) cm, usually sparsely pilose and grayish to whitish on both surfaces, margins undulate. **Inflorescences** narrowly cymose, open, 10–35(–40) × 5–20 cm; branches not fistulose, glabrous, glaucous; bracts 3, scalelike, 0.3–0.8(–1) × 0.2–0.8 mm. **Peduncles** erect, straight, capillary to slender, 0.5–2.5(–3) cm, glabrous. **Involucres** turbinate, (0.8–)1–1.6(–2) × (0.8–)1–1.2(–1.5) mm, glabrous; teeth 5, erect, 0.3–0.8 mm. **Flowers** (1.2–)1.5–2.5 mm; perianth yellowish to yellowish red or reddish with greenish or reddish to reddish brown midribs, glabrous; tepals dimorphic, those of outer whorl obovate with slightly enlarged, auriculate bases, those of inner whorl obovate; stamens included, 1–1.5 mm; filaments sparsely pilose proximally or glabrous. **Achenes** dark brown, 3-gonous, 1.5–2 mm, glabrous.

Flowering Mar–Jun and Aug–Nov. Sandy flats and gravelly washes, mixed grassland, saltbush, creosote bush, and mesquite communities; of conservation concern; 600–1100 m; Ariz.

Eriogonum arizonicum is widely distributed in scattered and isolated populations across central Arizona from extreme eastern Mohave and southern Yavapai counties through northern Maricopa County to southern Gila and northern Pinal counties. It rarely is common in any location. Limited field observations suggest that some individuals in a population may be first-year flowering perennials.

161. **Eriogonum capillare** Small, Bull. Torrey Bot. Club 25: 51. 1898 · San Carlos wild buckwheat C E

Herbs, erect, annual, (1–)2–4 dm, glabrous. **Stems:** caudex absent; aerial flowering stems erect, solid, not fistulose, 0.5–1.5 dm, glabrous. **Leaves** basal; petiole 1–3 cm, sparsely villous; blade obovate to round, 1–3 × 1–3 cm, sparsely villous to hirsute and greenish to grayish on both surfaces, margins plane. **Inflorescences** cymose, usually dense, 10–30 × 5–35 cm; branches not fistulose, glabrous; bracts 3, scalelike, 0.5–3 × 0.5–2.5 mm. **Peduncles** erect, straight, slender, 1–3 cm, glabrous. **Involucres** campanulate, 1–1.5 × 1–1.5 mm, glabrous; teeth 5, erect, 0.2–0.4 mm. **Flowers** 1–1.6 mm; perianth white with greenish or reddish midribs, becoming pink to rose, glabrous; tepals dimorphic, those of outer whorl pandurate with swollen, auriculate bases, those of inner whorl oblanceolate; stamens included to exserted, 0.8–1.2 mm; filaments glabrous. **Achenes** shiny brown to black, 3-gonous, 1.3–1.6 mm, glabrous.

Flowering Sep–Oct. Sandy flats and washes, saltbush, greasewood, and mesquite communities; 500–1500 m; Ariz., N.Mex.

Eriogonum capillare is known from southeastern Arizona (Gila, Graham, Greenlee, and Pima counties) and extreme southwestern New Mexico, where it is rare in northwestern Hidalgo County.

162. **Eriogonum esmeraldense** S. Watson, Proc. Amer. Acad. Arts 24: 85. 1889 E

Herbs, erect to spreading, annual, (0.5–)1–5(–10) dm, glabrous or glandular proximally, mostly grayish. **Stems:** caudex absent; aerial flowering stems erect, solid, not fistulose, 0.5–1(–3) dm, glabrous or sparsely glandular proximally. **Leaves** basal; petiole 0.5–4 cm, strigose to hispid; blade obovate to round-ovate, 0.5–2.5 × 0.4–2 cm, sparsely strigose to pilose-hispid, often slightly glandular, grayish, greenish, or reddish on both surfaces, or reddish adaxially, margins plane. **Inflorescences** cymose, open to diffuse, 5–40(–70) × 5–50 cm; branches not fistulose, glabrous or sparsely glandular near proximal nodes; bracts 3, scalelike, 0.5–3 × 0.5–2 mm. **Peduncles** spreading to reflexed, straight, slender to filiform, 0.2–1.5 cm, glabrous. **Involucres** narrowly turbinate, 0.8–1.8 × 0.5–1.2 mm, glabrous; teeth 4–5, erect, 0.2–0.4 mm. **Flowers** 1–3 mm; perianth white to pinkish with greenish or reddish midribs, glabrous, minutely pustulose proximally; tepals monomorphic, oblanceolate to oblong; stamens included, 1–1.5(–1.8) mm; filaments glabrous. **Achenes** light brown, 3-gonous, 1.4–2.5 mm, glabrous.

Varieties 2 (2 in the flora): w United States.

1. Flowering stems and inflorescence branches glabrous . . . .
. . . . . 162a. *Eriogonum esmeraldense* var. *esmeraldense*
1. Flowering stems and inflorescence branches glandular to sparsely glandular proximally . . .
. . . . 162b. *Eriogonum esmeraldense* var. *toiyabense*

162a. Eriogonum esmeraldense S. Watson var. **esmeraldense** • Esmeralda wild buckwheat [E]

Eriogonum esmeraldense var. *tayei* S. L. Welsh

Plants (0.5–)1–5(–10) dm. **Aerial flowering stems** glabrous. **Inflorescence branches** glabrous. **Flowers** 1–2 mm. **Achenes** 1.4–1.8 mm.

Flowering May–Oct. Sandy flats and slopes, sagebrush communities, pinyon-juniper and montane conifer woodlands; 1700–3200 m; Calif., Nev., Utah.

Variety *esmeraldense* is locally common throughout most of its range. Its greatest concentration is along the eastern edge of the Sierra Nevada in Mono and Inyo counties, California, and adjacent desert ranges eastward into western Esmeralda and Mineral counties, Nevada. A second area of concentration is the high mesa region of southern Nye County, Nevada, north and east of Beatty. An isolated population occurs in northeastern Washoe and adjacent northwestern Humboldt counties in northwestern Nevada. The isolated population in the Tushar Mountains of Sevier County, Utah, recently named var. *tayei*, is likely a recent introduction, probably via shipments of livestock.

162b. Eriogonum esmeraldense S. Watson var. **toiyabense** J. T. Howell, Leafl. W. Bot. 6: 178. 1952 • Toiyabe wild buckwheat [C][E]

Plants (0.5–)1–3 dm. **Aerial flowering stems** sparsely glandular proximally. **Inflorescence branches** glandular prox-imally. **Flowers** 2–3 mm. **Achenes** 1.8–2.5 mm.

Flowering Jun–Sep. Sandy to gravelly flats and slopes, saltbush, sagebrush, and mountain mahogany communities, pinyon-juniper and montane conifer woodlands; 2100–3200 m; Nev.

Variety *toiyabense* is totally isolated from all known populations of var. *esmeraldense*. In Nye County, it is found in the Toiyabe, Toquima, and Monitor ranges, where it can be locally common. To the east and north, it is locally infrequent in the Shoshone (Lander County) and Independent mountains (Elko County). It may well be more widely distributed than reported here, and not worthy of its current "sensitive" status.

163. Eriogonum concinnum Reveal, Bull. Torrey Bot. Club 96: 476, fig. 1. 1969 • Darin's wild buckwheat [C][E]

Herbs, erect, annual, (0.5–)1–6 (–10) dm, glabrous, green. **Stems:** caudex absent; aerial flowering stems erect, usually hollow and fistulose, 0.5–2 dm, glabrous. **Leaves** basal or sheathing up stems 1–6 cm; petiole 1–10 cm, pilose-hispid; blade subcordate, (0.5–)1–4(–6) × (0.5–)1–5(–7) cm, pilose-hispid and bright green on both surfaces, margins plane. **Inflorescences** narrowly cymose, open, 5–80 × 5–35(–40) cm; branches often fistulose, glabrous; bracts 3, scalelike, 1–4 × 0.5–2 mm. **Peduncles** deflexed, straight, slender, (0.05–)0.1–0.25 cm, glabrous. **Involucres** turbinate, 1.2–1.6 × 1–1.4 mm, glabrous; teeth 5, erect, 0.4–0.6 mm. **Flowers** 1–2.5 mm; perianth greenish white to reddish, glabrous; tepals monomorphic, oblong; stamens included to exserted, 1–1.5 mm; filaments glabrous. **Achenes** dark greenish brown to black, 3-gonous, 1.4–1.7 mm, glabrous. $2n = 40$.

Flowering May–Sep. Sandy to gravelly flats, washes, and slopes, saltbush and sagebrush communities, pinyon-juniper woodlands; of conservation concern; 1400–2100 m; Nev.

Eriogonum concinnum is locally common on the mesas of the U.S. Department of Energy's Nevada Test Site east and north of Beatty, Nye County. Although the area is subject to considerable habitat destruction, the species seems to be surviving and is neither threatened nor endangered, although it is properly considered a "sensitive" species given its limited range and the potential for more profound habitat destruction in the area.

164. Eriogonum rubricaule Tidestrom, Proc. Biol. Soc. Wash. 36: 181. 1923 • Lahontan Basin wild buckwheat E

Eriogonum laetum S. Stokes; *E. trichopes* Torrey var. *rubricaule* (Tidestrom) S. Stokes

Herbs, erect, annual, (0.5–)1–4 (–8) dm, glabrous, green to yellowish green. **Stems:** caudex absent; aerial flowering stems erect, occasionally hollow and fistulose, 0.3–1.3(–2) dm, glabrous. **Leaves** basal; petiole 2–3(–4) cm, pilose-hispid; blade round to reniform, 1–2(–3) × (0.5–)1–3 (3.5) cm, pilose-hispid and bright green on both surfaces, margins plane. **Inflorescences** narrowly cymose, open, 5–30 (–50) × 5–35 cm; branches occasionally fistulose, glabrous; bracts 3, scalelike, 1–4 × 0.5–2 mm. **Peduncles** erect or nearly so, straight, slender, 0.2–0.3 cm, glabrous. **Involucres** campanulate, 2.5–3 × 2.5–3 mm, glabrous; teeth 5, erect, 0.5–1 mm. **Flowers** 1.5–2 mm; perianth pale yellow to yellow with light green or reddish midribs, becoming yellowish red, glabrous, papillate in fruit; tepals monomorphic, narrowly ovate; stamens included to exserted, 1–2 mm; filaments glabrous. **Achenes** light brown, 3-gonous, 2–2.5 mm, glabrous.

Flowering May–Oct. Clayey, sandy or gravelly, often volcanic outcrops, washes, flats, and slopes, saltbush, greasewood, and sagebrush communities, pinyon-juniper woodlands; 1100–1800 m; Nev.

Eriogonum rubricaule occurs on the eastern half of the pluvial Lahontan Lake basin of northwestern Nevada (Churchill, Esmeralda, Humboldt, Lander, Lyon, Mineral, Nye, Pershing, and Washoe counties). It tends to occur on or near the old lakeshore at the bases of the mountains, where it is infrequent to locally common.

165. Eriogonum lemmonii S. Watson, Proc. Amer. Acad. Arts 12: 266. 1877 (as lemmoni) • Lemmon's wild buckwheat E

Herbs, erect, annual, (0.1–)1.5–8(–10) dm, glabrous, green. **Stems:** caudex absent; aerial flowering stems erect, occasionally hollow and fistulose, 0.1–3 dm, glabrous. **Leaves** basal; petiole 2–5 cm, pilose; blade round, 1–3 × 1–3 cm, pilose and bright green on both surfaces, margins plane. **Inflorescences** narrowly cymose, open, 10–80 × 5–50 cm; branches infrequently fistulose, glabrous; bracts 3, scalelike, 0.5–2 × 0.3–0.8 mm. **Peduncles** absent or only at proximal node, erect, straight, slender, 0.1–0.5 cm, glabrous. **Involucres** campanulate, 2–3 × (1.5–)2–3 mm, sparsely villous; teeth 6–8, erect, 0.5–0.8 mm. **Flowers** 1.5–2 mm; perianth white or greenish with pink to red midribs, becoming pinkish, glabrous, papillate in fruit; tepals monomorphic, lanceolate; stamens included to exserted, 1–2 mm; filaments glabrous. **Achenes** dark brown to black, 3-gonous, 1.3–1.5 mm, glabrous.

Flowering May–Jun. Clayey or tuffaceous outcrops, flats, and slopes, saltbush and greasewood communities; 1300–1700 m; Nev.

Eriogonum lemmonii is rare and localized, only infrequently being locally occasional. It is restricted to the southwestern edge of the pluvial Lahontan Lake basin of northwestern Nevada. It is found mainly in the Silver Springs and Wadsworth areas of Washoe, Churchill, and Storey counties, with outlying populations at the southern end of the Sahwave Mountains (Pershing County) and in Wilson Canyon (Lyon County). It is on the Nevada "watch list" of narrowly distributed endemic species.

166. Eriogonum deflexum Torrey in J. C. Ives, Rep. Colorado R. 4: 24. 1861 • Skeleton weed F

Herbs, erect to spreading, annual, (0.5–)1–10(–20) dm, glabrous, occasionally glaucous, greenish to grayish. **Stems:** caudex absent; aerial flowering stems erect, solid or occasionally hollow and fistulose, 0.3–3(–4) dm, glabrous, occasionally glaucous. **Leaves** basal; petiole 1–7 cm, usually floccose; blade cordate to reniform or nearly orbiculate, 1–2.5(–4) × 2–4(–5) cm, densely white-tomentose abaxially, less so to floccose or subglabrous and grayish to greenish adaxially, margins entire. **Inflorescences** cymose, open to diffuse, flat-topped, spreading, hemispheric or narrowly erect and strict with whiplike branches, 10–90(–180) × 5–50 cm; branches glabrous, occasionally glaucous; bracts 3, scalelike, 1–3 × 0.5–1.5 mm. **Peduncles** absent or deflexed, rarely some ± erect distally, straight, slender to stout, 0.1–1.5 cm, glabrous. **Involucres** narrowly turbinate to turbinate, 1.5–2.5(–3) × 1–2.5 mm, glabrous; teeth 5, erect, (0.2–)0.5–1 mm. **Flowers** 1–2.5 mm; perianth white to pink, with greenish to reddish midribs, becoming pinkish to reddish, glabrous; tepals dimorphic, those of outer whorl oblong or cordate to ovate, those of inner whorl lanceolate to narrowly ovate; stamens included, 1–1.5 mm; filaments glabrous or sparsely pilose proximally. **Achenes** brown to dark brown, 3-gonous, (1.5–)2–3 mm, glabrous.

Varieties 3 (3 in the flora): w North America, including Mexico.

Eriogonum deflexum is common and often weedy throughout most of its range. It is an important source of small seed for birds. The reported use of the stem (M. L. Zigmond 1981, as *E. insigne*) by the Kawaiisu people as a smoking pipe is incorrect; the taxon in point was actually *E. deflexum* var. *baratum*. The desert metalmark butterfly (*Apodemia mormo deserti*) is found in association with *E. deflexum*.

1. Involucres (2–)2.5–3 mm, narrowly turbinate; peduncles (0.3–)0.5–1.5 cm; stems fistulose; s and ec California, s Nevada . . . . 166c. *Eriogonum deflexum* var. *baratum*
1. Involucres 1.5–2.5 mm, turbinate; peduncles absent or 0.1–0.5 cm; stems not fistulose; Arizona, s and ec California, Nevada, sw New Mexico, w Utah.
  2. Tepals cordate to ovate; Arizona, s California, s Nevada, sw New Mexico, w Utah . . . . 166a. *Eriogonum deflexum* var. *deflexum*
  2. Tepals oblong; ec California, Nevada . . . . 166b. *Eriogonum deflexum* var. *nevadense*

166a. Eriogonum deflexum Torrey var. deflexum • Flat-topped skeleton weed F

Eriogonum deflexum var. *turbinatum* (Small) Reveal; *E. deflexum* subsp. *insigne* (S. Watson) S. Stokes; *E. deflexum* var. *rectum* Reveal; *E. insigne* S. Watson

Plants (0.5–)1–5(–20) dm. **Flowering stems** not fistulose. **Inflorescences** flat-topped, spreading, hemispheric, rarely narrowly erect and strict with whiplike branches. **Peduncles** absent or deflexed, rarely some ± erect distally, (0.1–)0.2–0.5 cm. **Involucres** turbinate, 1.5–2.5 mm. **Flowers** 1–2 mm; outer tepals cordate to ovate. $2n = 40$.

Flowering year-round. Sandy to gravelly washes, flats, and slopes, saltbush, creosote bush, greasewood, blackbrush, sagebrush, and mesquite communities, pinyon-juniper woodlands; -50–1900(–2000) m; Ariz., Calif., Nev., N.Mex., Utah; Mexico (Baja California, Sonora).

Variety *deflexum* is widespread and variable, found mainly in the Mojave and Sonoran deserts of southeastern California, southern Nevada, Arizona, and New Mexico, southward into northern Baja California Norte and northwestern Sonora, Mexico. Plants in southern Arizona tend to have slightly longer peduncles (3–5 mm) and involucres (2–2.5 mm) than do those to the north and west (peduncles 0.1–3 mm, involucres 1.5–2 mm); the former have been recognized as var. *turbinatum*. Variety *deflexum* occurs also in the colder Great Basin desert in southwestern Utah, where a sharp distinction between it and var. *nevadense* is not always possible.

166b. Eriogonum deflexum Torrey var. nevadense Reveal, Phytologia 25: 206. 1973 • Nevada skeleton weed E

Plants 0.5–3(–5) dm. **Flowering stems** not fistulose. **Inflorescences** flat-topped, spreading. **Peduncles** absent or deflexed, 0.1–0.3 cm. **Involucres** turbinate, 1.5–2 mm. **Flowers** 1–1.5(–2) mm; tepals oblong. $2n = 40$.

Flowering Jun–Oct. Sandy to gravelly washes, flats, and slopes, saltbush, greasewood, and sagebrush communities, pinyon-juniper woodlands; 1000–2200 m; Calif., Nev.

Variety *nevadense* is the common expression of the species in the Great Basin. It occurs from northeastern Inyo and southern Mono counties, California, eastward into central Nevada (Churchill, Douglas, Esmeralda, Eureka, Humboldt, Lander, Lincoln, Lyon, Mineral, Nye, Pershing, Washoe, and White Pine counties) north of the Mojave Desert (as defined by presence of creosote bush).

166c. Eriogonum deflexum Torrey var. baratum (Elmer) Reveal, Brittonia 20: 24. 1968 • Tall skeleton weed E

Eriogonum baratum Elmer, Bot. Gaz. 39: 52. 1905; *E. deflexum* subsp. *baratum* (Elmer) Munz

Plants 3–10 dm. **Flowering stems** fistulose. **Inflorescences** usually narrowly erect and strict. **Peduncles** deflexed, (0.3–)0.5–1.5 cm. **Involucres** narrowly turbinate, (2–)2.5–3 mm. **Flowers** 2–2.5 mm; tepals cordate. $2n = 40$.

Flowering Jul–Oct. Sandy to gravelly washes, flats, and slopes, saltbush, creosote bush, blackbrush, and sagebrush communities, pinyon-juniper woodlands; 700–2900 m; Calif., Nev.

Variety *baratum* occurs from the Transverse Ranges of southwestern California (Kern, Los Angeles, and Ventura counties) northward onto the eastern slope of the Sierra Nevada (Kern, Tulare, and Inyo counties), onto the desert ranges of northern San Bernardino County north to Mono County, and east into southern Nevada (Esmeralda, Lincoln, and Nye counties).

167. Eriogonum rixfordii S. Stokes, Leafl. W. Bot. 1: 29. 1932 • Pagoda wild buckwheat C E

Eriogonum deflexum Torrey subsp. *rixfordii* (S. Stokes) Munz

Herbs, erect, annual, (1.5–)2–4 dm, glabrous, greenish to grayish. **Stems:** caudex absent; aerial flowering stems erect, solid, not fistulose, 0.3–2 dm, glabrous. **Leaves** basal; petiole 1–3(–5) cm, floccose; blade cordate to orbiculate, (0.5–)1–3 × (0.5–)1–3 cm, densely white-tomentose abaxially, less so to subglabrous and grayish to greenish adaxially, margins entire. **Inflorescences** cymose, diffuse, of many horizontal tiers of flat-topped branches, 10–35 × 5–35 cm; branches glabrous; bracts 3, scalelike, 1–3 × 0.5–1.5 mm. **Peduncles** absent. **Involucres** deflexed, narrowly campanulate, 1–1.5 × 1.5–2 mm, glabrous; teeth 5, erect, 0.4–0.7 mm. **Flowers** 1.3–1.5 mm; perianth white with greenish midribs, becoming reddish, glabrous; tepals dimorphic, those of outer whorl narrowly oval to cordate, those of inner whorl narrowly lanceolate; stamens included, 1–1.3 mm; filaments glabrous. **Achenes** dark brown, lenticular, 1.3–1.5 mm, glabrous. $2n = 40$.

Flowering Jun–Dec. Sandy to gravelly washes and slopes, saltbush communities; of conservation concern; (30–)100–1600 m; Calif., Nev.

Eriogonum rixfordii is occasional in the Death Valley area of Inyo County, California, and adjacent southwestern Nye County, Nevada. Its habit is so striking that it is often collected as a novelty. It is rarely common and never weedy.

168. Eriogonum hookeri S. Watson, Proc. Amer. Acad. Arts 14: 295. 1879 • Hooker's wild buckwheat E

Eriogonum deflexum Torrey var. *gilvum* S. Stokes; *E. deflexum* subsp. *hookeri* (S. Watson) S. Stokes

Herbs, erect, annual, 1–6 dm, glabrous, glaucous, grayish. **Stems:** caudex absent; aerial flowering stems erect, solid, not fistulose, 0.1–0.4 dm, glabrous, glaucous. **Leaves** basal; petiole 1–5 cm, tomentose; blade cordate to subreniform, (1–)2–5 × 2–6 cm, densely white felty-tomentose abaxially, tomentose and grayish adaxially, margins occasionally wavy. **Inflorescences** cymose, open to diffuse, spreading to subglobose or flat-topped to umbrella-shaped, 5–35 × 5–50 cm; branches glabrous, glaucous; bracts 3, scalelike, 1–3 × 0.5–1.5 mm. **Peduncles** absent. **Involucres** deflexed, broadly campanulate to hemispheric, 1–2 × 1.5–3(–3.5) mm, glabrous; teeth 5, erect, 0.5–0.7 mm. **Flowers** 1.5–2 mm; perianth yellow to reddish yellow, glabrous; tepals dimorphic, those of outer whorl orbiculate, those of inner whorl narrowly ovate; stamens mostly included, 1.3–1.5 mm; filaments glabrous. **Achenes** light brown, 3-gonous, 2–2.5 mm, glabrous. $2n = 40$.

Flowering Jun–Oct. Sandy washes, flats, and slopes, saltbush, greasewood, sagebrush, and mountain mahogany communities, pinyon-juniper woodlands; 1300–2500(–3000) m; Ariz., Calif., Colo., Idaho, Nev., N.Mex., Oreg., Utah, Wyo.

Eriogonum hookeri is basically a species of the Intermountain Region of Utah and Nevada, just entering portions of east-central California, southeastern Oregon, southeastern Idaho, southwestern Wyoming, western Colorado, northwestern New Mexico, and northern Arizona. Populations of the species usually are scattered, with the plants rarely common and even less frequently locally weedy.

P. A. Vestal (1940) stated that the Hopi boiled *Eriogonum hookeri* with mush for flavoring.

169. Eriogonum brachypodum Torrey & A. Gray, Proc. Amer. Acad. Arts 8: 180. 1870 • Parry's wild buckwheat E

Eriogonum deflexum Torrey subsp. *brachypodum* (Torrey & A. Gray) S. Stokes; *E. deflexum* var. *brachypodum* (Torrey & A. Gray) Munz; *E. deflexum* subsp. *parryi* (A. Gray) S. Stokes; *E. parryi* A. Gray

Herbs, spreading, annual, 0.5–4 dm, glandular, greenish. **Stems:** caudex absent; aerial flowering stems erect, solid, not fistulose, 0.2–0.7 dm, glandular. **Leaves** basal; petiole 1–4 cm, tomentose; blade orbiculate to cordate, 1–3(–5) × (1.5–)2–5 (5.5) cm, densely white-tomentose abaxially, less so to subglabrous and green adaxially, margins usually smooth. **Inflorescences** cymose, open to rather diffuse, often flat-topped, 3–40 × 3–100 cm; branches glandular; bracts scalelike, 1–3 × 0.5–1.5 mm. **Peduncles** absent or deflexed, straight, stoutish, 0.1–1.5 cm, glandular. **Involucres** turbinate to campanulate, 1–2.5 × 1.5–2.5 mm, glandular; teeth 5, erect, 0.3–1 mm. **Flowers** 1–2.5 mm; perianth white with greenish or reddish midribs, becoming reddish, glabrous; tepals dimorphic, those of outer whorl ovate to oblong and often auriculate proximally, those of inner whorl usually lanceolate; stamens included to exserted, 1.5–2.5 mm; filaments glabrous or pilose proximally. **Achenes** brown to black, lenticular to 3-gonous, 1.5–2 mm, glabrous. $2n = 40$.

Flowering year-round. Sandy to gravelly washes, flats, and slopes, saltbush, creosote bush, greasewood, mesquite, blackbrush, and sagebrush communities, pinyon-juniper woodlands; 100–2300 m; Ariz., Calif., Nev., Utah.

Eriogonum brachypodum is variable in overall shape, peduncle length, and shape and size of involucres and flowers. Spreading plants with longish peduncles and narrow involucres occur in southwestern Utah and adjacent portions of Nevada and Arizona; this is the "*parryi*" phase of the species. Typical *E. brachypodum* is low and flat-topped, with the inflorescence up to 10 dm across. The involucres are usually sessile and campanulate rather than turbinate. The two phases intergrade completely and a taxonomic distinction is not warranted. Dried plants of the typical phase often are used as decorations.

The species is common and even weedy in places. It occurs from Churchill and Pershing counties, Nevada, south through Mono and Inyo counties to eastern Kern and San Bernardino counties, California, then eastward through Nevada (Clark, Esmeralda, Lincoln, Mineral, and Nye counties) into southwestern Utah (Beaver and Washington counties) and northwestern Arizona (Mohave County). It is much more common in the Mojave Desert than in the Great Basin.

170. Eriogonum bifurcatum Reveal, Aliso 7: 357. 1971 • Pahrump Valley wild buckwheat [C][E]

Herbs, spreading, annual, 1–4 dm, glabrous, greenish to grayish. **Stems:** caudex absent; aerial flowering stems erect, solid, not fistulose, 0.05–0.3 dm, often obscured by leaves, glabrous. **Leaves** basal; petiole 1–4 cm, tomentose; blade round-cordate, 1–3 × 1–3 cm, densely white-tomentose abaxially, less so to floccose and greenish adaxially, margins plane. **Peduncles** absent or erect, straight, slender, 0.1–0.5 cm, glabrous. **Inflorescences** cymose, spreading and flat-topped, 5–30 × 10–50 cm; branches glabrous; bracts scalelike, 1–1.5 × 0.4–1 mm. **Involucres** turbinate, 2–2.5 × 1.3–2 mm; teeth 5, erect, 0.4–0.5 mm. **Flowers** 1.5–2 mm, glabrous; perianth white with greenish to reddish midribs, becoming faintly pinkish, glabrous; tepals dimorphic, those of outer whorl obovate to cordate, those of inner whorl lanceolate; stamens exserted, 2–3 mm; filaments pilose proximally. **Achenes** light brown, 3-gonous, 2–2.3 mm, glabrous. *2n* = 40.

Flowering May–Jun. Alkaline sandy flats and slopes, saltbush communities; of conservation concern; 600–800 m; Calif., Nev.

Eriogonum bifurcatum is restricted to the Mesquite, Pahrump, and Stewart valleys along the California-Nevada border in southeastern Inyo, northwestern Clark, and southwestern Nye counties. It is closely related to *E. exaltatum*, and spring-flowering specimens of the latter are impossible to distinguish from mature individuals of *E. bifurcatum*. The Las Vegas populations (last collected in 1941) formerly called *E. bifurcatum* are here tentatively assigned to *E. exaltatum*. The Pahrump Valley wild buckwheat is currently considered a "sensitive" species, and habitat protection is being considered.

171. Eriogonum exaltatum M. E. Jones, Contr. W. Bot. 15: 61. 1929 • Ladder wild buckwheat [E]

Herbs, erect, annual, (0.5–)3–6 (–10) dm, glabrous, glaucous, grayish. **Stems:** caudex absent; aerial flowering stems erect, solid, not fistulose, (0.5–)1–2 dm, glabrous. **Leaves** basal; petiole 1–10 cm, tomentose; blade subcordate to orbiculate, (1.5–)2–5 (–8) × (1.5–)2–5(–8) cm, densely white-tomentose abaxially, floccose to subglabrous and green adaxially, margins often wavy. **Inflorescences** cymose, spreading when immature, quickly becoming narrowly erect and strict with whiplike branches, (5–)10–50(–80) × 10–50 cm; branches glabrous, glaucous; bracts 3, scalelike, 1–1.5(–2) × 0.4–1 mm. **Peduncles** absent or erect, straight, slender, 0.1–0.2 cm, glabrous. **Involucres** turbinate, 2–2.5(–3) × 1.5–2.5 mm, glabrous; teeth 5, erect, 0.5–1.5 mm. **Flowers** 1.5–2 mm; perianth white with green or reddish midribs, becoming pinkish, glabrous; tepals dimorphic, those of outer whorl oblong, those of inner whorl lanceolate; stamens exserted, 1.5–2.5 mm; filaments pilose proximally. **Achenes** dark brown to blackish, 3-gonous, 2–2.5 mm, glabrous. *2n* = 40.

Flowering May–Oct. Sandy to gravelly flats and slopes, saltbush, creosote bush, greasewood, blackbrush, and mesquite communities, rarely in pinyon-juniper woodlands; 500–1400 m; Ariz., Calif., Nev., Utah.

Eriogonum insigne has been a troublesome taxon. The type, collected by E. Palmer in 1876, is a curious specimen from near Paragonah in Iron County, Utah. That expression has not been re-collected and, after another frustrating review of the type material, it is now referred to *E. deflexum*, along with other odd specimens from southern Nevada and California that have always been difficult to assign. As a result, the tall, upright plants with long, whiplike branches of northwestern

Arizona (Mohave County), southern Nevada (northeastern Clark and southern Lincoln counties), and southwestern Utah (southwestern Washington County), are now recognized under the name *E. exaltatum*. Some specimens remain problematic. Immature plants of *E. exaltatum* (*Goodding 2302*, GH, MIN, MO, NY, RM, UC) resemble *E. bifurcatum*. Specimens from Eureka Valley (*DeDecker 4741*, NY) and western Pahrump Valley (*Abrams 14248*, DS, GH, NY) in Inyo County, California, appear to belong to *E. exaltatum*.

Like *Eriogonum exaltatum*, some populations of *E. deflexum* var. *deflexum* have nearly erect involucres. The plants from the Buried Hills (Nye and Clark counties, Nevada) are particularly odd. Single plants from a few collections in Kane County, Utah, have sessile, seemingly erect involucres in the forks of inflorescence branches. None of these plants has the characteristic long, whip-like branches of *E. exaltatum*, and they are not included in *E. deflexum* var. *deflexum*. The type of *E. deflexum* var. *rectum* from San Bernardino County, California (*Reveal & Broome 6385*, CAS, NY, US, etc.), has individual specimens with both deflexed and somewhat erect involucres, along with whiplike branches. Collections from Imperial and San Diego counties, previously associated with what is here termed *E. exaltatum*, are now considered specimens of *E. deflexum*. Clearly, more work is required to understand these species fully.

172. Eriogonum hoffmannii S. Stokes, Leafl. W. Bot. 1: 23. 1932 E

Herbs, spreading to erect, annual, 0.5–10 dm, glabrous, greenish to grayish. **Stems:** caudex absent; aerial flowering stems erect, solid, not fistulose, 0.4–4 dm, glabrous. **Leaves** basal or sometimes sheathing up stems 1–3(–5) cm; petiole 1–10 cm, tomentose; blade suborbicular to subcordate, 1–5 × 2–8 cm, densely white-tomentose abaxially, floccose or glabrous and green adaxially, margins plane or wavy, entire or crisped. **Inflorescences** cymose, open, spreading to rather erect and somewhat strict, 5–70 × 10–60 cm; branches glabrous; bracts 3, scalelike, 1–1.5 × 0.5–1 mm. **Peduncles** absent or erect, straight, stout, 0.01–0.1 cm, glabrous. **Involucres** turbinate, 1–2 × 1–1.8 mm, glabrous; teeth 5, erect, 0.4–1.2 mm. **Flowers** 1.5–1.8 mm; perianth white with greenish or reddish midribs, becoming reddish, glabrous; tepals monomorphic,

lanceolate to spatulate or ovate; stamens mostly included, 1–1.5 mm; filaments glabrous. **Achenes** brown, 3-gonous, 2 mm, glabrous.

Varieties 2 (2 in the flora): California.

Eriogonum hoffmannii is known only from the Death Valley region (M. DeDecker 1974) of Inyo County. Nearly all of its known populations are within the boundaries of Death Valley National Park (a few plants occur just to the west along California Highway 190 southeast of Travertine Point). It is of special concern but is well managed and protected.

1. Leaf margins entire, blade 1–4 × 2–4 cm; petioles 1–5 cm; plants 0.5–5 dm . . . .
   . . . . 172a. *Eriogonum hoffmannii* var. *hoffmannii*
1. Leaf margins crisped, blade 2–5 × 3–8 cm; petioles (3–)5–10 cm; plants 4–10 dm . . . .
   . . . . 172b. *Eriogonum hoffmannii* var. *robustius*

172a. Eriogonum hoffmannii S. Stokes var. **hoffmannii** · Hoffmann's wild buckwheat E

Plants 0.5–5 dm. **Leaves** basal; petiole 1–5 cm; blade 1–4 × 2–4 cm, margins entire. 2*n* = 40.

Flowering Jul–Sep. Sandy to gravelly slopes, sagebrush communities, pinyon-juniper woodlands; 1000–1700 m; Calif.

Variety *hoffmannii* is restricted to the Panamint Mountains in the Towne Pass and Wildrose Canyon areas of Inyo County.

172b. Eriogonum hoffmannii S. Stokes var. **robustius** S. Stokes, Leafl. W. Bot. 3: 16. 1941 (as hoffmanni) · Furnace Creek Wash wild buckwheat E

Eriogonum hoffmannii subsp. *robustius* (S. Stokes) Munz

Plants 4–10 dm. **Leaves** basal or sheathing; petiole (3–)5–10 cm; blade 2–5 × 3–8 cm, margins crisped. 2*n* = 40.

Flowering Aug–Nov. Sandy to gravelly flats, slopes, and washes, saltbush and creosote bush communities; 100–1700 m; Calif.

Variety *robustius* is known only from the Funeral and Black mountains of Inyo County. It is more common than var. *hoffmannii*, being found in the Furnace Creek and Ryan washes, and near Dante's View.

173. Eriogonum watsonii Torrey & A. Gray, Proc. Amer. Acad. Arts 8: 182. 1870 • Watson's wild buckwheat E

Eriogonum deflexum Torrey var. *multipedunculatum* (S. Stokes) C. L. Hitchcock; *E. deflexum* subsp. *watsonii* (Torrey & A. Gray) S. Stokes

Herbs, spreading, annual, (0.5–) 1–4 dm, glabrous, glaucous, greenish to grayish. **Stems:** caudex absent; aerial flowering stems erect, solid, not fistulose, 0.4–0.8 dm, glabrous. **Leaves** basal; petiole 1–3 cm, mostly floccose; blade subcordate to orbiculate, (0.5–)1–2.5 × (0.5–)1–2.5 cm, densely white-tomentose abaxially, less so to subglabrous and grayish to greenish adaxially, margins plane. **Inflorescences** cymose, open, spreading, (5–)10–20 × 5–20 cm; branches glabrous, glaucous; bracts 3, scalelike, 1–2 × 0.5–1 mm. **Peduncles** deflexed or curving and ascending upward, straight or slightly curved, slender, 0.3–1.5 cm, glabrous. **Involucres** narrowly turbinate, (1.5–)2–3 × 0.5–1.5 mm, glabrous; teeth 5, erect, 0.4–0.8 mm. **Flowers** 2–2.5 mm; perianth white to pink with greenish to reddish midribs, becoming pinkish to reddish, glabrous; tepals dimorphic, those of outer whorl oblong or obovate, those of inner whorl ovate; stamens included, 1.5–2 mm; filaments glabrous. **Achenes** brown, 3-gonous, 1.5–2 mm, glabrous. $2n = 40$.

Flowering May–Sep. Sandy to gravelly flats, slopes, and washes, saltbush, greasewood, sagebrush, and mountain mahogany communities, pinyon-juniper woodlands; 1100–2500 m; Idaho, Nev., Oreg.

Eriogonum watsonii is found in the center of northern Nevada (Churchill, Eureka, Lander, and Pershing counties), with scattered populations in the Elko area of Elko County, and near Ione in Nye County. To the north is a second center, where the plants are found mainly on the bluffs and flats above the Snake River in southwestern Idaho (Canyon, Owyhee, and Twin Falls counties), westward through the Harper area of Malheur County, Oregon, to near Buchanan in eastern Harney County. They also occur in a series of disjunct populations in and around Rome in Malheur County. Most of the individuals in these populations have deflexed peduncles, but some have long, gracefully upward-ascending peduncles that are most distinctive. The species usually is common but only rarely weedy along roadsides.

174. Eriogonum scabrellum Reveal, Ann. Missouri Bot. Gard. 55: 74. 1968 • Westwater wild buckwheat E

Herbs, spreading, annual, 1–3(–5) dm, floccose and scabrellous, greenish to reddish or reddish brown. **Stems:** caudex absent; aerial flowering stems erect, solid, not fistulose, 0.5–1.5 dm, floccose and scabrellous. **Leaves** basal, sheathing up stems 1–3 cm; petiole 1–4(–5) cm, floccose; blade cordate, 1–3(–4) × 1–3(–4) cm, densely white-tomentose abaxially, sparsely floccose and greenish to green adaxially, margins crisped and wavy. **Inflorescences** cymose, open, spreading, 5–40 × 20–100 cm; branches scabrellous; bracts 3, scalelike, 1–1.5 × 0.4–0.9 mm. **Peduncles** absent. **Involucres** horizontal, turbinate, 1.5–2.5 × 1.5–2 mm, scabrellous; teeth 5, erect, 0.4–0.6 mm. **Flowers** 1–1.5 mm; perianth white with greenish midribs, becoming pink to rose or deep red, pustulose; tepals dimorphic, those of outer whorl obovate, those of inner whorl ovate; stamens exserted, 1–1.5 mm; filaments glabrous. **Achenes** light brown to brown, 3-gonous, 2 mm, glabrous. $2n = 40$.

Flowering Sep–Nov. Clayey to gravelly washes, flats, and slopes, saltbush, blackbrush, and sagebrush communities, pinyon-juniper woodlands; 1400–2300 m; Colo., N.Mex., Utah.

Eriogonum scabrellum is widely scattered along the Colorado and San Juan river systems. It occurs from Grand Valley of Utah (Grand County) and Colorado (Mesa County), southward to the Lake Powell region of Utah (Garfield and Kane counties), and then to the Four Corners area of Utah (San Juan County), Colorado (Montezuma County), and New Mexico (San Juan County). It recently was collected in Chaco Canyon National Park in New Mexico, and just south of the town of Green River in Emery County, Utah.

175. Eriogonum eremicola J. T. Howell & Reveal, Leafl. W. Bot. 10: 174. 1965 • Telescope Peak wild buckwheat C E

Herbs, spreading, annual, 0.8–2.5 dm, glandular, greenish to reddish. **Stems:** caudex absent; aerial flowering stems erect, solid, not fistulose, 0.3–1 dm, glandular. **Leaves** basal; petiole 1–3 cm, floccose; leaf blade rounded, 1–2.5 × 1–2.5 cm, densely white-tomentose abaxially, floccose or glabrous and greenish adaxially, margins plane. **Inflorescences** cymose, open to somewhat diffuse, 5–20 × 5–20 cm; branches glandular; bracts 3, scalelike, 1–2

× 0.5–1.5 mm. **Peduncles** absent or deflexed, straight, slender, (0.1–)0.5–1 cm, glandular. **Involucres** turbinate-campanulate, 1.8–2 × 1–1.5 mm, glandular; teeth 5, erect, 0.5–0.8 mm. **Flowers** 2–2.5 mm; perianth white with red midribs, becoming red, glabrous; tepals monomorphic, oblong-ovate; stamens exserted, 2–2.5 mm; filaments glabrous. **Achenes** brown, 3-gonous, 2–2.5 mm, glabrous.

Flowering Jun–Sep. Sandy to gravelly slopes, sagebrush and mountain mahogany communities, montane conifer woodlands; of conservation concern; 220–3100 m; Calif.

Eriogonum eremicola is rare to infrequent, known only from the New York Butte area of the Inyo Mountains and the Telescope Peak area of the Panamint Range in Inyo County. The Bureau of Land Management considers it a "sensitive" species, and it is well protected in Death Valley National Park.

176. Eriogonum tenellum Torrey, Ann. Lyceum Nat. Hist. New York 2: 241. 1827 F

Herbs, erect to spreading, perennial, 1–6 dm, glabrous, often glaucous, grayish. **Stems:** caudex compact to spreading; aerial flowering stems erect, solid, not fistulose, 0.4–4 dm, glabrous, often glaucous. **Leaves** basal or sheathing up stem 3–8 cm; petiole 0.4–4 cm, tomentose; blade elliptic to deltoid or ovate to suborbiculate or orbiculate, 0.3–3 × 0.2–3 cm, densely white-tomentose and grayish on both surfaces, margins plane. **Inflorescences** cymose, open, erect to spreading, 5–40 × 5–40 cm; branches glabrous; bracts 3, scalelike, 1–3 × 0.5–2 mm. **Peduncles** erect or spreading, straight, slender, 0.6–6 cm, glabrous. **Involucres** turbinate to turbinate-campanulate, 2–4 × 1.5–3(–4) mm, glabrous but tomentose adaxially; teeth 5, erect or nearly so, 0.5–1 mm. **Flowers** 1.5–3.5 mm; perianth white to pink with greenish to reddish midribs, becoming pinkish or orange-brown to red, glabrous; tepals dimorphic, those of outer whorl ovate to obovate or cordate to suborbiculate, those of inner whorl narrowly oblong to oblong; stamens included, 1–1.5 mm; filaments sparsely pilose proximally. **Achenes** light brown to brown, 3-gonous, 2–3 mm, glabrous.

Varieties 3 (3 in the flora): c North America, including Mexico.

1. Leaves basal; caudex stem blades 0.3–1.5 × 0.3–1 cm; compact; se Colorado, e New Mexico, w Oklahoma, n and w Texas . . . . 176a. *Eriogonum tenellum* var. *tenellum*
1. Leaves sheathing up stems; caudex stem blades 0.3–3 × 0.2–3 cm; compact to spreading; Texas.
  2. Leaf blades elliptic to deltoid, 0.3–1.2 × 0.2–0.7 cm; caudex stems spreading; c Texas . . . . 176b. *Eriogonum tenellum* var. *ramosissimum*
  2. Leaf blades ovate to suborbiculate or orbiculate, 1–3 × 1–3 cm; caudex stems compact to slightly spreading; sw Texas . . . . 176c. *Eriogonum tenellum* var. *platyphyllum*

176a. Eriogonum tenellum Torrey var. **tenellum**
• Tall wild buckwheat   F

Plants 1–5 dm. **Caudex stems** compact. **Leaves** basal; blade usually elliptic, 0.3–1.5 × 0.3–1 cm. $2n = 40$.

Flowering May–Nov. Sandy to gravelly flats and slopes, mixed grassland, saltbush, creosote bush, mesquite, and sagebrush communities, oak and conifer woodlands; 900–2200 m; Colo., N.Mex., Okla., Tex.; Mexico (Chihuahua, Coahuila).

Variety *tenellum* is common and widespread throughout its northern range in southeastern Colorado, New Mexico, western Oklahoma, and northern and western Texas. Outlying populations, extending northward from Mexico, enter the flora area in Dona Ana County, New Mexico, and Webb County, Texas.

176b. Eriogonum tenellum Torrey var. **ramosissimum** Bentham in A. P. de Candolle and A. L. P. P. de Candolle, Prodr. 14: 20. 1856 • Granite Mountain wild buckwheat   E

Plants 3–4 dm. **Caudex stems** spreading. **Leaves** sheathing up stems; blade elliptic to deltoid, 0.3–1.2 × 0.2–0.7 cm. $2n = 40$.

Flowering Apr–Dec. Granitic flat outcrops, ledges, and crevices, mixed grassland and mesquite communities, juniper and oak woodlands; 200–600 m; Tex.

Variety *ramosissimum* is restricted to the "Hill Country" in Burnet, Gillespie, Llano, and Mason counties of central Texas, where it can be locally common.

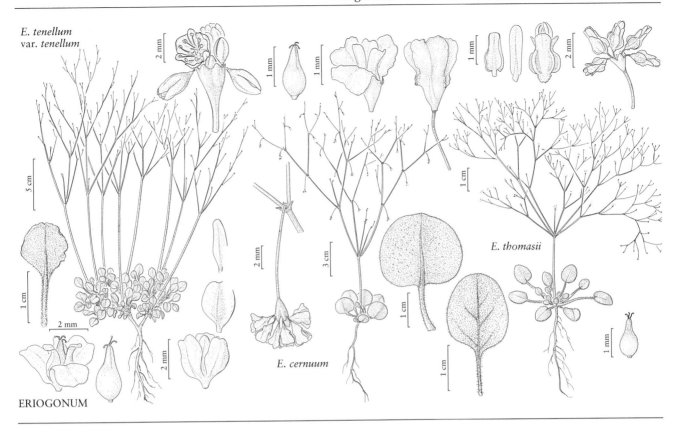

E. tenellum
var. tenellum

E. thomasii

E. cernuum

ERIOGONUM

176c. Eriogonum tenellum Torrey var. **platyphyllum** (Torrey ex Bentham) Torrey in W. H. Emory, Rep. U.S. Mex. Bound. 2(1): 176. 1859 • Broad-leaf wild buckwheat

Eriogonum platyphyllum Torrey ex Bentham in A. P. de Candolle and A. L. P. P. de Candolle, Prodr. 14: 20. 1856; *E. tenellum* subsp. *platyphyllum* (Torrey ex Bentham) S. Stokes

Plants 3–6 dm. **Caudex stems** compact to slightly spreading. **Leaves** sheathing up stems; blade ovate to suborbiculate or orbiculate, 1–3 × 1–3 cm.

Flowering May–Oct. Gravelly to rocky flats and slopes, saltbush, mesquite, and creosote bush communities, oak woodlands; 300–900 m; Tex.; Mexico (Coahuila, Nuevo León).

Variety *platyphyllum* is a Mexican taxon that just enters the flora area in Brewster, Presidio, Terrell, and Uvalde counties of western Texas. The type was gathered along the Nueces River in Uvalde County, but the variety has not been found in that area since 1849. Although this differs from the other varieties in terms of its leaf shap, a sharp distinction between this taxon and the others is not always possible in Texas.

177. Eriogonum cernuum Nuttall, Proc. Acad. Nat. Sci. Philadelphia 4: 14. 1848 • Nodding wild buckwheat [E] [F]

Eriogonum cernuum var. *psammophilum* S. L. Welsh; *E. cernuum* subsp. *tenue* (Torrey & A. Gray) S. Stokes; *E. cernuum* var. *tenue* Torrey & A. Gray; *E. cernuum* var. *umbraticum* Eastwood; *E. cernuum* subsp. *viminale* S. Stokes; *E. cernuum* var. *viminale* (S. Stokes) Reveal

Herbs, spreading to erect, annual, 0.5–6 dm, glabrous, grayish, greenish or reddish. **Stems:** caudex absent; aerial flowering stems erect, solid, not fistulose, 0.3–2 dm, glabrous. **Leaves** basal or sheathing up stems 2–10 cm; petiole 1–4 cm, tomentose; blade round-ovate to orbiculate, (0.5–)1–2(–2.5) × (0.5–)1–2(–2.5) cm, white- to grayish-tomentose abaxially, tomentose to floccose or glabrate and grayish or greenish adaxially, margins plane. **Inflorescences** cymose, open to diffuse, 5–50 × 5–40 cm; branches glabrous; bracts 3, scalelike, 1–2 × 1–2.5 mm. **Peduncles** spreading to ascending or deflexed to cernuous, infrequently absent, straight or curved, slender, 0.1–2.5 cm, glabrous. **Involucres** turbinate, (1–)1.5–2 × 1–1.5 mm, glabrous; teeth 5, erect, 0.4–0.7

mm. **Flowers** 1–2 mm, glabrous; perianth white to pinkish, becoming rose to red; tepals dimorphic, those of outer whorl pandurate, those of inner whorl obovate; stamens mostly exserted, 1–2 mm; filaments pilose proximally. **Achenes** light brown to brown, 3-gonous, 1.5–2 mm, glabrous.

Flowering Apr–Oct. Sandy to gravelly or clayey flats and slopes, mixed grassland, saltbush, sagebrush, and mountain mahogany communities, oak, pinyon-juniper, and conifer woodlands; 600–3100(–3300) m; Alta., Sask.; Ariz., Calif., Colo., Idaho, Mont., Nebr., Nev., N.Mex., Oreg., S.Dak., Utah, Wash., Wyo.

Eriogonum cernuum is widely distributed, being infrequent to common or even abundant and weedy. It is common throughout most of its range in southeastern Oregon, eastern California, southern Idaho, Nevada, Utah, northern Arizona, and New Mexico. The species is rare in southeastern Washington (Franklin County). It is less common and more widely scattered in Colorado, Wyoming, and Montana, and on the northern Great Plains in southern Alberta and Saskatchewan, western South Dakota, and western Nebraska. The northern Great Basin phase, with sessile involucres, has been called var. *viminale*, but this difference now appears to be ecologic rather than genetic.

Seeds of the nodding wild buckwheat were gathered by the Navajo (Diné) people, pounded into a meal, and eaten dry or made into a porridge (P. A. Vestal 1952; L. C. Wyman and S. K. Harris 1951).

178. Eriogonum rotundifolium Bentham in A. P. de Candolle and A. L. P. P. de Candolle, Prodr. 14: 21. 1856 • Round-leaf wild buckwheat

Eriogonum cernuum Nuttall subsp. *glaucescens* S. Stokes; *E. cernuum* subsp. *rotundifolium* (Bentham) S. Stokes; *E. rotundifolium* var. *angustius* Goodman

Herbs, spreading, annual, 0.5–4 dm, glabrous and often glaucous, greenish to grayish. **Stems:** caudex absent; aerial flowering stems erect, solid, not fistulose, 0.1–0.5(–0.7) dm, glabrous. **Leaves** basal; petiole 1.5–4 cm, floccose; blade cordate to orbiculate, 1–2(–3) × 1–2.5(–3) cm, densely white-tomentose abaxially, floccose or subglabrous and greenish adaxially, margins plane. **Inflorescences** cymose, open to diffuse, usually flat-topped, 5–35 × 5–35 cm; branches glabrous; bracts 3, scalelike, 1–2.5 × 0.5–2 mm. **Peduncles** erect, straight, stoutish, 0.3–1.5 cm, glabrous. **Involucres** turbinate to campanulate, 1–2 × 1.5–2.5 mm, glabrous; teeth 5, erect, 0.4–0.8 mm. **Flowers** 1–2.5 mm; perianth white to pink with greenish to reddish midribs, becoming rose to red, glabrous; tepals dimorphic, those

of outer whorl flabellate, those of inner whorl lanceolate; stamens included, 1.2–1.7 mm; filaments pilose proximally. **Achenes** dark brown, 3-gonous, 1.5–2 mm, glabrous. $2n = 40$.

Flowering May–Oct. Sandy to gravelly flats and slopes, mixed grassland, saltbush, creosote bush, and mesquite communities, juniper woodlands; 600–1800 m; Ariz., N.Mex., Tex.; Mexico (Chihuahua, Coahuila).

Eriogonum rotundifolium is the southern counterpart to *E. cernuum*, being common to abundant and occasionally even weedy. Its overall range, however, is significantly smaller. It occurs in Arizona only in Cochise County, but is found more widely in New Mexico, and is common in the trans-Pecos region of western Texas, with scattered populations in Dimmit, Ector, Foard, and Knox counties outside that region.

A sterile Edwin James specimen gathered in 1820 (NY) supposedly was collected near the Rocky Mountains and may be *Eriogonum rotundifolium*. Also seen at NY is an unattributed, redistributed collection of this species labeled only "Colorado." Until better documented material from that state is seen, the species is considered not to be a member of the Colorado flora.

F. A. Elmore (1943) reported that the round-leaf wild buckwheat was used by the Navajo (Diné) people as an emetic. My own consumption of a few seeds, as a self-experiment, produced no particular urge to vomit. Inasmuch as the treatment was taken after swallowing ants, it is difficult to know whether the ants or the seeds were the emetic. G. M. Hocking (1956) reported that the leaves were used for sore throats and the stems were eaten raw (the latter proving in the same self-experiment not to be particularly tasty, leaving a slightly sour aftertaste). Hocking also reported that the roots were used medicinally but mentioned no specific ailment.

179. Eriogonum nutans Torrey & A. Gray, Proc. Amer. Acad. Arts 8: 181. 1870   E

Eriogonum cernuum Nuttall var. *purpurascens* Torrey & A. Gray in War Department [U.S.], Pacif. Railr. Rep. 2(1): 124. 1857

Herbs, spreading to erect, annual, 0.5–3 dm, glandular or glabrous, occasionally glaucous, greenish, grayish, or reddish. **Stems:** caudex absent; aerial flowering stems erect, solid, not fistulose, 0.3–1.5 dm, glandular or glabrous, occasionally glaucous. **Leaves** basal; petiole 0.5–2.5 cm, tomentose; blade broadly reniform to round, 0.5–2.5 × 0.5–2.5 cm, densely white-tomentose abaxially, floccose to glabrate and greenish adaxially, margins plane. **Inflorescences** cymose, open to diffuse, 5–20 × 5–20 cm; branches glandular or glabrous, occasionally glaucous; bracts 3, scalelike, 0.5–2 × 1–2.5 mm. **Peduncles** recurved to cernuous, curved or sometimes straight, slender, 0.3–1 cm, glandular or glabrous. **Involucres** campanulate, 2–3 × 2–3.5 mm, glandular or glabrous; teeth 5, erect, 0.5–1 mm. **Flowers** 2–3 mm; perianth white with greenish or reddish midribs, becoming rose or red, glabrous; tepals dimorphic, those of outer whorl oblong to oval, those of inner whorl oblanceolate; stamens included, 2–3 mm; filaments pilose proximally. **Achenes** brown, 3-gonous, 1.7–2 mm, glabrous.

Varieties 2 (2 in the flora): w United States.

1. Peduncles and involucres glandular; nw Arizona, ec California, n and c Nevada, se Oregon, w and ec Utah . . . . 179a. *Eriogonum nutans* var. *nutans*
1. Peduncles and involucres glabrous; Elko County, Nevada, apparently introduced but not persisting in Nevada County, California . . . . 179b. *Eriogonum nutans* var. *glabratum*

179a. Eriogonum nutans Torrey & A. Gray var. **nutans**
• Dugway wild buckwheat ☒

Eriogonum deflexum Torrey subsp. *ultrum* S. Stokes; *E. nutans* var. *brevipedicellatum* S. Stokes

Peduncles glandular. **Involucres** glandular. $2n = 80$.

Flowering May–Sep. Sandy flats and slopes, saltbush, greasewood, sagebrush, and mountain mahogany communities, pinyon-juniper woodlands; 1200–2300(–2800) m; Ariz., Calif., Nev., Oreg., Utah.

Variety *nutans* is rare to infrequent throughout its range. It is most commonly encountered in Nevada and just enters Arizona (Mohave County), California (Mono County), Oregon (Harney County), and western Utah (Beaver, Juab, Millard, Piute, Sanpete, Sevier, and Tooele counties). It has been found east of Wellington, Carbon County, Utah (*Ripley & Barneby 8640*, CAS, NY), although it has not been seen there since 1947.

179b. Eriogonum nutans Torrey & A. Gray var. **glabratum** Reveal, Madroño 18: 172. 1966
• Deeth wild buckwheat ☒

Peduncles glabrous. **Involucres** glabrous. $2n = 80$.

Flowering Jun–Sep. Sandy flats and slopes, saltbush and sagebrush communities, and (as an introduction) in montane conifer woodlands; 1500–1900 m; Calif., Nev.

Variety *glabratum* occurs naturally only in Elko County, Nevada. It was

inadvertently introduced into the Truckee area of Nevada County, California, probably as a result of truckers stopping to put chains on their tires before crossing the Sierra Nevada. The variety was gathered there by Gordon True (*2588a, 2646, 6415*, CAS) from 1965 until 1970, but the plants apparently failed to persist.

180. Eriogonum thurberi Torrey in W. H. Emory, Rep. U.S. Mex. Bound. 2(1): 176. 1859 • Thurber's wild buckwheat

Eriogonum cernuum Nuttall subsp. *thurberi* (Torrey) S. Stokes; *E. cernuum* subsp. *viscosum* S. Stokes

Herbs, spreading, annual, 0.5–4 dm, glabrate, glabrous or sparsely glandular, greenish, grayish, or reddish. **Stems:** caudex absent; aerial flowering stems erect, solid, not fistulose, 0.3–1 dm, often sparsely tomentose and glandular proximally. **Leaves** basal; petiole 1–3 cm; blade oblong to narrowly ovate, 0.8–4.5 × 0.5–3 cm, densely white-tomentose abaxially, floccose or glabrous and greenish adaxially, margins often crenulate. **Inflorescences** cymose, mostly diffuse, 5–30 × 5–50 cm; branches sparsely glandular to glabrate or glabrous; bracts 3, scalelike, 1–2.5 × 1–2.5 mm. **Peduncles** erect, straight, capillary, 0.5–2.5 cm, glabrous and glandular-puberulent distally. **Involucres** broadly turbinate, 1.8–2 × 1.8–2 mm, minutely glandular-puberulent; teeth 5, erect, 0.4–0.6 mm. **Flowers** 1–1.7 mm; perianth white with greenish or reddish midribs, becoming red, glandular-puberulent with a tuft of long white hairs adaxially; tepals dimorphic, those of outer whorl broadly pandurate or flabellate, those of inner whorl oblanceolate; stamens included, 0.7–1.2 mm; filaments mostly glabrous. **Achenes** brown to black, usually lenticular, 0.6–0.8 mm, glabrous. $2n = 40$.

Flowering year-round. Sandy flats, washes, and slopes, saltbush, greasewood, and creosote bush communities, oak, pinyon and/or juniper woodlands, (montane conifer woodlands in Mexico); 100–1200 m; Ariz., Calif., N.Mex.; Mexico (Baja California, Sonora).

Eriogonum thurberi is common to abundant but rarely weedy in northwestern Mexico, southern California (Inyo, Los Angeles, Orange, Riverside, San Bernardino, and San Diego counties), and southern Arizona (Cochise, Gila, Graham, Maricopa, Mohave, Pima, Pinal, Santa Cruz, and Yavapai counties). It just enters New Mexico (Grant County). It and *E. thomasii* are the annual members of the *E. cernuum* complex typically found on the Mojave and Sonoran deserts, with *E. thurberi* extending farther to the east and *E. thomasii* farther to the north. The two occasionally grow together.

181. Eriogonum thomasii Torrey in War Department [U.S.], Pacif. Railr. Rep. 5(2): 364. 1857

• Thomas's wild buckwheat [F]

Herbs, spreading, annual, 0.5–3 dm, glabrous, greenish, sometimes grayish or reddish. **Stems:** caudex absent; aerial flowering stems erect, solid, not fistulose, 0.2–1 dm, glabrous except for few glands proximally. **Leaves** basal; petiole 0.5–3 cm; blade round to round-reniform, 0.5–2 × 0.5–2 cm, densely white-tomentose abaxially, floccose to glabrate and greenish adaxially, margins plane. **Inflorescences** cymose, open to diffuse, 5–25 × 5–25 cm; branches glabrous; bracts 3, scalelike, 1–2.5 × 1–2 mm. **Peduncles** capillary, spreading to slightly recurved, straight or curved, capillary, 0.5–3 cm, glabrous. **Involucres** turbinate-campanulate, 0.6–1.2 × 0.7–1.3 mm, glabrous; teeth 5, erect, 0.4–0.6 mm. **Flowers** 0.8–1 mm in early anthesis, becoming 1.2–2 mm; perianth yellow in early anthesis, becoming white to rose, short-hispidulous; tepals dimorphic, those of outer whorl cordate, becoming saccate-dilated proximally, those of inner whorl spatulate; stamens included, 0.5–0.9 mm; filaments mostly glabrous. **Achenes** brown to dark brown, usually lenticular, 0.8–1 mm, glabrous. $2n = 40.$

Flowering year-round. Sandy flats, washes, and slopes, saltbush, greasewood, creosote bush, and sagebrush communities, pinyon and/or juniper woodlands; -70–1200(–1400) m; Ariz., Calif., Nev., Utah; Mexico (Baja California, Sonora).

Eriogonum thomasii is common to abundant and even weedy throughout its range on the Mojave and Sonoran deserts in Arizona, southeastern California, southern Nevada, and southwestern Utah. The distinctive swollen bases of the outer tepals are easy to observe in fruiting material but not always obvious in early anthesis.

182. Eriogonum pusillum Torrey & A. Gray, Proc. Amer. Acad. Arts 8: 184. 1870 • Yellow turbans [E]

Eriogonum reniforme Torrey & Frémont var. *playanum* (M. E. Jones) S. Stokes; *E. reniforme* subsp. *pusillum* (Torrey & A. Gray) S. Stokes

Herbs, spreading, annual, 0.5–3 dm, glabrous, greenish, grayish, or reddish. **Stems:** caudex absent; aerial flowering stems erect, solid, not fistulose, 0.2–0.8 dm, glabrous except for few glands proximally. **Leaves** basal; petiole 1–3 cm; blade oblong-ovate to round, 0.5–2.5(–3) × 0.4–2(–2.5) cm, densely

white-tomentose abaxially, floccose to subglabrate and greenish adaxially, often with some glandular hairs, margins plane or rarely crenulate. **Inflorescences** cymose, mostly open, 5–25 × 05–25 cm; branches glabrous; bracts 3, scalelike, 1–3 × 1–2 mm. **Peduncles** erect, mostly straight, slender, 0.1–5(–7) cm, glabrous. **Involucres** campanulate, 1–1.5(–1.7) × 1.5–3 mm, glandular-puberulent; teeth 5, erect, 0.4–0.7 mm. **Flowers** 1–1.7 mm in early anthesis, becoming 1.5–3 mm; yellow in early anthesis, becoming reddish yellow to red, glandular; tepals dimorphic, those of outer whorl oblong-elliptic to obovate, those of inner whorl oblong; stamens included, 1–1.5 mm; filaments mostly glabrous. **Achenes** dark brown, usually lenticular, 0.6–0.8 mm, glabrous. $2n = 32.$

Flowering Feb–Aug. Sandy flats, washes, and slopes, saltbush, greasewood, creosote bush, and sagebrush communities, pinyon and/or juniper woodlands; 70–1300(–1600) m; Ariz., Calif., Nev., Oreg., Utah.

Eriogonum pusillum is common to abundant but rarely weedy. It is found mainly in the northern Mojave Desert and in the Great Basin of the Intermountain West of California and Nevada, extending just into southeastern Oregon, western Utah, and northwestern Arizona. It supposedly was found near Wickenburg in Maricopa County, California (*M. E. Jones s.n.*, 5 May 1903, POM), well out of the known range of the species today. The San Benito River canyon site in San Benito County, California (*Mason 5543*; UC), also is notably disjunct, and although the species has not been re-collected there, that location record is more reliable. The Plumas County record is based on an 1880 Austin collection (NY), and the location should be regarded as questionable.

Eriogonum pusillum is used as a food plant by the rare Mojave dotted-blue butterfly (*Euphilotes mojave*). Also, along with *E. reniforme*, it is an important food plant for juvenile desert tortoises (*Gopherus agassizii*). The Kawaiisu people of southern California gathered the achenes of *E. pusillum*, which were pounded into a powder and cooked into a mush or consumed dry (M. L. Zigmond 1981).

183. Eriogonum reniforme Torrey & Frémont in J. C. Frémont, Rep. Exped. Rocky Mts., 317. 1845

• Kidney-leaf wild buckwheat

Herbs, spreading, annual, 0.5–4 dm, glabrous, greenish, grayish, or reddish. **Stems:** caudex absent; aerial flowering stems erect, solid, not fistulose, 0.2–0.8 dm, glabrous except for few hairs proximally. **Leaves** basal; petiole 0.5–6 cm, tomentose to floccose; blade obovate or reniform to rounded, 0.5–2(–2.5) × 0.5–2 cm, densely white-tomentose abaxially, tomentose to subglabrous and grayish or greenish adaxially, margins plane. **Inflorescences** cymose, open, 5–35 × 5–35 cm; branches glabrous; bracts 3, scalelike, 1–3 × 1–2 mm. **Peduncles** erect, mostly curved, slender to capillary, 0.3–1.5 cm, glabrous. **Involucres** turbinate to turbinate-campanulate, 1.5–2(–2.2) × 1.5–2.5 mm, glabrous, glaucous; teeth 5, erect, 0.5–0.8 mm. **Flowers** 1–1.5 mm in early anthesis, becoming 1.5–2 mm; perianth yellowish to yellow in early anthesis, becoming reddish, glandular; tepals dimorphic, those of outer whorl broadly ovate, those of inner whorl oblong; stamens exserted, 1.5–2.5 mm; filaments pilose proximally. **Achenes** brown, lenticular, 0.8–1 mm, glabrous. $2n = 32$.

Flowering Feb–Aug. Sandy flats, washes, and slopes, saltbush, greasewood, creosote bush, and sagebrush communities, pinyon and/or juniper woodlands; 0–1700 m; Ariz., Calif., Nev.; Mexico (Baja California).

Eriogonum reniforme is infrequent to locally common throughout its range on the Mojave and Sonoran deserts in Arizona, southeastern California, and southern Nevada, where it extends northward along the Lahontan Trough onto the cold desert of the Great Basin. It is disjunct in the Tucson Mountains of Maricopa County, Arizona, and while specimens have not been collected there since 1906 (*Thornber 5886*, CAS, MO), the species recently was found north of Wickenburg (*Welnick s.n.*, undated, ASU). It just enters Mexico, where it is rare.

Eriogonum reniforme is a food plant for the rare Mojave dotted-blue butterfly (*Euphilotes mojave*). The small blue (*Philotiella speciosa*) also feeds on these plants.

184. **Eriogonum wetherillii** Eastwood, Proc. Calif. Acad. Sci., ser. 2, 6: 319. 1896 • Wetherill's wild buckwheat [E]

Eriogonum filiforme L. O. Williams; *E. sessile* S. Stokes ex M. E. Jones

Herbs, spreading, annual, 0.5–2.5 dm, glabrous, reddish green to reddish. **Stems:** caudex absent; aerial flowering stems erect, solid, not fistulose, 0.1–0.4 dm, glabrous except somewhat villous proximally. **Leaves** basal; petiole 1–5 cm, tomentose to floccose; blade oblong to orbicular, (0.5–)1–4 × (0.5–)1–3 cm, densely white-tomentose abaxially, floccose to glabrate and reddish or greenish adaxially, margins plane. **Inflorescences** cymose, diffuse, flat-topped to rounded, 5–20 × 10–35 cm; branches glabrous; bracts 3, scalelike, 0.5–2 × 1–2 mm. **Peduncles** erect, straight, filiform, (0.3–)0.5–1 cm, glabrous, rarely absent distally. **Involucres** turbinate, (0.3–)0.5–1 × (0.2–)0.5–1 mm, glabrous; teeth 4, erect, 0.1–0.3 mm. **Flowers** 0.5–1.2 mm in early anthesis, becoming 1–1.5 mm; perianth

yellowish to red in early anthesis, becoming pink to rose, glabrous; tepals monomorphic, elliptic to obovate; stamens included to slightly exserted, 0.7–1 mm; filaments glabrous. **Achenes** dark brown to black, lenticular, 0.6–1 mm, glabrous.

Flowering Apr–Oct. Sandy to clayey flats, washes, and slopes, saltbush, greasewood, blackbrush, and creosote bush communities, oak and pinyon and/or juniper woodlands; 1000–2100(–2700) m; Ariz., Colo., N.Mex., Utah.

Eriogonum wetherillii is a species of the Colorado Plateau in northern Arizona (Apache, Coconino, and Navajo counties), southwestern Colorado (Montezuma County), northwestern New Mexico (San Juan County), and southeastern Utah (Emery, Garfield, Grand, Kane, San Juan, eastern Sevier, and Wayne counties). It typically is locally common, especially on sandy soils, but only rarely weedy.

185. **Eriogonum subreniforme** S. Watson, Proc. Amer. Acad. Arts 12: 260. 1877 • Kidney-shaped wild buckwheat [E]

Eriogonum filicaule S. Stokes

Herbs, erect or slight spreading, annual, 1–5(–7) dm, mostly glabrous, usually greenish. **Stems:** caudex absent; aerial flowering stems erect, solid, not fistulose, (0.2–)0.5–1.5 dm, glabrous except floccose to glabrescent proximally. **Leaves** basal; petiole (1–) 2–6 cm, mostly floccose; blade reniform to orbiculate, (0.3–)1–3.5(–4) × (0.3–)1–4(–5) cm, densely white-tomentose abaxially, hirsute to floccose or glabrous and greenish adaxially, margins plane. **Inflorescences** cymose, open to somewhat diffuse, 5–40(–50) × 5–35 cm; branches glabrous; bracts 3, scalelike, 1–3 × 1.5–2.5 mm. **Peduncles** erect or nearly so, straight, filiform, 0.5–2.5 cm, glabrous or rarely floccose at or before anthesis. **Involucres** turbinate, 0.5–1 × 0.6–0.9 mm, glabrous; teeth 5, erect, 0.2–0.4 mm. **Flowers** 0.6–1.6 (–2) mm; perianth white to rose or rarely yellowish, glabrous or puberulent to sparsely hirsute; tepals monomorphic, lanceolate to spatulate or elliptic to ovate; stamens included to slightly exserted, 0.7–1.4 mm; filaments glabrous. **Achenes** light to dark brown, 3-gonous, 1.7–2 mm, glabrous.

Flowering Apr–Oct. Sandy to gypsophilous or clayey flats and slopes, saltbush, greasewood, blackbrush, and creosote bush communities, oak and pinyon-juniper woodlands; 800–2100 m; Ariz., N.Mex., Utah.

Eriogonum subreniforme is generally rare to infrequent throughout its scattered range in northern Arizona (Apache, Coconino, Mohave, and Navajo counties), northwestern New Mexico (McKinley and San

Juan counties), and southern Utah (Garfield, Kane, and Washington counties). Even in a "good" year, the plants are rarely common and never weedy.

186. **Eriogonum viscidulum** J. T. Howell, Leafl. W. Bot. 3: 138. 1942 • Sticky wild buckwheat C E

Herbs, erect or spreading, annual, 0.5–4(–5) dm, minutely viscid, yellowish green. Stems: caudex absent; aerial flowering stems erect, solid, not fistulose, 0.2–1 dm, minutely viscid. Leaves basal; petiole 0.5–4 cm, floccose; blade elliptic to broadly ovate, 0.5–3 × 0.5–3 cm, densely white-tomentose abaxially, thinly floccose to glabrate and greenish adaxially, margins plane. Inflorescences cymose, open, 3–35 × 3–30 cm; branches minutely viscid; bracts 3, scalelike, 1–2 × 1–2 mm. Peduncles erect or nearly so, straight, filiform, 0.5–1.5 cm, viscid. Involucres narrowly turbinate, 1–1.2 × 0.6–0.8 mm, viscid; teeth 4, erect, 0.3–0.5 mm. Flowers 1.3–1.5 mm in early anthesis, becoming 1.5–2 mm; perianth pale yellow to yellow in early anthesis, becoming tinged with red, glabrous; tepals monomorphic, oblong; stamens included, 0.9–1.1 mm; filaments glabrous. Achenes light to dark brown, 3-gonous, 0.8–1 mm, glabrous.

Flowering Apr–Jun. Sandy flats and slopes, saltbush and creosote bush communities; of conservation concern; 400–600 m; Ariz., Nev.

Eriogonum viscidulum is a rare and localized species known from a few locations in Clark County, Nevada, and adjacent Mohave County, Arizona. Its range extends from the Muddy River near Weiser Wash to the confluence with the Virgin River, and from Sandy Hollow Wash to Middle Point on the Colorado River. It is listed for protection by the state of Nevada and is considered a "sensitive" species by the Bureau of Land Management and the Nevada Heritage Program. In Arizona, it is a "special status species."

187. **Eriogonum collinum** S. Stokes ex M. E. Jones, Contr. W. Bot. 11: 15. 1903 • Juniper wild buckwheat E

Herbs, erect or spreading, annual, 1–5(–7) dm, glabrous or infrequently slightly floccose, greenish, grayish, or reddish. Stems: caudex absent; aerial flowering stems erect, solid, not fistulose, 0.3–1 dm, glabrous, infrequently slightly floccose proximally. Leaves basal; petiole 1–5 cm, floccose; blade elliptic to obovate or orbiculate, (0.5–)1–2.5(–3) ×

(0.5–)1–3 (3.5) cm, tomentose to hirsute abaxially, floccose or glabrous and greenish adaxially, margins plane. Inflorescences cymose to cymose-paniculate or racemose, open, 5–60 × 5–45 cm; branches glabrous; bracts 3, scalelike, 1–2(–4) × 1–2 mm. Peduncles ascending, curving, slender, 1–5 cm, glabrous. Involucres turbinate, (1.5–)2–3 × (1–)1.5–2.5 mm, glabrous; teeth 5, erect, 0.5–1 mm. Flowers 1–2.5 mm; perianth white to pale yellow, glabrous, pustulose proximally; tepals monomorphic, lanceolate to spatulate or ovate; stamens included, 1–1.8 mm; filaments glabrous. Achenes brown, lenticular, 2–2.5 mm, glabrous. 2*n* = 36.

Flowering May–Nov. Sandy to clayey or gravelly washes, flats, and slopes, saltbush, greasewood, sagebrush, and mountain mahogany communities, pinyon and/or juniper woodlands; 1300–2000 m; Calif., Nev., Oreg.

Eriogonum collinum is infrequent to locally common in the Lahontan Basin region of the Intermountain West in northeastern California (Lassen and Modoc counties), south-central Oregon (Lake County), and northwestern Nevada (Douglas, Humboldt, Lyon, Pershing, Storey, and Washoe counties).

188. **Eriogonum salicornioides** Gandoger, Bull. Soc. Roy. Bot. Belgique 42: 187. 1906 • Saltwort wild buckwheat E

Eriogonum demissum S. Stokes; *E. demissum* var. *romanum*; *E. vimineum* Douglas ex Bentham var. *salicornioides* (Gandoger) S. Stokes

Herbs, spreading, annual, (0.5–) 1–3 dm, glabrous, greenish to reddish. Stems: caudex absent; aerial flowering stems erect, solid, not fistulose, 0.2–0.6 dm, glabrous. Leaves basal; petiole 0.5–3 cm, floccose; blade ovate to orbiculate, 0.4–1.5 × 0.4–1.2 cm, white-tomentose abaxially, floccose or glabrous and greenish adaxially, margins plane. Inflorescences cymose, open, 5–25 × 5–25 cm; branches glabrous; bracts 3, scalelike, 1–2 × 1–2 mm. Peduncles absent or erect, straight, slender, 0.1–0.5 cm, glabrous, absent distally. Involucres turbinate, 1.5–2 × 1–1.7 mm, glabrous; teeth 5, erect, 0.4–0.8 mm. Flowers 1.2–1.7 mm; perianth white to pale yellow, glabrous, pustulose proximally; tepals monomorphic, ovate; stamens included, 1.2–1.8 mm; filaments glabrous. Achenes brown, lenticular, 1.6–2 mm, glabrous. 2*n* = 36.

Flowering May–Jul. Clayey flats and slopes, saltbush, greasewood, and sagebrush communities, juniper woodlands; 700–1300(–1500) m; Idaho, Nev., Oreg.

Eriogonum salicornioides is known from Harney and Malheur counties in southeastern Oregon and Elmore and Owyhee counties of southwestern Idaho. Its

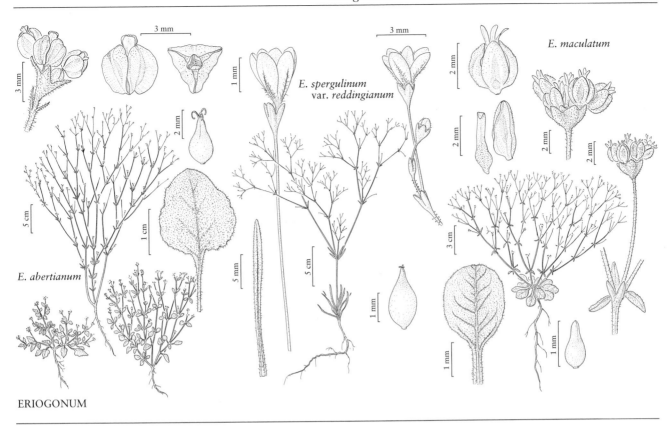

E. spergulinum var. reddingianum

E. maculatum

ERIOGONUM

attribution to Nevada is based on a Percy Train collection (Jun 1934, NY, PAC) from a dry desert valley in northern Humboldt County. Efforts to relocate the species in that state have been unsuccessful.

189. Eriogonum abertianum Torrey in W. H. Emory, Not. Milit. Reconn., 150. 1848 • Abert's wild buckwheat [F]

Eriogonum abertianum var. *cyclosepalum* (Greene) Fosberg; *E. abertianum* var. *lappulaceum* (Greene) S. Stokes; *E. abertianum* var. *neomexicanum* Gandoger; *E. abertianum* subsp. *pinetorum* (Greene) S. Stokes; *E. abertianum* var. *ruberrimum* Gandoger; *E. abertianum* var. *villosum* Fosberg; *E. cyclosepalum* Greene; *E. pinetorum* Greene

Herbs, erect or spreading, annual, 0.5–6(–7) dm, hirsute, greenish, grayish, tawny, or reddish. **Stems:** caudex absent; aerial flowering stems prostrate to erect, solid, not fistulose, 0.1–1 dm, appressed-hirsute. **Leaves** basal and cauline; basal: petiole 0.5–6 cm, villous to hoary, blade oblong to obovate, 1–4 × 1–3 cm, villous to hoary-tomentose and greenish, tawny, or reddish on both surfaces, margins plane, occasionally crenulate; cauline sessile, blade linear, lanceolate, or narrowly obovate, 1–4 × 0.3–2 cm, similar to basal blade. **Inflorescences** cymose, open to diffuse, 5–40(–60) × 5–50 cm; branches hirsute; bracts 3–6, semileaflike, 2–10 × 1–3 mm. **Peduncles** ascending to erect, mostly straight, slender, 0.5–6 cm, villous to hoary-tomentose. **Involucres** broadly campanulate, 2–3 × 2–3 mm, villous-canescent; teeth 5, lobelike, usually reflexed, 4–6 mm. **Flowers** 3–4.5 mm; perianth white to pale yellow in early anthesis, becoming reddish or rose, glabrous; tepals dimorphic, those of outer whorl orbiculate-cordate, those of inner whorl lanceolate to spatulate; stamens mostly exserted, 1.5–3.5 mm; filaments mostly pilose proximally. **Achenes** brown to dark brown, lenticular, 0.6–1 mm, glabrous. $2n = 40$.

Flowering year-round. Sandy, gravelly, or clayey flats, washes, and slopes, mixed grassland, saltbush, greasewood, creosote bush, blackbrush, and manzanita communities, oak and conifer woodlands; 400–2500 m; Ariz., N.Mex., Tex.; Mexico (Coahuila, Chihuahua, Sonora, San Luis Potosí).

Eriogonum abertianum is widespread and common to abundant or even locally weedy. It is basically a species of the Chihuahuan and Sonoran deserts, ranging from San Luis Potosí and Sonora in northern Mexico to western Texas and much of New Mexico and Arizona.

This species is exceedingly variable and can be differentiated into several geographic and seasonal phases (C. C. Baskin et al. 1993; G. A. Fox 1989, 1990, 1990b). Variety *abertianum* (including *E. pinetorum*) is a more northern and western, summer- and fall-flowering expression, with an erect habit and a cymose-paniculate inflorescence when the plants are in fruit. A vernal phase (var. *villosum*) of this northern form occurs throughout much of the same range (mainly Arizona, New Mexico, and the Sonoran Desert of Mexico). It is a hairier and more spreading expression, with elongated inflorescence branches bearing more floriferous involucres. Variety *ruberrimum* (including *E. cyclosepalum*) is the more southern, summer- to fall-flowering expression (mainly New Mexico, Texas, and the Chihuahuan Desert of Mexico) with a prostrate to spreading, compact habit and a racemose inflorescence at full maturity. There is a vernal expression of this as well; it differs from the vernal form of var. *abertianum* in having longer peduncles. There are no sharp morphologic or geographic boundaries for any of these expressions, and while the fruiting extremes are clearly distinguishable, far too many specimens are impossible to place satisfactorily.

The Navajo (Diné) people use these plants as a lotion for both themselves and their horses (P. A. Vestal 1952).

190. Eriogonum pharnaceoides　Torrey in L. Sitgreaves, Rep. Exped. Zuni Colorado Rivers, 167, plate 11. 1853

Herbs, erect or spreading, annual, 1–5 dm, villous, greenish, yellowish green, or reddish. **Stems:** caudex absent; aerial flowering stems erect or nearly so, solid, not fistulose, 0.1–0.8 dm, villous. **Leaves** basal and cauline; basal: petiole 0.1–0.5(–1) cm, blade linear-lanceolate to linear-oblanceolate, 1–4 × 0.1–0.4 cm, white-lanate abaxially, villous and greenish to yellowish green adaxially, margins plane or revolute; cauline sessile, blade linear, 0.5–2.5 × 0.05–0.3 cm, tomentose abaxially, thinly villous or glabrous and greenish adaxially. **Inflorescences** cymose, open, 5–45 × 3–40 cm; branches villous; bracts 3–8, semileaflike, 5–15 × 0.3–1.5(–2) mm. **Peduncles** erect or nearly so, straight, slender, (1–)2–5(–7) cm, sparsely villous or glabrous. **Involucres** campanulate, 1–2 × 2–3 mm, villous; teeth 5, mostly erect to spreading, 1–3 mm. **Flowers** 1–3 mm; perianth white to rose or yellow, sometimes becoming reddish, glabrous; tepals dimorphic, those of outer whorl oblong-ovate, 2-saccate-dilated proximally, those of inner whorl linear-oblong; stamens mostly included, 1–1.5 mm; filaments mostly pilose proximally. **Achenes** brown to blackish, lenticular, 1.8–2 mm, glabrous.

Varieties 2 (2 in the flora): w North America, including n Mexico.

190a. Eriogonum pharnaceoides　Torrey var. pharnaceoides　• Wire-stem wild buckwheat

Plants 1–5 dm. **Basal leaves:** petiole usually 0.3–0.5(–1) cm; blade 2–4 × 0.2–0.4 cm. **Inflorescences** (10–)20–45 cm. **Perianths** white to rose.

Flowering Jul–Nov. Sandy or gravelly slopes, sagebrush communities, oak, pinyon-juniper, and montane conifer woodlands; 1400–2500 m; Ariz., N.Mex.; Mexico (Chihuahua).

Variety *pharnaceoides* is found infrequently throughout its range in northern and eastern Arizona (Apache, Cochise, Coconino, Gila, Graham, Greenlee, Mohave, Navajo, Pima, Pinal, Santa Cruz, and Yavapai counties) and western New Mexico (Catron, Grant, Hidalgo, Luna, Sierra, Socorro, and Taos counties). Thurber (*773*, NY) gathered the only known collection from Mexico at Janos, Chihuahua, in 1852.

190b. Eriogonum pharnaceoides　Torrey var. cervinum　Reveal, Great Basin Naturalist 34: 246. 1974　• Deer Lodge wild buckwheat　E

Plants 1–3 dm. **Basal leaves:** petiole usually 0.1–0.3(–0.5) cm; blade 1–3(–4) × 0.1–0.25(–0.4) cm. **Inflorescences** 5–25 cm. **Perianths** yellow.

Flowering Jul–Sep. Sandy or gravelly slopes, sagebrush and mountain mahogany communities, oak, pinyon-juniper and montane conifer woodlands; (1400–)1800–2300 m; Ariz., Nev., Utah.

Variety *cervinum* is rare to infrequent in its limited range of eastern Lincoln County, Nevada, southwestern Iron and western Washington counties, Utah, and northeastern Mohave County, Arizona. It is totally disjunct from the typical variety.

191. Eriogonum spergulinum A. Gray, Proc. Amer. Acad. Arts 7: 389. 1868 [E] [F]

Oxytheca spergulina (A. Gray) Greene

Herbs, prostrate to spreading or erect, annual, 0.5–4 dm, glabrous or glandular and short-hispid, greenish, grayish, or reddish. **Stems:** caudex absent; aerial flowering stems prostrate to erect, solid, not fistulose, 0.1–0.5 dm, glabrous or glandular and short-hispid. **Leaves** basal and cauline; basal: petiole 0.05–0.3 cm, hispid, blade linear, (0.3–)1–3(–4) × 0.05–0.3 cm, short-hispid, margins plane or revolute, ciliate; cauline sessile, blade linear, 0.3–2.5 × 0.05–0.3 cm, similar to basal blade. **Inflorescences** cymose, open to diffuse, 4–25 × 5–35 cm; branches sparsely hispid to puberulent, internodes usually glandular; bracts 3–6, semileaflike, 2–10 × 0.5–2 mm. **Peduncles** erect, straight, filiform, 0.4–1.5 cm, glabrous. **Involucres** turbinate, 0.5–1 × 0.4–0.8 mm, glabrous; teeth 4, erect, 0.2–0.4 mm. **Flowers** 1.5–3.5 mm; perianth white with greenish to reddish midribs, becoming pinkish to rose, glabrous or sparsely pubescent; tepals monomorphic, oblong; stamens included, 0.5–2 mm; filaments usually glabrous. **Achenes** brown to blackish, lenticular, 1.5–2.3 mm, glabrous.

Varieties 3 (3 in the flora): w United States.

1. Branch internodes not glandular; plants prostrate to ascending; s Sierra Nevada, California . . . . 191c. *Eriogonum spergulinum* var. *pratense*
1. Branch internodes glandular; plants erect.
   2. Flowers 1.5–2.5 mm; Transverse, Sierra Nevada, and North Coast Ranges of nw, sw, and e California and w Nevada, also s Oregon, c Idaho . . . . 191a. *Eriogonum spergulinum* var. *reddingianum*
   2. Flowers 2.5–3.5 mm; c Sierra Nevada, California . . . . 191b. *Eriogonum spergulinum* var. *spergulinum*

191a. Eriogonum spergulinum A. Gray var. **reddingianum** (M. E. Jones) J. T. Howell, Leafl. W. Bot. 6: 79. 1950 • Redding's wild buckwheat [E] [F]

Oxytheca reddingiana M. E. Jones, Bull. Torrey Bot. Club 9: 32. 1882; *Eriogonum spergulinum* subsp. *reddingianum* (M. E. Jones) Munz ex Reveal

Plants erect, 0.8–4 dm. **Inflorescence branches:** internodes glandular. **Flowers** 1.5–2.5 mm, perianth glabrous or sparsely hairy.

Flowering Jun–Sep. Sandy to gravelly, often granitic or pumice flats and slopes, sagebrush communities, oak, montane, and subalpine conifer woodlands; 1300–3400 m; Calif., Idaho, Nev., Oreg.

Variety *reddingianum* is widespread and often common to even locally weedy. It occurs in the mountains and cold desert regions of California from the Transverse Ranges northward in the mountains of eastern California and western Nevada, and in California's North Coast Ranges. It is found also in less montane, mainly cold desert regions of southern Oregon, and disjunct at scattered sites in central Idaho.

191b. Eriogonum spergulinum A. Gray var. **spergulinum** • Spurry wild buckwheat [E]

Plants erect, 1–4 dm. **Inflorescence branches:** internodes glandular. **Flowers** 2.5–3.5 mm; perianth glabrous.

Flowering Jun–Sep. Sandy granitic flats and slopes, mixed grassland and sagebrush communities, oak and montane conifer woodlands; 1200–3000 m; Calif.

Variety *spergulinum* has a much more restricted range than var. *reddingianum*, being confined to the western side of the central Sierra Nevada in Calaveras, El Dorado, Fresno, Madera, Mariposa, Tulare, and Tuolumne counties. Nonetheless, the taxon can be locally abundant to even weedy. A sharp distinction between this variety and var. *reddingianum* is not always possible at the northern end of the latter's range. A collection from Bahia de la Angeles, Baja California, Mexico (*Palmer 196*, NY) is certainly mislabeled.

191c. Eriogonum spergulinum A. Gray var. **pratense** (S. Stokes) J. T. Howell, Leafl. W. Bot. 6: 80. 1950 • Mountain meadow wild buckwheat [E]

Eriogonum pratense S. Stokes, Leafl. W. Bot. 3: 201. 1943

Plants prostrate to ascending, 0.5–1 dm. **Inflorescence branches:** internodes not glandular. **Flowers** 1.8–2 mm; perianth hirsutulous.

Flowering Jul–Aug. Sandy granitic slopes, high-elevation meadow and sagebrush communities, subalpine to alpine conifer woodlands; (2500–)2800–3500 m; Calif.

Variety *pratense* is restricted to the southern Sierra Nevada in Inyo and Tulare counties.

192. Eriogonum hirtiflorum   A. Gray ex S. Watson, Proc. Amer. Acad. Arts 12: 259. 1877   • Hairy-flower wild buckwheat   E

Oxytheca hirtiflora (A. Gray ex S. Watson) Greene

Herbs, spreading, annual, 0.5–1.5 dm, glandular, greenish or reddish. **Stems:** caudex absent; aerial flowering stems mostly erect, solid, not fistulose, 0.2–0.5 dm, glandular. **Leaves** basal and cauline; basal: petiole 0.5–2 cm, glabrous, slightly winged, ciliate, blade obovate to spatulate, 1–2.5 × 0.4–0.8 cm, glabrous, margins ciliate; cauline sessile, blade lanceolate, 0.1–0.5 × 0.1–0.3 cm, similar to basal blade. **Inflorescences** cymose, usually diffuse, 5–13 × 5–25 cm; branches glandular; bracts 3, scalelike, 0.5–1.5 × 1–2 mm. **Peduncles** absent or erect, straight, filiform, 0.1–0.5 cm, glandular. **Involucres** turbinate, 0.8–1 × 0.7–0.9 mm, sparsely hispid; teeth 4, erect or nearly so, 0.4–0.8 mm. **Flowers** 2, 0.8–1.1 mm; perianth pink to reddish, hirsutulous with stiff, white, hooked hairs; tepals monomorphic, oblong; stamens included, 0.5–0.8 mm; filaments glabrous. **Achenes** brown, lenticular, 1–1.3 mm, glabrous; beak exserted beyond tepals.

Flowering May–Oct. Sandy to gravelly flats and slopes, mixed grassland and chaparral communities, oak and montane conifer woodlands; 200–2000 m; Calif.

Eriogonum hirtiflorum is found in scattered locations throughout much of California (Amador, Colusa, Fresno, Glenn, Kern, Lake, Los Angeles, Madera, Mariposa, Mendocino, Monterey, San Bernardino, San Benito, Santa Cruz, Siskiyou, Tulare, Tuolumne, Trinity, and Ventura counties) in the Coast Ranges and along the western slope of the Sierra Nevada. The species rarely is common at any single location and, being small and often obscure, it is frequently missed by collectors.

193. Eriogonum inerme   (S. Watson) Jepson, Fl. Calif. 1: 406. 1914   E

Oxytheca inermis S. Watson, Proc. Amer. Acad. Arts 12: 273. 1877

Herbs, spreading, annual, 0.5–3 dm, glandular, greenish or reddish. **Stems:** caudex absent; aerial flowering stems mostly erect, solid, not fistulose, 0.1–0.5 dm, glandular. **Leaves** basal and cauline; basal: petiole 0.5–2 cm, glabrous, not or indistinctly winged, ciliate, blade spatulate, 1–4(–4.5) × 0.4–1.5(–1.8) cm, glabrous, margins ciliate; cauline sessile, blade lanceolate, 0.3–1 × 0.1–0.3 cm, similar to basal blade. **Inflorescences**

cymose, open or diffuse, 5–25 × 5–30 cm; branches glandular; bracts 3, scalelike, 0.5–1.5 × 1–2 mm, occasionally semileaflike, 4–10(–12) × 1–5 mm. **Peduncles** absent or erect, straight, slender, 0.1–1 cm, glandular. **Involucres** turbinate, 1.5–1.9 × 1–1.2 mm, glabrous or hispidulous; teeth 4, erect or nearly so, 0.8–1.2 mm. **Flowers** 4–6, 1.2–1.8 mm; perianth pink to reddish, hirsutulous with stiff, white, hooked hairs; tepals monomorphic, oblong; stamens included, 1–1.5 mm; filaments glabrous. **Achenes** brown, lenticular, 1.5–1.9 mm, glabrous; beak not exserted beyond tepals.

Varieties 2 (2 in the flora): California, Idaho.

1. Involucres glabrous; Coast and Transverse ranges, California, disjunct in Boise County, Idaho . . . . 193a.   *Eriogonum inerme* var. *inerme*
1. Involucres hispidulous; Transverse Ranges and s Sierra Nevada, disjunct in Tuolumne County, California . . . . . . 193b.   *Eriogonum inerme* var. *hispidulum*

193a. Eriogonum inerme   (S. Watson) Jepson var. **inerme**   • Unarmed wild buckwheat   E

Eriogonum vagans S. Watson

Involucres glabrous.

Flowering May–Aug. Sandy to gravelly flats and slopes, mixed grassland and chaparral communities, oak and montane conifer woodlands; 600–2200 m; Calif., Idaho.

Variety *inerme* is rare to infrequent at scattered sites in Coast and Transverse ranges of California in Contra Costa, Kern, Lake, Monterey, San Benito, San Luis Obispo, Santa Clara, Tulare, and Ventura counties. The Idaho population (near New Centerville, Boise County) is a recent introduction, probably having come from California with mining equipment.

193b. Eriogonum inerme   (S. Watson) Jepson var. **hispidulum**   Goodman, Amer. Midl. Naturalist 39: 502. 1948   • Goodman's unarmed wild buckwheat   E

Eriogonum inerme subsp. *hispidulum* (Goodman) Munz

Involucres hispidulous.

Flowering May–Sep. Sandy to gravelly flats and slopes, mixed grassland and chaparral communities, oak and conifer woodlands; 800–2100 m; Calif.

Variety *hispidulum* is the Sierra Nevada component of the species. Like var. *inerme*, it is rare to infrequent at widely scattered sites. It is found in the Transverse Ranges (Santa Barbara, Ventura, southwestern Kern, Los Angeles, and San Bernardino counties) but is concentrated mainly in Fresno, Kern, and Tulare counties, with a disjunct series of populations in Mariposa County and in the Long Barn and Pinecrest areas of Tuolumne County. A sharp distinction between the two varieties is not always possible in the Transverse Ranges, where both may be encountered.

194. Eriogonum angulosum Bentham, Trans. Linn. Soc. London 17: 406, plate 18, fig. 1. 1836
• Angle-stem wild buckwheat E

Herbs, erect to spreading, annual, 1–5(–10) dm, tomentose to floccose or glabrous, usually grayish. **Stems:** caudex absent; aerial flowering stems erect, striated, angled, solid, not fistulose, 0.5–1 dm, tomentose to floccose. **Leaves** basal and cauline; basal: petiole 0.5–3 cm, mostly floccose, blade oblanceolate to oblong-lanceolate, 1–4(–4.5) × (0.2–)0.5–1(–1.3) cm, tomentose abaxially, floccose or glabrate and grayish or greenish adaxially, margins crenulate; cauline sessile, blade lanceolate to oblong, 0.5–2 × 0.3–0.8 cm, similar to basal blade. **Inflorescences** cymose, open, 5–80 × 10–60 cm; branches striated, angled, sparsely tomentose to glabrate; bracts 3, scalelike, 1–3 × 1–3 mm. **Peduncles** erect, straight, slender, 1–2 cm, sparsely tomentose to glabrous. **Involucres** turbinate-campanulate to campanulate, 1.5–2.5(–3) × 1.5–2.5(–3), sparsely puberulent; teeth 5, erect, 0.3–0.6 mm. **Flowers** 1.5–1.8 mm; perianth white to rose, without a conspicuous rose-purple spot on each outer tepal, minutely glandular-puberulent; tepals dimorphic, those of outer whorl elliptic to obovate, sometimes inflated proximally, those of inner whorl narrowly spatulate; stamens exserted, 2–3 mm; filaments pilose proximally. **Achenes** light brown to brown, 3-gonous, 1–1.5 mm, glabrous.

Flowering year-round. Clayey flats and slopes, mixed grassland, saltbush, and chaparral communities, oak and conifer woodlands; 0–800 m; Calif.

The name *Eriogonum angulosum* has been applied to all of the members of its species complex except *E. gossypinum*. Since the 1950s, the name consistently has been applied to plants with long, exserted stamens and strongly angled stems of the Inner Coast Ranges (Alameda and Contra Costa counties south), the western foothills of the southern Sierra Nevada (Tulare County south), and the Central Valley (San Joaquin County south). The southern edge of the range is the northern foothills of the Transverse Ranges (Ventura and Los Angeles counties). The species can be common and occasionally abundant but rarely is weedy. A mixed collection (with *E. gracillimum*) from Barstow, San Bernardino County (*K. Brandegee s.n.*, May 1913, UC), and two sheets of the species from San Diego gathered by Susan Stokes apparently in 1895 (B, SD) are discounted as to location.

In late fruit, the bractlets at the base of the pedicel inside the involucres of *Eriogonum angulosum* often elongate and broaden into oblanceolate segments that fill the involucre. As a result, the involucre appears to have several rows of teeth. This feature may be seen also in *E. viridescens*, but typically the involucres there appear to have only two or three rows of teeth. This feature is seen rarely in *E. maculatum*.

195. Eriogonum maculatum A. Heller, Muhlenbergia 2: 188. 1906 • Spotted wild buckwheat F

Eriogonum angulosum Bentham subsp. *maculatum* (A. Heller) S. Stokes; *E. angulosum* var. *maculatum* (A. Heller) Jepson; *E. angulosum* var. *rectipes* Gandoger

Herbs, erect to spreading, annual, 1–2(–3) dm, tomentose, greenish to reddish. **Stems:** caudex absent; aerial flowering stems erect, not striated or angled, solid, not fistulose, 0.1–0.5 dm, tomentose. **Leaves** basal and cauline; basal: petiole 0.3–1 cm, floccose, blade lanceolate to obovate, 1–3(–4) × 1–1.5(–2) cm, tomentose abaxially, sparsely floccose to glabrate and grayish to greenish adaxially, margins entire or infrequently crenulate; cauline sessile, blade lanceolate to oblanceolate, 0.5–2 × 0.3–1 cm, similar to basal blade. **Inflorescences** cymose, open, 5–25 × 10–30 cm; branches tomentose; bracts 3, scalelike, 0.5–2.5 × 1–2 mm. **Peduncles** spreading, straight or nearly so, filiform, (0.5–)1–3 cm, glandular-puberulent. **Involucres** campanulate, 1–1.5(–2) × 1.5–3(–3.5) mm, glandular-puberulent; teeth 5, erect, 0.4–0.8 mm. **Flowers** 1–2.5 mm; perianth white to yellow, becoming pink or red, with a conspicuous rose-purple spot on each outer tepal, glandular-puberulent; tepals dimorphic, those of outer whorl elliptic to roundish or obovate, inflated from base to middle, those of inner whorl lanceolate; stamens included, 1–1.5(–2) mm; filaments pilose proximally. **Achenes** brown, 3-gonous, 1–1.5 mm, glabrous. *2n* = 40.

Flowering Apr–Nov. Sandy to gravelly or clayey flats and slopes, mixed grassland, saltbush, creosote bush, and sagebrush communities, pinyon-juniper and montane conifer woodlands; 100–2500 m; Ariz., Calif., Idaho, Nev., N.Mex., Oreg., Utah, Wash.; Mexico (Baja California).

Eriogonum maculatum is the most common and widespread expression of the *E. angulosum* complex, being found in Arizona, California, southwestern Idaho,

Nevada, northwestern New Mexico, eastern Oregon, western Utah, and southeastern Washington. It is often common to abundant and may even be weedy, especially along roadsides intermixed with other annual wild buckwheats. The swollen bases of the outer tepals readily distinguish the species. The greatest concentrations of the spotted wild buckwheat are found in the Mojave Desert and the Great Basin. A specimen supposedly found on the "Laramie Plains" of Wyoming (*Parry s.n.*, 1882, ISC) is discounted as to location.

196. Eriogonum viridescens A. Heller, Muhlenbergia
2: 25. 1905 • Two-toothed wild buckwheat E

Eriogonum angulosum Bentham subsp. *bidentatum* (Jepson) S. Stokes; *E. angulosum* subsp. *viridescens*; *E. angulosum* var. *viridescens* (A. Heller) Jepson; (A. Heller) S. Stokes; *E. bidentatum* Jepson

Herbs, spreading, annual, (0.5–) 1–2(–3) dm, tomentose, greenish, grayish, or reddish. **Stems:** caudex absent; aerial flowering stems erect, not striated or angled, solid, not fistulose, 0.2–0.8 dm, tomentose. **Leaves** basal and cauline; basal: petiole 0.5–1 cm, floccose, blade lanceolate to obovate, (0.5–)2–3 × 1.5–2 cm, tomentose abaxially, floccose to glabrate and grayish to greenish adaxially, margins entire or occasionally crenulate; cauline sessile, blade lanceolate to oblanceolate, 0.5–2 × 0.3–1 cm, similar to basal blade. **Inflorescences** cymose, open, 4–25 × 5–30 cm; branches tomentose; bracts 3, scalelike, 0.5–2.5 × 1–2 mm. **Peduncles** spreading, straight, capillary, 1–2 cm, sparsely glandular. **Involucres** campanulate, 2–3 × 2–4 mm, finely glandular; teeth 5, erect, 0.4–0.8 mm. **Flowers** 1–2.5 mm; perianth white to rose, glandular; tepals dimorphic, those of outer whorl obovate to spatulate, broadly inflated above middle, those of inner whorl narrowly spatulate; stamens included, 1–2 mm; filaments pilose proximally. **Achenes** light brown, 3-gonous, 1–1.5 mm, glabrous.

Flowering Apr–Nov. Sandy to gravelly or clayey flats and slopes, mixed grassland, chaparral, saltbush, and creosote bush communities, oak and conifer woodlands; 100–1700 m; Calif.

Eriogonum viridescens is infrequent in the Inner Coast and the Transverse ranges (Fresno, Kern, Kings, Los Angeles, Merced, Monterey, Riverside, San Benito, western San Bernardino, San Luis Obispo, Santa Barbara, and Ventura counties). Unlike those of the more

widespread *E. maculatum*, the outer tepals of *E. viridescens* are swollen apically, rather than basally. A collection made at Crystal Springs in the Coso Mountains of Inyo County (*Coville & Funston 909*, DS, US) probably is accurately labeled but the location remains to be confirmed by modern collections.

197. Eriogonum gracillimum S. Watson in W. H.
Brewer et al., Bot. California 2: 480. 1880
• Rose-and-white wild buckwheat E

Eriogonum angulosum subsp. *gracillimum* (S. Watson) S. Stokes; *E. angulosum* Bentham var. *gracillimum* (S. Watson) M. E. Jones; *E. angulosum* subsp. *victorense* (M. E. Jones) S. Stokes

Herbs, erect to spreading, annual, 1–5 dm, thinly tomentose, greenish, grayish, or reddish. **Stems:** caudex absent; aerial flowering stems erect, not striated but sometimes angled, solid, not fistulose, 0.1–0.8 dm, tomentose. **Leaves** basal and cauline; basal: petiole 0.5–1 cm, floccose; basal blade oblanceolate to oblong, 2–4 × 0.3–1 cm, tomentose abaxially, floccose to glabrate and grayish or greenish adaxially, margins crenulate, often slightly revolute; cauline sessile, blade narrowly oblong, 0.5–2(–6) × 0.2–0.8(–1.5) cm, similar to basal blade. **Inflorescences** cymose, mostly open, 5–35 × 5–35 cm; branches thinly tomentose; bracts 3, scalelike, 1–3 × 1–2.5 mm. **Peduncles** spreading, straight, capillary, 0.8–2.5 cm, glabrous. **Involucres** campanulate, 1.8–2 × 2–3 mm, glandular-puberulent, not densely tomentose adaxially; teeth 5, erect, 0.4–0.8 mm. **Flowers** 2–2.5 mm; perianth white to rose, glandular-puberulent; tepals monomorphic, oblong to elliptic; stamens included, 1–2 mm; filaments pilose proximally. **Achenes** light brown, 3-gonous, 1–1.2(–1.5) mm, glabrous. $2n = 40$.

Flowering year-round. Sandy to gravelly or clayey flats and slopes, mixed grassland, chaparral, saltbush, and creosote bush communities, oak and conifer woodlands; 0–1100 m; Calif.

Eriogonum gracillimum is found in the Coast, Transverse, and scattered mountain desert ranges in Fresno, Inyo, Kern, Kings, Los Angeles, Madera, Merced, Monterey, Riverside, San Benito, San Bernardino, San Diego, San Luis Obispo, Santa Barbara, and Tulare counties. The species is often common and occasionally even weedy.

198. Eriogonum gossypinum Curran, Bull. Calif. Acad. Sci. 1: 274. 1885 • Cottony wild buckwheat E

Herbs, spreading, annual, 0.5–2 dm, tomentose, grayish to reddish. **Stems:** caudex absent; aerial flowering stems erect, not striated or angled, solid, not fistulose, 0.2–0.5 dm, tomentose. **Leaves** basal and cauline; basal: petiole 0.5–1 cm, floccose, blade broadly oblanceolate, 1.5–4 × 0.5–1 cm, tomentose abaxially, floccose or glabrate and grayish to green adaxially, margins crenulate, often slightly revolute; cauline sessile, blade lanceolate, 0.3–2 × 0.2–0.7 cm, similar to basal blade. **Inflorescences** cymose, diffuse, 3–17 × 5–25 cm; branches tomentose; bracts 3, scalelike, 1–2 × 0.5–1.5 mm. **Peduncles** spreading, straight, capillary, 0.2–1.5 cm, thinly tomentose to floccose or glabrous. **Involucres** turbinate, 2.7–3 × 2–2.5 mm, glandular-puberulent, densely tomentose adaxially; teeth 5, spreading, 0.8–1.2(–1.5). **Flowers** 1.5–1.7 mm; perianth white, glandular-puberulent; tepals monomorphic, narrowly oblong; stamens included, 1–1.5 mm; filaments pilose proximally. **Achenes** light brown, 3-gonous, 1.3–1.5 mm, glabrous. $2n = 40$.

Flowering year-round. Clayey gypsophilous flats and slopes, grassland and saltbush communities; 100–500 m; Calif.

Eriogonum gossypinum is rare to infrequent on the foothills of the Inner Coast Ranges in Kings (Kettleman Hills) and eastern San Luis Obispo counties, with two centers of concentration in Kern County, one in the Taft area and another east of Bakersfield. The Kern County populations are the most extensive and subject to the greatest habitat destruction. The species is protected at The Nature Conservancy's Sand Ridge Preserve near Edison.

1h. Eriogonum Michaux subg. Oregonium (S. Watson) Greene, Fl. Francisc., 146. 1891

Eriogonum sect. *Oregonium* S. Watson, Proc. Amer. Acad. Arts 12: 262. 1877

Herbs, spreading to erect or prostrate, rarely decumbent annuals, glabrous or floccose to tomentose or lanate, sometimes sericeous puberulent, or short-pilose; taproot not woody. **Stems** prostrate, ascending or erect, without persistent leaf bases, lanate, tomentose, floccose, or glabrous, sometimes sericeous, puberulent, or short-pilose; caudex absent; aerial flowering stems prostrate to ascending or erect, slender to stout, solid, not fistulose, not disarticulating in ringlike segments proximally, arising directly from root. **Leaves** persistent or marcescent, basal, cauline, or basal and cauline, 1 per node not fasciculate; blade tomentose to floccose or glabrous, sometimes sericeous, puberulent, or short-pilose, not glandular. **Inflorescences** cymose, open or diffuse; bracts 3, connate basally, scalelike. **Peduncles** usually absent, when present restricted to proximal nodes, erect, straight, slender. **Involucres** 1 per node, usually appressed to the inflorescence branches, cylindric to turbinate or rarely campanulate; teeth 5, rarely 6–8, erect or rarely spreading, rarely lobelike, spreading to somewhat reflexed and dividing involucral tube nearly to base. **Flowers** abruptly narrowing to acute base on slender pedicel, without stipelike base; perianth cream, white to pink or rose, yellow, or red, rarely ochroleucous, usually glabrous or glandular, sometimes hispid, hispidulous, pubescent, or hirtellous abaxially, occasionally papillose, mostly glabrous adaxially; tepals connate proximally 1/4–1/2 their length, monomorphic or dimorphic; stamens included or exserted; filaments glabrous or pilose proximally. **Achenes** light to dark brown or rarely nearly black, not winged, lenticular to more often 3-gonous, glabrous. **Seeds:** embryo curved.

Species 28 (26 in the flora): w North America including n Mexico, mainly California.

The species of *Eriogonum* subg. *Oregonium* not treated here are *E. foliosum* S. Watson and *E. hastatum* Wiggins; they occur in northern Mexico.

1. Involucres terminal at tips of short branchlets proximally, not appressed to branches; peduncles sometimes present.
    2. Flowers 0.7–1.5 mm; perianth yellow or cream; arid regions of e and se California, wc Nevada.

3. Perianths yellow; flowers 0.7–1 mm; inflorescences diffuse, spreading; Mojave Desert, se Kern, ne Los Angeles, and nw San Bernardino counties, California . . . . 199. *Eriogonum mohavense*
3. Perianths cream; flowers 1–1.5 mm; inflorescences narrow; Great Basin, s Mono and n Inyo counties, California, w Mineral and Esmeralda counties, Nevada . . . . 200. *Eriogonum ampullaceum*
2. Flowers 1–2.5(–3) mm; perianth yellow or white to pink, rose, or red; Coast Ranges and w Sierra Nevada, California.
    4. Stems glabrous; involucres sessile or, if short-pedunculate, plants from western foothills of Sierra Nevada, California.
        5. Basal leaf blades oblong-ovate, reniform, or rounded to cordate; involucres 3–4 mm, 5-toothed; n California from San Mateo, Santa Clara, and Mariposa counties n to sw and sc Oregon . . . . 208. *Eriogonum luteolum* (in part)
        5. Basal leaf blades suborbiculate or round to reniform; involucres 2–4 mm, 5–8-toothed; wc California from Santa Clara and Stanislaus counties s to Santa Barbara and Ventura counties.
            6. Involucres 2–2.5 mm; 5-toothed; flowers 2–2.5(–3) mm, perianth minutely puberulent, white to rose or yellow in fruit; Santa Clara and Stanislaus counties s to w Kern and nw Ventura counties, California . . . . 206. *Eriogonum covilleanum*
            6. Involucres 3–4 mm; 6–8-toothed; flowers 1–2 mm, perianth glabrous, white to rose in fruit; c Monterey and w San Benito counties, California . . . . 207. *Eriogonum nortonii*
    4. Stems tomentose or, if glabrous, involucres pedunculate at proximal nodes and plants from Coast Ranges of California.
        7. Involucres glabrous; e Monterey, w San Benito, and s Santa Clara (extirpated) counties, California . . . . 205. *Eriogonum argillosum*
        7. Involucres tomentose; Contra Costa, Fresno, Kern, Merced, Monterey, San Benito, San Luis Obispo, and Solano counties, California.
            8. Leaves strictly basal, blade suborbiculate; styles 0.1–0.3 mm; sw Fresno County, California. . . . . 201. *Eriogonum eastwoodianum*
            8. Leaves basal and subbasal, or basal and cauline, suborbiculate or narrowly oblong to ovate; styles 0.2–1 mm; Contra Costa, Fresno, Kern, Merced, Monterey, San Benito, San Luis Obispo, and Solano counties, California.
                9. Involucres 2.5–3.5(–4) mm; styles 0.2–0.3 mm; e Contra Costa and s Solano counties, California (extinct) . . . . 204. *Eriogonum truncatum*
                9. Involucres (1.5–)1.8–2.5 mm; styles 0.2–1 mm; Fresno, Kern, Merced Monterey, San Benito, and San Luis Obispo counties, California.
                    10. Perianths not pustulose; achene beaks granular; involucres 2–2.5 mm; leaves basal and subbasal, rarely cauline; nw Kern, se Monterey, and ne San Luis Obispo counties, California . . . . 202. *Eriogonum temblorense*
                    10. Perianths pustulose; achene beaks papillose; involucres (1.5–)1.8–2 mm; leaves basal and often cauline; w Fresno, sw Merced, and se San Benito counties, California . . . . 203. *Eriogonum vestitum*
1. Involucres not terminal at tips of short branchlets proximally, usually appressed to branches; peduncles absent.
  11. Leaf blades puberulent to short-pilose or sericeous; flowering stems and inflorescence branches puberulent to short-pilose or sericeous.
     12. Outer tepals oblong-lanceolate to oblong-ovate; plants puberulent to short-pilose; n Arizona, nw and sw Colorado, nw New Mexico, e and wc Utah, sw Wyoming . . . . 223. *Eriogonum divaricatum*
     12. Outer tepals fan-shaped and hooded; plants sericeous; nw Arizona, se Nevada, sc Utah . . . . 224. *Eriogonum darrovii*

[11. Shifted to left margin.—Ed.]

11. Leaf blades tomentose at least abaxially; flowering stems and inflorescence branches glabrous or tomentose to floccose or lanate.

    13. Perianths densely pubescent, white to rose; nw California . . . . 222. *Eriogonum dasyanthemum*

    13. Perianths glabrous or sparsely glandular, not densely hairy, white, creamy white, yellow, rose, or red; widespread.

        14. Involucres (1.8–)2–5(–7) mm; outer tepals not fan-shaped.

            15. Leaf blades oblong-obovate to oblanceolate or oblong; flowering stems and inflorescence branches lanate to tomentose or floccose or, if glabrous, plants not of central Sierra Nevada.

                16. Involucres (3.5–)4–5 mm, cylindric; outer tepals narrowly obovate to oblong; achenes 1.8–2(–2.2) mm; California and sw Oregon . . . . 212. *Eriogonum roseum*

                16. Involucres (1.8–)2–3 mm, turbinate; outer tepals lanceolate to oblong or oblong-obovate; achenes 1–2 mm; California.

                    17. Inflorescence branches straight; basal leaf blades oblanceolate to oblong, (0.8–)1–4(–6) cm, petiole not winged; involucres (1.8–)2–3 mm; flowers 1.5–3 mm; perianth white to pink or yellow; California . . . . 213. *Eriogonum gracile*

                    17. Inflorescence branches upwardly curved; basal leaf blades oblanceolate or elliptic to ovate or nearly rounded, 1–2 cm, petiole occasionally winged; involucres 2.5–3 mm; flowers 1.5–2 mm, perianth white to rose; sw California . . . . 214. *Eriogonum cithariforme*

            15. Leaf blades round-ovate to rounded, reniform, or nearly so or, if not, inflorescence branches glabrous and plants of Sierra Nevada; flowering stems and inflorescence branches glabrous or tomentose to floccose.

                18. Inflorescence branches floccose, especially proximally; leaves basal; perianths white to rose or pale yellow; nw California, w Idaho, n Nevada, c and e Oregon, se Washington . . . . 209. *Eriogonum vimineum*

                18. Inflorescence branches glabrous or occasionally tomentose, or if floccose then leaves basal and cauline; stems usually 1; perianths white to pink, red, or yellow; California.

                    19. Involucres (3.5–)4–5(–7) mm; plants 4–10 dm; s California . . . . 211. *Eriogonum molestum*

                    19. Involucres 2–4 mm; plants 1–5 dm; California n to Oregon, e to Arizona and Utah.

                      20. Leaves basal and cauline, involucres 3–3.5 mm, and flowers 1.5–2 mm on floccose inflorescence branches or, if leaves strictly basal and stems glabrous, involucres 2–3.5 mm and plants of foothills and mid-elevation of central Sierra Nevada; perianths white to rose; n California . . . . 208. *Eriogonum luteolum* (in part)

                      20. Leaves basal; involucres 3–4 mm; flowers 1.5–2 mm; perianth white to pink or red, rarely yellow; flowering stems glabrous; w Arizona, s California, s Nevada, s Utah . . . . 210. *Eriogonum davidsonii*

        14. Involucres 1–1.5 mm or, if 2 mm, outer tepals fan-shaped or slightly hastate proximally and plants primarily of Great Basin and warm desert regions.

            21. Stems glabrous or, if tomentose, then flowers glandular and plants of eastern slope of Sierra Nevada and Great Basin; outer tepals oblong to oblong-obovate.

                22. Perianths yellow or pale yellowish, rarely whitish; flowers 0.6–0.8(–1) mm; e California, w Nevada, sc Oregon . . . . 217. *Eriogonum brachyanthum*

                22. Perianths white to rose; flowers 1–2 mm; e and s California, sc Idaho, Nevada, e Oregon, sw Utah, e Washington.

23. Flowers 1.5–2 mm; perianths usually minutely glandular, rarely glabrous; flowering stems and inflorescence branches glabrous or tomentose; e and s California, sc Idaho, Nevada, e Oregon, sw Utah, e Washington . . . . 215. *Eriogonum baileyi*
23. Flowers 1–1.5 mm; perianths glabrous; flowering stems and inflorescence branches glabrous; sw California . . . . 216. *Eriogonum elegans*
[21. Shifted to left margin.—Ed.]
21. Stems tomentose to floccose; outer tepals narrowly to broadly fan-shaped or slightly hastate proximally.
  24. Outer tepals slightly hastate proximally in fruit; flowers 0.8–1.2 mm; involucres campanulate, 1–2 mm; montane habitats; c Riverside, w San Bernardino, and nc San Diego counties, California (presumed extinct) . . . . 218. *Eriogonum evanidum*
  24. Outer tepals fan-shaped; flowers 1–3.5 mm; involucres 1–2.5 mm; desert habitats; not of c Riverside, w San Bernardino, and nc San Diego counties, California.
    25. Leaves cauline; leaf blades narrowly oblanceolate to broadly elliptic; plants narrowly erect, (0.5–)1–6 dm; perianths white, becoming pink or red; c and s Arizona, se California (presumed extirpated), w and s New Mexico, w Texas, s Utah. . . . . 221. *Eriogonum polycladon*
    25. Leaves basal; leaf blades suborbiculate or rounded to cordate; plants usually spreading, 0.5–1.5(–3) dm; perianths white to pink or pale yellow to yellow, rarely creamy white, becoming pink or red; widespread.
      26. Perianths pale yellow or yellow, rarely creamy white, becoming red; outer tepals broadly fan-shaped; inflorescences diffuse, branches inwardly curved distally; achenes 1–1.3 mm; nw Arizona, e California, sw Idaho, Nevada, se Oregon, sw Utah. . . . . 219. *Eriogonum nidularium*
      26. Perianths white to pink or pale yellowish, outer tepals narrowly fan-shaped; inflorescences open, branches not inwardly curved distally; achenes 1.5–1.8 mm; Arizona, e California, nw and sw Colorado, sw Idaho, Nevada, w New Mexico, Utah . . . . 220. *Eriogonum palmerianum*

199. Eriogonum mohavense S. Watson, Proc. Amer. Acad. Arts 12: 266. 1877 · Western Mojave wild buckwheat [C] [E] [F]

Herbs, erect to spreading, 1–3 dm, glabrous, greenish to yellowish green. **Stems:** aerial flowering stems erect, 0.2–1 dm, glabrous. **Leaves** basal; petiole 1–4 cm, tomentose; blade oblong to rounded, (0.4–)0.6–2 × (0.4–)0.6–2 cm, white-tomentose and grayish on both surfaces, or slightly less so adaxially. **Inflorescences** cymose, diffuse, spreading, 5–25 × 5–20 cm; branches glabrous; bracts 0.5–1.5 × 1–2 mm. **Peduncles** absent. **Involucres** terminal at tips of slender branchlets at least proximally, not appressed to branches, turbinate, 1.7–2 × 1–1.5 mm, glabrous; teeth 5, erect, 0.5–0.9 mm. **Flowers** 0.7–1 mm; perianth yellow, glabrous, infrequently glandular proximally; tepals monomorphic, narrowly oblong to elliptic; stamens included to slightly exserted, 0.8–1.2 mm; filaments pilose proximally. **Achenes** dark brown to nearly black, lenticular, 1–1.2 mm.

Flowering May–Sep. Sandy to infrequently clayey flats and slopes, saltbush and creosote bush communities; of conservation concern; 600–1200 m; Calif.

Eriogonum mohavense is typically local and infrequent, and only rarely locally common. It is restricted to the northwestern corner of the Mojave Desert in southeastern Kern, northeastern Los Angeles, and northwestern San Bernardino counties. It occurs also just inside Inyo County. An 1882 C. C. Parry collection (NY) supposedly from San Diego is discounted as to location.

200. Eriogonum ampullaceum J. T. Howell, Leafl. W. Bot 1: 179. 1935 · Mono wild buckwheat [E]

Eriogonum mohavense S. Watson subsp. *ampullaceum* (J. T. Howell) S. Stokes

Herbs, erect, 1–3 dm, glabrous, yellowish green to gray. **Stems:** aerial flowering stems erect, 0.2–1 dm, glabrous. **Leaves** basal; petiole 0.5–4 cm, tomentose; blade orbiculate to subcordate, 0.5–2(–2.5) × 0.5–2(–2.5) cm, white-tomentose abaxially, tomentose to floccose and grayish or greenish

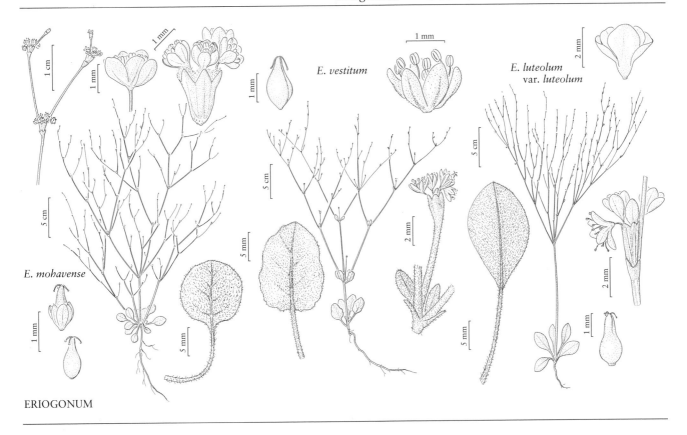

ERIOGONUM

adaxially. **Inflorescences** cymose, open, narrow, 5–25 × 5–10(–15) cm; branches glabrous; bracts 0.5–1.5 × 1–2 mm. **Peduncles** absent. **Involucres** terminal at tips of slender branchlets at least proximally, not appressed to branches, turbinate-campanulate, 1.5–2 × 1.5–2 mm, glabrous; teeth 5, erect, 0.5–0.8 mm. **Flowers** 1–1.5 mm; perianth cream, reddish green to reddish proximally, glabrous; tepals monomorphic, obovate; stamens exserted, 1–1.5 mm; filaments pilose proximally. **Achenes** brown, lenticular, 1–1.3 mm.

Flowering Jul–Sep. Sandy flats and slopes, sagebrush communities, pinyon-juniper, conifer woodlands; 1700–2200 m; Calif., Nev.

Eriogonum ampullaceum is the Great Basin counterpart to *E. mohavense* of the Mojave Desert. It is found along the eastern edge of the Sierra Nevada in Mono County, California, mainly in the Bridgeport, Mono Lake, and Long Valley areas (south of Mammoth). It occurs also in Alkali Valley in Mineral County, Nevada, and to the south in Fish Lake Valley of Esmeralda County. The species recently has been seen as a roadside weed along U.S. Highway 395 near Big Pine in Inyo County, California. The plants typically are infrequent and only occasionally locally common. The species is considered "sensitive" in Nevada.

201. Eriogonum eastwoodianum J. T. Howell, Leafl. W. Bot. 2: 133. 1938 • Eastwood's wild buckwheat E

Eriogonum covilleanum Eastwood subsp. *adsurgens* (Jepson) Abrams; *E. truncatum* Torrey & A. Gray var. *adsurgens* Jepson; *E. vimineum* Douglas ex Bentham subsp. *adsurgens* (Jepson) S. Stokes

Herbs, erect, 2–5 dm, tomentose, greenish gray to gray. **Stems:** aerial flowering stems erect, (0.4–)0.8–1.5 dm, tomentose. **Leaves** basal; petiole 2–8 cm, tomentose; blade suborbiculate, 1–3 × 1–3 cm, white-tomentose abaxially, subglabrous and greenish adaxially. **Inflorescences** cymose, open, 10–40 × 10–40 cm; branches tomentose; bracts 1–5(–6) × 1–2 mm. **Peduncles** erect, straight, slender, 1–3.5 cm, tomentose. **Involucres** terminal at tips of slender branchlets at least proximally, not appressed to branches, turbinate, 2–2.5 × 1.5–2 mm, tomentose; teeth 5, erect, 0.6–0.9 mm. **Flowers** 1.5–2.5(–3) mm; perianth white, glabrous; tepals monomorphic, elliptic to oblong-obovate; stamens included, 0.7–1.2 mm; filaments usually pilose proximally; styles 0.1–0.3 mm. **Achenes** brownish, 3-gonous, 1.6–2(–2.5) mm; beak granular. **2*n*** = 34.

Flowering May–Sep. Sandy shale outcrops and slopes, mixed grassland communities, oak and pine woodlands; (200–)500–1000 m; Calif.

Eriogonum eastwoodianum is rare and known only from two disjunct locations in Fresno County, one in the Silver Creek Canyon area and the other above Jacalitos Ranch on Parkfield Grade. Plants from a third site in the Cottonwood Pass area of San Luis Obispo County are now assigned to *E. temblorense*. That species and *E. vestitum* are closely related to *E. eastwoodianum*, and only fully mature plants can be assigned unquestionably to one of those species.

202. Eriogonum temblorense   J. T. Howell & Twisselmann, Leafl. W. Bot. 10: 42. 1963 • Temblor wild buckwheat   E

Herbs, erect, 1–8 dm, tomentose, greenish gray to gray. **Stems:** aerial flowering stems erect, 0.5–1(–1.5) dm, tomentose. **Leaves** basal and subbasal, rarely cauline; petiole 2–5 cm, tomentose; basal blade initially suborbiculate, then elliptic, 1.5–4 × 1–1.5 cm, rarely suborbiculate and 2–4 × 2–4 cm when first formed, densely tomentose on both surfaces, or slightly less so and grayish to greenish abaxially; cauline blade similar to basal but not suborbiculate. **Inflorescences** cymose, open, 8–70 × 10–50 cm; bracts 1–2.5 × 1–2 mm. **Peduncles** erect, straight, slender, 1–4 cm, tomentose. **Involucres** terminal at tips of slender branchlets proximally, not appressed to branches, turbinate, 2–2.5 × 1.5–2 mm, tomentose; teeth 5, erect, 0.5–0.8 mm. **Flowers** 1.5–2.5 mm; perianth white, glabrous; tepals monomorphic, oblong; stamens included, 1–1.5 mm; filaments usually pilose proximally; styles 0.2–1 mm. **Achenes** light brown, 3-gonous, 2–2.8 mm; beak granular. **2***n* = 34.

Flowering May–Sep. Sandy to loamy slopes, mixed grassland communities, oak and juniper woodlands; 300–900 m; Calif.

Eriogonum temblorense is rare but much more widely distributed than the closely related *E. eastwoodianum*. In early anthesis, it is difficult to differentiate between the two species, and plants gathered in May at Cottonwood Pass are particularly troublesome, as the first basal leaf blades are often suborbiculate. Those leaves quickly fall away, and the subsequent basal and cauline leaf blades are elliptic. As circumscribed here, *E. temblorense* is known from six sites in Monterey (Stone Canyon), San Luis Obispo (Cottonwood Pass and Polonio Pass areas), and Kern (Chico Martinez Canyon, Shale Hills, and hills west of McKittrick) counties.

203. Eriogonum vestitum   J. T. Howell, Leafl. W. Bot. 2: 42. 1937 • Idria wild buckwheat   E   F

Herbs, erect, (0.5–)1–4(–5) dm, tomentose, greenish gray to gray. **Stems:** aerial flowering stems erect, 0.5–1 dm, tomentose. **Leaves** basal and cauline; basal: petiole 1–4 cm, thinly tomentose, blade narrowly oblong to obovate, (1–)2–4(–5) × 1–2(–3) cm, thinly tomentose and greenish abaxially, floccose to glabrate and greenish adaxially; cauline: petiole 0.5–2 cm, thinly tomentose, blade elliptic or rarely ovate, (0.7–)1–3(–4.5) ×0.5–1.5(–3.5) cm, similar to basal blade. **Inflorescences** cymose, somewhat diffuse, 10–40(–45) × 5–30 cm; branches tomentose; bracts 2–4 × 0.5–1.5 mm. **Peduncles** erect, straight, slender, 0.5–4(–6) cm, tomentose. **Involucres** terminal at tips of slender branchlets proximally, not appressed to branches, narrowly campanulate, (1.5–)1.8–2 × 1.5–2(–2.2) mm, tomentose; teeth 5, erect, 0.3–0.5 mm. **Flowers** 1.5–2(–2.5) mm; perianth white, glabrous but conspicuously pustulose; tepals monomorphic, narrowly elliptic to oblong; stamens included, 1–1.5 mm; filaments pilose proximally; styles 0.7–1 mm. **Achenes** brown, 3-gonous, 2–2.5 mm; beak papillose. **2***n* = 34.

Flowering Mar–Nov. Clayey outcrops and slopes, mixed grassland communities, oak and pine woodlands; 400–700 m; Calif.

Eriogonum vestitum is restricted to the Idria region of San Benito County, the Silver Creek Canyon and Mercey Hot Springs areas of Fresno County, and the Los Banos Hills in Merced County. It is closely related to *E. eastwoodianum*. Some individuals mimic *E. gracile*, and a clear distinction is not always possible with immature specimens.

204. Eriogonum truncatum   Torrey & A. Gray, Proc. Amer. Acad. Arts 8: 173. 1870 • Mt. Diablo wild buckwheat   C   E

Herbs, erect, 1–4.5 dm, tomentose to floccose, grayish. **Stems:** aerial flowering stems erect or nearly so, (0.1–)0.2–0.6(–0.8) dm, tomentose. **Leaves** basal and cauline; basal: petiole (1–)2–5(–7) cm, floccose, blade narrowly oblong to obovate, 1–5(–7) × (0.5–)1–2 (–3) cm, densely grayish-tomentose abaxially, floccose to glabrate and greenish adaxially; cauline: petiole (0.5–)1–2(–3) cm, floccose, blade narrowly oblong, (0.8–)1–3(–6) × 0.2–1(–2) cm, similar to basal blade. **Inflorescences** cymose, open, (5–)10–30(–40) × 5–30 cm; branches tomentose to floccose; bracts 0.5–2 × 0.5–1 mm. **Peduncles** absent. **Involucres** terminal at tips of slender branchlets proximally, not appressed to branches, turbinate, 2.5–3.5(–4) × 2–2.5 mm, tomentose; teeth 5, erect, 0.1–0.3 mm. **Flowers** (1.5–)1.7–2(–2.2) mm; perianth white to rose, glabrous; tepals essentially monomorphic, elliptic to oblong or obovate; stamens included, 1.5–2 mm; filaments glabrous; styles 0.2–0.3 mm. **Achenes** light brown, 3-gonous, 1.7–2 mm; beak smooth.

Flowering Apr–Aug. Sandy slopes, mixed grassland and chaparral communities, oak woodlands; of conservation concern; 200–400 m; Calif.

Eriogonum truncatum is presumed extinct. It is known only from seven collections, most made in the Marsh Creek and Mt. Diablo areas of Contra Costa County. It was found also at Suisun in Solano County (*Earle s.n.*, 3 Aug 1888, DUKE). Three collections are widely distributed in herbaria: *M. L. Baker 2833*, Marsh Creek, 28 Apr 1903 (B, CAS, F, MICH, MIN, MSC, NY, POM, RM, UC), *M. K. Curran s.n.*, Antioch, May 1886 (CAS, CS, DS, ISC, JEPS, NY, POM, UC), and *J. T. Howell 11816*, 10 miles from Clayton on Marsh Creek Road, 8 Apr 1934 (CAS, DS, F, MICH, MIN, MO, MONTU, NY, UC, RSA). *Eriogonum truncatum* may be quickly distinguished from the morphologically similar *E. roseum* by its consistently dichotomously branched inflorescences, with each new branch curving upward gracefully from the node. The ultimate inflorescence branches of *E. roseum* are typically composed of several racemosely arranged involucres along a single, elongated branch.

205. **Eriogonum argillosum** J. T. Howell, Leafl. W. Bot. 1: 13. 1932 • Coast Range wild buckwheat [E]

Herbs, erect, 1–3 dm, glabrous or rarely sparsely floccose, mostly greenish. **Stems:** aerial flowering stems erect, (0.3–)0.5–1.5(–2.5) dm, glabrous, rarely sparsely floccose. **Leaves** basal and cauline; basal: petiole (0.5–)1–4 (–5) cm, sparsely floccose, blade oblong, 1–3(–5) × 0.3–1.2(–1.6) cm, densely white-tomentose abaxially, floccose to glabrate and greenish adaxially; cauline: petiole 0.3–2 cm, sparsely floccose, blade elliptic to oblong, 0.5–1.5 × 0.2–0.8 cm, similar to basal blade. **Inflorescences** cymose, occasionally distally uniparous due to suppression of secondary branches, open, (2–)5–20(–25) cm; branches glabrous or rarely sparsely floccose; bracts 1–5(–7) × 0.5–1(–1.5) mm. **Peduncles** erect, straight, slender, 0.5–4.5(–6) cm, glabrous. **Involucres** terminal at tips of slender branchlets proximally, not appressed

to branches, turbinate, (2–)2.5–3 × 2–3 mm, glabrous; teeth 5, erect, 0.2–0.5 mm. **Flowers** 1.5–2 mm; perianth white to rose, glabrous; tepals monomorphic, oblong; stamens mostly included, 1–2 mm; filaments glabrous; styles 0.2–0.3 mm. **Achenes** brown, 3-gonous, 2–2.5 mm, glabrous; beak smooth.

Flowering Mar–Oct. Clay to serpentine outcrops and slopes, mixed grassland and chaparral communities, oak and pine woodlands; 100–600 m; Calif.

Eriogonum argillosum is rare to locally infrequent in coastal central California. It was last seen in Santa Clara County in 1895 and now is almost certainly extirpated there. Otherwise, it is found in the San Benito River Canyon (San Benito County) area and east of San Lucas (Monterey County). This is the most distinct of the five species in the *E. eastwoodianum* complex.

206. **Eriogonum covilleanum** Eastwood, Proc. Calif. Acad. Sci., ser. 4, 20: 138. 1931 • Coville wild buckwheat [E]

Eriogonum vimineum Douglas ex Bentham var. *covilleanum* (Eastwood) S. Stokes

Herbs, erect, 1–4 dm, glabrous, green to reddish. **Stems:** aerial flowering stems erect, 0.3–1 dm, glabrous. **Leaves** basal; petiole 1–4(–5) cm, tomentose; blade suborbiculate to reniform, (0.3–) 0.5–1.5 × 0.5–1.5(–1.8) cm, densely white-tomentose abaxially, glabrous and bright green adaxially. **Inflorescences** cymose, infrequently distally uniparous due to suppression of secondary branches, open, 5–35 × 5–30 cm; branches glabrous; bracts 1–2 × 0.5–1 mm. **Peduncles** absent. **Involucres** terminal at tips of slender branchlets proximally, not appressed to branches, turbinate, 2–2.5 × 1–1.5 mm, glabrous; teeth 5, erect, 0.4–0.8 mm. **Flowers** 2–2.5(–3) mm; perianth pink at early anthesis, becoming white to rose or yellow, minutely puberulent; tepals monomorphic, narrowly elliptic; stamens included, 1.7–2 mm; filaments glabrous. **Achenes** light brown, 3-gonous, 1.8–2 mm. $2n = 34$.

Flowering Apr–Aug. Shale or serpentine outcrops and slopes, chaparral communities, oak and pine woodlands; 200–1400 m; Calif.

Eriogonum covilleanum is encountered infrequently in the Inner Coast Ranges of central and southwestern California (Fresno, Kern, Merced, Monterey, San Benito, San Luis Obispo, Santa Barbara, Santa Clara, Stanislaus, and Ventura counties). On occasion it can be locally common but is never abundant or weedy. It is often found growing with other annual wild buckwheats, and care must be taken to sort the species.

207. Eriogonum nortonii Greene, Pittonia 2: 165. 1891 (as nortoni) • Pinnacles wild buckwheat C E

Eriogonum vimineum Douglas ex Bentham subsp. *nortonii* (Greene) S. Stokes

Herbs, spreading, 0.5–2 dm, glabrous, usually reddish. **Stems:** aerial flowering stems prostrate to ascending or weakly erect, 0.1–0.3 dm, glabrous. **Leaves** basal and cauline; basal: petiole 1–3 cm, tomentose, blade round to reniform, 0.5–1.5 × 0.5–1.5 cm, densely white-tomentose abaxially, subglabrous or glabrous and green adaxially; cauline: petiole and blade similar to basal leaves. **Inflorescences** cymose, not distally uniparous due to suppression of secondary branches, open, 4–18 × 10–30 cm; branches glabrous; bracts 1–2 × 1–2 mm. **Peduncles** absent. **Involucres** terminal at tips of slender branchlets proximally, not appressed to branches, broadly turbinate, 3–4 × 2.5–3.5 mm, glabrous; teeth 6–8, erect, 0.4–0.8 mm. **Flowers** 1–2 mm; perianth white to rose, glabrous; tepals monomorphic, obovate; stamens included, 1–1.5 mm; filaments glabrous. **Achenes** light brown, 3-gonous, 1–1.3 mm.

Flowering May–Aug. Sandy to gravelly slopes, mixed grassland and chaparral communities, oak and pine woodlands; of conservation concern; 300–1200 m; Calif.

Eriogonum nortonii is a rare and localized species restricted to the Gabilan Range of Monterey and San Benito counties, from Fremont Peak south to The Pinnacles. It is confined largely to two protected areas, the Hastings Natural History State Reservation and the Pinnacles National Monument. A disjunct population is found southeast of Carmel Highlands in the Santa Lucia Mountains of Monterey County.

208. Eriogonum luteolum Greene, Pittonia 3: 200. 1897 E F

Eriogonum vimineum Douglas ex Bentham var. *luteolum* (Greene) S. Stokes

Herbs, erect or prostrate to spreading, 0.5–6 dm, glabrous or occasionally tomentose, greenish to reddish. **Stems:** aerial flowering stems prostrate to erect, 0.2–2 dm, glabrous or occasionally tomentose. **Leaves** basal or basal and cauline; basal: petiole 1–8 cm, floccose, blade oblong-ovate, reniform, or rounded to cordate, 0.5–5 × 0.5–3.5 cm, densely white-tomentose abaxially, floccose or glabrous and reddish or greenish adaxially; cauline: petiole 0.5–3 cm, floccose,

blade cordate to reniform or orbiculate, 0.5–3 × 0.5–3 cm, similar to basal blade. **Inflorescences** cymose, occasionally distally uniparous due to suppression of secondary branches, open, 2–50 × 3–40 cm; branches glabrous or occasionally tomentose; bracts 1–3(–4) × 1–2 mm. **Peduncles** absent or erect, straight, slender, 0.1–0.5 cm, glabrous. **Involucres** terminal at tips of slender branchlets proximally or appressed to branches, cylindric or occasionally turbinate, 2–4 × 1.5–2.5 mm, glabrous or rarely floccose; teeth 5, erect, 0.3–0.5 mm. **Flowers** 1–2.5 mm; perianth white to rose or yellow, glabrous; tepals monomorphic, obovate; stamens included, 1–1.5 mm; filaments glabrous. **Achenes** light brown, 3-gonous, 1–2 mm.

Varieties 4 (4 in the flora): w North America.

1. Plants prostrate to spreading, 0.5–3 dm; involucres 3–4 mm, often terminal on elongated branchlets; leaves basal and cauline; serpentine soil; w Alameda and s Marin counties, California . . . . 208d. *Eriogonum luteolum* var. *caninum*
1. Plants erect, 2–6 dm; involucres 2–3.5 mm, appressed to inflorescence branches; leaves basal, sometimes also cauline; serpentine or granitic soils; San Mateo, Santa Clara, and Mariposa counties, California, n to sw and sc Oregon.
  2. Involucres 2–3 mm; flowers 2–2.5 mm; granitic soils; Sierra Nevada, Alpine and Tuolumne counties, California . . . . 208c. *Eriogonum luteolum* var. *saltuarium*
  2. Involucres 3–3.5 mm; flowers 1–2 mm; serpentine soils; Coast Ranges and lower foothills of Sierra Nevada, San Mateo, Santa Clara, and Mariposa counties, California, n to sw and sc Oregon.
    3. Flowers 1.8–2 mm; Coast Ranges, w California from San Mateo and Santa Clara counties n to sw and sc Oregon . . . . 208a. *Eriogonum luteolum* var. *luteolum*
    3. Flowers 1–1.5(–1.8) mm; e California from Mariposa County to Plumas County . . . . 208b. *Eriogonum luteolum* var. *pedunculatum*

208a. Eriogonum luteolum Greene var. **luteolum** • Golden carpet wild buckwheat E F

Plants erect, (2–)3–5(–6) dm, glabrous or occasionally tomentose. **Aerial flowering stems** erect, 0.5–1 dm. **Leaves** basal and cauline; blade oblong-ovate or rounded to cordate. **Inflorescences** 10–45 cm. **Involucres** sessile and appressed to branches, cylindric, 3–3.5 mm. **Flowers** 1.8–2 mm; perianth white to rose or yellow. **Achenes** 1.8–2 mm. $2n = 24$.

Flowering Jul–Nov. Sandy to gravelly serpentine flats and slopes, mixed grassland and chaparral communities, oak and pine woodlands; 50–1600(–2100) m; Calif., Oreg.

Variety *luteolum* is rather common in northern California, where it is seen primarily in the North Coast Ranges and across the northern tier of counties to the northern tip of the Sierra Nevada. It continues northward into southern Oregon. In some places, it can even be abundant or infrequently rather weedy. It is difficult to distinguish from *E. gracile* in the southern part of its range and from *E. vimineum* in the northeastern portion. A collection from Crook County, Oregon, (*Steward & Steward 6199*, KANU, IA, OKLA, PENN, UTC) seems to be this species rather than *E. vimineum*.

208b. Eriogonum luteolum Greene var. **pedunculatum**
(S. Stokes) Reveal, Great Basin Naturalist 40: 148. 1980 • Mokelumne Hill wild buckwheat E

Eriogonum pedunculatum S. Stokes, Leafl. W. Bot. 2: 48. 1937

Plants erect, 3–6 dm, glabrous. **Aerial flowering stems** erect, 0.5–1 dm. **Leaves** basal or basal and cauline; blade oblong to oblong-ovate. **Inflorescences** 10–50 cm. **Involucres** sessile and appressed to branches, rarely on erect peduncles 1–5 mm at proximal nodes, cylindric, 3–3.5 mm. **Flowers** 1–1.5(–1.8) mm; perianth white. **Achenes** 1–1.4 mm.

Flowering Jun–Oct. Sandy to gravelly serpentine flats and slopes, mixed grassland and chaparral communities, oak and pine woodlands; 100–1500(–1950) m; Calif.

Variety *pedunculatum* is encountered infrequently on the western foothills of the Sierra Nevada in Amador, Butte, Calaveras, El Dorado, Mariposa, Nevada, Placer, Plumas, Sierra, Tulare, and Tuolumne counties. The feature that was the basis for the varietal epithet, a pedunculate involucre, is seen only rarely.

208c. Eriogonum luteolum Greene var. **saltuarium**
Reveal, Phytologia 66: 264. 1989 • Jack's wild buckwheat E

Plants erect, 2–4 dm, glabrous. **Aerial flowering stems** erect, 0.2–2 dm. **Leaves** basal or basal and cauline; blade reniform. **Inflorescences** 2–35 cm. **Involucres** sessile and appressed to branches, cylindric, 2–3 mm. **Flowers** 2–2.5 mm; perianth white to rose. **Achenes** 1.8–2 mm.

Flowering Jul–Sep. Sandy granitic flats and slopes, sagebrush communities, montane conifer woodlands; 1700–2400 m; Calif.

Variety *saltuarium* is known from only two locations, one near Dardanelle in Tuolumne County and another on Luther Pass in Alpine County. At present, this rare taxon has received no conservation consideration. Only a few hundred individuals have been observed in the field.

208d. Eriogonum luteolum Greene var. **caninum**
(Greene) Reveal, Great Basin Naturalist 40: 148. 1980 • Tiburon wild buckwheat E

Eriogonum vimenum Douglas ex Bentham var. *caninum* Greene, Fl. Francisc., 150. 1891; *E. caninum* (Greene) Munz

Plants protrate to spreading, 0.5–3 dm, glabrous. **Aerial flowering stems** mostly prostrate to weakly erect, 0.2–1 dm. **Leaves** basal and cauline; blade oblong-ovate. **Inflorescences** 3–30 cm. **Involucres** sessile or terminal on branchlets, turbinate, 3–4 mm. **Flowers** 1.5–2.5 mm; perianth white to rose. **Achenes** 1.4–1.6 mm. $2n = 24$.

Flowering May–Oct. Sandy to gravelly serpentine flats and slopes, mixed grassland and chaparral communities, oak and pine woodlands; 0–700 m; Calif.

The elegant variety *caninum* is localized in the Tiburon area of Marin County and in the Oakland Hills of Alameda County. It is similar to *E. nortonii* in form and habit. North of Tiburon, and especially on Mt. Tamalpais, a sharp distinction between var. *luteolum* and var. *caninum* is not always possible. Variety *caninum* is protected at Tiburon near Old St. Hilary's Chapel in the John Thomas Howell Botanical Garden and in The Nature Conservancy's Ring Mountain Preserve, and near Oakland in the W. P. Coe State Park.

209. Eriogonum vimineum Douglas ex Bentham,
Trans. Linn. Soc. London 17: 416. 1836 • Wicker-stem wild buckwheat E

Eriogonum shoshonense A. Nelson; *E. vimenum* subsp. *shoshonense* (A. Nelson) S. Stokes

Herbs, erect to slightly spreading, 0.5–3(–5) dm, thinly tomentose to floccose or glabrous, grayish or reddish gray. **Stems:** aerial flowering stems erect, 0.5–1 dm, thinly tomentose to floccose, infrequently glabrous, straight or nearly so, infrequently inwardly curved distally. **Leaves** basal; petiole 1–4 cm, mostly floccose; blade round-ovate to rounded, 0.5–2 (–2.5) × 0.5–2 cm, white-tomentose abaxially, less so to nearly glabrous and grayish or greenish adaxially. **Inflorescences** cymose, often distally uniparous due to suppression of secondary branches, open to somewhat diffuse, 5–25 × 5–25 cm; branches floccose or glabrous; bracts 1–3 × 1–2 mm. **Peduncles** absent. **Involucres** appressed to branches, narrowly cylindric, 2–3.5(–4) × 1–2 mm, glabrous or infrequently floccose; teeth 5, erect, 0.3–0.6 mm. **Flowers** 2–2.5 mm; perianth white to rose or pale yellow, glabrous; tepals slightly dimorphic, those of outer whorl broadly spatulate to obovate, those of inner whorl oblanceolate; stamens included, 1–1.5 mm; filaments pilose proximally or glabrous. **Achenes** brown, 3-gonous, 2–2.5 mm. $2n = 24$.

Flowering May–Sep. Sandy to gravelly flats and slopes, mixed grassland, saltbush, and sagebrush communities, oak, juniper, and montane conifer woodlands; 50–2400 m; Calif., Idaho, Nev., Oreg., Wash.

Eriogonum vimineum is widespread and common to abundant or even locally weedy from southeastern Washington southward through central and eastern Oregon and western Idaho to northeastern California and northern Nevada. Except for occasional populations in northeastern California, where it can be confused with *E. luteolum*, this species is distinct, albeit variable, throughout its range.

210. Eriogonum davidsonii Greene, Pittonia 2: 295. 1892 · Davidson's wild buckwheat

Eriogonum molestum S. Watson var. *davidsonii* (Greene) Jepson; *E. vimineum* Douglas ex Bentham var. *davidsonii* (Greene) S. Stokes; *E. vimineum* var. *glabrum* S. Stokes; *E. vimineum* subsp. *juncinellum* (Gandoger) S. Stokes

Herbs, erect, 1–5 dm, glabrous, greenish to grayish. **Stems:** aerial flowering stems erect, 0.5–1.5(–2) dm, glabrous. **Leaves** basal; petiole 1–5 cm, floccose; blade round to reniform, (0.3–)1–2(–4) × (0.3–)1–2(–4) cm, densely white-tomentose abaxially, floccose to glabrate and mostly greenish adaxially. **Inflorescences** cymose, occasionally distally uniparous due to suppression of secondary branches, open, 5–40 × 5–35 cm; branches straight or nearly so, infrequently inwardly curved distally, glabrous; bracts 1–3 × 1–2 mm. **Peduncles** absent. **Involucres** appressed to branches, cylindric-turbinate, 3–4 × 2–2.5 mm, glabrous; teeth 5, erect, 0.2–0.3 mm. **Flowers** 1.5–2 mm; perianth white to pink or red, rarely yellow, glabrous; tepals monomorphic, oblong-obovate; stamens included, 1–1.5 mm; filaments pilose proximally. **Achenes** brown, 3-gonous, 2 mm. $2n = 40$.

Flowering May–Sep. Sandy to gravelly flats and slopes, mixed grassland, saltbush, chaparral, and sagebrush communities, oak and montane conifer woodlands; (400–)900–2600 m; Ariz., Calif., Nev., Utah; Mexico (Baja California).

Eriogonum davidsonii is widespread and mostly common to occasionally abundant or weedy in Arizona, California, southern Nevada, and southern Utah. It is exceedingly variable. In the northern part of California, its range approaches that of *E. luteolum* var. *luteolum*, and the two can be difficult to differentiate. To the south, in Tulare County, the distinction between *E. davidsonii* and *E. luteolum* var. *pedunculatum* also is difficult. Specimens of *Eriogonum davidsonii* with curved inflorescence branches resemble *E. cithariforme* in the mountains of southern California, and care must be taken to separate *E. davidsonii* from its more robust relative, *E. molestum* in the San Jacinto Mountains of Riverside County. The disjunct populations in Utah and Arizona are somewhat different in appearance but presently do not seem worthy of taxonomic separation. The epithet *juncinellum* is available should recognition be desired.

Seeds of Davidson's wild buckwheat were pounded into a meal and eaten dry by the Kawaiisu people of southern California (M. L. Zigmond 1981).

211. Eriogonum molestum S. Watson, Proc. Amer. Acad. Arts 17: 379. 1882 · Pineland wild buckwheat C E

Eriogonum vimineum Douglas ex Bentham subsp. *molestum* (S. Watson) S. Stokes

Herbs, erect, 4–10 dm, glabrous, greenish to grayish. **Stems:** aerial flowering stems erect, 1–4 dm, glabrous. **Leaves** basal; petiole 1–6 cm, floccose; blade round to reniform, (0.5–)1–4 × (0.5–)1–4 cm, densely white-tomentose abaxially, floccose to glabrate and mostly greenish adaxially. **Inflorescences** cymose, occasionally distally uniparous due to suppression of secondary branches, open, 30–80 × 10–50 cm; branches straight, not inwardly curved distally, glabrous; bracts 1–3 × 1–3 mm. **Peduncles** absent. **Involucres** appressed to branches, cylindric-turbinate, (3.5–)4–5(–7) × 2.5–3 (3.5) mm, glabrous; teeth 5, erect, 0.2–0.4 mm. **Flowers** 1.5–3 mm; perianth white to pink, rarely pale yellow, glabrous; tepals monomorphic, oblong-obovate; stamens included, 1–1.5 mm; filaments pilose proximally. **Achenes** brown, 3-gonous, 2–2.5 mm, glabrous. $2n = 40$.

Flowering May–Sep. Sandy flats and slopes, grassland and chaparral communities, oak and montane conifer woodlands; of conservation concern; 1100–2200 m; Calif.

Eriogonum molestum is infrequent to occasionally common in the San Bernardino (San Bernardino County) and San Jacinto (Riverside County) mountains, and in scattered mountain ranges of San Diego County, California. It is easily confused with the perennial *E. nudum* var. *pauciflorum*.

212. **Eriogonum roseum** Durand & Hilgard, J. Acad. Nat. Sci. Philadelphia, n. s. 3: 45. 1855 • Wand wild buckwheat

Eriogonum vimineum Douglas ex Bentham subsp. *virgatum* (Bentham) S. Stokes; *E. virgatum* Bentham

Herbs, erect, 1–8 dm, thinly tomentose to floccose, whitish to grayish or brownish to reddish brown. **Stems:** aerial flowering stems erect, 0.5–3 dm, thinly tomentose to floccose. **Leaves** basal and cauline; basal: petiole 1–4 cm, thinly tomentose to floccose, blade oblanceolate to narrowly oblong, 1–3(–5) × (0.3–)0.5–1(–2) cm, tomentose on both surfaces, sometimes merely floccose and grayish or brownish to reddish brown adaxially; cauline: petiole 0.3–1.5(–2) cm, mostly floccose, blade elliptic, 0.5–3 × 0.3–1 cm, mostly tomentose and whitish to grayish. **Inflorescences** cymose, often distally uniparous due to suppression of secondary branches, open, 10–70 × 10–45 cm; branches straight or nearly so, infrequently inwardly curved distally, thinly tomentose to floccose; bracts 1–3 × 1–3 mm. **Peduncles** absent. **Involucres** appressed to branches, cylindric, (3.5–)4–5 × 2–3 mm; teeth 5, erect, 0.2–0.4 mm. **Flowers** 1.5–2(–2.5) mm; perianth white to pink or red, occasionally yellow, glabrous; tepals monomorphic, narrowly obovate to oblong; stamens included, 1–1.5 mm; filaments pilose proximally. **Achenes** brown, 3-gonous, 1.8–2(–2.2) mm. $2n = 18$.

Flowering May–Nov. Sandy and gravelly flats and slopes, mixed grassland, chaparral, and sagebrush communities, oak and pine woodlands; 0–2200 m; Calif., Oreg.; Mexico (Baja California).

Eriogonum roseum is widespread and typically common; occasionally it will be locally abundant but only rarely can it be considered weedy. It occurs from southwestern Oregon south through much of California to northern Baja California Norte, Mexico. Morphologically this annual species approaches the perennial *E. elongatum*, and poorly prepared specimens sometimes are difficult to differentiate. A clear distinction between *E. roseum* and *E. gracile* appears to be consistently possible in the field, but some herbarium material can be difficult to assign. By and large, *E. roseum* is more robust and less branched than the decidedly more slender and graceful *E. gracile*.

Seeds of *Eriogonum roseum* were pounded into a powder and either mixed with water and used as a beverage or eaten raw by the Kawaiisu people (M. L. Zigmond 1981).

213. **Eriogonum gracile** Bentham, Bot. Voy. Sulphur, 46. 1844

Eriogonum vimineum Douglas ex Bentham subsp. *gracile* (Bentham) S. Stokes

Herbs, erect to spreading or somewhat decumbent, (0.7–)1.5–5(–7) dm, lanate to tomentose or floccose, rarely glabrous, greenish to whitish or grayish. **Stems:** aerial flowering stems erect, 0.5–2 dm, lanate to tomentose or floccose, rarely glabrous. **Leaves** basal and cauline; basal: petiole 1–4 cm, floccose, not winged, blade oblanceolate to oblong, (0.8–)1–4(–6) × 0.5–2 cm, tomentose abaxially, less so to floccose and greenish to grayish adaxially; cauline: petiole 0.3–2 cm, floccose, blade oblanceolate, 0.5–3 × 0.3–1.5 cm, similar to basal blade. **Inflorescences** cymose, occasionally distally uniparous due to suppression of secondary branches, open to somewhat diffuse, 0.5–4 × 0.5–3 cm; branches lanate to tomentose or floccose, rarely glabrous, straight or nearly so, not inwardly curved distally; bracts 0.5–3 × 1–3 mm. **Peduncles** absent. **Involucres** somewhat appressed to branches, turbinate, (1.8–)2–3 × 1.5–2 mm, floccose or glabrous; teeth 5, erect, 0.3–0.6 mm. **Flowers** 1.5–3 mm; perianth white to pink or yellow, glabrous; tepals monomorphic, lanceolate to oblong; stamens mostly included, 1–1.5 mm; filaments pilose proximally. **Achenes** brown, 3-gonous, 1–2 mm. $2n = 22$.

Varieties 2 (2 in the flora): California; Mexico.

Eriogonum gracile is a highly variable species. The low, spreading plants from the mountains of coastal central California with 2–3 mm involucres and (usually) yellowish flowers differ from the more erect plants represented by the type collection, which have 1.8–2 mm involucres and white flowers. Such plants are found mainly farther south, in the Mt. Diablo and Mt. Hamilton areas of Alameda, Contra Costa, and Stanislaus counties. Some populations of *E. gracile* from the Inner Coast Ranges of eastern San Luis Obispo County are slightly different from the norm and the epithet *leucocladon* has been adopted in some local floristic treatments to account for them. However, the type of *E. leucocladon* does not seem to be this phase.

1. Inflorescence branches lanate to tomentose or floccose; throughout California . . . . 213a. *Eriogonum gracile* var. *gracile*
1. Inflorescence branches glabrous; s California . . . . 213b. *Eriogonum gracile* var. *incultum*

213a. Eriogonum gracile Bentham var. **gracile**
  · Slender woolly wild buckwheat

Eriogonum roseum Durand &
Hilgard var. *leucocladon* (Bentham)
Hoover

Inflorescence branches lanate to
tomentose or floccose.

Flowering year-round. Sandy
and gravelly flats and slopes,
grassland and chaparral communities, oak and pine woodlands;
0–1400(–1600) m; Calif.; Mexico (Baja California).

Variety *gracile* is widespread and common to locally
weedy throughout much of California, southward to
northern Baja California, Mexico. Two collections
(*Schott 4*, F; *Thurber s.n.*, 1852, F, NY) made in the
early 1850s supposedly in Sonora, Mexico, and an 1865
Torrey collection (*454*, NY) from Nevada, are well
outside the known range of *E. gracile* and are discounted
as to location.

213b. Eriogonum gracile Bentham var. **incultum**
Reveal, Phytologia 66: 265. 1989 · Palomar
Mountain wild buckwheat

Inflorescence branches glabrous.
Flowering Jul–Oct. Sandy and
gravelly flats and slopes, chaparral
communities, oak and pine
woodlands; 700–1600 m; Calif.;
Mexico (Baja California).

Variety *incultum* is infrequently
encountered in the foothills and
mountains of southern California
(Orange, Riverside, and San Diego counties).
Occasionally it is locally common. This variety is much
more slender and graceful than *Eriogonum davidsonii*,
with which it might be confused.

214. Eriogonum cithariforme S. Watson, Proc. Amer.
Acad. Arts 23: 266. 1888 (as citharaeforme) [E]

Eriogonum gracile Bentham var.
cithariforme (S. Watson) Munz; *E.
vimineum* Douglas ex Bentham var.
cithariforme (S. Watson) S. Stokes

Herbs, spreading to erect, 2–3
(–5) dm, glabrous or tomentose,
greenish or reddish. **Stems:** aerial
flowering stems erect, 0.5–1 dm,
glabrous or tomentose. **Leaves**
basal and sometimes cauline; basal: petiole 1–5 cm,
tomentose to floccose, often slightly winged, blade
oblanceolate or elliptic to ovate or nearly rounded, 1–2
× 0.5–1.5(–2) cm, tomentose abaxially, floccose to
glabrate and grayish to greenish adaxially; cauline:

petiole 0.5–2 cm, floccose, blade elliptic, 0.3–1 cm ×
0.2–0.7 cm, similar to basal blade. **Inflorescences**
cymose, distally uniparous due to suppression of
secondary branches, open, 5–25 × 5–25 cm; branches
usually inwardly curved, glabrous or tomentose; bracts
0.5–2.5 × 0.5–2 mm. **Peduncles** absent. **Involucres**
somewhat appressed to branches, turbinate, 2.5–3 × 1.5–
2 mm, glabrous; teeth 5, erect, 0.2–0.4 mm. **Flowers**
1.5–2 mm; perianth white to rose, glabrous; tepals
monomorphic, oblong-obovate; stamens included, 1–1.5
mm; filaments pilose proximally. **Achenes** brown, 3-
gonous, 1–1.5 mm.

Varieties 2 (2 in the flora): California.

The graceful, upwardly curved branch segments of
Eriogonum cithariforme generally are distinctive in older
plants, although an occasional individual of *E. davidsonii*
may have this feature. A distinction between *E.
cithariforme* and *E. roseum* also is troublesome, as in
young material the curved branch segments are not
always obvious.

1. Flowering stems and inflorescence branches
   tomentose . . . . 214a. *Eriogonum cithariforme*
                                              var. *cithariforme*
1. Flowering stems and inflorescence branches
   glabrous . . . . 214b. *Eriogonum cithariforme*
                                              var. *agninum*

214a. Eriogonum cithariforme S. Watson var.
  cithariforme · Cithara wild buckwheat [E]

Leaves: petiole usually winged;
basal blade oblanceolate. **Flowering stems and inflorescence
branches** tomentose.

Flowering May–Oct. Sandy
flats and slopes, chaparral communities, oak and montane
conifer woodlands; (100–)500–
2200 m; Calif.

Variety *cithariforme* is usually uncommon throughout
its range in the mountains and foothills of southwestern
California (Kern, Los Angeles, San Bernardino, San Luis
Obispo, Santa Barbara, and Ventura counties). In
vigorous plants, the winged petiole is striking.

214b. Eriogonum cithariforme S. Watson var.
  agninum (Greene) Reveal, Aliso 7: 223. 1970
  · Santa Ynez wild buckwheat [E]

Eriogonum agninum Greene,
Pittonia 2: 165. 1891; *E. gracile*
Bentham var. *polygonoides*
(S. Stokes) Munz; *E. vimineum*
Douglas ex Bentham var.
agninum (Greene) S. Stokes;
E. vimineum subsp.
polygonoides S. Stokes

Eriogonum • POLYGONACEAE

Leaves: petiole not winged; basal blade elliptic to ovate or nearly rounded. **Flowering stems and inflorescence branches** glabrous.

Flowering May–Oct. Sandy flats and slopes, chaparral communities, oak and montane conifer woodlands; 500–1800(–2100) m; Calif.

Variety *agninum* is known primarily from the mountains of Santa Barbara and Ventura counties. A single collection (*Davidson 1617*, NY) found along Big Rock Creek in Los Angeles County in 1900, and the type of subsp. *polygonoides*, gathered in San Luis Obispo County, are from disjunct populations. This variety is sometimes difficult to differentiate from *E. davidsonii*.

215. Eriogonum baileyi S. Watson, Proc. Amer. Acad. Arts 10: 348. 1875 E F

Eriogonum vimineum Douglas ex Bentham subsp. *baileyi* (S. Watson) S. Stokes

Herbs, erect to spreading, 1–4(–5) dm, glabrous or tomentose, grayish. **Stems:** aerial flowering stems erect, 0.5–1 dm, glabrous or tomentose. **Leaves** basal; petiole 0.5–3 cm, tomentose; blade suborbiculate, 0.5–2 × 0.5–2 cm, densely white-to-mentose abaxially, mostly tomentose and grayish to greenish adaxially. **Inflorescences** cymose, infrequently distally uniparous due to suppression of secondary branches, open to diffuse, 5–35(–45) × 5–35(–4) cm; branches glabrous or tomentose; bracts 0.5–3 × 1–3 mm. **Peduncles** absent. **Involucres** somewhat appressed to branches, turbinate, 1–1.5 × 0.5–1 mm, glabrous or tomentose; teeth 5, erect, 0.2–0.3 mm. **Flowers** 1.5–2 mm; perianth white to rose, minutely glandular, rarely glabrous; tepals monomorphic, oblong to oblong-obovate, somewhat constricted near middle and flaring adaxially; stamens included, 1–1.5 mm; filaments pilose proximally. **Achenes** brown, 3-gonous, 1–1.5 mm.

Varieties 2 (2 in the flora): w United States.

Eriogonum baileyi is closely related to and sometimes difficult to distinguish from *E. elegans* of the Inner Coast Ranges of California, and *E. brachyanthum* along the eastern edge of the Sierra Nevada, especially in Inyo and Mono counties. However, these species do not seem to intergrade. *Eriogonum baileyi* often grows with other annual wild buckwheats (especially *E. ampullaceum* and *E. brachyanthum*), so care must be taken not to make mixed collections.

The Kawaiisu people of southern California pounded the seeds into a meal, which they ate dry, and also mixed with water to serve as a beverage (M. L. Zigmond 1981).

1. Flowering stems and inflorescence branches glabrous; e and s California, sc Idaho, Nevada, e Oregon, sw Utah, e Washington . . . . 215a. *Eriogonum baileyi* var. *baileyi*
1. Flowering stems and inflorescence branches tomentose; ne California, w and nc Nevada, c Idaho . . . . 215b. *Eriogonum baileyi* var. *praebens*

215a. Eriogonum baileyi S. Watson var. **baileyi**
 • Bailey's wild buckwheat E F

Eriogonum vimineum var. *multiradiatum* S. Stokes; *E. vimineum* var. *porphyreticum* (S. Stokes ex M. E. Jones) S. Stokes; *E. vimineum* var. *restioides* (Gandoger) S. Stokes

Flowering stems and inflorescence branches glabrous.

Flowering May–Oct. Sandy to gravelly washes, flats, and slopes, mixed grassland, saltbush, greasewood, and sagebrush communities, pinyon-juniper, oak, or montane conifer woodlands; (100–)500–2900 m; Calif., Idaho, Nev., Oreg., Utah, Wash.

Variety *baileyi* basically is a taxon of arid regions of the far West, being found primarily in California and Nevada northward through eastern Oregon to eastern Washington. Isolated populations are known from south-central Idaho and from Beaver County, Utah. In southern California the variety is found along the desert edges of the Transverse Ranges but always just beyond the mixed grasslands and oak woodlands where *E. elegans* is typically encountered.

215b. Eriogonum baileyi S. Watson var. **praebens** (Gandoger) Reveal, Taxon 32: 294. 1983
 • Bailey's woolly wild buckwheat E

Eriogonum praebens Gandoger, Bull. Soc. Roy. Bot. Belgique 42: 196. 1906; *E. baileyi* var. *divaricatum* (Gandoger) Reveal; *E. commixtum* Greene ex Tidestrom

Flowering stems and inflorescence branches tomentose.

Flowering Jun–Oct. Sandy to gravelly washes, flats, and slopes, saltbush and sagebrush communities, pinyon and/or juniper woodlands; 1300–2500 m; Calif., Idaho, Nev.

Variety *praebens* is encountered infrequently over its scattered range in California (Inyo, Lassen, Los Angeles, Mono, Plumas, and Sierra counties), Idaho (Blaine and Elmore counties), and Nevada (Carson City, Douglas, Eureka, Humboldt, Lander, Storey, and Washoe counties).

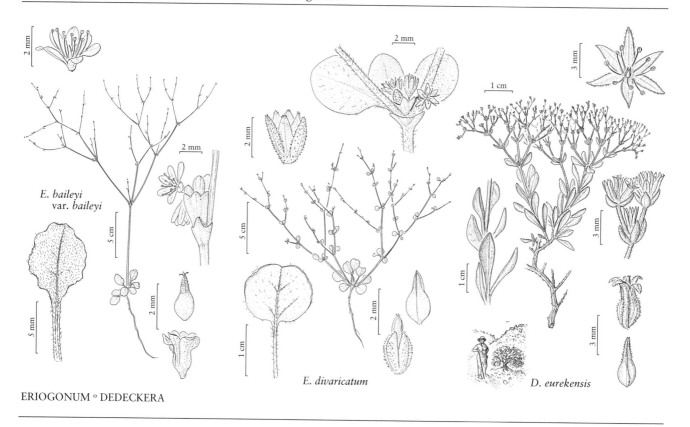

E. *baileyi*
var. *baileyi*

E. *divaricatum*

D. *eurekensis*

ERIOGONUM ∘ DEDECKERA

216. Eriogonum elegans   Greene, Pittonia 2: 173. 1891   • Elegant wild buckwheat   E

Eriogonum baileyi S. Watson subsp. *elegans* (Greene) Munz; *E. vimineum* Douglas ex Bentham var. *elegans* (Greene) S. Stokes

Herbs, erect, 1–4 dm; glabrous, grayish. **Stems:** aerial flowering stems erect, 0.5–1 dm, glabrous. **Leaves** basal; petiole 0.3–3 cm, thinly tomentose; blade oblong or rounded to subcordate, 0.3–1.5 × 0.3–1.5 cm, densely white-tomentose abaxially, mostly tomentose and grayish or greenish adaxially. **Inflorescences** cymose, occasionally distally uniparous due to suppression of secondary branches, open to somewhat diffuse, 5–3.5 × 5–30 cm; branches glabrous; bracts 0.5–2 × 1–2 mm. **Peduncles** absent. **Involucres** somewhat appressed to branches, turbinate, 1–1.5 × 0.6–0.8 mm, glabrous; teeth 5, erect, 0.2–0.3 mm. **Flowers** 1–1.5 mm; perianth white to rose, glabrous; tepals slightly dimorphic, those of outer whorl oblong-obovate, those of inner whorl oblong; stamens included, 1 mm; filaments pilose proximally. **Achenes** brown, 3-gonous, 1–1.5 mm.

Flowering May–Nov. Sandy to gravelly flats and slopes, mixed grassland communities, oak and pine woodlands; 200–1200 m; Calif.

Eriogonum elegans is found primarily on the coastal mountains (Fresno, Monterey, San Benito, San Luis Obispo, Santa Barbara, Santa Clara, and Ventura counties). It is more slender and elegant than *E. baileyi*. It is rather infrequent especially near the coast, but it can be locally common to even abundant on the Inner Coast Ranges.

217. Eriogonum brachyanthum   Coville, Contr. U.S. Natl. Herb. 4: 185. 1893   • Short-flower wild buckwheat   E

Eriogonum baileyi S. Watson var. *brachyanthum* (Coville) Jepson; *E. vimineum* Douglas ex Bentham var. *brachyanthum* (Coville) S. Stokes

Herbs, spreading, (0.5–)1–3 dm, glabrous, greenish. **Stems:** aerial flowering stems erect, 0.3–0.8 dm, glabrous. **Leaves** basal; petiole 0.5–3 cm, tomentose; blade ovate to rounded, 0.5–2 × 0.5–2 cm, densely white-tomentose on both surfaces, rarely glabrate and grayish or greenish adaxially. **Inflorescences** cymose, occasionally distally uniparous due to suppression of secondary branches, usually diffuse, 5–25 × 5–30 cm; branches glabrous; bracts 0.5–2 × 0.5–2 mm. **Peduncles** absent. **Involucres** somewhat appressed to branches, turbinate, 1–1.2 × 0.4–

0.6 mm, glabrous; teeth 5, erect, 0.2–0.3 mm. **Flowers** 0.6–0.8(–1) mm; perianth yellow or pale yellowish, rarely whitish, glabrous; tepals slightly dimorphic, those of outer whorl oblong to oblong-obovate, those of inner whorl narrowly oblong; stamens included, 0.5–0.8 mm; filaments pilose proximally. **Achenes** dark brown, 3-gonous, 0.8–1 mm.

Flowering Apr–Nov. Sandy to gravelly flats, washes, and slopes, saltbush, greasewood, and sagebrush communities, pinyon-juniper and montane conifer woodlands; 600–2300 m; Calif., Nev., Oreg.

Eriogonum brachyanthum basically is a species of the Mojave Desert, extending northward in eastern California (Inyo, Kern, Los Angeles, Mono, Riverside, and San Bernardino counties) and along the Lahontan Trough of Nevada (Churchill, Esmeralda, Eureka, Humboldt, Lander, Lyon, Mineral, Nye, Pershing, and Washoe counties) to just north of Denio in Harney County, Oregon. It is common to abundant and even locally weedy in the northern Mojave Desert. Elsewhere it is less common and much more widely scattered. It often is found with other annual wild buckwheats forming dense patches (especially along highways) composed of several (up to seven in places) species, and care must be taken to sort them properly when making herbarium specimens.

218. Eriogonum evanidum Reveal, Phytologia 86: 132. 2004 • Vanishing wild buckwheat C

Herbs, erect, 1–2 dm, tomentose to floccose, grayish. **Stems:** aerial flowering stems erect, 0.3–0.6 dm, tomentose to floccose. **Leaves** basal; petiole 0.5–1.5 cm, tomentose; blade broadly ovate to orbiculate or reniform, 0.7–1.2 × 0.7–1.2 cm, densely white-tomentose abaxially, floccose and greenish adaxially. **Inflorescences** narrowly cymose, infrequently distally uniparous due to suppression of secondary branches, open, 5–15 × 5–10 cm; branches tomentose to floccose; bracts 0.5–2 × 0.5–1 mm. **Peduncles** absent. **Involucres** somewhat appressed to branches, campanulate, 1–2 × 1–2 mm, glabrous; teeth 5, spreading, 0.4–0.8 mm. **Flowers** 0.8–1.2 mm; perianth ochroleucous, glabrous; tepals dimorphic, those of outer whorl ovate, slightly hastate proximally in fruit, those of inner whorl lanceolate to elliptic; stamens included, 0.4–0.6 mm; filaments glabrous. **Achenes** dark brown, 3-gonous, 1.3–1.5 mm.

Flowering Jul–Oct. Sandy to gravelly flats and slopes, sagebrush communities, oak and montane conifer woodlands; of conservation concern; 1100–2100 m; Calif.; Mexico (Baja California).

Eriogonum evanidum is presumed extinct. The last known collections were made in Bear Valley in the San Bernardino Mountains, San Bernardino County, in 1931 (*Templeton 1588*, BRY, KANS, NY); Hemet Valley in the San Jacinto Mountains, Riverside County (*Ziegler s.n.*, 10 Oct 1967, GH, RSA, UC, UTC); and the Pine Valley area, San Diego County, in 1938. It is related to two sprawling annual Mexican species with mostly elliptic basal leaf blades: *E. foliosum* S. Watson, with weakly hastate cauline leaves, and *E. hastatum* Wiggins, with strongly hastate cauline leaves.

219. Eriogonum nidularium Coville, Contr. U.S. Natl. Herb. 4: 186. 1893 • Bird nest wild buckwheat E

Eriogonum vimineum Douglas ex Bentham subsp. *nidularium* (Coville) S. Stokes

Herbs, erect to slightly spreading, 0.5–1.5(–2) dm, floccose to tomentose, greenish to tawny. **Stems:** aerial flowering stems mostly erect, 0.05–0.3 dm, floccose to tomentose. **Leaves** basal; petiole 1–3(–5) cm, tomentose; blade rounded to cordate, 1–2(–2.5) × 1–2(–2.5) cm, densely white-tomentose abaxially, tomentose to floccose and occasionally greenish adaxially. **Inflorescences** narrowly cymose, distally uniparous due to suppression of secondary branches, diffuse, 5–15(–20) × 5–20(–30) cm; branches inwardly curved distally, floccose to tomentose; bracts 0.5–3 × 1–3 mm. **Peduncles** absent. **Involucres** appressed to branches, cylindric-turbinate, 1–1.3 × 0.5–0.7 mm, floccose; teeth 5, erect, 0.1–0.2 mm. **Flowers** 1–3(–3.5) mm; perianth pale yellow or yellow, rarely creamy white, becoming red, glabrous; tepals dimorphic, those of outer whorl broadly fan-shaped, those of inner whorl oblanceolate; stamens included, 1–1.5 mm; filaments pilose proximally. **Achenes** light brown, 3-gonous, 1–1.3 mm.

Flowering Mar–Oct. Sandy to gravelly washes, flats, and slopes, saltbush, greasewood, creosote bush, blackbrush, and sagebrush communities, pinyon and/or juniper woodlands; (300–)500–2300(–2700) m; Ariz., Calif., Idaho, Nev., Oreg., Utah.

Eriogonum nidularium is common and widespread, sometimes becoming weedy, especially along roadsides. It is basically a species of the Great Basin and the Mojave Desert, found mainly in eastern California and Nevada, extending into northwestern Arizona, southwestern Idaho, southeastern Oregon, and southwestern Utah. The numerous, often distinctive, incurved inflorescence branches resemble a bird nest, as reflected in the common name.

220. Eriogonum palmerianum Reveal in P. A. Munz, Suppl. Calif. Fl., 58. 1968 • Palmer's wild buckwheat [E]

Eriogonum plumatella Durand & Hilgard var. *palmeri* Torrey & A. Gray, Proc. Amer. Acad. Arts 8: 180. 1870, not *E. palmeri* S. Watson 1877

Herbs, spreading to erect, (0.5–)1–3 dm, mostly floccose to tomentose, grayish to tawy. **Stems:** aerial flowering stems mostly erect, 0.3–0.8 dm, floccose to tomentose. **Leaves** basal; petiole 1–4 cm, floccose; blade suborbiculate to cordate, 0.5–1.5(–1.8) × 0.5–2 cm, densely white- to grayish-tomentose abaxially, less so to glabrate and often greenish adaxially. **Inflorescences** cymose, distally uniparous due to suppression of secondary branches, open, 5–2.5 × 10–30 cm; branches not inwardly curved distally, floccose to tomentose; bracts 0.5–3 × 1–3 mm. **Peduncles** absent. **Involucres** appressed to branches, campanulate, 1.5–2 × 1.5–2 mm, floccose to tomentose; teeth 5, erect, 0.2–0.3 mm. **Flowers** 1.5–2 mm; perianth white to pink or pale yellowish, becoming pink to red, glabrous; tepals dimorphic, those of outer whorl narrowly fan-shaped, those of inner whorl oblanceolate; stamens included, 1–1.5 mm; filaments pilose proximally. **Achenes** brown, 3-gonous, 1.5–1.8 mm. $2n = 40$.

Flowering Mar–Oct. Sandy to gravelly washes, flats, and slopes, saltbush, greasewood, creosote bush, blackbrush, and sagebrush communities, pinyon and/or juniper woodlands; (300–)600–2300(–2700) m; Ariz., Calif., Colo., Idaho, Nev., N.Mex., Utah.

Eriogonum palmerianum is common and widespread, sometimes becoming locally abundant and even weedy. It is found in the southern Great Basin and Mojave deserts of eastern California eastward through Nevada, southwestern Idaho, and Utah, on to the Colorado Plateau of western Colorado. In Arizona and New Mexico, it occurs mainly in the Sonoran Desert. Occasional collections are difficult to distinguish from *E. nidularium*, but such confusion usually is not encountered in the field.

221. Eriogonum polycladon Bentham in A. P. de Candolle and A. L. P. P. de Candolle, Prodr. 14: 16. 1856 • Sorrel wild buckwheat

Eriogonum densum Greene; *E. vimineum* var. *densum* (Greene) S. Stokes; *E. vimineum* Douglas ex Bentham subsp. *polycladon* (Bentham) S. Stokes

Herbs, narrowly erect, (0.5–)1–6 dm, tomentose, whitish to grayish. **Stems:** aerial flowering stems erect, (0.3–)1–3 dm, tomentose. **Leaves** cauline; petiole 0.3–1.5 cm; blade narrowly oblanceolate to broadly elliptic, (0.7–)1–3 × 0.5–1.5 cm, densely white-tomentose and whitish to grayish on both surfaces. **Inflorescences** narrowly cymose, distally uniparous due to suppression of secondary branches, open, (5–)10–50 × 10–25 cm; branches tomentose; bracts 1.5–3 × 1–2.5 mm. **Peduncles** absent. **Involucres** appressed to branches, turbinate, 1.5–2.5 × 1–2 mm, tomentose, rarely glabrous; teeth 5, erect, 0.4–1 mm. **Flowers** (1–)1.5–2 mm; perianth white, becoming pink or red, glabrous; tepals dimorphic, those of outer whorl broadly fan-shaped, those of inner whorl oblanceolate; stamens included, 1–1.5 mm; filaments pilose proximally. **Achenes** dark brown, 3-gonous, 1–1.3 mm. $2n = 26$.

Flowering year-round. Sandy to gravelly washes, flats, and slopes, saltbush, creosote bush, greasewood, blackbrush, and sagebrush communities, oak, pinyon and/or juniper, and montane conifer woodlands; (200–)500–2200(–2500) m; Ariz., Calif., N.Mex., Tex., Utah; Mexico (Chihuahua, Durango, Sonora).

Eriogonum polycladon is widely distributed from near Needles, San Bernardino County, California (where not found since the 1930s), eastward to western Texas, and from southern Utah throughout much of central and southern Arizona into western and southern New Mexico. It is found also in the Mexican states of Chihuahua, Durango, and Sonora. It can be locally common and occasionally weedy, especially in southeastern Arizona and southwestern New Mexico.

SELECTED REFERENCE Spellenberg, R., C. Leiva, and E. Lessa. 1988. An evaluation of *Eriogonum densum* (Polygonaceae). SouthW. Naturalist 33: 71–80.

222. Eriogonum dasyanthemum Torrey & A. Gray, Proc. Amer. Acad. Arts 8: 177. 1870 • Chaparral wild buckwheat [E]

Herbs, erect to spreading, 2–6 dm, floccose to tomentose, rarely glabrous, grayish to reddish. **Stems:** aerial flowering stems erect, 0.5–2 dm, floccose to tomentose, rarely glabrous. **Leaves** basal and occasionally cauline; basal: petiole (0.5–)1–3 cm, usually tomentose, blade broadly obovate to roundish, (0.3–)1–2 × (0.3–)1–2 cm, white-tomentose abaxially, floccose to glabrate or glabrous and greenish adaxially; cauline sessile, blade broadly obovate to roundish, rarely elliptic, 0.5–1.5 (–3) × 0.5–1(–2) cm, similar to basal blade. **Inflorescences** cymose, not distally uniparous due to suppression of secondary branches, mostly open, 10–50 × 10–50 cm; branches floccose to glabrate, rarely glabrous; bracts 0.5–2.5 × 1–3 mm. **Peduncles** absent. **Involucres** appressed to branches, cylindric, (3.8–)4–4.5 × 2–3 mm, glabrate; teeth 5, erect, 0.2–0.5 mm. **Flowers** (1.8–)2–2.5 mm; perianth white to rose, densely pubescent; tepals monomorphic, oblong-obovate; stamens included, 1.5–2 mm; filaments pilose proximally. **Achenes** brown, 3-gonous, 1.5–2 mm, glabrous; beak scabrous. $2n = 24$.

Flowering May–Oct. Sandy to gravelly flats and slopes, mixed grassland and chaparral communities, oak and pine woodlands; 50–1400 m; Calif.

Eriogonum dasyanthemum is common to locally abundant in the North Coast Ranges (Colusa, Glenn, Lake, Mendocino, Napa, Shasta, Solano, Tehama, Trinity, and Yolo counties). A single collection apparently made at Mendocino along the Pacific Ocean in Mendocino County (*Brown 942*, Aug 1898, B, BKL, F, GH, IND, MO, RM) requires modern confirmation. *Jepson 8962* (JEPS, UTC) supposedly was gathered "above Venado" in Colusa County, but Venado is a small town in Sonoma County. The densely pubescent flowers easily distinguish this species.

223. Eriogonum divaricatum Hooker, Hooker's J. Bot. Kew Gard. Misc. 5: 265. 1853 • Divergent wild buckwheat [E] [F]

Herbs, spreading, 1–2(–3) dm, puberulent to short-pilose, greenish to reddish. **Stems:** aerial flowering stems decumbent to spreading, 0.3–0.5 dm, puberulent to short-pilose. **Leaves** basal and cauline; basal: petiole 1–4 (–5) cm, puberulent to short-pilose, blade elliptic-oblong to orbiculate, 1–2(–2.5) × 1–2(–2.5) cm, puberulent to short-pilose and green on both surfaces; cauline: petiole (0–)0.1–2 cm, puberulent to short-pilose, absent distally, blades opposite, oblanceolate to oblong or elliptic, 0.3–1(–1.5) × 0.2–0.8(–1.2) cm, similar to basal blade. **Inflorescences** cymose, distally uniparous due to suppression of secondary branches, diffuse, 5–25 × 10–45 cm; branches puberulent; bracts 1–3(–5) × 1–2 mm. **Peduncles** absent. **Involucres** somewhat appressed to branches, campanulate, 1–2 × 1–2 mm, pilose; teeth 5, lobelike, spreading to somewhat reflexed, 0.7–1.5 mm. **Flowers** (1–)1.5–2 mm; perianth yellow, rarely pale yellow, hispidulous and glandular with yellowish-white hairs; tepals monomorphic, oblong-lanceolate to oblong-ovate; stamens included, 0.7–1.5 mm; filaments pilose proximally. **Achenes** light brown, 3-gonous, 1.5–2 mm.

Flowering May–Oct. Heavy clay flats and slopes, saltbush, greasewood, and sagebrush communities, pinyon-juniper woodlands; 1100–2300(–2500) m; Ariz., Colo., N.Mex., Utah, Wyo.

Eriogonum divaricatum basically is a species of the Colorado Plateau of southwestern Wyoming (Sublette, Sweetwater, and Uinta counties), eastern Utah (Carbon, Emery, Garfield, Grand, Kane, San Juan, Sevier, and Wayne counties), western Colorado (Mesa and Montezuma counties), northern Arizona (Apache, Coconino, and Navajo counties), and northwestern New Mexico (McKinley and San Juan counties), where it is infrequent to common but only rarely abundant. There are several disjunct populations well removed from that core area; the species occurs in Beaver and Millard counties of west-central Utah, and in Clayhole Valley, Mohave County, Arizona. In New Mexico, it was found around San Ysidro, Sandoval County, in 1926. The most extreme disjunctions were two populations found in Argentina during the growing season of 1899–1900. Those plants were named *Eriogonum ameghinoi* Spegazzini, which ultimately became the only species of the genus *Sanmartinia* M. Buchinger. Probably introduced there by migrating birds, the species did not persist in South America (J. L. Reveal 1981b).

The divergent wild buckwheat was used ceremonially by the Navajo (Diné) people in one of their snake dances, and portions of the plant were smoked in the treatment of snakebite (A. Clifford, pers. comm.; L. C. Wyman and S. K. Harris 1951).

224. Eriogonum darrovii Kearney, Leafl. W. Bot. 4: 267. 1946 • Darrow's wild buckwheat [C] [E]

Herbs, spreading, 0.5–1.5(–2) dm, sericeous, tawny to reddish. **Stems:** aerial flowering stems spreading, 0.3–0.5 dm, sericeous. **Leaves** basal and cauline; basal: petiole 0.5–2 cm, sericeous, blade obovate to rounded, 0.5–1.5 × 0.5–1.5 cm, sericeous and greenish on both surfaces; cauline: petiole (0–)0.1–1 cm, sericeous, absent distally, blades opposite, obovate to rounded, 0.2–1 × 0.2–1 cm, similar to basal blade. **Inflorescences** cymose, distally uniparous due to suppression of secondary branches, diffuse, 3–15 × 3–20 cm; branches sericeous; bracts 1–3 × 1–2 mm. **Peduncles** absent. **Involucres** somewhat appressed to branches, turbinate-campanulate, 2–2.5 × 1.5–2(–2.5) mm, sericeous; teeth 5, spreading, 0.8–1.5 mm. **Flowers** 1.5–2 mm; perianth pale yellow to pink or red, hirtellous with whitish hairs; tepals dimorphic, those of outer whorl fan-shaped, inflated and hooded distally, those of inner whorl lanceolate; stamens included, 1.5–2 mm; filaments pilose proximally. **Achenes** brown, 3-gonous, 1–5 mm.

Flowering May–Sep. Gravelly or clayey calcareous flats and slopes, saltbush and sagebrush communities, pinyon-juniper woodlands; of conservation concern; 1550–1900 m; Ariz., Nev., Utah.

Eriogonum darrovii is of special concern in Nevada and Arizona. In Arizona, it is known only from scattered locations in northern Mohave (west of Mt. Trumbull) and northwestern Coconino (Fredonia area) counties. In Nevada, it is known from White Valley (Nye County) and Spring Valley from Majors Place (White Pine County) to Pony Springs (Lincoln County). It is found also in the Buckskin Mountains of Kane County, Utah, and Coconino County, Arizona.

2. DEDECKERA   Reveal & J. T. Howell, Brittonia 28: 245, fig. 1. 1976   • Eureka or July gold [for Mary Caroline DeDecker, 1909–2000, noted California conservationist]

[C] [E]

James L. Reveal

Shrubs; taproot stout. **Stems** spreading, woody, hirsutulous; caudex absent; aerial flowering stems spreading, herbaceous, hirsutulous. **Leaves** usually persistent through anthesis, 1 per node, cauline, alternate; petiole present, blade narrow to broadly elliptic, margins entire. **Inflorescences** terminal, cymose; branches dichotomous, not brittle or disarticulating into segments, round, hirsutulous, each node with 1 short-pedicellate flower and 4–12 sessile or subsessile flowers; bracts 3–4 at proximal nodes and leaflike, 2–3 at distal nodes and scalelike, distinct, hirsutulous. **Peduncles** erect, slender. **Involucral bracts** obscure, in 2 whorls of 3 plus 2 lobes at proximal nodes, distinct, reduced to 2 lobes at distal nodes, not awn-tipped. **Flowers** 4–6 per involucral cluster, sessile or pedicellate; perianth yellow to reddish yellow, broadly campanulate when open, narrowly urceolate when closed, hispidulous abaxially; tepals 6, slightly dimorphic, those of outer whorl broader and longer than those of inner whorl, entire apically; stamens 9; filaments basally adnate, pilose basally; anthers yellow, oblong. **Achenes**, light reddish brown, not winged, 3-gonous, pubescent apically. **Seeds**: embryo curved. $x = 20$.

Species 1: California.

Dedeckera is rare but not endangered or threatened, primarily because most populations are in remote areas along sides of steep canyon walls. Most populations occur in areas managed by the Bureau of Land Management, the United States Forest Service, or the National Park Service. The plants rarely produce mature achenes (D. Wiens et al. 1989). The genus is allied to *Eriogonum* subg. *Eucycla*.

SELECTED REFERENCES   Nickrent, D. L. and D. Wiens. 1989. Genetic diversity in the rare California shrub *Dedeckera eurekensis* (Polygonaceae). Syst. Bot. 14: 245–253. Wiens, D. et al. 1989. Developmental failure and loss of reproductive capacity in the rare palaeoendemic shrub *Dedeckera eurekensis*. Nature 338: 65–67. Wiens, D., L. Allphin, D. H. Mansfield, and G. Thackray. 2004. Developmental failure and loss of reproductive capacity as a factor in extinction: A nine-year study of *Dedeckera eurekensis* (Polygonaceae). Aliso 21: 55–63. Wiens, D., C. I. Davern, and C. L. Calvin. 1988. Exceptionally low seed set in *Dedeckera eurekensis*: Is there a genetic component of extinction? In: C. A. Hall and V. Doyle-Jones, eds. 1988. Plant Biology of Eastern California. Los Angeles. Pp. 19–29. Wiens, D., M. DeDecker, and C. D. Wiens. 1986. Observations on the pollination of *Dedeckera eurekensis* (Polygonaceae). Madroño 33: 302–305.

1. Dedeckera eurekensis Reveal & J. T. Howell, Brittonia 28: 246, fig. 1. 1976 [C] [E] [F]

Plants 0.2–0.7(–1) × 0.5–2 m. **Leaves:** petiole 2–5 mm; blade 0.1–1.5 × (0.4–)0.5–0.8(–1.3) cm. **Inflorescences** 1–4(–6) cm; bracts (2–)5–12(–17) mm. **Peduncles** (0.1–)2–6(–7) mm. **Involucral bracts** 1–2 × 0.4–0.6 mm. **Flowers:** perianth 1.8–4 mm; filaments 1–2 mm; anthers 0.2–0.3 mm. **Achenes** 2–3(–3.5) mm. **2n** = 40.

Flowering Jun–Oct. Gravelly limestone talus slopes, saltbush communities; of conservation concern; 1200–2200 m; Calif.

Dedeckera eurekensis is known from small, disjunct populations along the western slope of the White Mountains from southern Mono County southward into the Last Chance Mountains of Inyo County.

3. STENOGONUM Nuttall, Proc. Acad. Nat. Sci. Philadelphia 4: 19. 1848 • Two-whorl buckwheat [Greek *stenos*, narrow, and *gonos*, seed, alluding to achene] [E]

James L. Reveal

Eriogonum Michaux sect. *Stenogonum* (Nuttall) Kuntze

Herbs, annual; taproot slender. **Stems** arising directly from the root, spreading to decumbent, solid, not fistulose or disarticulating into ringlike segments, glabrous or minutely strigose or glandular. **Leaves** usually persistent through anthesis, basal and rosulate or basal and cauline, alternate; petiole present (basal leaves); blade narrowly lanceolate to oblanceolate or spatulate to orbiculate, margins entire. **Inflorescences** terminal, cymose; branches mostly dichotomous, not brittle or disarticulating into segments, round, minutely strigose or glandular, glabrous; bracts 3 per node, connate basally, scalelike, triangular, not awn-tipped, glabrous or sparsely glandular. **Peduncles** straight or flexed, slender to filiform, sometimes absent. **Involucral bracts** in 2 whorls of 3, connate proximally, lanceolate, not awn-tipped. **Flowers** (6–)9–15 per involucral cluster at any single time during full anthesis; perianth yellow to reddish yellow, broadly campanulate when open, narrowly urceolate when closed, pilose abaxially; tepals 6, monomorphic, entire apically; stamens 9; filaments basally adnate, glabrous; anthers yellow, oval. **Achenes** usually included, light brown, not winged, 3-gonous, glabrous. **Seeds:** embryo curved. **x** = 20.

Species 2 (2 in the flora): w United States.

Stenogonum is clearly allied to *Eriogonum* subg. *Ganysma*, specifically *E. trichopes* and its close relatives. The genus is readily distinguished from *Eriogonum* by an involucre reduced to a series of two foliaceous whorls of three lanceolate bracts.

SELECTED REFERENCE Reveal, J. L. and B. Ertter. 1977. Re-establishment of *Stenogonum* (Polygonaceae). Great Basin Naturalist 36: 272–280.

1. Leaves basal; peduncles flexed; stems mostly erect, sparsely glandular . . . . 1. *Stenogonum flexum*
1. Leaves basal and cauline; peduncles straight; stems mostly spreading, glabrous . . . .
. . . . 2. *Stenogonum salsuginosum*

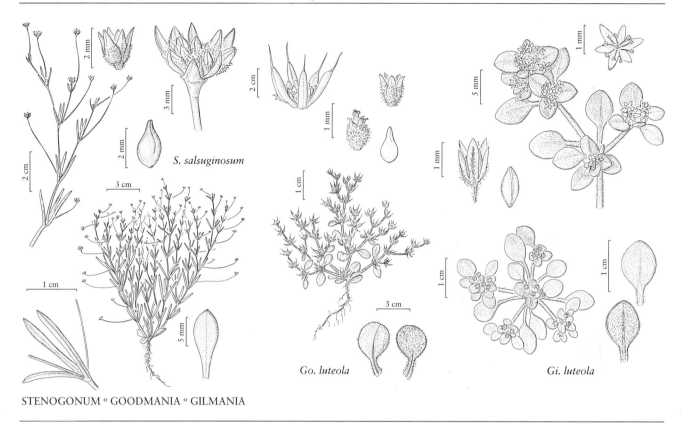

S. salsuginosum

Go. luteola

Gi. luteola

STENOGONUM ° GOODMANIA ° GILMANIA

1. Stenogonum flexum (M. E. Jones) Reveal & J. T. Howell, Brittonia 28: 249. 1976 • Bent two-whorl buckwheat E

Eriogonum flexum M. E. Jones, Zoë 2: 15. 1891

Stems mostly erect, (0.5–)1–3 dm, sparsely glandular. **Leaves** basal; petiole 1–4 cm; blade mostly orbiculate, 0.5–2 cm, minutely strigose when young, glabrous at maturity. **Inflorescences** erect or slightly spreading, 0.5–2.5 dm, glandular at nodes and proximal internode; bracts 0.5–2 mm, sometimes sparsely glandular. **Peduncles** flexed, filiform, 1–3 cm, glandular to middle. **Involucral bracts** 2–3 × 2–4 mm, glabrous, sometimes sparsely glandular. **Flowers:** perianth yellow to reddish yellow, 1.5–3.5 mm; tepals lanceolate; filaments 1.5–2 mm; anthers 0.2–0.3 mm. **Achenes** 2–2.5 mm. *2n* = 40.

Flowering Apr–Sep. Clay hills and flats, saltbush communities; 1300–2000 m; Ariz., Colo., N.Mex., Utah.

Stenogonum flexum is found in eastern Utah from Uintah County south to Kane and San Juan counties, with additional populations over the state lines in Arizona, Colorado, and New Mexico.

2. Stenogonum salsuginosum Nuttall, Proc. Acad. Nat. Sci. Philadelphia 4: 19. 1848 • Smooth two-whorl buckwheat E F

Eriogonum salsuginosum (Nuttall) Hooker

Stems mostly spreading, 0.5–2 dm, glabrous. **Leaves** basal and cauline; basal petiolate, petiole 0.5–2 cm, blade spatulate, (1–)2–4 × (0.5–)1–2.5 cm, glabrous; cauline sessile, blade narrowly lanceolate to oblanceolate, 0.5–4.5 × 0.2–1 cm. **Inflorescences** spreading, 0.5–2.5(–3) dm, glabrous, occasionally glandular at proximal nodes; bracts 0.5–4 mm. **Peduncles** straight, slender to filiform, 0.1–4 cm, glabrous, or absent. **Involucral bracts** 2–8 × 2–3 mm, glabrous. **Flowers:** perianth yellow, 1.5–3 mm; tepals lanceolate; filaments 1.5–2 mm; anthers 0.2–0.3 mm. **Achenes** 2–2.5 mm. *2n* = 40.

Flowering Apr–Sep. Clay hills and flats, saltbush communities; 1300–2200 m; Ariz., Colo., N.Mex., Utah, Wyo.

Stenogonum salsuginosum is common to abundant and even weedy on the Colorado Plateau. The species occurs from Washakie County, Wyoming, southward into eastern Utah and extreme western Colorado to northern Arizona (just entering Mohave County) and northwestern New Mexico.

4. GOODMANIA Reveal & Ertter, Brittonia 28: 427, fig. 1. 1977 • Yellow spinecape [For George Jones Goodman, 1904–1999, authority on *Chorizanthe*] E

James L. Reveal

Herbs, annual; taproot slender. **Stems** arising directly from the root, spreading to decumbent or prostrate, solid, not fistulose or disarticulating into ringlike segments, pubescent or glabrous. **Leaves** usually persistent through anthesis, basal and cauline, opposite; petiole present; blade broadly elliptic or round to reniform, becoming linear, margins entire, awn-tipped at proximal nodes. **Inflorescences** terminal, cymose; branches dichotomous, not brittle or disarticulating into segments, round, usually glabrous; bracts 2, distinct, somewhat leaflike and seemingly succulent, mucronate, pubescent to glabrate. **Peduncles** absent. **Involucral bracts** in 1 whorl of 5, distinct, narrowly lanceolate, awn-tipped. **Flowers** (6–)9–15 per involucral cluster at any single time during anthesis; perianth yellow, broadly campanulate when open, narrowly urceolate when closed, woolly-tomentose abaxially; tepals 6, connate proximally, monomorphic, entire apically; stamens 9; filaments basally adnate, glabrous; anthers yellow, ovate. **Achenes** included, light brown, not winged, 3-gonous, glabrous. **Seeds:** embryo curved.

Species 1: w United States.

Goodmania is allied to *Eriogonum* subg. *Ganysma* but its point of origin is obscure. The most logical point to suggest within subg. *Ganysma* is somewhere around the *E. inflatum* complex. The bright green color of the flowering stems and inflorescence branches, the pubescent yellow flowers, and the near-glabrous condition of the plant body are somewhat akin to those found in *Stenogonum*. Each of these segregate genera is confined to arid regions in the American West, all appear to have rather recent origins, and each seems to be exhibiting a type of variation different from the norm seen among the annual wild buckwheats.

1. Goodmania luteola (Parry) Reveal & Ertter, Brittonia 28: 427. 1977 E F

Oxytheca luteola Parry, Bull. Torrey Bot. Club 10: 23. 1883; *Eriogonum spinescens* S. Stokes

Plants (0.3–)0.5–1.5 × (0.1–)0.3–1.5 dm. **Leaves:** petiole (3–)10–20 mm; proximal blades 2–5(–6) × 2–5(–7) mm, acerose, distal blades 3–5 × 0.5–1.5 mm, awn-tipped. **Involucral bracts** 2–5 mm; awns 1–3 mm. **Flowers** 0.8–1 mm. **Achenes** 1–1.2 mm.

Flowering Apr–Aug. Dry lake-beds, flats and sinks, and meadows in mixed grassland, saltbush, creosote bush, and sagebrush communities; 70–2200 m; Calif., Nev.

Goodmania luteola occurs in the southern San Joaquin Valley and on the northwestern edge of the Mojave Desert in California northward to Mono County. It also occurs in Alkali Valley, Mineral County, Nevada. It rapidly is becoming rare and frequently locally extirpated.

5. GILMANIA Coville, J. Wash. Acad. Sci. 26: 210. 1936, name proposed for conservation • Golden carpet [for M. French Gilman, 1871–1944, Death Valley naturalist] C E

James L. Reveal

Phyllogonum Coville, Contr. U.S. Natl. Herb. 4: 190, plate 21. 1893, name proposed for rejection

Herbs, annual; taproot slender. **Stems** arising directly from the root, spreading to decumbent or prostrate, solid, not fistulose or disarticulating into ringlike segments, glabrous or sparsely pubescent. **Leaves** quickly deciduous, basal and cauline, in whorls of 3; petiole present only at proximal nodes; blade oblong to broadly elliptic or obovate, margins entire. **Inflorescences** terminal, cymose; branches mostly dichotomous, not brittle or disarticulating into segments, round, glabrous or sparsely pubescent; bracts absent. **Peduncles** absent. **Involucral bracts** absent. **Flowers** 3–9 per node at any single time during full anthesis; perianth yellow, broadly campanulate when open, narrowly urceolate when closed, thinly pubescent abaxially; tepals 6, connate proximally, monomorphic, entire apically; stamens 9; filaments basally adnate, pilose basally; anthers yellow, ovate. **Achenes** included, brown, not winged, 3-gonous, glabrous. **Seeds:** embryo curved.

Species 1: California.

Gilmania is restricted to the valley edges and low mountains surrounding Death Valley in Inyo County. In a "good" year, tens of millions of plants carpet the area, giving the edge of the valley a ring of golden yellow. Each individual produces thousands of flowers, and seed set in the plush years is enormous, thereby assuring long-term survival even in this harsh environment. The genus is allied to *Eriogonum* subg. *Ganysma*.

SELECTED REFERENCE   Reveal, J. L. 2004. Proposal to conserve the name *Gilmania* Coville against *Phyllogonum* Coville (Polygonaceae: Eriogonoideae)—A case of mistaken homonymy. Taxon 53: 573.

1. **Gilmania luteola** (Coville) Coville, J. Wash. Acad. Sci. 26: 210. 1936   C E F

Phyllogonum luteolum Coville, Contr. U.S. Natl. Herb. 4: 190, plate 21. 1893

Plants 0.3–1.2(–1.5) × 0.3–1.5 (–2) dm. **Leaves:** petiole 0.5–2 cm; blade 0.5–1.5 × 0.3–0.8(–1) cm. **Flowers:** perianth 1–2 mm. **Achenes** 1.5–2 mm.
  Flowering Feb–May. Alkaline, often barren slopes and flats, saltbush communities; of conservation concern; 10–500 m; Calif.

6. **OXYTHECA** Nuttall, Proc. Acad. Nat. Sci. Philadelphia 4: 18. 1848 • Puncturebract [Greek *oxys*, sharp, and *theke*, case, alluding to awned involucre]

James L. Reveal

Eriogonum Michaux sect. *Oxytheca* (Nuttall) Roberty & Vautier

Herbs, annual; taproot slender. **Stems** arising directly from the root, erect to spreading, slender, solid, not fistulose or disarticulating into ringlike segments, sparsely to densely glandular. **Leaves** persistent through anthesis or quickly deciduous, essentially basal, rosulate; petiole indistinct; blade linear to spatulate, margins entire. **Inflorescences** terminal, cymose; branches mostly dichotomous, not brittle or disarticulating into segments, round, glabrous or sparsely glandular; bracts 3–5, distinct or connate basally, opposite and linear to narrowly triangular, positioned to side of node, 3-lobed and broadly triangular, or forming a perfoliate disk, leaflike to scalelike, awned, occasionally glandular, glabrous to pubescent. **Peduncles** present or absent, erect or deflexed, slender or stout. **Involucres** 1 per node, not ribbed, tubular, narrowly turbinate; teeth

(3–)4, erect to spreading, awn-tipped. **Flowers** 2–10 per involucre at any single time during full anthesis; perianth white to rose or greenish yellow, broadly campanulate when open, narrowly urceolate when closed, glabrous or pubescent abaxially; tepals 6, connate ¼–⅓ their length, monomorphic or dimorphic, entire apically; stamens 9; filaments basally adnate, glabrous or minutely papillate basally; anthers cream to pink or red, ellipsoid to oval. **Achenes** usually included, light to dark brown or maroon, not winged, globose-lenticular, glabrous. **Seeds:** embryo curved. $x = 20$.

Species 3 (3 in the flora): w North America, s South America.

Oxytheca is allied to *Eriogonum* subg. *Ganysma*, specifically *E. spergulinum* and its close relatives. It is distinguished from *Eriogonum* by the awned involucral lobes. Several species of *Oxytheca* are food plants for the small dotted-blue butterfly (*Philotiella speciosa*).

SELECTED REFERENCE   Ertter, B. 1980. A revision of the genus *Oxytheca* (Polygonaceae). Brittonia 32: 70–102.

1. Inflorescence bracts connate into perfoliate disk; leaf blades glabrous, with ciliate margins
   . . . . 3. *Oxytheca perfoliata*
1. Inflorescence bracts not connate into perfoliate disk; leaf blades densely hirsute or strigose and sparsely glandular, without ciliate margins.
   2. Involucres pedunculate at proximal nodes; awns of inflorescence bracts 0.2–0.5 mm; leaf blades linear to linear-oblanceolate . . . . 1. *Oxytheca dendroidea*
   2. Involucres subsessile; awns of inflorescence bracts 1–3 mm; leaf blades broadly oblanceolate to spatulate . . . . 2. *Oxytheca watsonii*

1. Oxytheca dendroidea Nuttall, Proc. Acad. Nat. Sci. Philadelphia 4: 19. 1848 [F]

Eriogonum dendroideum (Nuttall) S. Stokes

Subspecies 2 (1 in the flora): w North America, s South America.

Subspecies *chilensis* (J. Rémy) Ertter is restricted to the Andes of Chile and Argentina. It, like another annual, *Chorizanthe commissuralis* J. Rémy, which is closely allied to our western *C. brevicornu*, is probably a recent introduction into South America (J. L. Reveal 1978; B. Ertter 1980; O. Shields and Reveal 1988).

1a. Oxytheca dendroidea Nuttall subsp. **dendroidea**
• Treelike puncturebract [E] [F]

Plants erect to spreading, 0.4–4 × 0.3–4.5 dm. **Stems** sparsely to densely glandular. **Leaf blades** linear to linear-oblanceolate, 1–4.5 × 0.1–0.7 cm, densely hirsute, sparsely glandular. **Inflorescences** open to diffuse, 0.5–4 dm; bracts (2–)3(–4) at first node, otherwise 2–3 and distinct or basally connate, linear to subulate or triangular, 1–18 × 0.5–4 mm, scalelike or sometimes leaflike, hirsute and glandular; awns 0.2–0.5 mm, often absent at distal nodes. **Peduncles** occasionally absent at distal nodes, erect or deflexed, slender, 0.5–1.5 cm. **Involucres** 1–2 mm, typically glabrous, rarely with few scattered hairs abaxially; teeth (3–)4; awns grayish, 0.5–3 mm. **Flowers** 2–6; perianth white to pink, 1–2 mm, glabrous or strigose and sparsely glandular abaxially; tepals dimorphic, margins essentially entire, those of outer whorl elliptic to ovate and pubescent adaxially, those of inner whorl elliptic or oblong to narrowly ovate and glabrous or sometimes strigose adaxially proximally; filaments 0.5–1.5 mm, glabrous; anthers cream to red, oval, 0.2–0.3 mm. **Achenes** yellow-brown to maroon, 2–2.5 mm. $2n = 40$.

Flowering Jun–Oct. Dry, sandy to rocky flats, washes, and slopes in mixed grassland, saltbush, sagebrush communities, pinyon and/or juniper and montane conifer woodlands; 300–3000 m; Calif., Idaho, Nev., Ore., Wash., Wyo.

Subspecies *dendroidea* is common and widespread in western North America from southeastern Oregon to southwestern Wyoming southward into eastern California (as far south as Inyo County), Nevada (to Nye County), but surprisingly unknown from northern Utah. Populations in Washington and Wyoming are extensions of the Snake River Plains populations found in Idaho.

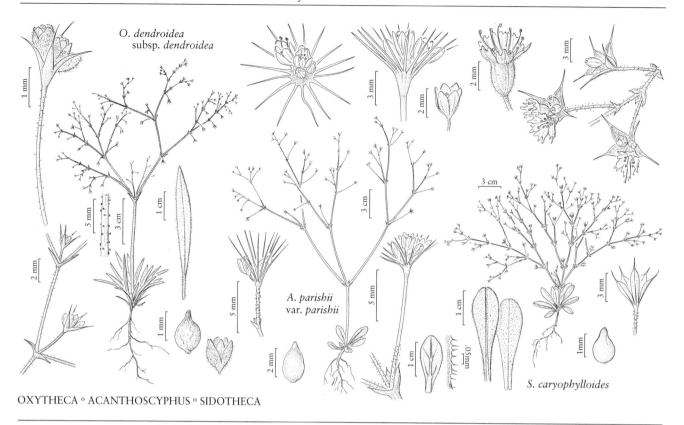

O. *dendroidea* subsp. *dendroidea*

A. parishii var. *parishii*

S. caryophylloides

OXYTHECA ° ACANTHOSCYPHUS ° SIDOTHECA

2. **Oxytheca watsonii** Torrey & A. Gray, Proc. Amer. Acad. Arts 8: 191. 1870 (as watsoni) · Watson's puncturebract E

Eriogonum cuspidatum S. Stokes

Plants erect to spreading, 0.5–2.5 × 0.4–4 dm. **Stems** glandular throughout. **Leaf blades** spatulate or obovate to oblanceolate, 0.7–4 × (0.1–)0.5–1.2 cm, strigose adaxially, less so abaxially, sparsely glandular on both surfaces. **Inflorescences** open to densely branched, 0.5–2 dm; bracts 1–5 × 0.5–3 mm, 3(–5) at first node and linear to ovate, otherwise distinct or basally connate and linear to triangular, ciliate and glandular; awns 1–3 mm. **Peduncles** deflexed, stout, 0.2–0.5 cm at proximal nodes, sometimes absent. **Involucres** 1.5–2 mm, typically glabrous, rarely with few scattered hairs abaxially; teeth 4; awns reddish, 2.5–3 mm. **Flowers** 2–4(–7); perianth white to pink, 1–1.5 mm, strigose and sparsely glandular abaxially; tepals dimorphic, entire, those of outer whorl oval to ovate and strigose abaxially, those of inner elliptic to oval or ovate and glabrous or sometimes pubescent adaxially at base; filaments 1–1.5 mm, glabrous; anthers cream to red, oval, 0.2 mm. **Achenes** dark brown to maroon, 1–1.5 mm. $2n = 40$.

Flowering Jun–Oct. Sandy flats and slopes, saltbush communities; 1200–2000 m; Calif., Nev.

Oxytheca watsonii is an uncommon species known only from scattered locations. The name was misapplied in pre-1980 California floras to *Acanthoscyphus parishii* var. *goodmaniana*.

3. **Oxytheca perfoliata** Torrey & A. Gray, Proc. Amer. Acad. Arts 8: 191. 1870 · Round-leaf puncturebract E

Eriogonum perfoliatum (Torrey & A. Gray) S. Stokes

Plants spreading, 0.6–2 × 0.5–4 dm. **Stems** glandular distally. **Leaf blades** spatulate to oblong or oblanceolate, 1–6 × 0.3–1.5 cm, margins ciliate, otherwise glabrous. **Inflorescences** open to densely branched, 0.8–1.7 dm; bracts at first node 4–5, triangular to lanceolate, 2–10 × 0.5–4(–8) mm, with awn 0.5–1 mm, sparsely glandular, bracts at remaining nodes 3, forming an orbiculate to somewhat triangular perfoliate disk mostly 1–2.5 cm across, glabrous or glandular, with awn 1–3 mm, terminal bracts 0.5–2 × 0.1–0.4 mm, sometimes merely acerose, with awn 0.5–2 mm. **Peduncles** erect, stout,

0.3–0.8 mm at proximal nodes, sometimes absent. **Involucres** 2–5 mm, glabrous or sparsely glandular abaxially; teeth 4; awns reddish, 2–3 mm. **Flowers** 5–10; perianth white or yellowish green to pink, 1.5–2.5 mm, echinulate and sparsely glandular abaxially; tepals monomorphic, lanceolate to ovate; filaments 1–1.5 mm, papillate basally; anthers pink to red, oval, 0.2–0.5 mm. **Achenes** dark brown to maroon, 1.5–2 mm. $2n = 40$.

Flowering Apr–Aug. Sandy to rocky flats, washes, and slopes mainly in saltbush communities; 600–1900 m; Ariz., Calif., Nev., Utah.

Oxytheca perfoliata is basically a plant of the Mojave Desert, with extensions along the Lahontan Trough into northwestern Nevada (to Humboldt and Washoe counties), and along the desert edge of California's Transverse Ranges and the more arid portions of the San Joaquin Valley. It also occurs in the Sonoran Desert south to Imperial County, California.

Oxytheca perfoliata is a food plant for the desert metalmark butterfly (*Apodemia mormo deserti*).

7. **ACANTHOSCYPHUS** Small, Bull. Torrey Bot. Club 25: 53. 1898 • Flowery puncturebract [Greek *acantha*, thorn, and *scyphos*, cup, alluding to awn on involucre] [E]

James L. Reveal

Oxytheca Nuttall sect. *Acanthoscyphus* (Small) Ertter

Herbs, annual; taproot slender. **Stems** arising directly from the root, erect to somewhat spreading, solid, not fistulose or disarticulating into ringlike segments, glabrous, partially glandular. **Leaves** usually persistent through anthesis, basal, rosulate; petiole indistinct; blade broadly obovate or spatulate to oblong or oblanceolate, margins ciliate-denticulate. **Inflorescences** terminal, cymose; branches dichotomous, not brittle or disarticulating into segments, round, glabrous except for glandular nodes and proximal ¹/₂ of internode; bracts 2–3, connate basally, scalelike, triangular to lanceolate or subulate, awned at proximal nodes, glabrous or sparsely glandular. **Peduncles** erect to deflexed. **Involucres** 1 per node, not ribbed, tubular, turbinate; teeth 4(–5) or 7–30(–36), ¹/₂ to completely connate, awn-tipped, glandular distally adaxially. **Flowers** 3–10(–20) per involucre at any single time during full anthesis; perianth white to rose, occasionally greenish white to cream, broadly campanulate when open, narrowly urceolate when closed, hirsute to strigose and sparsely glandular abaxially; tepals 6, connate ¹/₄–¹/₃ their length, monomorphic, entire or rarely irregularly divided apically; stamens 9; filaments basally adnate, glabrous or sparsely pubescent basally; anthers red to maroon, oblong to oval. **Achenes** usually included, dark brown to deep maroon, not winged, globose-lenticular, glabrous. **Seeds:** embryo curved. $x = 20$.

Species 1: s California.

Acanthoscyphus is allied to *Eriogonum* subg. *Ganysma*, specifically to *E. parishii*, *E. apiculatum*, and their close relatives. It is distinguished from *Eriogonum* by its awned involucres.

SELECTED REFERENCE Ertter, B. 1980. A revision of the genus *Oxytheca* (Polygonaceae). Brittonia 32: 70–102.

1. Acanthoscyphus parishii (Parry) Small, Bull. Torrey Bot. Club 25: 53. 1898 [E] [F]

Oxytheca parishii Parry, Proc. Davenport Acad. Nat. Sci. 3: 176. 1882

Plants (0.5–)1–6 × 0.5–4 dm. **Leaf blades** 1–7 × 0.5–2 cm. **Inflorescences** 1–5 dm; bracts 1–4(–10) × 0.5–3 mm, awns 0.2–1.5 mm. **Peduncles** (0.3–)1–5(–7.5) cm. **Involucres** 1.5–2 mm, awns ivory or reddish, 2–5 mm. **Flowers:** perianth 2–2.5 mm; filaments 2–2.5 mm; anthers 0.3–0.6 mm. **Achenes** 1.7–2 mm.

Varieties 4 (4 in the flora): s California.

1. Involucral awns 7–16, dark red, 3–4 mm; San Rafael Mountains, Topatopa Mountain, and Mount Pinos, Santa Barbara and Ventura counties, California . . . . 1d. *Acanthoscyphus parishii* var. *abramsii*
1. Involucral awns 4–36, ivory colored, 2–5 mm; San Bernardino and Ventura counties.
   2. Involucral awns 4(–5), 2–3 mm; ne side of San Bernardino Mountains, San Bernardino County, California . . . . 1c. *Acanthoscyphus parishii* var. *goodmaniana*
   2. Involucral awns 7–30(–36), 3–5 mm; se side of e San Bernardino County, and Ventura County, California.
      3. Involucral awns (10–)13–30(–36); inflorescence bracts triangular, awns 0.2–0.5 mm; Pine Mountain, Ventura County, e to c San Bernardino Mountains, San Bernardino County, California . . . . 1a. *Acanthoscyphus parishii* var. *parishii*
      3. Involucral awns 7–10; inflorescence bracts subulate, awns 0.5–1.5 mm; se side of e San Bernardino Mountains, San Bernardino County, California . . . . 1b. *Acanthoscyphus parishii* var. *cienegensis*

1a. Acanthoscyphus parishii (Parry) Small var. **parishii**
  • Parish's flowery puncturebract [E] [F]

Eriogonum abramsii (E. A. McGregor) S. Stokes subsp. *acanthoscyphus* S. Stokes

Plants erect, 1–6 dm. **Stems** usually stout, (1–)5–15 cm. **Leaf blades** 1–7 cm. **Inflorescences:** bracts (2–)3, triangular, awns 0.2–0.5 mm. **Peduncles** (1–)2–5(–7.5) cm. **Involucres:** awns (10–)13–30(–36), ivory colored, (3–)4–5 mm. **2n** = 40.

Flowering Jun–Oct. Sandy to gravelly flats and slopes, chaparral communities, montane conifer woodlands; 1900–2600 m; Calif.

Variety *parishii* occurs from the Pine Mountains, Ventura County, eastward through the San Gabriel Mountains to the central San Bernardino Mountains of San Bernardino County. It is the most common expression of the species.

1b. Acanthoscyphus parishii (Parry) Small var. **cienegensis** (Ertter) Reveal, Harvard Pap. Bot. 9: 144. 2004 • Cienega Seca flowery puncturebract [E]

Oxytheca parishii Parry var. *cienegensis* Ertter, Brittonia 32: 89, fig. 6G. 1980

Plants erect, 0.5–4 dm. **Stems** slender, 3–8 cm. **Leaf blades** 1–3.5 cm. **Inflorescences:** bracts 2(–3), subulate, awns 0.5–1.5 mm. **Peduncles** 0.5–2(–3.5) cm. **Involucres:** awns 7–10, ivory-colored, (3–)4–5 mm.

Flowering Jun–Sep. Sandy flats and slopes, montane conifer forests; 2100–2500 m; Calif.

Variety *cienegensis* is known only from the San Bernardino Mountains of San Bernardino County.

1c. Acanthoscyphus parishii (Parry) Small var. **goodmaniana** (Ertter) Reveal, Harvard Pap. Bot. 9: 144. 2004 • Cushenbury flowery puncturebract [E]

Oxytheca parishii Parry var. *goodmaniana* Ertter, Brittonia 32: 90, fig. 6H–J. 1980

Plants erect or spreading, 0.5–3 dm. **Stems** slender, (1.5–)2–6 cm. **Leaf blades** 1–3 cm. **Inflorescences:** bracts 2(–3), subulate, awns 0.5–1.5 mm. **Peduncles** 0.3–1.5(–2) cm. **Involucres:** awns 4(–5), ivory colored, 2–3 mm.

Flowering May–Sep. Sandy calcareous slopes, chaparral communities; 1300–2300 m; Calif.

Variety *goodmaniana* is known only from the northeastern slope of the San Bernardino Mountains in San Bernardino County. The name *Oxytheca watsonii* was misapplied to this taxon in the pre-1980 California literature.

1d. Acanthoscyphus parishii (Parry) Small var. **abramsii** (E. A. McGregor) Reveal, Harvard Pap. Bot. 9: 144. 2004 • Abrams's flowery puncturebract [E]

Oxytheca abramsii E. A. McGregor, Bull. Torrey Bot. Club 36: 605, figs. 1, 2. 1909; *Eriogonum abramsii* (E. A. McGregor) S. Stokes; *Oxytheca parishii* Parry subsp. *abramsii* (E. A. McGregor) Munz; *O. parishii* var. *abramsii* (E. A. McGregor) Munz

Plants erect or spreading, 1–3 dm. **Stems** slender, 1–3(–7) cm. **Leaf blades** 1–3(–5) cm. **Inflorescences:** bracts (2–)3, triangular, awns 0.5–1.5 mm. **Peduncles** 0.8–2 cm. **Involucres:** awns 7–16, dark red, 3–4 mm.

Flowering Jun–Aug. Sandy flats and slopes, chaparral communities, montane conifer woodlands; 1700–2000 m; Calif.

Variety *abramsii* is known only from the San Rafael Mountains, Topatopa Mountain, and Mount Pinos, in Santa Barbara and Ventura counties.

8. **SIDOTHECA** Reveal, Harvard Pap. Bot. 9: 211. 2004 • Starry puncturebract [Greek *sidus*, star, and *theke*, case, alluding to starlike involucres]

James L. Reveal

Oxytheca Nuttall sect. *Neoxytheca* Ertter, Brittonia 32: 92. 1980

Herbs, annual; taproots slender. **Stems** arising directly from the root, spreading or prostrate, sometimes erect, solid, not fistulose or disarticulating into ringlike segments, glabrous or sparsely glandular. **Leaves** persistent or quickly deciduous, basal, rosulate; petioles indistinct; blade broadly linear or spatulate to oblanceolate, margins entire, strigose and glandular. **Inflorescences** terminal, cymose; branches mostly dichotomous, not brittle or disarticulating into segments, round, glabrous or sparsely glandular; bracts (2–)3(–4) at first node, 2–3 at distal nodes, distinct or connate, often positioned to side of node, scalelike, triangular or linear to ovate and 3-lobed, awned, sparsely glandular. **Peduncles** present or absent, erect to spreading. **Involucres** 1 per node, not ribbed, tubular, narrowly turbinate to funnelform; teeth 5(–6), awn-tipped. **Flowers** 2–5(–10) per involucre at any single time during full anthesis; perianth white to rose or greenish yellow to red, funnelform when open, tubular when closed, hirsute and sparsely glandular abaxially; tepals 6, connate $^1/_4$–$^1/_3$ their length, monomorphic, 3–5-lobed or laciniate apically; stamens 9; filaments basally adnate, glabrous or minutely papillate basally; anthers red to maroon, ellipsoid or oblong to oval. **Achenes** usually included, golden- to red-brown, not winged, 3-gonous, glabrous. **Seeds:** embryo curved. *x* = 20.

Species 3 (3 in the flora): California, nw Mexico.

Sidotheca is allied to *Eriogonum* subg. *Ganysma*, approaching *E. inerme* in terms of foliar and overall habit. The trilobed to laciniate tepals resemble those of certain species of *Chorizanthe*. It is possible, as B. Ertter (1980) suggested, that the taxon was derived from *Acanthoscyphus*. In the 1950s, G. J. Goodman (1904–1999) proposed its recognition at generic rank, using the parahomonym "Neoxytheca."

SELECTED REFERENCE Ertter, B. 1980. A revision of the genus *Oxytheca* (Polygonaceae). Brittonia 32: 70–102.

1. Involucres funnelform, white-margined, bracts connate more than ³/₄ their length; tepals 3–5-lobed apically, lobes laciniate . . . . 3. *Sidotheca emarginata*
1. Involucres narrowly to broadly turbinate, concolored, bracts connate ca. ¹/₂ their length; tepals 3-lobed apically, lobes laciniate or not.
    2. Perianths 1–2 mm, greenish yellow to red; tepals 3-lobed apically ¹/₅ their length; peduncles present or absent; awns of involucral bracts 0.3–1 mm; awns of inflorescence bracts 0.2–0.5 mm . . . . 1. *Sidotheca caryophylloides*
    2. Perianths 2.5–4 mm, white to pink; tepals 3-lobed apically ¹/₃–¹/₂ their length; peduncles present; awns of involucral bracts 0.3–2 mm; awns of inflorescence bracts 0.8–1 mm
        . . . . 2. *Sidotheca trilobata*

1. **Sidotheca caryophylloides**   (Parry) Reveal, Harvard Pap. Bot. 9: 211. 2004   • Chickweed starry puncturebract   E F

Oxytheca caryophylloides Parry, Proc. Davenport Acad. Nat. Sci. 3: 175. 1882; *Eriogonum caryophylloides* (Parry) S. Stokes

Plants 1–2.5 × (0.3–)1–4(–5) dm. **Stems** spreading to prostrate. **Leaf blades** spatulate to oblanceolate, 1–6(–8) × 0.3–1.2(–1.8) cm. **Inflorescences** diffuse, 0.5–2 dm; bracts 3–8 × 0.5–1.5 mm, awns 0.2–0.5 mm. **Peduncles** erect, 0.1–1 cm, sparsely glandular or glabrous, otherwise absent. **Involucres** concolored, narrowly to broadly turbinate, 4–7 mm, sparsely glandular; teeth 5(–6), connate ca. ¹/₂ their length; awns greenish to reddish, 0.3–1 mm. **Flowers** 2–3; perianth greenish yellow to reddish, 1–2 mm; tepals broadly oblanceolate, 3-lobed apically ¹/₅ their length, lobes not laciniate; filaments 0.8–1.2 mm; anthers red, oval, 0.2–0.4 mm. **Achenes** golden to red-brown, 1.2–1.5 mm. $2n = 40$.

Flowering Jun–Sep. Sandy to gravelly flats, washes, and slopes, chaparral communities, montane conifer woodlands; 1300–2600 m; Calif.

Sidotheca caryophylloides grows in the Transverse Ranges and the southern Sierra Nevada.

2. **Sidotheca trilobata**   (A. Gray) Reveal, Harvard Pap. Bot. 9: 211. 2004   • Three-lobed starry puncturebract

Oxytheca trilobata A. Gray, Proc. Amer. Acad. Arts 12: 83. 1876; *Eriogonum trilobatum* (A. Gray) S. Stokes

Plants 0.7–5 × 0.7–6(–11) dm. **Stems** erect or more commonly spreading to prostrate. **Leaf blades** spatulate to oblanceolate or linear, 1–5(–9) × 0.2–0.7(–2) cm. **Inflorescences** open, 0.5–4 dm; bracts mostly 2–8 × 1–3 mm, awns 0.8–1 mm. **Peduncles** erect to spreading, 0.5–1.5 cm, glabrous except sparsely glandular proximally. **Involucres** concolored, broadly turbinate, 3–8 mm, sparsely glandular; teeth 5(–6), connate ¹/₃–¹/₂ their length; awns greenish to reddish, 0.3–2 mm. **Flowers** 3–5(–10); perianth white to pink or reddish, 2.5–4 mm; tepals oblong, 3-lobed apically ¹/₃–¹/₂ their length, lobes sometimes laciniate; filaments 1–4 mm; anthers red to maroon, ellipsoid to oval, 0.5–0.7 mm. **Achenes** golden-brown, 1.2–2 mm. $2n = 40$.

Flowering Apr–Sep. Sandy flats, washes, slopes, chaparral communities, montane coniferous woodlands; 700–2100 m; Calif.; Mexico (Baja California).

Sidotheca trilobata occurs in the Transverse Ranges of southern California from Los Angeles and San Bernardino counties southward into northern Baja California, Mexico. It is the most commonly encountered species of the genus.

3. **Sidotheca emarginata**   (H. M. Hall) Reveal, Harvard Pap. Bot. 9: 211. 2004   • White-margin starry puncturebract   C E

Oxytheca emarginata H. M. Hall, Univ. Calif. Publ. Bot. 1: 75, plate 14. 1902; *Eriogonum emarginatum* (H. M. Hall) S. Stokes

Plants 0.3–3 × 0.3–5 dm. **Stems** spreading to prostrate. **Leaf blades** spatulate to oblanceolate, 1.5–7.5 × 0.4–1.5 cm. **Inflorescences** open, 0.2–3 dm; bracts mostly 5–10 × 1–3(–5) mm, awns 1–2 mm. **Peduncles** erect, 0.5–3 cm, glandular. **Involucres** white-margined, broadly funnelform and laterally compressed, 4–8 × 9–12 mm, glabrous; teeth 5, connate more than ³/₄ their length, awns reddish, 1–1.5 mm. **Flowers** 3–6; perianth white to pink, (2–)3–4.5(–5) mm; tepals narrowly oblong, 3–5-lobed apically ¹/₃–¹/₂ their length, lobes laciniate; filaments 3–5 mm; anthers red, oval to oblong, 1–1.2 mm. **Achenes** golden brown, 1.8–2 mm.

Flowering Feb–Aug. Gravelly to rocky places, chaparral communities, montane coniferous woodlands; 1200–2500 m; of conservation concern; Calif.

Sidotheca emarginata grows in the San Jacinto and Santa Rosa mountains of Riverside County. The white-margined, papery involucre is a distinctive feature that might make the plant an attractive addition to an annual garden.

9. NEMACAULIS Nuttall, Proc. Acad. Nat. Sci. Philadelphia 4: 18. 1848 • Cottonheads [Greek *nema*, thread, and Greek *kaulos*, stem]

James L. Reveal

Eriogonum Michaux sect. *Nemacaulis* (Nuttall) Roberty & Vautier

Herbs annual; taproot thin. **Stems** arising from the root, prostrate to spreading or erect, glabrous to glandular or lanate. **Leaves** persistent through anthesis, essentially basal, rosulate; petiole indistinct; blade linear to spatulate, margins entire. **Inflorescences** terminal, cymose; branches mostly dichotomous, not brittle or disarticulating into segments, round, glabrous or sparsely lanate, sometimes glandular, secondary branches sometimes developing tardily; bracts 3, connate basally, triangular to obovate, leaflike or scalelike, not awned, glabrous or glandular. **Peduncles** erect, slender, or absent. **Involucral bracts** obscure to obvious, in tight spiral, linear to lanceolate or narrowly ovate, not awn-tipped. **Flowers** 5–30 per involucre, sessile or pedicellate; perianth white to rose, campanulate when open, narrowly urceolate when closed, glabrous or with gland-tipped hairs; tepals 6, connate proximally, slightly dimorphic, entire; stamens 3; filaments basally adnate, glabrous; anthers pink to red, oval. **Achenes** mostly included, brown to deep maroon or black, not winged, 3-gonous, glabrous. **Seeds:** embryo curved.

Species 1: sw United States, nw Mexico.

Nemacaulis is allied to *Eriogonum* subg. *Ganysma* and especially *E. gossypinum*. Like that species, *N. denudata* has flowers embedded in a dense mass of hairs, which is the only significant feature that the two have in common aside from being annuals, generally spreading in habit, and having rather densely pubescent stems and leaves. The flowers in *Nemacaulis* are arranged in glomerules, a feature unique in Eriogonoideae. The involucre is markedly different, being composed of several whorls of bracts, each of which subtends a flower. In this and all the other members of Eriogoneae, modification of the involucre appears to be a significant factor in the evolution of distinctive genera. In *Nemacaulis*, the entire glomerule is dispersed by on-shore breezes along the coast, an effective transport mechanism for the tiny seeds. Entire glomerules may be trapped in moist pockets of sand, thereby positioning several seeds in a favorable location. In the rapidly evolving derived annuals of Eriogonineae, any selective advantage is quickly adopted and survival enhanced in the rather harsh environment, where members of the subtribe abound.

SELECTED REFERENCE Reveal, J. L. and B. Ertter. 1980. The genus *Nemacaulis* Nutt. (Polygonaceae). Madroño 27: 101–109.

1. Nemacaulis denudata Nuttall, J. Acad. Nat. Sci. Philadelphia, n. s. 1: 168. 1848 [F]

Eriogonum nemacaulis S. Stokes

Plants 0.4–2.5(–4) × 0.2–8 dm. **Leaf blades** 1–8 × 0.1–1.5 cm. **Inflorescences** 0.3–3 × 2.8 dm; branches (2–)4–10 at first node, 1–2(–3) thereafter; bracts 1–5 × 0.5–2 mm. **Peduncles** 0.1–3 mm or absent. **Involucral bracts** (1.5–)2–4 × 0.5–2 mm. **Flowers:** perianth 0.5–1.5 mm; filaments 0.5–1 mm; anthers 0.2 mm. **Achenes** 1 mm.

Varieties 2 (2 in the flora): sw United States, nw Mexico.

1. Flowers (5–)12–30 per involucre; peduncles usually absent; involucral bracts dark red, white-tomentose; stems prostrate to decumbent; leaf blades usually spatulate, sometimes linear; outer tepals broadly obovate to ovate; coastal beaches . . . . 1a. *Nemacaulis denudata* var. *denudata*

1. Flowers 5(–12) per involucre; peduncles usually present; involucral bracts light brown to yellowish green, tawny-tomentose; stems ascending to erect; leaf blades usually linear or narrowly spatulate; outer tepals linear to oblong; coastal and inland deserts . . . . 1b. *Nemacaulis denudata* var. *gracilis*

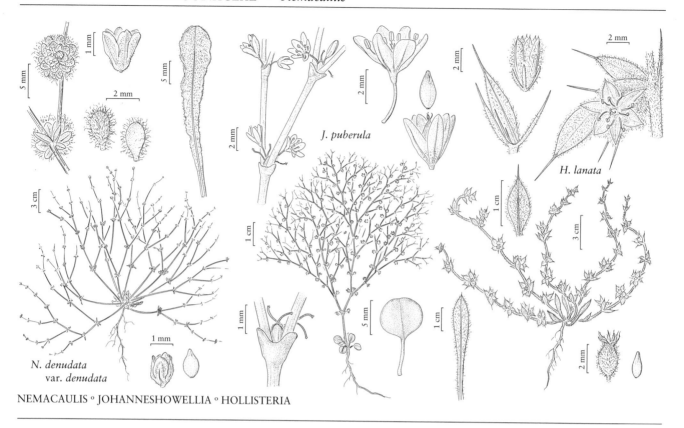

J. puberula

H. lanata

N. denudata
var. denudata

NEMACAULIS ∘ JOHANNESHOWELLIA ∘ HOLLISTERIA

1a. Nemacaulis denudata Nuttall var. **denudata** ☐F

Plants 0.4–2 × 0.2–8 dm. **Stems** prostrate to decumbent. **Leaf blades** usually spatulate, sometimes linear, 2–8 × 0.3–1.5 cm. **Inflorescences** with thick, dark red branches. **Peduncles** usually absent. **Involucral bracts** dark red, 1.5–3.5 × 1–2 mm, white-tomentose. **Flowers** (5–)12–30 per involucre, usually exposed in tomentum; perianth 1–1.5 mm; outer tepals broadly obovate to ovate.

Flowering Mar–Aug. Coastal sandy beaches, grassland communities; 0–100 m; Calif.; Mexico (Baja California).

Variety *denudata* occurs from Los Angeles County southward into Baja California.

1b. Nemacaulis denudata Nuttall var. **gracilis**

Goodman & L. D. Benson, Aliso 4: 89. 1958

• Slender cottonheads

Plants 0.4–2.5(–4) × 0.4–2 dm. **Stems** ascending to erect. **Leaf blades** usually linear or narrowly spatulate, 1–7 × 0.1–0.6 cm. **Inflorescences** with slender, light brown branches. **Peduncles** usually present, 0.5–3 mm. **Involucral bracts** light brown to yellowish green, 2–4 × 0.5–1 mm, tawny-tomentose. **Flowers** 5(–12) per involucre, usually obscured by tomentum; perianth 0.5–1.2 mm; outer tepals linear to oblong.

Flowering Jan–May. Coastal and desert sands, saltbush, creosote bush, and coastal grassland communities; -10–500 m; Ariz., Calif.; nw Mexico (Baja California, Sonora).

Variety *gracilis* is common in northwestern Mexico and somewhat less so in southwestern Arizona and southern California.

10. JOHANNESHOWELLIA Reveal, Brittonia 56: 299. 2004 • Howell's-buckwheat [for John Thomas Howell, 1903–1994, California botanist and *Eriogonum* scholar] E

James L. Reveal

Eriogonum Michaux [unranked] *Puberula* Rydberg, Fl. Rocky Mts., 211. 1917

Herbs, annual; taproot slender. **Stems** arising directly from the root, weakly erect to spreading, solid, not fistulose or disarticulating into ringlike segments, silky-puberulent. **Leaves** usually quickly deciduous, basal, rosulate; petiole present; blade obovate to rounded or somewhat reniform, margins entire. **Inflorescences** terminal, cymose; branches dichotomous or trichotomous at proximal node, otherwise dichotomous, not brittle or disarticulating into segments, round, silky-puberulent; bracts 3, connate proximally, triangular or linear to narrowly lanceolate, scalelike or somewhat leaflike, not awned, glabrous to puberulent. **Peduncles** absent. **Involucral bracts** obscure, in tight spiral of (3–)4(–7) distinct, oblanceolate to obovate lobes, bractlike, not awned. **Flowers** 4–7 per involucral cluster; perianth white or pale yellowish to rose or red, broadly campanulate when open, narrowly urceolate when closed, smooth or minutely pustulose abaxially, glabrous; tepals 6, connate ¼ their length, monomorphic or slightly dimorphic, entire apically; stamens 9; filaments basally adnate, glabrous; anthers white to pale pink, oblong. **Achenes** included, light brown, not winged, 3-gonous, glabrous. **Seeds:** embryo curved.

Species 2 (2 in the flora): w United States.

With recognition of *Johanneshowellia*, all members of *Eriogonum* have their involucral bracts fused into a distinct, turbinate to campanulate or hemispheric, tubular structure. In *Johanneshowellia*, the four to seven, distinct, involucral bracts are arranged in a tight spiral. Each 0.5–1.2 mm, awnless, entire to deeply lobed bract subtends a flower-bearing pedicel. These involucral bracts are themselves usually obscured by inflorescence bracts that subtend each node. As a result, care must be taken to observe the true nature of the involucral complex. The genus is allied with *Eriogonum* subg. *Oregonium*.

1. Perianths minutely pustulose basally and along midribs, 1.5–2 mm in fruit; outer tepals narrowly ovate, slightly auriculate basally with slightly undulate-crisped margins in fruit; plants grayish to greenish, mostly densely silky-puberulent . . . . 1. *Johanneshowellia puberula*
1. Perianths smooth, 2–2.5 mm in fruit; tepals lanceolate, not auriculate basally, margins entire; plants reddish, thinly silky-puberulent . . . . 2. *Johanneshowellia crateriorum*

1. Johanneshowellia puberula (S. Watson) Reveal, Brittonia 56: 302. 2004 • Red Creek Howell's-buckwheat E F

Eriogonum puberulum S. Watson, Proc. Amer. Acad. Arts 14: 295. 1879; *E. puberulum* var. *venosum* S. Stokes

Plants weakly erect, grayish to greenish, 0.5–3 × 1–2 dm, mostly densely silky-puberulent. **Stems** 0.3–0.8 dm. **Leaf blades** obovate to rounded, 0.5–1.5 × 0.5–1.5 cm. **Inflorescences** spreading, open, 5–25 cm; bracts linear to narrowly lanceolate, 2–5(–9) × 1–2(–2.5) mm, somewhat foliaceous, puberulent. **Involucral bracts** oblanceolate, 0.5–1.2 × 0.1–0.3 mm, villous abaxially, outermost one 2- or 3-lobed apically. **Flowers:** perianth white to pale yellow, becoming rose or red in fruit, 1.5–2 mm, minutely pustulose basally and along midribs; tepals slightly dimorphic, those of outer whorl narrowly ovate, slightly auriculate basally with undulate-crisped margins in fruit, those of inner whorl narrowly oblanceolate, often shorter than those of outer whorl; filaments 0.8–1.2 mm. **Achenes** 1–1.5 mm.

Flowering May–Sep. Sandy flats and slopes, saltbush and sagebrush communities, pinyon-juniper woodlands; (500–)800–2800 m; Calif., Nev., Utah.

Johanneshowellia puberula is infrequent to occasionally locally common. Populations are widely scattered from the Cottonwood Mountains of Inyo County, California, across northern Clark, southern Eureka, Lincoln, Nye, and White Pine counties of Nevada, into Beaver, Iron, Millard, and Washington counties in Utah.

2. **Johanneshowellia crateriorum** Reveal, Brittonia 56: 304, fig. 2. 2004 • Lunar Crater Howell's-buckwheat E

Plants spreading, reddish, 0.5–3 × 1–3(–5) dm, thinly silky-puberulent. **Stems** 0.3–0.5 dm. **Leaf blades** obovate to rounded or somewhat reniform, 0.5–1.5 × 0.5–1.5 cm. **Inflorescences** spreading, diffuse, 5–25 cm; bracts triangular, 1–2(–2.5) × 1–2(–2.5) mm, scalelike, glabrous but ciliate on margins. **Involucral bracts** oblanceolate to obovate, 0.5–0.8 × 0.1–0.3 mm, glabrous but ciliate on margins, outermost one deeply 2-lobed. **Flowers:** perianth white to rose, becoming red in fruit, 2–2.5 mm, smooth; tepals monomorphic, lanceolate, not auriculate, margins entire; filaments 1.2–1.8 mm. **Achenes** 1.3–1.8 mm.

Flowering May–Sep. Sandy, pumice flats and slopes, saltbush communities; 1700–1900 m; Nev.

Johanneshowellia crateriorum is known only from the Lunar Crater area of Nye County.

11. **HOLLISTERIA** S. Watson, Proc. Amer. Acad. Arts 14: 296. 1879 • False spikeflower [For William Welles Hollister, 1818–1886, California rancher] E

James L. Reveal

Eriogonum Michaux sect. *Hollisteria* (S. Watson) Roberty & Vautier

Herbs, annual; taproots slender. **Stems** arising directly from the root, spreading to decumbent or prostrate, solid, not fistulose or disarticulating into ringlike segments, tomentose. **Leaves** deciduous (basal) or persistent (cauline), basal and cauline, alternate; petiole indistinct at proximal node, absent at distal nodes; blade oblanceolate (proximal) or elliptic to ovate (distal), margins entire, apex mucronate. **Inflorescences** terminal, cymose, distally uniparous due to suppression of secondaries, each node bearing a flower; branches dichotomous, not brittle or disarticulating into segments, round, tomentose; bracts 3, typically situated at base of a cauline leaf and hidden by it, connate basally, linear to linear-lanceolate, somewhat foliaceous or scalelike, awned, thinly pubescent to densely tomentose. **Peduncles** absent. **Involucral bracts** obvious, in whorl of 3(–4), linear to lanceolate, awned. **Flowers** 1 per involucral cluster, sessile or pedicellate; perianth yellowish, broadly campanulate when open, tubular when closed, densely tomentose abaxially; tepals 6, connate proximally, monomorphic, mucronate apically; stamens 6 or 9; filaments basally adnate, glabrous; anthers yellow, oval to oblong. **Achenes** included, brown to black, not winged, 3-gonous, glabrous. **Seeds:** embryo curved. $x = 21$.

Species 1: California.

Hollisteria is the only member of subtribe Hollisteriineae H. Gross. This view differs from that presented by J. L. Reveal and C. B. Hardham (1989), who included *Chorizanthe* and its relatives also in Hollisteriineae. In an unpublished thesis, Adrienne Russell (pers. comm.) has suggested that *Hollisteria* is more closely related to *Eriogonum* than to *Chorizanthe*. Reveal and Hardham noted that the transfer of *Hollisteria* from a place near *Eriogonum* to a position with *Chorizanthe* was done with "some trepidation." There is no obvious place within *Eriogonum* subg. *Ganysma* to suggest as a point of origin for *Hollisteria*. Then, too, there is no place within *Chorizanthe* that one can comfortably suggest as such a point. By referring *Hollisteria* to its own subtribe, one acknowledges a relationship to *Eriogonum*, and recognizes that the single flower and base chromosome number of $x = 21$ are conditions unknown in the whole of Eriogonineae, although both features are common in Chorizanthineae.

1. Hollisteria lanata S. Watson, Proc. Amer. Acad. Arts 14: 296. 1879 [E] [F]

Eriogonum lanatum (S. Watson) Roberty & Vautier

Plants 0.3–0.8(–1) × (0.5–)0.8–3(–5) dm. **Leaves:** basal blades (1.5–)2–5(–6) × (0.2–)0.3–0.7(–0.9) cm; cauline blades (0.5–)1–3.5 × (0.1–)0.3–0.8(–1) cm. **Inflorescences:** bracts dimorphic, lateral 2 obvious, 2–5 mm, thinly pubescent, central 1 obscured, 1–2(–3) mm, densely tomentose. **Involucral bracts** 2 mm, densely tomentose. **Flowers:** perianth 1.5–2 mm. **Achenes** 1.7–2 mm. $2n = 42$.

Flowering Mar–Jul. Sandy to gravelly or clayey places, mixed grassland, desert scrub, chaparral communities, woodlands; 10–1000 m; Calif.

Hollisteria lanata is encountered occasionally in the San Joaquin Valley and the adjacent coastal ranges and eastward in the Transverse Ranges to Kern and Tulare counties. Plants are only rarely locally common.

12. CHORIZANTHE R. Brown ex Bentham, Trans. Linn. Soc. London 17: 416, plate 17, fig. 11; plate 19. 1836 • Spineflower [Greek *chorizo*, to divide, and *anthos*, flower, alluding to tepals]

James L. Reveal

Herbs [or subshrubs], annual [or perennial]; taproot slender to stout. **Stems** prostrate or decumbent to erect, pubescent; aerial flowering stems arising [at nodes of caudex branches, at distal nodes of aerial stems or] directly from the root, decumbent to erect, slender [to stout and solid, not disarticulating in ringlike segments], sometimes disarticulating at each node. **Leaves** persistent or quickly deciduous, basal and rosulate or basal and cauline, alternate; petiole present; blade linear to oblanceolate or spatulate, entire apically. **Inflorescences** terminal, cymose or capitate, uniparous due to suppression of secondaries; branches open and spreading or erect, typically trichotomously branched at proximal node, otherwise dichotomous, sometimes brittle and disarticulating into segments, round, pubescent [or rarely glabrous]; bracts mostly 2, opposite, sometimes numerous, whorled, distinct, leaflike to subulate or linear, occasionally awn-tipped, thinly pubescent (sometimes appressed), hirsute, villous, strigose, or tomentose, rarely woolly-floccose or minutely glandular. **Peduncles** absent. **Involucres** 1–6+ per node, 3–6-ribbed, tubular, cylindric to urceolate or turbinate to campanulate; teeth 3, 5, or 6, awn-tipped. **Flowers** bisexual, 1(–2) per involucre, pedicellate; perianth white to yellow or pink to rose-pink, red, maroon or purple, cylindric, funnelform, or campanulate when open, cylindric when closed, glabrous or pubescent abaxially; tepals (5–)6, connate $^2/_3$ their length, monomorphic or dimorphic, entire, emarginate, or lobed to laciniate apically; stamens 3, 6, or 9, or variously 3–9; filaments distinct or connate into staminal tube, sometimes adnate to floral tube, glabrous or pubescent; anthers maroon to red or cream to white or yellow, oval to oblong; styles erect to spreading. **Achenes** included, light brown to dark brown or black, not winged, lenticular, globose-lenticular, or 3-gonous, glabrous. **Seeds:** embryo straight or curved. $x = 10$.

Species 50 (33 in the flora): North America (including Mexico), South America.

Like *Eriogonum*, *Chorizanthe* is the basal element in its own subtribe, Chorizanthineae Reveal. Nonetheless, recent molecular data indicate that *Chorizanthe* is embedded within *Eriogonum* (A. S. Lamb Frye, pers. comm.), meaning that either all species of the Chorizanthineae should be moved to *Eriogonum*, or *Eriogonum* should be fragmented into several genera. Obviously, therefore, all of the segregate genera that follow could be merged into *Chorizanthe*, and this was common practice until 1989. A key factor still unresolved is the relationship between the perennial species of *Chorizanthe* (including the type of the genus) and *Eriogonum*. The traditional assumption is that *Eriogonum* and *Chorizanthe* represent independent lines of evolution from a basal, diploid ($n = 10$) ancestor that is now extinct. A corollary to this assumption is that this divergence occurred during the Eocene and the perennial spineflowers were successfully introduced into southern South America, whereas the wild buckwheats—lacking a ready means for long-distance dispersal—failed to make the trip. Thus, it is possible that the perennial members of *Chorizanthe* represent a genus distinct from the annuals treated here (33 of the 41 annuals occur in the flora area, the rest are in Mexico or in South America). In that case, our annual members (if not submerged into *Eriogonum*) would be called *Acanthogonum* Torrey.

The segregate genera allied to *Chorizanthe*, like those allied to *Eriogonum*, differ primarily in their involucres. *Aristocapsa* and *Dodecahema*, with haploid chromosome numbers of 14 and 17, respectively, are difficult to associate with any extant member of *Chorizanthe* (mainly $n = 19, 20, 38, 40$). *Centrostegia*, *Lastarriaea*, *Mucronea*, and *Systenotheca* all appear much more akin to *Chorizanthe*. A point of origin can be suggested only for *Lastarriaea*, namely *Chorizanthe interposita* Goodman of central Baja California, Mexico. C. D. Hardham (1989) reported a range of gametic numbers from single individuals. Until somatic counts are made, the primary chromosome numbers of some species remain uncertain.

J. L. Reveal and C. B. Hardham (1989b) divided the annual species of *Chorizanthe* genus into three subgenera: subg. *Amphietes* (39 species), and subg. *Eriogonella* and subg. *Quintaria* (one species each). The first is divided into four sections (*Ptelosepala*, *Acanthogonum*, *Fragile*, and *Clastoscapa* Reveal & Hardham), all but the last of which are found in our region. Some sections were divided into subsections, which are not treated here.

The approximately nine perennial species of subg. *Chorizanthe* occur only in arid regions of Chile and Argentina.

SELECTED REFERENCES Goodman, G. J. 1934. A revision of the North American species of the genus *Chorizanthe*. Ann. Missouri Bot. Gard. 21: 1–102. Reveal, J. L. and C. B. Hardham. 1989b. A revision of the annual species of *Chorizanthe* (Polygonaceae: Eriogonoideae). Phytologia 66: 98–198.

1. Involucres 3–5-toothed or 3–5-ribbed.
   2. Involucral teeth (4–)5.
      3. Involucral awns straight; perianths glabrous; filaments adnate at base of tepals; deserts, s California . . . . 1. *Chorizanthe spinosa*
      3. Involucral awns uncinate; perianths pubescent abaxially; filaments adnate near top of floral tube; mountains, coastal mesas, deserts, w North America.
         4. Involucres campanulate, 1.5–2.5 mm, anterior tooth not leaflike; perianths white to rose, densely pubescent abaxially, 1.5–1.8(–2) mm; mountains and coastal mesas of California . . . . 28. *Chorizanthe polygonoides* (in part)

4. Involucres cylindric, 3–4.5 mm, anterior tooth leaflike; perianths yellow, thinly pubescent abaxially, 1.5–2.5 mm; deserts of w North America . . . . 31. *Chorizanthe watsonii*
2. Involucral teeth 3.
  5. Involucres cylindric, markedly transversely corrugate; perianths thinly pubescent abaxially; stamens 6 . . . . 32. *Chorizanthe corrugata*
  5. Involucres urceolate to campanulate; perianths densely pubescent abaxially; stamens 9.
    6. Plants erect; involucres urceolate, anterior tooth 5–10 mm with straight awn; achenes 3-gonous; warm deserts of sw North America . . . . 29. *Chorizanthe rigida*
    6. Plants prostrate; involucres campanulate, anterior tooth 1.8–2 mm with uncinate awn; achenes lenticular; coastal mesas of extreme sw California . . . . 30. *Chorizanthe orcuttiana*
1. Involucres 6-toothed (teeth sometimes vestigial or 6-ribbed).
  7. Involucral teeth with scarious or membranous margins.
    8. Involucral teeth margins membranous, continuous across sinuses.
      9. Tepals obcordate to 2-lobed apically.
        10. Tepals not denticulate apically; n, c California . . . . 3. *Chorizanthe stellulata* (in part)
        10. Tepals denticulate apically; wc California . . . . 4. *Chorizanthe douglasii* (in part)
      9. Tepals entire apically.
        11. Involucres tomentose to floccose or glabrate, 3–4 mm; perianths densely pubescent abaxially, (1.5–)2.5–3 mm; sc Oregon and California . . . . 2. *Chorizanthe membranacea*
        11. Involucres sparsely pubescent and hispid at least along ridges, 3–5 mm; perianths slightly pubescent abaxially, 3.5–4(–4.5) mm; wc California . . . . 4. *Chorizanthe douglasii*
    8. Involucral teeth margins scarious, parted or divided at sinuses, sometimes with distinct margin between erect teeth.
      12. Involucral awns straight.
        13. Involucres congested in small bracteated clusters; leaf blades hirsute; tepals monomorphic; Coast Ranges of California . . . . 3. *Chorizanthe stellulata* (in part)
        13. Involucres 1; leaf blades villous; tepals dimorphic; coastal dunes and grasslands of California.
          14. Perianths (3–)3.5–4.5 mm, pubescent nearly throughout; plants spreading or decumbent to somewhat erect . . . . 10. *Chorizanthe howellii*
          14. Perianths (4–)5–6 mm, hairy on proximal 1/2; plants erect to spreading . . . . 11. *Chorizanthe valida* (in part)
      12. Involucral awns uncinate.
        15. Perianths glabrous; tepals entire apically; floral tubes lemon-yellow; coastal California . . . . 5. *Chorizanthe diffusa*
        15. Perianths pubescent abaxially; floral tubes white.
          16. Involucres 2.5–4 mm, thinly pubescent abaxially; perianths 2.5–4 mm; wc California . . . . 9. *Chorizanthe robusta*
          16. Involucres 1.5–2.5(–3) mm, villous abaxially; perianths 2–3 mm; coastal California.
            17. Tepals cuspidate apically . . . . 8. *Chorizanthe cuspidata* (in part)
            17. Tepals erose apically.
              18. Involucral margins white to pink or purple; involucres 2–2.5(–3) mm; perianths 2–3.5 mm; stamens 9 . . . . 6. *Chorizanthe pungens*
              18. Involucral margins pinkish; involucres 1.5–2(–2.5) mm; perianths 2–3 mm; stamens 3 or 6–9 . . . . 7. *Chorizanthe angustifolia* (in part)
  7. Involucral teeth without scarious or membranous margins.
    19. Involucral awns unequal, anterior one greatly elongated.

20. Anterior awns uncinate; c and n California .... 18. *Chorizanthe clevelandii*
20. Anterior awns straight; c California.
    21. Outer tepals obovate to nearly orbiculate, 2-lobed apically; inner tepals erose apically; stamens 9; Coast Ranges, c California .... 17. *Chorizanthe rectispina*
    21. Outer tepals narrowly oblong, with minute cusp or 3 teeth apically; inner tepals entire apically; stamens 3; Coast Ranges, Transverse Ranges, Tehachapi Ranges, s Sierra Nevada, c California .... 19. *Chorizanthe uniaristata*
[19. Shifted to left margin.—Ed.]
19. Involucral teeth of equal or alternating lengths but anterior tooth not greatly elongated.
  22. Tepals all, or at least inner ones, fimbriate or 2-lobed apically.
    23. Outer and inner tepals fimbriate to laciniate apically; montane s California .... 27. *Chorizanthe fimbriata*
    23. Outer tepals entire or if fimbriate to variously divided then inner tepals not fimbriate to laciniate; coastal ranges of California.
      24. Tepals white to pinkish.
        25. Inner tepals fimbriate apically .... 15. *Chorizanthe obovata*
        25. Inner tepals 2-lobed apically .... 16. *Chorizanthe blakleyi*
      24. Tepals red, maroon, or dark purple.
        26. Outer tepals entire, erect; involucres 3.5–4 mm; involucral awns 0.5–1 mm .... 12. *Chorizanthe palmeri*
        26. Outer tepals shallowly lobed, obcordate, 2-lobed, erose, or at least wavy apically, spreading to recurved; involucres 4–6 mm; involucral awns 0.5–2 mm.
          27. Outer tepals 2-lobed or obcordate apically; inner tepals fimbriate apically; perianths (4.5–)5–6 mm; involucres 4–6 mm, slightly ventricose basally .... 13. *Chorizanthe biloba*
          27. Outer tepals erose or at least wavy apically; inner tepals fimbriate or somewhat 2-lobed apically; perianths 4–4.5 mm; involucres 4–4.5 mm, strongly ventricose basally......................... 14. *Chorizanthe ventricosa*
  22. Tepals entire, cuspidate, or erose apically.
    28. Tepals monomorphic.
      29. Plants spreading to erect; stems disarticulating at nodes; stamens 3; filaments adnate at top of floral tube........................... 33. *Chorizanthe brevicornu*
      29. Plants prostrate or decumbent to ascending or erect; stems not disarticulating at nodes; stamens 3 or 6–9 and filaments adnate at base of floral tube or stamens 6 or 9 and filaments adnate at top of floral tube.
        30. Filaments adnate at top of floral tube; tepals obtuse to truncate or minutely emarginate apically; achenes 3-gonous; mountains and coastal mesas of California........................ 28. *Chorizanthe polygonoides* (in part)
        30. Filaments adnate at base of floral tube; tepals entire, erose, or cuspidate apically; achenes lenticular or globose-lenticular; sw and wc California.
          31. Tepals entire apically; filaments connate basally; sw California .... 21. *Chorizanthe procumbens*
          31. Tepals erose or cuspidate apically; filaments distinct; coastal c California.
            32. Tepals erose apically; flowers slightly exserted .... 7. *Chorizanthe angustifolia* (in part)
            32. Tepals cuspidate apically; flowers included to slightly exserted .... 8. *Chorizanthe cuspidata* (in part)
    28. Tepals dimorphic, sometimes monomorphic, inner ones usually narrower and shorter than outer ones.
      33. Involucral awns straight; tepals erose to denticulate.
        34. Plants erect to spreading; involucres 3–4(–4.5) mm; wc California .... 11. *Chorizanthe valida* (in part)
        34. Plants prostrate to spreading; involucres 1.5–2 mm; sw California .... 20. *Chorizanthe parryi* (in part)

[33. Shifted to left margin.—Ed.]
33. Involucral awns uncinate; tepals entire or denticulate.
 35. Involucres 1.5–2 mm; s California. . . . . 20. *Chorizanthe parryi* (in part)
 35. Involucres 2–6 mm; California.
  36. Proximal leaflike bracts soon deciduous or absent.
   37. Perianths 3–4(–5) mm; flowers mostly included; involucres usually congested
    terminally; coastal to montane sw California . . . . 22. *Chorizanthe staticoides*
   37. Perianths 4.5–6 mm; flowers long-exserted; involucres 1; montane s California
    . . . . 23. *Chorizanthe leptotheca*
  36. Proximal leaflike bracts persistent.
   38. Perianths 4.5–6 mm; flowers long-exserted; inland California . . . . 24. *Chorizanthe xanti*
   38. Perianths 2.5–3.5 mm; flowers exserted; coastal and insular sw California.
    39. Perianths 3–3.5 mm; involucres 2.5–3 mm, hairs slender, curly; stamens 9;
    coastal sw California . . . . 25. *Chorizanthe breweri*
    39. Perianths 2.5–3 mm; involucres 2–2.5 mm, hairs stoutish, recurved; sta-
    mens 6; insular sw California. . . . . 26. *Chorizanthe wheeleri*

12a. CHORIZANTHE R. Brown ex Bentham subg. QUINTARIA Reveal & Hardham, Phytologia 66: 108. 1989 E

Plants prostrate to spreading, thinly pubescent. **Leaf blades** oblong. **Stems** not disarticulating at each node. **Inflorescences:** bracts 3, whorled, connate proximally, leaflike, awn-tipped. **Involucres** urceolate, not ventricose basally, 5-toothed, without membranous or scarious margins; teeth erect, connate more than ³/₄ their length, shallow, unequal, 1 prominent and 4 shorter and equal. **Flowers** 1; perianth white, sometimes with yellowish tube, glabrous; stamens 9; filaments distinct, adnate at base of floral tube. **Achenes** black, 3-gonous. **Seeds:** embryo curved.

Species 1: California.

1. **Chorizanthe spinosa** S. Watson in W. H. Brewer et al., Bot. California 2: 481. 1880 • Mojave spineflower E

Eriogonella spinosa (S. Watson) Goodman

Plants spreading to prostrate, 0.3–0.8(–1) × 0.5–8 dm. **Leaves** basal; petiole 0.5–2 cm; blade (0.3–)0.5–1.5(–2) × (3–)5–10(–12) mm, thinly pubescent adaxially, more densely so to tomentose abaxially. **Inflorescences** greenish to reddish, mostly flat-topped and open to dense; bracts 3, whorled, short-petiolate, linear-lanceolate to lanceolate, acerose, 0.5–1.5 cm × 3–8(–10) mm, awns straight, 1–3.5 mm. **Involucres** usually congested in small terminal clusters of 1–3 at node of dichotomies, (4–)5-ribbed, weakly 3-angled, 2–2.5 mm, not corrugate, densely canescent; teeth (4–)5, essentially erect with longer, prominent, and thickened anterior one 2–4 mm, with straight awn 1–

2.5 mm, remaining teeth smaller, 0.5–1 mm, with straight awns 0.3–0.8 mm. **Flowers** 1, exserted; perianth, cylindric, 2.5–3.5 mm; tepals connate ¹/₂–²/₃ their length, dimorphic, entire, those of outer whorl spreading, broadly obovate and rounded apically, those of inner whorl erect, narrowly oblanceolate, ¹/₂ length of outer ones, acute apically; stamens slightly exserted; filaments 2.5–3 mm, glabrous; anthers yellowish, oblong, 0.5–0.7 mm. **Achenes** 2.5–3 mm. $2n = (40), 44, (46)$.

Flowering Apr–Jul. Sandy to gravelly flats and slopes, saltbush communities; 600–1300 m; Calif.

G. J. Goodman (1934) referred *Chorizanthe spinosa* to *Eriogonella*, but *C. spinosa* and *C. membranacea*, the type of *Eriogonella*, are well isolated from one another, and both are well removed from the remainder of the annual spineflowers. The Mojave spineflower is local and uncommon from southeastern Kern and southern Inyo counties, south into adjacent northeastern Los Angeles and northwestern San Bernardino counties to Antelope and Lucerne valleys.

12b. CHORIZANTHE R. Brown ex Bentham subg. **ERIOGONELLA** (Goodman) Reveal & Hardham, Phytologia 66: 110. 1989 [E]

Eriogonella Goodman, Ann. Missouri Bot. Gard. 21: 90. 1934

Plants erect, densely pubescent. **Leaf blades** linear to narrowly oblanceolate. **Stems** not disarticulating at each node. **Inflorescences:** bracts usually 2, opposite, rarely in whorls of 3–5, connate, proximally leaflike, awn-tipped. **Involucres** urceolate, ventricose basally, 6-toothed, with broad, continuous, membranous margins; teeth connate throughout their length, shallow, spreading, equal. **Flowers** 1(–2); perianth white to rose, densely pubescent; stamens 9, adnate at base of floral tube; filaments distinct. **Achenes** brown, 3-gonous, glabrous. **Seeds:** embryo curved.

Species 1: w United States.

2. Chorizanthe membranacea Bentham, Trans. Linn. Soc. London 17: 419, plate 17, fig. 11. 1836 • Pink spineflower [E] [F]

Eriogonella membranacea (Bentham) Goodman

Plants 1–6(–10) × 0.5–3(–5) dm, woolly-floccose. **Leaves** basal and cauline; petiole 0.1–0.5(–0.8) cm; blade linear to narrowly oblanceolate, (1–)1.5–5 × 0.1–0.3 cm, thinly to densely floccose adaxially, densely tomentose abaxially. **Inflorescences** strict, white to greenish, open; bracts usually 2, opposite, rarely in whorls of 3–5, short-petiolate, acerose, similar to proximal leaf blades only reduced, 0.3–3 cm × 1–3 mm, awns straight, 0.5–1 mm. **Involucres** usually congested in small terminal clusters of 1–3 at node of dichotomies, urceolate, ventricose basally, 3-angled, 6-ribbed, 3–4 mm, not corrugate, with conspicuous, white margins extending across sinuses, tomentose to floccose or glabrate with age, greenish to brownish; teeth 6; awns uncinate, 0.7–1.5 mm. **Flowers** 1(–2), slightly exserted; perianth white to rose, subcylindric, (1.5–)2.5–3 mm, densely pubescent abaxially; tepals connate $^2/_3$ their length, slightly dimorphic, entire and rounded apically, those of outer whorl obovate, those of inner whorl spatulate; stamens slightly exserted; filaments 1.5–2.5 mm, glabrous; anthers pink to red, oval, 0.2–0.3 mm. **Achenes** 2.5–3 mm. $2n$ = 38, 40, (42), 80, 82, 84.

Flowering Apr–Jul. Sandy to gravelly or rocky flats and slopes, mixed grassland and chaparral communities, oak-pine woodlands; 40–1400(–1600) m; Calif., Oreg.

Chorizanthe membranacea has long been considered an isolated element among the spineflowers. The strict, upright habit, numerous basal and cauline leaves, and broad, continuous, membranous margins of the involucre all reflect that isolation. Pink spineflower is widespread and often locally common in the Coast Ranges of southwestern Oregon and California and on the western foothills of the Sierra Nevada southward to the Transverse Ranges and the Tehachapi Mountains of Ventura and Kern counties, California.

12c. CHORIZANTHE R. Brown ex Bentham subg. **AMPHIETES** Reveal & Hardham, Phytologia 66: 113. 1989

Plants prostrate to spreading or erect, mostly thinly pubescent. **Leaf blades** linear to lanceolate, obovate, round, or spatulate. **Stems** sometimes disarticulating at each node. **Inflorescences:** bracts mostly 2, opposite, scalelike or if leaflike then similar to basal leaves only reduced, occasionally deciduous in early anthesis, with or without awns. **Involucres** cylindric to narrowly turbinate, campanulate, or urceolate, occasionally ventricose basally, 3-, 5-, or 6-toothed, with or without membranous or scarious margins; teeth erect to spreading or divergent, connate at least ¹/₂ their length, typically shallow, mostly unequal, with alternating long and short awns, often with anterior one longest. **Flowers** 1(–2), white to pink or rose, maroon or purple, or yellow, thinly pubescent at least along midribs abaxially; stamens 3–9; filaments adnate at base

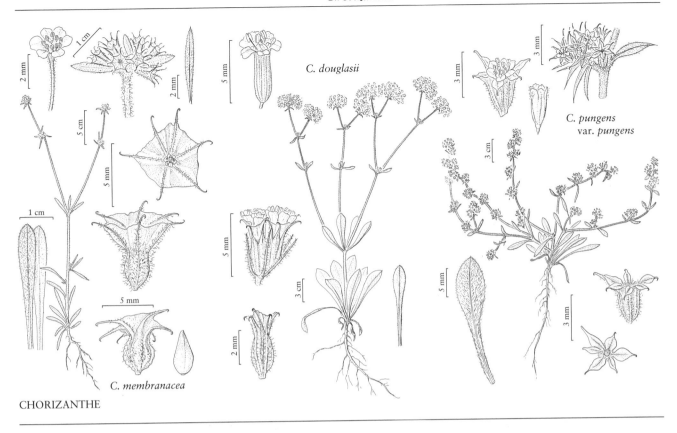

C. douglasii

C. pungens
var. pungens

C. membranacea

CHORIZANTHE

of floral tube or faucially; filaments sometimes connate into short tube. **Achenes** brown, lenticular or globose-lenticular, or 3-gonous. **Seeds:** embryo straight or rarely curved.

Species 39 (31 in the flora): w United States, nw Mexico, sw South America.

Most species of subg. *Amphietes* are found in California. Of the others, one is known only from southernmost Peru to central Chile (*Chorizanthe commissuralis* J. Rémy), while the rest are known only from Baja California, Mexico. Those include *C. inequalis* S. Stokes, *C. turbinata* Wiggins, *C. mutabilis* Brandegee, *C. rosulenta* Reveal, *C. pulchella* Brandegee, *C. flava* Brandegee, and *C. interposita* Goodman. The latter is the only member of sect. *Clastoscapa*, the only section of subg. *Amphietes* not found in our flora.

12c.1. CHORIZANTHE R. Brown ex Bentham (subg. AMPHIETES) sect. PTELOSEPALA Nuttall, Proc. Acad. Nat. Sci. Philadelphia 4: 17. 1848

Chorizanthe sect. *Anisogonum* Roberty & Vautier; *C.* sect. *Eriogonellopsis* Roberty & Vautier; *C.* sect. *Herbaceae* Bentham ex Goodman

Plants prostrate to spreading or somewhat erect. **Leaf blades** lanceolate to ovate or spatulate. **Stems** not disarticulating at each node. **Inflorescences:** bracts with or without awns. **Involucres** cylindric, campanulate, or urceolate, occasionally ventricose basally, 3-angled, 6-ribbed, 6-toothed, with or without membranous or scarious margins continuous across sinuses; awns unequal, typically with anterior one longest. **Flowers** 1; perianth white, yellow, rose, red, maroon, dark purple, or lavender, thinly pubescent; stamens 3–9; filaments adnate at base of floral tube. **Achenes** light brown to dark brown, lenticular or globose-lenticular.

Species 31 (25 in the flora): w United States, nw Mexico.

3. Chorizanthe stellulata Bentham in A. P. de Candolle and A. L. P. P. de Candolle., Prodr. 14: 26. 1856

• Starlite spineflower [E]

Plants erect, 0.5–2.5(–3) × 0.5–3 dm, hirsute. **Leaves** basal; petiole 0.1–0.5 cm; blade narrowly lanceolate to oblanceolate, 0.5–2 × 0.8–2(–2.2) cm, hirsute. **Inflorescences** cymose, dichotomously branched throughout, white to greenish or reddish; bracts usually 2, similar to leaves at proximal nodes only reduced, typically with whorl of 3–5 ca. midstem, short-petiolate, becoming linear and aciculate at distal nodes, acerose, 0.5–2(–3) cm × 10–30(–40) mm, awns absent. **Involucres** congested in small bracteated terminal clusters of 2–4 at node of dichotomies, tannish, cylindric, slightly ventricose basally, 3–4 mm, with conspicuous, white, broad, membranous margins typically extending up tooth to awn, finely corrugated, hispid at least along ridges, otherwise sparsely pubescent; teeth spreading, equal, 1–1.5 mm, awns straight, 0.5–1 mm. **Flowers** exserted; perianth cream to creamy white or rose, cylindric, 4–4.5(–5) mm, slightly pubescent abaxially; tepals connate ²/₃ their length, monomorphic, obovate, obcordate to 2-lobed apically, sometimes slightly irregular but not distinctly erose; stamens 9, slightly exserted; filaments distinct, 4–5 mm, glabrous; anthers pink to red, oblong, 0.5–0.6 mm. **Achenes** light brown, globose-lenticular, 3.5–4.5 mm. $2n$ = 38, 40, 44.

Flowering Apr–Jul. Sandy to gravelly flats and slopes, mixed grassland and chaparral communities, oak-pine woodlands; 30–900 m; Calif.

Chorizanthe stellulata can be locally common in the foothills bordering the Central Valley from Shasta County south to Stanislaus County on the western side, and to Tulare County on the eastern side. Post-flowering specimens of starlite spineflower and Douglas's spineflower are sometimes difficult to distinguish. The margins of the involucre in the former are always white; those of *C. douglasii* are purple.

4. Chorizanthe douglasii Bentham, Trans. Linn. Soc. London 17: 418. 1836 • Douglas's spineflower [E] [F]

Chorizanthe nortonii Greene

Plants erect, 1–4(–5) × 0.5–3 dm, villous. **Leaves** basal; petiole 1–3(–6) cm; blade oblanceolate, 0.5–2(–4) × 0.1–0.4(–1) cm, villous. **Inflorescences** cymose, dichotomously branched throughout, white to greenish or reddish; bracts usually 2, similar to proximal leaf blades, typically with whorl of 3–5 ca. midstem, short-petiolate, becoming linear and aciculate at distal nodes, acerose, 0.5–2(–3) cm × 1–5(–10) mm, awns absent. **Involucres** congested in small leafy terminal clusters of 2–4 at nodes of dichotomies, greenish, cylindric, slightly ventricose basally, 3–5 mm, with conspicuous, purple, broad, membranous margins typically extending across sinuses, finely corrugated, hispid at least along ridges, otherwise sparsely pubescent; teeth spreading, equal, (0.7–)1–1.5 mm, awns straight, 0.5–1 mm. **Flowers** exserted; perianth white to rose, cylindric, 3.5–4(–4.5) mm, slightly pubescent abaxially; tepals connate ²/₃ their length, monomorphic, obovate, 2-lobed or denticulate apically, infrequently inner whorl entire; stamens 9, slightly exserted; filaments distinct, 3–4 mm, glabrous; anthers pink to red, oblong, 0.5–0.6 mm. **Achenes** light brown, globose-lenticular, 3.5–4 mm. $2n$ = 40.

Flowering Apr–Jul. Sandy to gravelly flats and slopes, mixed grassland communities, oak and pine woodlands; (200–)300–1600 m; Calif.

Chorizanthe douglasii is restricted to the Santa Lucia Mountains and to the San Gabilan and La Panza ranges of west-central California. The species is infrequent but can be locally common. A single collection made in the Santa Cruz Mountains (*Rowntree s.n.*, 16 Jun 1929, CAS) may have been made in Santa Cruz County, but the location is uncertain and no other collection is known from that region.

5. Chorizanthe diffusa Bentham in A. P. de Candolle and A. L. P. P. de Candolle, Prodr. 14: 26. 1856

• Diffuse spineflower [E]

Chorizanthe andersonii Parry; *C. diffusa* var. *nivea* (Curran) Hoover; *C. nivea* (Curran) A. Heller; *C. pungens* Bentham var. *diffusa* (Bentham) Parry; *C. pungens* var. *nivea* Curran

Plants spreading, decumbent, prostrate, or rarely slightly erect, 0.3–1(–1.5) × 0.5–2(–10) dm, villous. **Leaves** basal; petiole 0.2–1.8(–2) cm; blade oblanceolate, 0.3–2 × 0.1–0.4 cm, villous. **Inflorescences** rather dense with secondary branches suppressed, white to greenish; bracts 2, similar to leaf blades at proximal nodes only reduced, short-petiolate, becoming linear and aciculate at distal nodes, acerose, 0.3–2 cm × 1–4 mm, awns absent. **Involucres** 1, mostly greenish, cylindric, not ventricose, 2–2.5 mm, with thin to broad and then conspicuous white or pinkish to purple, scarious margins extending nearly full length of awn, not corrugate, villous-hirsute; teeth spreading to divergent, equal, 0.5–1 mm, awns uncinate with longer ones 1–2 mm, anterior one mostly 2 mm, alternating with shorter (0.5–1 mm) ones. **Flowers** exserted; perianth bicolored with floral tube lemon-yellow and tepals white, campanulate, 2.5–3 mm, glabrous; tepals connate ca. ¹/₃ their length, monomorphic, oblong, acute to obtuse and entire apically; stamens 3–9, slightly exserted; filaments distinct, 1.5–2 mm, glabrous; anthers yellow, oval, 0.3–0.4 mm. **Achenes** dark brown, globose-lenticular, 2–2.5 mm. $2n = 38, 40, 42$.

Flowering Apr–Jul. Sandy to gravelly flats and slopes, coastal scrub communities, pine-oak woodlands; 30–800 m; Calif.

The involucral margins of *Chorizanthe diffusa* vary greatly. In some individuals, the white margins are barely visible. Plants with the margins of the involucre are predominantly white have been designated var. *nivea*. A full gradation between the extremes may be observed in most populations, although in coastal sands var. *nivea* is often the dominant expression. Diffuse spineflower occurs near the coast and in the Coast Ranges of central California.

Chorizanthe diffusa has been shown to inhibit carcinogen-induced preneoplastic lesions in a mouse mammary-organ culture model. This inhibitory activity is known to correlate with cancer chemopreventive effects in full-term models of tumorigenesis (see H. S. Chung et al. 1999).

6. **Chorizanthe pungens** Bentham, Trans. Linn. Soc. London 17: 419, plate 19, fig. 2. 1836 Ⓒ Ⓔ Ⓕ

Plants prostrate to ascending or erect, 0.5–2(–2.5) × 0.5–10 dm, grayish-villous. **Leaves** basal; petiole (0.5–)1–3(–4) cm; blade oblanceolate, (0.5–)1–5(–7) × (0.3–)0.4–0.7(–1) cm, villous. **Inflorescences** rather dense with secondary branches suppressed, grayish; bracts 2, similar to leaf blades at proximal nodes only reduced, short-petiolate, becoming linear and acicular at distal nodes, acerose, 0.5–7 cm × 2–7 mm, awns 0.5–1.2 mm. **Involucres** 1, grayish, cylindric, often ventricose basally, 2–2.5(–3) mm, with distinct, white to pink or purple, scarious margins extending nearly full length of awn, corrugate, villous abaxially; teeth spreading, equal, 0.5–1.5 mm; awns uncinate with longer ones 2–3 mm and alternating with shorter (1–1.5 mm) ones. **Flowers** exserted; perianth bicolored with floral tube white and tepals white to rose, cylindric, 2–3.5 mm, pubescent abaxially; tepals connate less than ¹/₄ their length, monomorphic, obovate to oblong, acute to truncate and erose apically; stamens

9, slightly exserted; filaments distinct, 2–3 mm, glabrous; anthers cream to rose, ovate, 0.3–0.4 mm. **Achenes** dark brown, globose-lenticular, 2–2.5 mm.

Varieties 2 (2 in the flora): California.

1. Margins of involucres white (rarely pinkish), scarious; plants prostrate to slightly ascending; coastal areas and adjacent inland valleys . . . . 6a. *Chorizanthe pungens* var. *pungens*
1. Margins of involucres dark pinkish to purple, scarious; plants slightly ascending to erect; coastal mountains . . . . 6b. *Chorizanthe pungens* var. *hartwegiana*

6a. **Chorizanthe pungens** Bentham var. **pungens**
• Monterey spineflower Ⓒ Ⓔ Ⓕ

Plants prostrate to slightly ascending, 0.5–1.5 × 0.5–10 dm. **Involucres:** margins white (rarely pinkish), scarious. $2n = 40$.

Flowering Apr–Jul. Sandy places, coastal scrub communities, oak woodlands; of conservation concern; 0–70 m; Calif.

Variety *pungens* is restricted to coastal regions of west-central California. It is rare and infrequent. A collection made by William Gambel in the 1840s suggests that the variety once occurred at San Simeon in extreme northern San Luis Obispo County; it is not known whether this is a labeling error or an indication of past distribution. Monterey spineflower is federally listed as threatened.

6b. **Chorizanthe pungens** Bentham var. **hartwegiana** Reveal and Hardham, Phytologia 66: 125. 1989
• Ben Lomand spineflower Ⓒ Ⓔ

Plants slightly ascending to erect, 0.5–2.5 × 0.5–10 dm. **Involucres:** margins dark pinkish to purple, scarious. $2n = 40$.

Flowering Apr–Jul. Sandy places, chaparral communities, pine-oak woodlands; of conservation concern; 90–500 m; Calif.

Variety *hartwegiana* is locally common but restricted mainly to the Ben Lomand sand hill area of Santa Cruz County. Until 1989, the name *Chorizanthe pungens* var. *hartwegii* was misapplied to this taxon.

7. Chorizanthe angustifolia Nuttall, Proc. Acad. Nat. Sci. Philadelphia 4: 17. 1848 • Narrow-leaf spineflower [C] [E]

Chorizanthe angustifolia var. *eastwoodiae* Goodman

Plants decumbent or prostrate, 0.3–1 × 0.5–10(–13) dm, villous. **Leaves** basal; petiole 1–4 cm; blade oblanceolate, (0.5–)1–4(–5) × (0.2–)0.3–0.6 cm, villous. **Inflorescences** rather dense with secondary branches suppressed, grayish to reddish; bracts 2, similar to proximal leaf blades only reduced, short-petiolate, becoming linear and aciculate at distal nodes, acerose, 1–4 cm × 2–8(–10) mm, awns absent. **Involucres** 1, reddish, cylindric, not ventricose, 1.5–2(–2.5) mm, without scarious margins or if so then pinkish, thin, and restricted to basal portion of teeth, not corrugate, villous abaxially; teeth spreading, equal, 0.5–1.5(–2) mm; awns uncinate with longer ones 1.5–2.5 mm and anterior one mostly 2–2.5 mm, these alternating with shorter 1–1.5 mm ones. **Flowers** slightly exserted; perianth bicolored with floral tube white and tepals white to rose, campanulate, 2–3 mm, pubescent abaxially; tepals connate ⅓ their length, monomorphic, oblong, usually rounded and erose apically; stamens 3 or 6–9, slightly exserted; filaments distinct, 2–2.5 mm, glabrous; anthers cream to rose, ovate, 0.2–0.3 mm. **Achenes** light brown, globose-lenticular, 2–2.5 mm. *2n* = 38, 40, (42, 44, 46).

Flowering Apr–Jul. Sandy places, coastal scrub communities, pine-oak woodlands; of conservation concern; 10–500 m; Calif.

Chorizanthe angustifolia is common along the immediate coast and mesas mainly in west-central California. Plants with slightly scarious, pink involucral margins have been distinguished as var. *eastwoodiae*, but as both this and the nonscarious var. *angustifolia* occur together, no distinction is made here. William Gambel obtained the only collection known from Los Angeles County in the 1840s; it has not been found there since. Narrow-leaf spineflower often grows with *C. diffusa* in intermingled populations and care must be taken to avoid mixed collections.

8. Chorizanthe cuspidata S. Watson, Proc. Amer. Acad. Arts 17: 379. 1882 [C] [E]

Chorizanthe pungens Bentham var. *cuspidata* (S. Watson) Parry

Plants decumbent to prostrate or ascending, 0.5–2(–2.5) × 0.5–10 dm, villous. **Leaves** basal; petiole (0.5–)1–3 cm; blade oblanceolate, (0.5–)1–5 × (0.3–)0.4–0.7(–1) cm, villous. **Inflorescences** rather dense with secondary branches suppressed, greenish to reddish; bracts 2, similar to proximal leaf blades only reduced, short-petiolate, becoming narrowly elliptic to linear-lanceolate and aciculate at distal nodes, acerose, 0.5–5 cm × 2–7 mm, awns 0.5–1.2 mm. **Involucres** 1, greenish, cylindric, often ventricose basally, 1–3 mm, without scarious margins or if so then white to pink, thin, and restricted to basal portions of teeth, corrugate, villous abaxially; teeth spreading, equal, 0.5–2 mm; awns uncinate or straight with longer ones 2–3 mm and anterior one mostly 2.5–3 mm, these alternating with shorter 1–1.5(–1.7) mm ones. **Flowers** included to slightly exserted; perianth bicolored with floral tube white and tepals white to rose, cylindric, 2–3 mm, pubescent abaxially; tepals connate less than ¼ their length, monomorphic, oblong, truncate to 3-lobed and distinctly cuspidate apically; stamens 9, slightly exserted; filaments distinct, 2–3 mm, glabrous; anthers cream to rose, narrowly oblong, 0.3–0.4 mm. **Achenes** light brown, globose-lenticular, 2–3 mm.

Varieties 2 (2 in the flora): California.

1. Involucral awns uncinate apically; San Francisco area . . . . 7a. *Chorizanthe cuspidata* var. *cuspidata*
1. Involucral awns straight, rarely curved or with 1 awn uncinate; Marin and Sonoma counties . . . . 7b. *Chorizanthe cuspidata* var. *villosa*

8a. Chorizanthe cuspidata S. Watson var. **cuspidata** • San Francisco spineflower [C] [E]

Plants decumbent to prostrate, 0.5–1.5 × 0.5–10 dm. **Leaf blades** 0.5–2(–2.5) × 0.3–0.7 cm. **Involucres** 1–2 mm, with or without pinkish, thin, scarious margins, awns uncinate apically. **Flowers** 2–2.5 mm. **Achenes** 2–2.5 mm.

Flowering Apr–Jul. Sandy places, coastal scrub communities, oak woodlands; of conservation concern; 0–300 m; Calif.

Variety *cuspidata* is localized in the greater San Francisco area, where it is rapidly becoming rare due to increased development. No collections of San Francisco spineflower have been made in Marin or Alameda counties since 1881. Populations may still be seen in San Mateo County, but their numbers have decreased greatly since the 1950s.

8b. Chorizanthe cuspidata S. Watson var. **villosa** (Eastwood) Munz, Aliso 4: 89. 1958 • Villose spineflower [C][E]

Chorizanthe villosa Eastwood, Bull. Torrey Bot. Club 30: 485. 1903

Plants ascending to slightly erect, 0.5–2 × 0.5–10 dm. **Leaf blades** 1–5 × 0.4–1 cm. **Involucres** (2–) 2.5–3 mm, without scarious margins, awns straight, rarely curved or with 1 awn uncinate. **Flowers** 2.5–3 mm. **Achenes** 2.5–3 mm. $2n = 38, 40, 42$.

Flowering May–Aug. Sandy places, coastal scrub communities; of conservation concern; 0–60 m; Calif.

Variety *villosa* is more common than var. *cuspidata*, due in part to habitat protection provided at Point Reyes National Seashore. Villose spineflower occurs from Point Reyes in Marin County to Bodega Bay in Sonoma County. Some of the awns of var. *villosa* may become bent due to pressing so as to resemble those of var. *cuspidata*. In the field, some truly uncinate awns can be found occasionally on individual involucres in var. *villosa*, but the condition is never uniform throughout the plant.

9. Chorizanthe robusta Parry, Proc. Davenport Acad. Nat. Sci. 5: 176. 1889 [C][E]

Plants erect to spreading or decumbent, 0.5–3 × 0.1–6 dm, villous. **Leaves** basal or nearly so; petiole 1–4(–7) cm; blade oblanceolate, 1.5–5 × 0.2–0.7(–1) cm, villous. **Inflorescences** with secondary branches not suppressed except in terminal clusters of involucres, green to reddish; bracts 2, similar to proximal leaf blades only reduced, short-petiolate, becoming linear and aciculate at distal nodes, acerose, 1–5 cm × 2–5(–7) mm, awns absent. **Involucres** 1, greenish, cylindric, not ventricose, 2.5–4 mm, with white to pinkish, thin scarious margins restricted to basal portion of teeth, not corrugate, thinly pubescent abaxially; teeth spreading, equal, 0.3–0.8 (–1) mm; awns uncinate with longer ones 0.7–1.3 mm

and anterior one mostly 1–1.3 mm, these alternating with shorter (0.3–0.7 mm) ones. **Flowers** slightly exserted; perianth bicolored with floral tube white and tepals white to rose, cylindric, 2.5–4 mm, pubescent abaxially; tepals connate ¼ their length, monomorphic, oblanceolate to narrowly oblong, usually truncate to rounded and erose or denticulate apically, occasionally distinctly cuspidate; stamens 9, included; filaments distinct, 2–3.5 mm, glabrous; anthers pink to red or maroon, oblong, 0.6–0.8 mm. **Achenes** light brown, globose-lenticular, 3.5–4 mm.

Varieties 2 (2 in the flora): wc California.

1. Margins of involucres white; plants spreading or decumbent; sandy to gravelly places; Alameda and San Mateo counties s in mountains and near coast to n Monterey County . . . . 9a. *Chorizanthe robusta* var. *robusta*
1. Margins of involucres rose-pink; plants erect; annual grasslands near Scotts Valley, Santa Cruz Mountains, Santa Cruz County . . . . 9a. *Chorizanthe robusta* var. *hartwegii*

9a. Chorizanthe robusta Parry var. **robusta** • Robust spineflower [C][E]

Chorizanthe pungens Bentham var. *robusta* (Parry) Jepson

Plants spreading or decumbent, 1–2 × 1–6 dm. **Leaf blades** 1.5–5 × 0.3–0.7(–1) cm. **Involucres** 2.5–4 mm, margins white.

Flowering May–Sep. Sandy to gravelly places, coastal scrub communities, oak woodlands; of conservation concern; 10–300 m; Calif.

Variety *robusta* is known only from the immediate coast in southern Santa Cruz County and extreme northern Monterey County. Nineteenth-century collections were made in Alameda County, the last in 1891, and in Santa Clara County, the last in 1888. The last collection from San Mateo County was made in 1913. Inland populations were known from Santa Cruz County in the 1930s, but none has been seen there since. Nearly all extant populations are known only from coastal state parks.

The type of var. *robusta* is a more upright, inland expression compared to the decumbent plants found along the coast. It is possible that the coastal form deserves recognition but until an extant population of the inland expression is found, no name is proposed. Robust spineflower is federally listed as endangered.

9b. Chorizanthe robusta Parry var. **hartwegii** (Bentham) Reveal & Rand. Morgan, Phytologia 67: 358. 1989 • Scott Valley spineflower C E

Chorizanthe douglasii Bentham var. *hartwegii* Bentham in A. P. de Candolle and A. L. P. P. de Candolle, Prodr. 14: 26. 1856 (as hartwegi); *C. pungens* Bentham var. *hartwegii* (Bentham) Goodman

Plants erect, (0.5–)1–3 × (0.1–)1–2.5 dm. **Leaf blades** 1.5–5 × 0.2–0.5(–0.7) cm. **Involucres** 2.5–3.5 mm, margins rose-pink.

Flowering Apr–Jul. Sandy outcrops, annual grassland communities; of conservation concern; 200–300 m; Calif.

Variety *hartwegii* is known from a few populations in Scott Valley, Santa Cruz County. The epithet "hartwegii" was misapplied to *Chorizanthe pungens* var. *hartwegiana* until 1989; see B. Ertter (1996) for details. Scott Valley spineflower is federally listed as endangered.

10. Chorizanthe howellii Goodman, Ann. Missouri Bot. Gard. 21: 44, plate 3, fig. 1. 1934 • Howell's spineflower C E

Plants spreading or decumbent to somewhat erect, 0.3–1 × 1–5 dm, villous. **Leaves** basal or nearly so; petiole 1–4 cm; blade spatulate to broadly obovate, 1–3 × 0.5–1.5 (–1.8) cm, villous. **Inflorescences** with secondary branches suppressed, greenish to grayish; bracts 2, similar to proximal leaf blades only reduced, short-petiolate, becoming linear and aciculate at distal nodes, acerose, 1–5 cm × 5–15 mm, awns absent. **Involucres** 1, greenish to grayish, broadly cylindric, not ventricose, 3–4 mm, with conspicuous, white, scarious margins between teeth and extending up awn, not corrugate, pubescent; teeth spreading, equal, 0.5–1 mm, awns straight with longer ones 1–2 mm and anterior one mostly 2 mm, these alternating with shorter (0.5–1 mm) ones. **Flowers** exserted; perianth bicolored with floral tube white and tepals white to rose, cylindric, (3–)3.5–4.5 mm, pubescent nearly throughout; tepals connate ¼ their length, dimorphic, oblong, truncate and erose to denticulate apically, those of outer lobes longer and wider than inner ones; stamens 9, included; filaments distinct, 3–4 mm, glabrous; anthers pink to red, oblong, 0.6–0.8 mm. **Achenes** light brown, globose-lenticular, 3–4.5 mm. *2n* = (72, 74, 76, 78), 80, (82, 84, 86, 88, 90).

Flowering May–Jul. Sandy places in coastal dunes and grassland communities; of conservation concern; 0–20 m; Calif.

Chorizanthe howellii is an octoploid probably derived from an ancient hybrid event involving *C. cuspidata* var. *villosa* and perhaps *C. valida*. It is known only from a dune area north of Fort Bragg in Mendocino County, and is federally listed as endangered.

11. Chorizanthe valida S. Watson, Proc. Amer. Acad. Arts 12: 271. 1877 • Sonoma spineflower C E

Plants erect to spreading, 1–3 × 1–6 dm, villous. **Leaves** basal or nearly so; petiole 1–3 cm; blade broadly oblanceolate, 1–2.5(–5) × 0.4–0.8(–1.2) cm, usually villous. **Inflorescences** with secondary branches suppressed, grayish; bracts 2, similar to proximal leaf blades only reduced, short-petiolate, becoming linear and aciculate at distal nodes, acerose, 1–3 cm × 6–10 mm, awns absent. **Involucres** 1, grayish, cylindric, not ventricose, 3–4(–4.5) mm, with white, scarious margins between teeth, finely corrugate, thinly pubescent; teeth erect, equal, 0.3–0.7(–1) mm; awns straight, with longer ones 0.7–1.3 mm and anterior one mostly 1.3 mm, these alternating with shorter, 0.5–1(–1.2) mm ones. **Flowers** exserted; perianth bicolored with floral tube white and tepals white to lavender or rose, cylindric, (4–)5–6 mm, pubescent on proximal ½; tepals connate ¼ their length, dimorphic, oblong, truncate and erose to denticulate, sometimes individual lobes entire, 2-lobed or even cuspidate apically, those of outer lobes longer and wider than inner ones; stamens 9, included; filaments distinct, 2–4.5 mm, glabrous; anthers pink to red or maroon, oblong, 0.6–0.8(–1) mm. **Achenes** light brown, lenticular-globose, 3–4.5 mm.

Flowering Jun–Aug. Sandy places, coastal grassland communities; of conservation concern; 10–100 m; Calif.

Chorizanthe valida may be distinguished by the highly colored involucre. The teeth and bases of awns are bright red. The awns then quickly transform to a bright ivory and this color dominates nearly the length of each awn. In the more inland populations (now extirpated), the awns observed in the old collections appear to be a straw color. It is not known if this is an artifact of age or potentially significant. Sonoma spineflower is now known only from grassy fields south of Abbott's Lagoon in the Point Reyes area of Marin County (L. Davis and R. J. Sherman 1990, 1992). The last collection from Sonoma County was made at Sebastopol in 1907. The type, collected in 1841, apparently was gathered near Fort Ross, also in Sonoma County. This species is federally listed as endangered.

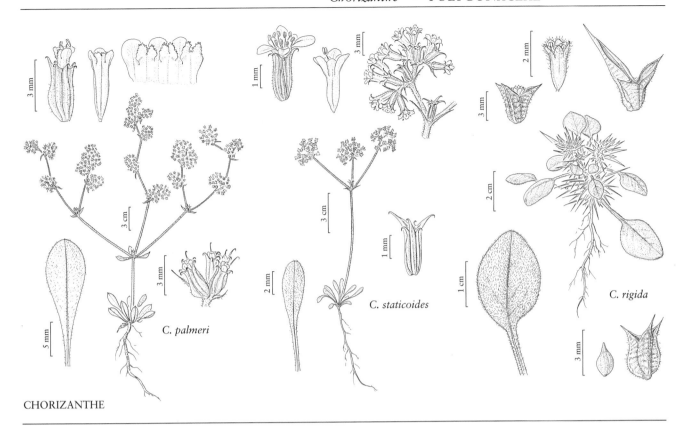

C. palmeri

C. staticoides

C. rigida

CHORIZANTHE

12. Chorizanthe palmeri S. Watson, Proc. Amer. Acad. Arts 12: 271. 1877 • Palmer's spineflower [E] [F]

Plants erect to spreading, (0.5–)1–3(–4) × 1–3 dm, appressed-pubescent. Leaves basal; petiole 1–3 cm; blade oblanceolate, 1–3 × 0.4–0.8 cm, thinly pubescent. Inflorescences with involucres in dense clusters 2–4 cm diam., greenish or reddish to purple; bracts 2–3 at proximal node, usually leaflike, often with whorl of sessile bracts about midstem, elliptic, 0.5–1.5 cm × 2–6 mm, becoming gradually lanceolate to elliptic, 0.2–1 cm × 1.5–5 mm, at distal nodes scalelike, linear and aciculate, acerose, awns straight, 1–3 mm. Involucres 3–10 or more, reddish to purplish, urceolate, slightly ventricose basally, 3.5–4 mm, without scarious or membranous margins, slightly corrugate, thinly pubescent with slender, curly hairs; teeth erect to spreading, unequal, 1–2 mm; awns uncinate, 0.5–1 mm with longer anterior one mostly 1 mm. Flowers exserted; perianth bicolored with floral tube white to yellow and tepals red, maroon, or dark purple, cylindric, 4–5 mm, glabrous or with few scattered hairs along midrib ca. midlength; tepals erect, connate 1/2 their length, dimorphic, obovate, those of outer whorl slightly longer than inner whorl, entire, rounded apically, those of inner whorl fimbriate and truncate or somewhat 2-lobed; stamens 9, exserted; filaments distinct, 4–5 mm, glabrous; anthers pink to red or maroon, oblong, 0.9–1 mm. Achenes brown, globose-lenticular, 3–3.5 mm. *2n* = 38, 40, (48).

Flowering May–Aug. Gravelly to rocky serpentine and serpentinized igneous outcrops, mixed grassland communities, pine-oak woodlands; 60–700 m; Calif.

Chorizanthe palmeri is known only from the Santa Lucia Mountains, the San Luis Range, and the Huasna area. Populations differ slightly both morphologically and ecologically, but recognition of variants is not suggested. At full anthesis, the reddish stems, involucres, and tepals, plus the localized concentrations of individuals provide for splashes of purplish red on the otherwise grass-brown slopes. Set against the often blackish green of serpentine barrens, the plants can be spotted even from a high-flying aircraft! This species would make an excellent addition to the garden border.

13. Chorizanthe biloba Goodman, Ann. Missouri Bot. Gard. 21: 73. 1934 [E]

Chorizanthe palmeri S. Watson var. *biloba* (Goodman) Munz

Plants erect or infrequently spreading, (0.5–)1–3(–4) × 1–3 (–4) dm, pubescent. **Leaves** basal; petiole 0.5–3 cm; blade oblance-olate, 1–3(–5) × 0.4–1(–1.3) cm, thinly pubescent. **Inflorescences** with involucres in open clusters 2–4(–6) cm diam., greenish or reddish to purplish; bracts 2–3 at proximal node, usually leaflike, often with whorl of sessile bracts ca. midstem, elliptic, 0.5–1.5 cm × 2–6 mm, gradually becoming lanceolate to elliptic, 0.2–2 cm × 1.5–8 mm, at distal nodes scalelike, linear and aciculate, acerose, awns straight, 1–3 mm. **Involucres** 3–10 or more, grayish or reddish, urceolate and slightly ventricose basally, 4–6 mm, without scarious or membranous margins, slightly corrugate, strigose; teeth erect to spreading, unequal, 1–2 mm; awns mostly uncinate, 0.5–2 mm, with longer anterior one straight, mostly 2 mm. **Flowers** exserted; perianth bicolored with floral tube white to yellow and tepals red, maroon, or dark purple, cylindric, (4.5–)5–6 mm, sparsely pubescent; tepals connate 1/2 their length, dimorphic, obovate, those of outer whorl spreading, slightly longer than those of inner whorl, 2-lobed, emarginate, or subacute apically, those of inner whorl erect, obtuse, fimbriate apically; stamens 9, exserted; filaments distinct, 4–5 mm, glabrous; anthers yellow to golden, oblong, 1.2–1.8 mm. **Achenes** brown, globose-lenticular, 4–4.5 mm. $2n = (34, 36, 38), 40, (42, 44, 46)$.

Varieties 2 (2 in the flora): California.

Unlike the serpentine species *Chorizanthe palmeri* and *C. ventricosa*, *C. biloba* occurs on gravelly or clayey soils. Those species and *C. obovata* can be quickly differentiated by carefully examining the colors, shapes, and modifications of the tepals. Immature specimens can be difficult to place, but a combination of geographic location and edaphic features can often enable accurate identification of even fragmentary material.

1. Outer tepals deeply 2-lobed, occasionally erose apically . . . . 13a. *Chorizanthe biloba* var. *biloba*
1. Outer tepals emarginate or subcordate apically . . . . 13b. *Chorizanthe biloba* var. *immemora*

13a. Chorizanthe biloba Goodman var. **biloba**

• Two-lobe spineflower [E]

Outer tepals deeply 2-lobed, occasionally erose apically. $2n = 40$.

Flowering May–Aug. Sandy, gravelly or clay soils, grassland communities, pine-oak woodlands; 200–700 m; Calif.

Variety *biloba* is found from the eastern foothills of the Santa Lucia Mountains of Monterey and San Luis Obispo counties eastward to the western foothills of the Diablo, La Panza, and Temblor ranges, and in extreme western Fresno County.

13b. Chorizanthe biloba Goodman var. **immemora** Reveal & Hardham, Phytologia 66: 138. 1989

• Hernandez's spineflower [E]

Outer tepals emarginate or subcordate apically. $2n = 34, 36, 38, 40, 42, 44, 46$.

Flowering May–Sep. Sandy to gravelly soils, mixed grassland communities, pine-oak woodlands; 600–800 m; Calif.

Variety *immemora* is found on the eastern slope of the Diablo Range in southern San Benito County and in adjacent Monterey County. Hernandez spineflower is restricted to five small populations of a few individuals each. Plants of this taxon were variously identified as *C. palmeri*, *C. biloba*, or *C. obovata* prior to 1989. C. B. Hardham (1989) believed that the variety may be a product of past hybridization, given the variation in chromosome number. She also found meiotic chromosomes of three size classes, numerous nucleoli, and atypical pollen grains.

14. Chorizanthe ventricosa Goodman, Leafl. W. Bot. 2: 193, figs. 1, 2. 1939 • Priest Valley spineflower [E]

Chorizanthe palmeri S. Watson var. *ventricosa* (Goodman) Munz

Plants spreading and diffuse, (0.5–)1–5 × 1–5(–7) dm, pubescent. **Leaves** basal; petiole 0.3–1(–1.5) cm; blade oblance-olate, (0.5–)1–3(–4) × (0.2–)0.4–1(–1.2) cm, thinly pubescent. **Inflorescences** with involucres in open clusters 2–6 cm diam., greenish or reddish; bracts 2–3 at proximal node, usually leaflike, often with whorl of sessile bracts about midstem, oblanceolate to elliptic,

0.5–1.5 cm × 1–4 mm, gradually becoming reduced, linear-lanceolate, 0.4–1.2(–1.5) cm × 1.5–5 mm, at distal nodes scalelike, linear and aciculate, acerose, awns straight, 1–3 mm. **Involucres** 3–10+, greenish or reddish, urceolate, strongly ventricose basally, 4–4.5 mm, without scarious or membranous margins, corrugate, thinly pubescent; teeth spreading, unequal, 1–3 mm; awns straight or uncinate with longer anterior one straight, mostly 2 mm, others uncinate, 0.5–1 mm. **Flowers** exserted; perianth bicolored with floral tube white to greenish yellow and tepals red to maroon, cylindric, 4–4.5 mm, sparsely pubescent; tepals connate ¹/₂ their length, dimorphic, oblong, those of outer whorl spreading and recurved, slightly longer than those of inner whorl, broadly obcordate, slightly erose or at least wavy and rounded apically, those of inner whorl erect, narrower, fimbriate and truncate or somewhat 2-lobed apically, erect; stamens 9, exserted; filaments distinct, 3.5–4 mm, glabrous; anthers pink to red or maroon, oblong, 1–1.3 mm. **Achenes** brown, globose-lenticular, 3–3.5 mm. $2n = 40, 42, 44$.

Flowering May–Sep. Serpentine outcrops, mixed grassland communities, oak-pine woodlands; 500–1000 m; Calif.

Chorizanthe ventricosa is restricted to isolated outcrops of serpentine in the coastal mountain ranges of southeastern Monterey County and southern San Benito County south in western Fresno County to the Parkfield Grade area and in Cottonwood Pass of San Luis Obispo County.

15. Chorizanthe obovata Goodman, Ann. Missouri Bot. Gard. 21: 70. 1934 • Spoon-sepal spineflower
E

Plants erect to prostrate, (0.5–)1–3(–4) × 1–4(–5) dm, pubescent. **Leaves** basal; petiole 0.5–2(–3) cm; blade oblanceolate, 0.5–2.5 × 0.3–1 cm, thinly pubescent adaxially, soft-hirsute abaxially. **Inflorescences** with involucres in open clusters 2–4(–6) cm diam., greenish or reddish; bracts 2–3 at proximal node, usually leaflike, without whorl of sessile bracts about midstem, elliptic, 0.5–1.5 cm × 2–6(–8) mm, abruptly reduced above proximal node, becoming scalelike, linear, aciculate, acerose, 0.2–1 cm × 1–3 mm, awns straight, 1–2 mm. **Involucres** 3–10+, grayish, urceolate, slightly ventricose basally, 3–4 mm, slightly corrugate, without scarious or membranous margins, thinly to densely pubescent; teeth erect to spreading, unequal, 1–2 mm; awns straight or uncinate with longer anterior one straight, mostly 1 mm, others uncinate, 0.5–1 mm. **Flowers** exserted; perianth bicolored with floral tube greenish white to white and tepals white to pink,

cylindric, 4–4.5(–5) mm, sparsely pubescent; tepals connate ¹/₂ their length, dimorphic, obovate, those of outer whorl spreading, 2 times longer than those of inner whorl, rounded or slightly obcordate apically, those of inner whorl erect, narrower, fimbriate apically; stamens (6–)9, mostly included; filaments distinct, 4–4.5 mm, glabrous; anthers yellow to golden, oblong, 0.9–1.1 mm. **Achenes** brown, globose-lenticular, 3–3.5 mm. $2n = 38, 40, 42$.

Flowering May–Jul. Sandy or calcareous soils, mixed grassland, coastal scrub, or chaparral communities, pine-oak woodlands; 10–1300 m; Calif.

Chorizanthe obovata is found in the Coast Ranges. The whitish flowers quickly distinguish it from *C. palmeri* and the other reddish-flowered members of this complex. Immature plants can be confused with *C. staticoides*; the floral features readily separate the two species.

16. Chorizanthe blakleyi Hardham, Leafl. W. Bot. 10: 95. 1964 • Blakley's spineflower C E

Plants spreading to ascending, 0.5–1.5 × 0.5–3 dm, thinly pubescent. **Leaves** basal; petiole 0.5–2 cm; blade oblanceolate, 0.5–2.5 × 0.3–0.8 cm, thinly pubescent. **Inflorescences** with involucres in dense clusters 1–2 cm diam., yellowish green; bracts 2, without whorl of sessile bracts about midstem, usually leaflike, oblanceolate, 0.5–1.5 cm × 1.5–3 mm, gradually reduced and becoming scalelike at distal nodes, linear, aciculate, acerose, 0.3–0.8 cm × 1–2 mm, awns straight, 1–2.5 mm. **Involucres** 3–10+, yellowish green, urceolate, slightly ventricose basally, 3–4.5 mm, slightly corrugate, without scarious or membranous margins, thinly pubescent; teeth spreading, unequal, 1–3 mm; awns straight or uncinate with longer anterior one straight or slightly curved, mostly 2 mm, others uncinate, 0.5–1.5 mm. **Flowers** exserted; perianth bicolored with floral tube greenish white to white and the lobes white to pinkish, cylindric, 5–6 mm, sparsely pubescent; tepals connate ²/₃ their length, dimorphic, obovate, those of outer whorl erect, slightly longer than those of inner whorl, 2-lobed apically, those of inner whorl erect, 2-lobed, erose apically; stamens 9, included; filaments distinct, 5–5.5 mm, glabrous; anthers yellow to golden, oblong, 1–1.2 mm. **Achenes** brown, globose-lenticular, 3–3.5 mm. $2n = $ ca. 38.

Flowering May–Jul. Sandy or gravelly flats and slopes, chaparral communities, oak woodlands; of conservation concern; 600–1600 m; Calif.

Chorizanthe blakleyi is known only from north-facing slopes and foothills of the Sierra Madre. The species is rare and localized. The yellowish green stem, branches, and involucres readily distinguish it. The white flowers align it with *C. obovata*.

17. **Chorizanthe rectispina** Goodman, Ann. Missouri Bot. Gard. 21: 72. 1934 · Prickly spineflower C E

Plants spreading to decumbent, 0.3–0.8(–1) × 0.5–4(–5) dm, appressed-pubescent. **Leaves** basal; petiole 0.5–2 cm; blade oblanceolate to spatulate, 0.5–1.5(–2) × 0.2–0.6 cm, thinly pubescent. **Inflorescences** with involucres in small, open clusters 0.5–1.5 cm diam., greenish to grayish; bracts 2, without whorl of sessile bracts about midstem, usually leaflike, oblanceolate to elliptic, 0.5–1.5 cm × 1.5–5 mm, gradually reduced and becoming scalelike at distal nodes, linear, aciculate, acerose, 0.3–0.8 cm × 1–2 mm, awns straight, 0.5–1.5 mm. **Involucres** 3–10+, grayish to reddish, urceolate, slightly ventricose basally, 2–2.5(–3) mm, slightly corrugate, without scarious or membranous margins, densely pubescent; teeth spreading, unequal, 1–2 mm; awns straight or uncinate, unequal, with longer anterior one straight, mostly 1.5–2.5 mm, others uncinate, 0.3–0.6 mm. **Flowers** exserted; perianth bicolored with floral tube yellow and tepals yellow or white, cylindric, 3.5–4 mm, sparsely pubescent; tepals connate ¹/₂ their length, dimorphic, obovate, those of outer whorl white, obovate to nearly orbiculate, 3–4 times longer than those of inner whorl, , truncate to slightly 2-lobed apically, those of inner lobes erect, yellow, broadly obovate, truncate and erose apically; stamens 9, included; filaments distinct, 1–1.5 mm, glabrous; anthers yellow to golden, oblong, 0.5–0.6 mm. **Achenes** brown, globose-lenticular, 3–3.5 mm. $2n$ = (36), 40, (44).

Flowering May–Jul. Sandy to gravelly flats and slopes, mixed grassland communities, pine-oak woodlands; of conservation concern; 200–600 m; Calif.

Chorizanthe rectispina is infrequent and localized in the Coast Ranges of west-central California.

18. **Chorizanthe clevelandii** Parry, Proc. Davenport Acad. Nat. Sci. 4: 62. 1884 (as clevelandi) · Cleveland's spineflower E

Plants spreading to decumbent, 0.2–0.8(–1) × 0.5–5(–7) dm, appressed-pubescent. **Leaves** basal; petiole 0.5–2 mm; blade oblanceolate, 0.5–1.5(–2) × 0.3–0.6(–0.8) cm, thinly pubescent. **Inflorescences** with involucres in small, open clusters 0.5–1.5 cm diam., greenish or grayish to reddish; bracts 2, sessile, usually leaflike, oblanceolate

to elliptic, 0.5–1.5 cm × 1.5–5 mm, gradually reduced and becoming scalelike at distal nodes, linear, aciculate, acerose, 0.4–1 cm × 1–2(–3) mm, awns straight, 1–3 mm. **Involucres** 3–10, grayish to reddish, urceolate, slightly ventricose basally, 3–3.5 mm, slightly corrugate, without scarious or membranous margins, densely pubescent; teeth widely spreading to divergent, unequal, 0.3–0.6 mm or 3–6 mm; awns uncinate, unequal, with longer anterior one 1.5–2.5 mm, others spreading, 0.3–0.6 mm. **Flowers** included or only slightly exserted; perianth bicolored with floral tube greenish white and tepals white, cylindric, 2.5–3 mm, sparsely pubescent; tepals connate ²/₃ their length, dimorphic, linear-oblong, those of outer whorl spreading, 1.5 times longer than those of inner whorl, rounded, entire or emarginate to slightly 2-lobed apically, those of inner whorl erect, acute, entire to erose, slightly fimbriate or 2-lobed apically; stamens 3, included; filaments distinct, 2–2.5 mm, glabrous; anthers white, ovate, 0.3–0.4 mm. **Achenes** brown, globose-lenticular, 2.5–3 mm. $2n$ = 42.

Flowering May–Sep. Sandy to gravelly flats and slopes, mixed grassland and chaparral communities, pine-oak woodlands; 400–2000 m; Calif.

Chorizanthe clevelandii is locally infrequent to common in scattered locations in the Coast Ranges from Mendocino and Lake counties south to Santa Barbara County, and across the Transverse and Tehachapi ranges of Ventura and Kern counties to the southern Sierra Nevada in Tulare County. It is the most widely distributed of the spineflowers endemic to California. The involucres stick to fur, clothing, and fingers, aiding dispersal of the achenes.

19. **Chorizanthe uniaristata** Torrey & A. Gray, Proc. Amer. Acad. Arts 8: 195. 1870 · One-awn spineflower E

Plants spreading or ascending, 0.2–0.6(–0.8) × 0.5–4(–5) dm, appressed-pubescent. **Leaves** basal; petiole 0.5–2 cm; blade oblanceolate, 0.5–1.5(–2) × 0.2–0.8 cm, thinly pubescent. **Inflorescences** with involucres in small open clusters 0.5–1.5 cm diam., greenish to grayish or reddish; bracts 2, sessile, usually leaflike, oblanceolate to elliptic, 0.5–1.5 cm × 1.5–5 mm, gradually reduced and becoming scalelike at distal nodes, linear, aciculate, acerose, 0.4–1.2 cm × 1–2(–3) mm, awns straight, 1.5–4 mm. **Involucres** 3–10, grayish to reddish, urceolate, slightly ventricose basally, 2–3 mm, without scarious or membranous margins, slightly corrugate, densely grayish-pubescent; teeth widely spreading to divergent, unequal, 0.3–0.5 or 3–6 mm; awns straight or uncinate, unequal, with longer anterior one straight, 2.5–5.5 mm, others spreading, uncinate, 0.3–0.5 mm. **Flowers**

included or only slightly exserted; perianth bicolored with floral tube greenish white and tepals white, cylindric, 2–3 mm, sparsely pubescent; tepals connate ²/₃ their length, dimorphic, linear-oblong, those of outer whorl spreading, narrowly oblong, 1.5 times longer than those of inner whorl, rounded but with minute cusp or 3 teeth apically, those of inner whorl erect to slightly spreading, acute, entire apically; stamens 3, included; filaments distinct, 1–2 mm, glabrous; anthers white, ovate, 0.4–0.5 mm. **Achenes** brown, globose-lenticular, 2–3 mm. **2n** = (78), 80, (82).

Flowering Apr–Jul. Sandy to gravelly talus or clay flats and slopes, mixed grassland and chaparral communities, pine-oak woodlands; 800–1900 m; Calif.

Chorizanthe uniaristata is scattered in the Inner Coast Ranges and across the Transverse and Tehachapi ranges to the southern Sierra Nevada.

One-awn spineflower is a polyploid, but whether an autopolyploid or an autoallopolyploid has not been determined. It has the smallest meiotic chromosomes observed by C. B. Hardham (1989).

20. **Chorizanthe parryi** S. Watson, Proc. Amer. Acad. Arts 12: 271. 1877 C E

Plants prostrate to spreading, 0.2–0.8(–1) × 0.5–4(–6) dm, strigose. **Leaves** basal; petiole 0.5–2(–3.5) cm; blade oblanceolate to narrowly oblong, 0.5–2.5(–4) × 0.2–0.6(–1.2) cm, thinly pubescent. **Inflorescences** with involucres in small, open clusters 0.3–1 cm diam., greenish or grayish to reddish; bracts 2, sessile, usually leaflike, oblanceolate to elliptic, 0.5–1.5 cm × 1.5–7 mm, gradually reduced and becoming scalelike at distal nodes, linear, aciculate, acerose, 0.1–0.5 cm × 1–2 mm, awns straight, 0.4–1 mm. **Involucres** 3–5, greenish or grayish to reddish, urceolate, slightly ventricose basally, 1.5–2 mm, corrugate, without scarious or membranous margins, pubescent; teeth widely spreading to divergent or recurved, equal, 0.5–1.5 mm or 1–3 mm; awns uncinate or straight, unequal, alternating 0.5–1.5 mm and 0.2–0.5 mm. **Flowers** slightly exserted; perianth bicolored with floral tube greenish white and tepals white, cylindric, 2.5–3 mm, sparsely pubescent; tepals connate ²/₃ their length, slightly dimorphic, those of outer whorl oblong to oblong-ovate, 1.5 times longer than those of inner whorl, rounded, erose or rarely some entire to denticulate apically, those of inner whorl linear-

oblanceolate, acute, entire or denticulate apically; stamens 9, included; filaments distinct, 2–2.5 mm, glabrous; anthers white, ovate, 0.2–0.3 mm. **Achenes** brown, globose-lenticular, 2.5–3 mm.

Varieties 2 (2 in the flora): California.

1. Involucral awns uncinate . . . . 20a. *Chorizanthe parryi* var. *parryi*
1. Involucral awns straight . . . . 20b. *Chorizanthe parryi* var. *fernandina*

20a. **Chorizanthe parryi** S. Watson var. **parryi**
• Parry's spineflower C E

Involucral awns uncinate.

Flowering May–Jun. Sandy soil on flats and foothills, mixed grassland and chaparral communities; of conservation concern; 90–800(–1300) m; Calif.

Variety *parryi* is known today only from scattered populations in the foothills of the San Gabriel, San Bernardino, and San Jacinto mountains. Much of its native habitat was destroyed by development in the twentieth century.

20b. **Chorizanthe parryi** S. Watson var. **fernandina** (S. Watson) Jepson, Man. Fl. Pl. Calif., 298. 1923
• San Fernando spineflower C E

Chorizanthe fernandina S. Watson in W. H. Brewer et al., Bot. California 2: 481. 1880

Involucral awns straight.

Flowering Apr–Jun. Sandy places on foothills, mixed grassland and chaparral communities; of conservation concern; 90–500 m; Calif.

Variety *fernandina* was thought to be extinct until two small populations were found on Laskey Mesa in Ventura County and on Grapevine Mesa in Los Angeles County. The first site consists of some 14 subpopulations in a sparsely vegetated, undisturbed area. The second site, found in 2000, has a single population. The species otherwise was last collected in Los Angeles County in 1929. A single collection (*Geis 541*, DS) was supposedly gathered near Santa Ana, Orange County, in 1902. San Fernando spineflower has been proposed for federal listing as an endangered species.

21. Chorizanthe procumbens Nuttall, Proc. Acad. Nat. Sci. Philadelphia 4: 17. 1848 • Prostrate spineflower

Chorizanthe chaetophora Goodman; *C. jonesiana* Goodman; *C. procumbens* var. *albiflora* Goodman; *C. procumbens* var. *mexicana* Goodman; *C. uncinata* Nuttall

Plants prostrate to decumbent, 0.2–0.8 × 0.5–4(–5) dm, thinly pubescent. **Leaves** basal; petiole 0.5–2(–3) cm; blade oblanceolate, (0.5–)1–3(–4) × 0.1–0.7(–1.2) cm, thinly pubescent. **Inflorescences** with involucres in small, open clusters 0.3–1 cm diam., greenish yellow to green or reddish green; bracts 2, sessile, leaflike and similar to proximal leaf blades only reduced, linear-oblanceolate to elliptic, 0.3–1(–1.5) cm × 1.5–5(–8) mm, rapidly reduced and scalelike at distal nodes, linear, acicular, often acerose, 0.1–0.5 cm × (0.3–)0.5–3 mm, awns straight, 0.2–1 mm. **Involucres** 3–10, rarely more, greenish yellow to reddish green, cylindric or narrowly to broadly campanulate, not ventricose, 1.5–3 mm, faintly corrugate, without scarious or membranous margins, thinly pubescent with spreading hairs, longest hairs on ribs and at base; teeth spreading, equal, 1–2.5 mm, or divergent, thickened basally, unequal, 1–2 mm or 2.5–5 mm with hyaline margins between teeth; awns uncinate, 0.2–0.5 mm. **Flowers** exserted; perianth yellow or sometimes white, cylindric, (1.7–)2–3 mm, pubescent; tepals connate ca. $^2/_3$ their length, essentially monomorphic, narrowly oblong to narrowly obovate, occasionally with outer lobes slightly broader and longer than inner ones, entire apically; stamens 9, exserted; filaments connate basally into 0.2–1 mm tube, (0.3–)0.5–2.5 mm, pilose-ciliate; anthers cream to pale yellow, oblong, (0.2–)0.5–0.7 mm. **Achenes** brown, lenticular, 1.5–2.5 mm. $2n = (38)$, 40, (42, 44, 46).

Flowering Apr–Jun. Sandy to gravelly flats and slopes, coastal grassland, coastal sage, chaparral, and desert scrub communities; (0–)10–1300 m; Calif.; Mexico (Baja California).

Chorizanthe procumbens is a variable complex of widely scattered, locally infrequent to common populations that occur from the Santa Monica, San Gabriel, and San Bernardino mountains southward through western Riverside and Orange counties to San Diego County. In a strict sense, prostrate plants from San Diego southward belong to *C. procumbens* (including *C. jonesiana*) while decumbent plants to the north are *C. uncinata* (*C. procumbens* sensu G. J. Goodman 1934), if such a distinction is considered taxonomically useful. Plants with a grayish hue south of our range in Baja California have been described as *C. chaetophora*. All of our plants have a greenish yellow cast.

22. Chorizanthe staticoides Bentham, Trans. Linn. Soc. London 17: 418. 1836 • Turkish rugging [E] [F]

Chorizanthe chrysacantha Goodman; *C. chrysacantha* var. *compacta* Goodman; *C. discolor* Nuttall; *C. nudicaulis* Nuttall; *C. staticoides* var. *brevispina* Goodman; *C. staticoides* subsp. *chrysacantha* (Goodman) Munz; *C. staticoides* var. *elata* Goodman; *C. staticoides* var. *latiloba* Goodman; *C. staticoides* var. *nudicaulis* (Nuttall) Jepson

Plants erect to spreading or ascending, 0.5–6 × 0.5–3 (–5) dm, thinly pubescent. **Leaves** essentially basal; petiole 1–3(–4) cm; blade oblong to oblong-ovate, 0.5–3(–8) × 0.3–1(–2.5) cm, thinly pubescent or glabrous adaxially, usually densely tomentose abaxially. **Inflorescences** mostly flat-topped and open to densely branched, green to gray or reddish; bracts soon deciduous, 2, occasionally leaflike at proximal nodes and similar to proximal leaf blades only short-petiolate, more reduced and usually sessile, obovate, others linear and acicular, often acerose, (0.1–)0.2–0.5(–1) cm × 1–3(–6) mm, awns straight, 0.5–2 mm. **Involucres** usually congested terminally with 1 at node of dichotomies, reddish to purplish, cylindric, not ventricose, 3–4(–5) mm, often irregularly corrugate, without scarious or membranous margins, thinly pubescent; teeth spreading, unequal, 0.7–1.3(–1.5) mm with longest of 3 longer ones more erect than others, these alternating with 3 shorter and less-prominent ones; awns uncinate, 0.5–1 mm. **Flowers** mostly included; perianth rose to red, cylindric, 3–4(–5) mm, pubescent; tepals connate $^1/_2$ their length, monomorphic to slightly dimorphic, oblong to obovate, rounded to obtuse or truncate apically, occasionally irregularly denticulate, those of outer whorl usually slightly broader and longer than those of inner whorl; stamens 9, mostly included; filaments distinct, 2.5–4 (–4.5) mm, glabrous; anthers pink to red, oblong, 0.5–0.6 mm. **Achenes** brown, lenticular, 3–4 mm. $2n = 38$, (40, 42).

Flowering Apr-Jul. Sandy to gravelly or rocky places, coastal scrub, mixed grassland and chaparral communities, pine-oak woodlands; 300–1700(–1900) m; Calif.

Chorizanthe staticoides is found in the foothills and mountains of the Coast Ranges from Monterey County southward into San Bernardino, Riverside, and Orange counties. A more insular series of populations occurs on Santa Catalina Island and along the coast and immediately adjacent foothills in Orange and San Diego counties.

Turkish rugging is a highly variable taxon. G. J. Goodman (1934) attempted to parse some of the variants but J. L. Reveal and C. B. Hardham (1989b) merged all of his segregates under a single name. Some of the variation is noteworthy. The insular phase, *C. discolor* (including *C. chrysacantha*), is a relatively rare expression restricted to the immediate (*C. chrysacantha*) and near coastal (*C. discolor*) mesas and bluffs. Such plants tend to have somewhat longer awns than the inland expression (*C. staticoides*). Also, the *C. chrysacantha* phase can be extremely depauperate, resulting in a compact mass overhanging the Pacific Ocean (var. *compacta*). While the extremes can be noted, useful taxonomic distinctions cannot be made because of intergradation in all features.

E. W. Voegelin (1938) noted the use of an infusion of this plant to treat pimples.

23. Chorizanthe leptotheca Goodman, Ann. Missouri Bot. Gard. 21: 61. 1934 • Ramona spineflower

Plants erect to spreading, 0.5–3 (–3.5) × 0.5–3(–5) dm, thinly pubescent. Leaves basal; petiole 1–3(–4) cm; blade oblong to oblong-ovate, 0.5–2(–3) × 0.3–0.5(–0.7) cm, thinly pubescent adaxially, usually densely tomentose adaxially. Inflorescences mostly flat-topped and openly branched, usually reddish; bracts soon deciduous, 2, occasionally leaflike at proximal nodes and similar to proximal leaf blades only more reduced, short-petiolate, ovate, 0.3–0.4 cm × 2–3 mm, otherwise sessile, linear and acicular, often acerose, 0.1–0.3 cm × 0.7–1 mm, awns straight, 0.5–1 mm. Involucres in congested clusters with 1 at node of dichotomies, reddish, cylindric, not ventricose, 3–4 mm, not corrugate, without scarious or membranous margins, thinly pubescent; teeth spreading, unequal, 0.7–1.5 mm with longer of 3 longest ones more erect than 3 other shorter and less-prominent ones, awns uncinate, 0.5–1 mm. Flowers long-exserted; perianth rose to red, infrequently with white lobes, cylindric, 4.5–6 mm, pubescent; tepals connate ca. ¹⁄₂ their length, dimorphic or sometimes monomorphic, narrowly oblanceolate, apex rounded, those of outer whorl slightly broader and occasionally longer than those of inner whorl; stamens 9, mostly included; filaments distinct, 4–6 mm, glabrous; anthers pink to red, ovate to oblong, 0.5–0.6 mm. Achenes brown, lenticular, 3–4 mm. $2n = 38$.

Flowering May–Aug. Sandy to gravelly flats and slopes, grassland and chaparral communities, pine-oak woodlands; (300–)600–1600(–1900) m; Calif.; Mexico (Baja California).

Chorizanthe leptotheca is found in the foothills of the San Bernardino Mountains of San Bernardino County southward along the eastern edge of the Santa Ana Mountains, and through the San Jacinto and Santa Rosa mountains of Riverside County into the mountains of central San Diego County. The species is also found in north-central Baja California.

Ramona spineflower is clearly related to *Chorizanthe staticoides*, but that species occurs to the west of the range of *C. leptotheca* and the two are not known to be sympatric.

24. Chorizanthe xanti S. Watson, Proc. Amer. Acad. Arts 12: 272. 1877 E

Plants erect to infrequently spreading, (0.3–)0.5–2.5(–3) × 0.5–3(–5) dm, thinly pubescent. Leaves basal or nearly so; petiole 1–2(–3) cm; blade oblong or oblong-ovate to ovate, 0.3–1 (–1.5) × 0.3–0.8(–1) cm, thinly pubescent adaxially, densely tomentose abaxially. Inflorescences mostly flat-topped and openly branched, usually reddish; bracts persistent, 2, usually leaflike at proximal nodes and similar to proximal leaf blades only more reduced, short-petiolate, oblong-ovate to ovate, 0.3–0.8 cm × 2–6 mm, becoming sessile, reduced and scalelike at distal nodes, linear, acicular, often acerose, 0.1–0.4 cm × 0.5–1 mm, awns straight, 0.5–1 mm. Involucres in open clusters with 1 at node of dichotomies, reddish, cylindric, not ventricose, 3–4.5 mm, not corrugate, without scarious or membranous margins, thinly to densely pubescent; teeth spreading, unequal, 0.7–1.5 mm, 3 longer ones more erect than 3 shorter and less-prominent ones; awns uncinate, 0.5–1 mm. Flowers long-exserted; perianth rose to red, infrequently with white lobes, cylindric, 4.5–6 mm, pubescent; tepals connate ca. ²⁄₃ their length, monomorphic to slightly dimorphic, narrowly oblanceolate, rounded apically, those of outer whorl occasionally slightly broader and longer than those of inner whorl; stamens 9, mostly included; filaments distinct, 4–6 mm, glabrous; anthers pink to red, oblong, 0.5–0.6 mm. Achenes brown, lenticular, 4–4.5 mm.

Varieties 2 (2 in the flora): California.

1. Involucres thinly pubescent ....
   .... 24a. *Chorizanthe xanti* var. *xanti*
1. Involucres densely white-pubescent ....
   .... 24b. *Chorizanthe xanti* var. *leucotheca*

24a. Chorizanthe xanti S. Watson var. xanti

• Pinyon spineflower   E

Involucres thinly pubescent. $2n$ = 38, 42.

Flowering Apr–Jul. Sandy to gravelly places in mixed grassland, saltbush and chaparral communities, pine-oak woodlands; (60–)300–1600 m; Calif.

Variety *xanti* is found in the Coast Ranges from western Merced, eastern Monterey, and San Benito counties south to the Transverse Ranges of northern Santa Barbara and Ventura counties, thence eastward along the northern foothills of the San Gabriel and San Bernardino mountains to extreme southwestern San Bernardino County, and then northward through the Tehachapi Mountains of central and eastern Kern County onto the foothills of the Sierra Nevada, going as far north as Madera County. In 1935, a disjunct population was sampled near Benton Station in southern Mono County (*Robinson & Lindner 20*, RSA) but it has not been recollected.

24b. Chorizanthe xanti S. Watson var. leucotheca

Goodman, Ann. Missouri Bot. Gard. 21: 60. 1934

• Whitewater spineflower   E

Chorizanthe xanti subsp. *leucotheca* (Goodman) Munz

Involucres densely white-pubescent.

Flowering Apr–Jun. Sandy to gravelly places in saltbush communities, pinyon-juniper and pine-oak woodlands; (60–)400–1300 m; Calif.

Variety *leucotheca* is infrequent and localized in the eastern San Bernardino Mountains of San Bernardino County and on the eastern slopes of the San Jacinto Mountains in Riverside County. The dense tomentum on the involucres completely obscures the reddish hue of the involucral surface.

25. Chorizanthe breweri S. Watson, Proc. Amer. Acad.

Arts 12: 270. 1877 • Brewer's spineflower   C E

Plants ascending to decumbent, (0.3–)0.5–1.5(–2) × 1–5(–7) dm, thinly pubescent. **Leaves** basal; petiole 1–3 cm; blade spatulate to ovate, 0.5–2 × 0.3–1.2(–1.5) cm, thinly pubescent to densely tomentose at least abaxially. **Inflorescences** mostly flat-topped, open to rather densely branched, greenish to reddish; bracts persistent, 2, usually leaflike at proximal nodes and similar to leaf blades, short-petiolate, oblong to ovate, 0.3–0.8 cm × 2–6 mm, becoming sessile, reduced and scalelike at distal nodes, linear, acicular, often acerose, 0.1–0.5 cm × 0.5–1 mm, awns straight, 0.5–1 mm. **Involucres** in open clusters with 1 at node of dichotomies, reddish, cylindric, not ventricose, 2.5–3 mm, corrugate, without scarious or membranous margins, thinly pubescent with slender curly hairs; teeth spreading, unequal, 0.4–1.2 mm with 3 longer ones more erect than 3 shorter and less-prominent ones; awns uncinate, 0.3–0.6 mm. **Flowers** exserted; perianth white to rose or red, usually with white lobes, cylindric, 3–3.5 mm, pubescent; tepals connate ca. 1/2 their length, monomorphic to slightly dimorphic, narrowly oblong to obovate, rounded apically, those of outer whorl usually slightly broader and longer than those of inner whorl; stamens 9, mostly included; filaments distinct, 2.5–3 mm, glabrous; anthers pink to red, oblong, 0.4–0.5 mm. **Achenes** brown, lenticular, 2.5–3 mm. $2n$ = 38.

Flowering Mar–Jul. Gravelly or rocky places, serpentine outcrops, mixed grassland and chaparral communities, pine-oak woodlands; of conservation concern; 60–800 m; Calif.

Chorizanthe breweri is found infrequently in the Coast Ranges of southwestern California.

26. Chorizanthe wheeleri S. Watson, Proc. Amer. Acad.

Arts 12: 272. 1877 • Wheeler's spineflower   E

Chorizanthe insularis R. Hoffmann

Plants erect to spreading, 0.5–2 (–2.5) × 1–2 dm, thinly pubescent. **Leaves** basal; petiole 0.5–3 cm; blade elliptic to oblong, 0.5–2 × 0.2–0.6 cm, thinly pubescent adaxially, tomentose abaxially. **Inflorescences** mostly flat-topped, openly branched, greenish to reddish; bracts persistent, 2, usually leaflike at proximal nodes and similar to leaf blades, short-petiolate, oblong, 0.5–1.2 cm × 2–4 mm, sessile, reduced and scalelike at distal nodes, linear, acicular, often acerose, 0.1–0.5 cm × 0.5–1 mm, awns straight, 0.5–1 mm. **Involucres** in dense terminal clusters with 1 at node of dichotomies, reddish, cylindric, not ventricose, 2–2.5 mm, corrugate, without scarious or membranous margins, thinly pubescent with stoutish, recurved hairs; teeth spreading, unequal, 0.3–0.8(–1) mm, with 3 longer ones more erect than 3 shorter and less-prominent ones; awns uncinate, 0.3–0.5 mm. **Flowers** exserted; perianth white or rose or red with white lobes, cylindric, 2.5–3 mm, glabrous except for few scattered hairs ca. midlength along midrib abaxially; tepals connate 1/2 their length, monomorphic to slightly dimorphic, oblong, rounded apically, those of outer whorl usually slightly broader and longer than those of inner whorl; stamens 6, included; filaments distinct, 1.5–2 mm, glabrous; anthers pink to red, oblong, 0.3–0.4 mm. **Achenes** brown, lenticular, 2.5–3 mm.

Flowering Apr–Jun. Gravelly to rocky slopes, coastal scrub communities; 0–400(–600) m; Calif.

Chorizanthe wheeleri is a rare insular endemic known only from Santa Cruz and Santa Rosa islands.

27. Chorizanthe fimbriata Nuttall, Proc. Acad. Nat. Sci. Philadelphia 4: 17. 1848

Plants erect to spreading, 1–3(–3.5) × 1–2.5 dm, pubescent and minutely glandular. **Leaves** basal; petiole 0.5–3(–5) cm; blade elliptic to obovate or spatulate, 1–3(–3.5) × 0.2–1(–2.5) cm, thinly pubescent adaxially, sparsely tomentose abaxially. **Inflorescences** open, reddish; bracts 3 at proximal node, otherwise 2, sessile, scalelike, linear, acicular, often acerose, 0.1–0.5 cm × 0.5–1 mm, awns straight, 1–2 mm. **Involucres** 1(–5+), reddish or greenish, cylindric, not ventricose, 4–6(–7) mm, finely corrugate, with thin hyaline margins between teeth, sparsely to densely pubescent; teeth divergent, unequal, 3 longer ones 1–3 mm, alternating with 3 shorter ones 0.3–1 mm; awns straight, unequal, 3 longer ones 1–2.5(–3) mm, shorter one (0.3–)0.5–1.5 mm. **Flowers** exserted; perianth bicolored, with floral tube yellow to yellowish white and tepals white to rose, becoming dark rose to red with age, cylindric, 6–9(–10) mm, glabrous abaxially except for few to several scattered hairs ca. midlength along midribs; tepals connate ¹/₂ their length, monomorphic, oblong, fimbriate to laciniate apically; stamens 9, included; filaments distinct, 3–8 mm, glabrous; anthers pink to red, oblong, 0.5–0.7 mm. **Achenes** brown, lenticular, 3–4 mm.

Varieties 2 (2 in the flora): California, nw Mexico.

Chorizanthe fimbriata is our only representative of subsect. *Flava*, a taxon of six species otherwise confined to Baja California. These are the most elegant of the annual species in their remarkable flowers. The flower color, the fine divisions of the tips of the tepals, plus the handsome habit make them of potential horticultural interest for the "need-to-be-challenged" gardeners. In addition to *C. fimbriata* var. *laciniata*, *C. pulchella* Brandegee would be a worthy introduction. *Chorizanthe flava* Brandegee has bright yellow flowers that contrast dramatically with its reddish mature inflorescence

branches. The large (7–9 mm) flowers of *C. mutabilis* Brandegee are a wonder to behold, the yellow floral tube contrasting with the white to red of the tepals themselves.

1. Tepals fimbriate apically, distal segments linear to lanceolate, broader than lateral segments; perianths 6–7(–8) mm . . . . 27a. *Chorizanthe fimbriata* var. *fimbriata*
1. Tepals laciniate apically, distal segments linear, scarcely broader than lateral segments; perianths 8–9(–10) mm . . . . 27b. *Chorizanthe fimbriata* var. *laciniata*

27a. Chorizanthe fimbriata Nuttall var. **fimbriata**
• Fringed spineflower

Plants 1–3(–3.5) × 1–2.5 dm. **Leaf blades** elliptic to obovate or spatulate, 1–3.5 × 0.2–1(–2.5) cm. **Perianths** 6–7(–8) mm; tepals fimbriate apically with distal segments linear to lanceolate, distinctly broader than lateral segments. $2n = 40, 44$.

Flowering Mar–Jul. Sandy to gravelly or rocky places, mixed grassland and chaparral communities, pine-oak woodlands; 30–1700 m; Calif.; Mexico (Baja California).

Variety *fimbriata* is found from eastern Orange and western Riverside counties southward through San Diego County into Baja California.

27b. Chorizanthe fimbriata Nuttall var. **laciniata** (Torrey) Jepson, Fl. Calif. 1: 394. 1914 • Lacinate spineflower

Chorizanthe laciniata Torrey in War Department [U.S.], Pacif. Railr. Rep. 7(3): 19. 1857

Plants 1–2(–2.5) × 1–2 dm. **Leaf blades** obovate or spatulate, 1–2(–3) × 0.2–1(–1.5) cm. **Perianths** 8–9(–10) mm; tepals laciniate apically with distal segments linear, scarcely broader than lateral ones. $2n = 40$.

Flowering Mar–Jul. Sandy to gravelly or rocky places, mixed grassland and chaparral communities, pine-oak woodlands; 400–1500 m; Calif.; Mexico (Baja California).

Variety *laciniata* is found from San Diego and western Imperial counties southward into Baja California. It can be weedy, especially in chaparral.

12c.2. CHORIZANTHE   R. Brown ex Bentham (subg. AMPHIETES) sect. ACANTHOGONUM   (Torrey) Torrey & A. Gray, Proc. Amer. Acad. Arts 8: 197. 1870

Acanthogonum Torrey in War Department [U.S.], Pacif. Railr. Rep. 4(5): 132. 1857; *Chorizanthe* sect. *Chorizanthella* Parry

Plants prostrate to spreading or erect. **Leaf blades** oblanceolate to elliptic or round. **Stems** not disarticulating at each node. **Inflorescences:** bracts awned or awnless. **Involucres** cylindric, campanulate, or urceolate, not ventricose basally, 3-angled, 3-ribbed, 3-, 5-, or occasionally 6-toothed, without membranous or scarious margins; awns equal, sometimes anterior one longest. **Flowers** 1(–2); perianth white to rose or yellow, thinly to densely pubescent; stamens 3, 6, or 9; filaments adnate faucially at or near top of floral tube. Achenes brown to black, lenticular or 3-gonous.

Species 5 (5 in the flora): w United States, nw Mexico.

G. J. Goodman (1934) assigned *Chorizanthe polygonoides* to *Acanthogonum*, which previously included *C. rigida* and *C. corrugata* (see below). J. L. Reveal and C. B. Hardham (1989b) considered that taxon to represent a section within *Chorizanthe*, expanded it further to include both *C. orcuttiana* and *C. watsonii*, and divided it into five subsections. Section *Acanthogonum* differs from sect. *Ptelosepala* by having involucral tubes that are not both 6-ribbed and 6-toothed at the same time.

28. Chorizanthe polygonoides Torrey & A. Gray, Proc. Amer. Acad. Arts 8: 197. 1870

Plants prostrate, 0.1–0.5 × 0.3–2(–2.5) dm, villous. **Leaves** basal; petiole 0.5–1 cm; blade oblanceolate to elliptic, 0.3–1 × 0.2–0.3 cm, thinly pubescent. **Inflorescences** with involucres in small clusters 0.5–1 cm across even at dichotomies, greenish or reddish; bracts 2, petiolate or sessile, similar to proximal leaf blades, oblanceolate or narrowly elliptic to linear-lanceolate, 0.3–1 cm × 1–3 mm, awns absent. **Involucres** 1, greenish or reddish, campanulate, 5–6-ribbed, 1.5–2.5 mm, corrugate, thinly pubescent; teeth 5–6, unequal, 3 longer ones 1.5–3 mm, these alternating with 2 or 3 shorter ones 1–2 mm, anterior tooth not leaflike; awns of prominent teeth uncinate and 1.5–3 mm, those of shorter teeth straight and 1–2 mm. **Flowers** 1, included to slightly exserted; perianth white to rose, cylindric, 1.5–1.8(–2) mm, densely pubescent abaxially; tepals connate $^2/_3$ their length, monomorphic, oblong, obtuse to truncate or minutely emarginate apically; stamens (6) 9, slightly exserted; filaments distinct, 0.8–1 mm, glabrous; anthers reddish, ovate, 0.2–0.3 mm. **Achenes** dark brown to black, 3-gonous, 2–2.5 mm.

Varieties 2 (2 in the flora): California, nw Mexico.

1. Involucres 2–2.5 mm; awns of prominent involucral teeth 1.5–2 mm . . . .
   . . . . . 28a. *Chorizanthe polygonoides* var. *polygonoides*
1. Involucres 1.5–2 mm; awns of prominent involucral teeth 2–3 mm . . . .
   . . . . . . 28b. *Chorizanthe polygonoides* var. *longispina*

28a. Chorizanthe polygonoides Torrey & A. Gray var. **polygonoides** • Knotweed spineflower E

Acanthogonum polygonoides (Torrey & A. Gray) Goodman

Plants generally greenish. **Involucres** 2–2.5 mm; prominent teeth with awns 1.5–2 mm. $2n = 40$.

Flowering Apr–Jun. Sandy to gravelly mostly volcanic soils, mixed grassland communities, oak-pine or montane conifer woodlands; 100–1500 m; Calif.

Variety *polygonoides* occurs from Modoc County to Calaveras County in the Sierra Nevada, in the Sacramento Valley, and in the Coast Ranges from Lake County to Santa Barbara County. In San Luis Obispo and Monterey counties, knotweed spineflower is on sandstones or conglomerates rather than volcanic substrates. All populations are widely scattered and the plants are mostly rare to infrequent.

28b. Chorizanthe polygonoides Torrey & A. Gray var. **longispina** (Goodman) Munz, Aliso 4: 89. 1958 • Long-awned spineflower

Acanthogonum polygonoides (Torrey & A. Gray) Goodman var. *longispinum* Goodman, Leafl. W. Bot. 7: 236. 1955; *Chorizanthe polygonoides* subsp. *longispina* (Goodman) Munz

Plants generally reddish. **Involucres** 1.5–2 mm; prominent teeth with awns 2–3 mm. $2n = 40$.

Flowering Apr–Jun. Sandy to gravelly soil, mixed grassland and chaparral communities, pine-oak woodlands; 30–1500 m; Calif.; Mexico (Baja California).

Variety *longispina* is localized and uncommon in the mountains from western Riverside County south through San Diego County into northern Baja California. There are scattered and even rarer disjunct populations of long-awned spineflower along the coast in the San Diego area.

29. Chorizanthe rigida (Torrey) Torrey & A. Gray, Proc. Amer. Acad. Arts 8: 198. 1870 • Devil's spineflower F

Acanthogonum rigidum Torrey in War Department [U.S.], Pacif. Railr. Rep. 4(5): 133. 1857

Plants erect, 0.2–0.8(–1.5) × 0.1–0.7(–1) dm, pubescent. **Leaves** basal and cauline; petiole 0.5–3 (–4) cm; basal blade broadly elliptic to obovate, 0.5–2.5 × (0.3–)0.5–2 cm, thinly pubescent adaxially, more densely so to tomentose abaxially; proximal cauline leaf soon deciduous, 1, blade similar to basal leaf blades only 1–2(–2.5) × 0.5–1.5 cm, mucronate to awn-tipped, awn mostly 2–4 mm; distal cauline leaf blade persistent, 1 per node, sessile, blade linear to linear-lanceolate, 0.1–1.5 × 0.05–0.15 cm, becoming hard and thornlike with age. **Inflorescences** with involucres in dense clusters in axils of bracts, these on short shoots and each subtended by cauline leaves; bracts 2, subopposite to opposite, linear, 0.5–1(–1.2) cm × 1–2 mm, awns straight, 2–4 mm. **Involucres** 1, greenish, urceolate, 3-ribbed, 2–3 mm, corrugate, pubescent, rarely villous near base in some; teeth 3, unequal, with thickened anterior tooth toward base, 5–10 mm, sometimes expanding and becoming lanceolate to narrowly elliptic, others 0.5–1.2 mm; awns straight. **Flowers** 1–2, included to slightly exserted; perianth yellow, cylindric, 1.5–1.8 mm, densely pubescent abaxially; tepals connate ca. $^2/_3$ their length, monomorphic, oblong, rounded, entire apically; stamens 9, slightly exserted; filaments distinct, 0.5–1 mm, glabrous;

anthers yellowish, ovate, 0.2–0.3 mm. **Achenes** brown, 3-gonous, (1.5–)1.8–2.2 mm. $2n = 38, 40$.

Flowering Feb–Jun. Sandy to gravelly or rocky flats and slopes, desert scrub; -60–1900 m; Ariz., Calif., Nev., Utah; Mexico (Baja California, Sonora).

Anyone with the misfortune to step bare-footed on *Chorizanthe rigida* after the plant has dried instantly appreciates its common name. The species is widespread on the Mojave and Sonoran deserts but only occasionally is it locally abundant or weedy. It is found also along the Lahontan Trough in western Nevada, a well-known biogeographic extension route north of the Mojave Desert (J. L. Reveal 1980). The exceedingly compact and dense inflorescences with suppressed secondary branches result in a series of leaves and bracts that subtend a closely arranged series of bracteated and involucrated flowers.

30. Chorizanthe orcuttiana Parry, Proc. Davenport Acad. Nat. Sci. 4: 54. 1884 • Orcutt spineflower C E

Plants prostrate, 0.1–0.5 × 0.3–2(–2.5) dm, villous. **Leaves** basal; petiole 1–2 cm; blade narrowly oblanceolate, 0.5–1.5 × 0.2–0.35(–0.5) cm, thinly pubescent. **Inflorescences** with involucres in small clusters 0.5–1 cm diam., greenish; bracts 2, sessile, unequal, 1 laminar and oblanceolate, 0.3–1 cm × 1–3 mm, awnless, this opposite linear, acicular, greatly reduced, 0.1–0.2 cm × 0.3–0.6 mm bract terminated by short, straight awn 0.6–1 mm. **Involucres** 1, greenish, campanulate, 3-ribbed, 0.8–2 mm, faintly corrugate, pubescent; teeth 3, equal, 1.8–2 mm; awns uncinate, 0.6–1 mm. **Flowers** 1, included to slightly exserted; perianth yellow, cylindric, 1.5–1.8 mm, densely pubescent abaxially; tepals connate ca. $^1/_2$ their length, monomorphic, narrowly oblanceolate, obtuse to truncate, entire apically, slightly spreading; stamens 9, slightly exserted; filaments distinct, 0.5–0.8 mm, glabrous; anthers reddish, ovate, 0.2–0.3 mm. **Achenes** dark brown, lenticular, 2–2.2 mm. $2n = (76, 78), 80, (84)$.

Flowering Mar–May. Sandy soil, mesas and hills near coast, coastal scrub communities; of conservation concern; 60–200 m; Calif.

Chorizanthe orcuttiana is known from a few populations on coastal mesas and hills near San Diego, San Diego County. It is federally listed as endangered. The species is an octoploid that may well have resulted from an ancient hybridization and doubling of chromosomes involving *C. procumbens* and *C. polygonoides* var. *longispina*. The Orcutt spineflower

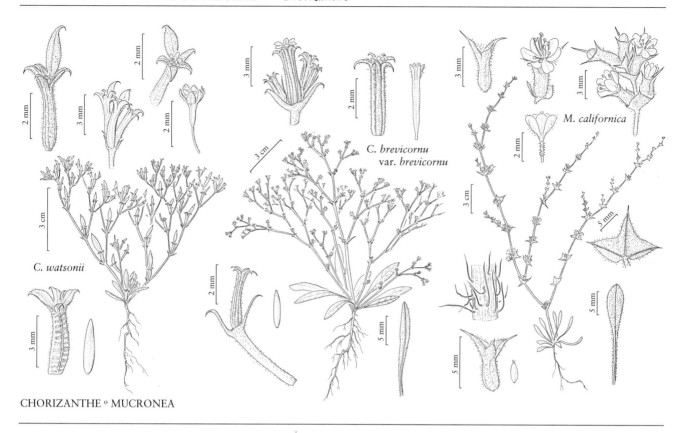

CHORIZANTHE ∘ MUCRONEA

grows in soft, white sand; *C. procumbens* and *C. polygonoides* var. *longispina* are restricted to gravelly sites.

31. Chorizanthe watsonii Torrey & A. Gray, Proc. Amer. Acad. Arts 8: 199. 1870 (as watsoni)

· Watson's spineflower   [E] [F]

Plants spreading to erect, 0.2–1 (–1.5) × 0.2–1(–1.5) dm, densely canescent-strigose. **Leaves** basal or nearly so; petiole (0.5–)1–2.5 (–3) cm; blade oblanceolate, (0.3–)0.5–1.5(–2) × 0.2–0.4(–0.5) cm, thinly floccose to sparsely tomentose. **Inflorescences** with involucres in small clusters 0.5–1 cm diam., greenish to reddish; bracts 2, with laminar ones oblanceolate, (0.5–)0.8–1.5(–2) cm × (1–)2–4 mm, those at distal nodes becoming sessile, reduced and scalelike, linear-lanceolate, acicular, awns slightly curved, 0.5–1 mm. **Involucres** 1, green, cylindric, 5-ribbed, 3–4.5 mm, finely corrugate, pubescent; teeth 5, erect, unequal, with leaflike, narrowly lanceolate, 2–6 mm anterior tooth, others linear, 1–2 mm; awn uncinate, 0.4–0.8(–1) mm. **Flowers** 1, included to slightly exserted; perianth yellow, cylindric, 1.5–2.5 mm, thinly pubescent abaxially; tepals connate ca. $^2/_3$ their length, monomorphic, oblong, acute, entire apically, mostly erect;

stamens 3 or 9, slightly exserted; filaments distinct, 0.8–1 mm, glabrous; anthers yellow, ovate, 0.2–0.3 mm. **Achenes** brown, lenticular, 2.5–3 mm.

Flowering Apr–Aug. Sandy to gravelly flats and slopes, mixed grassland, saltbush and sagebrush communities, pinyon-juniper woodlands; 300–2400 m; Ariz., Calif., Idaho, Nev., Oreg., Utah, Wash.

Chorizanthe watsonii is widely distributed in the cold desert of the Great Basin and in the northern part of the warmer Mojave Desert. Plants in the northern part of the range (especially on the Palouse Prairie of southeastern Washington) usually have three stamens.

32. Chorizanthe corrugata (Torrey) Torrey & A. Gray, Proc. Amer. Acad. Arts 8: 198. 1870 · Wrinkled spineflower

Acanthogonum corrugatum Torrey in War Department [U.S.], Pacif. Railr. Rep. 5(2): 364. 1857

Plants erect, 0.3–1.5 × 0.3–1 dm, thinly tomentose. **Leaves** basal or nearly so; petiole 0.5–2(–3) cm; blade round-ovate, (0.5–)0.8–1.5(–2) × (0.3–)0.5–1.5(–2) cm, thinly floccose to tomentose. **Inflorescences** with involucres in small clusters 0.5–1 cm diam., green to tan or reddish; bracts 2, linear to linear-lanceolate, acicular, 2–7 cm × 1–2.5 mm, awns slightly curved, 0.5–1 mm. **Involucres** 1, green to tan, cylindric, 3-angled but 3-ribbed, 3–4 mm, markedly transverse corrugate, glabrate; teeth 3, equal, 2–4.5 mm; awns uncinate, 0.6–1 mm. **Flowers** 1, included to slightly exserted; perianth white, cylindric, 2–2.5 mm, thinly pubescent abaxially; tepals connate ca. ²/₃ their length, monomorphic, oblong, acute, entire apically; stamens 6, slightly exserted; filaments distinct, 0.8–1 mm, glabrous; anthers cream, ovate, 0.4–0.5 mm. **Achenes** brown, lenticular, 2.5–3 mm. $2n = 38$.

Flowering Feb–May. Sandy to gravelly flats and slopes, mixed grassland, saltbush, creosote bush, and sagebrush communities; -70–1000 m; Ariz., Calif., Nev.; Mexico (Baja California, Sonora).

Chorizanthe corrugata is found mainly in the Mojave and Sonoran deserts. The narrow, transversely corrugated involucral tube is diagnostic. Some anomalous flowers with four or eight stamens have been seen but this condition was always associated with other flowers bearing the normal number.

12c.3. CHORIZANTHE R. Brown ex Bentham (subg. AMPHIETES) sect. FRAGILE Reveal & Hardham, Phytologia 66: 188. 1989

Plants spreading to erect. **Leaf blades** oblanceolate to narrowly elliptic or spatulate. **Stems** disarticulating at each node. **Inflorescences:** bracts awned; involucres cylindric, slightly ventricose basally, 3-angled, 6-ribbed, 6-toothed, without membranous or scarious margins; awns equal. **Flowers** 1; perianth greenish white to white or pale yellowish white, glabrous; stamens 3, adnate at top of perianth tube. **Achenes** dark brown, lenticular.

Species 2 (1 in the flora): w United States, nw Mexico.

Chorizanthe brevicornu of western North America, and its counterpart in western South America, *C. commissuralis* J. Rémy, are the most widely distributed members of the genus on their respective continents. The flowering stems and branches easily break apart, often with involucres still firmly attached. Even in this disarticulated condition, young achenes will continue to live and mature. No doubt this is a significant factor in their successful distribution.

33. **Chorizanthe brevicornu** Torrey in W. H. Emory, Rep. U.S. Mex. Bound. 2(1): 177. 1859 [F]

Plants spreading to erect, 0.5–3(–5) × 0.5–3 dm, thinly pubescent, often with appressed hairs, infrequently somewhat strigose or glabrate. **Leaves** basal; petiole 0.5–2 cm; blade oblanceolate to narrowly elliptic or spatulate, (1–)1.5–3(–4) × 0.1–1 cm, pubescent. **Inflorescences** green; bracts 2, similar to proximal leaf blades only more reduced, 0.3–1(–1.5) cm × 1–2.5 mm, becoming sessile and scalelike at distal nodes, linear, acicular, awns 0.2–0.5 mm. **Involucres** 1, green, 3–5 mm, not corrugate, thinly strigose; teeth divergent, 0.4–1.2 mm; awns uncinate, 0.2–0.5 mm. **Flowers** included; perianth greenish white to white or pale yellowish white, cylindric, 2–4 mm; tepals connate ³/₄ their length, monomorphic, linear to narrowly oblanceolate, acute, entire apically; stamens slightly exserted; filaments distinct, 2–3.5 mm, glabrous; anthers white to pale yellow, ovate, 0.3–0.4 mm. **Achenes** dark brown, lenticular, 3–4 mm.

Varieties 2 (2 in the flora): w United States, nw Mexico.

Chorizanthe brevicornu has stems and branches that easily disarticulate at the nodes. Dried specimens often are reduced to a mere jumble without careful handling. The vegetative fragments will not regenerate new plants, but the involucres (each with a single flower bearing a single achene) easily disarticulate from the parent plant, and with the aid of the awns on the teeth of the involucre, may be readily distributed.

1. Leaf blades oblanceolate to narrowly elliptic, 0.1–0.3(–0.5) cm wide; involucres prominently ribbed at maturity; Mojave and Sonoran deserts . . . . 33a. *Chorizanthe brevicornu* var. *brevicornu*
1. Leaf blades broadly oblanceolate to broadly spatulate, 0.5–1 cm wide; involucres obscurely ribbed at maturity; Great Basin Desert and Snake River Plains . . . . 33b. *Chorizanthe brevicornu* var. *spathulata*

33a. Chorizanthe brevicornu Torrey var. **brevicornu**
• Brittle spineflower   F

Plants 0.5–3(–5) × 0.5–3 dm. **Leaf blades** oblanceolate to narrowly elliptic, 1.5–3(–4) × 0.1–0.3(–0.5) cm, apex acute. **Involucres** distinctly and prominently ribbed at maturity. **2n** = 38, 40, 42, (46).

Flowering Feb–Jul. Sandy places, mixed grassland, saltbush, creosote bush, and sagebrush communities, pinyon-juniper woodlands; -60–3000 m; Ariz., Calif., Nev.; Mexico (Baja California, Sonora).

Variety *brevicornu* is known only from the warm Mojave and Sonoran deserts.

33b. Chorizanthe brevicornu Torrey var. **spathulata** (Small ex Rydberg) C. L. Hitchcock in C. L. Hitchcock et al., Vasc. Pl. Pacif. N.W. 2: 103. 1964
• Great Basin brittle spineflower   E

Chorizanthe spathulata Small ex Rydberg, Bull. Torrey Bot. Club 39: 309. 1912; *C. brevicornu* subsp. *spathulata* (Small ex Rydberg) Munz

Plants 0.5–2(–3) × 0.5–2 dm. **Leaf blades** broadly oblanceolate to broadly spatulate, 1–2 × 0.5–1 cm, apex round. **Involucres** obscurely ribbed at maturity. **2n** = 38.

Flowering Apr–Jul. Sandy to gravelly places, mixed grassland, saltbush, blackbrush, and sagebrush communities, pinyon and/or juniper woodlands; 700–2900(–3100) m; Calif., Idaho, Nev., Oreg.

Variety *spathulata* is known only from the cold deserts of the Intermountain West.

13. MUCRONEA Bentham, Trans. Linn. Soc. London 17: 405, 419, plate 20. 1836
• California spineflower [Latin *mucronis*, sharp point, alluding to awns of bracts and involucres]   E

James L. Reveal

Chorizanthe R. Brown ex Bentham sect. *Mucronea* (Bentham) A. Gray

Herbs, annual; taproot slender. **Stems** arising directly from the root, erect to spreading, solid, not fistulose or disarticulating into ringlike segments, sparsely glandular-pubescent. **Leaves** usually quickly deciduous, basal, rosulate; petiole absent; blade spatulate to obovate, margins entire. **Inflorescences** terminal, cymose, uniparous due to suppression of secondaries; branches dichotomous, not brittle or disarticulating into segments, round, sparsely glandular-pubescent; bracts 3(–5), positioned to side of node or perfoliate and completely but unequally surrounding node, connate nearly completely, triangular to ovate or oblong, awned, sparsely glandular-pubescent. **Peduncles** absent. **Involucres** 1–3 per node, tubular, cylindric, (2–)3–4-angled, slightly ventricose on the angles in some; teeth (2–)3–4, awn-tipped. **Flowers** 1(–2) per involucre; perianth white to pink, campanulate when open, cylindric when closed, pubescent abaxially; tepals 6, connate ca. ¹/₃ their length, monomorphic, entire or erose to fimbriate apically; stamens 6–9; filaments free, glabrous; anthers pink to red, oblong. **Achenes** mostly included, brown to black, not winged, globose-lenticular, glabrous. **Seeds:** embryo straight. $x = 19$.

Species 2 (2 in the flora): California.

1. Inflorescence bracts positioned to side of node, involucres 1–3 per node, 2.5–5(–7) mm, bracts (2–)3(–4), awns (0.5–)1–2.5 (3) mm; flowers 1–2; tepals entire apically . . . . . 1. *Mucronea californica*
1. Inflorescence bracts perfoliate; involucres 1 per node, 3–5(–6) mm, bracts 4, awns (0.3–) 0.5–1.2 mm; flowers 1; tepals entire to fimbriate apically . . . . 2. *Mucronea perfoliata*

1. **Mucronea californica** Bentham, Trans. Linn. Soc. London 17: 419, plate 20. 1836 E F

Chorizanthe californica (Bentham) A. Gray; *C. californica* var. *suksdorfii* J. F. Macbride; *Mucronea californica* var. *suksdorfii* (J. F. Macbride) Goodman

Plants (0.3–)0.5–3(–5) ×1–6(–8) dm. **Leaves:** petiole 0.5–3 cm; blade narrowly spatulate to obovate, (0.5–)1–5 × (0.1–)0.2–0.8(–1.2) cm. **Inflorescences:** bracts 3(–5), unilateral, spreading to nearly erect, connate for ¹/₂ their length, triangular to ovate or oblong, becoming acerose only at terminal nodes and then linear to linear-lanceolate, 0.5–1(–2) cm, apex acute to obtuse; awns 1–2.5(–3) mm. **Involucres** 1–3, (2–)3(–4)-angled, obscurely ribbed and not corrugate, not ventricose, 2.5–5(–7) mm; teeth (2–)3(–4), spreading to strongly divergent, glandular or slightly hirsute; awns divergent, (0.5–)1–2.5(–3) mm. **Flowers** 1(–2); perianth 1.5–2.5(–3) mm, pubescent at least near base abaxially; tepals oblong, entire apically; stamens 6–9; filaments 1–2(–2.5) mm; anthers 0.5–0.7 mm. **Achenes** 2–3 mm. $2n = 38$.

Flowering Mar–Aug. Sandy flats and slopes, coastal grassland, coastal sage and chaparral communities, pine-oak woodlands; 0–1000(–1400) m; Calif.

Mucronea californica is found mainly along the Pacific Coast from San Luis Obispo County south to San Diego County and inland in the southern Coast and Transverse ranges from Monterey County to Los Angeles and western Riverside County. Given the extensive urbanization in southern California, California spineflower is now uncommon there. Along the coast, and especially in San Luis Obispo and Santa Barbara counties, it can be weedy in deep, moving sands. *Mucronea californica* and *M. perfoliata* sometimes are found growing together without indication of intergradation.

2. **Mucronea perfoliata** (A. Gray) A. Heller, Muhlenburgia 2: 23. 1905 · Perfoliate spineflower E

Chorizanthe perfoliata A. Gray, Proc. Boston Soc. Nat. Hist. 7: 148. 1859; *Mucronea perfoliata* var. *opaca* Hoover

Plants (0.2–)0.3–2(–3) ×0.5–5 dm. **Leaves:** petiole inconspicuous; blade spatulate, (1–)2–5 × (0.2–)0.3–1.2(–2) cm. **Inflorescences:** bracts 3, spreading to nearly erect, perfoliate, becoming 1-sided and weakly perfoliate apically, connate for nearly their entire length, orbiculate, 0.5–1(–2) cm, apex acute to obtuse; awns 0.3–1.2(–1.5) mm. **Involucres** 1, 4-angled, distinctly ribbed and usually corrugate, often strongly ventricose, 3–5(–6) mm; teeth 4, spreading to strongly divergent, glandular or strongly hirsute; awns straight, (0.3–)0.5–1.2 mm. **Flowers** 1; perianth 1.5–3(–3.5) mm, pubescent abaxially; tepals narrowly oblanceolate, erose or fimbriate apically, infrequently entire or 2-lobed; stamens 9; filaments 1–2.5(–3) mm; anthers 0.5–0.8(–0.9) mm. **Achenes** 2–3 mm. $2n = 38, 40$.

Flowering Mar–Jul. Sandy to gravelly flats and slopes, mixed grassland, saltbush, creosote bush, and chaparral communities, pine-oak woodlands; 100–1600(–1900) m; Calif.

Mucronea perfoliata is found in the Coast and Transverse ranges from Stanislaus County south to Ventura County, and in the San Joaquin Valley from Kings County south to the Tehachapi Mountains and the northwestern edge of the Mojave Desert in Kern County.

14. SYSTENOTHECA Reveal & Hardham, Phytologia 66: 85. 1989 · Vortriede's spinyherb [Greek *systenos*, tapering to a point, and *theke*, case, alluding to involucre teeth] E

James L. Reveal

Herbs, annual, polygamodioecious; taproot slender. **Stems** arising directly from the root, spreading, solid, not fistulose or disarticulating into ringlike segments, sparsely glandular. **Leaves** usually quickly deciduous, basal, rosulate; petiole present; blade spatulate, margins entire. **Inflorescences** terminal, cymose; branches dichotomous, not brittle or disarticulating into segments, round, sparsely glandular; bracts 3–4, perfoliate and connate, spreading to recurved, usually 4-lobed proximally, 3-lobed distally, oblong to triangular, mucronate, sparsely glandular. **Peduncles** absent. **Involucres** 1 per node, tubular, turbinate, 4-angled, ventricose and sharply

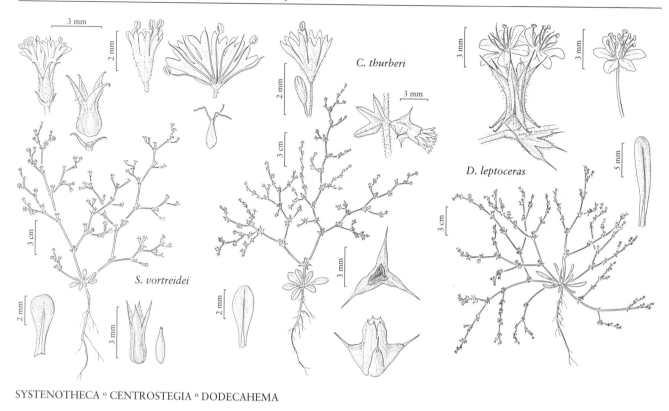

SYSTENOTHECA ° CENTROSTEGIA ° DODECAHEMA

angled; teeth 4, mucronate. **Flowers** bisexual or unisexual, distal shorter one bisexual, proximal longer one pistillate, 2 per involucre; perianth bicolored, with floral tube yellow and lobes white to pink or rose, funnelform when open, tubular when closed, glabrous but densely papillate abaxially; tepals 6, connate ⅓ their length, monomorphic, 2-lobed apically; stamens 9; filaments free, glabrous; anthers maroon to red, oblong. **Achenes** exserted (bisexual) or included and aborted (pistillate), dark brown to black, not winged, 3-gonous, glabrous. **Seeds:** embryo curved. $x = 19$.

Species 1: California.

Systenotheca is the only member of Chorizanthineae that is polygamodioecious, a feature otherwise found in some members of *Eriogonum* (Eriogonineae). Among the eriogonoid genera, *Systenotheca* may be readily distinguished also by the 4-angled, 4-bracted, boxlike involucres. No pistillate flower has been observed to produce a mature achene.

1. **Systenotheca vortriedei** (Brandegee) Reveal & Hardham, Phytologia 66: 85. 1989   E F

Chorizanthe vortriedei Brandegee, Zoë 4: 158. 1893; *Centrostegia vortriedei* (Brandegee) Goodman

Plants 0.2–1.5 × 0.2–1 cm. **Leaf blades** (1–)2–5 × (0.4–)0.5–1 cm. **Involucres** 2.5–4 mm. **Pedicels** to 1 mm (bisexual) or less than 0.3 mm (pistillate). **Flowers:** perianth of distal flower rose to pink, 1–1.5 mm; perianth of proximal flower white, 2–2.5 mm; filaments 2–3 mm; anthers 0.2–0.3 mm. **Achenes** 2–2.5 mm. $2n = 38, 40$.

Flowering May–Jul. Sandy flats and slopes, chaparral communities, pine-oak woodlands; 700–1500 m; Calif.

Systenotheca vortriedei is known only from the Santa Lucia Mountains.

15. CENTROSTEGIA A. Gray ex Bentham in A. P. de Candolle and A. L. P. P. de Candolle, Prodr. 14: 27. 1856 • Red triangles [Greek *kentron*, spur and *stegion*, roof, alluding to arched saccate spurs at base of involucre]

James L. Reveal

Chorizanthe R. Brown ex Bentham sect. *Centrostegia* (A. Gray) Parry

Herbs, annual; taproot slender. **Stems** arising directly from the root, erect to spreading, solid, not fistulose or disarticulating into ringlike segments, sparsely glandular. **Leaves** usually quickly deciduous, basal, rosulate; petiole present; blade oblong to broadly spatulate, margins entire. **Inflorescences** terminal, cymose, uniparous due to suppression of secondaries; branches dichotomous, not brittle or disarticulating into segments, solid, sparsely glandular; bracts 3, positioned to side of node, connate ca. ¼ their length, linear to linear-lanceolate, commonly acerose, awned, sparsely glandular. **Peduncles** absent. **Involucres** 1 per node, tubular, prismatic, 3-angled, 3-awned basally on saccate lobes; teeth 5, awn-tipped, awns straight. **Flowers** 2 per involucre; perianth white to pink, campanulate when open, cylindric when closed, pubescent proximally abaxially; tepals 6, connate proximally, monomorphic, 2-lobed apically; stamens 9; filaments free, glabrous; anthers pink to red, oblong. **Achenes** mostly included, brown, not winged, 3-gonous, glabrous. **Seeds:** embryo curved. *x* = 19.

Species 1: sw United States, nw Mexico.

The involucres of *Centrostegia* are highly modified and unique within Chorizanthineae. First, the involucres have basal and terminal awns. Second, the tube of the involucre is composed of five connate bracts—as in other genera with tubular involucres—but here four bracts are connate their entire length as two pairs with each bract terminated by a short, flatteened, keeled, awned tooth. The fifth bract, positioned opposite the point of attachment on the branch, is terminated by a short, flattened, awned tooth that is equal in size to the others.

SELECTED REFERENCE Goodman, G. J. 1957. The genus *Centrostegia*, tribe Eriogoneae. Leafl. W. Bot. 8: 125–128.

1. **Centrostegia thurberi** A. Gray in A. P. de Candolle and A. L. P. P. de Candolle, Prodr. 14: 27. 1856 F

Centrostegia thurberi var. *macrotheca* (J. T. Howell) Goodman; *Chorizanthe thurberi* (A. Gray) S. Watson; *C. thurberi* var. *macrotheca* J. T. Howell

Plants 0.3–2(–3) × (0.6–)1–4(–5) dm. **Leaf blades** (0.5–)1–3.5(–4) × 0.3–0.8(–1) cm. **Inflorescences:** bracts mostly spreading, (1–)2–6(–10) mm; awns 1–2 mm. **Involucres** (2–)3–6(–8) mm; basal spurs with awns 0.2–2 mm; teeth flattened and keeled distally, awn erect, short, 0.3–1 mm. **Flowers:** perianth 2–3(–3.5) mm; tepals oblanceolate; filaments 1–3 mm; anthers 0.4–0.5 mm. **Achenes** 2–2.5 mm. *2n* = 38.

Flowering Mar–Jul. Sandy to gravelly flats and slopes, mixed grassland, saltbush, creosote bush, and chaparral communities, pine-oak and montane conifer woodlands; 300–2400 m; Ariz., Calif., Nev., Utah; Mexico (Baja California, Sonora).

Centrostegia thurberi is fairly consistent morphologically throughout its range except for individuals in some populations in the mountains of the southern Coast Ranges that tend to have slightly longer and more robust involucres. The name var. *macrotheca* is available for such plants.

16. **DODECAHEMA** Reveal & Hardham, Phytologia 66: 86. 1989 • Spinyherb [Greek *dodeka*, twelve, and *hema*, dart or javelin, alluding to involucral awns] C E

James L. Reveal

Centrostegia A. Gray ex Bentham sect. *Diplostegia* Roberty & Vautier, Boissiera 10: 91. 1964

Herbs, annual; taproot slender. **Stems** arising directly from the root, spreading, solid, not fistulose or disarticulating into ringlike segments, sparsely glandular. **Leaves** usually quickly deciduous, basal, rosulate; petiole present; blade narrowly oblanceolate, margins entire. **Inflorescences** terminal, cymose, uniparous due to suppression of secondaries; branches dichotomous, not brittle or disarticulating into segments, round, sparsely glandular; bracts 3–4, positioned to side and opposite next distal subsequent branch or involucre, connate ca. ¹/₂ their length, lanceolate, awned, sparsely glandular. **Peduncles** spreading, peglike. **Involucres** 1 per node, tubular, cylindric, 6-angled; teeth 6, awn-tipped, awns uncinate. **Flowers** 3 per involucre; perianth white to pink, campanulate when open, cylindric when closed, pubescent abaxially; tepals 6, connate proximally, monomorphic, entire apically; stamens 9; filaments free, glabrous; anthers red to maroon, oblong. **Achenes** included, dark brown to black, not winged, 3-gonous, glabrous. **Seeds:** embryo curved. x = ca. 17.

Species 1: California.

Each of the 12 awns on each involucre of *Dodecahema leptoceras* is long and slender. They are more like those seen in *Sidotheca* (Eriogonineae) than those of any member of the Chorizanthineae. J. L. Reveal and C. B. Hardham (1989) reported a consistent chromosome count of n = 15 plus (seemingly) additional chromosome fragments that bring the number to approximately n = 17. More work is required to resolve the exact chromosome number for the species. A specimen (*Meiere s.n.*, 4 May 1917, CAS) supposedly from Needles, San Bernardino County, is discounted.

1. **Dodecahema leptoceras** (A. Gray) Reveal & Hardham, Phytologia 66: 87. 1989 C E F

Centrostegia leptoceras A. Gray, Proc. Amer. Acad. Arts 8: 192. 1870; *Chorizanthe leptoceras* (A. Gray) S. Watson

Plants 0.3–0.8(–1) × 0.2–0.8 dm. **Leaf blades** (1–)1.5–4(–6) × 0.2–0.4(–0.7) cm. **Peduncles** 0.5–1 mm. **Flowers:** perianth 1.2–1.8(–2) mm; filaments 0.6–1 mm; anthers 0.4–0.5 mm. **Achenes** 1.7–2 mm. $2n$ = ca. 34.

Flowering May–Jun. Sandy to gravelly flats and slopes, chaparral communities, pine-oak woodlands; of conservation concern; 200–700 m; Calif.

Dodecahema leptoceras is known only from the foothills of the San Gabriel Mountains of Los Angeles County, the San Bernardino Mountains of San Bernardino County, and the San Jacinto Mountains of western Riverside County. The urbanization of the greater Los Angeles area has significantly reduced the number of known northern populations. Slender-horned spinyherb is federally listed as endangered.

Dodecahema leptoceras is in the Center for Plant Conservation's National Collection of Endangered Plants.

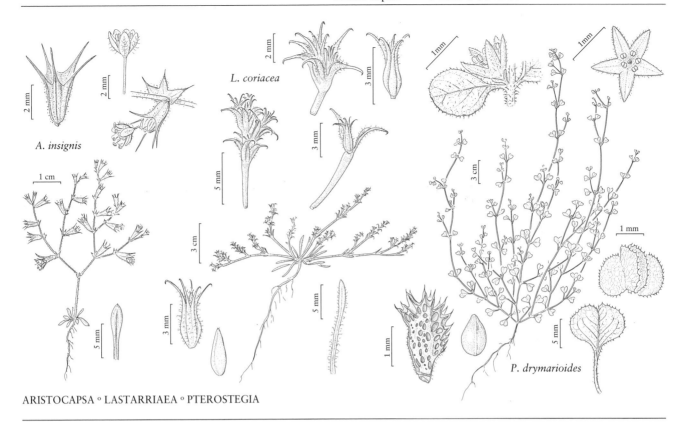

A. insignis

L. coriacea

P. drymarioides

ARISTOCAPSA ° LASTARRIAEA ° PTEROSTEGIA

17. ARISTOCAPSA Reveal & Hardham, Phytologia 66: 84. 1989 • Valley spinycape

[Latin *arista*, awn, and *capsa*, box, alluding to awned involucres] C E

James L. Reveal

Herbs, annual; taproot slender. **Stems** arising directly from the root, erect, solid, not fistulose or disarticulating into ringlike segments, glandular. **Leaves** usually quickly deciduous, basal, rosulate; petiole present; blade linear-spatulate. **Inflorescences** terminal, cymose; branches dichotomous, not brittle or disarticulating into segments, round, glandular; bracts 3, positioned to side of node opposite involucre, connate basally, oblong to linear-acicular, long-awned, glandular. **Peduncles** erect, peglike. **Involucres** 1 per node, 5-ribbed, tubular, narrowly turbinate; teeth 5, awn-tipped. **Flowers** (4–)6 per involucre; perianth white to pink or rose, campanulate when open, cylindric when closed, pubescent abaxially; tepals 6, connate proximally, monomorphic, entire apically; stamens 9; filaments free, glabrous; anthers red to maroon, oblong. **Achenes** mostly included, light greenish brown to tan, not winged, 3-gonous, glabrous. **Seeds:** embryo curved. $x = 14$.

Species 1: wc California.

Aristocapsa may be distinguished from other Eriogonoideae by the combination of five-awned involucres that are slightly corrugated, three-parted inflorescence bracts that are awn-tipped, and (4–)6 flowers per involucre. The base chromosome number is unique in tribe Eriogoneae. *Pterostegia* (tribe Pterostegieae) is the only other genus of the subfamily with that number. Among Chorizanthineae, only *Centrostegia* and *Aristocapsa* have curved embryos, the condition usually found in Eriogonineae.

1. **Aristocapsa insignis** (Curran) Reveal & Hardham, Phytologia 66: 84. 1989 [C] [E] [F]

Chorizanthe insignis Curran, Bull. Calif. Acad. Sci. 1: 275. 1885; *Centrostegia insignis* (Curran) A. Heller; *Oxytheca insignis* (Curran) Goodman

Plants 0.2–1 × 0.2–0.8 dm. **Leaf blades** (0.3–)0.5–1.5 × (0.1–)0.2–0.4 cm. **Peduncles** 1–2 mm. **Involucres** 3–5 mm, faintly corrugate; awns divergent, (1–)2–3 mm. **Flowers** 1.5 mm; filaments 1.5–2 mm. **Achenes** 1.5 mm. $2n = 28$.

Flowering May–Jun. Sandy soil in grassland communities, and in pine-oak or juniper woodlands; of conservation concern; 300–600 m; Calif.

18. **LASTARRIAEA** J. Rémy in C. Gay, Fl. Chil. 5: 289. 1851 or 1852 • Spineflower [For José Victorino Lastarria Santander, 1817–1888, lawyer and founder of the Liberal Party in Chile]

James L. Reveal

Herbs, annual; taproot slender. **Stems** arising directly from the root, prostrate to ascending, solid, not fistulose or disarticulating into ringlike segments, thinly pubescent. **Leaves** quickly deciduous, basal, rosulate; petiole present; blade linear, margins entire, hirsute. **Peduncles** absent. **Inflorescences** terminal, cymose, often uniparous due to suppression of secondaries; branches dichotomous, brittle and disarticulating into segments, round, thinly pubescent; bracts 2, opposite, connate proximally, lanceolate to narrowly ovate, mucronate or awned, thinly pubescent. **Involucral bracts** obvious, in 1 whorl of 3, linear to broadly lanceolate, awn-tipped. **Flowers** 1 per involucral cluster; perianth light green to greenish white, cylindric, thinly pubescent abaxially; tepals 5, connate ½–¾ their length, slightly dimorphic, coriaceous, acute and awn-tipped apically; stamens 3; filaments basally adnate, glabrous; anthers cream to white, oval. **Achenes** mostly included, brownish, not winged, lenticular to slightly 3-gonous, glabrous. **Seeds:** embryo straight. $x = 20$.

Species 3 (1 in the flora): w North America (including Mexico), s South America.

The awn-tipped, coriaceous tepals easily distinguish *Lastarriaea* from other genera of the Eriogonoideae. The mostly prostrate habit with brittle stems and inflorescence branches is seen elsewhere in *Chorizanthe*, most notably *C. brevicornu* and in the Baja California, Mexico, endemic *C. interposita* (the only species of sect. *Clastoscapa*), and because of these features early authors assigned the North and South American species of *Lastarriaea* to *Chorizanthe*. The names *L. chilensis* J. Rémy and *C. lastarriaea* Parry have been misapplied to the California species.

1. **Lastarriaea coriacea** (Goodman) Hoover, Leafl. W. Bot. 10: 342. 1966 • Leather spineflower [F]

Chorizanthe coriacea Goodman, Leafl. W. Bot. 3: 230. 1943; *Lastarriaea chilensis* J. Rémy subsp. *californica* H. Gross, Bot. Jahrb. Syst. 49: 345. 1913, not *Chorizanthe californica* (Bentham) A. Gray; *C. lastarriaea* Parry var. *californica* (H. Gross) Goodman

Plants 0.2–1.5 × 0.5–3(–5) dm. Leaf blades 0.5–3 cm × 0.2–0.8(–1) mm. **Inflorescences** yellowish green to green or red; bracts erect or nearly so, narrowly lanceolate, 0.4–1.5(–2) cm × 0.8–1.5(–2) mm; awns uncinate, (0.5–)1–2(–2.5) mm. **Involucral bracts** erect, linear, 0.3–1(–1.2) cm × 0.3–1 mm; awns uncinate, 0.5–2 mm. **Flowers:** perianth 2–3.5 mm (including awns); tepals connate ca. $^2/_3$ their length, narrowly lanceolate; filaments 0.5–1 mm; anthers 0.2–0.3 mm. **Achenes** 2.5–3 mm. $2n = 42, 60$.

Flowering Feb–Jun. Sandy to gravelly soils in coastal and inland grassland, coastal scrub and chaparral communities, pine-oak woodlands; 0–800 m; Calif.; Mexico (Baja California).

Lastarriaea coriacea grows along the coast from Sonoma County to San Diego County, and in the foothills and mountains from Calaveras County to western Riverside County.

19. PTEROSTEGIA Fischer & C. A. Meyer, Index Seminum (St. Petersburg) 2: 48. 1836 • Woodland threadstem [Greek *pteron*, wing, and *stege*, covering, alluding to winged bract]

James L. Reveal

Herbs, annual; taproot slender. **Stems** arising directly from the root, sprawling and spreading, solid, not fistulose or disarticulating into ringlike segments, thinly pubescent. **Leaves** persistent, cauline, opposite; petiole present; blade broadly elliptic to fan-shaped, margins entire or lobed. **Inflorescences** terminal, cymose, uniparous due to suppression of secondaries distally; branches dichotomous, not brittle or disarticulating into segments, round, thinly pubescent; bracts absent. **Peduncles** absent. **Involucral bracts** 1, erect, 2-winged, reticulately veined, lobed or notched, slightly gibbous with age, invaginated adaxial surface hyaline. **Flowers** 2–3 per involucral cluster; perianth pale yellow to pink or rose, campanulate when open, urceolate when closed, sparsely pubescent abaxially; tepals (5–)6, connate for ca. $^1/_3$ their length, monomorphic, entire apically; stamens 6; filaments adnate to perianth, glabrous; anthers yellow, oval. **Achenes** included, yellowish brown to brown, winged, globose, glabrous. **Seeds:** embryo straight. $x = 14$.

Species 1: w United States, nw Mexico.

Pterostegia and the Baja California, Mexico, endemic *Harfordia* are the only genera of the tribe Pterostegieae. In both, the involucre is highly modified, becoming a slightly to markedly gibbous, reticulated, winged structure that encloses the mature achene. In *Pterostegia*, the two wings are slightly enlarged, but in *Harfordia* the wings are greatly inflated, which apparently aids in achene dispersal.

SELECTED REFERENCE Reveal, J. L. 1989c. Remarks on the genus *Pterostegia* (Polygonaceae: Eriogonoideae). Phytologia 66: 228–235.

1. **Pterostegia drymarioides** Fischer & C. A. Meyer, Index Seminum (St. Petersburg) 2: 48. 1836 [F]

Plants 0.1–10(–12) dm across. **Leaves:** petiole 0.2–0.6(–1) cm; blade 0.3–2 × 0.5–2.5(–3) cm. **Involucral bracts** 1–1.5(–3) × (1.5–)2-3(–3.5) mm. **Flowers:** perianth 0.9–1.2 mm; filaments 0.5–0.6 mm; anthers 0.2–0.3 mm. **Achenes** 1.2–1.5 mm. $2n = 28$.

Flowering Mar–Jul. Sandy to gravelly soils in shady places in grassland and chaparral communities, and in oak-pine or occasionally montane conifer woodland; 0–1600 m; Ariz., Calif., Nev., Utah; Mexico (Baja California).

The only presumed Oregon collection of *Pterostegia drymarioides* (*Tolmie s.n.*, GH) was gathered along the Columbia River sometime in the 1830s. The only collection reportedly from New Mexico (*Parry et al. 1171*, US) supposedly was found in the Rio Grande Valley in the 1850s during the United States-Mexico Boundary Survey. The species likely will be found in southern Oregon (not northern), but it is unlikely that it ever occurred in New Mexico. A problem is that this species is often difficult to notice, as the plants tend to be in the shade under shrubs. Specimens often are misidentified as *Parietaria* (Urticaceae)—See *Flora of North America* volume 3, pages 406–408.

44b. Polygonaceae Jussieu subfam. Polygonoideae Eaton, Bot. Dict. ed. 4, 30. 1836 (as Polygoneae) • Knotweed

Craig C. Freeman

Trees, shrubs, vines, or herbs, perennial or annual, homophyllous (heterophyllous in some species of *Polygonum*); root fibrous or a solid taproot, rarely tuberous. **Stems** usually prostrate to erect, sometimes scandent, not scapose, rarely with recurved spines (some species of *Persicaria*), glabrous or pubescent, sometimes glandular; nodes usually swollen; branches free (adnate to stems distal to nodes and appearing to arise internodally in *Polygonella*); tendrils absent (except in *Antigonon* and *Brunnichia*). **Leaves** deciduous (persistent in *Coccoloba* and sometimes more than 1 year in *Antigonon* and *Polygonella*), basal or basal and cauline, rarely cauline only, mostly alternate; ocrea present, persistent or deciduous, cylindric to funnelform, chartaceous, membranous, coriaceous, or, rarely, foliaceous or partly so; petiole present or absent, rarely articulate basally (*Fagopyrum, Fallopia, Polygonella, Polygonum*), rarely with extrafloral nectaries (*Fallopia, Muehlenbeckia*); blade simple with entire margins, rarely undulate or lobed. **Inflorescences** terminal or terminal and axillary, spikelike, racemelike, paniclelike, cymelike, or, rarely, capitate, comprising simple or branched clusters of compound inflorescences; bracts absent; peduncle spreading to erect, sometimes absent; clusters of flowers subtended by connate bracteoles forming persistent membranous tube (ocreola), awnless. **Flowers** usually bisexual, sometimes bisexual and unisexual on same plant, rarely unisexual only, 1–20+ per ocreate fascicle, often with stipelike base distal to articulation; perianth often accrescent in fruit, often greenish, white, pink, yellow, red, or purple, usually unwinged and unkeeled (winged or, sometimes, keeled in *Fallopia*, rarely keeled in *Polygonum*), campanulate or urceolate, sometimes membranous, indurate, or fleshy in fruit, rarely developing raised tubercles proximally (*Rumex*), glabrous or pubescent, sometimes glandular or glandular-punctate; tepals 2–6, usually in 2 whorls, distinct or connate proximally and forming tube, petaloid or sepaloid, monomorphic or dimorphic; nectary a disk at base of ovary or glands associated with bases of filaments; stamens usually (1–)6–9, staminodes rarely present; filaments distinct, or connate basally and sometimes forming staminal tube, free or adnate to perianth tube; pistils (2–)3(–4)-carpellate; ovary 1-locular (sometimes with vestigial partitions proximally); ovule 1, orthotropous or, rarely, anatropous, placentation basal or free-central; styles 1–3, erect to spreading or recurved, distinct or connate proximally; stigmas peltate, capitate, fimbriate, or penicillate. **Achenes** yellowish, brown, red, or black, homocarpic (sometimes heterocarpic in *Polygonum*), winged or unwinged, usually 2–3-gonous, sometimes discoid, biconvex, or spheroidal, rarely 4-gonous. **Seeds:** embryo usually straight or curved, rarely folded.

Genera 28, species ca. 850 (16 genera, 160 species in the flora): mainly temperate regions of North America.

Morphological (K. Haraldson 1978; L.-P. Ronse Decraene and J. R. Akeroyd 1988; Ronse Decraene et al. 2000; Hong S. P. et al. 1998) and molecular (A. S. Lamb Frye and K. A. Kron 2003) data provide support for separation of *Persicaria* from *Polygonum*. Further studies are needed to elucidate the relationships of allied genera, particularly *Aconogonon, Bistorta,* and *Koenigia* with *Persicaria,* and *Fallopia* and *Polygonella* with *Polygonum*.

SELECTED REFERENCES Haraldson, K. 1978. Anatomy and taxonomy in Polygonaceae subfam. Polygonoideae Meisn. emend. Jaretzky. Acta Univ. Upsal., Symb. Bot. Upsal. 22: 1–93. Hong, S. P., L.-P. Ronse Decraene, and E. Smets. 1998. Systematic significance of tepal surface morphology in tribes Persicarieae and Polygoneae (Polygonaceae). Bot. J. Linn. Soc. 127: 91–116.

Ronse Decraene, L.-P. and J. R. Akeroyd. 1988. Generic limits in *Polygonum* and related genera (Polygonaceae) on the basis of floral characters. Bot. J. Linn. Soc. 98: 321–371. Ronse Decraene, L.-P., Hong S. P., and E. Smets. 2000. Systematic significance of fruit morphology and anatomy in tribes Persicarieae and Polygoneae (Polygonaceae). Bot. J. Linn. Soc. 134: 301–377.

1. Tendrils present; plants vines.
    2. Perianths pink to purple or, rarely, white or yellow, membranous; pedicels not 3-winged
    .... 20.   *Antigonon*, p. 481
    2. Perianths green to greenish yellow, indurate; pedicels 3-winged, 1 wing more prominent
    and becoming greatly expanded in fruit .... 21.   *Brunnichia*, p. 482
1. Tendrils absent; plants trees, shrubs, or herbs, rarely vinelike shrubs.
    3. Plants trees or shrubs; tubes of pistillate flowers becoming fleshy in fruit .... 22.   *Coccoloba*, p. 483
    3. Plants herbs, subshrubs, or, rarely, vinelike shrubs; tubes of pistillate flowers rarely
    becoming fleshy in fruit.
        4. Tepals 6.
            5. Flowers unisexual; outer 3 tepals of pistillate flowers each with apex ending in
            stout spine .... 24.   *Emex*, p. 487
            5. Flowers bisexual or, rarely, unisexual; outer 3 tepals each without apex ending
            in stout spine.
                6. Achenes winged; inner tepals of fruiting perianths nonaccrescent; stamens
                (6–)9 .... 25.   *Rheum*, p. 488
                6. Achenes unwinged; inner tepals of fruiting perianths usually accrescent;
                stamens 6... 26.   *Rumex*, p. 489
        4. Tepals 3, 4, or 5.
            7. Herbs annual; tepals 3 [4]; stamens (1–)3[–5].................... 35.   *Koenigia*, p. 600
            7. Herbs perennial or annual, or shrubs; tepals 4–5; stamens 3–8.
                8. Tepals 4; achenes lenticular, winged; leaves mostly basal .... 27.   *Oxyria*, p. 533
                8. Tepals 4 or 5; achenes 3-gonous, discoid, biconvex, spheroidal, or 4-gonous,
                unwinged or essentially so; leaves cauline or basal and cauline, rarely mostly
                basal.
                    9. Branches adnate to stems, appearing to arise internodally ..... 28.   *Polygonella*, p. 534
                    9. Branches not adnate to stems, not appearing to arise internodally.
                        10. Plants shrubs, vinelike; flowers unisexual, tubes of pistillate flowers
                        becoming fleshy in fruit.......................... 23.   *Muehlenbeckia*, p. 485
                        10. Plants herbs or, if shrubs, not vinelike; flowers bisexual or, rarely,
                        unisexual, if unisexual then tubes of pistillate flowers not becoming
                        fleshy.
                            11. Outer tepals winged or keeled.
                                12. Outer tepals winged (keeled in *F. ciliondis* and, usually, *F.
                                convolvulus*); ocreae chartaceous, tan to brownish,
                                glabrous or scabrous to variously pubescent, never 2-lobed
                                distally .... 29.   *Fallopia*, p. 541
                                12. Outer tepals keeled; ocreae often hyaline, silvery, glabrous,
                                2-lobed distally .... 30.   *Polygonum* (in part), p. 547
                            11. Outer tepals unwinged and unkeeled.
                                13. Leaves mostly basal, some cauline; inflorescences terminal,
                                spikelike; stems simple .... 33.   *Bistorta*, pp. 594
                                13. Leaves cauline; inflorescences terminal and axillary or
                                axillary; stems usually branched, rarely simple.
                                    14. Achenes strongly exserted; perianths nonaccrescent;
                                    tepals distinct......................... 31.   *Fagopyrum*, p. 572
                                    14. Achenes included or exserted; perianths accrescent or
                                    nonaccrescent; tepals connate to ²⁄₃ their lengths.

[15. Shifted to left margin.—Ed.]

20. ANTIGONON Endlicher, Gen. Pl. 4: 310. 1837 • Coralvine, Mexican creeper, corona de la reina, Confederate-vine [etymology uncertain; perhaps Greek *anti-*, against, and *gony*, knee, alluding to angled stems, or Greek *anti-*, in place of, and genus *Polygonum*, alluding to affinity] ⊡

Craig C. Freeman

Vines, perennial [annual]; roots tuberous. **Stems** scandent, tendril-bearing, pubescent or glabrous; tendrils terminal and axillary, branched. **Leaves** deciduous or persistent, cauline, alternate, petiolate; ocrea usually deciduous, chartaceous; blade broadly ovate to deltate or truncate, margins entire, sometimes undulate. **Inflorescences** terminal and axillary, often clustered near tips of stems, racemelike, pedunculate. **Pedicels** present. **Flowers** bisexual, (1–)2–5 per ocreate fascicle, base stipelike; perianth accrescent in fruit, pink to purple or, rarely, white or yellowish, campanulate, glabrous; tepals 5, connate proximally, petaloid, dimorphic, outer 3 broader than inner 2; stamens (7–)8(–9); filaments connate ca. ¹/₂ their length, forming staminal tube, adnate to perianth tube, glandular-pubescent; anthers yellow to reddish, ovate to elliptic; styles 3, recurved, distinct; stigmas reniform-capitate. **Achenes** included in membranous perianth, brown, unwinged, subglobose to bluntly 3-gonous proximally, 3-gonous distally, glabrous or pubescent. **Seeds:** embryo straight. $x = 20$.

Species ca. 6 (1 in the flora): introduced; subtropical and tropical, s North America (including Mexico), West Indies, Central America, South America.

1. **Antigonon leptopus** Hooker & Arnott, Bot. Beechy Voy., 308, plate 69. 1838 • Mountain-rose coralvine, queen's-jewels F I

Plants herbaceous or base sometimes woody. **Stems** climbing or sprawling by tendrils, branched, angular, to 15 m, sparsely to densely brownish- or reddish-pubescent or glabrous. **Leaves:** ocrea 0.2–2 mm; petiole often winged distally, 1–2.5(–5) cm, glabrate or pubescent; blade 5–14 × (2–)4–10 cm, base usually cordate, margins ciliate, apex acute to acuminate, glabrous or pubescent, especially on veins. **Inflorescences** 4–20 cm, axes puberulent to pilose; peduncle angular, 1–5 cm, puberulent to pilose. **Pedicels** articulated proximally, 3–5(–10) mm, glabrous or pubescent. **Flowers:** tepals ovate to elliptic, 4–8 × 2–6 mm, 8–20 × 4–15 mm in fruit, margins entire, apex acute. **Achenes** 8–12 × 4–7 mm, shiny. $2n = 14, 40, 42–44, 48$.

Flowering year-round. Cultivated and often persisting after abandonment, rarely escaping; 0–600 m; introduced; Ala., Fla., Ga., La., Miss., S.C., Tex.; Mexico; Central America; introduced in West Indies, Asia, Africa.

Antigonon leptopus is cultivated widely as an ornamental in warmer parts of the world and is grown extensively in South America. In the flora region, it appears to have naturalized only in Florida and southern Texas; records from elsewhere probably represent plants that have persisted from cultivation. It propagates easily by cuttings and seeds, and the tubers are edible.

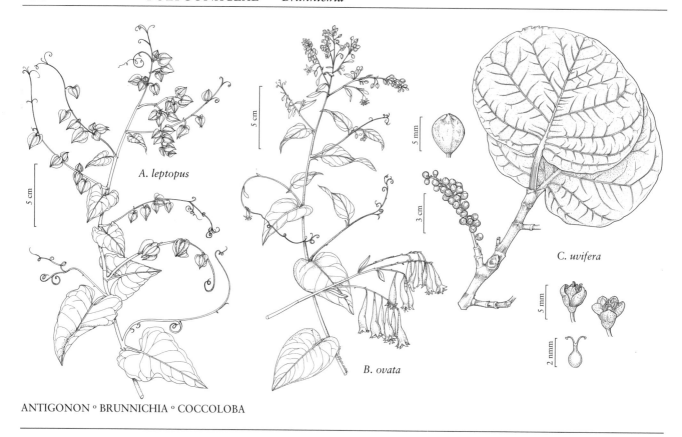

A. leptopus

B. ovata

C. uvifera

ANTIGONON ∘ BRUNNICHIA ∘ COCCOLOBA

21. BRUNNICHIA Banks ex Gaertner, Fruct. Sem. Pl. 1: 213, plate 45, fig. 2. 1788

• American buckwheat-vine, ladies'-eardrops, eardrop vine, anserine liana [for Morten Thrane Brunnich, 1737–1827, eighteenth-century Danish naturalist]

Walter C. Holmes

Vines, perennial; roots thin, fibrous. **Stems** scandent, tendril-bearing, glabrous or pubescent; tendrils terminating branches, simple to branched. **Leaves** cauline, alternate, petiolate; ocrea persistent, chartaceous; blade ovate to ovate-lanceolate, margins entire. **Inflorescences** terminal and axillary, paniclelike, usually pedunculate. **Pedicels** 3-winged, 1 wing more prominent and becoming greatly expanded in fruit, glabrous. **Flowers** bisexual, (1–)3–5 per ocreate fascicle, base stipelike; perianth accrescent, greenish to greenish yellow, campanulate, indurate, glabrous; tepals 5, connate proximally, petaloid, slightly dimorphic, subequal; stamens 8; filaments distinct, adnate proximally to perianth tube, glabrous; anthers pink to red, oval; styles 3, erect or spreading, distinct; stigmas 3, depressed-capitate. **Achenes** included in indurate perianth, tan to brown, unwinged, obscurely 3-gonous proximally, 3-gonous distally, glabrous. **Seeds:** embryo straight. $x = 24$.

Species 4 (1 in the flora): se United States, w Africa.

The African species of *Brunnichia* sometimes are segregated in the genus *Afrobrunnichia* Hutchinson & Dalziel.

1. Brunnichia ovata (Walter) Shinners, Sida 3: 115.
1967   E F W

Rajania ovata Walter, Fl. Carol., 247. 1788; *Brunnichia cirrhosa* Gaertner

Stems running on/in soil or water, rooting at nodes, younger ones climbing by tendrils, branched, terete, to 13 m, larger ones to 2 cm diam., soft-woody. **Leaves:** ocrea obscure, (0.1–)0.5–1 mm, usually fringed with reddish brown hairs; petiole dilated at base, 7–25 mm, glabrous or pubescent; blade 3–15 × 1.5–8 cm, base truncate to subcordate to broadly cuneate, apex acute to acuminate, glabrous or slightly pubescent abaxially, glabrous adaxially. **Inflorescences** 5–26 cm; peduncle (0.1–)5–35 mm. **Pedicels** jointed proximal to middle, 0.5–1 cm. **Flowers:** tepals linear-oblong to oblong, margins entire, apex acute; stamens prominently exserted. **Achenes** 8–10 × 3–5 mm, subglossy. $2n = 48$.

Flowering Jun–Jul. Riverbanks, margins of lakes, edges of wet woods and thickets; 0–200 m; Ala., Ark., Fla., Ga., Ill., Ky., La., Miss., Mo., Okla., S.C., Tenn., Tex., Va.

There has been some question regarding the valid name of this species, whether *Brunnichia ovata* or *B. cirrhosa*. The problem stems from the almost simultaneous publication of *Rajania ovata* Swartz (Dioscoreaceae) and *R. ovata* (Polygonaceae), the basionym of the now-accepted name of the only North American species of *Brunnichia*. The best evidence indicates that the Walter name was published between April and June 1788, while W. T. Stern (1980) put the date of publication of the Swartz name between 20 June and 31 July 1788, but suggested that July 1788 be accepted as the month of publication. Under that interpretation, the Swartz name is a later homonym, and the epithet of the Walter name is the earliest available for this species.

The most often used English common name for *Brunnichia ovata* alludes to the fruit's resemblance to an ear pendant. Fruits often persist on these interesting ornamentals until the following spring.

22. COCCOLOBA Browne, Civ. Nat. Hist. Jamaica, 209. 1756, name conserved (as Coccolobis) • Sea-grape [Greek, *coccos*, seed or berry, and *lobos*, capsule or pod, alluding to fleshy hypanthium surrounding fruit]

Craig C. Freeman

Trees or shrubs, evergreen; roots woody. **Stems** erect or spreading, glabrous or pubescent distally. **Leaves** persistent, cauline, alternate, petiolate; ocrea often deciduous, membranous to coriaceous; blade lanceolate to round or transversely elliptic, margins entire. **Inflorescences** terminal, racemelike, pedunculate. **Pedicels** present. **Flowers** functionally unisexual, some plants having only staminate flowers, others with only pistillate flowers, base stipelike; perianth white or greenish white, campanulate, glabrous; tepals 5, connate proximally, sepaloid, monomorphic. **Staminate flowers** 1–7 per ocreate fascicle, perianth nonaccrescent; stamens 8; filaments connate at base, adnate to perianth, glabrous; anthers white or bluish white, elliptic to round; pistil rudimentary. **Pistillate flowers** 1 per ocreate fascicle, perianth accrescent and fleshy in fruit; stamens rudimentary; styles 3, erect, distinct; stigmas capitate. **Achenes** usually included in fleshy perianth tube, brown to black, unwinged, bluntly 3-gonous, glabrous. **Seeds:** embryo straight. $x = 11$.

Species ca. 120 (2 in the flora): tropical, s North America (including Mexico), West Indies, Central America, South America.

The hypanthium usually completely invests the achene in both species of *Coccoloba* in the flora, becoming juicy and somewhat astringent at maturity. The fruits of *C. uvifera* are edible raw or are used to make jelly or wine (E. L. Little Jr. et al. 1969). Both species also enjoy some popularity in landscaping due to their attractive fruiting racemes and evergreen foliage, which on the two species in the flora is bronze colored when young (R. A. Howard 1958).

1. Leaf blades lanceolate, ovate, obovate, or elliptic, length usually 2–3 times width. . . 1. *Coccoloba diversifolia*
1. Leaf blades round to transversely elliptic, length equaling or less than width. . . . . 2. *Coccoloba uvifera*

1. Coccoloba diversifolia Jacquin, Enum. Syst. Pl., 19. 1760 • Pigeon-plum, dove-plum, uvilla, tie-tongue

Plants with branches spreading, to 10(–18) m. **Stems:** bark light gray, peeling off in short flakes, inner bark light brown; twigs green or grayish green when young, gray or whitish gray at maturity, glabrous or nearly so. **Leaves:** those of adventitious or juvenile shoots often much larger and of different shape from those of normal shoots; ocrea persistent proximally, deciduous distally, tan or brown, cylindric, 3–5 mm, coriaceous proximally, membranous distally, margins oblique, glabrous or puberulent; petiole 5–15 mm, glabrous or puberulent; blade pale green abaxially, green to dark green adaxially, lanceolate, ovate, obovate, or elliptic, (3–)5–10(–13) × (1–)3–5(–7) cm, length usually 2–3 times width, coriaceous, base acute to obtuse, margins often revolute, apex acuminate to obtuse or blunt, abaxial surface dull, adaxial surface shiny, minutely punctate, glabrous. **Inflorescences** (1.5–)3–10(–18) cm, glabrous, pistillate spreading or pendent in fruit; peduncle 1–6 cm, glabrous. **Pedicels** 1–3 mm, glabrous. **Flowers:** tepals round to broadly elliptic, margins entire, apex obtuse. **Staminate flowers** 1–3 per ocreate fascicle. **Pistillate flowers:** tube spherical to obpyriform, 9–14 × 6–10 mm, becoming fleshy. **Achenes** 6–10 × 6–9 mm, shiny. $2n = 22$ (West Indies).

Flowering year-round. Sandy coastal hummocks, limestone forests; 0–10 m; Fla.; s Mexico; West Indies; Central America (Belize, Guatemala).

The wood of *Coccoloba diversifolia* has a specific gravity of 0.8 and is strong and brittle (E. L. Little Jr. et al. 1969).

2. Coccoloba uvifera (Linnaeus) Linnaeus, Syst. Nat. ed. 10, 2: 1007. 1759 • Shore-grape, uva de playa

F

Polygonum uvifera Linnaeus, Sp. Pl. 1: 365. 1753

Plants with branches spreading or sprawling, 2–7(–15) m. **Stems:** bark gray, peeling off in small white, gray, or brown flakes, inner bark light brown; twigs green and puberulent when young, gray at maturity, glabrous or pubescent. **Leaves:** those of adventitious or juvenile shoots often much larger and of different shape from those of normal shoots; ocrea persistent proximally, deciduous distally, brown or reddish brown, cylindric to funnelform, 3–8 mm, coriaceous proximally, membranous distally, margins oblique, glabrous or densely puberulent; petiole 5–15 mm, puberulent to pilose; blade pale green abaxially, green to bluish green adaxially, round to transversely elliptic, (6–)10–20(–27) × 6–20(–27) cm, length equaling or less than width, coriaceous, base cordate, margins sometimes revolute, apex rounded to blunt or emarginate, abaxial surface dull, adaxial surface shiny or dull, minutely punctate, glabrous. **Inflorescences** 10–30 cm, puberulent or glabrous, pistillate pendent in fruit; peduncle 1–5 cm, glabrous. **Pedicels** 1–4 mm, glabrous. **Flowers:** tepals round to broadly elliptic, margins entire, apex obtuse. **Staminate flowers** 1–7 per ocreate fascicle. **Pistillate flowers:** tube obpyriform, 12–20 × 8–12 mm, becoming fleshy. **Achenes** 8–11 × 8–10 mm, shiny. $2n = 132$.

Flowering year-round. Sandy or rocky coastal hummocks, sand dunes. 0–10 m; Fla.; Mexico; West Indies; Central America; South America.

Coccoloba uvifera is an early colonizer of exposed, sandy shorelines. The wood has a specific gravity of 0.7, and a red sap obtained by cutting the bark has been used in commerce for tanning and dyeing (E. L. Little Jr. et al. 1969).

Pistillate inflorescences of some specimens of *Coccoloba uvifera* appear to bear clusters of up to five flowers at each node; all but one abort, leaving a single flower that produces a fruit. The pedicels of the abortive flowers usually are more slender than those of the fertile flowers.

23. MUEHLENBECKIA Meisner, Pl. Vasc. Gen. 1: 316; 2: 227. 1841, name conserved

• Wirevine [for H. G. Muehlenbeck, 1798–1845, Swiss physician] [1]

Craig C. Freeman

Shrubs, vinelike, perennial; rhizomatous. **Stems** suberect, prostrate, or scandent, glabrous or puberulent, sometimes papillose. **Leaves** deciduous, cauline, alternate, usually petiolate; ocrea usually deciduous, sometimes persistent, chartaceous; petiole base articulated, extrafloral nectaries present; blade linear to orbiculate, panduriform, or triangular-lanceolate, margins entire or irregularly wavy. **Inflorescences** terminal and axillary, spikelike, essentially not pedunculate. **Pedicels** present. **Flowers** bisexual and unisexual, with staminate, pistillate, or sometimes both sexes occurring with bisexual flowers on the same plant, 1–2(–5) per ocreate fascicle, base stipelike; perianth accrescent, white to greenish white, campanulate, glabrous; tepals 5, connate proximally, sepaloid, dimorphic, outer slightly larger than inner. **Staminate flowers:** stamens 8 (9); filaments distinct, adnate to base of perianth tube, glabrous; anthers yellow or pink to purple, ovate to elliptic; pistil rudimentary. **Pistillate flowers:** tube white or reddish purple to black in fruit, becoming fleshy; stamens rudimentary; styles 3, spreading, connate proximally; stigmas fimbriate. **Achenes** completely or partly included in fleshy perianth, black or dark brown, unwinged, 3-gonous to subglobose, glabrous. **Seeds:** embryo straight. $x = 10$.

Species 23 (2 in the flora); introduced; Central America, South America, Pacific Islands (New Zealand), Australia.

SELECTED REFERENCE Brandbyge, J. 1992. The genus *Muehlenbeckia* (Polygonaceae) in South and Central America. Bot. Jahrb. Syst. 114: 349–416.

1. Tubes of pistillate flowers white in fruit; distal branches of stems brownish-puberulent;
 leaf blades ovate-oblong to suborbiculate or panduriform . . . . 1. *Muehlenbeckia complexa*
1. Tubes of pistillate flowers reddish purple to black in fruit; distal branches of stems glabrous;
 leaf blades triangular-lanceolate . . . . 2. *Muehlenbeckia hastulata*

1. Muehlenbeckia complexa (A. Cunningham) Meisner, Pl. Vasc. Gen. 2: 227. 1841 • Maidenhair vine, lacy wirevine [F] [I]

Polygonum complexum A. Cunningham, Ann. Nat. Hist. 1: 455. 1838

Plants (0.5–)1–5 m. **Stems** prostrate to scandent, especially distally, sometimes rooting at nodes, striate, branched, glabrescent or puberulent, distal branches brownish-puberulent. **Leaves:** ocrea deciduous, brownish hyaline, cylindric, 2–3 mm, margins truncate, eciliate, faces puberulent along veins; petiole 3–10 mm, brownish-puberulent; blade ovate-oblong to suborbiculate or panduriform, 0.5–2.5 (–4) × 0.5–2.5(–4) cm, coriaceous, base truncate, margins entire, glabrous or scabrous, apex rounded to apiculate, glabrous adaxially and abaxially, or puberulent abaxially along midvein, obscurely punctuate abaxially. **Inflorescences** terminal and axillary, 5–30 mm. **Pedicels** ascending to spreading, 1.5–2 mm. **Flowers** 1–2(–5) per ocreate fascicle; perianth yellowish green or greenish; tepals connate ca. 1/4 their length, lanceolate-ovate to obovate, 2–4 mm, apex rounded to acute. **Staminate flowers:** anthers pink to purple, ovate to elliptic. **Pistillate flowers:** tube white in fruit. **Achenes** exserted or included, black or dark brown, 3-gonous, 3–4 × 2–3 mm, shiny, smooth. $2n = 20$ (New Zealand).

Flowering Jul–Sep. Sunny, disturbed sites, often in urban areas; 0–500 m; introduced; Calif.; Pacific Islands (New Zealand).

Muehlenbeckia complexa is cultivated as an ornamental and escapes rarely in the flora area.

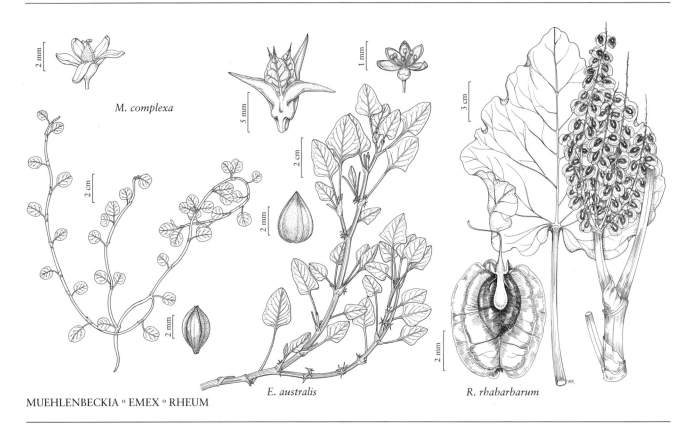

M. complexa

E. australis

R. rhabarbarum

MUEHLENBECKIA ∘ EMEX ∘ RHEUM

2. Muehlenbeckia hastulata (Smith) I. M. Johnston, Contr. Gray Herb. 81: 88. 1928 • Wirevine ⊡

Rumex hastulatus Smith in A. Rees, Cycl. 30: Rumex no. 29. 1815

Varieties 3 (1 in the flora): introduced; South America.

J. Brandbyge (1992) recognized two other varieties of *Muehlenbeckia hastulata* distinguished by leaf shape, inflorescence size, and achene morphology. Variety *rotundata* (Philippi) Brandbyge is endemic in Argentina and Chile; var. *fascicularis* (Meisner) Brandbyge is endemic in Chile.

2a. Muehlenbeckia hastulata (Smith) I. M. Johnston var. **hastulata** ⊡

Plants (0.5–)1–3 cm. **Stems** suberect to scandent or climbing, angular, striate, diffusely branched, glabrous, sometimes papillose, distal branches usually glabrous. **Leaves:** ocrea mostly persistent, brownish hyaline, cylindric, 3–5 mm, margins truncate to rounded, eciliate, faces glabrous; petiole (3–)6–12(–17) mm, glabrous; blade triangular-lanceolate, (2–)2.5–4(–5.5) × (0.8–)1.2–2.5 (–3) cm, subcoriaceous, base hastate, margins entire or irregularly wavy, glabrous or scabrous, apex acute, glabrous adaxially and abaxially, sometimes papillose abaxially, minutely punctate abaxially and adaxially. **Inflorescences** terminal and axillary, 3–5(–8) cm. **Pedicels** ascending to spreading, 1.5–2 mm. **Flowers** 1(–3) per ocreate fascicle; perianth white, greenish white, or greenish; tepals connate ca. ¼ their length, lanceolate-ovate to obovate, 2–3 mm, apex rounded to acute. **Staminate flowers:** anthers yellow or pink, ovate. **Pistillate flowers:** tube reddish purple to black in fruit. **Achenes** usually included, black, subglobose, 3–4 × 2.5–3.5 mm, shiny, smooth.

Flowering Jul–Aug. Sunny, disturbed sites, often in urban areas; 0–500 m; introduced; Calif.; South America (Argentina, Chile).

Variety *hastulata* is cultivated as an ornamental. It escapes rarely in the flora area and can be invasive.

24. EMEX Necker ex Campderá, Monogr. Rumex, 56, plate 1, fig. 1. 1819, name conserved • [Latin, *ex*, and *Rumex*, alluding to segregation from that genus] ⓘ ⓦ

Craig C. Freeman

Herbs, annual; taprooted. **Stems** prostrate to erect, glabrous. **Leaves** deciduous, cauline, alternate, petiolate; ocrea often deciduous, chartaceous; blade subhastate, triangular, or ovate to ovate-oblong, margins entire to obscurely crenulate or dentate, sometimes undulate. **Inflorescences** terminal and axillary, racemelike, pedunculate. **Pedicels** present on staminate flowers, absent or nearly so on pistillate flowers. **Flowers** unisexual, all plants having both staminate and pistillate flowers, the two types segregated in different inflorescences, base not stipelike; perianth accrescent (pistillate) or nonaccrescent (staminate), greenish, campanulate, glabrous; tepals (5–)6, sepaloid, dimorphic, especially in pistillate flowers. **Staminate flowers** 1–8 per ocreate fascicle, terminal and axillary; tepals distinct, subequal; stamens 4–6; filaments distinct, free, glabrous; anthers yellowish to reddish, elliptic to ovate. **Pistillate flowers** 1–7 per ocreate fascicle, axillary; tepals connate basally, outer 3 each with apex ending in a stout spine in fruit, inner 3 erect, tuberculate; styles 3, erect, distinct; stigmas penicillate. **Achenes** included in indurate perianth, brown, not winged, 3-gonous, glabrous. **Seeds:** embryo curved. *x* = 10.

Species 2 (2 in the flora): introduced; tropical to subtropical or temperate regions, South America, Eurasia, Africa, Atlantic Islands, Pacific Islands.

Both species of *Emex* are on the United States federal noxious weed list. *Emex australis* appears to be a more serious threat to agriculture and livestock production than does *E. spinosa*, based on experience in other parts of the world. The former is a serious agricultural pest in southern Australia, where it was introduced in the 1830s. *Emex spinosa* also was introduced there, apparently in the mid-1900s. The two species were formerly geographically isolated, but hybrids between them have recently been reported in Australia (E. Putievsky et al. 1980). Hybrids exhibit irregular meiosis and high sterility when self-pollinated; backcrosses with either parent often yield viable seeds.

1. Fruiting perianths 7–9 × 9–10 mm, spines 5–10 mm; inner tepals of pistillate flowers broadly triangular-ovate, apex mucronate . . . . 1. *Emex australis*
1. Fruiting perianths 5–6 × 5–6 mm, spines 2–4 mm; inner tepals of pistillate flowers linear-lanceolate, apex acute . . . . 2. *Emex spinosa*

1. Emex australis Steinheil, Ann. Sci. Nat., Bot., sér. 2, 9: 195, plate 7. 1838 • Doublegee, southern threecornerjack, spiny emex Ⓕ ⓘ ⓦ

Plants 1–4(–6) dm. **Stems** prostrate, decumbent, or ascending, base often reddish, branched proximally. **Leaves:** ocrea loose, glabrous; petiole (0.5–)1–8(–15) cm, glabrous; blade subhastate to elliptic or ovate, 1–10 × 0.5–6 cm, base truncate to cuneate, apex obtuse to acute. **Staminate flowers** 1–8 per ocreate fascicle; tepals narrowly oblong to oblanceolate, 1.5–2 mm. **Pistillate flowers** 1–4 per ocreate fascicle; outer tepals ovate to oblong, 4–6 mm in fruit, inner tepals broadly triangular-ovate, 5–6 mm in fruit, apex mucronate. **Fruiting perianths** 7–9 × 9–10 mm, spines ascending or spreading, 5–10 mm, base tapering. **Achenes** 4–6 × 2–3 mm, shiny. **2*n*** = 20.

Flowering year-round. Disturbed sites, especially in sandy soils; 0–200 m; introduced; Calif.; Africa (Republic of South Africa); introduced in West Indies (Trinidad), Europe, Asia (India, Pakistan, Taiwan), Africa (Kenya, Madagascar, Malawi, Tanzania, Zimbabwe), Pacific Islands (Hawaii), Australia.

2. Emex spinosa (Linnaeus) Campderá, Monogr. Rumex, 58. 1819 • Devil's-thorn, little jack, spiny threecornerjack, spiny emex ⊡ ⊡

Rumex spinosus Linnaeus, Sp. Pl. 1: 337. 1753

Plants 3–6(–8) dm. **Stems** ascending to erect, base often reddish, branched proximally. **Leaves:** ocrea loose, glabrous; petiole 2–29 cm, glabrous; blade ovate to ovate-oblong or triangular, 3–13 × 1.1–12 cm, base mostly truncate to subcordate, apex obtuse to acute. **Staminate flowers** 1–8 per ocreate fascicle; tepals narrowly oblong to oblanceolate, 1.5–2 mm. **Pistillate flowers** 2–7 per ocreate fascicle; outer tepals ovate to oblong, 4–6 mm in fruit, inner tepals linear-lanceolate, 5–6 mm in fruit, apex acute. **Fruiting perianths** 5–6 × 5–6 mm, spines spreading to reflexed, 2–4 mm, base broad. **Achenes** 4–5 × 2–3 mm, shiny. $2n = 20$.

Flowering year-round. Sandy shores, disturbed ground; 0–500 m; introduced; Calif., Fla., Mass., N.J., Tex.; South America (Argentina, Ecuador, Peru); Eurasia (Mediterranean region); n Africa (Mediterranean region); introduced in Eurasia (Cyprus, India, Iran, Iraq, Israel, Lebanon, Pakistan, Syria, Turkey), Africa (Algeria, Kenya, Kuwait, Mauritius, Morocco), Atlantic Islands (Canary Islands), Pacific Islands (Hawaii), Australia.

25. RHEUM Linnaeus, Sp. Pl. 1: 371. 1753: Gen. Pl. ed. 5, 174. 1754 • Rhubarb [Greek *rheon*, a name used by Dioscorides, probably for a plant in this genus] ⊡

Craig C. Freeman

Herbs, perennial; roots fleshy. **Stems** erect, glabrous or pubescent. **Leaves** deciduous, mostly basal, alternate, petiolate; ocrea persistent or deciduous, chartaceous; blade cordate-ovate to orbiculate or reniform, margins entire, undulate. **Inflorescences** terminal, paniclelike, pedunculate. **Pedicels** present. **Flowers** bisexual, 1–10 per ocreate fascicle, base stipelike; perianth nonaccrescent in fruit, whitish green or pinkish green, campanulate, glabrous; tepals 6, distinct, sepaloid, dimorphic, outer 3 narrower than inner 3; stamens (6–)9; filaments distinct, free, glabrous; anthers yellow or pinkish, elliptic; styles 3, erect or deflexed, distinct; stigmas capitate. **Achenes** exserted, dark brown, winged, 3-gonous, glabrous. **Seeds:** embryo straight or curved. $x = 11$.

Species ca. 60 (1 in the flora): introduced; temperate regions, Eurasia.

SELECTED REFERENCE Chin, T. C. and H. W. Youngken. 1947. The cytotaxonomy of *Rheum*. Amer. J. Bot. 34: 401–407.

1. Rheum rhabarbarum Linnaeus, Sp. Pl. 1: 372. 1753 • Garden rhubarb, pie-plant, wine-plant, rhubarbe ⊡ ⊡

Plants 3–20(–30) dm. **Stems** striate, often striped or suffused with red or pink, branched distally, hollow. **Leaves:** ocrea brown, loosely funnelform, 2–4 cm, margins strongly oblique, glabrous or puberulent with flattened, whitish hairs along veins; petioles of basal leaves pinkish green or reddish green, ca. equaling or longer than blade, thick, fleshy, those of cauline leaves absent distally; basal leaf blades palmately veined with 5–7 basal veins, 30–45(–60) × 10–30 cm, veins pubescent primarily along veins with flattened, whitish hairs. **Inflorescences** 250–500-flowered, 15–40 cm; peduncle 1–8 cm, glabrous. **Pedicels** articulated at or proximal to middle, 2–5 mm, glabrous. **Flowers:** tepals oblong-ovate, 2.3–4 × 1–2.5 mm, margins hyaline, apex obtuse. **Achenes** 6–10(–12) × 6–11 mm including wings, 4–8 times longer than perianth; wings tan, veined, 3–4 mm wide, membranous. $2n = 44$.

Flowering Jul–Aug, fruiting Aug–Sep. Cultivated and often persisting long after abandonment, escaping infrequently; 0–4000 m; introduced; Alta., Man., N.B., N.S., Ont., Que., Sask., Yukon; Alaska, Colo., Conn., Ill., Ind., Iowa, Maine, Mass., Mich., Minn., Miss., N.H., N.Y., N.C., Ohio, Pa., R.I., Utah, Vt., Va., Wis., Wyo.; Eurasia.

The name *Rheum rhaponticum* Linnaeus appears to have been misapplied widely to *R. rhabarbarum* in North America. *Rheum rhaponticum*, European rhubarb, is the only member of the genus confined to Europe. Rare in the wild but widely cultivated, it is a diploid (2*n* = 22); *R. rhabarbarum* is a tetraploid (B. Libert and R. Englund 1989). A chromosome count of 2*n* = 44 reported for *R. rhaponticum* from Wisconsin (N. A. Harriman 1981b) probably is from *R. rhabarbarum*.

Many rheums have culinary and medicinal uses, some of which originated in Asia more than 2 millennia ago. *Rheum rhabarbarum* is used alone and in combination with various other fruits to make pies, jellies, jams, and wine. All parts of the plant contain oxalic acid, which has been implicated in cases of poisoning. However, other potentially poisonous compounds also are produced, including citric acid and anthraquinone glycosides (W. H. Blackwell 1990). Raw or cooked leaf blades are poisonous to humans and livestock if ingested in sufficiently large quantities. The petioles typically are used as food and contain mostly malic acid, which is nontoxic. Plants traditionally are propagated and moved by taking cuttings from larger plants.

26. RUMEX Linnaeus, Sp. Pl. 1: 333. 1753; Gen. Pl. ed. 5, 156. 1754 • Dock, sorrel [classical Latin name for sorrel, probably derived from *rumo*, to suck, alluding to the practice among Romans of sucking the leaves to allay thirst]

Sergei L. Mosyakin

Herbs, perennial, biennial, or annual, synoecious (subg. *Rumex* and *Platypodium*) or dioecious (subg. *Acetosa* and *Acetosella*), occasionally polygamomonoecious, with taproots and usually short caudex, or sometimes rhizomatous and/or stoloniferous. **Stems** erect, ascending, or prostrate, glabrous or papillose-pubescent. **Leaves** basal (in some species) and cauline, alternate, petiolate; ocrea persistent or partially deciduous, membranous; petioles present on basal and proximal cauline leaves, absent on distal cauline leaves, bases not articulated; blades variable in shape, basal (if present) and proximal cauline leaves from broadly ovate or almost orbiculate to linear, becoming progressively smaller and narrower distally, margins entire (or basally lobate), flat, or occasionally undulate or crisped. **Inflorescences** terminal, sometimes terminal and axillary, paniclelike, rarely simple. **Pedicels** present. **Flowers** bisexual or unisexual, (1–)4–30 per ocreate fascicle, base stipelike; perianth green, pinkish, or red, campanulate, glabrous; tepals (5–)6, connate proximally, sepaloid, dimorphic, outer 3 remaining small, inner 3 usually enlarging, sometimes 1–3 with central vein transformed into tuberculate callosity (tubercle); stamens 6; filaments distinct, free, glabrous; anthers, yellow to brownish yellow, ovate to elongate; styles 3, spreading or reflexed, distinct; stigmas 3, fimbriate or plumose. **Achenes** included in accrescent and usually veiny perianth, tan to dark brown, unwinged to weakly winged, 3-gonous, sometimes compressed-3-gonous or nearly pyramidal, glabrous. **Seeds:** embryo straight. *x* = 7, 8, 9, 10 (polyploidy widespread in the genus).

Species 190–200 (63 in the flora): almost worldwide, but mostly in temperate regions of both hemispheres; some taxa occur in many regions of the world as naturalized or casual aliens.

Carefully collected mature specimens with well-developed inner tepals are desirable for reliable identification of *Rumex* species. Vegetative characters (in particular, growth habit, basal and proximal cauline leaves, and inflorescences) also are crucial.

In *Rumex* the distal part of a functional pedicel (below the articulation with a true pedicel) is formed by the narrowed connate basal parts of the outer tepals (also known as a stipelike hypanthium base or pseudopedicel). However, for simplification of the keys and descriptions, the whole functional pedicel (including pseudopedicel) is referred to simply as pedicel.

The genus *Rumex* in the broad sense may be divided into at least four segregate genera: *Rumex* in the narrow sense, *Acetosa*, *Acetosella*, and *Bucephalophora* (see e.g., Á. Löve 1983;

Löve and B. M. Kapoor 1967; N. N. Tzvelev 1987b, 1989b). These taxa probably represent distinct phylogenetic lineages; they have, however, not been generally accepted as separate genera by most taxonomists, including K. H. Rechinger (1937, 1949), monographer of *Rumex* in the broad sense. Moreover, in some cases they are connected by intermediate forms (especially *Acetosa* and *Acetosella*) and evidently are more closely related to each other than to any outgroup genus. In my opinion, this favors the retention of *Rumex* in the traditional broad sense, which is also nomenclaturally convenient.

Many Old World species of *Rumex* may be divided in their native areas of distribution into quite distinct subspecies or varieties. However, the same species occurring in North America as introduced aliens often are represented by atypical, intermediate specimens or even populations (as is true also for many native North American taxa occurring as aliens in Europe), which in many cases obscures those taxonomic distinctions.

Rumex rugosus Campderá, a commonly cultivated European species, was reported for North America by Á. Löve and D. Löve (1957) as a cultivated and occasionally escaped garden plant "in a few places in eastern Canada." No specimens from escaped plants in the flora area have been seen.

SELECTED REFERENCES Löve, Á. and B. M. Kapoor. 1968. A chromosome atlas of the collective genus *Rumex*. Cytologia 32: 328–342. Rechinger, K. H. 1937. The North American species of *Rumex*. (Vorarbeiten zu einer Monographie der Gattung *Rumex* 5.) Publ. Field Mus. Nat. Hist., Bot. Ser. 17: 1–150. Rechinger, K. H. 1949. *Rumices* Asiatici. (Vorarbeiten zu einer Monographie der Gattung *Rumex* 7.) Candollea 12: 9–152. Tolmatchew, A. I. 1966. *Rumex*. In: A. I. Tolmatchew, ed. 1960–1987. Flora Arctica URSS. 10 vols. Moscow and Leningrad. Vol. 5, pp. 143–161. Trelease, W. 1892. A revision of the American species of *Rumex* occurring north of Mexico. Rep. (Annual) Missouri Bot. Gard. 3: 74–98. Tzvelev, N. N. 1989b. *Rumex, Acetosella, Acetosa*. In: S. S. Kharkevich, ed. 1985+. Plantae Vasculares Orientis Extremi Sovetici. 7+ vols. Leningrad. Vol. 4, pp. 29–53.

1. Plants dioecious (rarely polygamomonoecious); flowers mostly unisexual; leaf blades in most species hastate or sagittate, with usually acute basal lobes (sometimes leaves not lobed, cuneate or narrowly cuneate at base, then pedicel articulated near base of tepals); pedicels in most cases with evident articulation.
    2. Pedicels articulated near base of tepals; outer tepals normally angled towards inner tepals; inner tepals not enlarged or slightly enlarged, normally 1.2–2.5(–3) mm, equaling to slightly wider than achenes; tubercles absent; leaf blades hastate at base, or in some species cuneate [25a. *Rumex* subg. *Acetosella*].
        3. At least basal leaf blades obovate-oblong, ovate-lanceolate, lanceolate-elliptic, or lanceolate (rarely linear-lanceolate), base hastate or at least broadly cuneate (almost truncate); inner tepals not enlarged, or rarely slightly enlarged at maturity, free wing absent to barely visible; widespread . . . . 1. *Rumex acetosella*
        3. Leaf blades narrowly linear, linear-oblanceolate, or distinctly spatulate, base usually not hastate (rarely some leaf blades with indistinct basal lobes), narrowly cuneate; inner tepals usually distinctly enlarged, free wing 0.2–0.6(–1) mm wide; arctic and subarctic.
            4. Plants with elongated underground stolons; inflorescences usually lax, with branches often reflexed; shoots not crowded, ± elongated, covered with whitish or silvery membranous ocreae at base . . . . 2. *Rumex graminifolius*
            4. Plants with thick, densely tufted underground stolons; inflorescences ± dense, with branches directed upward; shoots usually densely crowded, not elongated, covered with brownish or reddish brown membranous ocreae at base.
                5. Inflorescences interrupted at least at base, branched, occupying more than distal 1/2 of stem; inner tepals 1.6–2.3 × 1.8–2.5 mm (free wing 0.3–0.5 mm wide); achenes 1–1.5 mm . . . . 3. *Rumex beringensis*
                5. Inflorescences dense, simple or with few short branches, occupying distal 1/2 of stem; inner tepals 2.3–3 × 1.8–3 mm (free wing 0.4–0.8 mm wide); achenes 1.5–2 mm . . . . 4. *Rumex krausei*

2. Pedicels with articulation near middle, or in proximal part; outer tepals normally reflexed towards pedicel, or sometimes spreading; inner tepals distinctly enlarged, normally 2.5–5 mm (rarely more), always distinctly wider and longer than achene; tubercles small, recurved, developed only at base of inner tepals, occasionally absent; leaf blades in most species sagittate at base, sometimes hastate (normally in *R. hastatulus*, and occasionally in *R. thyrsiflorus*), cuneate (in *R. paucifolius*), or almost cordate (in *R. rugosus*) [25b. *Rumex* subgen. *Acetosa*].

  6. Leaf blades cuneate at base, usually not hastate or sagittate; Rocky Mountains region . . . . 5. *Rumex paucifolius*

  6. Leaf blades normally hastate or sagittate at base (rarely cuneate in underdeveloped plants); various regions (if in Rocky Mountains, then leaf blades sagittate at base).

    7. Leaf blades distinctly hastate at base, with spreading lobes (occasionally cuneate); inner tepals 2.7–3.2 mm wide; plants mostly annual or short-lived perennial . . . . 6. *Rumex hastatulus*

    7. Leaf blades sagittate at base, with lobes directed downward (towards petiole), sometimes also slightly incurved inward, or reflexed outward; inner tepals in (2.5–)3–4.5 mm wide; plants perennial.

      8. Rootstock thick, vertical or oblique (reaching deep into substrate), with remote 2d-order roots; 1st-order branches of inflorescence usually repeatedly branched, with numerous 2d-order branches; inflorescences broadly paniculate (in *R. lapponicus* occasionally simple), pyramidal, usually dense . . . . 9. *Rumex thyrsiflorus*

      8. Rootstock rather thin, horizontal or slightly oblique, short (not reaching deep into substrate), or plants rhizomatous; 1st-order branches of inflorescence usually simple or with few 2d-order branches; inflorescences narrowly paniculate, cylindric, usually lax.

        9. Ocreae entire (or sometimes laciniate only in distal parts); achenes dark brown to brownish yellow, usually dull; leaf blades broadly ovate (rarely almost rounded), oblong-ovate, or rarely oblong-lanceolate, normally less than 2.5 times as long as wide . . . . 8. *Rumex lapponicus*

        9. Ocreae laciniate (especially in middle and upper cauline leaves); achenes black to dark brown, shiny, smooth; leaf blades oblong-ovate to lanceolate, normally more than 2.5 times as long as wide . . . . 7. *Rumex acetosa*

1. Plants synoecious (rarely polygamodioecious or dioecious individuals in some species); flowers normally bisexual, sometimes bisexual and unisexual within same inflorescence; leaf blades never hastate or sagittate; pedicels with or without evident articulation [25c. *Rumex* subg. *Rumex* and 25d. subg. *Platypodium*].

10. Plants not developing basal rosette of leaves; stems erect, ascending, procumbent, or decumbent, normally with regular, leafy axillary shoots tending to develop 2d-order axillary inflorescences (often overtopping 1st-order ones); leaf blades mostly lanceolate, elliptic, ovate, ovate-lanceolate, or ovate-elliptic, base cuneate or almost rounded, or in some species broadly cuneate; inner tepal margins entire (rarely in some species minutely erose-denticulate) [25c.1. *Rumex* sect. *Axillares*].

  11. Inner tepals (20–)23–30 mm wide . . . . 10. *Rumex venosus*

  11. Inner tepals normally less than 15 mm wide.

    12. Pedicels 2.5–5 times as long as inner tepals, articulated in proximal part.

      13. Leaf blades ovate or ovate-elliptic, ca. 2 times as long as wide, lateral veins forming angle of 80° with midvein (especially near base) . . . . . 13. *Rumex fascicularis*

      13. Leaf blades linear-lanceolate, narrowly to broadly lanceolate (rarely ovate-lanceolate in *R. floridanus*), at least 3 times as long as wide, lateral veins forming angle of 4–60° with midvein.

        14. Leaf blades mostly linear-lanceolate, 5–7(–10) times as long as wide, thin; inflorescences normally interrupted (at least in basal ½); pedicels ca. 3–5 times as long as inner tepals; inner tepals longer than wide, or rarely as long as wide . . . . 11. *Rumex verticillatus*

14. Leaf blades mostly lanceolate to broadly lanceolate, 3–5(–6) times as long as wide, coriaceous and somewhat fleshy; inflorescences normally rather dense (sometimes interrupted only at base); pedicels ca. 2.5–3 times as long as inner tepals; inner tepals as wide as or wider than long . . . . 12. *Rumex floridanus*

[12. Shifted to left margin.—Ed.]

12. Pedicels usually not more than 2–2.5 times as long as inner tepals, articulated near middle or in proximal ¹/₂.

15. Leaf blades distinctly obovate or obovate-elliptic, widest in distal ¹/₂, coriaceous, apex obtuse, rounded; plants with long-creeping underground rhizomes and/or stolons, producing ascending or erect axillary shoots (5–)10–30(–40) cm . . . . 30. *Rumex cuneifolius*

15. Leaf blades ovate-lanceolate, elliptic-lanceolate, lanceolate, or linear-lanceolate, widest near middle or in proximal ¹/₂, subcoriaceous or coriaceous, apex acute (sometimes subobtuse, but never rounded); plants usually with vertical rootstock, occasionally with creeping rhizomes.

16. Inner tepals broadly cordate or broadly ovate-deltoid, 7–10 × 8–12 mm; leaf blades rounded or broadly truncate at base . . . . 14. *Rumex spiralis*

16. Inner tepals ovate, triangular, or deltoid, always less than 7 mm × 1–5 mm; leaf blades in most cases cuneate at base, rarely subtruncate.

17. Leaf blades ovate-lanceolate or elliptic-lanceolate, distinctly widest in proximal ¹/₂; inner tepals usually (4.5–)5–6 mm.

18. Inner tepals ovate or cordate-triangular (occasionally almost orbiculate), tubercles absent; stems normally ascending to decumbent . . . . 16. *Rumex ellipticus*

18. Inner tepals broadly triangular, ovate-triangular, or broadly ovate-deltoid, tubercles (2–)3; stems normally erect, rarely ascending . . . . 15. *Rumex altissimus*

17. Leaf blades in most cases lanceolate or linear-lanceolate, usually widest near middle; inner tepals usually 1.7–7 mm (in some species leaf blades ovate-elliptic or elliptic-lanceolate, but then inner tepals always less than 5 mm).

19. Inner tepals without tubercles (rarely 1 inner tepal with somewhat thickened midvein).

20. Inner tepals with margins minutely but distinctly denticulate, rarely subentire; inflorescences lax and broadly paniculate; stems ascending to suberect . . . . 19. *Rumex californicus*

20. Inner tepals with margins entire, sometimes margins indistinctly crenulate, but then inflorescences dense and paniculate or/and stems prostrate.

21. Inner tepals 2.5–3 × 2.5–3 mm; inflorescences dense, broadly paniculate, with crowded branches; leaf blade margins normally flat (sometimes slightly undulate); stems erect or ascending . . . . . . 20. *Rumex utahensis*

21. Inner tepals 3–4 × 3.2–4(–4.5) mm; inflorescences rather dense toward apex, usually broadly paniculate, interrupted in proximal ¹/₂, with remote branches almost perpendicular to main axis; leaf blade margins strongly undulate or crenulate; stems usually procumbent . . . . 28. *Rumex subarcticus*

19. At least 1 inner tepal with distinct tubercle, or all inner tepals with tubercles.

22. All or at least 1 tubercle large, subequal to inner tepals, or slightly narrower than inner tepals (then free margins of inner tepals distinctly narrower than tubercle).

23. Leaf blades thick, coriaceous, ovate-lanceolate, elliptic-lanceolate, or ovate-elliptic, not more than 2–3.5 times as long as wide; inner tepals (3–)4–5 × (2.5–)3–4 mm; tubercle 1 . . . . 21. *Rumex crassus*

23. Leaf blades thin, or in some species thick, occasionally coriaceous or subcoriaceous, oblanceolate, lanceolate, or linear-lanceolate, usually more than (3–)3.5 times as long as wide; inner tepals (1.7–)2–4 × (1.5–)1.8–3(–3.5) mm; tubercles 1–3.

24. Inner tepals (1.8–)2–2.5(–3) mm; tubercle 1 . . . . 18.   *Rumex salicifolius*
24. Inner tepals 2.5–4 mm; tubercles 3 (sometimes 1 or 2 distinctly smaller).
     25. Inner tepals usually 2.5–3(–3.5) mm, distinctly longer than tubercles. . . . . 27.   *Rumex sibiricus*
     25. Inner tepals usually 3–4 mm, subequal to or slightly longer than tubercles.
         26. Leaf blades rather thick, coriaceous, margins flat or slightly undulate; inner tepals deltoid-ovate; tubercles 3, equal, often minutely verrucose . . . . 26.   *Rumex pallidus*
         26. Leaf blades thin (rarely subcoriaceous), often with undulate margins; inner tepals ovate to ovate-lanceolate; tubercles 3, unequal (1 distinctly larger), smooth . . . . 25.   *Rumex transitorius*
22. Tubercles much narrower than inner tepals (free margins of inner tepal wider than or at least as wide as tubercle).
     27. Inner tepals 2–2.5(–3) mm, ovate or elliptic; leaf blades of aquatic submerged forms usually ovate-lanceolate to lanceolate, glabrous or nearly so; those of terrestrial forms lanceolate, papillose-pubescent abaxially . . . . 22.   *Rumex lacustris*
     27. Inner tepals normally more than (2.5–)3 mm, elliptic-lanceolate to linear-lanceolate; leaf blades of terrestrial or riparian plants linear-lanceolate, lanceolate; or elliptic-lanceolate, glabrous or nearly so.
         28. Inflorescences lax, distinctly interrupted; leaf blades thick, coriaceous, deep olive green, with strongly prominent veins abaxially, apex subobtuse; inner tepals orbiculate to ovate-triangular . . . . 17.   *Rumex chrysocarpus*
         28. Inflorescences rather dense, not interrupted, or interrupted only in proximal part; leaf blades thin, not coriaceous, light green to yellowish green, with scarcely prominent veins abaxially, apex in most cases distinctly acute; inner tepals triangular.
             29. Leaf blades elliptic-lanceolate or lanceolate, ca. (2–)3(–5) times as long as wide; tubercle 1 . . . . 29.   *Rumex hesperius*
             29. Leaf blades narrowly lanceolate or linear-lanceolate, normally more than 5 times as long as wide; tubercles 3 (1 in some forms of *R. triangulivalvis*).
                 30. Inner tepals 3.5–4.5(–5) mm; achenes 2–3 mm . . . . 23.   *Rumex mexicanus*
                 30. Inner tepals (2–)2.5–3.5(–3.8) mm; achenes 1.7–2.2 mm . . . . 24.   *Rumex triangulivalvis*

[10. Shifted to left margin.—Ed.]
10. Plants developing basal rosette of leaves (sometimes, especially in annual species, not persistent at maturity); stems mostly erect, sometimes ascending, spreading, or almost prostrate, simple or several from base, not branching below terminal paniculate inflorescence (single racemose in *R. bucephalophorus*), without axillary shoots; leaf blades variable in shape, base cordate to cuneate (rarely rounded in *R. bucephalophorus*); inner tepal margins entire or variously dentate [25c.2. *Rumex* sect. *Rumex*].
     31. Inner tepals with tubercles absent (or sometimes 1 inner tepal with indistinct tubercle or slightly thickened midvein in *R. longifolius* and *R. pseudonatronatus*), margins entire, indistinctly erose or, rarely, minutely denticulate.
         32. Inner tepals 11–16 mm; ocreae prominent; roots distinctly tuberous . . . 31.   *Rumex hymenosepalus*
         32. Inner tepals usually less than 10 mm; ocreae less prominent; roots not tuberous.

[33. Shifted to left margin.—Ed.]

33. Pedicels with distinctly swollen articulation point.
    34. Plants with creeping rootstock; leaf blades to broadly orbiculate to ovate-orbiculate, base deeply cordate; inner tepals ovate to ovate-triangular . . . . 36. *Rumex alpinus*
    34. Plants with vertical rootstock; leaf blades oblong, lanceolate, or narrowly lanceolate, base cuneate, truncate, or occasionally indistinctly cordate; inner tepals orbiculate, reniform, or cordate-orbiculate.
        35. Leaf blades 15–30 × 1–4 cm, base narrowly cuneate; inner tepals 3– 5 mm wide; achenes usually reddish brown, less than 1–1.5 mm wide . . . 46. *Rumex pseudonatronatus* (in part)
        35. Leaf blades 25–50(–60) × 7–15 cm, base broadly cuneate, rounded-truncate, or slightly cordate; inner tepals (4.5–)5–7(–7.5) mm wide; achenes dark brown or brown, normally 1.5–2 mm wide . . . . 45. *Rumex longifolius* (in part)
33. Pedicels without swollen articulation point.
    36. Plants with creeping rootstock or rhizome; inner tepals with margins entire to minutely dentate (especially near base) [*Rumex densiflorus* group]
        37. Leaf blades elliptic or oblong, sometimes almost orbiculate, ca. 2(–2.5) times as long as wide, apex obtuse or subacute; base rounded or broadly cuneate; inflorescences occupying more than proximal ¹/₂ of stem . . . . 35. *Rumex praecox*
        37. Leaf blades usually oblong-lanceolate, sometimes oblong, usually more than 3 times as long as wide, apex acute or subacute rarely obtuse, base weakly cordate, truncate, or broadly cuneate; inflorescences occupying distal ¹/₂ of stem.
            38. Inner tepals ovate-triangular or ovate-deltoid, widest at base, apex narrowly acute, margins usually minutely erose or weakly serrate (at least near base) . . . . 33. *Rumex pycnanthus*
            38. Inner tepals ovate-deltoid, ovate-triangular, or subcordate, widest above base, apex acute to subacute, margins entire to indistinctly erose, rarely minutely denticulate near base.
                39. Inner tepals abruptly contracted at apex, widest near middle; leaf blades with large lateral veins alternating with short ones . . . . 32. *Rumex densiflorus*
                39. Inner tepals gradually narrowed into acute apex, widest in proximal ¹/₃; leaf blades with lateral veins ± equal in size . . . . 34. *Rumex orthoneurus*
    36. Plants with vertical rootstock (or sometimes with short oblique rootstock); inner tepals with margins entire or rarely weakly erose [*Rumex aquaticus* group]
        40. Plants densely tomentose and/or papillose-pubescent (especially abaxial sides of leaf blades, ocreae, and petioles); pedicel distinctly swollen at distal part (near base of tepals, not at articulation point) . . . . 41. *Rumex tomentellus*
        40. Plants glabrous or nearly so (rarely sparsely papillose-pubescent); pedicel not distinctly swollen.
            41. Pedicel 12–20 mm, 3(–4) times as long as inner tepals . . . . 40. *Rumex nematopodus*
            41. Pedicel 5–13(–17) mm, usually not more than 2–2.5 times as long as inner tepals.
                42. Leaf blades narrowly lanceolate, lanceolate, or oblong-lanceolate, base cuneate to broadly cuneate; inflorescences usually simple or with comparatively short branches less than 7–8 cm . . . . 39. *Rumex arcticus*
                42. Leaf blades ovate-triangular, ovate-lanceolate, or oblong-lanceolate, base distinctly to weakly cordate, occasionally rounded or truncate; inflorescences normally with comparatively long branches more than 7–8 cm . . . . 38. *Rumex occidentalis*

[31. Shifted to left margin.—Ed.]

31. Inner tepals with at least 1 distinct tubercle, margins entire, denticulate, or variously dentate (sometimes tubercle absent, or inner tepals with indistinctly swollen midvein, then margins prominently dentate, with hooked teeth).

    43. Inner tepals with margins entire or minutely and indistinctly erose-denticulate (teeth less than 0.2 mm); however, in 3 species (*R. stenophyllus*, *R. cristatus*, *R. kerneri*) often more distinctly dentate, then inner tepals reniform, orbiculate, broadly ovate, or broadly ovate-triangular (ca. as long as wide, or wider than long), base often cordate.

        44. Inner tepals oblong-lanceolate, oblong, lingulate, ca. 2 times as long as wide, margins entire, largest tubercle almost as wide as inner tepal.

            45. Tubercles 3, equal or subequal; inflorescences with almost all but distalmost flower whorls with subtending leaves (panicle leafy at least in proximal $^2/_3$ of length), dense; pedicels 1–4(–5) mm . . . . 50. *Rumex conglomeratus*

            45. Tubercle 1 (occasionally 3, then 1 much larger); inflorescences with only proximalmost flower whorls with subtending leaves (panicle leafless, or leafy only near base), lax; pedicels (2–)4–6(–8) mm . . . . 51. *Rumex sanguineus*

        44. Inner tepals orbiculate, broadly ovate, or broadly ovate-triangular (deltoid or triangular-deltoid only in *R. violascens*), ca. as long as wide (or at least always distinctly less than 2 times as long as wide), margins entire or denticulate, largest tubercles normally much narrow than inner tepals.

            46. Plants annual or biennial, sometimes short-lived perennials, native to sw United States and Mexico; inner tepals deltoid or triangular-deltoid, ca. 1.5 times as long as wide; branches of inflorescences usually distinctly flexuous . . . . 57. *Rumex violascens* (in part)

            46. Plants perennial, mostly introduced; inner tepals orbiculate to broadly ovate-triangular, in most species as long as wide or nearly so; branches of inflorescences usually straight or arcuate, rarely indistinctly flexuous.

                47. Inner tepal margins denticulate or dentate, at least proximally.

                    48. Inner tepals normally less than 6 mm, with 3 equal or subequal tubercles . . . . 48. *Rumex stenophyllus*

                    48. Inner tepals normally more than 6 mm, with 1 distinct tubercle, other inner tepals without tubercles or tubercles small.

                        49. Inner tepals usually with 1 tubercle, teeth to 0.5 mm; branches of inflorescences mostly simple or nearly so; leaf blades distinctly papillose abaxially (on veins) . . . . 44. *Rumex kerneri*

                        49. Inner tepals usually with 3 unequal tubercles, teeth 0.5–1 mm; branches of inflorescences mostly with 2d-order branches; leaf blades indistinctly papillose to glabrous abaxially (on veins) . . . 43. *Rumex cristatus*

                47. Inner tepal margins entire or subentire to or weakly erose.

                    50. Inner tepals with 1 indistinct tubercle less than 1(–1.3) mm, or some with tubercles absent (usually both types occur within same inflorescence).

                        51. Leaf blades 15–30 × 1–4 cm, base narrowly cuneate; inner tepals usually 3–5 mm wide; achenes reddish brown, usually 1–1.5 mm wide . . . . 46. *Rumex pseudonatronatus* (in part)

                        51. Leaf blades 25–50(–60) × 7–15 cm, base broadly cuneate; inner tepals (4.5–)5–7(–7.5 mm wide; achenes brown to dark brown, normally 1.5–2 mm wide . . . . 45. *Rumex longifolius* (in part)

                    50. Inner tepals normally with 3 tubercles, or at least with 1 distinct tubercle more than (1–)1.5 mm wide.

52. Inner tepals with 3 distinctly equal or subequal tubercles; leaf blades 20–55(–70) cm, lanceolate or oblong-lanceolate, base cuneate, occasionally or rounded or truncate . . . . 49. *Rumex britannica*
52. Inner tepals with 1 tubercle, or with 3 unequal tubercles, at least 1 tubercle distinctly larger; leaf blades variable (rarely tubercles subequal, then largest leaves smaller than 55 cm).
  53. Leaf blades broadly ovate, ovate-triangular, or ovate-elliptic, base deeply and broadly cordate, apex obtuse to subacute; tubercle usually 1 . . . . 37. *Rumex confertus*
  53. Leaf blades ovate-lanceolate, oblong-lanceolate, or lanceolate, base cuneate, truncate, or subcordate, apex acute or subacute; tubercles 1–3.
    54. Leaf blades ovate-lanceolate or oblong-lanceolate, margins flat or weakly undulate; inner tepals (5–)5.5–8(–10) mm, broadly ovate to orbiculate, base usually distinctly cordate; tubercles normally 1 (occasionally 2–3); stems usually 80–150(–200) cm. . . . . 42. *Rumex patientia*
    54. Leaf blades usually lanceolate, margins strongly undulate and crisped; inner tepals 3.5–6 mm, orbiculate-ovate or ovate-deltoid, base truncate, or subcordate; tubercles normally 3 (rarely 1–2); stems 40–100(–150) cm . . . . 47. *Rumex crispus*

[43. Shifted to left margin.—Ed.]
43. Inner tepal margins variously dentate (at least some teeth 0.3 mm or longer, almost always evidently longer than wide (excluding teeth), base variable but normally not cordate.
  55. Inner tepals triangular, with 3–5 distinctly hooked teeth on each side, apex hooked, tubercles absent, or midveins indistinctly swollen . . . . 62. *Rumex brownii*
  55. Inner tepals various in shape, with straight apex and teeth (only in *R. bucephalophorus* sometimes with hooked, slender lateral teeth, but without hook at apex); at least 1 inner tepal with tubercle or (in *R. bucephalophorus*) tubercles absent.
    56. Plants annual (rarely biennial); leaf blades lanceolate or spatulate; flowers 2–3(–4) in whorls; pedicels usually distinctly heteromorphic, some swollen, curved, or clavate, others not swollen . . . . 63. *Rumex bucephalophorus*
    56. Plants perennial, annual, or biennial; leaf blade variable; flowers normally more than 3–4 in whorls; pedicels homomorphic.
      57. Leaf blades lanceolate-linear or lanceolate (rarely oblong-lanceolate), at least 4 times as long as wide; inner tepal margins with long bristlelike or subulate-filiform teeth longer than or equaling width of inner tepals (very rarely teeth shorter, or even absent); inner tepals (excluding teeth) narrowly triangular or narrowly rhombic-triangular, normally ca. 2 times as long as wide; plants annual (less commonly biennial or short-lived perennial).
        58. Leaf blades narrowly cuneate (rarely broadly cuneate) at base; inflorescence branches and leaf blades glabrous or indistinctly papillose; tubercles in most cases smooth (occasionally finely striate or indistinctly pitted in herbarium specimens); uncommon introduced species.
          59. Inner tepals (2.5–)3–3.5(–4) mm, teeth (bristles) ca. as long as width of inner tepals, tubercles obtuse at apex; inflorescences normally reddish brown, with flower whorls distinctly interrupted in proximal ½ or ⅔ . . . . 59. *Rumex palustris*
          59. Inner tepals 2.5–3(–3.5) mm, teeth (bristles) usually 1.5–2 times as long as width of inner tepals, tubercles acute or subacute at apex; inflorescences normally golden or greenish yellow, with flower whorls usually rather dense or interrupted in proximal part . . . . 58. *Rumex maritimus*
        58. Leaf blades slightly cordate, abruptly truncate, or broadly cuneate at base; inflorescence branches and leaf blades usually distinctly papillose-pubescent abaxially; tubercles usually distinctly reticulate-pitted; native species.

      60. Tubercles straw-colored, oblong-ovate, apex obtuse, almost as wide as inner tepals; inner tepal margins with teeth (bristles) ± equal to width of inner tepals . . . . 60. *Rumex persicarioides*
      60. Tubercles brownish or reddish, linear-lanceolate to fusiform, apex normally acute to subacute, distinctly narrower than inner tepals (ca. 0.5 times width of inner tepals); inner tepal margins with teeth (bristles) variable, but normally 1.5–2.5(–4) times as long as width of inner tepals . . . . 61. *Rumex fueginus*

    [57. Shifted to left margin.—Ed.]

57. Leaf blades ovate, obovate to elongate, occasionally broadly oblong-lanceolate, less than 4 times as long as wide; inner tepal margins with short-subulate or triangular-subulate (not bristlelike) teeth equaling or shorter than width of inner tepals (occasionally longer in some forms of *R. dentatus* and *R. pulcher*); inner tepals (excluding teeth) usually deltoid or broadly triangular (occasionally ligulate) normally ca. 1.5 times as long as wide (occasionally ca. 2 times as long as wide in some forms of perennial *R. obtusifolius*); plants annual, biennial, or perennial.
  61. Plants perennial; leaf blades oblong to ovate-oblong, sometimes panduriform in *R. pulcher*, base usually distinctly cordate (rarely truncate or rounded).
    62. Stems 20–60(–70) cm; leaf blades 4–10(–15) cm; inflorescence branches divaricately spreading (forming angle of 60–90° with 1st-order stem); tubercles usually verrucose (warty) . . . . 53. *Rumex pulcher*
    62. Stems 60–120(–150) cm; leaf blades 20–40 cm; inflorescence branches less spreading (normally forming angle of 30–45° with 1st-order stem); tubercles smooth . . . . 52. *Rumex obtusifolius*
  61. Plants annual or biennial (rarely short-lived perennial); leaf blades variable, base cuneate, truncate, or subcordate.
    63. Leaf blades distinctly obovate (rarely panduriform), usually coriaceous, apex obtuse, base cuneate (rarely rounded).
      64. Inner tepals 4–5(–5.5) mm; tubercles distinctly verrucose (warty), apex obtuse (rarely subacute) . . . . 55. *Rumex obovatus*
      64. Inner tepals 3–4 mm; tubercles smooth or minutely punctate, apex acute or subacute . . . . 56. *Rumex paraguayensis*
    63. Leaf blades elongate, ovate, oblong-lanceolate, obovate-elliptic, normally not coriaceous (occasionally subcoriaceous in *R. violascens*), apex obtuse or subacute, base subcordate, rounded, truncate, or broadly cuneate.
      65. Plants annual or biennial (occasionally short-lived perennial); leaf blades oblong-lanceolate to obovate-elliptic, base broadly cuneate to rounded; inner tepal margins with longest tooth less than 0.5 mm (rarely some inner tepal margins subentire) . . . . 57. *Rumex violascens* (in part)
      65. Plants annual (rarely biennial); leaf blades oblong, elliptic-lanceolate, or ovate-elliptic, base truncate or subcordate to weakly cordate; inner tepal margins with teeth 1–3(–5) mm . . . . 54. *Rumex dentatus*

26a. RUMEX Linnaeus subg. **ACETOSELLA** (Meisner) Rechinger f., Publ. Field Mus. Nat. Hist., Bot. Ser. 17: 6. 1937

Rumex sect. *Acetosella* Meisner in C. F. P. von Martius et al., Fl. Bras. 5(1): 10. 1855; *Acetosella* (Meisner) Fourreau

Plants dioecious (rarely polygamomonoecious). **Leaf blades** hastate with spreading basal lobes, or sometimes narrowly linear, unlobed, and narrowly cuneate at base. **Pedicels** filiform, articulation distinct near base of tepals, slightly swollen. **Flowers** unisexual, staminate and pistillate on different plants; outer tepals angled towards inner tepals; inner tepals not enlarged or slightly to distinctly enlarged, 1.2–2.5(–3) mm, equaling or slightly wider than achenes, margins entire; tubercles absent.

Species 7–8 (4 in the flora): mostly in temperate regions of both hemispheres; as naturalized almost worldwide.

SELECTED REFERENCE Löve, Á. 1983. The taxonomy of *Acetosella*. Bot. Helv. 93: 145–168.

1. **Rumex acetosella** Linnaeus, Sp. Pl. 1: 338. 1753
  • Sheep or field sorrel, oseille, petite oseille, sûrette

F I W

Acetosa acetosella (Linnaeus) Miller; *A. hastata* Scopoli; *A. vulgaris* (W. D. J. Koch) Fourreau; *Rumex acetosella* var. *vulgaris* W. D. J. Koch

Plants perennial, glabrous, with vertical rootstock and/or creeping rhizomes. **Stems** erect or ascending, several from base, branched in distal 1/2 (in inflorescence), 10–40(–45) cm; shoots variable. **Leaves:** ocrea brownish at base, silvery and lacerated in distal 1/2; blade normally obovate-oblong, ovate-lanceolate, lanceolate-elliptic, or lanceolate, occasionally linear-lanceolate to almost linear, 2–6 × 0.3–2 cm, base hastate (with spreading, entire or sometimes multifid, dissected lobes), occasionally without evident lobes, then base broadly cuneate, margins entire, flat or nearly so, apex acute or obtuse. **Inflorescences** terminal, usually occupying distal 1/2–2/3 of stem, usually lax and interrupted to top, broadly or narrowly paniculate. **Pedicels** 1–3 mm. **Flowers** (3–)5–8(–10) in whorls; inner tepals not or slightly enlarged, normally 1.2–1.7(–2) × 0.5–1.3 mm (free wing absent or barely visible), base cuneate, apex obtuse or subacute. **Achenes** brown or dark brown, 0.9–1.5 × 0.6–0.9 mm. $2n = 14, 28, 42$.

Flowering spring–summer. Roadsides, cultivated fields, waste places, disturbed areas, lawns, meadows, railroad gravels, sandy and muddy shores: usually in acidic soils; 0–2700 m; introduced; Greenland; St. Pierre and Miquelon; Alta., B.C., Man., N.B., Nfld. and Labr. (Nfld.), N.S., Ont., P.E.I., Que., Sask., Yukon; Ala., Alaska, Ariz., Ark., Calif., Colo., Conn., Del., D.C., Fla., Ga., Idaho, Ill., Ind., Iowa, Kans., Ky., La., Maine, Md., Mass., Mich., Minn., Miss., Mo., Mont., Nebr., Nev., N.H., N.J., N.Mex., N.Y., N.C., N.Dak., Ohio, Okla., Oreg., Pa., R.I., S.C., S.Dak., Tenn., Tex., Utah, Vt., Va., Wash., W.Va., Wis., Wyo.; Europe; w Asia; introduced almost worldwide.

Rumex acetosella in the broad sense is an extremely variable and taxonomically complicated polyploid complex, which includes diploids, tetraploids, hexaploids, and octoploids. This complex (excluding more distantly related arctic-montane *R. graminifolius* and its allies) probably originated and developed mostly in southern Europe and southwestern Asia. Some races

of *R. acetosella* now are distributed almost worldwide as introduced and often completely naturalized aliens.

Á. Löve (1941, 1983) assumed that in this group chromosome numbers are strictly correlated with morphology. In his opinion, every chromosome race represents a distinct species: diploid *Rumex angiocarpus* Murbeck [= *Acetosella angiocarpa* (Murbeck) Á. Löve]; tetraploid *R. multifidus* Linnaeus [= *R. tenuifolius* (Wallroth) Á. Löve = *Acetosella multifida* (Linnaeus) Á. Löve]; hexaploid *R. acetosella* in the narrow sense [= *A. vulgaris* (W. D. J. Koch) Fourreau, with gymnocarpous *A. vulgaris* subsp. *vulgaris* and angiocarpous *A. vulgaris* subsp. *pyrenaica* (Pourret ex Lapeyrouse) Á. Löve]; hexaploid *R. graminifolius* Rudolph ex Lambert [= *A. graminifolia* (Rudolph ex Lambert) Á. Löve]. However, the distribution given by Löve for these taxa seems unnatural. Studies by J. C. M. den Nijs and collaborators (den Nijs 1974, 1976, 1984; den Nijs and T. Panhorst 1980; den Nijs et al. 1980, 1985; see also W. Harris 1969, 1973) indicate that the situation is more complicated. They postulated the development of two major evolutionary lines into two ploidy complexes: a primary western Mediterranean one and a secondary eastern Mediterranean one. According to this scheme, polyploid races independently and spontaneously emerged (and still are emerging) within different ancestral populations.

The most widespread, almost cosmopolitan race, presumably native to the southwestern Mediterranean region, including southwestern and Atlantic Europe, which is common in North America, is characterized by a hexaploid chromosome set ($2n = 42$), nonmultifid lateral lobes of basal leaves, and angiocarpy (fruits are not easily separable from accrescent inner tepals). It was commonly and erroneously referred to as *Rumex angiocarpus* Murbeck, or *R. acetosella* subsp. *angiocarpus* (Murbeck) Murbeck. According to J. R. Akeroyd (1991), who in general followed the taxonomic revision of the group by J. C. M. den Nijs (1984), the correct name for this taxon is *R. acetosella* subsp. *pyrenaicus* (Pourret ex Lapeyrouse) Akeroyd (=*Acetosella vulgaris* subsp. *pyrenaica* (Pourret ex Lapeyrouse) Á. Löve). Gymnocarpous nonmultifid and multifid forms (*R. acetosella* subsp. *acetosella* and *R. acetosella* subsp. *acetselloides* (Balansa) den Nijs, respectively) also occur in North America, but evidently rarely. The distributions of subspecies of *R. acetosella* in North America are poorly known. Keys and detailed descriptions for the subspecies were provided by den Nijs

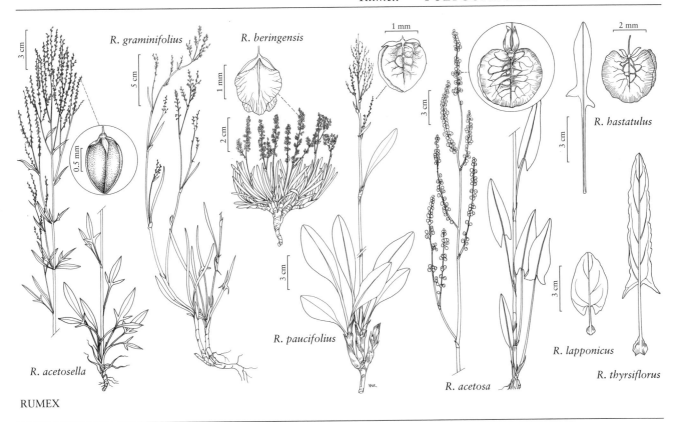

R. graminifolius

R. beringensis

R. hastatulus

R. paucifolius

R. acetosella

R. acetosa

R. lapponicus

R. thyrsiflorus

RUMEX

and Akeroyd. However, the tempting simplicity of the keys is somewhat suspicious. The alternative point of view (and an alternative key) may be found in Á. Löve (1983).

Rumex acetosella subsp. *arenicola* Mäkinen ex Elven was recently described from Greenland and reported for Scandinavia and arctic Russia (R. Elven et al. 2000). This entity seems to be morphologically transitional toward *Rumex graminifolius* (see discussion under that species below). According to Elven et al., it differs from other infraspecific entities of *R. acetosella* in having the following characters: leaves usually without basal lobes (as in *R. graminifolius*), with revolute margins; inflorescence sparsely branched; tepals and pedicels densely covered with red papillae (as in *R. graminifolius*). From *R. graminifolius* and related taxa (*R. beringensis* and *R. krausei*) it can be distinguished by narrower inner tepals (similar in size to those in other subspecies of *R. acetosella*). The distribution of subsp. *arenicola* and its relations to other taxa are in need of further study.

2. **Rumex graminifolius** Rudolph ex Lambert, Trans. Linn. Soc. London 10: 264, plate 10. 1811 • Grass-leaved or grassleaf sorrel [F]

Acetosella graminifolia (Rudolph ex Lambert) Á. Löve; *Rumex acetosella* Linnaeus var. *graminifolius* (Rudolph ex Lambert) Schrenk; *R. angustissimus* Ledebour

Plants perennial, glabrous, with creeping rhizomes and elongated underground stolons. **Stems** erect or ascending, rarely almost prostrate, branched at base and in distal $^1/_2$ (in inflorescence), 7–30(–40) cm; shoots not crowded, ± elongated. **Leaves:** ocrea whitish or silvery, membranous; blade normally narrowly linear, or occasionally linear-lanceolate, usually not hastate, rarely some with indistinct basal lobes, 3–10 × 0.1–0.2(–0.4) cm, base narrowly cuneate, margins entire, flat or occasionally slightly revolute, apex acute or obtuse. **Inflorescences** terminal, occupying distal $^2/_3$ of stem, usually lax and interrupted to top, paniculate, with branches often reflexed. **Pedicels** 1–4 mm. **Flowers** (3–)4–6(–8) in whorls; inner tepals distinctly enlarged, normally 2–2.6 × 1.5–2(–2.2) mm (free wing 0.3–0.5 mm wide), base cuneate, apex obtuse or subacute. **Achenes** brown or yellowish brown, 1.5–2 × 1–1.5 mm. $2n = 56$.

Flowering late spring–summer. Sandy and gravelly shores and slopes; 0–400 m; Greenland; Alaska; n Eurasia.

Records of *Rumex graminifolius* from Alaska in most cases refer to *R. beringensis* and *R. krausei*. The occurrence of typical *R. graminifolius* in northwestern North America remains uncertain. Some literature records of *R. acetosella* from northeastern North America (Greenland, Labrador, Newfoundland) may refer to *R. graminifolius* or *R. acetosella* subsp. *arenicola*. *Rumex graminifolius* was reported from Newfoundland also by M. L. Fernald (1950), but that record requires confirmation.

Some plants from northeastern Eurasia (northeastern Russian Far East and northern Siberia) are known in Russian literature as *Rumex aureostigmaticus* Komarov [*Acetosella aureostigmatica* (Komarov) Tzvelev], *R. acetosella* var. *subspathulatus* Trautvetter, or *R. graminifolius* var. *subspathulatus* (Trautvetter) Tolmatchew (see A. I. Tolmachew 1966; N. N. Tzvelev 1989b). They differ from *R. graminifolius* in having narrower inner tepals and wider spatulate leaves, usually without basal lobes. I have seen only one North American collection approaching this entity. Some specimens (mostly immature or staminate plants) from western Alaska differ from both *R. graminifolius* and *R. beringensis* in their habit; they need additional study. Some chromosome counts different from the most typical number ($2n = 56$) that have been reported for *R. graminifolius* in the broad sense from northeastern Russian Far East by several Russian authors (see references in Tzvelev) most probably also refer to *R. aureostigmaticus*. It is also possible that arctic and subarctic plants identified by various authors as *R. aureostigmaticus*, *R. acetosella* var. *subspathulatus*, *R. graminifolius* var. *subspathulatus*, and *R. acetosella* subsp. *arenicola* belong to one polymorphic complex of plants intermediate between *R. acetosella* and *R. graminifolius*.

3. Rumex beringensis Jurtzev & V. V. Petrovsky, Bot. Zhurn. (Moscow & Leningrad) 58: 1745. 1973

• Beringia or Bering Sea sorrel   F

Acetosella beringensis (Jurtzev & V. V. Petrovsky) Á. Löve & D. Löve

Plants perennial, glabrous, with thick, densely tufted underground stolons. **Stems** erect, rarely ascending, several from base, branched in inflorescence, 5–15 (–20) cm; shoots usually densely crowded, not elongated. **Leaves:** ocrea brownish or reddish brown, membranous; blade linear or spatulate-lanceolate, not hastate (without basal lobes), 1.5–5 × 0.1–0.3 cm, base narrowly cuneate (gradually narrowing into petiole), margins entire, flat or slightly convolute, apex obtuse or subacute. **Inflorescences** terminal, occupying more than distal $^1/_2$ of stem, ± dense, usually interrupted at least near base, narrowly paniculate with branches directed upward. **Pedicels** 1–4 mm. **Flowers** (3–)4–7 in whorls; inner tepals distinctly enlarged, 1.6–2.3 × 1.8–2.5 mm (free wing 0.3–0.5 mm wide), base cuneate, apex obtuse or subacute. **Achenes** brown or reddish brown, 1–1.5 × 0.8–1.2 mm. $2n = 14$.

Flowering summer. Sandy and gravelly soil, shores, limestone outcrops; 0–300 m; Yukon; Alaska; ne Asia (ne Russian Far East).

The name *Rumex graminifolius* was commonly misapplied to this species in both northwestern North America and northeastern Eurasia.

4. Rumex krausei Jurtzev & V. V. Petrovsky, Bot. Zhurn. (Moscow & Leningrad) 58: 1745. 1973 (as krausii) • Krause or Cape Krause sorrel   C

Acetosella krausei (Jurtzev & V. V. Petrovsky) Á. Löve & D. Löve

Plants perennial, glabrous, with thick, densely tufted underground stolons. **Stems** erect or slightly ascending, 1 or several from base, sparsely branched in inflorescence, occasionally inflorescence simple or nearly so, 8–20(–25) cm; shoots usually densely crowded, not elongated. **Leaves:** ocrea brownish, membranous; blade narrowly linear or spatulate-lanceolate, not hastate (without basal lobes), 2.5–6 × 0.15–0.3(–0.4) cm, base narrowly cuneate (gradually narrowing into petiole), margins entire, flat or slightly revolute, apex obtuse or subacute. **Inflorescences** terminal, occupying distal $^1/_2$ of stem, usually dense, narrowly paniculate with branches directed upward, or simple. **Pedicels** 1–4 mm. **Flowers** 3–7 in whorls; inner tepals distinctly enlarged, 2.3–3 × 1.8–3 mm (free wing 0.4–0.8 mm wide), base cuneate, apex obtuse or subacute. **Achenes** light brown, 1.5–2 × 1.2–1.9 mm. $2n = 21$.

Flowering summer. Clay and argillaceous soil, silty sand, rocky outcrops; 0–300 m; of conservation concern; Alaska; ne Asia (ne Russian Far East).

The name *Rumex graminifolius* was commonly misapplied to this species in northwestern North America and northeastern Eurasia.

Rumex krausei is closely related to *R. beringensis* and probably represents the latter's triploid race. *Rumex krausei* differs from *R. beringensis* in having larger flowers, fruiting inner tepals, and fruits, and shorter and less-branched or unbranched inflorescences. *Rumex krausei* occurs in eastern Chukotka, Russia, together with *R. beringensis*; however, it is believed to be confined mostly to clay soils and limestones. It has been reported from the Ogotoruk River on the northwestern coast of Alaska (B. A. Jurtzev et al. 1975) and some other Alaskan regions (western Seward Peninsula, capes Dyer and Thompson, and the Squirrel River). It may be expected anywhere within the range of *R. beringensis*.

26b. RUMEX Linnaeus subg. ACETOSA (Miller) Rechinger f., Publ. Field Mus. Nat. Hist., Bot. Ser. 17: 6. 1937

Acetosa Miller, Gard. Dict. Abr. ed. 4, vol. 1. 1754

Plants dioecious (rarely polygamomonoecious). **Leaf blades** in most species sagittate at base, with lobes directed downward, sometimes hastate, cuneate, or rarely almost cordate. **Pedicels** with articulation near middle or in proximal part. **Flowers** normally unisexual, staminate and pistillate on different plants; outer tepals normally reflexed towards pedicel, or sometimes spreading; inner tepals distinctly enlarged, normally longer than (2.5–)3 mm, always distinctly wider and longer than achene, margins entire; tubercles absent or small and indistinct, recurved, developed only at base of inner tepals.

Species ca. 45 (5 in the flora): mostly in the Old World, as naturalized nearly worldwide.

5. Rumex paucifolius Nuttall, J. Acad. Nat. Sci. Philadelphia 7: 49. 1834 • Alpine sheep sorrel E F

Acetosa gracilescens (Rechinger f.) Á. Löve & Everson; *A. paucifolia* (Nuttall) Á. Löve; *Acetosella gracilescens* (Rechinger f.) Á. Löve; *A. paucifolia* (Nuttall) Á. Löve; *Rumex engelmannii* Meisner var. *geyeri* Meisner; *R. geyeri* (Meisner) Trelease; *R. paucifolius* var. *gracilescens* Rechinger f.

Plants perennial, glabrous, with vertical rootstock and densely tufted underground stolons. **Stems** erect, rarely ascending, tufted at base and branched only in inflorescence, occasionally inflorescences simple or nearly so, 10–40(–60) cm. **Leaves:** blade normally broadly lanceolate or ovate-lanceolate, usually not hastate (without basal lobes), 3–7(–10) × (0.6–)1–3(–4) cm, base narrowly cuneate (gradually narrowing into petiole), margins entire, flat, apex obtuse or subacute. **Inflorescences** terminal, occupying distal 2/3 of stem, usually lax and interrupted at least near base, narrowly paniculate, rarely simple. **Pedicels** articulated in proximal 1/3, filiform, 1–3(–5) mm, articulation slightly swollen. **Flowers** 3–10(–12) in whorls; inner tepals broadly ovate or almost orbiculate, 2.8–3.8 × 2.7–3.6 mm, base cordate or rounded-truncate, apex obtuse or subacute; tubercles absent. **Achenes** brown, 1.2–1.8 × 0.8–1 mm. *2n* = 14, 28.

Flowering spring–fall. Meadows, gravelly and grassy slopes, banks of rivers and streams in alpine, subalpine, and montane zones; 2000–3000 m; Alta., B.C.; Calif., Colo., Idaho, Mont., Nev., Oreg., Utah, Wash., Wyo.

Rumex paucifolius is a montane species represented by two chromosome races (diploid and tetraploid) and several ecotypes. Smaller plants from California have been described as var. *gracilescens*; they are tetraploids and sometimes were regarded as a separate species (Á. Löve and V. Everson 1967; Löve 1986). B. W. Smith (1968) showed that both diploids and tetraploids (and even exceptional spontaneous triploids and individuals with higher polyploid chromosome numbers) occur in many other localities within the range of the species; the differences in chromosome number are not strictly correlated with distribution or morphology. Narrow-leaved ecotypes of *R. paucifolius* reported by Smith sometimes resemble other narrow-leaved taxa of subg. *Acetosella*, especially *R. beringensis*. *Rumex paucifolius* and *R. beringensis* may be regarded as morphologically and karyologically transitional between subg. *Acetosella* and subg. *Acetosa*. *Rumex paucifolius* was placed in the monotypic subsect. *Paucifoliae* Á. Löve & N. Sarkar. Later, Löve transferred it to the segregate genus *Acetosella*, based mostly on the chromosome number of the species, but morphology suggests it is a member of subg. *Acetosa*. Probably the best solution of this problem was proposed by Smith, who noted that "the composite range of vegetative, reproductive, and karyotypic characteristics of the forty-odd species now included in the diversified subgenus *Acetosa* would be only slightly extended by the addition of the five species now classified as *Acetosella*" (p. 683).

6. Rumex hastatulus Baldwin in S. Elliott, Sketch Bot. S. Carolina 1: 416. 1817 • Heartwing or wild sorrel E F W

Acetosa hastatula (Baldwin) Á. Löve; *Rumex engelmannii* Meisner

Plants annual or short-lived perennial, glabrous, with vertical rootstock. **Stems** solitary or several from base, erect or ascending, branched in distal $^2/_3$ (in inflorescence), 10–40(–45) cm. **Leaves:** blade obovate-oblong, ovate-lanceolate, oblong-lanceolate, or lanceolate; 2–6 (–10) × 0.5–2 cm, base hastate (with spreading lobes), auriculate, or occasionally without evident lobes, margins entire, flat, apex obtuse or subacute. **Inflorescences** terminal, occupying distal $^2/_3$ of stem, usually lax and interrupted, narrowly paniculate. **Pedicels** articulated in proximal part, filiform, 1.5–2.5 (–3) mm, articulation indistinct or slightly swollen. **Flowers** 3–6(–8) in whorls; inner tepals orbiculate or broadly ovate, 2.5–3.2 × 2.7–3.2 mm, base broadly cordate or rounded, apex obtuse or subacute; tubercles absent or some inner tepals with slightly swollen central veins. **Achenes** brown or dark brown, 0.9–1.2 × 0.6–0.8 mm. $2n$ = 8 (pistillate plants), 9 (staminate plants), 10 (both sexes).

Flowering spring–summer. Dry to moist alluvial and ruderal habitats, river valleys, sandy plains, meadows, waste places; 0–500 m; Ala., Ark., Fla., Ga., Ill., Ind., Kans., La., Md., Miss., Mo., N.Y., N.C., Okla., Pa., S.C., Tenn., Tex., Va.

Rumex hastatulus is distinct in subg. *Acetosa* and belongs to the monotypic subsect. *Americanae* Á. Löve & N. Sarkar. It is represented by at least two chromosome races: populations occurring from North Carolina to Florida and Mississippi normally have $2n$ = 8 in pistillate plants and $2n$ = 9 in staminate plants; populations from Louisiana to Texas and Oklahoma predominantly have $2n$ = 10 in both sexes. *Rumex hastatulus* has been reported from New Mexico (W. C. Martin and C. R. Hutchins 1980, vol. 1), but those records need confirmation. When fruiting, *R. hastatulus* has large inner tepals that distinguish it from *R. acetosella*, with which it is occasionally confused.

7. Rumex acetosa Linnaeus, Sp. Pl. 1: 337. 1753 • Common or garden sorrel, sourdock, grande oseille F I W

Acetosa pratensis Miller; *Rumex acetosa* subsp. *pratensis* (Miller) A. Blytt & O. C. Dahl

Plants perennial, glabrous or nearly so, with short and relatively thin, horizontal or slightly oblique rootstock (usually not reaching deep into substrate) and ± crowded 2d-order roots. **Stems** erect or rarely ascending, 1 to several from base, branched in distal $^1/_2$ (in inflorescence), (25–)30–90 (–110) cm. **Leaves:** ocrea normally laciniate; blade oblong-ovate, ovate-lanceolate, to lanceolate, 4–10(–15) × 1–4(–6) cm, normally more than 2.5 times as long as wide, base sagittate (with acute lobes directed downward, ± parallel to petiole), margins entire, normally flat, apex acute or subacute. **Inflorescences** terminal, occupying distal $^1/_3$ of stem, usually lax and interrupted especially in proximal part, narrowly paniculate, cylindric (with 1st-order branches simple, or with few 2d-order branches). **Pedicels** articulated near middle, filiform, 2–5(–6) mm, articulation distinct. **Flowers** (2–)4–8(–10) in whorls; inner tepals orbiculate, occasionally broadly ovate, 3–4(–5) × 3–4 mm, base rounded or cordate, apex obtuse; tubercles small or occasionally absent. **Achenes** black to dark brown, 1.8–2.5 × 1.2–1.5 mm, shiny, smooth. $2n$ = 14 (pistillate plants), 15 (staminate plants).

Flowering spring–early summer. Waste places, meadows, cultivated fields, alluvial habitats; 0–1000 m; introduced; St. Pierre and Miquelon; Alta., B.C., Man., N.B., Nfld. and Labr. (Nfld)., N.S., Ont., Que., Sask.; Alaska, Conn., Maine., Mass., Mich., Minn., N.H., N.Y., Oreg., Pa., Vt.; Europe; nw Africa; Asia.

Rumex acetosa is morphologically uniform in North America. It sometimes is misidentified as *R. hastatulus* or *R. acetosella*. Collections from North America are few in herbaria, and this species probably is not as common in the flora area as has been generally assumed. Some literature reports for *R. acetosa* may refer to other taxa of the species group.

8. **Rumex lapponicus** (Hiitonen) Czernov in B. N. Gorodkov and A. I. Pojarkova, Fl. Murmansk. Obl. 3: 154. 1956 • Lapland or Lapland mountain sorrel [F]

Rumex acetosa Linnaeus subsp. *lapponicus* Hiitonen, Suom. Kasvio, 300. 1933; *Acetosa alpestris* (Jacquin) Á. Löve subsp. *lapponica* (Hiitonen) Á. Löve; *A. lapponica* (Hiitonen) Holub; *Rumex alpestris* Jacquin subsp. *lapponicus* (Hiitonen) Jalas

Plants perennial, glabrous or nearly so, with short and relatively thin, horizontal or slightly oblique rootstock (usually not reaching deep into substrate) and ± crowded 2d-order roots (occasionally with short-creeping rhizome). **Stems** erect, rarely ascending, several from base or often solitary, branched in distal ¹/₂ (in inflorescence), (10–)20–60(–100) cm. **Leaves:** ocrea of at least middle and distal leaves with margins entire, not fringed, sometimes laciniate but only in distal parts; blade broadly ovate, rarely almost round, oblong-ovate, rarely oblong-lanceolate, 3–10(–14) × 1–4(–5) cm, normally less than 2.5 times as long as wide, base sagittate (with acute or subacute lobes directed downward, ± parallel to petiole, or slightly incurved inward), margins entire, normally flat, apex subacute or obtuse. **Inflorescences** terminal, occupying distal ¹/₃ of stem, usually lax and interrupted especially in proximal part, narrowly paniculate or occasionally simple, cylindric (with 1st-order branches simple, or with few 2d-order branches). **Pedicels** articulated near middle, filiform, 2–5 mm, articulation distinct. **Flowers** (2–)4–8 in whorls; inner tepals orbiculate, occasionally broadly ovate, 3.5–4.5 × 3.5–4.5 mm, base rounded or cordate, apex obtuse; tubercles small or occasionally absent. **Achenes** brown or dark brown to brownish yellow, 1.7–2.5 × 0.9–1.3 mm, dull. $2n = 14$ (pistillate plants), 15 (staminate plants).

Flowering late spring–summer. Meadows, rock outcrops, alluvial habitats along rivers and streams in tundra and montane and subalpine zones; 0–2500 m; Greenland; Alta., B.C., N.W.T., Yukon; Alaska, Mont., Wyo; n Eurasia.

The name *Rumex acetosa* has been commonly misapplied to *R. lapponicus*. Native North American montane plants of the *R. acetosa* aggregate from the Rocky Mountains southward to Beartooth Plateau in Montana and Wyoming usually were referred to as *R. alpestris* [= *Acetosa pratensis* Miller subsp. *alpestris* (Jacquin) Á. Löve; *A. pratensis* subsp. *arifolia* (A. Blytt & O. C. Dahl) Á. Löve; *R. arifolius* Allioni, not Linnaeus f.; *R. acetosa* subsp. *alpestris* (Jacquin) Á. Löve; *R. acetosa* subsp. *arifolius* A. Blytt & O. C. Dahl]. Recent nomenclatural studies demonstrated that *R. alpestris* is an ambiguous name, which was probably based on plants belonging to *R. scutatus* Linnaeus (see I. O. Pestova 1998), and accepted the name *R. arifolius* for the predominantly European montane taxon. It differs from arctic plants, as well as from montane forms of the *R. acetosa* aggregate from southern Siberia and temperate North America, by its more robust habit, more branched inflorescence (similar to that of *R. thyrsiflorus*), and larger and more acute triangular-sagittate leaves (see A. I. Tolmachew 1966; N. N. Tzvelev 1989b; Pestova; R. Elven et al. 2000). Montane races possibly developed independently from *R. lapponicus*-like or *R. acetosa*-like ancestors, and they are still unclear taxonomically. Because of that, I prefer to keep those forms within *R. lapponicus*. The whole aggregate needs careful study; however, some authors prefer to include all arcto-montane Holarctic races of this aggregate in the collective and rather polymorphic *R. alpestris* in the broad sense (see Á. Löve 1944; Löve and D. Löve 1957).

Some arctic plants from western Alaska may be conspecific with *Rumex pseudoxyria* (Tolmatchew) Khokhrjakov [= *R. acetosa* subsp. *pseudoxyria* Tolmatchew; *Acetosa pseudoxyria* (Tolmatchew) Tzvelev], a taxon described from arctic eastern Siberia (A. I. Tolmachew 1966). This entity is evidently closely related to *R. lapponicus* but differs from all other members of the *R. acetosa* group in having basal leaves less than two times as long as wide, almost hastate or at least rounded-truncate at base, resembling those of *Oxyria digyna* (Linnaeus) Hill, cauline leaves small or completely reduced, and inflorescence occupying more than one-half of the stem; the plant itself is also somewhat similar in appearance to the European alpine species *R. nivalis* Hegetschweiler.

According to E. Hultén (1973), his *Rumex arcticus* var. *perlatus*, described from a single specimen collected at Tin City, Seward Peninsula, Alaska, agrees perfectly with the original description of *R. acetosa* subsp. *pseudoxyria*. It is unlikely that Hultén would confuse two rather distantly related groups. There is also a possibility that var. *perlatus* is identical with *R. arcticus* var. *latifolius* Tolmatchew (see discussion under *R. arcticus*). Additional collections are needed to confirm the occurrence of *R. pseudoxyria* in northwestern North America.

9. Rumex thyrsiflorus Fingerhuth, Linnaea 4: 380. 1829 • Narrow-leaved sorrel, grande oseille thyrsiflore F I

Acetosa thyrsiflora (Fingerhuth) Á. Löve & D. Löve; *Rumex acetosa* Linnaeus subsp. *auriculatus* (Wallroth) A. Blytt & O. C. Dahl; *R. acetosa* var. *auriculatus* Wallroth; *R. acetosa* var. *crispus* (Roth) Čelakovský; *R. acetosa* var. *haplorhizus* (Czernjaev ex Turczaninow) Trautvetter; *R. acetosa* subsp. *thyrsiflorus* (Fingerhuth) Čelakovský; *R. auriculatus* (Wallroth) Murbeck; *R. haplorhizus* Czernjaev ex Turczaninow

Plants perennial, glabrous or nearly so, with thick, vertical or oblique rootstock (reaching deep into substrate) and remote 2d-order roots. **Stems** usually erect, several from base, or occasionally solitary, branched in distal ¹/₂ (in inflorescence), (30–)40–100 (–130) cm. **Leaves:** ocrea often with fringed margins; blade oblong-lanceolate to lanceolate, 3–12(–15) × 1–3(–5) cm, usually more than 4 times as long as wide, base sagittate or sometimes hastate (with acute lobes directed downward, ± parallel to petiole, or often reflexed outward), margins entire to obscurely and irregularly repand, usually crisped and undulate, occasionally flat, apex acute. **Inflorescences** terminal, occupying distal ¹/₃ of stem, usually dense, or interrupted in proximal part, broadly paniculate, pyramidal (1st-order branches usually repeatedly branched, with numerous 2d-order branches). **Pedicels** articulated near middle, filiform, 2–6(–7) mm, articulation distinct. **Flowers** (3–)4–8(–12) in whorls; inner tepals orbiculate, occasionally broadly ovate, 2.5–3.5(–4) × 2.5–3.5 mm, base rounded, truncate, or slightly cordate, apex obtuse; tubercles small or occasionally absent. **Achenes** black or dark brown, 1.5–1.8 × 0.8–1.2 mm, normally smooth. $2n = 14$ (pistillate plants), 15 (staminate plants).

Flowering late spring–early summer. Meadows, alluvial habitats, waste places, roadsides, edges of woods; 0–1400 m (in Europe); introduced; N.B., N.S., Ont., Que., Sask.; Mich.; c, e Europe; c Asia (s Siberia); introduced elsewhere.

Rumex thyrsiflorus is commonly misidentified as *R. acetosa*. The growth habit (stout, vertical rootstock), narrower, often undulate leaves with often slightly spreading basal lobes (however, some European specimens have the lobes of distal and middle cauline leaves curved inward), and pyramidal, usually much-branched panicle of *R. thyrsiflorus* are traits especially useful for field identification. In addition, the inner tepals of *R. thyrsiflorus* are distinctly smaller than those of *R. acetosa*. The southern European (Mediterranean) race of *R. thyrsiflorus*, characterized by narrower leaves with more spreading, almost hastate basal lobes and fruiting inner tepals less cordate at the base, is sometimes recognized as *R. intermedius* de Candolle [= *Acetosa thyrsiflora* subsp. *intermedia* (de Candolle) Á. Löve]. The same forms occasionally occur in North America (Á. Löve 1986).

26c. RUMEX Linnaeus subg. RUMEX

Lapathum Miller; *Rumex* sect. *Lapathum* (Miller) Campderá; *Rumex* subg. *Lapathum* (Miller) Rechinger f.

Plants synoecious (rarely polygamodioecious or dioecious individuals occur in some species). **Leaf blades** variable, cuneate, truncate, rounded, or cordate at base, but never hastate or sagittate, unlobed or in some species with rounded basal lobes. **Pedicels** articulated near middle or in proximal part (sometimes articulation indistinct). **Flowers** normally bisexual or, sometimes, bisexual and unisexual (staminate and pistillate on same plant) within same inflorescence; outer tepals normally angled towards inner tepals, or sometimes spreading; inner tepals distinctly enlarged, always distinctly wider and longer than achene; tubercles present or absent.

Species at least 150 (53 in the flora): nearly wordwide.

26c.1. RUMEX Linnaeus (subg. RUMEX) sect. AXILLARES Rechinger f., Publ. Field Mus. Nat. Hist., Bot. Ser. 17: 6. 1937

Plants not developing basal rosette of leaves, sometimes with long-creeping rhizomes. **Stems** erect, ascending, procumbent, or decumbent, normally with regular, leafy axillary shoots that tend to develop 2d-order axillary inflorescences (often overtopping 1st-order ones). **Leaf blades**

mostly lanceolate, elliptic, ovate-lanceolate, or ovate-elliptic, base cuneate, or in some species broadly cuneate, rounded, truncate-cuneate, or indistinctly cordate. **Inner tepals** with margins entire, rarely minutely erose-denticulate.

Species ca. 40 (21) in the flora: North America, South America, e Asia, Pacific Islands (Hawaii).

North american representatives of this section (except *Rumex venosus*, a distinctive species) form a polymorphic complex consisting of often interfertile races (N. M. Sarkar 1958). In most cases those races are separated morphologically and restricted geographically. They may be arranged into at least three aggregates (groups) that approximate the subsections described by K. H. Rechinger (1937): *R. verticillatus* aggr. (subsect. *Verticillati* Rechinger f.), *R. altissimus* aggr. (subsect. *Salicifolii* Rechinger f., in part), and *R. salicifolius* aggr. (subsect. *Salicifolii*, in part). In case of difficulty with identification, inevitable with immature or intermediate specimens, determination to the aggregate level is recommended. Those preferring more broadly circumscribed species may use existing infrageneric combinations (see J. C. Hickman 1984). Attempts have been made to submerge various "microspecies" (especially in the *R. salicifolius* group) into more broadly circumscribed taxa, usually *R. salicifolius* and *R. mexicanus*, and various combinations at subspecific and varietal ranks have been proposed (see Hickman). In my opinion, results of this approach are inconsistent (e.g., some taxa are treated as varieties while others, not less distinct, are accepted as species or subspecies) and in most cases less convincing than the original treatment by Rechinger and the thorough study by Sarkar. I agree with Sarkar (p. 993) that "any drastic taxonomic revision of the species of the *Axillares* section of *Rumex* should be postponed until more complete cytogenetic data have accumulated concerning the interrelationships of all the taxa in this section."

SELECTED REFERENCE Sarkar, N. M. 1958. Cytotaxonomic studies on *Rumex* sect. *Axillares*. Canad. J. Bot. 36: 947–996.

10. Rumex venosus Pursh, Fl. Amer. Sept. 2: 733. 1813

  • Veined or veiny dock, wild-begonia, rumex veine

E F W

Plants perennial, glabrous or nearly so, with creeping rhizomes. **Stems** ascending or, rarely, erect, usually producing axillary shoots near base, (10–)15–30(–40) cm. **Leaf blades** ovate-elliptic, ob-ovate-elliptic, or ovate-lanceolate, (2–)4–12(–15) × 1–5(–6) cm, subcoriaceous, base narrowly to broadly cuneate, margins entire, flat or slightly undulate, apex acute or acuminate. **Inflorescences** terminal and axillary, usually occupying distal 2/3 of stem/shoot, usually dense, or interrupted in proximal part, broadly paniculate. **Pedicels** articulated near middle, filiform or slightly thickened, (8–)10–16 mm, articulation distinct, slightly swollen. **Flowers** 5–15 in whorls; inner tepals distinctly double-reticulately veined, orbiculate or reniform-orbiculate, 13–18(–20) × (20–)23–30 mm, base deeply emarginate or cordate, margins entire, apex rounded, obtuse, rarely subacute, with short, broadly triangular tip; tubercles absent, occasionally very small. **Achenes** brown or dark brown, 5–7 × 4–6 mm. *2n* = 40.

Flowering spring–early summer. Sand dunes, sandy and gravelly riverbanks and slopes, deserts, grasslands 200–1500 m; Alta., Man., Sask.; Ariz., Calif., Colo., Idaho, Iowa, Kans., Minn., Mont., Nebr., Nev., N.Mex., N.Dak., Okla., Oreg., S.Dak., Tex., Utah, Wash., Wis., Wyo.

Rumex venosus is a distinctive species rarely confused with any other members of the genus. However, I have seen herbarium specimens of it misidentified as *R. hymenosepalus*, and vice versa.

11. Rumex verticillatus Linnaeus, Sp. Pl. 1: 334. 1753

  • Swamp dock   E F W

Plants perennial, glabrous or nearly so, with vertical rootstock. **Stems** erect or, rarely, ascending, simple or producing axillary shoots below 1st-order inflorescence or at proximal nodes, 40–100(–150) cm. **Leaf blades** with lateral veins forming angle of 45–60° with midvein, linear-lanceolate or lanceolate, 5–30(–40) × 1–5 cm, usually 5–7(–10) times as long as wide, normally rather thin or at most subcoriaceous, base narrowly cuneate, margins entire, flat or slightly undulate, apex acute or acuminate. **Inflorescences** terminal and axillary, terminal usually occupying distal 1/3–1/2 of stem, usually lax, interrupted at least in basal 1/2, narrowly paniculate. **Pedicels**

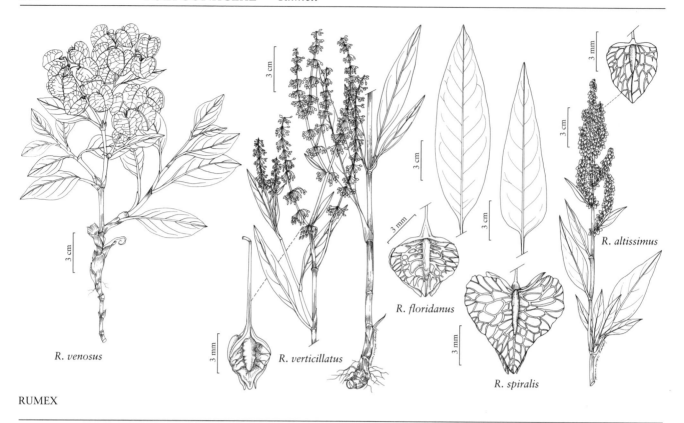

R. venosus

R. verticillatus

R. floridanus

R. spiralis

R. altissimus

RUMEX

articulated in proximal part, distinctly thickened distally, 10–17 mm, (2.5–)3–5 times as long as inner tepals, articulation distinctly or slightly swollen. **Flowers** 10–15(–25) in remote whorls; inner tepals ovate-triangular or ovate-deltoid, 3.5–5 × 2.5–4 mm, longer than wide or, very rarely, as long as wide, base truncate or rounded, margins entire or, rarely, very indistinctly erose, apex acute or subacute (then with broadly triangular-lingulate tip); tubercles 3, equal or subequal, minutely punctate and/or transversely rugose (wrinkled) in proximal part. **Achenes** brown or dark brown, 2.3–3.1 × 1.6–2.2 mm. $2n = 60$.

Flowering spring–early summer. Swamps, bogs, marshes, wet meadows, irrigation ditches, wet alluvial woods; 0–800 m; Ont., Que.; Ala., Ark., Conn., Del., Fla., Ga., Ill., Ind., Iowa, Kans., Ky., La., Md., Mass., Mich., Minn., Miss., Mo., Nebr., N.J., N.Y., N.C., Ohio, Okla., Pa., R.I., S.C., S.Dak., Tenn., Tex., Vt., Va., W.Va., Wis.

Reports of *Rumex verticillatus* for New Mexico (W. C. Martin and C. R. Hutchins 1980) need confirmation. The species was reported erroneously from Colorado (S. L. O'Kane et al. 1988) as a result of misidentification of *R. fueginus* (see W. A. Weber and R. C. Wittmann 1992).

I have not seen specimens of *Rumex verticillatus* from Delaware, Maine, New Hampshire, and New Jersey, but the species probably occurs in those states.

The following two species are closely related to *Rumex verticillatus* and sometimes treated as subspecies of it.

12. **Rumex floridanus** Meisner in A. P. de Candolle and A. L. P. P. de Candolle, Prodr. 14: 46. 1856
    • Florida dock   E F

Rumex verticillatus Linnaeus subsp. *floridanus* (Meisner) Á. Löve

Plants perennial, glabrous, with vertical rootstock. **Stems** erect, rarely ascending, especially axillary shoots, usually producing axillary shoots below 1st-order inflorescence, or at proximal nodes, 40–80(–120) cm. **Leaf blades** with lateral veins forming angle of 40–60° with midvein, lanceolate or broadly lanceolate, rarely ovate-lanceolate, 7–20(–30) × 3–5 cm, usually 3–5(–6) times as long as wide, normally rather fleshy, coriaceous or subcoriaceous when dry, base narrowly to broadly cuneate, margins entire, flat, apex acute or acuminate. **Inflorescences** terminal and axillary, terminal usually occupying distal 1/3–1/2 of stem, usually rather dense, interrupted only near base, narrowly to broadly paniculate. **Pedicels** articulated in proximal part, distinctly thickened distally, 7–15 mm, usually 2.5–3 times as long as inner tepals, articulation distinctly swollen. **Flowers** 10–20(–30) in whorls; inner tepals, broadly ovate-deltoid or deltoid, (3.5–)4–5.5 × 4–6 mm, usually as wide as or wider than long, base truncate, margins entire or, rarely, very indistinctly erose, apex acute or acuminate; tubercles 3, equal or subequal, often verrucose and/or transversely rugose (wrinkled) in proximal part. **Achenes** brown or dark brown, 2.5–3.5 × 2–3 mm. $2n = 60$.

Flowering late spring–early summer. Swamps, marshes, bogs, riverbanks, alluvial woods; 0–200 m; Ala., Del., Fla., Ga., La., Miss., N.J., N.C., S.C.

Rumex floridanus is closely related to and sometimes treated as a subspecies of *R. verticillatus*. Distribution of *R. floridanus* is not known sufficiently because of frequent confusion with *R. verticillatus*. Moreover, the name was partly misapplied by W. D. Trelease (1892) to *R. chrysocarpus*.

13. **Rumex fascicularis** Small, Bull. Torrey Bot. Club 22: 367, plate 246. 1895 [E]

Rumex verticillatus Linnaeus subsp. *fascicularis* (Small) Á. Löve

Plants perennial, glabrous, with vertical rootstock [and, according to Rechinger f. (1937), with fusiform-incrassate root fibers]. **Stems** ascending or decumbent, usually producing axillary shoots below 1st-order inflorescence or at proximal nodes, 50–60(–70) cm. **Leaf blades** with lateral veins forming angle of ca. 80° with midvein especially near base, ovate or ovate-elliptic, 10–25 × 4–12 cm, usually ca. 2 times as long as wide, fleshy, coriaceous, base rounded or truncate-cuneate, occasionally indistinctly cordate, margins entire, flat, apex acute. **Inflorescences** terminal and axillary, terminal usually occupying distal 1/3–1/2 of stem, usually lax, interrupted in proximal part, broadly paniculate. **Pedicels** articulated in proximal part, distinctly thickened distally 8–13 mm, (2.5–)3–4 times as long as inner tepals, articulation slightly swollen. **Flowers** 10–20 in whorls; inner tepals orbiculate or rounded-triangular, 4–5 × 4–5 mm, base truncate or subcordate, margins entire, or rarely indistinctly erose, apex acute or acuminate (with broadly triangular tip); tubercles 3, equal or subequal, minutely punctate and/or rugose in proximal part. **Achenes** brown or dark brown, 2–2.5(–3) × 1.8–2.5 mm. $2n = 60$.

Flowering spring–early summer. Swamps, marshes, wet meadows, shores of lakes and rivers; 0–100 m; Fla.

Rumex fascicularis was mentioned for North Carolina (Á. Löve 1986). It is closely related to and sometimes treated as a subspecies of *R. verticillatus*.

14. **Rumex spiralis** Small, Bull. Torrey Bot. Club 22: 44, plate 228. 1895 • Spiral tall dock [E] [F]

Plants perennial, glabrous, with creeping rhizomes. **Stems** ascending or erect, usually producing axillary shoots below 1st-order inflorescence or at proximal nodes, 50–90 cm. **Leaf blades** ovate-lanceolate, oblong-lanceolate, or lanceolate, 10–15 × 3–5.5 cm, usually 2.5–3.5 times as long as wide, widest in proximal 1/3, thick, usually not coriaceous, base broadly cuneate, truncate, or rounded, margins entire, flat or slightly undulate-crisped, apex acute or attenuate. **Inflorescences** terminal and axillary, terminal usually occupying distal 1/2 of stem, dense, narrowly to broadly paniculate (branches usually simple). **Pedicels** articulated in proximal 1/3, thin but slightly thickened distally, (2–)3–7(–8) mm, usually as long as or shorter than inner tepals, articulation slightly swollen. **Flowers** 12–20 in whorls; inner tepals broadly cordate or broadly ovate-deltoid, 7–10 × 8–12 mm, base deeply and broadly cordate, margins entire, apex acuminate; tubercles 3, equal or subequal, usually minutely to distinctly rugose. **Achenes** brown or dark reddish brown, 2.5–3.5 × 2–2.5 mm. $2n = 20$.

Flowering spring. Sandy and gravelly shores; 0–200 m; Tex.

Rumex spiralis is related to *R. altissimus*; however, it is geographically restricted and morphologically distinct. It has inner tepals larger than those of any other member of subsect. *Salicifolii* and distinctly wider leaves.

15. **Rumex altissimus** Alph. Wood, Class-book Bot. ed. 2, 477. 1847 • Tall or pale dock [F] [W]

Plants perennial, glabrous, with vertical rootstock. **Stems** erect, rarely ascending, usually producing axillary shoots below 1st-order inflorescence or at proximal nodes, 50–90(–120) cm. **Leaf blades** ovate-lanceolate, elliptic-lanceolate, or lanceolate, 10–15 × 3–5.5 cm, usually ca. 2.5–4 times as long as wide, widest in proximal 1/2, thick, often subcoriaceous, base broadly cuneate, rarely almost rounded, margins entire, flat, apex acute or attenuate. **Inflorescences** terminal and axillary, terminal usually occupying distal 1/5–1/3 of stem, rather dense, normally broadly paniculate. **Pedicels** articulated in proximal 1/3, sometimes almost near base, thick, (2–)3–7(–8) mm, usually approximately as long as inner tepals, occasionally slightly longer or shorter, articulation swollen. **Flowers** 12–20 in whorls; inner tepals with broadly triangular, ovate-triangular, or broadly ovate-deltoid, 4.5–6 × 3–4.5(–5) mm, base truncate or indistinctly cordate, margins entire, apex acute; tubercles (2–)3, equal or subequal, glabrous or minutely rugose. **Achenes** brown or dark reddish brown, 2.5–3.5 × 1.8–2.3 mm. $2n = 20$.

Flowering late spring–summer. Swamps, marshes, wet shores, alluvial woods, other wet habitats; 0–1800 m; Ont.; Ala., Ariz., Ark., Colo., Conn., Del., D.C., Fla., Ga., Ill., Ind., Iowa, Kans., Ky., La., Maine, Md., Mass., Mich., Minn., Miss., Mo., Nebr., N.H., N.J., N.Mex., N.Y., N.C., Ohio, Okla., Pa., R.I., S.C., S.Dak., Tenn., Tex., Vt., Va., W.Va., Wis., Wyo.; n Mexico; Europe (introduced in Denmark, Sweden, United Kingdom.

The name *Rumex britannica* Linnaeus was misapplied to this species by C. F. Meisner (1856) and some North American authors.

Some records of *Rumex altissimus* from Arizona and New Mexico may refer to *R. ellipticus*. Two reports from New Mexico were based on misidentification of *R. hymenosepalus*.

16. **Rumex ellipticus** Greene, Pittonia 4: 234. 1901

• Elliptic tall dock E

Rumex altissimus Alph. Wood subsp. *ellipticus* (Greene) Á. Löve

Plants perennial, glabrous, with vertical rootstock. **Stems** ascending or decumbent, usually producing axillary shoots below 1st-order inflorescence or at proximal nodes, 40–70 cm. **Leaf blades** lanceolate or broadly lanceolate, 5–10(–15) × 2–3(–4) cm, usually ca. 3–4 times as long as wide, widest in proximal 1/2, rarely near middle, thick, often subcoriaceous, base cuneate, margins entire, flat, apex acute or attenuate. **Inflorescences** terminal and axillary, terminal usually occupying distal 1/5–1/3 of stem, rather dense or interrupted in proximal 1/2, usually narrowly paniculate (branches simple and comparatively short). **Pedicels** articulated in proximal 1/2 almost near base, thickish, 3–6(–7) mm, usually approximately as long as or slightly shorter than inner tepals, articulation indistinctly swollen. **Flowers** 12–20 in whorls; inner tepals ovate or cordate-triangular, occasionally almost orbiculate, 5–6 × 4–5 mm, base truncate or indistinctly cordate, margins entire, apex obtuse or subacute; tubercles absent or 1 inner tepal with indistinctly swollen midvein. **Achenes** brown, 2.7–3.2 × 1.8–2.5 mm. $2n = 20$.

Flowering spring–early summer. Sandy, gravelly, and muddy shores of rivers and streams; 10–1000 m; Ariz., N.Mex., Tex.

Rumex ellipticus is closely related to *R. altissimus* and is sometimes regarded as a subspecies of it (Á. Löve 1986).

17. **Rumex chrysocarpus** Moris, Mem. Reale Accad. Sci. Torino 38: 46. 1835 (as chrysocarpos)

• Amamastla dock

Rumex berlandieri Meisner; *R. langloisii* Small

Plants perennial, glabrous, with vertical rootstock. **Stems** ascending or erect, usually producing axillary shoots below 1st-order inflorescence or at proximal nodes, 40–60(–80) cm. **Leaf blades** deep olive green, with strongly prominent veins abaxially, linear-lanceolate, occasionally lanceolate, 5–12 × 1.5–4 cm, usually ca. 3.5–5 times as long as wide, widest near or below middle, thick, coriaceous, base cuneate or rounded-cuneate, margins entire to crenulate, undulate or crisped, flat, apex subobtuse or broadly acute. **Inflorescences** terminal and axillary, terminal usually occupying distal 1/3 of stem, lax, interrupted almost to top, usually broadly paniculate (branches simple or nearly so). **Pedicels** articulated in proximal 1/3 or almost near middle, filiform or slightly thickened, especially distally 3–6(–7) mm, not more than 2–2.5 times as long as inner tepals, articulation indistinctly swollen. **Flowers** 5–15 in whorls; inner tepals orbiculate, ovate-deltoid, or ovate-triangular, 3.5–4.5(–5) × 3–4(–4.5) mm, base truncate or rarely indistinctly cordate, margins entire, apex subacute; tubercles 3, equal or subequal, much narrower than inner tepals, free margins of inner tepals wider than or at least as wide as tubercle, verrucose to subglabrous. **Achenes** brown or dark reddish brown, 2.5–3 × 1.5–2 mm. $2n = 20$.

Flowering spring–summer. Swamps, marshes, shores, wet alluvial forests; 0–200 m; La., Tex.; ne Mexico (Tamaulipas).

The name *Rumex floridanus* was misapplied (in part) to this species by W. D. Trelease (1892) and other North American authors. *Rumex chrysocarpus* is distinctive and rarely confused with other species of the *R. salicifolius* aggregate.

18. **Rumex salicifolius** Weinmann, Flora 4: 28. 1821

• Willow or willow-leaved dock

Plants perennial, glabrous, with vertical rootstock. **Stems** erect or ascending, usually producing axillary shoots below 1st-order inflorescence or at proximal nodes, 30–60(–90) cm. **Leaf blades** linear-lanceolate, occasionally almost linear, 5–13 × (0.5–)1–2.5 cm, usually ca. 7 or more times as long as wide, widest near middle, thin, occasionally subcoriaceous, base cuneate, margins entire, flat or rarely slightly crenulate, apex acute or attenuate.

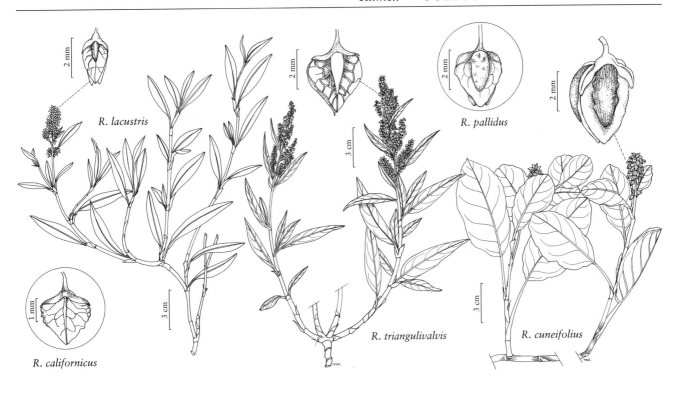

R. lacustris

R. pallidus

R. californicus

R. triangulivalvis

R. cuneifolius

RUMEX

Inflorescences terminal and axillary, terminal usually occupying distal ¹/₅–¹/₃ of stem, rather lax, interrupted in proximal ¹/₂, or almost to top, usually narrowly paniculate (branches normally simple and short). **Pedicels** articulated in proximal ¹/₃ or almost near base, filiform (slightly thickened towards base of tepals), 3–5 mm, not more than 2–2.5 times as long as inner tepals, articulation indistinctly swollen. **Flowers** 7–20 in whorls; inner tepals broadly triangular, (1.8–)2–2.5(–3) × 1.5–2.1 mm, base broadly cuneate or truncate, margins entire or indistinctly erose, apex acute; tubercle 1, large, subequal or slightly narrower than inner tepals (then free margins of inner tepal distinctly narrower than tubercle), smooth or indistinctly verrucose. **Achenes** dark reddish brown, 1.8–2 × 1.1–1.3 mm. $2n = 20$.

Flowering late spring–early summer. Shores of streams and rivers, wet mountain meadows, and rocky slopes; 0–3000 m; Ariz., Calif., Nev.; n Mexico (Sonora).

Rumex salicifolius occurs mostly in southern and central California; it has been reported also from adjacent parts of Arizona (N. M. Sarkar 1958) and Nevada (J. T. Kartesz 1987, vol. 1). The name *R. salicifolius* has been applied in a broad sense to nearly all species of subsect. *Salicifolii*, including even mostly Asian *R. sibiricus*. *Rumex salicifolius* appears to be most closely related to *R. californicus* and *R. utahensis*.

19. Rumex californicus Rechinger f., Repert. Spec. Nov. Regni Veg. 40: 297. 1936 • California willow dock F

Rumex salicifolius Weinmann var. *denticulatus* Torrey

Plants perennial, glabrous, with vertical rootstock. **Stems** usually ascending, rarely decumbent-ascending or suberect, usually producing axillary shoots below 1st-order inflorescence or at proximal nodes, 30–60 cm. **Leaf blades** linear-lanceolate or linear-oblanceolate, 5–10 × 1–3 cm, usually ca. (3–)5–7 times as long as wide, usually widest near middle, thin or, occasionally, subcoriaceous, base cuneate, margins entire, flat or, occasionally, undulate near base, apex acute or attenuate. **Inflorescences** terminal and axillary, terminal usually occupying distal ¹/₅–¹/₃ of stem, lax, interrupted at least in proximal ¹/₂, usually broadly paniculate (branches simple or with few 2d-order branches). **Pedicels** articulated in proximal ¹/₃ or almost near base, filiform, 3–8 mm, not more than 2–2.5 times as long as inner tepals, articulation indistinctly swollen. **Flowers** 10–15 (–20) in whorls; inner tepals usually broadly triangular or deltoid, 2.5–3.5 × 2.2–3.3 mm, base truncate, margins minutely but distinctly denticulate, rarely subentire, apex obtuse or subacute; tubercles absent, or only 1 midvein slightly swollen. **Achenes** brown or dark reddish brown, 2 × 1.3 mm. **2*n* = 20.**

Flowering late spring–summer. Moist coastal, alluvial, and montane habitats; 0–3000 m; Ariz., Calif., Nev., Oreg.; possibly n Mexico.

Rumex californicus is closely related to and often is regarded as a variety of *R. salicifolius.*

In Oregon this species has been reported only as a ballast waif in the Albina neighborhood of Portland (K. H. Rechinger 1937). It has been reported also from northeastern Nevada (J. T. Kartesz 1987, vol. 1), New Mexico (W. C. Martin and C. R. Hutchins 1980), and Wyoming (N. M. Sarkar 1958), but these records require confirmation.

20. **Rumex utahensis** Rechinger f., Repert. Spec. Nov. Regni Veg. 40: 298. 1936 • Utah willow dock [E]

Plants perennial, glabrous, with vertical rootstock. **Stems** usually erect, occasionally ascending, usually producing axillary shoots below 1st-order inflorescence or at proximal nodes, 15–40(–60) cm. **Leaf blades** lanceolate or linear-lanceolate, 6–15 × 2–3 cm, usually ca. 4–5 times as long as wide, widest near middle or slightly towards base, usually thin, base cuneate, margin entire, flat, rarely indistinctly undulate, apex acute. **Inflorescences** terminal and axillary, terminal usually occupying distal $^1/_5$–$^1/_3$ of stem, dense or occasionally slightly interrupted at base, usually broadly paniculate (branches normally simple and crowded). **Pedicels** articulated in proximal $^1/_3$ or almost near base, filiform (but thickened distally, 4–7 mm, not more than 2–2.5 times as long as inner tepals, articulation indistinctly or evidently swollen. **Flowers** 10–25 in whorls; inner tepals deltoid or broadly ovate-deltoid, 2.5–3 × 2.5–3 mm, base truncate or indistinctly cordate, margins entire, apex acute, rarely subacute; tubercles absent. **Achenes** dark reddish brown or almost black, 1.8–2 × 1–1.3 mm. **2*n* = 40.**

Flowering late spring–summer. Shores of rivers and streams, wet meadows, rocky slopes; 1000–3500 m; Alta.; Calif., Colo., Idaho, Mont., Nev., Oreg., Utah, Wyo.

The names *Rumex mexicanus* and *R. salicifolius* in the broad sense often have been applied to *R. utahensis.*

Records of "narrow-leaved forms" of *Rumex utahensis* from Yukon (E. Hultén 1968) probably refer to *R. hultenii* Tzvelev (see comments under 27. *R. sibiricus*) or *R. sibiricus* in the narrow sense.

21. **Rumex crassus** Rechinger f., Repert. Spec. Nov. Regni Veg. 40: 295. 1936 • Fleshy willow dock [E]

Plants perennial, glabrous, with vertical rootstock. **Stems** procumbent, occasionally ascending, usually producing axillary shoots below 1st-order inflorescence or at proximal nodes, 20–50(–60) cm. **Leaf blades** ovate-lanceolate, elliptic-lanceolate, or ovate-elliptic, 3–12 × 1–5 cm, usually not more than 2–3.5 times as long as wide, widest near middle or occasionally in proximal $^1/_2$, thick, coriaceous, base broadly cuneate, occasionally almost truncate or cuneate, margins entire, flat, apex acute. **Inflorescences** terminal and axillary, terminal usually occupying distal $^1/_3$–$^1/_2$ of stem, dense or interrupted only near base, usually broadly paniculate (branches simple or nearly so). **Pedicels** articulated in proximal $^1/_3$ or almost near base, usually rather thick, 4–9 mm, not more than 2–2.5 times as long as inner tepals, articulation distinctly swollen. **Flowers** 10–25 in whorls; inner tepals broadly ovate, ovate, or lingulate-deltoid, (3–)4–5 × (2.5–)3–4 mm, base rounded or subtruncate, margins entire or indistinctly erose, apex acute or subacute; tubercle 1, large, subequal to inner tepals, or slightly narrower than inner tepals (then free margins of inner tepal distinctly narrower than tubercle), subglabrous to verrucose. **Achenes** brown or dark reddish brown, 2–2.5 × 1.7–2.1 mm. **2*n* = 20.**

Flowering early spring–summer. Coastal dunes, sandy shores and marshes; 0–100 m; Calif., Oreg.

22. **Rumex lacustris** Greene, Erythea 3: 63. 1895 • Lake willow dock [E] [F]

Rumex salicifolius Weinmann var. *lacustris* (Greene) J. C. Hickman

Plants perennial, glabrous or papillose-pubescent especially on abaxial surface of leaf blades in terrestrial forms, with vertical or creeping rootstock, occasionally with creeping rhizomes. **Stems** erect, ascending, or submerged and/or floating (in aquatic forms), usually producing numerous axillary shoots below 1st-order inflorescence or at proximal nodes (especially in aquatic plants), 40–70(–90) cm. **Leaf blades** very variable, in aquatic submerged forms usually ovate-lanceolate to lanceolate, glabrous or nearly so; in terrestrial forms lanceolate, papillose-pubescent abaxially, (2–)3–7 × (0.5–)1–3 cm, usually ca. 3.5–5 times as long as wide, widest near middle, thin or occasionally subcoriaceous in terrestrial plants, base cuneate, margins entire, flat or undulate, apex acute or subobtuse. **Inflorescences** terminal and axillary, terminal usually occupying distal ¹/₅ or less of stem, dense or interrupted in proximal ¹/₂ usually narrowly paniculate (branches simple and comparatively short). **Pedicels** articulated in proximal ¹/₂, or almost near middle, filiform (slightly thickened distally), 2.5–4(–6) mm, subequal to inner tepals or at most 2 times longer, articulation indistinctly swollen. **Flowers** 10–20 in whorls; inner tepals ovate or elliptic, occasionally elliptic-triangular, 2–2.5(–3) × 1–1.5(–2) mm, base truncate or broadly rounded-cuneate, margins entire or very indistinctly erose, apex subacute or obtuse; tubercles 3, equal or subequal (much narrower than inner tepals), smooth or slightly verrucose. **Achenes** brown or dark reddish brown, 1.5–2.2 × 0.8–1.2 mm. $2n = 20$.

Flowering spring–early summer. Shores of slightly saline lakes, often aquatic; 1000–2500 m; Calif., Oreg.

Rumex lacustris is represented by two ecotypes: submerged aquatic forma *lacustris* (= forma *aquatilis* Rechinger f.), and terrestrial forma *terrestris* Rechinger f. The latter is unique among representatives of the *R. salicifolius* aggregate, which are normally glabrous, in having distinctly papillose-pubescent leaves. The species may occur also in the adjacent part of northern Nevada.

23. Rumex mexicanus Meisner in A. P. de Candolle and A. L. P. P. de Candolle, Prodr. 14: 45. 1856

• Mexican willow or dock

Plants perennial, glabrous; with vertical rootstock, occasionally with short, creeping rhizomes. **Stems** erect or ascending, usually producing axillary shoots below 1st-order inflorescence or at proximal nodes, 30–60(–90) cm. **Leaf blades** light green to yellowish green, linear-lanceolate, occasionally lanceolate, 6–14 × 1–3.5(–4) cm, usually ca. 5–7 times as long as wide, widest near middle, thin, not coriaceous, base cuneate, margins entire, flat or undulate, apex acute or attenuate. **Inflorescences** terminal and axillary, terminal usually occupying distal ¹/₅–¹/₃ of stem, rather dense or interrupted in proximal ¹/₂, usually broadly paniculate (branches simple or with few 2d-order branches). **Pedicels** articulated in proximal ¹/₃ or almost near base, filiform (thickened distally), 4–7 mm, not more than 2–2.5 times as long as inner tepals, articulation indistinctly swollen. **Flowers** 10–20 in whorls; inner tepals broadly ovate-triangular, occasionally broadly triangular, 3.5–4.5(–5) × 3.5–4 (–5) mm, base truncate or indistinctly cordate, margins

entire or indistinctly erose, apex obtuse or subacute; tubercles 3, equal or subequal (much narrower than inner tepals). **Achenes** brown or dark reddish brown, 2–3 × 1.5–2 mm. $2n = 40$.

Flowering spring–early summer. Shores of streams and rivers, wet meadows; 1000 m; N.Mex.; Mexico.

Some authors recognize *Rumex mexicanus* in the broad sense, including in it many other taxa treated here as separate entities. For consistency, the entities of the *R. salicifolius* aggregate that are recognized herein are kept separate pending additional taxonomic research.

24. Rumex triangulivalvis (Danser) Rechinger f., Repert. Spec. Nov. Regni Veg. 40: 297. 1936

• White, white willow, or triangular-valved dock

E F

Rumex salicifolius Weinmann subsp. *triangulivalvis* Danser, Ned. Kruidk. Arch. 1925: 415, plate 1, figs. 1, 2. 1926; *R. salicifolius* Weinmann var. *triangulivalvis* (Danser) J. C. Hickman

Plants perennial, glabrous, with vertical rootstock, occasionally with short-creeping rhizomes. **Stems** ascending or erect, usually producing axillary shoots below 1st-order inflorescence or at proximal nodes, (30–)40–100 cm. **Leaf blades** light or yellowish green, veins scarcely prominent abaxially, linear-lanceolate, 6–17 × 1–4(–5) cm, usually ca. 5–6 times as long as wide, widest near middle, thin, not coriaceous, base cuneate, margins entire, flat or undulate, apex acute. **Inflorescences** terminal and axillary, terminal usually occupying distal ¹/₅–¹/₃ of stem, rather dense or interrupted in proximal ¹/₂, broadly to narrowly paniculate (branches usually with 2d-order branches, rarely simple). **Pedicels** articulated in proximal ¹/₃ or almost near base, filiform (but slightly thickened distally), 4–8 mm, usually ca. 1.5 times as long as inner tepals, articulation indistinctly swollen. **Flowers** 10–25 in whorls; inner tepals broadly triangular, (2–)2.5–3.5 (–3.8) × (2–)2.5–3(–3.5) mm, base truncate or rounded, margins entire or indistinctly erose only near base, apex acute, occasionally subobtuse-triangular; tubercles usually 3, (1 in some forms, then large, occupying at least 0.5 width of inner tepal), equal or subequal, much narrower than inner tepals glabrous or minutely verrucose. **Achenes** brown or dark reddish brown, 1.7–2.2 × 1–1.5 mm. $2n = 20$.

Flowering late spring–summer. Many types of ruderal and alluvial habitats: waste places, roadsides, railroad embankments, cultivated fields, meadows, sandy and gravelly shores, ditches; 0–2500 m; Alta., B.C., Man., N.B., N.S., N.W.T.,Ont., P.E.I., Que., Sask., Yukon; Ariz., Calif., Colo., Conn., Idaho, Ill., Ind., Iowa, Kans., Ky., ?La., Maine., Mass., Mich., Minn., Mo., Mont., Nebr., Nev., N.H., N.J., N.Mex., N.Y., N.Dak., Ohio, Oreg., Pa., R.I., S.Dak., Tex., Utah, Vt., Wash., W.Va., Wis., Wyo.; Europe; introduced in Czech Republic, Denmark, Finland, Germany, Latvia, Norway, Russia, Switzerland, the Netherlands, Ukraine, United Kingdom, and elsewhere.

Rumex triangulivalvis is the most common and widespread species of the *R. salicifolius* group. It often occurs in ruderal habitats and may be expected outside its present range.

The names *Rumex salicifolius* and *R. mexicanus* (in the broad sense) were commonly applied to this species by many North American and European authors.

25. Rumex transitorius Rechinger f., Repert. Spec. Nov. Regni Veg. 40: 296. 1936 • Pacific willow dock

E

Rumex salicifolius Weinmann var. *transitorius* (Rechinger f.) J. C. Hickman

Plants perennial, glabrous, with vertical rootstock. **Stems** ascending, ascending-decumbent, or erect, simple or producing axillary shoots below 1st-order inflorescence or at proximal nodes, 25–70 cm. **Leaf blades** linear-lanceolate or lanceolate, 6–15(–17) × 2–4 cm, usually ca. 3.5–6 times as long as wide, widest near middle or nearly so, thin or rarely subcoriaceous, base cuneate, margins entire, undulate or slightly crisped, apex acute. **Inflorescences** terminal and axillary, terminal usually occupying distal $^1/_5$–$^1/_3$ of stem, dense or occasionally interrupted near base, usually broadly paniculate (branches simple or with few 2d-order branches). **Pedicels** articulated in proximal $^1/_3$, filiform, 3–7 mm, equaling or 1.5–2 times as long as inner tepals, articulation indistinctly swollen. **Flowers** 10–20(–25) in whorls; inner tepals, broadly ovate to ovate-lanceolate, occasionally almost triangular, (2.5–)3–3.5 × 2–2.5 mm, base truncate or rounded, margins entire or indistinctly erose, apex obtuse or subacute; tubercles 3 (occasionally 1 in var. *monotylos* Rechinger f., then very large, subequal or only slightly narrower than inner tepal), distinctly unequal (1 larger tubercle subequal or slightly narrower than inner tepal), usually smooth. **Achenes** dark reddish brown, 1.8–2.4 × 1–1.5 mm. *2n* = 20.

Flowering late spring–early summer. Coastal dunes and marshes, shores of rivers and streams, wet meadows; 0–2000 m; B.C.; Alaska, Calif., Oreg., Wash.

J. T. Kartesz (1987, vol. 1) reported *Rumex transitorius* from Washoe County, Nevada; the morphological characters mentioned in his description suggest another taxon of the *R. salicifolius* aggregate. Records from Idaho also need confirmation.

26. Rumex pallidus Bigelow, Fl. Boston. ed. 3, 153. 1840 • Seaside or pale willow dock E F

Plants perennial, glabrous, with vertical rootstock, occasionally with short-creeping rhizomes. **Stems** ascending or erect occasionally almost procumbent, usually producing axillary shoots below 1st-order inflorescence or at proximal nodes, 30–60(–80) cm. **Leaf blades** linear-lanceolate, 10–20(–22) × 1–3.5(–4.5) mm, usually ca. 7–10 times as long as wide, widest near middle, thick, coriaceous, base cuneate, margins entire, flat or slightly undulate, apex acute. **Inflorescences** terminal and axillary, terminal usually occupying distal $^1/_5$–$^1/_3$ of stem, rather dense or interrupted in proximal $^1/_2$, usually broadly paniculate (distal branches simple, proximal ones usually with few 2d-order branches). **Pedicels** articulated in proximal $^1/_3$ or almost near base, filiform, 4–6 mm, not more than 2–2.5 times as long as inner tepals, articulation indistinctly swollen. **Flowers** 10–20 in whorls; inner tepals, broadly ovate-triangular or almost deltoid, occasionally broadly triangular, (2.5–)3–4 × 2–3.5 mm, base truncate or round, occasionally indistinctly cordate, margins entire or indistinctly erose, apex obtuse or subacute; tubercles 3, equal or subequal, all tubercles, or at least largest tubercle pale, subequal to inner tepals or slightly narrower than inner tepals, often minutely verrucose. **Achenes** brown or dark reddish brown, 2–3 × 1–1.5 mm. *2n* = 20.

Flowering late spring–summer. Coastal marshes and dunes, sandy and rocky sea beaches; 0 m; N.B., Nfld. and Labr. (Nfld.), N.S., Ont., P.E.I., Que.; Maine, Mass., N.H., Vt.

Rumex pallidus is restricted to northeastern North America. Specimens with more "strict habit, larger leaves, and more erect fruiting branches" reported by K. H. Rechinger (1937) from Alaska and Yukon as this species belong instead to *R. sibiricus* (E. Hultén 1941–1950, vol. 4). Those species, as well as *R. subarcticus* and *R. hultenii* Tzvelev, represent a rather natural northern group of the *R. salicifolius* aggregate. The hybrid *R. pallidus* × *R. triangulivalvis* was reported from the western part of New York (R. S. Mitchell 1986).

27. Rumex sibiricus Hultén, Kongl. Svenska Vetensk. Acad. Handl., ser. 3, 5(2): 48. 1928 • Siberian or Siberian willow dock

Rumex salicifolius Weinmann var. *angustifolius* Meisner

Plants perennial, glabrous, with vertical rootstock. **Stems** usually ascending, occasionally erect, usually producing axillary shoots below 1st-order inflorescence, especially at proximal nodes, 15–40(–70) cm. **Leaf blades** linear-lanceolate or narrowly linear-lanceolate, 6–15 × 1–2 (–2.5) mm, usually ca. 6–10 times as long as wide, widest near middle, thin, not coriaceous, base cuneate, margins entire, flat or undulate, apex acute. **Inflorescences** terminal and axillary, terminal usually occupying distal $^1/_5$–$^1/_3$ of stem, rather dense or occasionally interrupted in proximal $^1/_2$, usually broadly paniculate (branches simple or with few 2d-order branches). **Pedicels** articulated in proximal $^1/_3$, filiform (but thickened distally), 1.5–3.5(–4) mm, not more than 2–2.5 times as long as inner tepals, articulation indistinctly swollen. **Flowers** 10–20(–25) in whorls; inner tepals ovate or occasionally ovate-lanceolate, 2.5–3(–3.5) × 1.6–2.5 mm, base broadly cuneate or truncate, margins entire or indistinctly erose, apex obtuse or subacute; tubercles 3, equal or subequal, usually distinctly narrower and much shorter than inner tepals, smooth or weakly rugose. **Achenes** dark brown, occasionally dark reddish brown, 2–2.5 × 1–1.2 mm. $2n = 20$.

Flowering late spring–summer. Sandy, gravelly, and clayey shores of rivers and streams, wet rocky and grassy slopes; 0–1500 m; N.W.T., Yukon; Alaska; Asia (n Russian Far East including Kamchatka, Siberia).

Rumex hultenii Tzvelev, a species closely related to *R. sibiricus* (it differs from the latter mostly by inner tepals without tubercles), was reported from Alaska by N. N. Tzvelev (1987b, 1989b). *Rumex hultenii* is reported to differ from *R. utahensis* by its narrower leaves, and from *R. subarcticus* by its smaller inner tepals and erect or ascending (not prostrate) habit, as well as by smaller inner tepals. Records of *R. utahensis* from Alaska may refer to *R. hultenii*. I have seen no reliable specimens from North America of *R. hultenii* and cannot confirm that it occurs in Alaska.

28. Rumex subarcticus Lepage, Naturaliste Canad. 82: 191. 1955 • Subarctic or subarctic willow dock E

Rumex pallidus Bigelow subsp. *subarcticus* (Lepage) Á. Löve

Plants perennial, glabrous, with vertical rootstock. **Stems** usually procumbent, rarely ascending, usually producing axillary shoots below 1st-order inflorescence or at proximal nodes, 30–60 cm. **Leaf blades** narrowly linear-lanceolate, 6–17 × 1–3 cm, usually ca. 7–10 times as long as wide, widest near middle, usually thick, not coriaceous or subcoriaceous, base cuneate, margins entire, usually strongly undulate and/or crenulate, apex acute. **Inflorescences** terminal and axillary, terminal usually occupying distal $^1/_5$–$^1/_3$ of stem, rather dense towards apex, distinctly interrupted in proximal $^1/_2$, usually broadly paniculate (branches almost at right angles to main axis, simple or with few 2d-order branches). **Pedicels** articulated in proximal $^1/_3$, filiform, 4–7 mm, not more than 2–2.5 times as long as inner tepals, articulation slightly swollen. **Flowers** 10–20 in whorls; inner tepals, broadly deltoid or deltoid-ovate, 3–4 × 3.2–4(–4.5) mm, base truncate, margins entire or indistinctly crenulate, apex obtuse or subacute; tubercles absent, rarely small and indistinct. **Achenes** brown or dark reddish brown, 2–3 × 1.5–2 mm. $2n = 20$.

Flowering early summer. Mostly coastal and alluvial habitats: sea beaches, shores of rivers and streams, wet meadows; 0–200 m; Nunavut., Que.

Some specimens of *Rumex subarcticus* have well-developed tubercles similar to those of *R. pallidus* (N. M. Sarkar 1958), to which it is closely related and of which it may be regarded as a northwestern subspecies or variety (see Á. Löve 1986).

29. Rumex hesperius Greene, Pittonia 4: 234. 1901 • Western willow dock E

Plants perennial, glabrous, with vertical rootstock. **Stems** usually ascending, usually producing axillary shoots below 1st-order inflorescence or at proximal nodes, 30–50(–60) cm. **Leaf blades** light or yellowish green, veins scarcely prominent abaxially, elliptic-lanceolate, occasionally lanceolate, 6–12 × (2–)3–6 cm, usually (2–)3(–5) times as long as wide, widest near middle, thin, not coriaceous, base cuneate, margins entire, undulate, apex acute, warty in appearance. **Inflorescences** terminal and axillary, terminal usually occupying distal $^1/_5$–$^1/_3$ of stem, dense or interrupted near base, usually broadly paniculate (branches simple or with few 2d-order branches). **Pedicels** articulated in proximal $^1/_3$, filiform (but thickened distally), 3–6 mm, not more than 2–2.5 times as long as inner tepals, articulation indistinctly swollen. **Flowers** 10–20 in whorls; inner tepals, triangular, broadly triangular, or deltoid, 3–4 × 3.5–4 (–5) mm, base truncate, margins entire, apex acute or subacute; tubercle 1, small, narrow, much narrower than inner tepals, smooth to weakly rugose. **Achenes** not seen. $2n = 20$.

Flowering late spring–early summer. Wet, alluvial habitats: shores, ditches; 500–1000 m; Wash.

Rumex hesperius is a little-known species reported only from a localized area in Washington. According to N. M. Sarkar (1958) it is "quite distinct from other species" in its general appearance—a small plant with large, broad leaves and compact inflorescences.

30. Rumex cuneifolius Campderá, Monogr. Rumex, 95. 1819 • Argentine or wedgeleaf dock ⸏F⸏ ⸏I⸏

Plants perennial, glabrous or slightly papillose-pubescent especially on leaf blades abaxially, with long-creeping rhizomes and/or stolons. **Stems** and axillary shoots ascending or erect, branching in distal $^2/_3$, (5–) 10–30(–40) cm. **Leaf blades** distinctly obovate or obovate-elliptic, 5–8(–12) × (2–)3–5(–7) cm, widest in distal $^1/_2$, rather fleshy, coriaceous, base cuneate, margins entire to obscurely repand, crisped or occasionally flat, apex obtuse, rounded. **Inflorescences** terminal and axillary, occupying distal $^1/_2$–$^2/_3$ of shoots, usually dense at least in distal part, narrowly to broadly paniculate (but branches in most cases simple or nearly so). **Pedicels** articulated near middle or in proximal $^1/_3$, thickened (especially distally), 3–5 mm, not more than 2–2.5 times as long as inner tepals, articulation distinct, swollen. **Flowers** 5–20 in whorls; inner tepals ovate-deltoid or ovate-triangular, 4–5 × (2.5–)3–3.5 mm, base broadly cuneate or truncate, margins entire, apex obtuse or subacute; tubercles 3, equal or unequal, finely punctate. **Achenes** glossy brown or dark reddish brown, 2.5–3 × 1.8–2.5 mm. $2n = 40$.

Flowering spring–summer. Waste places, sandy shores; 0–500 m; introduced; Ala., Fla., Oreg., South America; introduced in Europe, Australia.

Rumex cuneifolius, a representative of the predominantly South American subsection *Cuneifolii* Rechinger f., is an uncommon alien known from a few localities in North America. It may have become naturalized in the southwestern part of the United States, especially in coastal regions. The species is known definitely from Portland, Oregon, and was reported as "apparently well established in s.c. Oregon" (C. L. Hitchcock et al. 1955–1969, vol. 2).

This species may have been reported from Provo, Utah, by Á. Löve (1986), as *Rumex frutescens*. However, it is not mentioned by S. L.Welsh et al. (1993).

The name *Rumex frutescens* Thouars has been misapplied to *R. cuneifolius* in both Europe and North America. According to K. H. Rechinger (1990), *R. frutescens* in the narrow sense is an endemic species of the remote South Atlantic islands Tristan de Cuhna and Gough; it differs from *R. cuneifolius* in having thinner rhizomes, shorter petioles and leaf blades, and smaller inner tepals. *Rumex cuneifolius* can hybridize with some other species (J. E. Lousley 1953).

26c.2. Rumex Linnaeus (subg. Rumex) sect. Rumex

Rumex sect. *Simplices* Rechinger f.

Plants developing basal rosette of leaves (sometimes, especially in annual species, not persistent at maturity); rootstock usually vertical (in some species creeping). **Stems** mostly erect, solitary or several from base, not branching below terminal paniculate inflorescence, without axillary shoots. **Leaf blades** variable in shape, base cordate to narrowly cuneate. **Inner tepals** with margins entire or variously dentate.

Species ca. 110 (33 in the flora): nearly worldwide.

SELECTED REFERENCE Dawson, J. E. 1979. A Biosystematic Study of *Rumex* Section *Rumex* in Canada and the United States. Ph.D. thesis. Carleton University.

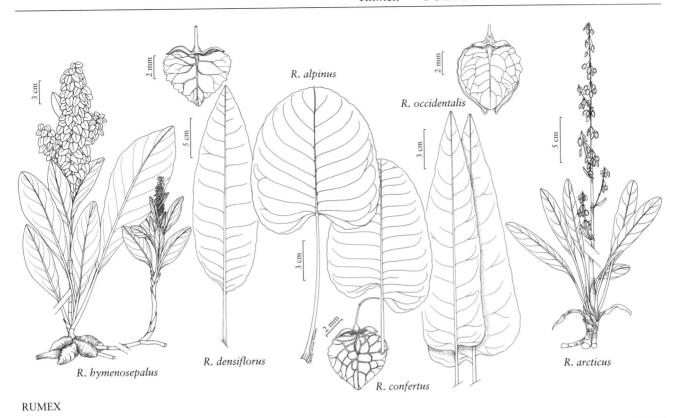

R. alpinus

R. occidentalis

R. densiflorus

R. confertus

R. hymenosepalus

R. arcticus

RUMEX

31. Rumex hymenosepalus Torrey in W. H. Emory, Rep. U.S. Mex. Bound. 2(1): 177. 1859 • Arizona or canaigre dock, wild-rhubarb F

Rumex arizonicus Britton; *R. hymenosepalus* var. *salinus* (A. Nelson) Rechinger f.; *R. salinus* A. Nelson; *R. saxei* Kellogg

Plants perennial, glabrous or indistinctly papillose-pubescent, with distinctly tuberous roots and short rhizomes. **Stems** usually erect, rarely ascending, branched above middle, 25–90(–100) cm. **Leaves:** ocrea prominent and persistent at maturity, whitish or silvery white, membranous; blade oblong, oblong-elliptic, or obovate-lanceolate, (5–)8–30 × 2–8(–12) cm, base cuneate or narrowly cuneate, margins entire, flat or indistinctly crisped, apex acute or acuminate, rarely obtuse. **Inflorescences** terminal, occupying distal 1/2 of stem, narrowly paniculate, rarely simple. **Pedicels** articulated near middle or in proximal 1/3, filiform, 5–15(–20) mm, articulation indistinct. **Flowers** 5–20 in whorls; inner tepals oblong-cordate or orbiculate-cordate, 11–16 × 9.5–14 mm, base sinuate or emarginate, margins entire, rarely with few extremely small denticles at base, apex obtuse or subacute; tubercles absent. **Achenes** brown or reddish brown, 4–5(–7) × 2.5–4.5(–5) mm. *2n* = 40.

Flowering spring. Sandy and rocky places: plains, slopes, stream beds, alkaline soils; 0–1700(–2000) m; Ariz., Calif., Colo., Mont., Nev., N.Mex., Okla., Tex., Utah, Wyo.; Mexico (Baja California, Chihuahua).

Rumex hymenosepalus is the only species of subsect. *Hymenosepali* Rechinger f.

Two varieties have been recognized. The typical variety has achenes 5 mm and ovate-elliptic or oblong-cordate inner tepals with a subacute apex. Variety *salinus* (A. Nelson) Rechinger f. has larger achenes (to 7 mm) and almost orbiculate inner tepals with an obtuse apex.

Rumex hymenosepalus was reported also from Montana (J. E. Dawson 1979), but no exact localities were given.

32. Rumex densiflorus Osterhout, Erythea 6: 13. 1898 • Dense-flowered dock E F

Rumex polyrrhizus Greene

Plants perennial, glabrous or indistinctly papillose-pubescent, with creeping horizontal rhizome. **Stems** erect, branched above middle (only in inflorescence), 50–100 cm. **Leaves:** ocrea deciduous or partially persistent at maturity; blade with large lateral veins alternating with short ones, oblong or oblong-lanceolate,

30–40(–50) × 10–12 cm, more than 3 times as long as wide, base broadly cuneate, truncate, or weakly cordate, margins entire or indistinctly repand, flat, apex obtuse or broadly acute. **Inflorescences** terminal, occupying distal $^1/_2$ of stem, usually dense, narrowly paniculate. **Pedicels** articulated in proximal $^1/_3$, filiform, 6–16 mm, articulation indistinct. **Flowers** 10–20 in whorls; inner tepals ovate-triangular or subcordate, 5–6 × 4.5–6 mm, widest at or near middle, base weakly emarginate, margins entire, erose, or indistinctly denticulate mostly at base, apex abruptly narrowed, acute or subacute; tubercles absent. **Achenes** deep brown to reddish brown, 2.5–4(–4.5) × 1.8–2.5 mm. **2*n*** = 120.

Flowering late spring–early summer. Along streams and rivers in montane, subalpine, and alpine zones; 1500–3000(–3500) m; Colo., N.Mex., Wyo.

The following three species are closely related to *Rumex densiflorus*, all belonging to subsect. *Densiflori* Rechinger f., and possibly form one polymorphic "macrospecies" (K. H. Rechinger 1937). Á. Löve (1986) treated *R. orthoneurus* and *R. pycnanthus* as subspecies of *R. densiflorus*. However, the variability of this aggregate is insufficiently known, and I prefer to treat it as consisting of four "microspecies."

Rumex densiflorus is reported from northwestern New Mexico (W. C. Martin and C. R. Hutchins 1980), where it most probably occurs; records for southern Idaho (R. J. Davis 1952) and Arizona (J. H. Lehr 1978) need confirmation.

33. **Rumex pycnanthus** Rechinger f., Repert. Spec. Nov. Regni Veg. 38: 372. 1935   E

Rumex subalpinus M. E. Jones, Proc. Calif. Acad. Sci., ser. 2, 5: 720. 1895 (as subalpina), not (Schur) Simonkai 1886; *R. densiflorus* Osterhout subsp. *pycnanthus* (Rechinger f.) Á. Löve

Plants perennial, glabrous or indistinctly papillose-pubescent, with fusiform or creeping horizontal rhizome. **Stems** erect, branched above middle (only in inflorescence), 60–100 cm. **Leaves:** ocreae deciduous or partially persistent at maturity; blade normally oblong-lanceolate, 20–45 × 8–10 cm, normally more than 3 times as long as wide, base broadly cuneate, truncate, or weakly cordate, margins entire, flat or indistinctly crisped, apex obtuse or broadly acute. **Inflorescences** terminal, occupying distal $^1/_2$ of stem, often dense, narrowly paniculate. **Pedicels** articulated in proximal $^1/_2$, occasionally almost near base, filiform, 3–11 mm, articulation indistinct or weakly evident. **Flowers** 10–20 in whorls; inner tepals ovate-deltoid or

ovate-triangular, 4–7 × 3–5 mm, widest at or near base, base truncate or weakly emarginate, margins erose to minutely dentate at least near base, apex narrowly acute; tubercles absent. **Achenes** deep brown to reddish brown, 3–4 × 1.5–2.2 mm. **2*n*** = 120.

Flowering late spring–summer. Along streams and rivers in montane, subalpine, and alpine zones; 1700–3000 m; Colo., Utah.

Rumex pycnanthus was reported (as *R. subalpinus* M. E. Jones) from White Pine County, Nevada (Mont E. Lewis 1973) and may occur in that state; according to J. T. Kartesz (1987, vol. 1), that record was based on misidentification of *R. californicus*.

34. **Rumex orthoneurus** Rechinger f., Repert. Spec. Nov. Regni Veg. 40: 294. 1936   • Chiricahua or Blumer's dock

Rumex densiflorus Osterhout subsp. *orthoneurus* (Rechinger f.) Á. Löve

Plants perennial, glabrous or indistinctly papillose-pubescent especially on leaf blade veins abaxially, with creeping or fusiform rhizomes. **Stems** erect, branched above middle (only in inflorescence), 60–100 cm. **Leaves:** ocrea deciduous or partially persistent at maturity; blade with lateral veins ± equal in size, oblong or oblong-lanceolate, 20–40 (–50) × 8–15(–18) cm, more than 3 times as long as wide, base broadly cuneate, obtuse, or weakly cordate, margins entire, flat, apex acute, subacute, or acuminate. **Inflorescences** terminal, occupying distal $^1/_2$ of stem, often dense, narrowly paniculate. **Pedicels** articulated in proximal $^1/_2$, filiform, (5–)12–15(–17) mm, articulation indistinct, scarcely visible. **Flowers** 10–20 in whorls; inner tepals ovate-deltoid, 4.5–7 × 3.5–7 mm, widest in proximal $^1/_3$, base truncate or weakly emarginate, margins erose to weakly serrate or indistinctly denticulate in basal part, apex acute to acuminate; tubercles absent. **Achenes** brown, 2.5–4 × 1.5–2.5 mm. **2*n*** = 120.

Flowering late spring–summer. Along streams; 2500 m; Ariz.; N.Mex., Mexico (Sonora).

Rumex orthoneurus has been reported from northern Mexico (M. Fishbein 1993). The species is in the Center for Plant Conservation's National Collection of Endangered Plants.

35. Rumex praecox Rydberg, Bull. Torrey Bot. Club 33: 137. 1906 • Early dock [E]

Plants perennial, glabrous (occasionally indistinctly papillose on veins of leaf blades abaxially), with incrassate creeping rhizome. **Stems** erect, branched from proximal ¹/₂ of stem, 20–45 cm. **Leaves:** ocrea deciduous or partially persistent at maturity; blade oblong or elliptic, sometimes almost orbiculate, 3–7(–14) × 1.5–5(–7) cm, ca. 2(–2.5) times as long as wide, base broadly cuneate or rounded, margins entire or weakly crisped, apex obtuse or subacute. **Inflorescences** terminal, occupying more than proximal ¹/₂ of stem, narrowly paniculate. **Pedicels** articulated in proximal ¹/₂, filiform, 3–7 mm, articulation indistinct, scarcely visible. **Flowers** 10–20 in whorls; inner tepals ovate or oblong, 4.5 × 2.5 mm, widest at or near middle, base emarginate or cordate, margins entire or weakly serrate, apex commonly obtuse; tubercles absent. **Achenes** greenish brown, 2.8–4 × 1.5–2 mm.

Flowering late spring. Along streams and river valleys; 2400–3000 m; Colo., Wyo.

36. Rumex alpinus Linnaeus, Sp. Pl. 1: 334. 1753, name proposed for conservation • Monk's-rhubarb, butter or Alpine dock [F][I]

Plants perennial, glabrous or minutely papillose-pubescent, with stout, creeping rhizome. **Stems** erect, branched above middle, 60–100 cm. **Leaves:** ocrea mostly deciduous or sometimes partially persistent at maturity; blade usually ovate-orbiculate, sometimes orbiculate or broadly ovate, 20–40 × 20–35 cm, base deeply and widely cordate, margins entire, flat or slightly undulate, apex obtuse. **Inflorescences** terminal, occupying distal ¹/₂ of stem, rather dense, widely paniculate to fusiform. **Pedicels** articulated at middle or in proximal ¹/₃, filiform, 3–9 mm, articulation distinctly swollen. **Flowers** 10–20 in whorls; inner tepals ovate or ovate-triangular, 4–5 (–6) × 3–5 mm, base truncate or slightly cordate, margins entire or subentire, apex obtuse or subobtuse; tubercles absent. **Achenes** brown to brownish green, 2.5–3.5 × 1–2 mm. **2n** = 20.

Flowering late spring–summer. Waste places: roadsides, old fields and gardens, disturbed meadows; 0–1500 m; introduced; N.S; Maine, Vt.; c, s Europe; w Asia.

Rumex alpinus belongs to subsect. *Alpini* Rechinger f. The name *R. alpinus* has been proposed for nomenclatural conservation (S. Cafferty and S. Snogerup 2000).

This species was first reported from North America in Nova Scotia (M. L. Fernald 1921; K. H. Rechinger 1937). It remains uncommon in the United States and Canada. *Rumex alpinus* never has been reported as being a serious invasive weed; however, it may persist at a site for a very long time. Previously, the species was cultivated widely, mostly for medicinal and veterinary purposes.

37. Rumex confertus Willdenow, Enum. Pl., 397. 1809 • Russian dock [F][I]

Rumex alpinus Linnaeus var. *subcalligerus* Boissier

Plants perennial, glabrous or weakly papillose-pubescent especially when young, with fusiform, vertical to oblique rootstock or short rhizomes. **Stems** erect, branched above middle, 50–100 (–130) cm. **Leaves:** ocrea mostly deciduous or rarely partially persistent at maturity; blade ovate-triangular, broadly ovate, or ovate-elliptic, 20–30 × 15–25 cm, base deeply and broadly cordate, margins entire to obscurely repand, usually slightly crisped or undulate, apex obtuse to subacute. **Inflorescences** terminal, occupying distal ¹/₂ of stem (branches often slightly arcuate at base), rather dense, widely paniculate. **Pedicels** articulated in proximal ¹/₃, filiform, 4–10 mm, articulation distinctly swollen. **Flowers** 15–30 in whorls; inner tepals orbiculate-reniform or broadly scutate, 6–9 × 6–11 mm, as long as wide or nearly so, base cordate to subcordate, margins entire or subentire, occasionally irregularly erose near base, apex abruptly acute to acute; tubercles usually 1, small, 1–2 mm, normally less than 2 times as wide as inner tepals, rarely absent or indistinct. **Achenes** reddish brown, 3–3.5 × 1.7–2.5 mm. **2n** =40.

Flowering late spring–summer. Roadsides, waste places, meadows, river valleys; 300–700 m; introduced; Alta., Man.; N.Dak.; e, ec Europe; w Asia (the Caucasus, Siberia); introduced elsewhere.

Rumex confertus was placed in subsect. *Conferti* Rechinger f. This species is common and ecologically successful in central and eastern Europe; it may be expected elsewhere in temperate regions of North America.

38. Rumex occidentalis S. Watson, Proc. Amer. Acad.
Arts 12: 253. 1877 • Western dock, rumex
occidental E F W

Rumex aquaticus Linnaeus subsp.
fenestratus (Greene) Hultén;
R. *aquaticus* subsp. *occidentalis*
(S. Watson) Hultén; R. *bakeri*
Greene; R. *fenestratus* Greene;
R. *fenestratus* var. *labradoricus*
Rechinger f.; R. *gracilipes* Greene;
R. *occidentalis* S. Watson subsp.
fenestratus (Greene) Hultén;
R. *occidentalis* var. *labradoricus* (Rechinger f.) Lepage

Plants perennial, glabrous or very indistinctly papillose, especially on veins of leaf blades abaxially, with fusiform, vertical or oblique rootstock. **Stems** usually erect, branched from above middle or in distal 2/3, 50–100 (–140) cm. **Leaves:** ocrea deciduous or partially persistent at maturity; blade narrowly ovate-triangular, ovate-lanceolate, or oblong-lanceolate, normally 10–35 × 5–12 cm, base weakly to distinctly cordate, truncate, or rounded, margins entire, undulate or indistinctly crisped, apex acute or subacute, rarely obtuse. **Inflorescences** terminal, occupying distal 2/3 of stem, dense to interrupted, narrowly paniculate, often repeatedly branched (branches usually more than 7–8 cm). **Pedicels** articulated in proximal 1/3, filiform, 5–13(–17) mm, normally not more than 2–2.5 times as long as inner tepals, articulation weakly evident, not swollen. **Flowers** mostly 12–25 in whorls; inner tepals orbiculate, ovate, or broadly ovate-triangular, 5–10 (–12) × 5–8(–11) mm, base truncate to weakly cordate, margins entire or subentire to very weakly erose, apex obtuse or subacute; tubercles absent. **Achenes** reddish brown, 3–4.5(–4.8) × 1.5–2.5 mm. $2n = 120$.

Flowering late spring–summer. Wet meadows, bogs, marshes, river banks, shallow water, other wet habitats; 0–2500(–3000) m; Alta., B.C., Man., Nfld. and Labr. (Nfld.), N.W.T., N.S., Ont., Que., Sask., Yukon; Alaska, Ariz., Calif., Colo., Idaho, Iowa, Maine, Minn., Mont., Nev., N.Mex., N.Dak., Oreg., S.Dak., Utah, Vt., Wash., Wyo.

In the nineteenth century, *Rumex occidentalis* commonly was misidentified as R. *aquaticus*, R. *longifolius*, or R. *domesticus*.

All of the species of subsect. *Aquatici* Rechinger f., represented in North America by *Rumex occidentalis*, R. *arcticus*, R. *nematopodus*, and R. *tomentellus*, form a taxonomically complex aggregate with poorly delimited, often intergrading species. Extremes are evidently distinct (e.g., R. *arcticus* and R. *tomentellus*). The taxonomy and distribution of members of this aggregate are still insufficiently known. Some authors prefer to treat all or most of these taxa as subspecies or varieties of R. *aquaticus* in the broad sense. From my

point of view, this does not promote a better understanding of their variability and relationships.

A number of segregate species have been described and recognized in regional floras in North America. In most cases the features upon which these species are based intergrade. One of the most widely recognized segregates is *Rumex fenestratus* Greene emend. Rechinger f. [R. *aquaticus* subsp. *fenestratus* (Greene) Hultén, R. *occidentalis* S. Watson subsp. *fenestratus* (Greene) Hultén], which, according to K. H. Rechinger (1937), may be distinguished mostly by larger and more cordate fruiting inner tepals (more than 7 mm in R. *fenestratus*, usually less than 7 mm in R. *occidentalis*), and larger achenes (3 mm, and more than 3.5 mm, respectively). The morphotype of R. *fenestratus* occurs mostly along the Pacific coast from central western California to Alaska. Plants with large fruiting inner tepals [known as R. *fenestratus* var. *labradoricus* Rechinger f. or R. *occidentalis* var. *labradoricus* (Rechinger f.) Lepage] occur also in eastern Canada (Newfoundland and Quebec). In this treatment, I follow the taxonomic decision by J. E. Dawson (1979), who carefully analyzed the clinal variability of the R. *occidentalis* aggregate. However, R. *fenestratus* probably deserves recognition at least as a subspecies of R. *occidentalis*, but its taxonomic status needs additional investigation.

Rumex occidentalis was reported also from New Brunswick (which seems to be a rather natural extension of its range); however, the present status of the species in that province is uncertain.

39. Rumex arcticus Trautvetter in A. T. von
Middendorff, Reise Siber. 1(2,1): 29. 1847
 • Arctic dock

Rumex aquaticus Linnaeus subsp.
arcticus (Trautvetter) Hiitonen;
R. *arcticus* var. *kamtschadalus*
(Komarov) Rechinger f. ex
Tolmatchew; R. *arcticus* var.
latifolius Tolmatchew; R.
domesticus Hartman var. *nanus*
Hooker; R. *kamtschadalus*
Komarov; R. *longifolius* de
Candolle var. *nanus* (Hooker)
Meisner; R. *ursinus* M. M. Maximova

Plants perennial, glabrous or nearly so, with fusiform, oblique rootstock, occasionally with horizontal, short-creeping rhizome. **Stems** erect, simple or branched in distal 2/3 (then with few, comparatively short branches), 10–70(–100) cm. **Leaves:** ocrea deciduous or partially persistent at maturity; blade narrowly lanceolate, lanceolate, or oblong-lanceolate, normally 5–15(–20) × 1.5–5 cm, base cuneate to broadly cuneate, rarely truncate or very weakly cordate, margins entire or rarely indistinctly repand, flat, apex acute or subacute. **Inflorescences** terminal, occupying distal ¹/₂–²/₃ of stem, interrupted, paniculate, simple or nearly so (branches, when present usually less than 7–8 cm). **Pedicels** articulated in proximal ¹/₃, filiform, 5–13(–17) mm, usually not more than 2–2.5 times as long as inner tepals, articulation weakly evident, not swollen. **Flowers** 7–15 in whorls; inner tepals ovate, 4.5–7.5(–8) × 4–6(–7) mm, base truncate to weakly cordate, margins entire, apex obtuse or subacute; tubercles absent. **Achenes** reddish brown, 3–4 × 1.5–2 mm. *2n* = 40, 120, ca. 170, ca. 200.

Flowering late spring–summer. Moist tundra, marshes, river valleys, sandy and gravelly shores and slopes; 0–1000 m; B.C., N.W.T., Nunavut, Yukon; Alaska; ne Europe, n Asia (arctic and subarctic zones).

Rumex arcticus is polymorphic, as are *R. aquaticus* in the narrow sense and *R. occidentalis*. However, unlike *R. aquaticus* and *R. occidentalus*, it is represented by at least two chromosome races (G. A. Mulligan and C. Frankton 1972; Á. Löve 1986). Plants morphologically transitional between *R. arcticus* and *R. aquaticus* were described from Kamchatka as *R. kamtschadalus* (= *R. arcticus* var. *kamtschadalus*). The same forms occasionally occur in northwestern North America. According to Á. Löve and D. Löve (1975b) and Á. Löve (1986), they are usually tetraploids (2*n* = 40) and deserve recognition at the species level. However, they are not always morphologically distinct from *R. arcticus*. The group needs additional study, and at present I prefer to keep the tetraploid plants provisionally within *R. arcticus*, regarding them as var. *kamtschadalus*.

A few highly sterile specimens with mostly abortive flowers, which I have seen in Alaskan herbarium material, most probably represent hybrids between tetraploid and 12-ploid races of the *R. aquaticus* group.

Plants with unusually wide, triangular-oblong, or almost ovate leaves were described as var. *latifolius* Tolmatchew. This seems to be a predominant variety on the Beringian coast of Chukotka and Wrangel and Ratmanov islands (A. I. Tolmachew 1966). I also have seen at least two collections of this variety from the western coast of Alaska. The enigmatic var. *perlatus* Hultén may belong here (see discussion under 8. *R. lapponicus*).

Rumex arcticus has been reported from Churchill in northeastern Manitoba (H. J. Scoggan 1978–1979, part 3). That record needs confirmation because some northern forms of *R. arcticus* and *R. occidentalis* are similar.

40. Rumex nematopodus Rechinger f., Leafl. W. Bot. 7: 134. 1954

Rumex aquaticus Linnaeus subsp. *nematopodus* (Rechinger f.) Á. Löve

Plants perennial, glabrous or nearly so, with fusiform, vertical rootstock. **Stems** erect, branched in distal ¹/₂, 40–85(–100) cm. **Leaves:** ocrea deciduous or partially persistent at maturity; blade oblong-lanceolate to lanceolate, 20–35 × 5–12 cm, base cordate to broadly cuneate, margins entire or nearly so, flat, apex acute or attenuate. **Inflorescences** terminal, occupying distal ¹/₂–²/₃ of stem, interrupted in proximal ¹/₂, paniculate, branched. **Pedicels** articulated in proximal ¹/₃, filiform, 12–20 mm, 3–4 times as long as inner tepals, articulation weakly evident, not swollen. **Flowers** 10–20 in whorls; inner tepals ovate-triangular, 4–5(–6) × 3–4(–5) mm, base truncate to subcordate, margins entire, apex acute to subacute; tubercles absent. **Achenes** brown, 3 × 1.5 mm. *2n* = 120.

Flowering spring–early summer. Seasonally wet habitats along rivers and streams; 2500–3000 m; Ariz., N.Mex.; n Mexico (Chihuahua).

Rumex nematopodus has been reported from several localities in Arizona, New Mexico, and northern Mexico (K. H. Rechinger 1954). It is closely related to *R. occidentalis*, differing from that species in having longer pedicels and smaller inner tepals with acute apices. Those characters are highly variable in almost all species of the *R. aquaticus* aggregate, and the taxonomic status of *R. nematopodus* remains unclear.

41. Rumex tomentellus Rechinger f., Leafl. W. Bot. 7: 133. 1954

Rumex aquaticus Linnaeus subsp. *tomentellus* (Rechinger f.) Á. Löve

Plants perennial, distinctly and often densely tomentose and/or papillose-pubescent, especially abaxial sides of leaf blades, ocreae, and petioles, with fusiform, vertical rootstock. **Stems** erect, branched in distal ¹/₂–²/₃, 30–70 (–100) cm. **Leaves:** ocrea deciduous or partially persistent at maturity; blade oblong-lanceolate, 20–30 × 5–10 cm, base cordate to broadly cuneate, margins entire, repand, flat or crisped, apex acute or attenuate. **Inflorescences** terminal, occupying distal ¹/₂–²/₃ of stem, interrupted in proximal ¹/₂, paniculate, branched. **Pedicels** articulated in proximal ¹/₃, filiform, distinctly swollen in distal part (near base of tepals, but not at articulation point), 4–14 mm, articulation weakly

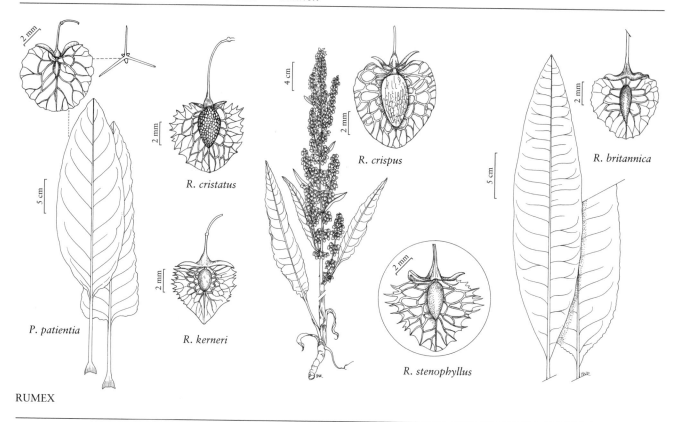

P. patientia

R. cristatus

R. kerneri

R. crispus

R. stenophyllus

R. britannica

RUMEX

evident, not swollen. **Flowers** 10–20 in whorls; inner tepals oblong-ovate, 4–6 × 3–5 mm, base rounded to subcordate, margins entire, apex obtuse or subacute; tubercles absent. **Achenes** mature specimens not seen. $2n = 120$.

Flowering spring–summer. Seasonally wet habitats along streams; 2500 m; N.Mex.

Rumex tomentellus is known only from New Mexico and needs additional study. It may represent one of the southern North American elements of the *R. aquaticus*– *R. occidentalis* aggregate.

42. **Rumex patientia** Linnaeus, Sp. Pl. 1: 333. 1753

· Patience dock F I W

Lapathum hortense Lamarck; *Rumex lonaczevskii* Klokov; *R. patientia* subsp. *orientalis* Danser

Plants perennial, glabrous or very indistinctly papillose normally only on veins of leaf blades abaxially, with fusiform, vertical rootstock. **Stems** erect, branched from above middle, 80–150 (–200) cm. **Leaves:** ocrea deciduous or partially persistent at maturity; blade ovate-lanceolate or oblong-lanceolate, normally 30–45(–50) × 10–15 cm, base truncate, broadly cuneate, or weakly cordate, margins entire, flat or weakly undulate, apex acute or subacute.

Inflorescences terminal, occupying distal ¹/₂ of stem, normally dense, narrowly to broadly paniculate, branches usually straight or arcuate, rarely indistinctly flexuous. **Pedicels** articulated in proximal ¹/₃, sometimes almost near base, filiform, 5–13(–17) mm, articulation usually distinctly swollen. **Flowers** 10–20(–25) in whorls; inner tepals broadly ovate, suborbiculate, or orbiculate, (5–)5.5–8(–10) × 5–9(–10) mm, base usually distinctly cordate, margins entire or subentire to very weakly erose, apex obtuse or occasionally subacute; tubercles normally 1, more than 1 mm wide, normally less than 2 times as wide as inner tepals, occasionally 3, then 2 much smaller. **Achenes** brown, 3–3.5 × 1.5–2.5 mm. $2n = 60$.

Flowering late spring–summer. Waste places, roadsides, old fields, gardens, disturbed meadows, occasionally in alluvial habitats; 0–2300 m; introduced; Ont., Que.; Conn., Idaho, Ill., Ind., Iowa, Kans., Ky., Maine, Md., Mass., Mich., Minn., Mo., Mont., Nebr., N.H., N.J., N.Y., N.C., N.Dak., Ohio, Okla., Pa., R.I., S.Dak., Tenn., Utah, Vt., Va., Wash., W.Va., Wis., Wyo.; e, s Europe; w Asia; introduced elsewhere.

Some North American specimens of *Rumex patientia* appear to belong to subsp. *orientalis* (= *R. orientalis* Bernhardi 1830, not Campderá 1819; *R. lonaczevskii*), which differs from subsp. *patientia* in having larger inner tepals (6–10 × 8–10 mm, not 4–8 × 4–8 mm).

A predominantly Asian variety with three tubercles sometimes is recognized as subsp. *callosus* (Fr. Schmidt ex Maximowicz) Rechinger f. [= var. *callosus* Fr. Schmidt ex Maximowicz; *Rumex callosus* (Fr. Schmidt ex Maximowicz) Rechinger f.]. However, the distribution of infraspecific taxa of *R. patientia* in North America has not been studied in detail.

Rumex patientia may be expected in southern Canada, especially the prairie regions of Alberta, Manitoba, and Saskatchewan, as well as in Colorado and other states. According to J. T. Kartesz (1987, vol. 1), a record from Nevada was based on misidentification of *R. crispus*.

Rumex patientia is the lectotype of the genus. It and the following two species belong to subsect. *Rumex*.

43. Rumex cristatus de Candolle, Cat. Pl. Hort. Monsp., 139. 1813 • Greek or crested dock [F] [I]

Rumex graecus Boissier & Heldreich; *R. patientia* Linnaeus subsp. *graecus* (Boissier & Heldreich) Lindberg

Plants perennial, glabrous or indistinctly papillose exclusively on veins of leaf blades abaxially, with fusiform, vertical rootstock. **Stems** erect, branched from above middle or in distal $^2/_3$, 70–150(–200) cm. **Leaves:** ocrea deciduous or partially persistent at maturity; blade broadly lanceolate to oblong-lanceolate, normally 15–25(–35) × 5–7(–12) cm, base truncate, rounded, or slightly cordate, margins entire, undulate or weakly crisped, occasionally flat, apex acute or acuminate. **Inflorescences** terminal, occupying distal $^1/_2$–$^2/_3$ of stem, normally dense or interrupted near base, broadly paniculate (branches of inflorescence mostly with 2d-order branches), branches usually straight or arcuate, rarely indistinctly flexuous. **Pedicels** articulated near middle, filiform, 6–14 mm, articulation distinctly swollen. **Flowers** 15–20 in whorls; inner tepals orbiculate, 6–8(–9) × 6–7.5(–8) mm, base usually distinctly cordate, margins entire or subentire near apex, distinctly dentate in basal $^1/_2$, apex acute or subacute, teeth 0.5–1 mm; tubercles normally 3, rarely 1, distinctly unequal. **Achenes** dark brown or brown, 2.8–3.5 × 2–2.5 mm. **2n** = 80.

Flowering late spring–early summer. Waste places, roadsides, along railroads, other ruderal habitats; 100–300 m; introduced; Ill., Kans., Mo.; se Europe (Balkans, Greece), naturalized elsewhere in s Europe.

Rumex cristatus is known from Missouri, Illinois, and Kansas; it may occur also in adjacent states (P. Shildneck et al. 1981). Although separable by the characters given in the key, *R. cristatus* sometimes is misidentified as *R. patientia*. The morphological distinctions between these related species are obscured by possible hybridization. Such hybrids are known as *R.* ×*xenogenus* Rechinger f. (G. D. Kitchener 2002); they may be expected in North America in areas where the parental species grow together.

44. Rumex kerneri Borbás, Fl. Comit. Temesiensis, 60. 1884 • Kerner's dock [F] [I]

Rumex confertoides Bihari; *R. cristatus* de Candolle subsp. *kerneri* (Borbás) Akeroyd & D. A. Webb

Plants perennial, distinctly papillose especially or almost exclusively on veins of leaf blades abaxially, with fusiform, vertical rootstock. **Stems** erect, branched from above middle or in distal $^2/_3$, 50–100(–150) cm. **Leaves:** ocrea deciduous or partially persistent at maturity; blade broadly lanceolate to oblong-lanceolate, normally 15–25 × 5–9 cm, base truncate or slightly cordate, margins entire, undulate or weakly crisped, occasionally flat, apex acute or attenuate. **Inflorescences** terminal, occupying distal $^1/_2$–$^2/_3$ of stem, normally dense or interrupted near base, broadly paniculate (branches mostly simple or nearly so), branches usually straight or arcuate, rarely indistinctly flexuous. **Pedicels** articulated near middle, filiform, 6–12 mm, articulation distinctly swollen. **Flowers** 15–20 in whorls; inner tepals orbiculate, 6–8(–9) × 6–7.5(–8) mm, base usually distinctly cordate, margins entire or subentire near apex, denticulate or dentate near base, apex acute or subacute, teeth to 0.5 mm; tubercle nor-mally 1 (rarely 3, then distinctly unequal), less than 2 times as wide as inner tepals. **Achenes** dark brown or brown, 2.5–3.3 × 1.8–2.3 mm. **2n** = 80.

Flowering late spring–early summer. Ruderal habitats: roadsides, waste places; 0–500 m; introduced; Calif.; se Europe.

Rumex kerneri is known in North America only from California, where it occurs mostly in Santa Barbara and San Luis Obispo counties. It is closely related to and has been regarded as a subspecies of *R. cristatus*.

45. Rumex longifolius de Candolle in J. Lamarck and A. P. de Candolle, Fl. Franç. ed. 3, 6: 368. 1815 • Long-leaved or northern or dooryard dock, rumex a feuilles longues ⬚I⬚ ⬚W⬚

Rumex domesticus Hartman; *R. hippolapathum* Fries

Plants perennial, glabrous or very indistinctly papillose normally only on branches of inflorescence, or on veins of leaf blades abaxially, with fusiform, vertical rootstock. **Stems** erect, branched distal to middle, 50–120(–160) cm. **Leaves:** ocrea deciduous or partially per-sistent at maturity; blade broadly lanceolate to ovate-lanceolate or oblong-lanceolate, normally 25–50(–60) × 7–15 cm, ca. 3–4 times as long as wide, base broadly cuneate, rounded-truncate, or slightly cordate, margins entire, undulate or weakly crisped, occasionally flat, apex acute or subacute. **Inflorescences** terminal, occupying distal 1/2 of stem, normally dense, narrowly paniculate, branches usually straight or arcuate. **Pedicels** articulated in proximal 1/3, filiform, 4–9 mm, articulation distinctly swollen. **Flowers** 10–20 in whorls; inner tepals broadly orbiculate or reniform, (4.5–)5–6(–7) × (4.5–)5–7(–7.5) mm, base usually distinctly cordate, margins entire or subentire to very weakly erose, flat, apex obtuse or, rarely, subacute; tubercles normally absent, sometimes with 1 indistinct tubercle or slightly thickened midvein less than 1–1.3 mm wide. **Achenes** dark brown or brown, (2.5–)3–3.5(–4) × 1.5–2 mm. 2*n* = 60.

Flowering late spring–summer. Waste places, roadsides, cultivated fields, river valleys, meadows; 0–1000 m; introduced; Greenland; Alta., B.C., Man., N.B., Nfld. and Labr. (Nfld.), N.S., Ont., P.E.I., Que., Sask., Yukon; Alaska, Conn., Maine, Mass., Mich., Minn., N.H., N.Y., N.Dak., R.I., Vt., Wis.; n temperate Europe; w temperate Asia; introduced elsewhere.

Most records of *Rumex longifolius* from the Great Plains (Great Plains Flora Association 1977; R. B. Kaul 1986) refer to *R. pseudonatronatus* (J. E. Dawson 1979), which can be distinguished from *R. longifolius* by its narrower leaves, smaller and more distinctly triangular inner tepals, and purplish or reddish brown stems at maturity. The two species are closely related; they were placed by K. H. Rechinger (1949) in subsect. *Longifolii* Rechinger f.

46. Rumex pseudonatronatus (Borbás) Murbeck, Bot. Not. 1899: 16. 1899 • Finnish or field dock, rumex de Finlande ⬚I⬚

Rumex domesticus Hartman var. *pseudonatronatus* Borbás, Értek. Term. Köréb. Magyar Tud. Akad. 11: 21. 1880; *R. fennicus* (Murbeck) Murbeck; *R. pseudonatronatus* subsp. *fennicus* Murbeck

Plants perennial, glabrous or very indistinctly papillose normally only on branches of inflorescences, or on veins of leaf blades abaxially, with fusiform, vertical rootstock. **Stems** erect, branched from above middle, 50–120(–150) cm. **Leaves:** ocrea deciduous or partially persistent at maturity; blade lanceolate to narrowly lanceolate or lanceolate-linear, normally 15–30 × 1–4 cm, apex acute. **Inflorescences** terminal, occupying distal 1/2 of stem, normally dense in distal part and interrupted at base, narrowly paniculate, branches usually straight or slightly arcuate. **Pedicels** articulated in proximal 1/3, filiform, 4–9 mm, articulation distinctly swollen. **Flowers** 15–25 in whorls, inner tepals orbiculate, ovate-orbiculate, or indistinctly reniform, 3.5–5(–6) × 3–5 mm, base slightly cordate, margins entire or rarely subentire to very weakly erose, undulate or nearly flat, apex obtuse or rounded, occasionally subacute; tubercles normally absent, sometimes with 1 indistinct tubercle or slightly thickened midvein less than 1–1.3 mm wide, normally less than 2 times as wide as inner tepals. **Achenes** usually reddish brown, 2–2.5 × 1–1.5 mm. 2*n* = 40.

Flowering late spring–summer. Ruderal and alluvial habitats, slightly saline soil, waste places, roadsides, shores of rivers and lakes, meadows, cultivated fields; 10–1000 m; introduced; Alta., B.C., Man., Ont., Sask., Yukon; Minn., N.Dak., S.Dak.; e Europe; c Asia (Siberia).

Rumex pseudonatronatus often is confused with *R. longifolius* and *R. crispus.*

47. Rumex crispus Linnaeus, Sp. Pl. 1: 335. 1753 • Curly or yellow dock, patience crépue, rumex crépu, reguette ⬚F⬚ ⬚I⬚ ⬚W⬚

Lapathum crispum (Linnaeus) Scopoli

Plants perennial, occasionally biennial, glabrous or very indistinctly papillose normally only on veins of leaf blades abaxially, with fusiform, vertical rootstock. **Stems** erect, branched distal to middle, 40–100(–150) cm. **Leaves:** ocrea deciduous, rarely partially persistent at maturity; blade lanceolate to narrowly lanceolate or lanceolate-linear, normally 15–30(–35) × 2–6 cm, base cuneate, truncate, or weakly cordate, margins entire to subentire, strongly crisped and undulate, apex acute. **Inflorescences** terminal, occupying distal ¹/₂ of stem, dense or interrupted at base, narrowly to broadly paniculate, branches usually straight or arcuate. **Pedicels** articulated in proximal ¹/₃, filiform, (3–)4–8 mm, articulation distinctly swollen. **Flowers** 10–25 in whorls; inner tepals orbiculate-ovate or ovate-deltoid, 3.5–6 × 3–5 mm, base truncate or subcordate, margins entire or subentire to very weakly erose, flat, apex obtuse or subacute; tubercles normally 3, rarely 1 or 2, unequal, at least 1 distinctly larger, more than (1–)1.5 mm wide. **Achenes** usually reddish brown, 2–3 × 1.5–2 mm. *2n* = 60.

Flowering late spring–early fall. Very broad range of ruderal, segetal, and seminatural habitats, disturbed soil, waste places, cultivated fields, roadsides, meadows, shores of water bodies, edges of woods; 0–2500 m; introduced; St. Pierre and Miquelon; Alta., B.C., Man., N.B., Nfld. and Labr. (Nfld.), N.W.T., N.S., Ont., P.E.I., Que., Sask., Yukon; Ala., Alaska, Ariz., Ark., Calif., Colo., Conn., Del., D.C., Fla., Ga., Idaho, Ill., Ind., Iowa, Kans., Ky., La., Maine, Md., Mass., Mich., Minn., Miss., Mo., Mont., Nebr., Nev., N.H., N.J., N.Mex., N.Y., N.C., N.Dak., Ohio, Okla., Oreg., Pa., R.I., S.C., S.Dak., Tenn., Tex., Utah, Vt., Va., Wash., W.Va., Wis., Wyo.; Eurasia; introduced almost worldwide.

Rumex crispus (belonging to subsect. *Crispi* Rechinger f.; see K. H. Rechinger 1937) is the most widespread and ecologically successful species of the genus, occuring almost worldwide as a completely naturalized and sometimes invasive alien. It has not been reported from Greenland, but it probably occurs there.

Rumex crispus hybridizes with many other species of subg. *Rumex*. Hybrids with *R. obtusifolius* (*Rumex* ×*pratensis* Mertens & Koch) are the most common in the genus, at least in Europe, and have been reported for several localities in North America. *Rumex crispus* × *R. patientia* (*Rumex* ×*confusus* Simonkai) was reported from New York. According to R. S. Mitchell (1986, p. 47), "this hybrid is now spreading along highway shoulders, and it has replaced *R. crispus* in some local areas." However, that information should be confirmed by more detailed studies since spontaneous hybrids between species of sect. *Rumex* usually are much less fertile and ecologically successful than the parental species. Hybrids of *Rumex* occuring in North America need careful revision.

Numerous infraspecific taxa and even segregate species have been described in the *Rumex crispus* aggregate. Many seem to represent minor variation of little or no taxonomic significance, but some are geographically delimited entities that may deserve recognition as subspecies or varieties. The typical variety has inner tepals with three well-developed tubercles; the less common var. *unicallosus* Petermann, with one tubercle, occurs sporadically in North America.

Some eastern Asian plants differ from typical *Rumex crispus* is having somewhat smaller inner tepals, longer pedicels, lax inflorescences with remote whorls, and narrower leaves that are almost flat or indistinctly undulate at the margins. These plants, originally described as *R. fauriei* Rechinger f., are now treated as *R. crispus* subsp. *fauriei* (Rechinger f.) Mosyakin & W. L. Wagner; the subspecies was recently reported from Hawaii (S. L. Mosyakin and W. L. Wagner 1998) and may be expected as introduced in western North America.

48. **Rumex stenophyllus** Ledebour, Fl. Altaica 2: 58. 1830 • Narrow-leaved or narrowleaf dock

F I W

Rumex alluvius F. C. Gates & McGregor; *R. crispus* Linnaeus var. *dentatus* Schur; *R. obtusifolius* Linnaeus var. *cristatus* Neilreich; *R. odontocarpus* Sandor ex Borbás

Plants perennial, glabrous or very indistinctly papillose normally only on veins of leaf blades abaxially, with fusiform, vertical rootstock. **Stems** erect, branched distal to middle, 40–80(–130) cm. **Leaves:** ocreae usually deciduous, rarely partially persistent at maturity; blade oblong-lanceolate, lanceolate, or narrowly lanceolate, normally 15–25(–30) × 2–7 cm, base cuneate or truncate, margins entire or irregularly denticulate, usually crisped and undulate, or, occasionally, flat, apex acute or subobtuse. **Inflorescences** terminal, occupying distal ¹/₂ of stem, dense or interrupted at base, narrowly paniculate, branches usually straight or occasionally arcuate. **Pedicels** articulated in proximal ¹/₃, filiform, 3–8 mm, articulation distinctly swollen. **Flowers** 20–25 in whorls; inner tepals orbiculate-ovate or occasionally ovate-deltoid, 3.5–5 × 3–5 mm, base truncate or slightly cordate, margins denticulate, apex acute or subacute, teeth 4–10 at each side, 0.2–1.5 mm; tubercles normally 3, equal or subequal, less than 2 times as wide as inner tepals. **Achenes** usually reddish brown or dark brown, 2–2.5(–3) × 1–1.5 mm. *2n* = 60.

Flowering late spring–early fall. Waste places, roadsides, fields, meadows, swamps and marshes, shores, saline soils; 0–1600 m; introduced; Alta., Man., Ont., Que., Sask.; Calif., Colo., Iowa, Kans., Minn., Mo., Mont., Nebr., Nev., N.Mex., N.Dak., Okla., S.C., S.Dak., Utah, Wash., Wyo.; c, se Europe; c Asia (s Siberia).

Within its native range *Rumex stenophyllus* is mostly confined to slightly saline coastal and alluvial (riparian) habitats. It has successfully colonized a wide range of ruderal and segetal habitats in both Europe and North America. Further spread of this species in the central and southwestern United States and southern Canada may be expected (D. Löve and J.-P. Bernard 1958). It was placed by K. H. Rechinger (1949) in subsect. *Stenophylli* Rechinger f.

According to J. K. Morton and J. M. Venn (1990), reports of *Rumex stenophyllus* from Ontario refer to the hybrid *R. crispus* × *R. obtusifolius*, but *R. stenophyllus* may be found in the province in the future. *Rumex stenophyllus* may be distinguished from that hybrid by its fertile fruits and more uniform inner tepals.

49. Rumex britannica Linnaeus, Sp. Pl. 1: 334. 1753 • Greater water or British dock   E   F

Rumex britannica var. *borealis* Rechinger f.; *R. hydrolapathum* Hudson var. *americanum* A. Gray; *R. orbiculatus* A. Gray

Plants perennial, normally glabrous or very indistinctly papillose on veins of leaf blades abaxially, with fusiform, vertical rootstock. **Stems** erect, branched distal to middle, 80–150(–200) cm. **Leaves:** ocrea deciduous or partially persistent at maturity; blade lanceolate or oblong-lanceolate, normally 20–55(–70) × 2–7 cm, base cuneate, occasionally rounded or truncate, margins entire, flat to slightly crisped, apex acute. **Inflorescences** terminal, occupying distal 1/2 of stem, dense or interrupted in proximal 1/2, broadly paniculate, branches usually straight or arcuate. **Pedicels** articulated in proximal 1/3, filiform, 5–13 mm, articulation barely evident, not distinctly swollen. **Flowers** 15–25 in whorls; inner tepals orbiculate or orbiculate-ovate, rarely ovate-deltoid, 4–7(–7.5) × 3.5–7 mm, base truncate or slightly cordate, margins entire or weakly erose, flat, apex obtuse to subacute; tubercles 3, equal or subequal, normally less than 2 times as wide as inner tepals. **Achenes** usually reddish brown, 3–4.5 × 1.5–2.5 mm. $2n = 20$.

Flowering late spring–summer. Marshes, shores, wet meadows, other damp areas; 0–1500 m; Alta., Man., N.B., Nfld. and Labr. (Nfld.), N.W.T., N.S., Ont., P.E.I., Que., Sask.; Calif., Conn., D.C., Ill., Ind., Iowa, Ky., Maine, Md., Mass., Mich., Minn., Nebr., N.H., N.J., N.Y., N.Dak., Pa., R.I., S.Dak., Vt., Va., Wis.

The name *Rumex orbiculatus* commonly was applied to this North American species. After study of the Linnaean type of *R. britannica*, J. E. Dawson (1979) concluded that that name is the earliest valid one for this taxon.

In early North American floristic literature, *Rumex britannica* commonly was misidentified as *R. hydrolapathum* Hudson, a closely related European species also belonging to subsect. *Hydrolapatha* Rechinger f. (K. H. Rechinger 1937); that species differs from *R. britannica* in having more triangular inner tepals with an acute apex. The name *R. acutus* Linnaeus was misapplied to *R. britannica* by W. J. Hooker ([1829–] 1833–1840, vol. 2) and other botanists.

Disjunct populations have been reported from California and Louisiana. The California record (from Plumas County) was confirmed by J. E. Dawson (1979); the records from Louisiana need confirmation.

50. Rumex conglomeratus Murray, Prodr. Stirp. Gott., 52. 1770 • Clustered or clustered green dock   F   I   W

Rumex acutus Smith

Plants perennial, normally glabrous, rarely very indistinctly papillose on veins of leaf blades abaxially, with fusiform, vertical rootstock. **Stems** erect, branched in distal 2/3 (sometimes with few flowering stems from rootstock), 30–80(–120) cm. **Leaves:** ocrea deciduous to partially persistent at maturity; blade oblong-lanceolate, obovate-lanceolate, or lanceolate, normally (5–)10–30 × 2.5–6 cm, base broadly cuneate, rounded, or truncate, rarely subcordate, margins entire, flat to very weakly undulate, apex subacute, occasionally obtuse. **Inflorescences** terminal, occupying distal 2/3 of stem, lax, interrupted, broadly paniculate, branches simple or nearly so, almost all but distalmost whorls with subtending leaves; panicle leafy at least in proximal 2/3 of length. **Pedicels** articulated in proximal 1/3 or occasionally near middle, filiform, 1–4(–5) mm, ca. as long as inner tepals or slightly longer, articulation distinctly swollen. **Flowers** 10–20 in dense remote whorls; inner tepals oblong-lanceolate, oblong, or lingulate, 2–3 × 1–1.6(–2) mm, ca. 2 times as long as wide, base cuneate or truncate, margins entire, apex obtuse; tubercles 3, equal or subequal, largest tubercle almost as wide as inner tepal. **Achenes** usually dark reddish brown, 1.5–1.8 × 1–1.4 mm. $2n = 20$.

Flowering early summer–early fall. Marshes, wet meadows, shores, alluvial woods, ditches, wet waste places; 0–1500 m; introduced; B.C.; Ala., Ariz., Ark., Calif., Ga., Ill., Ind., Ky., La., Mass., Miss., Mo., N.Y., N.C., Okla., Oreg., Pa., S.C., Tenn., Tex., Va., Wash., W.Va.; Europe; w, sw Asia; n Africa; introduced elsewhere.

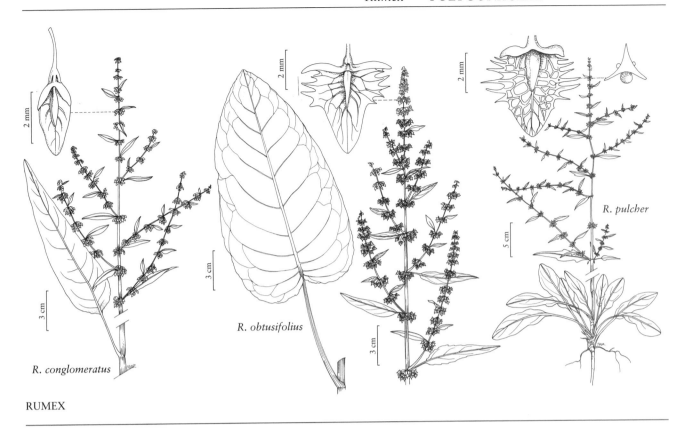

R. conglomeratus

R. obtusifolius

R. pulcher

RUMEX

Rumex conglomeratus often is confused with immature specimens of *R. obtusifolius*, as well as with other species (e.g., *R. sanguineus*). Its distribution in North America is insufficiently known, and some literature records may refer to *R. obtusifolius*.

Rumex conglomeratus and *R. sanguineus* were placed in subsect. *Conglomerati* Rechinger f. (K. H. Rechinger 1937).

51. Rumex sanguineus Linnaeus, Sp. Pl. 1: 334. 1753

• Wood or redvein or bloodwort dock [I]

Lapathum sanguineum (Linnaeus) Lamarck; *Rumex condylodes* M. Bieberstein; *R. nemorosus* Schrader ex Willdenow

Plants perennial, normally glabrous, rarely very indistinctly papillose on veins of leaf blades abaxially, with fusiform, vertical rootstock. **Stems** erect, branched in distal ²/₃, sometimes branched almost from base or with few flowering stems from rootstock, 30–70(–90) cm. **Leaves:** ocrea deciduous to partially persistent at maturity; blade oblong-lanceolate, obovate-lanceolate, or lanceolate, normally (5–)10–30 × 2.5–6 cm, base rounded, truncate, or subcordate, rarely cuneate, margins entire to obscurely repand, flat to slightly undulate, apex acute or subacute, occasionally attenuate. **Inflorescences** terminal, occupying distal ²/₃ of stem, lax, interrupted, broadly paniculate, branches simple or nearly so; panicle leafless or leafy only near base. **Pedicels** articulated in proximal ¹/₃ or rarely near middle, filiform, (2–)4–6(–8) mm, normally distinctly longer than inner tepals, articulation distinctly swollen. **Flowers** 10–20 in lax, remote whorls; inner tepals oblong-lanceolate, oblong, or lingulate, 2–3 × 0.8–1.3(–1.8) mm, ca. 2 times as long as wide, base cuneate or subtruncate, margins entire, apex obtuse; tubercle 1 (occasionally 3, then 1 much larger, almost as wide as inner tepals). **Achenes** usually dark reddish brown to almost black, 1.25–1.5 (–1.8) × 1–1.3 mm. $2n = 20$.

Flowering summer. Moist alluvial and riparian habitats, ruderal places, ballast grounds; 0–500 m; introduced; B.C., Ont., Que.; Ala., Oreg., Pa.; Europe; sw Asia; n Africa.

Distribution of *Rumex sanguineus* in North America is known insufficiently. Most reports from California, Washington, New Brunswick, Nova Scotia, Ontario, and Quebec were based on misidentified specimens of *R. conglomeratus* or immature *R. obtusifolius*.

Rumex sanguineus is represented in Europe by at least two varieties. The uncommon, cultivated, and occasionally escaped var. *sanguineus* (redvein dock or bloodwort) has bright red or purple venation of leaves. It probably arose as a mutant from the common, wild var. *viridis* Sibthorp.

52. Rumex obtusifolius Linnaeus, Sp. Pl. 1: 335. 1753

• Broad-leaved or broadleaf or bitter dock, patience a feuilles obtuses F I W

Rumex crispatulus Michaux; *R. rugelii* Meisner

Plants perennial, glabrous or ± papillose especially on veins of leaf blades abaxially, with fusiform, vertical rootstock. **Stems** erect, branched distal to middle or occasionally in distal $2/3$, often with few flowering stems from rootstock, 60–120(–150) cm. **Leaves:** ocrea deciduous to partially persistent at maturity; blade oblong to ovate-oblong, sometimes broadly ovate, 20–40 × 10–15 cm, usually less than 4 times as long as wide, base normally distinctly cordate, occasionally rounded, rarely truncate, margins normally entire, flat or undulate, rarely slightly crisped, apex obtuse or subacute. **Inflorescences** terminal, occupying distal $2/3$ of stem, usually lax and interrupted, narrowly or broadly paniculate, branches usually forming angle of 30–45° with 1st-order stem. **Pedicels** articulated in proximal $1/3$ or rarely near middle, filiform, 2.5–8.5(–10) mm, articulation distinctly swollen. **Flowers** 10–25 in lax whorls; inner tepals ovate-triangular, deltoid or, occasionally, lingulate, 3–6 × 2–3.5 mm (excluding teeth), ca. 1.5–2 times as long as wide, base truncate, margins usually distinctly dentate, rarely subentire, apex obtuse to subacute, straight, teeth 2–5, normally at each side of margin, short-subulate or triangular-subulate, straight, 0.5–1.8 mm, or shorter than width of inner tepals; tubercle usually 1, sometimes 3, then 1 distinctly larger, smooth. **Achenes** brown to reddish brown, 2–2.7 × 1.2–1.7 mm. $2n = 40$.

Flowering late spring–early fall. Waste places, roadsides, fields, shores, meadows, wet woods, swamps; 0–2300 m; introduced; Greenland; St. Pierre and Miquelon; B.C., N.B., Nfld. and Labr. (Nfld.), N.S., Ont., P.E.I., Que.; Ala., Alaska, Ariz., Ark., Calif., Colo., Conn., Del., D.C., Fla., Ga., Idaho, Ill., Ind., Iowa, Kans., Ky., La., Maine, Md., Mass., Mich., Minn., Miss., Mo., Mont., Nebr., N.H., N.J., N.Mex., N.Y., N.C., Ohio, Okla., Oreg., Pa., R.I., S.C., Tenn., Tex., Utah, Vt., Va., Wash., W.Va., Wis.; Europe; w Asia; introduced elsewhere.

Rumex obtusifolius, a member of subsect. *Obtusifolii* Rechinger f. (K. H. Rechinger 1937), is a polymorphic species represented in Eurasia by three or four rather distinct races often treated by European authors as subspecies or varieties. These taxa differ mostly in inner tepal dentation and geographic distribution. In North America the morphotypes often intergrade. In Eurasia this species is differentiated into predominantly western subsp. *obtusifolius* [including *R. obtusifolius* subsp. *agrestis* (Fries) Danser], eastern subsp. *sylvestris* (Wallroth) Rechinger f., intermediate central European subsp. *transiens* (Simonkai) Rechinger f., and montane subsp. *subalpinus* (Schur) Simonkai. Only subspp. *obtusifolius* and *sylvestris* occur in North America; the former seems to be more common. Subspecies *obtusifolius* differs from subsp. *sylvestris* in having larger and more prominently dentate inner tepals with one tubercle, or with three distinctly unequal tubercles; in subsp. *sylvestris* the teeth are usually less than 0.6 mm, developing only near the base of the inner tepals, and the tubercles often almost subequal.

Rumex obtusifolius may be expected elsewhere in the Great Plains region of the United States and Canada.

53. Rumex pulcher Linnaeus, Sp. Pl. 1: 336. 1753

• Fiddle dock F I W

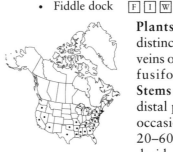

Plants perennial, glabrous or distinctly papillose especially on veins of leaf blades abaxially, with fusiform, vertical rootstock. **Stems** erect, often flexuous in distal part, branched in distal $2/3$, occasionally almost from base, 20–60(–70) cm. **Leaves:** ocrea deciduous or partially persistent at maturity; blade oblong to ovate-oblong, sometimes broadly lanceolate or panduriform, contracted near middle or proximally, 4–10(–15) × (2–)3–5 cm, less than 4 times as long as wide, base normally truncate or weakly cordate, occasionally rounded, margins entire, flat or undulate, rarely slightly crisped, apex obtuse or subacute. **Inflorescences** terminal, occupying distal $2/3$ of stem or more, usually lax and interrupted, broadly paniculate, branches usually divaricately spreading, forming angle of 60–90° with 1st-order stem. **Pedicels** articulated in proximal $1/3$ or occasionally near middle, thickened, not filiform, 2–5(–6) mm, articulation distinctly swollen. **Flowers** 10–20 in rather dense whorls; inner tepals ovate-triangular, deltoid, or oblong-deltoid, 3–6 × 2–3 mm (excluding teeth), normally ca. 1.5 times as long as wide, base truncate, margins usually distinctly dentate, rarely subentire, apex obtuse to subacute, straight, teeth 2–5(–9), normally on margins at each side, narrowly triangular, 0.3–2.5 mm, longer or shorter than width of inner tepals; tubercles (1–)3, equal or unequal, usually verrucose (warty). **Achenes** dark reddish brown to almost black, 2–2.8 × 1.3–2 mm. $2n = 20$.

Flowering late spring–summer. Waste places, roadsides, shores, fields, meadows, moist to dry habitats; 0–1500 m; introduced; Ala., Ariz., Ark., Calif., Fla., Ga., Ky., La., Md., Mass., Miss., Mo., Nev., N.J., N.Y., N.C., Okla., Oreg., Pa., R.I., S.C., Tenn., Tex., Va., W.Va.; s, w Europe; sw Asia; n Africa; introduced elsewhere.

Rumex pulcher is an extremely polymorphic species consisting of five or six more or less distinct subspecies (K. H. Rechinger 1949, 1964). Three of these were reported by Rechinger (1937) from North America: subsp. *pulcher*; subsp. *woodsii* (De Not.) Arcangeli [= *Rumex divaricatus* Linnaeus]; and subsp. *anodontus* (Haussknecht) Rechinger f. Judging from herbarium specimens, subsp. *woodsii* seems to be the most common. However, J. E. Dawson (1979) noted that many North American specimens cannot easily be assigned to any subspecies.

Some records require confirmation, especially from the midwestern states, since *Rumex pulcher* often is confused with other species with dentate inner tepals. Records from Colorado (H. D. Harrington 1954) belong to *R. stenophyllus* (W. A. Weber and R. C. Wittmann 1992).

54. Rumex dentatus Linnaeus, Mant. Pl., 226. 1771

• Toothed or dentate or Indian dock [1]

Plants annual, rarely biennial, glabrous or indistinctly papillose especially on veins of leaf blades abaxially, with fusiform, vertical rootstock. **Stems** erect, often flexuous in inflorescence, branched distal to middle, occasionally almost from base, 20–70(–80) cm. **Leaves:** ocrea deciduous or partially persistent at maturity; blade oblong, elliptic-lanceolate, or ovate-elliptic, 3–8(–12) × 2–5 cm, normally less than 4 times as long as wide, not coriaceous, base normally truncate or subcordate to weakly cordate, margins entire, flat to weakly undulate, occasionally slightly crisped, apex obtuse or subacute. **Inflorescences** terminal, occupying distal 1/2 of stem, usually lax and interrupted, broadly paniculate, branches usually ascending and straight. **Pedicels** articulated in proximal 1/3, filiform, 2–5 mm, articulation distinctly swollen. **Flowers** 10–20 in rather dense remote whorls; inner tepals ovate-triangular or deltoid, 3–5.5(–6) × 2–3 mm (excluding teeth), ca. 1.5 times as long as wide, base truncate, margins in most cases distinctly dentate, very rarely subentire, apex acute to subacute, straight, teeth 2–4(–5), normally at each side of margins, narrowly triangular, straight, 1–3(–5) mm, equaling or shorter than width of inner tepals; tubercles (1–)3, equal or subequal. **Achenes** dark reddish brown, 2–2.8 × 1.4–1.8 mm. **2n** = 40.

Flowering late spring–summer. Waste places, shores, cultivated fields; 0–300 m; introduced; Alta., Ont.; Ariz., Calif., Mo., Oreg., Tex.; se Europe; tropical and subtropical Asia; n Africa; introduced elsewhere.

Rumex dentatus, belonging to subsect. *Dentati* Rechinger f. (K. H. Rechinger 1937), is an extremely variable species. In Eurasia and northern Africa it is represented by several distinct races, usually regarded as subspecies. Rechinger reported from North America (California and Oregon, mostly as a casual alien occuring in ballast grounds) only subsp. *klotzschianus* (Meisner) Rechinger f., which is native in southern and eastern Asia (China, India, Japan, Korea). J. E. Dawson (1979) also regarded this as the most common subspecies in North America. However, I believe most North American representatives of this aggregate belong to subsp. *halacsyi* (Rechinger f.) Rechinger f., which is native in the eastern Mediterranean region (Asia Minor), the Caucasus, southeastern Europe, and western and central Asia. This subspecies, sometimes recognized as *R. halacsyi* Rechinger f., differs from subsp. *klotzschianus* by its broader triangular (not rounded) inner tepals and longer teeth (to 3 mm). Unfortunately, subspecies of *R. dentatus* still are insufficiently understood even in Eurasia. It would be premature to assign most North American specimens to any infraspecific entity.

55. Rumex obovatus Danser, Ned. Kruidk. Arch. 1920:

241, figs. 1–3. 1921 • Tropical or obovate-leaved dock [1]

Plants annual [sometimes biennial or perennial in tropics], glabrous or nearly so, with fusiform, vertical rootstock. **Stems** erect, branched distal to middle or in distal 2/3, 20–40(–70) cm. **Leaves:** ocrea deciduous or partially persistent at maturity; blade usually distinctly obovate, widest distal to middle, rarely weakly panduriform, 4–7(–11) × 2–5 cm, less than 4 times as long as wide, coriaceous, base cuneate, rarely almost truncate, margins entire, flat, rarely slightly undulate, apex obtuse or rounded. **Inflorescences** terminal, occupying distal 1/2 of stem, usually lax, occasionally rather dense, interrupted in basal 1/2, broadly paniculate, branches usually straight, unbranched, forming angle of 45° with 1st-order stem, leafy almost to top. **Pedicels** articulated in proximal 1/3, thickened, 3–5 mm, articulation distinctly swollen. **Flowers** 10–20 in rather dense whorls; inner tepals ovate-triangular or triangular-deltoid, 4–5(–5.5) × 2.4–3 mm (excluding teeth), ca. 1.5 times as long as wide, base truncate, margins distinctly dentate, apex acute to subacute, straight, teeth 3–5, normally at each side of margins, narrowly triangular or subulate, straight, 0.4–2.5 mm, equaling or shorter than width of inner tepals; tubercles usually 3, equal or subequal, apex obtuse, distinctly verrucose. **Achenes** brown, 2.3–2.8 (–3) × 1.4–1.8 mm.

Flowering summer. Mostly coastal habitats: sea shores, river deltas; 0–50 m; introduced; Fla., La.; South America; introduced in Europe.

Rumex obovatus is morphologically uniform. Individuals with panduriform leaves may be mistaken for *R. pulcher,* which is distinguished from *R. obovatus* by its less leathery leaves with subcordate to truncate (not distinctly cuneate) bases, usually more-spreading inflorescence branches, and perennial habit. *Rumex obovatus* was reported for Louisiana by J. W. Thieret (1969c), who followed determinations by K. H. Rechinger. It is known also from Florida and may be expected in adjacent coastal states. Thieret mentioned that one of his collections was made "within 150 feet of the Texas border."

56. **Rumex paraguayensis** D. Parodi, Anales Soc. Ci. Argent. 5: 160. 1878 • Paraguayan or Paraguay dock [I]

Plants annual [sometimes biennial], glabrous or nearly so, with fusiform, vertical rootstock. **Stems** erect, branched above middle or in distal 2/3, 20–40(–70) cm. **Leaves:** ocrea deciduous or partially persistent at maturity; blade distinctly obovate or sometimes panduriform, 4–6(–7) × 2–4(–5) cm, less than 4 times as long as wide, coriaceous, base cuneate or truncate, margins entire, flat or crisped, apex obtuse. **Inflorescences** terminal, occupying distal 1/2 of stem, usually lax, occasionally rather dense in distalmost part, interrupted in basal 1/2, broadly paniculate, branches spreading. **Pedicels** articulated in proximal 1/3, filiform or indistinctly thickened, 2.5–6 mm, articulation distinctly swollen. **Flowers** 10–20 in rather dense whorls; inner tepals obovate-triangular to deltoid, 3–4 × 1.8–2.5 mm (excluding teeth), ca. 1.5 times as long as wide, base truncate, margins distinctly dentate, apex acute to subacute, straight, teeth 2–3(–4), normally at each side of margins, triangular, straight, 0.4–1.5 mm, equaling or shorter than width of inner tepals; tubercles usually 3, equal or subequal, apex acute or subacute, smooth or minutely punctate. **Achenes** reddish brown, 1.7–1.9 (–2) × 1–1.4 mm.

Flowering summer. Coastal and riparian sites; 0 m; introduced; Fla., La.; South America.

Rumex paraguayensis is closely related to *R. obovatus* and may be confused with it or, less frequently, with other species having dentate inner tepals (e.g., *R. dentatus, R. obtusifolius,* and *R. pulcher*). *Rumex paraguayensis* and *R. obovatus* appear to belong to a separate, undescribed subsection.

This species was first reported for North America from Saint Tammany Parish, Louisiana, by J. W. Thieret (1969c). Its status in North America is uncertain; J. H. Horton (1972) excluded it from the list of Polygonaceae of the southeastern United States. I have seen only two immature specimens from Florida.

57. **Rumex violascens** Rechinger f., Repert. Spec. Nov. Regni Veg. 39: 171. 1936 • Violet dock

Plants annual or biennial, sometimes short-lived perennial, glabrous, with fusiform, vertical or oblique rootstock. **Stems** erect, branched above middle, 25–75 cm. **Leaves:** ocrea deciduous or partially persistent at maturity; blade oblong-lanceolate to obovate-elliptic, 6–12(–15) × (2–)3–4(–5) cm, less than 4 times as long as wide, not coriaceous, occasionally subcoriaceous, base broadly cuneate or rounded, occasionally truncate, margins entire, flat or, rarely, weakly undulate, apex obtuse, rarely subacute. **Inflorescences** terminal, occupying distal 1/2 of stem, usually lax, interrupted, broadly paniculate; branches usually distinctly arcuate and flexuous. **Pedicels** articulated in proximal 1/3, filiform, 3–8 mm, articulation distinctly swollen. **Flowers** 10–20 in rather dense remote whorls; inner tepals deltoid or triangular-deltoid, 2.5–3.7(–4) × 2–2.8(–3) mm (excluding teeth), ca. 1.5 times as long as wide, base truncate, margins dentate or rarely subentire, apex acute to subacute, teeth 2–3, normally at each side of margins, triangular, 0.2–0.5 mm; tubercles 3, unequal or, occasionally, subequal, normally less than 2 times as wide as inner tepals. **Achenes** brown or reddish brown, 1.8–2.3 × 1.2–1.5 mm. $2n = 20$.

Flowering late spring–summer. Wet habitats along streams, river valleys; 0–500 m; Ariz., Calif., Mass., N.Mex., Tex.; n Mexico.

The name *Rumex berlandieri* was misapplied to this species by some botanists who followed W. D. Trelease (1892).

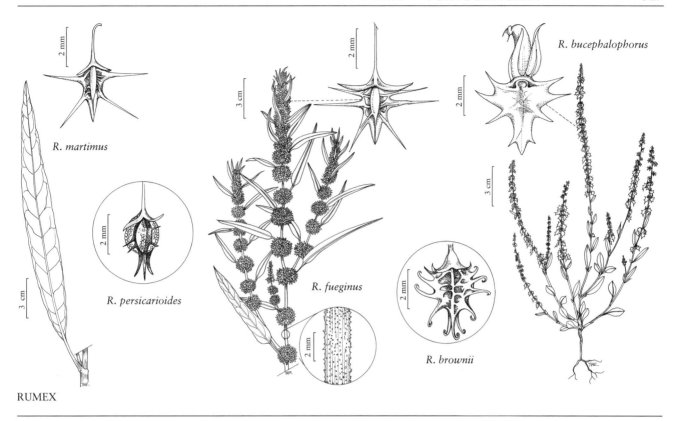

R. martimus

R. persicarioides

R. fueginus

R. brownii

R. bucephalophorus

RUMEX

This native species is poorly known taxonomically; it probably is distantly related to *Rumex obovatus* and *R. paraguayensis*. *Rumex flexicaulis* Rechinger f., a species possibly conspecific with *R. violascens*, occurs in Mexico.

58. Rumex maritimus Linnaeus, Sp. Pl. 1: 335. 1753

• Golden or maritime dock, rumex maritime F I

Lapathum minus Lamarck; *Rumex aureus* Miller

Plants annual, rarely biennial [perennial], glabrous or very weakly short-papillose, mostly in inflorescence and on leaf blades, with fusiform, vertical rootstock. **Stems** erect (some dwarf alluvial forms may be ascending or almost prostrate), branched in distal ²/₃, occasionally almost near base, (5–)15–75(–100) cm. **Leaves:** ocrea deciduous to partially persistent at maturity; blade lanceolate or lanceolate-linear, rarely oblong-lanceolate, usually very gradually narrowed at both ends, (4–)7–25(–40) × (1–)1.5–4(–5) cm, at least 4 times as long as wide, base narrowly cuneate, very rarely broadly cuneate, margins entire, flat or rarely weakly undulate, apex acute, very rarely subobtuse. **Inflorescences** terminal, occupying distal ¹/₂ of stem, occasionally most of stem, golden or greenish yellow, usually rather dense or interrupted in proximal part, broadly paniculate, branches spreading. **Pedicels** articulated near base or at least in proximal ¹/₃, filiform, 3–8 mm, articulation weakly evident. **Flowers** 15–30 (occasionally more) in rather dense whorls; inner tepals narrowly triangular or narrowly rhombic-triangular, 2.5–3(–3.5) × 0.75–1.2(–1.5) mm (excluding teeth), normally ca. 2 times as long as wide, base truncate or broadly cuneate, margins prominently dentate, apex acute, very rarely subacute, straight, teeth 2–3(–4), normally at each side of margins, subulate-filiform, bristlelike, 1–3.5 mm, usually 1.5–2 times as long as width of inner tepals; tubercles 3, equal or subequal, apex acute or subacute, smooth. **Achenes** light brown, small, 0.9–1.75 × 0.6–1 mm. $2n = 40$.

Flowering summer–early fall. Alluvial, riparian, and coastal habitats, mostly as a ruderal species; 0–500 m; introduced; Yukon; Alaska, Mass., N.J.; Europe.

This Eurasian species is known as a casual alien from several localities in North America. Its distribution is poorly known due to confusion with native American species of this aggregate. Plants from Alaska and Yukon reported by E. Hultén (1968) as *Rumex maritimus* need additional study; they may be conspecific with some eastern Asian races of the *R. maritimus* aggregate. It is rare or almost absent in eastern Asia, where it is replaced by closely related taxa.

Species of the *Rumex maritimus* aggregate can be placed in a separate subsection *Maritimi* Rechinger f. (K. H. Rechinger 1937) or even section *Orientales* A. I. Baranov & B. V. Skvortzov (see A. E. Borodina 1977).

In addition to characters mentioned in the key and descriptions, additional distinctive features of *Rumex maritimus* are the smooth tubercles (occasionally finely striate or indistinctly pitted in herbarium specimens), and golden yellow or greenish yellow mature inflorescences.

59. Rumex palustris Smith, Fl. Brit. 1: 394. 1800
· Marsh dock [I]

Plants annual or biennial, glabrous or very weakly short-papillose mostly in inflorescence and on leaf blades, with fusiform, vertical rootstock. **Stems** erect, normally branched in distal ⅔, occasionally almost near base, 10–60(–100) cm. **Leaves:** ocrea normally deciduous, rarely partially persistent at maturity; blade lanceolate or oblong-lanceolate, gradually narrowed at both ends, (10–)15–30(–35) × 1.5–6 cm, more than 4 times as long as wide, base narrowly to broadly cuneate, margins entire, flat or, rarely, weakly undulate, apex acute, rarely subobtuse. **Inflorescences** terminal, occupying distal ½ of stem, rarely more, reddish brown at maturity, usually lax, interrupted in proximal ½ or ⅔, broadly paniculate; branches spreading. **Pedicels** articulated near base or at least in proximal ⅓, filiform, 3–6 mm, articulation weakly evident. **Flowers** 15–25(–30) in rather dense whorls; inner tepals narrowly triangular, narrowly rhombic-triangular, or linguliform, (2.5–)3–3.5(–4) × 1.2–1.5(–2) mm (excluding teeth), normally ca. 2 times as long as wide, base truncate or broadly cuneate, margins prominently dentate, apex acute very rarely subacute, straight, teeth (1–)2–3, normally at each side of margins, subulate-filiform, bristlelike, straight, 1–2(–3) mm, usually as long as width of inner tepals; tubercles 3, equal or subequal, apex obtuse, smooth. **Achenes** light brown, 0.9–1.75 × 0.6–1 mm. $2n = 60$.

Flowering summer. Wet meadows, shores, marshes, ballast grounds, wet ruderal habitats; 0 m; introduced; Calif., N.J.; Europe.

Rumex palustris was reported for the first time from North America by J. E. Dawson (1979) based on specimens misidentified as *R. maritimus* and *R. pulcher*, collected in 1877 and 1959, respectively. *Rumex palustris* is not known to be invasive, and these collections probably represent only occasional, chance introductions.

60. Rumex persicarioides Linnaeus, Sp. Pl. 1: 335.
1753 · Rumex fausse-persicaire [E] [F]

Rumex maritimus Linnaeus var. *persicarioides* (Linnaeus) R. S. Mitchell

Plants annual, rarely biennial, usually distinctly papillose-pubescent mostly in inflorescence and on leaf blades abaxially, very rarely almost glabrous, with fusiform, vertical rootstock. **Stems** erect, branched in distal ⅔, occasionally almost near base, especially in dwarf littoral plants, (5–)15–70(–90) cm. **Leaves:** ocrea deciduous to partially persistent at maturity; blade lanceolate or oblong-lanceolate, 5–25(–30) × 1.5–5 cm, at least 4 times as long as wide, base abruptly truncate, slightly cordate or, rarely, broadly cuneate, margins entire, weakly undulate or crisped, apex acute or occasionally subobtuse. **Inflorescences** terminal, occupying distal ½ of stem (occasionally most of stem), usually dense in distal part, interrupted in proximal part, broadly paniculate. **Pedicels** articulated near base or at least in proximal ⅓, filiform, 3–7 mm, articulation weakly evident. **Flowers** 15–25 (occasionally more) in rather dense whorls; inner tepals narrowly triangular or narrowly rhombic-triangular, 2–2.5 × 0.75–1(–1.5) mm (excluding teeth), normally ca. 2 times as long as wide, base truncate or broadly cuneate, margins prominently dentate, apex acute, very rarely subacute, straight, teeth 2–3, normally at each side of margins, subulate-filiform, bristlelike, straight, 1–1.5(–1.7) mm, usually as long as width of inner tepals; tubercles 3, straw colored, oblong-ovate, ± equal, ca. as wide as inner tepals excluding teeth, apex obtuse, normally reticulate-pitted. **Achenes** brown, 0.9–1.5 × 0.5–0.8(–1) mm. $2n = 40$.

Flowering summer. Mostly coastal and slightly saline riparian habitats: shores, marshes. 0(–50) m; B.C., N.B., N.S., P.E.I., Que.; Calif., Conn., Mass., N.Y., Oreg.

Rumex persicarioides often has been treated by American botanists as a variety or synonym of *R. maritimus* (see R. S. Mitchell 1978). It and *R. fueginus* differ from Eurasian *R. maritimus* in many respects and are as distinct as many widely recognized Eurasian taxa of this aggregate (e.g., *R. palustris*, *R. rossicus* Murbeck, *R. ucranicus* Fischer ex Sprengel, *R. marschallianus* Reichenbach, *R. amurensis* Fr. Schmidt ex Maximowicz, *R. evenkiensis* Elisarjeva). When submerging *R. persicarioides* as a variety of *R. maritimus*, Mitchell noted: "Taxonomic treatment of the group from a Eurasian point of view would undoubtedly shed light on the minor problems which we face in North and South

America." However, from a Eurasian point of view (see e.g., K. H. Rechinger 1937, 1949; J. E. Lousley and D. H. Kent 1981; N. N. Tzvelev 1989b), all North American native taxa of subsect. *Maritimi* are evidently specifically different from any native Eurasian ones (with the only possible exception of Pacific plants, which are discussed below).

Plants similar to *Rumex persicarioides*, but with bigger tubercles and occuring along the Pacific coast from northern California to British Columbia, are, in my opinion, closer to *R. fueginus* in their habit and vegetative characters. K. H. Rechinger (1937) provisionally determined such specimens as *R. persicarioides*. J. E. Dawson (1979) noted that the Pacific plants differ from Atlantic ones in having bigger tubercles (more than 1.9 × 0.7–1 mm in western plants; less than 1.9 × 0.7 mm in eastern *R. persicarioides* in the narrow sense), and described these large-tubercled plants as a distinct variety, "*R. maritimus* var. *pacificus*", unfortunately, an invalid name. However, that taxon seems to be extremely closely related to or possibly conspecific with the northeastern Asian species, *R. ochotskius* Rechinger f., which is known in eastern Asia from northern Japan to the Okhotsk Sea region of Russian Far East (especially Sakhalin and Kuril islands). The latter species also has large (to 2–2.5 mm) botuliform tubercles with obtuse apices. In the original description Rechinger stated: "...*foliorum* forma *R. maritimo* simillimus...," but N. N. Tzvelev (1989b) in his recent treatment of the genus in the Russian Far East noted that most of the specimens of *R. ochotskius* seen by him had leaf blades rotundate-truncate or broadly cuneate at the base. The *R. persicarioides*-like plants from the Pacific coast of the United States and Canada (as well as their most probable allies from eastern Asia) need additional study. At present I prefer to place them provisionally into *R. persicarioides*, following Rechinger's treatment.

61. **Rumex fueginus** Philippi, Anales Univ. Chile 91: 493. 1895 • American golden or Tierra del Fuego dock F

Rumex maritimus Linnaeus subsp. *fueginus* (Philippi) Hultén; *R. maritimus* Linnaeus var. *fueginus* (Philippi) Dusen

Plants annual, rarely biennial, usually distinctly papillose-pubescent mostly in inflorescence and on leaf blades abaxially, or occasionally at most weakly papillose-pubecent, with fusiform, vertical rootstock. **Stems** erect (some dwarf alluvial forms may be with ascending or almost prostrate branches), branched in distal ²/₃, occasionally almost near base, (4–)15–60 (–70) cm. **Leaves:** ocrea mostly partially persistent at

maturity; blade lanceolate or lanceolate-linear, rarely oblong-lanceolate, (3–)5–25(–30) × (1–)1.5–3(–4) cm, more than 4 times as long as wide, base abruptly truncate, slightly cordate, or rarely broadly cuneate, margins entire or subentire to obscurely repand, normally undulate and crisped, apex acute very rarely subobtuse. **Inflorescences** terminal, occupying distal ¹/₂ of stem (occasionally most of stem), usually reddish brown or red (greenish yellow when mature), usually rather dense, interrupted in proximal part, broadly paniculate. **Pedicels** articulated near base or at least in proximal ¹/₃, filiform, 3–7(–9) mm, articulation weakly evident, occasionally indistinctly swollen. **Flowers** 15–30 (occasionally more) in rather dense whorls; inner tepals narrowly triangular or narrowly rhombic-triangular, 1.5–2.5 × 0.7–0.9(–1.2) mm (excluding teeth), normally ca. 2 times as long as wide, base truncate or broadly cuneate, margins usually prominently dentate, rarely with shorter teeth, or almost entire, apex acute, straight, teeth 2–3, at each side of margins, subulate-filiform, bristlelike, straight, 1–3 mm, usually 1.5–2.5(–4) times as long as width of inner tepals; tubercles 3, brownish or reddish, linear-lanceolate to fusiform, equal or subequal, rarely unequal, distinctly narrower than inner tepals, ca. 0.5 times as wide as inner tepals or less, apex acute or subacute, usually distinctly reticulate-pitted (especially in herbarium specimens). **Achenes** light brown, 1.1–1.4 × 0.6–0.8 mm. $2n = 40$.

Flowering late spring–early fall. Alluvial, riparian, and ruderal habitats, shores, marshes, bogs, wet meadows, dry streambeds; 0–2500 m; Alta., B.C., Man., N.B., N.W.T., N.S., Ont., P.E.I., Que., Sask., Yukon; Alaska, Ariz., Ark., Calif., Colo., Conn., Del., Idaho, Ill., Iowa, Kans., Ky., Maine, Md., Mass., Mich., Minn., Mo., Mont., Nebr., Nev., N.H., N.Mex., N.Y., N.Dak., Ohio, Oreg., Pa., R.I., S.Dak., Tex., Utah, Vt., Wash., Wis., Wyo.; Mexico; South America (s and mountains); Europe.

Rumex fueginus, in spite of its similarities to *R. maritimus*, is more closely related to *R. persicarioides*. Specimens of *R. fueginus* often are misidentified as *R. maritimus*, and the name *R. persicarioides* has been applied to *R. fueginus*. This confusion obscures distribution patterns among members of the aggregrate.

Several varieties have been described based mostly on teeth variation. These taxa appear to have little taxonomic significance, with the possible exception of var. *athrix* (St. John) Rechinger f., which has entire or subentire inner tepals and occurs in arid regions of the southwestern United States (H. St. John 1915; K. H. Rechinger 1937).

Rumex fueginus is known in Europe as an uncommon, casual alien.

62. Rumex brownii Campderá, Monogr. Rumex, 81. 1819 • Brown's dock F I

Rumex fimbriatus R. Brown, Prodr., 421. 1810, not Poiret 1804

Plants perennial, glabrous or very indistinctly papillose on veins of leaf blades abaxially, with vertical rootstock. **Stems** erect or ascending, divaricately branched in distal ¹/₂–²/₃, 30–80(–100) cm. **Leaves:** ocrea deciduous or partially persistent; blade shape variable, oblong to lanceolate or lanceolate-linear, often panduriform, contracted near or proximal middle, (3–)5–15(–17) × 1–3(–5) cm, base truncate, slightly cordate, or broadly cuneate, margins entire, normally slightly undulate and crisped, apex acute or subobtuse. **Inflorescences** terminal, occupying distal ²/₃ or most of stem, lax, interrupted in proximal part or throughout, broadly paniculate, branches spreading. **Pedicels** articulated in proximal ¹/₃, filiform (occasionally thickened distally), 2.5–5 mm, usually as long as or slightly longer than inner tepals, articulation distinctly swollen. **Flowers** 5–8(–10) in rather lax remote whorls; inner tepals broadly to narrowly triangular, 2.5–4 × 1.5–2.3 mm (excluding teeth), base truncate or broadly cuneate, margins prominently dentate, apex acute and ending in hooked tooth, teeth 3–5, at each side of margins, hooked, 0.5–1.5 mm; tubercles absent, or midveins indistinctly swollen. **Achenes** reddish brown, 1.8–2.3 × 1.2–1.5 mm. $2n = 20$.

Flowering summer. Waste places, near wool-combing mills; 0 m; introduced; S.C.; Europe; Australia; Pacific Islands (Java, New Guinea, New Zealand, Timor).

Rumex brownii, an uncommon "wool alien" in North America, was collected in South Carolina in the late 1950s (J. E. Dawson 1979). It is uncertain if it persists there. It occasionally occurs in Europe as a casual alien. This species is a member of subsect. *Acrancistron* Rechinger f. (see K. H. Rechinger 1984), which includes two Australian species.

26d. RUMEX Linnaeus subg. **PLATYPODIUM** (Willkomm) Rechinger f., Bot. Not. Suppl., 3(3): 106. 1954 I

Rumex sect. *Platypodium* Willkomm in H. M. Willkomm and J. M. C. Lange, Prodr. Fl. Hispan. 1: 284. 1862; *Bucephalophora* Pau; *Rumex* sect. *Heterolapathum* Nyman

Plants synoecious, with basal rosette of leaves usually not persistent at maturity. **Stems** ascending, spreading, or almost prostrate, occasionally erect. **Leaf blades** spatulate, lanceolate, or ovate-lanceolate, rarely rounded, base cuneate. **Pedicels** articulated in distal part, usually heteromorphic: some curved, swollen, and clavate, others straight, not swollen, and cylindric. **Flowers** normally bisexual; outer tepals not spreading; inner tepals enlarged, often heteromorphic, (1.5–)2–4(–5) mm, equaling or slightly wider and longer than achenes, margins usually dentate; tubercles very small, or absent.

Species 1: introduced; s Europe, sw Asia, n Africa; occasionally introduced in other regions.

Sometimes subg. *Platypodium* is recognized as a separate genus, *Bucephalophora*, together with other segregates from *Rumex* in the broad sense (see Á. Löve and B. M. Kapoor 1967).

63. Rumex bucephalophorus Linnaeus, Sp. Pl. 1: 336. 1753 • Horned, red, or ruby dock F I

Bucephalophora aculeata Pau; *Lapathum bucephalophorum* (Linnaeus) Lamarck

Plants annual, rarely biennial [perennial], glabrous or nearly so, with fusiform vertical root. **Stems** branched from base or near base, occasionally simple, slender, 5–50 cm. **Leaves:** ocrea deciduous or partially persistent; blade spatulate, lanceolate, or ovate-lanceolate, 1–5(–8) × (0.5–)1–2.5 cm, base cuneate, rarely rounded, margins normally entire, flat, apex obtuse. **Inflorescences** terminal, simple, racemose, occupying most of stem, interrupted, linear. **Pedicels** usually distinctly heteromorphic (much thickened distally), 4–7(–10) mm; others less than 4 mm. **Flowers** 2–3(–4) in lax clusters (reduced whorls); inner tepals variable, often heteromorphic, triangular, narrowly triangular, ligulate, or ovate-oblong, (1.5–)2–4(–5) × (0.5–)1–3 mm (excluding teeth), base truncate, margins usually dentate, sometimes entire, apex obtuse or acute, usually not hooked, teeth 2–4(–8), at each side of margins, slender, straight or hooked, 0.3–1 mm; tubercles absent, or 3, usually represented by minute swellings barely recognizable as tubercles. **Achenes** brown to dark brown, 1.3–2.3 × 0.7–1.4 mm. $2n = 16$.

Flowering summer. Ruderal habitats, ballast grounds; 0 m; introduced; La.; s Europe; w Asia; n Africa; occasionally introduced in other regions.

Rumex bucephalophorus is a polymorphic species, especially within its native range. K. H. Rechinger (1939, 1964) and J. R. Press (1988) recognized several subspecies, but no attempt has been made to distinguish infraspecific taxa among the limited North American materials. This species occurs in the flora area as an uncommon, casual alien. It has the potential to naturalize in the southern United States, especially in coastal regions from the Carolinas to Texas, and in California.

27. OXYRIA Hill, Veg. Syst. 10: 24, plate 24, fig. 2. 1765 • Mountain-sorrel, oxyrie
[Greek *oxys*, sour, and *-aria*, possession, alluding to acidic leaves]

Craig C. Freeman

John G. Packer

Herbs, perennial; roots fibrous; rhizomes present or not. **Stems** erect, glabrous. **Leaves** deciduous, basal, rarely also cauline, alternate, petiolate; ocrea sometimes deciduous, chartaceous; blade reniform to orbiculate-cordate, margins entire to obscurely wavy. **Inflorescences** terminal, paniclelike or racemelike, pedunculate. **Pedicels** present. **Flowers** bisexual, (1–)3–7 per ocreate fascicle, base stipelike; perianth nonaccrescent, greenish to reddish brown, campanulate, glabrous; tepals 4, in 2 whorls of 2, distinct, sepaloid, dimorphic, outer 2 narrower than inner 2; stamens (2–)6; filaments distinct, free, glabrous; anthers white to red or deep purple, elliptic to ovate; styles 2, spreading, distinct; stigmas penicillate. **Achenes** exserted, yellowish to tan, winged, lenticular, glabrous. **Seeds:** embryo straight or curved. $x = 7$.

Species ca. 4 (1 in the flora): arctic and alpine, n, w North America, Europe, e Asia.

SELECTED REFERENCES Chrtek, J. and M. Šourková. 1992. Variation in *Oxyria digyna*. Preslia 64: 207–210. Mooney, H. A. and W. D. Billings. 1961. Comparative physiological ecology of arctic and alpine populations of *Oxyria digyna*. Ecol. Monogr. 31: 1–29.

1. Oxyria digyna (Linnaeus) Hill, Hort. Kew., 158. 1768 • Alpine mountain-sorrel, oxyrie de montagne

Rumex digynus Linnaeus, Sp. Pl. 1: 337. 1753

Plants (3–)5–50 cm. **Stems** 1–4 (–8), often reddish, simple or branched distally. **Leaves** rarely 1–2 on stems, somewhat fleshy; ocrea hyaline or brownish hyaline, 2.5–10 mm, glabrous; petiole 1–15 cm; blade palmately veined with (5–)7(–9) basal veins, 0.5–6.5 × 0.5–6 cm, base cordate, apex rounded. **Inflorescences** (1–)2–20 cm; peduncle 1–17 cm. **Pedicels** spreading or reflexed, jointed proximal to middle, (1–)3–5 mm. **Flowers** 2–6 per ocreate fascicle; perianth 1–2.5 mm; outer 2 tepals spreading in fruit, navicular, 1.2–1.7 × 0.5–1 mm, inner 2 tepals appressed in fruit, broadly elliptic to orbiculate or obovate, 1.4–2.5 × 0.7–1.6 mm; stamens 1.5–2 mm; anthers 0.3–0.8(–1.1) mm; stigmas conspicuously exserted at anthesis, red. **Achenes** 3–4.5 × 2.5–5 mm including 2 wings, apex notched; wings reddish or pinkish, veiny. $2n = 14$.

Flowering Jun–Sep, fruiting Jul–Oct. Early melting snowbeds and zones of snow accumulation, gravel bars, mudflats, tundra, scree slopes, crevices in rock outcrops, talus slopes; 0–4200 m; Greenland; Alta., B.C., Nfld. and Labr., N.W.T., N.S., Nunavut, Que., Yukon; Alaska, Ariz., Calif., Colo., Idaho, Mont., Nev., N.H., N.Mex., Oreg., S.Dak., Utah, Wash., Wyo.; Europe; Asia.

Morphological and physiological differences between arctic and alpine populations of *Oxyria* in North America have been documented (H. A. Mooney and W. D. Billings 1961). Arctic plants (Alaska, northern Canada, and Greenland) taken from the field and grown in controlled environments tend to bear inflorescences with more branches, leaves with blades that are wider, and flowers with a more stable number of stamens as compared to alpine plants from populations in the south (California, Colorado, Montana, and Wyoming).

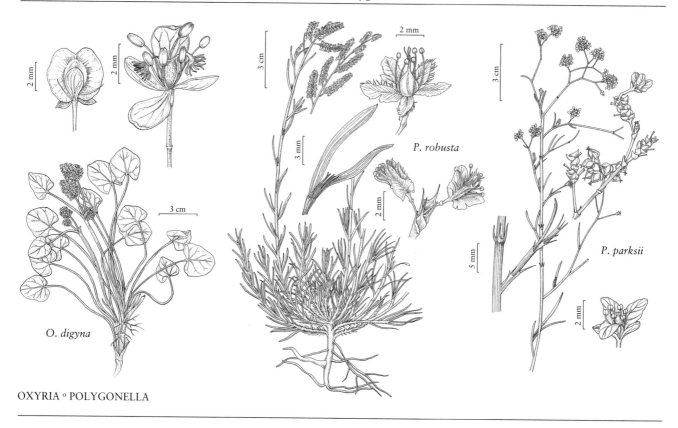

P. robusta

P. parksii

O. digyna

OXYRIA ∘ POLYGONELLA

Northern plants also have a greater tendency to reproduce asexually, often producing rhizomes and exhibiting relatively lower seed production.

Inuits consume the raw or cooked leaves and stems as a green or mixed with seal blubber or seal oil. Native American tribes in the Rocky Mountains also are reported to use the leaves as a salad (D. E. Moerman 1998). Caribou, muskoxen, and geese are reported to eat the leaves and stems, and arctic hares and lemmings consume the fleshy rhizomes (A. E. Porsild 1957).

28. POLYGONELLA Michaux, Fl. Bor.-Amer. 2: 240. 1803 • Jointweed, wireweed, polygonelle [genus name *Polygonum* and Latin *-ella*, diminutive]   E

Craig C. Freeman

Shrubs, subshrubs, or herbs, perennial or annual, synoecious, dioecious, gynodioecious, or gynomonoecious; taproots woody. **Stems** erect, decumbent, or prostrate, glabrous or scabrous. **Branches** adnate to stems, appearing to arise internodally. **Leaves** deciduous or, rarely, with leaves persisting more than 1 year, sometimes fugacious, cauline, alternate; ocrea usually persistent, sometimes disintegrating with age and deciduous distally, chartaceous or coriaceous; petiole apparently absent, articulate basally; blade filiform to broadly obovate, margins entire. **Inflorescences** terminal, racemelike, pedunculate. **Pedicels** present. **Flowers** bisexual, or some or all functionally unisexual, 1 per ocreate fascicle, base stipelike; perianth nonaccrescent, white, pink, red, greenish, or yellowish, campanulate, glabrous; tepals 5, distinct, petaloid, dimorphic, in 2 whorls with 2 outer and 3 inner or 2 outer and 2 inner plus 1 transitional; stamens 8, in 2 series with 5 outer and 3 inner; filaments distinct, free, dilated proximally, dimorphic, inner 3 dilated more abruptly than outer 5, with toothed or horned shoulders, or monomorphic (in *P. fimbriata* and *P. robusta*), glabrous (pubescent basally in *P. basiramia*); anthers white, yellow, orange, pink, or dark red, elliptic to ovate or round; styles (2–)3, erect, distinct; stigmas (2–)3, capitate. **Achenes** included or exserted, yellow-brown, brown, or reddish brown, wingless or narrowly winged, (2–)3(–4)-gonous, glabrous. **Seeds:** embryo straight or slightly curved. $x =$ 11.

Species 11 (11 in flora): e, sc United States.

Polygonella is distinct from other genera of Polygonaceae in having branches adnate to the stem and thus appearing to arise internodally. Palynological, anatomical, and morphological evidence suggests *Polygonella* is closely related to *Polygonum* sect. *Duravia* (L.-P. Ronse Decraene et al. 2004; Hong S. P. et al. 1998; P. O. Lewis 1991).

Within-population allozyme diversity is lower in the two most widespread species of the genus as compared to their narrowly endemic congeners (P. O. Lewis and D. J. Crawford 1995). High levels of selfing or depletion of diversity due to Pleistocene glaciation have been suggested as possible explanations for the lower allozyme diversity within populations of *Polygonella americana* and *P. articulata*.

SELECTED REFERENCES   Horton, J. H. 1963. A taxonomic revision of *Polygonella* (Polygonaceae). Brittonia 15: 177–203. Lewis, P. O. 1991. Allozyme Variation and Evolution in *Polygonella* (Polygonaceae). Ph.D. dissertation. Ohio State University. Lewis, P. O. and D. J. Crawford. 1995. Pleistocene refugium endemics exhibit greater allozymic diversity than widespread congeners in the genus *Polygonella* (Polygonaceae). Amer. J. Bot. 82: 141–149.   Nesom, G. L. and V. M. Bates. 1984. Reevaluation of infraspecific taxonomy in *Polygonella* (Polygonaceae). Brittonia 36: 37–44.   Nesom, G. L. and V. M. Bates. 1984b. A phylogenetic reconstruction of *Polygonella*. [Abstract.] A. S. B. Bull. 31: 74.   Ronse Decraene, L.-P., Hong S. P., and E. F. Smets. 2004. What is the taxonomic status of *Polygonella*? Evidence from floral morphology. Ann. Missouri Bot. Gard. 91: 320–345.

1. Margins of inner tepal deeply fringed; filaments of stamens monomorphic.
    2. Margins of leaf blade not hyaline; stems scabrous or, sometimes, glabrous proximally
       . . . . 1. *Polygonella fimbriata*
    2. Margins of leaf blade hyaline; stems glabrous or sparingly scabrous on angles distally
       . . . . 2. *Polygonella robusta*
1. Margins of inner tepal entire to erose; filaments of stamens dimorphic.
    3. Ocreae margins ciliate.
        4. Ocreolae not encircling rachis of raceme, their sides and bases adnate to rachis;
           inflorescences 2–6(–8) mm; outer tepals 0.5–0.9 mm in anthesis. . . . . 8. *Polygonella parksii*
        4. Ocreolae encircling rachis of raceme, only their bases adnate to rachis; inflorescences
           10–45 mm; outer tepals 0.7–1.5 mm in anthesis.
            5. Stems branched at or below ground level . . . . 3. *Polygonella basiramia*
            5. Stems simple, sometimes branched distally, if present, well above ground
               . . . . 4. *Polygonella ciliata*
    3. Ocreae margins not ciliate.
        6. Leaf blades (3–)9–30 mm wide, oblanceolate to obovate or broadly spatulate
           . . . . 6. *Polygonella macrophylla*
        6. Leaf blades 0.3–6(–8) mm wide, filiform to broadly spatulate.
            7. Styles and stigmas 0.4–1 mm in anthesis; plants perennial.
                8. Stems erect; outer tepals sharply reflexed early in anthesis and in fruit
                   . . . . 10. *Polygonella americana*
                8. Stems prostrate; outer tepals loosely appressed in anthesis, sometimes
                   spreading in fruit . . . . 11. *Polygonella myriophylla*
            7. Styles and stigmas ca. 0.1 mm or less in anthesis; plants annual or perennial.
                9. Plants perennial; margins of leaf blade hyaline at least along distal ¹/₂;
                   outer tepals usually reflexed in fruit. . . . . 7. *Polygonella polygama*
                9. Plants annual; margins of leaf blade not hyaline; outer tepals loosely
                   appressed to spreading in fruit.
                    10. Pedicels 0.9–3 mm, longer than subtending ocreolae; leaf blades 0.4–
                        1.2 mm wide . . . . 9. *Polygonella articulata*
                    10. Pedicles 0.1–0.3 mm, as long as subtending ocreolae; leaf blades 0.8–
                        5(–8) mm wide . . . . 5. *Polygonella gracilis*

1. Polygonella fimbriata (Elliott) Horton, Brittonia 15: 190. 1963 · Sandhill jointweed [E]

Polygonum fimbriatum Elliott, Sketch Bot. S. Carolina 1: 583. 1821; *Thysanella fimbriata* (Elliott) A. Gray

Herbs, annual, gynomonoecious, 1–6 dm. **Stems** erect, simple or sparingly branched proximally, scabrous or, sometimes, glabrous proximally. **Leaves** persistent; ocrea margins ciliate; blade linear to falcate, (10–)19–36(–50) × (0.6–)1–1.5(–3) mm, base tapered, margins not hyaline, apex acuminate, minutely scabrous. **Inflorescences** (5–)10–24(–30) mm; ocreola encircling rachis, only the base adnate to rachis, apex acuminate. **Pedicels** spreading in anthesis and fruit, 0.1–0.7 mm, as long as or much longer than subtending ocreola. **Flowers** bisexual or some pistillate, these usually distal; outer tepals loosely appressed in anthesis and fruit, pink with white margins, often drying orange, ovate, 1.1–2.3 mm in anthesis, margins erose; inner and transitional tepals loosely appressed in anthesis and fruit, pink with white margins, often drying orange, oblong, 1.2–2.2 mm in anthesis, margins deeply fringed; filaments monomorphic; anthers pink; styles and stigmas 0.6–1.3 mm in anthesis. **Achenes** included or exserted, yellow-brown, 3-gonous, 1.5–2.5 × 1–1.5 mm, shiny and smooth proximally, dull to shiny and minutely roughened distally.

Flowering Jul–Oct. Sandy pine-oak forests and sandhills; 10–200 m; Ala., Fla., Ga.

The chromosome number of 2*n* = 32 listed for *Polygonella fimbriata* by P. O. Lewis and D. J. Crawford (1995) appears to be in error. They cited J. H. Horton (1963) as the source; Horton did not count this species.

2. Polygonella robusta (Small) G. L. Nesom & V. M. Bates, Brittonia 36: 43. 1984 · Stout jointweed [E][F]

Thysanella robusta Small, Bull. Torrey Bot. Club 36: 159. 1909; *Polygonella fimbriata* (Elliott) Horton var. *robusta* (Small) Horton

Herbs, annual or perennial, gynomonoecious, 5–11 dm. **Stems** erect, usually branched at base, sometimes simple, glabrous or sparingly scabrous on angles distally. **Leaves** persistent; ocrea margins ciliate; blade linear, (10–)25–43(–69) × 1.2–2.5(–3) mm, base barely tapered, margins hyaline, apex acute to acuminate, glabrous. **Inflorescences** (10–)20–50(–60) mm; ocreola encircling rachis, only the base adnate to rachis, apex

acuminate. **Pedicels** spreading in anthesis and fruit, 0.3–1 mm, much longer than subtending ocreola. **Flowers** bisexual or some pistillate, these usually distal; outer tepals loosely appressed in anthesis and fruit, pink to white, drying orange, ovate, 1.7–4.5 mm in anthesis, margins erose; inner and transitional tepals pink to white, drying orange, loosely appressed in anthesis and fruit, oblong, 1.8–3.8 mm in anthesis, margins deeply fringed; filaments monomorphic; anthers pink; styles and stigmas 0.7–1.5 mm in anthesis. **Achenes** usually included, sometimes exserted distally, yellow-brown, 3-gonous, 1.8–2.6 × 1–1.5 mm, shiny and smooth proximally, dull to shiny and minutely roughened distally. **2***n* = 32.

Flowering Jul–Dec. Sandy roadsides, waste places, scrub; 0–60 m; Fla.

Polygonella robusta is closely related to *P. fimbriata* and has been treated as a variety thereof. Differences in morphology and distribution provide a basis for treating the two taxa as separate species (G. L. Nesom and V. M. Bates 1984).

3. Polygonella basiramia (Small) G. L. Nesom & V. M. Bates, Brittonia 36: 40. 1984 · Hairy wireweed [C][E]

Delopyrum basiramia Small, Bull. Torrey Bot. Club 51: 380. 1924; *Polygonella ciliata* Meisner var. *basiramia* (Small) Horton

Herbs, perennial, gynodioecious, 3–8 dm. **Stems** erect, branched or below ground level, glabrous or scabrous. **Leaves** fugacious; ocrea margins ciliate; blade linear, (3.5–)7–19(–28) × 0.2–0.5(–0.7) mm, base barely tapered, margins not hyaline, apex acuminate, minutely scabrous. **Inflorescences** (10–)14–22(–30) mm; ocreola encircling rachis, only the base adnate to rachis, apex acuminate. **Pedicels** spreading to reflexed in anthesis, sharply reflexed in fruit, 0.3–0.6 mm, as long or much longer than subtending ocreola. **Flowers** bisexual or pistillate; outer tepals loosely appressed in anthesis and fruit, white to pinkish, distal portion of midrib often inconspicuously greenish, narrowly oblong, 0.7–1.5 mm in anthesis, margins entire; inner tepals loosely appressed in anthesis and fruit, white to pinkish, narrowly oblong, 0.8–1.5 mm in anthesis, margins entire; filaments dimorphic, pubescent basally; anthers deep red; styles and stigmas ca 0.1 mm in anthesis. **Achenes** exserted, reddish brown to yellow-brown, 3-gonous, 2.2–2.8 × 0.5–0.7 mm, shiny, smooth. **2***n* = 22.

Flowering Sep–Dec. Xeric, white sand in rosemary scrub; of conservation concern; 30–60 m; Fla.

Polygonella basiramia is known only from the Lake Wales, Winter Haven, Lake Henry, and Bombing Range ridges of central peninsular Florida (S. P. Christman and W. S. Judd 1990). Habitat loss is a serious threat to this species, which disperses and colonizes new scrub habitat better than other regional endemics. C. V. Hawkes and E. S. Menges (1995) have shown a significant positive correlation between the amount of open-sand habitat and both plant density and seed production at sites where this species grows.

P. O. Lewis and D. J. Crawford (1995) considered *Polygonella basiramia* to be an annual; C. V. Hawkes and E. S. Menges (1995) treated it as a short-lived perennial. It is closely related to *P. ciliata* and was treated as a variety of that species by J. H. Horton (1963). G. L. Nesom and V. M. Bates (1984) advocated recognition of both taxa at the species level.

Polygonella basiramia is in the Center for Plant Conservation's National Collection of Endangered Plants.

SELECTED REFERENCE Hawkes, C. V. and E. S. Menges. 1995. Density and seed production of a Florida endemic, *Polygonella basiramia*, in relation to time since fire and open sand. Amer. Midl. Naturalist 133: 138–149.

4. Polygonella ciliata Meisner in A. P. de Candolle and A. L. P. P. de Candolle, Prodr. 14: 81. 1856 • Fringed jointweed E

Herbs, annual, gynodioecious, 2.5–11 dm. **Stems** erect, usually simple, sometimes branched distally, glabrous or minutely pubescent. **Leaves** fugaceous; ocrea margins ciliate; blade linear, larger ones falcate, (3–)11–33 (–51) × 0.3–1.3(–3) mm, base barely tapered, margins not hyaline, apex acuminate, glabrous. **Inflorescences** (10–)20–35(–45) mm; ocreola encircling rachis, only the base adnate to rachis, apex acuminate. **Pedicels** spreading to slightly reflexed in anthesis, sharply reflexed in fruit, 0.3–1.1 mm, as long or much longer than subtending ocreola. **Flowers** bisexual or pistillate; outer tepals appressed in anthesis and fruit, white, distal portion of midrib often inconspicuously greenish, narrowly oblong, 0.9–1.5 mm in anthesis, margins entire; inner tepals appressed in anthesis and fruit, white, narrowly oblong, 0.8–1.5 mm in anthesis, margins entire; filaments dimorphic; anthers deep red; styles and stigmas ca. 0.1 mm in anthesis. **Achenes** exserted, reddish brown to yellow-brown, 3-gonous, 1.7–3.4 × 0.6–0.9 mm, shiny, smooth. *2n* = 22.

Flowering Sep–Dec. Sandy pinelands and pine barrens; 0–50 m; Fla.

5. Polygonella gracilis Meisner in A. P. de Candolle and A. L. P. P. de Candolle, Prodr. 14: 80. 1856 • Slender wireweed

Polygonum gracile Nuttall, Gen. N. Amer. Pl. 1: 255. 1818, not Salisbury 1796

Herbs, annual, dioecious, 8–16 dm. **Stems** erect, simple or branched proximally and distally, glabrous. **Leaves** mostly fugaceous; ocrea margins not ciliate; blade filiform to broadly spatulate or, rarely, obovate, (9–)19–39(–65) × 0.8–5(–8) mm, base attenuate, margins not hyaline, apex obtuse, glabrous. **Inflorescences** (13–)20–40(–70) mm; ocreola encircling rachis, only the base adnate to rachis, apex acute. **Pedicels** spreading in anthesis, sharply reflexed in fruit, 0.1–0.3 mm, as long as subtending ocreola. **Flowers** functionally unisexual; filaments dimorphic; anthers deep red; styles and stigmas less than 0.1 mm in anthesis. **Staminate flowers:** outer tepals loosely appressed in anthesis and fruit, white, elliptic, 1.1–2 mm in anthesis, margins entire; inner tepals loosely appressed in anthesis and fruit, white, elliptic, 1–2 mm in anthesis, margins entire. **Pistillate flowers:** outer tepals loosely appressed in anthesis, usually spreading in fruit, white to pink, often drying red and streaked or dotted orange or brown, elliptic to oblong, 0.8–1.4 mm in anthesis, margins entire or obscurely erose; inner tepals loosely appressed in anthesis and fruit, white to pink, often drying yellow, broadly elliptic to ovate, 1.8–3.6 mm in flower, margins entire or obscurely erose. **Achenes** included or exserted, reddish brown to yellow-brown, (2–)3-gonous, 2–3.4 × 1–1.4 mm, shiny, smooth. *2n* = 22, 24.

Flowering Aug–Nov. Pinelands, sandhills, sandy roadsides; 0–60 m; Ala., Fla., Ga., Miss., N.C., S.C.

6. Polygonella macrophylla Small, Bull. Torrey Bot. Club 23: 407. 1896 • Large-leaf wireweed E

Subshrubs, perennial, gynodioecious, 8–11 dm. **Stems** erect, simple proximally, sometimes branched distally, glabrous. **Leaves** persistent; ocrea margins not ciliate; blade oblanceolate to obovate or broadly spatulate, (9–)26–56(–68) × (3–)9–30 mm, base attenuate to obtuse, margins hyaline, apex obtuse, glabrous. **Inflorescences** (14–)30–50(–70) mm; ocreola encircling rachis, only the base adnate to rachis, apex acute. **Pedicels** spreading in anthesis, reflexed in fruit, 1.2–2.1 mm, much longer than subtending ocreola. **Flowers** bisexual or pistillate; outer tepals loosely appressed in anthesis, reflexed in fruit, white, pink, or red, obovate, 1.6–2.4 mm in anthesis, margins entire; inner tepals loosely appressed in anthesis, reflexed in fruit, white, red, or pink, obovate, 1.7–2.8 mm in anthesis, margins entire; filaments dimorphic; anthers white or yellow; styles and stigmas 0.3–0.6 mm in anthesis. **Achenes** included, yellow-brown, 3-gonous, 3.1–4.1 × 1.5–1.9 mm, shiny, smooth. $2n = 28$.

Flowering Oct. Sand pine-oak scrub ridges; 0–70 m; Ala., Fla.

Polygonella macrophylla is known only from the Gulf Coast from Franklin County, Florida, westward to Baldwin County, Alabama. Two flower-color morphs exist. Two populations in the vicinity of Carrabelle, Florida, are red-flowered. All other populations produce white or pink flowers. P. O. Lewis (1991b) showed that expected levels of genetic diversity are much lower in white-flowered populations than in red-flowered ones, possibly due to lack of gene flow among populations and high levels of inbreeding. Morphological data suggest that *P. macrophylla* is most closely related to the more widespread *P. polygama* (Lewis and D. J. Crawford 1995).

Polygonella macrophylla is in the Center for Plant Conservation's National Collection of Endangered Plants.

SELECTED REFERENCE Lewis, P. O. 1991b. Allozyme variation in the rare Gulf Coast endemic *Polygonella macrophylla* Small (Polygonaceae). Pl. Spec. Biol. 6: 1–10.

7. Polygonella polygama (Ventenat) Engelmann & A. Gray, Boston J. Nat. Hist. 5: 231. 1845 · October-flower [E]

Polygonum polygamum Ventenat, Descr. Pl. Nouv., 65. 1800

Subshrubs, perennial, dioecious, 1.5–7 dm. **Stems** erect or decumbent, usually branched proximally and distally, glabrous or minutely pubescent. **Leaves** persistent; ocrea margins not ciliate; blade linear to narrowly clavate or broadly spatulate, (3–)4–16(–36) × (0.3–)0.6–3.6(–5) mm, base barely tapered to attenuate, margins hyaline at least along distal ¹/₂, apex obtuse, glabrous. **Inflorescences** (2–)4–20(–33) mm; ocreola encircling rachis, only the base adnate to rachis, apex acute to acuminate. **Pedicels** spreading in anthesis, spreading to reflexed in fruit, 0.2–0.9 mm, as long or much longer than subtending ocreola. **Flowers** functionally unisexual; filaments dimorphic; anthers pink, orange, or yellow; styles and stigma ca. 0.1 mm in anthesis. **Staminate flowers:** outer tepals loosely appressed in anthesis, reflexed in fruit, white, broadly elliptic, 0.9–1.8 mm in anthesis, margins entire; inner tepals appressed in

anthesis and fruit, white, elliptic, 0.9–1.7 mm in anthesis, margins entire. **Pistillate flowers:** outer tepals loosely appressed in anthesis, usually reflexed in fruit, white to pink, often drying yellow, broadly elliptic to ovate, 0.5–1.3 mm in anthesis, margins entire; inner tepals appressed in anthesis and fruit, white to pink, often drying yellow or red, broadly elliptic to ovate, 0.6–1.5 mm in anthesis, margins entire. **Achenes** mostly included, brown to yellow-brown, 3-gonous, 1.3–2.1 × 0.7–1.2 mm, shiny, smooth. $2n = 28$.

Varieties 3 (3 in the flora): se United States.

J. H. Horton (1963) found considerable intergradation among characters used by J. K. Small (1933) to separate *Polyonella polygama*, *P. brachystachya*, and *P. croomii*. He included the latter two taxa in *P. polygama*. R. P. Wunderlin (1981) as well as G. L. Nesom and V. M. Bates (1984) discussed morphological variation among these geographically distinct entities and recognized three varieties.

1. Ocreae with apex acuminate, tip 1–1.5 mm; leaf blades usually 0.5–1 mm wide . . . .
. . . . 7b. *Polygonella polygama* var. *croomii*
1. Ocreae with apex obtuse to acute, tip 0.1–0.5 mm; leaf blades 0.5–6 mm wide.
    2. Leaf blades 0.5–1 mm wide . . . .
    . . . . . 7a. *Polygonella polygama* var. *brachystachya*
    2. Leaf blades (2–)3–6 mm wide . . . .
    . . . . 7c. *Polygonella polygama* var. *polygama*

7a. Polygonella polygama (Ventenat) Engelmann & A. Gray var. brachystachya (Meisner) Wunderlin, Florida Sci. 44: 79. 1981 [E]

Polygonella brachystachya Meisner in A. P. de Candolle and A. L. P. P. de Candolle, Prodr. 14: 80. 1856

Leaves: ocrea with apex obtuse to acute, tip 0.1–0.5 mm; blade 0.5–1 mm wide.

Flowering Jul–Oct. Pinelands; 0–50 m; Fla.

7b. Polygonella polygama (Ventenat) Engelmann & A. Gray var. croomii (Chapman) Fernald, Rhodora 39: 406. 1937 [E]

Polygonella croomii Chapman, Fl. South. U.S., 387. 1860

Leaves: ocrea with apex acuminate, tip 1–1.5 mm; blade 0.5–1 mm wide. $2n = 28$.

Flowering Jul–Oct. Pinelands; 0–100 m; N.C., S.C.

7c. Polygonella polygama (Ventenat) Engelmann & A. Gray var. **polygama** [E]

Leaves: ocrea with apex obtuse to acute, tip 0.1–0.5 mm; blade (2–)3–6 mm wide. **2*n*** = 28.

Flowering Jul–Oct. Pinelands; 0–100 m; Ala., Fla., Ga., Miss., N.C., S.C., Tex., Va.

8. Polygonella parksii Cory, Rhodora 39: 417. 1937

• Parks's jointweed   [C] [E] [F]

Herbs, annual, gynodioecious, 5.5–15 dm. **Stems** erect, usually branched proximally and distally, sometimes simple, glabrous. **Leaves** fugaceous; ocrea margins ciliate; blade linear, (0.9–)4–10.5(–15) × 0.3–0.8 mm, base barely tapered, margins hyaline, apex obtuse, glabrous. **Inflorescences** 2–6(–8) mm; ocreola not encircling rachis, sides and bases adnate to rachis, apex acuminate. **Pedicels** spreading in anthesis, spreading to reflexed in fruit, 0.4–1.3 mm, much longer than subtending ocreola. **Flowers** bisexual or pistillate; outer tepals loosely appressed to spreading in fruit, white, distal portion of midrib often greenish, obovate to elliptic, 0.5–0.9 mm in anthesis, margins obscurely erose; inner tepals loosely appressed in anthesis and fruit, white, obovate to elliptic, 0.6–1.4 mm in anthesis, margins entire to erose; filaments dimorphic; anthers yellow or pink; styles and stigmas 0.1–0.3 mm in anthesis. **Achenes** exserted, yellow-brown, 3-gonous, 1.7–2.1 × 1.3–1.6 mm, shiny, smooth. **2*n*** = 36.

Flowering Jul–Oct. Deep, loose sand in oak woodlands, sandy rangeland, disturbed sites; of conservation concern; 100–200 m; Tex.

Polygonella parksii is known from eight counties in south and east-central Texas. It appears to belong to a clade including *P. articulata*, *P. americana*, and *P. myriophylla* (P. O. Lewis and D. J. Crawford 1995).

9. Polygonella articulata (Linnaeus) Meisner, Pl. Vasc. Gen. 2: 228. 1841 • Northern jointweed   [E]

Polygonum articulatum Linnaeus, Sp. Pl. 1: 363. 1753

Herbs, annual, synoecious, 1–5 dm. **Stems** erect, branched proximally and distally, glabrous. **Leaves** persistent or fugaceous; ocrea margins not ciliate; blade linear to narrowly clavate, (1.5–)5–20(–35) × 0.4–1.2 mm, base barely tapered, margins not hyaline, apex obtuse, glabrous. **Inflorescences** (13–)20–33(–40) mm; ocreola encircling rachis, only the base adnate to rachis, apex acute. **Pedicels** spreading in anthesis, reflexed in fruit, 0.9–2.1 mm, longer than subtending ocreola. **Flowers** bisexual; outer tepals loosely appressed in anthesis and fruit, pink to white, occasionally red, distal portion of midrib often inconspicuously greenish, elliptic to obovate, 1.3–2.3 mm in flower, margins entire to erose; inner tepals loosely appressed in anthesis and fruit, white to pink, occasionally red, elliptic, 1.3–2.3 mm in anthesis, margins mostly entire; filaments dimorphic; anthers yellow or pink; styles and stigmas ca. 0.1 mm in anthesis. **Achenes** included or exserted, reddish brown, 3(–4)-gonous, 2–2.8 × 1–1.3 mm, shiny, smooth. **2*n*** = 32.

Flowering Sep–Nov. Pine barrens, sandhills, and sandy sites along lake shores, beaches, river banks, roadsides, and railroads; 0–500 m; Ont., P.E.I., Que.; Conn., Del., Ga., Ill., Ind., Iowa, Maine, Md., Mass., Mich., Minn., N.H., N.J., N.Y., N.C., Pa., R.I., Wis., Vt., Va.

10. Polygonella americana (Fischer & C. A. Meyer) Small, Mem. Torrey Bot. Club 5: 141. 1894

• American jointweed, southern jointweed   [E] [F]

Gonopyrum americanum Fischer & C. A. Meyer, Mém. Acad. Imp. Sci. Saint-Pétersbourg, Sér. 6, Sci. Math., Seconde Pt. Sci. Nat. 4: 144. 1840

Subshrubs, perennial, synoecious, 5.5–9 dm. **Stems** erect, branched proximally and basally, glabrous. **Leaves** persistent; ocrea margins not ciliate; blade linear to spatulate, (4–)5–12(–19) × 0.5–0.9(–1.2) mm, base barely tapered, margins not hyaline, apex obtuse to acute, often erose, glabrous. **Inflorescences** (7–)18–30(–60) mm; ocreola encircling rachis, only the base adnate to rachis, apex acuminate. **Pedicels** spreading in anthesis and fruit, 0.4–2.3 mm, usually much longer than subtending ocreola. **Flowers** bisexual; outer tepals sharply reflexed early in anthesis and in fruit, white to pink, elliptic to ovate, 1.2–2.2 mm

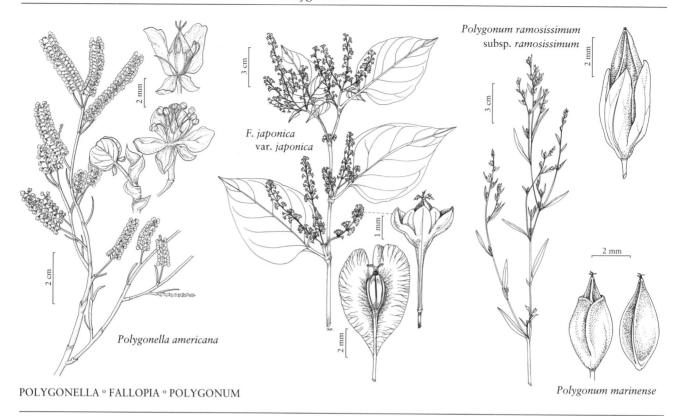

Polygonum ramosissimum subsp. *ramosissimum*

F. japonica var. *japonica*

Polygonella americana

POLYGONELLA ∘ FALLOPIA ∘ POLYGONUM

Polygonum marinense

in anthesis, margins erose; inner tepals loosely appressed in anthesis and fruit, white to pink, ovate to suborbiculate, 1.7–2.9 mm in anthesis, margins erose; filaments dimorphic; anthers yellow or pink; styles and stigmas 0.5–1 mm in anthesis. **Achenes** included, reddish brown, 3(–4)-gonous, 2.5–4 × 1.3–2.3 mm, shiny, smooth. $2n = 36$.

Flowering Jun–Oct. Sandy roadsides, fields, riverbanks, scrub forests, waste places; 10–200 m; Ala., Ark., Ga., La., Mo., N.Mex., Okla., S.C., Tex.

11. Polygonella myriophylla (Small) Horton, Brittonia 15: 196. 1963 · Woody jointweed, sandlace C E

Dentoceras myriophylla Small, Bull. Torrey Bot. Club 51: 389. 1924

Subshrubs, perennial, synoecious, 2–20 dm. **Stems** prostrate, creeping, usually with branches that form dense mats, glabrous. **Leaves** persistent; ocrea margins not ciliate; blade clavate to spatulate, (1.5–)2–8(–8.5) × 0.4–0.9 mm, base attenuate, margins not hyaline, apex obtuse, glabrous. **Inflorescences** 5–9(–13) mm; ocreola encircling rachis, only the base adnate to rachis, apex acute. **Pedicels** spreading in anthesis, spreading the reflexed in fruit, 0.7–1.9 mm, much longer than subtending ocreola. **Flowers** bisexual; outer tepals loosely appressed in anthesis, sometimes spreading in fruit, white, pink or yellow, often drying yellow, elliptic to oblong, 1.4–2.2 mm in anthesis, margins entire; inner tepals loosely appressed in anthesis and fruit, white, pink, or yellow, elliptic to suborbiculate, 1.7–2.5 mm in anthesis, margins entire; filaments dimorphic; anthers pink or yellow; styles and stigmas 0.4–0.9 mm in anthesis. **Achenes** mostly included, reddish brown, 3-gonous, 2.2–3.4 × 1.6–2.4 mm, dull to shiny, minutely roughened. $2n = 36$.

Flowering Apr–Nov. Xeric, white sand in sand pine scrub; of conservation concern; 30–60 m; Fla.

Polygonella myriophylla is restricted to 119 sand pine scrub sites in the Lake Wales, Lake Henry, and Winter Haven ridges of central peninsular Florida (S. P. Christman and W. S. Judd 1990). Loss of habitat to residential, commercial, and agricultural development is the most serious threat to this regional endemic.

29. FALLOPIA Adanson, Fam. Pl. 2: 277, 557. 1763, name conserved • False-buckwheat [for Gabriello Fallopio, 1532–1562, Italian anatomist]

Craig C. Freeman

Harold R. Hinds†

Bilderdykia Dumortier

Vines or herbs, annual or perennial; roots fibrous or woody; sometimes rhizomatous. **Stems** erect to scandent, rarely procumbent, glabrous or pubescent. **Leaves** deciduous, cauline, alternate, petiolate; ocrea persistent or deciduous, chartaceous; petiole base articulated, extrafloral nectaries sometimes present; blade broadly ovate to triangular, margins entire or wavy. **Inflorescences** terminal and spikelike, or terminal and axillary and paniclelike or racemelike, pedunculate or not. **Pedicels** present. **Flowers** bisexual, or bisexual and unisexual, some plants with bisexual flowers, other plants with only pistallate flowers 1–5 per ocreate fascicle, base stipelike; perianth usually accrescent in fruit, pale green or white to pink, campanulate, glabrous or, rarely, with blunt, hyaline hairs; tepals 5, connate nearly completely or only basally, petaloid, dimorphic, outer 3 winged or keeled, larger than inner 2; stamens 6–8; filaments distinct, free, glabrous or pubescent proximally; anthers yellow to pink or red, ovate to elliptic; styles 3, spreading, connate basally or nearly completely; stigmas capitate, fimbriate, or peltate. **Achenes** included or exserted, brown to dark brown or black, not winged, 3-gonous, glabrous. **Seeds:** embryo straight. $x = 10, 11$.

Species ca. 12 (8 in the flora): North America (including Mexico), South America, Europe, Asia, Africa.

Chromosome number and habit traditionally have been used to separate *Fallopia* ($x = 10$; climbing or sprawling, fibrous-rooted annuals and perennials) from *Reynoutria* ($x = 11$; erect, rhizomatous perennials). J. P. Bailey and C. A. Stace (1992) presented evidence to the contrary. *Fallopia* often is included in a broader concept of *Polygonum* but is distinguished by a syndrome of anatomical and morphological characters (K. Haraldson 1978; Hong S. P. et al. 1998; L.-P. Ronse Decraene and J. R. Akeroyd 1988; Ronse Decraene et al. 2000). Molecular data confirm its close relationship to *Polygonum* in the narrow sense (A. S. Lamb Frye and K. A. Kron 2003).

SELECTED REFERENCES Bailey, J. P. and C. A. Stace. 1992. Chromosome number, morphology, pairing, and DNA values of species and hybrids in the genus *Fallopia* (Polygonaceae). Pl. Syst. Evol. 180: 29–52. Kim, J. Y. and C. W. Park. 2000. Morphological and chromosomal variation in *Fallopia* section *Reynoutria* (Polygonaceae) in Korea. Brittonia 52: 34–48. Kim, M. H., J. H. Park, and C. W. Park. 2000. Flavonoid chemistry of *Fallopia* section *Fallopia* (Polygonaceae). Biochem. Syst. & Ecol. 28: 433–441. Kim, S. T., M. H. Kim, and C. W. Park. 2000. A systematic study on *Fallopia* section *Fallopia* (Polygonaceae). Korean J. Pl. Taxon. 30: 35–54.

1. Stigmas fimbriate; stems erect.
    2. Leaf blades with hairs along veins on abaxial face distinctly multicellular, 0.2–0.6 mm, bases of blades cordate . . . . 1. *Fallopia sachalinensis*
    2. Leaf blades with hairs along veins on abaxial face unicellular or appearing so, shorter than 0.1 mm, or veins scabrous, bases of blades truncate to attenuate or cordate.
        3. Leaf blades obscurely puberulent along some veins abaxially, tips of hairs acute, bases of blades truncate to cordate . . . . 2. *Fallopia* ×*bohemica*
        3. Leaf blades obscurely scabrous along some veins abaxially, tips of hairs blunt, bases of blades truncate to attenuate . . . . 3. *Fallopia japonica*
1. Stigmas capitate or peltate; stems usually trailing, twining, scandent, sprawling, or climbing, rarely erect.
    4. Plants perennial; stems woody, climbing . . . . 4. *Fallopia baldschuanica*
    4. Plants annual or perennial; stems herbaceous, scandent or sprawling.

[5. Shifted to left margin.—Ed.]

5. Ocreae bases fringed with reflexed hairs and slender bristles . . . . 5. *Fallopia cilinodis*
5. Ocreae bases glabrous or scabrid.
   6. Achenes minutely granular-tuberculate, dull; fruiting perianths glabrous or with blunt, hyaline hairs, wings absent or, rarely, 0.4–0.9 mm wide; plants annual . . . . 6. *Fallopia convolvulus*
   6. Achenes smooth, shiny; fruiting perianths glabrous, wings (0.7–)1.5–2.1 mm wide; plants perennial or annual.
      7. Plants perennial or annual; fruiting perianth wings decurrent on stipelike base, undulate or crinkled, rarely flat, margins wavy-crenulate to incised or lacerate, rarely entire . . . . 7. *Fallopia scandens*
      7. Plants annual; fruiting perianth wings usually truncate to attenuate-decurrent on stipelike base, flat or, less often, undulate or crinkled, margins entire or rarely undulate-crenate . . . . 8. *Fallopia dumetorum*

1. **Fallopia sachalinensis** (F. Schmidt) Ronse Decraene, Bot. J. Linn. Soc. 98: 369. 1988 · Giant knotweed

I W

Polygonum sachalinense F. Schmidt, Mém. Acad. Imp. St.-Pétersbourg Divers Savans 9: 233. 1859; *Reynoutria sachalinensis* (F. Schmidt) Nakai; *Tiniaria sachalinensis* (F. Schmidt) Janchen

Herbs, perennial, rhizomatous, 2–4(–5) m. **Stems** usually clustered, erect, sparingly branched, herbaceous, stiff, glabrous, glaucous. **Leaves:** ocrea persistent or deciduous, brownish, cylindric, 6–12 mm, margins oblique, face without reflexed and slender bristles at base, otherwise glabrous or puberulent; petiole 1–4 cm, glabrous; blade ovate-oblong, 15–30(–40) × 7–25 cm, base cordate, margins entire, glabrous or scabrous to ciliate, apex obtuse to acute, abaxial face minutely dotted, glaucous, with hairs along veins distinctly multicellular, 0.2–0.6 mm, tips acute to acuminate, adaxial face glabrous. **Inflorescences** axillary, mostly distal, erect or spreading, paniclelike, 3–8 cm, axes puberulent to pubescent; peduncle 0.1–4 cm or absent, puberulent to reddish-pubescent. **Pedicels** ascending or spreading, articulated proximal to middle, 2–4 mm, glabrous. **Flowers** bisexual or pistillate, 4–7 per ocreate fascicle; perianth accrescent in fruit, greenish, 4.5–6.5 mm including stipelike base, glabrous; tepals obovate to elliptic, apex obtuse to acute, outer 3 winged; stamens 6–8; filaments flattened proximally, glabrous; styles connate basally; stigmas fimbriate. **Achenes** included, brown, 2.8–4.5 × 1.1–1.8 mm, shiny, smooth; fruiting perianth glabrous, wings flat to undulate, 1.8–2.2 mm wide at maturity, decurrent on stipelike base to articulation, margins entire. $2n$ = 44, 66, 102, 132 (Japan, Korea).

Flowering Jul–Oct. Disturbed places; 0–500 m; introduced; B.C., N.B., Nfld. and Labr. (Nfld.), N.S., Ont., P.E.I., Que.; Calif., Conn., Del., Idaho, Ill., Ky., La., Maine, Md., Mass., Mich., Mont., N.J., N.Y., N.C., Ohio, Oreg., Pa., R.I., Tenn., Vt., Va., Wash., W.Va., Wis.; Asia (Japan); introduced in Europe.

Fallopia sachalinensis was introduced as a soil binder and garden ornamental. Like *F. japonica*, it spreads aggressively and has been declared noxious in California, Oregon, and Washington. It hybridizes with *F. japonica*, yielding *F.* ×*bohemica*. The mid-stem inflorescences of *F. sachalinensis* usually are shorter than the subtending leaves.

2. **Fallopia ×bohemica** (Chrtek & Chrtková) J. P. Bailey, Watsonia 17: 443. 1989 · Bohemian knotweed I

Reynoutria ×bohemica Chrtek & Chrtková, Čas. Nár. Muz. Praze Rada Přír. 152: 120. 1983; *Polygonum ×bohemicum* (Chrtek & Chrtková) Zika & Jacobson

Herbs, perennial, rhizomatous, 1.5–2.5 m. **Stems** usually clustered, erect, profusely branched, herbaceous, stiff, glabrous, glaucous. **Leaves:** ocrea usually deciduous, brownish, cylindric, 4–6(–10) mm, margins oblique, face without reflexed hairs and slender bristles at base, otherwise glabrous or minutely puberulent; petiole 1–3 cm, glabrous; blade ovate, 5–25(–30) × 2–10 cm, base truncate to cordate, margins entire, glabrous or scabrous to ciliate, apex cuspidate, abaxial face minutely dotted, glaucous, with hairs along veins unicellular or appearing so, obscure, shorter than 0.1 mm, tips acute, adaxial face glabrous. **Inflorescences** terminal and axillary, erect or spreading, paniclelike or sometimes racemelike, 4–12 cm, axes puberulent; peduncle 0.1–3.5 cm or absent, puberulent. **Pedicels** ascending or spreading, articulated proximal to middle, 3–5 mm, glabrous. **Flowers** bisexual or pistillate, 3–8(–15) per ocreate fascicle; perianth accrescent in fruit, white or greenish white to pink, 4–6 mm including stipelike base, glabrous; tepals obovate to elliptic, apex obtuse to acute, outer 3 winged; stamens 8; filaments flattened proximally, glabrous; styles connate basally; stigmas fimbriate. **Achenes** included, dark brown, 2.6–3.2 × 1.4–1.8 mm, shiny, smooth; fruiting perianth glabrous, wings flat to undulate, 1.5–2.1 mm wide at maturity, decurrent on stipelike base nearly to articulation, margins entire. **2n** = 44, 66, 88 (Korea).

Flowering Jul–Oct. Disturbed places; 0–800 m; introduced; B.C., N.S., Que.; Conn., Idaho, Ill., Iowa, Kans., Ky., La., Maine, Md., Mass., Minn., Nebr., N.Y., N.C., Oreg., Pa., Tenn., Vt., Va., Wash., W.Va., Wis.; Europe.

Fallopia ×bohemica is a widespread hybrid between *F. japonica* and *F. sachalinensis*. It went unnoticed in the North American flora until recently (P. F. Zika and A. L. Jacobson 2003). It exhibits morphological intermediacy with its parents and is distinguished most reliably by the pubescence along the veins on the abaxial surface of leaves produced early in the growing season. Hairs of *F. sachalinensis* are multicellular, usually twisted, acute to acuminate at the tip, and 0.2–0.6 mm; those of *F. japonica* are unicellular, blunt, and barely raised, making the veins appear scabrous. *Fallopia ×bohemica* has hairs that are unicellular or appearing so (actually often obscurely multicelluar), acute at the tip, and shorter than 0.1 mm. The hairs are easiest to find on fresh leaves; older leaves often are glabrescent and diagnostic hairs are hard to find. Hybrid specimens are most often misidentified as *F. japonica*.

Like its parents, *Fallopia ×bohemica* is gynodioecious. J. P. Bailey et al. (1996) reported it to be partially to fully fertile. In North America, reproduction appears to be largely vegetative by rhizomes.

SELECTED REFERENCE Zika, P. F. and A. L. Jacobson. 2003. An overlooked hybrid Japanese knotweed (*Polygonum cuspidatum × sachalinense*; Polygonaceae) in North America. Rhodora 105: 143–152.

3. Fallopia japonica (Houttuyn) Ronse Decraene, Bot. J. Linn. Soc. 98: 369. 1988 • Japanese knotweed [F] [I] [W]

Reynoutria japonica Houttuyn, Nat. Hist. 2: 640, plate 51, fig. 1. 1777

Varieties 4 (1 in the flora): introduced; Asia, introduced in Europe.

Fallopia japonica is planted widely as a garden ornamental; it has a proclivity to escape and spread aggressively. Once established, plants can be difficult to eradicate because of their extensive, woody rhizomes. The species has been declared noxious in Alabama, California, Oregon, Vermont, and Washington.

SELECTED REFERENCE Beerling, D. J., J. P. Bailey, and A. P. Conolly. 1994. *Fallopia japonica* (Houtt.) Ronse Decraene. J. Ecol. 82: 959–979.

3a. Fallopia japonica (Houttuyn) Ronse Decraene var. **japonica** [F] [I] [W]

Polygonum cuspidatum Siebold & Zuccarini

Herbs, perennial, rhizomatous, 1.5–2(–3) m. **Stems** usually clustered, erect, profusely branched, herbaceous, stiff, glabrous, glaucous. **Leaves:** ocrea usually deciduous, brownish, cylindric, 4–6(–10) mm, margins oblique, face not fringed with reflexed hairs and slender bristles at base, otherwise glabrous or puberulent; petiole 1–3 cm, glabrous; blade ovate, 5–15 × 2–10 cm, base truncate to attenuate, margins entire, glabrous or scabrous to ciliate, apex abruptly cuspidate, abaxial face minutely dotted, glaucous, with hairs along veins unicellular, shorter than 0.1 mm, tips blunt, veins obscurely scabrous, adaxial face glabrous. **Inflorescences** terminal and axillary, erect or spreading, paniclelike or, sometimes, racemelike, 4–12 cm, axes puberulent; peduncle 0.1–2.5 cm or absent, puberulent. **Pedicels** ascending or spreading, articulated proximal to middle, 3–5 mm, glabrous. **Flowers** bisexual or pistillate, 3–8 (–15) per ocreate fascicle; perianth accrescent in fruit, white or greenish white to pink, 4–6 mm including stipelike base, glabrous; tepals obovate to elliptic, apex obtuse to acute, outer 3 winged; stamens 8; filaments flattened proximally, glabrous; styles connate basally; stigmas fimbriate. **Achenes** included, dark brown, 2.3–3.6 × 1.4–1.9 mm, shiny, smooth; fruiting perianth glabrous, wings flat to undulate, 1.4–2 mm wide at maturity, decurrent on stipelike base nearly to articulation, margins entire. **2n** = 44, 66, 88 (Korea).

Flowering Aug–Sep. Disturbed sites, waste places; 0–1800 m; introduced; St. Pierre and Miquelon; B.C., Man., N.B., Nfld. and Labr. (Nfld.), N.S., Ont., P.E.I., Que.; Alaska, Ark., Calif., Colo., Conn., Del., D.C., Ga., Idaho, Ill., Ind., Iowa, Kans., Ky., La., Maine, Md., Mass., Mich., Minn., Miss., Mo., Mont., Nebr., N.H., N.J., N.Y., N.C., Ohio, Okla., Oreg., Pa., R.I., S.C., S.Dak., Tenn., Utah, Vt., Va., Wash., W.Va., Wis.; e Asia (China, Japan, Korea, Taiwan); introduced in Europe.

Variety *hachidyoensis* (Makino) Yonekura & H. Ohashi and var. *uzenensis* (Honda) Yonekura & H. Ohashi are endemic to Japan and are distinguished by differences in leaf size and pubescence. A dwarf alpine form of *Fallopia japonica* found in Korea also is sometimes recognized as var. *compacta* (Houttuyn) J. P. Bailey. It is shorter (to 8 dm), has reddish perianths, and often is planted as a ground cover.

Fallopia ×bohemica is a naturally occurring hybrid between var. *japonica* and *F. sachalinensis*. *Fallopia japonica* also hybridizes with *F. baldschuanica*.

4. Fallopia baldschuanica (Regel) Holub, Folia Geobot. Phytotax. 6: 176. 1971 • Bukhara fleeceflower, mile-a-minute vine I

Polygonum baldschuanicum Regel, Trudy Imp. S.-Petersburgsk Bot. Sada 8: 684. 1884; *Bilderdykia aubertii* (L. Henry) Moldenke; *B. baldschuanica* (Regel) D. A. Webb; *Fallopia aubertii* (L. Henry) Holub; *Polygonum aubertii* L. Henry; *Reynoutria baldschuanica* (Regel) Moldenke

Vines, perennial, not rhizomatous, 3–10 m. **Stems** climbing, branched from near base, woody, glabrous, not glaucous. **Leaves:** ocrea usually deciduous, hyaline or brownish, cylindric, 3–8 mm, margins truncate to oblique, face glabrous throughout; petiole 1–4 cm, glabrous or scabrid; blade narrowly ovate to ovate-oblong, 3–10 × 1–5 cm, base subcordate or cordate to sagittate, margins entire or wavy, glabrous or scabrid, apex obtuse to acuminate, abaxial face glabrous or scabrid along midvein, rarely minutely dotted, not glaucous, adaxial face glabrous. **Inflorescences** axillary and terminal, spreading or drooping, paniclelike, 3–15 cm, axes glabrous or papillate to scabrid in lines; peduncle 1–3 cm, glabrous or scabrid. **Pedicels** ascending or spreading, articulated proximal to middle, 1.5–4 mm, glabrous or scabrid. **Flowers** bisexual, 3–6 per ocreate fascicle; perianth accrescent in fruit, greenish white with white wings or mostly pink, sometimes bright pink in fruit, 5–8 mm including stipelike base, glabrous; tepals elliptic, apex obtuse to rounded, outer 3 winged; stamens 6–8; filaments flattened proximally, pubescent proximally; styles connate basally; stigmas peltate. **Achenes** included, dark brown to black, 2–4 × 1.8–2.2 mm, shiny, smooth; fruiting perianth glabrous, wings flat to undulate, 2–4 mm wide at maturity, decurrent on stipelike base nearly to articulation, margins entire. *2n* = 20 (Korea).

Flowering Aug–Sep. Disturbed sites; 0–1600 m; introduced; Calif., Colo., Md., Mass., Mich., N.J., N.Mex., N.Y., Pa., Utah, Va., Wash.; c Asia; introduced in Central America (Costa Rica), Europe.

Fallopia baldschuanica is cultivated as a trellis and garden plant; it escapes infrequently in the flora area. Plants with white or greenish white flowers and papillate or scabrid inflorescence axes have been recognized as *F. aubertii.*

5. Fallopia cilinodis (Michaux) Holub, Folia Geobot. Phytotax. 6: 176. 1971 (as cilinode) • Fringed black bindweed E W

Polygonum cilinode Michaux, Fl. Bor.-Amer. 1: 241. 1803; *Bilderdykia cilinodis* (Michaux) Greene; *B. cilinodis* var. *laevigata* (Fernald) C. F. Reed; *Polygonum cilinode* var. *laevigatum* Fernald; *Reynoutria cilinodis* (Michaux) Shinners; *Tiniaria cilinodis* (Michaux) Small

Herbs, perennial, not rhizomatous, 1–5 m. **Stems** usually scandent or sprawling, rarely erect, freely branched, herbaceous, pilose-hispid or, rarely, subglabrous, not glaucous. **Leaves:** ocrea usually deciduous, light brown, cylindric, 3–4 mm, margins oblique, base fringed with reflexed hairs and slender bristles, face glabrous or puberulent; petiole 1–6 cm, retrorsely pubescent; blade cordate-ovate, cordate-hastate, or cordate-sagittate, 2–6(–12) × 2–5(–10) cm, base cordate, margins wavy, often reddish-ciliate, apex acute to acuminate, abaxial face pilose-hispid, not minutely dotted, not glaucous, adaxial face glabrous. **Inflorescences** terminal and axillary, erect or spreading, paniclelike, 4–10(–15) cm, axes reddish-pilose; peduncle 1–12 cm, retrorsely pubescent. **Pedicels** ascending or spreading, articulated near middle or distally, 3–4 mm, glabrous or puberulent. **Flowers** bisexual, 4–7 per ocreate fascicle; perianth nonaccrescent, greenish white to white, 1.5–2 mm including stipelike base, glabrous; tepals elliptic, apex obtuse to acute, outer 3 obscurely keeled; stamens 6–8; filaments flattened proximally, pubescent proximally; styles connate basally; stigmas capitate. **Achenes** included or exserted, brownish black to black, 3–4 × 1.8–2.4 mm, shiny, smooth; fruiting perianth glabrous, wings absent. *2n* = 22.

Flowering Jun–Oct. Dry woods, thickets, clearings; 0–900 m; Man., N.B., Nfld. and Labr. (Nfld.), N.S., Ont., P.E.I., Que., Sask.; Conn., Ga., Ill., Ind., Ky., Maine, Md., Mass., Mich., Minn., N.H., N.J., N.Y., N.C., Ohio, Pa., R.I., Tenn., Vt., Va., W.Va., Wis.

Á. Löve and D. Löve (1982) reported a chromosome count of *2n* = 20 for *Fallopia cilinodis.* All other counts summarized by J. P. Bailey and C. A. Stace (1992) and counts by M. H. Kim et al. (2000) are *2n* = 22. It is not known if the *2n* = 20 count is an error.

6. Fallopia convolvulus (Linnaeus) Á. Löve, Taxon 29: 300. 1970 • Black bindweed [I] [W]

Polygonum convolvulus Linnaeus, Sp. Pl. 1: 364. 1753; *Bilderdykia convolvulus* (Linnaeus) Dumortier; *Fallopia convolvulus* var. *subalata* (Lejeune & Courtois) D. H. Kent; *Reynoutria convolvulus* (Linnaeus) Shinners; *Tiniaria convolvulus* (Linnaeus) Webb & Moquin-Tandon ex Webb & Berthelot

Herbs, annual, not rhizomatous, 0.5–1 m. **Stems** scandent or sprawling, branched proximally, herbaceous, puberulent, sometimes mealy, not glaucous. **Leaves:** ocrea persistent or deciduous, tan or greenish brown, cylindric, 2–4 mm, margins oblique, face not fringed with reflexed hairs and slender bristles at base, otherwise glabrous or scabrid; petiole 0.5–5 cm, puberulent in lines; blade cordate-ovate, cordate-hastate, or sagittate, 2–6 (–15) × 2–5(–10) cm, base cordate, margins wavy, scabrid, apex acuminate, abaxial face usually mealy and, rarely, minutely dotted, not glaucous, adaxial face glabrous. **Inflorescences** axillary, erect or spreading, spikelike, 2–10(–15) cm, axes puberulent; peduncle 0.1–10 cm or absent, glabrous or scabrid distally in lines. **Pedicels** ascending or spreading, articulated distally, 1–3 mm, glabrous or, rarely, scabrid. **Flowers** bisexual, 3–6 per ocreate fascicle; perianth nonaccrescent, greenish white, often with pinkish or purplish base, 3–5 mm including stipelike base, glabrous or outer 3 with blunt, hyaline hairs; tepals elliptic to obovate, apex obtuse to acute, outer 3 obscurely keeled; stamens 8; filaments flattened proximally, glabrous; styles connate distally; stigmas capitate. **Achenes** included, black, 4–5(–6) × 1.8–2.3 mm, dull, minutely granular-tuberculate, especially on faces; fruiting perianth glabrous or with blunt, hyaline hairs, wings absent or, rarely, flat to undulate, 0.4–0.9 mm wide at maturity, scarcely decurrent on stipelike base, margins entire. *2n* = 40.

Flowering May–Oct. Cultivated ground, waste places; 0–2700 m; introduced; Greenland; St. Pierre and Miquelon; Alta., B.C., Man., N.B., Nfld. and Labr., N.S., Ont., P.E.I., Que., Sask., Yukon; Ala., Alaska, Ariz., Ark., Calif., Colo., Conn., Del., D.C., Fla., Ga., Idaho, Ill., Ind., Iowa, Kans., Ky., La., Maine, Md., Mass., Mich., Minn., Miss., Mo., Mont., Nebr., Nev., N.H., N.J., N.Mex., N.Y., N.C., N.Dak., Ohio, Okla., Oreg., Pa., R.I., S.C., S.Dak., Tenn., Tex., Utah, Vt., Va., Wash., W.Va., Wis., Wyo.; Eurasia; introduced in South America (Argentina, Chile), Africa (Algeria, Morocco, Republic of South Africa), Pacific Islands (Hawaii, New Zealand), Australia.

Fallopia convolvulus can be an aggressive weed in crop fields. Rare plants with winged fruiting perianths have been named var. *subalata*; that characteristic often varies within populations.

7. Fallopia scandens (Linnaeus) Holub, Folia Geobot. Phytotax. 6: 176. 1971 • Climbing false-buckwheat [E] [W]

Polygonum scandens Linnaeus, Sp. Pl. 1: 364. 1753; *Bilderdykia cristata* (Engelmann & A. Gray) Greene; *B. scandens* (Linnaeus) Greene; *B. scandens* var. *cristata* (Engelmann & A. Gray) C. F. Reed; *Fallopia cristata* (Engelmann & A. Gray) Holub; *Polygonum cristatum* Engelmann & A. Gray; *P. dumetorum* var. *scandens* (Linnaeus) A. Gray; *P. scandens* var. *cristatum* (Engelmann & A. Gray) Gleason; *Reynoutria scandens* (Linnaeus) Shinners; *R. scandens* var. *cristata* (Engelmann & A. Gray) Shinners; *Tiniaria cristata* (Engelmann & A. Gray) Small; *T. scandens* (Linnaeus) Small

Herbs, perennial or annual, not rhizomatous, 1–5 m. **Stems** scandent or sprawling, freely branched, herbaceous, glabrous or papillose to scabrid, not glaucous. **Leaves:** ocrea usually deciduous, tan or brown, cylindric to funnelform, 1–6 mm, margins oblique, face not fringed with reflexed hairs and slender bristles at base, otherwise glabrous or scabrid; petiole 0.5–10 cm, glabrous or scabrid in lines; blade cordate, truncate-deltate, or hastate, 2–14 × 2–7 cm, base cordate, margins wavy, scabrid, apex acuminate, abaxially and adaxially faces glabrous or papillose to scabrid, not glaucous, the abaxial rarely minutely dotted. **Inflorescences** axillary, erect or spreading, racemelike, 1–28 cm, axes scabrid; peduncle 0.1–7 cm or absent, scabrid. **Pedicels** ascending or spreading to reflexed, articulated distally, 4–8 mm, glabrous. **Flowers** bisexual, 3–6 per ocreate fascicle; perianth accrescent in fruit, green to white or pinkish, 3.8–8 mm including stipelike base, glabrous; tepals elliptic to obovate, apex obtuse to acute, outer 3 winged; stamens 8; filaments flattened proximally, pubescent proximally; styles connate; stigmas capitate. **Achenes** included, dark brown to black, 2–6 × 1.4–3.5 mm, shiny, smooth; fruiting perianth glabrous, wings undulate or crinkled, rarely flat, (0.7–)1.5–2.1 mm wide, decurrent on stipelike base nearly to articulation, margins wavy-crenate to incised or lacerate, rarely entire. *2n* = 20.

Flowering Aug–Nov. Low habitats; 0–1800 m; Alta., Man., N.B., N.S., Ont., P.E.I., Que., Sask.; Ala., Ark., Conn., Del., D.C., Fla., Ga., Ill., Ind., Iowa, Kans., Ky., La., Maine, Md., Mass., Mich., Minn., Miss., Mo., Nebr., N.H., N.J., N.Y., N.C., N.Dak., Ohio, Okla., Pa., R.I., S.C., S.Dak., Tenn., Tex., Vt., Va., W.Va., Wis., Wyo.

Fallopia scandens has a complex nomenclatural history, which in North America usually involves three taxonomic elements: *F. scandens* and *F. cristata*, both native in North America, and *F. dumetorum*, which is native in Europe. Achene and perianth characters have been used to distinguish these elements, but variable and intergrading morphologies have caused taxonomists to combine them variously. Morphometric (S. T. Kim et al. 2000) and flavonoid (M. H. Kim et al. 2000) studies suggest that *F. scandens* and *F. dumetorum* are distinct species. Where *F. scandens* is absent, European specimens of *F. dumetorum* are distinctive. This distinction is far less clear in North America, where both species occur. Experience suggests that many North American herbarium specimens attributed to *F. dumetorum* are misidentified.

Fallopia cristata has been distinguished from *F. scandens* and *F. dumetorum* by its smaller fruiting perianths (5–7[–9] mm) bearing narrower (1.2–1.7 mm wide), undulate-crenate or lacerate wings, and smaller achenes (2.1–2.7 mm). Extreme forms are easily identified; some specimens grade gradually into *F. scandens*, making recognition of *F. cristata* of questionable utility. S. T. Kim et al. (2000) used morphometric studies as a basis for recommending that *F. cristata* is best treated as a variety of *F. scandens*.

8. **Fallopia dumetorum** (Linnaeus) Holub, Folia Geobot. Phytotax. 6: 176. 1971 • Corpse-bindweed, renouée des haies [I]

Polygonum dumetorum Linnaeus, Sp. Pl. ed. 2, 1: 522. 1762; *Bilderdykia scandens* (Linnaeus) Greene var. *dumetorum* (Linnaeus) Dumortier; *Polygonum scandens* Linnaeus var. *dumetorum* (Linnaeus) Gleason; *Reynoutria scandens* (Linnaeus) Shinners var. *dumetorum* (Linnaeus) Shinners; *Tiniaria dumetorum* (Linnaeus) Opiz

Herbs, annual, not rhizomatous, to 3 m. **Stems** scandent or climbing, rarely prostrate, freely branched, herbaceous, glabrous to papillose or scabrid, not glaucous. **Leaves:** ocrea usually deciduous, tan or brown, cylindric to funnelform, 1.5–3.5 mm, margins oblique, face not fringed with reflexed hairs and slender bristles at base, otherwise glabrous or scabrid; petiole 0.3–2.5 cm, glabrous or scabrid in lines; blade triangular to hastate, 2–8 × 1–5 cm, base cordate to truncate, margins wavy, scabrid, apex acute to acuminate, abaxially and adaxially faces glabrous or papillose to scabrid, not glaucous, the abaxial rarely minutely dotted. **Inflorescences** axillary, erect or spreading, racemelike, 2–20 cm, axes glabrous or scabrid; peduncle 0.1–6 cm or absent, glabrous or scabrid. **Pedicels** ascending or spreading to deflexed, articulated distally, 4–8 mm, glabrous. **Flowers** bisexual, 2–6 per ocreate fascicle; perianth accrescent in fruit, greenish white or pinkish, 3.5–7 mm including stipelike base, glabrous; tepals elliptic to obovate, apex obtuse to acute, outer 3 winged; stamens 8; filaments flattened proximally, pubescent proximally; styles connate; stigmas capitate. **Achenes** included, black, 2–4 × 1.8–2.4 mm, shiny, smooth; fruiting perianth glabrous, wings flat or, less often, undulate or crinkled, 1.5–2 mm wide, usually truncate or attenuate-decurrent on stipelike base nearly to articulation, margins entire or rarely undulate-crenate. $2n = 20$.

Flowering Jul–Oct. Hedges, wood borders, fields, waste ground; 0–300 m; introduced; Ont., Que.; Ala., Ark., Conn., Del., D.C., Fla., Ga., Ill., Ind., Iowa, Ky., La., Maine, Md., Mass., Mich., Miss., Mo., N.H., N.J., N.Y., N.C., Ohio, Okla., Pa., R.I., S.C., Tenn., Tex., Vt., Va., W.Va., Wis.; Europe; Asia.

Because of similarity to *Fallopia scandens*, specimens of *F. dumetorum* often are misidentified. Consequently, the range of *F. dumetorum* in North America is unclear; it is probably exaggerated in most floras.

Mature fruiting perianths provide the most reliable characters distinguishing *Fallopia dumetorum* from *F. scandens*. European specimens of *F. dumetorum*, have fruiting perianths that are consistently orbiculate in outline, with wings usually flat and abruptly contracted on the stipelike perianth bases. *Fallopia scandens* has fruiting perianths that are obovate in outline, with wings undulate and gradually decurrent on the stipelike perianth bases. Additional characteristics reported to distinguish *F. dumetorum* from *F. scandens* are its annual habit (*F. scandens* also is reported to be annual in some of the literature) and more triangular leaf blades with cordate or sagittate bases. In recognizing *F. dumetorum*, we follow S. T. Kim et al. (2000) and M. H. Kim et al. (2000), who concluded from limited morphological and flavonoid data that the taxa are best treated as distinct species. More detailed studies may prove the two to be conspecific.

30. POLYGONUM Linnaeus, Sp. Pl. 1: 359. 1753; Gen. Pl. ed. 5, 170. 1754

• Knotweed [Greek *poly*, many, and *gony*, knee joint (traditional interpretation), or *gone*, seed (grammatically correct interpretation)]

Mihai Costea

François J. Tardif

Harold R. Hinds†

Herbs, shrubs, or subshrubs, annual (perennial in *P. striatulum*), homophyllous or heterophyllous, sometimes heterocarpic; roots fibrous or woody. **Stems** prostrate to erect, glabrous, smooth or sometimes papillous-scabridulous. **Leaves** cauline, alternate (opposite in *P. humifusum*), petiolate or sessile; ocrea with distal part persistent, often hyaline, white or silvery, 2-lobed, chartaceous, glabrous, disintegrating into fibers, or disintegrating completely; petiole base articulated with ocrea or not; blade linear, lanceolate, elliptic, ovate, or subround, margins entire. **Inflorescences** axillary or axillary and terminal, spikelike, or flowers solitary; peduncle absent. **Pedicels** present or absent. **Flowers** bisexual, 1–7(–10) per ocreate fascicle, base not stipelike; perianth nonaccrescent, white or greenish white to pink, campanulate to urceolate, glabrous; tepals 5, connate 3–70% of their length, petaloid or sepaloid, monomorphic or, rarely, dimorphic, the inner usually flat, the outer flat or sometimes keeled and cucullate distally, sometimes of different length than the inner; stamens 3–8 (some may be reduced to staminodes); filaments distinct, free or adnate to perianth tube, glabrous; anthers whitish yellow, pink to purple or orange-pink, elliptic to oblong; styles (2–)3, mostly spreading, distinct or connate proximally; stigmas 2–3, capitate. **Achenes** included or exserted, yellow-green, brown, or black, unwinged, (2–)3-gonous, glabrous. **Seeds:** embryo curved. $x = 10$.

Species ca. 65 (33 in the flora): nearly worldwide.

Two sections of *Polygonum* are recognized here. Section *Polygonum* is nearly cosmopolitan and best represented in north-temperate regions; sect. *Duravia* comprises species restricted to North America. K. Haraldson (1978) recognized both sections based on differences in stem morphology, petiole structure, and pollen morphology. J. C. Hickman (1984) described sect. *Monticola* and included in it species of sect. *Duravia* occurring mostly in montane habitats, with leaves articulated to the ocreae, one-veined, and not mucronate, proximal leaves lanceolate to round, and styles connate at their bases and neither hardened nor persistent. L.-P. Ronse Decraene and J. R. Akeroyd (1988) and L.-P. Ronse Decraene et al. (2000) included sect. *Duravia* in sect. *Polygonum* based on floral and fruit characters.

Similarities in floral structure, fruit anatomy, and pollen morphology have been noted between *Polygonella* with *Polygonum* (L.-P. Ronse Decraene et al. 2000). Based on evidence from comparative morphological studies, Ronse Decraene et al. (2004) included *Polygonella* in sect. *Duravia* of *Polygonum*.

Four introduced taxa of sect. *Polygonum* that were collected in the flora area at the end of the nineteenth century and beginning of the twentieth century appear not to have persisted here and are not included in the keys. *Polygonum arenarium* Waldstein & Kitaibel and *P. bellardii* Allioni were reported by B. L. Robinson (1902) from Rhode Island and Massachusetts, respectively. The former resembles *P. patulum* but has open flowers. *Polygonum bellardii* is discussed below under *P. ramosissimum*. *Polygonum polycnemoides* Jaubert & Spach and *P. humifusum* C. Merck ex K. Koch subsp. *humifusum* were reported by J. F. Brenckle (1941).

The former was collected in New York City in 1894 and in Idaho in 1940. It differs from all other *Polygonum* species in having a tube 55–70% of the perianth length. *Polygonum humifusum* subsp. *humifusum* is discussed below under *P. humifusum* subsp. *caurianum*.

SELECTED REFERENCE   Jones, D. M. and T. R. Mertens. 1970. A taxonomic study of genus *Polygonum* employing chromatographic methods. Proc. Indiana Acad. Sci. 80: 422–430.

Key to the Sections of *Polygonum*

1. Stems distinctly and ± regularly 8–16-ribbed; leaf blade venation pinnate, secondary veins conspicuous; anthers whitish yellow; nearly worldwide . . . .   30a. *Polygonum* sect. *Polygonum*, p. 548
1. Stems 4-gonous, ribs obscure or absent; leaf blade venation parallel, secondary veins not conspicuous; anthers pink to purple; mostly w North America . . . . . .   30b. *Polygonum* sect. *Duravia*, p. 561

30a. POLYGONUM Linnaeus sect. POLYGONUM

Herbs, annual (perennial and rhizomatous in *P. striatulum*), not compact, not cushionlike, homophyllous or heterophyllous, frequently heterocarpic, rarely subsucculent (in *P. marinense* and *P. fowleri*). **Stems** prostrate, decumbent, or ascending to erect, straight (zigzag in *P. fowleri* and *P. humifusum*), distinctly and ± regularly 8–16-ribbed, smooth (papillose-scabridulous in *P. plebeium*). **Leaves:** ocrea 4–12-veined (1-veined in *P. plebeium*), proximal part not pruinose (pruinose in *P. glaucum* and *P. oxyspermum*); petiole articulated to proximal part of ocrea, when absent, blade is directly articulated to ocrea; blade linear to elliptic or obovate, rarely coriaceous (in *P. glaucum*), smooth (papillose in *P. plebeium*, rugulose and glaucous in *P. glaucum*); venation pinnate, secondary veins conspicuous. **Inflorescences** usually axillary, less commonly axillary and terminal; cymes 1–7(–10)-flowered. **Pedicels** erect to spreading, 0.5–7 mm. **Flowers** usually closed, sometimes semi-open; tepals ± monomorphic, outer tepals equaling or somewhat larger than inner, apices of outer tepals rounded; anthers whitish yellow. **Achenes** ovate to ovate-lanceolate, 2–3-gonous, shiny or dull, smooth to roughened or tubercled.

Species ca. 45 (13 in the flora): nearly worldwide.

Plants of sect. *Polygonum* produce two types of fruits (heterocarpy). Summer achenes are brown, ovate, and tubercled to smooth; late-season (autumn) achenes are hypertrophied, olivaceous, two to five times as long as the summer achenes, lanceolate, and smooth. The two types differ in their germination biology. Late-season achenes possess a low, innate dormancy and can germinate immediately after being produced, at 20–25°C (O. V. Yurtseva et al. 1999). Lower temperatures will delay germination until the following spring. Summer achenes are dormant when produced and undergo a cyclical dormant/nondormant pattern in the soil (A. D. Courtney 1968; J. M. Baskin and C. C. Baskin 1990).

Mature, early-season plants bearing leaves, flowers, and achenes are necessary for accurate determinations. Late-autumn specimens with hypertrophied achenes often are difficult or impossible to identify. In the descriptions, measurements of leaves, ocreae, and petioles are from the middle of the main stem. Leaf length includes the petiole. Heterophyllous taxa have cauline leaves at least three times as long as the branch leaves; homophyllous plants have stem and branch leaves about equal in size. Heterophylly is easily detected in most cases. However, some taxa show considerable variability in leaf morphology; identification should not rely on heterophylly alone. The character states "flowers closed" and "flowers semi-open" should be observed on herbarium specimens. Perianth description refers to the fruiting perianth measured from the pedicel joint to the apex of the tepals. Tepal descriptions refer to the outer three tepals (except for *P. heterosepalum*). Trigonous achenes have one face broader than the other two. Descriptions refer only to the two narrower faces, which can be subequal or distinctly unequal. Surface of achenes is best observed at magnifications of (or higher than) 100×. The achene surface may be: *smooth*, when no ornamentation is visible; *roughened*, when the tubercles are not discernible but the surface appears roughened; *striate-tubercled*, when conspicuous tubercles are oriented in rows; *uniformly tubercled*, when the tubercles are dense and rows difficult to distinguish; and *obscurely tubercled*, when the tubercles are weakly marked, inconspicuous, or restricted to a few (three to 15) rows on the central parts of faces. Late-season achenes in all species are hypertrophied, olivaceous, lanceolate, exserted, and smooth. They have little taxonomic significance.

SELECTED REFERENCES Mertens, T. R. and P. H. Raven. 1965. Taxonomy of *Polygonum*, section *Polygonum* (*Avicularia*) in North America. Madroño 18: 85–92. Wolf, S. J. and J. McNeill. 1986. Synopsis and achene morphology of *Polygonum* section *Polygonum* (Polygonaceae) in Canada. Rhodora 88: 457–479.

1. Plants perennial, rhizomatous . . . . 1. *Polygonum striatulum*
1. Plants annual, not rhizomatous.
  2. Leaves in distal part of inflorescence reduced, not overtopping flowers (shorter than or equaling flowers); inflorescences axillary and terminal, spikelike.
    3. Achenes striate-tubercled . . . . 12. *Polygonum patulum*
    3. Achenes smooth or roughened.
      4. Margins of tepals pink, rarely red or white; achenes 1.3–2.3 mm . . . . 13. *Polygonum argyrocoleon*
      4. Margins of tepals greenish yellow or yellow, rarely pink or white; achenes (2.3–)2.5–3.5 mm . . . . 4. *Polygonum ramosissimum* (in part)
  2. Leaves in distal part of inflorescence overtopping flowers; inflorescences entirely axillary.
    5. Proximal parts of ocreae pruinose.
      6. Distal parts of distal ocreae silvery, persistent; leaf blades glaucous, rugulose; perianth 2–3(–4) mm . . . . 6. *Polygonum glaucum*
      6. Distal parts of ocreae hyaline, soon disintegrating into fibers or deciduous; leaf blades not glaucous, not rugulose; perianth 3.5–5.5 mm . . . . . 7. *Polygonum oxyspermum*
    5. Proximal parts of ocreae not pruinose.
      7. Achenes striate-tubercled, uniformly tubercled, or obscurely tubercled.
        8. Plants green to bluish green; margins of tepals white, pink, or red.
          9. Achenes coarsely striate-tubercled . . . . 11. *Polygonum aviculare* (in part)
          9. Achenes obscurely tubercled.
            10. Plants dark brown to black after drying; distal part of ocreae disintegrating into persistent fibers, brown . . . . 4. *Polygonum ramosissimum* (in part)
            10. Plants green after drying (sometimes whitish from powdery mildew); distal part of ocreae persistent, silvery . . . . 11. *Polygonum aviculare* (in part)
        8. Plants light green or yellowish; margins of tepals green to yellow.
          11. Perianth tube 40–55% of perianth length; tepals ± keeled; pedicels 1.3–1.8(–2) mm, enclosed in ocreae . . . . 3. *Polygonum achoreum*
          11. Perianth tube 20–38% of perianth length; tepals not keeled; pedicels 2–7 mm, exserted from ocreae.
            12. Leaf blades elliptic to obovate; distal parts of ocreae ± persistent, silvery; achenes striate-tubercled . . . . 2. *Polygonum erectum*
            12. Leaf blades narrowly elliptic to lanceolate, rarely ovate; distal parts of ocreae soon disintegrating into persistent brown fibers; achenes uniformly tubercled . . . . 4. *Polygonum ramosissimum* (in part)
      7. Achenes smooth to roughened.
        13. Stems and leaf blades papillose-scabridulous; ocreae 1-veined . . . . 10. *Polygonum plebeium*
        13. Stems and leaf blades glabrous; ocreae 4–12-veined.

[14. Shifted to left margin.—Ed.]

14. Leaves often opposite at proximal nodes; achenes 1.4–1.6(–2.2) mm; Yukon, N.W.T., Nunavut, Alaska . . . . 9 *Polygonum humifusum*
14. Leaves all alternate; achenes 1.6–4.5 mm; broad distribution (including Yukon, N.W.T., Nunavut, Alaska).
  15. Perianth tube 40–57% of perianth length . . . . 11. *Polygonum aviculare* (in part)
  15. Perianth tube 18–38% of perianth length.
    16. Achenes beaked at apex, edges strongly concave; stems sometimes zigzagged . . . . 8. *Polygonum fowleri*
    16. Achenes not beaked at apex, edges straight; stems not zigzagged.
      17. Plants homophyllous; pedicels 1–2 mm, enclosed in ocreae . . . . 4. *Polygonum ramosissimum* (in part)
      17. Plants heterophyllous; pedicels 2–6 mm, mostly exserted from ocreae.
        18. Plants subsucculent; leaf blade apices rounded; proximal parts of ocreae funnelform; flowers semi-open; cymes in axils of most leaves; California . . . . 5. *Polygonum marinense*
        18. Plants not subsucculent; leaf blade apices acute to acuminate; proximal parts of ocreae cylindric; flowers closed; cymes crowded toward tips of branches; broad distribution in United States and Canada (including California) . . . . 4. *Polygonum ramosissimum* (in part)

1. **Polygonum striatulum** B. L. Robinson, Proc. Boston Soc. Nat. Hist. 31: 263. 1904 • Texas knotweed [E]

Polygonum striatulum var. *texense* (M. C. Johnston) Costea & Tardif; *P. texense* M. C. Johnston

Plants perennial, light green, heterophyllous; rhizomes brown, 0.7–3 cm diam. **Stems** erect to ascending, sparingly branched in distal ¹/₂, not wiry, 25–60 cm. **Leaves:** ocrea 6–12 mm, proximal part cylindric, distal part soon disintegrating into brown fibers, later leaving almost no fibrous remains; petiole 0–2 mm; blade light green, linear-lanceolate to oblanceolate, 8–35 × 2–8 mm, margins flat, apex acute or obtuse; stem leaves 2.1–4 times as long as adjacent branch leaves; distal leaves sharply reduced, not overtopping flowers (shorter than or equaling flowers). **Inflorescences** axillary and terminal, spikelike; cymes in distal nodes, 2–6-flowered. **Pedicels** exserted from ocreae, 2–4 mm. **Flowers** semi-open; perianth 2–3.5 mm; tube 18–25% of perianth length; tepals overlapping, green with white or sometimes pink margins, petaloid, not keeled, oblong to obovate, cucullate; midveins usually unbranched; stamens 7–8. **Achenes** enclosed in perianth, brown, ovate, 3-gonous, (1.6–)1.8–2.6(–3) mm, faces subequal or unequal, apex not beaked, edges concave, shiny, smooth or roughened; late-season achenes common, 4–6 mm.

Flowering Dec–Mar or Jun–Oct. Seasonal moist places, sterile prairies, granitic soils; 100–700 m; Tex.

Plants with leaves less-conspicuously veined, flowering August to October, and growing in seasonally moist habitats in central and western Texas are recognized by some authors as *Polygonum texense* or *P. striatulum* var. *texense* (M. Costea and F. J. Tardif 2003).

2. **Polygonum erectum** Linnaeus, Sp. Pl. 1: 363. 1753 • Erect knotweed, renouée dressée [E] [W]

Plants light green or yellowish, heterophyllous. **Stems** erect to ascending, sparingly branched in distal ¹/₂, not wiry, 15–75 cm. **Leaves:** ocrea 7–12 mm, proximal part cylindric, distal part usually persistent, with strong veins, margins entire or lacerate, silvery, later disintegrating into ± persistent brown fibers; petiole 1–5 mm; blade light green or yellowish, elliptic to obovate, 30–60(–80) × (8–)10–25 mm, margins flat, apex obtuse; stem leaves 1.5–3.5 (–4) times longer than branch leaves; distal leaves overtopping flowers in distal part of inflorescence. **Inflorescences** axillary; cymes in axils of most leaves and toward tips of stems and branchs, 1–5-flowered. **Pedicels** mostly exserted from ocreae, 3–7 mm. **Flowers** closed; perianth 2.8–3.8(–4.2) mm; tube 20–37% of perianth length; tepals overlapping, green with yellowish, rarely whitish green, margins sepaloid, not keeled, oblong to obovate, cucullate; midveins branched, moderately to heavily thickened; stamens 7–8. **Achenes** enclosed in perianth, brown to tan, ovate, 3-gonous, 2.3–3.5 mm, faces subequal, ± concave, apex not beaked, edges concave, dull, striate-tubercled; late-season achenes uncommon, 4–5 mm.

Flowering May–Oct. Dry, waste ground; 10–300 m; Alta., Ont., Que., Sask.; Ala., Ark., Colo., Conn., Del., D.C., Ga., Ill., Ind., Iowa, Kans., Ky., La., Maine, Md.,

Mass., Mich., Minn., Miss., Mo., Mont., Nebr., Nev., N.H., N.J., N.Mex., N.Y., N.C., Ohio, Okla., Oreg., Pa., R.I., S.C., Tenn., Vt., Va., Wash., W.Va., Wis., Wyo.

Polygonum erectum was cultivated in the midwest by Native Americans for its starchy seeds (C. M. Scarry 1993). It was formerly confused with *P. achoreum* (T. R. Mertens and P. H. Raven 1965).

3. Polygonum achoreum S. F. Blake, Rhodora 19: 232. 1917 • Blake's knotweed, renouée coriace [E]

Polygonum erectum Linnaeus subsp. *achoreum* (S. F. Blake) Á. Löve & D. Löve

Plants light green (often covered with whitish powdery mildew), homophyllous or, sometimes, heterophyllous. **Stems** erect when young, decumbent or prostrate later, moderately branched especially from base, not wiry, 50–70 cm. **Leaves:** ocreae 5–12 mm, proximal part cylindric, distal part soon disintegrating into brown fibers; petiole 0.3–1.5 mm; blade light yellowish green, elliptic to obovate, 8–35 × 3–15 mm, margins flat, apex rounded; stem leaves 1–2.1(–3) times longer than branch leaves; distal leaves overtopping flowers. **Inflorescences** axillary, cymes in axils of most leaves and toward tips of stems and branchs, 1–3(–5)-flowered. **Pedicels** enclosed in ocreae, 1.3–1.8(–2) mm. **Flowers** closed; perianth 2.6–4 mm; tube 40–55% of perianth length; tepals incurved, yellow-green with yellow to green, rarely pinkish, margins, sepaloid, ± keeled, narrowly oblong, cucullate; midveins unbranched, moderately to heavily thickened, tepals appearing keeled; stamens 5–8. **Achenes** enclosed in perianth, yellow-green to tan, ovate, 3-gonous, 2.4–3.5 mm, faces unequal, apex not beaked, edges concave or nearly straight, dull, uniformly tubercled; late-season achenes common, 3–5 mm. $2n = 40, 60$.

Flowering Jul–Sep. Disturbed areas, roadsides, sidewalks, edges of cultivated fields; 10–800 m; Alta., B.C., Man., N.B., N.W.T., N.S., Ont., Que., Sask., Yukon; Alaska, Colo., Conn., Idaho, Ill., Ind., Iowa, Kans., Maine, Mich., Minn., Mo., Mont., Nebr., Nev., N.Y., N.Dak., Ohio, Oreg., S.Dak., Utah, Vt., Wash., W.Va., Wis., Wyo.

Polygonum achoreum frequently is confused with *P. erectum*. It can be distinguished by its usually homophyllous leaves, its perianth, which is enlarged at the base and constricted above the fruit, its longer perianth tube, and its yellow-green to tan, tubercled achenes.

4. Polygonum ramosissimum Michaux, Fl. Bor.-Amer. 1: 237. 1803 • Bushy knotweed [E][F][W]

Plants yellowish green or bluish green, (dark brown to black after drying in subsp. *prolificum*), heterophyllous or homophyllous. **Stems** erect, usually profusely branched in distal 1/2, not wiry, 10–100(–200) cm. **Leaves:** ocrea 6–12(–15) mm, proximal part cylindric, distal part silvery, soon disintegrating into persistent brown fibers; petiole 2–4 mm; blade variable, light yellowish green to bluish green, proximal often caducous, narrowly elliptic, lanceolate, or oblanceolate, rarely ovate, 8–70 × 4–18(–35) mm, margins flat, apex acute to acuminate or obtuse; distal leaves, either overtopping or shorter than or equaling flowers. **Inflorescences** axillary or axillary and terminal, spikelike; cymes uniformly distributed or crowded toward tips of branches, 2–5-flowered. **Pedicels** enclosed in or exserted from ocreae, 1–6 mm. **Flowers** closed; perianth (2–)2.2–3.6(–4) mm; tube 20–38% of perianth length; tepals overlapping, greenish yellow with greenish yellow or yellow, rarely pink or white, margins, petaloid or sepaloid, not keeled, elliptic to oblong, cucullate; midveins thickened or not; stamens 3–6(–8). **Achenes** enclosed in or exserted from perianth, dark brown, ovate, 3-gonous, 1.6–3.5 mm, faces subequal, concave, apex not beaked, edges straight, shiny or dull, usually smooth to roughened, sometimes uniformly or obscurely tubercled; late-season achenes common, 4–15 mm.

Subspecies 2 (2 in the flora): North America.

Polygonum ramosissimum exhibits considerable morphological complexity and is similar in difficulty to the *P. aviculare* complex. Further research is necessary to understand the infraspecific variability of this species (M. Costea and F. J. Tardif 2003b).

1. Plants heterophyllous, usually yellowish green when fresh or dried; apices of leaf blades acute to acuminate; pedicels 2.5–6 mm . . . . 4a. *Polygonum ramosissimum* subsp. *ramosissimum*
1. Plants homophyllous, bluish green when fresh, dark brown or black after drying; apices of leaf blades rounded or obtuse; pedicels 1–2 mm . . . . 4b. *Polygonum ramosissimum* subsp. *prolificum*

4a. Polygonum ramosissimum Michaux subsp. **ramosissimum** • Renouée à fleurs jaunes

E F W

Polygonum atlanticum (B. L. Robinson) E. P. Bicknell; *P. exsertum* Small; *P. interior* Brenckle; *P. latum* Small ex Rydberg; *P. leptocarpum* B. L. Robinson; *P. stevensii* Brenckle; *P. triangulum* E. P. Bicknell

Plants usually yellowish greenish when fresh or dried, heterophyllous. **Stems** 30–200 cm. **Leaf blades** light yellowish green, narrowly elliptic to lanceolate, rarely ovate, 35–70 × 7–18(–35) mm, apex acute to acuminate; stem leaves 2.1–3.5(–4.2) times as long as branch leaves; distal leaves usually 1–2 mm, not overtopping flowers, or 5–15 mm, surpassing flowers. **Inflorescences** axillary and terminal, spikelike; cymes crowded toward tips of branches, 2–5-flowered. **Pedicels** exserted from ocreae, 2.5–6 mm. **Flowers:** perianth (2.6–)3–4 mm; tepal margins greenish yellow or yellow, rarely white or pink; stamens 3–6(–8). **Achenes** enclosed in or exserted from perianth, (2.3–)2.5–3.5 mm, shiny or dull, smooth to roughened, sometimes uniformly or obscurely tubercled; late-season achenes 5–12 mm. $2n = 60$.

Flowering Jul–Oct. Disturbed places, sandy shores or saline soils and shores; 0–1000 m; Alta., B.C., Man., N.B., N.W.T., N.S., Ont., P.E.I., Que., Sask.; Ariz., Ark., Calif., Colo., Conn., Del., Ga., Idaho, Ill., Ind., Iowa, Kans., Ky., La., Maine, Mass., Mich., Minn., Mo., Mont., Nebr., Nev., N.H., N.J., N.Mex., N.Y., N.C., N.Dak., Ohio, Okla., Pa., R.I., S.C., S.Dak., Tenn., Tex., Utah, Vt., Va., Wash., Wis., Wyo.

Subspecies *ramosissimum* is heterogeneous; some additional elements may deserve recognition. It is closely related to European *Polygonum bellardii* Allioni, which was collected in south Boston in 1785 (B. L. Robinson 1902). The latter species has semi-open flowers, petaloid tepals with white or pink margins, and eight stamens. A distinct form of *P. ramosissimum* growing in saline marshes from California has been mistakenly identified as *P. patulum* Bieberstein (M. Costea and F. J. Tardif 2003b). The morphology of late-season achenes and the branching patterns, which have been emphasized by some authors, appear to have little taxonomic value.

4b. Polygonum ramosissimum Michaux subsp. **prolificum** (Small) Costea & Tardif, Sida 20: 995. 2003 • Renouée prolifique   E W

Polygonum ramosissimum var. *prolificum* Small, Bull. Torrey Bot. Club 21: 171. 1894; *P. prolificum* (Small) B. L. Robinson; *P. prolificum* var. *autumnale* (Brenckle) Brenckle; *P. prolificum* var. *profusum* Brenckle

Plants bluish green when fresh, dark brown or black after drying, homophyllous. **Stems** 10–80 cm. **Leaf blades** bluish green, oblanceolate, 8–30(–35) × 4–6 mm, apex rounded or obtuse; stem leaves 1–2.5(–3.5) times as long as branch leaves; distal leaves overtopping flowers. **Inflorescences** axillary; cymes uniformly distributed or crowded toward tips of branches, 2–4-flowered. **Pedicels** enclosed in ocreae, 1–2 mm. **Flowers:** perianth 2–2.8(–3) mm; tepal margins white or pink; stamens 3. **Achenes** enclosed in or exserted from perianth, 1.6–2(–2.4) mm, shiny or dull, smooth, roughened, or obscurely tubercled; late-season achenes 4–10 mm. $2n = 60$.

Flowering Jul–Nov. Wet, saline places; 0–2000 m; Alta., B.C., Man., Ont., P.E.I., Que., Sask.; Ark., Calif., Conn., Del., Idaho, Ill., Iowa, Kans., Ky., La., Maine, Md., Mass., Minn., Miss., N.H., N.J., N.Mex., N.Y., N.C., N.Dak., Okla., Oreg., Pa., R.I., S.Dak., Tenn., Tex., Utah, Vt., Va., Wash., W.Va., Wyo.

5. Polygonum marinense T. R. Mertens & P. H. Raven, Madroño 18: 87. 1965 • Marin knotweed   C E F

Plants often reddish tinged, heterophyllous, subsucculent. **Stems** prostrate to ascending, branching from base, not wiry, 15–40 cm. **Leaves:** ocrea 4–6 mm, proximal part funnelform, distal part silvery hyaline, soon disintegrating, leaving almost no fibrous remains; petiole 2–5 mm; blade often reddish tinged, elliptic to obovate or oblanceolate; 20–35 × 9–16 mm, margins flat, apex rounded; stem leaves (1.3–)2–2.6(–3.5) times as long as branch leaves; distal leaves overtopping flowers in distal part of inflorescence. **Inflorescences** axillary; cymes in most leaf axils, 1–4-flowered. **Pedicels** mostly exserted from ocreae, 2–4 mm. **Flowers** semi-open; perianth 3–3.5(–4) mm; tube 18–25% of perianth length; tepals overlapping, green, margins white or pink, petaloid, not keeled, broadly rounded, cucullate; midveins unbranched; stamens 8. **Achenes** exserted from perianth, brown, ovate, 3-gonous, 2.8–3.4(–4) mm, faces subequal or evidently unequal, apex not beaked, edges straight, shiny, minutely roughened; late-season achenes uncommon, 4.5–5 mm. $2n = 60$.

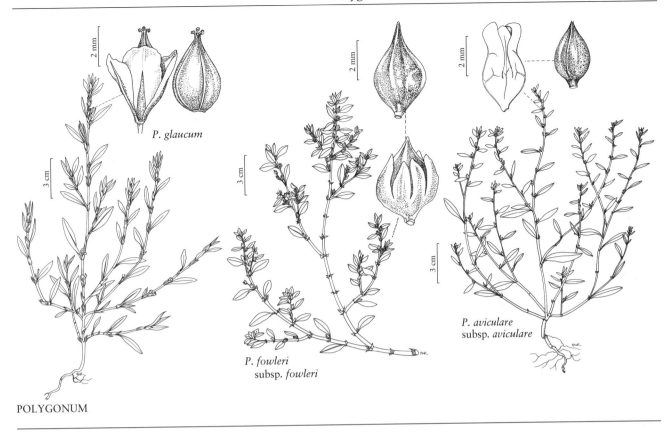

P. glaucum

P. fowleri
subsp. *fowleri*

P. aviculare
subsp. *aviculare*

POLYGONUM

Flowering Apr–Oct. Coastal salt and brackish marshes, swamps; of conservation concern; 0–10 m; Calif.

The origin and taxonomic affinities of *Polygonum marinense* are uncertain. T. R. Mertens and P. H. Raven (1965) suggested a relationship with *P. oxyspermum* C. A. Meyer & Bunge or the Mediterranean *P. robertii* Loiseleur-Deslongchamps. *Polygonum marinense* may be confused with *P. ramosissimum*. It can be distinguished by its subsucculent texture, funnelform ocreae, leaves rounded at the apices, and semi-open flowers. Marin knotweed is known from fewer than 15 locations in Marin, Napa, Solano, and Sonoma counties; it is threatened by coastal development.

6. Polygonum glaucum Nuttall, Gen. N. Amer. Pl. 1: 254. 1818 • Seabeach knotweed [E] [F]

Plants silvery, homophyllous. **Stems** prostrate to ascending, branched from base, not wiry, 20–70 cm. **Leaves:** ocrea persistent, 7–15 mm, proximal part cylindric, pruinose, distal parts silvery, margins overlapping, entire or lacerate; petiole 0.5–3 mm; blade bluish green, lanceolate, 10–30 × 2–8 mm, coriaceous, margins revolute, apex obtuse or acute, rugulose when fresh, markedly rugulose when dried, glaucous; middle stem leaves slightly larger than adjacent branch leaves, distal leaves overtopping flowers they subtend. **Inflorescences** axillary; cymes uniformly distributed, 1–3-flowered. **Pedicels** enclosed in ocreae, 3–5 mm. **Flowers** semi-open; perianth (2–)3–4 mm; tube 12–26% of perianth length; tepals ± recurved, overlapping, white, margins white or pink, petaloid, not keeled, oblong-obovate to spatulate, not cucullate; midvein usually unbranched; stamens 8. **Achenes** exserted from perianth, reddish brown to dark brown, ovate, 3-gonous, 2.5–3(–4) mm, faces subequal, apex not beaked, edges straight, shiny, smooth; late-season achenes common, 3–5 mm. $2n = 40$.

Flowering Jul–Nov. Coastal beaches, sand dunes, margins of salt ponds; 0–10 m; Conn., Del., D.C., Fla., Ga., Md., Mass., N.J., N.Y., N.C., R.I., S.C., Va.

Polygonum glaucum is restricted to maritime beaches along the Atlantic coast from Massachusetts south to Georgia. Over most of its range it is rare and declining; populations on coastal islands of Massachusetts and along the shore of Long Island appear to be secure. Seabeach knotweed appears to be related to *P. oxyspermum*.

7. Polygonum oxyspermum C. A. Meyer & Bunge ex Ledebour, Index Seminum (Dorpat) 1824: 5. 1824 ☐

Plants green to blue-green, homophyllous. **Stems** prostrate to ascending, branched at proximal and middle nodes, not wiry, 20–100 cm. **Leaves:** ocrea 6.5–12 mm, proximal part cylindric, pruinose, distal part hyaline, soon disintegrating into brown fibers or nearly completely deciduous; petiole 0–1 mm; blade green to bluish green, elliptic-lanceolate to linear, 10–35 × 1.5–15 mm, margins flat or narrowly revolute, apex acute; middle stem leaves slightly larger than adjacent branch leaves, distal leaves overtopping flowers. **Inflorescences** axillary, cymes uniformly distributed, 2–7-flowered. **Pedicels** exserted from ocreae, 2.5–5(–7) mm. **Flowers** semi-open; perianth 3.5–5.5 mm; tube 28–39% of perianth length; tepals slightly overlapping, green, margins white to pink, petaloid, not keeled, oblong to obovate, not cucullate; veins branched; stamens 8. **Achenes** exserted from perianth, pale brown to dark brown, ovate, 3-gonous, (3.5–)4.1–5.5(–6.5) mm, faces subequal, apex not beaked, edges straight, shiny, smooth or with fine tubercles especially toward apex; late-season achenes unknown.

Subspecies 3 (2 in the flora): introduced; e North America, Europe.

The treatment by D. A. Webb and A. O. Chater (1963) of the *Polygonum oxyspermum* complex is followed here. T. Karlsson (2000) accepted *P. raii* and *P. oxyspermum* as distinct species.

1. Ocreae with prominent veins, distal part disintegrating, veins persistent; leaf blades 5–8 times as long as wide; achenes 4.1–5.5(–6.5) mm, exserted 1–2.5 mm from perianth . . . .
    . . . . . 7a. *Polygonum oxyspermum* subsp. *oxyspermum*
1. Ocreae without prominent veins, distal part disintegrating, nearly completely deciduous; leaf blades 3.5–5 times as long as wide; achene (3.5–)4–5 mm, exserted 0.5–1.3(–1.8) mm from perianth . . . . 7b. *Polygonum oxyspermum* subsp. *raii*

7a. Polygonum oxyspermum C. A. Meyer & Bunge ex Ledebour subsp. **oxyspermum** ☐

Polygonum acadiense Fernald

Leaves: ocrea prominently veined, (6.5–)8–12 mm, distal part disintegrating, veins persistent, brown; blade green, narrowly elliptic to lanceolate, 5–8 times as long as wide. **Flowers:** perianth 3.5–5.5 mm; tepals slightly overlapping, green with pink margins, oblong to obovate, apex flat, flared. **Achenes** with apex exserted 1–2.5 mm from perianth, 4.1–5.5 (–6.5) mm, 1.5–2.6 times as long as wide, ± smooth.

Flowering Jul–Oct. Coastal beaches, dunes, and shores; 0–10 m; introduced; N.S.; Europe.

7b. Polygonum oxyspermum C. A. Meyer & Bunge ex Ledebour subsp. **raii** (Babington) D. A. Webb & Chater, Feddes Repert. Spec. Nov. Regni Veg. 68: 188. 1963 • Ray's knotweed, renouée de Ray ☐

Polygonum raii Babington, Trans. Linn. Soc. London 17: 458. 1836

Leaves: ocrea not prominently veined, 7–10 mm, distal part disintegrating, nearly completely deciduous; blade bluish green, elliptic, 3.5–5 times as long as wide. **Flowers:** perianth (3.5–)4–5 mm; tepals overlapping, green with white or pink margins, obovate, apex flat, flared. **Achenes** with apex exserted 0.5–1.3(–1.8) mm from perianth, (3.5–)4–5 mm, 1.3–1.6 times as long as wide, ± smooth or finely striate-tubercled. $2n = 40$.

Flowering Jul–Oct. Coastal beaches, dunes, and shores; 0–10 m; introduced; St. Pierre and Miquelon; N.B., Nfld. and Labr. (Nfld.), N.S., P.E.I., Que., Maine; Europe.

Some sources consider subsp. *raii* to be native in North America.

8. Polygonum fowleri B. L. Robinson, Rhodora 4: 67, plate 35, figs. 14, 15. 1902 ☐ ☐

Plants green, sometimes purple tinged, homophyllous or heterophyllous, sometimes subsucculent. **Stems** prostrate to ascending, sometimes zigzagged, branched from base, not wiry, 5–50 cm. **Leaves:** ocrea 2.5–12 mm, proximal part funnelform, distal part soon disintegrating, nearly completely deciduous or fibers persistent; petiole 2–7 mm; blade light green, sometimes purple tinged, elliptic to elliptic-obovate or obovate, 8–30(–50) × 4–15(–25) mm, margins flat, apex acute to obtuse; middle stem leaves 1.1–2.1(–3.4) times as long as adjacent branch leaves, distal leaves overtopping flowers. **Inflorescences** axillary; cymes uniformly distributed, 1–7(–10)-flowered. **Pedicels** enclosed in or sometimes exserted from ocreae, 1–2.5 mm. **Flowers** closed; perianth (2.2–)2.5–4.5 mm; tube 23–38% of perianth length; tepals initially overlapping, pushed apart as achene develops, green, margins white to pink, petaloid, not keeled, oblong, cucullate; midveins branched, sometimes not visible; stamens 6–8. **Achenes** exserted from perianth, brown to dark brown, broadly ovate to ovate-lanceolate, (2–)3-gonous, (1.8–)2–3.7(–4.5) mm, faces subequal or unequal, flat to concave, apex beaked, edges strongly concave, shiny to dull, roughened, rarely obscurely tubercled; late-season achenes common, 4–6 mm.

Subspecies 2 (2 in the flora): n North America.

1. Achenes broadly ovate, (2.5–)3–3.7(–4.5) mm; perianth 3–4.5 mm . . . .
. . . . 8a. *Polygonum fowleri* subsp. *fowleri*
1. Achenes ovate-lanceolate, (1.8–)2–2.5 (–3.1) mm; perianth (2.2–)2.5–3.3(–3.5) mm . . . .
. . . . 8b. *Polygonum fowleri* subsp. *hudsonianum*

8a. Polygonum fowleri B. L. Robinson subsp. **fowleri**
  • Fowler's knotweed, renouée de Fowler [E] [F]

Polygonum allocarpum S. F. Blake

Plants rarely purple tinged, homophyllous or heterophyllous, subsucculent. **Stems** prostrate to ascending, 15–50 cm. **Leaves:** ocrea 4–12 mm; petiole 5–7 mm; blade green, rarely purple tinged, elliptic to elliptic-obovate, 20–35(–50) × 10–15(–25) mm, apex acute or obtuse; middle stem leaves 1.1–2.1(–3.4) times as long as adjacent branch leaves. **Cymes** 1–5-flowered. **Pedicels** 1–2.5 mm. **Perianths** 3–4.5 mm. **Achenes** broadly ovate, 3-gonous, (2.5–)3–3.7(–4.5) mm, faces subequal, concave, roughened, rarely obscurely tubercled. $2n = 40, 60$.

Flowering Jul–Oct. Sandy, gravelly seashores; 0–30 m; St. Pierre and Miquelon; B.C., Man., N.B., Nfld. and Labr., N.W.T., N.S., Nunavut, Ont., P.E.I., Que., Yukon; Alaska, Calif., Maine, Oreg., Wash.

8b. Polygonum fowleri B. L. Robinson subsp. **hudsonianum** (S. J. Wolf & McNeill) Costea & Tardif, Sida 20: 994. 2003 • Hudsonian knotweed, renouée de la baie d'Hudson [E]

Polygonum caurianum B. L. Robinson subsp. *hudsonianum* S. J. Wolf & McNeill, Rhodora 88: 469, fig. 3. 1986; *P. hudsonianum* (S. J. Wolf & McNeill) H. R. Hinds

Plants often purplish tinged, homophyllous, rarely subsucculent. **Stems** prostrate, 5–30 cm. **Leaves:** ocrea 2.5–5 mm; petiole 2–5 mm; blade purple tinged, narrowly obovate or oblanceolate, 8–30 × 4–12(–15) mm, apex rounded to obtuse; middle stem leaves 1.1–1.8(–2.2) times as long as adjacent branch leaves. **Cymes** 2–7(–10)-flowered. **Pedicels** 1–1.5 mm. **Perianths** (2.2–)2.5–3.3(–3.5) mm.

Achenes ovate-lanceolate, (2–)3-gonous, (1.8–)2–2.5 (–3.1) mm, faces often unequal, ± flat to concave, roughened, rarely obscurely tubercled.

Flowering Jul–Sep. Gravelly seashores; 0–30 m; Man., Nfld. and Labr., N.W.T., N.S., Nunavut, Ont., Que..

Wolf and McNeill described Hudsonian knotweed as a subspecies of *Polygonum caurianum* (*P. humifusum* subsp. *caurianum* in this treatment). In raising it to the rank of species, H. R. Hinds (1995) noted that "separation between *P. fowleri* and *P. hudsonianum*, however, is more difficult." M. Costea and F. J. Tardif (2003) concluded that the taxon is best treated as a subspecies of *P. fowleri*, based on the existence of intermediate plants.

9. Polygonum humifusum C. Merck ex K. Koch, Linnaea 22: 205. 1849

Subspecies 2 (1 in the flora): nw North America, e Asia.

9a. Polygonum humifusum C. Merck ex K. Koch subsp. **caurianum** (B. L. Robinson) Costea & Tardif, Sida 20: 995. 2003 • Alaska knotweed, renouée du Nord-Ouest [E]

Polygonum caurianum B. L. Robinson, Proc. Boston Soc. Nat. Hist. 31: 264. 1904

Plants frequently reddish or purplish tinged; homophyllous or, rarely, heterophyllous. **Stems** prostrate, often zigzagged, branched mostly from base, wiry, 2–20(–40) cm. **Leaves** often opposite at proximal nodes; ocrea 2–3(–4) mm, proximal part funnelform, distal part soon lacerate, nearly completely deciduous; petiole 1–3.4 mm; blade usually reddish or purple tinged, obovate to oblanceolate, 3–12(–25) × (1.5–)2.5–4.5(–8) mm, margins flat, apex rounded to obtuse; stem leaves 1–1.5(–2) times as long as adjacent branch leaves, distal leaves overtopping flowers. **Inflorescences** axillary; cymes ± uniformly distributed, 2–6-flowered. **Pedicels** enclosed in ocreae, 0.5–1.5 mm. **Flowers** closed or semi-open; perianth 1.5–2.3(–3) mm; tube 28–49% of perianth length; tepals partially overlapping, green, margins pink, rarely white, petaloid, not keeled, oblong, ± outcurved, usually not cucullate; midveins unbranched or branched; stamens 5. **Achenes** exserted from perianth, dark brown to purple, ovate-lanceolate, 2–3-gonous, 1.4–1.6(–2.2) mm, faces unequal, apex not beaked or obscurely beaked, edges straight or concave, shiny or dull, smooth to roughened; late-season achenes common, 2–3.5 mm.

Flowering Jun–Aug. Gravel bars, waste places; 20–700 m; N.W.T., Nunavut, Yukon; Alaska.

Polygonum humifusum often has opposite proximal leaves. In North America, the closest relative of subsp. *caurianum* is *P. fowleri*, rare individuals of which produce opposite leaves.

Three specimens of the Asian subsp. *humifusum* were collected near Nanaimo, Vancouver Island, by J. Macoun in 1883 and 1887 (J. F. Brenckle 1941); the subspecies has not been recollected there. It has stems and leaves that are green, and achenes 2.1–2.7 mm, ± beaked, and exserted 0.9–1.3 mm from the perianth at maturity.

10. **Polygonum plebeium** R. Brown, Prodr., 420. 1810

I

Plants bluish green, homophyllous. **Stems** prostrate, much-branched from base, not wiry, 10–40 cm, papillose-scabridulous. **Leaves:** ocrea 1-veined, 2.5–3 mm, proximal part cylindric, distal part laciniate; petiole 0–1 mm; blade bluish green, narrowly elliptic or oblanceolate, 5–16 × 1–4 mm, margins flat, apex obtuse or acute, papillose-scabridulous; stem leaves 1–1.5(–2) times longer than adjacent branch leaves, distal leaves overtopping flowers. **Inflorescences** axillary; cymes uniformly distributed, 3–6-flowered. **Pedicels** exserted from ocreae, 3–5 mm. **Flowers** closed; perianth 1–1.5 mm; tube 12–27% of perianth length; tepals overlapping, green with white or pink margins, petaloid, not keeled, elliptic, cucullate; midveins unbranched, thickened; stamens 5. **Achenes** enclosed in perianth, black-brown, broadly ovate, 2–3-gonous, 1.5–2 mm, faces subequal, apex not beaked, edges concave, shiny, smooth; late-season achenes common, 2–4 mm. **2*n*** = 20.

Flowering May–Oct. Disturbed sites; 50–500 m; introduced; Idaho; e Africa; s Asia; Australia.

11. **Polygonum aviculare** Linnaeus, Sp. Pl. 1: 362. 1753 • Doorweed, knotgrass, renouée des oiseaux

F W

Plants green or bluish green, green after drying, sometimes whitish from powdery mildew, homophyllous or heterophyllous. **Stems** prostrate to erect, branched, flexuous, 5–200 cm. **Leaves:** ocrea 3–15 mm, proximal part cylindric or ± funnelform, distal part silvery, hyaline, soon disintegrating into persistent fibers or nearly completely deciduous; petiole 0.3–9 mm; blade green to gray-green, narrowly elliptic, lanceolate, elliptic, obovate, or spatulate, 6–50(–60) × 0.5–22 mm, margins flat, apex acute, obtuse, or rounded; stem leaves 1–4 times as long as adjacent branch leaves;

distal leaves overtopping flowers. **Inflorescences** axillary; cymes uniformly distributed or aggregated at tips of stems and branches, 1–6(–8)-flowered. **Pedicels** enclosed in or exserted from ocreae, 1.5–5 mm. **Flowers** closed or semi-open; perianth 1.8–5.5 mm; tube 20–57% of perianth length; tepals overlapping or not, green or reddish brown with white, pink, or red margins, petaloid, not keeled, oblong to obovate, often cucullate in fruit; midveins branched or unbranched, thickened or not; stamens 5–8. **Achenes** enclosed in or exserted from perianth, light to dark brown, ovate, (2–)3-gonous, 1.2–4.2 mm, faces subequal or unequal, apex not beaked, edges slightly concave, dull, usually coarsely striate-tubercled, sometimes obscurely tubercled; late-season achenes common or not, 2–5 mm.

Subspecies 7+ (6 in the flora): nearly worldwide.

Polygonum aviculare is a taxonomically controversial polyploid complex of selfing annuals. Although members of the complex have been considered inbreeders, they possess some structures that make cross pollination possible. Cleistogamous and chasmogamous flowers, heterostyly, protandry, and the capacity to secrete nectar suggest an ancestral mixed-mating system. Isoenzyme studies showed that the complex has an allopolyploid origin (P. Meerts et al. 1998) and has evolved as a swarm of inbreeding lines ("Jordanons") (J. Gasquez et al. 1978). The six subspecies included here have been treated variously (T. Karlsson 2000; M. Costea and F. J. Tardif 2003). Complex intergradation patterns among them make their recognition at the species level impractical. Multivariate analysis and isoenzyme studies show that populations with intermediate characteristics may occur (Meerts et al. 1990, 1998). Except for subsp. *boreale*, which occurs in Greenland and Labrador, all subspecies are partially sympatric and their distributions have been influenced greatly by humans.

SELECTED REFERENCES Meerts, P., T. Baya, and C. Lefèbvre. 1998. Allozyme variation in the annual weed species complex *Polygonum aviculare* (Polygonaceae) in relation to ploidy level and colonizing ability. Pl. Syst. Evol. 211: 239–256. Meerts, P., J.-P. Briane, and C. Lefèbvre. 1990. A numerical taxonomic study of the *Polygonum aviculare* complex (Polygonaceae) in Belgium. Pl. Syst. Evol. 173: 71–90. Styles, B. T. 1962. The taxonomy of *Polygonum aviculare* and its allies in Britain. Watsonia 5: 177–214.

1. Perianth tubes 40–57% of perianth length.
   2. Tepals green or reddish brown, margins white, veins unbranched . . . . 11e. *Polygonum aviculare* subsp. *depressum*
   2. Tepals green, margins usually pink or red, rarely white, veins branched . . . . 11d. *Polygonum aviculare* subsp. *neglectum* (in part)
1. Perianth tubes (15–)20–40(–42)% of perianth length.

[3. Shifted to left margin.—Ed.]

3. Perianths 3.3–5.5 mm; achenes 2.5–4.2 mm.
    4. Plants heterophyllous; leaf blades elliptic to oblanceolate; tepals oblong, cucullate in fruit; cymes aggregated at tips of stems and branchs; broad distribution in North America . . . . 11a. *Polygonum aviculare* subsp. *aviculare* (in part)
    4. Plants homophyllous or subheterophyllous; leaf blades obovate-spatulate or oblanceolate; tepals obovate, flat or curved outward in fruit; cymes ± uniformly distributed; Greenland, Newfoundland and Labrador . . . . . 11b. *Polygonum aviculare* subsp. *boreale*
3. Perianths 1.9–3.6 mm; achenes 1.2–2.8(–3) mm.
    5. Ocreae with distal parts relatively persistent, silvery; perianths 0.9–1.3(–1.5) times as long as wide, outer tepals pouched at base . . . . 11c. *Polygonum aviculare* subsp. *buxiforme*
    5. Ocreae soon disintegrating into persistent fibers or leaving almost no fibrous remains; perianths 1.5–2.9 times as long as wide; outer tepals not pouched at base.
        6. Leaf blades (6–)10–20 mm wide, 2–4.5 times as long as wide; cymes 3–8-flowered, aggregated at tips of stems and branches; achenes enclosed in or barely exserted from perianth . . . . 11a. *Polygonum aviculare* subsp. *aviculare* (in part)
        6. Leaf blades 0.5–6.8(–8) mm wide, (3.4–)4.2–15(–19) times as long as wide; cymes 1–3(–5)-flowered, uniformly distributed along stems and branches; achenes usually exserted from perianth.
            7. Ocreae 4–8 mm, veins inconspicuous, distal parts leaving almost no fibrous remains; lateral veins of leaf blades visible but not raised adaxially . . . . 11d. *Polygonum aviculare* subsp. *neglectum* (in part)
            7. Ocreae (6–)8–12 mm, veins conspicuous, distal parts disintegrating into persistent fibers; lateral veins of leaf blades raised adaxially . . . 11f. *Polygonum aviculare* subsp. *rurivagum*

11a. Polygonum aviculare Linnaeus subsp. **aviculare**

• Common knotweed   F I W

Polygonum aviculare var. *eximium* (Lindman) Ascherson & Graebner; *P. aviculare* subsp. *heterophyllum* Ascherson & Graebner; *P. aviculare* subsp. *monspeliense* (Thiébaut-de-Berneaud ex Persoon) Arcangeli; *P. aviculare* var. *vegetum* Ledebour; *P. heterophyllum* Lindman; *P. monspeliense* Thiébaut-de-Berneaud ex Persoon

Plants green, heterophyllous (often lacking larger proximal stem leaves at maturity). Stems 1–3, ascending or erect, sometimes decumbent, basal branches divaricate, (10–)25–75(–100) cm. Leaves: ocrea (5.5–)6–10 (–14) mm, proximal part cylindric, distal part silvery, hyaline, soon lacerate and disintegrating into fibers; petiole 1–6(–8) mm; blade green, lateral veins visible but not raised abaxially, elliptic to oblanceolate, 18–50 (–60) × (6–)10–20 mm, 2–4.5 times as long as wide, apex acute or obtuse; stem leaves 1.4–4 times as long as branch leaves. Cymes aggregated at tips of stems and branches, 3–8-flowered. Pedicels mostly exserted from ocreae, 2–5 mm. Flowers: perianth (2.3–)2.8–4.7 (–5) mm, 1.8–2.8 times as long as wide; tube (15–)20–37% of perianth length; tepals overlapping, green with pink, red, or white margins, oblong, flat (cucullate in fruit), outer tepals not pouched at base; veins branched, thickened; stamens 7–8. Achenes enclosed in or barely exserted from perianth, brown to dark brown, ovate, 3-gonous, (2.1–)2.7–3.7 mm, faces subequal, concave, apex straight, striate-tubercled; late-season achenes uncommon, 3.5–5 mm. $2n = 40, 60$.

Flowering May–Oct. Fields, uncultivated areas, waste places, roadsides; 0–3500 m; introduced; St. Pierre and Miquelon; Alta., B.C., Man., N.B., Nfld. and Labr. (Nfld.), N.W.T., N.S., Ont., P.E.I., Que., Sask., Yukon; Ala., Alaska, Ariz., Ark., Calif., Colo., Conn., Del., D.C., Fla., Ga., Idaho, Ill., Ind., Iowa, Kans., La., Maine, Md., Mass., Mich., Minn., Mont., Nebr., Nev., N.H., N.J., N.Mex., N.Y., N.C., N.Dak., Okla., Oreg., Pa., R.I., S.C., S.Dak., Tenn., Tex., Utah, Vt., Va., Wash., Wyo.; Eurasia.

11b. Polygonum aviculare Linnaeus subsp. **boreale** (Lange) Karlsson, Svensk Bot. Tidskr. 91: 249. 1998

• Northern knotweed   W

Polygonum aviculare var. *boreale* Lange, Consp. Fl. Groenland. 1: 105. 1880; *P. arenastrum* Boreau subsp. *boreale* (Lange) Á. Löve; *P. boreale* (Lange) Small; *P. heterophyllum* Lindman subsp. *boreale* (Lange) Á. Löve & D. Löve

Plants green, homophyllous or subheterophyllous. Stems 1–7, prostrate to ascending, mostly branching from base, 6–50(–90) cm. Leaves: ocrea 3.5–7 mm, proximal part cylindric to funnelform, distal part soon disintegrating, nearly completely deciduous; petiole 7–9 mm; blade green, lateral veins visible but not raised abaxially, obovate-spatulate or oblanceolate, (12.5–)16–44(–55) × (4–)6–18(–22) mm, 2–4(–5.5) times as long as wide, apex obtuse to rounded; stem leaves (1.1–)1.3–2.5(–3) times as long as branch leaves. Cymes uniformly distributed, rarely crowded at tips of stems and branches, (3–)4–7-flowered. Pedicels mostly exserted from ocreae,

2–5 mm. **Flowers:** perianth 3.3–5.5 mm, 1.6–2.8 times as long as wide; tube 25–35(–39)% of perianth length; tepals overlapping, green with white or pink margins, obovate, flat or curved outward in fruit, outer tepals not pouched at base; veins branched, thickened; stamens 6–8. **Achenes** enclosed in perianth, dark brown, ovate, 3-gonous, (2.5–)2.7–4(–4.2) mm, faces subequal, concave, apex straight, coarsely striate-tubercled; late-season achenes uncommon, 4–5 mm. $2n = 40$.

Flowering Aug–Oct. Sandy, gravelly, or rocky coastal areas; 0–500 m; Greenland; Nfld. and Labr. (Labr.); nw Europe.

11c. Polygonum aviculare Linnaeus subsp. **buxiforme** (Small) Costea & Tardif, Sida 20: 988. 2003

• American knotweed, renouée faux-buis [E] [W]

Polygonum buxiforme Small, Bull. Torrey Bot. Club 33: 56. 1906

Plants gray-green or bluish green, rarely green, homophyllous or subheterophyllous. **Stems** 1–3, erect to ascending, ± unbranched, 5–15 cm, or numerous, procumbent, mat-forming, extensively branched, 20–70(–200) cm. **Leaves:** ocrea 3.5–6.5(–8) mm, proximal part cylindric, distal part silvery, relatively persistent, with inconspicuous veins, leaving almost no fibrous remains after disintegrating; petiole 0.3–2(–3.5) mm; blade green or gray-green, lateral veins visible but not raised abaxially, lanceolate to elliptic, oblanceolate, or obovate, 6–30(–45) × 3–6(–13) mm, 2.5–5.6(–10) times as long as wide, apex acute to obtuse; stem leaves 1–2.5 times as long as branch leaves. **Cymes** mostly uniformly distributed, but also aggregated at tips of stems and branches, 2–6-flowered. **Pedicels** mostly enclosed in ocreae, 1–2.5 mm. **Flowers:** perianth (2–)2.3–3.4(–3.6) mm, 0.9–1.3(–1.5) times as long as wide; tube 20–36% of perianth length; tepals overlapping, green with white or sometimes pink margins, oblong, apex cucullate, outer tepals pouched at base; veins branched, moderately to strongly thickened; stamens 7–8. **Achenes** usually enclosed in perianth, light brown to brown, ovate, 3-gonous, (1.8–)2–2.8(–3) mm, faces subequal, concave to flat, apex straight, coarsely striate-tubercled to obscurely tubercled; late-season achenes common, 2.5–5 mm. $2n = 60$.

Flowering Jul–Nov. Roadsides, vacant lots, sidewalks, packed and nondrifting sands, borders of marshes and dunes; 0–3500 m; Alta., B.C., Man., N.B., Nfld. and Labr., N.W.T., N.S., Ont., Que., Sask., Yukon; Ala., Alaska, Ariz., Ark., Calif., Colo., Conn., Del., D.C., Idaho, Ill., Ind., Iowa, Kans., Ky., La., Maine, Mich., Minn., Miss., Mo., Mont., Nebr., Nev., N.H., N.J., N.Mex., N.Y., N.C., N.Dak., Ohio, Okla., Oreg., Pa.,

R.I., S.C., S.Dak., Tenn., Tex., Utah, Vt., Va., Wash., W.Va., Wis., Wyo.

Although apparently it has a North American origin, subsp. *buxiforme* is considered part of the *Polygonum aviculare* complex because it intergrades with subsp. *aviculare* (M. Costea and F. J. Tardif 2003).

11d. Polygonum aviculare Linnaeus subsp. **neglectum** (Besser) Arcangeli, Comp. Fl. Ital., 583. 1882

• Narrow-leaf knotweed, renouée négligée [I] [W]

Polygonum neglectum Besser, Enum. Pl., 45. 1822; *P. aequale* Lindman subsp. *oedocarpum* Lindman; *P. aviculare* subsp. *rectum* Chrtek

Plants green, homophyllous or sometimes heterophyllous. **Stems** usually 1–7, procumbent to ascending, sometimes erect, mostly branched from base, (5–)15–60 cm. **Leaves:** ocrea (3–)4–8 mm, proximal part cylindric, distal part with inconspicuous veins, eventually disintegrating and leaving few or no fibrous remains; petiole (0.3–)1–3 (–5.2) mm; blade green, lateral veins visible but not raised abaxially, narrowly elliptic or oblanceolate, (7.6–)12.2–34(–40) × 1.5–6.8(–8) mm, (3.4–)4.2–9.2 times as long as wide, apex acute or obtuse; stem leaves 1–2.7(–3.3) times as long as branch leaves. **Cymes** uniformly distributed along stems and branches, 1–3(–5)-flowered. **Pedicels** mostly enclosed in ocreae, 1.5–3 mm. **Flowers:** perianth (1.9–)2.3–3.4 mm, 1.6–2.6 times as long as wide; tube 28–48% of perianth length; tepals overlapping, spreading slightly as achene matures, green usually with pink or red, rarely white, margins, oblong, ± cucullate, outer tepals not pouched at base; veins branched, moderately to strongly thickened; stamens 7–8. **Achenes** exserted from perianth, dark brown, ovate, 3-gonous, 1.2–1.8 mm, faces unequal or, less often, subequal, flat to concave, apex with straight edges or somewhat bent toward narrow face, striate-tubercled or, rarely, obscurely so; late-season achenes uncommon, 2–3.7 mm. $2n = 40, 60$.

Flowering Jun–Nov. Disturbed places; 0–1500 m; introduced; Alta., B.C., Man., N.B., Nfld. and Labr. (Nfld.), N.S., Ont., P.E.I., Que., Sask.; Ala., Alaska, Ariz., Ark., Calif., Conn., Ga., Ill., Ind., Kans., Ky., La., Maine, Md., Mass., Mich., Minn., Miss., Mo., Mont., Nebr., Nev., N.H., N.J., N.Mex., N.Y., N.C., N.Dak., Ohio, Okla., Oreg., Pa., R.I., S.C., S.Dak., Tenn., Tex., Utah, Vt., Va., Wash., W.Va., Wyo.; Europe.

Subspecies *neglectum* has been reported from Colorado, Idaho, Iowa, and Wisconsin; those reports have not been confirmed.

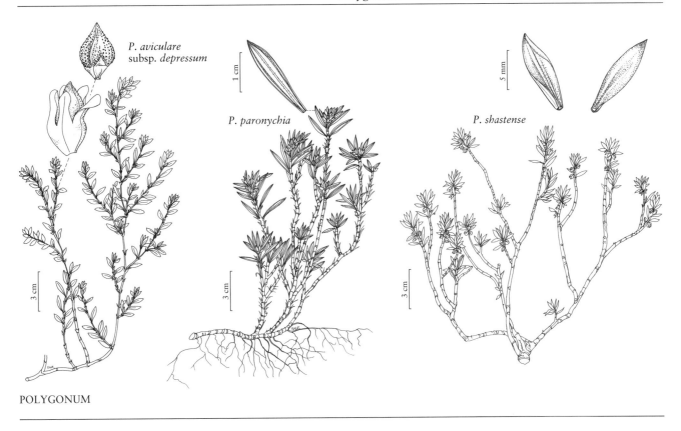

P. aviculare
subsp. *depressum*

P. paronychia

P. shastense

POLYGONUM

11e. Polygonum aviculare Linnaeus subsp. **depressum** (Meisner) Arcangeli, Comp. Fl. Ital., 583. 1882

• Common or oval-leaf knotweed, renouée à petits fruits [F] [I] [W]

Polygonum aviculare var. *depressum* Meisner in A. P. de Candolle and A. L. P. P. de Candolle, Prodr. 14: 98. 1856; *P. aequale* Lindman; *P. arenastrum* Boreau; *P. aviculare* subsp. *aequale* (Lindman) Ascherson & Graebner; *P. aviculare* subsp. *calcatum* (Lindman) Thellung; *P. aviculare* subsp. *microspermum* (Jordan ex Boreau) Berher; *P. calcatum* Lindman; *P. microspermum* Jordan ex Boreau; *P. montereyense* Brenckle

Plants green, homophyllous or subheterophyllous. **Stems** 3–15, prostrate to ascending, mat-forming, branched at most nodes, proximal branches divaricate, (5–)10–50 (–100) cm. **Leaves:** ocrea with proximal part cylindric to funnelform, either short, 3–5.5 mm, with distal part soon disintegrating and leaving almost no fibrous remains, or long, 7–12 mm, with distal part with veins inconspicuous, membranous, margins ± lacerate, overlapping towards apices of stems and branches; petiole 0.5–3 mm; blade green, lateral veins visible but not raised abaxially, elliptic to narrowly elliptic or oblanceolate, (6.2–)8–27(–35) × (1.4–)2–7(–10) mm, 2.8–5.7(–6.5) times as long as wide, apex obtuse or acute, rarely longitudinally striate; stem leaves 1–2.3(–3.4) times as long as branch leaves. **Cymes** uniformly distributed or, sometimes, crowded at tips of branches, 2–7-flowered. **Pedicels** enclosed in ocreae, 1–2.5 mm. **Flowers:** perianth (1.8–)2–3.4(–4) mm, 1.5–2.9 times as long as wide; tube 40–57% of perianth length; tepals overlapping but spreading slightly in fruit, green or reddish brown with white margins, oblong, flat or obscurely cucullate in fruit, outer tepals not pouched at base; midveins unbranched, thin to moderately thickened; stamens 5–7. **Achenes** usually slightly exserted from perianth, dark brown, ovate, (2–)3-gonous, 1.5–2.7(–3) mm, faces evidently unequal, flat to concave, apex straight or slightly bent toward narrow face, almost smooth, roughened, or coarsely striate-tubercled; late-season achenes common, 2.5–4.5 mm. $2n = 40, 60$.

Flowering May–Nov. Fields, uncultivated areas, waste places, roadsides; 0–2500 m; introduced; Alta., B.C., Man., N.B., Nfld. and Labr., N.W.T., N.S., Ont., P.E.I., Que., Sask.; Ala., Alaska, Ariz., Ark., Calif., Colo., Conn., Del., Fla., Ga., Idaho, Ill., Ind., Iowa, Kans., Ky., La., Maine, Md., Mass., Mich., Minn., Mo., Mont., Nebr., Nev., N.H., N.J., N.Mex., N.Y., N.C., N.Dak., Ohio, Okla., Oreg., Pa., R.I., S.C., S.Dak., Tenn., Tex., Utah, Vt., Va., Wash., W.Va., Wis., Wyo.; Europe.

Polygonum montereyense is a distinctive morphotype that may deserve infraspecific recognition. Plants referable to it have ocreae 8–13 mm, with distal parts silvery, persistent, entire or slightly lacerate, overlapping, and few-veined. Apparently, this type is restricted to California. Plants referable to *P. arenastrum* in the narrow sense are the most commonly encountered form of the subspecies in North America.

Subspecies *depressum* has been reported from Greenland and St. Pierre and Miquelon; those reports have not been confirmed.

11f. **Polygonum aviculare** Linnaeus subsp. **rurivagum** (Jordan ex Boreau) Berher, Fl. Vosges, 195. 1887

• Narrow-leaf knotweed [I] [W]

Polygonum rurivagum Jordan ex Boreau, Fl. Centre France ed. 3, 2: 560. 1857; *P. aviculare* var. *angustissimum* Meisner; *P. heterophyllum* var. *angustissimum* (Meisner) Lindman; *P. heterophyllum* subsp. *rurivagum* (Jordan ex Boreau) Lindman

Plants green, homophyllous to heterophyllous. **Stems** 1–10, procumbent to ascending, branched at most nodes, proximal branches divaricate, not wiry, 10–40 cm. **Leaves:** ocrea (6–)8–12 mm, proximal part cylindric, distal part with veins conspicuous, soon disintegrating into persistent fibers; petiole 0.3–2 mm; blade green, lateral veins strongly raised adaxially, narrowly elliptic to linear-lanceolate, (10–)15–27(–30) × 0.5–4.8(–8) mm, (4.5–)5–15(–19) times as long as wide, apex acute; stem leaves 1.2–2.4 times as long as branch leaves. **Cymes** uniformly distributed along stems and branches, 1–3-flowered. **Pedicels** enclosed in or exserted from ocreae, 1.5–3 mm. **Flowers:** perianth 2.2–3.1 mm, 1.6–2.6 times as long as wide; tube 26–40(–42)% of perianth length; tepals often not overlapping, green with pink or red margins, oblong, ± cucullate, outer tepals not pouched at base; veins branched, moderately to strongly thickened; stamens 7–8. **Achenes** usually exserted from perianth, blackish brown, ovate, 3-gonous, 2.1–2.6(–3) mm, faces subequal or unequal, flat to concave, apex straight, coarsely striate-tubercled; late-season achenes uncommon, 2.5–4 mm. $2n = 60$.

Flowering Jul–Oct. Disturbed places; 0–1000; introduced; Alta., Man., N.B., Nfld. and Labr. (Nfld.), Que., Sask.; Calif., Ind., Iowa, Maine, Mass., N.Y., R.I., S.Dak., W.Va.; Europe.

12. **Polygonum patulum** M. Bieberstein, Fl. Taur.-Caucas. 1: 304. 1808 [I]

Plants green or bluish green, heterophyllous. **Stems** mostly ascending or erect, branched from base, not wiry, 20–80 cm. **Leaves:** ocrea 7–9 mm, proximal part cylindic, distal part disintegrating into straight fibers; petiole 0.2–1 mm; blade green or bluish green, linear to lanceolate, 25–40 × 4–8 mm, margins flat, apex acute; stem leaves 2–4 times as long as branch leaves; distal leaves abruptly reduced and not overtopping flowers (shorter than or equaling flowers). **Inflorescences** axillary and terminal, spikelike; cymes aggregated at tips of stems and branches, 1–3-flowered. **Pedicels** enclosed in ocreae, 1.5–2 mm. **Flowers** closed or semi-open; perianth 2.2–3 mm; tube 15–30% of perianth length; tepals overlapping, green with white to pink margins, petaloid, not keeled, oblong, cucullate; veins branched; stamens 8. **Achenes** slightly exserted from perianth, light brown to dark brown, ovate, 3-gonous, 2–2.3(–2.8) mm, faces subequal, concave, apex not beaked, edges concave, dull, striate-tubercled; late-season achenes uncommon, 2.5–4 mm.

Flowering Jul–Oct. Waste places; 0–800 m; introduced; Ala., Ill., Wash.; Eurasia; n Africa.

The name *Polygonum patulum* has been misapplied to a distinctive form of *P. ramosissimum* in saline marshes in California (M. Costea and F. J. Tardif 2003b).

13. **Polygonum argyrocoleon** Steudel ex Kunze, Linnaea 20: 17. 1847 • Persian or silversheath knotweed [I] [W]

Plants green, heterophyllous. **Stems** erect or decumbent, branched mostly from base, not wiry, 15–100 cm. **Leaves:** ocrea 4–8 mm, proximal part cylindric, distal part soon disintegrating into curly or straight fibers; petiole 0–1.5 mm; blade bluish green, lanceolate or linear-lanceolate, 15–50 × 2–8 mm, margins flat, apex acute; stem leaves 2.1–4 times as long as branch leaves; distal leaves abruptly reduced and not overtopping flowers (shorter than or equaling flowers). **Inflorescence** axillary and terminal, spikelike; cymes aggregated at tips of stems and branches, 4–6-flowered, bracts inconspicuous. **Pedicels** enclosed in ocreae, 1–2 mm. **Flowers** closed; perianth 1.8–2.4 mm; tube 10–22% of perianth length; tepals overlapping, green or white, usually with pink, rarely red or white, margins, petaloid, not keeled, oblong to obovate, cucullate; midveins usually unbranched; stamens 7–8. **Achenes** enclosed in perianth, brown, ovate, 3-gonous, 1.3–2.3 mm, faces subequal, slightly concave, apex not beaked, with concave edges, shiny, smooth; late-season achenes unknown.

Flowering May–Oct. Fields, gardens, disturbed sites, often in saline soils; 0–1200 m; introduced; Ariz., Calif., Colo., Fla., Idaho, La., Mass., Mo., Nev., N.Mex., N.C., Tex., Utah, Vt., W.Va.; c Asia.

30b. **Polygonum** Linnaeus sect. **Duravia** S. Watson, Amer. Naturalist 7: 665. 1873   E

Duravia (S. Watson) Greene; *Polygonum* sect. *Monticola* J. C. Hickman

Shrubs, subshrubs, or annual **herbs,** sometimes compact and cushionlike (in *P. parryi, P. hickmanii,* and *P. heterosepalum*), homophyllous, homocarpic, not succulent, not glaucous. **Stems** erect or ascending, rarely prostrate, straight or zigzagged, 4-gonous, ribs obscure or absent, smooth or papillose-scabridulous. **Leaves:** ocrea 4–8-veined, proximal part not pruinose; petiole (or blade when petiole absent) articulated or not to ocrea; blade narrowly linear to subround, rarely coriaceous (*P. paronychia, P. shastense,* and *P. utahense*), margins smooth or papillose-denticulate, not rugulose, rarely glaucous adaxially (*P. nuttallii*); 1-veined or venation parallel (3-veined), secondary veins not conspicuous. **Inflorescences** usually axillary and terminal, sometimes entirely axillary; cymes 1–6-flowered. **Pedicels** erect to spreading or reflexed, 0.1–6 mm, sometimes absent. **Flowers** open, semi-open, or closed; tepals ± monomorphic, outer tepals equaling or somewhat larger than inner, (dimorphic, with outer tepals shorter than inner in *P. heterosepalum*), apices of outer tepals rounded or acute to acuminate; anthers pink to purple (orange-pink in *P. hickmanii*). **Achenes** ovate to elliptic or lanceolate, 3-gonous, shiny or dull, smooth, tubercled, or longitudinally reticulate.

Species 20 (20 in the flora): mostly w North America.

Section *Duravia* is heterogeneous and future studies may reveal the necessity of further subdivision. Seven species referred by J. C. Hickman (1984) to sect. *Monticola* (*Polygonum cascadense, P. heterosepalum, P. hickmanii, P. majus, P. polygaloides, P. sawatchense,* and *P. utahense*) are here included in sect. *Duravia* based on leaf, anther, achene, and pollen morphology. Heterocarpy does not occur in sect. *Duravia*.

SELECTED REFERENCE Costea, M. and F. J. Tardif. 2005. Taxonomy of the *Polygonum douglasii* (Polygonaceae) complex with a new species from Oregon. Brittonia 57: 1–27.

1. Plants shrubs or subshrubs.
  2. Stems erect, wiry; perianth 2.6–3.2 mm . . . . 19. *Polygonum bolanderi*
  2. Stems prostrate or ascending, not wiry; perianth (4.5–)5–10 mm.
    3. Ocreae 15–20 mm, distal parts persistent; w coastal dunes . . . . 14. *Polygonum paronychia*
    3. Ocreae 3–5 mm, distal parts deciduous; montane habitats . . . . 15. *Polygonum shastense*
1. Plants annual herbs.
  4. Leaves not articulated to ocreae, blades mostly linear, rarely linear-lanceolate.
    5. Distal parts of ocreae entire or shallowly dentate . . . . 21. *Polygonum bidwelliae*
    5. Distal parts of ocreae disintegrating into fibers.
      6. Plants not compact, not cushionlike; stems simple or divaricately branched, 4–40 cm . . . . 20. *Polygonum californicum*
      6. Plants compact, often cushionlike; stems simple or branched from base, 1.5–5(–8) cm.
        7. Distal parts of ocreae with curly fibers . . . . 16. *Polygonum parryi*
        7. Distal parts of ocreae with straight fibers.

8. Tepals ± monomorphic, outer longer than inner . . . . 17. *Polygonum hickmanii*
8. Tepals dimorphic, outer shorter than inner . . . . 18. *Polygonum heterosepalum*
[4. Shifted to left margin—Ed.]
4. Leaves articulated to ocreae, blades subulate, linear-lanceolate to subround.
  9. Apices of tepals acute to acuminate; achenes light yellow, light brown, or greenish to
    dark brown, smooth or with longitudinal ridges . . . . 22. *Polygonum polygaloides*
  9. Apices of tepals rounded; achenes black, smooth or minutely tubercled.
    10. Pedicels reflexed.
      11. Ocreae 5–12 mm; perianths and achenes 3–5 mm.
        12. Flowers open or semi-open; pedicels 0.5–1 mm; perianth tubes 9–17% of
          perianth lengths . . . . 31. *Polygonum majus*
        12. Flowers closed; pedicels 2–6 mm; perianth tubes 20–28% of perianth lengths
          . . . . 24. *Polygonum douglasii*
      11. Ocreae 3–5 mm; perianths and achenes 1.2–2.6 mm.
        13. Achenes usually exserted from perianth; leaf blades linear-oblanceolate,
          margins smooth . . . . 26. *Polygonum engelmannii*
        13. Achenes enclosed in perianth; leaf blades ovate to elliptic or ovate, margins
          papillose-denticulate . . . . 27. *Polygonum austiniae*
    10. Pedicels erect or erect to spreading.
      14. Leaf blades with 1 pleat on each side of midrib . . . . 23. *Polygonum tenue*
      14. Leaf blades without pleats.
        15. Ocreae 4–12 mm, distal parts disintegrating into persistent fibers; peri-
          anths (2.8–)3–5 mm; achenes 2.5–5 mm.
          16. Inflorescences dense, cymes ± overlapping at branch tips. . . . .
            . . . . 30. *Polygonum spergulariiforme*
          16. Inflorescences elongate, cymes widely spaced along branches . . . . .
            . . . . 25. *Polygonum sawatchense*
        15. Ocreae 1–5 mm, distal parts entire to lacerate; perianths 1.8–2.5 mm;
          achenes 1.5–2.3 mm.
          17. Stems 1.5–3.5 cm; leaf blades with margins revolute, touching along
            midrib . . . . 33. *Polygonum utahense*
          17. Stems 2–35 cm; leaf blades with margins flat, or if revolute then never
            touching along midrib.
            18. Flowers open . . . . 32. *Polygonum cascadense*
            18. Flowers semi-open or closed.
              19. Leaf blades narrowly elliptic, elliptic, ovate, obovate, or sub-
                round, green adaxially, margins flat . . . . 28. *Polygonum minimum*
              19. Leaf blades linear to narrowly oblong-elliptic, ± glaucous
                adaxially, margins revolute . . . . 29. *Polygonum nuttallii*

14. Polygonum paronychia Chamisso & Schlechtendal, Linnaea 3: 51. 1828 • Beach or black or dune knotweed E F

Shrubs or subshrubs. Stems prostrate or ascending, brown, branched, rooting at nodes, not wiry, 10–100 cm, glabrous, covered with remains of lacerate, hyaline ocreae. **Leaves** crowded at branch tips, articulated to ocreae, basal leaves caducous or persistent, distal leaves not reduced in size; ocrea 15–20 mm, glabrous, proximal part cylindric to funnelform, distal part silvery, entire or slightly lacerate, disintegrating into persistent white-gray curly fibers; petiole 0–0.5 mm; blade 1-veined, without pleats, linear to oblanceolate, (5–)10–20(–33) × 3–8 mm, coriaceous, margins revolute, smooth, apex acute or mucronate. **Inflorescences** axillary; cymes crowded in distal axils, 2–5-flowered. **Pedicels** enclosed in ocreae, erect to spreading, 2–5 mm. **Flowers** semi-open or open; perianth (4.5–)6–10 mm; tube 22–48% of perianth length; tepals partially overlapping, uniformly pink or white, reddish brown when dried, petaloid, oblong-ovate to ± lanceolate, apex rounded; midveins pinnately branched; stamens 8. **Achenes** enclosed in or slightly exserted from perianth, black, ovate, 4–5 mm, faces subequal, shiny, smooth.

Flowering Mar–Sep. Coastal sands, scrub along coast; 0–50 m; B.C.; Calif., Oreg., Wash.

Polygonum paronychia may be cultivated in rock gardens in open sites with sandy soil.

15. Polygonum shastense W. H. Brewer, Proc. Amer. Acad. Arts 8: 400. 1872 • Shasta knotweed E F

Subshrubs. Stems prostrate to ascending, brown, branched, gnarled, not wiry, 5–40 cm, glabrous. **Leaves** ± uniformly distributed, articulated to ocreae, basal leaves caducous or persistent, distal leaves not reduced in size; ocrea 3–5 mm, glabrous, proximal part cylindric, distal part membranous, deciduous with leaves; petiole 0–0.5 mm; blade 1-veined, without pleats, lanceolate to elliptic, 5–25 × 2–5 mm, coriaceous, margins revolute, smooth, apex acute. **Inflorescences** axillary; cymes in distal axils, 2–6-flowered. **Pedicels** enclosed in or slightly exserted from ocreae, erect, 2–4 mm. **Flowers** open or semi-open; perianth 5–9 mm; tube 7–15% of perianth length, tepals partially overlapping, uniformly pink or white, petaloid, ovate to ovate-round, apex rounded; midveins with numerous branched lateral veins; stamens 8. **Achenes** enclosed in or slightly exserted from perianth, brown, narrowly ovate, 3–4 mm, faces subequal, shiny, smooth.

Flowering Jul–Sep. Rocky or gravelly slopes; 2100–3400 m; Calif., Nev., Oreg.

Polygonum shastense may be cultivated in rock gardens in rock crevices with favorable water regime and shaded in summer.

16. Polygonum parryi Greene, Bull. Torrey Bot. Club 8: 99. 1881 • Parry's or prickly knotweed E

Herbs, compact, often cushion-like. **Stems** erect, green-brown, simple or branched from base, not wiry, 2–5(–8) cm, glabrous. **Leaves** ± uniformly distributed, dense, not articulated to ocreae, basal leaves ± persistent, distal leaves gradually reduced to bracts; ocrea 2–4(–5) mm, glabrous, proximal part cylindric, distal part deeply lacerate, disintegrating into white, curled fibers; petiole absent; blade 3-veined, without pleats, linear-lanceolate, subulate, 5–13(–20) × 0.4–1 mm, margins revolute, smooth, apex spine-tipped. **Inflorescences** axillary; cymes in most axils, 1-flowered. **Pedicels** absent. **Flowers** closed; perianth 1.5–2(–2.5) mm; tube 6–15% of perianth length; tepals overlapping, usually reddish with white margins, petaloid, oblong, navicular, apex acute; midveins unbranched; stamens 8. **Achenes** slightly exserted from perianth at maturity, dark brown, ovate, 1.2–1.6(–2) mm, faces subequal, shiny, smooth.

Flowering May–July. Vernally moist, open, sandy or rocky places; 500–2000 m; Calif., Wash.

17. Polygonum hickmanii H. R. Hinds & Rand. Morgan, Novon 5: 336. 1995 • Hickman's knotweed C E F

Herbs, compact, often cushion-like. **Stems** erect, color unknown, simple to profusely branched from near base, not wiry, 2–5 cm, glabrous. **Leaves** persistence and crowded at brance tips, not articulated to ocreae; ocrea 4–6 mm, glabrous, proximal part cylindric, distal part silvery, disintegrating nearly to base into straight fibers; petiole absent; blade 3-veined, without pleats, linear, 5–35 × 1–1.5 mm, margins revolute, smooth, apex acuminate. **Inflorescences** axillary; cymes insertion unknown, 1-flowered. **Pedicels** absent. **Flowers** closed; perianth 2–3 mm; tube 6–18 % of perianth length, tepals imbrication unknown, white with whitish or pink margins, petaloid, oblong, apex acute, mucronate; midveins unbranched; stamens 8; anthers orange-pink. **Achenes** enclosed in perianth, olive brown, ovate, 2–2.3 mm, faces subequal, shiny, smooth.

Flowering May–Oct. Open, seasonally dry grasslands; of conservation concern; 200–300 m; Calif.

The above description is based on the original one by Hinds and Morgan. *Polygonum hickmanii* is known only from the northern end of Scotts Valley in Santa Cruz County.

18. Polygonum heterosepalum M. Peck & Ownbey, Madroño 10: 250. 1950 • Dwarf desert knotweed C E

Herbs, compact, often cushion-like. **Stems** erect, green or reddish, simple or branched near base, not wiry, 1.5–5 cm, glabrous. **Leaves** uniformly distributed, dense, not articulated to ocreae, basal leaves persistent or caducous, distal leaves gradually reduced to bracts; ocrea 3–6 mm, glabrous, proximal part cylindric, distal part disintegrating almost to base, with whitish, straight, rigid fibers; petiole absent; blade 3-veined, without pleats, linear to lanceolate, 10–20 × 0.6–2.7 mm, margins revolute, smooth, apex spine-tipped. **Inflorescences** axillary; cymes in most axils, 2–3-flowered. **Pedicels** enclosed in ocreae, erect, 0.1–1 mm. **Flowers** closed; perianth 2.3–2.7 mm; tube 3–7% of perianth length; tepals overlapping, whitish with whitish or pink margins, petaloid, oblong, navicular, dimorphic, outer 2 shorter than inner 3, outer 2 0.8–1.2 mm, inner 2.3–2.7 mm, papillose at base, apex acute or acuminate; midveins unbranched; stamens 5–6. **Achenes** enclosed

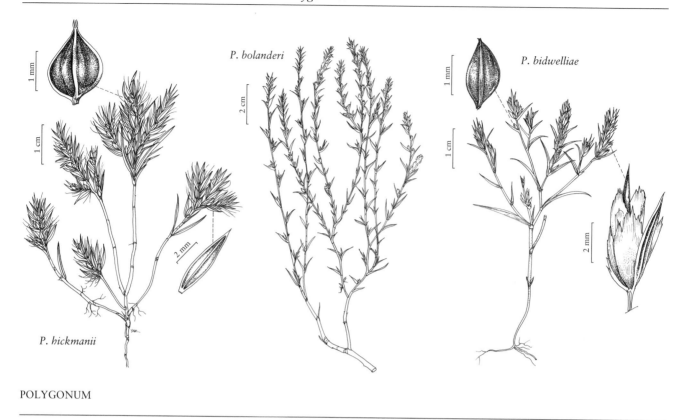

P. bolanderi

P. bidwelliae

P. hickmanii

POLYGONUM

in perianth, olive brown, narrowly ovate to ovate-lanceolate, 1.5–2 mm, faces subequal, shiny, smooth.

Flowering Jun–Aug. Dry waste ground, open flats in sagebrush plains, ponderosa pine forests; of conservation concern; 1000–1500 m; Idaho, Nev., Oreg.

19. Polygonum bolanderi W. H. Brewer, Proc. Amer. Acad. Arts 8: 400. 1872 • Bolander's knotweed E F

Duravia bolanderi (W. H. Brewer) Greene

Shrubs. Stems erect, brown, simple, wiry, gnarled with age, 20–60 cm, glabrous. **Leaves** crowded at branch tips, not articulated to ocreae, basal leaves caducous or persistent, distal leaves abruptly reduced to bracts; ocrea 6–10 mm, glabrous or papillose-scabridulous, proximal part cylindric, distal part silvery, deeply fringed, disintegrating; petiole absent; blade 3-veined, without pleats, linear to subulate, 3–15(–25) × 0.4–1.5 mm, margins flat, smooth, apex acuminate to spine-tipped. **Inflorescences** axillary; cymes in distal axils, 1–2-flowered. **Pedicels** absent. **Flowers** semi-open; perianth 2.6–3.2 mm; tube 18–33% of perianth length; tepals overlapping, ± recurved, uniformly white to pink, petaloid, elliptic-oblong to obovate, navicular, apex

rounded; midveins unbranched; stamens 8. **Achenes** enclosed in perianth, light brown, lanceolate to oblong-ovate, 2.5–3 mm, faces subequal, shiny, smooth.

Flowering Jun–Nov. Open, dry, gravelly, rocky places; 300–1500 m; Calif.

Polygonum bolanderi is known from several counties in northwestern California (Butte, Napa, Shasta, and Sonoma).

20. Polygonum californicum Meisner in A. P. de Candolle and A. L. P. P. de Candolle, Prodr. 14: 100. 1856 E

Duravia californica (Meisner) Greene; *Polygonum greenei* S. Watson

Herbs. Stems erect, green, simple or divaricately branched, ± wiry, 4–40 cm, papillose-scabridulous. **Leaves** uniformly distributed, not articulated to ocreae, basal leaves usually caducous, distal leaves abruptly reduced to bracts; ocrea 5–10 mm, glabrous or papillose-scabridulous, proximal part cylindric, distal part white or tawny, disintegrating into ± bristly-fringed fibers; petiole absent; blade 3-veined, without pleats, linear, 5–25(–30) × 0.5–2 mm, margins revolute, papillose-denticulate or smooth, apex mucronate or weakly spine-tipped. **Inflorescences** axillary; cymes in distal axils, 1-flowered. **Pedicels** absent. **Flowers** open or closed; perianth 2.5–3.5 mm; tube 10–20% of perianth length; tepals overlapping, uniformly white to pink, petaloid, elliptic, navicular, apex acute to acuminate; midveins unbranched; stamens 8. **Achenes** enclosed in or slightly exserted from perianth, brown, narrowly elliptic, 1.8–2.2 mm, faces subequal, shiny, smooth.

Flowering May–Sep. Open places, including serpentine; 40–1200 m; Calif., Oreg.

21. **Polygonum bidwelliae** S. Watson, Proc. Amer.
 Acad. Arts 14: 294. 1879 • Bidwell's knotweed
 E F

Herbs. Stems erect, green, simple or divaricately branched, ± wiry, 2–20 cm, minutely papillose-scabridulous. **Leaves** crowded at branch tips, not articulated to ocreae, basal leaves caducous or persistent, distal leaves abruptly reduced to bracts; ocrea 8–13 mm, papillose-scabridulous, proximal part cylindric, distal part overlapping and obscuring leaves and flowers, silvery, entire or shallowly dentate; petiole absent; blade 3-veined, without pleats, linear, 5–15(–20) × 0.5–1.5(–2) mm, margins revolute, papillose-denticulate, apex spine-tipped. **Inflorescences** axillary; cymes in distal axils, 1-flowered. **Pedicels** absent. **Flowers** closed; perianth 2–3 mm; tube 10–18% of perianth length; tepals overlapping, pink with pink or white margins, petaloid, elliptic, navicular, apex acute to acuminate; midveins unbranched; stamens 8. **Achenes** enclosed in perianth at maturity, light brown to brown, ovate-elliptic, 1.8–2.3 mm, faces subequal, shiny, smooth.

Flowering May–Jun. Thin volcanic soils, chaparral, montane woodland valleys, grasslands; 60–1200 m; Calif.

Polygonum bidwelliae occurs in the Cascade Range and northeastern Sacramento Valley in Butte, Shasta, and Tehama counties.

22. **Polygonum polygaloides** Meisner in A. P. de
 Candolle and A. L. P. P. de Candolle, Prodr. 14: 101.
 1856 • Polygala knotweed E F

Herbs. Stems erect, green, usually divaricately branched, rarely simple, ± wiry, (2–)6–20(–25) cm, glabrous. **Leaves** uniformly distributed, articulated to ocreae, basal leaves often caducous, distal leaves abruptly reduced to bracts; ocrea 4–8 mm, glabrous, proximal part cylindric, distal part silvery, with inconspicuous veins, lacerate; petiole absent;

blade 3-veined, lateral veins sometimes inconspicuous, without pleats, narrowly linear, 10–40 × 1–2.5 mm, margins ± revolute, smooth, apex acute or mucronate. **Inflorescences** axillary and terminal, spikelike, subglobose to cylindric; cymes in most axils or crowded distally, 1–3-flowered. **Pedicels** enclosed in ocreae, erect, 0.1–2 mm, sometimes absent. **Flowers** mostly closed; perianth 1.5–3 mm; tube 19–40% of perianth length; tepals overlapping, uniformly white, pink, or red, petaloid, oblong-lanceolate, ± navicular, apex acute to acuminate; midveins usually unbranched or with 2 lateral branches proximally, moderately to strongly thickened, tepals appearing ± keeled; stamens 3–8. **Achenes** enclosed in perianth, light yellow, light brown, or greenish brown to dark brown, ovate to lanceolate, 1.3–2.5 mm, faces subequal, shiny or dull, smooth or reticulate with longitudinal ridges.

Subspecies 4 (4 in the flora): w North America.

J. C. Hickman's (1993c) treatment of the *Polygonum polygaloides* complex is provisionally accepted here. Most of the intermediate specimens occur between subspp. *confertiflorum*, *esotericum*, and *kelloggii*. Alternatively, *P. polygaloides* could be recognized in the narrow sense and the three other taxa could be treated as subspecies of a separate *P. kelloggii*, the earliest available binomial.

1. Margins of bracts green, if white then scarious border less than 0.2 mm wide.
  2. Achenes lanceolate, 2–2.5 mm; bracts elliptic-lanceolate, appressed, rigid; stamens 5–8 . . . . 22c. *Polygonum polygaloides* subsp. *esotericum* (in part)
  2. Achenes ovate, 1.3–1.7 mm; bracts linear to linear-lanceolate, ± patent, soft; stamens 3 . . . . . 22d. *Polygonum polygaloides* subsp. *kelloggii*
1. Margins of bracts white, scarious border (0.2–) 0.25–1 mm wide.
  3. Inflorescences narrowly cylindric, continuous from bases of branches or stems; achenes 2–2.5 mm, lanceolate . . 22c. *Polygonum polygaloides* subsp. *esotericum* (in part)
  3. Inflorescences round to ovate or cylindric, confined to tips of branches, rarely continuous from bases of branches or stems; achenes 1.3–2.1 mm, ovate-lanceolate to ovate.
    4. Inflorescences round to ovate; scarious borders of bracts 0.5–1 mm wide; stamens 8 . . . . 22a. *Polygonum polygaloides* subsp. *polygaloides*
    4. Inflorescences ovate to cylindric; scarious borders of bracts 0.25–0.4 mm wide; stamens 3 . . . . 22b. *Polygonum polygaloides* subsp. *confertiflorum*

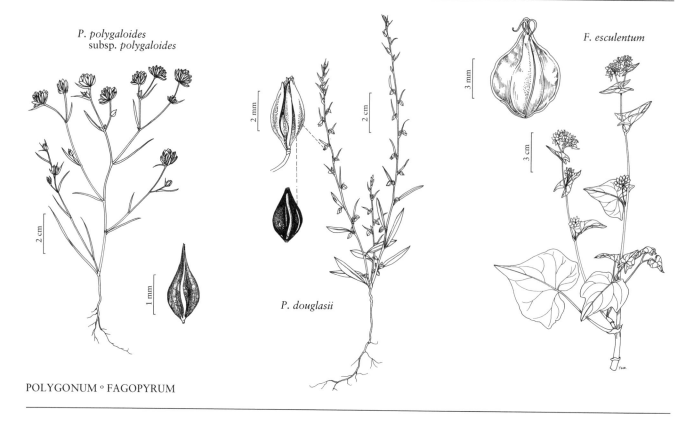

P. polygaloides
subsp. *polygaloides*

P. douglasii

F. esculentum

POLYGONUM ○ FAGOPYRUM

22a. Polygonum polygaloides Meisner subsp. **polygaloides** E F

Polygonum polygaloides var. *montanum* Brenckle

Stems 5–25 cm. **Inflorescences** mostly confined to tips of branches, globose to ovoid, 7–14 × 8–15 mm; bracts ascending or appressed, oblong to broadly ovate or elliptic, 4–9 mm, ± rigid, margins flat, undulate, or sometimes revolute, white, scarious portion 0.5–1 mm wide; midveins thin, lateral veins thickened and prominent adaxially. **Flowers:** perianth 2.5–3 mm; tube 19–30% of perianth length; tepals white or pink; perianth tube and base of tepals often ± crisped or undulate; stamens 8. **Achenes** dark brown, ovate-lanceolate, 1.3–2.1 mm, reticulate with longitudinal ridges, dull.

Flowering May–Aug. Dry beds of ponds and watercourses, roadsides, drainage pits; 500–1800 m; Alta.; Idaho, Mont., Oreg., Wash., Wyo.

22b. Polygonum polygaloides Meisner subsp. **confertiflorum** (Nuttall ex Piper) J. C. Hickman, Madroño 31: 251. 1984   E

Polygonum confertiflorum Nuttall ex Piper, Contr. U.S. Natl. Herb. 11: 228. 1906; *P. kelloggii* Greene var. *confertiflorum* (Nuttall ex Piper) Dorn; *P. watsonii* Small

Stems 2–15 cm. **Inflorescences** mostly confined to tips of branches, rarely interrupted and continuous from bases of branches or stems, ovoid to cylindric, 7–40 × 5–10 mm; bracts ascending to appressed, linear to linear-lanceolate, 4–11 mm, moderately rigid to rigid, margins revolute, white, scarious portion 0.25–0.4 mm wide; midvein moderately to heavily thickened, lateral veins inconspicuous. **Flowers:** perianth 1.8–2.2(–2.4) mm; tube 22–30% of perianth length; tepals white or red; perianth tube and base of tepals smooth or, sometimes, papillose; stamens 3. **Achenes** brown to dark brown, ovate-lanceolate to ovate, 1.5–2 mm, reticulate with longitudinal ridges, dull.

Flowering May–Aug. Vernal pools, wet meadows; 500–1900 m; Alta., B.C., Sask.; Ariz., Calif., Colo., Idaho, Mont., Nev., N.Mex., Oreg., Utah, Wash., Wyo.

22c. Polygonum polygaloides Meisner subsp. **esotericum** (L. C. Wheeler) J. C. Hickman, Madroño 31: 251. 1984 • Modoc County knotweed E

Polygonum esotericum L. C. Wheeler, Rhodora 40: 310, fig. 1. 1938

Stems 5–12 cm. **Inflorescences** continuous from bases of branches or stems, narrowly cylindric, 2–10 × 5–7 mm; bracts appressed, elliptic-lanceolate, 4–8 mm, rigid, margins flat, green, if white then scarious portion 0.1–0.2 mm wide; lateral and midveins prominent adaxially. **Flowers:** perianth 2.1–2.9 mm; tube 29–40% of perianth length; tepals white or pink; perianth tube and base of tepals smooth; stamens 5–8. **Achenes** dark brown, lanceolate, 2–2.5 mm, reticulately ridged to almost smooth, dull.

Flowering Jun–Aug. Vernal pools, seasonally wet places, desert scrub, pinyon and juniper woodlands; 1400–1600 m; Calif., Oreg.

22d. Polygonum polygaloides Meisner subsp. **kelloggii** (Greene) J. C. Hickman, Madroño 31: 251. 1984 • Kellogg's knotweed E

Polygonum kelloggii Greene, Fl. Francisc., 134. 1891; *P. minutissimum* L. O. Williams; *P. unifolium* Small ex Rydberg

Stems 2–15 cm. **Inflorescences** mostly terminal, prolonged almost to base of stems, ovoid, 3–15 × 5–15 mm, or ± continuous; bracts ± patent, linear to linear-lanceolate, 7–25 mm, soft, margins flat or revolute, green, not scarious; lateral and midveins inconspicuous. **Flowers:** perianth 1.5–2.3 mm; tube 19–34% of perianth length; tepals pink or white; perianth tube and base of tepals smooth; stamens 3. **Achenes** light yellow to greenish brown, ovate, 1.3–1.7 mm, smooth to obscurely longitudinally reticulate, shiny.

Flowering Jun–Sep. Mountain meadows, seeps, vernal pools, dry subalpine slopes; 1300–3300 m; B.C.; Ariz., Calif., Colo., Idaho, Mont., Nev., N.Mex., Oreg., Utah, Wash., Wyo.

23. Polygonum tenue Michaux, Fl. Bor.-Amer. 1: 238. 1803 • Slender or pleat-leaf knotweed E

Polygonum tenue var. *protrusum* Fernald

Herbs. Stems erect, green or brown-ish, simple or branched from below middle, not wiry, 5–50 cm, glabrous or papillose-scabridulous. **Leaves** uniformly distributed, articulated to ocreae, basal leaves caducous or persistent, distal leaves abruptly reduced to bracts; ocrea 6–15 mm, glabrous or papillose-scabridulous, proximal part cylindric, distal part soon disintegrating into a few brown fibers; petiole 0.1–1 mm; blade 1-veined, with 1 pleat on each side of midrib, narrowly lanceolate to linear, 25–40 × 1–8 mm, margins usually flat, papillose-denticulate, apex mucronate or cuspidate. **Inflorescences** axillary and terminal, spikelike, slender, elongate; cymes spaced along branches, 1–2(–3)-flowered. **Pedicels** enclosed in ocreae, erect, 1–1.5 mm. **Flowers** closed; perianth 2.5–4.2 mm; tube 15–22% of perianth length; tepals overlapping, green, often brownish when dried, with pink or white margins, petaloid or sepaloid, elliptic, cucullate, navicular, apex rounded; midveins usually unbranched, rarely branched; stamens 8. **Achenes** enclosed in or slightly exserted from perianth, black, elliptic to oblong, 2.3–4 mm, faces subequal, shiny, smooth or minutely striate-tubercled near edges and apex. $2n = 20$.

Flowering Jun–Oct. Dry acid soils in exposed sites; 100–1000 m; Ont.; Ala., Ark., Conn., Del., D.C., Ga., Ill., Ind., Iowa, Kans., Ky., La., Maine, Md., Mass., Mich., Minn., Mo., Nebr., N.H., N.J., N.Y., N.C., Ohio, Okla., Pa., R.I., S.Dak., Tenn., Tex., Vt., Va., W.Va., Wis., Wyo.

24. Polygonum douglasii Greene, Bull. Calif. Acad. Sci. 1: 125. 1885 • Douglas's knotweed, renouée de Douglas E F W

Polygonum douglasii var. *latifolium* (Engelmann) Greene; *P. emaciatum* A. Nelson; *P. montanum* (Small) Greene; *P. tenue* Michaux var. *commune* Engelmann; *P. tenue* var. *latifolium* Engelmann

Herbs. Stems erect, green, simple or branched, not wiry, 5–80 cm, glabrous or sparsely papillose-scabridulous. **Leaves** uniformly distributed, articulated to ocreae, basal leaves caducous, distal leaves abruptly reduced to bracts; ocrea 6–12 mm, glabrous or minutely papillose-scabridulous, proximal part cylindric, distal part hyaline, lacerate; petiole 0.1–2 mm; blade 1-veined, not pleated, linear, narrow-oblong, or oblanceolate, 15–55 × 2–8(–12) mm, margins revolute, smooth or papillose-denticulate; apex acute to mucronate. **Inflorescences** axillary and terminal, spikelike, elongate; cymes widely spaced along branches, 2–4-flowered. **Pedicels** mostly exserted from ocreae, reflexed, 2–6 mm. **Flowers** closed; perianth 3–4.5 mm; tube 20–28% of perianth length; tepals overlapping, green to tannish with white or pink margins, petaloid, oblong, cucullate, navicular, apex rounded; midveins usually branched, rarely unbranched; stamens 8. **Achenes** enclosed in perianth, black, elliptic or oblong to ovate, 3–4(–4.5) mm, faces subequal, shiny or dull, smooth or minutely striate-tuberculed.

Flowering Jun–Oct. Dry, often disturbed places, rock outcrops, sandy ground; 300–3000 m; Alta., B.C., Man., Ont., Que., Sask., Yukon; Ariz., Calif., Colo., Idaho, Iowa, Mich., Minn., Mont., Nebr., Nev., N.H., N.Mex., N.Y., Oreg., S.Dak., Utah, Vt., Va., Wash., Wyo.

Five taxa that have been included in *Polygonum douglasii* (E. Murray 1982; J. C. Hickman 1984; J. T. Kartesz and K. N. Gandhi 1990) are treated here as distinct species: *P. austiniae*, *P. majus*, *P. nuttallii*, *P. sawatchense*, and *P. spergulariiforme*. Hickman noted extensive intergradation and numerous intermediate specimens among those sympatric elements, but qualitative or quantitative characters allow reliable discrimination in most cases (M. Costea and F. J. Tardif 2005), and species are here circumscribed similar to C. L. Hitchcock (1964).

Greene described var. *latifolium* as having leaf blades and achenes broader than those of var. *douglasii*. C. L. Hitchcock (1964) recognized the former, but the characters used to distinguish it appear to vary continuously, and reliable separation is not possible.

25. **Polygonum sawatchense** Small, Bull. Torrey Bot. Club 20: 213, plate 156. 1893 [E]

Herbs. Stems erect, green to brownish, simple or branched from base, not wiry, 4–50 cm, glabrous or papillose-scabridulous. **Leaves** uniformly distributed, articulated to ocreae, basal leaves persistent or caducous, distal leaves abruptly or gradually reduced to bracts; ocrea 4–10 mm, glabrous or papillose-scabridulous, proximal part cylindric, distal part lacerate or disintegrating into a few persistent fibers; petiole 0.1–2 mm; blade 1-veined, not pleated, linear or narrowly oblong to oblanceolate, 8–45 × 2–8(–12) mm, margins revolute or flat, smooth or papillose-denticulate, apex acute, mucronate, rarely obtuse. **Inflorescences** axillary or axillary and terminal, spikelike, elongate; cymes widely spaced along branches, 2–4-flowered. **Pedicels** enclosed in or exserted from ocreae, erect, 1–4 mm. **Flowers** closed or at least some open; perianth (2.5–)4 mm; tube 20–40% of perianth length; tepals overlapping, greenish or reddish, sometimes flushed with purple, with white or pink margins, sepaloid or petaloid, oblong or oblong-elliptic, cucullate, navicular, apex rounded; midveins unbranched or with few branches at base; stamens 3–8. **Achenes** enclosed in perianth, black, elliptic or ovate, 2.5–3.3 mm, faces subequal, shiny, smooth.

Subspecies 2 (2 in the flora): w North America.

1. Stems and ocreae glabrous; leaf blades linear-lanceolate to narrowly oblong or oblanceolate, margins smooth; flowers closed . . . .
. . . . . 25a. *Polygonum sawatchense* subsp. *sawatchense*
1. Stems and ocreae papillose-scabridulous; leaf blades elliptic or oblong-elliptic, margins papillose-denticulate; at least some flowers open . . . .
. . . . 25b. *Polygonum sawatchense* subsp. *oblivium*

25a. **Polygonum sawatchense** Small subsp. **sawatchense** [E]

Polygonum douglasii Greene subsp. *johnstonii* (Munz) J. C. Hickman; *P. douglasii* var. *johnstonii* Munz; *P. exile* Eastwood; *P. triandrum* Coolidge

Stems green to brownish, usually branched from base, 4–50 cm, glabrous. **Leaves** basal leaves caducous or persistent, distal leaves abruptly reduced to bracts, rarely much longer than flowers; ocrea glabrous; blade linear-lanceolate to narrowly oblong or oblanceolate, 15–45 × 2–8(–12) mm, margins usually revolute, smooth. **Inflorescences** axillary or axillary and terminal, long, open, racemes, 5–15 cm; cymes (1–)2–4-flowered. **Pedicels** 1–4 mm. **Flowers** closed; perianth (2.5–)3–3.5 mm; tube 25–40% of perianth length; tepals greenish or reddish with white or pink margins, oblong; midveins greenish or brown, unbranched or with few branches at base; stamens 3–8. **Achenes** ovate or elliptic, 2.5–3 mm.

Flowering Jun–Aug. Dry meadows, pastures, sagebrush and forests on sandy, gravelly, or rocky substrates; 800–3500 m; Alta., Sask.; Ariz., Calif., Colo., Idaho, Mont., Nebr., Nev., N.Mex., N.Dak., S.Dak., Utah, Wash., Wyo.

25b. Polygonum sawatchense Small subsp. **oblivium**
Costea & Tardif, Sida 20: 1637, figs. 1b, 2b, d, f.
2003 [E]

Stems green, simple or with few branches from base, 5–15(–25) cm, papillose-scabridulous. **Leaves** persistent, gradually reduced in size distally; ocrea papillose-scabridulous; blade elliptic to oblong-elliptic, 8–20 (–25) × 5–10 mm, margins flat, papillose-denticulate. **Inflorescences** mostly axillary, if also terminal then less than 5 cm; cymes 1–3(–4)-flowered. **Pedicels** 1–3 mm. **Flowers:** at least some open; perianth 3–4 mm; tube 20–30% of perianth length; tepals greenish white or greenish yellow, sometimes flushed with purple, with white margins, oblong-elliptic; midveins greenish or brown, usually branched at base; stamens 8. **Achenes** elliptic to elliptic-ovate, 2.6–3.3 mm.

Flowering Jun–Aug. Dry or moist meadows, pastures, sagebrush, and forests, on sandy, gravelly, or rocky substrates; 1000–3500 m; B.C.; Calif., Idaho, Nev., Oreg. Wash.

26. Polygonum engelmannii Greene, Bull. Calif. Acad.
Sci. 1: 126. 1885 (as engelmanni) • Engelmann's knotweed [E]

Polygonum douglasii Greene subsp. *engelmannii* (Greene) Kartesz & Gandhi

Herbs. Stems erect, green or purplish brown, branched from base, not wiry, 4–30 cm, glabrous or minutely papillose-scabridulous. **Leaves** uniformly distributed, articulated to ocreae, basal leaves persistent, distal leaves abruptly reduced to bracts; ocrea 3–5 mm, papillose-scabridulous or glabrous, proximal part funnelform, distal part becoming lacerate with age; petiole 0.1–2 mm; blade 1-veined, not pleated, linear-oblanceolate, 10–20(–25) × 1–3(–4) mm, margins revolute, smooth, apex acute to mucronate. **Inflorescences** axillary and terminal, spikelike, loosely floriferous nearly to base, elongate; cymes spaced along branches, (1–)2–4-flowered. **Pedicels** exserted from ocreae, reflexed, 1–3 mm. **Flowers** closed; perianth 1.5–2(–2.5) mm; tube 18–26% of perianth length; tepals initially overlapping and cucullate, later forced apart by developing achene, greenish or sometimes purple, with white margins, petaloid or sepaloid, oblong, ± flat or navicular, apex rounded; midveins unbranched; stamens 5–8. **Achenes** exserted from perianth, black, elliptic, 1.2–2.3 mm, faces subequal, shiny, smooth.

Flowering Jun–Sep. Dry to moist sandy or well-drained soils, sagebrush desert to lower mountains; 1000–1500 m; Alta., B.C.; Colo., Idaho, Mont., Nev., S.Dak., Utah, Wyo.

27. Polygonum austiniae Greene, Bull. Calif. Acad. Sci.
1: 212. 1885 (as austinae) • Mrs. Austin's knotweed [E]

Polygonum douglasii Greene subsp. *austiniae* (Greene) E. Murray

Herbs. Stems ascending to erect, green to purplish, much-branched from base, not wiry, 5–10(–20) cm, papillose-scabridulous. **Leaves** uniformly distributed, articulated to ocreae, basal leaves persistent, distal leaves abruptly reduced to bracts, articulated to ocreae; ocrea 3–5 mm, papillose-scabridulous or glabrous, proximal part funnelform, distal part lacerate; petiole 0.1–2 mm; blade 1-veined, not pleated, ovate to elliptic or obovate, 5–15 × 4–7 mm, margins flat or narrowly revolute, papillose-denticulate, apex acute or mucronate. **Inflorescences** axillary and terminal, spikelike, slender; cymes spaced along branches, 1–4-flowered. **Pedicels** exserted from ocreae, reflexed, 1–2.5 mm. **Flowers** closed; perianth (1.8–)2–2.6 mm; tube 20–28% of perianth length; tepals overlapping, green or purple, with narrow whitish margins, petaloid or sepaloid, oblong, cucullate, navicular, apex rounded; midveins unbranched; stamens 8. **Achenes** enclosed in perianth, black, elliptic to obovate, 2–2.5 mm, faces subequal, shiny, ± smooth.

Flowering Jun–Sep. Dry to moist flats on banks, sagebrush plains, and ponderosa pine forests; 1300–1600 m; Alta., B.C.; Calif., Idaho, Mont., Nev., Oreg., Wash., Wyo.

28. Polygonum minimum S. Watson, Botany (Fortieth Parallel), 315. 1871 • Broad-leaf or leafy dwarf knotweed [E]

Polygonum torreyi S. Watson

Herbs. Stems prostrate to erect, often zigzagged, reddish brown, simple or branched from base, wiry, 2–30 cm, papillose-scabridulous. **Leaves** evenly distributed or crowded at branch tips, articulated to ocreae, basal leaves persistent, hardly reduced distally; ocrea 1–4 mm, papillose-scabridulous, proximal part cylindric, distal part entire or dentate-lacerate; petiole 0.1–3 mm; blade 1-veined, not pleated, narrowly elliptic, elliptic, ovate, obovate, or subround, 6–27 × 3–8 mm, margins flat, smooth, irregularly thickened or papillose-denticulate, apex apiculate, green adaxially. **Inflorescences** axillary; cymes from near stem and branch bases, sometimes also crowded at branch apices, 1–3-flowered. **Pedicels** enclosed in ocreae, erect to spreading, 2–3 mm. **Flowers** semi-open or closed; perianth 1.8–2.5 mm; tube 22–29% of perianth length; tepals overlapping, greenish with narrow white or pink margins, almost sepaloid, oblong, cucullate, ± navicular, apex rounded; midveins thickened, unbranched; stamens 8. **Achenes** enclosed in perianth or tip exserted, black, elliptic to ovate, 1.8–2.3 mm, faces subequal, smooth, shiny.

Flowering Jul–Sep. Alpine to subalpine sites, open or semibarren soil; 1500–3300 m; Alta., B.C.; Alaska, Calif., Colo., Idaho, Mont., Nev., Oreg., Utah, Wash., Wyo.

29. Polygonum nuttallii Small, Mongr. Amer. Sp. Polygonum, 132, plate 53. 1895 • Nuttall's knotweed E

Polygonum intermedium Nuttall ex S. Watson, Proc. Amer. Acad. Arts 17: 378. 1882, not Ehrhart 1791; *P. douglasii* Greene subsp. *nuttallii* (Small) J. C. Hickman

Herbs. Stems spreading to erect, sometimes zigzagged, purplish, simple or branched, wiry, (5–)10–35 cm, papillose-scabridulous. **Leaves** evenly distributed or crowded at branch tips, articulated to ocreae, basal leaves persistent, distal leaves gradually reduced to bracts; ocrea 3–4 mm, papillose-scabridulous, proximal part funnelform, distal part finely lacerate; petiole 0.1–2 mm; blade 1-veined, not pleated, linear to narrowly oblong-elliptic, 8–30 × 1–4(–7) mm, margins narrowly revolute, never touching along midrib, smooth, apex acute, mucronate, ± glaucous adaxially. **Inflorescences** axillary and terminal, spikelike, dense; cymes mostly congested toward tips of branches, 2–3-flowered. **Pedicels** enclosed in ocreae, erect to spreading, 2–3 mm. **Flowers** semi-open or closed; perianth 1.8–2.4 mm; tube 25–33% of perianth length; tepals overlapping, greenish, white, or pink with pink margins, petaloid, oblong, cucullate, navicular in distal ¼, apex rounded; midveins unbranched; stamens 8. **Achenes** enclosed in perianth, black, elliptic to ovate, 1.8–2.3 mm, faces subequal, shiny, smooth.

Flowering May–Oct. Dry prairies, open knolls in lower mountains, open sites in lowland and montane zones, sandy soil; 800–1100 m; B.C.; Oreg., Wash.

C. L. Hitchcock (1964) suggested that *Polygonum nuttallii* is but a small-flowered form of *P. spergulariiforme*. Although morphologically similar, *P.*

nuttallii differs from *P. spergulariiforme* in some respects, including its wiry, purplish stems, short and funnelform ocreae, adaxially glaucous leaves, longer bracts, shorter fruiting perianth, and achenes.

30. Polygonum spergulariiforme Meisner ex Small, Bull. Torrey Bot. Club 19: 366. 1892 (as spergulariaeforme) • Spurry or fall knotweed E

Polygonum douglasii Greene subsp. *spergulariiforme* (Meisner ex Small) J. C. Hickman; *P. emaciatum* A. Nelson

Herbs. Stems erect, green, divaricately branched, not wiry, 5–50 cm, usually papillose-scabridulous, rarely glabrescent. **Leaves** uniformly distributed, articulated to ocreae, basal leaves usually caducous, distal leaves abruptly reduced to bracts, articulated to ocreae; ocrea 8–12 mm, papillose-scabridulous, proximal part cylindric, distal part disintegrating into a few persistent fibers; petiole 0.1–2 mm; blade 1-veined, not pleated, linear to lanceolate, 35–60 × 1–3 mm, margins flat or narrowly revolute, smooth or papillose-denticulate, apex acute, mucronate. **Inflorescences** axillary and terminal, spikelike, dense; cymes crowded, ± overlapping at branch tips, 2–5-flowered. **Pedicels** enclosed in or slightly exserted from ocreae, erect to spreading, 0.5–2 mm. **Flowers** open or semi-open; perianth 3–5 mm; tube 9–17% of perianth length; tepals overlapping, uniformly pink or white, petaloid, oblong-obovate, cucullate, navicular in distal ¼, apex rounded; midveins branched; stamens 8. **Achenes** enclosed in perianth, black, narrowly elliptic or elliptic-lanceolate to narrowly ovate, 3–5 mm, faces subequal, shiny, smooth to striate-tubercled.

Flowering Jun–Oct. Moist to dry, open rocky places, including serpentine; 10–2000 m; B.C.; Calif., Idaho, Mont., Oreg., Wash., Wyo.

31. Polygonum majus (Meisner) Piper, Fl. Palouse Reg., 63. 1901 • Wiry knotweed E

Polygonum coarctatum Douglas ex Meisner var. *majus* Meisner in A. P. de Candolle and A. L. P. P. de Candolle, Prodr. 14: 101. 1856; *P. douglasii* Greene subsp. *majus* (Meisner) J. C. Hickman

Herbs. Stems erect, green, simple or branched, ± wiry, 15–60 cm, usually papillose-scabridulous. **Leaves** uniformly distributed, articulated to ocreae, basal leaves often caducous, distal leaves abruptly reduced to bracts, articulated to ocreae; ocrea 5–12 mm, glabrous or papillose-scabridulous, proximal part cylindric, distal part lacerate or disintegrating into fibers; petiole 0.1–2 mm; blade 1-veined, not pleated, linear to narrowly oblong or lanceolate, 15–70 × 2–8 mm, margins revolute, papillose-denticulate, apex acute or mucronate. **Inflorescences** axillary and terminal, spikelike, elongate; cymes spaced along branches, 2–5-flowered. **Pedicels** exserted from ocreae, reflexed, 0.5–1 mm. **Flowers** open or semi-open; perianth (3.5–)4–5 mm; tube 9–17% of perianth length; tepals overlapping, uniformly white to pink, petaloid, oblong to oblong-obovate, cucullate, navicular in distal ¹/₄, apex rounded; midveins unbranched or with short lateral branches; stamens 8. **Achenes** enclosed in perianth, black, elliptic, 3.5–5 mm, faces subequal, shiny or dull, smooth or striate-tubercled.

Flowering May–Aug. Dry plains, meadows, sometimes on serpentine; 500–2000 m; B.C.; Calif., Colo., Idaho, Nev., Oreg., Utah, Wash., Wyo.

32. **Polygonum cascadense** W. H. Baker, Madroño 10: 62, plate 1, fig. 1. 1949 • Cascade knotweed E

Herbs. Stems spreading to erect, zigzagged, green, simple or branched from base, wiry, 5–12 (–15) cm, glabrous. **Leaves** uniformly distributed, articulated to ocreae, basal leaves persistent, distal leaves abruptly reduced to bracts; ocrea 2–5 mm, glabrous, proximal part funnelform, distal part lacerate; petiole essentially absent; blade 1-veined, not pleated, oblanceolate to obovate, 5–20 × 2–5 mm, margins revolute, never touching along midrib, sparsely papillose-denticulate, apex rounded or apiculate. **Inflorescences** axillary and terminal, spikelike, dense; cymes congested at tips of stems and branches, 3–5-flowered. **Pedicels** enclosed in ocreae, erect to spreading, 2–3 mm. **Flowers** open; perianth 2–2.5 mm; tube 12–25% of perianth length; tepals overlapping, uniformly white, petaloid, oblong to obovate, cucullate, navicular in distal ¹/₄, apex rounded; midveins unbranched;

stamens 8. **Achenes** enclosed in or exserted from perianth, black, ovate to ovate-oblong, 1.8–2.1 mm, faces subequal, shiny, smooth.

Flowering Jun–Sep. Dry, usually rocky slopes, often on serpentine; 1600–1800 m; Oreg.

33. **Polygonum utahense** Brenckle & Cottam, Bull. Univ. Utah, Biol. Ser. 4(4): 3, plate 1. 1940 • Utah knotweed E

Polygonum douglasii Greene var. *utahense* (Brenckle & Cottam) S. L. Welsh

Herbs. Stems erect, green, simple or with few branches, not wiry, 1.5–3.5 cm, papillose-scabridulous. **Leaves** uniformly distributed, articulated to ocreae, basal leaves persistent, distal leaves abruptly reduced to bracts; ocrea 2–3.5 mm, papillose-scabridulous, proximal part funnelform, distal part hyaline, lacerate; petiole absent; blade 1-veined, not pleated, linear-subulate, 6–14 × 1 mm, coriaceous, margins revolute, touching along midrib, apex acute, mucronate, papillose. **Inflorescences** axillary and terminal, spikelike, dense; cymes overlapping, starting almost from stem base, 2–4-flowered. **Pedicels** enclosed in ocreae, erect to spreading, 0.4–1 mm. **Flowers** open; perianth 2–2.5 mm; tube 20–25% of perianth length; tepals overlapping, uniformly white, petaloid, oblong to widely obovate, cucullate, navicular in distal ¹/₄, apex rounded; midveins branched; stamens 8. **Achenes** enclosed in perianth, black, elliptic, 1.5–2 mm, faces subequal, shiny, smooth.

Flowering Jul–Sep. Dry, sandy ravines, rocky Navajo sandstone spur; 2300 m; Utah.

Polygonum utahense has been treated as a synonym of *P. sawatchense* by some authors (e.g., S. L. Welsh et al. 1987; J. T. Kartesz and C. A. Meacham 1999). Its perianth morphology suggests a relationship with *P. cascadense*, from which it differs in its dwarf habit and uniformly papillose leaves with margins revolute, touching along midribs.

31. FAGOPYRUM Miller, Gard. Dict. Abr. ed. 4, vol. 1. 1754, name conserved • Buckwheat, sarrasin [Latin *fagus*, beech, and Greek *pyrus*, wheat, alluding to resemblance of the achene to a beech-nut] ☐

Harold R. Hinds†

Craig C. Freeman

Herbs, annual; taprooted. **Stems** erect or ascending, glabrous or puberulent. **Leaves** deciduous, cauline, alternate, petiolate (proximal leaves) or sessile (distal leaves); ocrea persistent or deciduous, chartaceous; petiole base articulated; blade cordate, triangular, hastate, or sagittate, margins entire to sinuate. **Pedicels** present. **Inflorescences** axillary, or terminal and axillary, racemelike or paniclelike, pedunculate. **Flowers** bisexual or, rarely, bisexual and staminate on same plant, 2–6 per ocreate fascicle, heterostylous or homostylous, base stipelike; perianth nonaccrescent, white, pale pink, or green, broadly campanulate, glabrous; tepals 5, distinct, petaloid, dimorphic, outer smaller than inner; stamens 8; filaments distinct, free, glabrous; anthers white, pink, or red, oval to elliptic; styles 3, reflexed, distinct; stigmas capitate. **Achenes** strongly exserted, brown to dark brown or gray, sometimes mottled black, unwinged or essentially so, bluntly to sharply 3-gonous, glabrous. **Seeds:** embryo folded. *x* = 8.

Species 16 (2 in the flora): introduced; Eurasia, e Africa; introduced elsewhere, cultivated in temperate regions worldwide.

Fagopyrum esculentum and *F. tataricum* are cultivated widely. In North America, they often escape, but populations generally are ephemeral.

Archaeological evidence for the cultivation of buckwheat dates to 4600 BP in China and 3500 BP in Japan (O. Ohnishi 1998). Molecular studies indicate that *Fagopyrum* comprises two major clades, with *F. esculentum* and *F. tataricum* in the large-fruited "cymosum" group (O. Ohnishi and Y. Matsuoka 1996; Y. Yasui and O. Ohnishi 1998, 1998b; O. Ohsako and O. Ohnishi 2000).

SELECTED REFERENCES Ohnishi, O. 1998. Search for the wild ancestor of buckwheat III. The wild ancestor of cultivated common buckwheat, and of Tartary buckwheat. Econ. Bot. 52: 123–133. Ohnishi, O. and Y. Matsuoka. 1996. Search for the wild ancestor of buckwheat II. Taxonomy of *Fagopyrum* (Polygonaceae) species based on morphology, isozymes and cpDNA variability. Genes Genet. Systems 71: 383–390. Ohsako, T. and O. Ohnishi. 2000. Intra- and interspecific phylogeny of wild *Fagopyrum* (Polygonaceae) species based on nucleotide sequences of noncoding regions of chloroplast DNA. Amer. J. Bot. 87: 573–582. Yasui, Y. and O. Ohnishi. 1998. Phylogenetic relationships among *Fagopyrum* species revealed by the nucleotide sequences of the ITS region of the nuclear rRNA gene. Genes Genet. Systems 73: 201–210. Yasui, Y. and O. Ohnishi. 1998b. Interspecific relationships in *Fagopyrum* (Polygonaceae) revealed by the nucleotide sequences of the *rbc*L and *acc*D genes and their intergenic region. Amer. J. Bot. 85: 1134–1142.

1. Achene faces smooth, angles smooth; tepals (2.5–)3–5 mm; perianths, creamy white to pale pink; inflorescences paniclelike, 1–4 cm, terminal and axillary . . . . 1. *Fagopyrum esculentum*
1. Achene faces irregularly rugose, angles often sinuate-dentate; tepals 1.5–3 mm; perianths, green with whitish margins; inflorescences racemelike, 2–10 cm, axillary . . . . 2. *Fagopyrum tataricum*

1. Fagopyrum esculentum Moench, Methodus, 290. 1794 • Common buckwheat, sarrasin commun F I

Polygonum fagopyrum Linnaeus, Sp. Pl. 1: 364. 1753; *Fagopyrum sagittatum* Gilibert; *F. vulgare* T. Nees

Stems ascending or erect, green or striped with pink or red, branched, (7–)15–90 cm. **Leaves:** ocrea brownish hyaline, loose, funnelform, 2–8 mm, margins truncate, eciliate, glabrous or puberulent proximally; petiole 1.5–6(–9) cm, usually puberulent adaxially; blade palmately veined with 7–9 primary basal veins, hastate-triangular, sagittate-triangular, or cordate, 2.5–8 × 2–8 cm, base truncate or cordate to sagittate, margins ciliolate, apex acute to acuminate. **Inflorescences** terminal and axillary, paniclelike, 1–4 cm, usually crowded at stem apices; peduncle 0.5–4 cm, puberulent in lines. **Pedicels** ascending or recurved, 2.5–4 mm. **Flowers** chasmogamous, heterostylous [homostylous]; perianths creamy white to pale pink; tepals elliptic to obovate, (2.5–)3–5 mm, margins entire, apex obtuse to acute; stamens ca. $\frac{1}{2}$ as long as or slightly longer than perianth; styles 1.5–2 mm or 0.5–1 mm; stigmas purplish. **Achenes** uniformly light brown or streaked with dark brown or black, sharply 3-gonous, 4–6 × 4–6 mm, faces smooth, angles prominent, unwinged or essentially so, smooth or occasionally with blunt tooth in proximal $\frac{1}{3}$. $2n = 16$ (China).

Flowering Jun–Sep; fruiting Jun–Nov. Cultivated as crop plant, waif along railroads, roadsides, fields, waste places, occasionally weedy; 0–2200 m; introduced; Alta., Man., Nfld. and Labr. (Nfld.), N.S., Ont., P.E.I., Que., Sask., Yukon; Ala., Alaska, Ariz., Calif., Colo., Conn., Del., D.C., Fla., Ga., Idaho, Ill., Ind., Iowa, Kans., Ky., La., Maine, Md., Mass., Mich., Minn., Mo., Mont., Nebr., N.H., N.J., N.Mex., N.Y., N.C., N.Dak., Ohio, Okla., Oreg., Pa., R.I., S.C., S.Dak., Tenn., Tex., Vt., Va., Wash., W.Va., Wis., Wyo.; Asia (China); introduced in Central America, South America, Europe, Asia, Africa.

Fagopyrum esculentum is a heterostylous, obligate out-crosser. Morphological, allozyme, and molecular data suggest that the cultivated plants are most closely related to wild ones in northwestern Yunnan, China.

Common buckwheat is an important pseudocereal crop in China, the Russian Federation, Ukraine, Kazakhstan, and Poland; it is grown in many other countries. It is planted frequently in wildlife food plots, as a catch or cover crop, and as a honey plant in North America. Hulls from the achenes are used for pillow filling, which manufacturers claim has health benefits over traditional foam, polyester, or down fillings.

2. Fagopyrum tataricum (Linnaeus) Gaertner, Fruct. Sem. Pl. 2: 182. 1790 • Tartary or green buckwheat, India-wheat, sarrasin de Tartarie I

Polygonum tataricum Linnaeus, Sp. Pl. 1: 364. 1753

Stems ascending or erect, yellowish green, sometimes red-tinged, sparingly branched, (10–)30–80(–100) cm. **Leaves:** ocrea brownish hyaline, loose, funnelform, 5–11 mm, margins truncate to obtuse, eciliate, glabrous or puberulent proximally; petiole (0.5–)1–7 cm, usually puberulent adaxially; blade palmately veined with 7–9 primary basal veins, broadly triangular to broadly hastate, 2–7 × 2–8 cm, base truncate or cordate to sagittate, margins ciliolate, apex acute to acuminate. **Inflorescences** axillary, racemelike, 2–10 cm, not crowded at stem apices; peduncle 1–6 cm, puberulent in lines. **Pedicels** ascending or recurved, 1–3 mm. **Flowers** often cleistogamous, homostylous; perianths green with whitish margins; tepals triangular to ovate, 1.5–3 mm, margins entire, apex obtuse to acute; stamens ca. $\frac{1}{2}$ as long as perianth; styles 0.1–0.4 mm; stigmas purplish. **Achenes** uniformly gray or, infrequently, mottled with blackish spots medially, bluntly 3-gonous, 5–6 × 3–5 mm, faces irregularly rugose, angles usually obscure in proximal $\frac{1}{2}$, more conspicuous in distal $\frac{1}{2}$, unwinged, often sinuate-dentate. $2n = 16$ (China).

Flowering Jun–Sep; fruiting Jul–Nov. Cultivated as grain crop and green manure, waif in waste places, disturbed ground, and field margins, rarely persisting; 0–1000; introduced; Alta., N.B., Nfld. and Labr. (Nfld.), N.S., Ont., Que., Sask.; Maine, Mass., Mich., N.H., N.Y., Pa., R.I., Vt., W.Va.; Asia (China); introduced in Europe.

Fagopyrum tataricum is homostylous and self-pollinating. Cultivated plants appear to be most closely related to the wild ones in southwestern Sichuan, China. Tartary buckwheat is a less important crop plant and is encountered less frequently in the flora area than is *F. esculentum*. It is cultivated in mountainous areas of Asia and elsewhere (C. G. Campbell 1997).

32. PERSICARIA (Linnaeus) Miller, Gard. Dict. Abr. ed. 4, vol. 3. 1754 • Smartweed [Latin, *persica*, peach, and *-aria*, pertaining to, alluding to resemblance of leaves of some species]

Harold R. Hinds†

Craig C. Freeman

Polygonum [unranked] *Persicaria* Linnaeus, Sp. Pl. 1: 360. 1753

Herbs, perennial or annual (sometimes suffrutescent in *P. wallichii*); taprooted or fibrous-rooted; sometimes rhizomatous or stoloniferous. **Stems** erect or, sometimes, prostrate or scandent, simple or branched, glabrous or pubescent, rarely with recurved prickles. **Leaves** deciduous, mostly cauline, alternate, petiolate or sessile; ocrea persistent or disintegrating with age and deciduous entirely or distally, usually tan, brown, or reddish, chartaceous or partially to entirely foliaceous, rarely coriaceous proximally and chartaceous distally, glabrous or scabrous to variously pubescent, never 2-lobed distally; blade lanceolate or ovate to hastate or sagittate, margins entire or, rarely, hastately lobed. **Inflorescences** terminal or terminal and axillary, spikelike, paniclelike, or capitate; peduncle present. **Pedicels** present or absent. **Flowers** bisexual (often functionally unisexual in *P. amphibia* and *P. hydropiperoides*), 1–14 per ocreate fascicle, base not stipelike; perianth white, greenish white, roseate, red, or purple, campanulate or urceolate, rarely rotate, rarely becoming fleshy in fruit, glabrous, sometimes glandular-punctate, accrescent or nonaccrescent; tepals 4–5, connate 1/4–2/3 their lengths (less than 1/5 their lengths in *P. wallichii*), petaloid, dimorphic, outer larger than inner; stamens 5–8, filaments distinct or connate basally, outer ones sometimes adnate to perianth tube, glabrous; anthers yellow, pink, or red, elliptic to ovate; styles 2–3, erect to spreading or reflexed, distinct or connate; stigmas capitate. **Achenes** included or exserted, brown or dark brown to black, not winged, discoid, biconvex, 2–3-gonous, or spheroidal, glabrous. **Seeds:** embryo curved. x = 10, 11, 12.

Species ca. 100 (26 in the flora): nearly worldwide.

Opinions vary widely about the circumscription and infrageneric classification of *Persicaria*. The concept employed here generally follows L.-P. Ronse Decraene et al. (2000) and K. Haraldson (1978), with five sections recognized in the flora. *Aconogonon* and *Bistorta*, which often are included in *Persicaria* or in *Polygonum* in the broad sense, are treated here as separate genera.

Key to the Sections of *Persicaria*

1. Styles exserted, persistent on achenes; inflorescences spikelike, interrupted . . . . 32a. *Persicaria* sect. *Tovara*, p. 575
1. Styles included, rarely exserted, deciduous; inflorescences capitate, paniclelike, or spikelike, uninterrupted or interrupted.
   2. Stems with recurved prickles, scandent or, rarely, ascending to erect . . . . 32b. *Persicaria* sect. *Echinocaulon*, p. 575
   2. Stems unarmed, usually erect or ascending, rarely prostrate or decumbent.
      3. Inflorescences capitate; petioles usually winged, auriculate . . . . 32c. *Persicaria* sect. *Cephalophilon*, p. 579
      3. Inflorescence spikelike or paniclelike; petioles not winged, not auriculate.
         4. Inflorescences paniclelike; perianths rotate; tepals connate less than 1/5 their lengths . . . . 32d. *Persicaria* sect. *Rubrivena*, p. 580
         4. Inflorescences spikelike; perianths campanulate; tepals connate 1/4–2/3 their lengths . . . . 32e. *Persicaria* sect. *Persicaria*, p. 581

32a. PERSICARIA Miller sect. **TOVARA** (Adanson) H. Gross, Bull. Acad. Int. Géogr. Bot. 4: 27. 1913

Tovara Adanson, Fam. Pl. 2: 276, 612. 1763, name rejected; *Antenoron* Rafinesque

Stems erect, simple or branched distally, unarmed. **Leaves:** ocrea chartaceous, margins ciliate; petiole not winged, not auriculate; blade often with dark triangular or lunate blotch adaxially on leaves produced early in growing season, broadly lanceolate to ovate, base rounded to acute, margins entire. **Inflorescences** terminal and axillary, spikelike, interrupted. **Flowers** homostylous, articulation swollen; perianth narrowly campanulate to urceolate; tepals 4, connate ca. ¹/₂ their length; stamens 5; styles persistent on achenes, 2, exserted, reflexed distally.

Species 3 (1 in the flora): e North America, e Asia.

SELECTED REFERENCES Li, H. L. 1952b. The genus *Tovara* (Polygonaceae). Rhodora 50: 19–25. Mun, J. H. and C. W. Park. 1995. Flavonoid chemistry of *Polygonum* sect. *Tovara* (Polygonaceae): A systematic survey. Pl. Syst. Evol. 196: 153–159. Youngbae, S., S. Kim, and C. W. Park. 1997. A phylogenetic study of *Polygonum* sect. *Tovara* (Polygonaceae) based on ITS sequences of nuclear ribosomal DNA. J. Plant Biol. 40: 47–52.

1. Persicaria virginiana (Linnaeus) Gaertner, Fruct. Sem. Pl. 2: 180. 1790 • Jumpseed, renouée de Virginie F

Polygonum virginianum Linnaeus, Sp. Pl. 1: 360. 1753; *Antenoron virginianum* (Linnaeus) Roberty & Vautier; *Tovara virginiana* (Linnaeus) Rafinesque

Plants perennial, 4.5–6(–13) dm; rhizomatous. **Stems** ribbed, glabrous or strigose. **Leaves:** ocrea brownish hyaline, cylindric, 10–20 mm, base inflated or not, margins truncate, ciliate with bristles 0.5–4 mm, surface strigose to tomentose; petiole (0.1–)1–2 cm, leaves sometimes sessile; blade 5–17.5 × 2–10 cm, apex acute to acuminate, faces pubescent abaxially, strigose and scabrous adaxially. **Inflorescences** (50–)100–350 × 7–15 mm; peduncle 10–70 mm, pubescent or glabrous distally; ocreolae not overlapping, margins ciliate with bristles to 3 mm. **Pedicels** ascending to spreading, 0.5–1 mm. **Flowers** 1–3 per ocreate fascicle; perianth white, greenish white, or rarely pink, glabrous, accrescent; tepals elliptic to obovate, 2.5–3.5 mm, apex acute to acuminate; filaments distinct, outer ones sometimes adnate to perianth tube; anthers yellow or pink, ovate; styles distinct. **Achenes** included except for apex and styles, brown to dark brown, biconvex, 3.5–4 × 2–2.8 mm, dull to shiny, smooth to rugose. $2n = 44$.

Flowering Jul–Oct. Rich deciduous forests, floodplain forests, dry woodlands, thickets; 0–500 m; Ont., Que.; Ala., Ark., Conn., Del., D.C., Fla., Ga., Ill., Ind., Iowa, Kans., Ky., La., Md., Mass., Mich., Minn., Miss., Mo., Nebr., N.H., N.J., N.Y., N.C., Ohio, Okla., Pa., R.I., S.C., Tenn., Tex., Va., W.Va., Wis.; c Mexico.

Tension in the articulation of the pedicels is sufficient to throw mature achenes 3–4 m when the inflorescence is bumped, and the persistent, hooked styles aid in the dispersal of achenes in the fur of animals (H. S. Reed and I. Smoot 1906). A hot infusion of leaves with bark of honey-locust (*Gleditsia triacanthos* Linnaeus) was used by the Cherokee to treat whooping cough (D. E. Moerman 1998).

32b. PERSICARIA Miller sect. **ECHINOCAULON** (Meisner) H. Gross, Bull. Acad. Int. Géogr. Bot. 4: 27. 1913

Polygonum Linnaeus sect. *Echinocaulon* Meisner in N. Wallich, Pl. Asiat. Rar. 3: 58. 1832

Stems usually scandent or, rarely, ascending or erect, simple or branched, with recurved prickles. **Leaves:** ocrea chartaceous or foliaceous, margins eciliate or ciliate; petiole, if present, rarely winged, not auriculate; blade without dark triangular or lunate blotch adaxially, broadly lanceolate or elliptic to triangular, base truncate to sagittate to cordate, margins entire or hastately lobed. **Inflorescences** terminal and axillary, spikelike, paniclelike, or capitate, uninterrupted or, rarely, interrupted proximally. **Flowers** homostylous, articulation swollen or

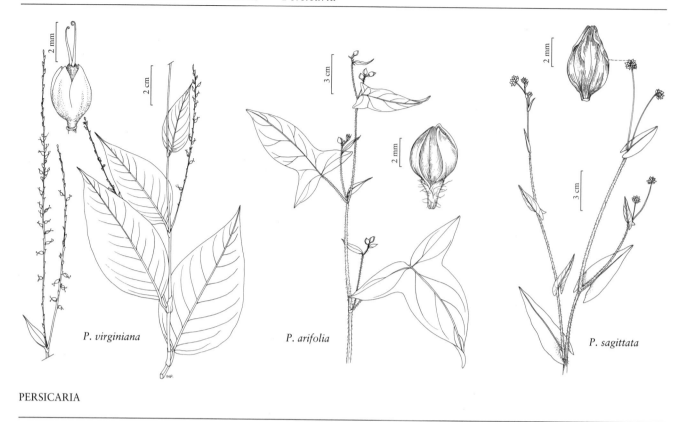

P. virginiana

P. arifolia

P. sagittata

PERSICARIA

not; perianth campanulate; tepals 4–5, connate $^{1}/_{4}$–$^{1}/_{2}$ their length; stamens 5–8; styles deciduous, 2–3, included or, rarely, exserted, erect or spreading.

Species 21 (5 in the flora): e North America, South America, Asia, se Africa, e Australia.

SELECTED REFERENCES Park, C. W. 1986. Nomenclatural typifications in *Polygonum* section *Echinocaulon* (Polygonaceae). Brittonia 38: 394–406. Park, C. W. 1987. Flavonoid chemistry of *Polygonum* sect. *Echinocaulon*: A systematic survey. Syst. Bot. 12: 167–179. Park, C. W. 1988. Taxonomy of *Polygonum* section *Echinocaulon* (Polygonaceae). Mem. New York Bot. Gard. 47: 1–82.

1. Ocreae, at least some, foliaceous, green; perianths fleshy and blue in fruit . . . . 2. *Persicaria perfoliata*
1. Ocreae chartaceous, brownish; perianths not fleshy and blue in fruit.
   2. Ocreae margins truncate, ciliate with bristles 2–4 mm . . . . 6. *Persicaria bungeana*
   2. Ocreae margins oblique, eciliate or ciliate with bristles to 2.5 mm.
      3. Leaf blades broadly hastate to hastate-cordate or triangular; tepals 4 . . . . 3. *Persicaria arifolia*
      3. Leaf blades linear-lanceolate to oblong; tepals 5.
         4. Peduncles usually glabrous, sometimes with retrorse prickles proximally; leaves petiolate; bases of leaf blades sagittate to cordate; stamens 8 . . . . 4. *Persicaria sagittata*
         4. Peduncles usually stipitate-glandular; leaves sessile or subsessile; bases of leaf blades cordate to truncate or cuneate; stamens 5 . . . . 5. *Persicaria meisneriana*

2. Persicaria perfoliata (Linnaeus) H. Gross, Beih. Bot.

Centralbl. 37(2): 113. 1919 • Devil's-tail or giant climbing tearthumb, mile-a-minute weed

Polygonum arifolium Linnaeus var. *perfoliatum* Linnaeus, Syst. Nat. ed. 10, 2: 1006. 1759; *P. perfoliatum* (Linnaeus) Linnaeus

Plants annual, 10–20(–70) dm; roots not also arising from proximal nodes. **Stems** scandent, ribbed, glabrous, often glaucous; prickles 0.5–1 mm. **Leaves:** ocrea green, plane to broadly funnelform, 9–14 mm, at least some foliaceous, base inflated or not, without prickles, margins oblique, eciliate, surface glabrous, glaucous; petiole 4.5–8 cm; blade triangular, 4–7 × 4.5–9 cm, base truncate to cordate, usually peltate, margins entire, sparsely retrorsely prickly, apex acuminate, faces glabrous, usually glaucous abaxially. **Inflorescences** capitate or spikelike, uninterrupted, 5–12 × 5–10 mm; peduncle 10–50 mm, retrorsely prickly; ocreolae overlapping, margins eciliate. **Pedicels** mostly ascending, 1–3 mm. **Flowers** 1–3 per ocreate fascicle; perianth greenish white, glabrous, accrescent, becoming fleshy and blue in fruit; tepals 5, connate to ca. 1/3 their length, broadly elliptic, 2–3.5 mm, apex acute to obtuse; stamens (6–)8, filaments distinct, free; anthers pinkish, ovate; styles 3, connate proximally. **Achenes** included, black or reddish black, spheroidal, 3–3.5 × 3–3.5 mm, shiny, smooth.

Flowering Jun–Oct. Thickets, streams banks, pastures, forest edges, roadsides, railroad embankments, other moist, disturbed sites; 0–300 m; introduced; Conn., Del., D.C., Md., Miss., N.J., N.Y., Ohio, Oreg., Pa., R.I., Va., W.Va.; Asia.

Persicaria perfoliata is an aggressive, fast-growing pest in its native range and in North America. At least some introductions appear to be through the nursery trade (J. C. Hickman and C. S. Hickman 1978; R. E. Riefener 1982). It was collected once in 1954 in British Columbia, but that population did not persist.

3. Persicaria arifolia (Linnaeus) Haraldson, Acta Univ.

Upsal., Symb. Bot. Upsal. 22: 72. 1978 • Halberd-leaf tearthumb, renouée à feuilles d'arum E F

Polygonum arifolium Linnaeus, Sp. Pl. 1: 364. 1753; *P. arifolium* var. *lentiforme* Fernald & Griscom; *P. arifolium* var. *pubescens* (R. Keller) Fernald; *P. sagittatum* Linnaeus var. *pubescens* R. Keller; *Tracaulon arifolium* (Linnaeus) Rafinesque; *Truellum arifolium* (Linnaeus) Soják

Plants annual, 2–15 dm; roots also often arising from proximal nodes. **Stems** scandent,

ribbed, glabrous; prickles 0.5–1 mm. **Leaves:** ocrea tan or brownish, cylindric, 8–15 mm, chartaceous, base inflated or not, with prickles, margins oblique, ciliate with bristles 0.5–2.5 mm, surface glabrous or appressed- to spreading-pubescent; petiole 1–7 cm; blade broadly hastate to hastate-cordate or triangular, (2–)6.5–13(–18) × (1–)6–11(–16) cm, base truncate to truncate-cordate, margins broadly hastate with lobes divergent, ciliate, sometimes also retrorsely prickly, apex acuminate, faces appressed-pubescent or, rarely, glabrous adaxially, stellate-pubescent or, rarely, glabrous abaxially, major veins often bearing prickles. **Inflorescences** capitate or paniclelike, uninterrupted, 5–12 × 3–8 mm; peduncle 10–80 mm, retrorsely prickly proximally, stellate-pubescent and stipitate-glandular distally, glands red or pink; ocreolae usually overlapping, sometimes not overlapping proximally, margins eciliate or ciliate with bristles to 0.5 mm. **Pedicels** mostly ascending, 2–3 mm. **Flowers** 2–4 per ocreate fascicle; perianth pink or red, often whitish green proximally, glabrous, accrescent, not becoming blue and fleshy in fruit; tepals 4, connate 1/3–1/2 their length, broadly elliptic, 5–6 mm, apex acute to obtuse; stamens (6–)8, filaments distinct, free; anthers pink, elliptic; styles 2, distinct. **Achenes** included, dark brown to black, biconvex, 3.5–6 × 3–4 mm, shiny, smooth.

Flowering Jul–Oct. Shaded swamps, ponds, tidal marshes along rivers, wet ravines in forests; 0–600 m; N.B., N.S., Ont., P.E.I., Que.; Conn., Del., D.C., Ga., Ill., Ind., Ky., La., Maine, Md., Mass., Mich., Minn., Mo., N.H., N.J., N.Y., N.C., Ohio, Pa., R.I., S.C., Tenn., Vt., Va., Wash., W.Va., Wis.

4. Persicaria sagittata (Linnaeus) H. Gross, Beih. Bot.

Centralbl. 37(2): 113. 1919 • Arrow-leaf tearthumb, arrow-vine, renouée sagittée F

Polygonum sagittatum Linnaeus, Sp. Pl. 1: 363. 1753; *P. sagittatum* var. *gracilentum* Fernald; *Tracaulon sagittatum* (Linnaeus) Small; *Truellum sagittatum* (Linnaeus) Soják

Plants annual, 3–20 dm; roots also often arising from proximal nodes. **Stems** scandent, ribbed,

glabrous; prickles 1–1.5 mm. **Leaves:** ocrea brownish, cylindric, (3–)5–13 mm, chartaceous, base inflated or not, without prickles, margins oblique, glabrous or ciliate at tip with bristles 0.2–1 mm, surface glabrous; petiole 0.5–4 cm; blade broadly lanceolate to oblong, 2–8.5 × 1–3 cm, base sagittate to cordate, margins entire, ciliate or eciliate, apex obtuse to acute, faces glabrous or densely appressed-pubescent, usually with retrorse prickles along midvein abaxially. **Inflorescences** capitate or paniclelike, uninterrupted, 5–15 × 4–10 mm; peduncle 10–80 mm, usually glabrous, sometimes with retrorse prickles proximally; ocreolae overlapping, margins eciliate. **Pedicels** mostly ascending, 1–1.5 mm. **Flowers** 2–3 per ocreate fascicle; perianth white or greenish white, often tinged pink or red, sometimes entirely pink, glabrous, accrescent, not becoming blue and fleshy in fruit; tepals 5, connate $\frac{1}{3}$–$\frac{1}{2}$ their length, broadly elliptic, 3–5 mm, apex obtuse; stamens 8, filaments distinct, free; anthers pink, ovate; styles 3, connate to middle. **Achenes** included or styles exserted, light or dark brown to black, 3-gonous, 2.5–4 × 1.8–2.5 mm, dull to shiny, smooth to minutely punctate. $2n = 40$.

Flowering Jun–Oct. Moist shaded sites, meadows, pastures, fens, swamps, shorelines of ponds and streams; 0–1000 m; Man., N.B., Nfld. and Labr. (Nfld.), N.S., Ont., P.E.I., Que.; Ala., Ark., Colo., Conn., Del., D.C., Fla., Ga., Ill., Ind., Iowa, Kans., Ky., La., Maine, Md., Mass., Mich., Minn., Miss., Mo., Nebr., N.H., N.J., N.Y., N.C., N.Dak., Ohio, Okla., Pa., R.I., S.C., S.Dak., Tenn., Tex., Vt., Va., W.Va., Wis.; e Asia.

Persicaria sagittata is an extremely variable species. Achene and leaf characters have been used by some authors to separate North American and Asian populations, but these characters show weak geographic variation (C. W. Park 1988).

5. **Persicaria meisneriana** (Chamisso & Schlechtendal) M. Gómez, Anales Inst. Segunda Enseñ. 2: 278. 1896 • Mexican tearthumb

Polygonum meisnerianum Chamisso & Schlechtendal, Linnaea 3: 40. 1828

Varieties 2 (1 in the flora): se United States, Mexico, West Indies, Central America, South America, se Africa.

Variety *meisneriana* has leaves with petioles 0.3–1 cm, blades prominently sagittate to hastate at bases, and ocreae usually moderately to densely pubescent. It is found only in South America (Argentina, Brazil, Paraguay).

5a. **Persicaria meisneriana** (Chamisso & Schlechtendal) M. Gómez var. **beyrichiana** (Chamisso & Schlechtendal) C. C. Freeman, Sida 21: 291. 2004

Polygonum beyrichianum Chamisso & Schlechtendal, Linnaea 3: 42. 1828; *P. meisnerianum* (Chamisso & Schlechtendal) M. Gómez var. *beyrichianum* (Chamisso & Schlechtendal) Meisner

Plants annual, 3–10 dm; roots not also arising from proximal nodes. **Stems** erect or scandent, ribbed, glabrous or pubescent with tufted hairs and glandular hairs; prickles 1–1.5 mm. **Leaves:** ocrea brownish, cylindric, 8–20 mm, chartaceous, base inflated or not,

sometimes with prickles, margins oblique, eciliate or ciliate with bristles 0.2–0.5 mm, surface glabrous or, rarely, glandular-pubescent; petiole 0.1–0.5 cm, leaves sometimes sessile; blade usually linear-lanceolate to lanceolate, rarely hastate, 5.5–14 × 0.8–1.5 cm, base cordate to truncate or cuneate, margins entire or, rarely, hastate, ciliate, apex acuminate, faces abaxially and adaxially usually glabrous, rarely appressed-pubescent with eglandular hairs, sometimes also with glandular hairs, major veins retrorsely prickly. **Inflorescences** capitate or racemelike, uninterrupted, 4–15 × 3–8 mm; peduncle 10–80 mm, stipitate-glandular at least distally, often with retrorse prickles proximally; ocreolae overlapping, margins eciliate or ciliate with bristles to 0.5 mm. **Pedicels** mostly ascending, 2–3.5 mm. **Flowers** 2–3 per ocreate fascicle; perianth white to pink, glabrous, accrescent, not becoming blue and fleshy in fruit; tepals 5, connate ca. $\frac{1}{2}$ their length, broadly elliptic to ovate, 2.3–3 mm, apex obtuse; stamens 5, filaments distinct, free; anthers yellow or pink, ovate; styles 3, connate to middle. **Achenes** included, dark brown to black, 3-gonous, 3–4 × 2–3 mm, shiny, smooth.

Flowering Jun–Oct. Swamps, wet ditches, thickets; 0–80 m; Fla., Ga., La., S.C., Tex.; Mexico; West Indies; Central America; South America; se Africa.

6. **Persicaria bungeana** (Turczaninow) Nakai in T. Mori, Enum. Pl. Corea, 131. 1922 • Prickly smartweed I W

Polygonum bungeanum Turczaninow, Bull. Soc. Imp. Naturalistes Moscou 13: 77. 1840

Plants annual, 3–8 dm; roots not also arising from proximal nodes. **Stems** ascending to erect, ribbed or obscurely so, glabrous or glandular-pubescent distally; prickles 1–1.5 mm. **Leaves:** ocrea brownish, cylindric, 8–14 mm, chartaceous, base inflated or not, without prickles, margins truncate, ciliate with bristles 2–4 mm, surface with appressed bristles along veins; petiole 0.5–1.5 cm; blade lanceolate to narrowly elliptic, 5–12.5 × 1.5–3.5 cm, base acute, margins entire, antrorsely ciliate, apex acute acuminate, rarely obtuse, faces glabrous or pubescent and, usually, with antrorse prickles along midvein abaxially and adaxially. **Inflorescences** racemelike, uninterrupted or interrupted proximally, 20–45 × 5–10 mm; peduncle 20–40 mm, usually stipitate-glandular at least proximally; ocreolae usually overlapping, sometimes not overlapping proximally, margins eciliate. **Pedicels** mostly ascending, 2–3 mm. **Flowers** 2–4 per ocreate fascicle; perianth pale green, often tinged red, glabrous, accrescent, not becoming blue and fleshy in fruit; tepals 5, connate $\frac{1}{4}$–$\frac{1}{3}$ their length, petaloid, elliptic to broadly elliptic, 3–4 mm, apex obtuse; stamens 8, filaments distinct, free; anthers pink, ovate; styles 2, connate to middle. **Achenes** included, black, biconvex, 2.5–3 × 2.3–2.8 mm, dull, rugose.

Flowering Jul–Sep. Cultivated fields; 300–400 m; introduced; Ill., Iowa, Minn.; e Asia (n China, Japan, Korea, Manchuria).

Persicaria bungeana is a weed of soybean fields (R. N. Andersen et al. 1985). It is not known how or when it was introduced into the midwestern United States.

32c. **PERSICARIA** Miller sect. **CEPHALOPHILON** (Meisner) H. Gross, Bull. Acad. Int. Géogr. Bot. 4: 27. 1913 ☐

Polygonum Linnaeus sect. *Cephalophilon* Meisner in N. Wallich, Pl. Asiat. Rar. 3: 59. 1832

Stems prostrate or decumbent to ascending or erect, branched, unarmed. **Leaves:** ocrea usually chartaceous, rarely coriaceous proximally and chartaceous distally, margins eciliate or ciliate; petiole usually winged, auriculate; blade usually with dark triangular or lunate blotch adaxially, broadly elliptic to ovate, base cordate, rounded, truncate, or cuneate, margins entire. **Inflorescences** terminal or terminal and axillary, capitate, uninterrupted. **Flowers** homostylous, articulation swollen or not; perianth urceolate or campanulate; tepals (4–)5, connate ¹/₃–¹/₂ their length; stamens (5–)8; styles deciduous, 2–3, included, spreading.

Species ca. 16 (3 in the flora): introduced; Asia; introduced also in Europe, Africa, Pacific Islands.

1. Achenes biconvex; abaxial surface of leaf blades glandular-punctate . . . . 7. *Persicaria nepalensis*
1. Achenes 3-gonous; abaxial surface of leaf blades not glandular-punctate.
    2. Leaf blade margins ciliate with reddish, multicellular hairs; ocreae lanate; peduncles glabrous or stipitate-glandular distally . . . . 8. *Persicaria capitata*
    2. Leaf blade margins glabrous or antrorsely scabrous with whitish hairs; ocreae glabrous or pubescent; peduncles stipitate-glandular along entire length . . . . 9. *Persicaria chinensis*

7. **Persicaria nepalensis** (Meisner) H. Gross, Bot. Jahrb. Syst. 49: 277. 1913 ☐

Polygonum nepalense Meisner, Monogr. Polyg., 84, plate 7, fig. 2. 1826

Plants annual, 3–5 dm; roots also often arising from proximal nodes. **Stems** decumbent to ascending, glabrous except for fleshy, retrorse, whitish hairs at nodes. **Leaves:** ocrea brownish or hyaline, cylindric to funnelform, 4–10 mm, chartaceous, base inflated or not, margins oblique, eciliate, surface glabrous or with bristlelike hairs proximally; petiole 0.1–3 cm, winged to base, leaves sometimes sessile; blade ovate-deltate, 1.5–5 × 1–4 cm, base rounded to truncate, margins glabrous or scabrous, apex acute, faces pilose and glandular-punctate abaxially, glabrous adaxially. **Inflorescences** terminal and axillary, 5–10 × 5–10 mm; peduncle 2–20 mm, apex stipitate-glandular; ocreolae overlapping, margins eciliate. **Pedicels** mostly ascending, 0.1–1 mm, flowers sometimes sessile. **Flowers** 1–2 per ocreate fascicle; perianth white to pink or lavender, urceolate, glabrous, scarcely accrescent; tepals 4(–5), oblong to broadly elliptic, 2.5–3 mm, apex acute to obtuse; stamens (5–)8, filaments distinct, free; anthers purplish black, elliptic; styles 2, connate proximally. **Achenes** included, dark brown to black, biconvex, 1.5–2 × 1–1.5 mm, dull, minutely punctate.

Flowering Jul–Oct. Disturbed sites, gravel bars in lowland zone; 0–900 m; introduced; B.C.; Conn., Fla., Mass., N.Y., Pa.; Asia; introduced also in Europe, Africa.

8. **Persicaria capitata** (Buchanan-Hamilton ex D. Don) H. Gross, Bot. Jahrb. Syst. 49: 277. 1913 • Pink-head knotweed ☐

Polygonum capitatum Buchanan-Hamilton ex D. Don, Prodr. Fl. Nepal., 73. 1825

Plants annual or perennial, 0.5–5 dm; roots also often arising from proximal nodes. **Stems** prostrate, glabrous or glandular-pubescent. **Leaves:** ocrea brown or reddish brown, cylindric to funnelform, 5–12 mm, chartaceous, base inflated or not, margins oblique, eciliate or ciliate with bristles to 1.5 mm, surface lanate, sometimes also glandular-pubescent; petiole 2–5 mm, winged distally; blade ovate to elliptic, 1.5–4(–6) × 0.6–2.5(–3.3) cm, base cuneate or tapering, margins ciliate with reddish, multicellular hairs, apex acute, faces glandular-pubescent abaxially and adaxially, not glandular-punctate. **Inflorescences** terminal, 5–20 × 7–18 mm; peduncle 10–40 mm, glabrous or stipitate-glandular in distal ¹/₅; ocreolae overlapping, margins eciliate. **Pedicels** spreading, 0.5–1 mm. **Flowers** 1–5 per ocreate fascicle; perianth greenish white proximally, pinkish distally, urceolate, glabrous, nonaccrescent; tepals 5, elliptic, 2–3 mm, apex acute to obtuse; stamens 8, filaments distinct, free; anthers pink to red, elliptic; styles 3, connate to middle or distally. **Achenes** included, reddish brown to brownish black, 3-gonous, 1.5–2.2 × 1–1.5 mm, shiny, smooth or minutely punctate.

Flowering Jun–Sep. Disturbed, urban places; 0–500 m; introduced; Calif., La., Oreg.; Asia (Bhutan, w China, n India, Nepal); introduced also in the Pacific Islands (Hawaii).

Persicaria capitata is planted as a garden groundcover. It escapes infrequently in the flora area; once established outside of cultivation it can be difficult to eradicate.

9. Persicaria chinensis (Linnaeus) H. Gross, Bot. Jahrb. Syst. 49: 269. 1913 · Chinese knotweed [I]

Polygonum chinense Linnaeus, Sp. Pl. 1: 363. 1753

Plants perennial, 7–10 dm; roots not also arising from proximal nodes; rhizomes present. **Stems** ascending to erect, sometimes scandent, glabrous or retrorsely hispid. **Leaves:** ocrea brownish, cylindric, 15–25(–50) mm, coriaceous proximally, chartaceous distally, base often inflated, margins oblique, eciliate, surface glabrous or pubescent; petiole 1–2.5 cm, winged at least distally; blade lanceolate to ovate or elliptic, 4–16 × 1.5–8 cm, base truncate to broadly cordate, margins glabrous or antrorsely scabrous with whitish hairs, apex acuminate, faces glabrous or hispid abaxially and adaxially, sometimes pubescent only along veins abaxially, not glandular-punctate but often minutely reddish-punctate abaxially. **Inflorescences** terminal or terminal and axillary, 3–6 × 3–6 mm; peduncle 10–30 mm, stipitate-glandular along entire length; ocreolae overlapping, margins eciliate. **Pedicels** mostly ascending, 2–3 mm. **Flowers** 1–3 per ocreate fascicle; perianth white to pink, campanulate, glabrous, accrescent; tepals 5, ovate, 3–4 mm, apex acute to obtuse; stamens 8, filaments distinct, free; anthers red or purple, elliptic; styles 3, connate proximally. **Achenes** included in fleshy, bluish black perianth, black, 3-gonous, 2.8–4 × 2–3 mm, dull, minutely punctuate.

Flowering Jul–Oct. Disturbed places; 0–100 m; introduced; Md., Mass., N.J.; Asia; introduced also in the Pacific Islands (Hawaii).

Varieties of *Persicaria chinensis* have been distinguished on the basis of stem pubescence, leaf shape, and leaf size. Whether those taxa merit recognition in the flora area is uncertain.

32d. Persicaria Miller sect. **Rubrivena** (M. Král) S. P. Hong, Pl. Syst. Evol. 186: 112. 1993 [I]

Rubrivena M. Král, Preslia 57: 65, fig. 1. 1985

Stems erect, branched distally, unarmed. **Leaves:** ocrea chartaceous or coriaceous proximally and chartaceous distally, margins eciliate; petiole not winged, not auriculate; blade without dark triangular or lunate blotch adaxially, lanceolate to elliptic-lanceolate, base cordate to truncate, margins entire. **Inflorescences** mostly terminal, sometimes also axillary, paniclelike, interrupted. **Flowers** heterostylous, articulation not swollen; perianth rotate; tepals 5, connate less than ¹/₅ their length; stamens 8; styles deciduous, 3, included, spreading.

Species 2 (1 in the flora): introduced; Asia.

The status of sect. *Rubrivena* has been controversial; its members have been included in six different genera. Herbarium specimens often are identified as *Polygonum polystachyum* or *Aconogonon polystachyum*. Hong S. P. (1993) cited morphological and anatomical evidence for including *Rubrivena* in *Persicaria* as sister to sect. *Cephalophilon*, also noting pollen and inflorescence characters that distinguish it from *Aconogonon*.

SELECTED REFERENCE Hong, S. P. 1993. Reconsideration of the generic status of *Rubrivena* (Polygonaceae, Persicarieae). Pl. Syst. Evol. 186: 95–122.

10. **Persicaria wallichii** Greuter & Burdet, Willdenowia 19: 41. 1989 • Himalayan knotweed, Kashmir plume [I]

Polygonum polystachyum Wallich ex Meisner in N. Wallich, Pl. Asiat. Rar. 3: 61. 1832, not *Persicaria polystachya* Opiz 1852; *Aconogonon polystachyum* (Wallich ex Meisner) M. Král; *Pleuropteropyrum polystachyum* (Wallich ex Meisner) Munshi & G. N. Javied; *Reynoutria polystachya* (Wallich ex Meisner) Moldenke; *Rubrivena polystachya* (Wallich ex Meisner) M. Král

Varieties 2 (1 in the flora): introduced; Asia.

Persicaria wallichii is an ornamental that escapes infrequently in the flora area. A population in Nova Scotia apparently was ephemeral. Plants with leaf blades sparsely to densely pubescent abaxially and pedicels glabrous are var. *wallichii*, to which naturalized North American plants appear to be referable. Plants with leaf blades brownish-tomentose abaxially and pedicels usually pubescent are var. *tomentosa* S. P. Hong, which may be in cultivation in North America.

10a. **Persicaria wallichii** Greuter & Burdet var. **wallichii** [I]

Herbs, perennial, sometimes suffrutescent, 7–12(–25) dm; roots not also arising from proximal nodes; rhizomes present. **Stems** ribbed, glabrous or densely pubescent. **Leaves:** ocrea reddish brown, cylindric, 10–40 mm, base inflated, surface glabrous or densely pubescent; petiole 0.3–2(–3) cm, glabrous or densely pubescent; blade (7.5–)9–22(–27) × 2.8–7.8 cm, base cordate to truncate, margins ciliate, apex acuminate to caudate-acuminate, faces pubescent abaxially, glabrous or pubescent adaxially. **Inflorescences** 40–110 × 10–55 mm; peduncle 10–80 mm, glabrous or pubescent; ocreolae not overlapping, margins eciliate. **Pedicels** ascending to spreading, 1–4 mm, glabrous. **Flowers** 3–9 per ocreate fascicle; perianth white or pinkish, often with scattered reddish glands, glabrous, nonaccrescent; tepals oblong to obovate, dimorphic, outer 3 larger than inner 2, 2.2–3.8 mm, margins entire, apex obtuse to rounded; filaments inserted at bases of tepals, distinct, glabrous; anthers red to purple, ovate; styles distinct. **Achenes** included, brown, 3-gonous, 2.1–2.5 × 1.3–1.8 mm, dull, minutely roughened. *2n* = 22.

Flowering Aug–Oct. Moist disturbed places, marshes; 0–500 m; introduced; St. Pierre and Miquelon; B.C.; Calif., Mass., Oreg.; Asia (Bhutan, China, India, Nepal, Tibet).

32e. PERSICARIA Miller sect. PERSICARIA

Stems ascending to erect or, rarely, prostrate, usually branched, sometimes simple, unarmed. **Leaves:** ocrea chartaceous, rarely foliaceous distally, margins eciliate or ciliate; petiole not winged, not auriculate; blade sometimes with dark triangular or lunate blotch adaxially, lanceolate to narrowly elliptic, base tapered, acute, rounded, or cordate, margins entire. **Inflorescences** terminal or terminal and axillary, spikelike, uninterrupted or interrupted. **Flowers** homostylous or heterostylous, articulation swollen or not; perianth campanulate; tepals 4–5, connate ¼–⅔ their length; stamens 5–8; styles deciduous, 2–3, included or exserted (exserted syles and stamens in heterostylous species), spreading.

Species ca. 60 (16 in the flora): nearly worldwide, especially n temperate regions.

Members of sect. *Persicaria* frequently grow in moist or inundated habitats. Many have evolved responses to submergence that allow them to survive extended periods of inundation

(R. S. Mitchell 1976). Hybridization often is implicated for the taxonomic difficulties in the section. However, J. Timson (1965) concluded that, at least among annual European species, hybridization is rare due to autogamous breeding systems. Mitchell (1971) demonstrated the usefulness of leaf morphology in distinguishing major groups among the native North American smartweeds.

SELECTED REFERENCES Consaul, L. L., S. I. Warwick, and J. McNeill. 1991. Allozyme variation in the *Polygonum lapathifolium* complex. Canad. J. Bot. 69: 2261–2270.   Dalci, M. 1972. The Taxonomy of Section *Persicaria* (Tourn.) L. in the Genus *Polygonum* (Tourn.) L. (Polygonaceae) in the United States East of the Rocky Mountains. Ph.D. dissertation. University of Nebraska.   McDonald, C. B. 1980. A biosystematic study of the *Polygonum hydropiperoides* (Polygonaceae) complex. Amer. J. Bot. 67: 664–670.   Timson, J. 1963. The taxonomy of *Polygonum lapathifolium* L., *P. nodosum* Pers., and *P. tomentosum* Schrank. Watsonia 5: 386–395.   Yang, J. and Wang J. W. 1991. A taxometric analysis of characters of *Polygonum lapathifolium* L. Acta Phytotax. Sin. 29: 258–263.

1. Some or all ocreae foliaceous and green distally.
  2. Plants annual; rhizomes and stolons absent; leaf blades ovate, 3–17 cm wide . . . 22. *Persicaria orientalis*
  2. Plants perennial; rhizomes or stolons usually present; leaf blades ovate-lanceolate to elliptic or oblong-lanceolate, 1–6(–8) cm wide . . . . 11. *Persicaria amphibia* (in part)
1. All ocreae chartaceous and hyaline, tan, brown, or reddish brown throughout, never foliaceous and green distally.
  3. Perianths glandular-punctate.
    4. Achenes minutely roughened, dull; axillary inflorescences sometimes enclosed in ocreae . . . . 21. *Persicaria hydropiper*
    4. Achenes smooth, shiny; inflorescences never enclosed in ocreae.
      5. Outer tepals with anchor-shaped veins; achenes discoid . . . . 20. *Persicaria lapathifolia* (in part)
      5. Outer tepals without anchor-shaped veins; achenes 3-gonous or biconvex.
        6. Ocreae margins eciliate; achenes biconvex; styles 2 . . . . 14. *Persicaria glabra* (in part)
        6. Ocreae margins ciliate with bristles 2–12 mm; achenes usually 3-gonous, rarely biconvex; styles 2–3.
          7. Punctae confined to bases of perianths and sometimes on inner tepals . . . . 13. *Persicaria hydropiperoides* (in part)
          7. Punctae ± uniformly distributed over perianths.
            8. Inflorescences interrupted; ocreolae mostly not overlapping, margins mostly ciliate with hairs to 2 mm; leaf blades 0.6–2.4 cm wide . . . . 15. *Persicaria punctata*
            8. Inflorescences uninterrupted; ocreolae usually overlapping, margins mostly eciliate or proximal sometimes with hairs to 1 mm; leaf blades 2–4.5 cm wide . . . . 12. *Persicaria robustior*
  3. Perianths not glandular-punctate.
    9. Peduncles stipitate-glandular.
      10. Plants perennial; rhizomes or stolons usually present; inflorescences terminal . . . . 11. *Persicaria amphibia* (in part)
      10. Plants annual; rhizomes and stolons absent; inflorescences terminal and axillary.
        11. Stems hirsute proximally; margins of ocreae ciliate with bristles 2–7 mm . . . . 23. *Persicaria careyi*
        11. Stems glabrous proximally; margins of ocreae eciliate or ciliate with bristles less than 1 mm.
          12. Outer tepals with anchor-shaped veins; tepals 4(–5); inflorescences mostly arching or nodding . . . . 20. *Persicaria lapathifolia* (in part)
          12. Outer tepals without anchor-shaped veins; tepals 5; inflorescences erect or, rarely, nodding.
            13. Flowers homostylous; achenes without central hump on 1 side . . . . 18. *Persicaria pensylvanica* (in part)
            13. Flowers heterostylous; achenes usually with central hump on 1 side . . . . 19. *Persicaria bicornis* (in part)

[9. Shifted to left margin.—Ed.]

9. Peduncles not stipitate-glandular.
    14. Plants perennial; rhizomes or stolons usually present.
        15. Achenes biconvex; styles 2.
            16. Perianth greenish white or white to pink; surfaces of ocreae glabrous, usually obscurely glandular-punctate . . . . 14. *Persicaria glabra* (in part)
            16. Perianth roseate to red; surfaces of ocreae glabrous or appressed-pubescent to hirsute, not glandular-punctate . . . . 11. *Persicaria amphibia* (in part)
        15. Achenes 3-gonous; styles 3.
            17. Proximal part of ocreae strigose, or ocreae glabrous . . . 13. *Persicaria hydropiperoides* (in part)
            17. Proximal part of ocreae hirsute, or strigose and with loosely ascending to spreading hairs at least proximally.
                18. Bases of leaf blades rounded to cordate; stems brownish-hirsute on internodes . . . . 16. *Persicaria hirsuta*
                18. Bases of leaf blades tapered to truncate; stems glabrous or loosely appressed-to spreading-hirsute near nodes . . . . 17. *Persicaria setacea*
    14. Plants annual; rhizomes and stolons absent.
        19. Margins of ocreae without bristles or with bristles to 1 mm; ocreolae mostly overlapping; achenes discoid, rarely 3-gonous.
            20. Outer tepals with anchor-shaped veins; tepals 4(–5); inflorescences mostly arching or nodding . . . . 20. *Persicaria lapathifolia* (in part)
            20. Outer tepals without anchor-shaped veins; tepals 5; inflorescences mostly erect, rarely nodding.
                21. Flowers homostylous; achenes without central hump on 1 side . . . . 18. *Persicaria pensylvanica* (in part)
                21. Flowers heterostylous; achenes usually with central hump on 1 side . . . . 19. *Persicaria bicornis* (in part)
        19. Margins of ocreae ciliate with bristles (0.2–)1–12 mm, if bristles less than 1 mm then ocreolae not overlapping; achenes discoid, biconvex, or 3-gonous.
            22. Inflorescences not dense; ocreolae not overlapping proximally, usually overlapping distally; leaf blades linear to linear-lanceolate . . . . 25. *Persicaria minor*
            22. Inflorescences dense; ocreolae mostly overlapping; leaf blades narrowly ovate or ovate-lanceolate to linear-lanceolate.
                23. Bristles of ocreolae 0.2–1.3(–2) mm; achenes discoid, biconvex, or 3-gonous; styles 2–3 . . . . 26. *Persicaria maculosa*
                23. Bristles of ocreolae (0.5–)1–4(–6) mm; achenes 3-gonous; styles 3 . . . 24. *Persicaria longiseta*

11. Persicaria amphibia (Linnaeus) Gray, Nat. Arr. Brit. Pl. 2: 268. 1821 • Water smartweed, renouée amphibie W

Polygonum amphibium Linnaeus, Sp. Pl. 1: 361. 1753; *Persicaria amphibia* (Linnaeus) Gray var. *emersa* (Michaux) J. C. Hickman; *P. amphibia* var. *stipulacea* (N. Coleman) H. Hara; *P. coccinea* (Muhlenberg ex Willdenow) Greene; *P. hartwrightii* (A. Gray) Greene; *P. muhlenbergia* (S. Watson) Small; *Polygonum amphibium* var. *emersum* Michaux; *P. amphibium* subsp. *laevimarginatum* Hultén; *P. amphibium* var. *natans* Michaux; *P. amphibium* var. *stipulaceum* N. Coleman; *P. coccineum* Muhlenberg ex Willdenow; *P. coccineum* var. *pratincola* (Greene) Stanford; *P. coccineum* var. *rigidulum* (E. Sheldon) Stanford; *P. emersum* (Michaux) Britton; *P. hartwrightii* A. Gray; *P. natans* (Michaux) Eaton

Plants perennial, 2–12 dm in terrestrial plants, to 30 dm in some aquatic plants; roots also sometimes arising from proximal nodes; rhizomes or stolons usually present. **Stems** prostrate to ascending or erect, simple or branched, ribbed, glabrous or strigose to hirsute. **Leaves:** ocrea tan to dark brown, cylindric or flared distally, 5–50 mm, chartaceous or, sometimes, foliaceous distally, base inflated, margins truncate to oblique, glabrous or ciliate with hairs 0.5–4.5 mm, surface glabrous or appressed-pubescent to hirsute, not glandular-punctate; petiole 0.1–3(–7) cm, glabrous or appressed-pubescent to hirsute, leaves sometimes sessile; blade without dark triangular or lunate blotch adaxially, ovate-lanceolate to elliptic or oblong-lanceolate, 2–15(–23) × 1–6(–8) cm, base usually tapered to acute or rounded, rarely cordate, margins antrorsely scabrous, apex acute to acuminate, faces glabrous or sparingly strigose, midveins glabrous or strigose, not glandular-punctate. **Inflorescences** terminal, ascending to erect, uninterrupted or interrupted proximally, 10–150 × 8–20 mm; peduncle 10–50 mm, glabrous or strigose to hirsute, often stipitate-glandular; ocreolae overlapping except sometimes proximal ones, margins ciliate with bristles to 1 mm. **Pedicels** ascending, 0.5–1.5 mm. **Flowers** bisexual or functionally unisexual, some plants having only staminate flowers, others with only pistillate flowers, 1–3(–4) per ocreate fascicle, heterostylous; perianth roseate to red, glabrous, not glandular-punctate, slightly accrescent; tepals 5, connate ca. ¹/₃ their lengths, obovate to elliptic, 4–6 mm, veins prominent, not anchor-shaped, margins entire, apex rounded to acute; stamens 5, included or exserted; anthers pink or red, elliptic; styles 2, included or exserted, connate ¹/₂–²/₃ their length. **Achenes** included, dark brown, biconvex, (2–)2.2–3 × (1.5–)1.8–2.6 mm, shiny or dull, smooth or minutely granular. $2n = 66, 132$.

Flowering Jun–Sep. Shallow water, shorelines of ponds and lakes, banks of rivers and streams, moist prairies and meadows; 0–3000 m; St. Pierre and Miquelon; Alta., B.C., Man., N.B., Nfld. and Labr., N.W.T., N.S., Ont. , P.E.I., Que., Sask., Yukon; Alaska, Ariz., Ark., Calif., Colo., Conn., Del., D.C., Idaho, Ill., Ind., Iowa, Kans., Ky., La., Maine, Md., Mass., Mich., Minn., Miss., Mo., Mont., Nebr., Nev., N.H., N.J., N.Mex., N.Y., N.C., N.Dak., Ohio, Okla., Oreg., Pa., R.I., S.C., S.Dak., Tenn., Tex., Utah, Vt., Va., Wash., W.Va., Wis., Wyo.; Mexico; South America; Eurasia; Africa.

Persicaria amphibia is widespread in the Northern Hemisphere and naturalized in Mexico, South America, and southern Africa. It is highly polymorphic and the most hydrophytic of the native North American smartweeds (R. S. Mitchell 1976). In recent decades, botanists have tended to follow Mitchell (1968) in recognizing two endemic, intergrading North American varieties. Studies by G. Turesson (1961) and Mitchell (1968, 1976) have shown that phenotypic extremes in the species are part of a cline of nearly continuous morphological variation that is strongly correlated with submergence, but also with some genetic integrity. Formal recognition of varieties is even less tenable when Eurasian elements also are considered.

Aquatic-adapted plants, which bloom in water or are sometimes stranded on land, have been called var. *stipulacea* (although that epithet may not be the oldest one available for the taxon). They produce ovoid-conic to short-cylindric inflorescences 10–40(–60) mm, prostrate aerial stems, and leaf blades that are glabrous with acute to rounded apices. Terrestrial forms of this ecotype usually are spreading-pubescent and often bear ocreae that are foliaceous, green, and flared distally, characters found only in North American plants (R. S. Mitchell 1968).

Terrestrial-adapted plants, referred to var. *emersa*, bloom on moist soil and produce short- to elongate-

cylindric inflorescences 40–110(–150) mm, spreading or erect aerial stems, and leaf blades that are appressed-pubescent with acute to acuminate apices. They produce ocreae that are entirely chartaceous and not flared distally. Emergent and terrestrial plants of this ecotype exhibit less phenotypic plasticity and a lower frequency of heterostyly than do plants of the aquatic ecotype (R. S. Mitchell 1968).

R. S. Mitchell and J. K. Dean (1978) and H. R. Hinds (2000) recognized var. *amphibia*, the Eurasian element, as introduced in New York and New Brunswick, respectively. These plants are morphologically intermediate between the North American ecotypes and often indistinguishable from North American plants (Mitchell and Dean).

The Meskwaki and Ojibwa used leaves, stems, and roots as a drug to treat a variety of maladies, the Potawatomi used roots to treat unspecified ailments, and the Lakota and Sioux used plants for food (D. E. Moerman 1998).

SELECTED REFERENCE Mitchell, R. S. 1968. Variation in the *Polygonum amphibium* complex and its taxonomic significance. Univ. Calif. Publ. Bot. 45: 1–65.

12. Persicaria robustior (Small) E. P. Bicknell, Bull. Torrey Bot. Club 36: 455. 1909 • Stout smartweed, renouée robuste W

Polygonum punctatum Elliott var. *robustius* Small, Bull. Torrey Bot. Club 21: 477. 1894; *Persicaria punctata* (Elliott) Small var. *robustior* (Small) Small; *Polygonum punctatum* var. *majus* (Meisner) Fassett; *P. robustius* (Small) Fernald

Plants perennial, 3–20 dm; roots also sometimes arising from proximal nodes; rhizomes present, stolons sometimes present. **Stems** ascending, usually branched proximally, scarcely ribbed, glabrous, glandular-punctate; branches sometimes creeping and rooting at nodes. **Leaves:** ocrea light brown, cylindric, 10–15 mm, chartaceous, base inflated, margins truncate, ciliate with bristles 3–12 mm, surface strigose, glandular-punctate; petiole 0.2–2 cm, glandular-punctate; blade without dark triangular or lunate blotch adaxially, lanceolate to elliptic-lanceolate, 4–20 × 2–4.5 cm, base tapered, margins antrorsely strigose, apex acute to acuminate, faces glabrous or main veins scabrous, glandular-punctate. **Inflorescences** terminal and axillary, erect, uninterrupted, 20–80 × 5–10 mm; peduncle 5–40 mm, glandular-punctate; ocreolae usually overlapping, margins eciliate or proximal ones sometimes ciliate with bristles to 1 mm. **Pedicels** ascending to spreading, 2–5 mm. **Flowers** 2–4 per ocreate fascicle, homostylous; perianth greenish proximally, white distally, glandular-punctate with punctae ± uniformly distributed, slightly

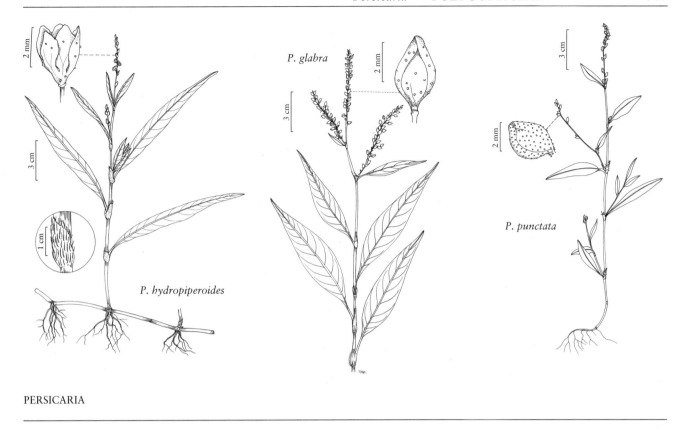

P. glabra

P. punctata

P. hydropiperoides

PERSICARIA

accrescent; tepals 5, connate ca.¹/₃ their length, obovate, 3.2–4.2 mm, veins prominent or not, not anchor-shaped, margins entire, apex obtuse to rounded; stamens 6–8, included; anthers pink or red, elliptic; styles 3, connate proximally. **Achenes** included or apex exserted, dark brown to brownish black, 3-gonous, 2.7–3.6 × 2–2.5 mm, shiny, smooth.

Flowering Jul–Oct. Peaty shores, often in water on coastal plain or near coast; 0–1500 m; N.S., Ont., Que.; Conn., Del., Fla., Ind., Ky., Maine, Md., Mass., Mich., Mo., N.H., N.J., N.Y., Ohio, Pa., R.I., Tex., Va.; West Indies; Central America; South America.

Persicaria robustior often is synonymized with *P. punctata.*

13. Persicaria hydropiperoides (Michaux) Small, Fl. S.E. U.S., 378. 1903 • Swamp smartweed, renouée faux-poivre-d'eau F W

Polygonum hydropiperoides Michaux, Fl. Bor.-Amer. 1: 239. 1803; *Persicaria opelousana* (Riddell ex Small) Small; *P. paludicola* Small; *Polygonum hydropiperoides* var. *adenocalyx* (Stanford) Gleason; *P. hydropiperoides* var. *asperifolium* Stanford; *P. hydropiperoides* var. *breviciliatum* Fernald; *P. hydropiperoides* var. *bushianum* Stanford; *P. hydropiperoides* var. *digitatum* Fernald; *P. hydropiperoides* var. *opelousanum* (Riddell ex Small) W. Stone; *P. hydropiperoides* var. *psilostachyum* H. St. John; *P. opelousanum* Riddell ex Small; *P. opelousanum* var. *adenocalyx* Stanford

Plants perennial, 1.5–10 dm; roots also often arising from proximal nodes; rhizomes often present. **Stems** decumbent to ascending, usually branched, without noticeable ribs, glabrous or obscurely strigose distally. **Leaves:** ocrea brown, cylindric, 5–23 mm, chartaceous, base inflated, margins truncate, ciliate with bristles (2–)4–10 mm, surface glabrous or strigose, not glandular-punctate; petiole 0.2–2 cm, glabrous or strigose; blade without dark triangular or lunate blotch adaxially, broadly lanceolate to linear-lanceolate, 5–25 × 0.4–3.7 cm, base tapered or acute, margins antrorsely appressed-pubescent, apex acuminate, faces glabrous or appressed-pubescent along midveins and sometimes on faces, usually punctate abaxially. **Inflorescences** terminal, sometimes also axillary, erect, uninterrupted or interrupted proximally, 30–80 × 2–5 mm; peduncle 10–30 mm, glabrous or strigose; ocreolae overlapping distally, often not overlapping proximally, margins ciliate with bristles to 2(–3) mm. **Pedicels** ascending, 1–1.5 mm. **Flowers** bisexual or unisexual and staminate, 2–6 per ocreate fascicle, homostylous; perianth roseate proximally, roseate, white, or greenish white distally, not glandular-punctate or sometimes glandular-punctate with punctae on tubes and inner tepals, scarcely accrescent; tepals 5, connate ca. ¹/₃–¹/₂ their length, obovate, 2.5–4 mm in bisexual flowers, 1.5–2.5 mm in staminate flowers, veins prominent or not, not anchor-shaped, margins entire, apex obtuse to rounded; stamens 8, included or exserted in staminate flowers; anthers pink or red, elliptic to ovate; styles 3, connate near middle. **Achenes** included or apex exserted, brown to brownish black or black, 3-gonous, 1.5–3 × 1–2.3 mm, shiny, smooth.

Flowering Jun–Nov. Wet banks and clearings, shallow water, marshes, moist prairies, ditches; 0–1500 m; B.C., N.B., N.S., Ont., Que.; Ala., Alaska, Ariz., Ark., Calif., Conn., Del., D.C., Fla., Ga., Idaho, Ill., Ind., Iowa, Kans., Ky., La., Maine, Md., Mass., Mich., Minn., Miss., Mo., Nebr., N.H., N.J., N.Y., N.C., N.Dak., Ohio, Okla., Oreg., Pa., R.I., S.C., S.Dak., Tenn., Tex., Vt., Va., Wash., W.Va., Wis.; Mexico; Central America; South America.

The extreme variability in *Persicaria hydropiperoides* is reflected in its extensive synonymy. Among the segregates most often recognized in floras and checklists is *P. opelousana*, which C. B. McDonald (1980) showed to be broadly sympatric and highly interfertile with *P. hydropiperoides*. Consistent with this conclusion, R. S. Mitchell (1971) found that *P. hydropiperoides* and *P. opelousana* are unique among native North American smartweeds in consistently possessing multicellular plate-glands on the abaxial surface of their leaves. Such glands also are found on *P. maculosa*, an introduced European species.

Herbarium specimens of *Persicaria hydropiperoides* sometimes are misidentified as *P. maculosa*, especially when the roots are missing. The former species may be distinguished reliably by its achenes all trigonous (trigonous and biconvex achenes are mixed in the inflorescences of *P. maculosa*) and bristles on the margins of the ocreae that average longer. M. L. Fernald (1922c) reported hybrids with *P. robustior* from Nova Scotia.

14. **Persicaria glabra** (Willdenow) M. Gómez, Anales Inst. Segunda Enseñ. 2: 278. 1896 • Smooth smartweed [F]

Polygonum glabrum Willdenow, Sp. Pl. 2: 447. 1799; *Persicaria densiflora* (Meisner) Moldenke; *P. portoricensis* (Bertero ex Small) Small; *Polygonum densiflorum* Meisner; *P. portoricense* Bertero ex Small

Plants perennial, 3–15 dm; roots also often arising from proximal nodes; rhizomes present. **Stems** decumbent to erect, usually branched distally, without noticeable ribs, glabrous or, rarely, pubescent distally, sometimes glandular-punctate. **Leaves:** ocrea light brown, cylindric, 12–23 mm, chartaceous, base inflated, margins truncate, eciliate, surface glabrous, usually obscurely glandular-punctate; petiole 0.2–2 cm, scabrous; blade without dark triangular or lunate blotch adaxially, lanceolate, (10–)15–30 ×

(1.5–)2–5.4 cm, base tapered, margins glabrous or antrorsely strigose, apex acute to acuminate, faces glabrous or scabrous along midveins, sometimes glandular-punctate. **Inflorescences** mostly terminal, sometimes also axillary, erect to slightly nodding, usually uninterrupted, 30–100 × 5–9 mm; peduncle 10–50 mm, glabrous or scabrid, glandular-punctate; ocreolae usually overlapping, margins eciliate. **Pedicels** erect to spreading, 2–5 mm. **Flowers** (1–)3–8 per ocreate fascicle, homostylous; perianth greenish white to white or pink, glabrous, not glandular-puncate or glandular-punctate with punctae ± uniformly distributed, scarcely accrescent; tepals 5, connate ca. $^{1}/_{3}$ their length, obovate, 3–3.6 mm, veins not prominent, not anchor shaped, margins entire, apex obtuse to rounded; stamens 5–7, included; anthers pink or red, ovate; styles 2, connate proximally. **Achenes** included, dark brown to brownish black, biconvex, 2–2.2 × 1.3–1.6 mm, shiny, smooth.

Flowering Aug–Nov. Swamps, wet thickets, marshy shores, frequently in water, mostly on coastal plain of e North America; 0–300; Ala., Ark., Del., Fla., Ga., Ky., La., Md., Miss., Mo., N.J., N.C., S.C., Tenn., Tex., Va.; Central America; South America; Asia; ne Africa; Pacific Islands (Hawaii, Philippines).

American plants here included in *Persicaria glabra* often have been treated as distinct and called *P. densiflora*. The morphological differences between them and Asian and Pacific *P. glabra* are minor. Regional tendencies exist but do not appear sufficient to warrant separation of the species (K. L. Wilson 1990b).

An infusion made from pounded whole plants was used by the Hawaiians as a blood medicine (D. E. Moerman 1998).

15. **Persicaria punctata** (Elliott) Small, Fl. S.E. U.S., 379. 1903 • Dotted smartweed, renouée ponctuée [F][W]

Polygonum punctatum Elliott, Sketch Bot. S. Carolina 1: 455. 1817; *P. acre* Kunth var. *leptostachyum* Meisner; *P. punctatum* var. *confertiflorum* (Meisner) Small; *P. punctatum* var. *ellipticum* Fassett; *P. punctatum* var. *leptostachyum* (Meisner) Small; *P. punctatum* var. *parviflorum* Fassett; *P. punctatum* var. *parvum* Marie-Victorin & Rousseau

Plants annual or perennial, 1.5–12 dm; roots also often arising from proximal nodes; rhizomes often present. **Stems** ascending to erect, branched, without noticeable ribs, glabrous, glandular-punctate. **Leaves:** ocrea brown, cylindric, (4–)9–18 mm, chartaceous, base inflated, margins truncate, ciliate with bristles 2–11 mm, surface glabrous or strigose, glandular-punctate; petiole 0.1–1 cm, glandular-punctate, leaves sometimes sessile; blade without dark triangular or lunate blotch adaxially, lanceolate to lanceolate-ovate or subrhombic, 4–10(–15) × 0.6–2.4 cm, base tapered or cuneate, margins antrorsely strigose, apex acute to acuminate, faces glabrous or scabrous along midveins, glandular-punctate. **Inflorescences** mostly terminal, sometimes also axillary, erect, interrupted, 50–200 × 4–8 mm; peduncle 30–60 mm, glabrous, glandular-punctate; ocreolae mostly not overlapping, margins mostly ciliate with bristles to 2 mm. **Pedicels** ascending, 1–4 mm. **Flowers** 2–6 per ocreate fascicle, homostylous; perianth greenish proximally, white distally, rarely tinged pink, glandular-punctate with punctae ± uniformly distributed, scarcely accrescent; tepals 5, connate ca. ¹/₃ their length, obovate, 3–3.5 mm, veins prominent or not, not anchor-shaped, margins entire, apex obtuse to rounded; stamens 6–8, included; anthers pink or red, elliptic to ovate; styles 2–3, connate proximally. **Achenes** included or apex exserted, brownish black, usually 3-gonous, rarely biconvex, (1.8–)2.2–3.2 × 1.5–2.2 mm, shiny, smooth. $2n = 44$.

Flowering Jun–Nov. Shallow water, shores, marshes, floodplain forests; 0–1500 m; B.C., Man., N.B., N.S., Ont., P.E.I., Que., Sask.; Ala., Ariz., Ark., Calif., Colo., Conn., Del., D.C., Fla., Ga., Idaho, Ill., Ind., Iowa, Kans., Ky., La., Maine, Md., Mass., Mich., Minn., Miss., Mo., Mont., Nebr., N.H., N.J., N.Mex., N.Y., N.C., N.Dak., Ohio, Okla., Oreg., Pa., R.I., S.C., S.Dak., Tenn., Tex., Vt., Va., Wash., W.Va., Wis., Wyo.; Mexico; West Indies (Puerto Rico); Central America (Guatemala); South America (Brazil); Pacific Islands (Hawaii).

N. C. Fassett (1949) proposed a complicated classification for *Persicaria punctata* with 12 varieties in North America and South America. He also identified numerous specimens that he considered to be morphologically intermediate between various varieties. M. Dalci (1972) documented a wide range of phenotypic and genotypic variation throughout the range of *P. punctata* and extensive overlap in many of the features used by Fassett to distinguish varieties. Consequently, recognition of varieties does not seem warranted. *Persicaria punctata* and its close relatives *P. robustior* and *P. glabra* are unique among native North American smartweeds in possessing complex glands called valvate chambers in their epidermises. *Persicaria punctata* is confused most frequently with *P. hydropiper*; the achenes are diagnostic.

The Chippewa, Houma, and Iroquois prepared decoctions from leaves, flowers, and roots for use as analgesics as well as gastrointestinal, orthopedic, and psychological aids (D. E. Moerman 1998).

SELECTED REFERENCE Fassett, N. C. 1949. The variations of *Polygonum punctatum*. Brittonia 6: 369–393.

16. **Persicaria hirsuta** (Walter) Small, Fl. S.E. U.S., 379. 1903 • Hairy smartweed [E]

Polygonum hirsutum Walter, Fl. Carol., 132. 1788

Plants perennial, 3–9 dm; roots also often arising from proximal nodes; rhizomes present. **Stems** decumbent to ascending or erect, branched, without noticeable ribs, brownish-hirsute on internodes. **Leaves:** ocrea brown to reddish brown, cylindric, 6–12 mm, chartaceous, base sometimes inflated, margins truncate, eciliate or ciliate with bristles 4–7.5 mm, surface hirsute, not glandular-punctate; petiole 0.1(–0.3) cm, hirsute, leaves sometimes sessile; blade without dark triangular or lunate blotch adaxially, ovate to lanceolate, (2–)4–8 × (0.5–)1–2.5 cm, base rounded to cordate, margins strigose to hirsute, apex acute to acuminate, faces sparingly hirsute abaxially and adaxially, midvein usually hirsute abaxially. **Inflorescences** mostly terminal, erect, interrupted proximally, usually uninterrupted distally, 20–80 × 4–8 mm; peduncle 30–60 mm, hirsute or, sometimes, nearly glabrous distally; ocreolae overlapping distally, usually not overlapping proximally, margins ciliate with bristles 0.4–1.5(–2) mm. **Pedicels** ascending, 1–2 mm. **Flowers** 1–3 per ocreate fascicle, homostylous; perianth white to pink, glabrous, not glandular-punctate, nonaccrescent; tepals 5, connate in proximally ¹/₃, obovate, 1.5–2 mm, veins not prominent, not anchor-shaped, margins entire, apex obtuse to rounded; stamens 5, included; anthers red, elliptic to ovate; styles 3, connate proximally. **Achenes** included or apex exserted, dark brown to brownish black, 3-gonous, 2–2.5 × 1.3–1.8 mm, shiny, smooth. $2n = 20$.

Flowering Jun–Oct. Sandy soils, open areas in savannahs, pond margins, ditches, often in shallow water; 0–100 m; Ala., Fla., Ga., Miss., N.C., S.C.

C. B. McDonald (1980) showed that *Persicaria hirsuta* is closely related to *P. setacea* and *P. hydropiperoides*. Hybrids between *P. hirsuta* and *P. setacea* have been produced experimentally but appear to be rare in the wild. Although geographically sympatric, the two species generally occupy different habitats. Experimental crosses between *P. hirsuta* and *P. hydropiperoides* were unsuccessful (McDonald).

17. Persicaria setacea (Baldwin) Small, Fl. S.E. U.S., 379. 1903 • Bog smartweed E

Polygonum setaceum Baldwin in S. Elliott, Sketch Bot. S. Carolina 1: 455. 1817; *P. hydropiperoides* Michaux var. *setaceum* (Baldwin) Gleason; *P. setaceum* var. *interjectum* Fernald; *P. setaceum* var. *tonsum* Fernald

Plants perennial, 5–15 dm; roots also often arising from proximal nodes; rhizomes present, stolons sometimes produced on plants in water. **Stems** ascending or erect, branched distally, slightly ribbed, glabrous or loosely appressed- to spreading-hirsute near nodes. **Leaves:** ocrea brown, cylindric, 10–20 mm, chartaceous, base usually inflated, margins truncate, ciliate with bristles 6–12 mm, surface strigose and with loosely ascending to spreading hairs at least proximally, not glandular-punctate; petiole 0.1–0.5 cm, spreading-hirsute, leaves sometimes sessile; blade without dark triangular or lunate blotch adaxially, lanceolate, (3–)6–15(–18) × (1.5–)2–3.2(–4.8) cm, base tapered to truncate, margins appressed-ciliate, apex acute to acuminate, faces sparsely hirsute to loosely appressed hirsute abaxially and adaxially. **Inflorescences** mostly terminal, erect, uninterrupted, 20–80 × 4–8 mm; peduncle 10–70 mm, strigose; ocreolae overlapping, margins ciliate with bristles (0.6–)1–3(–5) mm. **Pedicels** ascending, 1–3 mm. **Flowers** (1–)2–4(–5) per ocreate fascicle, homostylous; perianth greenish proximally, creamy or tan distally, occasionally tinged pink, glabrous, not glandular-punctate, nonaccrescent; tepals 5, connate ca. 1/3 their length, obovate, 2–3 mm, veins not prominent, not anchor-shaped, margins entire, apex obtuse to rounded; stamens 5, included; anthers pink or red, elliptic; styles 3, connate proximally. **Achenes** included or apex exserted, brown to black, 3-gonous, (1.5–)2–2.5 × 1.2–1.7 mm, shiny, smooth. *2n* = 20.

Flowering Jul–Oct. Alluvial woods, swamp forests; 0–300 m; Ala., Ark., Del., Fla., Ga., Ill., Ind., Ky., La., Md., Mass., Mich., Miss., Mo., N.J., N.Y., N.C., Ohio, Okla., Pa., R.I., S.C., Tenn., Tex., Va., Wash.

C. B. McDonald (1980) showed that *Persicaria setacea* is closely related to *P. hirsuta* and *P. hydropiperoides*. Hybrids between *P. setacea* and *P. hirsuta* have been produced experimentally but appear to be rare in the wild. *Persicaria setacea* and *P. hydropiperoides* occasionally occur in mixed populations but do not hybridize (McDonald). *Persicaria setacea* sometimes intergrades morphologically with *P. hydropiperoides*, especially in New England. Specimens of *P. setacea* without the characteristic ascending or spreading hairs on the ocreae usually can be distinguished from *P. hydropiperoides* by the extent of adnation of the hairs to the ocreae—up to one-third their lengths in *P. setacea*, but one-third to two-thirds their lengths in *P. hydropiperoides*.

18. Persicaria pensylvanica (Linnaeus) M. Gómez, Anales Inst. Segunda Enseñ. 2: 278. 1896 • Pennsylvania smartweed, renouée de Pennsylvanie W

Polygonum pensylvanicum Linnaeus, Sp. Pl. 1: 362. 1753; *Persicaria mississippiensis* (Stanford) Small; *P. pensylvanica* var. *dura* (Stanford) C. F. Reed; *Polygonum omissum* Greene; *P. pensylvanicum* var. *durum* Stanford; *P. pensylvanicum* var. *eglandulosum* Myers; *P. pensylvanicum* var. *laevigatum* Fernald; *P. pensylvanicum* var. *nesophilum* Fernald; *P. pensylvanicum* var. *rosiflorum* Norton

Plants annual, 1–20 dm; roots also occasionally arising from basal nodes; rhizomes and stolons absent. **Stems** ascending to erect, simple or branched, ribbed, glabrous or appressed-pubescent distally, eglandular or stipitate-glandular distally. **Leaves:** ocrea brownish, cylindric, 5–20 mm, chartaceous, base inflated, margins truncate, eciliate or ciliate with bristles to 0.5 mm, surface glabrous or appressed-pubescent, eglandular; petiole 0.1–2(–3) cm, glabrous or appressed-pubescent; blade sometimes with dark triangular or lunate blotch adaxially, narrowly to broadly lanceolate, 4–17(–23) × (0.5–)1–4.8 cm, base tapered to cuneate, margins antrorsely scabrous, apex acuminate, faces glabrous or appressed-pubescent, eglandular or glandular-punctate abaxially and occasionally adaxially. **Inflorescences** terminal and axillary, erect or rarely nodding, uninterrupted, 5–50 × 5–15 mm; peduncle 10–55(–70) mm, glabrous or pubescent, usually stipitate-glandular; ocreolae overlapping, margins eciliate or ciliate with bristles to 0.5 mm. **Pedicels** ascending, 1.5–4.5 mm. **Flowers** 2–14 per ocreate fascicle, homostylous; perianth greenish white to roseate, glabrous, not glandular-punctate, accrescent; tepals 5, connate ca. 1/4–1/3 their length, obovate to elliptic, 2.5–5 mm, veins prominent, not anchor-shaped, margins entire, apex obtuse to rounded; stamens 6–8, included; anthers yellow, pink, or red, elliptic; styles 2(–3), connate at bases. **Achenes** included or apex exserted, brown to black, discoid or, rarely, 3-gonous, without central hump on 1 side, 2.1–3.4 × 1.8–3 mm, shiny, smooth. *2n* = 88.

Flowering May–Dec. Moist, disturbed places, ditches, riverbanks, cultivated fields, shorelines of ponds and reservoirs; 0–1800 m; N.B., Nfld. and Labr. (Nfld.), N.S., Ont., Que.; Ala., Alaska, Ariz., Ark., Calif., Colo., Conn., Del., D.C., Fla., Ga., Ill., Ind., Iowa, Kans., Ky., La., Maine, Md., Mass., Mich., Minn., Miss., Mo., Mont., Nebr., Nev., N.H., N.J., N.Mex., N.Y., N.C., N.Dak., Ohio, Okla., Pa., R.I., S.C., S.Dak., Tenn., Tex., Utah, Vt., Va., W.Va., Wis., Wyo.; South America (Ecuador); Europe (England, Spain).

Persicaria pensylvanica is a morphologically variable allotetraploid, with *P. lapathifolia* probably one of the parents (L. L. Consaul et al. 1991). Three or four varieties (under *Polygonum*) often have been accepted in North American floras; the characters on which these are based vary greatly within and among populations. J. W. Taylor-Lehman (1987) concluded that *Polygonum pensylvanicum* is best treated as a polymorphic species without infraspecific taxa, based on specimens primarily from Ohio. The heterostylous *Persicaria bicornis* often is included in *P. pensylvanica*. A single chromosome count of 2*n* = 22 reported by Á. Löve and D. Löve (1982), which could not be confirmed by Consaul et al. because the voucher could not be found, is excluded. Flowers with three styles and trigonous achenes are produced; they are exceedingly rare and probably mostly overlooked. Several Native American tribes prepared infusions and decoctions from *P. pensylvanica*, which they used as drugs for humans and horses (D. E. Moerman 1998).

SELECTED REFERENCE Taylor-Lehman, J. W. 1987. Variation in *Polygonum pensylvanicum* L. (Polygonaceae) with an Emphasis on Variety *eglandulosum* Myers. M.S. thesis. Ohio State University.

19. **Persicaria bicornis** (Rafinesque) Nieuwland, Amer. Midl. Naturalist 3: 201. 1914 • Pink smartweed E

Polygonum bicorne Rafinesque, Fl. Ludov., 29. 1817; *Persicaria longistyla* (Small) Small; *Polygonum longistylum* Small

Plants annual, 2–18 dm; roots also rarely arising from basal nodes; rhizomes and stolons absent. **Stems** ascending to erect, rarely decumbent, branched, ribbed, glabrous or appressed-pubescent to spreading-pubescent distally, stipitate-glandular or, rarely, without stipitate-glands. **Leaves:** ocrea brownish, cylindric, 6–20 mm, chartaceous, base inflated, margins truncate, eciliate or ciliate with bristles to 1 mm, surface glabrous or scabrous proximally, eglandular; petiole 0.1–1.5(–2.3) cm, glabrous or appressed-pubescent; blade sometimes with dark triangular or lunate blotch adaxially, linear-lanceolate to ovate-lanceolate, 2.3–13(–18) × (0.4–)1–2.3 cm, base tapered to cuneate, margins antrorsely scabrous, apex acute to acuminate, faces glabrous or appressed-pubescent along midveins, glandular-punctate abaxially. **Inflorescences** terminal and axillary, erect, uninterrupted, 8–60 × 10–18 mm; peduncle 8–60(–70) mm, glabrous or pubescent, usually stipitate-glandular; ocreolae overlapping, margins eciliate or ciliate with bristles to 0.8 mm. **Pedicels** ascending, 1.5–5 mm. **Flowers** 2–11 per ocreate fascicle, heterostylous; perianth pink, glabrous, not glandular-punctate, accrescent; tepals 5, connate ca. 1/4–1/3 their length, obovate to elliptic, 3–4.6 mm, veins

prominent, not anchor-shaped, margins entire, apex obtuse to rounded; stamens 6–8, included or exserted; anthers pink or red, elliptic; styles 2(–3), included or exserted, connate at bases. **Achenes** included or apex exserted, brownish black to black, discoid or, rarely, 3-gonous, 1 side usually slightly concave and other with central hump, (2–)2.2–2.9 × 2–2.8(–3) mm, shiny, smooth.

Flowering Jun–Oct. Moist, disturbed places, permanent and ephemeral wetlands, ditches, cultivated fields, shorelines of ponds and reservoirs; 50–1600 m; Ark., Colo., Ill., Iowa, Kans., La., Mo., Nebr., N.Mex., Okla., S.Dak., Tex., Wyo.

Persicaria bicornis is a characteristic smartweed of permanent and ephemeral wetlands and moist, disturbed sites in the Great Plains. It often has been included in *P. pensylvanica* but can be distinguished readily by its heterostylous flowers. The achenes, which usually bear an obscure or prominent hump in the center of one face, also are diagnostic. This hump often ruptures the side of the perianth on fruiting herbarium specimens. *Persicaria bicornis* also has leaf blades that are on average narrower than are those of *P. pensylvanica*, and populations exhibit less variation in perianth color. As in *P. pensylvanica*, flowers with three styles and trigonous achenes are produced very rarely.

20. **Persicaria lapathifolia** (Linnaeus) Gray, Nat. Arr. Brit. Pl. 2: 270. 1821 • Pale smartweed, renouée à feuilles de patience F W

Polygonum lapathifolium Linnaeus, Sp. Pl. 1: 360. 1753; *P. incarnatum* Elliott; *P. lapathifolium* var. *ovatum* A. Braun; *P. lapathifolium* var. *salicifolium* Sibthorp; *P. linicola* Sutulov; *P. nodosum* Persoon; *P. pensylvanicum* Linnaeus var. *oneillii* (Brenckle) Hultén; *P. scabrum* Moench; *P. tomentosum* Willdenow

Plants annual, (0.5–)1–10 dm; roots also sometimes arising from proximal nodes; rhizomes and stolons absent. **Stems** ascending to erect, simple or branched, scarcely ribbed, glabrous or, rarely, appressed-pubescent distally, sometimes glandular-punctate or stipitate-glandular distally. **Leaves:** ocrea brownish, cylindric, 4–24(–35) mm, chartaceous, base inflated, margins truncate, eciliate or ciliate with bristles to 1 mm, surface glabrous, rarely strigose, eglandular; petiole 0.1–1.6 cm, usually strigose, sometimes glabrous; blade sometimes with dark triangular or lunate blotch adaxially, narrowly to broadly lanceolate, 4–12(–22) × (0.3–)0.5–4(–6) cm, base tapering to cuneate, margins antrorsely scabrous, apex acuminate, faces strigose on main veins, glabrous or tomentose abaxially, glandular-punctate abaxially. **Inflorescences** mostly terminal, sometimes also axillary, mostly arching

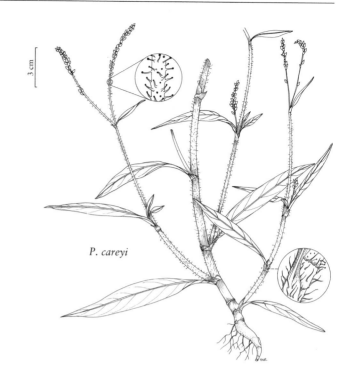

P. lapathifolia

P. careyi

PERSICARIA

or nodding, usually uninterrupted, 30–80 × 5–12 mm; peduncle 2–25 mm, often stipitate-glandular; ocreolae usually overlapping, margins eciliate or ciliate with bristles to 0.4 mm. **Pedicels** ascending, 0.5–2.3 mm. **Flowers** 4–14 per ocreate fascicle, homostylous; perianth greenish white to pink, glabrous, not glandular-punctate or glandular-punctate with punctae mostly on tubes and inner tepals, scarcely accrescent; tepals 4(–5), connate ca. ¹/₄–¹/₃ their length, obovate to elliptic, 2.5–3 mm, veins prominent, those of 2 or 3 outer tepals prominently bifurcate distally, anchor-shaped, margins entire, apex obtuse to rounded; stamens 5–6, included; anthers pink or red, elliptic; styles 2(–3), connate at bases. **Achenes** included or apex exserted, brown to black, discoid or, rarely, 3-gonous, 1.5–3.2 × 1.6–3 mm, shiny or dull, smooth. $2n = 22$.

Flowering (Apr–)Jul–Nov. Moist places, roadsides, floodplains, waste places, cultivated fields; 0–1500 (–1800) m; Greenland; St. Pierre and Miquelon; Alta., B.C., Man., N.B., Nfld. and Labr., N.W.T., N.S., Ont., P.E.I., Que., Sask., Yukon; Ala., Alaska, Ariz., Ark., Calif., Colo., Conn., Del., D.C., Fla., Ga., Idaho, Ill., Ind., Iowa, Kans., Ky., La., Maine, Md., Mass., Mich., Minn., Miss., Mo., Mont., Nebr., Nev., N.H., N.J., N.Mex., N.Y., N.C., N.Dak., Ohio, Okla., Oreg., Pa., R.I., S.C., S.Dak., Tenn., Tex., Utah, Vt., Va., Wash., W.Va., Wis., Wyo.; Mexico; South America; Europe; Asia; Africa; Pacific Islands (New Zealand).

Persicaria lapathifolia is a morphologically variable complex with more than two-dozen infraspecific taxa described in the New World and Old World. An allozyme study by L. L. Consaul et al. (1991) did not support recognition of elements often referred to *Polygonum lapathifolium* var. *salicifolium* or *P. scabrum*, which are synonymized here. Yang J. and Wang J. W. (1991) reached a similar conclusion regarding var. *salicifolium* and *P. nodosum* based on their morphometric analysis.

The Keres, Navajo, and Potawatomi prepared medicinal infusions with *Persicaria lapathifolia*, and the Zuni used decoctions made from the plants as cathartic and emetic drugs (D. E. Moerman 1998).

21. **Persicaria hydropiper** (Linnaeus) Spach, Hist. Nat. Vég. 10: 536. 1841 • Marsh-pepper smartweed, water-pepper, renouée poivre-d'eau Ⓘ Ⓦ

Polygonum hydropiper Linnaeus, Sp. Pl. 1: 361. 1753; *P. hydropiper* var. *projectum* Stanford

Plants annual, 2–8(–10) dm; roots also often arising from proximal nodes; rhizomes and stolons absent. **Stems** decumbent to ascending or erect, branched, without noticeable ribs, glabrous, glandular-punctate. **Leaves:** ocrea brown, cylindric or funnelform, (8–)10–15 mm, chartaceous, base inflated, margins truncate, ciliate with bristles 1–4 mm, surface glabrous or strigose, usually glandular-punctate; petiole 0.1–0.8 cm, glandular-punctate, leaves sometimes sessile; blade without dark triangular or lunate blotch adaxially, lanceolate to narrowly rhombic, (1.5–)4–10(–15) × 0.4–2.5 cm, base tapered or cuneate, margins antrorsely strigose, apex acute to acuminate, faces glabrous or scabrous along midveins, glandular-punctate, sometimes obscurely so adaxially. **Inflorescences** terminal and axillary, erect or nodding, interrupted or uninterrupted distally, 30–180 × 5–9 mm; peduncle (0–)10–50 mm, sometimes absent on axillary inflorescences and flowers thus enclosed in ocreae, glabrous, glandular-punctate; ocreolae not overlapping or overlapping distally, margins eciliate or ciliate with bristles to 1 mm. **Pedicels** ascending, 1–3 mm. **Flowers** 1–3(–5) per ocreate fascicle, homostylous; perianth greenish proximally, white or pink distally, glandular-punctate with punctae ± uniformly distributed, scarcely accrescent; tepals 4–5, connate ca. $^1/_3$ their length, petaloid, obovate, 2–3.5 mm, veins prominent or not, not anchor-shaped, margins entire, apex obtuse to rounded; stamens 6–8, included; anthers pink or red, elliptic to ovate; styles 2–3, connate proximally. **Achenes** included or apex exserted, brownish black, biconvex or 3-gonous, 1.9–3 × 1.5–2 mm, dull, minutely roughened. $2n = 20$.

Flowering May–Nov. Shorelines of lake and ponds, banks of streams and rivers, fens, forested wetlands, pastures, occasionally waste ground; 0–1500 m; introduced; St. Pierre and Miquelon; B.C., Man., N.B., Nfld. and Labr. (Nfld.), N.S., Ont., P.E.I., Que.; Ala., Alaska, Ark., Calif., Colo., Conn., Del., D.C., Fla., Ga., Idaho, Ill., Ind., Iowa, Kans., Ky., La., Maine, Md., Mass., Mich., Minn., Miss., Mo., Mont., Nebr., Nev., N.H., N.J., N.Mex., N.Y., N.C., N.Dak., Ohio, Okla., Oreg., Pa., R.I., S.C., S.Dak., Tenn., Tex., Utah, Vt., Va., Wash., W.Va., Wis., Wyo.; Europe; introduced also in Asia; nw Africa; Pacific Islands (Hawaii, New Zealand); Australia.

All parts of *Persicaria hydropiper* have an acrid, pepperlike taste. The plants have a long history of medicinal use in Europe, and the oily exudate produced in multicellular glands can cause skin irritation, hence the common name smartweed (R. S. Mitchell and J. K. Dean 1978). Some Native American tribes used *P. hydropiper* as a drug to treat a variety of ailments, and the Cherokee and Iroquois consumed it as food (D. E. Moerman 1998).

Herbarium specimens of *Persicaria hydropiper* often are misidentified as *P. punctata*. In addition to its minutely roughened and dull achenes, *P. hydropiper* differs from *P. punctata* frequently in bearing flowers enclosed in the ocreae, the inflorescences thus appearing somewhat leafy. By contrast, inflorescences of *P. punctata* generally appear terminal and leafless.

22. **Persicaria orientalis** (Linnaeus) Spach, Hist. Nat. Vég. 10: 537. 1841 • Kiss-me-over-the-garden-gate, princess-feather, renouée orientale [I][W]

Polygonum orientale Linnaeus, Sp. Pl. 1: 362. 1753

Plants annual, 6–25 dm; roots not also arising from proximal nodes; rhizomes and stolons absent. **Stems** erect, simple or branched distally, usually ribbed, strigose or glabrescent proximally, pilose to hirsute distally. **Leaves:** ocrea brownish proximally, green distally, narrowly funnelform, 10–20 mm, chartaceous proximally, foliaceous distally, rarely chartaceous throughout, base inflated or not, margins truncate, ciliate with bristles 1–3 mm, surface densely strigose to hispid, not glandular-punctate; petiole 1–8.5 (–14) cm, densely pilose to hirsute; blade without dark triangular or lunate blotch adaxially, ovate, 6–25(–30) × 3–17 cm, base cordate to truncate, margins scabrous to ciliate, apex acuminate, faces minutely strigose to densely hirsute, especially along veins abaxially, not glandular-punctate. **Inflorescences** mostly terminal, nodding or erect, uninterrupted, 10–150 × 8–18 mm; peduncle 20–100 mm, hirsute; ocreolae overlapping, margins ciliate with bristles 0.2–1 mm. **Pedicels** ascending to spreading, 1–4 mm. **Flowers** (1–)2–5 per ocreate fascicle, homostylous; perianth roseate to red, glabrous, not glandular-punctate, slightly accrescent; tepals 5, connate in proximal $^1/_3$, obovate, 3–4.5 mm, veins prominent or not, not anchor-shaped, margins entire, apex obtuse to rounded; stamens 6–8, included or exserted; anthers pink or red, elliptic; styles 2, connate proximally. **Achenes** included, dark brown to black, discoid, 2.5–3.5 × 3–3.5 mm, shiny to dull, smooth to minutely granulate.

Flowering Jun–Oct. Moist waste places; 0–500 m; introduced; N.B., Ont., Que.; Ala., Ark., Calif., Conn., Del., D.C., Fla., Ga., Ill., Ind., Iowa, Kans., Ky., La., Maine, Md., Mass., Mich., Minn., Miss., Mo., Nebr., N.H., N.J., N.Y., N.C., Ohio, Okla., Oreg., Pa., R.I., S.C., Tenn., Tex., Vt., Va., W.Va., Wis.; s Asia (India).

Persicaria orientalis was introduced as a garden ornamental. It often persists around homesteads and barnyards, and occasionally escapes and becomes weedy in moist waste places. A collection made in 1853 by F. V. Hayden at Fort Pierre, South Dakota (MO), is assumed to have come from a cultivated plant.

23. Persicaria careyi (Olney) Greene, Leafl. Bot. Observ. Crit. 1: 24. 1904 • Carey's smartweed, renouée de Carey [E] [F]

Polygonum careyi Olney, Proc. Providence Franklin Soc. 1: 29. 1847

Plants annual, 3–15(–20) dm; roots also rarely arising from proximal nodes; rhizomes and stolons absent. **Stems** erect, branched distally, ribbed distally, hirsute proximally, stipitate-glandular distally, usually smooth proximally. **Leaves:** ocrea brownish to reddish brown, cylindric, 8–20 mm, chartaceous, base inflated or not, margins truncate, ciliate with bristles 2–7 mm, surface strigose to hirsute, not glandular-punctate, rarely stipitate-glandular; petiole (0.1–)0.5–1.5 cm, hirsute, leaves sometimes sessile; blade without dark triangular or lunate blotch adaxially, narrowly lanceolate, 6–18 × 1–3.5 cm, base tapering, margins antrorsely strigose, apex acuminate to attenuate, faces sparingly hirsute abaxially and adaxially, veins often hirsute, sometimes stipitate-glandular. **Inflorescences** terminal and axillary, nodding or drooping, usually interrupted, 10–100 × 5–10 mm; peduncle 20–50 mm, stipitate-glandular; ocreolae overlapping or not overlapping proximally, margins eciliate or ciliate with bristles to 1.3 mm. **Pedicels** ascending to spreading, 1–4 mm. **Flowers** (1–)2–8 per ocreate fascicle, homostylous; perianth roseate or purple, glabrous, not glandular-punctate, scarcely accrescent; tepals 5, connate in proximal 1/3, obovate, 2.4–3.2 mm, veins prominent or not, not anchor-shaped, margins entire, apex obtuse to rounded; stamens 5 (or 8), included; anthers pink, elliptic; styles 2, connate to middle. **Achenes** included, dark brown to black, biconvex, 1.8–2.5 × 1.5–2 mm, shiny, smooth.

Flowering Jul–Oct. Low thickets, swamps, bogs, moist shorelines, clearings, recent burns, cultivated ground; 0–300 m; N.B., Ont., Que.; Conn., Del., Fla., Ill., Ind., Ky., Maine, Md., Mass., Mich., Minn., N.H., N.J., N.Y., Ohio, Pa., R.I., Vt., Va., Wis.

An infusion made from entire plants of *Persicaria careyi* was used by the Potawatomi as a cold remedy and febrifuge (D. E. Moerman 1998).

24. Persicaria longiseta (Bruijn) Kitagawa, Rep. Inst. Sci. Res. Manchoukuo 1: 322. 1937 • Bristly lady's-thumb [I]

Polygonum longisetum Bruijn in F. A. W. Miquel, Pl. Jungh. 3: 307. 1854; *Persicaria caespitosa* (Blume) Nakai var. *longiseta* (Bruijn) C. F. Reed; *Polygonum caespitosum* Blume var. *longisetum* (Bruijn) Steward

Plants annual, 3–8 dm; roots also often arising from proximal nodes; rhizomes and stolons absent. **Stems** decumbent to ascending, branched, without noticeable ribs, glabrous. **Leaves:** ocrea hyaline to brownish, cylindric, 5–12 mm, chartaceous, base sometimes inflated, margins truncate, ciliate with bristles 4–12 mm, surface glabrous or strigose, not glandular-punctate; petiole 0.1–0.3(–0.6) cm, glabrous, leaves sometimes sessile; blade without dark triangular or lunate blotch adaxially, ovate-lanceolate to linear-lanceolate, 2–8 × 1–3 cm, base tapering to cuneate, margins antrorsely strigose, apex acute to acuminate, faces glabrous or sparingly strigose along veins abaxially, glabrous or strigose along midvein and margins adaxially, not glandular-punctate. **Inflorescences** terminal, sometimes also axillary, erect, uninterrupted, 10–40(–80) × 3–7 mm; peduncle 10–50 mm, glabrous; ocreolae overlapping, margins ciliate with bristles (0.5–)1–4(–6) mm. **Pedicels** ascending, 1–2 mm. **Flowers** 1–5 per ocreate fascicle, homostylous; perianth pinkish green proximally, roseate distally, glabrous, not glandular-punctate, scarcely accrescent; tepals 5, connate ca. 1/3 their length, obovate, 2.2–2.8 mm, veins not prominent, not anchor-shaped, margins entire, apex obtuse to rounded; stamens 5, included; anthers yellow, elliptic to ovate; styles 3, connate proximally. **Achenes** included, dark brown to black, 3-gonous, 1.6–2.3 × 1.1–1.6 mm, shiny, smooth.

Flowering May–Oct. Floodplain forests and woodlands, shorelines of ponds, moist roadsides, waste places; 0–300 m; introduced; B.C., N.B., Ont.; Ala., Conn., Del., Fla., Ga., Ill., Ind., Iowa, Kans., Ky., La., Maine, Md., Mass., Mich., Minn., Miss., Mo., Nebr., N.J., N.Y., N.C., Ohio, Pa., S.C., Tenn., Tex., Vt., Va., W.Va., Wis.; e Asia; introduced also in Europe.

Persicaria longiseta is morphologically similar to another Asian species, *P. posumbu* (Buchanan-Hamilton ex D. Don) H. Gross (= *P. caespitosa*). Its spread in the United States since its introduction near Philadelphia in 1910 was summarized by A. K. Paterson (2000).

25. Persicaria minor (Hudson) Opiz, Seznam, 72. 1852 • Small water-pepper, petite renouée [I]

Polygonum minus Hudson, Fl. Angl., 148. 1762; *P. minus* var. *subcontinuum* (Meisner) Fernald

Plants annual, 0.5–3(–4) dm; roots also sometimes at from proximal nodes; rhizomes and stolons absent. **Stems** decumbent or ascending, branched proximally, scarcely ribbed, glabrous or scabrous distally. **Leaves:** ocrea brownish, cylindric, 3–10 mm, chartaceous, base not inflated, margins truncate, ciliate with bristles (0.3–)1–3(–5) mm, surface glabrous or strigose, not glandular-punctate; petiole 0.1–0.2 cm, glabrous or strigose, leaves sometimes sessile; blade without dark triangular or lunate blotch adaxially, linear to linear-lanceolate, (1–)2–7.5(–10) × (0.2–)0.4–1(–2.3) cm, base tapered to cuneate, margins antrorsely scabrous, apex acute to acuminate, faces glabrous or sparingly strigose, especially along midveins, not glandular-punctate. **Inflorescences** terminal and axillary, ascending to erect, usually interrupted proximally, uninterrupted distally, 10–50 × 2–4 mm; peduncle (0–)2–25 mm, sometimes absent on axillary inflorescences and flowers thus enclosed in ocreae, glabrous; ocreolae not overlapping proximally, usually overlapping distally, margins ciliate with bristles (0.1–)0.6–2(–2.7) mm. **Pedicels** ascending, 0.5–1 mm. **Flowers** 1–3(–4) per ocreate fascicle, homostylous; perianth roseate to red, rarely white, glabrous, not glandular-punctate, scarcely accrescent; tepals 5, connate ca. 1/3 their length, obovate to elliptic, 2.5–3 mm, veins not prominent, not anchor-shaped, margins entire, apex obtuse to rounded; stamens 5(–6), included; anthers yellow to pink, elliptic; styles 2(–3), connate at bases. **Achenes** included, brownish black to black, biconvex or, rarely, 3-gonous, (1.5–)1.8–2.3(–2.7) × (1.1–)1.3–1.5(–1.8) mm, shiny, smooth.

Flowering Jul–Oct. Damp, open places; 0–100 m; introduced; N.B., Ont., Que.; Conn., Ind., La., Mass., Nebr., Pa., Vt., Va.; Europe.

Persicaria minor is synonymized with *P. maculosa* in most North American floras; its distribution in the flora area is poorly known. Hybrids between *P. minor* and *P. maculosa* have been documented in Europe (R. H. Roberts 1977).

26. Persicaria maculosa Gray, Nat. Arr. Brit. Pl. 2: 269. 1821 • Spotted lady's-thumb, redshank, renouée persicaire [I] [W]

Polygonum persicaria Linnaeus, Sp. Pl. 1: 361. 1753; *Persicaria fusiformis* (Greene) Greene; *P. vulgaris* Webb & Moquin-Tandon; *Polygonum fusiforme* Greene; *P. persicaria* var. *ruderale* (Salisbury) Meisner; *P. puritanorum* Fernald

Plants annual, (0.5–)1–7(–13) dm; roots also often arising from proximal nodes; rhizomes and stolons absent. **Stems** procumbent, decumbent, ascending, or erect, simple or branched, without obvious ribs, glabrous or appressed-pubescent. **Leaves:** ocrea light brown, cylindric, 4–10(–15) mm, chartaceous, base inflated, margins truncate, ciliate with hairs 1–3.5(–5) mm, surface glabrous or strigose, rarely with spreading hairs, not glandular-punctate; petiole 0.1–0.8 cm, glabrous or strigose, leaves sometimes sessile; blade often with dark triangular or lunate blotch adaxially, lanceolate to narrowly ovate, (1–)5–10(–18) × (0.2–)1–2.5(–4) cm, base tapered or cuneate, margins antrorsely strigose, apex acute to acuminate, faces glabrous or strigose, especially along midveins, sometimes glandular-punctate abaxially. **Inflorescences** terminal and axillary, erect, usually uninterrupted, 10–45(–60) × 7–12 mm; peduncle 10–50 mm, glabrous or, rarely, pubescent; ocreolae overlapping or sometimes interrupted proximally, margins ciliate with bristles 0.2–1.3(–2) mm. **Pedicels** ascending, 1–2.5 mm. **Flowers** 4–14 per ocreate fascicle, homostylous; perianth greenish white proximally and roseate distally or entirely roseate, not glandular-punctate, scarcely accrescent; tepals 4–5, connate ca. 1/3 their length, obovate, 2–3.5 mm, veins prominent, not anchor-shaped, margins entire, apex obtuse to rounded; stamens 4–8, included; anthers yellow or pink, ovate; styles 2–3, connate proximally. **Achenes** included or apex exserted, brownish black to black, discoid or biconvex to 3-gonous, (1.9–)2–2.7 × (1.5–) 1.8–2.2 mm, shiny, smooth. $2n = 44$.

Flowering Mar–Nov. Weedy, moist semiwaste to cultivated areas; 0–2500 m; introduced; Greenland; St. Pierre and Miquelon; Alta., B.C., Man., N.B., Nfld. and Labr., N.S., Ont., P.E.I., Que., Sask., Yukon; Ala., Alaska, Ariz., Ark., Calif., Colo., Conn., Del., D.C., Fla., Ga., Idaho, Ill., Ind., Iowa, Kans., Ky., La., Maine, Md., Mass., Mich., Minn., Miss., Mo., Mont., Nebr., Nev., N.H., N.J., N.Mex., N.Y., N.C., N.Dak., Ohio, Okla., Oreg., Pa., R.I., S.C., S.Dak., Tenn., Tex., Utah, Vt., Va., Wash., W.Va., Wis., Wyo.; Eurasia; Africa; Pacific Islands (New Zealand).

An allozyme study by L. L. Consaul et al. (1991) provided evidence of the allotetraploid origin of *Persicaria maculosa*, with *P. lapathifolium* as one of the parents.

Plants with stems spreading-hairy and peduncles stipitate-glandular have been named *P. maculosa* subsp. *hirsuticaulis* (Danser) S. Ekman & Knutsson. Material referable to this subspecies has not been seen among North American specimens. Hybrids between *P. maculosa* and *P. minor* have been documented in Europe (R. H. Roberts 1977).

The Cherokee, Chippewa, and Iroquois prepared simple or compound decoctions of *Persicaria maculosa*, which they used as dermatological, urinary, gastrointestinal, and veterinary aids, for heart medicine, and as an analgesic (D. E. Moerman 1998).

33. BISTORTA (Linnaeus) Scopoli, Meth. Pl., 24. 1754 • Bistort [Latin, *bi*-, twice, and *tortus*, twisted, alluding to the rhizomes of some species]

Craig C. Freeman

Harold R. Hinds†

Polygonum [unranked] *Bistorta* Linnaeus, Sp. Pl. 1: 360. 1753

Herbs, perennial; roots fibrous, rhizomatous. **Stems** erect, simple, glabrous. **Leaves** mostly basal, some cauline, alternate, petiolate or sessile; ocrea persistent or disintegrating with age and deciduous entirely or distally, chartaceous; blade linear or lanceolate to elliptic, oblong-ovate, or ovate, margins entire or obscurely and irregularly repand. **Inflorescences** terminal, spikelike. **Pedicels** present. **Flowers** bisexual, 1–2 per ocreate fascicle, base not stipelike; perianth nonaccrescent, white, greenish white, pink, or purplish pink, rarely red, campanulate, glabrous; tepals 5, connate proximally ca. $^1/_5$ their length, petaloid, monomorphic or slightly dimorphic, outer larger than inner; stamens 5–8, sometimes poorly developed; filaments distinct or connate basally, outer ones sometimes adnate to perianth tube, glabrous; anthers yellow, pink, red, purple, or blackish, ovate to elliptic; styles 3, erect or spreading, distinct or connate proximally; stigmas 2–3, capitate. **Achenes** included or exserted, brown to dark brown, unwinged, 3-gonous, glabrous. **Seeds:** embryo curved. $x = 11, 12$.

Species ca. 50 (4 in the flora): arctic and temperate North America, Europe, Asia.

Bistorta often is included in *Polygonum* in the broad sense or in *Persicaria*. It is accepted here as a distinct genus based on habit, morphology, and anatomy (K. Haraldson 1978; L.-P. Ronse Decraene and J. R. Akeroyd 1988). In the species of the flora area, the base of the petiole forms a long, tubular sheath distal to the node from which the leaf arises and proximal to the point of divergence of the petiole. Distal to the sheath is the ocrea, which usually is darker and thinner.

1. Inflorescences narrowly elongate-cylindric, (15–)20–90 × 4–10 mm, usually bearing pyriform, pink to brown or purple bulblets proximally . . . . 1. *Bistorta vivipara*
1. Inflorescences short-cylindric to ovoid, 10–70 × 8–25 mm, bulblets absent.
    2. Leaf blade bases abruptly contracted, truncate to cuneate; petioles prominently winged distally; perianths pink . . . . 4. *Bistorta officinalis*
    2. Leaf blade bases usually tapered to rounded, rarely abruptly truncate or cuneate; petioles wingless or rarely winged distally; perianths white to pale pink, bright pink, or purplish pink.
        3. Perianths bright pink or purplish pink; basal leaf blades with apices rounded to acute; plants (8–)10–40(–50) cm; n Canada, Alaska . . . . 3. *Bistorta plumosa*
        3. Perianths white or pale pink; basal leaf blades with apices usually acute to acuminate, rarely obtuse; plants (10–)20–70(–75) cm; sw Canada, w United States . . . . 2. *Bistorta bistortoides*

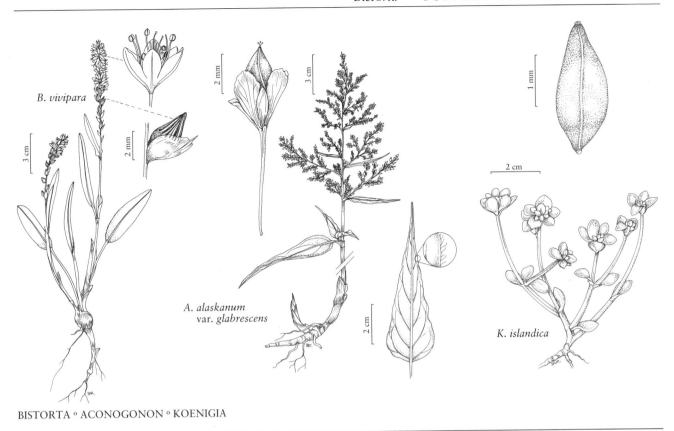

B. vivipara

A. alaskanum var. *glabrescens*

K. islandica

BISTORTA ° ACONOGONON ° KOENIGIA

1. Bistorta vivipara (Linnaeus) Delarbre, Fl. Auvergne ed. 2, 2: 516. 1800 • Alpine bistort F

Polygonum viviparum Linnaeus, Sp. Pl. 1: 360. 1753; *Bistorta vivipara* subsp. *macounii* (Small ex J. M. Macoun) Soják; *Persicaria vivipara* (Linnaeus) Ronse Decraene; *Polygonum viviparum* var. *macounii* (Small ex J. M. Macoun) Hultén

Plants (2–)8–30(–45) cm; rhizomes sometimes contorted. **Stems** 1–2(–3). **Leaves:** ocrea brown, cylindric, 4–22 (–27) mm, margins strongly oblique, glabrous; petiole attached to sheath 6–20(–45) mm, unwinged distally, 5–110(–200) mm; blade linear to lanceolate or oblong-ovate, 1–8(–10) × 0.5–1.7(–2.3) cm, base cuneate to rounded or cordate, often asymmetric, margins entire, usually revolute, not wavy, apex obtuse to acute, abaxial face pubescent with whitish or brownish hairs, glaucous, adaxial face glabrous, not glaucous; cauline leaves 2–4, petiolate proximally, sessile distally, gradually reduced distally, blade linear-lanceolate to linear. **Inflorescences** 1, narrowly elongate-cylindric, (15–)20–90 × 4–10 mm, usually bearing pink to brown or purple pyriform bulblets proximally and sterile flowers distally; peduncle 1–5 cm. **Pedicels** ascending or spreading, (1–)2–5 mm. **Flowers** 1–2 per ocreate fascicle; perianth greenish proximally, usually white or pink distally, rarely red; tepals obovate, 2.1–4 mm, apex obtuse to acute; stamens included or exserted, some or all often poorly developed; anthers reddish to purple. **Achenes** rarely produced, dark brown, 2.2–3.3 × 0.9–1.5 mm, dull, granular. $2n = 96, 120.$

Flowering Jun–Sep. Moist to wet spruce or mixed woods along shorelines, moist subalpine woods and meadows, alpine meadows, heaths, nutrient-rich sites; 0–4000 m; Greenland; St. Pierre and Miquelon; Alta., B.C., Man., N.B., Nfld. and Labr. (Labr.), N.W.T., Nunavut, Ont., Que., Sask., Yukon; Alaska, Ariz., Colo., Idaho, Maine, Mich., Minn., Mont., Nev., N.H., N.Mex., Oreg., S.Dak., Utah, Vt., Wash., Wyo.; Europe; Asia.

Bistorta vivipara is highly variable morphologically and cytologically. Robust plants with large leaves, compact spikes, and persistent bulblets have been named subsp. *macounii*. Abortion of stamens, production of bulblets, and the rarity of fruits suggest that reproduction is largely asexual; fruits and seedlings are produced rarely (N. Söyrinki 1989). B. Jonsell and T. Karlsson (2000+, vol. 1) summarized chromosome numbers that include $2n = 66$, ca. 77, ca. 80, 88, 99, ca. 100, 110, 120, and ca. 132.

A. E. Porsild and W. J. Cody (1980) reported that indigenous peoples of the circumpolar region eat the starchy, slightly astringent rootstocks raw or cooked, and preserve them in seal oil or by freezing. E. Hultén (1968) reported that the rootstocks taste like almonds.

2. Bistorta bistortoides (Pursh) Small, Bull. Torrey Bot. Club 33: 57. 1906 • Western or American bistort, smokeweed E

Polygonum bistortoides Pursh, Fl. Amer. Sept. 1: 271. 1813; *Bistorta bistortoides* var. *oblongifolia* (Meisner) Moldenke; *Persicaria bistortoides* (Pursh) H. R. Hinds; *Polygonum bistortoides* var. *linearifolium* (S. Watson) Small; *P. bistortoides* var. *oblongifolium* (Meisner) H. St. John

Plants (10–)20–70(–75) cm; rhizomes contorted. **Stems** 1–3. **Leaves:** ocrea brown, cylindric, 9–25(–32) mm, margins oblique, glabrous; petiole attached to sheath 10–35(–50) mm, usually wingless, rarely winged distally, (10–)30–70(–110) mm; blade elliptic to oblong-lanceolate or oblong-oblanceolate, (3.5–)5–22 × 0.8–4.8 cm, base tapered to rounded, rarely abruptly truncate or cuneate, often asymmetric, margins entire, sometimes wavy, apex usually acute to acuminate, rarely obtuse, abaxial face glabrous or pubescent with whitish or brownish hairs, glaucous, adaxial face glabrous, not glaucous; cauline leaves 2–6, petiolate proximally, sessile distally, gradually reduced distally, blade elliptic or lanceolate to linear-lanceolate. **Inflorescences** 1(–2), short-cylindric to ovoid, (10–)20–40(–50) × (8–)12–25 mm, bulblets absent; peduncle 1–10 cm. **Pedicels** ascending or spreading, 2–8(–11) mm. **Flowers** 1–2 per ocreate fascicle; perianth white or pale pink; tepals oblong, 4–5 mm, apex obtuse to acute; stamens exserted; anthers yellow, elliptic. **Achenes** yellowish brown or olive-brown, 3.2–4.2 × 1.3–2 mm, shiny, smooth. *2n* = 24.

Flowering Jul–Sep. Streambanks, moist or swampy meadows, alpine slopes; 1300–3800 m; Alta., B.C.; Ariz., Calif., Colo., Idaho, Mont., Nev., N.Mex., Oreg., Utah, Wash., Wyo.

Infrequent specimens of *Bistorta bistortoides* have basal leaf blades that are lance-ovate and abruptly contracted at the bases, and petioles distinctly winged distally, similar to those of *B. officinalis*.

Roots of western bistort were used in soups and stews by the Blackfoot, boiled with meat by the Cherokee, and used in a poultice that was applied to sores and boils by the Miwok (D. E. Moerman 1998).

3. Bistorta plumosa (Small) Greene, Leafl. Bot. Observ. Crit. 1: 18. 1904

Polygonum plumosum Small, Bull. New York Bot. Gard. 2: 166. 1901; *Bistorta major* Gray subsp. *plumosum* (Small) H. Hara; *Polygonum bistorta* Linnaeus subsp. *plumosum* (Small) Hultén; *P. bistorta* var. *plumosum* (Small) B. Boivin

Plants (8–)10–40(–50) cm; rhizomes contorted. **Stems** 1(–2). **Leaves:** ocrea brown, cylindric, 5–40 mm, margins oblique, glabrous; petiole attached to sheath 10–45 mm, unwinged or scarcely winged distally, 30–220 mm; blade lanceolate-elliptic to ovate, 2–12 × 0.5–3 cm, base tapered, rarely truncate, often asymmetric, margins entire, sometimes wavy, apex rounded to acute, abaxial face pubescent with whitish or brownish hairs, glaucous, adaxial face glabrous, not glaucous; cauline leaves 1–3, petiolate proximally, sessile distally, gradually reduced distally, blade triangular-lanceolate to linear. **Inflorescences** 1, short-cylindric to ovoid, 15–70 × 10–15 mm, bulblets absent; peduncle 1–8 cm. **Pedicels** ascending or spreading, 2–7 mm. **Flowers** 1–2 per ocreate fascicle; perianth bright pink or purplish pink; tepals oblong to elliptic, 3–4 mm, apex obtuse to acute; stamens included or exserted; anthers purple to blackish. **Achenes** brown, 2.5–4 × 1.2–2 mm, shiny, smooth. *2n* = 72.

Flowering May–Jun. Fields, meadows, arctic and alpine tundra, heathlands; 0–2000 m; B.C., N.W.T., Yukon; Alaska; e Asia.

Leaves of *Bistorta plumosa* are used as a dietary aid and consumed as a vegetable, and roots are boiled and added to stews by Alaskan Native Americans (D. E. Moerman 1998).

4. Bistorta officinalis Delarbre, Fl. Auvergne ed. 2, 2: 516. 1800 • European bistort I

Polygonum bistorta Linnaeus, Sp. Pl. 1: 360. 1753; *Bistorta major* Gray; *Persicaria bistorta* (Linnaeus) Sampaio

Plants (20–)30–100(–120) cm; rhizomes contorted. **Stems** 1(–2). **Leaves:** ocrea brown, cylindric, 15–50(–70) mm, margins oblique, glabrous; petiole attached proximally to sheath 10–45(–70) mm, prominently winged distally, 100–260 mm, wing 1–6(–12) cm; blade lanceolate to ovate, 6–15(–21) × 2.5–9(–10) cm, base abruptly contracted, truncate to cuneate, often asymmetric, margins entire or obscurely and irregularly repand, usually wavy, apex acute to acuminate, rarely obtuse, abaxial face glabrous or pubescent with whitish or brownish hairs, glaucous, adaxial face glabrous, not glaucous; cauline leaves 3–6, petiolate proximally, sessile distally, gradually reduced distally, blade triangular-lanceolate to linear. **Inflorescences** 1, short-cylindric to ovoid, 35–65 × 12–20 mm, bulblets absent; peduncle 1–15 cm. **Pedicels** ascending, 1.5–7 mm. **Flowers** 1–2 per ocreate fascicle; perianth pink; tepals elliptic to ovate, 3–4 mm, apex obtuse; stamens exserted; anthers pink to purple. **Achenes** brown, 3.3–4.5 × 2.5–3 mm, shiny, smooth. $2n = 48$.

Flowering May–Aug. Fields, meadows; 10–200 m; introduced; Nfld. and Labr. (Nfld.), N.S.; Maine, Mass., Vt.; Europe; Asia.

Chromosome numbers of $2n = 24, 44, 46, 48$, and ca. 50 have been reported for *Bistorta officinalis* (B. Jonsell and T. Karlsson 2000+, vol. 1). A count of $2n = 48$ made by Á. Löve and D. Löve (1988) was from a plant from California; it probably was cultivated.

34. ACONOGONON (Meisner) Reichenbach, Handb. Nat. Pfl.-Syst., 236. 1837 (as Aconogonum) • [Greek *acon*, whetstone, and *gone*, seed, perhaps alluding to rough seeds]

Harold R. Hinds†

Craig C. Freeman

Polygonum Linnaeus sect. *Aconogonon* Meisner, Monogr. Polyg., 55. 1826

Herbs, perennial; roots woody. **Stems** ascending to erect, glabrous or pubescent to pilose or tomentose. **Leaves** deciduous, mostly cauline, alternate, petiolate or sessile; ocrea persistent or deciduous, chartaceous; blade narrowly lanceolate to ovate, margins entire, sometimes irregularly undulate. **Inflorescences** terminal, subterminal, or axillary, racemelike or paniclelike; peduncle present or essentially absent. **Pedicels** present. **Flowers** bisexual, 1–5 per ocreate fascicle, base stipelike or not; perianth nonaccrescent, creamy or greenish to yellowish white or pink, rotate, glabrous; tepals 5, connate ca. ¼ their length, petaloid, slightly to distinctly dimorphic, outer 2 smaller than inner 3; stamens 8; filaments distinct, free or adnate to perianth tube, glabrous; anthers yellow to pink or reddish purple, ovate to elliptic; styles 3, erect or spreading, distinct or connate proximally; stigmas capitate. **Achenes** includedor exserted, yellowish or dark brown, unwinged, 3-gonous, glabrous. **Seeds:** embryo usually curved. $x = 8, 10, 11$.

Species ca. 25 (3 in the flora): w North America, Europe, Asia.

The orthography of and author citation for the genus name have been matters of confusion. Meisner published the taxon as *Polygonum* sect. *Aconogonon*. Reichenbach elevated it to generic rank and changed the orthography, apparently deliberately, to *Aconogonum*. Debate centers on whether Reichenbach's name is new (*Aconogonum* Reichenbach) or should be treated as Meisner's section name elevated to generic rank [*Aconogonon* (Meisner) Reichenbach]. Following established custom, we here use *Aconogonon*.

Aconogonon often is treated as a section of *Persicaria* (L.-P. Ronse Decraene and J. R. Akeroyd 1988); pollen morphology, inflorescence type, and seed and stem anatomy have been used by some taxonomists for segregation at the generic level (K. Haraldson 1978; Hong S. P. 1991).

Aconogonon campanulatum (Hooker f.) H. Hara is planted as an ornamental. C. L. Hitchcock and A. Cronquist (1973) reported it as escaping and probably established west of the Cascade Mountains, especially in Seattle, Washington. P. Zika (pers. comm.) searched numerous herbaria in the Pacific Northwest. He found no voucher indicating that *A. campanulatum* has escaped there, although it is cultivated occasionally in the Seattle area; it therefore is excluded here.

SELECTED REFERENCES   Hinds, H. 1995. Nomenclatural changes in *Polygonum, Persicaria,* and *Aconogonon* (Polygonaceae). Novon 5: 165–166. Hong, S. P. 1991. A revision of *Aconogonon* (= *Polygonum* sect. *Aconogonon,* Polygonaceae) in North America. Rhodora 93: 322–346.

1. Inflorescences axillary, racemelike; plants 12–42(–50) cm; leaf blades oblong-ovate to ovate, rarely broadly lanceolate, 2.1–7.5(–9.7) × 1.1–5 cm, often glaucous . . . . 3. *Aconogonon davisiae*
1. Inflorescences usually terminal or subterminal, sometimes also axillary, paniclelike; plants (30–)50–150(–200) cm; leaf blades narrowly lanceolate to ovate, 5–20 × 1.4–8 cm, not glaucous.
    2. Achenes 2.6–3.8 mm, included or exserted, tan to grayish, faces usually not concave; inflorescences terminal, sometimes also axillary; Alaska, n Yukon, nw Northwest Territories . . . . 1. *Aconogonon alaskanum*
    2. Achenes (3–)3.8–7 mm, usually exserted, yellowish brown, faces concave; inflorescences usually terminal, subterminal, and axillary; n California, Oregon, Washington, Idaho, w Montana, n Nevada . . . . 2. *Aconogonon phytolaccifolium*

1. Aconogonon alaskanum (Small) Soják, Preslia 46: 150. 1974 • Alaska wild-rhubarb [F]

Polygonum alpinum Allioni var. *alaskanum* Small, Monogr. Amer. Sp. Polygonum, 33. 1895; *P. alaskanum* (Small) W. Wight ex Harshberger; *P. alpinum* subsp. *alaskanum* (Small) S. L. Welsh

Plants (30–)50–150(–200) cm. **Stems** erect, glabrous or densely and retrorsely pubescent. **Leaves:** ocrea reddish brown, funnelform, 1–2.2 cm, margins oblique, glabrous or pilose; petiole 0.8–3.5(–4) mm; blade narrowly lanceolate to ovate, 5–20 × 2–8 cm, not subcoriaceous, base truncate or obtuse to subcordate, margins sometimes irregularly undulate, ciliate or, rarely, glabrous, apex acuminate, faces not glaucous, glabrous or pubescent. **Inflorescences** terminal, sometimes also axillary, paniclelike; peduncle 0–4 cm, glabrous or pubescent. **Pedicels** 0.5–0.1 mm. **Flowers** 2–4 per ocreate fascicle, all flowers with stamens well developed; perianth white or greenish white, 2–4 mm; tepals dimorphic, rarely slightly so, ovate, apex obtuse; anthers yellow to pink. **Achenes** included or exserted, tan to grayish brown, not beaked distally, 2.6–3.8 × 2–3 mm, shiny, faces usually not concave. $2n = 20$.

Varieties 2 (2 in the flora): nw North America, Europe (e Russia).

Aconogonon alaskanum has been considered conspecific with the morphologically similar Eurasian *A. alpinum* (Allioni) Schur, but the taxa differ in leaf size and achene characters. Nomenclatural issues were clarified by J. T. Kartesz and K. N. Gandhi (1990) and K. L. Chambers (1992).

1. Leaf blades pubescent; stems usually densely and retrorsely pubescent . . . . . . 1a. *Aconogonon alaskanum* var. *alaskanum*
1. Leaf blades glabrous except for margins; stems mostly glabrous, rarely pubescent proximal to nodes . . . . 1b. *Aconogonon alaskanum* var. *glabrescens*

1a. Aconogonon alaskanum (Small) Soják var. **alaskanum**

Aconogonon hultenianum (Yurtzev) Tsvelev var. *lapathifolium* (Chamisso & Schlechtendal) S. P. Hong; *Polygonum alpinum* Allioni var. *lapathifolium* Chamisso & Schlechtendal

Stem usually densely and retrorsely pubescent. **Leaf blades** 5–16.8 × 2–6 cm, margins ciliate, faces pubescent.

Flowering Jun–Aug. Montane slopes above treeline, steep hillsides, steep cut banks or sandy loam of rivers; 100–1300 m; Yukon; Alaska; Europe (e Russia).

1b. Aconogonon alaskanum (Small) Soják var. **glabrescens** (Hultén) H. R. Hinds, Novon 5: 166. 1995 [E] [F]

Polygonum alaskanum (Small) W. Wight ex Harshberger var. *glabrescens* Hultén, Acta Univ. Lund, n. s. 40: 612. 1944

Stem glabrous, rarely or pubescent proximal to nodes. **Leaf blades** 6–17(–20) × 2–8 cm, margins glabrous or ciliate, faces glabrous. $2n = 20$.

Flowering Jun–Aug. Moist hillsides, waste places, streambanks, sandy lake shores, talus slopes above treeline; 300–1700 m; N.W.T, Yukon; Alaska.

2.   **Aconogonon phytolaccifolium** (Meisner ex Small) Small in P. A. Rydberg, Fl. Rocky Mts., 1061. 1917 (as Aconogonum phytolaccaefolium) • Poke knotweed [E]

Polygonum phytolaccifolium Meisner ex Small, Bull. Torrey Bot. Club 19: 360. 1892

Plants (50–)70–150(–200) cm. **Stems** erect, glabrous or pubescent. **Leaves:** ocrea reddish brown, funnelform, 1–3 cm, margins oblique, glabrous or pubescent; petiole 5–20 mm; blade lanceolate to ovate-lanceolate, 5–15(–19) × 1.4–7.5 cm, not subcoriaceous, base rounded, margins entire, usually scabrous to ciliate, rarely glabrous, apex obtuse to acuminate, faces not glaucous, glabrous or densely pubescent. **Inflorescences** usually terminal, subterminal, and axillary, paniclelike; peduncle 0.5–12 cm, glabrous or pubescent. **Pedicels** 0.9–4 mm. **Flowers** 1–3 per ocreate fascicle, rarely some flowers with stamens poorly developed; perianth white to greenish white, 2.5–3.5 mm; tepals slightly dimorphic, ovate to obovate, apex obtuse; anthers yellow to pink. **Achenes** usually exserted, yellowish brown, not beaked distally, (3–)3.8–7 × 2.3–4.8 mm, shiny, faces concave.

Varieties 2 (2 in the flora): w United States.

1. Stems glabrescent or pubescent; leaf blade margins scabrous to ciliate . . . .
. . . . 2a. *Aconogonon phytolaccifolium* var. *phytolaccifolium*
1. Stems glabrous; leaf blade margins glabrous . . .
. . . . . . 2b. *Aconogonon phytolaccifolium* var. *glabrum*

2a.   **Aconogonon phytolaccifolium** (Meisner ex Small) Rydberg var. **phytolaccifolium** [E]

Stems glabrescent or pubescent. **Leaves:** ocrea glabrescent or pubescent; blade 5–15(–19) × 1.4–7.5 cm, margins scabrous to ciliate, faces glabrous or pubescent. **Pedicels** 0.9–3 mm.

Flowering Jun–Aug. Alpine and subalpine meadows, talus slopes, streambanks; 1200–3100 m; Calif., Idaho, Mont., Nev., Oreg., Wash.

2b.   **Aconogonon phytolaccifolium** (Meisner ex Small) Rydberg var. **glabrum** S. P. Hong, Rhodora 93: 339, fig. 4. 1991 [E]

Stems glabrous. **Leaves:** ocrea glabrous; blade 5.4–12.5 × 1.6–5 cm, margins glabrous, faces glabrous. **Pedicels** (1.1–)1.6–4 mm.

Flowering Jun–Aug. Sandy to rocky spruce-fir forests; 1800–3400 m; Calif, Idaho, Nev., Oreg.

3.   **Aconogonon davisiae** (W. H. Brewer ex A. Gray) Soják, Preslia 46: 151. 1974 • Davis's knotweed [E]

Polygonum davisiae W. H. Brewer ex A. Gray, Proc. Amer. Acad. Arts 8: 399. 1872; *Aconogonon newberryi* (Small) Soják; *P. newberryi* Small

Plants 12–42(–50) cm. **Stems** ascending to erect, glabrous or pubescent. **Leaves:** ocrea reddish brown, funnelform, 0.3–3 cm, margins oblique, glabrous or finely pubescent to pilose; petiole 0.3–10(–15) mm; blade oblong-ovate to ovate, rarely broadly lanceolate, 2.1–7.5(–9.5) × 1.1–5 cm, subcoriaceous, base truncate or, rarely, cordate, margins entire, glabrous or ciliate, apex obtuse to acute, faces often glaucous, glabrous or pubescent. **Inflorescences** axillary, racemelike; peduncle essentially absent. **Pedicels** 0.5–1.9(–2.4) mm. **Flowers** 1–4 per ocreate fascicle, rarely some flowers with stamens poorly developed; perianth greenish to pinkish white, 2–4.5 mm; tepals slightly dimorphic or, rarely, distinctly so, oblong-ovate to obovate, apex obtuse; anthers yellow or pink. **Achenes** exserted, yellowish brown, not beaked distally, 3.2–6(–8.3) × 2.1–5 mm, shiny, faces not concave. $2n = 20$.

Varieties 2 (2 in the flora): w United States.

The varieties are largely sympatric and often intergrade. The treatment here follows Hong S. P. (1991).

1. Faces of leaf blades pubescent, rarely glabrescent
. . . . 3a. *Aconogonon davisiae* var. *davisiae*
1. Faces of leaf blades glabrous . . . .
. . . . 3b. *Aconogonon davisiae* var. *glabrum*

3a. Aconogonon davisiae (W. H. Brewer ex A. Gray) Soják var. **davisiae** [E]

Stems pubescent, rarely glabrescent. **Leaf blades** 2.1–7.5(–9.5) × 1.1–5 cm, base truncate or, rarely, cordate, faces pubescent, rarely glabrescent. **Pedicels** 0.5–1.7 (–2.4) mm, glabrous. **Perianths** 2–4.5 mm. **2n** = 20.

Flowering Jun–Aug. Steep subalpine slopes, volcanic fell fields, often on talus or pumice; 1200–3100 m; Calif., Idaho, Oreg., Wash.

3b. Aconogonon davisiae (W. H. Brewer ex A. Gray) Soják var. **glabrum** (G. N. Jones) S. P. Hong, Rhodora 93: 343. 1991 [E]

Polygonum newberryi Small var. *glabrum* G. N. Jones, Rhodora 40: 359. 1938

Stems glabrous. **Leaf blades** 2.4–5.8 × 1.2–3.2 cm, base cordate or truncate, faces glabrous. **Pedicels** 0.6–1.4 mm, glabrous. **Perianths** 3–3.8 mm.

Flowering Jun–Aug. Mountain slopes, talus slopes, edges of serpentine outcrops; 1000–2200 m; Calif., Wash.

35. KOENIGIA Linnaeus, Mant. Pl. 1: 3, 35. 1767; Syst. Nat. ed. 12, 2: 71, 104. 1767

- [For Johann Gerhard König, 1827–1785, pupil of Linnaeus]

John G. Packer

Craig C. Freeman

Herbs, annual; taprooted. **Stems** decumbent, ascending, or erect, glabrous. **Leaves** cauline, alternate or subopposite, petiolate; ocrea persistent, chartaceous; blade spatulate-ovate to orbiculate, margins entire. **Inflorescences** terminal, paniclelike or cymelike, not pedunculate. **Pedicels** absent or present. **Flowers** bisexual, 3–10 per ocreate fascicle, bases not stipelike; perianth nonaccrescent, greenish, often tinged white or pink distally, narrowly campanulate, glabrous or occasionally with scattered glands; tepals 3 [4], distinct, sepaloid, monomorphic; stamens (1–)3[–5]; filaments distinct, free, glabrous; anthers white or yellowish, ovate to elliptic; styles 2(–3), erect, distinct; stigmas capitate. **Achenes** included or barely exserted, light brown or brown to black, unwinged, unevenly 2-gonous, rarely 3-gonous, glabrous. **Seeds:** embryo curved. *x* = 7.

Species 6 (1 in the flora): alpine, arctic, and circumpolar, n North America, s South America, n Europe, e Asia.

The five other species of *Koenigia* are endemic to high mountains of southeastern Asia, primarily the Himalayas (O. Hedberg 1997). K. Haraldson (1978) and L.-P. Ronse Decraene and J. R. Akeroyd (1988) placed *Koenigia* close to *Aconogonon* based on morphological data. Preliminary molecular data seem to support that relationship (A. S. Lamb Frye and K. A. Kron 2003).

SELECTED REFERENCES Hedberg, O. 1997. The genus *Koenigia* L. emend. Hedberg (Polygonaceae). Bot. J. Linn. Soc. 124: 295–330. Löve, Á. and P. Sarkar. 1957. Chromosomes and relationships of *Koenigia islandica*. Canad. J. Bot. 35: 507–514.

1. **Koenigia islandica** Linnaeus, Mant. Pl. 1: 35. 1767; Syst. Nat. ed. 12, 2: 71, 104. 1767 • Island koenigia, koenigia d'Islande [F]

Plants (0.5–)1–8(–20) cm. **Stems** often reddish, branched or simple, rooting adventitiously from proximal nodes. **Leaves:** ocrea brownish, broadly funnelform, 1–1.5 mm, margins oblique; petiole (0.1–)2–10 mm; blade 1–6.5 × 1–5 mm, base tapering to truncate, apex obtuse to acute, glabrous. **Pedicels** 0.1–1.5 mm or absent. **Flowers:** perianth 0.9–1.8 mm; tepals ovate to elliptic, margins entire, apex obtuse; stamens ca. ¹/₂ as long as tepals. **Achenes** 1–1.8 × 0.7–0.8 mm, dull. $2n = 14$ (Mongolia, Norway), 28.

Flowering Jul–Aug, fruiting Aug–Sep. Arctic tundra and alpine meadows with permanently moist gravel, especially around persistent snow patches near streams, ponds, and lakes; 0–4400 m; Greenland; Alta., B.C., Man., Nfld. and Labr. (Labr.), N.W.T., Nunavut, Ont., Que., Yukon; Alaska, Colo., Mont., Utah, Wyo.; s South America (Argentina, Chile); n Europe; c, e Asia.

Koenigia islandica is among the smallest of terrestrial flowering plants and one of few annual species in arctic and alpine floras. Some other species exhibit a similar bipolar distribution (e.g., *Anemone multifida* Poiret, *Osmorhiza chilensis* Hooker & Arnott, and *Carex macloviana* D'Urville).

45. PLUMBAGINACEAE Jussieu

• Leadwort Family

Nancy R. Morin

Herbs or shrubs [lianas], perennial or, rarely, annual; taprooted or rhizomatous. **Stems** woody stocks, acaulescent, or erect to prostrate, nodes swollen; indument of simple hairs, capitate glands that may secrete water or calcium salts, or multicelled glandlike structures. **Leaves** often basal, alternate, spiralled; stipules absent; petiole present or absent; blade linear to broadly obovate, ovate, or round, margins entire or lobed. **Inflorescences** terminal or axillary cymes, panicles, racemes, or corymbs, or solitary heads; bracts herbaceous, scarious, sometimes absent; involucral bracteoles (epicalyces) immediately subtending calyces usually present. **Pedicels** absent or present (short). **Flowers** bisexual, radially symmetric; perianth and androecium hypogynous; sepals persistent in mature fruits, 5, connate into 5- or 10-ribbed tube, mostly dry and membranous, sometimes petaloid, toothed or with distinct simple or lobed limbs; petals 5, nearly distinct, connate at bases or for most of their length (corolla salverform); blade clawed or claw absent, margins entire; corona absent; stamens 5; filaments adnate to bases of petals or free; ovary superior, 1-locular, placentation basal; ovules 1 per ovary, anatropous, bitegmic, crassinucellate; styles 1 with apically lobed stigma, or 5, each with linear stigma. **Fruits** utricles, achenes, or capsules. **Seeds** 1, embryo straight, endosperm present or absent.

Genera 24, species ca. 775 (3 genera, 11 species in the flora): worldwide, especially maritime areas.

A report of *Ceratostigma plumbaginioides* Bunge from Missouri is based on a single specimen collected in an alley in Columbia, Boone County (D. B. Dunn 1982) and probably is not naturalized, according to George Yatskievych (pers. comm.), who considers that it probably has not persisted. J. H. Schaffner (1932) reported the same species as a waif in Lake County, Ohio. T. C. Cooperrider (1995) cited that report and indicated that he had not seen a specimen. Cultivated *Ceratostigma* seems to have the potential for becoming naturalized. *Ceratostigma* resembles *Plumbago* but has stamens adnate to the corolla tube and a nonglandular calyx.

Plumbaginaceae may be a sister group to Polygonaceae (M. D. Lledó et al. 1998). It includes some plants of horticultural value, including *Ceratostigma*, *Armeria*, *Limonium*, and *Plumbago*. Some species of *Plumbago* and *Limonium* have medicinal uses. Plumbaginaceae often occur in saline habitats; basal leaves may have glands that excrete calcareous or chalklike salts. Some species of *Armeria* occur on soils rich in lead or on mine tailings. The family's Latin and common names derive from an early belief that the plants could cure lead poisoning.

SELECTED REFERENCES Carlquist, S. and C. J. Biggs. 1996. Wood anatomy of Plumbaginaceae. Bull. Torrey Bot. Club 123: 135–147. Lledó, M. D. et al. 1998. Systematics of Plumbaginaceae based upon cladistic analysis of *rbc*L sequence data. Syst. Bot. 23: 21–29. Luteyn, J. L. 1990. The Plumbaginaceae in the flora of the southeastern United States. Sida 14: 169–178.

1. Inflorescences dense hemispheric heads terminal on leafless scape; leaf blades mostly linear to lanceolate . . . . 1. *Armeria*, p. 603
1. Inflorescences terminal or axillary racemes, panicles, or corymbs; leaf blades elliptic to oblong to round, rarely linear.
    2. Plants acaulescent; petals nearly distinct; stamens adnate to bases of petals . . . . . . 2. *Limonium*, p. 606
    2. Plants with stems erect, prostrate, or climbing; petals connate for most of their length, corollas salverform; stamens free from petals . . . . 3. *Plumbago*, p. 610

1. ARMERIA Willdenow, Enum. Pl. 1: 333. 1809 • Thrift [Celtic *ar mor*, at seaside, alluding to habitat]

Claude Lefèbvre

Xavier Vekemans

Plants herbs, perennial, scapose, acaulescent; taprooted, rootstocks branched, woody. **Leaves** in basal rosettes, sessile; blade linear to linear-spatulate [lanceolate], narrowed or straight to base, margins entire. **Scapes** glabrous or densely pubescent, sometimes rugose, enclosed by tubular leafless sheath at apex. **Inflorescences** solitary, apical, dense hemispheric heads of scorpioid cymes, each surrounded by involucre of scarious bracts. **Pedicels** absent or present (short). **Flowers** monomorphic or dimorphic (in pollen and stigma characteristics); calyx 10-ribbed, funnel-shaped; tube usually pubescent on ribs only or all around, rarely glabrous, limbs membranaceous, awned or not; petals slightly connate basally, white to deep purple; filaments adnate to base of corolla; anthers included; styles 5, free, hairy proximally; stigmas linear, papillate or smooth. **Fruits** dry, enclosed in persistent calyces, dehiscing transversely. *x* = 9.

Species ca. 50 (1 in the flora): North America, s South America, Europe, w Asia (n Siberia), n Africa.

Armeria is known to be taxonomically difficult. Species concepts vary among authors. About 50 species can be recognized according to A. R. Pinto da Silva (1972).

SELECTED REFERENCES Bernis, F. 1952. Revisión del género *Armeria* Willd. con especial referencia a los grupos Ibéricos. Anales Inst. Bot. Cavanilles 11(2): 5–287. Lawrence, G. H. M. 1940. Armerias, native and cultivated. Gentes Herb. 4: 391–418. Lawrence, G. H. M. 1947. The genus *Armeria* in North America. Amer. Midl. Naturalist 37: 751–779. Lefèbvre, C. 1974. Population variation and taxonomy in *Armeria maritima* with special reference to heavy-metal tolerant populations. New Phytol. 73: 209–219. Lefèbvre, C. and X. Vekemans. 1995. A numerical taxonomic study of *Armeria maritima* (Plumbaginaceae) in North America and Greenland. Canad. J. Bot. 73: 1583–1595.

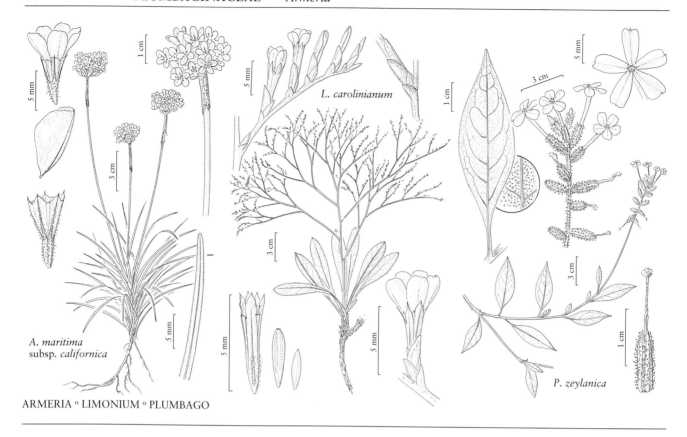

ARMERIA ○ LIMONIUM ○ PLUMBAGO

1. Armeria maritima (Miller) Willdenow, Enum. Pl. 1: 333. 1809 [F]

Statice maritima Miller, Gard. Dict. ed. 8, Statice no. 3. 1768

Rootstock erect. **Leaf blades** 1–15 cm × 0.5–3 mm, base 1- or ± 3-veined, faces glabrous or hairy. **Scapes** erect, 2–60 cm, glabrous or hairy. **Inflorescences:** involucral sheath 5–32 mm; outermost involucral bract ovate to triangular-lanceolate, 4–14 mm, shorter than, equaling, or exceeding head, mucronate or not; heads 13–28 mm diam. **Flowers** monomorphic, with all stigmas papillate and pollen reticulate, or dimorphic, with papillate stigmas and finely reticulate pollen or smooth stigmas and coarsely reticulate pollen; calyx tube hairy on and between ribs (holotrichous), on ribs only (pleurotrichous), or glabrous (atrichous); teeth triangular to shallowly triangular, awned or not; corolla pink to white; petals showy and exceeding calyx or reduced and included in calyx. $2n = 18$.

Subspecies 10 (4 in the flora): North America, Eurasia.

1. Flowers dimorphic, papillate stigmas associated with finely reticulate pollen and smooth stigmas associated with coarsely reticulate pollen . . . . 1a. *Armeria maritima* subsp. *maritima*
1. Flowers monomorphic, stigmas all papillate with coarsely reticulate pollen.
  2. Inflorescence sheath lengths usually 0.75 times diam. of flower heads; outer involucral bracts almost equaling or exceeding flower head; scapes glabrous; calyces hairy on ribs . . . . 1b. *Armeria maritima* subsp. *californica*
  2. Inflorescence sheath lengths usually 0.5–0.75 times diam. of flower heads; outer involucral bracts 0.5–0.75 times flower head; scapes glabrous or hairy; calyces hairy throughout, on ribs only, or glabrous.
    3. Calyces hairy throughout or on ribs only; scapes glabrous or hairy; leaf blades hairy or glabrous . . . 1c. *Armeria maritima* subsp. *sibirica*
    3. Calyces glabrous; scapes glabrous; leaf blades glabrous . . . . 1d. *Armeria maritima* subsp. *interior*

1a. Armeria maritima (Miller) Willdenow subsp. **maritima**

Leaf blades hairy or glabrous. **Scapes** hairy. **Inflorescences:** sheath length 0.33–0.75 times diam. of flower head; outer involucral bracts much shorter than flower head. **Flowers** dimorphic: papillate stigmas associated with finely reticulate pollen; smooth stigmas associated with coarsely reticulate pollen; calyx hairy throughout or on ribs only. $2n = 18$.

Flowering late spring–early summer. Maritime rocks, cliffs; 0 m; Greenland; Oreg.; Europe.

Populations of subsp. *maritima* occur mainly in coastal salt marshes, cliffs, rocks, and pastures in northwestern Europe and have their most western extension in southern Greenland, below 62°N latitude. It is introduced and naturalized at Yaquina Head.

1b. Armeria maritima (Miller) Willdenow subsp. **californica** (Boissier) A. E. Porsild, Bull. Natl. Mus. Canada 135: 174. 1955 [E] [F]

Armeria andina Poeppig ex Boissier var. *californica* Boissier in A. P. de Candolle and A. L. P. P. de Candolle, Prodr. 12: 682. 1848; *A. arctica* (Chamisso) Wallroth subsp. *californica* (Boissier) Abrams; *A. maritima* var. *californica* (Boissier) G. H. M. Lawrence

Leaf blades glabrous or hairy. **Scapes** glabrous. **Inflorescences:** sheath length usually 0.75 times diam. of flower head; outer involucral bracts almost equaling or exceeding flower head. **Flowers** monomorphic: stigmas papillate, pollen coarsely reticulate; calyx hairy on ribs. $2n = 18$.

Flowering mid spring–mid summer. Maritime rocks, cliffs, sandy bluffs, sandy dunes; 0–200 m; B.C.; Calif., Oreg., Wash.

In northern Washington and on Vancouver Island, populations with hairy leaves have been called *Armeria maritima* var. *purpurea* (Koch) G. H. M. Lawrence, a dimorphic-flowered taxon from central Europe. The American monomorphic-flowered specimens thought to belong to var. *purpurea* are not distinct from subsp. *californica*, except for their hairy leaves. We include hairy-leaved specimens in subsp. *californica*.

1c. Armeria maritima (Miller) Willdenow subsp. **sibirica** (Turczaninow) Nyman, Consp. Fl. Eur., 616. 1881

Armeria sibirica Turczaninow ex Boissier in A. P. de Candolle and A. L. P. P. de Candolle, Prodr. 12: 678. 1848; *A. arctica* (Chamisso) Wallroth; *A. labradorica* Wallroth; *A. labradorica* var. *submutica* (S. F. Blake) H. F. Lewis; *A. maritima* subsp. *arctica* (Chamisso) Hultén; *A. maritima* subsp. *labradorica* (Wallroth) Hultén; *A. maritima* var. *labradorica* (Wallroth) G. H. M. Lawrence; *A. maritima* var. *sibirica* (Turczaninow) G. H. M. Lawrence

Leaf blades hairy or glabrous. **Scapes** glabrous or hairy. **Inflorescences:** sheath length usually 0.5–0.75 times diam. of flower head; outer involucral bracts 0.5–0.75 times length of flower head. **Flowers** monomorphic: stigmas papillate, pollen coarsely reticulate; calyx hairy throughout or on ribs only. $2n = 18$.

Flowering summer. Gravelly tundras, flood plains, lakes and seashores, alpine meadows; 0–1200 m; Greenland; Man., Nfld. and Labr., N.W.T., Nunavut, Ont., Yukon; Alaska, Colo.; Eurasia.

The population of subsp. *sibirica* in Colorado is found on Hoosier Ridge. Plants with calyces hairy throughout and hairy scapes appear in eastern North America east of approximately 100°W longitude mixed with plants with hairy calyx ribs and glabrous scapes. Plants from this area have been called var. *labradorica*; there is no taxonomic reason for this recognition (C. Lefèbvre and X. Vekemans 1995).

1d. Armeria maritima (Miller) Willdenow subsp. **interior** (Raup) A. E. Porsild, Bull. Natl. Mus. Canada 135: 172. 1955 [E]

Statice interior Raup, J. Arnold Arbor. 17: 289, plate 198. 1936; *Armeria maritima* var. *interior* (Raup) G. H. M. Lawrence

Leaf blades glabrous. **Scapes** glabrous. **Inflorescences:** sheath length usually 0.5–0.75 times diam. of flower head; outer involucral bracts 0.5–0.75 times length of flower head. **Flowers** monomorphic: stigmas papillate, pollen coarsely reticulate; calyx glabrous. $2n = 18$.

Flowering summer. Sand dunes, gravel pavements; 200–300 m; Sask.

Subspecies *interior* is known only from the south shore of Lake Athabasca.

2. LIMONIUM Miller, Gard. Dict. Abr. ed. 4, vol. 2. 1754, name conserved • Sea lavender, statice, marsh rosemary [Greek *leimon*, meadow, referring to frequent occurrence of some species on salt meadows]

Alan R. Smith

Plants herbs, usually perennial, scapose, acaulescent; taprooted or rhizomatous. **Leaves** basal (sometimes also on inflorescence axes), sessile or petiolate; blade often punctate, elliptic to obovate, oblanceolate, spatulate, oblong, or round, usually coriaceous, base usually long-attenuate, margins entire or toothed to pinnatifid, apex rounded to apiculate or retuse. **Inflorescences** usually of terminal panicles or corymbs, ultimate branch tips bearing secund, usually 1–3(–5)-flowered spikelets. **Pedicels** absent or present (very short, subtended by 3 or 4 sheathing bracts). **Flowers** homostylous; calyx tubular to funnelform, 5-ribbed, glabrous or pubescent, plicate, lobes oblong to triangular, sometimes with smaller intervening lobes, or lobes ± connate and calyx mouth erose; petals nearly distinct, white, lavender, or yellow, long-clawed; filaments adnate to base of corolla; anthers included; styles 5, distinct to base; stigmas linear-clavate, papillate. **Fruits** utricles, usually exserted from persistent calyx, brownish green, usually capped by marcescent corolla and style bases. *x* = 8, 9.

Species ca. 300 (8 in the flora): worldwide, especially from Mediterranean region east to c Asia.

The greatest diversity in *Limonium* is found in Europe (ca. 100 species and many subspecies; see S. Pignatti 1972) and in Mediterranean and central Asian regions, often on saline or calcareous soils and cliffs near the coasts; other species are found in saline marshlands. The showiest species (*L. arborescens* and *L. perezii*), with a persistent blue-purple to lavender calyx, have their origin in the Canary Islands; they are often cultivated for ornament or their inflorescences are air-dried for floral arrangements under their Linnaean name "Statice." Other species have been used in rock gardens. Six species are locally naturalized in California.

Limonium vulgare Miller (*Statice limonium* Linnaeus), similar morphologically to *L. carolinianum*, has been reported by H. J. Scoggan (1978–1979, part 4) from central Saskatchewan and southern Ontario ("in a weedy...cemetery...York Co., where 'growing without cultivation'"). It is doubtful that the species persists or is spreading. Recent revisitation of the site in Ontario by J. E. Eckenwalder (pers. comm.) suggests that *Limonium vulgare* is no longer extant there. *Limonium leptostachyum* (Boissier) Kuntze (*S. leptostachya* Boissier) has been reported from New York by R. S. Mitchell and G. C. Tucker (1997); it is doubtful that this central Asian species is naturalized in the flora area. It differs from all other species in the flora area by having small (10–30 × 5 mm), deeply pinnatifid leaves and narrow, spikelike inflorescences.

Some species of *Limonium*, e.g., *L. sinuatum*, have dimorphic pollen and stigmas that result in self-incompatibility, although the native species in the flora area have been shown to be self-compatible (H. G. Baker 1953b). Agamospermy is also common in some extraterritorial species, and this may account, in part, for the taxonomic difficulty in some groups of *Limonium*.

SELECTED REFERENCES Baker, H. G. 1953b. Dimorphism and monomorphism in the Plumbaginaceae II. Pollen and stigmata in the genus *Limonium*. Ann. Bot. (Oxford), n. s. 17: 433–445. Luteyn, J. L. 1976. Revision of *Limonium* (Plumbaginaceae) in eastern North America. Brittonia 28: 303–317.

1. Leaf blade margins pinnately lobed; inflorescence axes with 3–5 wings, these with linear leaflike appendages 2–8 × 0.2–0.5 cm and stiff, stout hairs to ca. 1.5 mm . . . . 7. *Limonium sinuatum*
1. Leaf blade margins ± entire or obscurely undulate; inflorescence axes not winged, or if winged, then wings 2(–3), linear leaflike appendages absent, glabrous.
   2. Inflorescences with some nonflowering branches . . . . 6. *Limonium otolepis*
   2. Inflorescences with all (or nearly all) branches bearing flowers.
      3. Inflorescences greater than 1 m, axes strongly winged, wings leaflike, veined . . . . 5. *Limonium arborescens*
      3. Inflorescences less than 1 m, axes not winged.
         4. Calyces blue-purple distally; leaf blades round to broadly ovate or subcordate, bases abruptly narrowed, nearly as broad as long; floral bracts ciliate or fimbriate on margins; coastal s California . . . . 4. *Limonium perezii*
         4. Calyces whitish distally; leaf blades obovate to oblong or oblanceolate, bases gradually tapered; floral bracts glabrous at margins; widespread.
            5. Leaf blades less than 4 × 1.5 cm, each with single midrib and sometimes 2 lesser, ± parallel veins; petals pink to lavender, exserted 2–3 mm from calyx; calyx ribs glabrous; coastal salt marshes of s California . . . . . . 8. *Limonium ramosissimum*
            5. Leaf blades more than 4 × 1.5 cm, pinnately veined; petals lavender or whitish, exserted 0–2 mm from calyx; calyx ribs often pilose, sometimes glabrous; widespread.
               6. Calyx lobes spreading at maturity; spikelets always densely aggregated at tips of inflorescence branches; inland alkaline areas . . . . 2. *Limonium limbatum*
               6. Calyx lobes ascending or erect at maturity; spikelets not aggregated, or loosely to moderately or densely aggregated along inflorescence branches; coastal salt marshes or flats.
                  7. Spikelets loosely to moderately densely aggregated along inflorescence branches; leaf blade apices cuspidate, cusps 1–3 mm, soon falling; Atlantic and Gulf coastal plains . . . . 1. *Limonium carolinianum*
                  7. Spikelets moderately to densely aggregated along inflorescence branches; leaf blade apices rounded or, occasionally, retuse at tips, rarely cuspidate (cusps less than 0.5 mm); California and Oregon coasts, Arizona, s Nevada . . . . 3. *Limonium californicum*

1. **Limonium carolinianum** (Walter) Britton, Mem. Torrey Bot. Club 5: 255. 1894 • Canker or ink or marsh root, lavender or American or seaside thrift [F]

Statice caroliniana Walter, Fl. Carol., 118. 1788; *Limonium angustatum* (A. Gray) Small; *L. carolinianum* var. *angustatum* (A. Gray) S. F. Blake; *L. carolinianum* var. *compactum* Shinners; *L. carolinianum* var. *nashii* (Small) B. Boivin; *L. carolinianum* var. *obtusilobum* (S. F. Blake) H. E. Ahles; *L. carolinianum* var. *trichogonum* (S. F. Blake) B. Boivin; *L. nashii* Small; *L. nashii* var. *angustatum* (A. Gray) H. E. Ahles; *L. obtusilobum* S. F. Blake; *L. trichogonum* S. F. Blake

Leaves all in basal rosettes, living at anthesis, 5–25(–40) cm; petiole often narrowly winged distally, 0.1–20 cm, usually shorter than blade; blade usually elliptic, spatulate, or obovate to oblanceolate (rarely linear), 5–15(–30) × 0.5–5(–7.5) cm, leathery, base gradually tapered, margins usually entire, sometimes undulate, apex rounded or acute to retuse, cuspidate, cusp 1–3 mm, soon falling; main lateral veins ascending, obscurely pinnate. **Inflorescences:** axes not winged, 10–60(–95) cm × 1–5 mm, glabrous; nonflowering branchlets absent; spikelets loosely to moderately densely aggregated along branches, internodes 0.5–10 mm; subtending bracts 2–6 mm, obtuse, surfaces and margins glabrous; flowers solitary or 2–3(–5) per spikelet. **Flowers:** calyx whitish, obconic, 4–6.5(–7.5) mm; tube 2.5–5 mm, glabrous or densely pilose along ribs; lobes erect, to ca. 2 × 1 mm; petals lavender (rarely white), slightly exceeding calyx. **Utricles** 3–5.5 mm. $2n = 36$.

Flowering Jun–Dec. Salt marshes and salt flats along Atlantic and Gulf seacoasts; 0 m; N.B., Nfld. and Labr. (Nfld.), N.S., P.E.I., Que.; Ala., Conn., Del., Fla., Ga., La., Maine, Md., Mass., Miss., N.H., N.J., N.Y., N.C., R.I., S.C., Tex., Va.; Mexico (Tamaulipas); Bermuda.

J. L. Luteyn (1976, 1990) discussed the more or less continuous variation in this polymorphic species. He noted that seedling establishment is rare, and that populations spread primarily by vegetative means from horizontal rhizomes.

2. Limonium limbatum Small, Bull. Torrey Bot. Club 25: 317. 1898 E

Limonium limbatum var. *glabrescens* Correll; *Statice limbata* (Small) K. Schumann

Leaves all in basal rosettes, living at anthesis, 10–25 cm; petiole narrowly winged distally, 0.1–9 cm, shorter than blade; blade oblong-spatulate, obovate, or elliptic, 4–16 × 1.5–6.5 cm, leathery, base gradually tapered, margins entire, apex rounded or retuse, often short-cuspidate, cusp less than 1 mm; main lateral veins ascending, obscurely pinnate. **Inflorescences:** axes not winged, 30–60(–100) cm × 2–3 mm, glabrous; nonflowering branchlets absent; spikelets densely aggregated at tips of branchlets, internodes 0.5–3 mm; subtending bracts 1–5 mm, apex obtuse, surfaces and margins glabrous; flowers 1–3 per spikelet. **Flowers:** calyx whitish distally, with reddish brown ribs, obconic to slightly funnelform, 3.5–5 mm, ribs usually densely pubescent; tube ca. 3 mm; lobes spreading at maturity, 0.5–1.5 × 1–1.5 mm; petals blue to nearly white, not exceeding calyx. **Utricles** 2.5–3 mm.

Flowering Jun–Aug. Wet meadows, gypsum soils, salt flats, alkaline depressions in the interior; 400–1800 m; Ariz., N.Mex., Okla., Tex.

3. Limonium californicum (Boissier) A. Heller, Cat. N. Amer. Pl., 6. 1898 • Marsh rosemary

Statice californica Boissier in A. P. de Candolle and A. L. P. P. de Candolle, Prodr. 12: 643. 1848; *Limonium californicum* var. *mexicanum* (S. F. Blake) Munz; *L. commune* Gray var. *californicum* (Boissier) Greene; *L. mexicanum* S. F. Blake

Leaves all in basal rosettes, living at anthesis, 10–30 × 1–6 cm; petiole often very narrowly winged, 0.1–12 cm, usually shorter than blade; blade spatulate to oblanceolate or obovate, 7–20 × 1–6 cm, leathery, base gradually tapered and long-decurrent, margins entire to undulate, apex obtuse or rounded, sometimes retuse, rarely cuspidate, if so, cusp less than 0.5 mm; main lateral veins strongly ascending, obscurely pinnate. **Inflorescences:** axes not winged, 15–60 cm × 2–5 mm, glabrous; nonflowering branches absent; spikelets moderately to densely aggregated, internodes 1–2 mm; subtending bracts 3–6 mm, apex usually acute or apiculate, surfaces and margins glabrous; flowers 1–2 per spikelet. **Flowers:** calyx whitish distally, with brownish ribs, obconic, ribs glabrous or pilose; tube 4–

6 mm; lobes erect at maturity, triangular, ca. 1 mm; petals lavender to whitish, only slightly exceeding calyx. **Utricles** not seen. $2n = 18$.

Flowering Jul–Dec. Coastal strand, salt marshes, sand hills, beaches, bays, alkaline flats; 0–50(–600) m; Calif., Nev., Oreg.; Mexico (Baja California).

Limonium mexicanum (or *L. californicum* var. *mexicanum*) has been distinguished on the basis of having glabrous calyces. Plants with glabrous (or nearly glabrous) calyces occur throughout the species range, from Humboldt to San Diego counties, and so the character state seems of dubious taxonomic significance. The two variants seem otherwise indistinguishable.

The sole collection seen from Nevada (*Fosberg 14278*, UC) was collected at the highest elevation known for the species, on dried alkaline mud flats. J. Morefield (pers. comm.) reported that it has been established in southern Nevada since at least 1898, and so may be native there. Morefield also reported a collection from the Salt River drainage, Gila County, Arizona, but I have not seen that specimen.

4. Limonium perezii (Stapf) F. T. Hubbard, Rhodora 18: 158. 1916 I

Statice perezii Stapf, Ann. Bot. (Oxford) 22: 116. 1908

Leaves all in basal rosettes, living at anthesis, to 30 cm; petiole winged distally, to 18 cm, usually exceeding blade; blade round to broadly ovate or subcordate, to 15 × 9 cm, leathery, base subtruncate (abruptly narrowed) and then decurrent, margins entire, apex cuspidate, cusp to 5 mm, soon falling; main lateral veins pinnate. **Inflorescences:** axes not winged, to 100 cm × 7 mm, glabrous to puberulent (hairs ca. 0.1 mm); nonflowering branches absent; spikelets moderately to densely aggregated at tips of branches, internodes mostly 2–4 mm; subtending bracts 3–6 mm, acute or aristate (outer) to truncate (inner), ciliate or fimbriate at margins, surfaces glabrous or minutely appressed-pubescent; flowers 1–2 per spikelet. **Flowers:** calyx blue-purple in distal ½, with reddish brown, glabrous ribs, funnelform; tube ca. 5 mm, minutely pubescent along proximal end of ribs (hairs less than 0.1 mm); lobes spreading, ca. 5 mm (5 main lobes with shallower lobes between larger lobes), or lobes indistinct and calyx appearing erose or irregularly lobed at mouth; petals whitish, barely exceeding calyx. **Utricles** 4–5 mm. $2n = 14$.

Flowering Mar–Sep. Disturbed coastal areas, cliffs, sand dunes, roadsides (where it is sometimes planted); 0–100 m; introduced; Calif.; Atlantic Islands (Canary Islands).

5. Limonium arborescens Kuntze, Revis. Gen. Pl. 2: 395. 1891 I

Leaves all in basal rosettes, living at anthesis, sessile or with very short petiole; blade entire, oblanceolate, 15–35 × 5–10 cm, leathery, base tapered gradually to narrow or broad wing, margins entire, apex rounded, cuspidate, cusp 1–2 mm, soon falling; main lateral veins pinnate. **Inflorescences:** axes broadly 2–3-winged, to 100+ cm × 8 mm, glabrous or pubescent (hairs ca. 0.2 mm), wings to ca. 7 mm wide, leaflike, net-veined, variable in width, interrupted at branch points, leaflike appendages absent; nonflowering branches absent; spikelets moderately to densely aggregated at tips of branches, internodes mostly 3–8 mm; subtending bracts 5–7 mm, acute to truncate at tips, margins and surfaces glabrous or often densely pubescent or fimbriate; flowers 1–2 per spikelet. **Flowers:** calyx blue-purple in distal ½, with reddish brown, short-pubescent ribs, funnelform; tube 5–6 mm, expanded portion 5–7 mm, lobes indistinct, glabrous; petals whitish, barely exceeding calyx.

Flowering Mar–Oct. Disturbed urban areas, coastal lagoons, roadsides, dunes, vacant lots; 0 m; introduced; Calif.; Atlantic Islands (Canary Islands).

6. Limonium otolepis (Schrenk) Kuntze, Revis. Gen. Pl. 2: 396. 1891 I

Statice otolepis Schrenk, Bull. Cl. Phys.-Math. Acad. Imp. Sci. Saint-Pétersbourg 1: 362. 1843

Leaves in basal rosettes and on inflorescence axes, leaves in rosettes dead before anthesis; petiole 3–8 cm, ± equaling blade; blade obovate to oblong, 3–8 × 1.5–3 cm, base tapered, margins ± entire, apex unknown, venation not seen; leaves on inflorescence axes sessile, clasping stems, blade ± round, less than 3 cm. **Inflorescences:** axes not winged, sometimes angled, 40–80+ cm × 2–3 mm, glabrous; nonflowering branches present, especially in proximal part, slender; spikelets densely aggregated at branch tips, internodes 1–2 mm; subtending bracts whitish, 1–2 mm, truncate, surfaces and margins glabrous; flowers 1–2 (–3) per spikelet. **Flowers:** calyx whitish distally, with reddish brown ribs, obconic, proximal ½ pilose between and on ribs (hairs 0.2–0.4 mm); tube ± 1.5–2 mm; lobes 0.5–0.7 × 0.5–0.7 mm; petals blue to whitish, ca. 0.5 mm, exceeding tube. **Utricles** 1–2 mm. $2n = 18$.

Flowering Sep–Feb. Disturbed coastal and urban areas, especially salt marshes, roadsides; 0–100 m; introduced; Calif.; w, c Asia.

The name *Limonium perfoliatum* (Karelin ex Boissier) Kuntze, usually now treated as a synonym of *L. reniforme* (Girard) Linczevski, was misapplied to this species by J. T. Howell et al. (1958) and P. A. Munz (1968).

7. Limonium sinuatum (Linnaeus) Miller, Gard. Dict. ed. 8, Limonium no. 6. 1768 I

Statice sinuata Linnaeus, Sp. Pl. 1: 276. 1753

Leaves all in basal rosettes, living at anthesis, 6–16 × 1.5–3 cm; petiole to ca. 5 cm, shorter than blade; blade oblanceolate, 6–12 cm × 1.5–36 mm, herbaceous to chartaceous, base tapered to a sinuate wing, margins pinnately lobed to 1–3 mm from midrib (lobes mostly 4–6 per side, broadest near apex), apex cuspidate, cusp 1–3 mm, soon falling; main lateral veins pinnate. **Inflorescences:** axes narrowly 3–5-winged, 20–40(–50) cm × 3–5 mm, hispid (hairs to 1.5 mm), wings to 3 mm wide, each with ± leaflike, linear, hispid appendage 2–8 × 0.2–0.5 cm usually at branch points; nonflowering branches absent, spikelets moderately to densely aggregated at branch tips; internodes mostly 5–10 mm; subtending bracts 5–10 mm, narrowly acuminate or often awned at tips, surfaces and margins hispid; flowers 1–3 per spikelet. **Flowers:** calyx blue to lavender distally, funnelform, glabrous or minutely hairy on tube, lobes not distinct, expanded portion spreading, 5–7 mm, erose; petals pale yellow, exceeding calyx 2–4 mm. **Utricles** ca. 5 mm. $2n = 16, 18$.

Flowering Mar–Oct. Disturbed coastal areas, vacant lots, old fields, roadsides; 0-300 m; introduced; Calif.; Mediterranean region; w Asia.

8. Limonium ramosissimum (Poiret) Maire, Bull. Soc. Hist. Nat. Afrique N. 27: 244. 1936 I

Statice ramosissima Poiret, Voy. Barbarie 2: 142. 1789; *Limonium psilocladon* (Boissier) Kuntze; *Statice psiloclada* Boissier

Leaves more than 10, all in basal rosettes, living at anthesis, essentially sessile; blade obovate, 2–4 × 0.7–1.4 cm, base tapered, margins ± entire, apex rounded; main veins 1–3 per leaf, ± parallel, not obviously pinnate. **Inflorescences:** axes not winged or angled, 15–25 cm ×

1–1.5 mm, glabrous; nonflowering branches absent; spikelets moderately to densely aggregated at branch tips, internodes 2–3 mm; subtending bracts 1.5–5 mm, rounded to broadly acute, surfaces and margins glabrous; flowers 1–2 per spikelet. **Flowers:** calyx whitish distally with reddish brown ribs, funnelform, 4–5 mm, glabrous; tube 3–4 mm, lobes spreading, 1 × 1 mm; petals lavender to pink, exceeding calyx ca. 2–3 mm. **Utricles** unknown. $2n = 24, 27$.

Flowering Jun. Coastal salt marshes; 0 m; introduced; Calif.; Mediterranean region.

Limonium ramosissimum is abundantly naturalized in Carpenteria Salt Marsh, Santa Barbara County (C. F. Smith 1998). Its identification is somewhat problematic, for lack of comparative material in American herbaria and the immense size and complexity of the genus in Mediterranean areas; our plants appear to match specimens identified and keyed as *Limonium ramosissimum* from southern Europe and northern Africa, but further study is needed. S. Pignatti (1972) recognized five subspecies in that polymorphic species, and E. McClintock (as reported by Smith) identified our adventive as subsp. *provinciale* (Pignatti) Pignatti; however, it seems premature to assign our taxon to any of the subspecies without detailed comparison with European material. Plants have been seen in California nurseries and gardens under the name *L. psilocladon* (Boissier) Kuntze (as "psiloclada"), generally regarded as a synonym of *L. ramosissimum*. Another very similar species is *L. hyblaeum* Brullo, native to Sicily, which is thought to be naturalized around harbors and coastal marshes in southern Australia (D. B. Foreman et al. 1993–1999, vol. 3; J. Edmondson, pers. comm.). At its present naturalized location, *L. ramosissimum* may be a threat to the endangered *Cordylanthus maritimus* Nuttall, with which it grows (W. R. Ferren Jr., pers. comm.).

Another introduced and naturalized *Limonium*, as yet unidentified to species and probably from Mediterranean regions, has recently been collected in salt marshes in San Diego County, (*Lawhead 32*, SD, UC). It is similar in stature and inflorescence characters to *L. ramosissimum* but differs in having longer, thinner-textured leaves to 8 × 1 cm, with more gradually attenuate bases and apiculate blades, each having a single medial vein. It appears that non-native species of *Limonium* are being grown by the cut-flower industry in the area, escaping, and establishing, perhaps to the detriment of native species.

3. PLUMBAGO Linnaeus, Sp. Pl. 1: 151. 1753; Gen. Pl. ed. 5, 75. 1754 • Leadwort

[Latin *plumbago*, a leadlike ore, alluding to historical use as a cure for lead poisoning]

Alan R. Smith

Plants perennial shrubs or suffrutescent herbs; roots not known. **Stems** erect, prostrate, or climbing, ribbed. **Leaves** cauline, sessile or short-petiolate (petiole usually less than 1.5 cm); blade elliptic to oblanceolate or spatulate, base narrowed, margins entire, apex acute, acuminate, or obtuse, membranaceous. **Inflorescences** terminal or axillary spikelike racemes or panicles. **Pedicels** 2-bracteolate, short. **Flowers** sometimes heterostylous, short-pedicellate; bracts absent; calyx persistent, 5-ribbed, tubular, with stalked, capitate-glandular protuberances along ribs; lobes triangular, 1–2 mm; corolla salverform, evenly to somewhat unevenly 5-lobed, lobes spreading, obovate, round, or truncate, mucronate; stamens included or exserted, free from corolla; style 1 included or exserted; stigmas 5, linear. **Fruits** capsules, included, brownish, long-beaked; valves coherent at apex. $x = 7$.

Species 12 (2 in the flora); tropical and subtropical regions, North America, Central America, South America, Europe, Asia, Africa.

Several species of *Plumbago* are cultivated, including *P. auriculata*. The entire plant of that species, especially the root, contains plumbagin, a toxic naphthoquinone derivative (oil of plumbago), which may cause severe skin irritation or blistering in humans and may also be toxic to other animals (T. C. Fuller and E. McClintock 1986).

The remarkable glands on the calyces of *Plumbago* are often called "glandular hairs," but they are not true hairs, being much more massive and multicellular structures with enlarged, capitate apices.

1. Corollas pale blue, tube 2 or more times length of calyx; calyces with stipitate, glandlike protuberances and hairs; inflorescences compact, 2.5–3(–5) cm; plants cultivated and locally naturalized in Florida . . . . 1. *Plumbago auriculata*
1. Corollas white, tube mostly less than 2 times length of calyx; calyces with stipitate, glandlike protuberances, true hairs absent; inflorescences elongate, 3–15(–30) cm; plants native . . . . 2. *Plumbago zeylanica*

1. **Plumbago auriculata** Lamarck in J. Lamarck et al., Encycl. 2: 270. 1786 • Cape leadwort [I]

Plumbago capensis Thunberg

Plants evergreen shrubs. **Stems** erect, trailing, or climbing, diffusely branched, to 3+ m, glabrous or pubescent on youngest shoots. **Leaves** usually sessile, sometimes short-petiolate; blade elliptic, oblanceolate, or spatulate, (1–)2.5–9 × 0.5–2.5 cm, base usually long-attenuate, sometimes auriculate, apex acute or obtuse, mucronate. **Inflorescences** 2.5–3(–5) cm, rachises short-pilose (hairs ca. 0.1 mm), eglandular; floral bracts lanceolate, 3–9 × 1–2 mm. **Flowers** 3-stylous; calyx 10–13 mm, tube usually short-pilose and with stalked, capitate, glandlike protuberances ca. 1 mm along distal $^1/_2$–$^3/_4$ of ribs; corolla pale blue, 37–53 mm, tube 28–40 mm (more than 2 times length of calyx), lobes 10–16 × 6–15 mm; stamens included or exserted. **Capsules** 8 mm. **Seeds** brown, 7 mm. $2n = 14 + 0$–1B.

Flowering year-round. Hummocks, thickets, disturbed sites in dry soil; 0–50 m; introduced; Fla.; s Africa.

Plumbago auriculata is frequently cultivated in Mediterranean-type warmer climates, especially in California, Arizona, and Texas.

2. **Plumbago zeylanica** Linnaeus, Sp. Pl. 1: 151. 1753 • Doctorbush [F]

Plumbago scandens Linnaeus

Plants herbaceous. **Stems** prostrate, climbing, or erect, glabrous. **Leaves** petiolate (to 1.5 cm) or sessile; blade ovate, lance-elliptic, or spatulate to oblanceolate, (3–)5–9(–15) × (1–)2.5–4(–7) cm, base attenuate, apex acute, acuminate, or obtuse. **Inflorescences** 3–15 (–30) cm, rachises glandular, viscid; floral bracts lanceolate, 3–7 × 1–2 mm. **Flowers** heterostylous; calyx 7–11(–13) mm, tube glabrous but with stalked glands along length of ribs; corolla white, 17–33 mm, tube 12.5–28 mm (less than 2 times length of calyx), lobes 5–12 × 3–3.5 mm; stamens included. **Capsules** 7.5–8 mm. **Seeds** reddish brown to dark brown, 5–6 mm.

Flowering year-round. Palm groves, thickets, shady hummocks, shell mounds, rocky places in open areas; 0–50 m; Ariz., Fla., Tex.; Mexico; Central America; South America; Asia; Africa; Pacific Islands.

Plumbago zeylanica and *P. scandens*, both Linnaean species, have heretofore been treated as distinct, the former name applied exclusively to Old World plants, the latter to New World specimens. John Edmondson (pers. comm.) indicates that he believes this "could be a classic case of New World and Old World taxonomists each doing their own thing." Plants in herbaria under these two names appear indistinguishable.

Literature Cited

Robert W. Kiger, Editor

This is a consolidated list of all works cited in volume 5, whether as selected references, in text, or in nomenclatural contexts. In citations of articles, both here and in the taxonomic treatments, and also in nomenclatural citations, the titles of serials are rendered in the forms recommended in G. D. R. Bridson and E. R. Smith (1991). When those forms are abbreviated, as most are, cross references to the corresponding full serial titles are interpolated here alphabetically by abbreviated form. In nomenclatural citations (only), book titles are rendered in the abbreviated forms recommended in F. A. Stafleu and R. S. Cowan (1976–1988) and F. A. Stafleu and E. A. Mennega (1992+). Here, those abbreviated forms are indicated parenthetically following the full citations of the corresponding works, and cross references to the full citations are interpolated in the list alphabetically by abbreviated form. Two or more works published in the same year by the same author or group of coauthors will be distinguished uniquely and consistently throughout all volumes of *Flora of North America* by lower-case letters (b, c, d, ...) suffixed to the date for the second and subsequent works in the set. The suffixes are assigned in order of editorial encounter and do not reflect chronological sequence of publication. The first work by any particular author or group from any given year carries the implicit date suffix "a"; thus, the sequence of explicit suffixes begins with "b". There may be citations in this list that have dates suffixed "b" but that are not preceded by citations of "[a]" works for the same year, or that have dates suffixed "c," "d," "e," or "f" but that are not preceded by citations of "[a]," "b," "c," "d," and/or "e" works for that year. In such cases, the missing "[a]," "b," "c," "d," and/or "e" works are ones cited (and encountered first from) elsewhere in the *Flora* that are not pertinent in this volume.

A. S. B. Bull. = A S B Bulletin.

Abrams, L. and R. S. Ferris. 1923–1960. Illustrated Flora of the Pacific States: Washington, Oregon, and California. 4 vols. Stanford. (Ill. Fl. Pacific States)

Abuhadra, M. N. 2000. Taxonomic studies on the *Arenaria serpyllifolis* group (Caryophyllaceae). Fl. Medit. 10: 185–190.

Acta Bot. Neerl. = Acta Botanica Neerlandica.

Acta Phytotax. Sin. = Acta Phytotaxonomica Sinica. [Chih Wu Fen Lei Hsüeh Pao.]

Acta Univ. Lund. = Acta Universitatis Lundensis. Nova Series. Sectio 2, Medica, Mathematica, Scientiae Rerum Naturalium. [Lunds Universitets Årsskrift N.F., Avd. 2.]

Acta Univ. Upsal., Symb. Bot. Upsal. = Acta Universitatis Upsaliensis. Symbolae Botanicae Upsalienses.

Adanson, M. 1763[–1764]. Familles des Plantes. 2 vols. Paris. [Vol. 1, 1764; vol. 2, 1763.] (Fam. Pl.)

Aiton, W. and W. T. Aiton. 1810–1813. Hortus Kewensis; or a Catalogue of the Plants Cultivated in the Royal Botanic Garden at Kew. 5 vols. London. (Hortus Kew.)

Akeroyd, J. R. 1991. A new subspecific combination in *Rumex acetosella* L. Bot. J. Linn. Soc. 106: 97–99.

Akeroyd, J. R. 1993. *Herniaria*. In: T. G. Tutin et al., eds. 1993+. Flora Europaea, ed. 2. 1+ vol. Cambridge and New York. Vol. 1, pp. 182–184.

Alph. Aufz. Gew.—See: G. Heynhold 1846[–1847]

Amer. Anthropol. = American Anthropologist.

Amer. J. Bot. = American Journal of Botany.

Amer. J. Sci. Arts = American Journal of Science, and Arts.

Amer. Midl. Naturalist = American Midland Naturalist; Devoted to Natural History, Primarily That of the Prairie States.

Amer. Naturalist = American Naturalist....

Anales Inst. Bot. Cavanilles = Anales del Instituto Botánico A. J. Cavanilles.

Anales Inst. Segunda Enseñ. = Anales del Instituto de Segunda Enseñanza de la Habana.

Anales Jard. Bot. Madrid = Anales del Jardín Botánico de Madrid.

Anales Soc. Ci. Argent. = Anales de la Sociedad Científica Argentina.

Anales Univ. Chile, I, Mem. Ci. Lit. = Anales de la Universidad de Chile. I. Memorias Cientificas i Literarias.

Andersen, R. N., W. E. Lueschen, and J. R. Zaremba. 1985. Prickly smartweed (*Polygonum bungeanum*), a new weed in North America. Weed Sci. 33: 805–806.

Anderson, L. C. 1991. *Paronychia chartacea* ssp. *minima* (Caryophyllaceae): A new subspecies of a rare Florida endemic. Sida 14: 435–441.

Animadv. Bot. Spec. Alt.—See: P. Arduino 1764

Ann. A. C. F. A. S. = Annales de l'A C F A S.

Ann. Amélior. Pl. = Annales de l'Amélioration des Plantes.

Ann. Bot. (Oxford) = Annals of Botany. (Oxford.)

Ann. Lyceum Nat. Hist. New York = Annals of the Lyceum of Natural History of New York.

Ann. Missouri Bot. Gard. = Annals of the Missouri Botanical Garden.

Ann. Nat. Hist. = Annals of Natural History; or, Magazine of Zoology, Botany and Geology.

Ann. Sci. Nat., Bot. = Annales des Sciences Naturelles. Botanique.

Ann. Wiener Mus. Naturgesch. = Annalen des Wiener Museums der Naturgeschichte.

Annuaire Conserv. Jard. Bot. Genève = Annuaire du Conservatoire et Jardin Botaniques de Genève.

Anthropol. Rec. = Anthropological Records.

Arcangeli, G. 1882. Compendio della Flora Italiana.... Turin. (Comp. Fl. Ital.)

Arch. Naturwiss. Landesdurchf. Böhmen. = Archiv für die naturwissenschaftliche Landesdurchforschung von Böhmen.

Archibald, J. K., P. G. Wolf, V. J. Tepedino, and J. Bair. 2001. Genetic relationships and population structure of the endangered Steamboat buckwheat, *Eriogonum ovalifolium* var. *williamsiae* (Polygonaceae). Amer. J. Bot. 88: 608–615.

Arduino, P. 1764. Animadversionum Botanicarum Specimen Alterum. Venice. (Animadv. Bot. Spec. Alt.)

Ark. Bot. = Arkiv för Botanik Utgivet av K. Svenska Vetenskapsakademien.

Ascherson, P. F. A. et al. 1896–1939. Synopsis der mitteleuropäischen Flora. 12 vols., some in parts. Leipzig. (Syn. Mitteleur. Fl.)

Atlantic J. = Atlantic Journal, and Friend of Knowledge.

Austral. J. Bot. = Australian Journal of Botany.

Austral. Syst. Bot. = Australian Systematic Botany.

Baad, M. F. 1969. Biosystematic Studies of the North American species of *Arenaria*, Subgenus *Eremogone* (Caryophyllaceae). Ph.D. dissertation. University of Washington.

Bailey, J. P., L. E. Child, and A. P. Conolly. 1996. A survey of the distribution of *Fallopia* ×*bohemica* (Chrtek & Chrtková) J. Bailey (Polygonaceae) in the British Isles. Watsonia 21: 187–198.

Bailey, J. P. and C. A. Stace. 1992. Chromosome number, morphology, pairing, and DNA values of species and hybrids in the genus *Fallopia* (Polygonaceae). Pl. Syst. Evol. 180: 29–52.

Baileya = Baileya; a Quarterly Journal of Horticultural Taxonomy.

Baker, H. G. 1953b. Dimorphism and monomorphism in the Plumbaginaceae II. Pollen and stigmata in the genus *Limonium*. Ann. Bot. (Oxford), n. s. 17: 433–445.

Bakker, K. 1957. Revision of the genus *Polycarpaea* (Caryoph.) in Malaysia. Acta Bot. Neerl. 6: 48–53.

Ball, P. W. and V. H. Heywood. 1964. A revision of the genus *Petrorhagia*. Bull. Brit. Mus. (Nat. Hist.), Bot. 3: 121–172.

Barkoudah, Y. I. 1962. A revision of *Gypsophila, Bolanthus, Ankyropetalum* and *Phryna*. Wentia 9: 1–203.

Barneby, R. C. and E. C. Twisselman. 1970. Notes on *Loeflingia* (Caryophyllaceae). Madroño 20: 398–408.

Barrett, S. A. and E. W. Gifford. 1933. Miwok material culture. Bull. Public Mus. Milwaukee 2(4): 119–376.

Barrows, D. P. 1900. The Ethno-botany of the Coahuilla Indians of Southern California. Chicago. [Reprinted 1978, New York.]

Bartonia = Bartonia; a Botanical Annual.

Baskin, C. C., P. L. Chesson, and J. M. Baskin. 1993. Annual seed dormancy cycles in two desert winter annuals. J. Ecol. 81: 551–556.

Baskin, J. M. and C. C. Baskin. 1990. The role of light and alternating temperatures on germination of *Polygonum aviculare* seeds exhumed on various dates. Weed Res. 30: 397–402.

Bastard, T. 1812. Supplément à l'Essai sur la Flore du Département de Maine-et-Loire. Angers. (Suppl. Fl. Maine-et-Loire)

Baumgarten, J. C. G. 1816–1846. Enumeratio Stirpium Magno Transsilvaniae.... 4 vols. + suppl. + index. Vienna and Hermannstadt. (Enum. Stirp. Transsilv.)

Beena, K. R., K. Ananda, and K. R. Sridhar. 2000. Fungal endophytes of three sand dune species of west coast of India. Sydowia 52: 1–9.

Beerling, D. J., J. P. Bailey, and A. P. Conolly. 1994. *Fallopia japonica* (Houtt.) Ronse Decraene. J. Ecol. 82: 959–979.

Behnke, H.-D. 1982. *Geocarpon minimum*: Sieve-element plastids as additional evidence for its inclusion in the Caryophyllaceae. Taxon 31: 45–47.

Behnke, H.-D. and T. J. Mabry, eds. 1994. Caryophyllales: Evolution and Systematics. Berlin.

Beih. Bot. Centralbl. = Beihefte zum Botanischen Centralblatt. Original Arbeiten.

Beitr. Naturk.—See: J. F. Ehrhart 1787–1792

Bentham, G. 1839[–1857]. Plantas Hartwegianas Imprimis Mexicanas.... London. [Issued by gatherings with consecutive signatures and pagination.] (Pl. Hartw.)

Bentham, G. 1844[–1846]. The Botany of the Voyage of H.M.S. Sulphur, under the Command of Captain Sir Edward Belcher...during the Years 1836–1842. 6 parts. London. [Parts paged consecutively.] (Bot. Voy. Sulphur)

Berher, E.-L. 1887. La Flore des Vosges. Épinal. [Vol. 2 of L. Louis, Le Département des Vosges: Description, Histoire, Statistique.] (Fl. Vosges)

Bernis, F. 1952. Revisión del género *Armeria* Willd. con especial referencia a los grupos Ibéricos. Anales Inst. Bot. Cavanilles 11(2): 5–287.

Besser, W. S. J. G. von. 1822. Enumeratio Plantarum.... Vilna. (Enum. Pl.)

Bigelow, J. 1824. Florula Bostoniensis. A Collection of Plants of Boston and Its Vicinity..., ed. 2. Boston. (Fl. Boston. ed. 2)

Bigelow, J. 1840. Florula Bostoniensis. A Collection of Plants of Boston and Its Vicinity..., ed. 3. Boston. (Fl. Boston. ed. 3)

Biochem. Syst. & Ecol. = Biochemical Systematics and Ecology.

Biol. J. Linn. Soc. = Biological Journal of the Linnean Society.

Biol. Skr. = Biologiske Skrifter.

Bishop Mus. Occas. Pap. = Bishop Museum Occasional Papers.

Bittrich, V. 1993. Caryophyllaceae. In: K. Kubitzki et al., eds. 1990+. The Families and Genera of Vascular Plants. 4+ vols. Berlin etc. Vol. 2, pp. 206–236.

Blackwell, W. H. 1990. Poisonous and Medicinal Plants. Englewood Cliffs.

Bluff, M. J., C. A. Fingerhuth, and C. F. W. Wallroth. 1825–1833. Compendium Florae Germaniae. 4 vols. Nuremberg. (Comp. Fl. German.)

Bocek, B. R. 1984. Ethnobotany of Costanoan Indians, based on collections by John P. Harrington. Econ. Bot. 38: 240–255.

Böcher, T. W. 1963. Experimental and cytological studies on plant species. VIII. Racial differentiation in amphi-Atlantic *Viscaria alpina*. Biol. Skr. 11(6).

Böcher, T. W. 1977. *Cerastium alpinum* and C. *arcticum*, a mature polyploid complex. Bot. Not. 130: 303–309.

Bocquet, G. 1969. Revisio *Physolychnidum* (*Silene* Sect. *Physolychnis*).... Lehre.

Bogle, A. L., T. Swain, R. D. Thomas, and E. D. Kohn. 1971. *Geocarpon*: Aizoaceae or Caryophyllaceae? Taxon 20: 473–477.

Boissiera = Boissiera; Mémoires des Conservatoire et de l'Institut de Botanique Systématique de l'Université de Genève (later Mémoires des Conservatoire et Jardin Botaniques de la Ville de Genève). Supplement de Candollea.

Boivin, B. 1954. Les variations du *Stellaria humifusa* Rottboell. Ann. A. C. F. A. S. 20: 97–98.

Boivin, B. 1956. *Stellaria* sectio *Umbellatae* Schischkin (Caryophyllaceae). Svensk Bot. Tidskr. 50: 113–114.

Borbás, V. von. 1884. Temes Megye Vegetiója (Flora Comitatus Temesiensis).... Temesvárott. (Fl. Comit. Temesiensis)

Boreau, A. 1857. Flore du Centre de la France..., ed. 3. 2 vols. Angers. (Fl. Centre France ed. 3)

Borodina, A. E. 1977. O vidakh roda *Rumex* L. Evropeyshoy chasti SSSR. 1. Subgen. *Rumex* sect. *Rumex* subsect. *Maritimi* Rech. f. (De generis *Rumex* L. speciebus in parte Europaea URSS crescentibus. 1. Subgen. *Rumex* sect. *Rumex* subsect. *Maritimi* Rech. f.) Novosti Sist. Vyssh. Rast. 14: 64–72.

Boston J. Nat. Hist. = Boston Journal of Natural History.

Bot. Beechey Voy.—See: W. J. Hooker and G. A. W. Arnott [1830–]1841

Bot. Cab. = Botanical Cabinet; Consisting of Coloured Delineations of Plants from All Countries.... [Edited by G. Loddiges.]

Bot. California—See: W. H. Brewer et al. 1876–1880

Bot. Dict. ed. 4—See: A. Eaton 1836

Bot. Gaz. = Botanical Gazette; Paper of Botanical Notes.

Bot. Helv. = Botanica Helvetica.

Bot. J. Linn. Soc. = Botanical Journal of the Linnean Society.

Bot. Jahrb. Syst. = Botanische Jahrbücher für Systematik, Pflanzengeschichte und Pflanzengeographie.

Bot. Mag. = Botanical Magazine; or, Flower-garden Displayed.... [Edited by Wm. Curtis.] [With vol. 15, 1801, title became Curtis's Botanical Magazine; or....]

Bot. Mag. (Tokyo) = Botanical Magazine. [Shokubutsu-gaku Zasshi.] (Tokyo.)

Bot. Mus. Leafl. = Botanical Museum Leaflets. [Harvard University.]

Bot. Not. = Botaniska Notiser.

Bot. Not., Suppl. = Botisker Notiser. Supplement.

Bot. Tidsskr. = Botanisk Tidsskrift.

Bot. Voy. Sulphur—See: G. Bentham 1844[–1846]

Bot. Zhurn. (Moscow & Leningrad) = Botanicheskii Zhurnal. (Moscow and Leningrad.)

Botanical Society of America 2001. Botany 2001 Abstracts. Columbus.

Botanical Society of America 2002. Botany 2002 Abstracts. Columbus.

Botany (Fortieth Parallel)—See: S. Watson 1871

Bowerman, M. L. 1944. The Flowering Plants and Ferns of Mount Diablo, California; Their Distribution and Association into Plant Communities. Berkeley. (Fl. Pl. Ferns Mt. Diablo)

Bowlin, W. R., V. J. Tepedino, and T. L. Griswold. 1993. The reproductive biology of *Eriogonum pelinophilum* (Polygonaceae). In: R. C. Sivinski and K. Lightfoot, eds. 1993. Southwestern Rare and Endangered Plants: Proceedings of the...Conference. Santa Fe. Pp. 296–300.

Brandbyge, J. 1992. The genus *Muehlenbeckia* (Polygonaceae) in South and Central America. Bot. Jahrb. Syst. 114: 349–416.

Brandbyge, J. 1993. Polygonaceae. In: K. Kubitzki et al., eds. 1990+. The Families and Genera of Vascular Plants.

4+ vols. Berlin etc. Vol. 2, pp. 531–544.

Brenckle, J. F. 1941. Notes on *Polygonum* (*Avicularia*). Bull. Torrey Bot. Club 68: 491–495.

Brewer, W. H. et al. 1876–1880. Geological Survey of California.... Botany.... 2 vols. Cambridge, Mass. (Bot. California)

Bridson, G. D. R. and E. R. Smith. 1991. B-P-H/S. Botanico-Periodicum-Huntianum/Supplementum. Pittsburgh.

Britten, J. 1890. *Buda* v. *Tissa*. J. Bot. 28: 295–297.

Britton, N. L. et al., eds. 1905+. North American Flora.... 47+ vols. New York. [Vols. 1–34, 1905–1957; ser. 2, parts 1–13+, 1954+.]

Brittonia = Brittonia; a Journal of Systematic Botany....

Brown, L. E. and S. J. Marcus. 1998. Notes on the flora of Texas with additions and other significant records. Sida 18: 315–324.

Brown, R. 1810. Prodromus Florae Novae Hollandiae et Insulae van-Diemen.... London. (Prodr.)

Browne, P. 1756. The Civil and Natural History of Jamaica.... London. (Civ. Nat. Hist. Jamaica)

Brummitt, R. K. and C. E. Powell, eds. 1992. Authors of Plant Names. A List of Authors of Scientific Names of Plants, with Recommended Standard Forms of Their Names, Including Abbreviations. Kew.

Brysting, A. K. and R. Elven. 2000. The *Cerastium alpinum–C. arcticum* complex (Caryophyllaceae): Numerical analyses of morphological variation and a taxonomic revision of *C. arcticum* Lange s.l. Taxon 49: 189–216.

Bull. Acad. Int. Géogr. Bot. = Bulletin de l'Académie Internationale de Géographie Botanique.

Bull. Bot. Surv. India = Bulletin of the Botanical Survey of India.

Bull. Brit. Mus. (Nat. Hist.), Bot. = Bulletin of the British Museum (Natural History). Botany.

Bull. Calif. Acad. Sci. = Bulletin of the California Academy of Sciences.

Bull. Cl. Phys.-Math. Acad. Imp. Sci. Saint-Pétersbourg = Bulletin de la Classe Physico-mathématique de l'Académie Impériale des Sciences de Saint-Pétersbourg.

Bull. Herb. Boissier = Bulletin de l'Herbier Boissier.

Bull. Natl. Mus. Canada = Bulletin of the National Museum of Canada.

Bull. New York Bot. Gard. = Bulletin of the New York Botanical Garden.

Bull. New York State Mus. Sci. Serv. = Bulletin of the New York State Museum and Science Service.

Bull. Ohio Biol. Surv. = Bulletin of the Ohio Biological Survey.

Bull. Public Mus. Milwaukee = Bulletin of the Public Museum of Milwaukee.

Bull. S. Calif. Acad. Sci. = Bulletin of the Southern California Academy of Sciences.

Bull. Soc. Hist. Nat. Afrique N. = Bulletin de la Société d'Histoire Naturelle de l'Afrique du Nord.

Bull. Soc. Imp. Naturalistes Moscou = Bulletin de la Société Impériale des Naturalistes de Moscou.

Bull. Soc. Roy. Bot. Belgique = Bulletin de la Société Royale de Botanique de Belgique.

Bull. Torrey Bot. Club = Bulletin of the Torrey Botanical Club.

Bull. Univ. Utah, Biol. Ser. = Bulletin of the University of Utah. Biological Series.

Burleigh, J. G. and T. P. Holtsford. 2003. Molecular systematics of the eastern North American *Silene* (Caryophyllaceae): Evidence from nuclear ITS and chloroplast *trn*L intron sequences. Rhodora 105: 76–90.

Burnat, E., J. I. Briquet, and F. G. Cavillier. 1892–1931. Flore des Alpes Maritimes.... 7 vols. Geneva, Basel, and Lyon. (Fl. Alpes Marit.)

Byull. Moskovsk. Obshch. Isp. Prir., Otd. Biol. = Byulleten' Moskovskogo Obshchestva Ispytatelei Prirody. Otdel Biologicheskii.

Cafferty, S. and S. Snogerup. 2000. Proposal to conserve the name *Rumex alpinus* L. (Polygonaceae) with a conserved type. Taxon 49: 571–572.

Callihan, R. H., S. L. Carson, and R. T. Dobbins. 1995. NAWEEDS, Computer-aided Weed Identification for North America. Illustrated User's Guide plus Computer Floppy Disk. Moscow, Idaho.

Camazine, S. and R. A. Bye. 1980. A study of the medical ethnobotany of the Zuni Indians of New Mexico. J. Ethnopharmacol. 2: 365–388.

Campbell, C. G. 1997. Buckwheat. *Fagopyrum esculentum* Moench. Gatersleben and Rome. [Promoting the Conservation and Use of Underutilized and Neglected Crops 19.]

Campderá, F. 1819. Monographie des *Rumex*.... Paris etc. (Monogr. Rumex)

Canad. Field-Naturalist = Canadian Field-Naturalist.

Canad. J. Bot. = Canadian Journal of Botany.

Canad. J. Genet. Cytol. = Canadian Journal of Genetics and Cytology.

Canad. J. Pl. Sci. = Canadian Journal of Plant Science.

Candolle, A. P. de. 1813b. Catalogus Plantarum Horti Botanici Monspeliensis.... Monpellier, Paris, and Strasbourg. (Cat. Pl. Hort. Monsp.)

Candolle, A. P. de and A. L. P. P. Candolle, eds. 1823–1873. Prodromus Systematis Naturalis Regni Vegetabilis.... 17 vols. Paris etc. [Vols. 1–7 edited by A. P. de Candolle, vols. 8–17 by A. L. P. P. de Candolle.] (Prodr.)

Candollea = Candollea; Organe du Conservatoire et du Jardin Botaniques de la Ville de Genève.

Caratt. Nuov. Gen.—See: C. S. Rafinesque 1810

Carlquist, S. 2003. Wood anatomy of Polygonaceae: Analysis of a family with exceptional wood diversity. Bot. J. Linn. Soc. 141: 25–51.

Carlquist, S. and C. J. Biggs. 1996. Wood anatomy of Plumbaginaceae. Bull. Torrey Bot. Club 123: 135–147.

Carlson, G. G. and V. H. Jones. 1940. Some notes on uses of plants by the Comanche Indians. Pap. Michigan Acad. Sci. 25: 517–542.

Čas. Nár. Muz. Praze Rada Přír. = Časopis Národního Muzea v Praze. Rada Přírodovědný.

Castanea = Castanea; Journal of the Southern Appalachian Botanical Club.

Cat. N. Amer. Pl.—See: A. A. Heller 1898

Cat. Pl. Hort. Monsp.—See: A. P. de Candolle 1813b

Cat. Pl. Upper Louisiana—See: T. Nuttall 1813

Cattaneo, C. 1844. Notizie Naturali e Civili su la Lombardia. Milan. (Not. Nat. Civ. Lombardia)

Cavanilles, A. J. 1791–1801. Icones et Descriptiones Plantarum, Quae aut Sponte in Hispania Crescunt, aut in Hortis Hospitantur. 6 vols. Madrid. (Icon.)

Chamberlin, R. V. 1911. The ethno-botany of the Gosiute Indians of Utah. Mem. Amer. Anthropol. Assoc. 2: 331–405.

Chambers, K. L. 1992. Choosing the correct name for *Aconogonon* (*Polygonum* sect. *Aconogonon*) in Alaska. Rhodora 94: 319–322.

Chapman, A. W. 1860. Flora of the Southern United States.... New York. (Fl. South. U.S.)

Chapman, A. W. 1897. Flora of the Southern United States..., ed. 3. Cambridge, Mass. (Fl. South. U.S. ed. 3)

Chase, M. W. et al. 1993. Phylogenetics of seed plants: An analysis of nucleotide sequences from the plastid gene *rbc*L. Ann. Missouri Bot. Gard. 80: 528–580.

Chater, A. O. and G. Halliday. 1993. *Arenaria*. In: T. G. Tutin et al., eds. 1993+. Flora Europaea, ed. 2. 1+ vol. Cambridge and New York. Vol. 1, pp. 140–148.

Chaudhri, M. N. 1968. A revision of the Paronychiinae. Meded. Bot. Mus. Herb. Rijks Univ. Utrecht 285: 306–397.

Chestnut, V. K. 1902. Plants used by the Indians of Mendocino County, California. Contr. U.S. Natl. Herb. 7: 295–408.

Chin, T. C. and H. W. Youngken. 1947. The cytotaxonomy of *Rheum*. Amer. J. Bot. 34: 401–407.

Chinnappa, C. C. 1985. Studies on the *Stelleria longipes* complex (Caryophyllaceae): Interspecific hybridization. I. Triploid meiosis. Canad. J. Genet. Cytol. 27: 318–321.

Chinnappa, C. C. 1992. *Stellaria porsildii*, sp. nov., a new member of the *S. longipes* complex (Caryophyllaceae. Syst. Bot. 17: 29–32.

Chinnappa, C. C. and J. K. Morton. 1974. The cytology of *Stellaria longipes*. Canad. J. Genet. Cytol. 16: 499–514.

Chinnappa, C. C. and J. K. Morton. 1976. Studies on the *Stellaria longipes* Goldie complex—Variation in wild populations. Rhodora 78: 488–501.

Chinnappa, C. C. and J. K. Morton. 1984. Studies on the *Stellaria longipes* Goldie complex (Caryophyllaceae): Biosystematics. Syst. Bot. 9: 60–73.

Chinnappa, C. C. and J. K. Morton. 1991. Studies on the *Stellaria longipes* complex (Caryophyllaceae): Taxonomy. Rhodora 93: 129–135.

Chowdhuri, P. K. 1957. Studies in the genus *Silene*. Notes Roy. Bot. Gard. Edinburgh 22: 221–278.

Christman, S. P. and W. S. Judd. 1990. Notes on plants endemic to Florida scrub. Florida Sci. 53: 52–73.

Chrtek, J. and M. Šourková. 1992. Variation in *Oxyria digyna*. Preslia 64: 207–210.

Chung, H. S. et al. 1999. Flavonoid constituents of *Chorizanthe diffusa* with potential cancer chemo-preventive activity. J. Agric. Food Chem. 47: 36–41.

Civ. Nat. Hist. Jamaica—See: P. Browne 1756

Clairville, J. P. de. 1811. Manuel d'Herborisation en Suisse et en Valais.... Winterthur. (Man. Herbor. Suisse)

Class-book Bot. ed. 2—See: A. Wood 1847

Class-book Bot. ed. s.n.(b)—See: A. Wood 1861

Clute, O. and O. Palmer. 1893. Spurry (*Spergula arvensis*). Michigan Agric. Exp. Sta. Bull. 91: 3–8.

Coker, P. D. 1962. *Corrigiola litoralis* L. J. Ecol. 50: 833–840.

Comp. Fl. German.—See: M. J. Bluff et al. 1825–1833

Comp. Fl. Ital.—See: G. Arcangeli 1882

Consaul, L. L., S. I. Warwick, and J. McNeill. 1991. Allozyme variation in the *Polygonum lapathifolium* complex. Canad. J. Bot. 69: 2261–2270.

Conservation Genet. = Conservation Genetics.

Consp. Fl. Eur.—See: C. F. Nyman 1878–1890

Consp. Fl. Groenland.—See: J. M. C. Lange 1880[–1894]

Consp. Regn. Veg.—See: H. G. L. Reichenbach 1828

Contr. Gray Herb. = Contributions from the Gray Herbarium of Harvard University. [Some numbers reprinted from (or in?) other periodicals, e.g. Rhodora.]

Contr. Univ. Michigan Herb. = Contributions from the University of Michigan Herbarium.

Contr. U.S. Natl. Herb. = Contributions from the United States National Herbarium.

Contr. W. Bot. = Contributions to Western Botany.

Cooperrider, T. S. 1995. The Dicotyledoneae of Ohio: Linaceae through Campanulaceae. Columbus. [The Vascular Flora of Ohio, vol. 2(2).]

Core, E. L. 1939. A taxonomic revision of the genus *Siphonychia*. J. Elisha Mitchell Sci. Soc. 55: 339–345.

Core, E. L. 1940. Notes on the mid-Appalachian species of *Paronychia*. Virginia J. Sci. 1: 110–116.

Core, E. L. 1941. The North American species of *Paronychia*. Amer. Midl. Naturalist 26: 369–397.

Correll, D. S. and M. C. Johnston. 1970. Manual of the Vascular Plants of Texas. Renner, Tex.

Costea, M. and F. J. Tardif. 2003. Nomenclatural changes in the genus *Polygonum*, section *Polygonum* (Polygonaceae). Sida 20: 987–997.

Costea, M. and F. J. Tardif. 2003b. Does *Polygonum patulum* (Polygonaceae) grow in North America? Sida 20: 1707–1708.

Costea, M. and F. J. Tardif. 2005. Taxonomy of the *Polygonum douglasii* (Polygonaceae) complex with a new species from Oregon. Brittonia 57: 1–27.

Courtney, A. D. 1968. Seed dormancy and field emergence in *Polygonum aviculare*. J. Appl. Ecol. 5: 675–684.

Coville, F. V. 1897. Notes on the plants used by the Klamath Indians of Oregon. Contr. U.S. Natl. Herb. 5: 87–110.

Crépin, F. 1866. Manuel de la Flore de Belgique..., ed. 2. Brussels. (Man. Fl. Belgique ed. 2)

Cronquist, A. 1981. An Integrated System of Classification of Flowering Plants. New York.

Crow, G. E. 1978. A taxonomic revision of *Sagina* (Caryophyllaceae) in North America. Rhodora 80: 1–91.

Crow, G. E. 1979. The systematic significance of seed morphology in *Sagina* (Caryophyllaceae) utilizing SEM. Brittonia 31: 52–63.

Cuénoud, P. et al. 2002. Molecular phylogenetics of Caryophyllales based on nuclear 18S rDNA and plastid *rbc*L, *atp*B, and *mat*K DNA sequences. Amer. J. Bot. 89: 132–144.

Curr. Topics Pl. Sci.—See: J. E. Gunckel 1969

Curtis, W. [1775–]1777–1798. Flora Londinensis.... 2 vols. in 6 fasc. and 72 nos. London. (Fl. Londin.)

Cusick, A. W. 1983. *Spergularia* (Caryophyllaceae) in Ohio. Michigan Bot. 22: 69–71.

Cycl.—See: A. Rees [1802–]1819–1820

Cytologia = Cytologia; International Journal of Cytology.

Dalci, M. 1972. The Taxonomy of Section *Persicaria* (Tourn.) L. in the Genus *Polygonum* (Tourn.) L. (Polygonaceae) in the United States East of the Rocky Mountains. Ph.D. dissertation. University of Nebraska.

Darwent, A. L. 1975. The biology of Canadian weeds. 14. *Gypsophila paniculata*. Canad. J. Pl. Sci. 55: 1049–1058.

Davis, L. and R. J. Sherman. 1990. The rediscovered Sonoma spineflower at Point Reyes National Seashore. Fremontia 18(1): 17–18.

Davis, L. and R. J. Sherman. 1992. Ecological study of the rare *Chorizanthe valida* (Polygonaceae) at Point Reyes National Seashore, California. Madroño 39: 271–280.

Davis, R. J. 1952. Flora of Idaho. Dubuque. (Fl. Idaho)

Dawson, J. E. 1979. A Biosystematic Study of *Rumex* Section *Rumex* in Canada and the United States. Ph.D. thesis. Carleton University.

DeDecker, M. 1974. Redescription of *Eriogonum hoffmannii*, a Death Valley endemic. Madroño 5: 265–267.

Delarbre, A. 1800. Flore d'Auvergne..., ed. 2. 2 vols. Riom and Clermont. (Fl. Auvergne ed. 2)

den Nijs, J. C. M. 1974. Biosystematic studies of the *Rumex acetosella* complex. I. Angiocarpy and chromosome numbers in France. Acta Bot. Neerl. 23: 655–675.

den Nijs, J. C. M. 1976. Biosystematic studies of the *Rumex acetosella* complex. II. The Alpine region. Acta Bot. Neerl. 25: 417–447.

den Nijs, J. C. M. 1984. Biosystematic studies of the *Rumex acetosella* complex. VIII. A taxonomic revision. Feddes Repert. 95: 43–66.

den Nijs, J. C. M., H. Hooghiemstra, and P. H. Schalk. 1980. Biosystematic studies of the *Rumex acetosella* complex. IV. Pollen morphology and the possibilities of identification of cytotypes in pollen analysis. Phyton (Horn) 20: 307–323.

den Nijs, J. C. M. and T. Panhorst. 1980. Biosystematic studies of the *Rumex acetosella* complex. III. A note on the systematic position of "*R. tenuifolius* (Wallr.) Löve." Acta Bot. Neerl. 29: 179–192.

den Nijs, J. C. M., K. Sorgdrager, and J. Stoop. 1985. Biosystematic studies of the *Rumex acetosella* complex. IX. Cytogeography of the complex in the Iberian Peninsula and taxonomic discussion. Bot. Helv. 95: 141–156.

Descr. Pl. Nouv.—See: É. P. Ventenat [1800–1803]

Desert Pl. = Desert Plants; a Quarterly Journal Devoted to Broadening Our Knowledge of Plants Indigenous or Adaptable to Arid and Sub-arid Regions.

Desfontaines, R. L. [1798–1799.] Flora Atlantica sive Historia Plantarum, Quae in Atlante, Agro Tunetano et Algeriensi Crescunt. 2 vols. in 9 parts. Paris. (Fl. Atlant.)

Deut. Excurs.-Fl.—See: K. F. W. Jessen 1879

Deut. Fl.—See: H. Karsten 1880–1883

Deutschl. Fl.—See: G. F. Hoffmann [1791–1804]

Deutschl. Fl. ed. 2—See: J. Sturm et al. 1900–1907

Deutschl. Fl. ed. 3—See: J. C. Röhling et al. 1823–1839

Dodge, C. K. 1915. The flowering plants, ferns and fern allies growing without cultivation in Lambton County, Ontario. Rep. (Annual) Michigan Acad. Sci. 16: 132–200.

Don, D. 1825. Prodromus Florae Nepalensis.... London. (Prodr. Fl. Nepal.)

Don, G. 1831–1838. A General History of the Dichlamydeous Plants.... 4 vols. London. (Gen. Hist.)

Dorn, R. D. 1988. Vascular Plants of Wyoming. Cheyenne. (Vasc. Pl. Wyoming)

Dorn, R. D. 2001. Vascular Plants of Wyoming, ed. 3. Cheyenne. (Vasc. Pl. Wyoming ed. 3)

Downie, S. R., D. S. Katz(-Downie), and K. J. Cho. 1997. Relationships in the Caryophyllales as suggested by phylogenetic analyses of partial chloroplast DNA ORF2280 homolog sequences. Amer. J. Bot. 84: 253–273.

Duke, J. A. 1961. Preliminary revision of the genus *Drymaria*. Ann. Missouri Bot. Gard. 48: 173–268.

Dumortier, B. C. J. 1827. Florula Belgica.... Tournay. (Fl. Belg.)

Dunn, D. B. 1982. Problems in "keeping up" with the flora of Missouri. Trans. Missouri Acad. Sci. 16: 95–98.

Eaton, A. 1836. Eaton's Botanical Grammar and Dictionary..., ed. 4. Albany. (Bot. Dict. ed. 4)

Ecol. Monogr. = Ecological Monographs.

Ecology = Ecology, a Quarterly Journal Devoted to All Phases of Ecological Biology.

Econ. Bot. = Economic Botany; Devoted to Applied Botany and Plant Utilization.

Edinburgh J. Bot. = Edinburgh Journal of Botany.

Edinburgh Philos. J. = Edinburgh Philosophical Journal.

Edwards's Bot. Reg. = Edwards's Botanical Register....

Ehrhart, J. F. 1787–1792. Beiträge zur Naturkunde.... 7 vols. Hannover and Osnabrück. (Beitr. Naturk.)

Elliott, S. [1816–]1821–1824. A Sketch of the Botany of South-Carolina and Georgia. 2 vols. in 13 parts. Charleston. (Sketch Bot. S. Carolina)

Elmore, F. A. 1943. Ethnobotany of the Navajo. Albuquerque. [Reprinted 1978, New York.]

Elven, R., Ö. Nilsson, and S. Snogerup. 2000. *Rumex*. In: B. Jonsell and T. Karlsson, eds. 2000+. Flora Nordica. 2+ vols. Stockholm. Vol. 1, pp. 281–318.

Emery, R. J. N. and C. C. Chinnappa. 1994. Morphological variation among members of the *Stellaria longipes* complex: *S. longipes*, *S. longifolia* and *S. porsildii* (Caryophyllaceae). Pl. Syst. Evol. 190: 69–78.

Emory, W. H. 1848. Notes of a Military Reconnoissance, from Fort Leavenworth, in Missouri, to San Diego, in California, Including Part of the Arkansas, Del Norte, and Gila Rivers.... Made in 1846–7, with the Advanced Guard of the "Army of the West." Washington. (Not. Milit. Reconn.)

Emory, W. H. 1857–1859. Report on the United States and Mexican Boundary Survey, Made under the Direction of the Secretary of the Interior. 2 vols. in parts. Washington. (Rep. U.S. Mex. Bound.)

Encycl.—See: J. Lamarck et al. 1783–1817

Encycl. Brit. ed. 7—See: M. Napier 1830–1842

Endlicher, S. L. 1836–1840[–1850] Genera Plantarum Secundum Ordines Naturales Disposita. 18 parts + 5 suppls. in 6 parts. Vienna. [Paged consecutively through suppl. 1(2); suppls. 2–5 paged independently.] (Gen. Pl.)

Engler, H. G. A. et al., eds. 1924+. Die natürlichen

Pflanzenfamilien..., ed. 2. 26+ vols. Leipzig and Berlin. (Nat. Pflanzenfam. ed. 2)

Enum. Pl.—See: W. S. J. G. von Besser 1822; C. L. Willdenow 1809–1813[–1814]

Enum. Pl. Corea—See: T. Mori 1922

Enum. Pl. Nov.—See: F. E. L. von Fischer and C. A. von Meyer 1841–1842

Enum. Stirp. Transsilv.—See: J. C. G. Baumgarten 1816–1846

Enum. Stirp. Vindob.—See: N. J. Jacquin 1762

Enum. Syst. Pl.—See: N. J. Jacquin 1760

Eriogonum—See: S. G. Stokes 1936

Értek. Term. Köréb. Magyar Tud. Akad. = Értekezések a Természettudományi Köréböl; Kiadja a Magyar Tudományos Akadémia.

Ertter, B. 1980. A revision of the genus *Oxytheca* (Polygonaceae). Brittonia 32: 70–102.

Ertter, B. 1996. Saga of the Santa Cruz spineflower. Fremontia 24(4): 8–11.

Erythea = Erythea; a Journal of Botany, West American and General.

Escamilla, M. and V. Sosa. 2000. Características de la semilla en las series del género *Drymaria* (Caryophyllaceae). Kurtziana 28: 259–274.

Evol. Trends Pl. = Evolutionary Trends in Plants.

Evolution = Evolution, International Journal of Organic Evolution.

Fam. Pl.—See: M. Adanson 1763[–1764]

Fassett, N. C. 1949. The variations of *Polygonum punctatum*. Brittonia 6: 369–393.

Favarger, C. and M. Krähenbühl. 1996. Nombre chromosomique et affinities probables d'un *Cerastium* (Caryophyllaceae) des Balkans. Anales Jard. Bot. Madrid 54: 155–165.

Feddes Repert. = Feddes Repertorium.

Feddes Repert. Spec. Nov. Regni Veg. = Feddes Repertorium Specierum Novarum Regni Vegetabilis.

Felger, R. S. 2000. Flora of the Gran Desierto and Río Colorado of Northwestern Mexico. Tucson.

Fenzl, E. 1833. Versuch einer Darstellung der geographischen Verbreitungs- und Vertheilungs-Verhältnisse der natürlichen Familie der Alsineen.... Vienna. (Vers. Darstell. Alsin.)

Fernald, M. L. 1919. The unity of the genus *Arenaria*. Rhodora 21: 1–7.

Fernald, M. L. 1921. The Gray Herbarium expedition to Nova Scotia, 1920. Rhodora 23: 89–111, 153–171, 223–245.

Fernald, M. L. 1922c. Notes on the flora of western Nova Scotia 1921. Rhodora 24: 157–180, 201–208.

Fernald, M. L. 1936c. Notes on *Paronychia, Anychia*. Rhodora 38: 416–421.

Fernald, M. L. 1940b. Some spermatophytes of eastern North America. Rhodora 42: 239–276.

Fernald, M. L. 1943f. Virginian botanizing under restrictions. Part I. Field studies of 1942 and 1943. Rhodora 45: 357–413.

Fernald, M. L. 1950. Gray's Manual of Botany, ed. 8. New York.

Fewkes, J. W. 1896. A contribution to ethnobotany. Amer. Anthropol. 9: 14–21.

Field & Lab. = Field & Laboratory.

Firbank, L. G. 1988. Biological flora of the British Isles. 165. *Agrostemma githago* L. (*Lychnis githago* (L.) Scop.). J. Ecol. 76: 1232–1246.

Fischer, F. E. L. von and C. A. von Meyer. 1841–1842. Enumeratio...Plantarum Novarum.... 2 parts. St. Petersburg and Leipzig. [Parts paged independently.] (Enum. Pl. Nov.)

Fishbein, M. 1993. Noteworthy collections: Mexico. *Rumex orthoneurus* Rech. f. (Polygonaceae). Madroño 40: 271.

Fl. Alaska Yukon—See: E. Hultén 1941–1950

Fl. Aleut. Isl.—See: E. Hultén 1937

Fl. Alpes Marit.—See: E. Burnat et al. 1892–1931

Fl. Altaica—See: C. F. von Ledebour 1829–1833

Fl. Amer. Sept.—See: F. Pursh [1813] 1814

Fl. Angl.—See: W. Hudson 1762

Fl. Arct. URSS—See: A. I. Tolmatchew 1960–1987

Fl. Atlant.—See: R. L. Desfontaines [1798–1799]

Fl. Auvergne ed. 2—See: A. Delarbre 1800

Fl. Belg.—See: B. C. J. Dumortier 1827

Fl. Bor.-Amer.—See: W. J. Hooker [1829–]1833–1840; A. Michaux 1803

Fl. Boston. ed. 2—See: J. Bigelow 1824

Fl. Boston. ed. 3—See: J. Bigelow 1840

Fl. Bras.—See: C. F. P. von Martius et al. 1840–1906

Fl. Bras. Merid.—See: A. St.-Hilaire et al. 1824[–1833]

Fl. Brit.—See: J. E. Smith 1800–1804

Fl. Calif.—See: W. L. Jepson 1909–1943

Fl. Carol.—See: T. Walter 1788

Fl. Čech.—See: J. S. Presl and C. B. Presl 1819

Fl. Centre France ed. 3—See: A. Boreau 1857

Fl. Chil.—See: C. Gay 1845–1854

Fl. Colorado—See: P. A. Rydberg 1906

Fl. Comit. Temesiensis—See: V. von Borbás 1884

Fl. Dan.—See: G. C. Oeder et al. [1761–]1764–1883

Fl. Env. Paris ed. 2—See: J. L. Thuillier 1799

Fl. Franç. ed. 3—See: J. Lamarck and A. P. de Candolle 1805[–1815]

Fl. Francisc.—See: E. L. Greene 1891–1897

Fl. Helv.—See: J. Gaudin 1828–1833

Fl. Idaho—See: R. J. Davis 1952

Fl. Ital.—See: F. Parlatore 1848–1896

Fl. Lapp.—See: G. Wahlenberg 1812

Fl. Londin.—See: W. Curtis [1775–]1777–1798

Fl. Ludov.—See: C. S. Rafinesque 1817

Fl. Maine et Loire—See: J. P. Guépin 1838

Fl. Medit. = Flora Mediterranea; Acta Herbarii Mediterranei Panormitani sub Auspiciis Societatis Botanicorum Mediterraneorum "Optima" Nuncupata Edita.

Fl. Montpellier—See: H. Loret and A. Barrandon 1876

Fl. Murmansk. Obl.—See: B. N. Gorodkov and A. I. Pojarkova 1953–1966

Fl. N. Amer.—See: J. Torrey and A. Gray 1838–1843

Fl. N. Mitt.-Deutschland ed. 9—See: C. A. F. Garcke 1869

Fl. N.W. Amer.—See: T. J. Howell 1897–1903

Fl. N.W. Coast—See: C. V. Piper and R. K. Beattie 1915

Fl. Palouse Reg.—See: C. V. Piper and R. K. Beattie 1901

Fl. Pl. Ferns Mt. Diablo—See: M. L. Bowerman 1944

Fl. Rocky Mts.—See: P. A. Rydberg 1917

Fl. Ross.—See: C. F. von Ledebour [1841]1842–1853

Fl. Scand. Prodr. ed. 2—See: A. J. Retzius 1795

Fl. S.E. U.S.—See: J. K. Small 1903

Fl. Sicul.—See: C. B. Presl 1826

Fl. Sicul. Prodr.—See: G. Gussone 1827–1843

Fl. South. U.S.—See: A. W. Chapman 1860

Fl. South. U.S. ed. 3—See: A. W. Chapman 1897

Fl. Taur.-Caucas.—See: F. A. Marschall von Bieberstein 1808–1819

Fl. URSS—See: V. L. Komarov et al. 1934–1964

Fl. Vosges—See: E.-L. Berher 1887

Fl. W. Calif.—See: W. L. Jepson 1901

Flexner, S. B. and L. C. Hauck, eds. 1987. The Random House Dictionary of the English Language, ed. 2 unabridged. New York.

Flora = Flora; oder (allgemeine) botanische Zeitung. [Vols. 1–16, 1818–33, include "Beilage" and "Ergänzungsblätter"; vols. 17–25, 1834–42, include "Beiblatt" and "Intelligenzblatt."]

Florida Sci. = Florida Scientist.

Folia Geobot. Phytotax. = Folia Geobotanica et Phytotaxonomica.

Foreman, D. B., N. G. Walsh, and T. J. Entwisle, eds. 1993–1999. Flora of Victoria. 4 vols. Melbourne and Sydney.

Fors. Oecon. Plantel. ed. 3—See: J. W. Hornemann 1821–1837

Fox, G. A. 1989. Consequences of flowering-time variation in a desert annual: Adaptation and history. Ecology 70: 1294–1306.

Fox, G. A. 1990. Components of flowering time variation in a desert annual. Evolution 44: 1404–1423.

Fox, G. A. 1990b. Perennation and the persistence of annual life histories. Amer. Midl. Naturalist 135: 829–840.

Franklin, J. et al. 1823. Narrative of a Journey to the Shores of the Polar Sea, in the Years 1819, 20, 21 and 22. London. [Richardson: Appendix VII. Botanical appendix, pp. [729]–778, incl. bryophytes by Schwägrichen, algae and lichens by Hooker. Brown: Addenda [to Appendix VII], pp. 779–784.] (Narr. Journey Polar Sea)

Frémont, J. C. 1843–1845. Report of the Exploring Expedition to the Rocky Mountains in the Year 1842, and to Oregon and North California in the Year 1843–44. 2 parts. Washington. [Parts paged consecutively.] (Rep. Exped. Rocky Mts.)

Fremontia = Fremontia: Journal of the California Native Plant Society.

Fries, E. M. 1832–1845. Novitiarum Florae Suecicae Mantissa.... 3 vols. Lund. (Novit. Fl. Suec. Mant.)

Fruct. Sem. Pl.—See: J. Gaertner 1788–1791[–1792]

Fuller, T. C. and E. McClintock. 1986. Poisonous Plants of California. Berkeley.

Gaertner, J. 1788–1791[–1792]. De Fructibus et Seminibus Plantarum.... 2 vols. Stuttgart and Tübingen. [Vol. 1 in 1 part only, 1788. Vol. 2 in 4 parts paged consecutively: pp. 1–184, 1790; pp. 185–352, 353–504, 1791; pp. 505–520, 1792.] (Fruct. Sem. Pl.)

Gaertner, P. G., B. Meyer, and J. Scherbius. 1799–1802. Oekonomisch-technische Flora der Wetterau. 3 vols. Frankfurt am Main. [Vols. 1 and 2 paged independently, vol. 3 in 2 parts paged independently.] (Oekon. Fl. Wetterau)

Gagnepain, F. 1908. Contribution à la connaissance du genre Polycarpaea Lamk. J. Bot. (Morot), sér. 2, 1: 275–280.

Garcke, C. A. F. 1869. Flora von Nord- und Mittel-Deutschland..., ed. 9. Berlin. (Fl. N. Mitt.-Deutschland ed. 9)

Gard. Dict. ed. 8—See: P. Miller 1768

Gard. Dict. Abr. ed. 4—See: P. Miller 1754

Gartenflora = Gartenflora; Monatsschrift für deutsche und schweizerische Garten- und Blumenkunde.

Gasquez, J. et al. 1978. Essai de taxonomie d'une espèce adventice annuelle: Polygonum aviculare L. Ann. Amélior. Pl. 28: 565–577.

Gaudin, J. 1828–1833. Flora Helvetica.... 7 vols. Zürich. (Fl. Helv.)

Gay, C. 1845–1854. Historia Física y Política de Chile.... Botánica [Flora Chilena]. 8 vols., atlas. Paris. (Fl. Chil.)

Gen. Hist.—See: G. Don 1831–1838

Gen. N. Amer. Pl.—See: T. Nuttall 1818

Gen. Pl.—See: S. L. Endlicher 1836–1840[–1850]; N. M. Wolf 1776

Gen. Pl. ed. 5—See: C. Linnaeus 1754

Gen. Sp. Pl.—See: M. Lagasca y Segura 1816b

Genes Genet. Systems = Genes and Genetic Systems.

Gentes Herb. = Gentes Herbarum; Occasional Papers on the Kinds of Plants.

Geobios (Jodhpur) = Geobios; an International (Bimonthly) Journal of Life Sciences.

Ges. Naturf. Freunde Berlin Mag. Neuesten Entdeck. Gesammten Naturk. = Der Gesellschaft naturforschender Freunde zu Berlin Magazin für die neuesten Entdeckungen in der gesammten Naturkunde.

Ges. Naturf. Freunde Berlin Neue Schriften = Der Gesellschaft naturforschender Freunde zu Berlin, neue Schriften.

Gifford, E. W. 1936. Northeastern and western Yavapai. Univ. Calif. Publ. Amer. Archaeol. Ethnol. 34: 247–345.

Gilbert, M. G. 1987. The taxonomic position of the genera Telephium and Corrigiola. Taxon 36: 47–49.

Gleason, H. A. 1952. The New Britton and Brown Illustrated Flora of the Northeastern United States and Adjacent Canada. 3 vols. New York.

Good, D. A. 1984. Revision of the Mexican and Central American species of Cerastium. Rhodora 86: 339–379.

Goodman, G. J. 1934. A revision of the North American species of the genus Chorizanthe. Ann. Missouri Bot. Gard. 21: 1–102.

Goodman, G. J. 1957. The genus Centrostegia, tribe Eriogoneae. Leafl. W. Bot. 8: 125–128.

Gorodkov, B. N. and A. I. Pojarkova, eds. 1953–1966. Flora Murmanskoi Oblasti. 5 vols. Moscow. (Fl. Murmansk. Obl.)

Graham, S. A. and C. E. Wood Jr. 1965. The genera of Polygonaceae in the southeastern United States. J. Arnold Arbor. 46: 91–113.

Gray, A. 1867. A Manual of the Botany of the Northern United States..., ed. 5. New York and Chicago. [Pteridophytes by D. C. Eaton.] (Manual ed. 5)

Gray, A., S. Watson, and J. M. Coulter. 1890. A Manual of the Botany of the Northern United States..., ed. 6. New York and Chicago. (Manual ed. 6)

Gray, S. F. 1821. A Natural Arrangement of British Plants.... 2 vols. London. (Nat. Arr. Brit. Pl.)

Great Basin Naturalist Mem. = Great Basin Naturalist Memoirs.

Great Plains Flora Association 1977. Atlas of the Flora of the Great Plains. Ames.

Great Plains Flora Association 1986. Flora of the Great Plains. Lawrence, Kans.

Greene, E. L. 1891–1897. Flora Franciscana. An Attempt to Classify and Describe the Vascular Plants of Middle California. 4 parts. San Francisco. [Parts paged consecutively.] (Fl. Francisc.)

Greuter, W. 1995. *Silene* (Caryophyllaceae) in Greece—A subgeneric and sectional classification. Taxon 44: 542–581.

Grinnell, G. B. 1923. The Cheyenne Indians: Their History and Ways of Life. 2 vols. New Haven and London. [Reprinted 1972, Lincoln, Nebr.]

Grønland—See: H. J. Rink 1852–1857

Guépin, J. P. 1838. Flore de Maine et Loire.... Paris. (Fl. Maine et Loire)

Gunckel, J. E., ed. 1969. Current Topics in Plant Science. New York. (Curr. Topics Pl. Sci.)

Gussone, G. 1827–1843. Florae Siculae Prodromus.... 2 vols. + suppl. Naples. (Fl. Sicul. Prodr.)

Gustafson, D. J., G. Romano, R. E. Latham, and J. K. Morton. 2003. Amplified fragment length polymorphism analysis of genetic relationships among the serpentine barrens endemic *Cerastium velutinum* Rafinesque var. *villosissimum* Pennell (Caryophyllaceae) and closely related *Cerastium* species. J. Torrey Bot. Soc. 130: 218–223.

Hall, C. A. Jr. and V. Doyle-Jones, eds. 1988. Plant Biology of Eastern California. Los Angeles.

Hammer, K., P. Hanelt, and H. Knupffer. 1982. Vorarbeiten zur monographischen Darstellung von Wildpflanzensortimenten: *Agrostemma* L. (Studies towards a monographic treatment of wild plant collections: *Agrostemma* L.) Kulturpflanze 30: 45–96.

Handb. Nat. Pfl.-Syst.—See: H. G. L. Reichenbach 1837

Handb. Skand. Fl.—See: C. J. Hartman 1820

Handbuch—See: J. H. F. Link 1829–1833

Hannover. Mag. = Hannoverisches Magazin worin kleine Abhandlungen...gesamlet (gesammelt) und aufbewahret sind.

Haraldsen, K. B. and J. Wesenberg. 1993. Population genetic analysis of an amphi-Atlantic species: *Lychnis alpina* (Caryophyllaceae). Nordic J. Bot. 13: 377–387.

Haraldson, K. 1978. Anatomy and taxonomy in Polygonaceae subfam. Polygonoideae Meisn. emend. Jaretzky. Acta Univ. Upsal., Symb. Bot. Upsal. 22: 1–93.

Hardham, C. B. 1989. Chromosome numbers of some annual species of *Chorizanthe* and related genera (Polygonaceae: Eriogonoideae). Phytologia 66: 89–94.

Harriman, N. A. 1981b. In: IOPB chromosome number reports LXX. Taxon 30: 68–80.

Harrington, H. D. 1954. Manual of the Plants of Colorado. Denver.

Harris, S. K. 1965. Notes on the flora of Coös County, New Hampshire. Rhodora 67: 195–197.

Harris, W. 1969. Seed characters and organ size in the cytotaxonomy of *Rumex acetosella* L. New Zealand J. Bot. 7: 125–141.

Harris, W. 1973. Leaf form and panicle height variability in *Rumex acetosella* L. New Zealand J. Bot. 11: 115–144.

Hart, J. A. 1981. The ethnobotany of the Northern Cheyenne Indians of Montana. J. Ethnopharmacol. 4: 1–55.

Hartman, C. J. 1820. Handbok i Skandinaviens Flora, Innefettande Sveriges och Norriges Vexter, till och med Mossorna. Stockholm. (Handb. Skand. Fl.)

Hartman, R. L. 1971. The Family Caryophyllaceae in Wyoming. M.S. thesis. University of Wyoming.

Hartman, R. L. 1972. [Flora of Wyoming] Caryophyllaceae. Res. J. Wyoming Agric. Exp. Sta. 64: 14–45.

Hartman, R. L. 1974. Rocky Mountain species of *Paronychia* (Caryophyllaceae): A morphological, cytological, and chemical study. Brittonia 26: 256–263.

Hartman, R. L. 1979. *Cerastium clawsonii* (Caryophyllaceae): A synonym of *Linum hudsonioides* (Linaceae). Rhodora 81: 283.

Hartman, R. L. 1993. *Arenaria*. In: J. C. Hickman, ed. 1993. The Jepson Manual. Higher Plants of California. Berkeley, Los Angeles, and London. Pp. 478–480.

Hartman, R. L. and R. K. Rabeler. 2004. New combinations in North American Caryophyllaceae. Sida 21: 753–754.

Harvard Pap. Bot. = Harvard Papers in Botany.

Hawkes, C. V. and E. S. Menges. 1995. Density and seed production of a Florida endemic, *Polygonella basiramia*, in relation to time since fire and open sand. Amer. Midl. Naturalist 133: 138–149.

Hedberg, O. 1997. The genus *Koenigia* L. emend. Hedberg (Polygonaceae). Bot. J. Linn. Soc. 124: 295–330.

Hedges, K. 1986. Santa Ysabel Ethnobotany. San Diego. [Ethnic Technol. Notes 20.]

Heide, O. M., K. Pederson, and E. Dahl. 1990. Environmental control of flowering and morphology in the high arctic *Cerastium regelii*, and the taxonomic status of C. *jenisejense*. Nordic J. Bot. 10: 141–147.

Heldreich, T. [1851.] Plantae Atticae, a. 1848. N.p. [Indelible autograph list.] (Pl. Atticae)

Heller, A. A. 1898. Catalogue of North American Plants North of Mexico, Exclusive of the Lower Cryptogams. [Lancaster, Pa.] (Cat. N. Amer. Pl.)

Herman, K. D. 1996. *Gypsophila paniculata*. Baby's-breath. In: J. M. Randall and J. Marinelli, eds. 1996. Invasive Plants: Weeds of the Global Garden. Brooklyn. P. 78.

Hermann, F. 1937. Übersicht über die *Herniaria*-Arten des Berliner Harbars. Repert. Spec. Nov. Regni Veg. 42: 203–224.

Hess, W. J. and J. L. Reveal. 1976. A revision of *Eriogonum* (Polygonaceae) subgenus *Pterogonum*. Great Basin Naturalist 36: 281–333.

Heynhold, G. 1846[–1847]. Alphabetische und synonymische Aufzählung der...Gewächse.... Dresden and Leipzig. (Alph. Aufz. Gew.)

Hickman, J. C. 1971. *Arenaria*, section *Eremogone* (Caryophyllaceae) in the Pacific Northwest: A key and discussion. Madroño 21: 201–207.

Hickman, J. C. 1984. Nomenclatural changes in *Polygonum*, *Persicaria* and *Rumex* (Polygonaceae). Madroño 31: 249–252.

Hickman, J. C., ed. 1993. The Jepson Manual. Higher Plants of California. Berkeley, Los Angeles, and London.

Hickman, J. C. 1993b. Geographic subdivisions of California. In: J. C. Hickman, ed. 1993. The Jepson Manual. Higher Plants of California. Berkeley, Los Angeles, and London. Pp. 37–46.

Hickman, J. C. 1993c. Polygonaceae. In: J. C. Hickman, ed. 1993. The Jepson Manual. Higher Plants of California. Berkeley, Los Angeles, and London. Pp. 854–895.

Hickman, J. C. and C. S. Hickman. 1978. *Polygonum perfoliatum*: A recent Asiatic adventive. Bartonia 45: 18–23.

Hiitonen, H. I. A. [1933.] Suomen Kasvio.... Helsinki. (Suom. Kasvio)

Hill, J. 1759–1775. The Vegetable System.... 26 vols. London. (Veg. Syst.)

Hill, J. 1768. Hortus Kewensis. London. (Hort. Kew.)

Hinds, H. R. 1995. Nomenclatural changes in *Polygonum*, *Persicaria*, and *Aconogonon* (Polygonaceae). Novon 5: 165–166.

Hinds, H. R. 2000. Flora of New Brunswick..., ed. 2. Fredericton.

Hinton, L. 1975. Notes on La Huerta Diegueno ethnobotany. J. Calif. Anthropol. 2: 214–222.

Hist. Nat. Vég.—See: E. Spach 1834–1848

Hist. Pl. Dauphiné—See: D. Villars 1786–1789

Hitchcock, C. L. 1964. Polygonaceae. In: C. L. Hitchcock et al. 1955–1969. Vascular Plants of the Pacific Northwest. 5 vols. Seattle. Vol. 2, pp. 102–182.

Hitchcock, C. L. and A. Cronquist. 1973. Flora of the Pacific Northwest: An Illustrated Manual. Seattle.

Hitchcock, C. L., A. Cronquist, M. Ownbey, and J. W. Thompson. 1955–1969. Vascular Plants of the Pacific Northwest. 5 vols. Seattle. [Univ. Wash. Publ. Biol. 17.] (Vasc. Pl. Pacif. N.W.)

Hitchcock, C. L. and B. Maguire. 1947. A Revision of the North American Species of *Silene*. Seattle. [Univ. Wash. Publ. Biol. 13.] (Revis. N. Amer. Silene)

Hocking, G. M. 1956. Some plant materials used medicinally and otherwise by the Navaho Indians in the Chaco Canyon, New Mexico. Palacio 56: 146–165.

Hoffmann, G. F. [1791–1804.] Deutschland Flora oder botanisches Taschenbuch.... 4 vols. Erlangen. (Deutschl. Fl.)

Hong, S. P. 1991. A revision of *Aconogonon* (= *Polygonum* sect. *Aconogonon,* Polygonaceae) in North America. Rhodora 93: 322–346.

Hong, S. P. 1993. Reconsideration of the generic status of *Rubrivena* (Polygonaceae, Persicarieae). Pl. Syst. Evol. 186: 95–122.

Hong, S. P., L.-P. Ronse Decraene, and E. Smets. 1998. Systematic significance of tepal surface morphology in tribes Persicarieae and Polygoneae (Polygonaceae). Bot. J. Linn. Soc. 127: 91–116.

Hooker, W. J. [1829–]1833–1840. Flora Boreali-Americana; or, the Botany of the Northern Parts of British America.... 2 vols. in 12 parts. London, Paris, and Strasbourg. (Fl. Bor.-Amer.)

Hooker, W. J. and G. A. W. Arnott. [1830–]1841. The Botany of Captain Beechey's Voyage; Comprising an Account of the Plants Collected by Messrs Lay and Collie, and Other Officers of the Expedition, during the Voyage to the Pacific and Bering's Strait, Performed in His Majesty's Ship Blossom, under the Command of Captain F. W. Beechey...in the Years 1825, 26, 27, and 28. 10 parts. London. [Parts paged and plates numbered consecutively.] (Bot. Beechey Voy.)

Hooker's J. Bot. Kew Gard. Misc. = Hooker's Journal of Botany and Kew Garden Miscellany.

Hornemann, J. W. 1821–1837. Forsøg til en Dansk Oeconomisk Plantelaere, ed. 3. 2 vols. Copenhagen. (Fors. Oecon. Plantel. ed. 3)

Hort. Kew.—See: J. Hill 1768

Horton, J. H. 1963. A taxonomic revision of *Polygonella* (Polygonaceae). Brittonia 15: 177–203.

Horton, J. H. 1972. Studies of the southeastern United States flora. IV. Polygonaceae. J. Elisha Mitchell Sci. Soc. 88: 92–102.

Hortus Kew.—See: W. Aiton and W. T. Aiton 1810–1813

Houttuyn, M. 1773–1783. Natuurlijke Historie of Uitvoerige Beschrijving der Dieren, Planten en Mineraalen, Volgens het Samenstel van den Heer Linnaeus. Met Naauwkeurige Afbeeldingen. Tweede Deels [Planten].... 14 fasc. Amsterdam. (Nat. Hist.)

Howard, R. A. 1958. A history of the genus *Coccoloba* in cultivation. Baileya 6: 204–212.

Howell, J. T. 1941. Notes on *Polycarpon*. Leafl. W. Bot. 3: 80.

Howell, J. T., P. H. Raven, and P. Rubtzoff. 1958. A flora of San Francisco, California. Wasmann J. Biol. 16: 1–157.

Howell, T. J. 1897–1903. A Flora of Northwest America. 1 vol. in 8 fasc. Portland. [Fasc. 1–7 (text) paged consecutively, fasc. 8 (index) independently.] (Fl. N.W. Amer.)

Hudson, W. 1762. Flora Anglica.... London. (Fl. Angl.)

Hultén, E. 1937. Flora of the Aleutian Islands and Westernmost Alaska Peninsula with Notes on the Flora of Commander Islands.... Stockholm. (Fl. Aleut. Isl.)

Hultén, E. 1941–1950. Flora of Alaska and Yukon. 10 vols. Lund and Leipzig. [Vols. paged consecutively and designated as simultaneous numbers of Lunds Univ. Årsskr. (= Acta Univ. Lund.) and Kungl. Fysiogr. Sällsk. Handl.] (Fl. Alaska Yukon)

Hultén, E. 1943. *Stellaria longipes* and its allies. Bot. Not. 1943: 251–270.

Hultén, E. 1956. The *Cerastium alpinum* complex. A case of world-wide introgressive hybridization. Svensk Bot. Tidskr. 50: 411–495.

Hultén, E. 1964. Remarkable range extension for *Minuartia groenlandica* (Retz.) Ostenf. Svensk Bot. Tidskr. 58: 432–435.

Hultén, E. 1968. Flora of Alaska and Neighboring Territories: A Manual of the Vascular Plants. Stanford.

Hultén, E. 1973. Supplement to Flora of Alaska and Neighboring Territories. A study in the flora of Alaska and the transberingian connections. Bot. Not. 126: 459–512.

Hutchinson, J. 1973. The Families of Flowering Plants, ed. 3. Oxford.

Icon.—See: A. J. Cavanilles 1791–1801

Ikonnikov, S. S. 1973. Zametki o gvozdichnykh

(Caryophyllaceae). I. O rode *Eremogone* Fenzl. Novosti Sist. Vyssh. Rast. 10: 136–142.

Ill. Fl. Pacific States—See: L. Abrams and R. S. Ferris 1923–1960

Index Kew.—See: B. D. Jackson et al. [1893–]1895+

Index Seminum (Berlin), App. = Index Seminum (Berlin), Appendix.

Index Seminum (Dorpat) = Index Seminum Horti Academici Dorpatensis.

Index Seminum (St. Petersburg) = Index Seminum, Quae Hortus Botanicus Imperialis Petropolitanus pro Mutua Commutatione Offert.

Ives, J. C. 1861. Report upon the Colorado River of the West, Explored in 1857 and 1858 by Lieutenant Joseph C. Ives.... 5 parts, appendices. Washington. (Rep. Colorado R.)

J. Acad. Nat. Sci. Philadelphia = Journal of the Academy of Natural Sciences of Philadelphia.

J. Agric. Food Chem. = Journal of Agricultural and Food Chemistry.

J. Appl. Ecol. = Journal of Applied Ecology.

J. Arnold Arbor. = Journal of the Arnold Arboretum.

J. Bot. = Journal of Botany, British and Foreign.

J. Bot. (Morot) = Journal de Botanique. [Edited by L. Morot.]

J. Calif. Anthropol. = Journal of California Anthropology.

J. Ecol. = Journal of Ecology.

J. Elisha Mitchell Sci. Soc. = Journal of the Elisha Mitchell Scientific Society.

J. Ethnopharmacol. = Journal of Ethnopharmacology; Interdisciplinary Journal Devoted to Bioscientific Research on Indigenous Drugs.

J. Hist. Nat. = Journal d'Histoire Naturelle.

J. Jap. Bot. = Journal of Japanese Botany.

J. Lepid. Soc. = Journal of the Lepidopterists' Society.

J. Linn. Soc., Bot. = Journal of the Linnean Society. Botany.

J. Pl. Biol. = Journal of Plant Biology.

J. Range Managem. = Journal of Range Management.

J. Torrey Bot. Soc. = Journal of the Torrey Botanical Society.

J. Wash. Acad. Sci. = Journal of the Washington Academy of Sciences.

Jackson, B. D. et al., comps. [1893–]1895+. Index Kewensis Plantarum Phanerogamarum.... 2 vols. + 21+ suppls. Oxford. (Index Kew.)

Jacquin, N. J. 1760. Enumeratio Systematica Plantarum, Quas in Insulis Caribaeis Vicinaque Americes Continente Detexit Novas.... Leiden. (Enum. Syst. Pl.)

Jacquin, N. J. 1762. Enumeratio Stirpium Plerarumque, Quae Sponte Crescunt in Agro Vindobonensi, Montibusque Confinibus. Vienna. (Enum. Stirp. Vindob.)

Jahrb. Wiss. Bot. = Jahrbücher für wissenschaftliche Botanik.

Jaretzky, R. 1928. Histologische und karyologische Studien an Polygonaceen. Jahrb. Wiss. Bot. 69: 357–490.

Jepson, W. L. 1901. A Flora of Western Middle California.... Berkeley. (Fl. W. Calif.)

Jepson, W. L. 1909–1943. A Flora of California.... 3 vols. in 12 parts. San Francisco etc. [Pagination consecutive within each vol.; vol. 1 page sequence independent of part number sequence (chronological); part 8 of vol. 1 (pp. 1–32, 579–index) never published.] (Fl. Calif.)

Jepson, W. L. [1923–1925.] A Manual of the Flowering Plants of California.... Berkeley. (Man. Fl. Pl. Calif.)

Jessen, K. F. W. 1879. Deutsche Excursions-Flora. Hannover. (Deut. Excurs.-Fl.)

Johnston, A. 1987. Plants and the Blackfoot. Lethbridge, Alta.

Jones, D. M. and T. R. Mertens. 1970. A taxonomic study of genus *Polygonum* employing chromatographic methods. Proc. Indiana Acad. Sci. 80: 422–430.

Jonsell, B. 2001. *Arenaria*. In: B. Jonsell and T. Karlsson, eds. 2000+. Flora Nordica. 2+ vols. Stockholm. Vol. 2, pp. 99–104.

Jonsell, B. and T. Karlsson, eds. 2000+. Flora Nordica. 2+ vols. Stockholm.

Judd, W. S. 1983. The taxonomic status of *Stipulicida filiformis* (Caryophyllaceae). Sida 10: 33–36.

Jurtzev, B. A. et al. 1975. Floristicheskie nakhodki v Chukotskoy tundre, 2. (Novitates pro flora Terrae Tschuktschorum, 2.) Novosti Sist. Vyssh. Rast. 12: 301–324.

Kalmia = Kalmia; Botanic Journal.

Kan, T. M. 1992. The Distribution and Ecology of the Narrow Endemic *Eriogonum umbellatum* var. *torreyanum*. M.S. thesis. University of California, Davis.

Karlsson, T. 2000. *Polygonum*. In: B. Jonsell and T. Karlsson, eds. 2000+. Flora Nordica. 2+ vols. Stockholm. Vol. 1, pp. 255–273.

Karsten, H. 1880–1883. Deutsche Flora. Pharmaceutisch-medicinische Botanik.... 13 Lieferungen. Berlin. (Deut. Fl.)

Kartesz, J. T. 1987. A Flora of Nevada. Ph.D. thesis. 3 vols. University of Nevada, Reno.

Kartesz, J. T. and K. N. Gandhi. 1990. Nomenclatural notes for the North American flora, II. Phytologia 68: 421–427.

Kartesz, J. T. and C. A. Meacham. 1999. Synthesis of the North American Flora, ver. 1.0. Chapel Hill. [CD-ROM.]

Kaul, R. B. 1986. Polygonaceae. In: Great Plains Flora Association. 1986. Flora of the Great Plains. Lawrence, Kans. Pp. 214–235.

Kearney, T. H. and R. H. Peebles. 1960. Arizona Flora, ed. 2. Berkeley.

Ketzner, D. M. 1996. *Crepis pulchra* (Asteraceae) and *Moenchia erecta* (Caryophyllaceae) in Illinois. Trans. Illinois State Acad. Sci. 89: 21–23.

Kew Bull. = Kew Bulletin.

Khalaf, M. K. and C. A. Stace. 2001. The distinction between *Cerastium tomentosum* Linnaeus and *C. biebersteinii* Candolle (Caryophyllaceae), and their occurrence in the wild in Britain. Watsonia 23: 481–491.

Kharkevich, S. S., ed. 1985+. Plantae Vasculares Orientis Extremi Sovietici. 7+ vols. Leningrad.

Kiger, R. W. and D. M. Porter. 2001. Categorical Glossary for the Flora of North America Project. Pittsburgh.

Kim, J. Y. and C. W. Park. 2000. Morphological and chromosomal variation in *Fallopia* section *Reynoutria* (Polygonaceae) in Korea. Brittonia 52: 34–48.

Kim, M. H., J. H. Park, and C. W. Park, C. W. 2000. Flavonoid chemistry of *Fallopia* section *Fallopia* (Polygonaceae). Biochem. Syst. & Ecol. 28: 433–441.

Kim, S. T., M. H. Kim, and C. W. Park. 2000. A systematic study on *Fallopia* section *Fallopia* (Polygonaceae). Korean J. Pl. Taxon. 30: 35–54.

Kitchener, G. D. 2002. *Rumex ×xenogenus* Rech. fil. (Polygonaceae), the hybrid between Greek and patience docks, found in Britain. Watsonia 24: 209–213.

Komarov, V. L., B. K. Schischkin, and E. Bobrov, eds. 1934–1964. Flora URSS.... 30 vols. Leningrad. (Fl. URSS)

Kongl. Svenska Vetensk. Acad. Handl. = Kongl[iga]. Svenska Vetenskaps Academiens Handlingar.

Kongl. Vetensk. Acad. Nya Handl. = Kongl[iga]. Vetenskaps Academiens Nya Handlingar.

Korean J. Pl. Taxon. = Korean Journal of Plant Taxonomy.

Kozhevnikov, Y. P. 1985. De Caryophyllaceis notae criticae. Novosti Sist. Vyssh. Rast. 22: 95–114.

Kral, R., ed. 1983. A Report on Some Rare, Threatened, or Endangered Forest-related Vascular Plants of the South. 2 vols. Washington. [U.S.D.A. Forest Serv., Techn. Publ. R8-TP 2.]

Kruckeberg, A. R. 1961. Artificial crosses of western North American silenes. Brittonia 13: 305–333.

Kruckeberg, A. R. 1962. Intergeneric hybrids in the Lychnideae (Caryophyllaceae). Brittonia 14: 311–321.

Kubitzki, K. et al., eds. 1990+. The Families and Genera of Vascular Plants. 4+ vols. Berlin etc.

Kulturpflanze = Kulturpflanze. Berichte und Mitteilungen aus dem Institut für Kulturpflanzenforschung der Deutschen Akademie der Wissenschaften zu Berlin in Gatersleben Krs. Aschersleben.

Kuntze, O. 1891–1898. Revisio Generum Plantarum Vascularium Omnium atque Cellularium Multarum.... 3 vols. Leipzig etc. (Revis. Gen. Pl.)

Kurtto, A. 2001. *Silene* [in part]. In: B. Jonsell and T. Karlsson, eds. 2000+. Flora Nordica. 2+ vols. Stockholm. Vol. 2, pp. 183–188.

Kurtto, A. 2001b. *Honckenya*. In: B. Jonsell and T. Karlsson, eds. 2000+. Flora Nordica. 2+ vols. Stockholm. Vol. 2, pp. 106–109.

Kuyper, K. F., U. Yandell, and R. S. Nowak. 1997. On the taxonomic status of *Eriogonum robustum* (Polygonaceae), a rare endemic in western Nevada. Great Basin Naturalist 57: 1–10.

Kuzmina, M. L. 2002. The sections *Macrolepides* (F. N. Williams) Klok. and *Barbulatum* F. N. Williams of the genus *Dianthus* L. (Caryophyllaceae) in East Europe and the Caucasus. Komarovia 2: 29–54.

Kuzmina, M. L. 2003. The sections *Carthusiani* and *Armerium* of the genus *Dianthus* (Caryophyllaceae) in East Europe and the Caucasus. Komarovia 3: 85–102.

Lagasca y Segura, M. 1816b. Genera et Species Plantarum.... Madrid. (Gen. Sp. Pl.)

Lakela, O. 1962. Occurrence of species of *Polycarpaea* Lam. (Caryophyllaceae) in North America. Rhodora 64: 179–182.

Lakela, O. 1963. Annotation of North American *Polycarpaea*. Rhodora 6: 35–44.

Lamarck, J. et al. 1783–1817. Encyclopédie Méthodique. Botanique.... 13 vols. Paris and Liège. [Vols. 1–8, suppls. 1–5.] (Encycl.)

Lamarck, J. and A. P. de Candolle. 1805[–1815]. Flore Française, ou Descriptions Succinctes de Toutes les Plantes Qui Croissent Naturellement en France..., ed. 3. 5 tomes in 6 vols. Paris. [Tomes 1–4(2), 1805; tome 5, 1815.] (Fl. Franç. ed. 3)

Lamarck, J. and J. Poiret. 1791–1823. Tableau Encyclopédique et Méthodique des Trois Règnes de la Nature. Botanique.... 6 vols. Paris. [Vols. 1–2 = tome 1; vols. 3–5 = tome 2; vol. [6] = tome 3. Vols. paged consecutively within tomes.] (Tabl. Encycl.)

Lamb Frye, A. S. and K. A. Kron. 2003. *rbc*L phylogeny and character evolution in Polygonaceae. Syst. Bot. 28: 326–332.

Lambinon, J. 1981. Proposition de rejet *Spergularia media* (L.) C. Presl, Fl. Sic. 1: 161. 1826 (Caryophyllaceae). Taxon 30: 364.

Landry, G. P., W. D. Reese, and C. M. Allen. 1988. *Stellaria parva* Pederson new to North America. Phytologia 64: 497.

Lange, J. M. C. 1880[–1894]. Conspectus Florae Groenlandicae. 3 parts. Copenhagen. [Parts paged consecutively.] (Consp. Fl. Groenland.)

Laubengayer, R. A. 1937. Studies in the anatomy and morphology of the polygonaceous flower. Amer. J. Bot. 24: 329–343.

Lawrence, G. H. M. 1940. Armerias, native and cultivated. Gentes Herb. 4: 391–418.

Lawrence, G. H. M. 1947. The genus *Armeria* in North America. Amer. Midl. Naturalist 37: 751–779.

Leafl. Bot. Observ. Crit. = Leaflets of Botanical Observation and Criticism.

Leafl. W. Bot. = Leaflets of Western Botany.

Ledebour, C. F. von. 1829–1833. Flora Altaica. 4 vols. Berlin. (Fl. Altaica)

Ledebour, C. F. von. [1841]1842–1853. Flora Rossica sive Enumeratio Plantarum in Totius Imperii Rossici Provinciis Europaeis, Asiaticis, et Americanis Hucusque Observatarum.... 4 vols. Stuttgart. (Fl. Ross.)

Lefèbvre, C. 1974. Population variation and taxonomy in *Armeria maritima* with special reference to heavy-metal tolerant populations. New Phytol. 73: 209–219.

Lefèbvre, C. and X. Vekemans. 1995. A numerical taxonomic study of *Armeria maritima* (Plumbaginaceae) in North America and Greenland. Canad. J. Bot. 73: 1583–1595.

Lehr, J. H. 1978. A Catalogue of the Flora of Arizona. Phoenix.

Leslie, A. C. 1983. The International *Dianthus* Register, ed. 2. London.

Lewis, Mont E. 1973. Wheeler Peak Area: Species List.... N.p.

Lewis, P. O. 1991. Allozyme Variation and Evolution in *Polygonella* (Polygonaceae). Ph.D. dissertation. Ohio State University.

Lewis, P. O. 1991b. Allozyme variation in the rare Gulf Coast endemic *Polygonella macrophylla* Small (Polygonaceae). Pl. Spec. Biol. 6: 1–10.

Lewis, P. O. and D. J. Crawford. 1995. Pleistocene refugium endemics exhibit greater allozymic diversity than widespread congeners in the genus *Polygonella* (Polygonaceae). Amer. J. Bot. 82: 141–149.

Lex. Gen. Phan.—See: T. E. von Post and O. Kuntze 1903

Li, H. L. 1952b. The genus *Tovara* (Polygonaceae). Rhodora 50: 19–25.

Libert, B. and R. Englund. 1989. Present distribution and ecology of *Rheum rhaponticum* (Polygonaceae). Willdenowia 19: 91–98.

Link, J. H. F. 1829–1833. Handbuch zur Erkennung der nutzbarsten und am häufigsten vorkommenden Gewächse.... 3 vols. Berlin. (Handbuch)

Linnaea = Linnaea. Ein Journal für die Botanik in ihrem ganzen Umfange.

Linnaeus, C. 1753. Species Plantarum.... 2 vols. Stockholm. (Sp. Pl.)

Linnaeus, C. 1754. Genera Plantarum, ed. 5. Stockholm. (Gen. Pl. ed. 5)

Linnaeus, C. 1758[–1759]. Systema Naturae per Regna Tria Naturae..., ed. 10. 2 vols. Stockholm. (Syst. Nat. ed. 10)

Linnaeus, C. 1762–1763. Species Plantarum..., ed. 2. 2 vols. Stockholm. (Sp. Pl. ed. 2)

Linnaeus, C. 1766–1768. Systema Naturae per Regna Tria Naturae..., ed. 12. 3 vols. Stockholm. (Syst. Nat. ed. 12)

Linnaeus, C. 1767[–1771]. Mantissa Plantarum. 2 parts. Stockholm. [Mantissa [1] and Mantissa [2] Altera paged consecutively.] (Mant. Pl.)

Little, E. L. Jr., R. O. Woodbury, and F. H. Wadsworth. 1969. Common Trees of Puerto Rico and the Virgin Islands. 2 vols. Washington. [Agric. Handb. 249.]

Lledó, M. D. et al. 1998. Systematics of Plumbaginaceae based upon cladistic analysis of *rbc*L sequence data. Syst. Bot. 23: 21–29.

Loret, H. and A. Barrandon. 1876. Flore de Montpellier.... 2 vols. Montpellier and Paris. [Vols. paged consecutively.] (Fl. Montpellier)

Lousley, J. E. 1953. *Rumex cuneifolius* and a new hybrid. Watsonia 2: 394–397.

Lousley, J. E. and D. H. Kent. 1981. Docks and Knotweeds of the British Isles. London.

Löve, Á. 1941. *Rumex tenuifolius* (Wallr.) Löve, spec. nova. Bot. Not. 1941: 99–101.

Löve, Á. 1944. The dioecious forms of *Rumex* subgenus *Acetosa* in Scandinavia. Bot. Not. 1944: 237–254.

Löve, Á. 1983. The taxonomy of *Acetosella*. Bot. Helv. 93: 145–168.

Löve, Á. 1986. In: IOPB chromosome number reports XCII. Taxon 35: 610–613.

Löve, Á. and V. Everson. 1967. The taxonomic status of *Rumex paucifolius*. Taxon 16: 423–425.

Löve, Á. and B. M. Kapoor. 1967. New combinations in *Acetosa* and *Bucephalophora*. Taxon 16: 519–522.

Löve, Á. and B. M. Kapoor. 1968. A chromosome atlas of the collective genus *Rumex*. Cytologia 32: 328–342.

Löve, Á. and D. Löve. 1957. *Rumex thyrsiflorus* new to North America. Rhodora 59: 1–5.

Löve, Á. and D. Löve. 1975. Nomenclatural notes on Arctic plants. Bot. Not. 128: 497–523.

Löve, Á. and D. Löve. 1975b. Cytotaxonomical Atlas of the Arctic Flora. Vaduz.

Löve, Á. and D. Löve. 1982. In: IOPB chromosome number reports LXXIV. Taxon 31: 119–128.

Löve, Á. and D. Löve. 1988. In: IOPB chromosome number reports XCIX. Taxon 37: 396–399.

Löve, Á. and P. Sarkar. 1957. Chromosomes and relationships of *Koenigia islandica*. Canad. J. Bot. 35: 507–514.

Löve, D. and J.-P. Bernard. 1958. *Rumex stenophyllus* in North America. Rhodora 60: 54–57.

Luteyn, J. L. 1976. Revision of *Limonium* (Plumbaginaceae) in eastern North America. Brittonia 28: 303–317.

Luteyn, J. L. 1990. The Plumbaginaceae in the flora of the southeastern United States. Sida 14: 169–178.

Macloskie, G. 1903–1914. Reports of the Princeton University Expeditions to Patagonia 1896–1899...Botany.... 2 vols. + suppl. Princeton and Stuttgart. [Vols. 1 and 2 paged consecutively; suppl. paged independently.]

Madroño = Madroño; Journal of the California Botanical Society [from vol. 3: a West American Journal of Botany].

Maguire, B. 1947. Studies in the Caryophyllaceae. III. A synopsis of the North American species of *Arenaria* sect. *Eremogone* Fenzl. Bull. Torrey Bot. Club 74: 38–56.

Maguire, B. 1950. Studies in the Caryophyllaceae. IV. A synopsis of the North American species of the subfamily Silenoideae. Rhodora 52: 233–245.

Maguire, B. 1951. Studies in the Caryophyllaceae. V. *Arenaria* in America north of Mexico. Amer. Midl. Naturalist 46: 493–511.

Maguire, B. 1958. *Arenaria rossii* and some of its relatives in America. Rhodora 60: 44–53.

Maguire, B. 1960. *Arenaria*. In: T. H. Kearney and R. H. Peebles. 1960. Arizona Flora, ed. 2. Berkeley. Pp. 295–298.

Man. Fl. Belgique ed. 2—See: F. Crépin 1866

Man. Fl. Pl. Calif.—See: W. L. Jepson [1923–1925]

Man. Herbor. Suisse—See: J. P. de Clairville 1811

Man. Higher Pl. Oregon—See: M. E. Peck 1941

Man. S. Calif. Bot.—See: P. A. Munz 1935

Mant. Pl.—See: C. Linnaeus 1767[–1771]

Manual ed. 5—See: A. Gray 1867

Manual ed. 6—See: A. Gray et al. 1890

Marschall von Bieberstein, F. A. 1808–1819. Flora Taurico-Caucasica.... 3 vols. Charkow. (Fl. Taur.-Caucas.)

Martin, W. C. and C. R. Hutchins. 1980. A Flora of New Mexico. 2 vols. Vaduz.

Martius, C. F. P. von, A. W. Eichler, and I. Urban, eds. 1840–1906. Flora Brasiliensis. 15 vols. in 40 parts, 130 fasc. Munich, Vienna, and Leipzig. [Volumes and parts numbered in systematic sequence, fascicles numbered independently in chronological sequence.] (Fl. Bras.)

Mattfeld, J. 1922. Geographisch-genetische Untersuchungen über die Gattung *Minuartia* (L.) Hiern. Repert. Spec. Nov. Regni Veg. Beih. 15.

McCormick, J. F., J. R. Bozeman, and S. A. Spongberg. 1971. A taxonomic revision of granite outcrop species of *Minuartia* (*Arenaria*). Brittonia 23: 149–160.

McDonald, C. B. 1980. A biosystematic study of the *Polygonum hydropiperoides* (Polygonaceae) complex. Amer. J. Bot. 67: 664–670.

McNeill, J. 1962. Taxonomic studies in the Alsinoideae: I. Generic and infra-generic groups. Notes Roy. Bot. Gard. Edinburgh 24: 79–155.

McNeill, J. 1978. *Silene alba* and *S. dioica* in North America and the generic delimitation of *Lychnis*, *Melandrium* and *Silene* (Caryophyllaceae). Canad. J. Bot. 56: 297–308.

McNeill, J. 1980b. The delimitation of *Arenaria* (Caryophyllaceae) and related genera in North America, with 11 new combinations in *Minuartia*. Rhodora 82: 495–502.

McNeill, J. and I. J. Bassett. 1974. Pollen morphology and the infrageneric classification of *Minuartia* (Caryophyllaceae). Canad. J. Bot. 52: 1225–1231.

McNeill, J. and J. N. Findlay. 1972. Introduced perennial species of *Stellaria* in Quebec. Naturaliste Canad. 99: 59–60.

Med. Repos. = Medical Repository.

Meddel. Grønland = Meddelelser om Grønland, af Kommissionen for Ledelsen af de Geologiske og Geografiske Undersølgeser i Grønland.

Meded. Bot. Mus. Herb. Rijks Univ. Utrecht = Mededeelingen van het Botanisch Museum en Herbarium van de Rijks Universiteit te Utrecht.

Meerts, P., T. Baya, and C. Lefèbvre. 1998. Allozyme variation in the annual weed species complex *Polygonum aviculare* (Polygonaceae) in relation to ploidy level and colonizing ability. Pl. Syst. Evol. 211: 239–256.

Meerts, P., J.-P. Briane, and C. Lefèbvre. 1990. A numerical taxonomic study of the *Polygonum aviculare* complex (Polygonaceae) in Belgium. Watsonia 5: 177–214.

Meinke, R. J. and P. F. Zika. 1992. A new annual species of *Minuartia* (Caryophyllaceae) from Oregon and California. Madroño 39: 288–300.

Meisner, C. F. 1826. Monographiae Generis *Polygoni* Prodromus. Geneva. (Monogr. Polyg.)

Meisner, C. F. 1837–1843. Plantarum Vascularium Genera.... 2 parts in 14 fasc. Leipzig. [Parts (1: "Tab. Diagn."; 2: "Commentarius") issued together incrementally within fascicles and paged independently, fascicles paged consecutively for each part.] (Pl. Vasc. Gen.)

Meisner, C. F. 1856. *Rumex*. In: A. P. de Candolle and A. L. P. P. de Candolle, eds. 1823–1873. Prodromus Systematis Naturalis Regni Vegetabilis.... 17 vols. Paris etc. Vol. 14, pp. 41–74.

Mém. Acad. Imp. Sci. Saint-Pétersbourg, Sér. 6, Sci. Math., Seconde Pt. Sci. Nat. = Mémoires de l'Académie Impériale des Sciences de Saint-Pétersbourg. Sixième Série. Sciences Mathématiques, Physiques et Naturelles. Seconde Partie: Sciences Naturelles.

Mém. Acad. Imp. Sci. St.-Pétersbourg Divers Savans = Mémoires Présentés à l'Académie Impériale des Sciences de St.-Pétersbourg par Divers Savans et Lus dans ses Assemblées.

Mém. Acad. Imp. Sci. St. Pétersbourg Hist. Acad. = Mémoires de l'Académie Impériale des Sciences de St. Pétersbourg. Avec l'Histoire de l'Académie.

Mém. Acad. Imp. Sci. St.-Pétersbourg, Sér. 6, Sci. Math. = Mémoires de l'Académie Impériale des Sciences de St.-Pétersbourg. Sixième Série. Sciences Mathématiques, Physiques et Naturelles.

Mem. Amer. Acad. Arts = Memoirs of the American Academy of Arts and Science.

Mem. Amer. Anthropol. Assoc. = Memoirs of the American Anthropological Association.

Mem. New York Bot. Gard. = Memoirs of the New York Botanical Garden.

Mem. Reale Accad. Sci. Torino = Memorie della Reale Accademia delle Scienze di Torino

Mem. Torrey Bot. Club = Memoirs of the Torrey Botanical Club.

Mem. Tour N. Mexico—See: F. A. Wislizenus 1848

Memoranda Soc. Fauna Fl. Fenn. = Memoranda Societatis pro Fauna et Flora Fennica.

Mentzelia = Mentzelia; Journal of the Northern Nevada Native Plant Society.

Merriam-Webster 1988. Webster's New Geographical Dictionary. Springfield, Mass.

Mertens, T. R. and P. H. Raven. 1965. Taxonomy of *Polygonum*, section *Polygonum* (*Avicularia*) in North America. Madroño 18: 85–92.

Meth. Pl.—See: J. A. Scopoli 1754

Methodus—See: C. Moench 1794

Meyer, C. A. von. 1831. Verzeichniss der Pflanzen...Reise im Caucasus und in den Provinzen am westlichen Ufer des Caspischen Meeres gefunden.... St. Petersburg. (Verz. Pfl. Casp. Meer.)

Michaux, A. 1803. Flora Boreali-Americana.... 2 vols. Paris and Strasbourg. (Fl. Bor.-Amer.)

Michigan Agric. Exp. Sta. Bull. = Michigan Agricultural Experiment Station. Bulletin.

Michigan Bot. = Michigan Botanist.

Middendorff, A. T. von. 1847–1867. Reise in den äussersten Norden und Osten Siberiens während der Jahre 1843 und 1844.... 4 vols. in parts and fasc. St. Petersburg. (Reise Siber.)

Miller, P. 1754. The Gardeners Dictionary.... Abridged..., ed. 4. 3 vols. London. (Gard. Dict. Abr. ed. 4)

Miller, P. 1768. The Gardeners Dictionary..., ed. 8. London. (Gard. Dict. ed. 8)

Miquel, F. A. W. [1851–]1853–1855[–1857]. Plantae Junghuhnianae. Enumeratio Plantarum, Quas in Insulis Java et Sumatra, Detexit Fr. Junghuhn.... 5 parts. Leiden and Paris. [Parts paged consecutively.] (Pl. Jungh.)

Mitchell, R. S. 1968. Variation in the *Polygonum amphibium* complex and its taxonomic significance. Univ. Calif. Publ. Bot. 45: 1–65.

Mitchell, R. S. 1971. Comparative leaf structure of aquatic *Polygonum* species. Amer. J. Bot. 58: 342–360.

Mitchell, R. S. 1976. Submergence experiments on nine species of semi-aquatic *Polygonum*. Amer. J. Bot. 63: 1158–1165.

Mitchell, R. S. 1978. *Rumex maritimus* L. versus *R. persicarioides* L. (Polygonaceae) in the Western Hemisphere. Brittonia 30: 293–296.

Mitchell, R. S. 1986. A checklist of New York State plants. Bull. New York State Mus. Sci. Serv. 458.

Mitchell, R. S. and J. K. Dean. 1978. Polygonaceae (buckwheat family) of New York State. Bull. New York State Mus. Sci. Serv. 431: 1–79.

Mitchell, R. S. and G. C. Tucker. 1991. *Sagina japonica* (Sw.) Ohwi (Caryophyllaceae), an overlooked adventive in the northeastern United States. Rhodora 93: 192–194.

Mitchell, R. S. and G. C. Tucker. 1997. Revised checklist of New York State plants. Bull. New York State Mus. Sci. Serv. 490.

Mitchell, R. S. and L. J. Uttal. 1969. Natural hybridization in Virginia *Silene* (Caryophyllaceae). Bull. Torrey Bot. Club 99: 544–549.

Mizushima, M. 1965. Critical studies on Japanese plants, II: The genus *Pseudostellaria* Pax in Japan. Bull. Bot. Surv. India 7: 62–72.

Moench, C. 1794. Methodus Plantas Horti Botanici et Agri Marburgensis.... Marburg. (Methodus)

Moerman, D. E. 1986. Medicinal Plants of Native America. 2 vols. Ann Arbor. [Univ. Michigan, Mus. Anthropol., Techn. Rep. 19.]

Moerman, D. E. 1998. Native American Ethnobotany. Portland.

Mohr, C. T. 1901. Plant life of Alabama. Contr. U.S. Natl. Herb. 6.

Monogr. Amer. Sp. Polygonum—See: J. K. Small 1895

Monogr. Polyg.—See: C. F. Meisner 1826

Monogr. Rumex—See: F. Campderá 1819

Monogr. Syst. Bot. Missouri Bot. Gard. = Monographs in Systematic Botany from the Missouri Botanical Garden.

Mooney, H. A. and W. D. Billings. 1961. Comparative physiological ecology of arctic and alpine populations of *Oxyria digyna*. Ecol. Monogr. 31: 1–29.

Mori, T. 1922. An Enumeration of Plants hitherto Known from Corea.... Seoul. (Enum. Pl. Corea)

Morton, J. K. 1972. On the occurrence of *Stellaria pallida* in North America. Bull. Torrey Bot. Club 99: 95–97.

Morton, J. K. 1973. Spontaneous hybrids between *Cerastium tomentosum* Linn. and *C. arvense* Linn. Rhodora 74: 519–521.

Morton, J. K. and R. K. Rabeler. 1989. Biosystematic studies on the *Stellaria calycantha* (Caryophyllaceae) complex. I. Cytology and cytogeography. Canad. J. Bot. 67: 121–127.

Morton, J. K. and J. M. Venn. 1990. A Checklist of the Flora of Ontario: Vascular Plants. Waterloo.

Moseley, R. K. and J. L. Reveal. 1995. The Taxonomy and Preliminary Conservation Status of *Eriogonum shockleyi* S. Wats. in Idaho. Boise. [U.S. Bur. Land Managem., Idaho, Techn. Bull. 96-4.]

Mosyakin, S. L. and W. L. Wagner. 1998. Notes on two alien taxa of *Rumex* L. (Polygonaceae) naturalized in the Hawaiian Islands. Bishop Mus. Occas. Pap. 55: 39–44.

Muhlenbergia = Muhlenbergia; a Journal of Botany.

Mulligan, G. A. and C. Frankton. 1972. Chromosome races in *Rumex arcticus* (Polygonaceae). Canad. J. Bot. 50: 378–380.

Mun, J. H. and C. W. Park. 1995. Flavonoid chemistry of *Polygonum* sect. *Tovara* (Polygonaceae): A systematic survey. Pl. Syst. Evol. 196: 153–159.

Munz, P. A. 1935. A Manual of Southern California Botany.... San Francisco. (Man. S. Calif. Bot.)

Munz, P. A. 1968. Supplement to A California Flora. Berkeley and Los Angeles. (Suppl. Calif. Fl.)

Murphey, E. V. A. 1959. Indian Uses of Native Plants. Fort Bragg, Calif.

Murray, E. 1982. Notae spermatophytae. Kalmia 12: 18–27.

Murray, J. A. 1770. Prodromus Designationis Stirpium Gottingensium.... Göttingen. (Prodr. Stirp. Gott.)

Napier, M., ed. 1830–1842. Encyclopedia Britannica..., ed. 7. 21 vols. Edinburgh. (Encycl. Brit. ed. 7)

Narr. Journey Polar Sea—See: J. Franklin et al. 1823

Nat. Arr. Brit. Pl.—See: S. F. Gray 1821

Nat. Hist.—See: M. Houttuyn 1773–1783

Nat. Pflanzenfam. ed. 2—See: H. G. A. Engler et al. 1924+

Naturaliste Canad. = Naturaliste Canadien. Bulletin de Recherches, Observations et Découvertes se Rapportant à l'Histoire Naturelle du Canada.

Nature = Nature; a Weekly Illustrated Journal of Science.

Ned. Kruidk. Arch. = Nederlandsch Kruidkundig Archief; Verslagen en Mededelingen der Nederlandsche Botanische Vereeniging.

Neel, M. C. and N. C. Ellstrand. 2003. Conservation of genetic diversity in the endemic endangered plant *Eriogonum ovalifolium* var. *vineum* (Polygonaceae). Conservation Genet. 4: 337–352.

Neel, M. C., J. Ross-Ibarra, and N. C. Ellstrand. 2001. Implications of mating patterns for conservation of the endangered plant *Eriogonum ovalifolium* var. *vineum* (Polygonaceae). Amer. J. Bot. 88: 1214–1222.

Nepokroeff, M. et al. 2001. Origin of the Hawaiian subfam. Alsinoideae and preliminary relationships in Caryophyllaceae inferred from *mat*K and *trn*L C-F sequence data. In: Botanical Society of America. 2001. Botany 2001 Abstracts. Columbus. P. 130.

Nepokroeff, M. et al. 2002. Relationships within Caryophyllaceae inferred from molecular sequence data. In: Botanical Society of America. 2002. Botany 2002 Abstracts. Columbus. P. 105.

Nesom, G. L. and V. M. Bates. 1984. Reevaluation of infraspecific taxonomy in *Polygonella* (Polygonaceae). Brittonia 36: 37–44.

Nesom, G. L. and V. M. Bates. 1984b. A phylogenetic reconstruction of *Polygonella*. [Abstract.] A. S. B. Bull. 31: 74.

Neues Mag. Aerzte = Neues Magazin für Aerzte.

New, J. K. 1961. Biological flora of the British Isles. *Spergula arvensis* L. (*S. sativa* Boenn., *S. vulgaris* Boenn.). J. Ecol. 49: 205–215.

New Phytol. = New Phytologist; a British Botanical Journal.

New Zealand J. Bot. = New Zealand Journal of Botany.

Nicholls, G. 2002. Alpine Plants of North America: An Encyclopedia of Mountain Flowers from the Rockies to Alaska. Portland.

Nickrent, D. L. and D. Wiens. 1989. Genetic diversity in the rare California shrub *Dedeckera eurekensis* (Polygonaceae). Syst. Bot. 14: 245–253.

Nilsson, Ö. 2001. *Minuartia*. In: B. Jonsell and T. Karlsson, eds. 2000+. Flora Nordica. 2+ vols. Stockholm. Vol. 2, pp. 108–113.

Nomencl. Bot. ed. 2—See: E. G. Steudel 1840–1841

Nordic J. Bot. = Nordic Journal of Botany.

Not. Milit. Reconn.—See: W. H. Emory 1848

Not. Nat. Civ. Lombardia—See: C. Cattaneo 1844

Notes Roy. Bot. Gard. Edinburgh = Notes from the Royal Botanic Garden, Edinburgh.

Nova Acta Phys.-Med. Acad. Caes. Leop.-Carol. Nat. Cur. = Nova Acta Physico-medica Academiae Caesareae Leopoldino-Carolinae Naturae Curiosorum Exhibentia Ephemerides, sive Observationes Historias et Experimenta....

Novit. Fl. Suec. Mant.—See: E. M. Fries 1832–1845

Novon = Novon; a Journal for Botanical Nomenclature.

Novosti Sist. Vyssh. Rast. = Novosti Sistematiki Vysshikh Rastenii.

Nowicke, J. W. and J. J. Skvarla. 1977. Pollen morphology and the relationship of the Plumbaginaceae,

Polygonaceae, and Primulaceae to the order Centrospermae. Smithsonian Contr. Bot. 37: 1–64.

Nuovo Giorn. Bot. Ital. = Nuovo Giornale Botanico Italiano.

Nuttall, T. 1813. A Catalogue of New and Interesting Plants Collected in Upper Louisiana.... London. (Cat. Pl. Upper Louisiana)

Nuttall, T. 1818. The Genera of North American Plants, and Catalogue of the Species, to the Year 1817.... 2 vols. Philadelphia. (Gen. N. Amer. Pl.)

Nuytsia = Nuytsia; Bulletin of the Western Australian Herbarium.

Nygren, A. 1951. Experimental studies in Scandinavian alpine plants. II. On the origin of the Greenlandic species *Melandrium triflorum* (R. Br.) J. Vahl. Hereditas (Lund) 37: 373–381.

Nyman, C. F. 1878–1890. Conspectus Florae Europaeae.... 4 parts + 2 suppls. Örebro. [Parts 1–4 and suppl. 1 paged consecutively; suppl. 2 paged independently.] (Consp. Fl. Eur.)

Nyt Mag. Naturvidensk. = Nyt Magazin for Naturvidenskaberne; Grundlagt af den Physiographiske Forening i Christiania.

Oeder, G. C. et al., eds. [1761–]1764–1883. Icones Plantarum...Florae Danicae Nomine Inscriptum. 17 vols. in 51 fasc. Copenhagen. [Fascicles paged independently and numbered consecutively throughout volumes.] (Fl. Dan.)

Oekon. Fl. Wetterau—See: P. G. Gaertner et al. 1799–1802

Ohnishi, O. 1998. Search for the wild ancestor of buckwheat III. The wild ancestor of cultivated common buckwheat, and of Tartary buckwheat. Econ. Bot. 52: 123–133.

Ohnishi, O. and Y. Matsuoka. 1996. Search for the wild ancestor of buckwheat II. Taxonomy of *Fagopyrum* (Polygonaceae) species based on morphology isozymes and cpDNA variability. Genes Genet. Systems 71: 383–390.

Ohsako, T. and O. Ohnishi. 2000. Intra- and interspecific phylogeny of wild *Fagopyrum* (Polygonaceae) species based on nucleotide sequences of noncoding regions of chloroplast DNA. Amer. J. Bot. 87: 573–582.

O'Kane, S. L., D. H. Wilken, and R. L. Hartman. 1988. Noteworthy collections: Colorado. Madroño 35: 72–74.

Opiz, P. M. 1852. Seznam Rostlin Květeny České.... Prague. (Seznam)

Opler, P. A. and A. B. Wright. 1999. A Field Guide to Western Butterflies, ed. 2. Boston.

Orcutt, C. R. 1886. A botanical trip. W. Amer. Sci. 2: 53–58.

Osmond, C. D., S. D. Smith, B. Gui-Ying, and T. D. Sharkey. 1987. Stem photosynthesis in a desert ephemeral, *Eriogonum inflatum*. Characterization of leaf and stem CO_2 fixation and H_2O vapor exchange under controlled conditions. Oecologia 72: 542–549.

Oxelman, B., B. Ahlgren, and M. Thulin. 2002. Circumscription and phylogenetic relationships of *Gymnocarpos* (Caryophyllaceae–Paronychioideae). Edinburgh J. Bot. 59: 221–227.

Oxelman, B. and M. Lidén. 1995. Generic boundaries in the tribe Sileneae (Caryophyllaceae) as inferred from nuclear rDNA sequences. Taxon 44: 525–542.

Oxelman, B., M. Lidén, and D. Berglund. 1997. Chloroplast

*rps*16 intron phylogeny of the tribe Sileneae (Caryophyllaceae). Pl. Syst. Evol. 206: 411–420.

Oxelman, B., M. Lidén, R. K. Rabeler, and M. Popp. 2000. A revised generic classification of the tribe Sileneae (Caryophyllaceae). Nordic J. Bot. 20: 743–748.

Pacif. Railr. Rep.—See: War Department 1855–1860

Packer, J. G., ed. 1995+. Flora of the Russian Arctic.... 3+ vols. Edmonton. [Translation by G. C. D. Griffiths of: A. I. Tolmatchev, ed. 1960–1987. Flora Arctica SSSR.... 10 vols. Moscow and Leningrad.]

Palacio = El Palacio.

Palmer, E. J. and J. A. Steyermark. 1950. Notes on *Geocarpon minimum* Mackenzie. Bull. Torrey Bot. Club 77: 268–273.

Pap. Michigan Acad. Sci. = Papers of the Michigan Academy of Sciences, Arts and Letters.

Pap. Peabody Mus. Amer. Archaeol. = Papers of the Peabody Museum of American Archaeology and Ethnology.

Parfitt, B. D. and W. C. Hodgson. 1985. *Drymaria viscosa* (Caryophyllaceae): Correct author citation and range extension to the United States. Sida 11: 96–98.

Park, C. W. 1986. Nomenclatural typifications in *Polygonum* section *Echinocaulon* (Polygonaceae). Brittonia 38: 394–406.

Park, C. W. 1987. Flavonoid chemistry of *Polygonum* sect. *Echinocaulon*: A systematic survey. Syst. Bot. 12: 167–179.

Park, C. W. 1988. Taxonomy of *Polygonum* section *Echinocaulon* (Polygonaceae). Mem. New York Bot. Gard. 47: 1–82.

Parlatore, F. 1848–1896. Flora Italiana.... 10 vols. + index. Florence. (Fl. Ital.)

Paterson, A. K. 2000. Range expansion of *Polygonum caespitosum* var. *longisetum* in the United States. Bartonia 60: 57–69.

Patterson, D. T. et al. 1989. Composite List of Weeds. Champaign.

Pax, F. A. and K. Hoffmann. 1934c. Caryophyllaceae. In: H. G. A. Engler et al., eds. 1924+. Die natürlichen Pflanzenfamilien..., ed. 2. 26+ vols. Leipzig and Berlin. Vol. 16c, pp. 275–364.

Peck, M. E. 1941. A Manual of the Higher Plants of Oregon. Portland. (Man. Higher Pl. Oregon)

Peck, M. E. 1961. A Manual of the Higher Plants of Oregon, ed. 2. Corvallis.

Perez, C. J. et al. 1998. Seedbank characteristics of a Nebraska sandhills prairie. J. Range Managem. 51: 55–62.

Persoon, C. H. 1805–1807. Synopsis Plantarum.... 2 vols. Paris and Tubingen. (Syn. Pl.)

Pestova, I. O. 1998. Nomenklaturno-taksonomichne doslidzhennya *Rumex arifolius* complex (Polygonaceae). [Taxonomic and nomenclatural study of the *Rumex arifolius* complex (Polygonaceae).] Ukrayins'k. Bot. Zhurn. 55: 169–173.

Petrovsky, V. V. 1971. *Honckenya*. In: A. I. Tolmatchew, ed. 1960–1987. Flora Arctica URSS. 10 vols. Moscow and Leningrad. Vol. 6, pp. 70–74.

Petrovsky, V. V. 2000. *Honckenya*. In: J. G. Packer, ed. 1995+. Flora of the Russian Arctic.... 3+ vols. Edmonton. Vol. 3, pp. 270–273.

Petrusson, L. and M. Thulin. 1996. Taxonomy and

biogeography of *Gymnocarpos*. Edinburgh J. Bot. 53: 1–26.

Physiogr. Sällsk. Tidskr. = Physiographiska Sällskapets Tidskrift.

Phytologia = Phytologia; Designed to Expedite Botanical Publication.

Phyton (Horn) = Phyton; Annales Rei Botanica.

Pignatti, S. 1972. *Limonium*. In: T. G. Tutin et al., eds. 1964–1980. Flora Europaea. 5 vols. Cambridge. Vol. 3, pp. 38–50.

Pinto da Silva, A. R. 1972. *Armeria*. In: T. G. Tutin et al., eds. 1964–1980. Flora Europaea. 5 vols. Cambridge. Vol. 3, pp. 30–38.

Piper, C. V. 1906. Flora of the State of Washington. Contr. U.S. Natl. Herb. 11: 1–637.

Piper, C. V. and R. K. Beattie. 1901. The Flora of the Palouse Region. Pullman. (Fl. Palouse Reg.)

Piper, C. V. and R. K. Beattie. 1915. Flora of the Northwest Coast.... Lancaster, Pa. (Fl. N.W. Coast)

Pl. Asiat. Rar.—See: N. Wallich [1829–]1830–1832

Pl. Atticae—See: T. Heldreich [1851]

Pl. Hartw.—See: G. Bentham 1839[–1857]

Pl. Jungh.—See: F. A. W. Miquel [1851–]1853–1855[–1857]

Pl. Spec. Biol. = Plant species Biology; an International Journal.

Pl. Syst. Evol. = Plant Systematics and Evolution.

Pl. Vasc. Gen.—See: C. F. Meisner 1837–1843

Poiret, J. 1789. Voyage en Barbarie.... 2 vols. Paris. (Voy. Barbarie)

Porsild, A. E. 1957. Illustrated flora of the Canadian Arctic Archipelago. Bull. Natl. Mus. Canada 146: 1–209.

Porsild, A. E. 1963. *Stellaria longipes* Goldie and its allies in North America. Bull. Natl. Mus. Canada 186: 1–35.

Porsild, A. E. and W. J. Cody. 1980. Vascular Plants of Continental Northwest Territories, Canada. Ottawa.

Porter, T. C. 1892. Ballast-plants at South Bethlehem, Penn. Bull. Torrey Bot. Club 19: 9–10.

Post, T. E. von and O. Kuntze. 1903. Lexikon Generum Phanerogamarum.... Stuttgart. (Lex. Gen. Phan.)

Pratt, G. F. and G. R. Ballmer. 1991. Three biotypes of *Apodemia mormo* (Riodinidae) in the Mojave Desert. J. Lepid. Soc. 45: 46–51.

Précis Découv. Somiol.—See: C. S. Rafinesque 1814

Presl, C. B. 1825–1835. Reliquiae Haenkeanae seu Descriptiones et Icones Plantarum, Quas in America Meridionali et Boreali, in Insulis Philippinis et Marianis Collegit Thaddeus Haenke.... 2 vols. in 7 parts. Prague. (Reliq. Haenk.)

Presl, C. B. 1826. Flora Sicula, Exhibens Plantas Vasculosas in Sicilia.... Prague. (Fl. Sicul.)

Presl, J. S. and C. B. Presl. 1819. Flora Čechica. Prague. (Fl. Čech.)

Preslia = Preslia. Věstník (Časopis) Československé Botanické Společnosti.

Press, J. R. 1988. Infraspecific variation in *Rumex bucephalophorus* L. Bot. J. Linn. Soc. 97: 344–355.

Pringle, J. S. 1976. *Gypsophila scorzonerifolia* (Caryophyllaceae), a naturalized species in the Great Lakes region. Michigan Bot. 15: 215–219.

Proc. Acad. Nat. Sci. Philadelphia = Proceedings of the Academy of Natural Sciences of Philadelphia.

Proc. Amer. Acad. Arts = Proceedings of the American Academy of Arts and Sciences.

Proc. Biol. Soc. Wash. = Proceedings of the Biological Society of Washington.

Proc. Boston Soc. Nat. Hist. = Proceedings of the Boston Society of Natural History.

Proc. Calif. Acad. Sci. = Proceedings of the California Academy of Sciences.

Proc. Davenport Acad. Nat. Sci. = Proceedings of the Davenport Academy of Natural Sciences.

Proc. Indiana Acad. Sci. = Proceedings of the Indiana Academy of Science.

Proc. Providence Franklin Soc. = Proceedings of the Providence Franklin Society.

Proc. Utah Acad. Sci. = Proceedings of the Utah Academy of Sciences.

Prodr.—See: R. Brown 1810; A. P. de Candolle and A. L. P. P. de Candolle 1823–1873

Prodr. Fl. Hispan.—See: H. M. Willkomm and J. M. C. Lange 1861–1880

Prodr. Fl. Nepal.—See: D. Don 1825

Prodr. Stirp. Gott.—See: J. A. Murray 1770

Publ. Bot. (Ottawa) = Publications in Botany, National Museum of Natural Sciences, Canada.

Publ. Field Mus. Nat. Hist., Bot. Ser. = Publications of the Field Museum of Natural History. Botanical Series.

Pugsley, H. W. 1930. The duration of *Herniaria glabra*. J. Bot. 68: 214–218.

Pursh, F. [1813] 1814. Flora Americae Septentrionalis; or, a Systematic Arrangement and Description of the Plants of North America. 2 vols. London. (Fl. Amer. Sept.)

Putievsky, E., P. W. Weiss, and D. R. Marshall. 1980. Interspecific hybridization between *Emex australis* and *E. spinosa*. Austral. J. Bot. 28: 323–328.

Rabeler, R. K. 1981. *Gypsophila muralis:* Is it naturalized in Michigan? Michigan Bot. 20: 21–26.

Rabeler, R. K. 1985. *Petrorhagia* (Caryophyllaceae) in North America. Sida 11: 6–44.

Rabeler, R. K. 1986. Revision of the *Stellaria calycantha* (Caryophyllaceae) Complex and Taxonomic Notes on the Genus. Ph.D. dissertation. Michigan State University.

Rabeler, R. K. 1991. *Moenchia erecta* (Caryophyllaceae) in eastern North America. Castanea 56: 150–151.

Rabeler, R. K. 1992. A new combination in *Minuartia* (Caryophyllaceae). Sida 15: 95–96.

Rabeler, R. K. 1993. The occurrence of anther smut, *Ustilago violacea* s.l., on *Stellaria borealis* (Caryophyllaceae) in North America. Contr. Univ. Michigan Herb. 19: 165–169.

Rabeler, R. K. 1996. *Sagina japonica* (Caryophyllaceae) in the Great Lakes region. Michigan Bot. 35: 43–44.

Rabeler, R. K. 2004. Caryophyllaceae (pink family). In: N. P. Smith et al., eds. 2004. Flowering Plants of the Neotropics. Princeton. Pp. 88–90.

Rabeler, R. K. and A. W. Cusick. 1994. Comments on the introduced Caryophyllaceae of Ohio and nearby states. Michigan Bot. 33: 87–100.

Rabeler, R. K. and R. E. Gereau. 1984. Eurasian introductions to the Michigan flora. I. Michigan Bot. 23: 39–47.

Rabeler, R. K. and R. R. Old. 1992. *Lepyrodiclis holosteoides*

(Caryophyllaceae), "new" to North America. Madroño 39: 240–242.

Rabeler, R. K. and J. W. Thieret. 1988. Comments on the Caryophyllaceae of the southeastern United States. Sida 13: 149–156.

Rabeler, R. K. and J. W. Thieret. 1997. *Sagina* (Caryophyllaceae) range extensions in Canada: *S. japonica* new to Newfoundland, *S. procumbens* new to the Northwest Territories. Canad. Field-Naturalist 111: 309–310.

Rafinesque, C. S. 1810. Caratteri di Alcune Nuovi Generi.... Palermo. (Caratt. Nuov. Gen.)

Rafinesque, C. S. 1814. Précis des Découvertes et Travaux Somiologiques.... Palermo. (Précis Découv. Somiol.)

Rafinesque, C. S. 1817. Florula Ludoviciana; or, a Flora of the State of Louisiana. Translated, Revised, and Improved, from the French of C. C. Robin.... New York. (Fl. Ludov.)

Randall, J. M. and J. Marinelli, eds. 1996. Invasive Plants: Weeds of the Global Garden. Brooklyn.

Ray, V. F. 1933. The Sanpoil and Nesplem: Salishan peoples of northeastern Washington. Univ. Wash. Publ. Anthropol. 5. [Reprinted 1954, New Haven; 1980, New York.]

Reagan, A. B. 1929. Plants used by the White Mountain Apache Indians of Arizona. Wisconsin Archaeologist 8: 143–161.

Rechinger, K. H. 1937. The North American species of *Rumex*. (Vorarbeiten zu einer Monographie der Gattung *Rumex* 5.) Publ. Field Mus. Nat. Hist., Bot. Ser. 17: 1–150.

Rechinger, K. H. 1939. Versuch einer natürlichen Gliedewrung des Formenkrieses von *Rumex bucephalophorus* L. Bot. Not. 1939: 485–504.

Rechinger, K. H. 1949. *Rumices* Asiatici. (Vorarbeiten zu einer Monographie der Gattung *Rumex* 7.) Candollea 12: 9–152.

Rechinger, K. H. 1954. Some new American species of *Rumex*. (Beitrage zur Kenntnis von *Rumex* 12.) Leafl. W. Bot. 7: 133–135.

Rechinger, K. H. 1964. *Rumex*. In: T. G. Tutin et al., eds. 1964–1980. Flora Europaea. 5 vols. Cambridge. Vol. 1, pp. 82–89.

Rechinger, K. H. 1984. *Rumex* (Polygonaceae) in Australia: A reconsideration. Nuytsia 5: 75–122.

Rechinger, K. H. 1990. *Rumex* subgen. *Rumex* sect. *Axillares* (Polygonaceae) in South America. Pl. Syst. Evol. 172: 151–192.

Reed, H. S. and I. Smoot. 1906. The mechanism of seed-dispersal in *Polygonum virginianum*. Bull. Torrey Bot. Club 33: 377–386.

Rees, A. [1802–]1819–1820. The Cyclopaedia; or, Universal Dictionary of Arts, Sciences, and Literature.... 39 vols. in 79 parts. London. [Pages unnumbered.] (Cycl.)

Reichenbach, H. G. L. 1828. Conspectus Regni Vegetabilis.... Leipzig. (Consp. Regn. Veg.)

Reichenbach, H. G. L. 1837. Handbuch des natürlichen Pflanzensystems.... Dresden and Leipzig. (Handb. Nat. Pfl.-Syst.)

Reise Siber.—See: A. T. von Middendorff 1847–1867

Reliq. Haenk.—See: C. B. Presl 1825–1835

Rep. (Annual) Bur. Amer. Ethnol. = Annual Report of the Bureau of American Ethnology.

Rep. (Annual) Michigan Acad. Sci. = Report (Annual) of the Michigan Academy of Science, (Arts, and Letters).

Rep. (Annual) Missouri Bot. Gard. = Report (Annual) of the Missouri Botanical Garden.

Rep. Colorado R.—See: J. C. Ives 1861

Rep. Exped. Rocky Mts.—See: J. C. Frémont 1843–1845

Rep. Exped. Zuni Colorado Rivers—See: L. Sitgreaves 1853

Rep. Inst. Sci. Res. Manchoukuo = Report of the Institute of Scientific Research, Manchoukuo.

Rep. U.S. Mex. Bound.—See: W. H. Emory 1857–1859

Repert. Bot. Syst.—See: W. G. Walpers 1842–1847

Repert. Spec. Nov. Regni Veg. = Repertorium Specierum Novarum Regni Vegetabilis.

Repert. Spec. Nov. Regni Veg. Beih. = Repertorium specierum novarum regni vegetabilis. Beihefte.

Res. J. Wyoming Agric. Exp. Sta. = Research Journal, Wyoming Agricultural Experiment Station.

Res. Stud. State Coll. Wash. = Research Studies of the State College of Washington.

Retzius, A. J. 1795. Florae Scandinaviae Prodromus..., ed. 2. Leipzig. (Fl. Scand. Prodr. ed. 2)

Rev. Bot. Recueil Mens. = Revue Botanique; Recueil Mensuel Renfermant l'Analyse des Travaux Publiés en France et à l'Étranger sur la Botanique, et sur Ses Applications à l'Horticulture, l'Agriculture, la Médecine, etc.

Reveal, J. L. 1968b. *Eriogonum* (part). In: P. A. Munz. 1968. Supplement to A California Flora. Berkeley and Los Angeles. Pp. 33–72.

Reveal, J. L. 1970b. *Eriogonum*. In: D. S. Correll and M. C. Johnston. 1970. Manual of the Vascular Plants of Texas. Renner, Tex. Pp. 510–516.

Reveal, J. L. 1973. *Eriogonum*. In: C. L. Hitchcock and A. Cronquist. 1973. Flora of the Pacific Northwest: An Illustrated Manual. Seattle. Pp. 79–84.

Reveal, J. L. 1973b. *Eriogonum* (Polygonaceae) of Utah. Phytologia 25: 169–217.

Reveal, J. L. 1976. *Eriogonum* (Polygonaceae) of Arizona and New Mexico. Phytologia 34: 409–484.

Reveal, J. L. 1978. Distribution and phylogeny of the Eriogonoideae (Polygonaceae). Great Basin Naturalist Mem. 2: 169–190.

Reveal, J. L. 1980. Intermountain biogeography—a speculative appraisal. Mentzelia 4: 1–92.

Reveal, J. L. 1981b. *Eriogonum divaricatum* Hook. (Polygonaceae), an intermountain species in Argentina. Great Basin Naturalist 41: 143–146.

Reveal, J. L. 1985. An annotated key to *Eriogonum* (Polygonaceae) of Nevada. Great Basin Naturalist 45: 493–519.

Reveal, J. L. 1989. The eriogonoid flora of California (Polygonaceae: Eriogonoideae). Phytologia 66: 295–416.

Reveal, J. L. 1989b. Notes on selected genera related to *Eriogonum* (Polygonaceae: Eriogonoideae). Phytologia 66: 236–245.

Reveal, J. L. 1989c. Remarks on the genus *Pterostegia* (Polygonaceae: Eriogonoideae). Phytologia 66: 228–235.

Reveal, J. L. 2004. Proposal to conserve the name *Gilmania* Coville against *Phyllogonum* Coville (Polygonaceae: Eriogonoideae)—A case of mistaken homonymy. Taxon 53: 573.

Reveal, J. L. 2004b. Nomenclatural summary of Polygonaceae subfamily Eriogonoideae. Harvard Pap. Bot. 9: 143–230.

Reveal, J. L. and B. Ertter. 1977. Re-establishment of *Stenogonum* (Polygonaceae). Great Basin Naturalist 36: 272–280.

Reveal, J. L. and B. Ertter. 1980. The genus *Nemacaulis* Nutt. (Polygonaceae). Madroño 27: 101–109.

Reveal, J. L. and C. B. Hardham. 1989. Notes on selected genera related to *Chorizanthe* (Polygonaceae: Eriogonoideae). Phytologia 66: 199–220.

Reveal, J. L. and C. B. Hardham. 1989b. A revision of the annual species of *Chorizanthe* (Polygonaceae: Eriogonoideae). Phytologia 66: 98–198.

Revis. Gen. Pl.—See: O. Kuntze 1891–1898

Revis. N. Amer. Silene—See: C. L. Hitchcock and B. Maguire 1947

Reznicek, A. A. 1980. Halophytes along a Michigan roadside with comments on the occurrence of halophytes in Michigan. Michigan Bot. 19: 23–30.

Rhodora = Rhodora; Journal of the New England Botanical Club.

Riefner, R. E. 1982. Studies on the Maryland flora VIII: Range extensions of *Polygonum perfoliatum* L., with notes on introduction and dispersal in North America. Phytologia 50: 152–159.

Rink, H. J. 1852–1857. Grønland Geografisk og Statistisk Beskrevet.... 2 vols. Copenhagen. [Vol. 1 in 2 parts, paged independently.] (Grønland)

Roberts, R. H. 1977. *Polygonum minus* Huds. × *P. persicaria* L. in Anglesey. Watsonia 11: 255–256.

Roberty, G. E. and S. Vautier. 1964. Les genres de Polygonacées. Boissiera 10: 7–128.

Robinson, B. L. 1902. The New England polygonums of the section *Avicularia*. Rhodora 40: 65–73.

Roché, C. T. 1991. Rupturewort (*Herniaria cinerea* DC.). Olympia etc.

Roemer, J. J., J. A. Schultes, and J. H. Schultes. 1817[–1830]. Caroli a Linné...Systema Vegetabilium...Editione XV.... 7 vols. Stuttgart. (Syst. Veg.)

Rogers, D. J. 1980. Lakota Names and Traditional Uses of Native Plants by Sicangu (Brulé) People in the Rosebud Area, South Dakota: A Study Based on Father Eugene Buechel's Collection of Plants of Rosebud around 1920. St. Francis, S.Dak.

Röhling, J. C., F. C. Mertens, and W. D. J. Koch. 1823–1839. Deutschlands Flora, ed. 3. 5 vols. Frankfurt am Main. (Deutschl. Fl. ed. 3)

Rohrer, J. R. 1998. Noteworthy collection. Wisconsin. *Gypsophila muralis* L. (Caryophyllaceae). Cushion baby's breath. Michigan Bot. 37: 113–114.

Romero, J. B. 1954. The Botanical Lore of the California Indians. New York.

Ronse Decraene, L.-P. and J. R. Akeroyd. 1988. Generic limits in *Polygonum* and related genera (Polygonaceae) on the basis of floral characters. Bot. J. Linn. Soc. 98: 321–371.

Ronse Decraene, L.-P., Hong S. P., and E. F. Smets. 2000. Systematic significance of fruit morphology and anatomy in tribes Persicarieae and Polygoneae (Polygonaceae). Bot. J. Linn. Soc. 134: 301–337.

Ronse Decraene, L.-P., Hong S. P., and E. F. Smets. 2004. What is the taxonomic status of *Polygonella*? Evidence from floral morphology. Ann. Missouri Bot. Gard. 91: 320–345.

Ross, T. S. and S. Boyd. 1996b. *Arenaria macradenia* var. *kuschei* (Caryophyllaceae), re-collection of an obscure California taxon, with notes on known habitat and morphological variation. Crossosoma 22: 65–71.

Rossbach, R. P. 1940. *Spergularia* in North and South America. Rhodora 42: 57–83, 105–143, 158–193, 203–213.

Ruprecht, F. 1850. Ueber die Verbreitung der Pflanzen im nördlichen Ural. St. Petersburg. [Alt. title: Beiträge zur Pflanzenkunde des Russischen Reiches.... Siebente Lieferung.] (Verbr. Pfl. Ural)

Rydberg, P. A. 1906. Flora of Colorado.... Fort Collins, Colo. (Fl. Colorado)

Rydberg, P. A. 1917. Flora of the Rocky Mountains and Adjacent Plains. New York. (Fl. Rocky Mts.)

Sargentia = Sargentia; Continuation of the Contributions from the Arnold Arboretum of Harvard University.

Sarkar, N. M. 1958. Cytotaxonomic studies on *Rumex* sect. *Axillares*. Canad. J. Bot. 36: 947–996.

Scarry, C. M., ed. 1993. Foraging and Farming in the Eastern Woodlands. Gainesville.

Schaffner, J. H. 1932. Revised catalog of Ohio vascular plants.... Bull. Ohio Biol. Surv. 5(2): 89–215.

Schenck, S. M. and E. W. Gifford. 1952. Karok ethnobotany. Anthropol. Rec. 13: 377–392.

Schischkin, B. K. 1936. *Cerastium*. In: V. L. Komarov et al., eds. 1934–1964. Flora URSS.... 30 vols. Leningrad. Vol. 6, pp. 430–466.

Schlising, R. A. and H. H. Iltis. 1961. Preliminary reports on the flora of Wisconsin, no. 46. Caryophyllaceae—pink family. Trans. Wisconsin Acad. Sci. 50: 89–139.

Sci. Bull. Brigham Young Univ., Biol. Ser. = Science Bulletin, Brigham Young University. Biological Series.

Sci. Stud. Montana Coll. Agric., Bot. = Science Studies, Montana College of Agriculture and Mechanic Arts. Botany.

Scoggan, H. J. 1978–1979. The Flora of Canada. 4 parts. Ottawa. [Natl. Mus. Nat. Sci. Publ. Bot. 7.]

Scopoli, J. A. 1754. Methodus Plantarum.... Vienna. (Meth. Pl.)

Seznam—See: P. M. Opiz 1852

Shields, O. and J. L. Reveal. 1988. Sequential evolution of *Euphilotes* (Lycaenidae: Scolitantidini) on their plant host *Eriogonum* (Polygonaceae: Eriogonoideae). Biol. J. Linn. Soc. 33: 51–93.

Shildneck, P. and A. G. Jones. 1986. *Cerastium dubium* (Caryophyllaceae) new for the eastern half of North America (a comparison with sympatric *Cerastium* species, including cytological data). Castanea 51: 49–55.

Shildneck, P., A. G. Jones, and V. Muhlenbach. 1981. Additions to the vouchered records of Illinois plants and a note on the occurrence of *Rumex cristatus* in North America. Phytologia 47: 265–290.

Shinners, L. H. 1962c. *Siphonychia* transferred to *Paronychia* (Caryophyllaceae). Sida 1: 101–103.

Shinners, L. H. 1965. *Holosteum umbellatum* (Caryophyllaceae) in the United States: Population explosion and fractionated suicide. Sida 2: 119–128.

Shults, V. A. 1989. Rod Myl'nyanka (*Saponaria* L. s.l.) vo Flore SSSR. Riga.

Shultz, L. M. 1998. A new species of *Eriogonum* (Polygonaceae: Eriogonoideae) from Utah and Nevada. Harvard Pap. Bot. 3: 49–52.

Sida = Sida; Contributions to Botany.

Sitgreaves, L. 1853. Report of an Expedition down the Zuni and Colorado Rivers.... Washington. [Botany by J. Torrey, pp. 153–178, plates 1–21.] (Rep. Exped. Zuni Colorado Rivers)

Sivinski, R. C. and K. Lightfoot, eds. 1993. Southwestern Rare and Endangered Plants: Proceedings of the...Conference. Santa Fe.

Sketch Bot. S. Carolina—See: S. Elliott [1816–]1821–1824

Skr. Kiøbenhavnske Selsk. Laerd. Elsk. = Skrifter, Som Udi det Kiøbenhavnske Selskab af Saerdoms og Videnskabers Elskere ere Fremlagte og Oplaeste.

Skr. Vidensk.-Selsk. Christiana, Math.-Naturvidensk. Kl. = Skrifter Udgivne af Videnskabs-Selskabet i Christiana. Mathematisk-Naturvidenskabelig Klasse.

Small, J. K. 1895. A Monograph of the American Species of the Genus *Polygonum*. New York. [Mem. Dept. Bot. Columbia Coll. 1.] (Monogr. Amer. Sp. Polygonum)

Small, J. K. 1903. Flora of the Southeastern United States.... New York. (Fl. S.E. U.S.)

Small, J. K. 1933. Manual of the Southeastern Flora, Being Descriptions of the Seed Plants Growing Naturally in Florida, Alabama, Mississippi, Eastern Louisiana, Tennessee, North Carolina, South Carolina and Georgia. New York.

Smissen, R. D., J. C. Clement, P. J. Garnock-Jones, and G. K. Chambers. 2002. Subfamilial relationships within Caryophyllaceae as inferred from 5' NDHF sequences. Amer. J. Bot. 89: 1336–1341.

Smissen, R. D. and P. J. Garnock-Jones. 2002. Relationships, classification and evolution of *Scleranthus* (Caryophyllaceae) as inferred from analysis of morphological characters. Bot. J. Linn. Soc. 140: 15–29.

Smissen, R. D., P. J. Garnock-Jones, and G. K. Chambers. 2003. Phylogenetic analysis of ITS sequences suggests a Pliocene origin for the bipolar distribution of *Scleranthus* (Caryophyllaceae). Austral. Syst. Bot. 16: 301–315.

Smith, B. W. 1968. Cytogeography and cytotaxonomic relationships of *Rumex paucifolius*. Amer. J. Bot. 55: 673–683.

Smith, C. F. 1998. A Flora of the Santa Barbara Region, California, ed. 2. Santa Barbara.

Smith, J. E. 1800–1804. Flora Britannica. 3 vols. London. [Vols. paged consecutively. Vols. 1 and 2, 1800; vol. 3, 1804.] (Fl. Brit.)

Smith, J. F. and T. A. Bateman. 2002. Genetic differentiation of rare and common varieties of *Eriogonum shockleyi* (Polygonaceae) in Idaho using ISSR variability. W. N. Amer. Naturalist 62: 316–326.

Smith, N. P. et al., eds. 2004. Flowering Plants of the Neotropics. Princeton.

Smithsonian Contr. Bot. = Smithsonian Contributions to Botany.

Smithsonian Contr. Knowl. = Smithsonian Contributions to Knowledge.

Snyder, D. B. 1987. Notes on some of New Jersey's adventive flora. Bartonia 53: 17–23.

SouthW. Naturalist = Southwestern Naturalist.

Söyrinki, N. 1989. Fruit production and seedlings in *Polygonum viviparum*. Memoranda Soc. Fauna Fl. Fenn. 65: 13–15.

Sp. Pl.—See: C. Linnaeus 1753; C. L. Willdenow et al. 1797–1830

Sp. Pl. ed. 2—See: C. Linnaeus 1762–1763

Spach, E. 1834–1848. Histoire Naturelle des Végétaux. Phanérogames.... 14 vols., atlas. Paris. (Hist. Nat. Vég.)

Spellenberg, R., C. Leiva, and E. Lessa. 1988. An evaluation of *Eriogonum densum* (Polygonaceae). SouthW. Naturalist 33: 71–80.

Spier, L. 1930. Klamath ethnography. Univ. Calif. Publ. Amer. Archaeol. Ethnol. 30.

Sprengel, K. [1824–]1825–1828. Caroli Linnaei...Systema Vegetabilium. Editio Decima Sexta.... 5 vols. Göttingen. [Vol. 4 in 2 parts paged independently; vol. 5 by A. Sprengel.] (Syst. Veg.)

St. John, H. 1915. *Rumex persicarioides* and its allies in North America. Rhodora 17: 73–83.

St.-Hilaire, A., J. Cambessèdes, and A. H. L. de Jussieu. 1824[–1833]. Flora Brasiliae Meridionalis.... 3 vols. in 24 parts. Paris. [Volumes paged independently, parts and plates numbered consecutively.] (Fl. Bras. Merid.)

Stafleu, F. A. and R. S. Cowan. 1976–1988. Taxonomic Literature: A Selective Guide to Botanical Publications and Collections with Dates, Commentaries and Types, ed. 2. 7 vols. Utrecht, Antwerp, The Hague, and Boston.

Stafleu, F. A. and E. A. Mennega. 1992+. Taxonomic Literature. A Selective Guide to Botanical Publications and Collections with Dates, Commentaries and Types. Supplement. 6+ vols. Königstein.

Stearn, W. T. 1980. Swartz's contributions to West Indian botany. Taxon 29: 1–13.

Steedman, E. V. 1930. Ethnobotany of the Thompson Indians of British Columbia, based on field notes by James A. Teit. Rep. (Annual) Bur. Amer. Ethnol. 45: 441–522. [Reprinted 1973, Seattle.]

Steudel, E. G. 1840–1841. Nomenclator Botanicus Enumerans Ordine Alphabetico Nomina atque Synonyma tum Generica tum Specifica.... 2 vols. Stuttgart and Tübingen. (Nomencl. Bot. ed. 2)

Steury, B. W. 2004. Noteworthy collections from the District of Columbia and Maryland. Castanea 69: 154–157.

Stevenson, M. C. 1915. Ethnobotany of the Zuñi Indians. Rep. (Annual) Bur. Amer. Ethnol. 30: 35–102. [Reprinted 1993, New York.]

Steward, A. N. 1930. The Polygoneae of eastern Asia. Contr. Gray Herb. 88: 1–129.

Steward, J. H. 1933. Ethnography of the Owens Valley Paiute. Univ. Calif. Publ. Amer. Archaeol. Ethnol. 33: 233–350.

Steyermark, J. A. 1941. A study of *Arenaria patula*. Rhodora 43: 325–333.

Steyermark, J. A. 1963. Flora of Missouri. Ames.

Steyermark, J. A., J. W. Voigt, and R. H. Mohlenbrock. 1959. Present biological status of *Geocarpon minimum* Mackenzie. Bull. Torrey Bot. Club 86: 228–235.

Stokes, S. G. 1936. The Genus *Eriogonum,* a Preliminary Study Based on Geographical Distribution. San Francisco. (Eriogonum)

Stone, A. M. and C. T. Mason. 1979. A study of stem inflation in wild buckwheat, *Eriogonum inflatum*. Desert Pl. 1: 77–81.

Sturm, J. et al. 1900–1907. J. Sturm's Flora von Deutschland..., ed. 2. 15 vols. Stuttgart. (Deutschl. Fl. ed. 2)

Styles, B. T. 1962. The taxonomy of *Polygonum aviculare* and its allies in Britain. Watsonia 5: 177–214.

Sundaramoorthy, S. and D. N. Sen. 1988. Ecology of Indian arid zone weeds. XI. *Polycarpaea corymbosa* (Linn.) Lamk. Geobios (Jodhpur) 15: 235–237.

Suom. Kasvio—See: H. I. A. Hiitonen [1933]

Suppl. Calif. Fl.—See: P. A. Munz 1968

Suppl. Fl. Maine-et-Loire—See: T. Bastard 1812

Svensk Bot. Tidskr. = Svensk Botanisk Tidskrift Utgifven af Svenska Botaniska Föreningen.

Svensson, L. 1988. Inbreeding, crossing and variation in stamen number in *Scleranthus annuus* (Caryophyllaceae), a selfing annual. Evol. Trends Pl. 2: 31–37.

Svensson, R. and M. Wigren. 1986. A changing flora—a matter of human concern. Acta Univ. Upsal., Symb. Bot. Upsal. 27: 241–251.

Swink, F. and G. S. Wilhelm. 1994. Plants of the Chicago Region: An Annotated Checklist of the Vascular Flora..., ed. 4. Indianapolis.

Sydowia = Sydowia; Annales Mycologici Editi in Notitiam Scientiae Mycologicae Universalis.

Syn. Mitteleur. Fl.—See: P. F. A. Ascherson et al. 1896–1939

Syn. Pl.—See: C. H. Persoon 1805–1807

Syst. Bot. = Systematic Botany; Quarterly Journal of the American Society of Plant Taxonomists.

Syst. Nat. ed. 10—See: C. Linnaeus 1758[–1759]

Syst. Nat. ed. 12—See: C. Linnaeus 1766[–1768]

Syst. Veg.—See: J. J. Roemer et al. 1817[–1830]; K. Sprengel [1824–]1825–1828

Tabl. Encycl.—See: J. Lamarck and J. Poiret 1791–1823

Takhtajan, A. L. 1997. Diversity and Classification of Flowering Plants. New York.

Taxon = Taxon; Journal of the International Association for Plant Taxonomy.

Taylor-Lehman, J. W. 1987. Variation in *Polygonum pensylvanicum* L. (Polygonaceae) with an Emphasis on Variety *eglandulosum* Myers. M.S. thesis. Ohio State University.

Thieret, J. W. 1969c. *Rumex obovatus* and *Rumex paraguayensis* (Polygonaceae) in Louisiana: New to North America. Sida 3: 445–446.

Thomas, S. M. and B. G. Murray. 1983. Chromosome studies in species and hybrids of *Petrorhagia* sect. *Kohlrauschia* (Caryophyllaceae). Pl. Syst. Evol. 141: 243–255.

Thuillier, J. L. 1799. Flore des Environs de Paris..., ed. 2. Paris. (Fl. Env. Paris ed. 2)

Timson, J. 1963. The taxonomy of *Polygonum lapathifolium* L., *P. nodosum* Pers., and *P. tomentosum* Schrank. Watsonia 5: 386–395.

Timson, J. 1965. A study of hybridization in *Polygonum* section *Persicaria*. J. Linn. Soc., Bot. 59: 155–161.

Tolmatchew, A. I., ed. 1960–1987. Flora Arctica URSS. 10 vols. Moscow and Leningrad. (Fl. Arct. URSS)

Tolmatchew, A. I. 1966. *Rumex*. In: A. I. Tolmatchew, ed. 1960–1987. Flora Arctica URSS. 10 vols. Moscow and Leningrad. Vol. 5, pp. 143–161.

Torrey, J. and A. Gray. 1838–1843. A Flora of North America.... 2 vols. in 7 parts. New York, London, and Paris. (Fl. N. Amer.)

Torrey, J. and A. Gray. 1870. A revision of the Eriogoneae. Proc. Amer. Acad. Arts 8: 145–200.

Torreya = Torreya; a Monthly Journal of Botanical Notes and News.

Train, P., J. R. Henrichs, and W. A. Archer. 1941. Medicinal Uses of Plants by Indian Tribes of Nevada. Washington. [Contr. Fl. Nevada 33; reprinted 1982, Lawrence, Mass.]

Trans. Amer. Philos. Soc. = Transactions of the American Philosophical Society Held at Philadelphia for Promoting Useful Knowledge.

Trans. Illinois State Acad. Sci. = Transactions of the Illinois State Academy of Science.

Trans. Linn. Soc. London = Transactions of the Linnean Society of London.

Trans. Missouri Acad. Sci. = Transactions of the Missouri Academy of Science.

Trans. Wisconsin Acad. Sci. = Transactions of the Wisconsin Academy of Sciences, Arts and Letters.

Transylvania J. Med. Assoc. Sci. = The Transylvania Journal of Medicine and the Associate Sciences.

Trelease, W. 1892. A revision of the American species of *Rumex* occurring north of Mexico. Rep. (Annual) Missouri Bot. Gard. 3: 74–98.

Trudy Bot. Muz. = Trudy Botanicheskogo Muzeya.

Trudy Imp. S.-Peterburgsk. Bot. Sada = Trudy Imperatorskago S.-Peterburgskago Botanicheskago Sada.

Tsukui, T. and T. Sugawara. 1992. Dioecy in *Honckenya peploides* var. *major* (Caryophyllaceae). Bot. Mag. (Tokyo) 105: 615–624.

Turesson, G. 1961. Habitat modifications in some widespread plant species. Bot. Not. 114: 435–452.

Turkington, R., N. C. Kenkel, and G. C. Franko. 1980. The biology of Canadian weeds. 42. *Stellaria media* (L.) Vill. Canad. J. Pl. Sci. 60: 981–982.

Turland, N. J. and M. Wyse Jackson. 1997. Proposals to reject the names *Cerastium viscosum* and *C. vulgatum* (Caryophyllaceae). Taxon 46: 775–778.

Turner, B. L. 1983b. The Texas species of *Paronychia* (Caryophyllaceae). Phytologia 54: 9–23.

Turner, B. L. 1995. *Cerastium texanum* does not occur in Texas. Phytologia 79: 356–363.

Turner, N. J., R. Bouchard, and D. I. D. Kennedy. 1980. Ethnobotany of the Okanagan-Coleville Indians of British Columbia and Washington. Victoria.

Tutin, T. G. et al., eds. 1964–1980. Flora Europaea. 5 vols. Cambridge.

Tutin, T. G. et al., eds. 1993+. Flora Europaea, ed. 2. 1+ vol. Cambridge and New York.

Tutin, T. G. and S. M. Walters. 1993. *Dianthus*. In: T. G. Tutin et al., eds. 1993+. Flora Europaea, ed. 2. 1+ vol. Cambridge and New York. Vol. 1, pp. 227–246.

Tzvelev, N. N. 1987b. Zametki o Polygonaceae vo flore Dal'nego Vostoka. (Notulae de Polygonaceis in flora Orientis Extremi.) Novosti Sist. Vyssh. Rast. 24: 72–79.

Tzvelev, N. N. 1989b. *Rumex, Acetocella, Acetosa*. In: S. S. Kharkevich, ed. 1985+. Plantae Vasculares Orientis Extremi Sovietici. 7+ vols. Leningrad. Vol. 4, pp. 29–53.

Ugborogho, R. E. 1974. North American *Cerastium arvense* Linnaeus. IV. Phenotypic variation. Phyton (Horn) 32: 89–97.

Ugborogho, R. E. 1977. North American *Cerastium arvense* Linnaeus: Taxonomy, reproductive system and evolution. Phyton (Horn) 35: 169–187.

Ukrayins'k. Bot. Zhurn. = Ukrayins'kyi Botanichnyi Zhurnal.

Univ. Calif. Publ. Amer. Archaeol. Ethnol. = University of California Publications in American Archaeology and Ethnology.

Univ. Calif. Publ. Bot. = University of California Publications in Botany.

Univ. Colorado Stud., Ser. Biol. = University of Colorado Studies. Series in Biology.

Univ. Wash. Publ. Anthropol. = University of Washington Publications in Anthropology.

University of Chicago Press 1993. The Chicago Manual of Style, ed. 14. Chicago.

Utah Fl. ed. 3—See: S. L. Welsh et al. 2003

Vasc. Pl. Pacif. N.W.—See: C. L. Hitchcock et al. 1955–1969

Vasc. Pl. Wyoming—See: R. D. Dorn 1988

Vasc. Pl. Wyoming ed. 3—See: R. D. Dorn 2001

Veg. Syst.—See: J. Hill 1759–1775

Ventenat, É. P. [1800–1803.] Description des Plantes Nouvelles et Peu Connues Cultivés dans le Jardin de J. M. Cels.... 10 parts. Paris. [Plates numbered consecutively.] (Descr. Pl. Nouv.)

Verbr. Pfl. Ural—See: F. Ruprecht 1850

Vers. Darstell. Alsin.—See: E. Fenzl 1833

Verz. Pfl. Casp. Meer.—See: C. A. von Meyer 1831

Vestal, P. A. 1940. Notes on a collection of plants from the Hopi Indian region of Arizona made by J. G. Owens in 1891. Bot. Mus. Leafl. 8: 153–168.

Vestal, P. A. 1952. The ethnobotany of the Ramah Navaho. Pap. Peabody Mus. Amer. Archaeol. 40(4).

Vestal, P. A. and R. E. Schultes. 1939. The Economic Botany of the Kiowa Indians. Cambridge, Mass.

Villars, D. 1786–1789. Histoire des Plantes de Dauphiné. 3 vols. Grenoble, Lyon, and Paris. (Hist. Pl. Dauphiné)

Virginia J. Sci. = Virginia Journal of Science.

Voegelin, E. W. 1938. Tübatulabal Ethnography. Berkeley. [Anthropol. Rec. 2(1).]

Vogel, A. 1997. Die Verbreitung, Vergesellschaftung und Populationsökologie von *Corrigiola litoralis, Illecebrum verticillatum* und *Herniaria glabra* (Illecebraceae). Berlin.

Voy. Barbarie—See: J. Poiret 1789

W. Amer. Sci. = West American Scientist.

W. N. Amer. Naturalist = Western North American Naturalist.

Wagner, W. L. and E. M. Harris. 2000. A unique Hawaiian *Schiedea* (Caryophyllaceae: Alsinoideae) with only five fertile stamens. Amer. J. Bot. 87: 153–160.

Wagstaff, S. J. and R. J. Taylor. 1988. Genecology of *Cerastium arvense* and *C. beeringianum* (Caryophyllaceae) in northwest Washington. Madroño 35: 266–277.

Wahlenberg, G. 1812. Flora Lapponica Exhibens Plantas Geographice et Botanice Consideratas.... Berlin. (Fl. Lapp.)

Wallich, N. [1829–]1830–1832. Plantae Asiaticae Rariores.... 3 vols. in 12 parts. London, Paris, and Strasburgh. [Parts paged consecutively within volumes, parts and plates numbered consecutively throughout.] (Pl. Asiat. Rar.)

Walpers, W. G. 1842–1847. Repertorium Botanices Systematicae.... 6 vols. Leipzig. (Repert. Bot. Syst.)

Walter, T. 1788. Flora Caroliniana, Secundum Systema Vegetabilium Perillustris Linnaei Digesta.... London. (Fl. Carol.)

War Department [U.S.]. 1855–1860. Reports of Explorations and Surveys, to Ascertain the Most Practicable and Economical Route for a Railroad from the Mississippi River to the Pacific Ocean. Made under the Direction of the Secretary of War, in 1853[–1856].... 12 vols. in 13. Washington. (Pacif. Railr. Rep.)

Ward, D. B. 1977c. Keys to the flora of Florida. 2. *Paronychia* (Caryophyllaceae). Phytologia 35: 414–418.

Ward, D. B. 2001. New combinations in the Florida flora. Novon 11: 360–365.

Wasmann J. Biol. = Wasmann Journal of Biology.

Watson, S. 1871. United States Geological Expolration [sic] of the Fortieth Parallel. Clarence King, Geologist-in-charge. [Vol. 5] Botany. By Sereno Watson.... Washington. [Botanical portion of larger work by C. King.] [Botany (Fortieth Parallel)]

Watsonia = Watsonia; Journal of the Botanical Society of the British Isles.

Weaver, R. E. 1970. The arenarias of the southeastern granitic flat-rocks. Bull. Torrey Bot. Club 97: 40–52.

Webb, D. A. and A. O. Chater. 1963. *Polygonum oxyspermum* subsp. *raii* (Bab.) D. A. Webb & A. O. Chater, stat. nov. Feddes Repert. Spec. Nov. Regni Veg. 68: 187–189.

Weber, S. A. and P. D. Seaman. 1985. Havasupai Habitat: A. F. Whiting's Ethnography of a Traditional Indian Culture. Tucson.

Weber, W. A. 1961. *Stellaria irrigua* Bunge in America. Univ. Colorado Stud., Ser. Biol. 7: 12–15.

Weber, W. A. and R. L. Hartman. 1979. *Pseudostellaria jamesiana*, comb. nov., a North American representative of a Eurasian genus. Phytologia 44: 313–314.

Weber, W. A., B. C. Johnston, and D. H. Wilken. 1979. Additions to the flora of Colorado–VI. Phytologia 41: 486–498.

Weber, W. A. and R. C. Wittmann. 1992. Catalog of the Colorado Flora: A Biodiversity Baseline. Niwot, Colo.

Weed Res. = Weed Research.

Weed Sci. = Weed Science; Journal of the Weed Science Society of America.

Welsh, S. L., N. D. Atwood, S. Goodrich, and L. C. Higgins,

eds. 1987. A Utah flora. Great Basin Naturalist Mem. 9.

Welsh, S. L., N. D. Atwood, S. Goodrich, and L. C. Higgins, eds. 1993. A Utah Flora, ed. 2. Provo.

Welsh, S. L., N. D. Atwood, S. Goodrich, and L. C. Higgins. 2003. A Utah Flora, ed. 3. Provo. (Utah Fl. ed. 3)

Whitehead, F. H. and R. P. Sinha. 1967. Taxonomy and taximetrics of *Stellaria media* (Linnaeus) Vill., *S. neglecta* Weihe and *S. pallida* (Dumort.) Piré. New Phytol. 66: 769–784.

Wiens, D. et al. 1989. Developmental failure and loss of reproductive capacity in the rare palaeoendemic shrub *Dedeckera eurekensis.* Nature 338: 65–67.

Wiens, D., L. Allphin, D. H. Mansfield, and G. Thackray. 2004. Developmental failure and loss of reproductive capacity as a factor in extinction: A nine-year study of *Dedeckera eurekensis* (Polygonaceae). Aliso 21: 55–63.

Wiens, D., C. I. Davern, and C. L. Calvin. 1988. Exceptionally low seed set in *Dedeckera eurekensis:* Is there a genetic component of extinction? In: C. A. Hall Jr. and V. Doyle-Jones, eds. 1988. Plant Biology of Eastern California. Los Angeles. Pp. 19–29.

Wiens, D., M. DeDecker, and C. D. Wiens. 1986. Observations on the pollination of *Dedeckera eurekensis* (Polygonaceae). Madroño 33: 302–305.

Willdenow, C. L. 1809–1813[–1814]. Enumeratio Plantarum Horti Regii Botanici Berolinensis.... 2 parts + suppl. Berlin. [Parts paged consecutively.] (Enum. Pl.)

Willdenow, C. L., C. F. Schwägrichen, and J. H. F. Link. 1797–1830. Caroli a Linné Species Plantarum.... Editio Quarta.... 6 vols. Berlin. [Vols. 1–5(1), 1797–1810, by Willdenow; vol. 5(2), 1830, by Schwägrichen; vol. 6, 1824–1825, by Link.] (Sp. Pl.)

Williams, F. N. 1896. A revision of the genus *Herniaria.* Bull. Herb. Boissier 4: 556–570.

Williams, F. N. 1896b. A revision of the genus *Silene* Linn. J. Linn. Soc., Bot. 32: 1–196.

Willkomm, H. M. and J. M. C. Lange. 1861–1880. Prodromus Florae Hispanicae.... 3 vols. in 9 parts. Stuttgart. (Prodr. Fl. Hispan.)

Wilson, K. L. 1990b. Some widespread species of *Persicaria* (Polygonaceae) and their allies. Kew Bull. 45: 621–636.

Wilson, P. 1934. Tetragoniaceae. In: N. L. Britton et al., eds. 1905+. North American Flora.... 47+ vols. New York. Vol. 21, pp. 267–277.

Wislizenus, F. A. 1848. Memoir of a Tour to Northern Mexico, Connected with Col. Doniphan's Expedition, in 1846 and 1847.... Washington. (Mem. Tour N. Mexico)

Wofford, B. E. 1981. External seed morphology of *Arenaria* (Caryophyllaceae) of the southeastern United States. Syst. Bot. 6: 126–135.

Wofford, B. E., D. H. Webb, and W. M. Dennis. 1977. State

records and other recent noteworthy collections of Tennessee plants II. Castanea 42: 190–193.

Wolf, N. M. 1776. Genera Plantarum.... [Danzig.] (Gen. Pl.)

Wolf, S. J. and J. McNeill. 1986. Synopsis and achene morphology of *Polygonum* section *Polygonum* (Polygonaceae) in Canada. Rhodora 88: 457–479.

Wolf, S. J., J. G. Packer, and K. E. Denford. 1979. The taxonomy of *Minuartia rossii* (Caryophyllaceae). Canad. J. Bot. 57: 1673–1686.

Wood, A. 1847. A Class-book of Botany..., ed 2. Boston and Claremont, N.H. (Class-book Bot. ed. 2)

Wood, A. 1861. A Class-book of Botany.... New York. [Class-book Bot. ed. s.n.(b)]

Wrightia = Wrightia; a Botanical Journal.

Wunderlin, R. P. 1981. *Polygonella polygama* (Polygonaceae) in Florida. Florida Sci. 44: 78–80.

Wyatt, R. 1984. The evolution of self-pollination in granite outcrop species of *Arenaria* (Caryophyllaceae). I. Morphological correlates. Evolution 38: 804–816.

Wyman, L. C. and S. K. Harris. 1951. The Ethnobotany of the Kayenta Navaho: An Analysis of the John and Louisa Wetherill Ethnobotanical Collection. Albuquerque.

Yang, J. and J. W. Wang. 1991. A taxometric analysis of characters of *Polygonum lapathifolium* L. Acta Phytotax. Sin. 29: 258–263.

Yasui, Y. and O. Ohnishi. 1998. Phylogenetic relationships among *Fagopyrum* species revealed by the nucleotide sequences of the ITS region of the nuclear rRNA gene. Genes Genet. Systems 73: 201–210.

Yasui, Y. and O. Ohnishi. 1998b. Interspecific relationships in *Fagopyrum* (Polygonaceae) revealed by the nucleotide sequences of the *rbc*L and *acc*D genes and their intergenic region. Amer. J. Bot. 85: 1134–1142.

Yatskievych, G. and J. Turner. 1990. Catalogue of the flora of Missouri. Monogr. Syst. Bot. Missouri Bot. Gard. 37.

Youngbae, S., S. Kim, and C. W. Park. 1997. A phylogenetic study of *Polygonum* sect. *Tovara* (Polygonaceae) based on ITS sequences of nuclear ribosomal DNA. J. Pl. Biol. 40: 47–52.

Yurtseva, O. V., N. D. Yakolevleva, and T. I. Ivanova-Radkevich. 1999. Heterocarpy in *Polygonum aviculare* L. s. str. and related species (*Polygonum,* subsect. *Polygonum*). Byull. Moskovsk. Obshch. Isp. Prir., Otd. Biol. 104: 13–20.

Zigmond, M. L. 1981. Kawaiisu Ethnobotany. Salt Lake City.

Zika, P. F. and A. L. Jacobson. 2003. An overlooked hybrid Japanese knotweed (*Polygonum cuspidatum* × *sachalinense;* Polygonaceae) in North America. Rhodora 105: 143–152.

Zoë = Zoë; a Biological Journal.

Index

Names in *italics* are synonyms, casually mentioned hybrids, or plants not established in the flora. Page numbers in **boldface** indicate the primary entry for a taxon. Page numbers in *italics* indicate an illustration. Roman type is used for all other entries, including author names, vernacular names, and accepted scientific names for plants treated as established members of the flora.